Outstanding Contributions to Logic

Volume 5

For further volumes:
http://www.springer.com/series/10033

Alexandru Baltag · Sonja Smets
Editors

Johan van Benthem on Logic and Information Dynamics

 Springer

Editors
Alexandru Baltag
Sonja Smets
Institute for Logic, Language
  and Computation
University of Amsterdam
Amsterdam
The Netherlands

ISSN 2211-2758   ISSN 2211-2766   (electronic)
ISBN 978-3-319-06024-8   ISBN 978-3-319-06025-5   (eBook)
DOI 10.1007/978-3-319-06025-5
Springer Cham Heidelberg New York Dordrecht London

Library of Congress Control Number: 2014946756

Printed on acid-free paper

Springer is part of Springer Science+Business Media (www.springer.com)

Contents

Part III   Games

Part VI   Styles of Reasoning

Preface by Johan van Benthem

This book appears in a series highlighting contributions to logic, as seen through the eyes of a congenial community of colleagues. It is a great honor to be a focus for the group of authors assembled here. Though public mirrors seldom reflect self-images, we learn most about people, not only by their own words, but also by the company they keep.

But enough said about people, let me turn to the topic. This book is about *logical dynamics*, a bundle of interests and a program that may not cover my whole work, but that definitely constitutes the largest chunk of what I have done over the last decades. Let me explain what it means to me. You may find what follows ideological, some people prefer context-free theorems—but I need such broader perspectives even for myself, to remind me of why I do the things I do—or even, why I do research at all.

The main idea of logical dynamics is the pervasive duality between information-related actions and their products. Standard logical systems emphasize notions like formula or proof in the sense of static objects that can be viewed or even manipulated externally. But these objects are produced in activities of communicating statements, engaging in reasoning, and many other intellectual skills. Interestingly, our natural language is often ambiguous in this respect between verbs or other activity-related expressions and static nouns. A dance is an activity I can engage in, but also an object that can be produced by dancing—and the same duality holds for many logical terms, like "statement" or "argument." The idea of logical dynamics is to take this duality seriously, and bring the core logical activities explicitly into formal systems that satisfy the same standards of rigor as the ones that we know and love. This is possible, since activities and events, too, have a formal structure that lends itself to logical analysis. In this way, to borrow a happy phrase, "logic can be more than it is."

Over time, my view of what are logical core activities has evolved from single-agent acts of inference and observation to social scenarios involving more agents, with asking a question, perhaps, as the major instance of a basic logical act. This took some time, since this social turn went against central tenets of my upbringing.

Dutch Calvinists like me were raised with the idea that there are two modes of life. The horizontal mode looks at other people and what they think, the vertical one looks only at one's relationship to God. Naturally, the latter, more lonely but also more heroic stance appealed much more to me, and logic seemed very much in that spirit, putting one in direct communion with the intellectual joints of the universe. By contrast, the horizontal stance is all about being influenced by and dependent on others, that is, the realm of human frailty and folly. But over time, I have come to appreciate that social behavior and the intricate network of dependencies that form our life may be the more exciting and challenging phenomenon—or at least, that it has equal importance to solitude in logic and intellectual life, just as the various interactions of particles that constitute our physical world. In fact, perhaps the original source for logic is argumentation between different parties, with formal systems coming only later as a methodological device. And again, logic can deal with all these perspectives at once.

Formal versions of these views in their various phases can be found in a sequence of my books: *Language in Action* (1991), *Exploring Logical Dynamics* (1996), *Logical Dynamics of Information and Interaction* (2011), and *Logic in Games* (2013). Another important source is the dissertations of my students since roughly 2000. What all these publications reflect are influences on my thinking from the worlds that meet in my academic environments at Amsterdam and Stanford: logics of action and processes in computer science, dynamic semantics of natural language, philosophical theories of knowledge and information, and interaction as studied in game theory. I see logic as lying at a crossroads of the university, absorbing many ideas that pass.

Still, being a logician also implies a certain modus operandi, and in my view, a unity of methods persists even when we expand the agenda: logical dynamics uses formal systems. In much of my work, systems of modal logic play an important role, as a convenient light formalism that allows us to see a lot of interesting structures without importing too much machinery. But I see exclusive allegiance to one formalism or school as an intellectual weakness, and I have in fact devoted a lot of time to seeing connections and parallels between different logical systems, as in my work on correspondence theory. Still, the main point is the formal slant in this kind of work per se. Even when I theorize about noisy "horizontal" social reality, the methodology is "vertical," the mathematical truth is absolute, and social strategizing would not help.

So much for my own take on the topic of this book. But a book like this is a risk, since it is a mirror in one's colleagues' eyes, who may see things quite differently. Sometimes you wish you were the person portrayed, sometimes the mirror confirms your worst suspicions. That is why so many people with books devoted to their work are engaged in frantic spin covering the entries with added responses, conclusions, and other types of cotton candy. I will try to minimize this spin mode, though I cannot refrain from making a few points about the book as I experience it—both the editorial process of producing it, and the product that now lies before us.

For a start, though one can have lots of soul-searching thoughts, at a most simple and immediate level, this book just consists of topics that I like! Many chapters represent some aspect of logic, information, and agency that I would like to understand better—and that is exactly what the authors have provided. Moreover, I admit to just liking abstract technical logic, and again many authors have done just that, stepping up their abstraction levels in those typical ways that please logicians. I will not even begin to enumerate all chapter topics here: a later separate piece will present some more detailed thoughts concerning what the authors have to say. But even so, it will be clear that this book contains many trails of happy hiking in the landscape of logical dynamics, very broadly conceived. Some of these trails start out in places where I have walked myself these past decades, such as dynamic epistemic logics, temporal logics, logics of games, or belief revision—while other trails in the book move out into relatively new territory for me, such as learning theory, social dynamics, database theory, proof theory, cognitive science, or probability theory.

But the material collected here defies easy description. What also appeals very much to me is chapters that remind me of my earlier interests in natural language, philosophical logic, and philosophy of science. They made me realize that there may be much more continuity of concerns than I have perhaps thought over the past period, and many more things to be learnt by returning there, than I had imagined. Likewise, there is material on my old and persistent technical interests in modal model theory and foundations of computation that I find extremely suggestive, especially, as I feel that applied agenda extensions for our discipline, as envisaged in logical dynamics, had better be accompanied by rigorous theoretical investigations from the start.

Of course, not every author approaches things the way I myself would do it: I guess this realization on my part is the process called learning. In fact, on a self-critical note, several chapters have taught me that logical dynamics is not such a clear concept as I would like to think. There are serious philosophical issues about its precise claims and its relationship to classical logics, and there are mathematical issues about a best understanding of how its dynamic systems had best be formulated and understood. Much of this has generated lively correspondence with authors, and I hope that some of this ongoing discussion will itself find its way into the literature.

In order to give this book its present focus, selections had to be made. Some loves from my earlier life do not occur, or not enough justice is done to them, such as the interfaces of logic with natural language, philosophy, and cognition described in my scientific autobiography. This restricted focus is the format of this series, and I think it is inevitable for any readable book. Still, several authors have made connections to these other topics that set me thinking. I now feel that natural language is much more important to logical dynamics than I had realized so far, while there is also a clear potential for revitalizing the interface of logic and philosophy. And even cognitive science is just around the corner: while my systems of logical dynamics remain normative, they can only function in the real world. On the sunny side, even gaps and loose ends that come to light tell me where I might be going from here. For instance, I find myself drawn increasingly to interfaces between logic and probability, and many chapters in this book whet that appetite. I find this a comforting thought. Although a book series like this new initiative might be considered a polite invitation to outstanding logicians to finally shut up and leave the field to a younger generation, I see some rays of future for me shining through its pages.

I find it hard to tell other people what sort of book this is. It is not a Festschrift, it is not just an anthology, it has no systematically enforced message or methodology. It is much more ambiguous than that, like life itself. The way I experience this book, it is a panorama of a world I enjoy. It demonstrates the broad interests and methods that have shaped my own work in logic. But I hope it does not do that too obtrusively. Even if you are not into logical dynamics (or Johan van Benthem), the pieces that follow should still be of interest. Their topics are important, and represent a future for logic. Moreover, the group of their authors itself conveys an important message. They come from many disciplines: mathematics, philosophy, computer science, artificial intelligence [the love child of computer science and philosophy], but also game theory and beyond. This diversity is my world where I feel comfortable, this is how I was educated, and how my academic environment functions. I deeply feel that the broad logic that is at stake here can only flourish in this sort of intellectual company.

Thanks to the authors for contributing what they did, and lending their presence to this book. And thanks to the editors Alexandru and Sonja for making it happen.

Johan van Benthem

Acknowledgements

The editors would like to thank the many people who contributed to the making of this book. First of all, we thank the invited authors for their insightful contributions, going way beyond a mere homage: these are fascinating new research papers, and collectively, they will push our field ahead in one single stroke. Just reading these papers has been a pleasurable and mind-opening experience for us. We would also like to thank the Springer publishing team for their great work. The general series editors Ryszard Wojcicki and Sven Ove Hansson were instrumental in inviting us, and reminding us, to realize this volume. We give our thanks to Ties Nijssen and Christi Lue for their long-standing and crucial involvement with this volume. We are grateful to Floor Oosting for providing essential help at a later stage, and also to the typesetting teams in Chennai for their speedy and professional work. We thank Nina Gierasimczuk for her drawing of all the "logic birds" perching on the tree of this volume. Finally, we are indebted to Johan van Benthem, the gathering point that made this book possible, for his quiet but persistent assistance in our editorial work.

On the Trails of Logical Dynamics: a bird's-eye view of this volume, by Alexandru Baltag and Sonja Smets[1]

1 The Main Theme

Reducing Johan van Benthem's vast interdisciplinary work, published over 40 years, to one single research theme, however broad, seems to us both impossible and counterproductive. However, as part of the Springer series "Outstanding Contributions", our volume is meant to be *thematic*, as well as being personally dedicated to a world-famous logician.

Indeed, "Outstanding Contributions" is a series of book profiles of major themes pursued by leading logicians today. The way we see it, the idea behind this book is two-fold: (a) to highlight van Benthem's contributions to the chosen topic, and (b) to develop the field further, by inviting other leading researchers to contribute papers exploring specific dimensions of the main theme. But these dimensions are often as important and as broad in their own right as the one chosen as the "main" topic. In this sense, our thematic focus could be deceptive: in reality, the chosen theme is itself only one perspective among others in a family of interrelated on-going research areas, lying at the cutting edge of contemporary logic.

Together with Johan himself, we decided on a theme that has been at the core of his research agenda over at least the last 23 years: *Logical-Informational Dynamics*. In fact, we will argue here that, if broadly conceived, this topic can be seen to be highly relevant for *most* of van Benthem's research, from its very beginning. Moreover, while reiterating our feeling that we cannot subsume or reduce all the many facets of van Benthem's work to one unique catchphrase, we nevertheless think that this body of work can best be interpreted *as a whole* only when seen from the vantage point of Logical Dynamics.

The reader might well think that such a thematic consistency of all van Benthem's lines of work sounds "too good to be true". But we definitely do not claim that Johan van Benthem had consciously and deliberately pursued one overall theme throughout all his life! We only argue that this theme *can* be used to provide a deeper, more unitary understanding of van Benthem's various research pursuits.

[1] Sonja Smets' contribution to this paper was funded by the European Research Council under the European Community's Seventh Framework Programme (FP7/2007-2013)/ERC Grant agreement no. 283963.

Such an understanding is typically unavailable in the heat of the actual pursuit. It becomes transparent only retroactively. So we do not feel that we are falsifying van Benthem's research history by seeking to uncover its underlying unity.[2]

The same can be said about the invited contributions that form this volume. While each of them fits very well with (at least one of) Johan's research interests, together they testify to the wide range of these interests. The life-like variety of the topics investigated in the chapters of this volume, and the seemingly uncontrolled and irreducible diversity (and sometimes apparent divergence) of their perspectives, may seem confusing to the reader at a first sight. But a discerning eye will recognize the deeper unity behind this friendly clash of paradigms: not an abstract, all-subsuming unity under the aegis of some dead formal deity, but the live, striving unity-in-competition that makes them part of a shared "eco-system". Together, these contributions map the landscape of a field in the making, locating the core issues and the most desirable spots on the map, filling the gaps where the dragons dwell, defining the borders and outlining the main shapes of the New World of logical-informational dynamics.

2 A Field in the Making

2.1 Dynamics in van Benthem's Work

Johan van Benthem's first publications with an explicitly "dynamic" bent date from around 1990 [42–45], though these are anticipated by his early work on temporal logics in the late 1970s and early 1980s [29], on logical games in late 1980s, and at a more abstract level they are connected to his earlier work on relational semantics and modal correspondence. More generally, dynamics was "in the air" in the late 80s, with concomitant work in computer science [147], logic [140], philosophy [156] and linguistics [148], with even earlier historical precursors in the philosophical literature [25, 154, 194], logic [168] and computer science [155].

The 'General Dynamics' program [45] is van Benthem's first full-fledged programmatic text on logical dynamics, followed in quick succession by his books *Language in Action* [43], *Exploring Logical Dynamics* [47], his Spinoza project "Logic in Action", a vast number of research papers on this topic and, more recently, his books *Logical Dynamics of Information and Interaction* [102] and *Logic in Games* [113], as well as his forthcoming book *The Music of Knowledge* [114] (with A. Baltag and S. Smets) on dynamic epistemology.

---

[2] But, of course, we may be *wrong*: the view presented here is our own, "creative" (hence, possibly inaccurate) interpretation.

2.2 What is Logical-Informational Dynamics?

We understand the theme of this volume very broadly as *the logical study of information flow, cognitive and computational processes, strategic interaction and rational agency*, study lying at the intersection of many different disciplines, and extending from more mathematical to more philosophical dimensions. We see Logic as closely connected to the concept of *information*, taken mainly in its *qualitative-cognitive* sense, though potentially also connecting to the more quantitative notions of information, such as the Bayesian account of belief change. In their chapter in the Handbook of Philosophy of Information [89], van Benthem and Martinez extensively discuss the various conceptions of information encountered in philosophy and science (information-as-range, information-as-correlation, information-as-code etc.), as well as their relevance for logic. This diversity of conceptions extends to dynamics: each notion of information comes with its own specific type of information flow.[3]

The Static-Dynamic Duality

As repeatedly stressed by van Benthem [47], informational-cognitive concepts often have a *dual* character: both "static" and "dynamic", or what may be called the duality between *product* versus *process*. van Benthem's favorite examples of such cognitive notions with dual meaning are the concepts of *reasoning*, *argument* and *judgment*, which denote both intellectual processes and their contents or products. This is very similar to the opposition "procedural"-"declarative" in Computer Science, as well as the ambiguity in natural language between activities and their end-products (e.g. words such as *dance*, *play* or *move*). The ambiguity suggests the existence of two complementary aspects of these concepts: a static, object-like, "finished" (and thus "well-defined") side, and a dynamic, processual, "free-flowing" and on-going (thus "unfinished") aspect. The *epistemic* version of the procedural-declarative duality is the distinction between "knowledge how" and "knowledge that". The same duality also occurs in Mathematics: in Model Theory, between *structures* (sets, graphs, algebras, models etc.) and structure-preserving (or structure-changing) *transformations* (morphisms, isomorphisms, permutations, automorphisms, bisimulations, operations on models such as ultraproducts etc.); in Category Theory, between "objects" and "arrows", or at a higher level between "categories" and "functors"; and in Proof Theory, between "definitions" and "theorems" on the one hand, and "constructions" or "proofs" (as algorithms, procedures for producing theorems or counterexamples) on the other hand.

---

[3] For instance, information-as-range naturally gives rise to the notion of update as *range-restriction*.

Logical Dynamics: The Search for Abstract Structures

According to van Benthem, the distinction static-dynamic is related to the existence of different *types of logical structures*. The static aspects are usually captured by *Boolean algebras* or other types of *lattices* (underlying "classical" logic, as well as its more traditional "non-classical" alternatives). As for the dynamic aspects, van Benthem [47] sees *relational algebras* (underlying arrow logics, multi-dimensional modal logics etc.) as the "dynamic" analogue of Boolean algebras.

The interplay between the static and the dynamic aspects is captured by "mixed" *two-sorted logical structures*, such as Kleene algebras, dynamic algebras and quantales (underlying e.g. dynamic logic or quantum logic), or by *higher-typed structures* (e.g. typed lambda calculus, or the Hilbert-space formalism for quantum mechanics, which use higher-typed objects or systems of higher dimension to encode the dynamics of lower-typed objects or lower-dimensional systems). More complex mixed algebras correspond to the more sophisticated two sorted Boolean-Algebraic logics introduced by van Benthem, such as the family of Dynamic Modal logics [40, 49].

But this is just a first stab. More generally, van Benthem's work throughout the years can be seem as a constant search for the best mathematical frameworks for dynamics: from relational algebras and abstract Dynamic Modal Logic, to the dynamic significance of other frameworks such as Barwise's Situation theory and Chu spaces [55, 89]; to logics such as DEL [102], that go beyond traditional model-theoretic semantics, by talking about potential updates of their own models; and more recently, to abstract "update universes" [115].

The Drive Towards Concreteness: Learning, Games, Agency, Language

As we'll see, the constant search for better abstract-mathematical perspectives on dynamics is paralleled in van Benthem's work by a dual move: a constant drive towards *concreteness*. This focuses on modeling, and reasoning about, the actual information flow via "real" channels between full-fledged agents. We call this move *informational* dynamics, distinguishing it from the "purely logical" one, since it goes beyond inference and reasoning, and comprises other types of informational events.

In particular, we may distinguish four aspects of informational dynamics, in increasing order of concreteness. The first refers to the dynamics of knowledge and belief over time: essentially, this is about *learning*, observation, communication and belief revision. The second aspect adds (static) preferences or goals that guide agents' actions: this is the "rational dynamics" underlying strategic interactions and forming the main topic of *game theory*. The third aspect concerns the more sophisticated features that characterize *agency*, going beyond the game-theoretic perspective: concepts such as choice, control and free will, activities such as deliberation, inquiry and preference-change, the dynamics of desires, intentions and norms, and the ensuing moral dilemmas. Finally, the fourth aspect concerns human *cognition and natural language*. This last aspect is the most "concrete" of all: here, logic faces the empirical "facts" about people's actual cognitive, psychological, social and linguistic practices in real life.

Back and Forth Between Abstract and Concrete: Styles of Inference

As we have seen, the drive towards concreteness is counterbalanced by an opposite one, that tends to recover the abstract structures and patterns of inference underlying concrete informational processes. Indeed, from an abstract perspective, logical dynamics is just one of many different *styles of reasoning*, and dynamic reasoning is just one among a wide variety of alternative forms of inference. But the dual perspective may also be applied: other styles of reasoning, usually formalized as non-classical consequence relations (e.g. substructural and non-monotonic logics, relevance logic etc.), can be given dynamic interpretations, by looking at the specific informational tasks (e.g. forms of learning, communication etc.) that they are meant to capture.

So we have a continuous back-and-forth move between abstract and concrete. On the one hand, we can go from concrete dynamics to consequence relations via a process of *abstraction*; on the other hand, we can go the other way around, from consequence relations to dynamics, via *dynamic representations or interpretations*. Concrete information flow can be described via abstract inferences, which themselves can be interpreted as concrete informational processes.

2.3   The Six Dimensions of Logical Dynamics

Putting together the conclusions of the above discussion, we have decomposed our main theme into six different "dimensions" or aspects: (1) *mathematical and computational perspectives on logical dynamics*; (2) *dynamics of knowledge and belief over time*; (3) *games and strategic interaction*; (4) full-fledged *dynamic agency*; (5) *dynamics in natural language and cognition*; (6) *dynamics and "styles of reasoning"*. The first dimension belongs to the search for abstract foundations for dynamics, the next four represent the different sides of concrete dynamics that we uncovered above (learning, games, agency and language). Finally, the sixth dimension captures in a sense the above-mentioned back-and-forth move between abstract and concrete: on the one hand, dynamics is understood at an abstract level as just one among many alternative styles of reasoning; while on the other hand, other reasoning styles are re-interpreted as forms of dynamic inference. But in reality this back-and-forth move also happens within each dimension: e.g. the usual models for Dynamic Epistemic Logic (in which an update is a very specific model-transformation) could be thought of as "concrete" models, while van Benthem's "update universes" [115] are more abstract models for the same language.

Not coincidentally, these six viewpoints correspond to the six parts of our volume, representing thematic groupings of our invited papers: indeed, each of these divisions group together papers that explore in more depth one of the above dimensions (while of course occasionally touching on other aspects, both within and beyond the above-mentioned six). So, as part of our investigation, each of the subsequent six sections of this Introduction will explore one of the viewpoints (1–6) above, by discussing the parts of van Benthem's work and the invited papers that are relevant for that specific dimension.

3 Mathematical and Computational Perspectives on Logical Dynamics

In a sense, all modal logic, especially when endowed with its relational semantics based on Kripke models, can be said to be "dynamic" in general: the truth of a modal formula at a given world depends on what happens at *other* worlds, related to the first one via the usual transition (or accessibility) relation. Such transition relations always involve a logical change or shift (from a given context, situation, world or state, to another), and hence they capture a form of abstract "dynamics", even when the intended interpretation of these relations is in fact static (e.g. epistemic indistinguishability, or world-similarity, or spatial nearness). This is the aspect that embodies "*logical* dynamics" proper, i.e. the dynamics of our logical manipulation of information.

3.1 Johan van Benthem on Mathematical-Computational Perspectives

Modalization as Dynamification

Any move towards the "modalization" of a specific logical area can thus be interpreted as a "dynamification" of that area. Examples are van Benthem's work on (a) the modal fine structure of classical predicate logic, and (b) the (non-classical) modal semantics of predicate logic.

The modal fine structure of first order logic. This area of research pertains to the investigation of fragments of predicate logic that are "modal-like", in the sense of sharing some of the desirable features of modal logic, in particular its *low complexity* (decidability). In [195], Vardi has famously asked the question "Why is modal logic so robustly decidable?" What was needed to answer this was to isolate some key features of modal logic, and then to show that the other logics having these features are also robustly decidable. A first attempt in this sense were the finite-variable fragments introduced earlier by Henkin, but this type of generalization proved to be a red herring as far as tractability is concerned: all fragments with at least 3 variables are undecidable. In fact, the best answer was suggested by van Benthem in [47, 52], and only truly developed in his joint work with Andréka and Németi [1]. This came in the form of the *bounded fragment* (in which the quantifiers $\exists \vec{y}(R\vec{x}\vec{y} \wedge \phi(\vec{y}))$ are always "bounded" by some relation R) and the larger *guarded fragment* (in which the quantifiers $\exists \vec{y}(R\vec{x}\vec{y} \wedge \phi(\vec{x}, \vec{y}))$ are "guarded" by relations R), later extended in [52] to the even larger *loosely guarded fragment* (in which the quantifiers are guarded by conjunctions of atomic formulae of certain forms). By teaming up with Andréka and Németi [1], van Benthem investigated in depth the meta-logical properties of these fragments. Even larger extensions in the same spirit were later proposed: the *packed fragment*, and the *clique-guarded fragment*. All these logics share the "locality" of modal logic, by restricting the scope of quantifiers to whatever is "locally" reachable via some accessibility relation. But if we think of R as an "action" or program (going from the string of input-values stored in the variables $\vec{x}$ to the output-values stored in $\vec{y}$), then guarded quantification is really the prototypical dynamic move: the statement $\exists \vec{y}(R\vec{x}\vec{y} \wedge \phi(\vec{x}, \vec{y}))$ says that program R is locally "correct" (on the current input $\vec{x}$), in the sense that it fulfills some desirable condition ϕ (holding between its input and output values). The "modal" nature of the guarded fragments is confirmed by the fact that they are decidable and have the finite model property. Moreover, their decidability is "robust", in the sense that it is inherited by their fixed-point extensions, as shown by Graedel and Walukiewicz [142]. As such, this line of research provided a clear conceptual answer to Vardi's question: it is exactly the locality or "dynamicity" of modal logic (shared with the above-mentioned fragments, in the form of guardedness) that is responsible for its robustly good behavior!

The modalization of first order logic. An alternative route pursued by van Benthem was the "generalization by modalization" of the standard semantics of predicate logic. A Kripke structure is assumed, with each possible world or "state" coming with its own variable assignment, while the existential quantifier $\exists x$ is a modal diamond along some binary "update" relation R_x between states. The complete logic of this setting is the *minimal polyadic logic*, which thus represents the "modal core" of first-order logic. The standard Tarskian semantics is recovered as a special case, namely the one satisfying three additional constraints: (1) states are completely identified with the corresponding variable assignments; (2) the "update" relation R_x is the standard one, holding between two assignments iff they agree on all variables different from x; (3) all possible assignment functions are represented in the model. The latter is a strong existence condition, which might reasonably be considered as set-theoretic rather than purely logical, and which is responsible for the undecidability of classical predicate logic. Another choice, somewhere in between the two extremes, is given by *generalized assignment models*, which keep the "logical" requirements (1) and (2) while dropping the existential condition (3). The corresponding logic, proven decidable by Németi [174, 175] is called CRS (from "cylindric relativized set algebras"). But more generally, each of the axioms and rules in any standard axiomatization of first-order logic corresponds to an additional semantic condition on the update relation R_x: in [47], van Benthem gives a detailed such analysis for the axiomatic system in Enderton's standard textbook [128].

The Further "Dynamification" of Modal Logic and Inference

If we consider (as argued above, following van Benthem) that the "dynamic" character of a logic is witnessed by the match between its underlying mathematical structure and Relational Algebras, then basic modal logic (with its standard Kripke semantics) is insufficiently dynamic: there is ample scope for further "dynamification", by internalizing more of the relational structure and its natural operators. A classical example in this sense is PDL (Propositional Dynamic Logic), but van Benthem went further in this direction, by introducing *arrow logic* and *dynamic modal logic*. In addition, one can also "dynamify" our logics' consequence relation, obtaining a variety of *dynamic styles of inference*.

Arrow Logic. This is a line of research initiated by van Benthem [43] and Venema [201, 202]. This is a "truly" dynamic logic, in the same sense as Dynamic Predicate Logic: the meaning of a logical formula is given by a type of informational change, i.e. a set of "arrows". Arrows are abstractions of state-transitions (or "arcs" in graph theory), but they are considered as objects in themselves, which can be composed and reversed, have an identity element, and play the role of possible worlds (so that arrow formulas are evaluated *at* arrows). A special case is given by *pair models*: in these, arrows are identified with pairs of states (s, t), representing transitions between these states, so that the meaning of an arrow formula in a pair model is a binary *relation* on states (a set of pairs). If composition, reversal and identity are interpreted in the natural way (as relational composition, converse and identity relation) and no further existential assumptions are made (so that not all state relations are necessarily represented), then the complete arrow logic of pair models is axiomatizable and decidable [170]. Various extensions of arrow logic have been investigated, especially by researchers in Amsterdam, Budapest and Sofia. In particular, it has been shown that adding the requirement that composition is transitive leads to undecidable arrow logics. "Dynamic arrow logic" is obtained by adding an infinitary operator ϕ^* denoting the reflexive-transitive closure of the relation ϕ: once again, if no further existence assumptions are made, dynamic arrow logic is axiomatizable and decidable on pair models.

Dynamic Modal Logic (DML). This is a powerful type of logical formalism, originating in van Benthem's thinking about AGM Belief Revision and in his search for an abstract approach that could fully internalize the AGM-style dynamics. Introduced by van Benthem [40] and extended by de Rijke [180–182], who also did a thorough investigation of its meta-theory, its complexity and its applications, DML extends both (the star-free fragment of) Propositional Dynamic Logic and Intuitionistic Logic, as well as forms of preferential dynamics, which include standard belief-revision operators (such as expansion, contraction and other types of preference upgrades). The semantics is a combination of intuitionistic semantics (with an information preorder $\leq$ on states) and standard multi-modal Kripke semantics (with binary accessibility relations on states, denoting atomic actions). Like PDL, the syntax has a *static* component ("formulas" ϕ, that are to be interpreted as *sets of* *states* in the usual way), and a *dynamic* component ("procedures" α, interpreted as binary *relations* on states). The static repertoire of operators includes the *usual propositional and modal connectives* of multi-modal logic, and the dynamic repertoire includes *converse, intersection* (and thus also union) and relational *composition* (denoting the sequential composition of procedures). In addition, there are a number of "projections" (operators taking procedures into formulas) and "modes" (operations taking formulas into procedures). The projections include a *domain* operator $do(\alpha)$ (capturing the domain of the relation α), a *range* operator (capturing the range of α) and a fixed-point operator $fix(\alpha)$ (capturing the set of reflexive points of the relation α). The modes include the standard *test* operator ϕ? as in PDL, an *updating* operator $upd(\alpha)$ (which is a kind of "loose", non-deterministic expansion, relating a state x to any state $y \geq x$ satisfying ϕ) and a *downdating* operator (a "loose" contraction, relating x to any state $y \leq x$ satisfying ϕ). Using this language, one can define the usual type of *expansion* ("strict updating", relating x to the *lowest* state $y \geq x$ satisfying ϕ) and the usual contraction ("strict downdating"), as abbreviations. Interestingly, DML may be seen as an early precursor to more recent abstract approaches to Belief Revision theory, such as Segerberg's Dynamic Doxastic Logic [190–192], or the abstract update universes recently proposed by van Benthem [115] as an abstraction of Dynamic Epistemic Logic.

Dynamic styles of inference. In the context of the program of Logical Dynamics, the "dynamification" of consequence relations is a natural step. Driven by the spirit of Logical Pluralism, logicians produced in fact a bewildering variety of types and styles of dynamic inference. We follow here van Benthem's classification of this multitude, that uses the above-mentioned DML projections (*domain, range* and *fixed-point* operators) and modes (*test* and *update* operators). Though not exhaustive, this classification gives us a high-ground vantage point, from which we can get a better, clearer perspective of the field of possibilities. The *Update-to-Update* Consequence, introduced by van Benthem [43], has the shape $P_1 \circ \ldots \circ P_n \subseteq C$ (where both the premises $P_1, \ldots, P_n$ and the conclusion C are binary relations). The *Update-to-Test* Consequence, due to Veltman [199], has the shape $range(P_1 \circ \ldots \circ P_n) \subseteq fix(C)$. The *Update-to-Domain* Consequence was the one proposed by Groenendijk and Stokhof [143]: $range(P_1 \circ \ldots \circ P_n) \subseteq dom(C)$. The variant proposed by van Eijck and de Vries [125] corresponds to *Domain-to-Domain* Consequence: $dom(P_1 \circ \ldots \circ P_n) \subseteq dom(C)$. The *dynamic presupposition* introduced by Beaver [24] corresponds to *Domain-to-Test* Consequence: $dom(P_1 \circ \ldots \circ P_n) \subseteq fix(C)$. Finally, classical (Tarskian) inference can be recovered as the *Test-to-Test* Consequence: $fix(P_1) \cap \ldots \cap fix(P_n) \subseteq fix(C)$. Only the last one satisfies all the classical structural rules of inference. The others correspond to various "substructural" logics. In his book [47], van Benthem gives complete axiomatizations of the system of structural rules governing each of the above types of dynamic consequence, studies the translations between these various consequence relations, as well as their translations into more standard dynamic logics such as PDL.

Process Equivalence, Invariance, Characterization, Safety

All dynamic-logical frameworks mentioned above come with an implicit "dynamic" ontology, embodied in their formal semantics in terms of various types of processes, procedures, events, actions or programs. For each framework and each type of process, one can pose a number of key conceptual problems: first, answering the question "when are two processes the same?", by finding an appropriate notion of *process equivalence*. Processes that are structurally equivalent differ only by the irrelevant details of their chosen representation. The second task is *invariance*: finding logics that do not distinguish between equivalent processes. Only such logics are truly "dynamic": they express only properties that are invariant under equivalence, and hence belong to the process itself, rather than being dependent on a specific representation. The third task is to circumscribe the dynamic expressivity of some given, canonical logic, by finding a nice *characterization* of the maximal (most expressive) dynamic fragment of that logic (encompassing, up to logical equivalence, all the formulas that are invariant under process equivalence). The final problem is *safety*, which can be seen as an analogue of invariance for the *relational* side of a dynamic language: a safe operation is one that adds new relations to the structure, while keeping unchanged the notion of structural equivalence.

Bisimulation. Johan van Benthem is one of the co-discoverers of the concept of *bisimulation* (which he called "modal p-relation"), one of the fundamental notions in Modal Logic and Theoretical Computer Science. The relation of *bisimilarity* (defined as the existence of a bisimulation) is one of the most natural answers to the first question mentioned above: "when are two processes equivalent?" (Here, a "process" is just a "pointed Kripke model", i.e. a Kripke model with a designated state.) Bisimilarity is an essentially dynamic notion, capturing a notion of *behavioral (or observational) equivalence* between processes: *two processes are the same when they can simulate each others' behavior* (i.e. every "move" to another state in either process can be "matched" by the other process, step by step, in such a way that matching states satisfy the same atomic formulas). Note that the notion of bisimilarity is relative to the basic "dynamic vocabulary" (the underlying transition relations, the types of "moves" that have to be matched). To be precise, we should in fact always talk about *bisimilarity with respect to a given family of binary relations*: this matches only the relations in the given family, while disregarding other types of "moves".

Invariance. Bisimilarity poses a fundamental upper limit to the expressivity of modal logic: indeed, *modal formulas are invariant under bisimulation*. So modal logic is truly "dynamic" in the above-mentioned sense, i.e. it cannot distinguish between bisimilar processes: if two processes are bisimilar then they satisfy exactly the same modal formulas. The converse is true only for *infinitary* modal logic.

However, there exists one partial converse for finitary modal logic, known as the Hennessy-Milner Theorem: *two finitely-branching*[4] *processes are bisimilar if and only if they satisfy the same modal formulas.* One of the major contributions of van Benthem's Ph.D. thesis [26] was to show that the key property here is not a cardinality restriction (e.g. finitely-branching), but a logical "saturation" property, saying that all the appropriate consistent types of formulas are satisfied in the given structure. Johan van Benthem's theorem generalizes Hennessy-Milner's result[5] to all modally saturated models. This is a real, wide-ranging generalization: finitely branching models are modally saturated, but the converse fails.

Characterization. Another partial converse to the bisimulation-invariance of modal formulas is van Benthem's celebrated Modal Characterization Theorem (also known as the Modal Invariance Theorem): *a first-order sentence is (equivalent to) a modal formula if and only if it is invariant under bisimulation.* This beautiful result shows that *modal logic is the largest "dynamic" fragment*[6] *of first-order logic.*

Over the years, logicians introduced other notions of bisimulation appropriate for other logics, and computer scientists investigated other interesting notions of process equivalence. Analogues and extensions of van Benthem's Characterization Theorem were proved for many of these logics, as well as many of the other notions of bisimulation. We only mention here two such results. First, Janin and Walukievicz [160] gave a similar characterization of modal mu-calculus (obtained from modal logic by adding fixed points of monotonic operators definable by positive modal formulas): *mu-calculus is the largest "dynamic" fragment of monadic second-order logic.* Second, the analogue of van Benthem's theorem for the guarded fragment was proved by Andréka et al. [1], and this was later extended by Graedel, Hirsch and Otto to an analogue of the Janin-Walukievicz theorem for the fixed-point extension of the guarded fragment [141].

Safety for Bisimulation. An analogue of the notion of invariance for *program operations* (i.e. operations on binary relations) is the concept of safety, also due to van Benthem [51, 54]. An operation $O(R_1, \ldots, R_n)$ on programs is *safe for bisimulation* if, whenever two processes or models are bisimilar with respect to the relations $R_1, \ldots, R_n$, they are also bisimilar with respect to the relation $O(R_1, \ldots, R_n)$. E.g. the standard regular operations of PDL are safe for bisimulation. van Benthem's Safety Theorem is a kind of analogue for programs of the Modal Characterization Theorem: *a first-order definable relational*

---

[4] A Kripke model is finite-branching (or "finite-image") if every state has only finitely many immediate successors. A process, seen as a pointed Kripke model, is finitely-branching if the underlying model is finitely-branching.

[5] And in fact it predates it! So, since this result is a just an obvious special case of van Benthem's earlier work, it might be more appropriate to call it the van Benthem-Hennessy-Milner Theorem.

[6] Here, we follow the terminology introduced above, by calling "dynamic" the logics which do not distinguish between bisimilar processes.

operation $O(R_1, \ldots, R_n)$ is safe for bisimulation if and only if it is definable using atomic relations, PDL tests, union and relational composition.

"Meta-Logical Dynamics": Correspondence and Transfer

There is more to abstract dynamics than the relational semantics and modalization. Johan van Benthem is especially well-known for taking a "bird's-eye view" of logical systems: looking, not so much at one system at a time in isolation from others, but at *families of logical systems*, the way these logics *relate* to each other, and the way in which concepts, techniques and results can be *imported and exported* between logics. This is what one may call Meta-Logical Dynamics. It starts with van Benthem's Ph.D. thesis [26], which (together with Sahlquist's work, but independently from it) laid the foundations for *Modal Correspondence Theory*. By developing systematic ways of identifying for many modal logics their "corresponding" fragments of first-order logic, and using this correspondence to study completeness, canonicity and other important meta-theoretical properties of these logics, Correspondence Theory remains one of the cornerstones of modern modal logic. Analogues of this theory have been recently developed for other classes of logics. We mention here in this sense van Benthem's work on fixed-point logics and their correspondent fragments of second-order logic [67, 72, 104]. More generally, the study of *systematic translations* [33] and of *transfer results* between logics is central to the contemporary model-theoretic and algebraic[7] approaches to Logic. So is the search for "characterization results" giving structural and meta-logical conditions that characterize the expressive power of a given language inside another logic, or characterize a given logic among a *class of logics*. Many of the results mentioned in the above sections (including van Benthem's Characterization and Safety Theorems, the Janin-Walukiewicz theorem etc.), as well as many of the results presented in the rest of the volume, are examples in this sense.

After following the work of Johan van Benthem on abstract logical dynamics, it is now the time to look at the same story from the multiple perspectives of our invited authors. The papers gathered in part I of our volume explore in depth the mathematical-computational perspective on dynamics, connecting in interesting ways with some of the above long-standing research strands pursued by van Benthem.

3.2   The Invited Contributions on Mathematical and Computational Perspectives

The paper by Erich Graedel and Martin Otto is an insightful survey of the different notions of *generalized bisimulation associated to the various guarded fragments and their fixed-point extensions*, with a particular stress on the *complexity results* (e.g. the Graedel-Walukiewicz theorem on the decidability of the fixed-point

---

[7] See e.g. the work on Duality Theory, in the framework of universal algebra.

guarded logic) and on the *expressive-completeness results* that provide the appropriate analogues for these logics of the van Benthem Modal Characterization Theorem (or of the Janin-Walukiewicz theorem, in the case of the fixed-point extensions). As such, this area brings together some of the key concepts and model-theoretic methods from van Benthem's work (modalization, bisimulation, invariance and characterization, fixed points, generalized tree property etc.) with other, new and very powerful methods (primarily from automata theory).

Yde Venema's paper continues this strand at a more abstract level, by bringing together model theory, automata and the theory of *coalgebras*, developed over the last 20 years by category-theorists, computer-scientists and logicians. Coalgebras are maybe the most general working model for dynamical systems today, and they successfully generalize most of the classical modal notions introduced by van Benthem: they come equipped with a general notion of *coalgebraic bisimulation*, general ways to construct *coalgebraic logics*, general invariance and character-ization results, as well as natural ways to extend these results to fixed point logics by using the appropriate notions of *coalgebraic automata*. By choosing various concrete functors, one can obtain as special instances classical modal logic (and mu-calculus), neighborhood and topological semantics, game logics, probabilistic modal logics etc. Moreover, taking such an abstract perspective gives us more insight into results that used to appear impenetrably hard: Venema's coalgebraic analysis of the Janin-Walukiewicz proof of their (mu-calculus analogue of van Benthem's characterization) theorem gives maybe for the first time an easily readable and conceptually simple rendering of this proof, while also showing that it is based on a much more general argument of a coalgebraic nature.

The paper by Balder ten Cate and Phokion Kolaitis is a fascinating exploration of the "adventures of logical dynamics" in *database theory*. Databases can themselves be considered as a very general model for information and knowledge representation, hence developing an abstract dynamics on databases seems to us an essential contribution towards a better understanding of information change and knowledge update. The notion of "schema mappings" embodies this form of abstract database dynamics, in the spirit of *abstract model theory* and of van Benthem's work on correspondence, structural mappings/relations and transfer of results between logics. The paper shows how these abstract-logical notions can be seen to play a key role in the understanding of down-to-earth manipulation of databases, such as data integration, data exchange and other data-interoperability tasks. Very much in the spirit of van Benthem's model-theoretic work is also the smooth interplay between the *structural properties* of schema mappings and their *syntactic specifications* in various schema-mapping languages. Once again, the model-theoretic aspects tie up well in this work with the *complexity-theoretic* ones, which are obviously of crucial importance to database theory.

The last three papers in this part form a bridge between the abstract-logical dynamics that is at the core of the mathematical-computational perspective and the more concrete informational dynamics explored in the next parts of our volume. The paper by Pietro Galliani and Jouko Vaananen investigates from a logical point of view the notion of *dependency*, essential for both database theory and mathematics at large. Dependency is an intrinsically dynamic notion, being deeply connected to van Benthem's vision of variable substitution as "update". In this context, dependency is expressed by correlated updates, or if you like by the "non-local" effect of a local update: changing the value of one variable affects other (dependent) variables. Indeed, this is exactly how dependency manifests itself as "entanglement" in quantum mechanics: a local measurement may induce non-local changes. The authors' axiomatic investigation of the abstract logic of dependency and independence is enthralling. But, beyond the abstractions, they also provide an interpretation in terms of beliefs and belief dynamics, that points in the direction of the concrete dynamics of multi-agent information flow explored in part II of this volume. Moreover, the game semantics of (in)dependence logic ties this up with the strategic dynamics in part III.

Samson Abramsky's paper is a conceptual exploration of a question that is central for Computer Science, as well as for logical dynamics: *what is a process?* In contrast to the purely extensional and well-established notions in Computability theory ("computable function", "recursively enumerable set"), the notion of algorithmic/computational process is *intensional*, and covers a bewildering multiplicity of process calculi and other approaches. The conceptual elucidation of this notion is closely related to the study of appropriate notions of *process equivalence*, study that as we saw was initiated by van Benthem (among others) and actively pursued in his subsequent work. As shown by Abramsky, this leads to the "full abstraction problem": finding a good representation of all relevant processes, that gives a common (invariant) representation to equivalent processes. The paper traces a success story, namely how Abramsky's *game semantics* provides an elegant resolution of this problem for *sequential functional processes*. At the same time, the *concrete* nature of game semantics and especially its multi-agent and strategic features lead us beyond the purely logical dynamics, connecting to the logics of game-theoretic interactions investigated in part III.

Finally, the contribution by Hajnal Andréka and Istvan Németi can be considered as a breathtaking exercise in Meta-Logical Dynamics, in the spirit of van Benthem's work on translations [33], transfer and correspondence between logics [26]. Indeed, the paper gives a systematic comparison between apparently very different first-order logic axiomatizations of Special Relativity Theory, each of which seems to talk about different kinds of objects: the logic SPECREL developed in Budapest is about reference frames, Ax's Signaling Theory is about particles and signals, Goldblatt's approach focuses on the geometry of orthogonality, while the key notion of Suppes' theory is the Minkowski metric. But the paper uncovers the deep underlying unity of these theories, and explores in detail their interconnections, using definability theory (in a version that allows defining new entities, not only new predicates). A byproduct of this investigation is the development of a *concrete operational semantics for special relativity theory.* Overall, this paper is an excellent example of an application of logical techniques to the Foundations of Physics as well as to the Philosophy and Methodology of Science. At the same time, this paper can be said to belong to the *concrete* informational dynamics that forms the topic of the next part of our volume.

4 Informational Dynamics: Time, Space, Knowledge and Belief

We move now to the part of Johan van Benthem's work, and to the contributions in our volume, that deal with the logical understanding of *"concrete" informational dynamics*: the way information flows between actual epistemic agents through "real" channels (sometimes spatially located), potentially generating a temporal succession of "informational events".

4.1 Johan van Benthem on Informational Dynamics

Dynamic Epistemic Logic

While originating in the work of Gerbrandy and Groeneveld [132] on public announcements (independent of, but anticipated by Plaza [177]), and having acquired its standard framework through the work of Baltag et al. [10] on epistemic action models, the field of Dynamic Epistemic Logic (DEL) has drawn inspiration from van Benthem's ideas from the very beginning (especially through his influence on the work of Jelle Gerbrandy, Alexandru Baltag and Hans van Ditmarsch). Johan van Benthem is also the most influential champion of the DEL approach, as well as one of the most active contributors to this field (through a long sequence of papers [56, 62, 64, 80, 86, 90], including some with many of his Ph.D. students [66, 82, 98, 92–94, 101, 105, 119, 134, 157, 158, 198, 203] and other collaborators [76, 89, 106, 114], as well as through his recent book [102]).

DEL is in fact, not one logic, but a family of logics, or rather a *general type of logical approach to information flow*, that subsumes many logical formalisms. In its most common form, Dynamic Epistemic Logic [10, 123] combines the syntax of epistemic and dynamic logic, having both knowledge modalities $K_a\phi$ (asserting that ϕ is known to agent a) and dynamic modalities $[e]\phi$ (asserting that ϕ becomes true after some epistemic event e). However, at a semantic level, DEL combines the standard relational semantics for static epistemic logic (based on epistemic Kripke models) with a *non-standard semantics for events or "actions"* (based on *model transformers*, i.e. relations or functions *between* models, rather than the usual dynamic relations within a given model). The simplest such epistemic action is the *public announcement* !P (also known as "update"), corresponding to the simple old idea of *learning by elimination of possibilities*: after P is learnt, all the non-P worlds are deleted. Known as "conditioning" in Probability Theory and Belief Revision theory, this move was really formalized as an *action* (i.e. model transformation) only in DEL. As a consequence, DEL is the first approach that uncovered the ambiguity underlying the old terminology, by distinguishing between "static" and "dynamic" conditioning. This simple distinction is at the basis of Gerbrandy's solutions to the puzzles posed by Moore sentences and the Muddy Children, as well as of van Benthem's solution to Fitch's Knowability Paradox and his study of "self-fulfilling" and "self-refuting" sentences.

More complex epistemic actions are the *fully private announcements* due to Gerbrandy [131], the fair game announcements studied in [122] and [9], and more generally the *epistemic action models* introduced in [10] and renamed "event models" by van Benthem. These last ones can represent complex multi-agent scenarios that are (partially) visible to some and (partially) opaque to other agents, or even potentially deceiving to some of them. An operation known as "product update" is used to combine the original (static) epistemic model and the event model into a new, updated (static) epistemic model, that accurately represents the agents' knowledge *after* the event.

Among the many other contributions of van Benthem to the development of DEL we mention here: the novel and conceptually fertile notion of *conditional common knowledge* [76], introduced initially by van Benthem as a tool for the complete axiomatization of the logic of public announcements and common knowledge; the study of the limit behavior of iterated public announcements and their application to game-theoretic notions [79]; the study of the dynamic-epistemic logic of distributed knowledge and of the "actualization" of distributed knowledge via public communication [73]; the investigation of the properties of the *dynamic inference* system induced by public announcements [63]; the introduction (in joint work with van Eijck and Kooi) of a hierarchy of "levels of conditional knowledge" known as *epistemic PDL*[8], and its use as a static logical basis for a new axiomatization of DEL [76]; the subsequent investigation of stronger logics that are "product-closed" (i.e. closed under product update with any epistemic actions): not only epistemic PDL, but also epistemic mu-calculus [88] and other logics; the extension of DEL to fact-changing events [76]; the exploration of games, strategies, rationality and game-theoretic solutions using DEL [56, 77–79]; analogues of DEL for preference change [82, 77]; the systematic comparison and merge of DEL with Epistemic Temporal Logic [92] and with other frameworks for interaction such as STIT logics [112]; the dynamic logic of questions and issues, leading to the development of "interrogative DEL" [105]; extensions of DEL dealing with the inferential dynamics and awareness [101], as well as the evidential dynamics and evidence-managing actions [106]; etc.

Finally, we want to stress one other line of research within DEL, to which van Benthem made key contributions: the development of *probabilistic versions of DEL*. Kooi [165] was the first to introduce a Probabilistic Public Announcement Logic, while other authors [2, 16] developed belief-revision-friendly versions dealing with 'surprise' events (of probability 0). Kooi's setting was later extended by van Benthem [60] by decisively enriching event models with *occurrence probabilities* (a probabilistic version of the usual notion of *precondition* of an action). A full-fledged Probabilistic Dynamic Epistemic Logic, obtained by adding to the above setting *observation probabilities* (as a probabilistic version of the doxastic accessibility relations on events in standard DEL), was developed by van Benthem et al. [93], and was later extended to infinite models by Sack [187].

---

[8] This hierarchy generalizes Rohit Parikh's "levels of knowledge".

Epistemology and Belief Revision

Johan van Benthem played a central role in the investigation of dynamic logics for Belief Revision Theory (BRT), and especially in the development of a "belief-revision-friendly" version of DEL: essentially, it seems to us that the two fields are currently in the process of merging. The first step in this direction (using a quantitative product update rule based on Spohn ordinals) was taken by Guillaume Aucher under van Benthem's supervision [2]. The next most significant step was the introduction of a qualitative logic of conditional beliefs and its use as a static basis for formalizing updates (public announcements) over belief-revision structures (called "plausibility models"): this was done independently at about the same time by van Benthem [80] and Baltag and Smets [13]. Moreover, van Benthem [80] formalized "softer", more "revisable" versions of learning ("upgrades"), by internalizing in the logic some of the most popular revision methods used in BRT (Spohn's lexicographic revision and Boutilier's minimal revision). Baltag and Smets went on to propose a qualitative product update construction for general belief-revision with arbitrary event models [14, 15, 17], construction that was later called "Priority Update" by van Benthem, who generalized this idea further to a conception of "belief revision as a special case of preference merge", using concepts from Social Choice Theory [81]. In a separate line of work with Liu [82], van Benthem proposed a different kind of generalization of belief upgrades, in the form of relational transformers expressible in PDL format; this idea was later combined with epistemic PDL and with fact-changing actions (substitutions) by van Eijck [76], and further generalized in the so-called GDDL by Girard et al. [138].

Over the years, van Benthem has applied dynamic-logical concepts and techniques to the elucidation of some of the core issues in Epistemology [71]. In [64], he proposed an analysis of Fitch's Knowability Paradox, based on a "dynamification" of the possibility operator used in Fitch's argument: interpreting "ψ is knowable" ($K\phi$) as "ϕ could become known by learning some more information" ($\exists P[!P]\phi$). On the technical side, this lead to the introduction and axiomatization of the so-called "arbitrary announcement" modality, by Balbiani et alia [4]. On the conceptual level, it allowed van Benthem to give a sophisticated dynamic-epistemic treatment of the Fitch Paradox. In other work [95] he dealt with the problem of logical omniscience, by developing a non-omniscient version of DEL, that can capture the dynamics of inferential actions, such as the deduction of a new fact by the application of an inference rule, or the learning of a new rule etc. In work with Pacuit [106], he developed a version of DEL that combines belief revision with evidence-management actions, getting a better hold on the philosophical concepts of justified belief by making explicit the evidential basis of our beliefs and the way in which evidence-gathering can lead to belief revision. In subsequent work [102], he took a new look at current epistemological debates from a belief-revision perspective: at the role played by informational correlations and dependencies in the establishment of knowledge; at the use of the notion of "safe belief" introduced by Baltag and Smets [17], as a first step towards a more general formalization of knowledge as "dynamic robustness" of belief (in the face of new informational events); at the importance of falsification and revision for epistemic logic. These themes came together in the so-called "Erlangen program for epistemology" proposed by Baltag [6], as a van Benthem-inspired epistemological version of Felix Klein's famous Erlangen program for mathematics. Recently, this line of investigation has led to a general conception of knowledge as a "dynamic-doxastic equilibrium", conception developed in van Benthem's forthcoming book [114]. The idea is that various types of "acceptance" (belief, strong belief, safe belief, various forms of knowledge) are characterized in terms of being invariant to different types of dynamics: they are "fixed points" of specific types of informational events. (You "know" something when it is redundant to learn it; you believe something when it is redundant to be persuaded of it.) Beyond this "fixed-point theory of acceptance" (with roots in classical epistemological texts by Lehrer, Klein, Hintikka, Stalnaker, Rott etc.), the book calls for a change of focus (from the "static" concepts of knowledge or acceptance) to the notion of *epistemic interaction*, as a self-correcting, self-testing, evidence-gathering, truth-seeking, reality-oriented, socially-involved process of iterated belief revision. Other lines of van Benthem's thinking on the dynamic-logical aspects of epistemology were further pursued by his Stanford students Wes Holliday, Tomohiro Hoshi and Thomas Icard [149, 157, 150, 151].

Long-Term Doxastic Dynamics: Time, Protocols, Learning

We arrive in this way at the notion of *belief-changing process* that unfolds over *time*. Instead of the one-step input-output perspective of standard BRT (and of standard DEL!), we are now looking at a temporal succession of informational events, or even a branching tree (or forest) of such events. Johan van Benthem, well-known in his youth for his contributions to temporal logic [29, 32] (especially to the logic of *time intervals*), has returned in the last years to the study of time after a long detour through dynamic logics of action. But now the focus is on temporal aspects of information flow. In a series of joint papers [85, 92, 96], van Benthem and his collaborators gave an explicit formalization of the implicit temporal processes generated by DEL-style event models; compared this setting with *epistemic-temporal logic* (ETL) as investigated by Parikh and Ramanujam [176] (or in a different version by Fagin et al. [129], under the name of "interpreted systems"), by showing how DEL-generated models can be embedded in ETL forests; characterized the forests corresponding to DEL-generated models (for both the classical version of DEL with "hard" information, and the belief-revision-friendly version with "soft" information), in terms of general and natural semantic constraints (e.g. Perfect Recall, "No Miracles" etc.); compared the complexity and the expressive power of the various DEL-related fragments of ETL; extended DEL by adding procedural information about the long-term constraints of the given informational processes, obtaining the notion of *DEL protocol*, and axiomatizing the resulting dynamic protocol logic; used both temporal and fixed-point extensions of these logics to express long-term properties of epistemic protocols; and applied this framework to the analysis of game trees, extensional games and strategies.

Formal Learning Theory originates in the computational investigation [139] of the long-term truth-tracking power of the various "learning methods" that are available to a single agent observing an infinite stream of incoming data. In recent years, it has been used for epistemological purposes by Kelly [163] and Hendricks [152, 153]. Its main advantage is that it doesn't make any "rationality" presuppositions about the learning methods (unlike DEL, the AGM approach to Belief Revision, or the Bayesian theory of credence update), thus being able to compare and evaluate various paradigms from a "neutral" point of view: the only criterion for success is... success! (At truth-tracking, of course.) A first learning-theoretic analysis of updates was done by van Benthem's students Gierasimczuk and Dégrémont [120], and a more thorough investigation of the learning-theoretic power of various (DEL-versions of) belief-revision methods over plausibility models was done by Gierasimczuk in her Ph.D. thesis [134] under van Benthem's supervision , as well as in her joint work with de Jongh [135] and with Baltag and Smets [21].

Generalized Structures for Evidence-Based Knowledge: Spatial, Topological and Neighborhood Models

From a conceptual perspective, the epistemic notion of "accessibility" can be understood at a first approximation in terms of a spatial-topological notion such as "closeness". From a purely mathematical perspective, epistemic logic can be considered as a special case of "spatial" logics: Kripke semantics for the epistemic logic S4 is just a type of *topological* semantics (itself a special case of the more general neighborhood semantics) and the "knowledge" modality is a special case of the topological *interior* operator [197].[9] Philosophically, the topological semantics models a notion of "evidence-based knowledge": the open sets represent the *agent's evidence*, and hence according to this interpretation a proposition (set of possible worlds) P is "known" if there exists some "true evidence" (i.e. an open set O containing the real world w) that *entails* P (i.e. $w \in O \subseteq P$).

Johan van Benthem's interest in spatial logics is in fact deeper and goes way beyond the needs of epistemic logic [58, 59], but he has been particularly interested in the topological interpretation of knowledge, and its generalization to neighborhood semantics. In joint work with his Stanford student Sarenac [70], he investigated the properties of common knowledge in a topological setting. Starting from the work of Barwise [22] on distinguishing the various concepts of common knowledge (the fixed-point notion and the countable iteration version) that are

---

[9] As for doxastic logic, the "belief" modality is, depending on one's favorite interpretation, either an instance of the topological co-derivative operator [196] or of the closure-of-interior operator [7].

lumped together by the Kripke semantics, they proposed a new topological concept of common knowledge (given by the standard product topology, induced by the topologies underlying each agents' knowledge), stronger than the others, and closer to Lewis' original conception of "having a shared situation". This line of inquiry was pursued further in Sarenac's Ph.D. thesis [189]. van Benthem's Amsterdam students Raul Leal and Jonathan Zvesper went on to propose a topological version of DEL, obtained by generalizing the update product operation using the product topology, and to axiomatize (the finite-event fragment of) this logic, as well as its even more general topological version [167, 203].

In recent joint work [106, 110, 116], van Benthem and his collaborators use neighborhood semantics to develop a very general model for *evidence, evidential dynamics* and *justified belief*. In fact, the authors show that the evidence structure induces in a natural way a plausibility order on possible worlds, which encodes, not only beliefs, but also a system of *belief revision*.[10] The neighborhoods are once again interpreted as pieces of "evidence" possessed by the agent, but now this can also be *false* evidence: unlike topological neighborhoods, general neighborhoods of a given point (world) may fail to contain that point. Moreover, the available pieces of evidence might be *mutually inconsistent*; but nevertheless van Benthem and Pacuit show how this will still give rise to *consistent beliefs* in a natural way.[11] Finally, unlike in a topological setting, the agent cannot always "combine" two pieces of evidence into one piece. This captures a form of bounded rationality. The authors study the dynamics of evidence induced by various actions: updates, upgrades, combining pieces of evidence, acquiring new evidence etc.

Once again, we see how the search for the right level of abstraction and for the best logical-mathematical framework for dynamics continues, now even in the context of "concrete" informational dynamics. Indeed, as we already saw in section 2.2, the back-and-forth shift of perspectives between abstract and concrete dynamics represents a constant feature of van Benthem's work across many decades, and so this theme will reoccur throughout this volume. But now is the time to look at informational dynamics from our invited authors' perspectives.

4.2   The Invited Contributions on Informational Dynamics

In his penetrating contribution, Jan van Eijck gives an excellent survey of the field of Dynamic Epistemic Logics, with a particular focus on epistemic PDL and fact-changing actions (substitutions), and their connections with logics for computer programs, such as Hoare logic and Propositional Dynamic Logic. This sweeping

---

[10] Interestingly enough, the resulting Belief Revision Theory does not necessarily satisfy all the standard AGM postulates. In recent work by Fiutek [130], this setting is generalized further, and ways to regain AGM-type revision are proposed.

[11] In this context, "having evidence for *P*" is not the same as "believing *P*". The agent only believes the propositions that are entailed by all the (conjunctions of) maximally consistent families of evidence. (Here, "maximal" is taken in the sense of inclusion).

bird's view of DEL is not limited to the purely theoretical work, but is complemented by some beautiful applications to practical computational tasks. Indeed, the author nicely illustrates DEL theory with applications to *navigation problems* (on a grid) and to *epistemic planning* using DEL protocols. He ends with presenting a long-standing open problem (finding a structural characterization of when two epistemic action models induce equivalent updates), some partial solutions to this problem proposed in his own joint work [126, 193], and a brief note on some of his very recent results on probabilistic DEL [127].

The paper by Patrick Girard and Hans Rott has two sides. On the one hand, the paper contains an insightful philosophical reflection on the epistemological significance and the interpretation of the main concepts and results from Belief Revision Theory (BRT) and Dynamic Epistemic Logic (DEL), including discussions of the negative introspection for knowledge, the mis-identification of Segerberg's irrevocable belief with knowledge, and the Limit Assumption. On the other hand, the paper also includes a sophisticated technical contribution to the dynamic logic of belief revision, based on having a static logic with doxastic modalities for *strict and non-strict plausibility* and for *epistemic indistinguishability*, and a dynamic logic using *doxastic PDL-transformations* (in the style of Girard et al. [138]), to simulate operations such as expansion, revision, contraction and the very interesting *two-dimensional belief change operations* introduced by Rott [183]), e.g. *bounded revision* and *revision by comparison*. This is overall a fascinating contribution to the on-going efforts towards building bridges between formal logic and Mainstream Epistemology.[12]

The next three papers are about the *long-term temporal* aspects of epistemic dynamics. First, the chapter by Valentin Goranko and Eric Pacuit is a comprehensive and very insightful overview of the various approaches to the *temporal logic of knowledge*, and of their mutual relations. After giving an introduction to epistemic logic on the one hand, and to (basic) temporal logic on the other hand, the authors discuss the logics obtained by simple *fusions* between the two. They move on to what is by now the *standard* approach to temporal-epistemic logic in Computer Science, the framework of *interpreted systems*, due to Halpern and Vardi and developed further in the classic [129]. They then present the protocol-based approach of *Epistemic Temporal Logic* (ETL), introduced by Rohit Parikh and Ram Ramanujam, after which they look at temporal frameworks with *uncertainties between histories* (closely related to the STIT models widely used in philosophical literature, and which form the topic of the chapter by Ciuni and Horty in this volume). Next, they take a look at *DEL and DEL protocols*, DEL's embedding into ETL, and van Benthem's abstract characterization of the ETL models that are generated by uniform DEL protocols. Then they briefly present logics (such as ATL or STIT) that investigate the *interaction between players'*

---

[12] Tensions between the two sides of the paper surface occasionally: e.g. world elimination and the use of the word 'knowledge' for an *S*5-modality are criticized in the philosophical part, then re-adopted later. But these tensions reflect precisely the intriguing challenges of working at the interface of logic and epistemology.

knowledge and their abilities to achieve their goals in games. They discuss the differences between players' *a priori information* (that they have prior to the actual play) and their *empirical information* (gained during the play). Finally, they present the work of Sack, Yap, Hoshi and Renne [159, 178, 179, 185, 186, 188] on Temporal DEL (TDEL). Overall, the amount of material covered in this chapter is truly breathtaking, and amazingly this is done without sacrificing in any way the rigour and clarity of exposition. On the contrary: the paper's systematic comparative discussion of all these approaches is extremely insightful, bringing much-needed clarity and order to (what at first sight looks like) the lush impenetrable jungle of temporal-epistemic logics.

In their contribution, Nina Gierasimczuk, Vincent F. Hendricks, and Dick de Jongh map the exciting new research area opened up in recent years at the interface between Formal Learning Theory and Logic. After presenting the work of Kelly [161, 162], Martin and Osherson [171, 172], Kelly et al. [164] on the connections between Belief Revision Theory, Logic and Learning Theory, the paper focuses on the ideas and results uncovered in the recent work done in Amsterdam by Gierasimczuk and her collaborators [21, 120, 121, 133, 133, 135, 136], under the influence of van Benthem's "dynamic turn" in Logic. This includes results on the learning-theoretic power of standard belief-revision methods (conditioning, lexicographic revision and minimal revision) over epistemic spaces, on computable learning by conclusive updates, on preset learning and fastest learning, and on capturing notions of learnability in both Dynamic Epistemic Logic and Temporal Epistemic Logic. The paper ends with a discussion of the significance of this line of research for scientific methodology and other key issues in Philosophy of Science.

Kevin Kelly's paper is a far-reaching attempt to give a new foundation for Epistemology, based on a *new semantics for inductive knowledge*. Solidly anchored within the learning-theoretical approach to Epistemology championed by Kelly in his previous papers and in his book, the chapter looks at an epistemic agent as a "learner", who is faced with a potentially infinite sequence of incoming data and who is continuously (re)computing her beliefs based on the available information. The beliefs are represented in a "hyper-intensional" way, so that there is no logical omniscience problem. The paper argues that the sensitivity and safety conditions proposed by Nozick, Sosa and others are too strict to be useful for empirically-based inductive knowledge, especially in their "negative" side (the requirement that, if the given belief were false, the agent wouldn't believe it). Instead, the author proposes a notion of knowledge that combines *true stable belief*[13] at the current moment in the actual world with *avoidance of error in the limit* (rather than at the current moment) in counterfactual worlds: if the belief were false, the agent would *eventually* stop believing it (possibly after a time lag, used for acquiring more data).

---

[13] This fits well with the "defeasibility" theory of knowledge, shown by Gierasimczuk to agree with the learning-Theoretic notion of identifiability in the limit. See the chapter by Gierasimczuk, Hendricks and de Jongh.

A formalization of this conception is proposed, using a complex multi-modal language (with no less than 10 modal operators!), whose semantics is given in terms of *computational learning models*. Kelly uses this setting to address an impressive number of well-known epistemological issues and paradoxes: the problem of inductive skepticism, Fitch's paradox, Duhem's problem, deductive closure of knowability, the questions whether inductive knowledge is deductively closed and/ or (positively, or negatively) introspective, etc.

The chapter by Wiebe van der Hoek and Nick Bezhanishvili is a much-needed survey of the vast *landscape of epistemic structures*. Starting with epistemic (and doxastic) Kripke models, as well as their multi-agent generalizations, going then to epistemic temporal models and interpreted systems, and finally moving to the more general *topological and neighborhood models for knowledge and belief*, the authors give a sweeping overview of the wide range of mathematical structures used in epistemic/doxastic logic. Among other things, they touch on the fascinating topics of the *spatial aspects of knowledge* and of the connections between *epistemic logic and spatial reasoning*. In addition, they give a very good introduction to the more technical aspects of epistemic logic: for each of these types of structures, they explain the appropriate notions of canonical models, completeness proofs and bisimulation, and present the appropriate results on definability, expressivity and characterization. They also give an insightful discussion of more conceptual issues, such as: the way in which topological structures can represent the subtle differences (first stressed by Barwise [22]) between various notions of common knowledge, including the work by van Benthem and Sarenac on modeling common knowledge using topological products [70]; the problems raised by the topological semantics of knowledge, etc.

The last two papers of this part are about the *logic of probabilistic beliefs*. Lorenz Demey and Barteld Kooi give a conceptually insightful and technically sophisticated overview of *probabilistic epistemic logics* and of their *dynamic extensions*. They later include Probabilistic Public Announcement Logic, as well as the full-fledged Probabilistic Dynamic Epistemic Logic of van Benthem, Gerbrandy and Kooi. The authors discuss subtle features, such as the problems posed by updating with higher-order information, which requires a distinction between *actual* public announcements (the "real thing") and *hypothetical* announcements (essentially equivalent to simple Bayesian conditioning). They explain the concept of *occurrence probability* and its use in the concrete applications of Probabilistic DEL, and discuss van Benthem's recent thoughts [109] about the lessons that the usual, qualitative DEL might still have to learn from these probabilistic notions. At the end they tackle a number of technical and conceptual issues, among which we mention the problems posed by learning surprising information (i.e. which has prior probability 0), the game-theoretic applications of probabilistic DEL (in particular, gaining a better understanding of Aumann's well-known "Agreeing to Disagree" theorem), and the so-called *Lockean thesis*, which identifies belief and "high probability" (above a certain threshold). They discuss the so-called *Lottery Paradox*, showing that the Lockean thesis in its original form is incompatible with the commonly accepted principle of *additivity of beliefs*.

Hannes Leitgeb's paper starts where the one of Demey and Kooi ends: with the problems posed by the Lockean thesis and by other attempts for *connecting the qualitative and the probabilistic accounts of belief*. In a series of recent papers, Leitgeb has developed and defended a principled solution: the *stability theory of belief*, requiring that the set of doxasticaly-possible worlds must be "stable", i.e. its subjective probability *stays high* after learning any new information which doesn't contradict it. In this paper, Leitgeb first pinpoints the fundamental intuition that lies behind the Lockean thesis and other natural attempts for unification: the intuition that *belief ought to be understood as a "simplification" of subjective probability*. He then proposes a precise mathematical formalization of the concept of "simplification", in terms of approximating probability by means of belief or disbelief. Finally, he shows that the stability theory of belief fits the bill: *the "best approximations" of the given subjective probability measure are exactly the stable sets*. It is worth noting that, even beyond the specific justification given here, Leitgeb's theory fits perfectly with van Benthem's Logical-Informational Dynamics and the above-mentioned "Erlangen program for epistemology" [6], underlying the conception developed in van Benthem's forthcoming book [114]. Recall that, according to this program, various types of "acceptance" are characterized in terms of *invariance* under various types of dynamics. Lehrer's defeasibility theory, as well as the simplified version identified by Rott as the "stability theory of knowledge", fit well within this general "fixed-point" conception. Though developed independently from the above-mentioned ideas and motivated by a very different research agenda, Leitgeb's theory of belief *could* be seen as one of the most beautiful and philosophically richest developments of this general program.

5 Games

We move on now to a more sophisticated kind of dynamics, obtained by adding to the picture the agents' *preferences* or payoffs. The assumption of "rationality" converts preferences into *intentions* or goals: a rational agent intends to maximize her payoff. The dynamics becomes goal-oriented: this is what van Benthem calls "rational dynamics". The interplay between information, actions and preferences leads to the notions of *decision problem* and *planning*, and in the multi-agent case to the concepts of *game* and *strategy*. Games are temporal-epistemic-payoff structures, in which strategic reasoning, competition, cooperation, equilibrium, the formation of coalitions and other complex dynamic-epistemic-intentional phe-nomena play key roles.

5.1   *Johan van Benthem's Work on Logic and Games*

We are thus now in the realm of Game Theory, but this doesn't necessarily mean that we must leave Logic behind: on the contrary, the focus is on the *interplay* between the two. Johan van Benthem's long-standing interest in the issues lying at the interface of Logic and Game Theory has manifested in a long series of papers [37, 42, 56, 57, 61, 65, 66, 68, 69, 74, 75, 78, 79, 83, 87, 98, 107], culminating in his latest book *Logic in Games* [113]. According to his view, "games provide a rich model for cognitive processes, carrying vivid intuitions" (op. cit, p. 7), which explains why they provide a crucial tool for the study of informational dynamics.

Logical games versus the logic of games. The first important distinction made by van Benthem is between on the one hand the view of "logic *as* games", view embodied in a multitude of *logical games*, and on the other hand the logic *of* games, approach that manifests itself in a variety of logics for reasoning about games. While the first is a game-theoretic approach to logic, the second provides a logic-based perspective on games.

Games in Logic. In his work [37, 38, 113] van Benthem has looked at various types of argumentation games (e.g. Lorenzen's dialogue games) and (related to this) proof-theoretic games; at semantical evaluation games (including non-classical versions such as the Hintikka-Sandu semantics of independence-friendly logic using games of imperfect information); at model-checking games; at model-comparison games (e.g. the bisimilarity game) and model-building games etc.

New games for new logics: from Sabotage to Learning. van Benthem is also an inventor of new games, usually each coming with its own logic. Besides a number of "knowledge games" [48, 56], the most interesting such example is van Benthem's Sabotage Game [68], which is a "gamification" of a known search problem (the Graph Reachability problem). The Traveler is trying to reach a certain node of a graph by navigating across the connecting arrows, while at each step the Saboteur can cut some connection (to try to prevent the traveler from reaching its destination). A more abstract interpretation (with changed payoffs) gives rise to the Learning Game (in which the Saboteur is replaced by a Teacher, who is trying to "help" the Student stay on the right path by cutting "wrong" connections). There is an associated logic (the Sabotage Modal Logic), and numerous generalizations: any algorithmic task can be "gamified" into a sabotage version by adding an obstructing player.

Logical foundations of games. In a number of papers, van Benthem investigates abstract logical foundations of game-related concepts. Crucial earlier themes, that dominated abstract logical dynamics, return in more concrete game-theoretic incarnations: notions of game equivalence (when are two games the same?); natural operations on games; connections between games and process-graphs; the use of fixed-point logics on game trees, for expressing notions of equilibrium, Backwards Induction or other solution concepts [56, 57, 78, 79, 98].

Towards a "Theory of Play": multi-modal dynamic logics for games. Going towards more concrete features of games, van Benthem and others have looked at multi-modal logics that combine various types of modalities to capture different aspects of game-theoretic interactions: game logics based on dynamic-epistemic logic [56, 87, 113]; the use of belief-revision conditionals to capture strategic reasoning; modal logics of preference and best action [77]; combinations of temporal and epistemic logics; logics that make explicit the role of strategies as first-class objects [108, 113]. In more recent work [103, 107, 113], van Benthem and his collaborators argue that the combination of Logic and Game Theory that has developed in the last decade is about to reach a critical mass, giving rise to a new field: "we are witnessing the birth of something new which is not just logic, nor just game theory, but rather a Theory of Play" [107].

We have now seen how van Benthem's conception about games and logic has unfolded, from logic games to game logics, from knowledge games to the "gamification" of algorithmic tasks, and from the logical foundations of traditional Game Theory to the new modal foundations of a Theory of Play. It is time to look at the story of games and logic from the perspective of our invited authors.

5.2   The Invited Contributions on Logic and Games

The chapter by Giacomo Bonanno and Cédric Dégremont is an excellent survey of the work on the "Theory of Play" by van Benthem and his collaborators [107], as well as a thorough comparison of this work with other logic-based approaches to games and strategic interaction. The paper contains insightful discussions of *key conceptual issues* at the interface of game theory and logic: the problem of identifying the best modal languages to reason about extensive games; the relations between logical characterizations of game-theoretic concepts and their computational analysis; how van Benthem's ideas give a new perspective on the question of under what conditions two games can be considered the same; when and how can the convergence of iterative solution concepts be analyzed in fixed-point modal languages, in the style of van Benthem in [79]; how the results of van Benthem and Geerbrand [98] on characterizing backward induction in fixed-point languages can be used to give different interpretations to this solution concepts; etc.

The paper by Thomas Ågotnes and Hans van Ditmarsch is a fascinating investigation in the spirit of van Benthem's Theory of Play. The authors present new results on the distinction between knowledge *de dicto* and *de re* for individual agents and for coalitions; they use dynamic modalities that *quantify* over various classes of public announcements to capture the group's *coalitional abilities in knowledge games*. They also isolate and study two interesting classes of such games: *public announcement games* and *question-answer games*, and finally they look at the "state of the art" in the dynamic-epistemic analysis of some "real" games (such as chess, bridge, cluedo, pit and sudoku).

Sergei Artemov's powerful piece gives a refreshingly different perspective on games: while most logics for games are developed in a semantic manner, the author adopts a *proof-theoretic* approach. More precisely, he proposes a *"Syntactic" Epistemic Logic* for games, in which the conclusions are deduced directly from the syntactic description of the game (rather than from the, typically huge, model of the game). In this setting, Artemov formalizes Nash's notion of "definitive solution" of a game (prescribing that "a rational prediction should be unique, that the players should be able to deduce and make use of it") and studies the existence of such definitive solutions, showing that their meaning depends on the underlying notion of rationality. For Aumann rationality, they are *not* equivalent to Nash equilibria: games with multiple Nash equilibria cannot have definitive solutions, and some games with a unique Nash equilibrium have definitive solutions, but others don't. However, each Nash equilibrium can be a definitive solution for an appropriate refinement of Aumann rationality. The conclusion, similar to the one of Aumann [3], is that *equilibrium is not the way to look at games*; the most basic concept should be: to maximize your utility *given your information*.

In his beautiful paper, Ramanujam proposes a new formal framework, based on *constructible player types*, realizable by *automata*. This requirement, adopted as a constructive implementation of van Benthem's slogan *"the players matter"* [113], gives a way to take into account the players' resource limitations, their finite memory structure and their selective process of observation and update. It also helps to keep the logical complexity under control, by avoiding the exponential explosion that is so typical for most game-theoretic analyses. Ramanujam goes on to present a logic for specifying player types, and uses it to study rationalizability and other solution concepts. Last but not least, one of the most original and far-reaching features of Ramanujam's approach is his conception of player types and strategies as *evolving entities*. So, while games were already dynamic objects, Ramanujam's frameworks amounts to a further, higher-level "dynamification" of Game Theory, in the spirit of van Benthem's Theory of Play.

In his provocative contribution, Gabriel Sandu starts from van Benthem's friendly criticism of the Game Theoretical Semantics (GTS), and in particular of the IF (Independence Friendly) Logic, proposed by Hintikka and Sandu. The objection was that GTS *presupposes* that the denotations of the basic lexical items are already shared knowledge between the players, but it does not seem to give us any insight into *how* these shared meanings came to be established. To address this, Sandu takes a fresh look at a different kind of games: the *signaling games*, proposed by Lewis precisely to explain the emergence of conventions (including shared meanings). Lewis' games had the defect of allowing for some undesirable equilibria: the so-called non-strict (noncommunicative) equilibria, which do not correspond to conventions. Sandu argues that Lewis' signaling games are a bad formalization of their intended scenarios, and he proposes a new formalization of signaling games, as win-lose extensive games of imperfect information. By looking then at the strategic form of these games, Sandu observes that they differ in an essential way from Lewis' games: the *Sender and the Receiver are now on the same side*, playing against Nature and trying to coordinate no matter what is the state of Nature. This move eliminates Lewis' noncommunicative equilibria. Moreover, these games can be encoded as sentences in IF logic. So game-theoretical semantics *does* in fact tell us something interesting about the emergence of shared meanings: the signaling games that lead to linguistic conventions are essentially IF games!

6 Dynamic Agency

Games are not the end of the story of Intelligent Agency. Full-fledged agency goes beyond multi-agent epistemic actions and even beyond strategic rationality: it is about the agents' *powers* and their forcing abilities; their *free will*, their choices and *control* of options over time; their *active inquiry* (via actions such as raising questions, or answering them); their capacity for *argumentation* and deliberation; their ever-changing *preferences*, their conflicting *desires*, their mixed motives and *intentions*; their evolving *norms* and their moral conflicts.

6.1   Johan van Benthem on Agency

In recent years, Johan van Benthem and his collaborators have been working on extensions of dynamic logics meant to deal with each of the more sophisticated aspects of agency mentioned above. We will now proceed to look at these features one by one from the perspective of van Benthem's recent work.

Agents' Powers to Choose and Act

Dynamic logics in the usual style (including PDL and DEL) may use linguistic expressions such as "agent a makes a public announcement that P", or "agent a performs action α", but in reality they can only express *changes affecting* the agents ("events"), rather than *actions performed* (in a deliberate manner) by the agents. DEL gives a nice semantics for informational events and their effects, but not for the agents' control over events, their freedom (or lack of freedom) to choose among the possible actions. In contrast, formalisms such as Belnap's See-to-it-that Logic (STIT) and Coalition Logic were designed precisely to reason about choices, abilities, actions and freedom (while Alternating Temporal Logic, though not designed for this explicit purpose, can also capture such features of agency).

In [112], Johan van Benthem and Eric Pacuit make a systematic comparison between these two apparently very different formal approaches, and make some concrete proposals towards combining the two. Their first observation is that, at a purely formal level, the semantics of the STIT operator is very similar to the one of the knowledge modality in a fully introspective ($S5$) epistemic logic. They show that this similarity is not a purely formal accident: the STIT operator really corresponds to a form of knowledge, namely what the authors call *ex interim* knowledge (the knowledge that a player has right after choosing her action or strategy, but before observing the other players' actions). They use this observation to derive a faithful embedding of STIT into the Matrix Game Logic introduced in [79]. This last formalism is a dynamic-epistemic-style logic of games, with Kripke models having as possible worlds all the strategy profiles in a given game (in strategic form), and with two accessibility relations for each agent i: an "epistemic" relation $\sim_i$, relating profiles that agree on i's strategy, and a "freedom" relation $\approx_i$, relating profiles that agree on all the other agents' strategies (though may disagree on i's strategy). The first describes what agent i knows (essentially, she only knows her own strategy), while the second describes i's range of freedom (i.e. her alternative choices of action, assuming fixed the others' choices). But on the other hand, the first modality $[\sim_i]\phi$ really captures the STIT operator: the properties that i ensures, no matter what the others do.[14]

In another line of investigation in the same paper, the authors extend the PDL-style labelled transition systems (in which transitions bear labels denoting events) with new labels indicating the *control* (or the range of choices) available to the agent who does the action. In this analysis, *Actions = events + control*.

The Dynamics of Inquiry: Question-Raising Actions and Issue Management

Real agents do not just learn. They seek the answers to specific *questions*: they always come endowed with their own "issues", their *interrogative agenda*, which guides their learning and their other interactions. They may also *raise* new questions, thus trying to interrogate Nature or other agents. They may *answer* others' questions (or their own).

Building on previous DEL-style work on questions by Baltag [5], van Benthem and Minica [105] develop an Interrogative version of DEL. Starting from the standard analysis of *questions as partitions* (grouping together in the same cell worlds in which the answer is the same), the authors introduce *epistemic issue models*. These are Kripke models endowed with two equivalence relations: an epistemic relation $\sim$ and an "issue" relation $\approx$. The last embodies (the partition induced by) all the agents' questions. In addition to the standard $S5$ operator for knowledge, the language has two interrogative operators: $Q\phi$ is the Kripke modality for the issue relation, expressing the fact that ϕ *is entailed by the (true) answer(s) to the agent's question(s)*; while $R\phi$ is the modality for the intersection

---

[14] To obtain a faithful embedding, the authors need to add an "Independence" Axiom to Matrix Logic, capturing the standard STIT assumption of *independence* (of an agent's choices from the others' choices). The authors discuss the complexity of this richer version of Matrix Logic, which leads them to suggest more general versions of STIT, that can model correlations, or dependencies, between agents' choices, by dropping the Independence axiom.

$\sim \cap \approx$, thus capturing a form of *conditional knowledge*. Indeed, $R\phi$ means that the agent *"knows" ϕ conditional on the (correct) answers to her questions*: if given these true answers, the agent would be able to derive ϕ. Various actions of "issue management" are considered: *resolution* is the action by which the agent's questions are answered (so that her new epistemic relation is given by $\sim \cap \approx$); while *refinement* is the action by which her issue relation is refined to incorporate her knowledge (so that her new issue relation is $\sim \cap \approx$, i.e. the strongest question that she can answer if she learns the answers to all her questions). The authors formalize the complete dynamic logic of these interrogative actions, then move on to the *multi-agent* case. They extend DEL event models with "issues" (about the current event), similarly to the way they extended epistemic models to issue models, and they introduce an interrogative version of the usual DEL product update.

In Minica's Ph.D. thesis [173], written under van Benthem's supervision, one can find a complete DEL-style logic for interrogative product update, as well as applications to more sophisticated scenarios (such as informative questions, whose preconditions give new information to the listener). A more thorough discussion and related open questions can be found in [102].

Agents' Preferences and Their Dynamics

Starting with his older work with van Eijck and Frolova [46], continuing with his work with Otterloo and Roy [77], Liu [82], Girard and Roy [94], as well as the Ph.D. theses by Liu [169], Roy [184], Girard [137] and Zvesper [203], written under his supervision, and then with his more recent papers [91, 117] and books [102, 113], Johan van Benthem has worked towards a dynamic approach to preference logics. The basic semantic setting is given by *modal betterness models*, i.e. Kripke models with a reflexive and transitive "betterness" relation $\leq$ between worlds. The basic language involves a *preference modality* $[\leq]$, in which $[\leq]\phi$ means that ϕ holds in all the worlds that are at least as good as the actual world (or, equivalently, that all non-ϕ worlds are worse than the actual one). This syntax is extended to richer languages in [137] and [94]. The preference modality is a very useful tool, since in combination with the standard dynamic modalities it can be used to express interesting features of the interplay between preference and action, such as various notions of *optimality* and rationality, including the backward induction solution [77].

To obtain a preference relation in the usual sense (between propositions, i.e. sets of worlds, rather than between individual worlds), one has to "lift" the betterness relation to the level of sets. In [94] and [169], various relation-lifting proposals are discussed, showing that von Wright's celebrated approach corresponds to a very specific choice among others. For many such choices, the resulting preference operator $P\phi\psi$ on sentences ("ψ is preferred to ϕ") can be defined in terms of the basic preference modality $[\leq]$ and a universal modality U.

In joint work with Liu [82], van Benthem introduces *relation transformers* in DEL-style to model specific preference-changing actions (such as "suggestions" and various kinds of "commands"), then generalize them to a very wide class of *PDL-format transformers* (in which the new relation can be defined from the old one using the regular PDL operations). This format allows one to "read off" reduction axioms (for the corresponding actions) from the PDL definitions in an automatic manner.

In [94], the authors introduce and study an interesting *"ceteris paribus* logic", meant to capture a different type of preference, namely equality-based ceteris-paribus preferences $P\phi\psi$ ("all things being equal, I prefer ψ to ϕ).

Deontics and the Dynamics of Norms

In a succession of papers with Grossi and Liu [99, 100] and a forthcoming paper with Liu [117], van Benthem applied the above-mentioned recent developments in the logic of preference to a number of topics in deontic logic. The approach in [99] looks at deontic logic as resulting from both a *betterness* ordering (deontic preference) on states, and a *priority* ordering on properties (meant to encode a law explicitly representing a standard of behavior). The authors use the correspondence between these two orderings to look at deontic scenarios and classical deontic paradoxes from a new perspective. The framework is extended to describe dynamics involving both orderings, thus leading to a new analysis of norm change as betterness change.

In [100], van Benthem et al. continue their investigation of priority structures in deontic logic, focusing on the so-called Hansson conditionals. They generalize these conditionals by pairing them with reasoning about syntactic priority structures. They test this approach, first against the usual scenarios involving contrary-to-duty obligations, then by applying it to the modelling of two intuitively different sorts of deontic dynamics of obligations (one based on information changes and the second based on genuine normative events). In this two-level setting, the authors also discuss the Chisholm paradox and the issue of modelling strong permission. Finally, the priority framework is shown to provide a unifying setting for the study of dynamic operations on norms (e.g. *adding* or *deleting* individual norms, *merging* normative systems etc.).

The story of van Benthem's past work on dynamic agency comes to a temporary stop here. But there is more to come: in our view, van Benthem's reflections on agency do not seem to have fully settled yet into a natural fixed point. In particular, his reflections on connections between logic and Argumentation Theory [48, 97, 111], as well as his recent thoughts on collective agency [102] and its relevance to Social Epistemology, Social Choice Theory [81] and the study of social networks, have still to crystalize in a definitive conception. The understanding of "super-agents" (corporate agents, such as artistic or religious groups, sportive teams, companies, states etc., but also socially-produced agents, such cultural icons, literary characters, memes, role models, imaginary being, mythical heroes and gods) is a great challenge both for philosophy and for logical dynamics. To paraphrase Shakespeare, *there are more things to agency in heaven and earth, than are dreamt of in any individual agent's mind.*

6.2 The Invited Contributions on Dynamic Agency

In their fascinating discussion of agency in Artificial Intelligence, Peter Millican and Michael Wooldridge address a central question: *what is an agent*? In this powerful piece, they discuss two main views of agents: agents as *actors* versus agents as *intentional systems*. The first view stresses the role of *action*, identifying agents as the originators of deliberate actions. The second view stresses the agents' *beliefs, desires*, intentions and plans. The authors show that these different views really make a difference when it comes to the actual task of *building* artificial agents.

The contribution by Hector Levesque and Yongmei Liu is a great technical achievement of fundamental conceptual and applied importance. They *reunite two very different paradigms* to reasoning about action: Situation Calculus (the long-standing dominant approach in AI) and Dynamic Epistemic Logic (which can be said to be rapidly becoming the dominant approach to informational dynamics in Logic). By importing into Situation Calculus the key DEL concept of *action models* (or "event models", in van Benthem's terminology), these authors give Situation Calculus a modern facelift, that not only contributes to its rejuvenation, but it hugely extends its capacity for modelling complex informational scenarios.

In their insightful paper, Wes Holliday and John Perry address some of the many conceptual problems and paradoxes that have plagued epistemic predicate logic from Frege to Quine to Kripke and Hintikka. Their solution constitutes a new philosophical approach to epistemic agency, based on the distinction between *agents' names* (and *object variables*) and their *epistemic roles*. This approach is formally implemented in a modified version of Melvin Fitting's First-Order Intensional Logic, in which Fitting's intensional variables are reinterpreted as role variables. While the names are rigid designators, and similarly the assignments of object variables are world-independent, the assignments of role variables depend on the world: the same role may be played by different agents in different worlds. This simple move from individual concepts to agent-relative roles allows the authors to deal in a natural and elegant way with a large number of philosophical puzzles. Moreover, this move opens up new ways of thinking about agency in a dynamic-epistemic context, that are only hinted here: as authors point at the end, one should look now not only at the dynamics of agents' epistemic (accessibility) relations, as in standard DEL, but also at the *dynamics of their epistemic roles*.

The elegant paper on STIT logic by Roberto Ciuni and John Horty goes right to the very essence of agency: the notions of *choice* and *freedom*, and their relationships with *knowledge*, both one's own and the others'. Belnap-style STIT logics (with "agent *a* sees to it that..." as their main operator) were especially created to analyze these concepts. On the other hand, the above-mentioned Matrix Game Logic introduced by van Benthem [79] is a dynamic-epistemic logic, endowed with epistemic modalities for "*ex interim* knowledge" and a "freedom operator". As we saw, a first comparison between this logic and STIT logics was made in [112]. Ciuni and Horty develop a group version of a particular version of STIT logic, called CSTIT (since it is based on the so-called "Chelas STIT", introduced by Horty and Belnap), and show that Matrix Game Logic is equivalent to a fragment of group CSTIT (namely the *anti-group CSTIT*, having stit modalities only for individual agents and their anti-groups): each of these can be embedded in the other. The "bad news" is that, as a consequence, they show that Matrix Game Logic is undecidable and not finitely axiomatizable. The good news is that this mutual embedding provides new insights into the nature and properties of van Benthem's freedom operator. The authors give several examples of such derived principles, of great philosophical interest. One of them encodes the interesting relation between one-agent's knowledge and the other agents' freedom. Another principle states that, even if we restrict ourselves (as the authors do) to situations in which different agents *are independent* with respect to their *individual choices*, those agents may still be *not* independent with respect to their *margins of freedom*!

In their innovative contribution, Oliver Roy, Albert Anglberger and Norbert Gratzl discuss the *normative* aspects of game-theoretic solution concepts and of their logical correlatives (the "best" action operators, used by van Benthem and other logicians to investigate rationality in games). The authors adopt a *systemic* view of the normative interpretation of solution concepts: even if each norm is plausible, the normative system as a whole may be counter-intuitive. They connect this to the older debate of normative conflicts in deontic logic. From the non-uniqueness of the "best action" in most games, they derive a "permissive" interpretation of solution concepts, as stated in their *O/P-Best Principle*: "A player ought to play *a* best action. All specific best actions are permitted, although none is obligatory." By looking at sets of actions as *action types*, the authors restate their proposed interpretation as saying that *obligations are the weakest permitted action type available to the agents* (where "weakest" means logically weakest). They argue that this interpretation with the often-assumed *normality* of the deontic operators, and they go on to propose a new, *non-normal* formalization of deontic logic, based on a *neighborhood semantics*. In this logic, permission and obligation are no longer definable in terms of each other (as in the standard systems), but are they are only weakly related by a number of connecting principles (such as the one relating obligations to the weakest permissions). As a side-effect, the resulting logic can deal with the usual deontic paradoxes: not being normal, the proposed deontic operators are not necessarily additive, thus allowing for conflicting obligations without automatically entailing inconsistent obligations. However, unlike in other such approaches, the move to a non-normal logic is not just an ad-hoc move with the express purpose of solving the deontic paradoxes. On the contrary, the non-normal, neighborhood-based semantics has independent conceptual motivations, of a game-theoretic nature, and it is rooted in a very interesting and original interpretation of obligation!

Dov Gabbay and Davide Grossi connect two of van Benthem's long-standing interests: *argumentation* on the one hand, as one of the informational processes that is among the most essential for rational agency [48, 97, 111], and on the other hand the general study of notions of *process equivalence* that lies at the core of van Benthem's program of Logical Dynamics. Even more intriguingly, this excellent paper is a formal-logical study of an approach that was for a long time understood as being fundamentally opposed to formal logic (at least in its traditional, abstract-inference incarnation): Argumentation Theory. By adopting the perspective on Argumentation Theory proposed by Grossi in [144–146], based on looking at attack graphs as Kripke models, the authors are able to overcome the apparent opposition, and pursue an unashamedly modal-logical analysis of argumentation. They develop a natural notion of *equivalence* between arguments, and compare it with the standard notion of *bisimulation*. Then they "dynamify" these notions by tying them up with the theory of *argument games* and with a notion of *strategic equivalence*.

7 Dynamics in Natural Language and Cognition

Our journey from abstract logical dynamics towards the concreteness of real-life informational processes is about to reach its ultimate limit: real agency meets real, empirical *facts*. The logics of agency, understood as theoretical investigations of intelligent interaction, are now facing the empirical sciences dedicated to the same topic: psychology, cognitive science, linguistics, social sciences. This is the final frontier of concrete dynamics, a frontier that Johan van Benthem, as well as our invited authors, do not hesitate to cross over.

7.1   Johan van Benthem on Natural Language and Cognition

"Do the Facts Matter?"

The traditional view of Logic is as a *normative* discipline: one that prescribes the "correct" ways of thinking. If human reasoning does not obey the rules of Logic, then too bad for the humans! According to this account, the "facts" about how people actually reason and how informational interactions actually happen are supposedly irrelevant for logic, and empirical sciences studying these facts (e.g. psychology and sociology) should have nothing to say to logicians.

As is well-known, this divide between formal logic, as a normative science, and the empirical disciplines, providing only descriptive accounts of human reasoning practices, was codified by Frege in his celebrated "Anti-Psychologism" stance, which became a cornerstone of modern philosophy of logic. From the other side, in psychology, cognitive science and behavioral economics there is a large body of experimental literature on the empirical "failures" of logic, purporting to show that formal-mathematical logic is useless for human practice.

In his paper [84], Johan van Benthem reawakens this dispute, by asking once again a key question: "Do the facts matter" (for logic)? His answer is positive: if logic were totally disjoint from human reasoning then it would have no practical use at all. While still admitting that logic is not psychology, that the correctness of an argument is not decided solely by practice and conversely that a logical theory is not useless just because people do not behave quite according to it, van Benthem argues that the recent cognitive turn in logic has lead to what he calls "the fall of Frege's wall" separating logical research from psychology and other empirical sciences. The twentieth century saw the explosion of the fortress of classical logic and the flourishing of a wild variety of alternative styles of formal inference: non-monotonic logics and default reasoning, substructural and relevant logics, linear logic and other resource-sensitive formalisms, constructive methods and intuitionistic logic etc. These developments were justified philosophically by the Logical Pluralism stance, but many of them were also motivated by intended applications to various forms of practical reasoning. As van Benthem argues, far from having any genuine clash between "Logic" and practice (as the above-mentioned body of literature aims to show), we may now have the opposite problem: almost *any* human reasoning practice can be explained by some combination of the above-mentioned classical and non-classical formalisms. "The resulting immunity for logic would not please Popper, and even worse than that: it seems boring at times." (van Benthem, op. cit., p. 5).

Instead, van Benthem prefers to let logic actually *learn* from direct confrontation with practice, as already argued in his older meditation on "logical semantics as an empirical science" [28]. He goes on to show that this is already happening, and in fact it has been happening for some time now: from Prior's temporal logics to Stalnaker's and Lewis' accounts of counterfactuals, inspired at least in part from actual ordinary usage; to the more recent developments in the logical semantics of natural language, which go beyond standard logical formalisms in an attempt to come closer to the reasoning patterns that are actually used by real-life speakers; to the many logics for "common-sense reasoning" proposed in the Artificial Intelligence literature; to the work of Veltman [200] using dynamic default logics to justify some of the "mistakes" in human reasoning high-lighted by Kahneman and Tversky; to the work on psychologically plausible models for revision of beliefs and goals [118], and on the formation and maintenance of collective intentions [124], etc.

Moreover, van Benthem argues that the influences between logic and cognitive science go *both ways*, and proposes a New Psychologism, as a codification of the actual recent practice of many logicians: avail ourselves of broad psychological insights about real human reasoning, in order to build richer logical theories, that are closer to the facts. An example in this sense is the dynamic turn itself: van Benthem argues that the dynamic view of logic is closer to empirical practice. Instead of seeing logic as the static guardian of "correctness", we should come to see it as an *immune system of the mind*, by focussing on its dynamic role of constantly *correcting* the mistakes that inevitably pop up in practical reasoning.

Natural Language as a Programming Language for Cognition

In van Benthem's vision, the study of *communication* (in natural language), *computation* and *cognition* go together. To use van Benthem's famous slogan, *"natural language is a programming language for cognition"*. But moreover, most of van Benthem's work on the syntax, semantics and pragmatics of Natural Language is "dynamic" in spirit, being in fact closely connected to his research on logical dynamics. Indeed, the original version of the above-mentioned quote is more precise in this sense: *"Natural language is a programming language for effecting cognitive transitions between information states and its users"* [43]. This points to a dynamic-relational understanding of natural language sentences in terms of *information updates*, a view that reflects the Amsterdam milieu of the time[15]: indeed, the independently emerging dynamic semantics for Predicate Logic was being developed around the same time by Groenendijk and Stokhof.

Relations to Dynamic Predicate Logic. Phenomena such as anaphora, presuppositions, non-commutativity of conjunction in natural language etc., are dealt with in a systematic manner in the so-called Dynamic Predicate Logic (DPL), due to Groenendijk and Stokhof [143] (though with roots going back all the way to Stalnaker and Barwise). This approach, very influential in the field of Natural Language Semantics, represents in a sense a more radical "dynamification" of first-order logic than van Benthem's "modalization" of this logic: the meaning of a sentence is now given by a *function* on information states, representing the information *update* induced in the listener's state when that sentence is uttered. Nevertheless, van Benthem showed that DPL can be presented in a more traditional style, namely it is essentially equivalent to the *propositional dynamic logic of atomic tests Pt? and actions of variable-value reassignment $\exists x$*. Put symbolically: $DPL = PDL(Pt?, \exists x)$.

Categorial Grammar

Johan van Benthem's work on categorial grammar [30, 41, 43] is in line with the above-mentioned dynamic view of language. In Categorial Grammar, a grammatical category is identified with a function type $A \rightarrow B$, taking arguments of type A into values of type B. This is an inherently *procedural* view: linguistic expressions denote procedures that change a state of some given type into a state of another type. The procedures corresponding to compound expressions are computed by deriving corresponding sequents in some appropriate categorial logic. Such logics are typically *sub-structural* and "resource-sensitive", since the usual structural rules of classical logic do not necessarily hold for natural language categories. One obtains a whole hierarchy of such substructural calculi (the so-called Categorial Hierarchy), of which the most famous is Lambek Calculus.

There are two kinds of "semantics" in categorial grammar. The first is the proof-theoretic "semantics", given by the derivations themselves ("proofs"). Various categorial calculi are mapped into different fragments of typed lambda calculus, via the Curry-Howard isomorphism between categorial derivations and lambda-calculus terms. Since lambda-terms denote functions, this "semantics" is

---

[15] Further details of contributions, sources and references for this period may be found in [53].

dynamic in character. But there exist also more standard, "logical" dynamics, in which derivable sequents are interpreted as valid ones, "forgetting" the derivation structure. The most interesting from the perspective of logical dynamics are of course the *relational* models. Such a relational semantics for Lambek Calculus was proposed by van Benthem [43], motivated by the above-mentioned dynamic view of language, in terms of binary update relations on information states. van Benthem [43] showed the soundness of Lambek Calculus for the relational interpretation, and posed the problem of its completeness, which was positively answered by Andréka and Mikulas (1993). van Benthem also gave a relational interpretation for the Non-Associative Lambek Calculus (in terms of *ternary* relations, analogue to Dunn's semantics for relevance logics), whose completeness was proved in [166]. In this last interpretation, Lambek Calculus becomes a fragment of van Benthem's Arrow Logic, and indeed the later is a faithful extension of the first.

Generalized Quantifiers

Johan van Benthem's early work on generalized quantifiers, polyadic quantifiers etc. [31, 39], may seem at first sight to be less relevant to logical dynamics. But this first impression is wrong: van Benthem's use of *semantic automata* to analyze quantifiers as model-checking procedure of various complexities [35], and his classification of quantifiers in terms of the computational complexity measured by both automata-theoretic and logical (definability) means [43], amount in effect to providing a *procedural semantics for generalized quantifiers*. In van Benthem's own words, linguistic expressions "denote certain 'procedures' performed within models for the language" [35]. This research uncovered deep connections with modal tree logics and fixed points. His work on changing of contexts or 'local domains of quantification' during the evaluation of sentences adds another level of "dynamicity" to the study of quantifiers.

Natural Logic

More generally, van Benthem's program of "natural logic" [35, 43, 44, 84] aims to identify the "logical core" of feasible, "easy", intuitive reasoning, core underlying the most common forms of inference used in natural language. One can see natural logic as a dynamic theme: a search for the abstract structures governing very concrete concrete and simple forms of informational flow, underlying the low-level fast inferential mechanism that constitutes an essential part of linguistic ability. According to van Benthem, any natural logical system for linguistic reasoning should contain at least the following modules: *monotonicity reasoning* (predicate replacement), *conservativity reasoning* (domain restriction) and *alge-braic laws for reasoning with specific lexical items* (e.g. for negation). The concept of *polarity* plays an important role in van Benthem's analysis of monotonicity reasoning, in the form of the so-called *monotonicity calculus*.

The Dynamics of Context in the Pragmatics of Natural Language

Many phenomena in linguistic pragmatics, like presupposition, anaphora, index-icality etc., are best explained in terms of processes that change utterance contexts. Such "context shifts" were investigated in terms of *context logics*, proposed in [50], in line with van Benthem's relational theory of meaning [36]. These are formalisms which make explicit the contexts and their dynamics, in the form of two-sorted first-order logics with variables ranging over contexts and over objects.

In [47], van Benthem went on to propose more "radical" type of context logics, obtained by both *relativizing* the original context logic (so that not all contexts are always available) and by "localizing" the contexts (so that a context may interpret in different ways each different occurrence of the same predicate symbol, or each occurrence of an existential quantifier).

In this section, we have followed the fascinating story of logic's encounter with the empirical facts about natural language and about the cognitive-interactive behavior of actual human beings, as this story unfolded in the work of Johan van Benthem. It is now time to look at the same topic from the multiple perspectives of our invited authors.

7.2   The Invited Contributions on Natural Language and Cognition

The paper by Larry Moss is a beautiful contribution to the "natural logic" line of research line inaugurated by van Benthem. Moss isolates three types of tractable fragments of natural language reasoning. The first is in fact a family of logical fragments, called *syllogistic logics*, some of which extend classical Aristotelian syllogistics with verbs, generalized quantifiers, cardinal comparisons, comparative adjective phrases etc. The second is a nice spatial-dynamic modal logic, with two modalities: a spatial *location-switching* operator $*\varphi$ ("φ holds in the *other* room") and a temporal/dynamic *and-then* operator $\varphi; \psi$ ("φ and then ψ"). The semantics is a very interesting combination of a classical Tarskian semantics (interpreting sentences as sets of worlds) and a non-classical dynamic-relational semantics in the style of van Benthem's Arrow Logic or of Groenendijk and Stokhof's Dynamic Predicate Logic (interpreting sentences as transition relations between states). Finally, the third is a modern formalization of van Benthem's Monotonicity Calculus, obtained by grafting monotonicity and polarity information into the sentences derived from categorial grammars. To use the author's expression, these three jewels are *etudes*: simple, enlightening versions of more complex phenomena, which give a flavor of the bigger picture but are also interesting in their own respect. This paper is really an excellent illustration of the exciting new research results that have recently reawakened the interest in the Natural Logic program.

The article by Sven Ove Hansson and Fenrong Liu is a powerful piece of original research into the *logic of value expressions*. There are two major classes of non-numerical value predicates: the monadic predicates expressing classificatory notions (e.g. "good", "bad", "best", "very bad", "fairly good", etc.) and the dyadic predicates expressing comparative notions (e.g. "better", "equally good as", and "at least as good as"). The second form the topic of preference logic, extensively explored by van Benthem, Hansson, Liu and many others. The first, i.e. the logic of monadic value predicates, taken by themselves, has not attracted much attention, since it was generally thought as being a relatively impoverished fragment of the second: the monadic value notions can be easily defined in terms of the dyadic ones, while definitions in the opposite direction are usually assumed to be impossible. In this paper, the authors show that the later direction is also feasible, by building on the work of van Benthem [27], who had shown that dyadic value terms can be reduced to *context-dependent* monadic value predicates (e.g. "good" interpreted contextually as "good among the Z's"). The authors develop two logical approaches to this issue: a contextual modal logic of preference, as well as a dynamic contextual logic (in the spirit of van Benthem's context logic) that directly formalizes context shifts as context-changing actions. In fact, they completely overturn the above-mentioned common-wisdom assumptions, showing that although dyadic preference orderings can be defined from context-indexed monadic notions, the monadic notions cannot be regained from the preference relation that they gave rise to. So, in a contextual setting, the logic of monadic values is actually *more* expressive than dyadic preference logic!

Martin Stokhof's paper is a fascinating meditation on what distinguishes the various approaches to the semantics of natural language, and in particular in what sense either of them can be said to be "dynamic", and why does this matter. Stokhof looks comparatively at three approaches to natural language semantics: the Discourse Representation Theory (DRT) of Kamp and Heim; the "dynamic semantics" approach illustrated by Veltman's Update Semantics as well as by the Dynamic Predicate Logic (DPL) of Groenendijk and Stokhof; and the approach taken in Stalnaker's work in the 1970s, based on the distinction between (dynamic) speech acts and the (static) content of an assertion. Stokhof characterizes the first approach as "dynamic assignment of static meanings", the second as "dynamic meaning as such" and the third as "dynamic employment of static meanings". He goes on to ask how real are these distinctions: is there really a deep conceptual difference here, or just different packaging? He takes as a starting point van Benthem's already mentioned thesis that relational algebras play a similar foundational role for dynamic theories to the one played by Boolean algebras for static logics, and his corollary that the difference static-dynamic can be spotted by looking at *the formal properties of the updates* permitted by a given theory. Stokhof mentions a number of increasingly more general such update-based characterizations, and uses them to conclude that these are really different conceptions. Finally, Stokhof asks whether this formal distinction matters at all for actual linguistic applications; he answers this question, by distinguishing between two different roles ("modelling" versus "describing") that a formal system can play in natural language semantics. The static-dynamic distinction is crucial when we insist on our systems fulfilling the second role, but it is not so relevant if we focus only on the first role.

Hans Kamp's deep-reaching contribution is also about the roles played by our model-theoretic accounts of the semantics of natural language, and more generally by our formal theories of various aspects of human behavior (including non-linguistic features, such as action or cognition). Are these theories *descriptive*, and so *testable* against human users' intuitions about their own behavior, or are they prescriptive, in which case the human users can *learn* from them how to better understand themselves and how to improve their own behavior? Kamp's conclusion is that formal models play *both* roles, but in *two stages*: first, logical theories should be tested against users' firm intuitions about those aspects of the formalized concepts about which they are confident; second, the theories and models which are successful in the first stage can then be used by humans as guides towards a better understanding of other, more obscure aspects, and hence towards improving their behavior. Of course, in practice these two stages will not be passed only once, but again and again, giving rise to looping feedbacks, leading to more and more accurate theories *and* to "better" (more consistent, more "rational", more effective) behavior. Kamp argues that these lessons apply to models of both linguistic and non-linguistic features of human life, and concludes that these similarities suggest a deeper connection than usually assumed between the formal logics for natural language and the logical models of action and cognition.

The contribution by Alistair Isaac, Jakub Szymanik and Rineke Verbrugge is an excellent survey of the current state of research into the *logical and computational complexity of cognitive tasks*, and how they relate to humans' actual performance in these tasks. The authors argue (in agreement with, but independently from, Kamp's similar argument in the previous chapter) that things go beyond the mere distinction between normative and prescriptive, showing how by first adopting a computational perspective one can generate feedbacks from empirical results back into the development of better computational models. Once again, van Benthem's models and methods, such as the use of semantic automata and natural logic in the study of language and cognition, play an essential role here, together with newer perspectives such as the one provided by the P-Cognition Thesis.

Finally, Peter Gärdenfors' paper is an insightful reflection on the limits of the computational (and the connectionist) perspective on cognition, arguing in favor of the *situated cognition* paradigm. Gärdenfors shows how *embodiment*, the interaction of the brain with the body and the world at large, can help reduce the complexity of cognitive tasks that would otherwise be intractable, and thus hugely improve the agent's learning performance. By an analogy to the way in which "tunas and dolphins swim *with* the water, not in the water", Gärdenfors suggests that our brains "think *with* the world, not in the world". He gives examples showing in particular how "the geometric structure of the external world reduces the complexity" of typical spatial-manipulation problems in robotics. He goes on to argue that the embodiment solves a number of other cognitive mysteries, including the difficulties encountered by the Chomskyan paradigm in explaining children's amazing capacity for learning grammatical structures:

natural language grammar is *not independent* of its intended, real-world semantics, and the fact that children live in (and interact with) the real world is in this view essential for their successful language acquisition.

8 Styles of Reasoning

Our movement towards concreteness has reached its limits, in the shape of empirical facts, natural language and cognitive realities. It is here where dynamics re-encounters other styles of reasoning. Indeed, many competing approaches have already been extensively used to explain and organize this empirical evidence, especially in the study of natural language and common-sense reasoning: alternative semantical frameworks for classical connectives, alternative conse-quence relations, situation theory etc. This encounter resumes the theme of searching for the right level of generality, that dominated the *abstract* Logical Dynamics in section 3. So the back-and-forth move between abstraction and concrete representation reappears here as a fertile debate between two polar attitudes: from an abstract perspective, logical dynamics is just one of many styles of reasoning, and dynamic operators are only one special type of non-classical inferential tools; but from a concrete perspective, all the various reasoning styles are best understood as talking in fact about dynamics, rather than about some non-classical consequence; and inference itself (both classical and non-classical) is just one among many other informational processes (e.g. communication, observation, introspection etc.).

8.1 Johan van Benthem on Styles of Reasoning

The above-mentioned polarity is reflected in van Benthem's ambivalent attitude towards Logical Pluralism. On the one hand he is one of its champions, continuously looking at *combinations of different styles* and paradigms (e.g. modal logic and information channels in the style of Barwise and Seligman, model theory and situation theory, etc.) [34, 55], and always stressing the *peaceful coexistence* of dynamics with other approaches. But on the other hand he usually adopts a *"dynamics-first" viewpoint*, taking information flow as fundamental, with inference as only one of its aspects. In this sense, the dynamic approach aims to gain a deeper understanding of non-classical forms of inference by looking at the concrete informational tasks underlying these formalisms.

So this is not only a debate within Logic at large, but also a debate within van Benthem's own work and thinking: a friendly logical conversation, as well as a fierce philosophical confrontation, between Johan van Benthem and himself.

Dynamics Versus Pluralism

The traditional attitude towards Logic was to see it as the science of *reasoning*, thus having *inference* as its central topic. Before the twentieth century, it was generally assumed that there was only one Logic, and hence one correct inference relation. Logical Pluralism has exploded this assumption, and so Logic became the art of charting the variety of possible *reasoning styles*, encoded as different *consequence relations*, each with its own rules. This is how the marvelous, bewildering diversity of *non-classical logics* came into being, starting with intuitionistic logic, and continuing with multi-valued logics, partial logics, fuzzy logics, paraconsistent logics, relevance logics, substructural logics (in particular, non-monotonic logics) etc. Johan van Benthem's early work on substructural logics and styles of inference fell within this paradigm, and by working within this line of research he and others first saw dynamic logics as just one family of logics among many others. As the reader may recall from section 2.2, "dynamic inference" was yet another style of inference, on a par with others.

But somewhere along the line, something happened. "I have changed my mind", confesses van Benthem in the section entitled "What is logic?" in one of his latest books [102], and he continues: "The Logical Dynamics of this book says that the main issue is not reasoning styles, but the *variety of informational tasks* performed by intelligent agents, of which inference is only one among many, including observation, memory, questions, answers, dialogue, and strategic interaction. Logical systems should deal with all of these, making information-carrying events first-class citizens." ([102], p. 295).

Already in [40], van Benthem was voicing two concerns about non-monotonic and non-classical inference: first, that these features are only *symptoms* of some underlying phenomenon. "Non-monotonicity is like fever: it does not tell you which disease is." ([102], p. 297). Second, most non-classical and substructural logics are radical-revolutionary with respect to semantics and proof theory, but conservative with respect to the language: they retain the classical repertoire (classical connectives plus the Gentzen consequence meta-relation), changing only their properties. But why so? What is so sacred about classical logical operators, but not about their interpretations? "Why not be radical with respect to the language as well, and reconsider what we want to say?" ([102], p. 297).

The logics of *informational dynamics* investigated by van Benthem in the last decades, with Dynamic Epistemic Logics at the forefront, address these concerns: they are, not just about pluralistic consequence relations, but also about many kinds of concrete informational tasks, and they use richer languages to talk about them. According to van Benthem, these logics set themselves a "more ambitious goal" than non-classical and substructural logics: they look for a *diagnosis* for non-classicality, instead of administering only a symptomatic treatment.

Dynamic Re-interpretations of Other Styles of Reasoning

Johan van Benthem was thus lead to propose a program involving a *dynamic (re)interpretation* of substructural and non-classical styles of reasoning. This aims at providing an *explanation* of these non-classical phenomena, by "deconstructing them into classical logic plus an explicit account of the underlying informational events" ([102], p. 296). As we saw, this is a move towards *concreteness* and *realism*, and away both from the "one true inference" obsession of traditional logic and from the warring abstractions of Logical Pluralism.

In [102], van Benthem exemplifies this program, by showing how some non-monotonic logics can be naturally embedded into dynamic logics of belief change; in particular, he gives a representation of McCarthy's *circumscription* in terms of acts of belief revision. By looking at the Kripke semantics for *intuitionistic logic*, he goes beyond Gödel's proof-theoretic embedding of this logic into the modal logic *S4* (usually misunderstood as a purely *epistemic* interpretation), showing that from a semantic point of view intuitionistic logic is really a *dynamic-epistemic* logic: it is all about persistent acceptance, or persistent non-acceptance, or (in the case of implication) persistent correlation.

The same program can be applied to many non-classical logics. As we've seen, Lambek calculus, relevance logic and other substructural logics can be given relational models (sometimes involving ternary relations), that reveal their true dynamic nature. (In fact, this dynamic flavor is already present in some of the intuitive explanations given by the founding fathers of relevance logic.) To deconstruct relevance logic into a classical and a dynamic component, one needs to make explicit, not only events such as information updates or upgrades, but also acts of information *merge* (fusion).

In our own work, we gave a similar deconstruction of *quantum logic* [11, 12, 18–20], showing that the so-called quantum implication (known as "Sasaki Hook") $P \rightarrow Q$ is best understood semantically as a PDL-style dynamic modality $[?P]Q$ for a "quantum-test" action $?P$, representing a successful measurement of the quantum property P. In contrast to classical PDL tests (and classical idealized measurements), quantum tests are "really dynamic": they change the state of the system under observation, which explains the non-classical behavior of quantum "implication". Dynamic interpretations can be similarly given to other quantum-logical connectives.

The general motto of this program is that *"natural" non-classical logics can be modeled as classical dynamic logics, by making explicit the informational event that causes the apparent non-classicality.* As van Benthem states, this observation seems quite general, but it is not proved. This remains still a program, for which van Benthem mentions important challenges (e.g. obtaining dynamic deconstructions of paraconsistent logics or of linear logic).

Alternative Consequence Relations for Dynamic-Epistemic Logics

One should *not* conclude from the above discussion that van Benthem aims at eliminating non-classical styles of inference! The move suggested above can also be applied *in reverse*: by "dynamifying" the inference systems for dynamic-epistemic logics, one can obtain non-classical, non-monotonic consequence relations that may provide a *more adequate* proof theory for these logics than classical inference. Johan van Benthem exemplifies this by proposing in [63] a dynamic inference relation based on updates: $P_1, \ldots, P_k \implies P$ is equivalent to $[!P_1]\ldots[!P_k]CkP$ (where $!Q$ is public announcement of Q and Ck is common knowledge). This is a version of the so-called Update-to-Test consequence relation, which is substructural: due to Moore-type sentences, *all* classical structural rules fail! However, some modified structural rules hold, and van Benthem proves an abstract completeness result for the system formed of these rules. Similar systems have been used [8] to axiomatize versions of DEL. In [102], van Benthem goes on to propose a similar semantics in terms of belief (rather than knowledge), and notices that in this case there are natural alternatives, obtained by replacing updates with "soft" doxastic upgrades of various kinds.

Back-and-Forth Between Dynamics and Styles of Inference

To conclude, Johan van Benthem's position towards styles of inference is in fact rather complex, going beyond the reductionist polar attitudes discussed above. He seeks a genuine compromise between Logical Pluralism and Dynamics, according to which one can go back and forth between the two via appropriate meta-logical transformations. We go from concrete dynamics to consequence relations via a process of *abstraction*; and we go the other way around, from consequence relations to dynamics, via *representations*.

The second direction is generally *harder*, since it involves proving a representation theorem. This is natural: intuitively, the dynamics gives us *more* information (or at least more *explicit* information), and recovering this additional information from the abstract inference patterns is not in general an easy task. But, according to van Benthem's view, the two programs live side by side in peaceful coexistence, and exchange interesting insights via the back-and-forth movement.

Nevertheless, while at a technical-formal level they can easily coexist, at a conceptual-philosophical level these programs are still based on two fundamentally different viewpoints: on the one hand, a "static" but non-classical logical world, focused on *consequence* relations; on the other hand, a classical view of inference combined with a shift to *dynamics* as the main focus of logic.

Is it possible to achieve a deep-level conceptual unification of these perspectives, going beyond a mere formal pluralism? Are these viewpoints complementary in some sense: are inference and dynamics just two faces of the same coin?

These questions, which remain open, touch the very core of van Benthem's debate with himself. Their answers will tell us something fundamental about Logic, giving us a deeper understanding of its role as an almost "magic" gateway between concrete informational processes and abstract inference patterns.

8.2 The Invited Contributions on Styles of Reasoning

The chapter of Denis Bonnay and Dag Westerståhl is an enlightening, systematic comparison between classical consequence and a *dynamic-type* consequence (namely, Update-to-Test consequence), in two different settings: an abstract setting given by *information frames*, and a concrete setting in which updates are given by public announcements. They show that the two notions of consequence diverge when applied to non-persistent information (e.g. epistemic information that something is not yet known). They also compare the condition of classicality for updates (i.e. the condition that an update can be classically represented) with classicality for (dynamic) consequence, showing that the second is a much weaker requirement.

The paper by Guillaume Aucher goes right at the heart of van Benthem's program mentioned above, showing how to go back and forth between a version of dynamic epistemic logic (with the standard product update semantics) and a certain substructural logic (with the sort of ternary relational semantics used both by Routley and Meyer for relevance logic and by van Benthem in his work on categorial grammar). Aucher shows that the product update semantics can be seen as a "concrete" implementation of the Routley-Meyer semantics, and that DEL itself can be seen as a two-sorted substructural logic. This helps with the study of a better-behaved (non-classical) proof theory for DEL, which turns out to be related to the dynamic consequence relation introduced by van Benthem. Conversely, Aucher argues that the DEL connection throws light on the meaning of substructural logics, including relevance logic: they are essentially *dynamic* in nature, being all based on (an abstract version of) an update operator. The fascinating abstract correspondences uncovered by Aucher represent a truly spectacular realization of van Benthem's program of going back-and-forth between substructural logics and dynamic logics of belief change.

In his insightful contribution, Mike Dunn looks at the connection between relevance logics and another dynamic logic: van Benthem's arrow logic. He provides a detailed comparison between arrow logics with van Benthem-Venema semantics, and relevance logics with Routley-Meyer semantics, showing how to go between the two logics, which operators are common and how to add the non-common operators to the side that's missing them. Then he proceeds to make an even more interesting comparison, between a version of arrow frames endowed with a number of additional requirements by van Benthem (to make them closer to relation algebras) and Dunn's own version of the Routley-Meyer semantics (meant to represent relation algebras). As a side benefit, he shows how van Benthem's conditions can be slightly modified to obtain a representation of relation algebras. In both spirit and implementation, Dunn's elegant work on these connections is indeed very close to van Benthem's above-mentioned program: on the one hand, this work can be understood as revealing in a new way the "dynamic", transition-like nature of relevance logic; and on the other hand, it can be seen as a way to solve problems in the field of dynamic arrow logics by importing technical insights from the semantics of relevance logic.

Jeremy Seligman's paper is a tribute, not only to Johan van Benthem, but at the same time to one of his best friends and collaborators, the late Jon Barwise. His Situation Theory remains to date one of the most ambitious and wide-ranging attempts to provide new mathematical foundations for philosophy and semantics, in which the notion of context, or "situatedness", would play a crucial role. One of the several specific implementations of this project was the work by Barwise and Seligman [23] on the theory of *classifications*,[16] as a general theory of situated information flow, based on the concepts of *info-morphisms*, information channels and "local logics". At the time seen as a rival to the Kripke semantics approach to information, Classification Theory was in fact of great interest to van Benthem, one of his dearest and most long-standing projects being the investigation of the relations between these two approaches, investigation started by him in [55]. In this fascinating paper, Seligman continues and widens this project, by first proposing a new formalization of Situation Theory within Classification Theory, based on the identification of "situations" with a type of local logics, and then comparing this approach with van Benthem's own "constraint-logic" proposal, based on modeling situations in terms of Kripke-style *constraint models*. This seems to us the most promising and far-reaching new approach to the topic to date, and has the potential of reawakening the interest in Situation Theory and bringing its core ideas into the twenty first century.

The last paper of this volume, by Willem Conradie, Silvio Ghilardi and Alessandra Palmigiano, can be seen as both a manifesto and a proof-of-concept for an ambitious algebraic project, that reconnects all the topics of this book and aims to give a uniform proof-theoretic treatment to both logical dynamics and non-classical styles of reasoning, from the high vantage point of a Unified Correspondence Theory. The unifying power of Algebra and its deep significance for the understanding of logical dynamics and its proof theory can be seen in all shining glory in this beautiful piece.

The circle is now full: we are back to van Benthem's early work in his Ph.D. thesis on modal correspondence, which is now seen to provide a "royal road" through all the new territories we have explored in this book.

We cannot imagine a better ending point for our volume.

---

[16] Also known as Chu spaces.

9 Conclusions

In this volume, we have followed the trajectory of Johan van Benthem's epic logical adventure, using as our guide the contributions of so many world-renowned scholars and friends of Johan. *We now want to thank Johan and his friends for giving us the opportunity to share this adventure with them and bask in their reflected light.*

Together, we have mapped van Benthem's itinerary, across the six dimensions of "Logical-Informational Dynamics". We see this itinerary as an *anabasis*, in the Greek tradition: an upward journey from the safety of the (by now) well-known, well-mapped coast of classical and non-classical logics into the uncharted interior highlands of a New Continent. As it became apparent in the last section, this is in fact a journey towards the *concreteness* and richness of "real life". It is a move towards full-fledged *agency* (and not just "logical agents"), towards *meeting others*, towards stepping out of the unending circles of reason and daring to actually *look* at the world and *interact* with it.

Logic in Johan's view is not only about reasoning and inference (in no matter how many styles). It is also about *acting intelligently*; about *asking questions* to Nature and to each other; about *experimentation* and *communication*; about changing your mind and imagining different perspectives; about learning from your own mistakes and from the testimony of others; about beneficial *social encounters* and sometimes tragic social *conflicts*; about choices, and goals, and norms, and desires; and about how to live with all these, despite their mutual inconsistency; about duty, and privacy, and freedom, and their limits.

As noticed in the last section, the opposite move also continues to happen, in parallel with the first one: a *katabasis*, a perpetual return back down to the coast, by which all those rich, concrete, "real-life" informational processes feed back into the abstraction of inferential logic, as so many rivers flowing into the sea. Anything that logicians touch becomes Logic, and so a subject of inference: as post-modern Midas kings, they convert all reality into formal proof systems. Life, evolution and learning, intelligence, interaction and agency: according to Johan van Benthem, these all are legitimate topics of logical investigation.

To paraphrase the last line of Darwin's magnum opus[17]: *There is logic in this view of life.* Dynamic logic, more precisely: the logic of living and acting, cooperation and competition, love and strife. Information highways and information wars: both are first-class citizens, with full rights, in Johan's logical society of informational processes.

And (to paraphrase once again) *there is grandeur in this view of logic.*

---

[17] "There is grandeur in this view of life, with its several powers, having been originally breathed into a few forms or into one." (Charles Darwin, *Origins of Species*).

References

1. Andréka H, van Benthem J, Németi I (1998) Modal languages and bounded fragments of predicate logic. J Philos Logic 27:217–274
2. Aucher G (2003) A combined system for update logic and belief revision. Master thesis, University of Amsterdam, ILLC Publications MoL-2003-03
3. Aumann R (2010) Interview. In: Hendricks V, Roy O (eds) Epistemic logic: 5 questions. Automatic Press/VIP, New York, pp 21–33
4. Balbiani P, Baltag A, van Ditmarsch H, Herzig A, de Lima T, Hoshi T (2008) Knowable" as "known after and announcement". Rev Symbol Logic 1(3):305–334
5. Baltag A (2001) Logics for insecure communication. In: Proceedings of TARK'01 (eighth conference on rationality and knowledge), pp 111–122
6. Baltag A (2008) An interview on epistemology. In: Hendricks V, Pritchard D (eds) Epistemology: 5 questions. Automatic Press/VIP, New York
7. Baltag A, Bezhanishvili N, Özgün A, Smets S (2013) The topology of belief, belief revision and defeasible knowledge. In: Proceedings of the LORI 2013 conference. Lecture notes in computer science, vol 8196, pp 27–40
8. Baltag A, Coecke B, Sadrzadeh M (2005) Algebra and sequent calculus for epistemic actions. Electron Notes Theor Comput Sci 126:27–52
9. Baltag A, Moss L (2004) Logics for epistemic programs. Synthese 139(2):165–224
10. Baltag A, Moss L, Solecki S (1998) The logic of public announcements, common knowledge and private suspicions. In: Proceedings TARK. Morgan Kaufmann Publishers, Los Altos, pp 43–56
11. Baltag A, Smets S (2005) Complete axiomatizations for quantum actions, in the proceedings of IQSA 2004. Int J Theor Phys 44(12):2267–2282
12. Baltag A, Smets S (2006) LQP: The dynamic logic of quantum information. Math Struct Comput Sci 16(3):491–525
13. Baltag A, Smets S (2006) Conditional doxastic models: a qualitative approach to dynamic belief revision. In:Mints G, de Queiroz R (eds) Proceedings of WOLLIC 2006. Electronic notes in theoretical computer science, vol 165, pp 5–21
14. Baltag A, Smets S (2006) The logic of conditional doxastic actions: a theory of dynamic multi-agent belief revision, In: Artemov S, Parikh R (eds) Proceedings of the workshop on rationality and knowledge, ESSLLI 2006, pp 13–30
15. Baltag A, Smets S (2006) Dynamic belief revision over multi-agent plausibility models. In: Bonanno G, van der Hoek W, Woolridge M (eds) Proceedings of the 7th conference on logic and the foundations of game and decision (LOFT 2006), University of Liverpool, pp 11–24
16. Baltag A, Smets S (2008) Probabilistic dynamic belief revision. Synthese 165(2):179–202
17. Baltag A, Smets S (2008) A qualitative theory of dynamic interactive belief revision. In: Bonanno G, van der Hoek W, Wooldridge M (eds) TLG 3: logic and the foundations of game and decision theory (LOFT 7). Texts in logic and games, vol 3, pp 11–58 Amsterdam University Press, Amsterdam
18. Baltag A, Smets S (2008) A dynamic—logical perspective on quantum behavior. In: Douven, I, Horsten, L (eds) Studia logica 89:185–209
19. Baltag A, Smets S (2011) Quantum logic as a dynamic logic. In: Kuipers T, van Benthem J, Visser H (eds) Synthese 179:285–306
20. Baltag A, Smets S (2012) The dynamic turn in quantum logic. Synthese 186(3): 753–773
21. Baltag A, Gierasimczuk G, Smets S (2011) Belief revision as a truth tracking process. In: Proceedings of theoretical aspects of rationality and knowledge, TARK
22. Barwise J (1989) On the model theory of common knowledge. In: The situation in logic, 201–221. CSLI Press, Stanford
23. Barwise J, Seligman J (1997) Information flow in distributed systems. Cambridge tracts in theoretical computer science. Cambridge University Press, Cambridge

24. Beaver D (1995) Presupposition and assertion in dynamic semantics. Doctoral dissertation, Centre for Cognitive Science, University of Edinburgh
25. Belnap N (1977) A useful four-valued logic. In: Dunn M, Epstein G (eds) Modern uses of multiple-valued logics. Reidel, Dordrecht, pp 8–37
26. van Benthem J (1977) Modal correspondence theory. Doctoral dissertation, Mathematical Institute, University of Amsterdam
27. van Benthem J (1982) Later than late: on the logical origin of the temporal order. Pac Philos Q 63:193–203
28. van Benthem J (1983) Logical semantics as an empirical science. Studia Log 42(2/3):299–313
29. van Benthem J (1991) The logic of time, Reidel, Dordrecht (Synthese library 156, 1983. Revised and expanded edition published in 1991)
30. van Benthem J (1983) The semantics of variety in categorical grammar. Technical report 83–26. Department of Mathematics, Simon Fraser University, Burnaby
31. van Benthem J (1984) Questions about quantifiers. J Symb Log 49(2):443–466
32. van Benthem J (1984) Tense logic and time. Notre Dame J Formal Log 25(1):1–16
33. van Benthem J, Pearce D (1984) A mathematical characterization of interpretation between theories. Studia Log 43(3):295–303
34. van Benthem J (1985) Situations and inference. Linguist Philos 8:3–9
35. van Benthem J (1986) Essays in logical semantics. Studies in linguistics and philosophy. Reidel, Dordrecht, p 29
36. van Benthem J (1986) The relational theory of meaning. Logique et Analyse 29:251–273
37. van Benthem J (1988) Games in logic. In: Hoepelman J (ed) Representation and reasoning. Niemeyer Verlag, Tubingen, pp 3–15
38. van Benthem J (1999) Logic in games. Lecture notes. University of Amsterdam, ILLC, Amsterdam
39. van Benthem J (1989) Polyadic quantifiers. Lingu Philos 12(4):437–464
40. van Benthem J (1989) Semantic parallels in natural language and computation. In: Ebbinghaus H-D et al (eds) Logic colloquium. Ganada 1987. North-Holland, Amsterdam, pp 331–375
41. van Benthem J (1990) Categorial grammar and type theory. J Philos Log 19:115–168
42. van Benthem J (1990) Computation versus play as a paradigm for cognition. Acta Philos Fennica 49:236–251
43. van Benthem J (1991) Language in action: categories, lambdas and dynamic logic. Studies in logic, vol 130. The MIT Press, North-Holland (Paperback reprint with new Appendix 1995)
44. van Benthem J (1991) Language in action. J Philos Log 20:1–39
45. van Benthem J (1991) General dynamics. Theor Ling 17:1(2/3):151–201
46. van Benthem J, van Eijck J, Frolova A (1993) Changing preferences. Report CS-93-10. CWI, Amsterdam
47. van Benthem J (1996) Exploring logical dynamics. Studies in logic, language and information. CSLI Publications and Cambridge University Press, Cambridge
48. van Benthem J (1996) Logic and argumentation theory. In: van Eemeren F, Grootendorst R, van Benthem J, Veltman F (eds) Proceedings colloquium on logic and argumentation. Royal Dutch Academy of Arts and Sciences, Amsterdam, pp 27–41
49. van Benthem J (1996) Modal logic as a theory of information. In: Copeland J (ed) Logic and reality. Essays on the Legacy of Arthur Prior. Oxford University Press, Oxford
50. van Benthem J (1986) Shifting contexts and changing assertions. In: Llera A, van Glabbeek A, Westerrstahl D (eds) Proceedings 4th CSLI workshop on logic, language and computation. CSLI Publications, Stanford
51. van Benthem J (1996) Programming operations that are safe for bisimulation. In: Logic colloquium 94. Clermont-Ferrand. Special issue of studia logica.

52. van Benthem J (1997) Dynamic bits and pieces. ILLC research report, University of Amsterdam

53. van Benthem J, Muskens R, Visser A (1997) Dynamics. In: van Benthem J, ter Meulen A (eds) Handbook of logic and language. Elsevier Science Publishers, Amsterdam, pp 587–648

54. van Benthem J (1998) Programming operations that are safe for bisimulation. Studia logica, vol 60, pp 2, 311–330 (Logic Colloquium. Clermont-Ferrand 1994)

55. van Benthem J (2000) Information transfer across chu spaces. Log J IGPL 8(6):719–731

56. van Benthem J (2001) Games in dynamic epistemic logic. In: Bonanno G, van der Hoek W (eds) Bulletin of economic research, vol 53, pp 4, 219–248 (Proceedings LOFT-4, Torino).

57. van Benthem J (2002) Extensive games as process models. In: Pauly M, Dekker P (eds) J Log Lang Inf 11:289–313 (Special issue)

58. Aiello M, van Benthem J, Bezhanishvili G (2003) Reasoning about space: the modal way. Log Comput 13(6):889–920

59. Aiello M, van Benthem J (2003) A modal walk through space. J Appl Non-Classical Log 12(3/4):319–363

60. van Benthem J (2003) Conditional probability meets update logic. J Log Lang Inf 12(4):409–421

61. van Benthem J (2003) Logic games are complete for game logics. Studia Log 75:183–203

62. van Benthem J (2003) Logic and the dynamics of information. In: Floridi L (ed) Minds and machines, vol 13, pp 4, 503–519 (Special issue)

63. van Benthem J (2003) Structural properties of dynamic reasoning. In: Peregrin J (ed) Meaning: the dynamic turn. Elsevier, Amsterdam, pp 15–31 (Final version (2008) Inference in action. Publications de L'Institut Mathématique, Nouvelle Série, 82, Beograd, pp 3–16)

64. van Benthem J (2004) What one may come to know. Analysis 64(282):95–105

65. van Benthem J (2004) Probabilistic features in logic games. In: Kolak D, Symons J (eds) Quantifiers, questions, and quantum physics. Springer, New York, pp 189–194

66. van Benthem J, Liu F (2004) Diversity of logical agents in games. Philos Sci 8(2):163–178

67. van Benthem J (2005) Minimal predicates, and definability. J Symb Log 70(3):696–712

68. van Benthem J (2005) An essay on sabotage and obstruction. In Hutter D (ed) Mechanizing mathematical reasoning, essays in honor of Jörg Siekmann on the occasion of his 60th birthday, Springer, New York, pp 268–276

69. van Benthem J (2005) Open problems in logic and games. In: Artemov S, Barringer H, d'Avila Garcez A, Lamb L, Woods J (eds) Essays in honour of Dov Gabbay. King's College Publications, London, pp 229–264

70. van Benthem J, Sarenac D (2005) The geometry of knowledge. In: Béziau JY, Costa Leite A, Facchini A (eds) Aspects of universal logic, Centre de Recherches Sémiologiques, université de Neuchatel, Neuchatel, pp 1–31

71. van Benthem J (2006) Epistemic logic and epistemology: the state of their affairs. Philos Stud 128(2006):49–76

72. van Benthem J (2006) Modal frame correspondences and fixed-points. Studia Log 83(1):133–155

73. van Benthem J (2006) One is a lonely number: on the logic of communication. In: Chatzidakis Z, Koepke P, Pohlers W (eds) Logic colloquium 02. ASL and A.K, Peters, Wellesley, pp 96–129

74. van Benthem J (2006) Logical construction Games. In: Aho T, Pietarinen AV (eds) Truth and games, essays in honour of Gabriel Sandu. Acta Philos Fennica, vol 78, pp 123–138

75. van Benthem J (2006) The epistemic logic of IF games. In: Auxier R, Hahn L (eds) The philosophy of Jaakko Hintikka (Schilpp series). Open Court Publishers, Chicago, pp 481–513

76. van Benthem J, van Eijck J, Kooi B (2006) Logics of communication and change. Inf Comput 204(11):1620–1662

77. van Benthem J, van Otterloo S, Roy O (2006) Preference logic, conditionals and solution concepts in games. In: Lagerlund H, Lindström S, Sliwinski R (eds) Modality matters. University of Uppsala, Uppsala, pp 61–76

78. van Benthem J (2007) Rationalizations and promises in games, philosophical trends, supplement 2006 on logic. Chinese Academy of Social Sciences, Beijing, pp 1–6

79. van Benthem J (2007) Rational dynamics and epistemic logic in games, Int Game Theory Rev 9:1, 13–45 (Erratum reprint 9:2, 377–409)

80. van Benthem J (2007) Dynamic logic of belief revision. J Appl Non-Classical Log 17(2):129–155

81. van Benthem J (2007) The social choice behind belief revision. Lecture presented at the workshop dynamic logic montreal

82. van Benthem J, Liu F (2007) Dynamic logic of preference upgrade. J Appl Non-Classical Log 17(2):157–182

83. van Benthem J (2008) Games that make sense: logic, language and multi-agent interaction. In: Apt K, van Rooij R (eds) New perspectives on games and interaction. Texts in logic and games 4. Amsterdam University Press, Amsterdam, pp 197–209

84. van Benthem J (2008) Logic and reasoning: do the facts matter? Studia Log 88:67–84

85. van Benthem J, Dégrémont C (2008) Multi-agent belief dynamics: bridges between dynamic doxastic and doxastic temporal logics. In: Bonanno G, van der Hoek W, Löwe B (eds) Proceedings LOFT

86. van Benthem J (2008) Logical pluralism meets logical dynamics? Aust J Log 6:28

87. van Benthem J, Ghosh S, Liu F (2008) Modeling simultaneous games in dynamic logic. Synthese (KRA) 165(2):247–268

88. van Benthem J, Ikegami D (2008) Modal fixed-point logic and changing models. In: Avron A, Dershowitz N, Rabinovich A (eds) Pillars of computer science: essays dedicated to Boris (Boaz) Trakhenbrot on the occasion of his 85th birthday. Springer, Berlin, pp 146–165

89. van Benthem J, Martinez M (2008) The stories of logic and information. In: Adriaans P, van Benthem J (eds) Handbook of the philosophy of information. Elsevier Science Publishers, Amsterdam, pp 217–280

90. van Benthem J (2009) The information in intuitionistic logic. Synthese 167(2):251–270

91. van Benthem J (2009) For better of for worse: dynamic logics of preference. In: Grüne-Yanoff T, Hansson S (eds) Preference change. Springer, Dordrecht, pp 57–84

92. van Benthem J, Gerbrandy J, Hoshi T, Pacuit E (2009) Merging frameworks for interaction. J Philos Log 38(5):491–526

93. van Benthem J, Gerbrandy J, Kooi B (2009) Dynamic update with probabilities. Studia Log 93(1):67–96

94. van Benthem J, Girard P, Roy O (2009) Everything else being equal: a modal logic for ceteris paribus preferences. J Philos Log 38(1):83–125

95. van Benthem J, Velázquez-Quesada F (2009) Inference, promotion, and the dynamics of awareness. Knowl Ration Action 177(1):5–27

96. Van Benthem J, Pacuit E (2010) Temporal logics of agency, editorial. J Log Lang Inf 19(4):1–5

97. van Benthem J (2010) A logician looks at argumentation theory. Cogency 1:2 (Universidad Diego Portales, Santiago de Chili)

98. van Benthem J, Gheerbrant A (2010) Game solution, epistemic dynamics, and fixed-point logics. Fundam Inf 100:19–41

99. van Benthem J, Grossi D, Liu F (2010) Deontics = betterness + priority. In: Governatori G, Sartor G (eds) Proceedings deontic logic in computer science, DEON 2010, Fiesole, Italy. Lecture notes in computer science 6181, Springer, New York, pp 50–65

100. van Benthem J, Grossi D, Liu F (2013) Priority structures in deontic logic. Theoria. doi:10.1111/theo.12028

101. van Benthem J, Velázquez-Quesada F (2010) The dynamics of awareness, knowledge, rationality and action. Synthese 177(1):5–27 (Springer on-line)

102. van Benthem J (2011) Logical dynamics of information and interaction. Cambridge University Press, Cambridge
103. van Benthem J (2011) Exploring a theory of play. In: Apt K (ed) Proceedings TARK 2011, Groningen, pp 12–16
104. van Benthem J, Bezhanishvili N, Hodkinson I (2011) Sahlqvist correspondence for modal mu-calculus. Studia Log 100:31–60
105. van Benthem J, Minica S (2011) Toward a dynamic logic of questions. J Philos Log 41(4): 633–669
106. van Benthem J, Pacuit E (2011) Dynamic logic of evidence-based beliefs. Studia Log 99(1): 61–92
107. van Benthem J, Pacuit E, Roy O (2011) Toward a theory of play: a logical perspective on games and interaction. Games 2011 2(1):52–86. doi:10.3390/g2010052
108. van Benthem J (2012) In praise of strategies. In: van Eijck J, Verbrugge R (eds) Games, actions, and social software, vol 7010. Lecture notes in computer science. Springer, Heidelberg, pp 96–116
109. van Benthem J (2010) A problem concerning qualitative probabilistic update. Unpublished manuscript, 10 p
110. van Benthem J, Fernández-Duque D, Pacuit E (2012) Evidence logic: a new look at neighborhood structures. In: Bolander T et al (eds) Advances in modal logic. College Publications, London, pp 97–118
111. van Benthem J (2012) The nets of reason. Arg Comput 3(2/3):83–86
112. van Benthem J, Pacuit E (2014) Connecting logics of choice and change. In: Outstanding contributions to logic: Nuel Belnap on indeterminism and free action, Springer, New York
113. van Benthem J (2014) Logic in games. The MIT Press, Cambridge
114. van Benthem J, Baltag A, Smets S (2014) The music of knowledge, dynamic-epistemic logic and epistemology (Forthcoming)
115. van Benthem J (2014) Two logical faces of belief revision. In: Trypuz R (ed) Outstanding contributions to logic: Krister Segerberg on logic of actions. vol 1, pp 281–300
116. van Benthem J, Fernández-Duque D, Pacuit E (2014) Evidence and plausibility in neighborhood structures. Ann Pure Appl Log 165(1):106–133
117. van Benthem J, Liu F (nd) Deontic logic and changing preferences. In: Gabbay D, Horty J, van der Meyden R, Parent X, van der Torrre L (eds) Handbook of deontic logic and normative systems
118. Castelfranchi C, Paglieri F (2007) The role of beliefs in goal dynamics: prolegomena to a constructive theory of intentions. Synthese 155:237–263
119. Dégrémont, C (2010) The temporal mind: observations on belief change in temporal systems. Dissertation, ILLC, University of Amsterdam
120. Dégrémont C, Gierasimczuk N (2009) Can doxastic agents learn? On the temporal structure of learning. In: He X, Horty J, Pacuit E (eds) Proceedings of LORI. Lecture notes in computer science, vol 5834, pp 90–104. Springer, New York
121. Dégrémont C, Gierasimczuk N (2011) Finite identification from the viewpoint of epistemic update. Inf Comput 209(3):383–396
122. van Ditmarsch H (2000) Knowledge games. Dissertation. ILLC, University of Amsterdam and Department of Informatics, University of Groningen
123. van Ditmarch H, van der Hoek W, Kooi B (2007) Dynamic epistemic logic. Synthese Library, vol 337. Springer, Berlin
124. Dunin-Keplicz B, Verbrugge R (2002) Collective intentions. Fundam Inf 51(3):271–295
125. van Eijck J, de Vries F-J (1991) Dynamic interpretation and hoare deduction. J Log Lang Inf 1:1–44
126. van Eijck J, Ruan J, Sadzik T (2012) Action emulation. Synthese 185:131–151
127. van Eijck J (2014) Learning about probability. Unpublished manuscript, available from homepages.cwi.nl: ~/jve/ software/prodemo. (to appear)
128. Enderton H (1972) A mathematical introduction to logic. Academic Press, New York

129. Fagin R, Halpern JY, Moses Y, Vardi M (1995) Reasoning about knowledge. MIT Press, New York
130. Fiutek V (2013) Playing with knowledge and belief. Ph.D. Dissertation, ILLC, University of Amsterdam
131. Gerbrandy J (1999) Bisimulations on planet kripke. Dissertation, ILLC, University of Amsterdam
132. Gerbrandy J, Groeneveld W (1997) Reasoning about information change. J Log Lang Inf 6:147–169
133. Gierasimczuk N (2009) Bridging learning theory and dynamic epistemic logic. Synthese 169(2):371–384
134. Gierasimczuk N (2010) Knowing one's limits: logical analysis of inductive inference. Dissertation, ILLC, University of Amsterdam
135. Gierasimczuk N, de Jongh D (2013) On the complexity of conclusive update. Comput J 56(3):365–377
136. Gierasimczuk N, Kurzen L, Velázquez-Quesada F (2009) Learning and teaching as a game: a sabotage approach. In: He X, Horty J, Pacuit E (eds) Proceedings of LORI. Lecture notes in computer science, vol 5834, pp 119–132
137. Girard P (2008) Modal logic for belief and preference change. Ph.D. Dissertation, Department of Philosophy, Stanford University (ILLC-DS-2008-04)
138. Girard P, Seligman J, Liu F (2012) General dynamic logic. In: Bolander T, Braüner T, Ghilardi S, Moss I (eds) Advances in modal logics, vol 9, pp 239–260
139. Gold EM (1967) Language identification in the limit. Inf Control 10:447–474
140. Golddblatt R (1987) Logics of time and computation. Lecture notes. CSLI Publications, Stanford
141. Graedel E, Hirsch C (2002) M. back and forth between guarded and modal logics. ACM Trans Comput Log 3:418–463
142. Graedel E, Walukiewicz I (1999) Guarded fixed point logic. In: Proceedings of the 14th IEEE symposium on logic and computer science
143. Groenendijk J, Stokhof M (1991) Dynamic predicate logic. Linguist Philos 14:39–100
144. Grossi D (2009) Doing argumentation theory in modal logic. ILLC prepublication series PP-2009-24, Institute for Logic, Language and Computation
145. Grossi D (2010) On the logic of argumentation theory. In: van der Hoek W, Kaminka G, Lespérance Y, Sen S (eds) Proceedings of the 9th international conference on autonomous agents and multiagent systems (AAMAS 2010), IFAAMAS, pp 409–416
146. Grossi D (2011) Argumentation theory in the view of modal logic. In: McBurney P, Rahwan I (eds) Post-proceedings of the 7th international workshop on argumentation in multi-agent systems, no. 6614 in LNAI, pp 190–208
147. Harel D (1984) Dynamic logic. In: Gabbay DM, Guenthner F (eds) Handbook of philosophical logic, vol 2. Reidel, Dorcrecht, pp 497–604
148. Harrah D (1984) The logic of questions. In: Gabbay DM, Guenthner F (eds) Handbook of philosophical logic, vol 2. Reidel, Dordrecht, pp 715–764
149. Icard T, Holliday HW, Hoshi T (2013) Information dynamics and uniform substitution. Synthese
150. Icard T, Holliday HW, Hoshi T (2012) A uniform logic of information dynamics. Adv Modal Log 9
151. Icard T, Holliday HW (2010) Moorean phenomena in epistemic logic. Adv Modal Log 8
152. Hendricks V (2001) The convergence of scientific knowledge: a view from the limit. Kluwer Academic Publishers, Dordrecht
153. Hendricks V (2007) Mainstream and formal epistemology. Cambridge University Press, New York
154. Hintikka J (1973) Logic, language games and information. Clarendon Press, Oxford
155. Hoare CAR (1969) An axiomatic basis for computer programming. Commun ACM 12(10):576–585

156. Harman G (1985) Change in view: principles of reasoning. The MIT Press/Bradford Books, Cambridge
157. Holliday HW (2012) Knowing what follows: epistemic closure and epistemic logic. Ph.D. Thesis, Department of Philosophy, Stanford University
158. Hoshi T (2009) Epistemic dynamics and protocol information. Ph.D. Thesis, Department of Philosophy, Stanford University (ILLC-DS-2009-08)
159. Hoshi T, Yap A (2009) Dynamic epistemic logic with branching temporal structures. Synthese 169(2):259–281
160. Janin D, Walukiewicz I (1996) On the expressive completeness of the propositional mu-calculus with respect to monadic second order logic. In: International conference on concurrency theory (CONCUR), pp 263–277
161. Kelly K (1998a) Iterated belief revision, reliability, and inductive amnesia. Erkenntnis 50:11–58
162. Kelly K (1998b) The learning power of belief revision. In: TARK98: Proceedings of the 7th conference on theoretical aspects of rationality and knowledge, Morgan Kaufmann Publishers Inc., San Francisco, CA, USA, pp 111–124
163. Kelly K (2004) Learning theory and epistemology. In: Niiniluoto I, Sintonen M, Smolenski J (eds) Handbook of epistemology. Kluwer, Dordrecht
164. Kelly K, Schulte O, Hendricks V (1995) Reliable belief revision. In: Proceedings of the 10th international congress of logic, methodology, and philosophy of science. Kluwer Academic Publishers, Dordrecht, pp 383–398
165. Kooi B (2003) Probabilistic dynamic epistemic logic. J Log Lang Inf 12:381–408
166. Kurtonina N (1995) Frames and labels. A modal analysis of categorial deduction. Doctoral dissertation, Onderzoeksinstituut voor Taal, Spraak, university of Utrecht and ILLC, University of Amsterdam
167. Leal RA (2011) Modalities through the looking glass: a study on coalgebraic modal logic and their applications. Ph.D. Disseration, ILLC, University of Amsterdam
168. Lorenzen P, Lorenz K (1979) Dialogische Logik. Wissenschaftliche Buchgesellschaft, Darmstadt
169. Liu F (2008) Changing for the better: preference dynamics and agent diversity, Dissertation DS-2008-02, ILLC, University of Amsterdam
170. Marx M (1995) Arrow logic and relativized algebras of relations. Doctoral Dissertation, CCSOM and ILLC, University of Amsterdam
171. Martin E, Osherson D (1997) Scientific discovery based on belief revision. J Symbol Log 62(4):1352–1370
172. Martin E, Osherson D (1998) Elements of scientific inquiry. MIT Press, Cambridge
173. Minica S (2011) Dynamic logic of questions. Dissertation, DS-2011-08 ILLC, University of Amsterdam
174. Németi I (1985) The equational theory of cylindric relativized set algebras is decidable. Preprint no 63/95. Mathematical Institute, Hungarian Academy of Sciences, Budapest
175. Németi I (1995) Decidability of weakened versions of first-order logic. In: Csimarcz L, Gabbay DM, de Rijke M (eds) Logic colloquium '92. Studies in logic, language and information. CSLI Publications, Stanford, pp 177–241
176. Parikh R, Ramanujam R (2003) A knowledge based semantics of messages. J Log Lang Inf 12:453–467
177. Plaza J (1989) Logics of public communications. In: Proceedings 4th international symposium on methodologies for intelligent systems, pp 201–216
178. Renne B, Sack J, Yap A (2009) Dynamic epistemic temporal logic. In: Proceedings of LORI, pp 263–277
179. Renne B, Sack J, Yap A (2010) Dynamic epistemic temporal logic (extended version to appear)
180. de Rijke M (1992) A system of dynamic modal logic. Technical report CSLI-92-170, Stanford University

181. de Rijke M (1993) Extending modal logic. Doctoral dissertation, Institute for Logic, Language and Computation. University of Amsterdam
182. de Rijke M (1994) Meeting some neighbours. In: van Eijck J, Visser A (eds) Logic and information flow. MIT Press, Cambridge, pp 170–195
183. Rott H (2012) Bounded revision: two-dimensional belief change between conservative and moderate revision. J Philos Log 41:173–200
184. Roy O (2008) Thinking before acting: intentions, logic, and rational choice. Dissertation DS-2008-03, Institute for Logic, Language and Computation, University of Amsterdam
185. Sack J (2007) Adding temporal logic to dynamic epistemic logic. Ph.D. thesis, Indiana University at Bloomington
186. Sack J (2008) Temporal languages for epistemic programs. J Logic, Lang Inform 17(2):183–216
187. Sack J (2009) Extending probabilistic dynamic epistemic logic. Synthese 169(2):241–257
188. Sack J (2010) Logic for update products and steps into the past. Ann Pure Appl Log 161(12):1431–1461
189. Sarenac D (2006) Products of topological modal logics. Ph.D. thesis, Stanford University
190. Segerberg K (1998) Irrevocable belief revision in dynamic doxastic logic. Notre Dame J Formal Log 39(3):287–306
191. Segerberg K (2001) The basic dynamic doxastic logic of AGM. In: Williams MA, Rott H (eds) Frontiers in belief revision. Kluwer, Dordrecht, pp 57–84
192. Segerberg K, Leitgeb H (2007) Dynamic doxastic logic—why, how and where to? Synthese 155:167–190
193. Sietsma F, van Eijck J (2013) Action emulation between canonical models. J Philos Log 42(6):905–925
194. Stalnaker R (1972) Pragmatics. In: Davidson D, Harman G (eds) Semantics of natural language. Reidel, Dordrecht, pp 380–397
195. Vardi M (1997) Why is modal logic so robustly decidable? In: Immerman N, Kolaitis P (eds) Descriptive complexity and finite models. DIMACS series in discrete mathematics and theoretical computer science, AMS, vol 31, pp 149–184
196. Steinsvold C (2006) Topological models of belief logics. Ph.D. thesis, City University of New York, New York, USA
197. McKinsey J, Tarski A (1944) The algebra of topology. Ann Math 45(2):141–191
198. Velázquez-Quesada F (2010) Small steps in the dynamics of information. Dissertation, ILLC, University of Amsterdam
199. Veltman F (1991) Defaults in update semantics. Technical report LP-91-02. ILLC, University of Amsterdam (Published in Journal of Philosophical Logic)
200. Veltman F (2001) Een zogenaamde denkfout. Institute for Logic, Language and Computation, University of Amsterdam
201. Venema Y (1991) Many-dimensional modal logic. Doctoral Dissertation, Department of Mathematics and Computer Science, University of Amsterdam
202. Venema Y (1994) A crash course in arrow logic. Technical report 107. Logic Group Preprint Series. Utrecht University
203. Zvesper J (2010) Playing with information. Dissertation, ILLC, University of Amsterdam

Part I
Mathematical and
Computational Perspectives

Chapter 1
The Freedoms of (Guarded) Bisimulation

Erich Grädel and Martin Otto

Abstract We survey different notions of bisimulation equivalence that provide flexible and powerful concepts for understanding the expressive power as well as the model-theoretic and algorithmic properties of modal logics and of more and more powerful variants of guarded logics. An appropriate notion of bisimulation for a logic allows us to study the expressive power of that logic in terms of semantic invariance and logical indistinguishability. As bisimilar nodes or tuples in two structures cannot be distinguished by formulae of the logic, bisimulations may be used to control the complexity of the models under consideration. In this manner, bisimulation-respecting model constructions and transformations lead to results about model-theoretic properties of modal and guarded logics, such as the tree model property of modal logics and the fact that satisfiable guarded formulae have models of bounded tree width. A highlight of the bisimulation-based analysis are the characterisation theorems: inside a classical level of logical expressiveness such as first-order or monadic second-order definability, these provide a tight match between bisimulation invariance and logical definability. Typically such characterisation theorems state that a modal or guarded logic is not only invariant under bisimulation but, conversely, also expressively complete for the class of all bisimulation invariant properties at that level. Finally, the bisimulation-based analysis of modal and guarded logics also leads to important insights concerning their algorithmic properties. Since satisfiable formulae always admit simple models, for instance tree-like ones, and since modal and guarded logics can be embedded or interpreted in monadic second-order logic on trees, powerful automata theoretic methods become available for checking satisfiability and for evaluating formulae.

E. Grädel (✉)
RWTH Aachen University, Aachen, Germany
e-mail: graedel@logic.rwth-aachen.de

M. Otto
TU Darmstadt, Darmstadt, Germany
e-mail: otto@mathematik.tu-darmstadt.de

A. Baltag and S. Smets (eds.), *Johan van Benthem on Logic*
and Information Dynamics, Outstanding Contributions to Logic 5,
DOI: 10.1007/978-3-319-06025-5_1, © Springer International Publishing Switzerland 2014

1.1 Introduction

Bisimulation equivalence is one of the leading themes in modal logic. As the quintessential back-and-forth notion for two-player combinatorial games it may not only be regarded as a special case in the model-theoretic tradition of Ehrenfeucht–Fraïssé games but may also be seen as their common backbone. Bisimulation equivalence (of game graphs or transition systems) grasps the complex equivalence between dynamic behaviours as a natural structural equivalence. The generalisation of this graph-based bisimulation concept to higher dimensions in the form of guarded bisimulation opened up one further branch in the rich world of model-theoretic games; the study of guarded bisimulation in the wake of the inception of the guarded fragment of first-order logic in [1] has led to a new conceptual understanding of well-behaved logics that are 'modal' in a more general sense. Guarded logics far transcend basic modal logics while retaining some of the key features of modal model theory precisely through the parallelism between the underlying notions of bisimulation equivalence. Guarded bisimulation can be seen as derived from a hypergraph version of ordinary (modal, graph-based) bisimulation. And just as preservation under ordinary bisimulation accounts for much of the good model-theoretic behaviour of modal logics, so hypergraph bisimulation and guarded bisimulation are the keys to understanding the model theory of guarded logics. Model constructions and transformations that are compatible with guarded bisimulation account for the malleability of models and the tractability of the finite and algorithmic model theory of various guarded logics. We here survey and summarise a number of model-theoretic techniques and results, especially in the light of bisimulation respecting model constructions, including some more recent developments. Results to be surveyed include finite and small model properties, decidability results, complexity and expressive completeness issues. Among the more recent developments are notions of guardedness that focus on the role of negation rather than on just the quantification pattern. Unary and guarded negation bisimulation and the corresponding unary and guarded negation fragments of first-order logic from [10] and [3] have contributed yet another aspect to our understanding of the good behaviour of 'modal' logics with a yet wider scope.

1.2 Bisimulation: Behavioural and Structural Equivalence

1.2.1 Ehrenfeucht–Fraïssé, Back-and-forth, Zig-zag, Pebble Games: Games Model-Theorists Play

Notions like 'behaviour' and 'strategies' seem to be quintessentially dynamic, while the analysis of structure and structural comparisons are mostly construed as static concerns. Yet modal logics, transition systems and game graphs bridge the apparent gap in a natural manner and typically allow us to understand behavioural comparisons as structural comparisons, and behavioural equivalences as structural equivalences. This is not even really surprising if we remind ourselves how, e.g., game graphs can be regarded as extensional (and static) descriptions of the possible plays (hence behaviours) of the game, so that, e.g., the existence of a winning strategy for one of the players can be determined by structural analysis. The dynamics and intuitive appeal of games can also be harnessed for the analysis of the semantics and expressive power of logics: model checking games account for the evaluation of logical formulae over structures, and model comparison games are used to account for distinctions and degrees of indistinguishability between structures w.r.t. properties expressible in a given logic. In the classical context of first-order logic the model comparison games are at the centre of the Ehrenfeucht–Fraïssé technique.

In the world of modal logics, the essential model comparison game is the *bisimulation* game. It is a typical model-theoretic back and forth game, played by two players over the two structures at hand (Kripke structures or transition systems). A position in the game is a pair of (similar) nodes, one from each of the two structures, marked by pebbles; players take turns to move the pebbles along available transitions in the respective structure; in each new round the first player is free to choose one of the structures and one of the available transitions to move the pebble across that transition, and the second player must respond likewise in the opposite structure. Overall, the game protocol ensures that the second player has a winning strategy in a position precisely if—recursively—every transition in the one structure can be matched by a transition in the opposite structure, ad infinitum. Bisimulation relations and bisimulation equivalence capture this notion of game equivalence by means of back&forth closure conditions on a (or the maximal) set of pairs that are winning positions for the second player.

Definition 1.1 For structures $\mathfrak{A} = (A, (R_i^{\mathfrak{A}}), (P_j^{\mathfrak{A}}))$ and $\mathfrak{B} = (B, (R_i^{\mathfrak{B}}), (P_j^{\mathfrak{B}}))$ with binary accessibility relations R_i and unary predicates P_j:

A binary relation $Z \subseteq A \times B$ between the nodes of $\mathfrak{A}$ and nodes of $\mathfrak{B}$ is a *bisimulation relation* if for all $(a, b) \in Z$:

   (i)   (*atom eq.*): for each P_j, $a \in P_j^{\mathfrak{A}}$ iff $b \in P_j^{\mathfrak{B}}$;
  (ii)   (R_i-*back*): for every b' with $(b, b') \in R_i^{\mathfrak{B}}$ there is some a' such that $(a, a') \in R_i^{\mathfrak{A}}$ and $(a', b') \in Z$;
 (iii)   (R_i-*forth*): for every a' with $(a, a') \in R_i^{\mathfrak{A}}$ there is some b' such that $(b, b') \in R_i^{\mathfrak{B}}$ and $(a', b') \in Z$.

As the union of bisimulation relations is again a bisimulation relation, there is a well-defined $\subseteq$-maximal *largest bisimulation* between $\mathfrak{A}$ and $\mathfrak{B}$. Pointed structures $\mathfrak{A}, a$ and $\mathfrak{B}, b$ are *bisimilar*, $\mathfrak{A}, a \sim \mathfrak{B}, b$, if (a, b) is in some (hence in the largest) bisimulation between $\mathfrak{A}$ and $\mathfrak{B}$.

Clearly $\sim$ captures a strong form of behavioural equivalence, if we think of 'behaviours' not just as traces of actions, but rather as the complex interactive and responsive patterns that can evolve in any step-wise alternating exploration of potential transitions. The conditions (R_i-*back*) and (R_i-*forth*) capture the challenge-response requirements posed for the second player by one additional round.

Correspondingly, the largest bisimulation on $A \times B$ forms a greatest fixed point w.r.t. the refinement operator induced by (*atom eq.*) and the (R_i-*back*) and (R_i-*forth*) conditions:

$$Z \longmapsto \mathscr{F}(Z),$$

where $\mathscr{F}(Z)$ consist of those pairs $(a, b) \in Z$ that satisfy (*atom eq.*) and the (R_i-*back*) and (R_i-*forth*) conditions w.r.t. Z. Locally, over every pair of structures, the bisimulation relation $\sim$ is the greatest fixed point of this operation $\mathscr{F}$ (which is guaranteed to exist since $\mathscr{F}$ is monotone w.r.t. $\subseteq$).

This direct—more static—description of the target equivalence as a greatest fixed point is typical for comparison games of this kind; in the case of bisimulation equivalence the typical back and forth conditions were introduced in the modal world under the name of *zig-zag* conditions by Johan van Benthem. The term *bisimulation* equivalence, which points to an intuition based on the behaviour of transition systems, was introduced by Milner and Park.

A more dynamic view is also extracted from the greatest fixed point characterisation, if we look at the refinement process that recursively generates the fixed point $\sim$ as a limit of relations $\sim^\alpha$:

$$\sim = \bigcap_\alpha \sim^\alpha, \quad \text{where}$$

$$\sim^0 = \text{atom equivalence,}$$
$$\sim^{\alpha+1} = \mathscr{F}(\sim^\alpha),$$
$$\sim^\lambda = \bigcap_{\alpha < \lambda} \sim^\alpha \quad \text{for limit ordinals } \lambda.$$

Formally, the intersection in the above definition of $\sim$ is over all ordinal levels α, but in restriction to any two concrete structures can be bounded by any infinite ordinal that is of cardinality greater than the structures at hand. Over all finite, and indeed over finitely branching structures and also over the class of all ω-saturated or the class of all modally saturated structures, the limit is reached by stage ω, i.e., coincides with the limit of the finite approximations $\sim^\ell$ for $\ell \in \mathbb{N}$,

$$\sim^\omega = \bigcap_{\ell \in \omega} \sim^\ell .$$

Over finite $\mathfrak{A}$ and $\mathfrak{B}$ of sizes $|A|$ and $|B|$, the natural game analysis even shows that full bisimulation is reached no later than by level $\sim^\ell$, where $\ell = \max(|A|, |B|)$.

The game counterpart of $\sim^\ell$ for $\ell \in \mathbb{N}$ is the ℓ-round bisimulation game, which is won by the second player if she does not lose during the first ℓ rounds. Bisimulation equivalence and its infinite game, and especially its finite approximations $\sim^\ell$ for $\ell \in \mathbb{N}$ in relation to the ℓ-round game, can be viewed as a special adaptation to the modal scenario of the classical back&forth games in the Ehrenfeucht–Fraïssé tradition.

We write $\mathfrak{A}, a \equiv^{\ell}_{\mathrm{ML}} \mathfrak{B}, b$ for the modal levels of elementary equivalence up to quantifier rank (modal nesting depth) ℓ: $\mathfrak{A}, a \equiv^{\ell}_{\mathrm{ML}} \mathfrak{B}, b$ if $\mathfrak{A}, a \models \varphi \Leftrightarrow \mathfrak{B}, b \models \varphi$ for all $\varphi \in \mathrm{ML}$ of nesting depth up to ℓ. Similarly, $\mathfrak{A}, a \equiv_{\mathrm{ML}} \mathfrak{B}, b$ stands for full modal equivalence, and $\mathfrak{A}, a \equiv^{\infty}_{\mathrm{ML}} \mathfrak{B}, b$ for equivalence w.r.t. the infinitary variant of modal logic which allows for infinite conjunctions and disjunctions.

Theorem 1.2 (Ehrenfeucht–Fraïssé and Karp theorems for ML) *In restriction to finite modal vocabularies, and for every $\ell \in \mathbb{N}$:*

$$\mathfrak{A}, a \sim^{\ell} \mathfrak{B}, b \quad \textit{if, and only if,} \quad \mathfrak{A}, a \equiv^{\ell}_{\mathrm{ML}} \mathfrak{B}, b.$$

Consequently, in restriction to finite modal vocabularies $\mathfrak{A}, a \sim^{\omega} \mathfrak{B}, b$ if, and only if, $\mathfrak{A}, a \equiv_{\mathrm{ML}} \mathfrak{B}, b$. Without any restriction on the size of the modal vocabulary,

$$\mathfrak{A}, a \sim \mathfrak{B}, b \quad \textit{if, and only if,} \quad \mathfrak{A}, a \equiv^{\infty}_{\mathrm{ML}} \mathfrak{B}, b.$$

Many other logics, and in particular other fragments of first-order logic besides the modal fragment, can be analysed via specifically associated Ehrenfeucht–Fraïssé games. The analysis of the guarded fragment GF of first-order logic in the light of its invariance under guarded bisimulation equivalence is a prime example to be discussed in Sect. 1.3. The very proposal of GF in [1] was inspired by considerations concerning the taming of first-order logic through variations that involve a generalised (or, depending on the point of view: restricted) semantics in 'general assignment models' in the sense of [6]. Returning to our opening remarks about 'behaviour' in terms of logic and games, different logics with distinct semantics may be obtained by admitting different *observable configurations* and different *modes of navigation* between these. (For classical modal semantics, think of possible worlds and accessibility relations.) It is in this view, that games and game graphs provide yet another link to bisimulation as the quintessential notion of behavioural equivalence. Bisimulation as the master game equivalence is adaptable to different logics if, instead of the usual structures, we look at the game graphs induced by the semantic games of those other logics. For suitable logics, the associated game graphs formalise the notion of observable configurations (or admissible assignments) and transitions between these (quantification patterns). Thus, levels of bisimulation equivalence between the associated game graphs correspond to levels of Ehrenfeucht–Fraïssé equivalence between the underlying structures, capturing the specific restrictions embodied in the semantics of the logic in question. Some correspondences of this kind are explored at first-order level in [20], and, with much greater generality in mind, in [6], in the terminology of *general assignment models*. In the same vein, suitable abstractions of the associated game graphs (intuitively akin to filtrations or bisimulation quotients) may serve as concise descriptions of structures up to equivalence, or as blue-prints for desired models (quasi-models) towards decidability and complexity arguments.

1.2.2 Bisimulation in Modal Model Theory

The essential observation for a view of bisimulation equivalences as specialisations of corresponding classical first-order Ehrenfeucht–Fraïssé equivalences is the manner in which its back&forth conditions precisely reflect the power of modal quantification. The existential diamond modality $\Diamond_i$, whose semantics in structure $\mathfrak{A}$ is defined in terms of the accessibility relation $R_i^{\mathfrak{A}}$, precisely captures the available moves in the game along R_i-transitions, and the back&forth clauses for R_i reflect potential distinctions w.r.t. properties of nodes accessible from the current nodes through R_i-edges in their respective structures.

On the other hand, the bisimulation games can be taken as the quintessential template for a large class of model-theoretic Ehrenfeucht–Fraïssé style comparison games: if we correctly abstract from the structures at hand a game graph that models the relevant configurations and transitions between them, then levels of bisimulation equivalence correspond to winning strategies for the second player in a game that reflects the expressive power and quantification pattern of some other target logic [20]. In some key examples, the relevant configurations correspond to the *admissible assignments* to first-order variables, and the transitions to their relative accessibility by means of basic quantification steps. In this vein, variations and especially restrictions to the admissible assignments in a first-order framework lead to fragments that can be analysed and understood in terms of bisimulation equivalences between derived game graphs. Among the most pertinent examples are the k-variable fragments FO^k of first-order logic, and the guarded fragment GF of first-order logic. The finite variable fragments FO^k work with a uniform restriction of assignments to size k. This purely quantitative restriction is contrasted in the seminal paper on the guarded fragment [1] by Andréka, van Benthem and Németi with a qualitative restriction of assignments to clusters that are 'guarded' by some relational hyperedge. The new fragment is proposed with a view to a 'dynamic' bounding of the available assignments—it is 'dynamic' in the sense of a position-dependent restriction familiar from modal logics; yet static in the sense of structural analysis. We shall discuss the guarded fragment and the associated ramification of bisimulation in Sect. 1.3. Before that, let us summarise some key features and uses of ordinary, modal bisimulation equivalence, which account for its pivotal role in modal model theory.

The first is a direct corollary of the modal Ehrenfeucht–Fraïssé theorem. If $\varphi \in$ ML has modal quantifier depth ℓ, then its semantics is invariant under $\sim^\ell$.

The essential feature of *bisimulation invariance* extends to more powerful logics that share the underlying modal quantification pattern, like the modal μ-calculus.

Corollary 1.3 *The semantics of basic modal logic* ML *is invariant under bisimulation equivalence: for* $\varphi \in$ ML, $\mathfrak{A}, a \sim \mathfrak{B}, b \implies \mathfrak{A}, a \models \varphi \Leftrightarrow \mathfrak{B}, b \models \varphi$.

Bisimulation invariance is *the* model-theoretic hallmark of modal logics; in fact so much so, that modal model theory could be equated with *model theory up to bisimulation equivalence*.

1.2.3 Tree Models and Robust Decidability of Modal Logics

The familiar process of tree-unfolding takes a pointed structure $\mathfrak{A}$, a to a tree structure $\mathfrak{A}_a^*$ with root a, built on the tree of all R_i-labelled paths from a in $\mathfrak{A}$.

Definition 1.4 Let $\mathfrak{A} = (A, (R_i^{\mathfrak{A}}), (P_j^{\mathfrak{A}}), a)$ be a pointed structure (Kripke structure or transition system). Its *tree unfolding* from a is the tree-like structure $\mathfrak{A}_a^* = (A_a^*, (R_i^{\mathfrak{A}^*}), (P_j^{\mathfrak{A}^*}))$ with root a, where A_a^* is the set of edge-labelled paths of the form $w = (a_0, i_0, a_1, \ldots, a_\ell, i_\ell, a_{\ell+1}, \ldots, a_n)$ where $a_0 = a$, i_ℓ such that $e_\ell = (a_\ell, a_{\ell+1}) \in R_{i_\ell}^{\mathfrak{A}}$, with the natural projection

$$\pi : A_a^* \longrightarrow A$$
$$(a_0, \ldots, a_n) \longmapsto a_n;$$

$(w, w') \in R_i^{\mathfrak{A}^*}$ if w' is an extension of w by one R_i-edge, $w = w\widehat{}(i, a')$; and $w \in P_j^{\mathfrak{A}^*}$ if $\pi(w) \in P_j^{\mathfrak{A}}$.

Clearly $\mathfrak{A}^*$, $a \sim \mathfrak{A}$, a. It follows that any bisimulation invariant logic has the *tree model property*. For the finite-depth approximation $\sim^\ell$ of $\sim$, even the truncation $\mathfrak{A}_a^\ell$ to paths of lengths $n \leqslant \ell$ from a satisfies $\mathfrak{A}^\ell$, $a \sim^\ell \mathfrak{A}$, a. For finite vocabulary (finitely many R_i and P_j), the equivalence relation $\sim^\ell$ has finite index. Therefore, $\mathfrak{A}_a^\ell$ can be pruned so as to retain at most one sibling of each $\sim^\ell$-type among the immediate children of any node, without affecting $\sim^\ell$-types. For basic modal logic, this pruning yields finite tree models.

Corollary 1.5 *Every satisfiable formula $\varphi \in$ ML (of modal quantifier depth ℓ) has a finite tree model (of depth ℓ).*

These observations are essential for decidability and complexity results for the satisfiability problem, and for what has been called the robust decidability of modal logics. Indeed, it is not just the basic propositional modal logic ML that is decidable for satisfiability. This property is shared by many extensions of ML to much stronger and practically more relevant logics, including linear or branching time temporal logics such as LTL, CTL, CTL*, dynamic logics of programs such as PDL, Parikh's game logic GL and the modal μ-calculus L_μ, the extension of ML by least and greatest fixed points. While basic modal logic ML can be seen as a fragment of first-order logic, this is not the case for these stronger logics; all of them can express properties based on reachability and on other non-local properties that are not first-order. However, it is easy to see that all these logics can be embedded into monadic second-order logic MSO. Among the extensions of modal logics, the modal μ-calculus occupies a special rôle. It encompasses the other logics mentioned (and many more) and it has a clean and interesting model theory. The modal μ-calculus remains decidable in the presence of backward modalities.

The tree model property provides powerful tools for proving decidability and complexity results and for constructing efficient decision procedures. For a quick proof of decidability one can translate formulae of these logics into monadic second-order formulae and invoke Rabin's famous theorem saying that SωS, the monadic theory of the ω-branching tree, is decidable [27]. However, the complexity of monadic logics on infinite trees (and words) is non-elementary. But recall that the proof of Rabin's Theorem is based on tree automata. A much more practical approach for constructing decision procedures for modal logics avoids the detour through monadic second-order logic and directly applies suitable variants of tree automata to modal logics. The theory of finite automata on trees is very well developed, with many different automata models tailored for specific applications, with efficient algorithms for manipulating automata and for reductions between different models, a good understanding of the complexity of the common reasoning tasks for automata (emptiness problems, word problems etc.), and sophisticated optimisation techniques. The tree model property paves the way to make tree automata applicable to the world of modal logics.

The typical complexity level of satisfiability problems for modal logics is EXPTIME. An exception is the basic modal logic ML for which satisfiability is PSPACE-complete. But the addition of rather modest features to ML, for instance a global modality, push up the complexity to EXPTIME; on the other hand, also rather strong extensions of ML such as the modal μ-calculus and even the modal μ-calculus with backward modalities remain EXPTIME-complete. Such results rely on efficient translations of formulae into, say, alternating tree automata, and the EXPTIME-completeness of the emptiness problem for such automata.

1.2.4 Expressive Completeness

As mentioned above, one of the highlights of modal model theory in this sense is the characterisation of basic modal logic as the bisimulation-invariant fragment of first-order logic.

Theorem 1.6 (van Benthem) *For every first-order formula $\varphi(x)$ in a vocabulary of binary relations R_i and unary predicates P_j as above, the following are equivalent:*

(i) φ is bisimulation invariant.
(ii) φ is logically equivalent to a formula of basic modal logic ML.

In shorthand notation, FO/$\sim$ $\equiv$ ML, where the left-hand side suggestively stands for the (syntactically undecidable) collection of bisimulation invariant first-order formulae.

By no means a direct consequence, not even via the finite model property, but rather yet another striking feature of bisimulation equivalence and of modal logic, the same characterisation holds also in the sense of finite model theory:

$$\text{FO}/\sim \ \equiv \ \text{ML} \quad \text{(FMT)}.$$

In its basic form this result is due to Rosen [28]; alternative proofs that yield strengthenings and lend themselves to further generalisations have been presented in [19]. We state a few of these generalisations from [11, 18]. *Global bisimulation equivalence*, $\mathfrak{A}, a \sim_\forall \mathfrak{B}, b$, refers to a bisimulation relation in which every $a \in A$ is matched to some $b \in B$ and vice versa; *modal logic with a global modality*, ML[$\forall$], is the extension of basic modal logic ML by a global modality, with the full binary relation as its accessibility relation. A *rooted structure* is a structure $\mathfrak{A}, a$ with a single binary accessibility relation R such that every node is reachable on a directed R-path from the root a. *Equivalence structures* are structures that interpret all the binary relations R_i as equivalence relations (*S5* models).

Theorem 1.7 *Bisimulation invariant fragments of first-order logic are captured by modal logics over some classes of structures, as follows.*

   (i) FO/$\sim$ $\equiv$ ML *over the class of all (finite) structures.*
  (ii) FO/$\sim_\forall$ $\equiv$ ML[$\forall$] *over the class of all (finite) structures.*
 (iii) FO/$\sim$ $\equiv$ ML *over the class of all (finite) equivalence structures.*
 (iv) FO/$\sim$ $\equiv$ ML[$\forall$] *over the class of all finite rooted structures.*
 (iv) FO/$\sim$ $\equiv$ ML *over the class of all finite irreflexive transitive trees.*

Here (i) is the van Benthem–Rosen characterisation from [5] and [28], respectively; the rest are due to [11, 18].

Several of the finite model theory results above make use of *finite* unfoldings of finite structures that produce *locally* tree-like and fully bisimilar finite models—which is not achievable by tree unfoldings since any globally acyclic bisimilar companion of any cyclic structure is necessarily infinite. Simple combinatorial constructions of finite locally acyclic bisimilar covers of finite graphs for this purpose are presented in [18]. They play a crucial role in the analysis of the expressiveness of first-order formulae that are bisimulation invariant over finite structures. Locally acyclic behaviour suffices due to Gaifman's locality theorem: the semantics of any first-order formula $\varphi(x)$ only depends on certain global multiplicities and the local neighbourhood around x; up to bisimulation, global multiplicities (Gaifman's basic local sentences) can be adjusted comparatively easily even when working in special classes of finite models; what remains is the necessity to control the local neighbourhoods and this is where local tree-likeness is useful.

The van Benthem characterisation of bisimulation invariant first-order logic, as FO/$\sim$ $\equiv$ ML, also has a an exciting extension to its monadic second-order counterpart:

Theorem 1.8 (Janin–Walukiewicz) MSO/$\sim$ $\equiv$ L_μ, *i.e., for every monadic second-order formula* $\varphi(x)$ *in a vocabulary of binary relations* R_i *and unary predicates* P_j, *the following are equivalent:*

   (i) φ *is bisimulation invariant.*
  (ii) φ *is logically equivalent to a formula of the μ-calculus* L_μ.

Whether this characterisation holds in the sense of finite model theory, remains one of the great challenges in modal model theory.

1.3 Guarded Bisimulation: A Systematic Lifting to Higher Dimension

The 'dynamic' behaviour of modal logics w.r.t. locally available transitions between single-node assignments is vastly generalised in the setting of guarded logics.

The generalisation manifests itself on various levels: as a liberalisation in the relational type of structures (from graph-like transition systems to relational structures with relations of any arity); a generalisation w.r.t. the restrictions on admissible assignments and quantification patterns (from modal $\square$ and $\Diamond$ to universal and existential quantification over guarded tuples); a generalisation w.r.t. the relevant notion of bisimulation (from modal to guarded bisimulation); and, in the wake of these generalisations, a shift form graph theory to hypergraph theory as the underlying combinatorial framework.

1.3.1 Guardedness and the Guarded Fragment

With a relational structure $\mathfrak{A} = (A, (R_i^{\mathfrak{A}})_{i \in I})$ with relation symbols R_i of arity r_i, we associate a hypergraph of guarded sets, and a notion of guarded tuples as follows. It will be convenient to use the notation $[\mathbf{a}] := \{a_1, \ldots, a_k\}$ to denote the set of components of the tuple $\mathbf{a} = (a_1, \ldots, a_k) \in A^k$.

Definition 1.9 A subset $s \subseteq A$ is *guarded* in $\mathfrak{A}$ if s is a singleton set or if there is some tuple $\mathbf{a} \in R_i^{\mathfrak{A}}$ for one of the R_i such that $s \subseteq [\mathbf{a}]$. The *hypergraph of guarded sets* of $\mathfrak{A}$ is the hypergraph $H(\mathfrak{A}) := (A, S[\mathfrak{A}])$ with the set S of all guarded subsets of $\mathfrak{A}$ as the set of hyperedges. A tuple $\mathbf{a} \in A^k$ is a *guarded tuple* if $[\mathbf{a}] \in S(\mathfrak{A})$.

The guarded fragment of first-order logic essentially restricts the relevant assignments of first-order variables to guarded tuples. The actual definition is in terms of the restriction of all quantification by means of an explicit relativisation to some guarded tuples. It thus allows only outermost free variables to be instantiated by unguarded assignments, but for many purposes this does not matter (since outer boolean combinations could be treated separately).

Definition 1.10 For arbitrary relational vocabularies, the guarded fragment GF $\subseteq$ FO is the syntactic fragment of FO generated from atomic formulae by the boolean connectives and quantifications of the form

$$\forall \mathbf{y}\big(\alpha(\mathbf{xy}) \to \varphi(\mathbf{xy})\big), \quad \text{and, dually,} \quad \exists \mathbf{y}\big(\alpha(\mathbf{xy}) \wedge \varphi(\mathbf{xy})\big),$$

where $\varphi(\mathbf{xy}) \in$ GF has free variables among those listed in $\mathbf{xy}$ and $\alpha(\mathbf{xy})$ is an atomic formula in which all the listed variables occur. The formula α is called the *guard* of this quantification.[1] The semantics of GF is that of FO.

---

[1] If $\mathbf{xy}$ consists of a single variable symbol z, α can be the equality $z = z$.

The definition generalises the relativised quantification of modal logic, so that it is clear that, w.r.t. expressiveness, ML $\subseteq$ GF $\subseteq$ FO, and in fact even the extension of basic modal logic by global and backward modalities is naturally covered by GF.

1.3.2 Guarded Bisimulation and Model Theory

Just as the model theory of modal logics is governed by (modal) bisimulation equivalence, the nice model-theoretic properties of the guarded fragment are closely related to its invariance under guarded bisimulation equivalence. Guarded bisimulation equivalence $\sim_g$ and its finite approximations $\sim_g^\ell$ exactly cover the same station for GF as do $\sim$ and $\sim^\ell$ for ML—also w.r.t. their nature as the appropriate specialisations of the first-order framework of back&forth games to the quantification pattern of GF.

The positions of the guarded bisimulation game on structures $\mathfrak{A}$ and $\mathfrak{B}$ are partial isomorphisms between A and B whose domain and image are guarded sets[2]; we use a tuple-based notation $p: \mathbf{a} \mapsto \mathbf{b}$ to indicate a partial map from A to B with domain $[\mathbf{a}]$ and image $[\mathbf{b}]$ where $b_i = p(a_i)$. One may also think of a placement of matched pebbles on $\mathbf{a}$ and $\mathbf{b}$; the requirements are that $\mathbf{a}$ and $\mathbf{b}$ are guarded and that $p: \mathfrak{A} \restriction [\mathbf{a}] \simeq \mathfrak{B} \restriction [\mathbf{b}]$ is an isomorphism of induced substructures (p a partial isomorphism, $\mathbf{a}$ and $\mathbf{b}$ atom equivalent). Then the available moves for the first player, e.g. on the $\mathfrak{A}$-side, are to guarded tuples $\mathbf{a}'$ together with some specified sub-tuple $\mathbf{a}_0$ of both $\mathbf{a}$ and $\mathbf{a}'$ that stay put—and the response by the second player needs to keep the sub-tuple $\mathbf{b}_0 := p(\mathbf{a}_0)$ fixed and produce an extension $\mathbf{b}'$ such that the new $p': \mathbf{a}' \mapsto \mathbf{b}'$ is again a partial isomorphism between $\mathfrak{A}$ and $\mathfrak{B}$.

An alternative set-based view has partial isomorphisms between guarded subsets as the positions; the moves correspond to transitions from one guarded subset to another, with a specified (possible empty) subset of their intersection to be respected by the second player's response. This view highlights the hypergraph-theoretic nature, and indeed can be cast as a notion of *hypergraph bisimulation* that additionally needs to respect relational content.

Definition 1.11 For two relational structures $\mathfrak{A}$ and $\mathfrak{B}$ (of the same vocabulary), a set of partial maps Z between $\mathfrak{A}$ and $\mathfrak{B}$ is a *guarded bisimulation* if it satisfies the following, for every $p: \mathbf{a} \mapsto \mathbf{b}$ in Z:

(i) *(atom eq.)*: $p: \mathfrak{A} \restriction \mathbf{a} \simeq \mathfrak{B} \restriction \mathbf{b}$ is a partial isomorphism;
(ii) *(back)*: for every guarded tuple $\mathbf{b}'$ of $\mathfrak{B}$ and $\mathbf{b}_0$ with $[\mathbf{b}_0] \subseteq [\mathbf{b}] \cap [\mathbf{b}']$, there is some guarded tuple $\mathbf{a}'$ of $\mathfrak{A}$ and $p': \mathbf{a}' \mapsto \mathbf{b}'$ in Z such that $p'^{-1}(\mathbf{b}_0) = p^{-1}(\mathbf{b}_0)$;
(iii) *(forth)*: for every guarded tuple $\mathbf{a}'$ of $\mathfrak{A}$ and $\mathbf{a}_0$ with $[\mathbf{a}_0] \subseteq [\mathbf{a}] \cap [\mathbf{a}']$, there is some guarded tuple $\mathbf{b}'$ of $\mathfrak{B}$ and $p': \mathbf{a}' \mapsto \mathbf{b}'$ in Z such that $p'(\mathbf{a}_0) = p(\mathbf{a}_0)$.

---

[2] One should except the initial position from the guardedness requirement in order to match the liberal treatment of (outermost) free variables in GF.

We write $\mathfrak{A}, \mathbf{a} \sim_g \mathfrak{B}, \mathbf{b}$ if there is a guarded bisimulation Z containing $p: \mathbf{a} \mapsto \mathbf{b}$. Finite approximations $\sim_g^\ell$ are introduced in complete analogy with the modal $\sim$ and $\sim^\ell$, and similarly correspond to the existence of winning strategies for ℓ rounds in the guarded bisimulation game. As in the modal case, we introduce $\sim_g^\omega$ as the common refinement of the finite levels $\sim_g^\ell$.

One obtains the natural variant of the first-order Ehrenfeucht–Fraïssé and Karp theorems for GF. The equivalence relations $\equiv_{GF}^\ell$ and $\equiv_{GF}$ are introduced as levels of elementary equivalence in GF, where the ℓ in $\equiv_{GF}^\ell$ refers to the nesting depth of guarded quantification (which is typically lower than the first-order quantifier rank, as guarded quantification may quantify over tuples in a single step). The relation $\equiv_{GF}^\infty$ similarly denotes equivalence w.r.t. the infinitary variant of GF, with infinite disjunctions and conjunctions.

Theorem 1.12 (Ehrenfeucht–Fraïssé and Karp theorems for GF) *In restriction to finite relational vocabularies, and for every $\ell \in \mathbb{N}$:*

$$\mathfrak{A}, \mathbf{a} \sim_g^\ell \mathfrak{B}, \mathbf{b} \quad \textit{if, and only if,} \quad \mathfrak{A}, \mathbf{a} \equiv_{GF}^\ell \mathfrak{B}, \mathbf{b}.$$

Consequently, in restriction to finite vocabularies $\mathfrak{A}, \mathbf{a} \sim_g^\omega \mathfrak{B}, \mathbf{b}$ if, and only if, $\mathfrak{A}, \mathbf{a} \equiv_{GF} \mathfrak{B}, \mathbf{b}$. Without any restriction on the size of the vocabulary,

$$\mathfrak{A}, \mathbf{a} \sim_g \mathfrak{B}, \mathbf{b} \quad \textit{if, and only if,} \quad \mathfrak{A}, \mathbf{a} \equiv_{GF}^\infty \mathfrak{B}, \mathbf{b}.$$

1.3.3 Guarded Bisimulation Invariance

The following is an immediate consequence of the guarded Ehrenfeucht–Fraïssé theorem.

Corollary 1.13 *The semantics of $\varphi \in$ GF is invariant under $\sim_g$.*

The expressive completeness assertion in the following characterisation theorem of Andréka–van Benthem–Németi rests on a non-trivial but canonical classical proof by means of compactness and saturation. It provides a beautiful analogue and generalisation of van Benthem's semantic characterisation of ML $\subseteq$ FO, Theorem 1.6.

Theorem 1.14 (Andréka–van Benthem–Németi) *The guarded fragment is semantically characterised as a fragment of first-order logic by its invariance under guarded bisimulation equivalence: $FO/\sim_g \equiv$ GF. In more detail, for every first-order formula $\varphi(\mathbf{x})$ in a relational vocabulary, the following are equivalent:*

(i) φ is invariant under guarded bisimulation.
(ii) φ is logically equivalent to a formula of GF.

Moreover, a guarded analogue of the Janin–Walukiewicz Theorem (Theorem 1.8) can also be obtained via a natural translation between the guarded and modal worlds. The logics involved are the following: *guarded second-order logic* GSO, which here takes the place of MSO, is the natural restriction of second-order logic that allows to quantify over *sets of guarded tuples*; *guarded fixpoint logic* μGF is the extension of GF by constructors for least and greatest fixed points.

Theorem 1.15 (Grädel–Hirsch–Otto) GSO/$\sim_g$ $\equiv$ μGF, *i.e., For every GSO-formula* $\varphi(\mathbf{x})$, *the following are equivalent:*

(i) φ *is invariant under guarded bisimulation equivalence.*
(ii) φ *is logically equivalent to a formula of* μGF.

The translations in [14] that directly reduce this assertion to Theorem 1.8 involve an interesting parallelism between modal and guarded tree unfoldings.

Guarded tree unfoldings of relational structures $\mathfrak{A} = (A, (R^{\mathfrak{A}}))$ can be constructed from a tree unfolding of the associated transition system $I(\mathfrak{A}) = (S[\mathfrak{A}] \cup \{\emptyset\}, E)$ where $S[\mathfrak{A}]$ is the set of guarded subsets of $\mathfrak{A}$ and $E = \{(s, s') : s \neq s', s = \emptyset \text{ or } s \cap s' \neq \emptyset\}$.[3] From a tree unfolding $I^* := I^*_\emptyset$ of $I(\mathfrak{A})$ from the root node $\emptyset$, with natural projection $\pi : I^* \to S(\mathfrak{A}) \cup \{\emptyset\}$ we reconstruct a relational structure

$$\hat{\mathfrak{A}} = (\hat{A}, (R^{\hat{\mathfrak{A}}}))$$

as follows. The universe $\hat{A}$ is the quotient of the disjoint union of copies of sets $\pi(\hat{s}) \subseteq A$,

$$\bigcup_{\hat{s} \in I^*} \{\hat{s}\} \times \pi(\hat{s})$$

w.r.t. the equivalence relation that identifies $a \in \pi(\hat{s}_1)$ with $a \in \pi(\hat{s}_2)$ if, and only if, $\hat{s}_2$ and $\hat{s}_1$ are connected in $I^*_\emptyset$ by a path whose π-projection involves just edges $e = (s, s') \in E$ for which $a \in s \cap s'$. We denote the equivalence class of $(\hat{s}, a)$ for $a \in \pi(\hat{s})$ by $[\hat{s}, a]$, and the set $\{[\hat{s}, a] : a \in \pi(\hat{s})\} \subseteq \hat{A}$ by $[\hat{s}]$. The map that sends the equivalence class $[\hat{s}, a]$ of $a \in \hat{s}$ to $a \in A$ is the natural projection associated with the unfolding, for simplicity also denoted $\pi : \hat{A} \to A$. Locally, in restriction to every $[\hat{s}] \subseteq \hat{A}$, this projection π is a bijection onto the corresponding guarded subset $s = \pi(\hat{s})$ of $\mathfrak{A}$. Relations R are interpreted in $\hat{\mathfrak{A}}$ such that precisely the sets $[\hat{s}] \subseteq \hat{A}$ are guarded subsets, and such that $\pi : \hat{\mathfrak{A}} \to \mathfrak{A}$ is a global relational homomorphism and a local isomorphism in restriction to every subset $[\hat{s}]$.

Definition 1.16 The *guarded tree unfolding* of a relational structure $\mathfrak{A} = (A, (R^{\mathfrak{A}}))$ is the structure $\hat{\mathfrak{A}} = (\hat{A}, (R^{\hat{\mathfrak{A}}}))$ as constructed from a tree unfolding of the intersection graph $I(\mathfrak{A})$ above, together with the natural homomorphic projection $\pi : \hat{\mathfrak{A}} \to \mathfrak{A}$, which bijectively associates the guarded subsets $[\hat{s}] \in S(\hat{\mathfrak{A}})$ with their underlying guarded subsets $s = \pi(\hat{s}) \in S[\mathfrak{A}]$.

---

[3] We attach the empty set as a root to $I(\mathfrak{A})$ and join it to every guarded set to obtain a natural tree unfolding for our purposes, rather than a forest.

It is straightforward to check that the restrictions of the projection homomorphism $\pi : \hat{\mathfrak{A}} \to \mathfrak{A}$ to the guarded subsets of $\hat{\mathfrak{A}}$ form a guarded bisimulation. Therefore, for any guarded subset $[\hat{s}]$ of $\hat{\mathfrak{A}}$ above the guarded subset $s = \pi(\hat{s})$ of $\mathfrak{A}$,

$$\hat{\mathfrak{A}}, [\hat{s}] \sim_g \mathfrak{A}, s,$$

where we allow ourselves to write just the guarded sets $[\hat{s}]$ and s, instead of π-compatible listings of their elements as tuples.

Tree unfoldings as just defined are tree-like also in the sense that their hypergraphs of guarded subsets $S[\mathfrak{A}]$ are *acyclic*. There are several equivalent characterisations of the relevant notion of hypergraph acyclicity (also called α-acyclicity in the literature, cf. [4, 9]): in terms of *tree decompositions* that use guarded subsets (hyperedges) as bags; in terms of reducibility by means of reduction steps that allow for

(i)   removal of a vertex (from the universe and every hyperedge) provided it is contained in at most one hyperedge, and
(ii) retraction of a hyperedge provided it is fully contained in some other hyperedge;

and in terms of the local criteria of *conformality* and *chordality* for the hypergraph and its associated Gaifman graph.

Definition 1.17 For a hypergraph $H = (A, S)$, define the associated *Gaifman graph* $G(H)$ to have vertex set A and an edge between distinct $a, a' \in A$ precisely if a and a' occur together in some hyperedge $s \in S$.

The hypergraph $H = (A, S)$ is *acyclic* if it is both

(i)   *conformal*: each clique in $G(H)$ is contained in a single hyperedge, and
(ii) *chordal*: every cycle in $G(H)$ of length greater than 3 has a chord, i.e., $G(H)$ has no induced subgraphs isomorphic to the k-cycle for $k > 3$.

Since every relational structure $\mathfrak{A}$ is guarded bisimulation equivalent to its guarded tree unfolding, and as GF is invariant under guarded bisimulation equivalence, we find that every satisfiable formula of GF has an acyclic model. This was first stated in [12] as the *generalised tree model property* of GF.

Corollary 1.18 (Grädel) *Every logic that is invariant under guarded bisimulation equivalence has this generalised tree model property: every satisfiable formula has a model whose hypergraph of guarded subsets is acyclic, i.e., a model that admits a tree-decomposition with guarded subsets as bags.*

For a relational vocabulary of width w, this further entails that every satisfiable formula of GF or μGF has a (countable) model of tree width $w - 1$.

1.3.4 Decidability and Complexity for GF and Its Extensions

As in the case of modal logics, the tree model property for guarded models paves the way to decidability and automata based decision procedures. These do not only work for the guarded fragment GF in its basic form, but also for guarded fixed-point logic μGF and for other variants of guarded logics based on more liberal notions of guarded sets.

Indeed, structures of bounded tree width can be uniformly represented by standard trees, in the graph-theoretic sense, with a bounded set of labels. More precisely, given a tree decomposition of width $k - 1$ of a relational τ-structure $\mathfrak{D}$ we fix a set K of $2k$ constants and assign to every element $d \in \mathfrak{D}$ a constant $a_d \in K$ such that distinct elements living at adjacent nodes in the tree decomposition are represented by distinct constants. On the tree T underlying the decomposition of $\mathfrak{D}$ we define monadic predicates $\mathscr{O}_a$ (for $a \in K$) and $R_\mathbf{a}$ (for m-ary $R \in \tau$ and $\mathbf{a} \in K^m$) where $\mathscr{O}_a$ is true at those nodes of T where an element represented by a occurs, and $R_\mathbf{a}$ is the set of nodes of T where a tuple $(d_1, \ldots, d_m) \in R$ occurs that is represented by $\mathbf{a}$. We thus obtain a tree structure $T(\mathfrak{D})$ which has (beyond the edge relation of the tree) only monadic predicates and which carries all structural information about $\mathfrak{D}$ and its tree decomposition.

On the other hand, a tree T with such monadic relations $\mathscr{O}_a$ and $R_\mathbf{a}$ is indeed a tree representation $T(\mathfrak{D})$ for some τ-structure $\mathfrak{D}$ if, and only if, it satisfies certain consistency axioms that turn out to be first-order definable.

There are several options to exploit this for proving decidability and complexity results. The simplest way to prove decidability of guarded fixed-point logic μGF is by an interpretation into SωS, the monadic logic of the countable branching tree. That is, with every formula $\varphi(x_1, \ldots x_m)$ of μGF and every tuple $\mathbf{a} \in K^m$ one can associate a monadic second-order formula $\psi_\mathbf{a}(z)$ that describes on the tree structure $\mathscr{T}(\mathfrak{D})$ the same properties of *guarded* tuples that $\varphi(\bar{x})$ does on $\mathfrak{D}$, in the following sense: if $\mathbf{d}$ is a guarded tuple of $\mathfrak{D}$ living at node v of the tree T, and if $\mathbf{a}$ represents $\mathbf{d}$ at v, then

$$\mathfrak{D} \models \varphi(\mathbf{d}) \Longleftrightarrow T(\mathfrak{D}) \models \psi_\mathbf{a}(v).$$

On the basis of this translation and of the facts that the consistency axioms for tree representations are first-order, that μGF (and least fixed point logic in general) has the Löwenheim-Skolem property, and that the monadic theory of countable trees is decidable, it is then not difficult to prove that the satisfiability problem for μGF is decidable.

Instead of the reduction to the monadic second-order theory of trees, one can define a similar reduction to the modal μ-calculus with backward modalities. The decidability (and EXPTIME-complexity) of this logic has been established by Vardi [30] by means of two-way alternating automata. To make such a reduction work, one has to observe that the consistency axioms for tree representations can be formulated in this logic (in fact, it is sufficient to use basic modal logic with a global modality and backward modalities) and that least and greatest fixed points in μGF on $\mathfrak{D}$ can be encoded by *simultaneous* modal fixed-point formulae on $T(\mathfrak{D})$.

It should be pointed out that the usual modal μ-calculus, without backward modalities, does not seem to be sufficient for such an approach. Indeed, besides the tree model property, the modal μ-calculus also has the finite model property, while one easily obtains formulae that have only infinite models in μGF and in the μ-calculus with backward modalities.

Finally, the satisfiability problem for guarded fixed-point logic can also be solved by direct application of suitably tailored automata-theoretic methods. The general idea is to associate with every sentence $\psi \in \mu$GF an alternating tree automaton $\mathscr{A}_\psi$ that accepts precisely the (tree descriptions of the) like-tree models of ψ. This reduces the satisfiability problem of ψ to the emptiness problem of the automaton, a problem that is solvable in exponential time with respect to the number of states of the automaton. This was the approach taken in [15] where the decidability of μGF had first been established. Instead of Vardi's two-way automata, Grädel and Walukiewicz use a different variant of alternating automata that work on trees of arbitrary, finite or infinite, degree and do not make use of the orientation of edges. The behaviour of such an automaton on a given tree structure is described by a parity game, and by means of the positional determinacy of these games one can reduce the input trees to trees of bounded branching (and the automata to those used by Vardi for the decidability of the μ-calculus with backward modalities). The size of the automaton $\mathscr{A}_\psi$ is bounded by $|\psi|^{2k \log k}$ where k is the width of ψ. For the following see [15].

Theorem 1.19 (Grädel–Walukiewicz) *The satisfiability problem for μGF is decidable, and complete for* 2EXPTIME. *For μGF-sentences of bounded width the satisfiability problem is* EXPTIME-*complete.*

It is worth pointing out that the same complexity bounds also hold for GF, the guarded fragment without fixed points [12]. The double exponential complexity of GF and μGF may seem high (and disappointing for practical applications). However, it is not really surprising, since these logics admit predicates of unbounded arity (whereas modal logics are evaluated on graph-like structures). Even a single predicate of arity n on a universe with just two elements admits 2^{2^n} types already at the atomic level, so one cannot really expect lower complexity bounds. In many practical applications, the underlying vocabulary will be fixed and the arity therefore bounded. In such cases the satisfiability problems for GF and μGF are in EXPTIME and thus on the same level as for most modal logics.

Beyond GF and μGF the general approach outlined here also works for other, more general, notions of guarded logics based on more liberal definitions of guardedness. This includes *loosely guarded*, *packed*, or *clique-guarded* logics. While the classical notion of a guarded set means that the entire set is covered by one atomic fact, the most liberal notion, of a clique-guarded set, just requires that any two elements of the set coexist in some atomic fact, which means that the set is a clique in the Gaifman graph of the structure. Most of the algorithmic results on GF and μGF can be extended to the clique-guarded extensions CGF and μCGF (with appropriate modifications, in particular for the notion of bisimulation). For details, see [13].

1.3.5 Guarded Model Constructions

Guarded tree unfoldings provide one example of a specific form of model construction, or in this case: model transformation, that is tailored for the model theoretic analysis of guarded logics. The requirements of acyclicity and finiteness will in general be incompatible; we shall return to the interesting question how much acyclicity can in general be achieved in finite models further below. For a start, however, we consider the finite model property for the guarded fragment, disregarding the issue of acyclicity. The following proof idea stems from [12] and uses a nice combinatorial result, about finite extension properties of partial isomorphisms due to Herwig [16].

Theorem 1.20 (Herwig) *Any finite relational structure $\mathfrak{A}$ admits a finite extension $\bar{\mathfrak{A}} \supseteq \mathfrak{A}$ ($\mathfrak{A}$ becomes an induced substructure of $\bar{\mathfrak{A}}$) with the property that every partial isomorphism $p: \mathfrak{A} \restriction \mathrm{dom}(p) \simeq \mathfrak{A} \restriction \mathrm{image}(p)$ extends to (is induced by) an automorphism $\bar{p}$ of $\bar{\mathfrak{A}}$.*

It is easy to see that any Herwig extension $\bar{\mathfrak{A}}$ of $\mathfrak{A}$ can be thinned out so that each $R^{\bar{\mathfrak{A}}}$ is generated by the orbit of $R^{\mathfrak{A}}$ under the automorphism group. Let us call such a Herwig extension *special*.

Special Herwig extensions of sufficiently rich finite substructures $\mathfrak{A} \subseteq \mathfrak{B}$ are $\sim_g^\ell$-equivalent to $\mathfrak{B}$ itself; this is the core of the finite model property for GF as proved in [12], see Theorem 1.22.

Lemma 1.21 *Let $\mathfrak{B}$ be a relational structure, $\mathfrak{A} = \mathfrak{B} \restriction A$ an induced finite substructure on a subset $A \subseteq B$ that is sufficiently rich to contain, for every guarded tuple $\mathbf{b}$ of $\mathfrak{B}$, at least one realisation of that $\sim_g^\ell$-type: there is $\mathbf{a} \in \mathfrak{A}$ such that $\mathfrak{B}, \mathbf{a} \sim_g^\ell \mathfrak{B}, \mathbf{b}$. Then any special Herwig extension $\bar{\mathfrak{A}} \supseteq \mathfrak{A}$ is $\sim_g^\ell$-equivalent to $\mathfrak{B}$ in the sense that*

(i) $\bar{\mathfrak{A}} \sim_g^\ell \mathfrak{B}$;
(ii) for every guarded tuple $\mathbf{a} \in \mathfrak{A}$: $\bar{\mathfrak{A}}, \mathbf{a} \sim_g^\ell \mathfrak{B}, \mathbf{a}$.

Proof Using the fact that every guarded tuple in $\bar{\mathfrak{A}}$ is in the orbit of some guarded tuple $\mathbf{a}$ of $\mathfrak{A}$ under an automorphism of $\bar{\mathfrak{A}}$ (because $\bar{\mathfrak{A}}$ is special), and that, up to $\sim_g^\ell$, every guarded tuple of $\mathfrak{B}$ is represented in $A \subseteq B$, claim (i) directly follows from claim (ii). For claim (ii) it essentially suffices to observe that every back&forth requirement for $\mathbf{a}$ that can be met in $\mathfrak{B}$ can also be met in $\bar{\mathfrak{A}}$, as follows.

Let $\mathbf{a} \in \mathfrak{A}$ be guarded, $\mathbf{b}$ guarded in $\mathfrak{B}$, and $\mathbf{c}$ a tuple in the intersection $[\mathbf{a}] \cap [\mathbf{b}]$. By the richness assumption on A, there is some $\mathbf{a}' \in \mathfrak{A}$ such that $\mathfrak{B}, \mathbf{a}' \sim_g^\ell \mathfrak{B}, \mathbf{b}$. This implies in particular that the tuple $\mathbf{c}'$ in $[\mathbf{a}']$ corresponding to $\mathbf{c}$ in $[\mathbf{a}] \cap [\mathbf{b}]$ is linked to $\mathbf{c}$ by a partial isomorphism p of $\mathfrak{A}$. The automorphism $\bar{p}$ of $\bar{\mathfrak{A}}$ then shows that $\bar{p}(\mathbf{a}')$ overlaps with $\mathbf{a}$ in the tuple $\mathbf{c}$ in $\bar{\mathfrak{A}}$ (just as $\mathbf{b}$ overlaps with $\mathbf{a}$ in $\mathbf{c}$ in $\mathfrak{B}$). By induction on ℓ for claim (ii), i.e. assuming claim (ii) at level $\ell - 1$, we find

$$\bar{\mathfrak{A}}, \bar{p}(\mathbf{a}') \simeq \bar{\mathfrak{A}}, \mathbf{a}' \sim_g^{\ell-1} \mathfrak{B}, \mathbf{a}' \sim_g^\ell \mathfrak{B}, \mathbf{b}.$$

This, for all available $\mathbf{b}$ in $\mathfrak{B}$, shows that $\bar{\mathfrak{A}}, \mathbf{a} \sim_g^\ell \mathfrak{B}, \mathbf{a}$ as required for (ii) at level ℓ. □

Claim (i) of the lemma directly yields the finite model property for GF, since any $\varphi \in$ GF of nesting depth ℓ is preserved under $\sim_g^\ell$, and since every $\mathfrak{B}, \mathbf{b} \models \varphi$ has a finite substructure $\mathfrak{A} \subseteq \mathfrak{B}$ that contains at least one realisation of each one of the finitely many $\sim_g^\ell$-types realised by guarded tuples of $\mathfrak{B}$.

Theorem 1.22 (Grädel) GF *has the finite model property: every satisfiable $\varphi \in$ GF has a finite model.*

Better bounds on the size of small models for a given satisfiable $\varphi \in$ GF are obtained by a more recent construction in [2], which builds a small model not directly from a given (infinite) model, but from a complete abstract description of the required $\sim_g^\ell$-type to be realised.

Proposition 1.23 (Bárány–Gottlob–Otto) *Every satisfiable $\varphi \in$ GF(σ), where σ is any relational vocabulary of width w, has a small finite model whose size can be bounded exponentially in the length of φ, for fixed w; the dependence on w, on the other hand, is doubly exponential.*

The core construction of [2], of which the above really is a technical corollary, yields finite guarded bisimilar covers that are weakly N-acyclic in the sense of the following definitions.

Definition 1.24 A *guarded bisimilar covering* of a relational structure $\mathfrak{A}$ is a homomorphism $\pi : \hat{\mathfrak{A}} \to \mathfrak{A}$ from some relational structure $\hat{\mathfrak{A}}$ (the cover) onto $\mathfrak{A}$, such that the restrictions of π to guarded subsets of $\hat{\mathfrak{A}}$ induce a guarded bisimulation.

Guarded tree unfoldings are natural examples in point; however, we are here mostly interested in coverings of finite $\mathfrak{A}$ by *finite covers* $\hat{\mathfrak{A}}$. The restrictions of the cover homomorphism π to guarded subsets must in particular be partial iso-morphisms. The *forth*-property is thus subsumed in the requirement that π is a homomorphism. The *back*-property corresponds to a lifting property familiar from topological or geometric notions of coverings.[4]

Guarded tree unfoldings provide fully acyclic coverings, albeit infinite ones. One useful approximation to acyclicity in finite covers is the following from [2].

Definition 1.25 A covering $\pi : \hat{\mathfrak{A}} \to \mathfrak{A}$ is *weakly N-acyclic* if every induced sub-structure of $\hat{\mathfrak{A}}$ of size up to N is tree-decomposable with bags that project onto guarded subsets of $\mathfrak{A}$ under π.

Proposition 1.26 (Bárány–Gottlob–Otto) *For every $N \in \mathbb{N}$, each finite relational $\mathfrak{A}$ admits weakly N-acyclic coverings by finite structures.*

---

[4] It may be worth to point out that, unlike the finite bisimilar coverings obtained for graph-like structures in [18], the bisimilar coverings of relational structures or of hypergraphs will necessarily be *branched coverings*, and do not provide unique liftings.

An analysis of homomorphisms $h: \mathfrak{C} \to \hat{\mathfrak{A}}$, from structures $\mathfrak{C}$ of size up to N into a weakly acyclic cover $\pi: \hat{\mathfrak{A}} \to \mathfrak{A}$, shows that $\mathfrak{A}$ must satisfy one of a finite list of potential GF-descriptions of all possible acyclic homomorphic images of $\mathfrak{C}$.[5] If $\mathfrak{A}$ does not satisfy this GF-expressible finite 'disjunction of acyclic conjunctive queries', then $\hat{\mathfrak{A}}$ cannot even admit cyclic homomorphic images of $\mathfrak{C}$. Together with existence of finite, weakly N-acyclic covers, this argument from [2] yields a considerable strengthening of the finite model property for GF, as well as natural applications to database issues regarding conjunctive queries under GF-definable constraints.

For the following, a class $\mathscr{C}$ of σ-structures is said to be defined in terms of *finitely many forbidden homomorphisms* if, for some finite list of finite σ-structures $\mathfrak{C}_1, \ldots, \mathfrak{C}_m$, the class $\mathscr{C}$ consists of precisely those σ-structures $\mathfrak{C}$ that admit no homomorphisms $h: \mathfrak{C}_i \to \mathfrak{C}$ for $1 \leqslant i \leqslant m$.

Corollary 1.27 (Bárány–Gottlob–Otto) GF *has the finite model property in restriction to any class $\mathscr{C}$ of relational structures that is defined in terms of finitely many forbidden homomorphisms: for any such class $\mathscr{C}$, φ has a model in $\mathscr{C}$ if, and only if, it has a finite model in $\mathscr{C}$.*

Interestingly, this strengthening of the finite model property for GF can also be obtained from a corresponding strengthening of Herwig's theorem. We briefly present this new alternative proof from [24], which may be of independent systematic interest.[6] The Herwig–Lascar theorem [17] asserts a finite model property for the extension task for partial isomorphisms over classes with finitely many forbidden homomorphisms. An alternative proof of the Herwig–Lascar theorem itself, which is inspired by hypergraph constructions related to the exploration of the finite model theory of GF, see Sect. 1.3.6, can be found in [22, 24, 25].

Theorem 1.28 (Herwig–Lascar) *Let the class of relational structures $\mathscr{C}$ be defined in terms of finitely many forbidden homomorphisms. Suppose that a finite structure $\mathfrak{A} \in \mathscr{C}$ has a possibly infinite extension $\mathfrak{B} \supseteq \mathfrak{A}$ in $\mathscr{C}$ that extends every partial isomorphism of $\mathfrak{A}$ to an automorphism of $\mathfrak{B}$. Then $\mathfrak{A}$ also possesses a finite extension with this property in $\mathscr{C}$.*

Just as Lemma 1.21 links Herwig's theorem to the basic finite model property for GF, the following links the Herwig–Lascar theorem to the stronger finite model property for GF expressed in Corollary 1.27.

A structure $\mathfrak{B}$ is $\sim_g^\ell$-*homogeneous* if any guarded tuples $\mathbf{b}, \mathbf{b}'$ in $\mathfrak{B}$ such that $\mathfrak{B}, \mathbf{b} \sim_g^\ell \mathfrak{B}, \mathbf{b}'$ are related by an automorphism of $\mathfrak{B}$.

Lemma 1.29 *Let $\mathscr{C}$ be a class of relational structures defined in terms of finitely many forbidden homomorphisms. Let $\mathfrak{B} \in \mathscr{C}$ be $\sim_g^\ell$-homogeneous. Let $\mathfrak{B}'$ be the*

---

expansion of $\mathfrak{B}$ *by a new relation for each one of the finitely many* $\sim_g^\ell$-*types realised in* $\mathfrak{B}$. *Let* $\mathfrak{A}' = \mathfrak{B}' \upharpoonright A$ *be large enough to contain, for every guarded tuple* $\mathbf{b}$ *of* $\mathfrak{B}$, *at least one realisation of that* $\sim_g^\ell$-*type.*

Then $\mathfrak{A}'$ *has a special Herwig extension* $\bar{\mathfrak{A}}' \supseteq \mathfrak{A}'$ *in* $\mathscr{C}$ *that is* $\sim_g$-*equivalent to* $\mathfrak{B}'$ *in the sense that* $\bar{\mathfrak{A}}' \sim_g \mathfrak{B}'$ *and* $\bar{\mathfrak{A}}', \mathbf{a} \sim_g \mathfrak{B}', \mathbf{a}$ *for every guarded tuple* $\mathbf{a} \in \mathfrak{A}$.

Proof In view of Lemma 1.21 and Theorem 1.28 it suffices to show that the extension task for $\mathfrak{A}'$ has some, possibly infinite, solution in $\mathscr{C}$. But $\mathfrak{B}'$, being homogeneous, is such an infinite solution.   □

Proof (of Corollary 1.27) Let $\mathscr{C}$ be defined by the condition that there are no homomorphic images of the finite structures $\mathfrak{C}_1, \ldots, \mathfrak{C}_m$. The class $\mathscr{C}_0 \supseteq \mathscr{C}$ of structures that admit no acyclically embedded homomorphic images of the $\mathfrak{C}_i$ is definable in GF by some $\gamma \in$ GF of guarded nesting depth ℓ, for some ℓ. To find finite models of $\varphi \in$ GF in $\mathscr{C}$, we moreover choose ℓ greater or equal to the nesting depth of φ. If φ has an infinite model in $\mathscr{C}$, then a $\sim_g^\ell$-homogeneous infinite model $\mathfrak{B}$ of φ in $\mathscr{C}$ can be obtained as a suitable regular tree-like model of $\varphi \wedge \gamma$ (which in turn could be obtained from an arbitrary finite model of $\varphi \wedge \gamma$). An application of the lemma then yields a finite model in $\mathscr{C}$.   □

Beside the notion of weakly N-acyclic coverings from [2], there is the stronger notion of N-acyclic coverings from [21], which rules out any small cyclic substructures in the cover. This yields an even stronger finite model property for GF and is essential for an expressive completeness proof for GF in finite model theory, as sketched in the next section. More canonical constructions of N-acyclic coverings and related hypergraph constructions have recently been explored in [22, 25]. But unlike the case of weakly N-acyclic covers, the known constructions of fully N-acyclic finite covers do not provide feasible size bounds.

Definition 1.30 A guarded bisimilar covering $\pi : \hat{\mathfrak{A}} \to \mathfrak{A}$ is *N-acyclic* if every induced substructure of size up to N of the cover $\hat{\mathfrak{A}}$ is acyclic.

Proposition 1.31 (Otto) *For every* $N \in \mathbb{N}$, *each finite relational* $\mathfrak{A}$ *admits* N-*acyclic coverings by finite structures.*

Corollary 1.32 (Otto) GF *has the finite model property in restriction to any class* $\mathscr{C}$ *of relational structures that is defined in terms of finitely many forbidden cyclic substructures.*

1.3.6 Expressive Completeness

The N-acyclic finite guarded bisimilar covers of Proposition 1.31 are also essential for the proof of the finite model theory version of Theorem 1.14. The issue at stake is the expressive completeness assertion, that a first-order definable property of guarded tuples in (finite) relational structures is expressible in GF (over all finite structures) if it is closed under guarded bisimulation equivalence (among finite structures). For both, the classical and the finite model theory reading, the Ehrenfeucht–Fraïssé theorem for GF shows that it suffices to prove the following, which may be read as a compactness property for $(\sim_g^\ell)_{\ell \in \mathbb{N}}$ versus $\sim_g$: for any $\varphi(\mathbf{x}) \in$ FO (in an explicitly guarded tuple $\mathbf{x}$ of free variables),

$$(*) \quad \begin{cases} \varphi(\mathbf{x}) & \text{invariant under } \sim_g \Rightarrow \\ \varphi(\mathbf{x}) & \text{invariant under } \sim_g^\ell \text{ for some } \ell \in \mathbb{N}. \end{cases}$$

The classical proof typically achieves this through

(i) a compactness argument that reduces $(*)$ to: invariance under $\sim_g$ implies invariance under $\sim_g^\omega$ (i.e., $\equiv_{GF}$); and

(ii) a proof of claim (i) through an upgrading argument involving saturated models: for $\mathfrak{A} \equiv_{GF} \mathfrak{B}$ there are $\mathfrak{A}^* \equiv_{FO} \mathfrak{A}$ and $\mathfrak{B}^* \equiv_{FO} \mathfrak{B}$ for which (by saturation) $\mathfrak{A}^* \equiv_{GF} \mathfrak{B}^*$ implies $\mathfrak{A}^* \sim_g \mathfrak{B}^*$; the claim is then apparent from this diagram:

$$
\begin{array}{ccc}
\mathfrak{A} & \!\!\!\!\!-\!\!-\!\! \equiv_{GF} -\!\!-\!\!\!\!\! & \mathfrak{B} \\
| & & | \\
\equiv_{FO} & & \equiv_{FO} \\
| & & | \\
\mathfrak{A}^* & \!\!\!\!\!-\!\!-\!\! \sim_g -\!\!-\!\!\!\!\! & \mathfrak{B}^*
\end{array}
$$

For the finite model theory version, a passage through the necessarily infinite companion structures, which are involved in both parts of this classical argument, is not supported by the assumptions.

Instead, the upgrading needs to be based on a more constructive approach to model transformations, and focuses on a concrete level ℓ in $(*)$ that is determined by the width of the vocabulary and the quantifier rank q of the given φ. It follows this pattern:

$$
\begin{array}{ccc}
\mathfrak{A} & \!\!\!\!\!-\!\!-\!\! \equiv_{GF}^\ell -\!\!-\!\!\!\!\! & \mathfrak{B} \\
| & & | \\
\sim_g & & \sim_g \\
| & & | \\
\hat{\mathfrak{A}} & \!\!\!\!\!-\!\!-\!\! \equiv_{FO}^q -\!\!-\!\!\!\!\! & \hat{\mathfrak{B}}
\end{array}
$$

Here $\hat{\mathfrak{A}}$ and $\hat{\mathfrak{B}}$ are obtained as (finite) guarded bisimilar covers of $\mathfrak{A}$ and $\mathfrak{B}$, respectively, that need to be sufficiently acyclic and finitely saturated w.r.t. multiplicities: a certain level of N-acyclicity is necessary because $\hat{\mathfrak{A}}$ and $\hat{\mathfrak{B}}$ may necessarily have cycles, and differences w.r.t. short cycles would be FO-expressible at low quantifier rank; similarly for differences w.r.t. small branching degrees between relational hyperedges, which can also not be controlled in GF.

Technically rather intricate arguments in [21] use Proposition 1.31 as a starting point to provide companions $\hat{\mathfrak{A}}$ and $\hat{\mathfrak{B}}$ that support this proof idea.

Theorem 1.33 (Otto) FO/$\sim_g$ $\equiv$ GF, *also in the sense of finite model theory: For every first-order formula $\varphi(\mathbf{x})$ in a relational vocabulary, the follwong are equivalent:*

(i) φ is invariant under guarded bisimulation among finite structures.
(ii) φ is logically equivalent over all finite structures to a formula of GF.

1.4 Guarded Negation Bisimulation

One natural decidable fragment of first-order logic that stands out because of its considerable algorithmic importance, is the positive existential fragment: $\exists$posFO $\subseteq$ FO is generated from atomic formulae by conjunction, disjunction and existential quantification. It is semantically characterised, as a fragment of FO, by preservation under homomorphisms. This characterisation is known as the Lyndon–Tarski theorem in classical model theory; for finite model theory, it was proved by Rossman in [29], with characteristically different techniques that also shed new light on the classical version. Any $\exists$posFO-formula can be equivalently re-written as a disjunction over existentially quantified conjunctions of atoms—so that it corresponds, in database terminology, to a *union of conjunctive queries*. And a conjunctive query asserts the existence of a homomorphism: consider a conjunctive query $\varphi = \varphi(\mathbf{x}) = \exists \mathbf{y} \bigwedge_i \alpha_i(\mathbf{z_i})$ with relational atoms $\alpha_i(\mathbf{z_i})$ for tuples of variables $\mathbf{z_i}$ from $[\mathbf{xy}]$. With the template $\bigwedge_i \alpha_i(\mathbf{z_i})$ associate a relational structure $\mathfrak{C}_\varphi$ whose universe is the set of variables $[\mathbf{xy}]$, and whose relations are interpreted by putting $\mathbf{z_i}$ into the relation involved in the atom α_i. Then $\mathfrak{A}, \mathbf{a} \models \varphi$ if, and only if, there is a homomorphism $h \colon \mathfrak{C}_\varphi \to \mathfrak{A}$ that maps $\mathbf{x}$ to $\mathbf{a}$. Interestingly, φ can equivalently be expressed in GF (i.e., is invariant under guarded bisimulation equivalence) if, and only if, $\mathfrak{C}_\varphi$ is acyclic.

$\exists$posFO $\subseteq$ FO or the formalism of (unions of) conjunctive queries are closed under nesting, but closure under (unconstrained) negation generates all of relational FO and becomes undecidable for satisfiability. The guarded fragment GF $\subseteq$ FO, on the other hand, is closed under negation, but not under (unconstrained) nesting.

The introduction of the *guarded negation fragment* GN $\subseteq$ FO in [3] combines the innocuous ingredients in GF and $\exists$posFO with the natural constraints to produce a common extension of GF and $\exists$posFO that retains many of the good features, most notably decidability.

We follow the pattern of the treatment so far and put the appropriate notions of back&forth equivalence centre-stage. The characteristic feature is the interleaving of (local, and possibly size-bounded) homomorphisms with modal or guarded bisimulation.

1.4.1 Homomorphisms and Bisimulation

We start with a back&forth equivalence that interleaves homomorphisms with modal bisimulation; this will provide the Ehrenfeucht–Fraïssé notion and semantic characterisation of the *unary negation fragment* UN $\subseteq$ FO of [10], a modal precursor to the *guarded negation fragment* GN $\subseteq$ FO of [3].

A *unary negation bisimulation* relation between relational structures $\mathfrak{A}$ and $\mathfrak{B}$ is a set $Z \subseteq A \times B$ of positions, which are just pairs of related vertices in $\mathfrak{A}$ and $\mathfrak{B}$ as in modal bisimulation, subject to atom equivalence and more complex back&forth conditions involving homomorphisms. For all $(a, b) \in Z$:

(i) *(atom eq.)*: $\mathfrak{A} \upharpoonright \{a\} \simeq \mathfrak{B} \upharpoonright \{b\}$;
(ii) (hom-*back*): for every $B_0 \subseteq B$ there is a homomorphism $h \colon \mathfrak{B} \upharpoonright B_0 \to \mathfrak{A}$ such that $(h(b), b) \in Z$ for all $b \in B_0$, and $h(b) = a$ if $b \in B_0$;
(iii) (hom-*forth*): for every $A_0 \subseteq A$ there is a homomorphism $h \colon \mathfrak{A} \upharpoonright A_0 \to \mathfrak{B}$ such that $(a, h(a)) \in Z$ for all $a \in A_0$, and $h(a) = b$ if $a \in A_0$.

We write $\mathfrak{A}, a \sim_{\mathsf{hom}} \mathfrak{B}, b$ if $(a, b) \in Z$ for some unary negation bisimulation relation Z between $\mathfrak{A}$ and $\mathfrak{B}$; $\mathfrak{A}, a \sim^{\ell}_{\mathsf{hom}} \mathfrak{B}, b$ for the finite approximation corresponding to a strategy for the second player for ℓ rounds in the natural bisimulation game associated with this back&forth scenario.

A generalisation of this idea leads from an equivalence between individual elements (as in modal bisimulation) to an equivalence based on guarded tuples (as in guarded bisimulation), similarly interleaving bisimulation with local homomorphisms: this is the notion of *guarded negation bisimulation* equivalence from [3].

A *guarded negation bisimulation* relation between relational structures $\mathfrak{A}$ and $\mathfrak{B}$ is a set Z of partial isomorphisms $\rho \colon \mathbf{a} \mapsto \mathbf{b}$ between guarded tuples or subsets, such that, for all $\rho \colon \mathbf{a} \mapsto \mathbf{b}$ in Z:

(i) *(atom eq.)*: $\rho \colon \mathfrak{A} \upharpoonright \mathbf{a} \simeq \mathfrak{B} \upharpoonright \mathbf{b}$ (isomorphism of guarded substructures);
(ii) (hom-*back*): for all $B_0 \subseteq B$ there is a homomorphism $h \colon \mathfrak{B} \upharpoonright B_0 \to \mathfrak{A}$ that is compatible with the restriction of ρ^{-1} to B_0, and such that $\rho' \colon h(\mathbf{b}') \mapsto \mathbf{b}'$ is in Z for all guarded tuples $\mathbf{b}'$ from B_0;
(iii) (hom-*forth*): for all $A_0 \subseteq A$ there is a homomorphism $h \colon \mathfrak{A} \upharpoonright A_0 \to \mathfrak{B}$ that is compatible with the restriction of ρ to A_0, and such that $\rho' \colon \mathbf{a}' \mapsto h(\mathbf{a}')$ is in Z for all guarded tuples $\mathbf{a}'$ from A_0.

We write $\mathfrak{A}, \mathbf{a} \sim_{\mathsf{ghom}} \mathfrak{B}, \mathbf{b}$ and $\mathfrak{A}, \mathbf{a} \sim^{\ell}_{\mathsf{ghom}} \mathfrak{B}, \mathbf{b}$ to denote guarded bisimulation equivalence and its finite approximations.

Simple size-bounded versions of $\sim_{\mathsf{hom}}$ and $\sim_{\mathsf{ghom}}$ and their finite approximations are technically useful: we restrict conditions (hom-*back*) and (hom-*forth*) to subsets $B_0 \subseteq B$ and $A_0 \subseteq A$ of size up to k, for some fixed $k \in \mathbb{N}$. We write e.g. $\mathfrak{A}, \mathbf{a} \sim_{\mathsf{ghom};k} \mathfrak{B}, \mathbf{b}$ and $\mathfrak{A}, \mathbf{a} \sim^{\ell}_{\mathsf{ghom};k} \mathfrak{B}, \mathbf{b}$ in connection with this restricted notion of *k-bounded guarded negation bisimulation*, and similarly, e.g., $\mathfrak{A}, a \sim_{\mathsf{hom};k} \mathfrak{B}, b$ for a corresponding notion of *k-bounded unary negation bisimulation*.

We discuss briefly the extensions of modal logic and the guarded fragment that are obtained by closure of the existential positive fragment of FO under negation in suitably restricted settings:

- negation of 'unary' formulae in a single free variable for the unary negation fragment [10];
- negation of 'guarded' formulae in an explicitly guarded tuple of free variables for the guarded negation fragment [3].

Definition 1.34 The formulae of the *unary negation fragment* UN $\subseteq$ FO are generated from the atomic formulae by positive boolean connectives, existential quantification, and negation on formulae in at most one free variable.

It is obvious that, for suitable modal vocabularies, ML $\subseteq$ UN and that generally ∃posFO $\subseteq$ UN; both inclusions are easily seen to be strict (for non-trivial vocabularies). It turns out that formulae of UN (in at most a single free variable) are preserved under unary negation bisimulation, and in fact this property characterises the unary negation fragment as a fragment of FO, classically. See [10] for this and many related model-theoretic results, also regarding the fixpoint extension of UN and including decidability for satisfiability and finite satisfiability.

Definition 1.35 The formulae of the *guarded negation fragment* GN $\subseteq$ FO are generated from the atomic formulae by positive boolean connectives, existential quantification, and negation on formulae in an explicitly guarded tuple of free variables.

It is not hard to see that UN $\subseteq$ GN and GF $\subseteq$ GN, and that these inclusions are strict in general. Formulae of GN (in an explicitly guarded tuple of free variables) are preserved under guarded negation bisimulation equivalence; this preservation property also characterises GN as a fragment of FO, in the sense of classical model theory, as shown in [3].

For useful Ehrenfeucht–Fraïssé correspondences, which rely on the natural notion of nesting depth in GN and UN and induce equivalence relations of finite index, we need to bound the size of the existential quantifications (conjunctive queries) by some width parameter. For the games and bisimulation notions this restriction leads to the size bounded equivalences like $\sim^\ell_{\text{ghom};k}$. For the logics, we correspondingly let GN[k] $\subseteq$ GN stand for those formulae that can be generated with existential quantifications over up to k variables at a time. To avoid pathologies, we shall always assume that k is no less than the width of the vocabulary.

It is then not hard to see that equivalence w.r.t. GN[k] up to nesting depth ℓ and $\sim^\ell_{\text{ghom};k}$ are related in an Ehrenfeucht–Fraïssé correspondence. The theorem gives an indicative example; its variants for UN and also for infinitary versions of UN and GN in the style of Karp theorems are straightforward.

Theorem 1.36 (Ehrenfeucht–Fraïssé for GN[k]) *In restriction to finite relational vocabularies, fixed $k \in \mathbb{N}$, and for every $\ell \in \mathbb{N}$:*

$$\mathfrak{A}, \mathbf{a} \sim^\ell_{\text{ghom};k} \mathfrak{B}, \mathbf{b} \quad \text{if, and only if,} \quad \mathfrak{A}, \mathbf{a} \equiv^\ell_{\text{GN}[k]} \mathfrak{B}, \mathbf{b}.$$

1.4.2 Towards a (Finite) Model Theory of Guarded Negation

We summarise some key techniques and a few further results for the model theory of GN and GN[k], especially pertaining to the finite model property and to the expressive completeness concern in finite model theory. We concentrate on guarded negation rather than unary negation, since this is the richer of the two settings; technically it is, moreover, more directly related to one of our main themes, viz., to the interesting passage from graph-like structures to general relational structures with an emphasis on the hypergraph of guarded subsets.

Theorem 1.37 (Bárány–ten Cate–Segoufin) GN *has the finite model property.*

The argument from [3] is based on a reduction from GN-satisfiability to satisfiability of GF under constraints imposed by *forbidden homomorphisms*, and thus, essentially, a reduction to Corollary 1.27.

The semantics of a formula $\varphi(\mathbf{x}) \in$ GN (in explicitly guarded free variables $\mathbf{x}$) can be translated into a collection of auxiliary specifications that subject certain guarded tuples $\mathbf{a}$ in a prospective model $\mathfrak{A}$ to *positive* or *negative* requirements w.r.t. homomorphisms:

- (pos. hom.) requiring the existence of a homomorphism $h \colon \mathfrak{C}, \mathbf{c} \rightarrow \mathfrak{A}, \mathbf{a}$, for certain finite templates $\mathfrak{C}, \mathbf{c}$;
- (neg. hom.) ruling out the existence of any homomorphism $h \colon \mathfrak{C}, \mathbf{c} \rightarrow \mathfrak{A}, \mathbf{a}$, for certain finite templates $\mathfrak{C}, \mathbf{c}$.

In both cases, the templates $\mathfrak{C}, \mathbf{c}$ are abstracted from the underlying conjunctive queries or positive existential parts (in a suitable normal form). A standard process of relational Skolemisation thus translates $\varphi(\mathbf{x})$ into a positive boolean combination of requirements of the form (pos. hom.) and (neg. hom.) for all tuples in certain (auxiliary) relations. A further crude Skolemisation step serves to provide realisations of positive requirements in image substructures that are guarded as a whole by new auxiliary relations; this puts all (pos. hom.) requirements into GF, and leaves just the negative requirements of the form (neg. hom.) to cope with. But this is precisely the situation in which Corollary 1.27 yields finite models whenever there are any models.

The requirements for an expressive completeness proof for GN[k] in relation to all $\sim_{\mathrm{ghom};k}$-invariant FO-definable properties (of guarded tuples), which is meant to work in finite model theory, are considerable higher. The basic idea again is to use an upgrading through $\sim_{\mathrm{ghom};k}$-compatible model transformations that work in finite structures. I.e., we want to follow this pattern, presented without the guarded parameter tuples:

$$
\begin{array}{ccc}
\mathfrak{A} & \overset{\sim^{\ell}_{\mathrm{ghom};k}}{\rule{3cm}{0.4pt}} & \mathfrak{B} \\
\Big| \sim_{\mathrm{ghom};k} & & \Big| \sim_{\mathrm{ghom};k} \\
\hat{\mathfrak{A}} & \underset{\equiv^{q}_{\mathrm{FO}}}{\rule{3cm}{0.4pt}} & \hat{\mathfrak{B}}
\end{array}
$$

More precisely, given some first-order φ of quantifier rank q that is invariant under $\sim_{\mathrm{ghom};k}$, and finite structures $\mathfrak{A}$ and $\mathfrak{B}$ that are $\sim^{\ell}_{\mathrm{ghom};k}$-equivalent for sufficiently high level ℓ, we need to provide finite $\sim_{\mathrm{ghom};k}$-equivalent companion structures $\hat{\mathfrak{A}}$ and $\hat{\mathfrak{B}}$ for which $\sim^{\ell}_{\mathrm{ghom};k}$-equivalence implies $\equiv^{q}_{\mathrm{FO}}$-equivalence, so that

$$\hat{\mathfrak{A}} \models \varphi \quad \text{iff} \quad \hat{\mathfrak{B}} \models \varphi.$$

If this can generally be achieved, for a uniform level ℓ that only depends on φ, then the diagram shows that φ is preserved under $\sim^{\ell}_{\mathrm{ghom};k}$, and by the Ehrenfeucht–Fraïssé theorem for GN[k], Theorem 1.36, is equivalently expressible in GN[k].

The crucial features with respect to which $\hat{\mathfrak{A}}$ and $\hat{\mathfrak{B}}$ need to agree, even though these features are *not* GN-definable are

- presence of small cyclic configurations other than those explicitly ruled out by (neg. hom.) assertions;
- multiplicities (up to a threshold) and isomorphism types of realisations of (pos. hom.) assertions.

That $\mathfrak{A}$ and $\mathfrak{B}$ agree w.r.t. the relevant (pos. hom.) and (neg. hom.) assertions follows from their $\sim_{\mathrm{ghom};k}$-equivalence. Then agreement w.r.t. to the above features is relatively easy to achieve in infinite tree unfoldings of $\mathfrak{A}$ and $\mathfrak{B}$ that are simultaneously saturated w.r.t. *all* admissible isomorphism types of the relevant (pos. hom.) assertions. Relational Skolemisation and an application of the finite model property for GN, Theorem 1.37, yield finite companions $\hat{\mathfrak{A}}'_0$ and $\hat{\mathfrak{B}}'_0$. These further admit finite coverings by suitable $\hat{\mathfrak{A}}'$ and $\hat{\mathfrak{B}}'$ whose degree of acyclicity and saturation w.r.t. small multiplicities show them to be equivalent in the sense of $\equiv^{q}_{\mathrm{FO}}$ (this last part of the argument is as for Theorem 1.33). This yields the following result from [23].

Theorem 1.38 (Otto) FO/$\sim_{\mathrm{ghom};k} \equiv$ GN[k], *classically and in the sense of finite model theory.*

1.5 Summary

We have seen that bisimulation equivalence is a very flexible and powerful concept for the analysis of many logics. In its classical form it is one of the crucial tools in the study of modal logics, and its generalisations to various forms of guarded bisimulation provide indispensable methods for understanding the expressive power as well as the model-theoretic and algorithmic properties of more and more powerful variants of guarded logics.

First of all, an appropriate notion of bisimulation for a logic L characterises semantic invariance and logical indistinguishability: bisimilar nodes or tuples in two structures cannot be distinguished by formulae of L. In this sense, bisimulation is closely related to the characterisation of elementary equivalence via

Ehrenfeucht-Fraïssé games, and bisimulation games can indeed be viewed as special cases of these. The specific form of a bisimulation depends mostly on the nature of the quantification patterns that the associated logic provides. In game theoretic terms, the restrictions on the permitted forms of quantification are reflected by the rules in the associated bisimulation game. In modal and guarded bisimulation games the configurations at any position in a play are restricted in the sense that they may only contain elements that are, in a sense, 'close together'. As a consequence, bisimulation permits us to control the complexity of model constructions and leads to results about model-theoretic properties of modal and guarded logics such as the tree model property of modal logics and the fact that satisfiable guarded formulae have models of bounded tree width. While such results are usually not too difficult to establish for infinite models, corresponding constructions for finite models may be quite challenging and require intricate combinatorial arguments and sophisticated mathematical techniques.

A further highlight of the bisimulation-based analysis of logics are the characterisation theorems that provide, inside a classical level of logical expressiveness such as first-order or monadic second-order definability, a sort of converse of bisimulation invariance. Typically such characterisation theorems state that a modal or guarded logic is not only invariant under bisimulation, but is in fact (up to logical equivalence) precisely the bisimulation invariant part of that level. Again such theorems are, by means of compactness and model-theoretic notions such as saturation or by automata-theoretic methods, better understood and easier to prove for arbitrary (i.e. finite or infinite) models, and much more challenging, and in some cases open, on finite structures.

A related issue that we have not treated here concerns Lindström characterisations of modal and guarded logics. It is shown in [7, 8] that no logic that is bisimulation invariant, compact, and closed under relativisation can properly extend the basic modal logic ML. In this proof, a crucial role is played by a locality criterion (which is implied by compactness and relativisation for any bisimulation closed logic) saying that the truth of a formula at a given node only depends on a neighbourhood of points reachable in a bounded number of steps. For guarded logics, and even for modal logics with a global modality no such locality criterion is available. To obtain Lindström characterisations for GF and ML[∀], Otto and Piro [26] use instead the Tarski Union Property saying that the union of any elementary chain is itself an elementary extension of each structure in the chain. They show that ML[∀] and GF are the maximal compact logics that satisfy the Tarski Union Property and the corresponding bisimulation invariance. It is open whether there are Lindström characterisations of these logics that are not based on the Tarski Union Property but, say, on compactness and relativisation.

Finally the bisimulation-based analysis of modal and guarded logics also leads to important insights concerning their algorithmic properties. Since satisfiable formulae always admit simple models, for instance tree-like ones, and since modal and guarded logics, including the fixed-point variants such as the modal μ-calculus and the guarded fixed-point logic μGF can be embedded or interpreted in monadic second-order logic on trees, powerful automata theoretic methods become available for checking satisfiability and for evaluating formulae. It still remains to determine where the limits are for fragments of first-order logic (and fixed-point logic or even second-order logic) that are invariant under a suitable notion of (guarded) bisimulation that is sufficient to ensure similar model-theoretic and algorithmic properties as those that have been established for modal and guarded logic. In particular, can we find in this way stronger decidable fragments of first-order logic, fixed-point logic and second-order logic than those known so far?

References

1. Andréka H, van Benthem J, Németi I (1998) Modal languages and bounded fragments of predicate logic. J Philoso Logic 27:217–274
2. Bárány V, Gottlob G, Otto M (2014) Querying the guarded fragment. Logical Methods Comput Sci (to appear)
3. Bárány V, ten Cate B, Segoufin L (2011) Guarded negation. In: Proceedings of ICALP, pp 356–367
4. Beeri C, Fagin R, Maier D, Yannakakis M (1983) On the desirability of acyclic database schemes. J ACM 30:497–513
5. van Benthem J (1983) Modal logic and classical logic. Bibliopolis, Napoli
6. van Benthem J (2005) Guards, bounds, and generalized semantics. J Logic Lang Inform 14(3):263–279
7. van Benthem J (2007) A new modal Lindström theorem. Log Univers 1:125–138
8. van Benthem J, ten Cate B, Väänänen J (2007) Lindström theorems for fragments of first-order logic. In: Proceedings of 22nd IEEE symposium on logic in computer science, LICS 2007, pp 280–292
9. Berge C (1973) Graphs and hypergraphs. North-Holland, Amsterdam
10. ten Cate B, Segoufin L (2011) Unary negation. In: Proceedings of STACS, pp 344–355
11. Dawar A, Otto M (2009) Modal characterisation theorems over special classes of frames. Ann Pure Appl Logic 161:1–42
12. Grädel E (1999) On the restraining power of guards. J Symbolic Logic 64:1719–1742
13. Grädel E (2002) Guarded fixed point logics and the monadic theory of countable trees. Theoret Comput Sci 288:129–152
14. Grädel E, Hirsch C, Otto M (2002) Back and forth between guarded and modal logics. ACM Trans Comput Logics 3:418–463
15. Grädel E, Walukiewicz I (1999) Guarded fixed point logic. In: Proceedings of 14th IEEE symposium on logic in computer science, LICS 1999, pp 45–54
16. Herwig B (1995) Extending partial isomorphisms on finite structures. Combinatorica 15:365–371
17. Herwig B, Lascar D (2000) Extending partial isomorphisms and the profinite topology on free groups. Trans AMS 352:1985–2021
18. Otto M (2004) Modal and guarded characterisation theorems over finite transition systems. Ann Pure Appl Logic 164(12):1418–1453
19. Otto M (2006) Bisimulation invariance and finite models. In: Colloquium logicum 2002. Lecture notes in logic, pp 276–298, ASL
20. Otto M (2011) Model theoretic methods for fragments of FO and special classes of (finite) structures. In: Esparza J, Michaux C, Steinhorn C (eds) Finite and algorithmic model theory, volume 379 of LMS lecture notes, pp 271–341. CUP
21. Otto M (2012a) Highly acyclic groups, hypergraph covers and the guarded fragment. J ACM 59:1
22. Otto M (2012b) On groupoids and hypergraphs. arXiv:1211.5656 (preprint)

23. Otto M (2013a) Expressive completeness through logically tractable models. Ann Pure Appl Logic 164(12):1418–1453
24. Otto M (2013b) Groupoids, hypergraphs and symmetries in finite models. In: Proceedings of 28th IEEE symposium on logic in computer science, LICS 2013
25. Otto M (2014) Finite groupoids, finite coverings and symmetries in finite structures. arXiv:1404.4599 (preprint)
26. Otto M, Piro R (2008) A Lindström characterisation of the guarded fragment and of modal logic with a global modality. In: Areces C, Goldblatt R (eds) Advances in modal logic 7, pp 273–288
27. Rabin M (1969) Decidability of second-order theories and automata on infinite trees. Trans AMS 141:1–35
28. Rosen E (1997) Modal logic over finite structures. J Logic Lang Inform 6:427–439
29. Rossman B (2008) Homomorphism preservation theorems. J ACM 55:1–53
30. Vardi M (1998) Reasoning about the past with two-way automata. In: Automata, languages and programming ICALP 98. Lecture notes in computer science, vol 1443. Springer, pp 628–641

Chapter 2
Expressiveness Modulo Bisimilarity: A Coalgebraic Perspective

Yde Venema

Abstract One of van Benthem's seminal results is the Bisimulation Theorem characterizing modal logic as the bisimulation-invariant fragment of first-order logic. Janin and Walukiewicz extended this theorem to include fixpoint operators, showing that the modal μ-calculus μML is the bisimulation-invariant fragment of monadic second-order logic MSO. Their proof uses parity automata that operate on Kripke models, and feature a transition map defined in terms of certain fragments of monadic first-order logic. In this paper we decompose their proof in three parts: (1) two automata-theoretic characterizations, of MSO and μML respectively, (2) a simple model-theoretic characterization of the identity-free fragment of monadic first-order logic, and (3) an automata-theoretic result, stating that (a strong version of) the second result somehow propagates to the level of full fixpoint logics. Our main contribution shows that the third result is an instance of a more general phenomenon that is essentially coalgebraic in nature. We prove that if one set Λ of predicate liftings (or modalities) for a certain set functor T uniformly corresponds to the T-natural fragment of another such set Λ', then the fixpoint logic associated with Λ is the bisimulation-invariant logic of the fixpoint logic associated with Λ'.

2.1 Introduction

Johan van Benthem is one of the founders of *correspondence theory* [3] as a branch of modal logic where the expressiveness of modal logic as a language for describing Kripke structures is compared to that of more classical languages such as first-order logic. Perhaps his most important contribution to this area is the Bisimulation

Y. Venema (✉)
Institute for Logic, Language and Computation, Universiteit van Amsterdam,
Science Park 107, 1098 XG Amsterdam, The Netherlands
e-mail: y.venema@uva.nl

A. Baltag and S. Smets (eds.), *Johan van Benthem on Logic and Information Dynamics*, Outstanding Contributions to Logic 5, DOI: 10.1007/978-3-319-06025-5_2, © Springer International Publishing Switzerland 2014

Theorem stating that modal logic is the bisimulation-invariant fragment of first-order logic [2], in a slogan:

$$ML = FO/\underline{\leftrightarrow}. \tag{2.1}$$

More precisely, van Benthem showed that a formula $\varphi(x)$, in the language of first-order logic for Kripke models, is invariant under bisimulations iff it is equivalent to (the standard translation of) a modal formula. This observation fits the model-theoretic tradition of *preservation results*, characterizing fragments of (first-order) logic through a certain semantic property. What makes the result important is that in many applications of modal logic, it is natural to identify bisimilar states, and so properties that are not bisimulation invariant are irrelevant. From this perspective, the Bisimulation Theorem states an *expressive completeness* result: when it comes to relevant properties, modal logic has the same expressive power as first-order logic.

van Benthem's result has inspired many modal logicians, and over the years a wealth of variants of the Bisimulation Theorem have been obtained. Roughly, these can be classified as follows:

- Results showing that van Benthem's result still holds on *restricted classes* of models. In particular, Rosen proved that van Benthem's result is one of the few preservation results that transfers to the setting of finite models [19]; for a recent, rich source of van Benthem-style characterization results, see Dawar and Otto [8].
- Results characterizing *extensions* of basic modal logic as the bisimulation-invariant fragment of some extension of first-order logic. Here a key example is the theorem by Janin and Walukiewicz [14], characterizing the modal μ-calculus as the bisimulation-invariant fragment of monadic second-order logic.
- Results characterizing *variants* of modal logic as fragments of first-order logic that are invariant under some appropriate variant of the standard notion of bisimulation. Here we mention the result by Andréka, van Benthem and Németi, who characterized the guarded fragment [1] as the fragment of first-order logic that is invariant under guarded bisimulations; Otto [17] provides a overview of the results in this area, and of the (game-theoretic) methods used to prove these.
- Results on variants of modal logic where the modalities find their interpretation in different structures than the standard Kripke models. For instance, ten Cate et al. proved a van Benthem-style characterisation result for *topological* structures [4]. Recently, coalgebraic variations and generalizations of van Benthem's result have been obtained by Litak et al. [15].
- Clearly, researchers have been considering *combinations* of the above variations and generalizations; for example, Grädel et al. [11] proved a characterization result for guarded fixpoint logic. An outstanding open problem is whether the Janin-Walukiewicz theorem also holds for finite models, or equivalently, whether Rosen's result can be extended to the modal μ-calculus.

In this paper we will look in some detail at the result by Janin and Walukiewicz, which we can formulate as:

$$\mu ML = MSO/\underline{\leftrightarrow}. \tag{2.2}$$

Taking a *coalgebraic* perspective, we will show how the proof of (2.2) can be decomposed into three more or less independent parts:

1. a (non-trivial) result showing that both the modal μ-calculus and (on the class of tree models) monadic second-order logic can be characterized by certain *automata*,
2. a fairly simple model-theoretic characterisation result in monadic first-order logic, and
3. a general result on coalgebra automata.

Towards the end of this section we briefly discuss the relation of this chapter to Johan's work, and to the theme of this volume, viz., Logical Dynamics. First we turn to a fairly detailed explanation of the above decomposition, motivating our coalgebraic perspective. For this purpose we need to introduce automata. We fix a set Q of proposition letters. Elements of PQ will be called *colors*, and given a valuation $V : Q \to PS$, we define its associated *coloring* as the *transposed* map $V^\flat : S \to PQ$ given by $V^\flat : s \to \{p \in Q \mid p \in V(s)\}$.

The automata that we will consider here will be of the shape $\mathbb{A} = (A, \delta, \Omega)$, where A is a finite set of states and Ω is a parity map, $\Omega : A \to \mathbb{N}$. We will see a state $a \in A$ as a propositional variable, or, very much in the spirit of modal correspondence theory, as a monadic predicate. In this way, A provides a (monadic) first-order signature, which we will also denote as A. We define $\Phi^=(A)$ and $\Phi(A)$ as the sets of sentences (with and without equality, respectively) in this signature. We may *initialize* the automaton $\mathbb{A}$ by selecting an initial state $a \in A$. What shall interest us most is the *transition map* δ associating, with each state $a \in A$ and each color $c \in PQ$, a first-order sentence $\delta(a, c) \in \Phi(A)$.

Acceptance of a pointed Kripke model $(\mathbb{S}, s)$ by such an initialized automaton $(\mathbb{A}, a)$ is defined in terms of an infinite two-player *acceptance game* $\mathscr{A}(\mathbb{A}, \mathbb{S})$. A match of this game consists of the two players, $\forall$ and $\exists$, moving a token from one position to another. In a so-called *basic position*, which is of the form $(a, s) \in A \times S$, $\exists$ needs to define an A-valuation $M : A \to PS$ on S, with the proviso that M turns the set of successors of s into a *structure* for the signature A where the formula $\delta(a, V^\flat(s))$ is *true*. That, is we require that

$$(\sigma_R(s), M{\upharpoonright}_{\sigma_R(s)}) \models \delta(a, c),$$

where $c = V^\flat(s) \in PQ$ is the color of s in $\mathbb{S}$, $\sigma_R(s)$ is the set of successors of s, and $M{\upharpoonright}_{\sigma_R(s)}$ is the A-valuation M restricted to $\sigma_R(s)$. In other words, the pair $(\sigma_R(s), M{\upharpoonright}_{\sigma_R(s)})$ is an A-structure (in the sense of first-order model theory), and it is the aim of $\exists$ to make the sentence $\delta(a, V^\flat(s))$ true in this structure by chosing an appropriate A-valuation M. Given such a choice $M : A \to PS$, the game moves on with $\forall$ picking a next basic position from the set $\{(b, t) \mid t \in M(b)\}$. In this way, a match proceeds from one basic position (a_i, s_i) to the next (a_{i+1}, s_{i+1}). An infinite match of this game is won by $\exists$ if the highest parity $\Omega(a_i)$ occurring infinitely often during this match is even.

If ∃ has a winning strategy in the instantiation of the game that starts at the basic position (a, s), we say that the initialized automaton $(\mathbb{A}, a)$ *accepts* the pointed model $(\mathbb{S}, s)$. Initialized automata thus determine classes of pointed Kripke models, and we may compare the expressive power of such automata to that of a logic such as the modal μ-calculus or monadic second-order logic.

Definition 2.1.1 Given a fragment Θ of monadic first-order logic (in the sense that Θ assigns to each A a set of sentences $\Theta(A) \subseteq \Phi^=(A)$), we obtain an associated class Aut_Θ of initialized automata $(\mathbb{A}, a)$ by requiring that $\Theta(A)$ is the co-domain of the transition function of $\mathbb{A}$, that is, we have $\delta : A \times PQ \to \Theta(A)$. □

Definition 2.1.2 We let $\Pi^=(A)$ and $\Pi(A)$ denote the sets of sentences in $\Phi^=(A)$, with and without equality, respectively, where each occurrence of a monadic predicate is positive. □

In particular, we can now substantiate the claim that both monadic second-order logic (on the class of tree models) and the modal μ-calculus can be captured by automata-theoretic means. This link between logic and automata theory essentially goes back to the work of Rabin and Büchi on stream and tree automata. The two statements in Fact 2.1.3 below can be found in Walukiewicz [23] and Janin and Walukiewicz [13], respectively.

Fact 2.1.3 1. *On tree models, monadic second-order logic corresponds to* $\text{Aut}_{\Pi^=}$.
2. *The modal μ-calculus corresponds to* Aut_Π.

The main point of this paper is that, for any fragment Θ of $\Phi^=$, properties of Θ-automata are determined by properties of Θ. In particular, given two distinct fragments Θ and Θ', we will see how the question whether $\text{Aut}(\Theta)$ is the bisimulation invariant fragment of $\text{Aut}(\Theta')$, may already be determined at the level of Θ and Θ'.

For this purpose we introduce the notion of *P-invariance*. We say that two A-structures (D, V) and (D', V') are P-equivalent, notation: $(D, V) \equiv_P (D', V')$, if for all $d \in D$ there is a $d' \in D'$ with the same A-color, and vice versa. A first-order sentence $\alpha \in \Phi^=(A)$ is *P-invariant* if $(D, V) \models \alpha \iff (D', V') \models \alpha$, for all pairs of P-equivalent A-structures (D, V) and (D', V'). (As we will see, this property is equivalent to being preserved under surjective homomorphisms.) Given two fragments Θ, Θ' of $\Phi^=$, we say that Θ *corresponds to the P-invariant fragment of* Θ' if any sentence $\alpha \in \Theta'(A)$ is *P*-invariant iff it is equivalent to a formula $\alpha^* \in \Theta(A)$. The 'fairly simple result in monadic first-order logic' mentioned as the second item above, can now be stated as follows.

Proposition 2.1.4 *Π corresponds to the P-invariant fragment of $\Pi^=$.*

Our observation is that the Janin-Walukiewicz Theorem is a direct *corollary* of Fact 2.1.3 and Proposition 2.1.4. To be more precise, we will define a translation $(\cdot)^*$ mapping a $\Pi^=(A)$-sentence α to $\Pi(A)$-sentence α^* satisfying

$$\alpha \equiv \alpha^* \quad \text{iff } \alpha \text{ is } P\text{-invariant.} \tag{2.3}$$

On the basis of this, we can present the proof of Janin and Walukiewicz as follows.

First, given a *MSO*-formula φ, consider the equivalent initialized $\Pi^=$-automaton $(\mathbb{A}_\varphi, a_\varphi)$ given by Fact 2.1.3(1). Where $\mathbb{A}_\varphi = (A, \delta, \Omega)$, with $\delta : A \times PQ \to \Pi^=(A)$, define the Π-automaton $\mathbb{A}_\varphi^* := (A, \delta^*, \Omega)$, with $\delta^* : A \times PQ \to \Pi(A)$ by putting $\delta^*(a, c) := (\delta(a, c))^*$. Let φ^* be the μML-formula that is equivalent to $(\mathbb{A}_\varphi^*, a_\varphi)$, given by Fact 2.1.3(2). Using (2.3) one may then show that for any pointed Kripke model $(\mathbb{S}, s)$ there is a pointed Kripke model $(\mathbb{S}', s')$, and a bounded morphism $f : \mathbb{S}' \to \mathbb{S}$ such that $f s' = s$, while for any *MSO*-formula φ we have

$$\mathbb{S}, s \Vdash \varphi^* \quad \text{iff} \quad \mathbb{S}', s' \Vdash \varphi. \tag{2.4}$$

Now suppose that φ is a bisimulation-invariant *MSO*-formula. Then for any pointed Kripke model $(\mathbb{S}, s)$ we have that

$$\begin{aligned} \mathbb{S}, s \Vdash \varphi \quad &\text{iff} \quad \mathbb{S}', s' \Vdash \varphi && \text{(assumption on } \varphi) \\ &\text{iff} \quad \mathbb{S}, s \Vdash \varphi^* && \text{(2.4)} \end{aligned}$$

Clearly this shows that φ is equivalent to φ^*, and since φ^* is a formula in the modal μ-calculus, this suffices to prove the Janin-Walukiewicz theorem.

In fact, the argument just given can be generalized to prove the following result.

Theorem 2.1.5 *Let Θ and Θ' be fragments of monadic first-order logic. If Θ corresponds to the P-invariant fragment of Θ', then Aut_Θ corresponds to the bisimulation-invariant fragment of $\mathrm{Aut}_{\Theta'}$.*

The second and main contribution of this paper is the observation that Theorem 2.1.5 is itself an instance of a more general phenomenon that is essentially *coalgebraic* in nature. Universal Coalgebra [20] provides the notion of a *coalgebra* as the natural mathematical generalization of state-based evolving systems such as streams, (infinite) trees, finite state automata, Kripke frames and models, (probabilistic) transition systems, and many others. Formally, a coalgebra is a pair $\mathbb{S} = (S, \sigma)$, where S is the carrier or state space of the coalgebra, and $\sigma : S \to TS$ is its unfolding or transition map. This approach combines simplicity with generality and wide applicability: many features, including input, output, nondeterminism, probability, and interaction, can easily be encoded in the coalgebra type T (formally an endofunctor on the category Set of sets as objects with functions as arrows).

Logic enters the picture if one wants to specify and reason about *behavior*, one of the most fundamental notions admitting a coalgebraic formalization. With Kripke structures constituting key examples of coalgebras, it should come as no surprise that most coalgebraic logics are some kind of modification or generalization of *modal logic* [5]. Moss [16] introduced a modality ∇_T generalizing the so-called 'cover modality' from Kripke structures to coalgebras of arbitrary type. This approach is uniform in the functor T, but as a drawback only works properly if T satisfies a certain category-theoretic property (viz., it should preserve weak pullbacks); also the nabla modality is syntactically rather nonstandard. As an alternative, Pattinson [18] and others developed coalgebraic modal formalisms, based on a completely standard syntax, that work for coalgebras of arbitrary type. In this approach, the semantics of each modality is determined by a so-called *predicate lifting* (see Definition 2.2.15 below). Many well-known variations of modal logic in fact arise as the coalgebraic logic L_Λ associated with a set Λ of such predicate liftings; examples include both standard and (monotone) neighborhood modal logic, graded and probabilistic modal logic, coalition logic, and conditional logic.

In order to reason about *ongoing* coalgebraic behavior, modal logicians have introduced fixpoint extensions of coalgebraic logics [6, 22] and developed the corresponding automata theory [10]. For instance, each set Λ of predicate liftings comes with a modal logic L_Λ, a coalgebraic μ-calculus μL_Λ, and an equivalent class of automata Aut_Λ.

Kripke frames are coalgebras for the power set functor P, and each sentence α in monadic first-order logic induces a predicate lifting $\widehat{\alpha}$ for the power set functor. However, corresponding to the fact that we are looking at logics that are not bisimulation invariant, not all of these predicate liftings will be *natural* (in some technical sense to be defined below). In fact, we will introduce a coalgebraic novelty in this paper, in that we will consider non-natural predicate liftings for an arbitrary functor T. Generalizing the notion of P-invariance discussed above, we will define what it means for one set of predicate lifting to be the *natural fragment* of another set. Our coalgebraic generalization of Theorem 2.1.5 then roughly states the following:

if Λ provides the T-natural fragment of Λ', then μL_Λ is the bisimulation-invariant fragment of $\mu L_{\Lambda'}$.

For a more precise formulation, we refer to Theorem 2.5.1 below.

To conclude this introduction, we briefly discuss the relation of this chapter with van Benthem's work. First of all, it may have struck the reader's attention that while van Benthem's Bisimulation Theorem concerns the bisimulation-invariant fragment of *first-order* logic, our focus is on monadic *second-order* logic. We certainly believe that our coalgebraic perspective has some bearing on first-order logic as well, but we will leave this topic for later work. The main reason for this is that we wanted to give a detailed account of the coalgebraic perspective on fixpoint logics and automata theory. Note that in a general coalgebraic context, it is always clear how to define modal fixpoint logics and their associated automata. This is not necessarily the case with first-order logic, although recently some interesting proposals have been made, see for instance Litak et al. [15].

Another matter concerns the link between this chapter and the volume's theme, viz., *Logical Dynamics*. Here, again, *coalgebra* is the key word: as mentioned, universal coalgebra is a very natural mathematical framework for the kind of state-based evolving systems that play a fundamental role in the study of dynamics. In particular, many of the game-like processes that van Benthem is interested in, allow for a coalgebraic presentation. From this perspective coalgebraic modal logics, and in particular their fixed-point variants, provide natural logics for representing dynamic phenomena. The question of bisimulation invariance then makes us focus on the power of logical languages to express those properties that are *relevant* from the perspective of modelling dynamics. As such, our chapter not only connects with van Benthem's earliest technical work, but also with his foundational studies to the nature of the dynamics of information-related processes.

2.2 Coalgebra and Modal Logic

This section contains an introducton to coalgebra and coalgebraic modal logic.

We assume familiarity with basic notions from category theory, but not going beyond categories, functors, and natural transformations. We let Set denote the category with sets as objects and functions as arrows. Functors that feature prominently in this paper are the *co-* and the *contravariant power set functor*, P and $\breve{P}$, respectively. Both act on objects by mapping a set S to its power set $PS = \breve{P}S$; a function $f : S' \to S$ is mapped by P to the direct image function $Pf : PS' \to PS$ given by $(Pf)X' := \{fs' \in S \mid s' \in X'\}$, and by $\breve{P}$ to the inverse image function $\breve{P}f : PS \to PS'$ given by $(\breve{P}f)X := \{s' \in S' \mid fs' \in X\}$.

2.2.1 Coalgebra

We start with introducing coalgebras and their morphisms.

Definition 2.2.1 Let $T : \mathsf{Set} \to \mathsf{Set}$ be a (covariant) set functor. A *T-coalgebra* is a pair $\mathbb{S} = \langle S, \sigma \rangle$ where S is a set and σ is a function $\sigma : S \to TS$. Elements of S are called *states* of the coalgebra and σ is called the *transition map* of coalgebra map of $\mathbb{S}$. We may refer to T as the *type* of $\mathbb{S}$. A *pointed T-coalgebra* is a pair $(\mathbb{S}, s)$ consisting of a T-coalgebra $\mathbb{S}$ and a state $s \in S$.

If, for a function $f : S' \to S$, the following diagram commutes:

$$
\begin{array}{ccc}
S' & \overset{f}{\longrightarrow} & S \\
{\scriptstyle \sigma'}\downarrow & & \downarrow{\scriptstyle \sigma} \\
TS' & \underset{Tf}{\longrightarrow} & TS
\end{array}
\tag{2.5}
$$

we call f a *(coalgebra) morphism* from $\mathbb{S}' = \langle S', \sigma \rangle$ to $\mathbb{S} = \langle S, \sigma \rangle$, and write $f : \mathbb{S}' \to \mathbb{S}$.   □

Convention 2.2.2 Throughout this paper we will discuss an arbitrary but fixed (covariant) set functor that we denote as $T : \mathsf{Set} \to \mathsf{Set}$.

Many structures that are well-known from theoretical computer science or from modal logic admit a natural presentation as coalgebras.

Example 2.2.3 • Kripke frames are coalgebras for the power set functor P: a Kripke frame $\langle S, R \rangle$ with $R \subseteq S \times S$ can be represented as the coalgebra $\langle S, \rho_R \rangle$, where

ρ_R maps a state in S to the collection of its successors: $\rho_R : s \mapsto \{t \in S \mid Rst\}$. It is straightforward to verify that the notion of a coalgebra morphism for P-coalgebras coincides with that of a bounded morphism between Kripke frames. In other words, the category of P-coalgebras is isomorphic to that of Kripke frames (with bounded morphisms).

• Kripke models are coalgebras as well. Fix a set Q of proposition letters, and observe that the information given by a valuation $V : Q \to PS$ can just as well be provided by its *transpose* $V^\flat : S \to PQ$ given by $V^\flat : s \to \{p \in Q \mid p \in V(s)\}$. On the basis of this, we may represent a Kripke model $\langle S, R, V \rangle$ as a a coalgebra $\langle S, \sigma_{V,R} \rangle$, where $\sigma_{R,V} : S \to PQ \times PS$ is given by $\sigma_{R,V} : s \mapsto (V^\flat(s), \rho_R(s))$. In other words, Kripke models (over Q) are coalgebras for the functor $P_Q := PQ \times P-$.

• Recall that a deterministic finite state automaton (DFA) over a finite alphabet (or color set) C is a triple $\langle S, \delta, F \rangle$ with $\delta : S \times C \to S$ and $F \subseteq S$. Representing the transition map δ, through currying, by a function $\delta' : S \to S^C$, and the set F of accepting states by its characteristic function $\chi_F : S \to \{0, 1\}$, we may think of this DFA as a coalgebra $\langle S, (\chi_F, \delta') \rangle$ for the functor $D_C := \{0, 1\} \times (-)^C$. Here (as in subsequent examples) we omit to check that the *morphisms* induced by the coalgebra framework are the natural, standard ones.

• Given a set A of atomic actions, we can represent a transition system $(S, (R_a)_{a \in A})$, where each atomic action a is interpreted as a binary relation $R_a \subseteq S \times S$, as a coalgebra for the functor $(P-)^A$.

• Define the covariant set functor $N : \mathsf{Set} \to \mathsf{Set}$ as the composition of the contravariant power set with itself, $N := \breve{P} \circ \breve{P}$. Coalgebras for this functor correspond to the well-known *neighborhood* models in modal logic.

Restricting this example somewhat, we may obtain various interesting classes of structures. For instance, take the functor M given by $MS := \{\mathscr{U} \in NS \mid \mathscr{U}$ is upward closed with respect to $\subseteq \}$ and $Mf = Nf$. M-coalgebras are known in modal logic as monotone neighborhood frames.

• For a slightly more involved example, consider the finitary *multiset* or *bag* functor B_ω. This functor takes a set S to the collection $B_\omega S$ of maps $\mu : S \to \mathbb{N}$ of finite support (that is, for which the set $Supp(\mu) := \{s \in S \mid \mu(s) > 0\}$ is finite), while its action on arrows is defined as follows. Given an arrow $f : S \to S'$ and a map $\mu \in B_\omega S$, we define $(B_\omega f)(\mu) : S' \to \mathbb{N}$ by putting $(B_\omega f)(\mu)(s') := \sum \{\mu(s) \mid f(s) = s'\}$. Coalgebras for this functor are *weighted* transition systems, where each transition from one state to another carries a weight given by a natural number. Observe that a Kripke frame $\langle S, R \rangle$ can be seen as a B_ω-coalgebra $\langle S, \rho'_R \rangle$ by putting $\rho'_R(s)(t) = 1$ if Rst, and $\rho_R(s)(t) = 0$ otherwise.

• As a variant of B_ω, consider the finitary probability functor D_ω, where $D_\omega S = \{\delta : S \to [0, 1] \mid Supp(\delta)$ is finite and $\sum_{s \in S} \delta(s) = 1\}$, while the action of D_ω on arrows is just like that of B_ω. Coalgebras for this functor are known as *Markov chains*.

The connection between Kripke frames and Kripke models can be generalized to coalgebras of arbitrary type.

Definition 2.2.4 Let T be a set functor and let Q be a set of proposition letters. We define the set functor $T_Q := PQ \times T$. A T-*model over* Q is a pair $(\mathbb{S}, V)$ consisting of a T-coalgebra $\mathbb{S} = \langle S, \sigma \rangle$ and a Q-*valuation* V on S, that is, a function $V : Q \to PS$. The *coloring* associated with V is the *transpose* map $V^\flat : S \to PQ$ given by

$$V^\flat(s) := \{p \in Q \mid s \in V(p)\}.$$

Hence the pair $(\mathbb{S}, V)$ induces a T_Q-coalgebra $\langle S, (V^\flat, \sigma) \rangle$. □

Convention 2.2.5 In the remainder of this paper we will identify T-models over Q with the T_Q-coalgebras they induce. For instance, morphisms between T-models are implicitly defined as coalgebra morphisms between the induced T_Q-coalgebras. That is, a map $f : S' \to S$ is a morphism from $(\mathbb{S}', V')$ to $(\mathbb{S}, V)$ if (1) $f : \mathbb{S}' \to \mathbb{S}$ and (2) $s' \in V'(p)$ iff $fs' \in V(p)$, for all $s' \in S'$ and all $p \in Q$.

The key coalgebraic notion of equivalence is that of two pointed coalgebras being *behaviorally equivalent*. In case the functor T admits a coalgebra $\mathbb{Z} = \langle Z, \zeta \rangle$ which is *final* (in the sense that for every T-coalgebra $\mathbb{S}$ there is a unique coalgebra morphism $!_\mathbb{S} : \mathbb{S} \to \mathbb{Z}$), the elements of Z often provide an intuitive encoding of the notion of *behaviour*, and the unique coalgebra morphism $!_\mathbb{S}$ can be seen as a map that assigns to a state x in $\mathbb{S}$ its behaviour. In this case we call two pointed coalgebras, $(\mathbb{S}, s)$ and $(\mathbb{S}', s')$, *behaviorally equivalent* if $!_\mathbb{S}s = !_{\mathbb{S}'}s'$. In the general case, when we may not assume the existence of a set-based final coalgebra, we define the notion as follows.

Definition 2.2.6 Let $(\mathbb{S}, s)$ and $(\mathbb{S}', s')$ be two pointed coalgebras. If there are coalgebra morphisms f, f' with a common codomain such that $f(s) = f'(s')$, we call the two pointed coalgebras *behaviorally equivalent*, notation: $\mathbb{S}, s \simeq \mathbb{S}', s'$. We will often apply this notion to the states s and s'. □

Remark 2.2.7 In many cases, including those of Kripke frames and models, behavioral equivalence is the same as *bisimilarity*, but in cases where the two notions diverge, behavioral equivalence is the more natural notion. For the purpose of this paper it suffices to work with behavioral equivalence, and we do not need to discuss generalisations of the notion of a bisimulation to coalgebras of arbitrary type, referring the reader to [21] for more information.

2.2.2 Coalgebraic Logics

It will be convenient to have a rather abstract notion of coalgebraic languages and logics for T (so that for instance, we can think of automata as proper formulas).

Definition 2.2.8 An abstract coalgebraic logic is a pair $(L, \Vdash^L)$ such that L is a set and $\Vdash$ is a collection of relations associating with each T-coalgebra $\mathbb{S} = \langle S, \sigma, V \rangle$ a binary relation $\Vdash^L_\mathbb{S} \subseteq S \times L$. The set L is called the *language* of the logic, and its elements will be called *formulas*. If $s \Vdash^L_\mathbb{S} \varphi$ we say that the formula φ is *true* or *satisfied* at s in $\mathbb{S}$, and we will often write $\mathbb{S}, s \Vdash \varphi$.

The satisfaction relation $\Vdash^L_\mathbb{S}$ induces a *meaning function* $[\![\cdot]\!]^\mathbb{S} : L \to PS$ given by

$$s \in [\![\varphi]\!]^\mathbb{S} \quad \text{iff} \quad s \Vdash^L_\mathbb{S} \varphi. \tag{2.6}$$

$\square$

Example 2.2.9 Let us see how monadic second-order logic fits as a coalgebraic logic for $P_\mathbf{Q}$-coalgebras (Kripke models over some fixed set $\mathbf{Q}$ of proposition letters). Clearly we may also see elements of $\mathbf{Q}$ as monadic predicate symbols.

To define the syntax of this logic, let $\mathsf{IVar} = \{u, v, \ldots\}$ be a set of individual (first-order) variables, and let $\mathsf{Var} = \{x, y, \ldots\}$ be a set of objects that one may think of alternatively as propositional variables or monadic predicate (that is, second-order) variables. Define the set of *MSO(**Q**)-formulas* by the following grammar:

$$\varphi ::= p(v) \mid x(v) \mid Ruv \mid \bot \mid \neg\varphi \mid \varphi_0 \vee \varphi_1 \mid \exists v.\varphi \mid \exists x.\varphi,$$

where $p \in \mathbf{Q}, v \in \mathsf{IVar}$, and $x \in \mathsf{Var}$. The interpretation of this language on a Kripke model $\langle S, R, V \rangle$ is standard.

Finally, we define $MSO_v(\mathbf{Q})$ as the set of $MSO(\mathbf{Q})$-formulas $\varphi(v)$ that contain a single free individual variable v, and no free variables in Var. (We need the free variable v in order to interpret formulas in *pointed* models.) Thus we obtain a coalgebraic logic $(MSO_v(\mathbf{Q}), \Vdash^{MSO})$ by putting $\mathbb{S}, s \Vdash \varphi$ iff $\mathbb{S} \models \varphi(s)$.

Coalgebraic logics naturally induce equivalence relations between formulas, and between pointed coalgebras.

Definition 2.2.10 Let $(L, \Vdash)$ be a coalgebraic logic. Two formulas φ and ψ are called *equivalent*, notation: $\varphi \equiv_{L, \Vdash} \psi$, if for all pointed coalgebras $(\mathbb{S}, s)$ we have $\mathbb{S}, s \Vdash \varphi \iff \mathbb{S}, s \Vdash \psi$.

Similarly, two pointed coalgebras $(\mathbb{S}, s)$ and $(\mathbb{S}', s')$ are called *equivalent*, notation: $\mathbb{S}, s \equiv^{L, \Vdash} \mathbb{S}', s'$, if $\mathbb{S}, s \Vdash \varphi \iff \mathbb{S}', s' \Vdash \varphi$, for all $\varphi \in L$. $\square$

Since the satisfaction relation is usually determined by the language, in practice we will often blur the distinction between logics and their languages. For instance, we will write $\equiv^L$ rather than $\equiv^{(L, \Vdash)}$, etc.

Generally, in abstract model theory, an abstract logic is required not to distinguish isomorphic structures. Clearly such a condition would make sense here as well, but it is not relevant to our story. On the other hand, a condition that features crucially in our story is the requirement that coalgebraic logics cannot distinguish behaviorally equivalent states. Since we are generally much more interested in the behavior of a system then in its precise representation, this so-called *adequacy* property is a very natural one.

Definition 2.2.11 A formula φ is *behaviorally invariant* if for all pairs of behaviorally equivalent pointed coalgebras $\mathbb{S}, s \simeq \mathbb{S}', s'$ it holds that $\mathbb{S}, s \Vdash \varphi \iff$

$\mathbb{S}', s' \Vdash \varphi$. A coalgebraic language is *behaviorally invariant* or *adequate* if all its formulas are behaviorally invariant, or equivalently, if $\simeq_T \subseteq \equiv^L$. $\qquad\square$

We have now arrived at the central notion in this paper, namely, that of one logic corresponding to the behaviorally invariant fragment of another.

For its definition, observe that if $(L, \Vdash)$ is a coalgebraic logic, then any set $L' \subseteq L$ induces a logic $(L', \Vdash_{L'})$, where $\Vdash_{L'}$ is the obviously defined restriction of the relation $\Vdash$ to L'. In the sequel we will simply write $\Vdash$ rather than $\Vdash_{L'}$.

Definition 2.2.12 Let $(L, \Vdash)$ be a coalgebraic logic, and let $L' \subseteq L$. We say that L' *corresponds to the behaviorally invariant fragment of* L, notation: $L' \equiv L/T$, if (1) $(L', \Vdash)$ is behaviorally invariant, and (2) every behaviorally invariant formula $\varphi \in L$ is equivalent to some formula $\varphi' \in L'$. $\qquad\square$

The technical work in this paper will be based on a stronger, somewhat more 'constructive' version of this notion.

Definition 2.2.13 Let $(L, \Vdash)$ be a coalgebraic logic, and let $L' \subseteq L$. We say that L' *strongly corresponds to the behaviorally invariant fragment of* L, notation: $L' \equiv^s L/T$, if (1) $(L', \Vdash)$ is behaviorally invariant, and $(2')$ there is a translation $(\cdot)^* : L \to L'$ and a map associating with each pointed coalgebra $(\mathbb{S}, s)$ a pointed coalgebra $(\mathbb{S}', s')$, together with a morphism $f : (\mathbb{S}', s') \to (\mathbb{S}, s)$ such that

$$\mathbb{S}, s \Vdash \varphi^* \quad \text{iff} \quad \mathbb{S}', s' \Vdash \varphi. \tag{2.7}$$

for all formulas $\varphi \in L$. $\qquad\square$

The following proposition justifies our terminology.

Proposition 2.2.14 *If* $L' \equiv^s L/T$ *then* $L' \equiv L/T$.

Proof Assume that L' strongly corresponds to the behaviorally invariant fragment of L, via the translation $(\cdot)^* : L \to L'$, and let φ be an arbitrary behaviorally invariant formula in L. In order to prove the Proposition, it suffices to show that $\varphi \equiv \varphi^*$. For this purpose, take an arbitrary pointed coalgebra $(\mathbb{S}, s)$. By our assumption there is a morphism $f : \mathbb{S}' \to \mathbb{S}$ and a state s' in $\mathbb{S}'$ with $fs' = s$, and satisfying

$$\mathbb{S}, s \Vdash \varphi \quad \text{iff} \quad \mathbb{S}', s' \Vdash \varphi \quad \text{iff} \quad \mathbb{S}, s \Vdash \varphi^*. \tag{2.8}$$

Here the first equivalence is by our assumption on φ, and the second equivalence is by (2.7). The equivalence of φ and φ^* is immediate from (2.8). $\qquad$ QED

2.2.3 Predicate Liftings

Many coalgebraic logics are induced by a set of so-called predicate liftings. In this subsection we will be interested in T-models; we fix a set Q of proposition letters.

Definition 2.2.15 An *n*-ary *predicate lifting for* T is a collection λ of maps, associating a function

$$\lambda_S : (PS)^n \to PTS$$

with each set S.   □

In other words, an *n*-ary predicate lifting λ associates, with each set S, a map that yields a subset $\lambda_S(X_1, \ldots, X_n) \subseteq TS$ for each *n*-tuple $X_1, \ldots, X_n$ of subsets of S.

Note that our definition deviates from the usual one in that we do not require predicate liftings to be *natural* (see Definition 2.2.23).

Example 2.2.16 Here are some predicate liftings for the functors discussed in Example 2.2.3.

• Given a set S, a unary predicate lifting λ for the power set functor yields a map $\lambda_S : PS \to PPS$. Here are three examples, $\Diamond$, $\Box$ and ∞:

$$\Diamond_S : X \mapsto \{D \in PS \mid D \cap X \neq \varnothing\},$$
$$\Box_S : X \mapsto \{D \in PS \mid D \subseteq X\},$$
$$\infty_S : X \mapsto \{D \in PS \mid |D \cap X| \geq \omega\}.$$

For an example of a binary predicate lifting, consider the following definition, for a set S:

$$\star_S : \quad (X, Y) \mapsto \{D \in S \mid X \subseteq D \subseteq Y\}.$$

• A *nullary* predicate lifting λ assigns to each set S, a function λ_S from $(PS)^0$ to PTS; such a function can be identified with a subset of TS that we will also denote as λ_S. As a particularly interesting example, consider the functor T_Q. With each proposition letter $p \in Q$ we may associate a nullary predicate lifting $\underline{p}$ by defining, for each set S, the following subset of $T_Q S$:

$$\underline{p}_S := \{(\Pi, \tau) \in PQ \times TS \mid p \in \Pi\}.$$

• Regarding the functor D_C corresponding to finite state automata over C, consider the nullary predicate lifting $\sqrt{}$ and the unary $\copyright$ (for any $c \in C$), defined, for a set S, by

$$\sqrt{}_S := \quad \{(i, f) \in 2 \times S^C \mid i = 1\},$$
$$\copyright_S : X \mapsto \{(i, f) \in 2 \times S^C \mid f(c) \in X\}.$$

• For the functor P^A of *A*-labelled transition systems, consider the unary lifting $\langle a \rangle$ given by

$$\langle a \rangle_S : X \mapsto \{D \in (PS)^A \mid D(a) \cap X \neq \varnothing\},$$

• With respect to the neighborhood functor N, we define a unary predicate lifting $\Diamond$ by putting, for a set S:

$$\Diamond_S : X \mapsto \{\mathscr{A} \in NS \mid X \in \mathscr{A}\}.$$

• Finally, consider the functor B_ω. Given a natural number $k \in \omega$, we define the predicate lifting $\underline{k}$ by putting

$$\underline{k}_S : X \mapsto \{\mu \in B_\omega S \mid \sum_{x \in X} \mu(x) \geq k\}.$$

Definition 2.2.17 With each predicate lifting λ we associate a modality $\heartsuit_\lambda$ with the same arity as λ. Given a set Λ of predicate liftings, we obtain the modal language $L_\Lambda(Q)$ by defining its set of formulas by the following grammar:

$$\varphi ::= p \mid \bot \mid \neg\varphi \mid \varphi_0 \vee \varphi_1 \mid \heartsuit_\lambda(\varphi_1, \ldots, \varphi_n)$$

where $p \in Q$, and $\lambda \in \Lambda$ is n-ary. □

Definition 2.2.18 Let Λ be a set of predicate liftings. For any T-model $\mathbb{S} = \langle S, \sigma, V \rangle$, by induction on the complexity of L_Λ-formulas, we define the *meaning function* $[\![\cdot]\!]^{\mathbb{S}} : L_\Lambda \to PS$:

$$
\begin{aligned}
[\![\bot]\!]^{\mathbb{S}} &:= \varnothing \\
[\![p]\!]^{\mathbb{S}} &:= V(p) \\
[\![\neg\varphi]\!]^{\mathbb{S}} &:= S \setminus [\![\varphi]\!]^{\mathbb{S}} \\
[\![\varphi_0 \vee \varphi_1]\!]^{\mathbb{S}} &:= [\![\varphi_0]\!]^{\mathbb{S}} \cup [\![\varphi_1]\!]^{\mathbb{S}} \\
[\![\heartsuit_\lambda(\varphi_1, \ldots, \varphi_n)]\!]^{\mathbb{S}} &:= (\breve{P}\sigma)(\lambda_S([\![\varphi_1]\!]^{\mathbb{S}}, \ldots, [\![\varphi_n]\!]^{\mathbb{S}})).
\end{aligned}
$$

The meaning function $[\![\cdot]\!]^{\mathbb{S}}$ induces a satisfaction relation $\Vdash_{\mathbb{S}}$ given by (2.6). □

In terms of the satisfaction relation $\Vdash$, the meaning of the modality $\heartsuit_\lambda$ is given by

$$\mathbb{S}, s \Vdash \heartsuit_\lambda(\varphi_1, \ldots, \varphi_n) \quad \text{iff} \quad \sigma(s) \in \lambda_S([\![\varphi_1]\!], \ldots, [\![\varphi_n]\!]).$$

Example 2.2.19 • It is easy to see that, for a Kripke model $\mathbb{S} = \langle S, R, V \rangle$, we have

$$
\begin{aligned}
\mathbb{S}, s \Vdash \heartsuit_\Diamond \varphi &\iff \mathbb{S}, t \Vdash \varphi \text{ for some } t \in R(s) \\
\mathbb{S}, s \Vdash \heartsuit_\Box \varphi &\iff \mathbb{S}, t \Vdash \varphi \text{ for all } t \in R(s) \\
\mathbb{S}, s \Vdash \heartsuit_\infty \varphi &\iff \mathbb{S}, t \Vdash \varphi \text{ for infinitely many } t \in R(s) \\
\mathbb{S}, s \Vdash \heartsuit_\star(\varphi, \psi) &\iff \mathbb{S}, t \Vdash \varphi \text{ for all } t \in R(s), \text{ and } Rsu \text{ for all } u \text{ with } \mathbb{S}, u \Vdash \psi.
\end{aligned}
$$

The first two examples shows in particular that the well-known diamond and box operator from modal logic are coalgebraic modalities indeed.

• The definition of the predicate lifting $\underline{p}$, for a proposition letter $p \in Q$, ensures that for any T-model $\mathbb{S} = (S, \sigma, V)$ we have

$$
\begin{aligned}
\mathbb{S}, s \Vdash \underline{p} \quad &\text{iff} \quad p \in V^\flat(p) && (\text{semantics } \underline{p}) \\
&\text{iff} \quad s \in V(p) && (\text{definition } V^\flat) \\
&\text{iff} \quad \mathbb{S}, s \Vdash p && (\text{semantics } p)
\end{aligned}
$$

- For a model $\mathbb{S}$ based on a finite state automaton $\langle S, F, \delta \rangle$, we have

$$\mathbb{S}, s \Vdash \heartsuit_{\surd}\varphi \iff s \in F,$$
$$\mathbb{S}, s \Vdash \heartsuit_{\copyright}\varphi \iff \mathbb{S}, \delta(c, s) \Vdash \varphi.$$

- If $\mathbb{S}$ is a model based on an A-labelled transition system, we find

$$\mathbb{S}, s \Vdash \heartsuit_{\langle a \rangle}\varphi \iff \mathbb{S}, t \Vdash \varphi \text{ for some } t \in R_a(s)$$

- For a neighborhood model $\mathbb{S}$ we obtain

$$\mathbb{S}, s \Vdash \heartsuit_{\diamond}\varphi \iff [\![\varphi]\!]^{\mathbb{S}} \in \sigma(s),$$

showing that classical modal logic is a coalgebraic logic indeed.

- Finally, suppose that we consider a Kripke frame as a coalgebra $\mathbb{S}$ for the functor B_ω. Then for any natural number $k \in \omega$ we obtain

$$\mathbb{S}, s \Vdash \heartsuit_{\underline{k}}\varphi \iff s \text{ has at least } k \text{ successors } t \text{ such that } \mathbb{S}, t \Vdash \varphi.$$

In other words, *graded modal logic* can be presented as a coalgebraic logic too.

We now turn to *coalgebraic μ-calculi*, that is, extensions of coalgebraic logics with fixpoint operators. In order to guarantee well-definedness of the semantics, we need to restrict attention to monotone predicate liftings.

Definition 2.2.20 An n-ary predicate lfiting λ is *monotone* if for every set S, the map $\lambda_S : (PS)^n \to PTS$ is order-preserving in each coordinate (with respect to the subset order). The predicate lifting $\overline{\lambda} : (P-)^n \to PT-$, given by

$$\overline{\lambda}_S(X_1, \ldots, X_n) := TS \setminus \lambda_S(S \setminus X_1, \ldots, S \setminus X_1),$$

is called the *(Boolean) dual* of Λ. □

Since we are working with a fixed set Q of proposition letters, we need to introduce a set $\mathsf{Var} = \{x, y, z, x_0, \ldots\}$ of propositional *variables* in our formal set-up.

Definition 2.2.21 Let Λ be a set of monotone predicate liftings. The modal language $\mu L_\Lambda(\mathsf{Q})$ is defined by the following grammar:

$$\varphi ::= p \mid x \mid \bot \mid \neg\varphi \mid \varphi_0 \vee \varphi_1 \mid \heartsuit_\lambda(\varphi_1, \ldots, \varphi_n) \mid \mu x.\varphi$$

where $p \in \mathsf{Q}$, $x \in \mathsf{Var}$, and the application of the fixpoint operator μx is subject to the proviso that all occurrences of x in φ are positive (that is, under an even number of negations).

The sets of free and bound variables in a formula is defined as usual, and we define a $\mu L_\Lambda(\mathsf{Q})$-*sentence* as a formula with no bound variables. □

The semantics of this language contains no surprises.

Definition 2.2.22 Let Λ be a set of monotone predicate liftings, and let $\mathbb{S}$ be a T-model. An *assignment* is a map $h : \mathsf{Var} \to PS$ assigning a meaning to each variable in Var. By induction on the complexity of $\mu L_\Lambda(\mathsf{Q})$-formulas we define, for each assignment h, a meaning function $[\![\cdot]\!]^{\mathbb{S},h} : \mu L_\Lambda(\mathsf{Q}) \to PS$. Here we only give the following two clauses:

$$[\![x]\!]^{\mathbb{S},h} := h(x)$$
$$[\![(\mu x.\varphi)]\!]^{\mathbb{S},h} := \bigcap_{X \subseteq S} [\![\varphi]\!]^{\mathbb{S},h[x \mapsto X]},$$

where $h[x \mapsto X]$ is the assignment sending every $y \in \mathsf{Var}$ to $V(y)$, except for x which is sent to X. □

It is a routine exercise to verify that with this definition, the formula $\mu x.\varphi$ is interpreted as the least fixed point of the formula $\varphi(x)$.

Returning to the notion of adequacy, for logics generated by predicate liftings this property follows from naturality of the predicate liftings.

Definition 2.2.23 A predicate lifting λ is *natural for T* if it is a natural transformation $\lambda : (\breve{P}-)^n \dot{\to} (\breve{P}T-)$, i.e. if for each function $f : S' \to S$, the following diagram commutes:

$$
\begin{array}{ccc}
S & (PS)^n \xrightarrow{\lambda_S} PTS & (2.9) \\
f \uparrow & (\breve{P}f)^n \downarrow \qquad \downarrow \breve{P}Tf & \\
S' & (PS')^n \xrightarrow{\lambda_{S'}} PTS' &
\end{array}
$$

A set of predicate liftings is natural for T if this property applies to each of its members. □

Proposition 2.2.24 *If Λ is natural for T, then $L_\Lambda(\mathsf{Q})$ and $\mu L_\Lambda(\mathsf{Q})$ are behaviorally invariant for each set Q.*

Proof By definition of behavioral equivalence it suffices to prove that for every morphism $f : \mathbb{S}' \to \mathbb{S}$, every formula φ, and every state $s' \in S'$, it holds that

$$\mathbb{S}', s' \Vdash \varphi \quad \text{iff} \quad \mathbb{S}, fs' \Vdash \varphi. \tag{2.10}$$

In terms of the meaning function $[\![\cdot]\!]$, we may prove equivalently that for all formulas φ

$$[\![\varphi]\!]^{\mathbb{S}'} = (\breve{P}f)[\![\varphi]\!]^{\mathbb{S}}. \tag{2.11}$$

We prove (2.11) by induction on φ. For the key inductive clause, assume that $\varphi = \Diamond_\lambda \psi$ (for notational simplicity we assume that λ is unary). Then

$$\llbracket\varphi\rrbracket^{\mathbb{S}'} = (\check{P}\sigma')\lambda_{S'}(\llbracket\psi\rrbracket^{\mathbb{S}'}) \qquad \text{(semantics of } \heartsuit_\lambda)$$
$$= (\check{P}\sigma')\lambda_{S'}(\check{P}f)(\llbracket\psi\rrbracket^{\mathbb{S}}) \qquad \text{(induction hypothesis)}$$
$$= (\check{P}\sigma')(\check{P}Tf)\lambda_S(\llbracket\psi\rrbracket^{\mathbb{S}}) \qquad \text{(naturality of } \lambda \text{ (2.9))}$$
$$= (\check{P}f)(\check{P}\sigma)\lambda_S(\llbracket\psi\rrbracket^{\mathbb{S}}) \qquad (f \text{ a morphism)}$$
$$= (\check{P}f)\llbracket\varphi\rrbracket^{\mathbb{S}} \qquad \text{(semantics of } \heartsuit_\lambda)$$

QED

We leave it as an exercise for the reader to check that all predicate liftings of Example 2.2.16 are natural, except ∞ and $\star$.

For concreteness, we define (basic) modal logic and the modal μ-calculus as follows.

Definition 2.2.25 We define $ML(\mathsf{Q})$, *basic modal logic* over Q, as the coalgebraic logic $L_{\Diamond,\Box}(\mathsf{Q})$, and $\mu ML(\mathsf{Q})$, the *modal μ-calculus* over Q, as its fixpoint extension: $\mu ML(\mathsf{Q}) := \mu L_{\Diamond,\Box}(\mathsf{Q})$. $\qquad\Box$

2.3 Coalgebra Automata and MSO

As usual in the theory of fixpoint logics, it will be easier to work with *automata* rather than with formulas. In this section we define the notion of a *coalgebra automaton* associated with a set Λ of monotone predicate liftings (and a set Q of proposition letters). These devices will provide the automata-theoretic counterpart to the coalgebraic μ-calculus, and we will see how monadic second-order can be captured by automata (and thus correspond to a coalgebraic fixed point logic), when we restrict attention to the class of tree models. This is also a good place to introduce the one-step perspective on coalgebraic logic, a key coalgebraic concept.

2.3.1 One-Step Syntax and Semantics

Definition 2.3.1 Given a set A of propositional variables and a collection Λ of predicate liftings, we define the set $L^1_\Lambda(A)$ via the following grammar:

$$\varphi ::= \heartsuit_\lambda(a_1, \ldots, a_n) \mid \bot \mid \top \mid \varphi_0 \vee \varphi_1 \mid \varphi_0 \wedge \varphi_1$$

where $\lambda \in \Lambda$ is n-ary, and $a_i \in A$, for each i. Elements of $L^1_\Lambda(A)$ will be called *rank-1 Λ-formulas over A*, or simply: *rank-1 formulas*. $\qquad\Box$

Observe that we do not allow negations to occur in rank-1 formulas. Given a set S, we can interpret rank-1 formulas over A as subsets of TS, once we have been given a valuation assigning a meaning to the variables in A.

Definition 2.3.2 Given sets A and S, an A-*valuation* or A-*marking* on S is a map $V : A \to PS$. Given such a valuation, we inductively define the *one-step* satisfaction relation $\Vdash^1_V \subseteq TS \times L^1_\Lambda(A)$. For the basic formulas of the form $\heartsuit_\lambda(a_1, \ldots, a_n)$ we put, for $\tau \in TS$,

$$\tau \Vdash^1_V \heartsuit_\lambda(a_1, \ldots, a_n) \quad \text{iff} \quad \tau \in \lambda_S(V(a_1), \ldots, V(a_n)),$$

while inductively each Boolean connective receives its standard set-theoretic interpretation. Frequently we will write $TS, V, \tau \Vdash^1 \varphi$ rather than $\tau \Vdash^1_V \varphi$. □

The link with the ordinary semantics for coalgebraic logic is given by the coalgebra map. That is, given a T-coalgebra $\mathbb{S} = (S, \sigma : S \to TS)$ and a valuation $V : A \to PS$, we have

$$(\mathbb{S}, V), s \Vdash \heartsuit_\lambda(a_1, \ldots, a_n) \quad \text{iff} \quad TS, V, \sigma(s) \Vdash^1 \heartsuit_\lambda(a_1, \ldots, a_n).$$

2.3.2 Coalgebra Automata

We are now ready to introduce coalgebra automata, which will be parametrized by a set Λ of predicate liftings, and a set $\mathbf{Q}$ of proposition letters.

Definition 2.3.3 Let Λ be a set of monotone predicate liftings for T. A $(\Lambda, \mathbf{Q})$-automaton $\mathbb{A}$ is a triple $\mathbb{A} = (A, \delta, \Omega)$, where A is a finite set of states, $\delta : A \times P\mathbf{Q} \to L^1_\Lambda(A)$ is the transition map, and $\Omega : A \to \mathbb{N}$ is a parity map.

An *initialized* automaton is a pair $(\mathbb{A}, a_I)$ where $a_I \in A$. The class of initialized $(\Lambda, \mathbf{Q})$-automata is denoted as $\mathrm{Aut}_\Lambda(\mathbf{Q})$. □

The semantics of these automata is defined in terms of an infinite parity graph game. We assume that the reader has some familiarity with these games, and with associated notions such as matches, (positional) strategies, etc. (Details can be found in [12]).

Definition 2.3.4 Let $\mathbb{S} = \langle S, \sigma, V \rangle$ be a T-model and let $\mathbb{A} = (A, \delta, \Omega)$ be a $(\Lambda, \mathbf{Q})$-automaton. The associated *acceptance game* $\mathscr{A}(\mathbb{A}, \mathbb{S})$ is the parity game given by the table below.

Position	Player	Admissible moves	Priority
$(a, s) \in A \times S$	$\exists$	$\{M : A \to PS \mid TS, V, \sigma(s) \Vdash^1 \delta(a, M^\flat(s))\}$	$\Omega(a)$
$M \in (PS)^A$	$\forall$	$\{(b, t) \mid t \in M(b)\}$	0

A pointed coalgebra $(\mathbb{S}, r)$ is *accepted* by an initialized automaton $(\mathbb{A}, a_I)$, notation: $\mathbb{S}, r \Vdash (\mathbb{A}, a_I)$, if the pair (a_I, r) is a winning position for player $\exists$ in $\mathscr{A}(\mathbb{A}, \mathbb{S})$. □

Observe that the acceptance game of $(\Lambda, \mathbf{Q})$-automata proceeds in *rounds* moving from one basic position in $A \times S$ to another. In each round, at position (a, s) first $\exists$ picks an A-marking M on S that makes the depth-one formula $\delta(a, V^\flat(s))$ true at $\sigma(s)$. Looking at this $M : A \to PS$ as a binary relation $\{(b, t) \mid t \in M(b)\}$ between A and S consisting of *witnesses* picked by $\exists$, it is $\forall$ who closes the round by choosing a witness (b, t) from this relation, which will then serve as the starting position of the next round of the game.

Acceptance games feature both maps of the form $V : \mathbf{Q} \to PS$ (valuations that are part of the models $\mathbb{S}$ on which the automaton operates) and maps $M : A \to PS$ (that correspond to sets of witnesses picked by $\exists$). To emphasize the distinct roles that these two kinds of maps play, we will refer to the first ones as *valuations* and to the second ones as *markings* in situations where both types occur.

The following proposition instantiates the connection between fixpoint logics and parity automata in our setting of coalgebraic logic. Observe that we may think of the set $\text{Aut}_{\Lambda, \mathbf{Q}}$ of initialized $(\Lambda, \mathbf{Q})$-automata, together with the acceptance relation, as a coalgebraic logic.

Proposition 2.3.5 *Let Λ be a set of monotone predicate liftings for T which is closed under taking Boolean duals. Then $\mu L_\Lambda(\mathbf{Q}) \equiv \text{Aut}_\Lambda(\mathbf{Q})$.*

We omit the (completely routine) proof.

2.3.3 MSO As a Coalgebraic Fixpoint Logic

We will now see how we may think of MSO as a coalgebraic fixpoint logic, at least if we restrict our attention to tree models.

Definition 2.3.6 A *tree model* is a Kripke model $\langle S, R, V \rangle$ in which there is a unique path to each point from a certain, fixed, state called the *root* of the tree. $\qquad\square$

Throughout this subsection we fix a set A of syntactic objects that, as mentioned in the introduction, we may think of as either propositional variables or monadic *predicate symbols* of some first-order language. In other words, we will see A as a first-order signature (that we will also denote as A). The formulas in this language are given by the following grammar:

$$\alpha ::= x = y \mid a(x) \mid \neg\alpha \mid \alpha \vee \alpha \mid \exists x. \alpha$$

If we use the notation $\alpha(a_1, \ldots, a_n)$ for a sentence in this language, this indicates that the predicate symbols in α (not the first-order variables) are among $a_1, \ldots, a_n$.

Definition 2.3.7 We let $\Phi^=(A)$ and $\Phi(A)$ denote the sets of all first-order sentences over the signature A, respectively with and without identity. $\Pi^=(A)$ and $\Pi(A)$ are the positive fragments of $\Phi^=(A)$ and $\Phi(A)$, respectively, consisting of those sentences in which all occurrences of atomic formulas are positive. $\qquad\square$

We have now arrived at the key observation underlying this paper. In the case that our coalgebra functor T is the *power set functor*, given an A-valuation V on S and an element $D \in PS$, we may think of the pair $\langle D, V \rangle$ as a *structure* for the signature A (in the sense of first-order model theory), where the predicate symbol $a \in A$ is interpreted as the subset $V(a) \cap D \subseteq D$. Consequently, we will now see that each first-order sentence $\alpha(a_1, \ldots, a_n)$ of this signature induces a (not necessarily natural) n-ary predicate lifting $\widehat{\alpha}$ for the power set functor.

Definition 2.3.8 Let $\alpha(a_1, \ldots, a_n) \in \Phi^=(A)$. For any set S, α induces a map $\widehat{\alpha} : (PS)^n \to PPS$, given by

$$\widehat{\alpha}(X_1, \ldots, X_n) := \{D \in PS \mid \langle D, V_{\overline{X}} \rangle \models \alpha\},$$

where $V_{\overline{X}}$ is the A-valuation on S given by $V_{\overline{X}}(a_i) := X_i$. By a slight abuse of notation, for any fragment Θ of $\Phi^=$, we let Θ also denote the corresponding set of predicate liftings $\{\widehat{\alpha} \mid \alpha \in \Theta\}$. □

That this approach makes sense follows by the following Proposition which states that the coalgebraic and the first-order perspective coincide.

Proposition 2.3.9 *Let $\alpha(a_1, \ldots, a_n) \in \Phi^=(A)$. For any set S, any valuation $V : A \to PS$ and any subset $D \subseteq S$ we have*

$$PS, V, D \Vdash^1 \heartsuit_{\widehat{\alpha}}(a_1, \ldots, a_n) \quad \text{iff} \quad \langle D, V \rangle \models \alpha(a_1, \ldots, a_n). \tag{2.12}$$

Proof The proof of Proposition 2.3.9 consists of a four line unravelling of the definitions. Consider a first-order sentence $\alpha(a_1, \ldots, a_n) \in \Phi^=$. Then

$$
\begin{aligned}
PS, V, D \Vdash^1 \heartsuit_{\widehat{\alpha}}(a_1, \ldots, a_n) \quad &\text{iff} \quad D \in \widehat{a}(V(a_1), \ldots, V(a_n)) &&\text{(semantics } \heartsuit\text{)} \\
&\text{iff} \quad \langle D, V_{V(a_1), \ldots, V(a_n)} \rangle \models \alpha &&\text{(definition } \widehat{\alpha}\text{)} \\
&\text{iff} \quad \langle D, V \rangle \models \alpha &&\text{(†)}
\end{aligned}
$$

where the last equivalence (†) follows from the fact that the predicate symbols in α are among $a_1, \ldots, a_n$. QED

In Sect. 2.4 we will see that all predicate liftings in Φ and Π are *natural* for P, while this is definitely not the case for $\Phi^=$ and Π (cf. Theorem 2.4.9).

As we will see further on, the following theorem is the reason why our coalgebraic approach can be applied to the Janin-Walukiewicz theorem. It states that, on the class of tree models, monadic second order logic can be captured by automata-theoretic means. As a corollary of this result and Proposition 2.3.5, MSO (see Example 2.2.9) is in fact a coalgebraic fixpoint logic. Theorem 2.3.10 below can be seen as a more precise formulation of Fact 2.1.3; as mentioned there, the two statements can be found in Walukiewicz [23] and Janin and Walukiewicz [13], respectively.

Theorem 2.3.10   1. $MSO_v(\mathbf{Q}) \equiv_P \mu L_{\Pi=}(\mathbf{Q})$ *on tree models*[1];
2. $\mu ML(\mathbf{Q}) \equiv_P \mu L_\Pi(\mathbf{Q})$.

Unfortunately, a proof of this Theorem would go beyond the scope of this paper. For proof details, the reader is referred to the above-mentioned papers, or to Chap. 16 of [12].

Remark 2.3.11 For the interested reader, we give a very rough sketch of the proof for part 1, which consists of two parts. First, rather than working with $MSO(\mathbf{Q})$ one defines a variant MSO' with only second-order variables. For a definition of the set of MSO'-formulas, consider the following grammar:

$$\varphi ::= p \sqsubseteq q \mid p \lhd q \mid \Downarrow p \mid \neg\varphi \mid \varphi \vee \psi \mid \exists p.\varphi$$

where p, q belong to some set $\mathbf{Q}' \supseteq \mathbf{Q}$ of variables. Then we define $MSO'(\mathbf{Q})$ as the set of MSO'-formulas whose free variables belong to $\mathbf{Q}$.

Intuitively, it may be useful to think of MSO' as a *first-order* logic in which the variables are interpreted on the power set of (the state space of) a Kripke frame. The valuation of a Kripke model is then to be seen as a first-order assignment of an element $V(p) \in PS$ to an arbitrary letter $p \in \mathbf{Q}$. More precisely, the semantics of this language on a pointed Kripke model $(\mathbb{S}, r)$ is defined inductively—we only give the clause of the atomic formulas:

$$
\begin{aligned}
(\mathbb{S}, r) &\models p \sqsubseteq q && \text{iff } V(p) \subseteq V(q) \\
(\mathbb{S}, r) &\models p \lhd q && \text{iff for all } s \in V(p)\text{there is a}t \in V(q)\text{with } Rst. \\
(\mathbb{S}, r) &\models \Downarrow p && \text{iff } V(p) = \{r\}
\end{aligned}
$$

It is not too difficult to see why this language corresponds to standard MSO. To start with, it is easy to interpret MSO'-formulas in standard MSO; for a translation in the opposite direction, the key idea is to encode elements of S as the corresponding singleton sets, and define a formula $sing(p) \in MSO'$ characterizing the singleton subsets of S in the sense that $\mathbb{S}, r \models sing(p)$ iff $V(p)$ is a singleton.

In the second part of the proof of Theorem 2.3.10(1) one defines, by induction on the complexity of a formula $\varphi \in MSO'(\mathbf{Q}')$, an automaton $\mathbb{A}_\varphi \in Aut_{\Pi=,\mathbf{Q}'}$ which is equivalent to φ in the sense that for any tree model $\mathbb{T}$ with root r, we have $(\mathbb{T}, r) \models \varphi$ iff $\mathbb{A}_\varphi$ accepts $(\mathbb{T}, r)$. This part of the proof is nontrivial, involving closure properties of specific classes of automata $\langle A, \delta, \Omega \rangle$ that are defined by restricting the range of the transition map δ to fragments of $\Phi^=$ (such as, in particular, the set $N^{=,+}$ defined in the next section).

---

[1] Here the tacit understanding is that the variable v is interpreted as the *root* of the tree.

2.4 One-Step Adequacy

2.4.1 The General Case

In this section we define and compare various one-step versions of the notion of adequacy, and of the notion of one logic corresponding the one-step behaviorally invariant fragment of another. First we need a notion of one-step equivalence; to understand this notion, consider an A-valuation $V : A \to PS$. Lifting the associated coloring $V^\flat : S \to PA$, we obtain a map $TV^\flat : TS \to TPA$, which associates, with an element $\tau \in TS$ an object $TV^\flat(\tau)$ that one may think of as a 'T-color'.

Definition 2.4.1 Given two A-valuations $V_i : A \to PS_i$ ($i = 0, 1$), we define a relation $\sim_{V_0,V_1} \subseteq TS_0 \times TS_1$ by putting $\tau_0 \sim_{V_0,V_1} \tau_1$ iff τ_0 and τ_1 have the same T-color, that is,

$$\tau_0 \sim_{V_0,V_1} \tau_1 \quad \text{iff} \quad (TV_0^\flat)\tau_0 = (TV_1^\flat)\tau_1.$$

We call a rank-1 formula φ *one-step T-invariant* if for all pairs of valuations $V_i : A \to PS_i$, and all pairs of elements $\tau_i \in TS_i$ ($i = 0, 1$) such that $\tau_0 \sim_{V_0,V_1} \tau_1$ it holds that $TS_0, V_0, \tau_0 \Vdash^1 \varphi$ iff $TS_1, V_1, \tau_1 \Vdash^1 \varphi$. A coalgebraic logic is called *one-step behaviorally invariant* if each of its rank-1 formulas is one-step T-invariant. $\square$

In the sequel we will need a characterization of the notion of one-step adequacy that involves pairs of valuations on two sets that are linked by some function.

Definition 2.4.2 Fix two sets S, S' and a map $f : S' \to S$. Then with every valuation $V : A \to PS$ we may associate an A-valuation V_f on S' given by

$$V_f := \check{P}f \circ V,$$

while for a valuation $U : A \to PS'$ defining

$$U^f := Pf \circ U.$$

we obtain an A-valuation U^f on S. $\square$

Concerning these definitions we need the following fact, which can be proved via routine verification.

Proposition 2.4.3 *Let $f : S' \to S$ be some map, and let $V : A \to PS$ and $U : A \to PS'$ be two valuations. Then*

1. $V = (V_f)^f$ *and* $U \subseteq (U^f)_f$ *(in the sense that $U(a) \subseteq (U^f)_f(a)$ for all $a \in A$).*
2. $U = V_f$ *iff* $U^\flat = V^\flat \circ f$.

The following proposition provides a useful characterization of one-step invariance.

Proposition 2.4.4 *Let Λ and A be sets of predicate liftings and proposition letters, respectively. A formula $\varphi \in L_\Lambda^1(A)$ is one-step T-invariant iff for each map $f : S' \to S$, for each $V : A \to PS$, and for each $\sigma' \in TS'$:*

$$TS, V, (Tf)\sigma' \Vdash^1 \varphi \quad \text{iff} \quad TS', V_f, \sigma' \Vdash^1 \varphi. \tag{2.13}$$

Proof For the direction from left to right, assume that $\varphi \in L_\Lambda^1$ is one-step T-invariant, and let f, V, and σ' be as in the formulation of the proposition. It follows from Proposition 2.4.3(2) that $V_f^\flat = V^\flat \circ f$, and from this it is immediate that $(Tf)\sigma' \sim_{V,V_f} \sigma'$. But then (2.13) follows from the one-step T-invariance of φ.

Conversely, suppose that φ satisfies the condition on the right hand side of the Proposition. Let $V_i : A \to PS_i$ $(i = 0, 1)$ be two A-valuations, and let $\sigma_i \in TS_i$ be objects such that $\sigma_0 \sim_{V_0,V_1} \sigma_1$. Our aim is to prove that

$$TS_0, V_0, \sigma_0 \Vdash^1 \varphi \quad \text{iff} \quad TS_1, V_1, \sigma_1 \Vdash^1 \varphi. \tag{2.14}$$

For this purpose, consider the *natural* valuation $N : A \to PPA$ on PA given by $N(a) := \{B \in PA \mid a \in B\}$. We leave it as an exercise for the reader to verify that (i) $N^\flat = id_{PA}$, that (ii) $(\check{P}V^\flat)N(a) = V(a)$ for all $a \in A$, and that (iii) $V_{V_i^\flat} = V_i$. From this we conclude that, taking $S = PA$, $S' = S_i$, $f = V_i^\flat$ and $\sigma_i = \sigma'$, we may read Eq. (2.13) as follows:

$$TPA, N, (TV_i^\flat)\sigma_i \Vdash^1 \varphi \quad \text{iff} \quad TS_i, V_i, \sigma_i \Vdash^1 \varphi.$$

From this, (2.14) is immediate by the assumption that $(TV_0^\flat)\sigma_0 = (TV_1^\flat)\sigma_1$. QED

The next theorem makes a link between some of the notions we have been discussing.

Theorem 2.4.5 *The following are equivalent, for any set Λ of predicate liftings:*

1. Λ *is natural;*
2. Λ *is one-step behaviorally invariant;*
3. *for each set A of proposition letters, for each function $f : S' \to S$, for each $V : A \to PS$, and for each $\sigma' \in TS'$, (2.13) holds for each formula $\varphi \in L_\Lambda(A)$.*

Proof Since the equivalence of (2) and (3) is an immediate consequence of Proposition 2.4.4, it suffices to show that (1) $\iff$ (3).

For this purpose, first assume that (1) Λ is natural, and let A, S', S, and f be as in item 3. Clearly it suffices to prove (2.13) for an arbitrary atomic rank-1 formula $\heartsuit_\lambda(\overline{a})$:

$$TS, V, (Tf)\sigma' \Vdash^1 \heartsuit_\lambda(\overline{a})$$

$$\begin{aligned}
&\text{iff} \quad (Tf)\sigma' \in \lambda_S(V(a_1), \ldots, V(a_n)) && (\text{semantics } \heartsuit_\lambda)\\
&\text{iff} \quad \sigma' \in (\check{P}Tf)\lambda_S(V(a_1), \ldots, V(a_n)) && (\text{definition } \check{P}Tf)\\
&\text{iff} \quad \sigma' \in \lambda_{S'}\left((\check{P}f)V(a_1), \ldots, (\check{P}f)V(a_n)\right) && (\text{naturality of } \lambda)\\
&\text{iff} \quad \sigma' \in \lambda_{S'}\left(V_f(a_1), \ldots, V_f(a_n)\right) && (\text{definition } V_f)\\
&\text{iff} \quad TS', V_f, \sigma' \Vdash^1 \heartsuit_\lambda(\overline{a}) && (\text{semantics } \heartsuit_\lambda)
\end{aligned}$$

Conversely, assume (3) and consider an arbitrary n-ary predicate lifting $\lambda \in \Lambda$. In order to prove that λ is natural, take an arbitrary function $f : S' \to S$, and an arbitrary n-tuple $\overline{X} = (X_1, \ldots, X_n)$ of subsets of S. We need to show that

$$\lambda_{S'}\left((\check{P}f)X_1, \ldots, (\check{P}f)X_n\right) = (\check{P}Tf)\lambda_S(X_1, \ldots, X_n). \tag{2.15}$$

For this purpose, define $A := \{a_1, \ldots, a_n\}$, and consider the valuation $V : A \to PS$ such that $V(a_i) = X_i$; observe that by definition of V_f, this implies that $V_f(a_i) = (\check{P}f)X_i$. Hence it suffices to prove, for an arbitrary element $\sigma' \in S'$, that

$$\sigma' \in \lambda_{S'}\left(V_f(a_1), \ldots, V_f(a_n)\right) \quad \text{iff} \quad \sigma' \in (\check{P}Tf)\lambda_S(V(a_1), \ldots, V(a_n)).$$

This we prove as follows:

$$\sigma' \in \lambda_{S'}\left(V_f(a_1), \ldots, V_f(a_n)\right)$$

$$\begin{aligned}
&\text{iff} \quad TS', V_f, \sigma' \Vdash^1 \heartsuit_\lambda(\overline{a}) && (\text{semantics } \heartsuit_\lambda)\\
&\text{iff} \quad TS, V, (Tf)\sigma' \Vdash^1 \heartsuit_\lambda(\overline{a}) && (\text{assumption})\\
&\text{iff} \quad (Tf)\sigma' \in \lambda_S(V(a_1), \ldots, V(a_n)) && (\text{semantics } \heartsuit_\lambda)\\
&\text{iff} \quad \sigma' \in (\check{P}Tf)\lambda_S(V(a_1), \ldots, V(a_n)) && (\text{definition } \check{P}Tf)
\end{aligned}$$

QED

We now turn to the one-step version of one coalgebraic logic corresponding to the T-invariant fragment of another. On the basis of Theorem 2.4.5, we may nicely formulate this property in terms of predicate liftings.

Definition 2.4.6 Let Λ and Λ' be two sets of predicate liftings for the functor T. We say that Λ' *corresponds to the behaviorally invariant fragment of* Λ *at the one-step level*, notation: $\Lambda' \equiv_1 \Lambda/T$, if

1. $\Lambda' \subseteq \Lambda$,
2. Λ' is natural, and
3. every one-step T-invariant formula $\varphi \in L_\Lambda^1(A)$ is equivalent to a formula φ^* in $L_{\Lambda'}^1(A)$.

If Λ and Λ' satisfy the conditions 1, 2 and 3' below:

3'. there is a translation $(\cdot)^* : L_\Lambda^1(A) \to L_{\Lambda'}^1(A)$ and a construction associating, with each set S, a set S' together with a map $f : S' \to S$, such that for each $\sigma \in TS$ there is a $\sigma' \in TS'$ with $\sigma = (Tf)\sigma'$, and for each valuation $V : A \to PS$, and each formula $\varphi \in L_\Lambda^1(A)$ it holds that

$$TS, V, \sigma \Vdash^1 \varphi^* \quad \text{iff} \quad TS', V_f, \sigma' \Vdash^1 \varphi. \tag{2.16}$$

we say that Λ' *uniformly corresponds to the behaviorally invariant fragment of* Λ *at the one-step level*, notation: $\Lambda' \equiv_1^u \Lambda/T$.   □

Similarly to Proposition 2.2.14, the following Proposition states that uniform correspondence implies correspondence.

Proposition 2.4.7 *If* $\Lambda' \equiv_1^u \Lambda/T$, *then* $\Lambda' \equiv_1 \Lambda/T$.

Proof Similarly to the proof of Proposition 2.2.14, one may show that if $\varphi \in L_\Lambda^1$ is one-step T-invariant, then φ is equivalent to φ^*, where $(\cdot)^* : L_\Lambda^1(A) \to L_{\Lambda'}^1(A)$ is the translation given by uniform correspondence.   QED

Remark 2.4.8 There are many variants of the notion of uniform one-step correspondence that may be of interest as well. For instance, it makes sense to weaken the condition (1), stating that Λ' is an actual *subset* of Λ, to a condition requiring that there is a translation $(\cdot)^\dagger$ mapping any formula $\varphi \in L_{\Lambda'}^1(A)$ to an equivalent formula $\varphi^\dagger \in L_\Lambda^1(A)$. As a second example, all of the results in this paper still hold if we weaken condition 3' to a non-uniform version in which the set S' and the function $f : S' \to S$ depend on the object $\sigma \in TS$. In detail, this condition would read as follows

3''. there is a translation $(\cdot)^* : L_\Lambda^1(A) \to L_{\Lambda'}^1(A)$ and a construction associating, with each set S, and each $\sigma \in TS$, a set S', a map $f : S' \to S$, and an object $\sigma' \in TS'$ such that $\sigma = (Tf)\sigma'$, and for each valuation $V : A \to PS$, (2.16) holds for each formula $\varphi \in L_\Lambda^1(A)$.

Finally, in some cases it may be convenient to consider (one-step) languages in which we admit only a (not necessarily functionally complete) selection of Boolean connectives.

2.4.2 The Case of Kripke Models

Our key example of the notions just defined is given by the predicate liftings for Kripke structures, that are induced by monadic first-order sentences. Our main goal in this section will be to prove the following Theorem, which states that Π uniformly corresponds to the behaviorally invariant fragment of $\Pi^=$. This theorem will be crucial in our proof of the Janin-Walukiewicz theorem.

Theorem 2.4.9 $\Pi \equiv_1^u \Pi^= / P$.

Let us first see what one-step invariance means in this context. Recall from the introduction the notions of P-equivalence of structures, and of P-invariance of monadic sentences. We will say that a monadic first-order sentence $\alpha \in \Phi^=$ is *invariant under surjective homomorphisms* iff for each valuation $V : A \to PD$ and each surjection $f : D' \to D$,

$$\langle D, V \rangle \models \alpha \quad \text{iff} \quad \langle D', V_f \rangle \models \alpha.$$

Proposition 2.4.10 *For any first-order sentence* $\alpha(\overline{a}) \in \Phi^=$, *the following are equivalent:*

1. *the predicate lifting* $\widehat{a}$ *is natural;*
2. *the rank-1 formula* $\heartsuit_{\widehat{a}}(\overline{a})$ *is P-invariant;*
3. α *is P-invariant;*
4. α *is invariant under surjective homomorphisms.*

Proof This Proposition is a straightforward consequence of Propositions 2.3.9 and 2.4.4.   QED

As a first corollary of this, we obtain the following.

Corollary 2.1 Φ *and* Π *are P-invariant sets of predicate liftings.*

Proof By Proposition 2.4.10 it suffices to show that all identity-free sentences of monadic first-order logic are invariant under surjective homomorphisms. This is a routine exercise in first-order logic.   QED

As a more important consequence of Proposition 2.4.10, we may prove Theorem 2.4.9 as a corollary of the following result on monadic first-order logic.

Proposition 2.4.11 *There is a translation* $(\cdot)^* : \Pi^=(A) \to \Pi(A)$ *such that for all structures* (D, V) *and all sentences* $\alpha \in \Pi^=$ *we have*

$$(D, V) \models \alpha^* \text{ iff } (D \times \omega, V_\pi) \models \alpha, \tag{2.17}$$

where $\pi : D \times \omega \to D$ *is the first projection function,* $\pi : (d, n) \mapsto d$.

In order to prove Proposition 2.4.11, we will need certain normal forms for monadic first-order sentences. First we supply some preliminary definitions.

Definition 2.4.12 For a sequence $\overline{x} = x_1, \ldots, x_n$ of variables, write $\text{diff}(\overline{x}) := \bigwedge_{1<j} x_i \neq x_j$. Given a set $B \subseteq A$ and a variable x, abbreviate $\tau_B(x) := \bigwedge_{a \in B} a(x) \wedge \bigwedge_{a \notin B} \neg a(x)$ and $\tau_B^+(x) := \bigwedge_{a \in B} a(x)$. $\square$

In words, $\text{diff}(\overline{x})$ states that the variables $x_1, \ldots x_n$ denote *distinct* elements of the domain. The formulas $\tau_B(x)$ and $\tau_B^+(x)$ state, respectively, that the *type* of the element denoted by x is equal to (contains, respectively) B. Here the *type* of an element d in a structure $\langle D, V \rangle$ for A is the set $V^b(d)$.

Definition 2.4.13 Fix a set A of propositional variables. Let $\overline{B} = B_1, \ldots, B_n$ and $\overline{C} = C_1, \ldots, C_m$ be two sequences of subsets of A, respectively. We define the following formulas:

$$\chi^{=,+}(\overline{B}, \overline{C}) := \exists y_1 \cdots y_n \left(\mathrm{diff}(\overline{y}) \wedge \bigwedge_i \tau^+_{B_i}(y_i) \wedge \forall z \, (\mathrm{diff}(\overline{y}z) \rightarrow \bigvee_j \tau^+_{C_j}(z)) \right)$$

$$\chi^{=}(\overline{B}, \overline{C}) := \exists y_1 \cdots y_n \left(\mathrm{diff}(\overline{y}) \wedge \bigwedge_i \tau_{B_i}(y_i) \wedge \forall z \, (\mathrm{diff}(\overline{y}z) \rightarrow \bigvee_j \tau_{C_j}(z)) \right)$$

$$\chi^{+}(\overline{B}, \overline{C}) := \exists y_1 \cdots y_n \left(\bigwedge_i \tau^+_{B_i}(y_i) \wedge \forall z \bigvee_j \tau^+_{C_j}(z) \right)$$

$$\chi(\overline{B}, \overline{C}) := \exists y_1 \cdots y_n \left(\bigwedge_i \tau_{B_i}(y_i) \wedge \forall z \bigvee_j \tau_{C_j}(z) \right)$$

We let $N^{=,+}(A)$ denote the set of sentences of the form $\chi^{=,+}(\overline{B}, \overline{C})$, and proceed similarly for the sets $N^{=}(A)$, $N^+(A)$ and $N(A)$. $\qquad\square$

We need the following fact from first-order logic, which explains why we may think of (disjunctions of) χ-type formulas as providing *normal forms* for monadic first-order logic.

Proposition 2.4.14 *Every sentence in $\Phi^{=}$ is equivalent to a disjunction of sentences in $N^{=}$, and similarly for Φ and N, $\Pi^{=}$ and $N^{=,+}$ and Π and N^+, respectively.*

Proof The proof of this Proposition can be seen as an exercise in the theory of Ehrenfeucht-Fraïssé games. We confine ourselves to a sketch (rephrasing the proof of Lemma 16.23 in [12]), and we only consider the case of $\Phi^{=}$.

Given a set $B \subseteq A$, and a first-order structure $\mathbb{D} = \langle D, V \rangle$ for A, let $N_{B,\mathbb{D}}$ be the number of elements in D of type B. We say that two such structures $\mathbb{D}$ and $\mathbb{D}'$ are n-equivalent, notation $\mathbb{D} \sim_n \mathbb{D}'$, if for every $B \subseteq A$, either $N_{B,\mathbb{D}} = N_{B,\mathbb{D}'} \leq n$, or both $N_{B,\mathbb{D}} > n$ and $N_{B,\mathbb{D}'} > n$. Clearly $\sim_n$ is an equivalence relation of finite index, and each equivalence class of $\sim_n$ is described by a formula in $N^{=}$. Using Ehrenfeucht-Fraïssé games it is not difficult to show that $\mathbb{D} \sim_n \mathbb{D}'$ implies that $\mathbb{D}$ and $\mathbb{D}'$ satisfy the same sentences of quantifier rank at most n. From this it follows that the class of models of such a sentence is the union of a (finite) number of $\sim_n$-cells, and that the sentence itself is thus equivalent to the disjunction of the formulas asssociated with these $\sim_n$-cells. QED

Now we are ready for the proof of Proposition 2.4.11.

Proof of Proposition 2.4.11 Given a formula $\alpha \in \Pi^{=}(A)$, we need to come up with a translation $\alpha^* \in \Pi(A)$ such that (2.17) holds.

First assume that α is of the form $\chi^{=,+}(\overline{B}, \overline{C}) \in N^{=,+}$, and define

$$\alpha^* := \chi^+(\overline{B}, \overline{C}).$$

Proving (2.17) in this situation boils down to showing that

$$\langle D, V \rangle \models \chi^+(\overline{B}, \overline{C}) \quad \text{iff} \quad \langle D \times \omega, V_\pi \rangle \models \chi^{=,+}(\overline{B}, \overline{C}). \tag{2.18}$$

For this purpose, first observe that V_π satisfies

$$d \in V(a) \quad \text{iff} \quad (d, n) \in V_\pi(a) \tag{2.19}$$

for each $d \in S$, $a \in A$, and $n \in \omega$. Suppose that $\overline{B} = B_1, \ldots, B_n$ and $\overline{C} = C_1, \ldots, C_m$.

For the left-to-right direction of (2.18), assume that $\langle D, V \rangle \models \chi^+(\overline{B}, \overline{C})$. Let $d_1, \ldots, d_n$ be elements in D satisfying the existential part of $\chi^+(\overline{B}, \overline{C})$, that is, for each i we find $d_i \in \bigcap_{b \in B_i} V(b)$. From the universal part of the formula it follows that for each $d \in D$ there is a subset $C_d \subseteq A$ such that $d \in \bigcap_{c \in C_d} V(c)$. Now we move to $D \times \omega$; it is easy to see that its elements $(d_1, 1), \ldots, (d_n, n)$ provide a sequence of n distinct elements that satisfy $(d_i, i) \in \bigcap_{b \in B_i} V_\pi(b)$ for each i. In addition, every element (d, n) distinct from the ones in the mentioned tuple will satisfy $(d, n) \in \bigcap_{c \in C_d} V_\pi(c)$. From these observations it is immediate that $\langle D \times \omega, V_\pi \rangle \models \chi^{=,+}(\overline{B}, \overline{C})$.

For the opposite direction of (2.18), assume that $\langle D \times \omega, V_\pi \rangle \models \chi^{=,+}(\overline{B}, \overline{C})$. Let $(d_1, k_1), \ldots, (d_n, k_n)$ be the sequence of distinct elements of $D \times \omega$ witnessing the existential part of $\chi^{=,+}(\overline{B}, \overline{C})$ in $\mathbb{D}'$. Then clearly, $d_1, \ldots, d_n$ witness the existential part of $\chi^+(\overline{B}, \overline{C})$ in $\langle D, V \rangle$. In order to show that $\langle D, V \rangle$ also satisfies the universal part $\forall z \bigvee_j \tau_{C_j}^+(z)$ of χ^+, consider an arbitrary element $d \in D$. Take any $m \in \omega \setminus \{k_1, \ldots, k_n\}$, then (d, m) is distinct from each (d_i, k_i). It follows that for some j we have $(d, m) \in \bigcap_{c \in C_j} V_\pi(c)$, and so we obtain $d \in \bigcap_{c \in C_j} V(c)$. Since d was arbitrary this shows that indeed $\langle D, V \rangle \models \forall z \bigvee_j \tau_{C_j}^+(z)$. So we have proved that $\langle D, V \rangle \models \chi^+(\overline{B}, \overline{C})$.

Now consider the general case, where α is arbitrary. It follows from Proposition 2.4.14 that α is equivalent to a formula $\alpha \equiv \bigvee_i \alpha_i$, with each formula α_i belongs to $N^{=,+}$. With

$$\alpha^* := \bigvee_i \alpha_i^*$$

it is straightforward to verify (2.17). $\hfill$ QED

Both Proposition 2.1.4 and Theorem 2.4.9 are straightforward corollaries of Proposition 2.4.11.

Proof of Proposition 2.1.4 Assume that $\alpha \in \Pi^=$ is a P-invariant monadic sentence, and let $\alpha^* \in \Pi$ be the formula given by Proposition 2.4.11. Consider an arbitrary structure (D, V), and observe that $(D, V) \equiv_P (D \times \omega, V_\pi)$. But then we obtain the following equivalences:

$$
\begin{aligned}
(D, V) \models \alpha \quad &\text{iff} \quad (D \times \omega, V_\pi) \models \alpha && \text{(assumption on } \alpha\text{)} \\
&\text{iff} \quad (D, V) \models \alpha^* && (2.17)
\end{aligned}
$$

From this it is immediate that α and α^* are equivalent, which suffices to prove Proposition 2.1.4. $\hfill$ QED

Proof of Theorem 2.4.9 It is obvious that $\Pi \subseteq \Pi^=$, and Corollary 2.1 states the one-step P-invariance of Π. Hence we may focus on item $3'$ of Definition 2.4.6.

We need to define a translation $(\cdot)^* : L^1_{\Pi^=} \to L^1_{\Pi}$. By the definition of rank-1 formulas, it suffices to come up with a translation for *atomic* rank-1 formulas, that is, formulas of the form $\heartsuit_{\hat{a}}(\overline{a})$ for some sentence $\alpha \in \Pi^=$. But for such a formula, we can simply put

$$\left(\heartsuit_{\hat{a}}(\overline{a})\right)^* := \heartsuit_{\alpha^*}(\overline{a}).$$

We leave it for the reader to verify that this defines a formula of the right shape and with the right properties.   QED

2.5 Main Result

We are now ready for the main technical result of the paper. Intuitively, Theorem 2.5.1 states that, given two sets Λ, Λ' of monotone predicate liftings for a functor T, if Λ corresponds to the T-invariant fragment of Λ' at the one-step level, then the coalgebraic μ-calculus $\mu L_{\Lambda,Q}$ is the bisimulation-invariant fragment of $\mu L_{\Lambda',Q}$.

Theorem 2.5.1 *Let T be some set functor, and let Λ, Λ' be two sets of monotone predicate liftings for T such that $\Lambda \equiv^u_1 \Lambda'/T$. Then for any set Q,*

$$\mathrm{Aut}_{\Lambda,Q} \equiv^s \mathrm{Aut}_{\Lambda',Q}/{\simeq_T}.$$

As a corollary, if Λ' is closed under Boolean duals, then $\mu L_{\Lambda,Q} \equiv^s \mu L_{\Lambda',Q}/{\simeq_T}$.

Proof Fix a set Q of proposition letters, and assume that $\Lambda \equiv^u_1 \Lambda'/T$. It easily follows from this assumption that $\mathrm{Aut}_{\Lambda,Q} \subseteq \mathrm{Aut}_{\Lambda',Q}$ and that $\mathrm{Aut}_{\Lambda,Q}$ is invariant under behavioral equivalence. This leaves the following tasks:

1. define a translation from initialized Λ, Q-automata to initialized Λ', Q-automata,
2. outline a construction, that assocates with an arbitrary pointed T-model $(\mathbb{S}, r)$, a pointed T-model $(\mathbb{S}', r')$ and a morphism $f : \mathbb{S}' \to \mathbb{S}$, and
3. prove, for every initialized Λ, Q-automaton $(\mathbb{A}, a_i)$, and every pointed model $(\mathbb{S}, r)$ that

$$\mathbb{S}, r \Vdash (\mathbb{A}, a_I)^* \quad \text{iff} \quad \mathbb{S}', r' \Vdash (\mathbb{A}, a_I). \tag{2.20}$$

The first of these tasks is easy to accomplish. Given a Λ, Q-automaton $\mathbb{A} = \langle A, \delta, \Omega \rangle$, recall that $\delta(a, \Pi)$ is a rank-1 Λ-formula for each $a \in A$ and $\Pi \in PQ$. Hence we obtain a Λ', Q-automaton $\mathbb{A}^*$ by putting $\mathbb{A}^* := \langle A, \delta^*, \Omega \rangle$, where $\delta^* : A \times PQ \to L^1_{\Lambda'}(A)$ is given by Definition 2.4.6($3'$): $\delta^*(a, \Pi) := (\delta(a, \Pi))^*$. For the initialized automaton $(\mathbb{A}, a_I)$ we put $(\mathbb{A}, a_I)^* := (\mathbb{A}^*, a_I)$.

Concerning the second task, consider an arbitrary T-model $\mathbb{S} = \langle S, \sigma, W \rangle$. Take the set S' and the map $f : S' \to S$ provided by clause 3 of Definition 2.4.6. In order to endow the set S' with coalgebra structure, consider an arbitrary element $s' \in S'$. Applying the properties of S, S' and f (given by the mentioned clause) to the element $\sigma(fs') \in TS$, we obtain an element $\sigma's' \in TS'$ such that

$$(Tf)(\sigma's') = \sigma(fs') \tag{2.21}$$

and such that for every rank-1 formula $\varphi \in L^1_A(A)$ and every marking $V : A \to PS$ we have

$$TS, V, \sigma(fs') \Vdash^1 \varphi^* \quad \text{iff} \quad TS', V_f, \sigma's' \Vdash \varphi. \tag{2.22}$$

Clearly this procedure defines a coalgebra structure $\sigma' : S' \to TS'$. For the valuation W' on S' we take $W' := W_f$. It is immediate by (2.21) and the fact that $W^\flat_f = W^\flat \circ f$, that the map $f : S' \to S$ is in fact a T_Q-coalgebra morphism.

Finally, we need to come up with a designated point r' of $S' := \langle S', \sigma' \rangle$ which is mapped to r by f. Clearly if S' already contains such an element we are done; if not, then we can simply *adjoin* a fresh element r' to S'. We define $\sigma'r'$ so that $(Tf)(\sigma'r') = \sigma r$ (this is possible by the assumptions), adapt the valuation W' so that the type of r' is that of r in S, and add the pair (r', r) to (the graph of) f. Modulo some renaming, this ensures that we obtain a pointed coalgebra (S', r'), with a map $f : S' \to S$ satisfying (2.21) and (2.22), and such that $W' = W_f$ and $fr' = r$. (Formally, we define a model S'' based on the set $S'' := S' \uplus \{r'\}$, and in the sequel work with the pointed model (S'', r'). We omit the details of this construction which are coalgebraically obvious but somewhat tedious).

We are now ready to prove (2.20). Fix an initialized Λ, Q-automaton (A, a_I) and a pointed T-model (S, r). Clearly it suffices to show that

$$(a_I, r) \in \text{Win}_\exists(\mathscr{A}(A^*, S)) \quad \text{iff} \quad (a_I, r') \in \text{Win}_\exists(\mathscr{A}(A, S')). \tag{2.23}$$

Abbreviate $\mathscr{A}^* = \mathscr{A}(A^*, S)$ and $\mathscr{A}' = \mathscr{A}(A, S')$.

For the direction from left to right of (2.23), without loss of generality we may assume that in $\mathscr{A}^*$, $\exists$ has a *positional* strategy $\theta : A \times S \to (PS)^A$ which is winning when played from each position $(a, s) \in \text{Win}_\exists(\mathscr{A}^*)$. Now consider the following (positional) strategy θ_f for $\exists$ in $\mathscr{A}'$:

at position $(a, s')\exists$ picks the marking V_f,
where $V : A \to PS$ is the marking $V = \theta(a, fs')$provided in $\mathscr{A}'$ by θ at(a, fs').

The legitimacy of this move is immediate by (2.22).

In order to show that θ_f is in fact a winning strategy in $\mathscr{A}'@(a_I, r)$, consider an arbitary match

$$\pi = (a_I, r')U_1(a_1, s'_1)U_2(a_2, s'_2)\dots$$

in which $\exists$ plays the strategy θ_f just defined. The point is that there is an associated θ-conform $\mathscr{A}^*$-match

$$\pi^* = (a_I, r)V_1(a_1, s_1)V_2(a_2, s_2)\ldots$$

such that $U_i = (V_i)_f$ and $fs'_i = s_i$ for all $i < \omega$. To see this, consider a round of the game, starting at position $(a, s') \in A \times S'$ with $(a, fs') \in \mathrm{Win}_\exists(\mathscr{A}^*)$. If $\exists$ plays her strategy θ_f, picking the marking V_f with $V = \theta(a, fs')$, then for every pair (b, t') picked by $\forall$, by definition of V_f, the pair (b, ft') is a legitimate move for $\forall$ in $\mathscr{A}^*$.

By our assumptions on (a_I, r) and θ, the match π^* is won by $\exists$. But since π and π^* project to exactly the same sequence of A-states, and the winning conditions of $\mathscr{A}^*$ and $\mathscr{A}'$ are the same, this means that $\exists$ also wins π. Thus we conclude that θ_f is a winning strategy for $\exists$ in the game $\mathscr{A}'$ initialized at (a_I, r).

For the opposite direction '$\Leftarrow$' of (2.23), we may assume that in $\mathscr{A}^*$, $\exists$ has a *positional* strategy $\eta : A \times S \to (PS)^A$ which is winning from all positions $(a, s') \in \mathrm{Win}_\exists(\mathscr{A}')$. We will use this η to define a (partial) strategy η^f for $\exists$ in $\mathscr{A}^*$.

For the definition of η^f, consider a position in $\mathscr{A}^*$ of the form (a, fs') for some $s' \in S'$ such that $(a, s') \in \mathrm{Win}_\exists(\mathscr{A}')$. (Note that the position (a_I, r) has this shape.) Suppose that in $\mathscr{A}'$, at position (a, s'), $\exists$'s strategy η tells her to pick a marking $U : A \to PS'$. Our first claim is that the marking $U^f : A \to PS$ constitutes a legitimate move for $\forall$ in the game $\mathscr{A}^*$ at position (a, fs').

To see this we need to verify that $TS, U^f, \sigma(fs') \Vdash^1 \delta^*(a, W^\flat(fs'))$. But because U is a legitimate move at (a, s') in $\mathscr{A}'$, we know that $TS', U, \sigma's' \Vdash^1 \delta(a, W_f^\flat(s'))$. Observe that $U \subseteq (U^f)_f$ (Proposition 2.4.3), so that by monotonicity it follows that $TS, (U^f)_f, \sigma's' \Vdash^1 \delta(a, W_f^\flat(s'))$. From this it is immediate by (2.22) and the fact that $W^\flat(fs') = W_f^\flat(s')$ (Proposition 2.4.3), that $TS, U^f, \sigma(fs') \Vdash^1 \delta^*(a, W^\flat(fs'))$, as required. Now consider an arbitrary response (b, t) of $\forall$ to $\exists$'s move U^f at position (a, fs'). It follows from $t \in U^f(b)$ that t is of the form ft' for some $t' \in S'$ such that $t = ft'$. This means that in $\mathscr{A}'$, the move (b, t') is legitimate at position U. Furthermore, since we assumed that U was given by a winning strategy, the position (b, t') belongs to the set $\mathrm{Win}_\exists(\mathscr{A}')$. Summarizing, this shows that in any round of $\mathscr{A}^*$ starting at a position (a, fs') with $(a, s') \in \mathrm{Win}_\exists(\mathscr{A}')$, $\exists$ has the power to end the round at a position (b, t) of the same kind; and more specifically, she maintains an η-conform 'shadow round' of the game $\mathscr{A}'$ starting at (a, s') and ending at a position $(b, t') \in \mathrm{Win}_\exists(\mathscr{A}')$ with $ft' = t$.

On the basis of the above observations, we may easily equip $\exists$ with a (partial) strategy η^f with the property, that for any η^f-conform match

$$\pi = (a_I, r)V_1(a_1, s_1)V_2(a_2, s_2)\cdots$$

there is an η-conform 'shadow match'

$$\pi' = (a_I, r')U_1(a_1, s'_1)U_2(a_2, s'_2)\cdots$$

such that $s_i = fs'_i$ and $V_i = U_i^f$ for all $i \in \omega$. From this we may derive, using a similar argument as given before, that π is won by $\exists$. Thus in this case we conclude that η^f is a winning strategy for $\exists$ in the game $\mathscr{A}^*$ initialized at (a_I, r).   QED

Finally we show how to derive the Janin-Walukiewicz theorem, stating that μML is the bisimulation-invariant fragment of MSO, from the results obtained (or mentioned) above.

Corollary 2.2 (Janin and Walukiewicz) *For any set* Q *of proposition letters,* μML $(Q) \equiv^s MSO_v(Q)/P.$

Proof This result is a straightforward corollary of the Theorems 2.3.10, 2.4.9, and 2.5.1, together with Proposition 2.3.5.

To see this, fix a set Q of proposition letters, and note that by Theorem 2.3.10(1), there is an initialized automaton $(\mathbb{A}_\varphi, a_\varphi)$ in $\mathrm{Aut}_{\Pi^=}(Q)$ such that

$$\varphi \equiv (\mathbb{A}_\varphi, a_\varphi) \text{ on trees.} \tag{2.24}$$

By the Theorems 2.3.10(2), 2.4.9, and 2.5.1 and by Proposition 2.3.5, there is a translation $\xi : \mathrm{Aut}_{\Pi^=}(Q) \to \mu ML$ such that for all pointed Kripke models $(\mathbb{S}, s)$, there is a pointed Kripke model $(\mathbb{S}', s')$ and a morphism $f : (\mathbb{S}', s') \to (\mathbb{S}, s)$ such that for all initialized automata $(\mathbb{A}, a)$ it holds that

$$\mathbb{S}, s \Vdash \xi(\mathbb{A}, a) \quad \text{iff} \quad \mathbb{S}', s' \Vdash (\mathbb{A}, a). \tag{2.25}$$

Now let $\varphi(v) \in MSO_v(Q)$ be invariant under bisimilarity, or behavioral equivalence (these are the same for the power set functor P). We claim that $\varphi' := \xi(\mathbb{A}_\varphi, a_\varphi) \in \mu ML$ is equivalent to φ. To see this, let $(\mathbb{S}_0, s_0)$ be an arbitrary pointed Kripke model, and let $(\mathbb{S}_1, s_1)$ be a tree model bisimilar (or behaviorally equivalent) to $(\mathbb{S}_0, s_0)$.

Then we have the following chain of equivalences:

$$
\begin{array}{llr}
\mathbb{S}_0, s_0 \Vdash \varphi & \text{iff} \quad \mathbb{S}_1, s_1 \Vdash \varphi & \text{(assumption on } \varphi) \\
& \text{iff} \quad \mathbb{S}_1', s_1' \Vdash \varphi & \text{(assumption on } \varphi) \\
& \text{iff} \quad \mathbb{S}_1', s_1' \Vdash (\mathbb{A}_\varphi, a_\varphi) & \text{(2.24)} \\
& \text{iff} \quad \mathbb{S}_1, s_1 \Vdash \varphi' & \text{(2.25)} \\
& \text{iff} \quad \mathbb{S}_0, s_0 \Vdash \varphi' & \text{(adequacy of } \mu ML)
\end{array}
$$

which shows that $\varphi \equiv \varphi' \in \mu ML$ indeed. QED

2.6 Conclusion

We finish the paper with some general observations and questions for further research.

First of all, given the fact that it is an open problem whether the Janin-Walukiewicz theorem also holds in the setting of finite models, it may be interesting to note that both Proposition 2.1.4 and Theorem 2.1.5 can be proved in that setting, as can Fact 2.1.3(2). Hence, the 'only' hurdle to prove a finite model theory version of their result is the fact that the correspondence between monadic second-order logic and $\Pi^=$-automata is only proven for tree models (Fact 2.1.3(1)).

Second, commenting on an earlier version of this chapter, van Benthem asked some questions concerning the translation $(\cdot)^*$ from *MSO* to μML. His question concerning interpolation can be answered positively: given two *MSO*-formulas φ and χ, one may show that φ implies χ 'along bisimilarity' iff φ and χ have an interpolant ψ in the modal μ-calculus.

The analysis of fixed-point logics at the level of syntax for the transition functions of automata, which started with the work of Janin and Walukiewicz, has yielded some other basic results about the modal μ-calculus. For instance, it was used by d'Agostino and Hollenberg to prove uniform interpolation [7] and by Fontaine and Venema to obtain various preservation results, such as the characterization of the continuous fragment of μML [9].

Finally, it would be interesting to extend the coalgebraic analysis of the modal μ-calculus from model-theoretic aspects to axiomatics and proof theory.

Acknowledgments The research of this author has been made possible by VICI grant 639.073.501 of the Netherlands Organization for Scientific Research (NWO). Furthermore, the author is grateful to Facundo Carreiro for many comments on an earlier version of the paper.

References

1. Andréka H, van Benthem J, Németi I (1998) Modal languages and bounded fragments of predicate logic. J Philos Log 27:217–274
2. van Benthem J (1976) Modal correspondence theory. PhD thesis, Mathematisch Instituut & Instituut voor Grondslagenonderzoek, University of Amsterdam
3. van Benthem J (1984) Correspondence theory. In: Gabbay DM, Guenthner F (eds) Handbook of philosophical logic, vol 2. Reidel, Dordrecht, pp 167–247
4. ten Cate B, Gabelaia D, Sustretov D (2009) Modal languages for topology: expressivity and definability. Ann Pure App Log 159(1–2):146–170
5. Cîrstea C, Kurz A, Pattinson D, Schröder L, Venema Y (2011) Modal logics are coalgebraic. Comput J 54:524–538
6. Cîrstea C, Kupke C, Pattinson D (2009) EXPTIME tableaux for the coalgebraic μ-calculus. In: Grädel E, Kahle R (eds) Computer science logic 2009. Lecture notes in computer science, vol 5771. Springer, New York, pp 179–193
7. D'Agostino G, Hollenberg M (2000) Logical questions concerning the μ-calculus. J Symbol Log 65:310–332
8. Dawar A, Otto M (2009) Modal characterisation theorems over special classes of frames. Ann Pure Appl Log 161:1–42
9. Fontaine G, and Venema Y (2010) Some model theory for the modal mu-calculus: syntactic characterizations of semantic properties (Manuscript, submitted)
10. Fontaine G, Leal R, Venema Y (2010) Automata for coalgebras: an approach using predicate liftings. In: Abramsky s et al. (eds) Proceedings of ICALP 2010, part II. LNCS, vol 6199. Springer, New York
11. Grädel E, Hirsch C, Otto M (2002) Back and forth between guarded and modal logics. ACM Trans Comput Log 3:418–463
12. Grädel E, Thomas W, Wilke T (eds) (2002) Automata, logic, and infinite games. LNCS, vol 2500. Springer, New York

13. Janin D, Walukiewicz I (1995) Automata for the modal μ-calculus and related results. In: Proceedings of the twentieth international symposium on mathematical foundations of computer science, MFCS'95. LNCS, vol 969. Springer, New York, pp 552–562
14. Janin D, Walukiewicz I (1996) On the expressive completeness of the propositional μ-calculus w.r.t. monadic second-order logic. In: Proceedings CONCUR '96
15. Litak T, Pattinson D, Sano K, Schröder L (2012) Coalgebraic predicate logic. In: Czumaj A et al. (eds) ICALP 2012. Lecture notes in computer science, vol 7392. Springer, New York, pp 299–311
16. Moss L (1999) Coalgebraic logic. Ann Pure Appl Log 96:277–317 (Erratum published Ann P Appl Log 99:241–259)
17. Otto M (2011) Model theoretic methods for fragments of fo and special classes of (finite) structures. In: Esparza J, Michaux C, Steinhorn C (eds) Finite and algorithmic model theory. LMS Lecture notes series, vol 379. Cambridge University Press, Cambridge
18. Pattinson D (2003) Coalgebraic modal logic: soundness, completeness and decidability of local consequence. Theor Comput Sci 309(1–3):177–193
19. Rosen E (1997) Modal logic over finite structures. J Log Lang Inf 6:427–439
20. Rutten J (2000) Universal coalgebra: a theory of systems. Theor Comput Sci 249:3–80
21. Venema Y (2006) Algebras and coalgebras. In: van Benthem J, Blackburn P, Wolter F (eds) Handbook of modal logic. Elsevier, Amsterdam, pp 331–426
22. Venema Y (2006) Automata and fixed point logic: a coalgebraic perspective. Inf Comput 204:637–678
23. Walukiewicz I (2002) Monadic second-order logic on tree-like structures. Theor Comput Sc 275:311–346

Chapter 3
Schema Mappings: A Case of Logical Dynamics in Database Theory

Balder ten Cate and Phokion G. Kolaitis

Abstract A schema mapping is a high-level specification of the structural relationships between two database schemas. This specification is expressed in a schema-mapping language, which is typically a fragment of first-order logic or second-order logic. Schema mappings have played an essential role in the study of important data-interoperability tasks, such as data integration and data exchange. In this chapter, we examine schema mappings as a case of logical dynamics in action. We provide a self-contained introduction to this area of research in the context of logic and data-bases, and focus on some of the concepts and results that may be of particular interest to the readers of this volume. After a basic introduction to schema mappings and schema-mapping languages, we discuss a series of results concerning fundamental structural properties of schema mappings. We then show that these structural properties can be used to obtain characterizations of various schema-mapping languages, in the spirit of abstract model theory. We conclude this chapter by highlighting the surprisingly subtle picture regarding compositions of schema mappings and the languages needed to express them.

Database theory has been one of the most fruitful areas of application of logic to computer science. In fact, over the past four decades there has been an extensive exchange of ideas between logic and database theory that has benefitted both areas. This chapter is about *schema mappings*, a research topic in database theory that has been developed in the context of data inter-operability, which can be described as the problem of combining, managing, and querying data from heterogeneous sources.

Research on this paper was supported by NSF Grants IIS-0905276 and IIS-1217869.

B. ten Cate (✉)
UC Santa Cruz, Santa Cruz, USA
e-mail: btencate@ucsc.edu

P. G. Kolaitis
UC Santa Cruz and IBM Research, Almaden, USA
e-mail: kolaitis@cs.ucsc.edu

A. Baltag and S. Smets (eds.), *Johan van Benthem on Logic*   67
and Information Dynamics, Outstanding Contributions to Logic 5,
DOI: 10.1007/978-3-319-06025-5_3, © Springer International Publishing Switzerland 2014

One can view schema mappings as a case of "logical dynamics" in database theory: while traditional research on databases focuses on instances over a single schema, schema mappings are concerned with describing relationships across schemas, and with the access to and the transformation of data across different schemas. The aim of this chapter is to provide a "logician friendly" introduction to this area of research, focusing on some of the results that may be of particular interest to the readers of this volume. The chapter aims to be self-contained, and as such, it includes a brief introduction to the fundamentals of relational databases. In addition, at various points (marked in text by the '♣' sign), we take the opportunity to make short excursions into topics that we think may be of interest to the reader.

This chapter is organized as follows. After an introduction to relational database theory in Sect. 3.1, we introduce *schema mappings* in Sect. 3.2, where we also present the most important languages used for specifying schema mappings. In Sects. 3.3 and 3.4, we review the role of schema mappings in *data exchange* and *data integration*, two important data inter-operability tasks. Throughout these sections, we emphasize important structural properties of schema mappings that enable solving these tasks. Then, in Sect. 3.5, we turn the tables around and show that the main schema-mapping languages admit characterizations of abstract model-theoretic flavor in terms of the aforementioned important structural properties. Finally, in Sect. 3.6 we examine schema mapping from a dynamic perspective by highlighting some of the results concerning the composition of schema mappings.

3.1 Background: Relational Database Theory

One of the main ideas behind the relational model of databases is *physical data independence*, which amounts to a strict separation between the physical level at which data are stored and processed, and the logical level that describes how the data are organized and presented to the database user. At the logical level, all data are specified in terms of *relations* (informally, tables), and are accessed by means of queries in some declarative language, which is typically based on a logical formalism, such as first-order logic. The relational *Database Management System* (DBMS) takes care of choosing appropriate data structures and indexes for storing the data, and of translating the declarative queries into physical query plans that are then evaluated and the answers produced are returned to the database user.

Below, we recall briefly the basic ingredients of the relational database model. For a more detailed exposition, we refer the reader to [1].

3.1.1 Database Schemas and Instances

A *database schema* is a finite collection S of relation symbols, each of which has a specified arity and also names for its *attributes* (i.e., its columns). For example, the notation BOOK(ISBN, TITLE, PUBLISHER) indicates that BOOK is a ternary relation symbol having ISBN, TITLE, and PUBLISHER as the names of the three attributes. If $S = (R_1, \ldots, R_m)$ is a database schema, then a database instance over S is a sequence $I = (R_1^I, \ldots, R_m^I)$ of finite relations such that the arity of the relation R_i^I matches the arity of the relation symbol R_i it interprets, $1 \leq i \leq m$. If a tuple of values c belongs to R_i^I for some relation name $R_i \in S$, then we say that $R(c)$ is a *fact* of I. In what follows, we assume that the values occurring in relations come from some countably infinite domain D of data values (in practice, D may include integers, strings, dates, and other such data values; this, however, is not relevant for us here).

Note that there are striking similarities, but also differences, between database schemas and instances on the one had, and relational signatures and structures on the other. To begin with, a database schema $S = (\mathbf{R_1}, \ldots, \mathbf{R_m})$ can be thought of as a finite relational signature equipped with names for the columns of its relation symbols. Consequently, at first sight, a database instance over S can be thought of as a finite relational structure over the same schema. There is, however, an important difference between instances and structures. The specification of a relational structure includes the *domain* (*universe*) of the structure, which means that a structure over S is sequence $A = (V, R_1^A, \ldots, R_m^A)$ such that V is a set and each R_i^A is a relation on V. In contrast, the specification of a database instance does not include a domain. As we will soon see, the absence of an explicit domain in the specification of database instances will cause some difficulties in defining rigorously the semantics of queries asked against database instances. Given a database instance, however, it is possible to extract an implicit domain, called the *active domain*. In precise terms, the *active domain* of an instance I, denoted by $adom(I)$, is the (finite) set of values from D occurring in facts of I.

3.1.2 Database Queries

Queries are used to extract information from a database. When a query is evaluated on a database instance, it produces a finite relation. More formally, if $k \geq 1$, then a *k-ary query* over a database schema S is a function defined on instances over S and such that if I is an instance over S, then $q(I)$ is a k-ary relation on the active domain of I. The precise definition of queries as semantic objects involves one further condition called "genericity", according to which the function must behave in an isomorphism-invariant way. We say that a bijection $f : D \rightarrow D$ is an *isomorphism* between two relations R and R' of the same arity k, if for all tuples $a \in D^k$, we have that $a \in R$ if and only if $f(a) \in R'$; an isomorphism between two instances over the same schema is a function that is an isomorphism between every pair of corresponding relations. The genericity requirement, then, says that the query q is *isomorphism-invariant*, in the sense that each isomorphism f between instances I and I' is also an isomorphism between the relations $q(I)$ and $q(I')$, which means that $q(I') = f(q(I))$. A *zero-ary query*, also called a *Boolean query*, is a function from instances over S to the set

{0, 1} that is invariant under isomorphisms; thus, a Boolean queries is a decision problem (a "yes" or "no" question) about database instances up to isomorphism.

♣ A reader familiar with abstract model theory may observe an intuitive analogy between Boolean queries and generalized quantifiers. Both can be defined as isomorphism-invariant functions that take as input a relational structure or instance, yielding as output a truth value. The isomorphism-invariance condition can be seen as expressing a form of topic neutrality, and as such it features also prominently in the literature on generalized quantifiers and the demarcation of logical constancy (see for instance [8]).

3.1.3 First-Order Queries and Domain Independence

Queries are semantic objects. From a syntactic point of view, a k-ary query can often be specified by a first-order formula with k free variables, where $k \geq 1$. For example, the first-order formula

$$q_1(x_1, x_2) = \exists yzu \; \text{BOOK}(y, x_1, u) \wedge \text{BOOK}(z, x_2, u)$$

defines the query that returns a binary relation containing all pairs of titles of books that have the same publisher. Similarly,

$$q_2(x_1) = \exists yzuv \; \text{BOOK}(x_1, y, z) \wedge \text{BOOK}(x_1, u, v) \wedge y \neq u)$$

defines the query that returns the set of all authors with at least two books, and

$$q_3(x_1) = \forall yzuv(\text{BOOK}(x_1, y, z) \wedge \text{BOOK}(x_1, u, v) \rightarrow z = v)$$

defines the query that returns the set of all authors whose books are all published by the same publisher. When we specify a query by means of a formula ϕ, we assume that the free variables of ϕ are ordered, as in $x_1, \ldots, x_n$, hence the order of the attributes of the resulting relation is clear.

Boolean queries are often specified by sentences of first-order logic. For example, the first-order sentence

$$q_4 = \forall yzwy'z'w'(\text{BOOK}(y, z, w) \wedge \text{BOOK}(y', z', w') \rightarrow w = w')$$

defines the query that is true on an instance of BOOK precisely when all books in the database are published by the same publisher.

In general, the denotation of a first-order formula on a relational structure $A = (V, R_1, \ldots, R_m)$ depends not only on the relations $R_1, \ldots, R_m$ of the structure, but also on its domain V. In particular, a formula may have different denotations on two structures $A = (V, R_1, \ldots, R_m)$ and $A' = (V', R_1, \ldots, R_m)$ that have the same relations, but different domains. For example, this is the case for the first-order formula $\neg R_1(x)$, as well as for the first-order sentence $\forall x(R_1(x))$, which may be true on A, but false on A' in case the domain V' of A' contains properly the domain V of A.

This state of affairs implies that the denotation of a first-order formula on a database instance can be ambiguous, since, as discussed earlier, the specification of a database instance does not include an explicit domain. This is unsatisfactory, hence it is customary in database theory to restrict attention to *domain independent* first-order formulas. Formally, domain independence is defined as follows: if I is an instance and A is any superset of $adom(I)$, we write $\langle A, I \rangle$ for the relational structure with domain A in which all relation symbols are interpreted as in I. We also write $\phi^A(I)$ for the denotation of the formula ϕ on the structure $\langle A, I \rangle$. Now, we say that a first-order formula ϕ is *domain independent* if for all instances I and sets $A, B \supseteq adom(I)$, we have that $\phi^A(I) = \phi^B(I)$. In particular, this implies that $\phi^D(I) = \phi^{adom(I)}(I)$, and so, when evaluating a domain-independent formula, we can safely ignore values outside the active domain. From now on, whenever we speak about *first-order queries*, we will always mean queries that are definable by a domain-independent first-order formula.

♣ Not surprisingly, domain independence is an undecidable semantic property of first-order formulas [20]. This is an easy consequence of Trakhtenbrot's Theorem [52], which states that the satisfiability problem for first-order formulas on finite structures is undecidable. However, it is possible to define a syntactic fragment of first-order logic that captures the full domain independent fragment of first-order logic, up to logical equivalence. For instance, this can be done following the same general idea underlying the first-order fragment "F3" from [2], in which all quantifiers are required to be relativized by relational atomic formulas to make sure that they range over the active domain. Moreover, broader such syntactic fragments have been identified; for example, see [53].

♣ In reality, database queries may refer to fixed elements of the domain, in a way that is similar to the use of individual constants in first-order logic. Thus, for example, the first-order formula $\forall yz(\text{BOOK}(x_1, y, z) \rightarrow z \neq \text{'Springer'})$ defines the query that computes the set of authors who have never published a book with Springer. Allowing arbitrary values from the domain to be used as constants in queries does not have any fundamental implications for the results presented in this chapter (nor for other basic results in database theory), but it complicates many definitions, such as the above definitions of genericity and of domain independence. For this reason, we will not consider queries with constants here.

At its core, the industry-standard relational database query language SQL can be viewed as a friendly syntax for first-order queries. In practice, of course, SQL has many features that go beyond first-order logic, such as involving arithmetical operations, string operations, and aggregate operations.

3.1.4 Query Evaluation and Query Containment

The *query evaluation problem* is one of the most fundamental problems about databases; this is the problem of computing $q(I)$, for a query q and an instance I. The computational complexity of the query evaluation problem has been extensively studied for various query languages. It is common to distinguish between the *data complexity* and the *combined complexity* of query evaluation [54]. The *data complexity* is the complexity of query evaluation for fixed queries, where the complexity is measured only in terms of the size of the instance. The *combined complexity* is the complexity of query evaluation when both the query and the instance are considered to be part of the input. Often, data complexity is a more sensible measure of complexity, because the size of the instance tends to be many orders of magnitude larger than the size of the query; moreover, it is often the case that one is interested in only a small number of queries that stay fixed, while the instance changes frequently.

For first-order queries, the data complexity of query evaluation is in the complexity class PTIME. This means that, for any fixed first-order query, the query evaluation problem can be solved in time bounded by a polynomial in the size of the input instance (the degree of the polynomial, however, depends on the fixed query). In fact, the data complexity of evaluating first-order formulas is in LOGSPACE (logarithmic space), a complexity class that is contained in PTIME. The *combined complexity* of evaluating first-order queries, on the other hand, is complete for the complexity class PSPACE (polynomial space), which contains NP and other higher complexity classes.

Another fundamental problem is the *query containment problem*: given two queries q, q', decide if q is contained in q', meaning that, for all instances I, we have that $q(I) \subseteq q'(I)$. From the point of view of logic, this is of course precisely the entailment problem for (the formulas that define the) queries. For first-order queries, the query containment problem is undecidable. This follows from the aforementioned Trakhtenbrot's Theorem. There are, however, broad classes of frequently asked queries for which the containment problem is decidable and, in fact, has relatively low computational complexity. We shall discuss such a class next.

3.1.5 Conjunctive Queries and Homomorphisms

The query q_1 discussed earlier is an example of a *conjunctive query* (CQ), unlike the queries q_2, q_3, and q_4. Conjunctive queries form one of the most important classes of database queries in practice. They are the queries defined by first-order formulas of the form $q(\mathbf{x}) = \exists \mathbf{y} \phi(\mathbf{x}, \mathbf{y})$, where $\phi(\mathbf{x}, \mathbf{y})$ is a conjunction of atomic formulas; these atomic formulas may include equalities. In addition, it is required that each free variable $x_i \in \mathbf{x}$ actually occurs in a relational atomic formula of ϕ, in order to ensure that the query is domain-independent.

For arbitrary queries, the query evaluation problem and the query containment problem are very different algorithmic problems. In particular, for first-order queries, the query evaluation problem is decidable, while the query containment problem is undecidable. In the case of conjunctive queries, however, it turns out that these two problems are essentially the same problem and, moreover, are intimately connected to the *homomorphism problem*. A *homomorphism* from an instances I to an instance J over the same schema $\mathbf{S}$ is a function from the active domain of I to the active domain of J such that, for every fact of I, its h-image is a fact of J; in other words, for every relation symbol R_i of $\mathbf{S}$ and for every tuple $(a_1, \ldots, a_n) \in R_i^I$, we have that $(h(a_1), \ldots, h(a_n)) \in R_i^J$. The notation $h : I \to J$ indicates that h is a homomorphism from I to J. The existence of a homomorphism from I to J means that I is included in J, modulo some (not-necessarily bijective) substitution of values; thus, I is, in some sense, more "general" than J. The *homomorphism problem* is the following decision problem: given two instances I and J over the same schema **S**, is there a homomorphism from I to J?

♣ The existence-of-a-homomorphism relation between instances is a pre-order. It induces a partial order between homomorphism-equivalence classes; in fact, this preorder is a lattice, where the meet and join can be defined in terms of direct products and disjoint unions. The structure of this lattice has been the focus of an extensive study in graph theory, resulting in a rich theory [33].

The *Chandra-Merlin Theorem* is a basic result in database theory that establishes an intimate connection between conjunctive-query evaluation, conjunctive-query containment, and the homomorphism problem. To simplify the presentation, we will explain the connection for the special case of Boolean conjunctive queries. Given an instance I over a schema **S**, we can naturally associate with it a Boolean conjunctive query q_I over the same schema, which is called the *canonical Boolean conjunctive query* of I. This query has an existentially quantified variable for each value from $adom(I)$, and it contains one conjunct for each fact of I. For example, if I is the instance consisting of the facts BOOK(0-201-53771-0, Foundations of Databases, Addison-Wesley) and LOCATION(0-201-53771-0, A7.14), then the canonical Boolean conjunctive query of I would be $\exists xyzu\, \text{BOOK}(x, y, z) \wedge \text{LOCATION}(x, u)$. Conversely, if q is a Boolean conjunctive query, then, by choosing an arbitrary distinct value for each existentially quantified variable, we can associate with it a *canonical instance* I_q whose facts are, up to a renaming, the conjuncts of q. Together, these two transformations establish a one-to-one correspondence between instances and Boolean conjunctive queries, modulo renaming of values and variables; this is so because the canonical instance associated with the canonical query of an instance I is isomorphic to I.

Theorem 3.1 (Chandra-Merlin [17]) For every instance I and every Boolean conjunctive query q, the following statements are equivalent:

1. q is true on I.
2. There is a homomorphism $h : I_q \to I$
3. The canonical query q_I of I is contained in q.

The Chandra-Merlin Theorem, whose proof is not difficult, has played a pivotal role in the development of database theory, as it shows that, in a precise sense, conjunctive-query evaluation and conjunctive-query containment coincide with the homomorphism problem. This is remarkable given that, for arbitrary first-order queries, query evaluation and query containment are computationally very different problems (indeed, as we discussed earlier, the query containment problem is undecidable for first-order queries).

An immediate consequence of the Chandra-Merlin theorem is that the conjunctive-query containment problem is decidable. Moreover, both the conjunctive-query containment problem and the combined complexity of conjunctive-query evaluation are in NP, since these two problems amount to guessing a homomorphism between two instances. Actually, it is easy to see that both these problems are

NP-hard, hence they are NP-complete. Indeed, the well known NP-complete problem *Graph 3-Colorability* is a special case of the homomorphism problem, because a graph G is 3-colorable if and only if there is a homomorphism from G to the three-element clique K_3. Note also that, by the Chandra-Merlin Theorem, *Graph 3-Colorability* can be viewed as the following special case of the conjunctive-query evaluation problem: given a graph G, does K_3 satisfy the canonical conjunctive query q_G of G?

> ♣ There has been an extensive investigation of subclasses of the class of conjunctive queries for which the combined complexity of query evaluation is in polynomial time. One important such subclass is the class of *acyclic conjunctive queries* [55]. Various generalizations of acyclicity have been studied over the years, including queries of bounded tree-width and queries of bounded hypertree-width (see [27, 28] for a survey). This is an area of research that has enjoyed extensive interaction with constraint satisfaction, logic, and graph theory (see for instance [38]).

There is a second fundamental relationship between conjunctive queries and homomorphisms, namely the fact that conjunctive queries are preserved under homomorphisms. A query q is said to be *preserved by homomorphisms* if, for every homomorphism $h : I \to J$ and for every tuple $\mathbf{a} \in q(I)$, we have that the h-image $h(\mathbf{a})$ of $\mathbf{a}$ belongs to $q(J)$. Every conjunctive query is preserved under homomorphisms. Indeed, conjunctive queries are positive existential first-order formulas, and it is a well known fact in model theory that positive existential first-order formulas are preserved under homomorphisms.

A *union of conjunctive queries* (UCQ) is a query defined by a disjunction of conjunctive queries all of which have the same set of free variables. Clearly, every UCQ is, in particular, a positive existential first-order query. Conversely, it is not hard to show that every domain-independent positive existential first-order formula is equivalent to a union of conjunctive queries. Rossman [51] proved that, on finite structures, a first-order formula is preserved under homomorphism if and only if it is equivalent to a positive existential first-order formula. This implies that every first-order query (i.e., every query defined by a domain-independent first-order formula) preserved under homomorphism is a union of conjunctive queries.

Theorem 3.2 (Rossman [51]) *A first-order query is preserved under homomorphisms if and only if it is equivalent to a union of conjunctive queries.*

> ♣ Rossman's theorem is a rare example of a *preservation theorem* in model theory that holds true in the finite. A preservation theorem is a theorem that characterizes a fragment of first-order logic, up to logical equivalence, in terms of preservation under some model-theoretic operation or relation on structures (in this case, the relation of homomorphism). Well known examples of preservation theorems in classical model theory are the Łoś-Tarski Theorem (a first-order formula is preserved under extensions if and only if it is equivalent to an existential first-order formula) and Lyndon's Positivity Theorem (a first-order formula is preserved under surjective homomorphisms if and only if it is equivalent to a positive first-order formula). Most preservation theorems from classical model theory, including these two, fail in the finite (see [50] for a survey). It was a longstanding open problem to determine whether or not the homomorphism preservation theorem holds in the finite, and Rossman's remarkable theorem confirmed that it does. Its proof is highly sophisticated and required the

development of elaborate machinery. van Benthem's characterization of modal logic [7] is another notable example of the few preservation theorems that survive the passage to finite structures, as shown by Rosen [49].

3.1.6 Database Constraints

By organizing all data in relations, some inherent semantic information about the data may get lost. In the preceding example about books, the schema does not reflect the fact that ISBN numbers uniquely identify books. Thus, there is no a priori reason to exclude the existence of two entries with the same ISBN number but with different titles. To capture this semantic information, *database constraints* are used. Database constraints are statements concerning structural properties of the relations in a database schema (in this sense, they play a similar role as frame conditions in modal logic).

An example is the *key constraint*, customarily written as

$$\text{BOOK} : \text{ISBN} \rightarrow \text{TITLE}, \text{ PUBLISHER}$$

that expresses that the ISBN number functionally determines the title and the publisher of a book. The above key constraint can be equivalently expressed in first-order logic as $\forall xyzuv(\text{BOOK}(x, y, z) \wedge \text{BOOK}(x, u, v) \rightarrow y = u \wedge z = v)$.

Inclusion dependencies form another commonly used type of constraints. Suppose that the database schema contains, besides the BOOK relation, also a binary relation LOCATION(ISBN,STACK). Then we may wish to require that, for every entry in the BOOK relation, there is a corresponding entry in the LOCATION relation. This constrained is usually written as

$$\text{BOOK}[\text{ISBN}] \subseteq \text{LOCATION}[\text{ISBN}]$$

and, in first-order logic, the same constraint would be expressed as $\forall xyz$ (BOOK $(x, y, z) \rightarrow \exists u \text{ LOCATION}(x, u))$.

Constraints are included in the specification of the database, and the database management system may take these constraints into account when deciding on the physical layout for the stored data. For instance, knowing that ISBN numbers uniquely identify books, the system may decide to store the data using a hash table, instead of an array. Constraints have applications in other areas as well, including database design and query optimization.

In the early 1980s, *tuple-generating dependencies* and *equality-generating dependencies* emerged as suitable fragments of first-order logic that capture the most important types of database constraints in practice. A *tuple-generating dependency* (*tgd*) is a first-order sentence of the form

$$\forall \mathbf{x}(\phi(\mathbf{x}) \rightarrow \exists \mathbf{y}\psi(\mathbf{x}, \mathbf{y})),$$

where $\phi(\mathbf{x})$ and $\psi(\mathbf{x,y})$ are conjunctions of atomic relational formulas, and every variable in $\mathbf{x}$ occurs in ϕ. An *equality-generating dependency (egd)* is a first-order sentence of the form

$$\forall\mathbf{x}(\phi(\mathbf{x}) \rightarrow x_i = x_j),$$

where $\phi(\mathbf{x})$ is a conjunction of relational atomic formulas containing all the variables in $\mathbf{x}$, and $x_i, x_j \in \{\mathbf{x}\}$.

Clearly, every key constraint can be equivalently expressed as a conjunction of egds, and every inclusion dependency can be expressed as a tgd.

♣ There is another perspective on tgds, which is based on query containment. A tgd $\forall\mathbf{xy}$ $(\phi(\mathbf{x}, \mathbf{y}) \rightarrow \exists\mathbf{z}\psi(\mathbf{x}, \mathbf{z}))$ can be viewed as expressing the requirement that an instance satisfies the containment $q_1 \subseteq q_2$, where q_1 and q_2 are the conjunctive queries $\exists\mathbf{y}\phi(\mathbf{x}, \mathbf{y})$ and $\exists\mathbf{z}\psi(\mathbf{x}, \mathbf{z})$, respectively. Conversely, for any two given conjunctive queries q_1, q_2 of the same arity, the containment $q_1 \subseteq q_2$ can be expressed as a tgd.

Tgds and egds, as fragments of first-order logic, have been extensively studied from a computational, axiomatic, and model-theoretic point of view. In particular, the *implication problem* (does a given set of constraints logically imply another constraint) has been investigated in considerable depth. The implication problem for arbitrary tgds turned out to be undecidable; however, it is decidable for the restricted case of tgds without existential quantifiers (which are known as *full tgds*) and edgs. The computational complexity of the implication problem for various classes of dependencies has been studied, and axiom systems have been developed for the entailment problem of classes of dependencies. We refer the reader to [25] for a survey of results in this area. Also, some model theoretic characterizations of classes of dependencies have been established [45].

♣ It is worth mentioning an interesting open problem in the theory of database constraints. It is easy to see that every full tgd is logically equivalent to a finite conjunction of domain-independent *universal Horn sentences*; moreover, the converse holds as well. A well known result in classical model theory asserts that a first-order sentence is preserved under direct products and substructures on all (finite and infinite) structures if and only if it is equivalent to a finite conjunction of universal Horn sentences (see [18]). It is not known whether this result holds true in the finite. If it does, then it would yield a model-theoretic characterization of full tgds in the finite.

In recent years, database constraints have found new applications in the context of data exchange and data integration, as they provide a suitable language for specifying schema mappings. We discuss this next.

3.2 Schema Mappings

If you ask two different people to design a database schema for a particular application, they will most likely come up with different schemas. The differences may be innocuous, e.g., merely involving the name of a relation or of an attribute, or they may be substantial, e.g., one relation in one schema may contain the same information as two relations together in the other. As a result of such differences, combining related data from different sources can be a difficult task, a task that is known as the *data inter-operability problem* [12, 29]. The research community has investigated several different facets of the data inter-operability problem, including *data exchange* and *data integration*, which we will discuss in more detail in what follows. What all data inter-operability tasks have in common is that they require an understanding of the relationships between different database schemas. *Schema mappings* have emerged as an important tool for achieving this [46, 47].

A schema mapping is a high-level, declarative specification of the relationships between two database schemas. Formally, a schema mapping is a triple $M = (\mathbf{S}, \mathbf{T}, \Sigma)$, where $\mathbf{S}$ and $\mathbf{T}$ are disjoint database schemas, called the *source schema* and the *target schema*, and Σ is a finite collection of constraints over $\mathbf{S} \cup \mathbf{T}$, typically defined in some suitable logical language. On the face of this definition, a schema mapping is a syntactic object. We can, however, assign semantics to a schema mapping $M = (\mathbf{S}, \mathbf{T}, \Sigma)$ as follows. Let I be a source instance, i.e., an instance over the schema $\mathbf{S}$, and let J be a target instance, i.e., an instance over the schema $\mathbf{T}$. We say that J is a *solution* for I with respect to the schema mapping M if the pair (I, J) (viewed as an instance over the schema $\mathbf{S} \cup \mathbf{T}$) satisfies the constraints in Σ. Then, from a semantic point of view, the schema mapping M can be identified with the set Sem(M) of all pairs (I, J) such that I is a source instance, J is a target instance, and J is a solution for I with respect to M. In symbols,

$$\text{Sem}(M) = \{(I, J) : (I, J) \models \Sigma\}.$$

An example of a schema mapping is given in Fig. 3.1. The pair (I, J_1) consisting of the source relation I and the target relation J_1 satisfies both constraints; the same holds true for the pairs (I, J_2) and (I, J_3). In contrast, the pair (I, J_4) fails to satisfy the first constraint because the SALES relation does not contain the required quadruple (05-01-2009, UCSC, TFT-933SN-Wide, 100). Therefore, J_1, J_2, and J_3 are solutions for I, whereas J_4 is not.

What is a "good" language for specifying schema mappings? At first, one may think that first-order logic is a natural candidate. However, several important algorithmic problems about data inter-operability are undecidable, when the full expressive power of first-order logic is used to specify schema mappings. In particular, it is not hard to show that there is a schema mapping M defined by a finite set of first-order sentences for which the *existence-of-solutions problem* is undecidable: given a source instance I, is there a solution J for I with respect to M?

To motivate the choice of a "good" schema-mapping specification language, let us proceed in a bottom-up way by considering some basic constraints that every such language ought to be able to express.

(i) Copy (Nicknaming): $\forall xyz(P(x, y, z) \rightarrow R(x, y, z))$
(ii) Projection: $\forall xyz(P(x, y, z) \rightarrow R(x, y))$
(iii) Column Addition: $\forall xy(P(x, y) \rightarrow \exists z R(x, y, z))$

Source schema **S***:*

DIRECTCUSTOMER(CUST-ID,NAME,ADDRESS)
DIRECTORDER(CUST-ID, DATE, PROD, QUANT)
RETAIL(STORE-ID, DATE, PROD, QUANT)

Target schema **T***:*

SALES(DATE, CUST, PROD, QUANT)

Source instance I:

DIRECTCUSTOMER

cust-id	name	address
c1	UCSC	1156 High St, Santa Cruz

DIRECTORDER

cust-id	date	prod	quant
c1	05-01-09	Quadcore-9950-PC	100
c1	05-01-09	TFT-933SN-Wide	100

RETAIL

store-id	date	prod	quant
s1	05-03-09	Quadcore-9950-PC	1
s1	05-03-09	Quadcore-9800-PC	1

A target instance J_1

SALES

date	cust	prod	quant
05-01-09	UCSC	Quadcore-9950-PC	100
05-01-09	UCSC	TFT-933SN-Wide	100
05-03-09	N_1	Quadcore-9950-PC	1
05-03-09	N_2	Quadcore-9800-PC	1

A second target instance J_2

SALES

date	cust	prod	quant
05-01-09	UCSC	Quadcore-9950-PC	100
05-01-09	UCSC	TFT-933SN-Wide	100
05-03-09	N_1	Quadcore-9950-PC	1
05-03-09	UCSC	Quadcore-9800-PC	1

A third target instance J_3

SALES

date	cust	prod	quant
05-01-09	UCSC	Quadcore-9950-PC	100
05-01-09	UCSC	TFT-933SN-Wide	100
05-03-09	N_1	Quadcore-9950-PC	1
05-03-09	N_1	Quadcore-9800-PC	1

A fourth target instance J_4

SALES

date	cust	prod	quant
05-01-09	UCSC	Quadcore-9950-PC	100
05-03-09	N_1	Quadcore-9950-PC	1
05-03-09	N_2	Quadcore-9800-PC	1

Schema mapping

$\forall xyzuvw$ (DIRECTCUSTOMER(x,y,z) $\land$ DIRECTORDER(x,u,v,w) $\rightarrow$ SALES(u, y, v, w))

$\forall xyzvw$ (RETAIL(x, y, v, w) $\rightarrow$ $\exists N$ SALES(y, N, v, w))

Fig. 3.1 An example of a schema mapping

(iv) Join: $\forall xyz(E(x, y) \land F(y, z) \rightarrow T(x, y, z))$

(v) Decomposition: $\forall xyz(P(x, y, z) \rightarrow R(x, y) \land T(y, z))$

(vi) Combination of Join and Column Addition:
$\forall xyz(E(x, z) \land F(z, y) \rightarrow \exists w(R(x, y) \land T(x, y, z, w)))$.

Observe that the above constraints have something striking in common, namely, each of them is a tuple-generating dependency, whose antecedent consists of source relations, and whose consequent consists of target relations. Recall that a tuple-generating dependency (tgd) is a first-order sentence of the form

$$\forall \mathbf{x}(\phi(\mathbf{x}) \rightarrow \exists \mathbf{y}\psi(\mathbf{x}, \mathbf{y})).$$

A *source-to-target tuple-generating-dependency (s-t tgd)*, also known as a *GLAV (Global-and-Local-As-Views)* constraint, is tgd whose antecedent refers only to a source schema and whose consequent refers only to a target schema. Thus, the constraints in the preceding examples (i)–(vi) are GLAV constraints, and the same holds true for the two constraints that specify the schema mapping in Fig. 3.1.

As it turns out, schema mappings defined by GLAV constraints strike a good balance between expressive power and computational complexity. Indeed, they are powerful enough to specify interesting relationships between schemas, while at the same time they have tame algorithmic behavior, as we will soon see. For this reason, the language of GLAV constraints is the most extensively studied schema-mapping language to date.

Two important special cases of GLAV constraints are GAV ("Global-As-View") constraints and LAV ("Local-As-View") constraints. A GAV constraint is a GLAV constraint in which the consequent is a single atomic formula without existential quantifiers. The constraints in the preceding examples (i), (ii), (iv), as well as the first constraint in Fig. 3.1 are GAV constraints. Furthermore, the constraint in example (v) is logically equivalent to a conjunction of two GAV constraints. Dually, a LAV constraint is a GLAV constraint in which the antecedent is a single atomic formula. The constraints in the preceding examples (i), (ii), (iii), (v), as well as the second constraint in Fig. 3.1 are LAV constraints.

Besides GAV, LAV, and GLAV constraints, other classes of constraints have been used in the context of schema mapping specification. In particular, *second-order tgds*, which involve existential second-order quantification, turn out to be important in the context of schema mapping composition, as we shall see in Sect. 3.6.

This use of constraints in the specification of schema mappings can be described as *dynamic* as opposed to the more *static* traditional use of constraints. This is because constraints are now being used to describe relationships that hold across instances, as opposed to inside a single instance. Furthermore, there is another sense in which this use of constraints is dynamic: GLAV constraints can be interpreted not only declaratively, as specifying intended relationships between instances of the two schemas, but also procedurally, as providing a recipe for constructing a target instance on the basis of a source instance. In fact, this explains the suggestive names we gave to the constraints in the preceding examples (i)–(vi). We will discuss this in more detail in the next section.

In Sects. 3.3 and 3.4, we will discuss in depth two important data inter-operability tasks, namely, *data exchange* and *data integration* that are depicted in Fig. 3.2. Data exchange is the problem, given a source instance, of materializing a suitable target instance. Data integration is the problem of answering target queries on the basis of

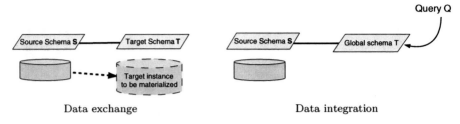

Fig. 3.2 Diagrammatic depiction of the data exchange and data integration tasks

a source instance. Schema mappings play a crucial role in formalizing and studying both these data inter-operability tasks.

3.3 Data Exchange: Moving Data from Source to Target

Data exchange, intuitively speaking is the problem, transforming data structured according to one schema into data structured according to another schema. Although it has a long history, the problem was first rigorously formalized and studied in [21], where the basic concepts and results were established that we discuss below.

Formally, data exchange via a schema mapping $M = (S, T, \Sigma)$ is the problem, given a source instance I, of constructing a target instance J that is a solution for I with respect to M [21]. In general, a source instance may have many solutions. Indeed, suppose that Σ is a finite set of GLAV constraints. An inspection of the syntax of GLAV constraints reveals that when more facts are added to a solution, the result is still a solution (see also Structural Property 2 below). Consequently, every source instance has infinitely many solutions. This raises the question: given a source instance I, which is a "suitable" solution to materialize when solving the data exchange problem? *Universal solutions* were introduced in order to capture the preferred solutions in data exchange. The idea behind universal solutions is simple: they contain no more and no less information than is necessary to satisfy the constraints of the schema mapping. This is formally defined using the notion of a homomorphism, that we encountered earlier.

3.3.1 Universal Solutions

Before we give the formal definition of universal solutions, we illustrate the idea with an example. Consider the schema mapping given in Fig. 3.1, and let I be a source instance consisting of the single fact

RETAIL(s1, 05-03-09, Quadcore-9950-PC, 1)

The target instance J consisting of the single fact

$$\text{SALES}(05\text{-}03\text{-}09, \text{xxx}, \text{Quadcore-9950-PC}, 1)$$

is a solution for I. Intuitively, in the target instance J, the values '05-03-09' and 'xxx' play a different role. The former has been copied from I, in order to satisfy the constraints of the schema mapping. The latter is used to witness an existentially quantified requirement imposed by the schema mapping, and the exact choice of value is arbitrary. In a sense, 'xxx' plays the role of a placeholder for an unknown value; such values are called *null* values. Another example is the target instance J_1 given in Fig. 3.1, where N_1 and N_2 are null values, while all other values are taken from $adom(I)$. The definition of universal solution takes this distinction into account.

In Fagin et al. [21], it was assumed that the domain D from which values are drawn consists of two types of values, *constant values* and *nulls values*. Here, to simplify things, we will identify constant values with those values that occurred already in the source instance, and null values with values that are fresh, i.e., that did not occur in the source instance (note that this distinction is then relative to a given source instance). A *universal solution* for a source instance I with respect to a schema mapping M is a solution J for I with respect to M, such that, for every solution J' of I with respect to M, there is a homomorphism $h : J \rightarrow J'$ that is constant on $adom(I)$. Note that, in the above example, the values '05-03-09', 'Quadcore-9950-PC', and '1' belong to $adom(I)$, but 'xxx' does not. In this case, J is in fact a universal solution for I with respect to M. Indeed, it is clearly a solution and it can be homomorphically mapped into any other solution by mapping 'xxx' to an appropriate value, while leaving all other values unchanged.

Recall that, intuitively, the existence of a homomorphism $h : I \rightarrow J$ indicates that I is "more general" than J or, equivalently, that J "contains more information" than I. Thus, the above definition of universal solutions essentially says that universal solutions are the most general solutions or, equivalently, that they contain a minimal amount of information. We illustrate this intuition with one more example.

Let I, J_1, J_2, J_3, J_4 be the source and target instance given in Fig. 3.1. Then J_1 is a universal solution for I with respect to the depicted schema mapping. While J_2 and J_3 are solutions as well, they are not universal solutions. Intuitively, this is because they contains the additional, unjustified information that the customer of one of the retail transaction was UCSC, or that both retail transactions involved the same customer. Finally, J_4 is not even a solution for I.

♣ Universal solutions are not unique up to isomorphism; they are, however, unique up to homomorphism equivalence. For example, consider the source instance I' consisting of the fact RETAIL(s1, 05-03-09, Quadcore-9950-PC, 1). Let K_1 be the target instance

$$\text{SALES}(05\text{-}03\text{-}09, \text{xxx}, \text{Quadcore-9950-PC}, 1)$$
$$\text{SALES}(05\text{-}03\text{-}09, \text{yyy}, \text{Quadcore-9950-PC}, 1)$$

and let K_2 be the target instance consisting of just the first of these two facts. Both K_1 and K_2 are universal solutions for I'; moreover, K_1 and K_2 are homomorphically equivalent, which means that there are homomorphisms from K_1 to K_2, and from K_2 to K_1.

In Fagin et al. [22], it was shown that if M is a schema mapping defined by a finite set of GLAV constraints, then every source instance has a unique-up-to-isomorphism *minimal* universal solution, which is known as the *core universal solution*. This follows from the fact every finite instance I has a *core*, that is, a unique-up-to-isomorphism minimal homomorphically equivalent instance [34]. In fact, the core of I is a subinstance of I.

The concept of a core of a finite structure can be seen as playing an analogous role to the concept of bisimulation contraction of a Kripke structure in modal logic (see [10]).

3.3.2 Constructing Universal Solutions Using the Chase

As we already mentioned earlier, there are two ways of thinking about GLAV constraints. First, GLAV constraints are declarative specifications of the relationships between source and target relations. Second, they can also be interpreted procedurally, as a recipe for computing a solution, or, in other words, as a call to action. For example, the GLAV constraint

$$\forall xyzvw(\text{RETAIL}(x, y, v, w) \rightarrow \exists N \text{ SALES}(y, N, v, w))$$

can be read procedurally as follows: "for each triple (x, y, v, w) in the RETAIL relation, choose a fresh value N and insert (y, N, v, w) in the SALES relation (if such a tuple was not already present)". This idea naturally leads to what is known as the *chase procedure*. In essence, the chase procedure consists of repeatedly adding facts as dictated by the constraints, until all constraints are satisfied. Several different variants of the chase procedure have been introduced and studied over the year. Originally, the chase procedure was used in the study of the implication problem for arbitrary tuple-generating and equality-generating dependencies [6, 44], in which case it is a recursive procedure that is not guaranteed to terminate. However, in the case of GLAV constraints and due to the separation of source and target relations, there is no genuine recursion. As a matter of fact, it turns out that, for every fixed schema mapping defined by a finite set of GLAV constraints, whenever given a source instance I, the chase terminates in polynomial time and yields a universal solution for I with respect to M (the degree of the polynomial, however, depends on the schema mapping).

Let us say that a schema mapping *admits universal solutions* if every source instance has a universal solution. The preceding discussion shows that GLAV schema mappings possess this structural property (and moreover, universal solutions can be computed in polynomial time):

Structural Property 1 *Every schema mapping defined by a finite set of GLAV constraints admits universal solutions.*

♣ The original use of the chase procedure as a technique for testing whether a given set of constraints implies another constraint bears a close resemblance to the proof method of semantic tableaux developed by Beth [13]. There, it is also the case that formulas are interpreted procedurally, as a call to action (e.g., when a tableau branch contains a conjunction, we add each conjunct; when a tableau branch contains an existentially quantified formula, we add a substitution instance with a fresh value witnessing the existential claim, and so on). Just as for semantic tableaux, sophisticated conditions have been identified that guarantee termination of the chase procedure for restricted classes of tuple-generating dependencies.

3.3.3 Closure Under Target Homomorphisms

A universal solution (with respect to a GLAV schema mapping) not only is a most general solution, but can also be viewed as a representation of the entire space of all solutions. This follows from the second structural property of GLAV schema mappings, which we will discuss next, namely, *closure under target homomorphisms*.

A schema mapping is *closed under target homomorphisms* if for every source instance I, every pair of target instances J, J', and every homomorphism $h : J \to J'$ that is constant on $adom(I)$, we have that if J is a solution for I, then also J' is a solution for I. In other words, closure under target homomorphisms means that when more information is added to a solution (by extending the solution with additional facts and/or by replacing null values by other values), then the result is still a solution. Note, however, that if more information is added to a universal solution, then the resulting instance need not be a universal solution.

Structural Property 2 *Every schema mapping defined by a finite set of GLAV constraints is closed under target homomorphisms.*

The above structural property, in combination with the property of admitting universal solutions, has three applications.

First, it implies that, for every source instance I and every universal solution J for I, the set of all solutions of I is precisely the set of all target instances J' for which there is a homomorphism $h : J \to J'$. In other words, the infinite set of all solutions for I is "captured" by a single solution.

Second, it implies that each source instance has a core universal solution (we omit the details).

Finally, it enables a natural approach to data exchange with multiple sources. Thus far, we have focused on the case of data exchange with a single source. In general, one may wish to combine data from different sources, and construct from it a single target instance over a unified schema. In this case, for each source schema, a schema mapping is needed that spells out the relationships between that source schema and the target schema. Once these schema mappings are available, we can apply the same techniques as before. To make this more precise, let us say that a target instance J is a *solution* for a collection of source instances $I_1, \ldots, I_n$ over disjoint schemas $S_1, \ldots, S_n$, if J is a solution for each source instance I_i with respect to the corresponding schema mapping between S_i and T. Let us say that J is a *universal solution* for the collection of source instances $I_1, \ldots, I_n$ if it is a solution and, for every solution J', there is a homomorphism $h : J \to J'$ that is constant on $adom(I_1) \cup \cdots \cup adom(I_n)$. If all the schema mappings involved are closed under target homomorphisms, then a universal solution for $I_1, \ldots, I_n$ can be constructed as follows: we construct a universal solution J_i for each source instance I_i (making sure to always use a disjoint set of null values, not occurring in $adom(I_k)$ for any $k \le n$). Next, we take the union $J = \bigcup_i J_i$ of all the resulting target instances. Then J is a solution for each source instance I_i, since it is a homomorphic extension of each target instance J_i. Furthermore, it is not difficult to show that J is in fact a universal solution for $I_1, \dots, I_n$.

3.4 Data Integration: Answering Target Queries Using Source Data

In the previous section, we discussed *data exchange*, the problem of transforming source data into target data. In this section, we discuss *data integration*, which is a different, but closely related, facet of data inter-operability. Here, the problem is to answer target queries using source data. In other words, given a schema mapping and a source instance, the problem is to compute answers to target queries (see [41] for a survey). Since a source instance can have many different solutions, we first have to make precise the semantics of data integration, that is, what the intended answers of a target query are. A natural way to define this semantics is via the notion of *certain answers*.

3.4.1 Certain Answers

Given source instance I, we can think of each solution of I as being, intuitively, a different possible world. The *certain answers* of a target query, are the tuples that are *necessarily* an answer to the query, no matter what possible world we choose. More precisely, given a schema mapping M, a source instance I, and a query q over the target schema, the set of certain answers of q in I with respect to M, denoted by $\text{CERT}_{q,M}(I)$, is the intersection $\bigcap \{q(J) \mid J \text{ is a solution of } I\}$.

Example 3.1 Returning to the schema mapping and the source instance I from Fig. 3.1, consider the following conjunctive query q over the target schema

$$q(x, y) = \exists uvwz \ \text{SALES}(u, v, x, w) \land \text{SALES}(u, v, y, z)$$

This query asks for all pairs of products (x, y), such that some customer bought x and y (in some quantities) on the same date. It is not hard to see that, for every solution J for I, the pair (*Quadcore-9950-PC,TFT-933SN-Wide*) belongs to $q(J)$. In other words, this tuple belongs to the certain answers of q in I with respect to the schema mapping.

♣ The concept of certain answers, which has a strong modal-logic flavor, originated in the study of *incomplete databases* [37]. An incomplete database can be thought of as a specification of a set of possible worlds, where each possible world is a complete database. Certain answers then naturally arise, and indeed they constitute the standard semantics for query answering over incomplete databases. Note that *possible answers* have also been considered, but they play a less prominent role than the certain answers do, as the certain answers provide the guarantee that they are returned by the query on every possible world.

Due to the inherent second-order quantification over solutions in this definition, the certain answers to a first-order query is in general not computable. Indeed, the satisfiability problem for first-order sentences on finite structures, which is known to be undecidable, coincides with the problem of testing whether (the negation of) a first-order sentence is true in all solutions of the empty instance with respect to the empty schema mapping. Fortunately, for conjunctive queries, and, more generally, for unions of conjunctive queries, the problem of computing certain answers is decidable. As shown in Fagin et al. [21], one way to compute certain answers is by first constructing universal solutions.

Theorem 3.3 *Let M be an arbitrary schema mapping, I a source instance, J a universal solution for I, and q a k-ary conjunctive query. Then*

$$\mathrm{CERT}_{q,M}(I) = q(J) \cap adom(I)^k.$$

In particular, if M is a schema mapping defined by a finite set of GLAV constraints and q is a conjunctive query, then, given an instance I, the certain answers $\mathrm{CERT}_{q,M}(I)$ can be computed in polynomial time.

Proof Recall that a universal solution of a source instance I is a solution J for I such that for every solution J' for I, there is a homomorphism $h : J \rightarrow J'$ that is constant on $adom(I)$. Also, recall that conjunctive queries are preserved under homomorphisms. Putting these two facts together, it is not hard to show that if J is a universal solution for I, then a tuple of values from $adom(I)$ is a certain answer to a conjunctive query q in I if and only if the tuple is an answer to q in J. Furthermore, using an isomorphism invariance argument, it can be shown that if a tuple of values is a certain answer to a query q, then the tuple must consist entirely of values from $adom(I)$.

It follows that the certain answers of a conjunctive query can be computed simply by evaluating the query on an arbitrary universal solution (and disregarding the answers that contain values outside $adom(I)$). Let M be a fixed schema mapping defined by a finite set of GLAV constraints. Since universal solutions with respect to M can be constructed in polynomial time and since every fixed conjunctive query can be evaluated over polynomial time, it follows that for every fixed conjunctive query q, the certain answers $\mathrm{CERT}_{q,M}(I)$ can be computed in time polynomial in I (and, in fact, a careful analysis shows that $\mathrm{CERT}_{q,M}(I)$ is even computable in logarithmic space). □

Thus, in Example 3.1, the certain answers of the query $q(x, y)$ can be obtained simply by evaluating the query on the universal solution J_1 and keeping only those tuples all of whose values are from the active domain of I.

Theorem 3.3 not only shows how to compute certain answers using universal solutions, but also provides additional justification for the definition of universal solutions. Indeed, using the Chandra-Merlin Theorem, a target instance J is a universal solution of a source instance I if and only if Proposition 3 holds [21].

The pre-processing step of computing a universal solution in order to compute certain answers can be avoided, however, by using another method that is based on the idea of query rewriting.

3.4.2 Computing Certain Answers Via Query Rewriting

We now discuss how to compute certain answers by rewriting the target query q to a source query q' that, on input I, yields precisely the certain answers of q in I. We say that a schema mapping M *allows for CQ rewriting* if for every conjunctive query q over the target schema, there is a union of conjunctive queries q' over the source schema such that, for all source instance I, we have that $q'(I) = \mathrm{CERT}_{q,M}(I)$.

Structural Property 3 *Every schema mapping defined by a finite set of GLAV constraints allows for CQ rewriting.*

Example 3.2 Continuing from Example 3.1, it can be shown that the certain answers of q on a source instance I are precisely the answers in $q'(I)$, where q' is the following union of conjunctive queries over the source schema:

$$
\begin{aligned}
q'(\mathrm{x,y}) = \big(&\exists \mathrm{cid}_1, \mathrm{cid}_2, \mathrm{name,addr}_1, \mathrm{addr}_2, \mathrm{date,n,m} \\
\big(\textsc{DirectOrder}&(\mathrm{cid}_1, \mathrm{date,x,n}) \wedge \textsc{DirectOrder}(\mathrm{cid}_2, \mathrm{date,y,m}) \wedge \\
\textsc{DirectCustomer}&(\mathrm{cid}_1, \mathrm{name,addr}_1) \wedge \\
\textsc{DirectCustomer}&(\mathrm{cid}_2, \mathrm{name,addr}_2) \big) \big) \\
\vee \big(x = y \wedge &\exists \mathrm{sid,date,n}\ \textsc{Retail}(\mathrm{sid,date,x,n}) \big)
\end{aligned}
$$

Note that the RETAIL relation does not provide information about the name of the buyer, and therefore, can only contribute identity pairs to the certain answers of q.

Structural Property 3 shows again, now via a different route, that, for fixed GLAV schema mappings M and conjunctive queries q, $\mathrm{CERT}_{q,M}(I)$ can be computed in polynomial time, and, in fact, in logarithmic space. Moreover, it shows that the certain answers of a conjunctive query can be obtained by evaluating a union of conjunctive queries using any off-the-shelf database management system.

Different techniques have been developed for computing the source query q' from the target query q. In the special case of GAV schema mappings, there is a very simple method which is known as *unfolding*. Essentially, this method consists of replacing each occurrence of a target relation in q by a union of many conjunctive queries, one for each left-hand side of a GAV constraints in which the relation R occurs on the right. We omit the details but illustrate the method by means of an example.

Example 3.3 Consider the schema mapping M defined by the GAV constraint

$$\forall xyz(R(x, y, z) \rightarrow T(x, y))$$

as well as the GAV constraint $\forall x(S(x, x) \rightarrow T(x, x))$, which, for convenience, we will write here using a non-standard syntax as

$$\forall xy(S(x, y) \wedge x = y \rightarrow T(x, y))$$

so as to make sure that there is no repetition of variables on in the right-hand side. Consider the unary conjunctive query $q(u) = \exists v \, T(u, v)$. Then unfolding q with respect to M will cause the subexpression $T(u, v)$ in q to be replaced by the disjunction of $\exists w \, R(u, v, w)$ (derived from the first GAV constraint) and $S(u, v) \wedge u = v$ (derived from the second GAV constraint). This yields $q'(u) = \exists v((\exists w \, R(u, v, w)) \vee (S(u, v) \wedge u = v))$, which is equivalent to the union of conjunctive queries $\exists vw \, Ruvw \vee \exists v(S(u, v) \wedge u = v)$.

In the LAV case, query rewriting is less straightforward. Intuitively, this is because it no longer suffices to treat each atomic formula in the query q independently. A single application of a LAV constraint may account for several atomic formulas in the query, and therefore it is necessary to consider all possible ways in which (i) the atomic formulas in q can be partitioned into groups, and (ii) each group is mapped, in an appropriate way, to the right-hand side of a single LAV constraint. In general, this results in a union of exponentially many conjunctive queries. We omit the details, but refer to [48], for the MiniCon algorithm, which is most well known query rewriting algorithm for LAV schema mappings.

Finally, in the case of GLAV schema mappings, a combination of the above two techniques can be used. In fact, every GLAV schema mapping can be "decomposed" into a LAV schema mapping and a GAV schema mapping, and query rewriting can be performed by successively applying, for instance, the MiniCon algorithm and the unfolding technique.

To summarize, we have discussed two approaches to query answering: via universal solutions and via query rewriting. Each has its own advantages. In scenarios where the source data is not likely to change anymore, computing a universal solution may be a sensible pre-processing step, enabling us to quickly evaluate target queries afterwards. In scenarios where the source data keeps changing, it may be better to use the query-rewriting approach instead.

♣ Query rewritings are intimately related to the concept of *weakest preconditions* in Hoare logic [36]. Recall that, in Hoare logic, the weakest precondition of a postcondition P with respect to a (possibly non-deterministic) program π is a necessary and sufficient condition that needs to hold *before* execution of the program π, in order to guarantee that P holds *after* the execution. For example, if π is the assignment statement $x := 4$ and P is the postcondition $x + y > 10$, then the weakest precondition of P with respect to π is $y > 6$. In dynamic logics, the notation $[\pi]\phi$ is used to denote the weakest precondition of the postcondition ϕ with respect to a program π. We can view a schema mapping as a non-deterministic program that takes us from a given source instance I to any target instance that is a solution for I.

If we view a target query as a postcondition, then Structural Property 3 can be viewed as stating that, for every GLAV schema mapping M and target conjunctive query q, the weakest precondition of q with respect to M is expressible by a union of conjunctive queries. As such, Structural Property 3 may be compared to the use of reduction axioms in dynamic epistemic logic [39].

3.5 Structural Characterizations of Schema Mapping Languages

In the previous two sections, we saw that schema mappings defined by a finite set of GLAV constraints satisfy a number of desirable structural properties that have applications to data exchange and data integration. In this section, we turn the tables around: we will present a number of results of abstract model theoretic flavor that characterize schema mapping languages in terms of the structural properties of schema mappings definable in these languages. These results are from [15, 16].

Before we can state these results, we need to give an abstract definition of a schema mapping, which is based on the semantics Sem(M) of a (syntactically defined) schema mapping $M = (\mathbf{S}, \mathbf{T}, \Sigma)$ introduced in Sect. 3.2.

Definition 3.1 An *(abstract)* schema mapping is a triple $M = (\mathbf{S}, \mathbf{T}, \mathcal{W})$, where $\mathbf{S}$ and $\mathbf{T}$ are disjoint schemas, and $\mathcal{W}$ is a binary relation between $\mathbf{S}$-instances and $\mathbf{T}$-instances, satisfying the following isomorphism-invariance condition: for every two pairs of instances (I, J) and (I', J'), if $I \cup J$ and $I' \cup J'$ are isomorphic (as $\mathbf{S} \cup \mathbf{T}$-instances) and $(I, J) \in \mathcal{W}$, then also $(I', J') \in \mathcal{W}$.

We say that J is a solution for I with respect to M if the pair (I, J) belongs to $\mathcal{W}$.

Clearly, every schema mapping defined by finitely many GLAV constraints is a schema mapping in the above abstract sense. The results from [15] that we will present below characterize when an schema mapping is definable by finitely many GLAV constraints, or finitely many GAV constraints, or finitely many LAV constraints. These and several other variations of these results were reported in [15, 16]. Here, we will limit ourselves to presenting the ones that are easier to prove and, in each case, we will include a hint about the proofs. Also, in what follows, we will use the term "GLAV schema mapping" for a schema mapping defined by a finite set of GLAV constraints; the terms "GAV schema mapping" and "LAV schema mapping" have an analogous meaning.

3.5.1 LAV Schema Mappings

We begin with a characterization of the class of LAV schema mappings. LAV schema mappings, being a special case of GLAV schema mappings, possess the three structural properties that we discussed earlier: admitting universal solutions, closure under target homomorphisms, and allowing for CQ rewriting. In addition, LAV schema mappings possess a structural properties that GLAV schema mappings, in general, lack, namely, *closure under union.*

A schema mapping M is said to be *closed under union* if, for all source instances I, I' and all target instances J, J', if J is a solution for I with respect to M and J' is a solution for I' with respect to M, then $J \cup J'$ is a solution for $I \cup I'$ with respect to M. Closure under union can be viewed as a form of *modularity*, since it allows for a *divide and conquer* approach that makes it possible to construct a solution for a large source instances out of solutions for smaller subinstances. It is not hard to show that every LAV schema mapping is closed under union. Furthermore, the following characterization holds.

Theorem 3.4 *A schema mapping is definable by a finite set of LAV constraints if and only if it admits universal solutions, is closed under target homomorphisms, allows for CQ rewritings, and is closed under union.*

The proof of this result makes extensive use of the Chandra-Merlin Theorem. In particular, we consider all possible facts over the source schema, of which there are only finitely many up to isomorphism, and for such fact F, we construct a LAV constraint, whose left-hand side is the fact F, and whose right-hand side is the canonical query of a universal solution of the source instance consisting of the single fact F. Using the various properties of the schema mapping (in particular, closure under union, which tells us that the behavior of a schema mapping on an arbitrary source instance is determined by its behavior on source instances consisting of a single fact), it can be shown that the resulting set of LAV constraints defines the schema mapping at hand.

3.5.2 GAV Schema Mappings

Next, let us consider the case of GAV schema mappings. GAV schema mappings, being a special case of GLAV schema mappings, also possess the three structural properties of admitting universal solutions, closure under target homomorphisms, and allowing for CQ rewriting. In addition, GAV schema mappings possess a structural properties that GLAV schema mappings, in general, lack, namely, namely *closure under intersection.* A schema mapping M is said to be *closed under intersection* if for every source instance I and every target instances J, J', if both J and J' are solutions of I with respect to M, then $J \cap J'$ is a solution of I with respect to M. It can be shown that every GAV schema mapping is closed under intersections (this follows from closure under target homomorphism, in combination with the existence of "null-free" universal solutions, i.e., universal solutions that only contain values from the active domain of the source instance). Furthermore, the following characterization holds.

Theorem 3.5 *A schema mapping is definable by a finite set of GAV constraints if and only if it admits universal solutions, is closed under target homomorphisms, allows for CQ rewriting, and is closed under intersection.*

The proof of this result makes extensive use of allowing for CQ rewriting. In particular, we consider all possible queries over the target instance consisting of a single atomic formula (up to isomorphism, there are only finitely many). For each such query $q(\mathbf{x}) = R(\mathbf{x})$, there is a union of $q_1(\mathbf{x}) \cup \cdots \cup q_n(\mathbf{x})$ of conjunctive queries over the source schema that computes the certain answers of q. For each of these conjunctive queries $q_i(\mathbf{x}) = \exists \mathbf{y}\phi_i(\mathbf{x}, \mathbf{y})$, we construct a GAV constraint of the form $\forall \mathbf{xy}(\phi_i(\mathbf{x}, \mathbf{y}) \to R(\mathbf{x}))$, for each $i \leq n$. Using the various properties of the schema mapping, it can be shown that the (finite) set of all GAV constraints obtained in this way defines the schema mapping at hand.

3.5.3 GLAV Schema Mappings

At this point, it may seem natural to expect that GLAV schema mappings are characterized by the properties of admitting universal solutions, closure under target homomorphisms, and allowing for CQ rewriting. Unfortunately, this turns out not to be quite the case.

Example 3.4 The schema mapping defined by the first-order sentence $\exists y \forall x$ $(Px \to Rxy)$ admits universal solutions, is closed under target homomorphisms, and allows for CQ rewriting, but is not definable by any finite set of GLAV constraints.

It should be pointed out that the schema mapping in the preceding Example 3.4 is definable by the infinite set of all GLAV constraints of the form $\forall x_1 \ldots x_n(\bigwedge_i P(x_i)$ $\to \exists y \bigwedge_i R(x, y))$. This is not an accident, because it can be shown that every schema mapping admitting universal solutions, closure under target homomorphisms and allowing for CQ rewriting is definable by a possibly infinite set of GLAV constraints. However, this is not a characterization, because not every schema mapping defined by an infinite set of GLAV constraints admits (finite) universal solutions.

To obtain a structural characterization of GLAV schema mappings, we introduce one further structural property, which can be viewed as a weakening of closure under union. A schema mapping M is said to be *n-modular*, where n is natural number, if whenever a target instance J is not a solution for a source instance I with respect to M, there is a sub-instance $I' \subseteq I$ such that $|adom(I')| \leq n$ and J is not a solution for I' with respect to M. Intuitively, this means that, if J is a solution for every small sub-instance of I, then J is a solution of I. It is not difficult to see that every GLAV schema mapping is *n*-modular for some *n*. Specifically, *n* can be taken to be the maximum number of existentially quantified variables in the GLAV constraints that define the schema mapping. The property of *n*-modularity can be naturally viewed as a generalization of closure under union, provided that *n* is at least as large as the maximum arity of a relation in the source schema.

Theorem 3.6 *A schema mapping is definable by a finite set of GLAV constraints if and only if it admits universal solutions, is closed under target homomorphisms, allows for CQ rewritings, and is n-modular for some n > 0.*

The proof is a generalization of the proof of Theorem 3.4, where, instead of considering source instance containing a single fact, we consider source instances whose active domain has size at most n.

From a model theoretic perspective, the structural property of *allowing for CQ rewriting* looks rather "syntactic". However, it is closely related to a semantic property of *reflecting source homomorphisms*, which roughly states that every homomorphism between two source instances extends to a homomorphism between corresponding universal solutions. We refer to [15] for details. In Theorem 3.4 and in Theorem 3.6, the condition of allowing for CQ rewriting can be replaced by reflecting source homomorphisms. For Theorem 3.5 also, the condition of allowing for CQ rewriting can be replaced by reflecting source homomorphisms, provided that the schema mapping in question is first-order definable; we note that the proof of this makes essential use of results proved by Rossman [51] to obtain his homomorphism preservation theorem.

♣ Example 3.4 suggests that we may be able to extend the language of GLAV constraints with a form of quantifier alternation, while, at the same time, preserving the structural properties of admitting universal solutions, closure under target homomorphisms, and allowing for CQ rewriting. Indeed, it is possible to define a language of *nested GLAV constraints* that possess all these properties.

Nested GLAV constraints are essentially GLAV constraints in which the consequent may contain conjuncts that are themselves again nested GLAV constraints (with free variables). Formally, we use two disjoint sets of variables, universal variables X and existential variables Y, and then a *nested GLAV constraint* is defined as a first-order sentence that is generated by the following grammar (which may generate intermediate formulas with free variables):

$$\chi := \alpha \mid \forall x_1 \ldots x_n (\beta_1 \wedge \cdots \wedge \beta_k \rightarrow \exists y_1 \ldots x_m \chi_1 \wedge \cdots \wedge \chi_\ell),$$

where α is an atomic formula over the target schema, each $x_i \in X$, each $y_i \in Y$, each β_i is an atomic formula over the source schema containing only variables from X, and each variable x_i occurs in some β_j. The first-order formula from Example 3.4 can be equivalently written in this form as

$$\forall x (Px \rightarrow \exists y\, Rxy \wedge \forall x' (Px' \rightarrow Rx'y)).$$

Nested GLAV constraints are in fact incorporated in the schema mapping language used in the data exchange prototype system Clio developed at the IBM Almaden Research Center [26, 30, 35].

Every schema mapping defined by a finite set of nested GLAV constraints admits universal solutions, is closed under target homomorphisms, and allows for CQ rewriting. In ten Cate [15], we had conjectured that these three properties characterize the language of nested tgds. This conjecture has been recently refuted [4]. It remains an open problem to find a structural characterization of schema mappings definable by a finite set of nested tgds.

The language of nested GLAV constraints bears a noticeable syntactic similarity to the language of *flow formulas*, which was introduced in van Benthem [9], in the study of information flow [5], to characterize the first-order formulas that are preserved under Chu-space transformations. One of the differences is that nested GLAV constraints do not contain disjunction. It remains an interesting question whether the apparent similarity between these languages is indicative of a more fundamental relationship between schema mappings and

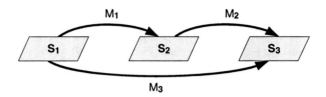

Fig. 3.3 Diagrammatic depiction of the composition of schema mappings

formal frameworks such as that of information flow [5] studied in the literature on philosophy of information.

3.6 Composing Schema Mappings

As seen in previous sections, schema mappings embody logical dynamics. In particular, GLAV schema mappings provide a declarative specification for data exchange that suggests an efficient procedural counterpart (namely, the chase procedure) for performing data exchange. Up to this point, however, our study of schema mappings has taken place in a static context, in the sense that each result presented is about some (arbitrary but) fixed schema mapping. In this section, we will examine schema mappings in a dynamic context in which schema mappings become mathematical objects that can be manipulated, operated on, and transformed.

Schemas and schema mappings can be thought of as *metadata*, since they contain information about data (for example, how data are organized and how they are related). In practice, schemas and, with them, schema mappings do not remain stationary, but, instead, evolve over time. Furthermore, in real-life applications, schemas and schemas mappings are large and complex objects. These and other related considerations served as the motivation for Bernstein [11] to introduce a *model management* framework for manipulating schemas, schema mappings, and other such metadata. The main ingredient of Bernstein's model management framework is a set of *operators* on schema mappings, that is, functions that take one or more schema mappings as arguments and return some other schema mapping as value. Out of these operators, the *composition* operator turned out to be the most fundamental and extensively studied one to date. In this section, we give a brief overview of some of the main results concerning the composition of schema mappings; most of these results are drawn from [23].

The composition of two schema mappings is depicted in Fig. 3.3. Suppose we are given two consecutive schema mappings M_1 and M_2, that is, two schema mappings of the form $M_1 = (S_1, S_2, \Sigma_1)$ and $M_2 = (S_2, S_3, \Sigma_2)$. Intuitively, the *composition* of M_1 and M_2 is a schema mapping M_3 between the schemas S_1 and S_3 that is "equivalent" to the sequential application of M_1 and M_2. The first task at hand is to make this intuition precise and give rigorous semantics for the composition operator on schema mappings.

We begin by considering the composition of abstract schema mappings. Let $M_1 = (S_1, S_2, W_1)$ and $M_2 = (S_2, S_3, W_2)$ be two consecutive abstract schema mappings. Thus, W_1 is a binary relation between S_1-instances and S_2-instances, while W_2 is a binary relation between S_2-instances and S_3-instances. Consequently, W_1 and W_2 can be composed in the standard sense of composition of binary relations. In other words, it is perfectly meaningful to form the (set-theoretic) composition $W_1 \circ W_2$, which is the set of all pairs (I, K) such that I is an S_1-instance, K is an S_3-instance, and there an S_2-instance J such that $(I, J) \in W_1$ and $(J, K) \in W_2$. Moreover, it is easy to see that the binary relation that $W_1 \circ W_2$ is invariant under isomorphisms, because so are the binary relations W_1 and W_2.

If $M_1 = (S_1, S_2, W_1)$ and $M_2 = (S_2, S_3, W_2)$ are two consecutive abstract schema mappings, then the *composition of M_1 and M_2* is defined to be the schema mapping $M_3 = (S_1, S_3, W_1 \circ W_2)$. In what follows, we will write $M_1 \circ M_2$ to denote the composition of M_1 and M_2. The preceding remarks show that $M_1 \circ M_2$ is indeed an abstract schema mapping with S_1 as its source schema and S_3 as its target schema.

♣ Bernstein [11] introduced the composition operator on schema mappings, but did not provide formal semantics for it. The first rigorous semantics for the composition operator were formulated by Madhavan and Halevy [43] and were based on the notion of the certain answers of queries. In particular, the resulting notion of composition depended on the class of queries considered, and was not unique. The semantics of composition we just presented were introduced in Fagin et al. [23] and became the most widely adopted and used ones.

Recall that a syntactically specified schema mapping $M = (S, T, \Sigma)$ can be identified with the abstract schema mapping $M = (S, T, \text{Sem}(M))$, where $\text{Sem}(M)$ is the set of all pairs (I, J) of source and target instances such that $(I, J) \models \Sigma$ (i.e., J is a solution for I with respect to M). Consequently, if $M_1 = (S_1, S_2, \Sigma_1)$ and $M_2 = (S_2, S_3, \Sigma_2)$ are two consecutive schema mappings specified syntactically, then the *composition of M_1 and M_2* is the abstract schema mapping $M_1 \circ M_2 = (S_1, S_3, \text{Sem}(M_1) \circ \text{Sem}(M_2))$. We also say that a syntactically specified schema mapping $M_3 = (S_1, S_3, \Sigma_3)$ is the *composition of M_1 and M_2)* if $\text{Sem}(M_3) = \text{Sem}(M_1) \circ \text{Sem}(M_2)$, which means that the abstract schema mapping associated with M_3 is the composition of M_1 and M_2. It is easy to see that, as a syntactic object, the composition of two schema mapping is unique up to logical equivalence, that is to say, if both $M_3 = (S_1, S_3, \Sigma_3)$ and $M_4 = (S_1, S_3, \Sigma_4)$ are compositions of M_1 and M_2, then Σ_3 and Σ_4 are logically equivalent.

The following questions arise naturally now concerning the interplay between abstract schema mappings and syntactically specified ones. Suppose that $M_1 = (S_1, S_2, \Sigma_1)$ and $M_2 = (S_2, S_3, \Sigma_2)$ are two consecutive schema mappings specified syntactically. What is a suitable language for expressing their composition? In particular, is the composition of two arbitrary GLAV schema mappings again definable by a finite set of GLAV constraints? Equivalently, are GLAV schema mappings *closed under composition*? Note that if the composition $M_1 \circ M_2$ of two GLAV schema mappings $M_1 = (S_1, S_2, \Sigma_1)$ and $M_2 = (S_2, S_3, \Sigma_2)$ is a GLAV schema mapping, then we can exchange data directly from S_1 to S_3 using the chase, and without having to first exchange data from S_1 to S_2, and then from S_2 to S_3.

♣ The study of composition permeates practically every area of mathematics. This has been eloquently expressed by Lawvere and Schanuel [40], who wrote: "The notion of composition of maps leads to the most natural account of fundamental notions of mathematics, from multiplication, addition, and exponentiation, through the basic notions of logic."

Closure under composition is a desirable property of classes of functions. For example, the composition of two continuous functions from the real numbers to the real numbers is a continuous function. Moreover, the composition of two differentiable functions from the real numbers to the real numbers is a differentiable function as well. These fundamental closure properties have played a key role in the development of real analysis. Recall also that the notion of a *category* is based on closure under composition.

Let M_1 be the schema mapping defined by the two GAV constraints:

$$\forall smc(\text{TAKES}(s, m, c) \rightarrow \text{STUDENT}(s, m))$$
$$\forall smc(\text{TAKES}(s, m, c) \rightarrow \text{ENROLLS}(s, c)),$$

where TAKES is a ternary relation symbol with information about students (identified by a unique student identification number), majors, and courses. Similarly, STUDENT is a binary relation symbol with information about students and majors, while ENROLLS is a binary relation symbol with information about students and courses. Let M_2 be the schema mapping defined by the GLAV constraint

$$\forall smc(\text{STUDENT}(s, m) \wedge \text{ENROLLS}(s, c) \rightarrow \exists g\text{RECORD}(s, m, c, g)),$$

where RECORD is a 4-ary relation symbol containing information about students, majors, courses, and grades. It is not too difficult to verify that the composition $M_1 \circ M_2$ is defined by the GLAV constraint:

$$\forall smcm'c'(\text{TAKES}(s, m, c) \wedge \text{TAKES}(s, m', c') \rightarrow \exists g\text{RECORD}(s, m, c', g)).$$

Intuitively, the above GLAV constraint is a correct specification of the composition of M_1 and M_2 because a student may have more than one majors (e.g., mathematics and music) and may take courses under either major. The relation TAKES need not list all combinations of courses and majors for each student, while the GLAV constraint that defines M_2 stipulates that all such combinations must appear in the relation RECORD. This explains the role of the self-join of TAKES in the antecedent of the GLAV constraint that defines $M_1 \circ M_2$, as well as the occurrence of m and c' in the atom in the conclusion of that GLAV constraint.

We have just seen an example of a GAV schema mapping M_1 and a GLAV schema mapping M_2 whose composition $M_1 \circ M_2$ is a GLAV schema mapping. It turns out that this is not an isolated example, but an illustration of a general result about the composition of a GAV schema mapping with a GLAV schema mapping. Moreover, if both schema mappings are GAV, then their composition is also a GAV schema mapping.

Theorem 3.7 Let $M_1 = (S_1, S_2, \Sigma_1)$ and $M_2 = (S_2, S_3, \Sigma_2)$ be two consecutive schema mappings.

1. *If both M_1 and M_2 are GAV schema mappings, then their composition $M_1 \circ M_2$ is also a GAV schema mapping.*
2. *If M_1 is a GAV schema mapping and M_2 is a GLAV schema mapping, then then their composition $M_1 \circ M_2$ is a GLAV schema mapping.*

An immediate consequence of Theorem 3.7 is that if $M_1, \ldots, M_n, M_{n+1}$ is a sequence of consecutive schema mappings such that $M_1, \ldots, M_n$ are GAV schema mappings and M_{n+1} is a GLAV schema mapping, then the composition $M_1 \circ \cdots \circ M_n \circ M_{n+1}$ is a GLAV schema mapping.

As it turns out, closure under composition does not extend to schema mappings belonging to some other combination of the classes of GAV, LAV, and GLAV schema mappings. We begin with an informative example of a LAV schema mapping $M_1 = (S_1, S_2, \Sigma_1)$ and a GAV schema mapping $M_1 = (S_2, S_3, \Sigma_1)$, whose composition is not even first-order definable. The construction entails encoding 3-COLORABILITY, a well known NP-complete problem on graphs, as the composition of a LAV schema mapping and a GAV schema mapping. Schema S_1 consists of a unary relation V denoting the nodes of a graph and a binary relation E denoting the edge relation of a graph $G = (V, E)$. Let M_1 be the schema mapping defined by the LAV constraints

$$\forall x(V(x) \rightarrow \exists u C(x, u))$$
$$\forall xy(E(x, y) \rightarrow F(x, y)),$$

and let M_2 be the schema mapping defined by the GAV constraint

$$\forall xy(F(x, y) \land C(x, u) \land C(y, v) \rightarrow D(u, v)).$$

Intuitively, $C(x, u)$ means that the vertex x of the graph is colored with color u, the relation F is simply used to copy the edge information from the source instance into the intermediate instance, and the relation D captures when two colors are required to be distinct. It can be shown that if I is the S_1-instance consisting of the nodes and the edges of a graph $G = (V, E)$, and if J is the S_3-instance consisting of the facts $D(a, b), D(b, a), D(a, c), D(c, a), D(b, c), D(c, b)$, then the graph G is 3-colorable if and only if J is a solution for I with respect to the composition $M_1 \circ M_2$ of M_1 and M_2. It is well known that no first-order formula defines 3-COLORABILITY on finite graphs; this can be proved using Ehrenfeucht-Fraïssé games. Actually, a much stronger inexpressibility result for 3-COLORABILITY is known. Specifically, Dawar [19] has shown that 3-COLORABILITY is not even expressible in *finite-variable infinitary logic*. In particular, 3-COLORABILITY is not expressible in *least fixed-point logic* LFP, a powerful extension of first-order logic that embodies recursion and also subsumes modal μ-calculus.

What is the "right" language for expressing the composition of two GLAV schema mappings? The preceding discussion shows that GLAV schema mappings are not closed under composition and that, in fact, we must go well beyond first-order logic to find such a language. It can be shown that the composition of two GLAV schema mappings is always expressible by a formula of *existential second-order logic*. As we are about to see, the "right" language for expressing the composition of two GLAV schema mappings is a certain fragment of existential second-order logic, which we will introduce by way of two examples.

To begin with, the composition $M_1 \circ M_2$ encoding 3-COLORABILITY can be defined by the existential second-order sentence

$$\exists f \forall xy(V(x) \land V(y) \land E(x, y) \rightarrow D(f(x), f(y))).$$

This sentence contains most of the constructs of the fragment of existential second-order logic needed for expressing the composition of two GLAV schema mappings: it involves existentially quantified function symbols and its first-order part resembles a (first-order) tuple-generating dependency, but also allows for terms as arguments. The next example will point at additional constructs.

Let S_1 be a schema containing a unary relation Emp listing employees of a company, let S_2 be a schema containing binary relation Mgr listing the manager of each employee, and let S_3 be a schema containing a similar binary relation Rep listing employees and managers, as well as a unary relation SelfMgr listing all employees who are their own manager. Let M_1 be the schema mapping defined by the LAV constraint

$$\forall e(\text{Emp}(e) \rightarrow \exists m\, \text{Mgr}(e, m)),$$

and let M_2 be the schema mapping defined by the LAV constraints

$$\forall em(\text{Mgr}(e, m) \rightarrow \text{Rep}(e, m))$$
$$\forall e(\text{Mgr}(e, e) \rightarrow \text{SelfMgr}(e)).$$

It can be shown that the composition $M_1 \circ M_2$ of M_1 and M_2 is not definable by any set of GLAV constraints, not even an infinite one; in particular, this implies that the composition of two LAV schema mappings need be a GLAV schema mapping. However, the composition $M_1 \circ M_2$ is defined by the existential second-order sentence

$$\exists f(\forall e(\text{Emp}(e) \rightarrow \text{Rep}(e, f(e))) \land \forall e(e = f(e) \rightarrow \text{SelfMgr}(e))).$$

Intuitively, this is a correct specification of the composition of $M_1 \circ M_2$ because if a source instance contains a fact Emp(e), then $M_1 \circ M_2$ must require that the target instance contains Rep(e, m) for a suitable value m. Moreover, either this must be the case for some value m distinct from e or the target instance must be required to contain the fact SelfMgr(e).

The preceding two existential second-order sentences are examples of *second-order tuple-generating dependencies* or, in short, *SO tgds*. The precise definition of an SO tgd can be found in [23], where this class of formulas was introduced and studied. The salient features of SO tgds are as follows: they are formulas of existential second-order logic with existentially quantified function symbols; moreover, their first-order part resembles a conjunction of (first-order) tgds, but allows also for terms as arguments and for atoms involving equalities between terms. Note that the first-order part of the SO tgd expressing 3-COLORABILITY had no equality between terms as a conjunct. In general, however, such equalities are indispensable in the syntax of SO tgds.

Note that every GLAV constraint is logically equivalent to an SO tgd. In fact, it can be shown that every nested GLAV constraint is logically equivalent to an SO tgd. Moreover, a finite conjunction of SO tgds is logically equivalent to a single SO tgd.

The next result asserts that SO tgds are the "right" language for expressing the composition of GLAV schema mappings, and that the theory of data exchange can be extended to SO tgds.

Theorem 3.8 *The following statements are true.*

1. *If M_1 and M_2 are two consecutive GLAV schema mappings, then their composition $M_1 \circ M_2$ is defined by an SO tgd.*
2. *If M is a schema mapping defined by an SO tgd, then there are two consecutive GLAV schema mappings M_1 and M_2 such that $M = M_1 \circ M_2$.*
3. *If M_1 and M_2 are two consecutive schema mappings defined by SO tgds, then their composition $M_1 \circ M_2$ is defined by an SO tgd.*
4. *The chase procedure can be extended to SO tgds. As a result, every schema mapping M defined by an SO tgd admits universal solutions. Moreover, given a source instance I, a universal solution for I with respect to M can be constructed in polynomial time in the size of I.*
5. *If M is a schema mapping defined by an SO tgd, then M allows for CQ rewriting.*

The first two parts of Theorem 3.8 reveal that SO tgds are precisely the closure of GLAV schema mappings under composition. The third part tells that SO tgds are themselves closed under composition, while the fourth part tells that SO tgds are chaseable, hence they admit universal solutions. Moreover, the chase procedure for SO tgds is a polynomial-time algorithm for constructing universal solutions. Consequently, if M is a schema mapping defined by an SO tgd and q is a target conjunctive query, then the certain answers of q with respect to M can be computed in polynomial time. This also follows from the last part of Theorem 3.8, which, in turn, is a consequence of the second part of this theorem and the fact that GLAV schema mappings allow for CQ-rewriting.

Schema mappings defined by SO tgds enjoy some, but, of course, not all structural properties of GLAV schema mappings. In particular, schema mappings defined by SO tgds are not, in general, closed under target homomorphisms. It is an interesting open problem to give a structural characterization of schema mappings defined by SO tgds.

It should be noted that SO tgds have been incorporated in the Clio data exchange system, which is now part of the IBM InfoSphere Data Architect enterprise data modeling and integration tool. Composition of schema mappings is a case study of not only logic *in* computer science, but also logic *from* computer science, since the class of SO tgds represents a well-behaved fragment of second-order logic that was identified during the pursuit of the "right" language for composing schema mappings.

As mentioned in the beginning of this section, composition is one of several key operators on schema mappings proposed by Bernstein [11] in the context of the model management framework. Several other such operators have been also investigated, including the *inverse* operator on schema mappings. Unlike for the composition operator, both the semantics and the language for the inverse operator are not completely settled. In particular, no definitive notion of inverse has emerged; instead, several competing notions, each with its own advantages and disadvantages, have been proposed. We refer the reader to the survey [3] and the chapter [24]. The latter discusses also applications of the composition operator and the inverse operator to *schema evolution*, which is one of the most challenging problems in data inter-operability.

3.7 Concluding Remarks

Schema mappings are the essential building blocks in formalizing and investigating challenging data inter-operability tasks, such as data exchange and data integration. In this chapter, we have attempted to demonstrate that schema mappings can be viewed as a case of logical dynamics in action. In doing so, we focused on individual schema mappings that relate two schemas. Schema mappings have also been used to model the flow of data in a network of peers, where different peers use different schemas to hold data [31, 32]. Different semantics have been explored for query answering in such settings, including semantics based on epistemic logic [14, 42]. We believe that there is room for more interaction between epistemic logic, logical dynamics, and database theory, and we look forward to this happening in the near future.

References

1. Abiteboul S, Hull R, Vianu V (1995) Foundations of databases. Addison-Wesley, Boston
2. Andréka H, van Benthem J, Németi I (1998) Modal languages and bounded fragments of predicate logic. J Philos Logic 27:217–274
3. Arenas M, Pérez J, Reutter JL, Riveros C (2009) Composition and inversion of schema mappings. SIGMOD Rec 38(3):17–28
4. Arenas M, Pérez J, Reutter JL, Riveros C (2013) The language of plain so-tgds: composition, inversion and structural properties. J Comput Syst Sci 79(6):763–784
5. Barwise J, Seligman J (1997) Information flow: the logic of distributed systems., Cambridge tracts in theoretical computer science, Cambridge University Press, Cambridge
6. Beeri C, Vardi MY (1984) A proof procedure for data dependencies. J ACM 31(4):718–741
7. van Benthem J (1983) Modal logic and classical logic. Bibliopolis, Berkeley
8. van Benthem J (1989) Logical constants across varying types. Notre Dame J Formal Logic 30(3):315–342
9. van Benthem J (2000) Information transfer across chu spaces. J Logic IGPL 8(6):719–731
10. van Benthem J (2010) Modal logic for open minds. CSLI lecture notes, Center for the Study of Language and Information
11. Bernstein PA (2003) Applying model management to classical meta data problems. In: Proceedings of the 1st Biennial conference on innovative data systems research (CIDR)
12. Bernstein PA, Haas LM (2008) Information integration in the enterprise. Commun ACM 51(9):72–79

13. Beth EW (1955) Semantic entailment and formal derivability. Meded van de KNAW, Afdeling Letterkunde 18(13):309–42 (Reprinted in 1969, Hintikka J (ed) The philosophy of mathematics, Oxford University Press)
14. Calvanese D, Giacomo GD, Lenzerini M, Rosati V (2004) Logical foundations of peer-to-peer data integration. In: Proceedings of the 23rd ACM SIGMOD-SIGACT-SIGART Symposium on Principles of Database Systems (PODS), pp 241–251
15. ten Cate B, Kolaitis PG (2009) Structural characterizations of schema-mapping languages. In: International conference on database theory, pp 63–72
16. ten Cate B, Kolaitis PG (2010) Structural characterizations of schema-mapping languages. Commun ACM 53(1):101–110
17. Chandra A, Merlin P (1977) Optimal implementation of conjunctive queries in relational databases. In: Proceedings of 9th ACM symposium on theory of computing, pp 77–90
18. Chang CC, Keisler J (1973) Model theory. Number 73 in Studies in Logic and the Foundations of Mathematics, North-Holland (3rd edn, 1990)
19. Dawar A (1998) A restricted second order logic for finite structures. Inf Comput 143(2): 154–174
20. Di Paola RA (1969) The recursive unsolvability of the decision problem for the class of definite formulas. J ACM 16(2):324–327
21. Fagin R, Kolaitis PG, Miller RJ, Popa L (2005) Data exchange: semantics and query answering. Theoret Comput Sci 336(1):89–124
22. Fagin R, Kolaitis PG, Popa L (2005) Data exchange: getting to the core. ACM Trans Database Syst 30(1):174–210
23. Fagin R, Kolaitis PG, Popa L, Tan W-C (2005) Composing schema mappings: second-order dependencies to the rescue. ACM Trans Database Syst 30(4):994–1055
24. Fagin R, Kolaitis PG, Popa L, Tan WC (2011) Schema mapping evolution through composition and inversion. In: Schema matching and mapping. Springer, pp 191–222
25. Fagin R, Vardi MY (1986) The theory of data dependencies—a survey. In: Anshel M, Gewirtz W (eds) Proceedings of symposia in applied mathematics, vol 34. Mathematics of Information Processing American Mathematical Society, Providence, pp 19–71
26. Fuxman A, Hernández MA, Ho CTH, Miller RJ, Papotti P, Popa L (2006) Nested mappings: schema mapping reloaded. In: Proceedings of VLDB, pp 67–78
27. Gottlob G, Leone N, Scarcello F (2001) Hypertree decompositions: a survey. In: Sgall J, Pultr A, Kolman P (eds) MFCS of lecture notes in computer science, vol 2136. Springer, pp 37–57
28. Gottlob G, Leone N, Scarcello F (2002) Hypertree decompositions and tractable queries. J Comput Syst Sci 64(3):579–627
29. Haas LM (2007) Beauty and the beast: the theory and practice of information integration. In: Schwentick T, Suciu D (eds) ICDT, lecture notes in computer science, vol 4353. Springer, pp 28–43
30. Haas LM, Hernández MA, Ho H, Popa L, Roth M (2005) Clio grows up: from research prototype to industrial tool. In: Özcan F (ed) SIGMOD conference, ACM, pp 805–810
31. Halevy AY, Ives ZG, Madhavan J, Mork P, Suciu D, Tatarinov I (2004) The piazza peer data management system. IEEE Trans Knowl Data Eng 16(7):787–798
32. Halevy AY, Ives ZG, Suciu D, Tatarinov I (2005) Schema mediation for large-scale semantic data sharing. VLDB J 14(1):68–83
33. Hell P, Nešetřil J (2004) Graphs and homomorphisms. Oxford lecture series in mathematics and its applications, Oxford University Press
34. Hell P, Nešetřil J (1992) The core of a graph. Discrete Math 109:117–126
35. Hernández MA, Miller RJ, Haas LM (2001) Clio: a semi-automatic tool for schema mapping. In: SIGMOD conference, p 607
36. Hoare CAR (1969) An axiomatic basis for computer programming. Commun. ACM 12(10):576–580
37. Imielinski T, Jr WL (1984) Incomplete information in relational databases. J ACM 31(4): 761–791

38. Kolaitis PG, Vardi MY (2000) Conjunctive-query containment and constraint satisfaction. J Comput Syst Sci 61(2):302–332
39. Kooi B, van Benthem J (2004) Reduction axioms for epistemic actions. In: Schmidt R, Pratt-Hartmann I, Reynolds M, Wansing H (eds) Preliminary proceedings of AiML-2004. Department of Computer Science, University of Manchester, pp 197–211
40. Lawvere FW, Schanuel SS (1997) Conceptual mathematics: a first introduction to category theory. Cambridge University Press, Cambridge
41. Lenzerini M (2002) Data integration: a theoretical perspective. In: Proceedings of principles of database systems, pp 233–246
42. Lenzerini M (2004) Principles of P2P data integration. In: Proceedings of the 3rd International Workshop on Data Integration Over the Web (DIWeb), pp 7–21
43. Madhavan J, Halevy AY (2003) Composing mappings among data sources. In: Proceedings of 29th International Conference on Very Large Data Bases (VLDB), pp 572–583
44. Maier D, Mendelzon AO, Sagiv Y (1979) Testing implications of data dependencies. ACM Trans Database Syst 4(4):455–469
45. Makowsky JA, Vardi MY (1986) On the expressive power of data dependencies. Acta Informatica 23(3):231–244
46. Miller RJ, Haas LM, Hernández MA (2000) Schema mapping as query discovery. In: Abbadi AE, Brodie ML, Chakravarthy S, Dayal U, Kamel N, Schlageter G, Whang K.-Y (eds) Proceedings of 26th International Conference on Very Large Data Bases (VLDB), Morgan Kaufmann, pp 77–88
47. Miller RJ, Hernández MA, Haas LM, Yan L-L, Ho CTH, Fagin R, Popa L (2001) The clio project: managing heterogeneity. SIGMOD Rec 30(1):78–83
48. Pottinger R, Halevy A (2001) Minicon: a scalable algorithm for answering queries using views. VLDB J 10(2–3):182–198
49. Rosen E (1997) Modal logic over finite structures. J Logic Lang Inform 6:427–439
50. Rosen E (2002) Some aspects of model theory and finite structures. Bull Symb Logic 8(3):380–403
51. Rossman B (2008) Homomorphism preservation theorems. J ACM 55(3):15:1–15:53
52. Trakhtenbrot B (1950) Impossibility of an algorithm for the decision problem on finite classes. Dokl Akad Nauk SSSR 70:569–572
53. Van Gelder A, Topor RW (1991) Safety and translation of relational calculus. ACM Trans Database Syst 16(2):235–278
54. Vardi MY (1982) The complexity of relational query languages (extended abstract). In: Proceedings of the 14th annual ACM symposium on theory of computing, STOC '82, ACM, New York, pp 137–146
55. Yannakakis M (1981) Algorithms for acyclic database schemes. In: Proceedings of 7th International Conference on Very Large Data Bases (VLDB), pp 82–94

Chapter 4
On Dependence Logic

Pietro Galliani and Jouko Väänänen

Abstract Dependence logic extends the language of first order logic by means of *dependence atoms* and aims to establish a basic theory of dependence and independence underlying such seemingly unrelated subjects as causality, random variables, bound variables in logic, database theory, the theory of social choice, and even quantum physics. In this work we summarize the setting of dependence logic and recall the main results of this rapidly developing area of research.

4.1 Introduction

The goal of dependence logic is to establish a basic theory of dependence and independence underlying such seemingly unrelated subjects as causality, random variables, bound variables in logic, database theory, the theory of social choice, and even quantum physics. There is an avalanche of new results in this field demonstrating remarkable convergence. The concepts of (in)dependence in the different fields of

J. Väänänen's research partially supported by grant 40734 of the Academy of Finland and the EUROCORES LogICCC LINT programme.
P. Galliani's research partially supported by the EUROCORES LogICCC LINT programme, by the Väisälä Foundation and by by Grant 264917 of the Academy of Finland.

P. Galliani (✉)
Department of Mathematics and Statistics, University of Helsinki, Helsinki, Finland
e-mail: pgallian@gmail.com

J. Väänänen
Department of Mathematics and Statistics, University of Helsinki, Helsinki, Finland
e-mail: jouko.vaananen@helsinki.fi

J. Väänänen
Institute for Logic, Language and Computation, University of Amsterdam, Amsterdam, The Netherlands

A. Baltag and S. Smets (eds.), *Johan van Benthem on Logic and Information Dynamics*, Outstanding Contributions to Logic 5, DOI: 10.1007/978-3-319-06025-5_4, © Springer International Publishing Switzerland 2014

humanities and sciences have surprisingly much in common and a common logic is starting to emerge.

Dependence logic [29] arose from the compositional semantics of Wilfrid Hodges [19] for the independence friendly logic [18, 25]. In dependence logic the basic semantic concept is *not* that of an assignment s satisfying a formula ϕ in a model $\mathfrak{M}$,

$$\mathfrak{M} \models_s \phi,$$

as in first order logic, but rather the concept of a *set* S of assignments satisfying ϕ in $\mathfrak{M}$,

$$\mathfrak{M} \models_S \phi.$$

Defining satisfaction relative to a *set* of assignments opens up the possibility to express dependence phenomena, roughly as passing in propositional logic from one valuation to a Kripke model leads to the possibility to express modality. The focus in dependence logic is not on truth values but on variable values. We are interested in dependencies between individuals rather than between propositions.

In [3] Johan van Benthem writes:

> "Sets of assignments S encode several kinds of 'dependence'
> between variables. There may not be one single intuition.
> 'Dependence' may mean functional dependence
> (if two assignments agree in S on x, they also agree on y),
> but also other kinds of 'correlation' among value ranges.
> ...
> Different dependence relations may have different mathematical
> properties and suggest different logical formalisms."   (4.1)

This is actually how things have turned out. For a start, using the concept of functional dependence it is possible, as Wilfrid Hodges [19] demonstrated, to define compositionally[1] the semantics of independence friendly logic, the extension of first order logic by the quantifier

$$\exists x/y\phi \quad \text{i.e. "there is an } x, \text{ independently of } y, \text{ such that } \phi\text{",}$$

as follows: Suppose S is a team of assignments, a "plural state", in a model $\mathfrak{M}$. Then

$$\mathfrak{M} \models_S \exists x/y\phi$$

if and only if there is another set S' such that

$$\mathfrak{M} \models_{S'} \phi$$

---

[1] Before [19] it was an open question whether a compositional semantics can be given to independence friendly logic.

and the following "transition"-conditions hold:

- If $s \in S$, then there is $s' \in S'$ such that if z is a variable other than x, then $s(z) = s'(z)$.
- If $s' \in S'$, then there is $s \in S$ such that if z is a variable other than x, then $s(z) = s'(z)$.
- If $s, s' \in S'$ and $s(z) = s'(z)$ for all variables other than y or x, then $s(x) = s'(x)$.

In a sense, independence friendly logic is a logical formalism suggested by the functional dependence relation, but its origin is in game theoretical semantics, not in dependence relations. With dependence logic the situation is different. It was directly inspired by the functional dependence relation introduced by Wilfrid Hodges.

Peter van Emde Boas pointed out to the second author in the fall of 2005 that the functional dependence behind dependence logic is known in database theory [2]. This led the second author to realize—eventually—that the dependence we are talking about here is not just about variables in logic but a much more general phenomenon, covering such diverse areas as algebra, statistics, computer science, medicine, biology, social science, etc.

As Johan van Benthem points out in (4.1), there are different dependence intuitions. Of course the same is true of intuitions about independence. For some time it was not clear what would be the most natural concept of *independence*. There was the obvious but rather weak form of independence of x from y as dependence of x on some variable z other than y. Eventually a strong form of independence was introduced in [15], which has led to a breakthrough in our understanding of dependence relations and their role.

We give an overview of some developments in dependence logic (Sect. 4.2) and independence logic (Sect. 4.3). This is a tiny selection, intended for a newcomer, from a rapidly growing literature on the topic. Furthermore, in Sect. 4.4 we discuss conditional independence atoms and we prove a novel result—that is, that conditional and non-conditional independence logic are equivalent. Finally, in Sect. 4.6 we briefly discuss an application of our logics to belief representation.

4.2 Functional Dependence

The approach of [29] is that one should look for the strongest concept of dependence and use it to define weaker versions. Conceivably one could do the opposite, start from the weakest and use it to define stronger and strong concepts. The weakest dependence concept—whatever it is—did not offer itself immediately, so the strongest was more natural to start with. The wisdom of focusing in the extremes lies in the hope that the extremes are most likely to manifest simplicity and robustness, which would make them susceptible to a theoretical study.

Let us start with the strongest form of dependence, functional dependence. We use the vector notation $\vec{x}$ for finite sequences $x_1, \ldots, x_n$ of variables.[2] We add to first order logic[3] new atomic formulas

$$=(\vec{y}, \vec{x}), \tag{4.2}$$

with the intuitive meaning

the $\vec{y}$ totally determine the $\vec{x}$.

In other words, the meaning of (4.2) is that the values of the variables $\vec{y}$ functionally determine the values of the variables $\vec{x}$. We think of the atomic formulas (4.2) on a par with the atomic formula $x = y$. In particular, the idea is that the formula (4.2) is a purely logical expression, not involving any non-logical symbols, in particular no function symbol for the purported function manifesting the functional dependence.

The best way to understand the concept (4.2) is to give it exact semantics: To this end, suppose $\mathfrak{M}$ is a model. Suppose S is a set of assignments into M (or a *team* as such sets are called). We define:

Definition 4.1 The team S satisfies $=(\vec{y}, \vec{x})$ in $\mathfrak{M}$, in symbols

$$\mathfrak{M} \models_S =(\vec{y}, \vec{x})$$

if

$$\forall s, s' \in S(s(\vec{y}) = s'(\vec{y}) \rightarrow s(\vec{x}) = s'(\vec{x})). \tag{4.3}$$

One may ask, why not define the meaning of $=(y, x)$ as "there is a function which maps y to x"? The answer is that if we look at the meaning of $=(y, x)$ under *one* assignment s, then there *always* is a function f mapping $s(y)$ to $s(x)$, namely the function $\{(s(y), s(x))\}$, and if we look at the meaning of $=(y, x)$ under *many* assignments, a team, then (4.3) is indeed equivalent to the statement that there is a function mapping $s(y)$ to $s(x)$ for all s in the team.

A special case of $=(\vec{y}, \vec{x})$ is $=(\vec{x})$, the *constancy atom*. The intuitive meaning of this atom is that the value of $\vec{x}$ is constant in the team. It results from $=(\vec{y}, \vec{x})$ when $\vec{y}$ is the empty sequence.

Functional dependence has been studied in database theory and some basic properties, called **Armstrong's Axioms** have been isolated [2]. These axioms state the following properties of $=(\vec{y}, \vec{x})$:

(A1) $=(\vec{x}, \vec{x})$. Anything is functionally dependent of itself.
(A2) If $=(\vec{y}, \vec{x})$ and $\vec{y} \subseteq \vec{z}$, then $=(\vec{z}, \vec{x})$. Functional dependence is preserved by increasing input data.

---

[2] Or attributes, something that has a value.

[3] The basic ideas can be applied to almost any logic, especially to modal logic.

(A3) If $\vec{y}$ is a permutation of $\vec{z}$, $\vec{u}$ is a permutation of $\vec{x}$, and $=(\vec{z}, \vec{x})$, then $=(\vec{y}, \vec{u})$.
Functional dependence does not look at the order of the variables.

(A4) If $=(\vec{y}, \vec{z})$ and $=(\vec{z}, \vec{x})$, then $=(\vec{y}, \vec{x})$. Functional dependences can be transitively composed.

The following result is well-known in the database community and included in textbooks of database theory[4]:

Theorem 4.2 [2] *The axioms* (A1)–(A4) *are complete in the sense that a relation* $=(\vec{y}, \vec{x})$ *follows by the rules* (A1)–(A4) *from a set* Σ *of relations of the same form if and only if every team which satisfies* Σ *satisfies* $=(\vec{y}, \vec{x})$.

Proof Suppose $=(\vec{y}, \vec{x})$ does not follow by the rules from a set Σ of atoms. Let V be the set of variables z such that $=(\vec{y}, z)$ follows by the rules from Σ. Let W be the remaining variables in $\Sigma \cup \{=(\vec{y}, \vec{x})\}$. Thus $\vec{x} \cap W \neq \emptyset$. Consider the model $\{0, 1\}$ of the empty vocabulary and the team

The variables in V				The variables in W			
0	0	...	0	0	...	...	0
0	0	...	0	1	1	...	1

The atom $=(\vec{y}, \vec{x})$ is not true in this team, because $\vec{y} \subseteq V$ and $\vec{x} \cap W \neq \emptyset$. Suppose then $=(\vec{v}, \vec{w})$ is one of the assumptions. If each v is in V, then so is each w so they all get value 0. On the other hand, if some v is in W, it gets in this team two values, so it cannot violate dependence.   □

We now extend the truth definition (Definition 4.1) to the full first order logic augmented by the dependence atoms $=(\vec{x}, \vec{y})$. To this end, let $s(a/x)$ denote the assignment which agrees with s except that it gives x the value a. We define for formulas which have negation in front of atomic formulas only:

$$
\left.
\begin{aligned}
\mathfrak{M} \models_S x = y \quad &\Longleftrightarrow \forall s \in S(s(x) = s(y)). \\
\mathfrak{M} \models_S \neg x = y \quad &\Longleftrightarrow \forall s \in S(s(x) \neq s(y)). \\
\mathfrak{M} \models_S R(x_1, \ldots, x_n) \quad &\Longleftrightarrow \forall s \in S((s(x_1), \ldots, s(x_n)) \in R^{\mathfrak{M}}). \\
\mathfrak{M} \models_S \neg R(x_1, \ldots, x_n) \quad &\Longleftrightarrow \forall s \in S((s(x_1), \ldots, s(x_n)) \notin R^{\mathfrak{M}}). \\
\mathfrak{M} \models_S \phi \wedge \psi \quad &\Longleftrightarrow \mathfrak{M} \models_S \phi \text{ and } \mathfrak{M} \models_S \psi. \\
\mathfrak{M} \models_S \phi \vee \psi \quad &\Longleftrightarrow \text{There are } S_1 \text{ and } S_2 \text{ such that} \\
&\quad S = S_1 \cup S_2, \mathfrak{M} \models_{S_1} \phi, \text{ and } \mathfrak{M} \models_{S_2} \psi. \\
\mathfrak{M} \models_S \exists x \phi \quad &\Longleftrightarrow \mathfrak{M} \models_{S'} \phi \text{ for some } S' \text{ such that} \\
&\quad \forall s \in S \, \exists a \in M(s(a/x) \in S') \\
\mathfrak{M} \models_S \forall x \phi \quad &\Longleftrightarrow \mathfrak{M} \models_{S'} \phi \text{ for some } S' \text{ such that} \\
&\quad \forall s \in S \, \forall a \in M(s(a/x) \in S')
\end{aligned}
\right\} \quad (4.4)
$$

It is easy to see that for formulas not containing any dependence atoms, that is, for pure first order formulas ϕ,

---

[4] See e.g. [26].

$$\mathfrak{M} \models_{\{s\}} \phi \iff \mathfrak{M} \models_s \phi$$

and

$$\mathfrak{M} \models_S \phi \iff \forall s \in S(\mathfrak{M} \models_s \phi),$$

where $\mathfrak{M} \models_s \phi$ has its usual meaning. This shows that the truth conditions (4.4) agree with the usual Tarski truth conditions for first order formulas. Thus considering the "plural state" S rather than individual "states" s makes no difference for first order logic, but it makes it possible to give the dependence atoms $=(\vec{x}, \vec{y})$ their intended meaning.

What about axioms for non-atomic formulas of dependence logic? Should we adopt new axioms, apart from the Armstrong Axioms [A1–A4]? There is a problem! Consider the sentence

$$\exists x \forall y \exists z (=(z, y) \land \neg z = x). \tag{4.5}$$

It is easy to see that this sentence is satisfied by a team in a model $\mathfrak{M}$ if and only M is infinite. As a result, by general considerations going back to Gödel's Incompleteness Theorem, the semantic consequence relation

$$\phi \models \psi \iff \forall \mathfrak{M} \forall S (\mathfrak{M} \models_S \phi \to \mathfrak{M} \models_S \psi)$$

is non-arithmetical. Thus there cannot be any completeness theorem in the usual sense. However, this does not prevent us from trying to find axioms and rules which are as complete as possible. This is what is done in [24], where a complete axiomatization is given for *first order* consequences of dependence logic sentences. The axioms are a little weaker than standard first order axioms when applied to dependence formulas, but on the other hand there are two special axioms for the purpose of dealing with dependence atoms as parts of formulas in a deduction. Rather than giving all details (which can be found in [24]) we give just an example of the use of both new rules.

Suppose we are given ϵ, x, y and f, and we have already concluded, in the middle of some argument, the following:

> *if $\epsilon > 0$, then there is $\delta > 0$ depending only on ϵ such that*
> *if $|x - y| < \delta$, then $|f(x) - f(y)| < \epsilon$.*

By merely logical reasons we should be able to conclude

> *There is $\delta > 0$ depending only on ϵ such that*
> *if $\epsilon > 0$ and $|x - y| < \delta$, then $|f(x) - f(y)| < \epsilon$.*

Note that "*depending only on ϵ*" has moved from inside the implication to outside of it. The new rule of dependence logic, isolated in [24], which permits this, is called

Dependence Distribution Rule. Neither first order rules nor Armstrong's Axioms give this because neither of them gives any clue of how to deal with dependence atoms as parts of bigger formulas.

Here is another example of inference in dependence logic: Suppose we have arrived at the following formula in the middle of some argument:

> *For every x and every $\epsilon > 0$ there is $\delta > 0$ depending only on ϵ*
> *such that for all y, if $|x - y| < \delta$, then $|f(x) - f(y)| < \epsilon$.*

On merely logical grounds we should be able to make the following conclusion:

> *For every x and every $\epsilon > 0$ there is $\delta > 0$*
> *such that for all y, if $|x - y| < \delta$, then $|f(x) - f(y)| < \epsilon$,*
> *and moreover, for any other x' and $\epsilon' > 0$ there is $\delta' > 0$*
> *such that for all y', if $|x' - y'| < \delta'$, then $|f(x') - f(y')| < \epsilon$*
> *and if $\epsilon = \epsilon'$, then $\delta = \delta'$.*

The new rule, isolated in [24] which permits this step is called *Dependence Elimination Rule*, because the dependence atom "*depending only on ϵ*" has been entirely eliminated. The conclusion is actually first order, that is, without any occurrence of dependence atoms.

The first author [11] has given an alternative complete axiomatization, not for first order consequences of dependence sentences, but for dependence logic consequences of first order sentences. Clearly, more results about partial axiomatizations of the logical consequence relation in dependence logic can be expected in the near future.

An important property of dependence logic is the *downward closure* [20]: If $\mathfrak{M} \models_S \phi$ and $S' \subseteq S$, then $\mathfrak{M} \models_{S'} \phi$. It is a trivial matter to prove this by induction on the length of the formula. Once the downward closure is established it is obvious that we are far from having a negation in the sense of classical logic. Intuitively, dependence is a restriction of freedom (of values of variables in assignments). When the team gets smaller there is even less freedom. This intuition about the nature of dependence prevails in all the logical operations of dependence logic. Since dependence formulas are easily seen to be representable in existential second order logic, the following result shows that downward closure is really *the* essential feature of dependence logic:

Theorem 4.3 [23] *Let us fix a vocabulary L and an n-ary predicate symbol $S \notin L$. Then:*

- *For every L-formula $\phi(x_1, ..., x_n)$ of dependence logic there is an existential second order $L \cup \{S\}$-sentence $\Phi(S)$, closed downward with respect to S, such that for all L-structures M and all teams X:*

$$\mathfrak{M} \models_X \phi(x_1, \ldots, x_n) \iff \mathfrak{M} \models \Phi(X). \qquad (4.6)$$

- *For every existential second order $L \cup \{S\}$-sentence $\Phi(S)$, closed downward with respect to S, there exists an L-formula $\phi(x_1, \ldots, x_n)$ of dependence logic such that (4.6) holds for all L-structures M and all teams $X \neq \emptyset$.*

This shows that dependence logic is maximal with respect to the properties of being expressible in existential second order logic and being downward closed. This theorem is also the source of the main model theoretical properties of dependence logic. The Downward Löwenheim-Skolem Theorem, the Compactness Theorem and the Interpolation Theorem are immediate corollaries. Also, when the above theorem is combined with the Interpolation Theorem of first order logic, we get the fact that dependence logic sentences ϕ for which there exists a dependence logic sentence ψ such that for all $\mathfrak{M}$

$$\mathfrak{M} \models \psi \iff \mathfrak{M} \not\models \phi$$

are first order definable. So not only does dependence logic not have the classical negation, the only sentences that have a classical negation are the first order sentences.

4.3 Independence Logic

Independence logic was introduced in [15]. Before going into the details, let us look at the following precedent:

In [3] Johan van Benthem suggested, as an example of an "other kind of correlation" than functional dependence, the following dependence relation for a team S in a model $\mathfrak{M}$:

$$\exists a \in M \exists b \in M(\{s(x) : s \in S, s(y) = a\} \neq \{s(x) : s \in S, s(y) = b\}). \quad (4.7)$$

The opposite of this would be

$$\forall a \in M \forall b \in M(\{s(x) : s \in S, s(y) = a\} = \{s(x) : s \in S, s(y) = b\}), \quad (4.8)$$

which is a kind of independence of x from y, for if we take $s \in S$ and we are told what $s(y)$ is, we have learnt nothing about $s(x)$, because for each $a \in M$ the set

$$\{s(x) : s \in S, s(y) = a\}$$

is the same. This is the idea behind the independence atom $\vec{x} \perp \vec{y}$: the values of $\vec{x}$ should not reveal anything about the values of $\vec{y}$ and vice versa. More exactly, suppose $\mathfrak{M}$ is a model and S is a team of assignments into M. We define:

Definition 4.4 A team S satisfies the atomic formula $\vec{x} \perp \vec{y}$ in $\mathfrak{M}$ if

$$\forall s, s' \in S \exists s'' \in S(s''(\vec{y}) = s(\vec{y}) \wedge s''(\vec{x}) = s'(\vec{x})). \qquad (4.9)$$

We can immediately observe that a constant variable is independent of every variable, including itself. To see this, suppose x is constant in S. Let y be any variable, possibly $y = x$. If $s, s' \in S$ are given, we need $s'' \in S$ such that $s''(x) = s(x)$ and $s''(y) = s'(y)$. We can simply take $s'' = s'$. Now $s''(x) = s(x)$, because x is constant in S. Of course, $s''(y) = s'(y)$. Conversely, if x is independent of every variable, it is clearly constant, for it would have to be independent of itself, too. So we have

$$= (\vec{x}) \iff \vec{x} \perp \vec{x}.$$

We can also immediately observe the symmetry of independence, because criterion (4.9) is symmetrical in x and y. More exactly, $s''(y) = s(y) \wedge s''(x) = s'(x)$ and $s''(x) = s'(x) \wedge s''(y) = s(y)$ are trivially equivalent.

Dependence atoms were governed by Armstrong's Axioms. Independence atoms have their own axioms introduced in the context of random variables in [14]:

Definition 4.5 The following rules are the **Independence Axioms**

1. $\vec{x} \perp \emptyset$ (Empty Set Rule).
2. If $\vec{x} \perp \vec{y}$, then $\vec{y} \perp \vec{x}$ (Symmetry Rule).
3. If $\vec{x} \perp \vec{y}\vec{z}$, then $\vec{x} \perp \vec{y}$ (Weakening Rule).
4. If $\vec{x} \perp \vec{x}$, then $\vec{x} \perp \vec{y}$ (Constancy Rule).
5. If $\vec{x} \perp \vec{y}$ and $\vec{x}\vec{y} \perp \vec{z}$, then $\vec{x} \perp \vec{y}\vec{z}$ (Exchange Rule).

Note that $xy \perp xy$ is derivable from $x \perp x$ and $y \perp y$, by means of the Empty Set Rule, the Constancy Rule and the Exchange Rule.

It may seem that independence must have much more content than what these four axioms express, but they are actually complete in the following sense[5]:

Theorem 4.6 (Completeness of the Independence Axioms, [14]) *If T is a finite set of independence atoms of the form $\vec{u} \perp \vec{v}$ for various $\vec{u}$ and $\vec{v}$, then $\vec{y} \perp \vec{x}$ follows from T according to the above rules if and only if every team that satisfies T also satisfies $\vec{y} \perp \vec{x}$.*

Proof We adapt the proof of [14] into our framework. Suppose $\vec{x} \perp \vec{y}$ follows semantically from Σ but does not follow by the above rules. W.l.o.g. Σ is closed under the rules. We may assume that $\vec{x}$ and $\vec{y}$ are minimal, that is, if $\vec{x}' \subseteq \vec{x}$ and $\vec{y}' \subseteq \vec{y}$ and at least one containment is proper, then if $\vec{x}' \perp \vec{y}'$ follows from Σ semantically, it also follows by the rules. It is easy to see that if $\Sigma \models u \perp u$, then $\Sigma \vdash u \perp u$.

Suppose $\vec{x} = (x_1, \ldots, x_l)$ and $\vec{y} = (y_1, \ldots, y_m)$. Let $\vec{z} = (z_1, \ldots, z_k)$ be the remaining variables. W.l.o.g., $l \geq 1$ and $m \geq 1$, $x_1 \perp x_1 \notin \Sigma$, and $x_1 \notin \{y_1, \ldots, y_m\}$.

---

[5] This was originally proved for random variables in [14] and then adapted for databases in [22].

We construct a team S in a 2-element model $M = \{0, 1\}$ of the empty vocabulary as follows: We take to S every $s : \vec{x}\vec{y}\vec{z} \to M$, which satisfies $s(u) = 0$ for u such that $u \perp u \in \Sigma$ and in addition

$$s(x_1) = \text{the number of ones in } s[\{x_2, \ldots, x_l, y_1, \ldots, y_m\}] \bmod 2$$

Claim 1 $\vec{x} \perp \vec{y}$ is not true in S. Suppose otherwise. Consider the following two assignments in S:

	x_1	other x_i	y_1	other y_i	other
s	1	0	1	0	0
s'	0	0	0	0	0

If s'' is such that $s''(\vec{x}) = s(\vec{x})$ and $s''(\vec{y}) = s'(\vec{y})$, then $s'' \notin S$. Claim 1 is proved.

Claim 2 S satisfies all the independence atoms in Σ. Suppose $\vec{v} \perp \vec{w} \in S$. If either $\vec{v}$ or $\vec{w}$ contains only variables in Z, then the claim is trivial, as then either $\vec{v}$ or $\vec{w}$ has in S all possible binary sequences. So let us assume that both $\vec{v}$ and $\vec{w}$ meet $\vec{x}\vec{y}$. If $\vec{v}\vec{w}$ does not cover all of $\vec{x}\vec{y}$, then S satisfies $\vec{v} \perp \vec{w}$, because we can fix parity on the variable in $\vec{x}\vec{y}$ which does not occur in $\vec{v}\vec{w}$. So let us assume $\vec{v}\vec{w}$ covers all of $\vec{x}\vec{y}$. Thus $\vec{v} = \vec{x}'\vec{y}'\vec{z}'$ and $\vec{w} = \vec{x}''\vec{y}''\vec{z}''$, where $\vec{x}'\vec{x}'' = \vec{x}$ and $\vec{y}'\vec{y}'' = \vec{y}$. W.l.o.g., $\vec{x}' \neq \emptyset$ and $\vec{x}'\vec{y}' \neq \vec{x}\vec{y}$. By minimality $\vec{x}' \perp \vec{y}' \in \Sigma$ and $\vec{x}'' \perp \vec{y} \in \Sigma$. Since $\vec{v} \perp \vec{w} \in \Sigma$, a couple of applications of the Exchange and Weakening Rules gives $\vec{x}'\vec{y}' \perp \vec{x}''\vec{y}'' \in \Sigma$. But then $\vec{x}'\vec{x}'' \perp \vec{y}'\vec{y}'' \in \Sigma$, contrary to the assumption. $\square$

We can use the conditions (4.4) to extend the truth definition to the entire *independence logic*, i.e. the extension of first order logic by the independence atoms. Can we axiomatize logical consequence in independence logic? The answer is again no, and for the same reason as for dependence logic: Recall that the sentence (4.5) characterizes infinity and ruins any hope to have a completeness theorem for dependence logic. We can do the same using independence atoms:

Lemma 4.7 *The sentence*

$$\exists z \forall x \exists y \forall u \exists v (xy \perp uv \wedge (x = u \leftrightarrow y = v) \wedge \neg v = z) \tag{4.10}$$

is true exactly in infinite models.

The conclusion is that the kind of dependence relation needed for expressing infinity can be realized either by the functional dependence relation or by the independence relation. Another such example is parity in finite models. The following two sentences, the first one with a dependence atom and the second with an independence atom, both express the evenness of the size of a finite model:

$$\forall x \exists y \forall u \exists v (=(u, v) \wedge (x = v \leftrightarrow y = u) \wedge \neg x = y)$$

$$\forall x \exists y \forall u \exists v (xy \perp uv \wedge (x = v \leftrightarrow y = u) \wedge \neg x = y)$$

The fact that we could express, at will, both infinity and evenness by means of either dependence atoms or independence atoms, is not an accident. Dependence logic and independence logic have overall the same expressive power:

Theorem 4.8 *The following are equivalent:*

(1) *K is definable by a sentence of the extension of first order logic by the dependence atoms.*

(2) *K is definable by a sentence of the extension of first order logic by the independence atoms.*

(3) *K is definable in existential second order logic.*

Proof The equivalence of (1) and (3), a consequence of results in [8] and [31], as observed in [20], is proved in [29]. So it suffices to show that (1) implies (2). We give only the main idea. Sentences referred to in (1) have a normal form [29]. Here is an example of a sentence in such a normal form

$$\forall x \forall y \exists v \exists w (=(x, v) \wedge =(y, w) \wedge \phi(x, y, v, w)),$$

where $\phi(x, y, v, w)$ is a quantifier free first order formula. This sentence can be expressed in terms of independence atoms as follows:

$$\forall x \forall y \exists v \exists w (xv \perp y \wedge yw \perp xv \wedge \phi(x, y, v, w)). \qquad \square$$

Note that independence, as we have defined it, is not the negation of dependence. It is rather a very strong denial of dependence. However, there are uses of the concepts of dependence and independence where the negation of dependence *is* the same as independence. An example is vector spaces.

There is an earlier common use of the concept of independence in logic, namely the independence of a set Σ of axioms from each other. This is usually taken to mean that no axiom is provable from the remaining ones. By Gödel's Completeness Theorem this means the same as having for each axiom $\phi \in \Sigma$ a model of the remaining ones $\Sigma \backslash \{\phi\}$ in which ϕ is false. This is not so far from the independence concept $\vec{y} \perp \vec{x}$. Again, the idea is that from the truth of $\Sigma \backslash \{\phi\}$ we can say nothing about the truth-value of ϕ. This is the sense in which Continuum Hypothesis (CH) is independent of ZFC. Knowing the ZFC axioms gives us no clue as to the truth or falsity of CH. In a sense, our independence atom $\vec{y} \perp \vec{x}$ is the familiar concept of independence transferred from the world of formulas to the world of elements of models, from truth values to variable values.

4.4 Conditional Independence

The independence atom $\vec{y} \perp \vec{x}$ turns out to be a special case of the more general atom $\vec{y} \perp_{\vec{x}} \vec{z}$, the intuitive meaning of which is that the variables $\vec{y}$ are totally independent of the variables $\vec{z}$ when the variables $\vec{x}$ are kept fixed (see [15]). Formally,

Definition 4.9 A team S satisfies the atomic formula $\vec{y} \perp_{\vec{x}} \vec{z}$ in $\mathfrak{M}$ if

$$\forall s, s' \in S(s(\vec{x}) = s'(\vec{x}) \rightarrow \exists s'' \in S(s''(\vec{x}\vec{y}) = s(\vec{x}\vec{y}) \wedge s''(\vec{z}) = s'(\vec{z}))).$$

Some of the rules that this "conditional" independence notion obeys are

Reflexivity:	$\vec{x} \perp_{\vec{x}} \vec{y}$,
Symmetry:	If $\vec{y} \perp_{\vec{x}} \vec{z}$, then $\vec{z} \perp_{\vec{x}} \vec{y}$,
Weakening:	If $\vec{y}y' \perp_{\vec{x}} \vec{z}z'$, then $\vec{y} \perp_{\vec{x}} \vec{z}$,
First Transitivity:	If $\vec{x} \perp_{\vec{z}} \vec{y}$ and $\vec{u} \perp_{\vec{z}\vec{x}} \vec{y}$, then $\vec{u} \perp_{\vec{z}} \vec{y}$,
Second Transitivity:	If $\vec{y} \perp_{\vec{z}} \vec{y}$ and $\vec{z}\vec{x} \perp_{\vec{y}} \vec{u}$, then $\vec{x} \perp_{\vec{z}} \vec{u}$,
Exchange:	If $\vec{x} \perp_{\vec{z}} \vec{y}$ and $\vec{x}\vec{y} \perp_{\vec{z}} \vec{u}$, then $\vec{x} \perp_{\vec{z}} \vec{y}\vec{u}$.

Are these axioms complete? More in general, is it possible to find a finite, decidable axiomatization for the consequence relation between conditional independence atoms?

The answer is negative. Indeed, in [16, 17] Hermann proved that the consequence relation between conditional independence atoms is undecidable; and as proved by Parker and Parsaye-Ghomi in [28], it is not possible to find a finite and complete axiomatization for these atoms. However, the consequence relation is recursively enumerable, and in [27] Naumov and Nicholls developed a proof system for it.

The logic obtained by adding conditional independence atoms to first order logic will be called in this paper *conditional independence logic*. It is clear that it contains (nonconditional) independence logic; and furthermore, as discussed in [15], it also contains dependence logic, since a dependence atom $=(\vec{x}, \vec{y})$ can be seen to be equivalent to $\vec{y} \perp_{\vec{x}} \vec{y}$. It is also easy to see that every conditional independence logic sentence is equivalent to some Σ_1^1 sentence, and therefore that conditional independence logic is equivalent to independence logic and dependence logic with respect to sentences.

But this leaves open the question of whether every conditional independence logic formula is equivalent to some independence logic one. In what follows, building on the analysis of the expressive power of conditional independence logic of [10],[6] we prove that independence logic and conditional independence logic are indeed equivalent.

In order to give our equivalence proof we first need to mention two other atoms, the inclusion atom $\vec{x} \subseteq \vec{y}$ and the exclusion atom $\vec{x} \mid \vec{y}$. These atoms correspond to the database-theoretic inclusion [4, 9] and exclusion [5] dependencies, and hold in a

---

[6] In that paper, conditional independence logic is simply called "independence logic". After all, the two logics are equivalent.

team if and only if no possible value for $\vec{x}$ is also a possible value for $\vec{y}$ and if every possible value for $\vec{x}$ is a possible value for $\vec{y}$ respectively. More formally,

Definition 4.10 A team S satisfies the atomic formula $\vec{x} \subseteq \vec{y}$ in $\mathfrak{M}$ if

$$\forall s \in S \exists s' \in S(s'(\vec{y}) = s(\vec{x}))$$

and it satisfies the atomic formula $\vec{x} \mid \vec{y}$ in $\mathfrak{M}$ if

$$\forall s, s' \in S(s(\vec{x}) \neq s'(\vec{y})).$$

As proved in [10],

1. Exclusion logic (that is, first order logic plus exclusion atoms) is equivalent to dependence logic;
2. Inclusion logic (that is, first order logic plus inclusion atoms) is not comparable with dependence logic, but is contained in (nonconditional) independence logic;
3. Inclusion/exclusion logic (that is, first-order logic plus inclusion and exclusion atoms) is equivalent to *conditional* independence logic (that is, first-order logic plus conditional independence atoms $\vec{y} \perp_{\vec{x}} \vec{z}$).

Thus, if we can show that exclusion atoms can be defined in terms of (nonconditional) independence atoms and of inclusion atoms, we can obtain at once that independence logic contains conditional independence logic (and, therefore, is equivalent to it). But this is not difficult: indeed, the exclusion atom $\vec{x} \mid \vec{y}$ is equivalent to the expression

$$\exists \vec{z}(\vec{x} \subseteq \vec{z} \land \vec{y} \perp \vec{z} \land \vec{y} \neq \vec{z}).$$

This can be verified by checking the satisfaction conditions of this formula. But more informally speaking, the reason why this expression is equivalent to $\vec{x} \mid \vec{y}$ is that it states that that every possible value of $\vec{x}$ is also a possible value for $\vec{z}$, that $\vec{y}$ and $\vec{z}$ are independent (and therefore, any possible value of $\vec{y}$ must occur together with any possible value of $\vec{z}$), and that $\vec{y}$ is always different from $\vec{z}$. Such a $\vec{z}$ may exist if and only if no possible value of $\vec{x}$ is also a possible value of $\vec{y}$, that is, if and only if $\vec{x} \mid \vec{y}$ holds.

Hence we may conclude at once that

Theorem 4.11 *Every conditional independence logic formula is equivalent to some independence logic formula.*

In [10] it was also shown the following analogue of Theorem 4.3:

Theorem 4.12 *Let us fix a vocabulary L and an n-ary predicate symbol $S \notin L$. Then:*

- *For every L-formula $\phi(x_1, \ldots, x_n)$ of conditional independence logic there is an existential second order $L \cup \{S\}$-sentence $\Phi(S)$ such that for all L-structures M and all teams X:*

$$\mathfrak{M} \models_X \phi(x_1, \ldots, x_n) \iff \mathfrak{M} \models \Phi(X). \tag{4.11}$$

- *For every existential second order $L \cup \{S\}$-sentence $\Phi(S)$ there exists an L-formula $\phi(x_1, \ldots, x_n)$ of conditional independence logic such that (4.11) holds for all L-structures M and all teams $X \neq \emptyset$.*

Due to the equivalence between independence logic and conditional independence logic, the same result holds if we only allow nonconditional independence atoms. In particular, this implies that over finite models independence logic captures precisely the NP properties of teams.

4.5 Further Expressivity Results

The results mentioned in the above section left open the question of the precise expressive power of inclusion logic. This was answered in [14], in which a connection was found between inclusion logic and positive greatest fixed point logic GFP⁺. In brief, GFP⁺ is the logic obtained by adding to the language of first order logic the operator

$$[\mathrm{gfp}_{R, \vec{x}} \psi(R, \vec{x})](\vec{t}),$$

which asserts that the value of $\vec{t}$ is in the greatest fixed point of the operator $O(R) = \{\vec{a} : \psi(R, \vec{a})\}$,[7] and further requiring that no such operator occurs negatively.

Over finite models, it is known by [21] that this logic is equivalent to the better-known *least fixed point logic* LFP, which captures PTIME over linearly ordered models [21, 30]. Thus, the same is true of inclusion logic: more precisely,

Theorem 4.13 *A class of linearly ordered finite models is definable in inclusion logic if and only if it can be recognized in polynomial time.*

As dependence logic is equivalent to existential second order logic over sentences, it follows at once that there exists a fragment of dependence logic which also captures PTIME. Which fragment may it be? This is answered—in a slightly different setting—by Ebbing, Kontinen, Müller and Vollmer in [7] by introducing the class of *D^*-Horn formulas* and proving that they capture PTIME over *successor structures*.[8]

The complexity-theoretic properties of further fragments of these logics have been studied in the papers [6] and [13], in which *hierarchy theorems* are developed for the restrictions of these logics to dependencies of certain maximum lengths or to maximum numbers of quantifiers. We will not however present here a complete summary of these results, and we refer the interested readers to these papers for the details.

---

[7] In order to guarantee that such a fixed point exist, R is required to appear only positively in ψ.

[8] That is, over finite structures with a built-in successor operator and two constants for the least and greatest elements.

Finally, yet another direction of investigation consists in the search for *weak* dependency notions, which if added to the language of first order logic do not increase its expressive power (with respect to sentences). This problem in studied in [12], in which a fairly general class of such dependencies is found. One particularly interesting result along these lines is that the contradictory negations of functional dependence, inclusion, exclusion and (conditional or non-conditional) independence atoms are all weak and do not increase the expressive power of first order logic. This is somewhat surprising, since as we saw the corresponding non-negated atoms greatly increase the expressivity of first order logic instead; and the study of the manner in which applying the contradictory negation acts on dependence notions and on the logics they generate is an intriguing and largely unexplored avenue of research.

4.6 Belief Representation and Belief Dynamics

Given a model $\mathfrak{M}$, a variable assignment s admits a natural interpretation as the representation of a possible *state of things*, where, for every variable v, the value $s(v)$ corresponds to a specific *fact* concerning the world. To use the example discussed in Chap. 7 of [11], let the elements of $\mathfrak{M}$ correspond to the participants to a competition: then the values of the variables x_1, x_2 and x_3 in an assignment s may correspond respectively to the first-, second- and the third-placed players.

With respect to the usual semantics for first order logic, a first order formula represents a *condition* over assignments. For example, the formula

$$\phi(x_1, x_2, x_3) := (\neg x_1 = x_2) \wedge (\neg x_2 = x_3) \wedge (\neg x_1 = x_3)$$

represents the (very reasonable) assertion according to which the winner, the second-placed player and the third-placed player are all distinct.

Now, a team S, being a set of assignments, represents a set of states of things. Hence, a team may be interpreted as the *belief set* of an agent α: $s \in S$ if and only if the agent α believes s to be possible. Moving from assignments to teams, it is possible to associate to each formula ϕ and model $\mathfrak{M}$ the family of teams $\{S : \mathfrak{M} \models_S \phi\}$, and this allows us to interpret formulas as *conditions over belief sets*: in our example, $\mathfrak{M} \models_S \phi(x_1, x_2, x_3)$ if and only if $\mathfrak{M} \models_s \phi(x_1, x_2, x_3)$ for all $s \in S$, that is, if and only if our agent α believes that the winner, the second-placed player and the third-placed player will all be distinct.

However, there is much that first order logic cannot express regarding the beliefs of our agent. For example, there is no way to represent the assertion that the agent α *knows* who the winner of the competition will be: indeed, suppose that a first order formula θ represents such a property, and let s_1 and s_2 be any two assignments with $s_1(x_1) \neq s_2(x_1)$, corresponding to two possible states of things which disagree with respect to the identity of the winner. Then, for $S_1 = \{s_1\}$ and $S_2 = \{s_2\}$, we should have that $\mathfrak{M} \models_{S_1} \theta$ and that $\mathfrak{M} \models_{S_2} \theta$: indeed, both S_1 and S_2 correspond to belief sets in which the winner is known to α (and is respectively $s_1(x_1)$ or $s_2(x_1)$). But since a team S satisfies a first order formula if and only if all of its assignments satisfy it, this implies that $M \models_{S_1 \cup S_2} \theta$; and this is unacceptable, because if our agent α believes both s_1 and s_2 to be possible then she does not know whether the winner will be $s_1(x_1)$ or $s_2(x_1)$.

How to represent this notion of knowledge? The solution, it is easy to see, consists in adding *constancy atoms* to our language: indeed, $\mathfrak{M} \models_S =(x_1)$ if and only if for any two assignments $s, s' \in S$ we have that $s(x_1) = s'(x_1)$, that is, if and only if all states of things the agent α consider possible agree with respect to the identity of the winner of the competition. What if, instead, our agent could infer the identity of the winner from the identity of the second- and third-placed participants? Then we would have that $\mathfrak{M} \models_S =(x_2 x_3, x_1)$, since any two states of things which the agent considered possible and which agreed with respect to the identity of the second- and third-placed participants would also agree with respect to the identity of the winner. More in general, a dependence atom $=(\vec{y}, \vec{x})$ describes a form of *conditional knowledge*: $\mathfrak{M} \models_S =(\vec{y}, \vec{x})$ if and only if S corresponds to the belief state of an agent who would be able to deduce the value of $\vec{x}$ from the value of $\vec{y}$.

On the other hand, independence atoms represent situations of *informational independence*: for example, if $\mathfrak{M} \models_S x_1 \perp x_3$ then, by learning the identity of the third-placed player, our agent could infer nothing at all about the identity of the winner. Indeed, suppose that, according to our agent, it is possible that A will win (that is, there is a $s \in S$ with $s(x_1) = A$) and it is possible that B will place third (that is, there is a $s' \in S$ such that $s'(x_3) = B$). Then, by the satisfaction conditions of the independence atom, there is also a $s'' \in S$ such that $s''(x_1) = A$ and $s''(x_3) = B$: in other words, it is possible that A will be the winner *and* B will place third, and telling our agent that B will indeed place third will not allow her to remove A from her list of possible winners.

Thus, it seems that dependence and independence logic, or at least fragments thereof, may be interpreted as *belief description languages*. This line of investigation is pursued further in [11]: here it will suffice to discuss the interpretation of the *linear implication*[9] $\phi \multimap \psi$, a connective introduced in [1] whose semantics is given by

$$\mathfrak{M} \models_S \phi \multimap \psi \Leftrightarrow \text{ for all } S' \text{ such that } \mathfrak{M} \models_{S'} \phi \text{ it holds that } \mathfrak{M} \models_{S \cup S'} \psi.$$

How to understand this connective? Suppose that our agent α, whose belief state is represented by the team S, interacts with another agent β, whose belief state is represented by the team S': one natural outcome of this interaction may be represented by the team $S \cup S'$, corresponding to the set of all states of things that α *or* β consider possible. Then stating that a team S satisfies $\phi \multimap \psi$ corresponds to asserting that whenever our agent α interacts with another agent β whose belief state satisfies ϕ, the result of the interaction will be a belief state satisfying ψ: in other words, using

---

[9] The name "linear implication" is due to the similarity between the satisfaction conditions of this connective and the ones of the implication of linear logic. Another similarity is the following Galois connection: $\theta \models \phi \multimap \psi \iff \theta \vee \phi \models \psi$ [1].

the linear implication connective allows us to formulate predictions concerning the future *evolution* of the belief state of our agent.

One can, of course, consider other forms of interactions between agents and further connectives; and quantifiers can also be given natural interpretations in terms of belief updates (the universal quantifier $\forall v$, for example, can be understood in terms of the agent α *doubting* her beliefs about v). But what we want to emphasize here, beyond the interpretations of the specific connectives, is that team-based semantics offers a very general and powerful framework for the representation of beliefs and belief updates, and that notions of dependence and independence arise naturally under such an interpretation. This opens up some fascinating—and, so far, relatively unexplored—avenues of research, such as for example a more in-depth investigation of the relationship between dependence/independence logic and dynamic epistemic logic (DEL) and other logics of knowledge and belief; and, furthermore, it suggests that epistemic and doxastic ideas may offer some useful inspiration for the formulation and analysis of further notions of dependence and independence.

4.7 Concluding Remarks

We hope to have demonstrated that both dependence and independence can be given a logical analysis by moving in semantics from single states s to plural states S. Future work will perhaps show that allowing limited transitions from one plural state to another may lead to decidability results concerning dependence and independence logic, a suggestion of Johan van Benthem.

Furthermore, we proved the equivalence between conditional independence logic and independence logic, thus giving a novel contribution to the problem of characterizing the relations between extensions of dependence logic.

Finally, we discussed how team-based semantics may be understood as a very general framework for the representation of beliefs and belief updates and how notions of dependence and independence may be understood under this interpretation. This suggests the existence of intriguing connections between dependence and independence logic and other formalisms for belief knowledge representation, as well as a possible application for this fascinating family of logics.

References

1. Abramsky S, Väänänen J (2009) From IF to BI. Synthese 167(2):207–230. doi:10.1007/s11229-008-9415-6
2. Armstrong WW (1974) Dependency structures of data base relationships. Inf Process 74
3. van Benthem J (1997) Modal foundations for predicate logic. Log J IGPL 5(2):259–286 (electronic). doi:10.1093/jigpal/5.2.259
4. Casanova MA, Fagin R, Papadimitriou CH (1982) Inclusion dependencies and their interaction with functional dependencies. In: Proceedings of the 1st ACM SIGACT-SIGMOD sympo-

sium on principles of database systems, PODS'82. ACM, New York, NY, USA, pp 171–176. doi:10.1145/588111.588141

5. Casanova MA, Vidal VMP (1983) Towards a sound view integration methodology. In: Proceedings of the 2nd ACM SIGACT-SIGMOD symposium on principles of database systems, PODS'83. ACM, New York, NY, USA, pp 36–47. doi:10.1145/588058.588065

6. Durand A, Kontinen J (2012) Hierarchies in dependence logic. ACM Trans Comput Log (TOCL) 13(4):31. doi:http://dx.doi.org/10.1145/2362355.2362359

7. Ebbing J, Kontinen J, Müller J-S, Vollmer H (2012) A fragment of dependence logic capturing polynomial time. arXiv:1210.3321. URL: http://arxiv.org/abs/1210.3321

8. Enderton HB (1970) Finite partially-ordered quantifiers. Math Log Q 16(8):393–397. doi:10.1002/malq.19700160802

9. Fagin R (1981) A normal form for relational databases that is based on domains and keys. ACM Trans Database Syst 6:387–415. doi:10.1145/319587.319592

10. Galliani P (2012) Inclusion and exclusion dependencies in team semantics: on some logics of imperfect information. Ann Pure Appl Log 163(1):68–84. doi:10.1016/j.apal.2011.08.005

11. Galliani P (2012) The dynamics of imperfect information. PhD thesis, University of Amsterdam, September 2012. URL: http://dare.uva.nl/record/425951

12. Galliani P (2013) Upwards closed dependencies in team semantics. In: Puppis G, Villa T (eds) Proceedings fourth international symposium on games, automata, logics and formal verification, vol 119 of EPTCS, pp 93–106. doi:http://dx.doi.org/10.4204/EPTCS.119

13. Galliani P, Hannula M, Kontinen J (2012) Hierarchies in independence logic. In: Rocca SRD (ed) Computer science logic 2013 (CSL 2013), Leibniz international proceedings in informatics (LIPIcs), vol 23. Schloss Dagstuhl-Leibniz-Zentrum fuer Informatik, Dagstuhl, Germany, pp 263–280. doi:http://dx.doi.org/10.4230/LIPIcs.CSL.2013.263

14. Galliani P, Hella L (2013) Inclusion logic and fixed point logic. In: Rocca SRD (ed) Computer science logic 2013 (CSL 2013), Leibniz international proceedings in informatics (LIPIcs), vol 23. Schloss Dagstuhl-Leibniz-Zentrum fuer Informatik, Dagstuhl, Germany, pp 281–295. doi:http://dx.doi.org/10.4230/LIPIcs.CSL.2013.281

15. Grädel E, Väänänen J (2013) Dependence and independence. Studia Logica 101(2):399–410. doi:10.1007/s11225-013-9479-2

16. Herrmann C (1995) Corrigendum: corrigendum to "on the undecidability of implications between embedded multivalued database dependencies" [Inf Comput 122:221–235 (1995)]. Inf Comput 204(12):1847–1851. doi:10.1016/j.ic.2006.09.002

17. Herrmann C (1995) On the undecidability of implications between embedded multivalued database dependencies. Inf Comput 122(2):221–235. doi:10.1006/inco.1995.1148

18. Hintikka J, Sandu G (1989) Informational independence as a semantic phenomenon. In: Fenstad JE, Frolov IT, Hilpinen R (eds) Logic, methodology and philosophy of science, pp. 571–589. Elsevier, Amsterdam. doi:10.1016/S0049-237X(08)70066-1

19. Hodges W (1997) Compositional semantics for a language of imperfect information. Log J IGPL 5(4):539–563 (electronic). doi:10.1093/jigpal/5.4.539

20. Hodges W (1997) Some strange quantifiers. In: Structures in logic and computer science, Lecture Notes in Computer Science, vol 1261. Springer, Berlin, pp 51–65. doi:10.1007/3-540-63246-8_4

21. Immerman N (1986) Relational queries computable in polynomial time. Inf Control 68(1):86–104

22. Kontinen J, Link S, Väänänen J (2013) Independence in database relations. In: Proceedings of the 20th International workshop (WOLLIC) on logic, language, information, and computation. Lecture notes in computer science, vol 8071. Springer, pp 152–156

23. Kontinen J, Väänänen J (2009) On definability in dependence logic. J Log Lang Inf 18(3):317–332. doi:10.1007/s10849-009-9082-0

24. Kontinen J, Väänänen J (2013) Axiomatizing first order consequences in dependence logic. Ann Pure Appl Log 164:1101–1117. URL: http://arxiv.org/abs/1208.0176

25. Mann AL, Sandu G, Sevenster M (2011) Independence-friendly logic: A game-theoretic approach. In: London Mathematical Society lecture note series, vol 386. Cambridge University Press, Cambridge. doi:10.1017/CBO9780511981418

26. Mannila H, Räihä KJ (1992) The design of relational databases. Addison-Wesley, Cambridge
27. Naumov P, Nicholls B (2013) Re axiomatization of conditional independence. In: TARK, 2013. http://www2.mcdaniel.edu/pnaumov/papers/2013.tark.nn.pdf
28. Parker D Jr, Parsaye-Ghomi K (1980) Inferences involving embedded multivalued dependencies and transitive dependencies. In: Proceedings of the 1980 ACM SIGMOD international conference on management of data, ACM, pp 52–57. doi:10.1145/582250.582259
29. Väänänen J (2007) Dependence logic. In: London Mathematical Society student texts, vol 70. Cambridge University Press, Cambridge. doi:10.1017/CBO9780511611193
30. Vardi MY (1982) The complexity of relational query languages. In: Proceedings of the fourteenth annual ACM symposium on theory of computing. ACM, pp 137–146
31. Walkoe WJ Jr (1970) Finite partially-ordered quantification. J Symbolic Log 35:535–555. doi:10.2307/2271440

Chapter 5
Intensionality, Definability and Computation

Samson Abramsky

Abstract We look at intensionality from the perspective of computation. In particular, we review how game semantics has been used to characterize the sequential functional processes, leading to powerful and flexible methods for constructing fully abstract models of programming languages, with applications in program analysis and verification. In a broader context, we can regard game semantics as a first step towards developing a positive theory of intensional structures with a robust mathematical structure, and finding the right notions of invariance for these structures.

5.1 Introduction

Our aim in this paper is to give a conceptual discussion of some issues concerning intensionality, definability and computation. Intensionality remains an elusive concept in logical theory, but actually becomes much more tangible and, indeed, inescapable in the context of computation. We will focus on a particular thread of ideas, leading to recent and ongoing work in game semantics. Technical details will be kept to a minimum, while ample references will be provided to the literature.

It is a pleasure and a privilege to contribute to a volume in honour of Johan van Benthem. Johan has been a friend and an inspiring and supportive colleague for well over a decade. He has encouraged me to broaden my intellectual horizons and to address wider conceptual issues. Although our technical languages and backgrounds are rather different, we have found a great deal of common ground in pursuing a broad and inclusive vision of logical dynamics. I look forward to continuing our interactions and discussions over the coming years.

S. Abramsky (✉)
Department of Computer Science, University of Oxford, Wolfson Building, Parks Road, Oxford, OX1 3QD, UK
e-mail: samson.abramsky@cs.ox.ac.uk

A. Baltag and S. Smets (eds.), *Johan van Benthem on Logic*
and Information Dynamics, Outstanding Contributions to Logic 5,
DOI: 10.1007/978-3-319-06025-5_5, © Springer International Publishing Switzerland 2014

5.1.1 Computability Versus Computer Science

Computability theory [21, 68] is concerned with the computability (or, most often, degree of **non**-computability) of **extensional objects**: numbers, sets, functions etc. These objects are inherited from mathematics and logic. To define when such objects are computable requires some additional structure: we say e.g. that a function is computable if there exists an **algorithmic process** for computing it. Hence the notion of computability relies on a characterization of algorithmic processes. This was, famously, what Turing achieved in his compelling analysis [70].

Computer Science asks a broader question:

> What is a process?

By contrast with the well-established extensional notions which provide the reference points for computability, there was no established mathematical theory of what processes are predating computer science.

5.1.2 Why Processes Matter in Computer Science

Let us pause to ask why Computer Science asks this broader question about the nature of informatic processes. The purpose of much of the software we routinely run is not to compute a function, but to **exhibit some behaviour**. Think of communication protocols, operating systems, browsers, iTunes, Facebook, Twitter, ... The purpose of these systems is not adequately described as the computation of some function.

Thus we are led ineluctably to questions such as:

> What is a process? When are two processes equivalent?

The situation is very different to that which we find in computability theory, where we have

- A confluence of notions, whereby many different attempts to characterize the notion of algorithmic process have converged to yield the same class of computable functions [21, 68].
- A definitive calculus of functions: the λ-calculus [17, 19].

There has been active research on concurrency theory and processes for the past five decades in Computer Science [31, 55, 56, 65, 66]. Many important concepts and results have emerged. However, one cannot help noticing that:

- Hundreds of different process calculi, equivalences, logics have been proposed.
- No λ-calculus for concurrency has emerged.
- There is no plausible Church-Turing thesis for processes.

This has been referred to as the 'next 700' syndrome, after Peter Landin's paper (from 1966!) on 'The Next 700 Programming Languages' [47]; for a discussion of this syndrome, see [4]. Finding an adequate characterization of informatic processes in general can plausibly be considered to be a much harder problem than that of characterizing the notion of computable set or function. Despite the great achievements of Petri, Milner, Hoare et al., we still await our modern-day Turing to provide a definitive analysis of this notion.

5.1.3 Prospectus

Our aim in this paper is to tell one limited but encouraging success story: the characterization of **sequential functional processes** using **game semantics**, solving in best possible terms the 'full abstraction problem' for PCF [54, 67]. This has led on to many further developments, notably:

- Full abstraction and full completeness results for a wide range of programming languages, type theories and logics [9–15, 32, 35, 44, 46].
- A basis for compositional program verification [8, 24].

There is much ongoing work [58–60], and this continues to be a flourishing field.[1]

In a broader context, we can regard game semantics as a first step towards developing a positive theory of intensional structures with a robust mathematical structure, and finding the right notions of invariance.

5.2 Intensionality Versus Extensionality

The notions of intensionality and extensionality carry symmetric-sounding names, but this apparent symmetry is misleading. Extensionality is enshrined in mathematically precise axioms with a clear conceptual meaning. Intensionality, by contrast, remains elusive. It is a "loose baggy monster"[2] into which all manner of notions may be stuffed, and a compelling and coherent general framework for intensional concepts is still to emerge.

Let us recall some basic forms of extensionality. For sets we have:

$$x = y \quad \leftrightarrow \quad \forall z. z \in x \leftrightarrow z \in y.$$

This says that a set is completely characterized by its **members**.

For functions we have:

---

[1] See e.g. https://sites.google.com/site/galopws/ for a workshop series devoted to this topic.

[2] Cf. Henry James on the Russian masters.

$$f = g \iff \forall x.\, f(x) = g(x).$$

This says that a function is completely characterized by the **input-output corre-spondence** (i. e. the set of input-output pairs) which it defines.

The common idea underlying these principles is that mathematical entities can be characterized in **operational** terms: mathematical objects should be completely determined by their **behaviour** under the operations which can be applied to them. In the case of sets, this operation is testing elements for membership of the set, while in the case of functions, it is applying functions to their arguments.

The basic question we are faced with in seeking to make room for intensional notions is this: is intensionality just a **failure** to satisfy such properties? Or is there some **positive story** to tell?

We shall focus on the case of functions. Here we can say that the modern, exten-sional view of functions as completely determined by their graphs of input-output correspondences over-rode an older, intensional notion, of a function being given by its **rule**. That older notion was never adequately formalized. Much of modern logic, in its concern with issues of definability, can be seen as providing tools for capturing the old intuitions in a more adequate fashion.

5.2.1 Intrinsic Versus Extrinsic Properties of Functions

We shall now draw a distinction which will be useful in our discussion. We say that a property of functions

$$f : A \to B$$

is **intrinsic** if it can be defined purely in terms of f (as a set of input-output pairs) and any structure pertaining to A and B.

This is, of course, not very precise. More formally, we could say: if it can be defined using only bounded quantification over the structures A and B. But we shall rest content with the informal rendition here. We believe that the distinction will be quite tangible to readers with some mathematical experience.

5.2.2 Examples

- A and B are groups. A function $f : A \to B$ is a **group homomorphism** if

$$f(xy) = f(x)f(y), \qquad f(e) = e.$$

In general, homomorphisms of algebraic structures are clearly intrinsic in the sense we intend.

- *A* and *B* are topological spaces. A function $f : A \to B$ is **continuous** if $f^{-1}(U)$ is open in the topology on *A* for every open subset *U* of *B*. Here the topology is viewed as part of the structure.

5.2.3 A Non-example: Computability

Computability is **not** an intrinsic property of partial functions

$$f : \mathbb{N} \rightharpoonup \mathbb{N}$$

in this sense. In order to define whether a function is computable, we need to refer to something **external**; a **process** by which *f* is computed. A function is computable if there is some **algorithmic process** which computes *f*. But what is an algorithmic process?

Here of course we can appeal to Turing's analysis [70], and to subsequent, axiomatic studies by Gandy et al. [23, 69]. But note that, not only do we have to appeal to some external notion of machine or algorithmic process, but there is **no single canonical form** of external structure witnessing computability. Rather, we have a **confluence**: all 'reasonable' notions lead to the same class of computable functions. But this confluence still leaves us with an extrinsic definition, and moreover one in which the external witness has no canonical form.

More concretely, suppose we are given some function

$$f : \{0, 1\}^* \longrightarrow \{0, 1\}$$

i. e. a predicate on binary strings. To say that whether such a function is computable is an extrinsic property of *f* simply means that we cannot say if *f* is computable just by looking at its input-output graph and properties relating to the structure of $\{0, 1\}^*$ and $\{0, 1\}$. Clearly, we need something more, typically either:

- A suitable notion of **machine**, or
- An inductive definition given by some 'function algebra' or logical theory.

5.2.4 A Comparison Point: Regular Languages

One might reasonably ask what it would even mean to have an intrinsic (or, at least, **more** intrinsic) way of defining computability. For this purpose, it is useful to consider a much simpler notion which pertains to (sub-)computability in a non-trivial fashion, and which **does** admit an intrinsic definition in our sense.

Such an example is provided by regular languages, i. e. those accepted by finite-state automata [33].

Let A be some finite alphabet. We write A^* for the set of finite words or strings over this alphabet. Given a language $L \subseteq A^*$, we define

$$s \equiv_L t \quad \leftrightarrow \quad \forall v \in A^*. sv \in L \leftrightarrow tv \in L.$$

Clearly, $\equiv_L$ is an equivalence relation on A^*.

Theorem 5.1 (Myhill-Nerode [63, 64]) *L is regular if and only if $\equiv_L$ is of finite index (i. e. has finitely many distinct equivalence classes).*

Of course, computability is a much richer notion than regularity, and one may well suppose that for metamathematical reasons, no intrinsic or quasi-intrinsic definition of computability can be achieved—although we are not aware of any specific formal result which implies this.

Still, the question seems worth asking: we suspect that a better understanding of this issue may be important in addressing some fundamental questions in computability and complexity. We shall return briefly to this point in the concluding section.

5.3 From Functions to Functionals

The issues become clearer if we include **functionals** (functions which take functions as arguments) in our discussion. This leads to the following hierarchy of types:

$$\text{Type } 0 : \mathbb{N}$$
$$\text{Type } 1 : \mathbb{N} \rightharpoonup \mathbb{N}$$
$$\text{Type } 2 : [\mathbb{N} \rightharpoonup \mathbb{N}] \rightharpoonup \mathbb{N}$$
$$\vdots$$

In general, a type $n + 1$ functional takes type n functionals as arguments. Of course, more general types can also be considered.

Functionals are not so unfamiliar: e.g. the quantifiers!

$$\forall, \exists : [\mathbb{N} \rightarrow \mathbb{B}] \rightarrow \mathbb{B}$$

Here $\mathbb{B}$ is the set of truth-values; the quantifiers over the natural numbers are seen as functionals taking natural number predicates to truth-values.

While it might seem that bringing functionals into the picture will merely complicate matters, in fact when we consider higher-order functions, some intrinsic structure emerges naturally, which is lacking when we only look at first-order functions over discrete data.

When we compute with a function (or procedure) parameter P, we can immediately distinguish two paradigms.

- Extensional paradigm: the only way we can interact with P is to call it with some arguments, and use the results:

$$\text{let } m = P(n) \text{ in} \dots$$

In any finite computation, we can only make finitely many such calls, and hence 'observe' a finite subset of the graph of the function defined by P:

$$\{m_1 = P(n_1), m_2 = P(n_2), \dots, m_k = P(n_k)\}$$

- Intensional paradigm: we have access to the **code** of P. So we can compute such things as the code itself (as a text string), or how many symbols appear in it, etc. Examples: interpreters, program analyzers, and other **programs which manipulate programs**. Usually, though, in computer science we keep programs as data (subject to manipulation) distinct from programs as code (performing manipulations); it is generally seen as bad practice to mingle these two modes. By contrast, computability theory does use codes of programs viewed as data in a pervasive and essential fashion. We shall return to this contrast when we discuss fixpoint theorems.

5.3.1 Intrinsic Structure of Computable Functionals

When we pass to functionals, some intrinsic structure begins to emerge. Firstly, there is a natural ordering on partial functions:

$$f \sqsubseteq g \quad \leftrightarrow \quad \mathsf{graph}(f) \subseteq \mathsf{graph}(g).$$

Here $\mathsf{graph}(f)$ is the set of input-output correspondences which—on the extensional view—uniquely characterize f. If P is a program code, we can define $\mathsf{graph}(P)$ to be the set of input-output correspondences defined by the (partial) function computed by P.

We can view a function which takes codes of programs and returns numbers as an (*a priori* intensional) functional. We say that such a function F is extensional if $F(P) = F(Q)$ whenever $\mathsf{graph}(P) = \mathsf{graph}(Q)$, and hence defines a functional $\hat{F} : [\mathbb{N} \rightharpoonup \mathbb{N}] \rightharpoonup \mathbb{N}$.

We have the following classical result:

Theorem 5.2 (Myhill-Sheperdson [62]) *An extensional computable functional* $\hat{F}$ *satisfies the following properties:*

Monotonicity : $f \sqsubseteq g \Rightarrow \hat{F}(f) \sqsubseteq \hat{F}(g)$.
Continuity : *For any increasing sequence of partial functions*

$$f_0 \sqsubseteq f_1 \sqsubseteq f_2 \sqsubseteq \cdots$$

we have

$$\hat{F}\left(\bigsqcup_i f_i\right) = \bigsqcup_i \hat{F}(f_i).$$

Moreover, there is a sort of converse (which still needs to appeal to the usual notion of computability for functions on the natural numbers).

5.3.2 Generalization to Domains

These ideas were elaborated into a beautiful mathematical theory of computation by Dana Scott and others. Domains are partial orders with least elements and least upper bounds of increasing sequences. The corresponding functions are the monotonic and continuous ones.

A function $f : D \to E$ is **monotonic** if, for all $x, y \in D$:

$$x \sqsubseteq y \implies f(x) \sqsubseteq f(y).$$

It is **continuous** if it is monotonic, and for all ω-chains $(x_n)_{n \in \omega}$ in D:

$$f\left(\bigsqcup_{n \in \omega} x_n\right) = \bigsqcup_{n \in \omega} f(x_n).$$

Continuity serves as an 'intrinsic approximation' to computability: an intrinsic property which is a necessary condition for computability. Indeed, one speaks of 'Scott's thesis', that computable functions are continuous [75]. Of course, this condition is not sufficient, and hence does not offer a complete analysis of computability.

5.3.2.1 Examples

We consider functions $f : \{0, 1\}^\infty \to \mathbb{B}_\perp$. Here $\{0, 1\}^\infty$ is the domain of finite and infinite binary sequences (or 'streams'), ordered by prefix; while $\mathbb{B}_\perp = \{\mathbf{tt}, \mathbf{ff}, \perp\}$ is the 'flat' domain of booleans with $\mathbf{tt} \sqsupseteq \perp \sqsubseteq \mathbf{ff}$, representing computations which either fail to halt, or return a boolean value.

We consider the following definitions for such functions:

1. $f(x) = \mathbf{tt}$ if x contains a 1, $f(x) = \perp$ otherwise.
2. $f(x) = \mathbf{tt}$ if x contains a 1, $f(0^\infty) = \mathbf{ff}$, $f(x) = \perp$ otherwise.
3. $f(x) = \mathbf{tt}$ if x contains a 1, $f(x) = \mathbf{ff}$ otherwise.

Of these: (1) is continuous, (2) is monotonic but not continuous, and (3) is not monotonic.

The conceptual basis for monotonicity is that the information in Domain Theory is **positive**; negative information is not regarded as stable observable information. That is, if we are at some information state s, then for all we know, s may still increase to t, where $s \sqsubseteq t$. This means that if we decide to produce information $f(s)$ at s, then we must produce all this information, and possibly more, at t, yielding $f(s) \sqsubseteq f(t)$. Thus we can only make decisions at a given information state which are stable under every possible information increase from that state. This idea is very much akin to the use of partial orders in Kripke semantics for Intuitionistic Logic [42], in particular in connection with the interpretation of negation in that semantics.

The continuity condition, on the other hand, reflects the fact that a computational process will only have access to a finite amount of information at each finite stage of the computation. If we are provided with an infinite input, then any information we produce as output at any finite stage can only depend on some finite observation we have made of the input. This is reflected in one of the inequations corresponding to continuity:

$$f\left(\bigsqcup_{n \in \omega} x_n\right) \sqsubseteq \bigsqcup_{n \in \omega} f(x_n)$$

which says that the information produced at the limit of an infinite process of information increase is no more than what can be obtained as the limit of the information produced at the finite stages of the process. Note that the "other half" of continuity

$$\bigsqcup_{n \in \omega} f(x_n) \sqsubseteq f\left(\bigsqcup_{n \in \omega} x_n\right)$$

follows from monotonicity.

5.3.3 The Fixpoint Theorem

Theorem 5.3 (The Fixpoint Theorem [49]) *Let D be an ω-cpo with a least element, and $f : D \to D$ a continuous function. Then f has a least fixed point* $\mathsf{lfp}(f)$. *Moreover,* $\mathsf{lfp}(f)$ *is defined explicitly by:*

$$\mathsf{lfp}(f) = \bigsqcup_{n \in \omega} f^n(\bot). \tag{5.1}$$

This is a central pillar of domain theory in its use in the semantics of computation, providing the basis for interpreting recursive definitions of all kinds [27, 76]. Note however that the key fixpoint result for computability theory is the **Kleene second recursion theorem** [36, 57], an **intensional** result, which refers to fixpoints for **programs** rather than the functions which they compute. This theorem is strangely absent from Computer Science, although it can be viewed as strictly stronger than the extensional theorem above. This reflects the fact which we have already alluded to, that while Computer Science embraces wider notions of processes than computability theory, it has tended to refrain from studying intensional computation, despite its apparent expressive potential. This reluctance is probably linked to the fact that it has proved difficult enough to achieve software reliability even while remaining within the confines of the extensional paradigm. Nevertheless, it seems reasonable to suppose that understanding and harnessing intensional methods offers a challenge for computer science which it must eventually address.

5.3.4 Trouble in Paradise

The semantic theory of higher-type functional computation we get from domain theory seems compelling. But there is a problem: a mismatch between what the semantic theory allows, and what we can actually compute in 'natural' programming languages.

5.3.4.1 Example: Parallel or

We consider a (curried) function of two arguments

$$\mathsf{por} : \mathbb{B}_\perp \to \mathbb{B}_\perp \to \mathbb{B}_\perp$$

$$\mathsf{por} \perp \mathbf{tt} = \mathbf{tt} = \mathsf{por}\,\mathbf{tt}\,\perp, \qquad \mathsf{por}\,\mathbf{ff}\,\mathbf{ff} = \mathbf{ff}$$

This is the 'strong or' of Kleene 3-valued logic.

This function lives in the semantic model, since it is monotonic and continuous, but it is not definable in realistic languages—or those arising from logical calculi (essentially, the λ-calculus). This is because a definable function of two arguments must examine its arguments in some definite order; whichever it examines first, the function will yield an undefined result if that argument is undefined. We can define second order functions which can only be distinguished by their values at parallel or; they will be **observationally equivalent**, i. e. equivalent in the operational sense, since no experiment we can perform within the language by applying them to inputs which can be defined in the language will serve to distinguish them. However, these second-order functions will have **different denotations** in a model which includes parallel or. Thus such a model—and in particular, the canonical domain-theoretic model—introduces operationally unjustified distinctions.

Note that **definability** and **higher types** are crucially important here. This mismatch between model and operational content is known as the **failure of full abstraction** [54, 67].

5.3.5 Sequentiality

Consider the following functions, which **are** definable:

$$\mathsf{lsor}(b_1, b_2) \equiv \text{ if } b_1 \text{ then } \mathbf{tt} \text{ else } b_2$$

$$\mathsf{rsor}(b_1, b_2) \equiv \text{ if } b_2 \text{ then } \mathbf{tt} \text{ else } b_1$$

Note that

$$\mathsf{lsor}(\mathbf{tt}, \bot) = \mathbf{tt}, \qquad \mathsf{lsor}(\bot, \mathbf{tt}) = \bot$$

$$\mathsf{rsor}(\mathbf{tt}, \bot) = \bot, \qquad \mathsf{rsor}(\bot, \mathbf{tt}) = \mathbf{tt}$$

To get **por**, we need to run the **processes** for evaluating b_1 and b_2 in parallel. For example, we could interleave the executions of these processes, running each one step at a time. As soon as one of these returned the result **tt**, we could return **tt** as the value of **por**. This means that we need to have access, not just to the purely extensional information about the arguments—i. e. their values—but to their **intensional descriptions**. This form of **dovetailing** is fundamental to many constructions in computability theory [68], but is not available in logical calculi and functional languages based on the λ-calculus. Indeed, contemporary functional programming languages take an extensional view of data as one of their cardinal virtues [34, 71, 74].

This raises the question: How can we capture the notion of **sequential functional**?

5.3.6 Sequentiality Is Extrinsic

We need additional information to characterize those functionals which are sequential. The following remarkable result, due to Ralph Loader [50], shows that this is unavoidable.

Theorem 5.4 (Loader) *The set of functionals definable in Finitary PCF, the typed λ-calculus over the booleans with conditionals and a term denoting $\bot$, is not recursive.*

Note that, since the base type here is just the flat domain of booleans, the set of functionals at any type is **finite**. Thus if we form the logical type theory over this structure by closing under cartesian product and powerset, then all the logical types over this structure will also be finite. If there were any form of intrinsic definition of sequentiality, given in terms of some structure defined at each type even in full higher-order logic, this finiteness would ensure that the notion of sequentiality was recursive. By Loader's theorem, we conclude that no such intrinsic definition can exist.[3] This rules out any hope of a reasonable intrinsic definition of the sequential functionals.

---

[3] This argument is due to Gordon Plotkin (personal communication).

So the best we can do is to characterize which are the **sequential functional processes**—i. e. an intensional notion.

Attempts at such characterizations coming from a computability theory perspective were made by Kleene in his series of papers [37–41], and by Gandy and Pani in unpublished work. The work of Berry and Curien on sequential algorithms on concrete data structures [18], directly inspired by the full abstraction problem for PCF, should also be mentioned.

A characterization of the sequential functional processes was eventually achieved using the newly available tools of Game Semantics, developed in the 1990's by the present author, Radha Jagadeesan and Pasquale Malacaria, and by Martin Hyland and Luke Ong [11, 35]. We shall now give a brief account of these ideas.

5.4 Game Semantics

We shall give a brief, informal introduction to game semantics, emphasizing concepts and intuitions rather than technical details, for which we refer to works such as [1, 11, 14, 35].

The traditional approach to denotational semantics, exemplified by domain theoretic semantics [27, 76], was to interpret the types of a programming language by (possibly structured) sets, and the programs as functions. Game semantics fundamentally revises this ontology:

- Types of a programming language are interpreted as 2-person games; the Player is the System (program fragment) currently under consideration, while the Opponent is the Environment or context.
- Programs are **strategies** for these games.

So game semantics is inherently a semantics of **open systems**; the meaning of a program is given by its potential interactions with its environment.

A key feature of game semantics as developed in computer science (and significantly differentiating it from previous work on games in logic [30, 51]) is its **compositionality**. The key operation is plugging two strategies together, so that each **actualizes** part of the environment of the other. The familiar game-theoretic idea of playing one strategy off against another is a special case of this, corresponding to a **closed** system, with no residual environment. This form of interaction exploits the game-theoretic P/O duality.

5.4.1 Types as Games

- A simple example of a basic datatype of natural numbers:

$$\mathbb{N} = \{q \cdot n \mid n \in \mathbb{N}\}$$

Fig. 5.1 Strategy for $\lambda f : \mathbb{N} \Rightarrow \mathbb{N}.\, \lambda x : \mathbb{N}.\, f(x) + 2$

Note a further classification of moves, orthogonal to the P/O duality;the O-move q is a **question**, the P-moves n are **answers**. This turns out to be important for capturing **control features** of programming languages.

- Forming function or procedure types $A \Rightarrow B$. We form a new game from disjoint copies of A and B, **with P/O roles in A reversed**. Thus we think of $A \Rightarrow B$ as a **structured interface** to the Environment; in B, we interact with the caller of the procedure, **covariantly**, while in A, we interact with the argument supplied to the procedure call, **contravariantly**.

5.4.2 Example

We consider the strategy corresponding to the term $\lambda f : \mathbb{N} \Rightarrow \mathbb{N}.\, \lambda x : \mathbb{N}.\, f(x) + 2$. This term defines the procedure $P(f, x)$ such that $P(f, x)$ returns $f(x) + 2$.

We show a typical play for the strategy corresponding to this term in Fig. 5.1.

This should be read as follows. Time flows downwards. Moves by Opponent alternate with those of Player. Each column corresponds to one of the occurrences of atomic types in the overall type of the term. Each of these will be a copy of the simple game for natural numbers. In effect, we are playing on several game boards, switching from one to another.

The play begins with the Opponent, or environment, requesting an output. The strategy must call the function argument f, so it requests an output from this argument. Note that the variance rules dictate that, since the type of f occurs negatively in the overall type, this opening move is indeed a Player move. The environment now has to respond to this request. Typically, it will do so by requesting its input. Since the strategy is realizing the function call $f(x)$, the value of this input will be obtained by the strategy from the natural number argument x, and thus the strategy requests the value of this argument. When the evironment responds to this request with some

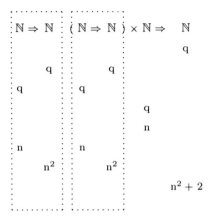

Fig. 5.2 Applying $\lambda f : \mathbb{N} \Rightarrow \mathbb{N}. \lambda x : \mathbb{N}. f(x) + 2$ to $\lambda x : \mathbb{N}. x^2$

number n, the strategy **copies** this value to answer the request of the function argument for its input. This copying of data is a key feature of game semantics, corresponding to the logical flow of information. Indeed, purely logical strategies, i. e. those corresponding to logical proofs, are typically made entirely from such 'copy-cat' processes [10].

After the function argument receives its input n, it will typically emit some output m; the strategy for the term will now answer the original request from the environment with the number $m + 2$.

We have simply described what the evident strategy corresponding to the above λ-term will do. Of course, the purpose of a compositional game semantics is precisely to allow this strategy to be constructed systematically as the denotation of the above term in a syntax-directed fashion.

5.4.3 Composition

We now show how the key operation of plugging one strategy together with another, corresponding to applying a procedure to its argument, or more generally to composing procedures, will proceed, again through an example. We apply the higher-order procedure $\lambda f : \mathbb{N} \Rightarrow \mathbb{N}. \lambda x : \mathbb{N}. f(x) + 2$ to the argument $\lambda x : \mathbb{N}. x^2$. A typical run of the corresponding strategy is shown in Fig. 5.2.

Here the common parts of the two strategies are shown in the dashed boxes. The key point is that type matching between function and argument guarantees that the Opponent moves for one are Player moves for the other, and vice versa. This intrinsic duality allows the interactive interpretation of composition to be defined without requiring any additional structure. Thus when the strategy for the functional plays its opening move in response to the environment request for an output, the other strategy sees this as the opening move of **its** environment, and responds accordingly. The functional now sees this response as the next move of the nvironment, and so on. With each of the strategies following their own scripts, we get a uniquely determined path through the common part of the type, where they interact.

Note that the **residual strategy**, after we hide the part where the interaction between function and argument occurs, is that corresponding to the function $\lambda x . x^2 + 2$, consistent with the result of performing β-reduction syntactically. This 'parallel composition plus hiding' paradigm for composition of strategies [10] is a fundamental component of game semantics.

5.4.4 Technical Notes

Once the ideas shown informally by example in the previous subsection are properly formalized, games and strategies organize themselves into a very nice mathematical structure—a cartesian closed category $\mathcal{G}$ [11, 35]. We can then use standard methods of denotational semantics to give a compositional semantics for functional languages such as PCF in $\mathcal{G}$ [27].

The key result is:

Theorem 5.5 (Abramsky-Jagadeesan-Malacaria and Hyland-Ong [11, 35]) *Every compact (even: recursive) strategy in $\mathcal{G}$ is definable by a PCF term.*

This can be understood as a **completeness theorem**. It is saying, not only that every sequential functional process can be interpreted as a strategy in the game semantics, but that the 'space' of strategies corresponds **exactly** to the class of sequential functional processes. In short, we have achieved a characterization, albeit at the intensional level.

Note that similar results can be achieved for proof calculi for various logical systems, leading to notions of **full completeness** [10, 15].

If we refer to strategies as in our examples in the previous sub-section, we can get some intuition for how a result such as this is proved. The first move (by Player) in the strategy corresponds to the head variable in the head-normal form of the λ-term, or the last rule in a proof. Decomposing the strategy progressively uncovers the defining term. This argument is formalized in [11, 35], and even axiomatized in Abramsky [2].

There is an important caveat, which actually leads to a major positive feature of game semantics. In order to achieve this completeness result, strategies must be **constrained**, as we shall now explain.

5.4.5 Constraints on Strategies

There are two main kinds of constraints which must be imposed on strategies in order to capture exactly the sequential functional processes:

1. Strategies do not have perfect information about the previous computation history (play of the game). For example, consider a call

$$f t_1 t_2$$

   When it is evaluated (called) by f, t_2 should not 'know' whether t_1 has already been evaluated—and vice versa. Note that if we had imperative state, the arguments could use this to pass such information to each other. Thus this constraint is a distinctive feature of purely functional computation.
2. Properly nested call-return flow of control. This is visibly a feature of the example we discussed previously. It is often referred to as the 'stack discipline' [11], for obvious reasons. It corresponds to the absence of non-local control features such as jumps or exceptions. Again, this constraint is a distinctive feature of purely functional computation.

Both these constraints can be formulated precisely (the first as 'innocence' or 'history-freedom', the second as 'well-bracketing'), and shown to be closed under composition and the other semantic constructions [11, 35]. The resulting cartesian closed categories of games and constrained strategies yield the completeness results as in Theorem 5.5.

5.4.6 Discussion

It should be noted that there is an important caveat to the result in Theorem 5.5. We get a definability result—but strategies are much finer grained than functions. They correspond to certain 'PCF evaluation trees' [11]. To get an equationally fully abstract model, we must quotient the games model. This is unavoidable by Loader's Theorem, from which it follows that the fully abstract model, even for Finitary PCF, is **not effectively presentable**.

What is the significance of the result? In retrospect, the real payoff is in other cases—but the PCF result is the keystone of the whole development. Relaxing the constraints which characterize functional computation leads to fully abstract models for languages with (locally scoped) state, or control operators, or both [14]. This picture of a semantic 'cube' or hierarchy of constraints has proved very fruitful as a paradigm for exploring a wide range of programming language features.

5.4.7 The Game Semantics Landscape

Game semantics has proved to be a flexible and powerful paradigm for constructing highly structured fully abstract semantics for languages with a wide range of computational features:

- (Higher-order) Functions and procedures [11, 35]
- Call by name and call by value [13, 32]
- Locally scoped state [12, 58]
- General reference types [9, 60]
- Control features (continuations, exceptions) [44, 45]
- Non-determinism, probabilities [22, 29]
- Concurrency [26, 43, 46]
- Names and freshness [6, 72].

In many cases, game semantics have yielded the first, and often still the only, semantic construction of a fully abstract model for the language in question. Moreover, where sufficient computational features (typically state or control) are present, then the observational equivalence is more discriminating, and game semantics captures the fully abstract model directly, without the need for any quotient. Intensions become extensions!

More generally: the point of conceptual interest is to find **positive reasons**—structural invariants—**for non-expressiveness**. This is a positive story for a form of intensionality. Indeed, we can see here the beginnings of a **structural theory of processes**.

5.4.8 Mathematical Aside

Categories of games and strategies have fascinating mathematical structure in their own right. They give rise to:

- Constructions of **free categories with structure** of various kinds.
- **Full completeness** results characterizing the "space of proofs" for various logical systems [10, 15].
- There are even connections with **geometric topology**, e.g. Temperley-Lieb and other diagram algebras [5].

5.4.9 Algorithmic Game Semantics

We can also take advantage of the **concrete nature** of game semantics. A play is a sequence of moves, so a strategy can be represented by the set of its plays, i.e. by a **language** over the alphabet of moves, and hence by an automaton.

There are significant finite-state fragments of the semantics for various interesting languages, as first observed by Ghica and McCusker [3, 25]. This means we can **compositionally construct** automata as (representations of) the meanings of open (incomplete) programs, giving a powerful basis for compositional software model-checking. There has been an extensive development of these ideas in the last few years, by Ghica [24], Ong [7], Murawski [48], Tzevelekos et al. [61].

5.4.10 Other Aspects

It should also be noted that there are other strands in game semantics, which has become a rich and diverse field. For example, there is work on clarifying the mathematical structure of game semantics itself, and its relation to syntax and to categorical structure. Examples include [20, 28, 52, 53, 77].

More broadly, we must emphasise that our subject in this paper has been the characterisation of sequential, functional processes. We have argued that a considerable measure of success has been achieved in answering this question. However, finding a compelling answer to the general question of "What is a process?" which we raised in the Introduction remains a major challenge of the field. It is likely that new conceptual ingredients will be needed to enable further advances towards this goal.

5.5 Conclusions: Some Questions and Dreams

- Can we use intensional recursion to give more realistic models of reflexive phenomena in biology, cognition, economics, etc.?
  Working with codes seems more like what biological or social mechanisms might plausibly do, rather than with abstract mathematical objects *in extenso*.
- How does this relate to current interest in higher categories, homotopy type theory etc. [16, 73], where equalities are replaced by **witnesses**? Do higher categories, and the intensional type theories they naturally support, provide the right setting for a systematic intensional view of mathematics and logic? What kind of novel applications will these structures support?
- Can we develop a positive theory of intensional structures, and find the right notions of invariance?
- The dream: to use this to give some (best-possible) **intrinsic** characterization of computability, and of complexity classes.
- Could this even be the missing ingredient to help us separate classes? Well, we can dream!

References

1. Abramsky S (1997) Semantics of interaction: an introduction to game semantics. In: Dybjer P, Pitts A (eds) Semantics and logics of computation. Publications of the Newton Institute, Cambridge University Press, pp 1–31
2. Abramsky S (1999) Axioms for definability and full completeness. In: Plotkin G, Tofte M, Stirling C (eds) Proof, language and interaction: essays in honour of Robin Milner. MIT Press, Cambridge, pp 55–75
3. Abramsky S (2001) Algorithmic game semantics. In: Schwichtenberg H, Steinbrüggen R (eds) Proof and system-reliability: proceedings of the NATO advanced study institute. Kluwer Academic Publishers, Marktoberdorf, pp 21–47 (24 July–5 Aug 2001)
4. Abramsky S (2006) What are the fundamental structures of concurrency?: we still don't know!. Electron Notes Theoret Comput Sci 162:37–41
5. Abramsky S (2007) Temperley-Lieb algebra: from knot theory to logic and computation via quantum mechanics. In: Chen G, Kauffman L, Lomonaco S (eds) Mathematics of quantum computing and technology, Taylor and Francis, New York, pp 415–458
6. Abramsky S, Ghica DR, Murawski AS, Ong CHL, Stark IDB (2004) Nominal games and full abstraction for the nu-calculus. In: Proceedings of the 19th annual IEEE symposium on logic in computer science, pp 150–159
7. Abramsky S, Ghica DR, Murawski AS, Ong CHL (2003) Algorithmic game semantics and component-based verification. In: SAVCBS 2003, specification and verification of component-based systems, p 66
8. Abramsky S, Ghica D, Murawski A, Ong CHL (2004) Applying game semantics to compositional software modeling and verification. In: Conference on tools and algorithms for the construction and analysis of systems. Lecture Notes in Computer Science, vol 2988. Springer, pp 421–435
9. Abramsky S, Honda K, McCusker G (1998) A fully abstract game semantics for general references. In: Proceedings of 13th annual IEEE symposium on logic in computer science, pp 334–344
10. Abramsky S, Jagadeesan R (1994) Games and full completeness for multiplicative linear logic. J Symb Log 59(2):543–574
11. Abramsky S, Jagadeesan R, Malacaria P (2000) Full abstraction for PCF. Inf Comput 163(2):409–470
12. Abramsky S, McCusker G (1997) Linearity, sharing and state: a fully abstract game semantics for idealized algol with active expressions. In: O'Hearn P, Tennent RD (eds) Algol-like languages, Birkhauser, pp 317–348
13. Abramsky S, McCusker G (1998) Call-by-value games. In: Proceedings of computer science logic, Springer, pp 1–17
14. Abramsky S, McCusker G (1999) Game semantics. In: Schwichtenberg H, Berger U (eds) Computational logic: proceedings of the 1997 Marktoberdorf summer school, Springer, pp 1–55
15. Abramsky S, Mellies PA (1999) Concurrent games and full completeness. In: Proceedings of 14th symposium on logic in computer science, pp 431–442
16. Awodey S, Warren MA (2009) Homotopy theoretic models of identity types. Math Proc Camb Phil Soc 146(1):45–55
17. Barendregt HP (1984) The lambda calculus: its syntax and semantics, vol 103. Studies in Logic and the Foundations of Mathematics, North Holland
18. Berry G, Curien PL (1982) Sequential algorithms on concrete data structures. Theoret Comput Sci 20(3):265–321
19. Church A (1941) The calculi of lambda-conversion, vol 6. Princeton University Press, Princeton
20. Curien PL, Faggian C (2012) An approach to innocent strategies as graphs. Inf Comput 214:119–155
21. Cutland N (1980) Computability: an introduction to recursive function theory. Cambridge University Press, Cambridge

22. Danos V, Harmer RS (2002) Probabilistic game semantics. ACM Trans Comput Log 3(3):359–382
23. Gandy R (1980) Church's thesis and principles for mechanisms. In: Barwise J, Keisler HJ, Kunen K (eds) The kleene symposium of studies in logic and the foundations of mathematics, vol 101. Elsevier, Amsterdam, pp 123–148
24. Ghica DR (2009) Applications of game semantics: from program analysis to hardware synthesis. In: Proceedings of 24th Annual IEEE Symposium on logic in computer science, pp 17–26
25. Ghica DR, McCusker G (2003) The regular-language semantics of second-order idealized ALGOL. Theoret Comput Sci 309(1):469–502
26. Ghica DR, Murawski AS (2008) Angelic semantics of fine-grained concurrency. Ann Pure Appl Log 151(2):89–114
27. Gunter CA (1992) Semantics of programming languages: structures and techniques. MIT press, Cambridge
28. Harmer R, Hyland H, Mellies PA (2007) Categorical combinatorics for innocent strategies. In: 22nd annual IEEE symposium on logic in computer science, pp 379–388
29. Harmer R, McCusker G (1999) A fully abstract game semantics for finite nondeterminism. In: 14th annual IEEE symposium on logic in computer science, pp 422–430
30. Hintikka J, Sandu G (1997) Game-theoretical semantics. In: van Benthem J, ter Meulen A (eds) Handbook of logic and language, Elsevier, Amsterdam, pp 361–410
31. Hoare CAR (1978) Communicating sequential processes. Commun ACM 21(8):666–677
32. Honda K, Yoshida N (1997) Game theoretic analysis of call-by-value computation: automata, languages and programming, Springer, Berlin, pp 225–236
33. Hopcroft JE, Ullman JD (1979) Introduction to automata theory, languages, and computation. Addison-Wesley, Cambridge
34. Hughes J (1989) Why functional programming matters. Comput J 32(2):98–107
35. Hyland JME, Ong CHL (2000) On full abstraction for PCF: I, II, and III. Inf Comput 163(2):285–408
36. Kleene SC (1938) On notation for ordinal numbers. J Symb Log 3(4):150–155
37. Kleene SC (1978) Recursive functionals and quantifiers of finite types revisited I. In: Fenstad JE, Gandy RO, Sacks GE (eds) Generalized recursion theory II: proceedings of the 1977 Oslo symposium of studies in logic and the foundations of mathematics, vol 94. Amsterdam, pp 185–222
38. Kleene SC (1980) Recursive functionals and quantifiers of finite types revisited II. In: Barwise J, Keisler HJ, Kunen K (eds) The Kleene symposium of studies in logic and the foundations of mathematics, vol 101. Elsevier, pp 1–29
39. Kleene SC (1982) Recursive functionals and quantifiers of finite types revisited III. In: Metakides G (ed) Patras logic symposion: proceedings of studies in logic and the foundations of mathematics, logic symposion held at Patras, Greece, vol 109. Elsevier, pp 1–40, Aug 18–22 1982
40. Kleene SC (1985) Unimonotone functions of finite types (recursive functionals and quantifiers of finite types revisited IV). Recur Theory 42:119–138
41. Kleene SC (1991) Recursive functionals and quantifiers of finite types revisited V. Trans Am Math Soc 325:593–630
42. Kripke S (1965) Semantical analysis of intuitionistic logic I. In: Formal systems and recursive functions, studies in logic and foundations of mathematics, North Holland, pp 92–130
43. Laird J (2001) A game semantics of idealized CSP. Electron Notes Theor Comput Sci 45:232–257
44. Laird J (2001) A fully abstract game semantics of local exceptions. In: 16th annual IEEE symposium on logic in computer science, pp 105–114
45. Laird J (2003) A game semantics of linearly used continuations. In: Foundations of software science and computation structures, Springer, pp 313–327
46. Laird J (2005) A game semantics of the asynchronous π-calculus. CONCUR 2005-concurrency theory, pp 51–65

47. Landin PJ (1966) The next 700 programming languages. Commun ACM 9(3):157–166
48. Legay A, Murawski A, Ouaknine J, Worrell J (2008) On automated verification of probabilistic programs: tools and algorithms for the construction and analysis of Systems, pp 173–187
49. Lassez JL, Nguyen VL, Sonenberg E (1982) Fixed point theorems and semantics: a folk tale. Inf Process Lett 14(3):112–116
50. Loader R (2001) Finitary PCF is not decidable. Theor Comput Sci 266(1):341–364
51. Lorenzen P (1960) Logik und agon. Atti Congr Int di Filosofia 4:187–194
52. Melliès PA (2006) Asynchronous games 2: the true concurrency of innocence. Theor Comput Sci 358(2):200–228
53. Mellies PA (2012) Game semantics in string diagrams. In: 27th annual IEEE symposium on logic in computer science, pp 481–490
54. Milner R (1977) Fully abstract models of typed λ-calculi. Theor Comput Sci 4(1):1–22
55. Milner R (1989) Communication and concurrency. Prentice-Hall, Upper Saddle River
56. Milner R (1999) Communicating and mobile systems: the pi calculus. Cambridge University Press, Cambridge
57. Moschovakis YN (2010) Kleene's amazing second recursion theorem. Bull Symb Log 16(2):189–239
58. Murawski A, Tzevelekos T (2009) Full abstraction for reduced ML: foundations of software science and computational structures, pp 32–47
59. Murawski A, Tzevelekos T (2011) Algorithmic nominal game semantics. In: Proceedings of the 20th European symposium on programming, Lecture Notes in Computer Science, vol 6602. Springer, pp 419–438
60. Murawski AS, Tzevelekos N (2011) Game semantics for good general references. In: 26th annual IEEE symposium on logic in computer science, pp 75–84
61. Murawski AS, Tzevelekos N (2012) Algorithmic games for full ground references. In: Proceedings of the 39th international colloquium on automata, languages and programming, Lecture Notes in Computer Science, vol 7392. pp 312–324
62. Myhill J, Shepherdson JC (1955) Effective operations on partial recursive functions. Math Log Quart 1(4):310–317
63. Myhill JR (1957) Finite automata and the representation of events. Technical Report WADD TR-57-624, Wright Patterson AFB
64. Nerode A (1958) Linear automaton transformations. Proc Am Math Soc 9(4):541–544
65. Petri CA (1962) Fundamentals of a theory of asynchronous information flow. In: IFIP Congress 1962, Amsterdam, pp 386–390
66. Petri CA (1966) Communication with automata, New York: Griffiss Air Force Base. vol. 1, suppl. no. 1. Tech Rep RADC-TR-65-377
67. Plotkin GD (1977) LCF considered as a programming language. Theor Comput Sci 5(3):223–255
68. Rogers H (1967) Theory of recursive functions and effective computability. McGraw Hill, New York
69. Sieg W (2002) Calculations by man and machine: mathematical presentation. In: Gärdenfors P, Wolenski J, Kijania-Placek K (eds) Proceedings of the 11th international conference on logic, methodology and philsophy of science, Vol 1. Kluwer Academic Publishers, pp 247–262
70. Turing AM (1937) On computable numbers, with an application to the entscheidungs problem. Proc Lond Math Soc 42(2):230–265
71. Turner D (1995) Elementary strong functional programming. In: Functional programming languages in education, Springer, pp 1–13
72. Tzevelekos N (2007) Full abstraction for nominal general references. In: 22nd annual IEEE symposium on logic in computer science, pp 399–410
73. Voevodsky V (2010) Univalent foundations project, NSF grant application
74. Wadler P (1992) The essence of functional programming. In: Proceedings of the 19th ACM SIGPLAN-SIGACT symposium on principles of programming languages, pp 1–14

75. Winskel G (1987) Event structures. In: Brauer W, Reisig W, Rozenberg G (eds) Proceedings of an advanced course on Petri nets: Applications and relationships to other models of concurrency, advances in Petri nets 1986, Part II, LNCS, Bad Honnef, September 1986, vol 255. Springer, pp 325–392
76. Winskel G (1993) The formal semantics of programming languages: an introduction. MIT press, Cambridge
77. Winskel G (2012) Bicategories of concurrent games. In Foundations of Software Science and Computational Structures, Springer Berlin Heidelberg, pp 26–41

Chapter 6
Comparing Theories: The Dynamics of Changing Vocabulary

Hajnal Andréka and István Németi

Abstract There are several first-order logic (FOL) axiomatizations of special relativity theory in the literature, all looking different but claiming to axiomatize the same physical theory. In this chapter, we elaborate a comparison between these FOL theories for special relativity. We do this in the framework of mathematical logic. For this comparison, we use a version of definability theory in which new entities can also be defined besides new relations over already available entities. In particular, we build an interpretation (in Alfred Tarski's sense) of the reference-frame oriented theory SPECREL developed in the Budapest Logic Group into the observationally oriented Signalling theory of James Ax published in Foundations of Physics. This interpretation provides SPECREL with an operational/experimental semantics. Then we make precise, "quantitative" comparisons between these two theories via using the notion of definitional equivalence. This is an application of mathematical logic to the philosophy of science and physics in the spirit of Johan van Benthem's work.

6.1 Introduction

This chapter is about an application of logic to the methodology of science in the spirit of van Benthem's [8, 9].

Research supported by the Hungarian grant for basic research OTKA No K81188.

H. Andréka (✉)
Alfréd Rényi Institute of Mathematics, Hungarian Academy of Sciences, Reáltanoda u. 13-15, Budapest 1053, Hungary
e-mail: andreka.hajnal@renyi.mta.hu

I. Németi
Alfréd Rényi Institute of Mathematics, Hungarian Academy of Sciences, Reáltanoda u. 13-15, Budapest 1053, Hungary
e-mail: nemeti.istvan@renyi.mta.hu

A. Baltag and S. Smets (eds.), *Johan van Benthem on Logic and Information Dynamics*, Outstanding Contributions to Logic 5, DOI: 10.1007/978-3-319-06025-5_6, © Springer International Publishing Switzerland 2014

There are several axiomatizations of special relativity theory available in the literature, all looking different but claiming to axiomatize the same physical theory. Such are, among many others, the ones in Andréka et al. [4], Ax [6], Goldblatt [17], Schelb [32], Schutz [33], Suppes [34]. These papers talk about very different kinds of objects: [4] talks about reference frames, [6] talks about particles and signals, [17] looks like a purely geometrical theory about orthogonality, the central notion of [34] is the so-called Minkowski-metric, etc. While, as usual, one gets a better picture of this area via a variety of different "eyeglasses", the following questions arise. What are the connections between these theories? Do they all talk about the same thing? If they do, do they capture it to the same extent, or is one axiomatization more detailed or accurate than some of the others? In this chapter we want to show how, in the framework of mathematical logic, a concrete, tangible comparison/connection can be elaborated between these theories for special relativity. We also want to show what we can gain from such an investigation.

For this comparison, we have to use a form of logical definability theory in which totally new kinds of entities can be defined as opposed to traditional definability theory where only new relations can be defined over already available entities. The existing methods of definability theory had to be modified and refined for the purposes of the present situation. Thus, definability theory, too, profits from such an application. In the present chapter, we elaborate in detail on one piece of comparing relativity theories: we construct an interpretation of the relativity theory in [4] talking about reference frames into the theory in [6] which talks about particles emitting and absorbing signals. Then we construct an inverse interpretation and we discuss which versions of the two theories are definitionally equivalent.[1] Since this is a case-study for applicability of the proposed method for connecting theories, we tried to give all the detail needed. This is why some sections of the chapter may look somewhat technical.

An insight of last century mathematical logic is that it is important to fix the vocabulary of a first-order logic theory and stay inside the so obtained language while working in a specific theory (see, e.g., [40]). The symbols in the vocabulary[2] are the concepts that are not analyzed further in the given theory, they are thus called basic (or primitive) concepts. But this is not a forever frozen state: we may decide to analyze further the basic concepts of this vocabulary and we can do this in the form of building an interpretation (in the sense of mathematical logic) into another language the vocabulary of which consists of new basic concepts, and the interpretation gives us the information of how the "old basic concepts" are built up from the "new basic concepts" as refined ones. The interpretation we construct in this chapter thus refines the basic concept of a reference frame in terms of just sending and receiving signals. To refine further the basic concepts of this Signalling theory, we can interpret it to, say, in a theory of electromagnetism, or in a quantum-mechanical theory.

---

[1] A similar investigation for Newtonian gravitation, but not in the framework of mathematical logic, can be found in [45].

[2] Other names for vocabulary are signature and set of nonlogical constants.

An interpretation of this kind may also be regarded as defining a so-called operational semantics for the basic concepts of the first theory. Starting with the Vienna Circle, several authors suggest that a physical theory is a more complex object than just a set of first-order logic (FOL for short) formulas. A physical theory, they propose, is a FOL theory together with instructions for how to interpret the basic symbols (or vocabulary) of this theory "in the real physical world". (Following Carnap [12], this is often called a "(partially) interpreted theory".) We want to show in Sect. 6.6 that such an "operational semantics" can be taken to be an interpretation in the sense of mathematical logic.

Returning to our concrete example, an operational semantics should say something about how we obtain or set up (in the real world) the reference frames for special relativity theory. Usually, rigid meter-rods and standard clocks are used for this purpose (e.g., [44]). However, as [36] points out, we cannot use these rigid meter-rods in astronomy or cosmology. The interpretation we give in detail in this chapter results also in an operational/experimental/observational definition for setting up a reference frame by just relying on sending and receiving light-signals. This method can be used, in principle, in the above mentioned astronomical scale.

Summing up, the first language in an interpretation has the theoretical concepts while the basic concepts of the second language are the observational ones (for the observational-theoretical duality see, e.g., [8, 13]). We can look at the same interpretation "from the other direction": In our example, we may imagine someone living in a space-time, exploring his surroundings by sending and receiving signals, and during this process, he devises so-called theoretical concepts which make thinking more efficient. In particular, he may devise the concept of a reference frame, and even the concept of quantities forming a field, as mental constructs having concrete definitions in terms of observations. The tools of mathematical logic, and more closely those of definability theory (interpretations are among them) can be used for modeling this emergence of theoretical concepts.

A further aim of the present approach of comparing theories is shifting the emphasis from working inside a single huge theory to working in a modularized hierarchy of smaller theories connected in many ways. Usually this approach is called theory-hierarchy. We note that this is not so much a hierarchy as rather a category of theories, technically the category of all FOL-theories as objects with interpretations as morphisms of the category. This direction of replacing a huge theory with a category of small theories is present in many parts of science [7]. In foundational thinking, [14] emphasizes this. In computer science, it is present in the form of structured programming. "Putting theories together" of Burstall and Goguen [11] refers to the act of computing/generating colimits of certain diagrams in this category. Even in such practical areas as using a huge medical data-base the need of modularizing arises: it is necessary to "break up" the given data-base and generate many smaller ones according to the query at hand [21, 22]. The interpretation going from special relativity as formalized in [4] into the more observational Signalling theory of [6] we build in the present chapter is but one morphism of this huge dynamic category of FOL theories.

The content of this chapter can also be viewed as preparing the ground for an application of algebraic logic to relativity theory, as follows. The cylindric algebra of a theory is an abstract representation of the structure of concepts expressible/definable in that theory and a homomorphism between two cylindric algebras corresponds to an interpretation between the corresponding theories. Hence the category of all FOL theories is basically the same as the category of cylindric algebras as objects and homomorphisms as morphisms. There are, for example, well known and understood methods for how to compute colimits in this category of algebras.

There are still many questions and phenomena to be understood in this area of application of logic. For example, what are the desirable or good properties of an interpretation for being informative about the theories in one or other respect? Consider for a second the definability/interpretability picture between scientific theories (in FOL) in two versions: (1) in the framework of traditional definability theory, and (2) in the new, extended theory of definability used in the present chapter. What are the characteristic differences? We think it is useful to keep this picture/issue in mind.

In Sect. 6.2 we briefly recall the relativity theory SPECREL from [4], in Sect. 6.3 we recall Signalling theory SIGTH from [6] and we try to give a basic feeling for it by sketching the proof of the completeness theorem in [6]. Section 6.4 is an important part of the present chapter, it contains an algorithm for how to set up a reference frame in Signalling theory, this is an "operational semantics" for setting up reference frames of Andréka et al. [4]. At the end of the section we outline how the same method could be used for space-times other than special relativistic, e.g., for the Schwarzshild space-time of a black hole. This algorithm is at the heart of the interpretation elaborated in Sect. 6.6. Section 6.5 recalls the features of the more refined definability theory that are needed for defining the interpretation of SPECREL$_0$ into SIGTH. Section 6.7 rounds up the picture between SPECREL and SIGTH by interpreting SIGTH in a slightly reinforced version of SPECREL and then giving more information about connections between various concrete theories of special relativity. We end the chapter with a conclusion.

6.2 Special Relativity

In this section we give a list of basic concepts and axioms of the FOL theory SPECREL in [3–5, 23, 38].

The basic notions not analyzed further in SPECREL are "observers" having reference frames in which they represent the world-lines of bodies (or test particles), of which signals (light-particles, or photons) are special ones. The world-line of a body represents its motion, it is a function that describes the location of the body at each instant. For representing "time" and "location", observers use quantities, quantities are endowed with addition and multiplication in order to be able to express whether a motion is "uniform" or not. To make life simpler, we treat also observers as special bodies. (Another, equivalent, option would be to treat them as entities of different "kind", or of different "sort", than bodies and quantities.) The reference frame or world-view of an observer o gives the information which bodies b are present at time t at location x, y, z; thus $W(o, b, t, x, y, z)$ expresses that body b is present at t in $\langle x, y, z \rangle$, according to observer o. We treat quantities as entities of a different nature, of a different kind, than bodies.

According to the above, the vocabulary of the language of SPECREL is the following: we have two sorts, bodies B and quantities Q, we have two unary relations Obs, Ph of sort B, we have two binary functions $+$, $\star$ of sort Q, and we have a six-place relation W the first 2 places of which are of sort B and the rest of sort Q.

Next, we list the five axioms of SPECREL. Concrete formulas and more intuition can be found in, e.g., [3–5, 23, 38].

AxPh   The world-lines of photons are exactly the straight lines of slope 1, in each reference frame.

AxEv   All observers coordinatize the same physical reality (i.e., the same set of events).

AxSelf   The "owner" of a reference frame sits tight (stays put) at the origin.

AxFd   The quantities form a Euclidean field w.r.t. the operations $+$, $\star$, this means that Q, $+$, $\star$ form an ordered field in which each positive quantity has a square root.

AxSym   All observers use the same units of measurement: if two events are simultaneous for observers o, o', then the spatial distance between them is the same according to o, o'.

$\text{SPECREL}_0 := \{\text{AxPh, AxEv, AxSelf, AxFd}\}$ and
$\text{SPECREL} := \{\text{AxPh, AxEv, AxSelf, AxFd, AxSym}\}$.

SPECREL may seem to be a rather weak axiom system. However, this is not so. All the well-known theorems/predictions of the (kinematics of) special relativity can be proved even from SPECREL_0. Below is a sample of theorems that can be proved from SPECREL_0 (for proofs, further theorems provable from SPECREL, and for extensions see the references given earlier as well as [24, 39]):

• Each observer moves uniformly and slower than light in any other observer's world-view (i.e., the world-line of an observer is a straight line with slope less than 1).
Assume that o, o' are moving relative to each other.
• Events that are separated in o's world-view in a direction of o''s motion and simultaneous according to o, are not simultaneous according to o'.
• Events that are simultaneous according to both o and o' are exactly the ones that are separated orthogonally to the direction of motion of o'.
• Assume that o and o' use the same units of measurement, i.e., the spatial distance between events that are simultaneous to both of them is the same according to them. Then a-synchronicity, time-dilation and length-contraction between o and o' are exactly according to the known formula of special relativity, see e.g., [3, p. 633].

- The world-view transformations between observers in SPECREL$_0$ are exactly the bijections that preserve Minkowski-equidistance; these bijections are the so-called Poincaré-transformations composed with dilations and field-automorphisms.
- The world-view transformations between observers in SPECREL are exactly the bijections that preserve Minkowski-distance; these bijections are the so-called Poincaré-transformations.

In SPECREL, a reference frame is a basic (or primitive) notion, just an "out-of-the-blue" assigning space-and-time coordinates to events, which all together have to satisfy some regularities (our axioms). The theory does not address the question of how an observer sets up his reference frame. As already outlined in the introduction, according to some authors, a physical theory (a theory about our physical reality), should say something about the meaning (in the "real" physical world) of the basic concepts, if not otherwise, then in natural language one could amend the theory with a set of so-called *operational rules* about how the basic concepts (the reference frames in our case) are set up (experimentally). Here usually meter-rods and wrist-watches, or standard clocks, are used, see e.g., Taylor and Wheeler [44, Fig. 9, 135], Szabo [36]. In Sect. 6.4 we give a more ambitious algorithm for setting up coordinate systems.

6.3 James Ax's Signalling theory

The intention of Ax's theory is to give an axiom system for special relativity so that its basic symbols and axioms are designed to be observational. The players of this theory are experimenters that can "communicate with each other" by sending signals to each other. Together, as a team the experimenters can "map" (or explore) space-time, without having rigid meter rods or clocks. A definition of an introduced (or defined) term in this first-order logic theory can be viewed as an experiment designed to establish whether the defined term holds or not. The basic terms of space and time are defined this way. (Indeed, in this theory one can define "rigid meter rods" and "clocks" from signalling experiments.) The results of the experiments we make can be built into axioms then (which are designed to be observational-oriented), and they can tell us in what kind of space-time we live in. Euclidean? Special relativistic? Hyperbolic space with relativistic time? Newtonian? General relativistic? Etc. All this amounts to an implementation of Leibnizian relational notion of space and time. We return to this subject in more detail in the next section.

We begin to describe Ax's theory which we call Signalling theory SIGTH. In the vocabulary of SIGTH we have two sorts, Par for "particles" (or experimenters, or agents) and Sig for "signals" (or light-signals); and we have two binary relations T, R between particles and signals. The intended meanings of $a\mathsf{T}\sigma$ and $a\mathsf{R}\sigma$ are "a transmits (or emits, or sends out) σ", and "a receives (or absorbs) σ", respectively. Ax [6] uses an impersonal terminology of particle physics, particles emit and absorb signals. We are more attracted to a terminology of communication between active experimenters. These experimenters (players) of SIGTH are somewhat analogous to the observers of SPECREL. In this chapter when talking about Ax's Signalling theory, we will use the terms experimenter and particle interchangeably.

The "standard" (or intended) model we have in mind is the following: Let us fix a Euclidean field F. Then Par is the set of all straight lines in F^4 with slope less than 1, and Sig is the set of all directed finite segments (including the segments of length zero) of straight lines with slope 1. A particle a transmits a signal σ iff the beginning point of σ lies on a, and a receiving the signal σ means that the endpoint of σ lies on a. Let us denote this structure by $\mathfrak{M}(F)$.

The main result of Ax [6] is a finite set Σ of axioms, our SIGTH, which characterizes the class of standard models, i.e., the models of Σ are exactly the standard models $\mathfrak{M}(F)$ over some Euclidean field F (Theorem 1 in [6]). SIGTH consists of three groups of axioms, altogether it has 23 elements. Instead of listing these 23 axioms, in this chapter we will use Ax's completeness theorem, since that implies that a formula ψ is provable from SIGTH iff ψ is true in all the standard models.

The question immediately arises: what does this theory SIGTH have to do with special relativity? Do we not lose much expressive power by using such meager resources? Where do a-synchronicity, time-dilation, length-contraction come into the picture in SIGTH? Some answers are in the proof of Theorem 6.1 which we briefly outline below. We will give more explicit answers to these questions in the coming Sects. 6.4 and 6.6. In particular, we will show that everything we can say in the language of SPECREL can be said in the Spartan language of SIGTH, too. One of the ideas for proving this can be traced back to Hilbert, as will be noted in Fig. 6.2.

To give a feeling for SIGTH and the expressive power of its language, we briefly outline the proof of Ax's completeness theorem. Let's begin by making a little elbow-room for working. We will need to express things such as "two signals are received by an experimenter at the same time", and "signal σ was received by an experimenter just when he transmitted signal γ". Since we have no notion of time in our language, we have to express these notions just by using the basic concepts of transmitting and receiving signals. Here comes how we can do this. The (open) formula "$\phi := \forall a\ aT\sigma \rightarrow aT\gamma$" is true in a standard model just when the beginning points of the segments σ, γ coincide, we say that ϕ expresses this fact.[3] Similarly, "$\forall a\ aR\sigma \rightarrow aT\gamma$" expresses that the endpoint of σ coincides with the beginning point of γ, etc. Now can we express that two particles/experimenters meet? Well, they meet if there is a signal that both of them transmit. From now on we will use similar statements without translating them to the language of SIGTH.

To begin outlining the idea of Ax's completeness proof, let $\mathfrak{M}$ be any model of SIGTH, and let e be any experimenter in this model. We will construct an isomorphism between $\mathfrak{M}$ and a standard model $\mathfrak{M}(F)$ which takes e to the time-axis in $\mathfrak{M}(F)$. From now on, in this section e denotes this fixed experimenter.

---

[3] In the formulas, the scope of a quantifier is till the end of the formula if not indicated otherwise. Lower case Roman and Greek letters denote variables of sorts Par and Sig, respectively. Instead of conjunction $\wedge$ we will simply write a comma.

We define the set Space of "places" or "locations" for experimenter e to consist of those particles which are motionless w.r.t. e. For expressing that two particles are motionless w.r.t. each other, any formula expressing this in the standard models will do. Ax uses the following formula: e' is motionless w.r.t. e exactly when e and e' do not meet (if they are not equal) and there are two other particles d, c which meet them and each other in 5 distinct events. Ax then expresses the betweenness relation Bw for such places as well as the equidistance relation Ed with suitable formulas. Having all this, the first group of axioms in SIGTH states Tarski's axioms for axiomatizing Euclidean geometry over the Euclidean fields (see [42]). From Tarski's theorem then Ax gets a Euclidean field F and an isomorphism between $\langle F^3, \text{Bw}, \text{Ed} \rangle$ and $\langle \text{Space}, \text{Bw}, \text{Ed} \rangle$.

Having Space for our experimenter e, what is "time" for him? What are the things that we mark with time? The events. And what are the events? In the present vocabulary we take them to be "particle b emits/receives a signal σ", more precisely we take the equivalence classes of them described when we made the elbow-room for this proof (e.g., particle b may send out signal σ in the same event when it sends out another signal γ or when it receives γ). Then our experimenter e's time will be the events that happened to e. For simplicity, we will represent events with special signals, as explained below.

In the standard models, there are special signals that are received by everyone who transmitted them, we call these signals *events*, we will denote them by variants of ε:

$$\text{Ev}(\varepsilon) \; :\Leftrightarrow \; \forall a \; a\text{T}\varepsilon \to a\,\text{R}\,\varepsilon.$$

In the standard models, events are the light-like segments of zero length, so they correspond to elements of F^4. (These zero-length signals may look counter-intuitive to some readers. It is just handy and not important that we use or have these at hand, everything works with a slight modification if we omit these short signals from the standard models.) We say that event ε happened to experimenter e, or in other words, experimenter e *participated in event* ε, if e transmitted (and then also received) ε. The events that happened to e will constitute e's *world-line*.

We can then express simultaneity of events by using that the speeds of light-signals are the same (see Fig. 6.1). Ax then states an axiom to the effect that signals make a one-to-one correspondence between e's world-line and the simultaneous events on any given line in Space. This makes e's world-line isomorphic to F, we take this to be the time-axis.From now on it is more or less straightforward what we have to include to SIGTH in order to make $\mathfrak{M}$ isomorphic to $\mathfrak{M}(F)$. E.g., we can state that for any event ε there is a simultaneous event ε' on the world-line of e, and there is a particle e' that participates in ε' and is motionless w.r.t. e. This concludes the proof-idea.

6.4 An Algorithm for Setting Up Coordinate Systems

The purpose of this section is twofold. Firstly, in Sect. 6.6 we want to give an interpretation of SPECREL$_0$ into SIGTH, and for this we need concrete formulas representing the proof-idea given in the previous section. For example, Ax used Tarski's theorem for getting the Euclidean field F, but we will need to exhibit concrete formulas defining this F. Secondly, we want to make the previous proof-idea into an algorithm for setting up coordinate systems (i.e., reference frames) with the use of just light-signals and freely moving particles. This could also be viewed as providing *operational semantics* to the basic notion of a coordinate system of SPECREL.

What we give in this section will not be an algorithm in the strict sense, it will be more like a recipe for how to design experiments/measurements for assigning coordinates to events. These experiments will also be suitable for finding out/confirming that we live in a special relativistic space-time (if we do). For this reason, we will try to make the formulas "executable" when possible. There will be plenty of room for improving on this aspect, the reader is invited to design more practical experiments.

Assume that we are given a model $\mathfrak{M}$ of SIGTH, and e is an experimenter in this model. Just as in the previous section, this experimenter e is fixed throughout this section. We are going to give e a recipe for defining a field F of quantities and for assigning four quantities to each event. Such an assignment is called a *coordinate system* (or reference frame). These coordinate systems will satisfy the axioms of SPECREL$_0$.

A *location* for e was defined as a particle that is motionless w.r.t. e. In the previous section we recalled a formula, from [6], expressing whether e' is motionless w.r.t. e (in symbols, $e' \| e$). However, the algorithm suggested by that formula is not very convenient since it involves deciding whether e meets e' or not, and for this e has to know all the events that happened and will happen to him. This is not very practical as an experiment, since e may need to "wait" for an infinity of time before he could know the result. Using the Affine Desargues Property (ADP for short, see, e.g., [17, p. 20]) one can design a more realistic experiment which decides $e' \| e$ "in a finite time", we are going to describe it now. We note that in the standard models $\mathfrak{M}(F)$ the ADP is true, because it is true in the affine space F^4, for any field F.

For a while, it will be easier to think in 4-dimensional space-time than tracing motion in 3-dimensional space. Geometrically, e' is motionless w.r.t. e iff the world-line of e' is parallel to that of e. The conclusion of the ADP is that two lines are parallel, but in the hypothesis part parallelism of two other sets of lines are used. We are lucky: we have light-signals and their speeds are the same in both directions, thus we can use parallelism of world-lines of two sets of light-signals in the hypothesis part of the ADP. The experiment is depicted in geometrical form in the left-hand part of Fig. 6.1. Here is the "non-geometrical" description of the experiment: Assume e wants to decide whether e' is motionless w.r.t. him or not. He asks a brother (another experimenter) to throw towards him three "test" particles ("balls") b_1, b_2, b_3 at once (in one event ε), b_1 faster than b_3 and b_3 faster than b_2 in such a way that when b_1

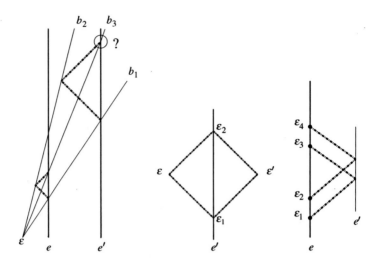

Fig. 6.1 On the left: Experiment for checking whether e' is motionless w.r.t. e. In the middle: Experiment to make sure that ε, ε' are simultaneous w.r.t. e. On the right: Time-equidistance of events $\varepsilon_1, \ldots, \varepsilon_4$

meets e, the latter sends out a signal towards b_2 that b_2 reflects back and the reflected signal reaches e just when b_3 reaches e. (The brother and e have to experiment a little while till finding the right velocities for such three particles.) After checking that b_1, b_2, b_3 have the desired property, e asks e' to do the same: when b_1 reaches a', he should send a signal towards b_2 that reflects this signal back towards e'. If the reflected signal reaches e' just when b_3 reaches e', then e' is motionless w.r.t. e; otherwise e' is not motionless w.r.t. e. It is best to imagine this experiment to take place in outer space, far from heavy heavenly objects so that gravity and friction do not bend the world-lines of the "balls".[4] From now on, we will use "locations" and "places" as being particles/experimenters motionless w.r.t. our fixed experimenter e.

Two events ε, ε' are defined to be *simultaneous* w.r.t. e iff there is a place e' such that from e' two signals can be sent at the same event towards the locations of ε and ε' respectively such that if these signals are sent back from ε and ε' right away, they will arrive back to e' at the same event, see middle of Fig. 6.1. Formally: $\varepsilon \equiv_e \varepsilon' :\Leftrightarrow \exists e' \| e, \sigma_1, \ldots, \sigma_4, \varepsilon_1, \varepsilon_2$ $\mathsf{Ev}(\varepsilon_1)$, $\mathsf{Ev}(\varepsilon_2)$, $(\varepsilon_1, \sigma_1, \varepsilon)$, $(\varepsilon_1, \sigma_2, \varepsilon')$, $(\varepsilon, \sigma_3, \varepsilon_2)$, $(\varepsilon', \sigma_4, \varepsilon_2)$, $e'\mathsf{T}\varepsilon_1$, $e'\mathsf{T}\varepsilon_2$, where $(\varepsilon, \sigma, \gamma)$ means that ε, γ are the events of sending and receiving σ, respectively, formally: $(\varepsilon, \sigma, \gamma) :\Leftrightarrow (\mathsf{Beg}(\sigma, \varepsilon), \mathsf{End}(\sigma, \gamma))$ where $\mathsf{Beg}(\sigma, \varepsilon)$ expresses that σ, ε are sent out at the same event, formally: $\mathsf{Beg}(\sigma, \varepsilon) :\Leftrightarrow \forall b\, b\mathsf{T}\sigma \rightarrow b\mathsf{T}\varepsilon$ and a similar definition for End. (Note that if we want a more experiment-friendly formula for Beg, then we can use the following: $\mathsf{Beg}(\sigma, \varepsilon) \Leftrightarrow (\exists b, c\ b \neq c, b\mathsf{T}\sigma, b\mathsf{T}\varepsilon, c\mathsf{T}\sigma, c\mathsf{T}\varepsilon)$.) We even can provide instructions for where to look for such a place e': it can be chosen to be the midpoint of the

---

[4] Or, if we are content with more approximate measurements, we can imagine all this happening on a big lake covered with smooth ice (but then we have to take space to be 2-dimensional).

line-segment connecting the locations of ε and ε'. (We can use this experiment for setting two clocks at the places of ε, ε' which "tick simultaneously").

We get an ordering on all the events from the fact that we send a signal earlier than receiving it, namely ε is *earlier* than ε' iff we can send a signal at ε to an event from where it bounces back to ε' ($\varepsilon \prec \varepsilon'$:$\Leftrightarrow$ [$\exists \varepsilon''$, σ_1, σ_2 (ε, σ_1, ε''), (ε'', σ_2, ε')]). For example, ε_1 is earlier than ε_2 in the middle part of Fig. 6.1. We note that, while two events being simultaneous or not depends on which experimenter makes the experiment deciding simultaneity, one event being earlier than another does not depend on any experimenter.

Let's see, what structure the set of events happening to e has. Let Time_e denote the world-line of e and let $\varepsilon_1, \ldots, \varepsilon_4 \in \mathsf{Time}_e$. Besides the ordering, we also have *time-equidistance* of events, since the speed of all signals is the same: the time elapsed between ε_1 and ε_2 is the same as that between ε_3 and ε_4, iff there is a place e' to which we can send signals from ε_1, ε_2 resp., these bounce from e' and arrive back to e at ε_3, ε_4 respectively. See the right-hand part of Fig. 6.1. More precisely, this is the definition when $\varepsilon_1 \prec \varepsilon_3$. When $\varepsilon_3 \prec \varepsilon_1$, we get the definition by interchanging the pairs ε_1, ε_2 and ε_3, ε_4. (Formally, $\mathsf{Edt}_e(\varepsilon_1, \varepsilon_2, \varepsilon_3, \varepsilon_4)$, $\varepsilon_1 \prec \varepsilon_3$:$\Leftrightarrow$ [$\exists e' \| e, \sigma_1, \ldots, \sigma_4, \varepsilon, \varepsilon'$ $\mathsf{Ev}(\varepsilon)$, $\mathsf{Ev}(\varepsilon')$, ($\varepsilon_1, \sigma_1, \varepsilon$), ($\varepsilon, \sigma_3, \varepsilon_3$), ($\varepsilon_2, \sigma_2, \varepsilon'$), ($\varepsilon', \sigma_4, \varepsilon_4$), $e'\mathsf{T}\varepsilon$, $e'\mathsf{T}\varepsilon'$], and $\mathsf{Edt}_e(\varepsilon_1, \varepsilon_2, \varepsilon_3, \varepsilon_4)$, $\varepsilon_3 \prec \varepsilon_1$:$\Leftrightarrow$ $\mathsf{Edt}_e(\varepsilon_3, \varepsilon_4, \varepsilon_1, \varepsilon_2)$, $\varepsilon_3 \prec \varepsilon_1$.) Note that $\mathsf{Edt}_e(\varepsilon_1, \ldots, \varepsilon_4)$ implies that ε_1 happens earlier than ε_2 iff ε_3 happens earlier than ε_4. By using time-equidistance, we can define *addition* by selecting a "*zero*" time $o \in \mathsf{Time}_e$ as parameter, namely $\tau = \tau_1 + \tau_2$:$\Leftrightarrow$ $+$ (τ, τ_1, τ_2, a, o) :$\Leftrightarrow$ $\mathsf{Edt}_e(o, \tau_1, \tau_2, \tau)$. Now that we have addition, we do not stop before having *multiplication*. For this we have to choose a *unit* time $\iota \in \mathsf{Time}_e$, distinct from o and happening later than o, as another parameter. For defining multiplication, we will need the *collinearity* relation on locations, we will get this by noticing that the space-trajectories of signals are (3-dimensional) straight lines in the standard models: $\mathsf{Col}(a_1, a_2, a_3)$ iff exist signals $\sigma_1, \sigma_2, \sigma_3$ and events $\varepsilon_1, \varepsilon_2, \varepsilon_3$ such that ($\varepsilon_i, \sigma_1, \varepsilon_j$), ($\varepsilon_j, \sigma_2, \varepsilon_k$), ($\varepsilon_i, \sigma_3, \varepsilon_k$) and $a_1\mathsf{T}\varepsilon_1$, $a_2\mathsf{T}\varepsilon_2$, $a_3\mathsf{T}\varepsilon_3$, for some permutation i, j, k of $1, 2, 3$.

We define $\tau_1 \star \tau_2$ for the case when τ_1 happened later than ι, and τ_2 happened later than o. See the left-hand part of Fig. 6.2. (The other cases are similar, we leave them out.) Here is how we find out whether τ is $\tau_1 \star \tau_2$: we find two places b_1 and b_2 collinear with e and we find a particle p such that if b_1 and b_2 send towards e, simultaneously, at time zero, a light-signal and p, and another light-signal and another particle q "with the same speed" as p, then these four arrive (to e) at times ι, τ_1, τ_2, τ, respectively. Formally: $\tau = \tau_1 \star \tau_2$:$\Leftrightarrow$ $\star$ (τ, τ_1, τ_2, e, o, ι) :$\Leftrightarrow$ $\exists b_1 \| e, b_2 \| e, \iota', \tau_2', \sigma_1, \sigma_2, p, q$ [$\mathsf{Ev}(\iota')$, $\mathsf{Ev}(\tau_2')$, $\mathsf{Col}(e, b_1, b_2)$, $o \equiv_e \iota'$, $o \equiv_e \tau_2'$, $b_1\mathsf{T}\iota'$, $b_2\mathsf{T}\tau_2'$, (ι', σ_1, ι), ($\tau_2', \sigma_2, \tau_2$), $p\mathsf{T}\iota'$, $p\mathsf{T}\tau_1$, $q\mathsf{T}\tau_2'$, $q\mathsf{T}\tau$, $p\|q$].

The reader will have noticed that the above definitions of addition and multiplication on Time_e are a special case of Hilbert's coordinatization procedure, see e.g., [17, pp. 23–28] or [23, pp. 296–308].

By the above, we have a structure $\mathsf{F}(e, o, \iota) = \langle \mathsf{Time}_e, +, \star \rangle$ which is isomorphic to our field F in the intended models $\mathfrak{M}(\mathsf{F})$. We define the above structure to be

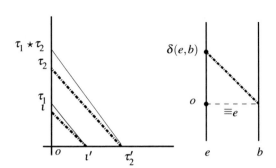
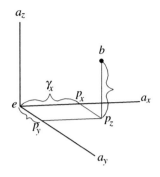

Fig. 6.2 On the left: Experiment for computing $\tau_1 \star \tau_2$. In the middle: distance between locations e, b. On the right: Spatial coordinates of location b. In the picture, $\gamma_x = \delta(e, p_x)$. In this part points represent (3-dimensional) locations, while in the previous pictures points represent (4-dimensional) events

the field of *quantities* of our fixed experimenter e. We define the *time-coordinate* of an arbitrary event ε as an element of this field, namely the unique event on e's world-line which is simultaneous with it (simultaneous according to e). Next, we define three coordinates, three elements of this field, for each location b. From now on, let Space_e denote the set of locations for e.

We begin by defining a geometric structure on Space_e, namely we will define distance of locations, parallelism and orthogonality of (3-dimensional) spatial lines.

We define the *distance* of any two locations. Let $b \in \mathsf{Space}_e$ be arbitrary. We define the distance of b from our fixed e as the event when a signal sent from b at time zero arrives to e, see the middle part of Fig. 6.2. This definition corresponds to a convention that we measure spatial distances in light-years (if we measure time in years). Having this, we get the distance between any two locations b_1, b_2 by measuring the distance between their parallel translated versions so that b_1 gets to e, i.e., $\delta(e, b) = \varepsilon :\Leftrightarrow \exists\varepsilon', \sigma \, [\mathsf{Ev}(\varepsilon'), \varepsilon' \equiv_e o, b\mathsf{T}\varepsilon', (\varepsilon', \sigma, \varepsilon), \varepsilon \in \mathsf{Time}_e]$, and $\delta(b_1, b_2) = \varepsilon :\Leftrightarrow [\varepsilon = \delta(e, b), \mathsf{pa}(b_1, b_2, e, b), \mathsf{pa}(b_1, e, b_2, b)]$ where $\mathsf{pa}(b_1, b_2, b_3, b_4)$ means that the spatial lines defined by b_1, b_2 and b_3, b_4 are *parallel*, we easily can express this by using the collinearity relation Col between locations as defined earlier in this section.

We also need the orthogonality relation which is definable from the equidistance of pairs of locations. We define orthogonality of two intersecting lines only. We call the lines going through a, b and a, c orthogonal, if $a \neq b, a \neq c$ and there is a $b' \neq b$ on the spatial line going through a, b such that the distances between a, b' and a, b equal, and also those between c, b' and c, b equal ($\mathsf{Ort}(a, b, a, c) :\Leftrightarrow \exists b'[\mathsf{Col}(b', a, b), \delta(a, b') = \delta(a, b), \delta(c, b') = \delta(c, b)]$). By now we defined a structure $\langle \mathsf{Space}_a, \mathsf{Col}, \mathsf{pa}, \mathsf{Ort} \rangle$ and we defined distance $\delta : \mathsf{Space}_e^2 \longrightarrow \mathsf{Time}_e$.

Setting up a coordinate system needs three more parameters, the three space-axes. Let $a_x, a_y, a_z \in \mathsf{Space}_e$ be such that e, a_x, e, a_y and e, a_z are pairwise orthogonal.

We have everything for defining the usual spatial coordinates of the place b. See the right-hand part of Fig. 6.2. The spatial coordinates of a location b are defined the usual way by "projecting" b to the three coordinate axes, along lines parallel with some of the axes, and measuring the distance of the projected points from the origin (our experimenter e in our case). See the formula cor below.

We can now round up the definition of the coordinate system our experimenter e is setting up. We already defined the time-coordinate of an event ε, and we define the space-coordinates of ε to be the spatial coordinates just defined for the "location of ε", the latter being the unique particle participating in ε and motionless w.r.t. our experimenter e. The formula $\mathrm{cor}(\varepsilon, \tau, \gamma_x, \gamma_y, \gamma_z, e, o, \iota, a_x, a_y, a_z)$ defined below expresses that the coordinates of the event ε are $\tau, \gamma_x, \gamma_y, \gamma_z$ in the coordinate system specified by $e, o, \iota, a_x, a_y, a_z$.

$$\mathrm{cor}(\varepsilon, \tau, \gamma_x, \gamma_y, \gamma_z, e, o, \iota, a_x, a_y, a_z) \ :\Leftrightarrow\ \varepsilon \equiv_e \tau, \exists b, p_x, p_y, p_z \in \mathsf{Space}_e \lfloor b \mathsf{T} \varepsilon,$$
$$\mathrm{pa}(b, p_z, e, a_z), \mathrm{pa}(p_z, p_x, e, a_y), \mathrm{pa}(p_z, p_y, e, a_x), \mathsf{Col}(e, p_x, a_x), \mathsf{Col}(e, p_y, a_y),$$
$$\delta(p_z, p_x) = \gamma_x, \delta(p_z, p_y) = \gamma_y, \delta(b, p_z) = \gamma_z. \rfloor.$$

Since it can be proved that the associated coordinates are unique, we will also use the functional form

$$\mathrm{cor}(\varepsilon, e, o, \iota, a_x, a_y, a_z) = (\tau, \gamma_x, \gamma_y, \gamma_z) \ :\Leftrightarrow\ \mathrm{cor}(\varepsilon, \tau, \gamma_x, \gamma_y, \gamma_z, e, o, \iota, a_x, a_y, a_z).$$

By the above, we have defined coordinate systems to each particle $e \in \mathrm{Par}$. Such a coordinate system is defined by six parameters: $e, o, \iota, a_x, a_y, a_z$. Before going on, we show that the relativistic (or, in other words, Minkowski-) distance between events can be defined in these coordinate systems. We call two events $\varepsilon, \varepsilon'$ *time-like separated* iff there is a particle participating in both. For simplicity, we will define relativistic distance between time-like separated events only. See Fig. 6.3.

The relativistic distance we are going to define will depend on experimenter e and on the chosen zero o of its coordinate system. Let first $\varepsilon \succ o$ be any event time-like separated from o. Then $\mu_{e,o}(o, \varepsilon) = \xi$ iff there is an event ε' which is simultaneous with o both according to e and according to the unique observer participating in o, ε, and there are signals from ε' to ε and from ε' to ξ, respectively. It can be checked that in any standard model $\mathfrak{M}(\mathsf{F})$, if o, ε, ξ are in the above described configuration, then the "standard" Minkowski-distances between o, ε and o, ξ are the same.Conversely, if these two distances agree then there exists an event ε' as in Fig. 6.3. Let now $\varepsilon_1, \varepsilon_2$ be any two time-like separated events, $\varepsilon_1 \prec \varepsilon_2$. Then the relativistic distance between $\varepsilon_1, \varepsilon_2$ is the same as that between the "parallel translations" o, ε of these, where the "parallel translation" happens according to Fig. 6.3 (where for ε'' it is important only that it is connected to both o and to ε_1 with a light-signal, e.g., it is not important that $o \prec \varepsilon''$). If $\varepsilon_1 \succ \varepsilon_2$ then we define $\mu_{e,o}(\varepsilon_1, \varepsilon_2) = -\mu_{e,o}(\varepsilon_2, \varepsilon_1)$. We note that, while this relativistic distance strongly depends on the parameters e, o, the relativistic *equi*distance relation we get from this does not depend on e, o any more. So, let us define relativistic equidistance, or 4-equidistance, as

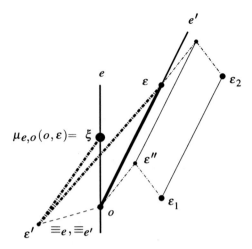

Fig. 6.3 Relativistic distance $\xi = \mu_{e,o}(o, \varepsilon) = \mu_{e,o}(\varepsilon_1, \varepsilon_2)$ between events $\varepsilon_1, \varepsilon_2$

$\mathsf{Edr}(\varepsilon_1, \varepsilon_2, \varepsilon_3, \varepsilon_4) :\Leftrightarrow \mu_{e,o}(\varepsilon_1, \varepsilon_2) = \mu_{e,o}(\varepsilon_3, \varepsilon_4)$, for any $e \in \mathsf{Par}$ and event o on e's world-line.

Having defined the desired coordinate systems in SIGTH, we conclude this section with some remarks on what this method can give us, what it can be used for.

We asked earlier, in Sect. 6.3, where the paradigmatic effects—a-synchronicity, time-dilation, length-contraction—of special relativity theory came into the picture in Signalling theory. One answer is the following. We defined natural coordinate systems to the particles. (These coordinate systems correspond to the observers in SPECREL, this correspondence will be made explicit in Sect. 6.6.) Now, the coordinate-transformations between these are so that the three paradigmatic effects of special relativity (mentioned in Sect. 6.2) hold in a version where we can recalibrate the units of measurement.

This section contains definitions only, definitions (with some parameters) in the language of SIGTH that in the standard models define coordinate systems for the particles/experimenters. We can get an axiom system characterizing the standard models (thus doing the job of SIGTH) via using these definitions. Namely, we can state as axioms that the coordinate systems defined for the experimenters have all the good properties we want (e.g., the beginning and end-points of light-signals are exactly those of the ordered segments of slope 1). This alternative axiom system would be more complicated and less natural than SIGTH of Ax [6], however, it would be the result of a clear-cut method that can be used in many other situations, as indicated below.

We can use the method of this section for exploring space-times other than the special relativistic one, and for using signals of various different nature, too. We mention some examples briefly, we think that elaborating these examples would be worthwhile.

We can use the method of setting up a coordinate system as described in this section, for example, for a particle moving faster-than-light (FTL) in a special relativistic space-time. So, let us take as standard models the standard models $\mathfrak{M}(\mathsf{F})$ modified so that the particles are the lines with slope more than 1 (and not the ones with slope less than 1). If we apply our method to these modified models, then the FTL experimenter e will find that its space Space_e is a 3-dimensional Minkowski-space $\mathsf{MS} \; := \; \langle \mathsf{F}^3, \mathsf{Bw}, \mathsf{Edr} \rangle$, and not a Euclidean space $\langle \mathsf{F}^3, \mathsf{Bw}, \mathsf{Ed} \rangle$. He can reach by signals directly, and check whether they are motionless w.r.t. him, only those places/brothers that are time-like separated from him in terms of MS, but he can get indirect information about the rest of places by communicating with these primarily reachable brothers. By working through the details, we can get an axiom system $\mathsf{SIGTH}^{\mathsf{ftl}}$ axiomatizing the signalling models of FTL experimenters that would be quite analogous to Ax's SIGTH. The main difference would be that the first group of axioms for 3-dimensional Euclidean space would be replaced by an analogous axiom system for 3-dimensional Minkowski space. For this we can use the one devised by Goldblatt in [17, Appendix A]. For a slightly different approach for including FTL observers in this setting see [20].

However, communicating with directed signals (as in SIGTH) between FTL experimenters is rather restricted if we want to take the experiments to be executable (FTL experimenters can get information this way only about the part of their space MS which is in their "past" in terms of MS as a Minkowski-space). We can change the nature of signals to be undirected (but otherwise letting their speed to be 1), imagining that if two events are connected with a signal, then the information this signal carries appears at both events "at once". This is connected somehow to time-travel, a subject strongly connected to FTL motion. The method given in the present section is suitable for exploring space-time with undirected signals, too.

The method given in this section can also be used for giving meaning to two-dimensional time. Time being 2-dimensional could simply mean that the events happening with the experimenters can be best described by, say, the structure $\langle \mathsf{F}^2, \prec \rangle$. For example, one could assume that our experimenter lives in a world characterized by the $2+2$-dimensional Minkowski-metric $\sqrt{t_1^2 + t_2^2 - x^2 - y^2}$ and then apply our method to see what kind of coordinate system he would set up for himself, and in general, what kind of responses he would get to his experiments.

Finally, we can imagine using signals of infinite velocity, this way we can explore the Newtonian space-time characterized by absolute time. Or, we can use bent signals of general relativity. For example, we can explore the outer part of the Schwarzshild black hole (the space-time outside the event horizon) with the same method. We would take as experimenters a team of densely placed suspended observers (spaceships in outer space using their drives to maintain their desired positions), constantly checking positions by communicating with photons (as light-signals), and using freely-falling spaceships (or astronauts) as messengers.

6.5 Defining New Entities, Interpretations

Our aim is to clarify the connections between SPECREL and SIGTH. Not only the vocabularies of these two theories are disjoint, even on the intuitive level they speak of different kinds of things. We can see that somehow photons and observers of SPECREL correspond to signals and particles of SIGTH, but what correspond to quantities in SIGTH? Quantities of SPECREL do not seem to enter the picture in SIGTH. Yet, in Sect. 6.4 we defined something that intuitively could correspond to quantities in SIGTH. In this section we recall some tools from mathematical definability theory by which we can make explicit the way quantities arise in SIGTH.

We briefly recall the tools that we will use in the next section for making connections between theories for special relativity in a very precise sense. We elaborated these tools in [1, 23] for the specific purpose of establishing a strong connection between two versions of special relativity theory, the so-called observer-independent geometrical and the reference-frame oriented ones. We only recall the syntactic form to be used in specifying a concrete interpretation together with some background intuition. We elaborated a more extensive definability theory for this kind of connecting theories that we do not recall here. We will say some words about it at the end of this section. For simplicity, we will treat function symbols as special relation symbols.

In "traditional", one-sorted definability theory, an interpretation of a theory Th' in language $\mathscr{L}'$ into a theory Th in another language $\mathscr{L}$ is the following. For each n-place relation symbol R of $\mathscr{L}'$ we assign a formula φ_R of $\mathscr{L}$ with at most n free variables. (We think of φ_R as the "definition of R" in $\mathscr{L}$.) This then defines a natural translation function $\mathsf{tr} : \mathscr{L}' \longrightarrow \mathscr{L}$ by replacing each atomic formula $R(v_1, \ldots, v_n)$ with $\varphi_R(v_1, \ldots, v_n)$. This is an interpretation of $\mathscr{L}'$ into $\mathscr{L}$. This interpretation is an interpretation of Th' into Th iff Th proves the translated theory Th', i.e.,

$$\star \quad \mathsf{Th} \models \mathsf{tr}(\psi) \text{ whenever } \mathsf{Th}' \models \psi, \quad \text{for all } \psi \in \mathscr{L}'.$$

On the semantic side, an interpretation of Th' into Th "constructs" a model of Th' inside each model of Th. Namely, it associates a model $\mathsf{tr}(\mathfrak{M})$ of $\mathscr{L}'$ to each model $\mathfrak{M}$ of $\mathscr{L}$ in such a way that the universe of $\mathsf{tr}(\mathfrak{M})$ is the same as that of $\mathfrak{M}$, and for each assignment k of the variables into this universe we have

$$\star\star \quad \mathsf{tr}(\mathfrak{M}) \models \psi[k] \quad \text{if and only if} \quad \mathfrak{M} \models \mathsf{tr}(\psi)[k], \quad \text{for each formula } \psi \text{ in } \mathscr{L}'.$$

In the new, "non-traditional" or "generalized" definability theory we will use a notion of interpretation that does the same thing, except that the universe of $\mathsf{tr}(\mathfrak{M})$ will not necessarily be a subset of the universe of $\mathfrak{M}$, therefore its definition and the property analogous to ($\star\star$) above will be more involved. We will define new entities as elements of new "sorts". Using many-sorted FOL is not an essential feature of this generalized definability theory, just it is convenient in many cases, as it is in our present task.

We illustrate the idea of defining new sorts with a simple example. The language of affine planes in, e.g., [17] is two-sorted, we have two sorts Points, Lines and we have a binary relation between them, the relation I of incidence (or membership) between a point and a line. Another language in use for the same is one-sorted, see, e.g., [43], we have one sort Points and we have a three-place relation Col of "collinearity" between three points. Everyone can connect the two ways of thinking about affine planes immediately: a line is the set of all points collinear with given two distinct points. Thus a line ℓ is a subset of the old universe, given two distinct points p, q the line ℓ going through them is defined by

$$\ell(p, q) \ := \ \{x : \mathsf{Col}(x, p, q)\}.$$

But the new sort Lines stands for the set of all these subsets! We can specify one line with the open formula $\mathsf{Col}(x, p, q)$ with one free variable x, but how can we define the set of all lines? Well, we will define the set of the parameters p, q specifying the individual lines: we identify the set of all lines with the set of pairs of distinct points. Thus the formula defining the new sort Lines will have two free variables p, q and it will state $p \neq q$. We are almost there, except that different pairs of distinct points may specify the same line, and we have to take this into account when talking about equality of lines, i.e., when interpreting the equality symbol on the sort Lines. We can do this again with a formula using 4 free variables p, q, p', q' stating when the lines specified by p, q and p', q' coincide. In our case this formula can be taken to be $\mathsf{Col}(p', p, q) \wedge \mathsf{Col}(q', p, q)$.

So far we have defined the universe of the new sort Lines and the equality relation of this new sort by two formulas in the "old" language, i.e., in the language talking about Points and Col. Having defined a universe means that we have variables ranging over this universe (and we can quantify over them). In other words, we have to introduce variables Var(Lines) of sort Lines. Then, in order to be able to use the definition of the new sort Lines, we need to connect Var(Lines) to variables used in the definition for Lines, i.e., to Var(Points). We can state this connection by matching a variable ℓ of sort Lines with variables denoting its "defining parameters", e.g., we can state that ℓ_p, ℓ_q denote parameters that define ℓ. After this we can define the incidence relation, too: $I(x, \ell) :\Leftrightarrow \mathsf{Col}(x, \ell_p, \ell_q)$, where x is a variable of sort Points and ℓ is a variable of sort Lines.

Summing up: defining the new sort Lines goes by defining the variables Var(Lines) of the new sort and matching them to the variables of the old sort Var(Points) occurring in the defining formula of the sort Lines, defining the equality on the sort Lines, and defining the non-logical symbol of incidence I which involves the sort Lines. Thus we can interpret the 2-sorted language of affine planes into the one-sorted one by the following data:

$$\mathsf{var} : \ell \mapsto \langle \ell_p, \ell_q \rangle \quad \text{for } \ell \in \mathsf{Var(Lines)},$$
$$\mathsf{Lines}(\ell) :\Leftrightarrow \ell_p \neq \ell_q,$$
$$\ell = h :\Leftrightarrow \mathsf{Col}(\ell_p, h_p, h_q), \mathsf{Col}(\ell_q, h_p, h_q),$$
$$I(x, \ell) :\Leftrightarrow \mathsf{Col}(x, \ell_p, \ell_q).$$

The above data then define a translation function tr from the 2-sorted language of affine planes to their one-sorted language as follows:

$$\mathsf{tr}(\exists \ell \psi) := \exists \ell_p, \ell_q \; \ell_p \neq \ell_q, \mathsf{tr}(\psi),$$
$$\mathsf{tr}(\ell = h) := \mathsf{Col}(\ell_p, h_p, h_q), \mathsf{Col}(\ell_q, h_p, h_q),$$
$$\mathsf{tr}(I(x, \ell)) := \mathsf{Col}(x, \ell_p, \ell_q),$$

the rest of the definition of tr is more or less straightforward.

The new feature in this translation function, over the traditional one, is that we translate the quantifiers according to the defining formula and variable-matching of the new sort and we translate equality on the new sort, too. Throughout, we will use the above variable matching $\mathsf{var} : \ell \mapsto \langle \ell_p, \ell_q \rangle$ without recalling it.

This translation is not only recursive and structural, it is also meaning preserving in the sense analogous to $(\star\star)$. In more detail: let $\mathfrak{M} = \langle P, \mathsf{Col} \rangle$ be a model of the one-sorted language. We will construct its "translation", a model $\mathsf{tr}(\mathfrak{M})$ of the two-sorted language. Let

$$U := \{\langle x, y \rangle \in P \times P : x \neq y\}, \text{ and let } E \subseteq U \times U \text{ be defined by}$$
$$E := \{\langle u, v \rangle \in U \times U : \mathsf{Col}(u_1, v_1, v_2), \mathsf{Col}(u_2, v_1, v_2)\}.$$

Assume that E is an equivalence relation on U, then define

$$\mathsf{tr}(\mathfrak{M}) := \langle P, L, I \rangle \quad \text{where}$$
$$L := U/E \quad \text{and}$$
$$I := \{\langle x, u \rangle \in P \times L : \mathsf{Col}(x, v_1, v_2) \text{ for some } v \in u/E\}.$$

Let Var_P, Var_L denote the sets of variables in the 2-sorted language of the affine planes and let $\mathsf{Var}'_P := \mathsf{Var}_P \cup (\mathsf{Var}_L \times \{1\}) \cup (\mathsf{Var}_L \times \{2\})$ be the variables of the one-sorted language. Now, let $k : \mathsf{Var}_P \cup \mathsf{Var}_L \longrightarrow \mathsf{tr}(\mathfrak{M})$ be any evaluation of the variables of the 2-sorted language, and let $\mathsf{tr}(k) : \mathsf{Var}'_P \longrightarrow \mathfrak{M}$ be an evaluation of the variables of the one-sorted language such that $\mathsf{tr}(k)(x) = k(x)$ if $x \in \mathsf{Var}_P$, and if $\ell \in \mathsf{Var}_L$ then $\langle \mathsf{tr}(k)(\ell, 1), \mathsf{tr}(k)(\ell, 2) \rangle$ is an arbitrary element of $k(\ell)$. Then the following is true for each formula ψ of the 2-sorted language:

$$(\star\star') \quad \mathsf{tr}(\mathfrak{M}) \models \psi[k] \quad \text{if and only if} \quad \mathfrak{M} \models \mathsf{tr}(\psi)[\mathsf{tr}(k)].$$

The above ($\star\star'$) expresses that the translation function preserves meaning when we talk about the 2-sorted model constructed inside the one-sorted model.

Now, such a translation tr is an interpretation from Th$'$ into Th iff, just as before,

$$(\star') \quad \text{Th} \models \text{tr}(\psi) \text{ whenever } \text{Th}' \models \psi, \quad \text{for all } \psi \in \mathscr{L}'.$$

Definitional equivalence of theories Th$'$, Th in different languages $\mathscr{L}'$, $\mathscr{L}$ is a strong connection between them, much stronger than mutual interpretability requiring that the two interpretations be inverses of each other, up to isomorphism. (Cf. [23, Ex. 4.3.46, p. 266]).

Two theories Th$'$ and Th are said to be *definitionally equivalent* if they have a common definitional extension. Here, two theories are said to be the same if they prove the same formulas. But what is a definitional extension? In the one-sorted case, definitional extension of Th is Th $\cup$ Δ where Δ is a union of definitions of the form $\Delta(R) := \{R(v_1, \ldots, v_n) \leftrightarrow \varphi_R(v_1, \ldots, v_n)\}$ with φ_R as above ($\star$) (see, e.g., [19, pp. 60–61]). For telling what definitional extension is in the many-sorted case, we return to our previous example of defining the sort Lines. Let us write $\delta(p, q)$ and $\varepsilon(p, q, p', q')$ for $p \neq q$ and Col(p', p, q), Col(q', p, q) respectively, for the formulas defining the "domain" and the "equality" on the new sort Lines. The explicit definition of the sort Lines will also involve a new relation π fixing the connection of the new sort to the old ones. Now, $\Delta($Lines$, \pi)$ is defined to be the set of the following sentences

$$\exists p, q \, (\pi(p, q, \ell), \pi(p, q, \ell')) \leftrightarrow \ell = \ell',$$
$$\exists \ell \, (\pi(p, q, \ell), \pi(p', q', \ell)) \leftrightarrow \varepsilon(p, q, p', q'),$$
$$\exists \ell \, (\pi(p, q, \ell)) \leftrightarrow \delta(p, q).$$

We note that the intuitive meaning of $\pi(p, q, \ell)$ is that "p, q are distinct points lying on ℓ", or, "p, q code, or represent, line ℓ". So far it was the variable matching that played this role and, intuitively, $\pi(p', q', \ell)$ is an explicit way of saying $\varepsilon(p', q', \ell_p, \ell_q)$.

After having defined the new sort Lines, the definition $\Delta(I)$ of the incidence relation is the same as in the one-sorted case:

$$I(p, \ell) \leftrightarrow \exists p', q' \, (\pi(p', q', \ell), \text{Col}(p, p', q')).$$

Now, Th $\cup$ $\Delta($Lines$, \pi) \cup \Delta(I)$ is a definitional extension of Th, where Th is the "one-sorted" theory of affine planes. A *definitional extension* of any theory Th is Th $\cup$ Δ where Δ is a union of definitions of the above form. Instead of describing the above in more detail, we refer to [1], [23, Sect. 4.3], [2, Sect. 6.3] where many examples can also be found.

The notion of definitional equivalence is important for our purposes, and we believe that it is an important one in understanding how we form our concepts. We try to illustrate this with an example. We will see that the theory EFD of Euclidean fields and the theory SIGTH of special relativity are mutually interpretable into each other. However, they are not definitionally equivalent.[5] Namely, SIGTH and EFD cannot have a common definitional extension because of the following two reasons. (i) SIGTH has to be an "information-losing" reduct of any definitional extension of EFD, and (ii) any theory is an "information-preserving" reduct of any of its own definitional extensions. We note that (ii) holds because the very idea of "definitional extension" is an extension based on "information" contained in the unextended theory; thus by forgetting this extra structure we lose nothing, we can recover it from the unextended theory. We explain (i): In a definitional extension of EFD of which SIGTH is a reduct, we will define the new sort Par of experimenters together with a projection function π_P which ties the behavior of Par to EFD. Such a projection function will single out the experimenter whose world-line is the time-axis, in other words, we can single out "the" motionless experimenter.[6] Absolute motion! However, the *essence* of relativity theory is that motion is relative. This is formalized in the so called Special Principle of Relativity, which states that all the experimenters are equivalent, we cannot tell which one is motionless and which one moves. Indeed, any experimenter can be taken to any other experimenter by an automorphism, in any model of SIGTH. Thus, when making the reduct of a definitional extension of EFD in order to obtain SIGTH, we have to forget π_P, otherwise we do not get the right concept of experimenter. This is "information-loss" since we cannot recover π_P from SIGTH. This shows that "forgetting" is an important part in forming the concept of experimenter in this case. "Less is more" in this case. Definitional equivalence keeps track of these kinds of "forgetting", while mutual interpretability may not do this.

We conclude this section with a few words about interpretations. We already wrote about the philosophical importance of interpretations between theories in the introduction. Here we write about more technical aspects. An interpretation tr from theory Th′ to Th is a connection between them, and this connection imports some properties of one theory to the other. For example, if Th is consistent, then Th′ is also consistent. If tr is faithful and Th′ is undecidable, then Th is also undecidable, and if Th′ and Th are mutually interpretable in each other, then an axiom system for Th can be imported to Th′ via any two mutual interpretations. Definitional equivalence induces a strong duality between Th′ and Th. For these kinds of application of interpretations see, e.g., [15, 23, 28, 41]. The present chapter intends to show the usefulness of interpretations in physical theories, e.g., defining operational semantics for a physical theory. We note that definability theory is quite extensively used in geometry, see, e.g., [17, Appendix B], [28–30, 43].

Versions of the general interpretability we use in this chapter appeared in various different forms as early as in 1969, see [10, 19, 25–27, 31, 37]. Almost all of these works use a syntactic device similar to ours, let's call it explicit definitions, but

---

[5] Similar observations apply to a slight variant SPECREL$_0$ + COMPL of SPECREL in place of SIGTH (cf. Theorem 6.2 in Sect. 6.7). This can be extended to the Newtonian theory in [2, Sect. 4.1, p. 423].

[6] The easiest way of making this precise is that there are fields with no automorphisms at all, e.g., the field of real numbers, and this means that the structure $\langle$Par, π_P, EFD$\rangle$ will have no automorphism, either.

they all elaborate on different semantical aspects of this general definability. For example, [10] characterizes when a functor of a given form is the semantical part of an interpretation. Makkai [27], Makkai and Reyes [25] recast model theory in a categorical form, where both the syntactical and semantical parts of an interpretation are functors between pretoposes, and it is proved that both functors are equivalences when one is. This theorem is called a conceptual completeness theorem. For the model theoretical forms, meaning and impacts of this completeness theorem we refer to [18]. (We refer specifically to [18, Sect. 6, item (3)] for connections with the notion of general interpretability.) In [1, 23], it is shown that our form of explicit definitions outlined in this section is not ad hoc in the sense that any sensible definition can be brought to this form. Namely, a notion of implicit definability suggests itself as a necessary condition for these new entities to be called "defined", see, e.g., [23, Sect. 4.3.1] and [19]. An analogue of the Beth definability theorem ([23, Sect. 4.3.48]) states that if a sort of new elements is implicitly definable, then it is explicitly definable, too. We note that the powerset of the universe of an infinite model is not implicitly definable in the sense of [23, Sect. 4.3.1], while, say, the set of two-element subsets of it is implicitly (and thus also explicitly) definable.

We hope that the content of this section is enough to give us a guiding intuition for what comes in the rest of this chapter.

6.6 Reducing SpecRel to Signalling Theory: An Interpretation

In this section we define in detail an interpretation of SPECREL_0 in SIGTH. We have to define (over SIGTH) the new sorts Q and B, and the new operations and relations $+$, $\star$, Obs, Ph, W that involve these new sorts.

We begin with defining the new sort Q. In Sect. 6.4 we already defined a field $F(e, o, \iota)$, that will provide the definition to our new sort Q and to $+$, $\star$. However, that definition had three parameters e, o, ι (the particle who was setting up his coordinate system, the "beginning of the era", and duration of one year). Up to isomorphism, we get the same field no matter how we choose these 3 parameters, but their universes strongly depend on the parameters (namely, the universe of $F(e, o, \iota)$ is the set of events on e's world-line). Which one should we take as the set of elements of sort Q? The answer is: take neither one, take all of them! Intuitively, this means that we take the disjoint union of all the fields belonging to the different parameters, and then we define an equivalence relation on this set that relates the isomorphic images of the same element. For this, in our explicit definition of the quantity sort we need a uniform formula that defines the isomorphisms between the fields $F(e, o, \iota)$. One such formula is given in [23, p. 305]. Here we give a simpler formula defining the isomorphisms between the various incarnations of our field. We can give this simpler formula because relativistic equidistance is available for us, while [23] used only the betweenness relation.

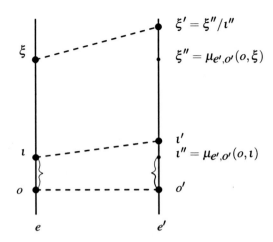

Fig. 6.4 The isomorphism $\varphi_{\mathsf{iso}}(e, o, \iota, e', o', \iota')$ between $\mathsf{F}(e, o, \iota)$ and $\mathsf{F}(e', o', \iota')$

We are going to define the isomorphisms sought for between the fields $\mathsf{F}(e, o, \iota)$. See Fig. 6.4. Let $e, o, \iota, e', o', \iota'$ be suitable parameters for defining the fields (as in Sect. 6.4). The isomorphism between them will take o to o', ι to ι' and it will take an arbitrary ξ on the world-line of e to $\xi' := \xi''/\iota''$ where ξ'', ι'' are events on e''s world-line such that $\mathsf{Edr}(\xi, o, \xi'', o')$ and $\mathsf{Edr}(\iota, o, \iota'', o')$, further $/$ denotes the division operation of the field belonging to e', o', ι'. Let $\varphi_{\mathsf{iso}}(\xi, \xi', e, o, \iota, e', o', \iota')$ denote the formula expressing the above. We denote the isomorphism as $\varphi_{\mathsf{iso}}(e, o, \iota, e', o', \iota')$, and we denote the unique ξ with the property $\varphi_{\mathsf{iso}}(\xi, \xi', e, o, \iota, e', o', \iota')$ as $\xi = \varphi_{\mathsf{iso}}(\xi', e, o, \iota, e', o', \iota')$.

Let $\mathsf{Fp}(e, o, \iota)$ express that e, o, ι are appropriate parameters for a field $\mathsf{F}(e, o, \iota)$, let U be the disjoint union of the universes of all the fields $\mathsf{F}(e, o, \iota)$, and let E denote the binary relation relating isomorphic elements, i.e.,

$$\mathsf{Fp}(e, o, \iota) :\Leftrightarrow \mathsf{Ev}(o), \mathsf{Ev}(\iota), o \neq \iota, o \prec \iota, eTo, eT\iota,$$
$$U := \{\langle \xi, e, o, \iota \rangle : \mathsf{Fp}(e, o, \iota), \xi \in \mathsf{F}(e, o, \iota)\},$$
$$E := \{\langle \langle \xi, e, o, \iota \rangle, (\xi', e', o', \iota') \rangle : \varphi_{\mathsf{iso}}(\xi, \xi', e, o, \iota, e', o', \iota')\}.$$

It can be shown that E is an equivalence relation on U, in each standard model of SIGTH. Our quantity sort will be U/E.

Recall that we are in the process of defining SPECREL$_0$ over SIGTH.

We are ready to define the quantity sort Q explicitly, by using the tools we introduced in the previous section. If q is a variable of the (new) sort Q, then q_ξ, q_e, q_o, q_ι denote the corresponding variables of the (old) sorts Sig and Par. We can think of this variable matching as q denotes an equivalence block of E (i.e., an element of U/E), and $\langle q_\xi, q_e, q_o, q_\iota \rangle$ denotes an arbitrary (unknown) element in the equivalence block q. Intuitively, q denotes an "abstract" quantity, and $\varphi_{\mathsf{iso}}(q_\xi, q_e, q_o, q_\iota, e, o, \iota)$ is the corresponding "concrete" quantity in the field $\mathsf{F}(e, o, \iota)$. Let us denote this last thing as

$$\mathsf{rep}(q, e, o, \iota) \; := \; \varphi_{\mathsf{iso}}(q_\xi, e, o, \iota, q_e, q_o, q_\iota).$$

This situation is somewhat analogous to the concept of a manifold in general relativity theory, the elements of the manifold are the "observer-independent" entities, and the charts/observers associate concrete values to these. Below comes the definition of the sort Q:

$$\mathsf{var} : q \mapsto \langle q_\xi, q_e, q_o, q_\iota \rangle \quad \text{for } q \in \mathsf{Var}_{\mathsf{Q}}.$$
$$\mathsf{Q}(q) \; :\Leftrightarrow \; q_e \mathsf{T} q_\xi, \mathsf{Ev}(q_\xi), \mathsf{Fp}(q_e, q_o, q_\iota),$$
$$q = q' \; :\Leftrightarrow \; \varphi_{\mathsf{iso}}(q_\xi, q'_\xi, q_e, q_o, q_\iota, q'_e, q'_o, q'_\iota).$$

Note that this definition of the sort Q is analogous to the one given for the new sort Lines in the example of affine planes in the previous section.

We get the definitions for $+$, $\star$ from writing up the definitions given in Sect. 6.4, as follows. Recall the formula $+(\tau, \tau_1, \tau_2, e, o)$ from Sect. 6.4.

Now, here is the definition of addition of sort Q:

$$+(q, q_1, q_2) \; :=$$
$$+(q_\xi, \varphi_{\mathsf{iso}}(q_{1\xi}, q_e, q_o, q_\iota, q_{1e}, q_{1o}, q_{1\iota}), \varphi_{\mathsf{iso}}(q_{2\xi}, q_e, q_o, q_\iota, q_{2e}, q_{2o}, q_{2\iota}), q_e, q_o).$$

The formula defining multiplication of sort Q is obtained analogously.

The rest of this section (interpreting SPECREL$_0$ in SIGTH) will be relatively straightforward.

We turn to defining the sort B. We will define the sort B of bodies as the union of observers and photons. So first we define the entities that we will call photons. A photon will be defined just as a signal σ that is not an event. The world-line of this photon will be defined as the set of all events that lie on the 4-dimensional line defined by the beginning and end points of σ. This way, many photons will share the same world-line, just as in the case of affine planes many pairs of distinct points define the same line, and we will define two photons to be equal if they share the same world-line. An observer will be defined to be a coordinate system. We recall from Sect. 6.4 that six parameters are required for defining a coordinate system, namely the experimenter e, a "zero" o and a time-unit ι, and three locations a_x, a_y, a_z specifying the space coordinate axes. These parameters have to satisfy the conditions below, which we will denote by Op (Op refers to "observer parameters"):

$$\mathsf{Op}(e, o, \iota, a_x, a_y, a_z) \; :\Leftrightarrow \; \mathsf{Fp}(e, o, \iota), e \| a_x, e \| a_y, e \| a_z, \mathsf{Ort}(e, a_x, e, a_y),$$
$$\mathsf{Ort}(e, a_x, e, a_z), \mathsf{Ort}(e, a_y, e, a_z).$$

Two observers will be defined equal if they assign the same coordinates to all events.

We are ready to formalize these definitions by using the tools we introduced in Sect. 6.5. Let $\mathsf{Var}_{\mathsf{B}}$ denote the set of variables of sort B. If b is a variable of sort B, then $b_\sigma, b_e, b_o, b_\iota, b_x, b_y, b_z$ will denote the corresponding variables of "old" sorts. Intuitively, this body will be b_σ if this is a "real", non-degenerate signal (i.e., if b_σ is not an event), and if b_σ is "degenerate" (i.e., if it is an event), then the body b will be the observer $\langle b_e, b_o, b_\iota, b_x, b_y, b_z \rangle$. We are ready to define the new sort B together with the unary formulas $\mathsf{Ph}(b)$ and $\mathsf{Obs}(b)$:

$$\mathsf{var} : b \mapsto \langle b_\sigma, b_e, b_o, b_\iota, b_x, b_y, b_z \rangle \quad \text{for } b \in \mathsf{Var_B}.$$
$$\mathsf{Ph}(b) :\Leftrightarrow \neg\mathsf{Ev}(b_\sigma),$$
$$\mathsf{Obs}(b) :\Leftrightarrow \mathsf{Ev}(b_\sigma), \mathsf{Op}(b_e, b_o, b_\iota, b_x, b_y, b_z),$$
$$\mathsf{B}(b) :\Leftrightarrow \mathsf{Ph}(b) \vee \mathsf{Obs}(b),$$

We are going now to define the equality relation on this new sort B. For stating equality of photons, first we express that three events are on one light-like line ($\lambda(\varepsilon_1, \varepsilon_2, \varepsilon_3)$), then we express that an event is on the world-line of a signal ($\mathsf{wl}(\varepsilon, \sigma)$).

$$\lambda(\varepsilon_1, \varepsilon_2, \varepsilon_3) :\Leftrightarrow \bigwedge \{\exists \sigma [(\varepsilon_i, \sigma, \varepsilon_j) \vee (\varepsilon_j, \sigma, \varepsilon_i)] : i, j \in \{1, 2, 3\}\},$$
$$\mathsf{wl}(\varepsilon, \sigma) :\Leftrightarrow \exists \varepsilon_1, \varepsilon_2 \ \lambda(\varepsilon, \varepsilon_1, \varepsilon_2), \mathsf{Beg}(\sigma, \varepsilon_1), \mathsf{End}(\sigma, \varepsilon_2).$$

Recall from Sect. 6.4 that the formula $\mathsf{cor}(\varepsilon, e, o, \iota, a_x, a_y, a_z) = (\tau, \gamma_x, \gamma_y, \gamma_z)$ expresses that the coordinates of the event ε are $\tau, \gamma_x, \gamma_y, \gamma_z$, in the coordinate system specified by $e, o, \iota, a_x, a_y, a_z$.

$$b = b' :\Leftrightarrow$$
$$(\neg\mathsf{Ev}(b_\sigma), \neg\mathsf{Ev}(b'_\sigma), \forall \varepsilon \ \mathsf{wl}(\varepsilon, b_\sigma) \leftrightarrow \mathsf{wl}(\varepsilon, b'_\sigma)) \vee$$
$$(\mathsf{Ev}(b_\sigma), \mathsf{Ev}(b'_\sigma), \forall \varepsilon \ \mathsf{cor}(\varepsilon, b_e, b_o, b_\iota, b_x, b_y, b_z) = \mathsf{cor}(\varepsilon, b'_e, b'_o, b'_\iota, b'_x, b'_y, b'_z)).$$

It remains to define the world-view relation W. The intuitive meaning of the formula $\mathsf{W}(m, b, t, x, y, z)$ will be that m is an observer, and the event at place t, x, y, z in m's coordinate system is on the world-line of b. Let m, b be variables of sort B and let t, x, y, z be variables of sort Q. Assume that m is an observer, i.e., $\mathsf{Ev}(m_\sigma)$. Let us denote the concrete value of an abstract quantity q in m's coordinate system by

$$m(q) := \mathsf{rep}(q, m_e, m_o, m_\iota).$$

We can now define W as follows:

$$\mathsf{W}(m, b, t, x, y, z) :\Leftrightarrow \exists \varepsilon \ \mathsf{cor}(\varepsilon, m(t), m(x), m(y), m(z), m_e, m_o, m_\iota, m_x, m_y, m_z),$$
$$(\neg\mathsf{Ev}(b_\sigma) \to \mathsf{wl}(\varepsilon, b_\sigma)), (\mathsf{Ev}(b_\sigma) \to b_e \mathsf{T} \varepsilon), \mathsf{Ev}(m_\sigma).$$

By the above, we gave definitions for all the sort and relation symbols of the language of SPECREL in the language of SIGTH. This defines a translation function tr between the two languages. Let $=_\mathsf{Q}$ and $=_\mathsf{B}$ stand for the equality relations between terms of sort Q and B, respectively. In the next theorem we state, without proof, that we indeed obtained an interpretation.

Theorem 6.1 tr *as given in this section is an interpretation of* SPECREL$_0$ *into* SIGTH, *that is, the following are true:*

SIGTH $\models$ "$=_Q$ *and* $=_B$ *are equivalence relations*",

SIGTH $\models$ "*the formulas defining* $+$, $\star$, Ph, Obs, W *are invariant under* $=_Q$, $=_B$",

SIGTH $\models$ tr(ψ) *for all* $\psi \in$ SPECREL$_0$.

Having defined the desired interpretation of SPECREL$_0$ into SIGTH, in the next section we extend this interpretation to a definitional equivalence between a slightly stronger version of SPECREL$_0$ and SIGTH.

6.7 Definitional Equivalence Between SpecRel and Signalling Theory

In this section we investigate interpretability and definitional equivalence between some of the FOL theories formalizing special relativity. We show that a slightly reinforced version of SPECREL$_0$ is definitionally equivalent to SIGTH. We mean interpretability and definitional equivalence in the sense of the generalized definability theory of Andréka et al. [1, 2], Madarász [23] outlined in Sect. 6.5.

The interpreted theory tr(SPECREL$_0$) is stronger than the original one in the sense that there are sentences ψ in the language of SPECREL$_0$ such that SIGTH $\models$ tr(ψ) while SPECREL$_0$ $\not\models \psi$. Such a sentence is, e.g., "all lines of slope less than 1 are world-lines of observers". We can express exactly how much more is true in the translated models by amending SPECREL$_0$ with some existence, extensionality, and time-orientation axioms (see below) and showing that the so obtained theory is definitionally equivalent with SIGTH. This is what we are going to do now.

The formulas describing the "difference" between SPECREL$_0$ and SIGTH are as follows. Formulas expressing that we have all kinds of possible observers (from each point, in each direction, for each velocity less than the speed of light there is an observer moving in that direction with that speed, each observer can re-coordinatize its coordinate-system with any space-isometry, each observer can set the unit of its clock arbitrarily), and otherwise we are as economic as possible (at most one photon through any two distinct events, only one observer with the same coordinate-system, only photons and observers as bodies, only one time-orientation for each observer).

These additional axioms, except the one about setting the clocks, are denoted as AxThEx, AxCoord, AxExtOb, AxExtPh, AxNobody, Ax$\uparrow$ in [3, Sect. 2.5]. Let AxClock formulate that each observer can set the unit of its clock arbitrarily (in the spirit of the above axioms). Let COMPL denote the set of these axioms and let SPECREL$_0^+$ denote the theory SPECREL$_0$ amended with these formulas:

COMPL := {AxThEx, AxCoord, AxClock, AxExtOb, AxExtPh, AxNobody, Ax↑},

SPECREL_0^+ := SPECREL_0 + COMPL.

To state definitional equivalence between SPECREL_0^+ and SIGTH, we now define an interpretation Tr of SIGTH into SPECREL_0^+. We have to define the universes Par, Sig of particles and signals and the relations T, R of transmitting and receiving, inside SPECREL_0. Intuitively, particles are defined to be observers, with two particles being equal if their world-lines coincide:

$$\text{var} : a \mapsto a_b \quad \text{for } a \in \text{Var}_{\text{Par}}, \text{ where } a_b \in \text{Var}_{\text{B}}.$$
$$\text{Par}(a) :\Leftrightarrow \text{Obs}(a_b),$$
$$a = a' :\Leftrightarrow \forall t, x, y, z \; W(a, a', t, x, y, z) \leftrightarrow x = y = z = 0.$$

Signals are defined to be photons with two events on their world-lines representing the beginning and end-points of the signal. We represent the two events with observers meeting the photon. The following formulae express in SPECREL_0 that "in b's world-view, p meets a at time t", and "a, p, e meet in one event", respectively:

$$\text{Meet}(b, p, a, t) :\Leftrightarrow \exists x, y, z \; W(b, p, t, x, y, z), W(b, a, t, x, y, z),$$
$$\text{meet}(a, p, e) :\Leftrightarrow \exists b, t \; \text{Meet}(b, a, p, t), \text{Meet}(b, a, e, t).$$

Now we are ready to interpret signals in SPECREL_0:

$$\text{var} : \sigma \mapsto \langle \sigma_b, \sigma_p, \sigma_e \rangle \quad \text{for } \sigma \in \text{Var}_{\text{Sig}}, \text{ where } \sigma_b, \sigma_p, \sigma_e \in \text{Var}_{\text{B}}.$$
$$\text{Sig}(\sigma) :\Leftrightarrow \text{Ph}(\sigma_p), \text{Obs}(\sigma_b), \text{Obs}(\sigma_e), \exists t \le t' \; \text{Meet}(\sigma_b, \sigma_p, \sigma_b, t), \text{Meet}(\sigma_b, \sigma_p, \sigma_e, t').$$
$$\sigma = \sigma' :\Leftrightarrow \text{meet}(\sigma_b, \sigma_b', \sigma_p), \text{meet}(\sigma_e, \sigma_e', \sigma_p), \neg\text{meet}(\sigma_b, \sigma_p, \sigma_e) \to \sigma_p = \sigma_p'.$$

Finally,

$$a \mathsf{T} \sigma :\Leftrightarrow \text{meet}(a_b, \sigma_b, \sigma_p),$$
$$a \mathsf{R} \sigma :\Leftrightarrow \text{meet}(a_b, \sigma_e, \sigma_p).$$

The above define a translation function Tr as indicated in Sect. 6.5. We state in the next theorem, without proof, that this Tr interprets SIGTH in SPECREL_0^+, and moreover, together with the interpretation tr defined in the previous section it forms a definitional equivalence between SPECREL_0^+ and SIGTH. This is the main theorem of this chapter.

Theorem 6.2 SIGTH *is definitionally equivalent to* SPECREL_0 + COMPL, *the pair* tr, Tr *of interpretations forms a definitional equivalence between them.*

We can read the above theorem as saying that what the theory SIGTH tells about special relativity is exactly what the theory SPECREL_0 + COMPL says. Since no axiom in COMPL follows from SPECREL_0, we can conclude that SIGTH tells more than SPECREL$_0$, the amount of "more" is exactly COMPL. However, we did not include the axioms of COMPL into SPECREL, because we do not need them in proving the main predictions of relativity theory; we feel that they do not belong to the core of the physical theory. Moreover, of the axioms of COMPL, we consider only Ax↑ as having a physically (or even philosophically) relevant content, namely it says that "time is oriented".

On the other hand, we will see that SPECREL has a content that SIGTH does not say about special relativity theory. This is the axiom AxSym of SPECREL. So, what is the connection between AxSym and SIGTH? Below we answer this question.

The interpretation tr we defined in the previous section does not interpret SPECREL in SIGTH, because tr(AxSym) does not follow from SIGTH (i.e., it is not true in the standard models $\mathfrak{M}(F)$ of SIGTH).[7] The reason for this is the following. AxSym states that any two observers use the same units of measurement. We can express in the language of SIGTH that "two observers use the same units of measurement", and this defines an equivalence relation on the set of all observers. For AxSym to be true, we should select any one of the blocks of this equivalence relation (since AxSym states that any two observers use the same units of measurements). But which one should we select? The question might sound familiar. In the previous analogous case (that concerned the various incarnations $F(e, o, \iota)$ of the field F) we took all the classes "up to isomorphism". However, in the present case there are no definable bijections between the blocks of this equivalence relation.

We can get around this problem by adding to the models $\mathfrak{M}(F)$ of SIGTH a "unit of measurement". We can do this, e.g., the following way. We add a new basic two-place relation symbol Tu (short for "Time unit") of sort Sig to the language of Signalling theory. In each standard model $\mathfrak{M}(F)$ we interpret Tu as the set of pairs of events with Minkowski-distance 1. (We note that these relations are not definable in $\mathfrak{M}(F)$ in the language of Signalling theory.) Let us denote the so expanded standard models by $\mathfrak{M}(F)^+$, and let SIGTH$^+$ denote an axiom system for their theory (in the extended language). Now, the interpretation we gave in this section can be extended to interpret SPECREL in SIGTH$^+$. Moreover, it also can be made into a definitional equivalence between a stronger version of SPECREL and SIGTH$^+$, that we obtain from SPECREL$_0^+$ by exchanging AxClock with AxSym:

$$\text{COMPL}^- := \{\text{AxThEx, AxCoord, AxExtOb, AxExtPh, AxNobody, Ax}\uparrow\},$$

To our minds, the following theorem clarifies the connection between AxSym and SIGTH. It says that the content of AxSym is to set the time-unit: the difference between SIGTH$^+$ and SIGTH is that in SIGTH$^+$ we can express Minkowski-distance, while in SIGTH we have only Minkowski equidistance.

Theorem 6.3 *(i)* SIGTH$^+$ *is definitionally equivalent to* SPECREL + COMPL$^-$.
(ii) SIGTH$^+$ *is not definitionally equivalent to* SIGTH.

---

[7] We note that SPECREL can be interpreted in SIGTH in the way that we interpret SPECREL in the field Q, +, $\star$.

The proof of part (i) of the above theorem goes by extending the interpretation Tr to SIGTH$^+$, this amounts to defining the new relation Tu in SPECREL; and also making some (minor) changes in the definition of tr. The proof of part (ii) of the above theorem goes by showing that the automorphism groups of members of SIGTH and SIGTH$^+$ differ from each other, this technique is elaborated in, e.g., [19, 23].

We included AxSym into SPECREL as a tool for convenience, it seems to carry no philosophical or physical importance. AxSym is only a simplifying assumption.

Concerning some of the other theories for special relativity, we mention that SIGTH is definitionally equivalent to Goldblatt's theory for special relativity in [17, Appendix A] amended with time-orientation. I.e., the two theories are almost the same, the only difference is that SIGTH assumes time-orientation while Goldblatt's theory does not. The proof of this last statement can be put together from the definitions and ideas in Sects. 6.4 and 6.6. Also, (a slight variant of) our SPECREL is definitionally equivalent to (a slight variant of) Suppes's axiomatization of special relativity in [34, 35].

6.8 Conclusion

We intended to show in this chapter some results the methods of mathematical logic can provide for other branches of science, in particular, for physics and the methodology of science. Using the tools of definability theory of first-order logic, we compared in detail two rather different axiom systems for special relativity theory. One of these, SPECREL of Andréka et al. [5], is coordinate-system-, or reference frame-oriented, while the other, SIGTH of James Ax [6], uses meager resources and talks about particles emitting and absorbing signals. The two theories use disjoint languages and talk about different kinds of entities. Yet, a precise comparison was made possible by using mathematical logic, and we obtained the following: SIGTH can express and states everything that SPECREL does, except for the relativistic (Minkowski) distance between events (implied by AxSym in SPECREL), while in addition it states time-orientation for space-time together with some auxiliary simplifying axioms (COMPL). Informally,

$$\text{SIGTH} = \text{SPECREL} - \text{relativistic distance} + \text{time-orientation} + \text{auxiliaries},$$

and a little more formally

$$\text{SIGTH} + \text{AxSym} = \text{SPECREL} + \text{COMPL}^-.$$

A byproduct of these investigations is a concrete operational semantics for special relativity theory. We believe that interpreting one theory in another is a flexible methodology for connecting physical theories with each other as well as with the "physical reality".

References

1. Andréka H, Madarász JX, Németi I (2001) Defining new universes in many-sorted logic. Mathematical Institute of the Hungarian Academy of Sciences, Budapest, p 93 (Preprint)
2. Andréka H, Madarász JX, Németi I (2002) On the logical structure of relativity theories. Alfréd Rényi Institute of Mathematics, Hungarian Academy of Sciences, Budapest. Research report, 5 July 2002, with contributions from Andai A, Sági G, Sain I, Tőke Cs, 1312 pp. http://www.math-inst.hu/pub/algebraic-logic/Contents.html
3. Andréka H, Madarász JX, Németi I (2007) Logic of space-time and relativity theory. In: Aiello M, Pratt-Hartmann I, van Benthem J (eds) Handbook of spatial logics. Springer, Berlin, pp 607–711
4. Andréka H, Madarász JX, Németi I, Németi P, Székely G (2011) Vienna circle and logical analysis of relativity theory. In: Máte A, Rédei M, Stadler F (eds) The Vienna circle in Hungary (Der Wiener Kreis in Ungarn). Veroffentlichungen des Instituts Wiener Kreis, Band 16, Springer, New York, pp. 247–268
5. Andréka H, Madarász JX, Németi I, Székely G (2012) A logic road from special relativity to general relativity. Synthese 186(3):633–649
6. Ax J (1978) The elementary foundations of spacetime. Found Phys 8(7/8):507–546
7. Balzer W, Moulines U, Sneed JD (1987) An architectonic for science. The structuralist program. D. Reidel Publishing Company, Dordrecht
8. van Benthem J (1982) The logical study of science. Synthese 51:431–472
9. van Benthem J (2012) The logic of empirical theories revisited. Synthese 186(2):775–792
10. van Benthem J, Pearce D (1984) A mathematical characterization of interpretation between theories. Studia Logica 43(3):295–303
11. Burstall R, Goguen J (1977) Putting theories together to make specifications. In: Proceeding of IJCAI'77 (Proceedings of the 5th international joint conference on artificial intelligence), vol 2. Morgan Kaufmann Publishers, San Francisco, pp 1045–1058
12. Carnap R (1928) Die Logische Aufbau der Welt. Felix Meiner, Leipzig
13. Friedman M (1983) Foundations of space-time theories: Relativistic physics and philosophy of science. Princeton University Press, Princeton
14. Friedman H (2004) On foundational thinking 1. FOM (foundations of mathematics) posting. http://www.cs.nyu.edu/pipermail/fom/. Posted 20 Jan 2004
15. Friedman H (2007) Interpretations of set theory in discrete mathematics and informal thinking. Lectures 1–3, Nineteenth annual Tarski lectures, Berkeley. http://u.osu.edu/friedman.8/
16. Gärdenfors P, Zenker F (2013) Theory change as dimensional change: conceptual spaces applied to the dynamics of empirical theories. Synthese 190:1039–1058
17. Goldblatt R (1987) Orthogonality and spacetime geometry. Springer, Berlin
18. Harnik V (2011) Model theory versus categorical logic: two approaches to pretopos completion (a.k.a. T^{eq}). In: Centre de Recherches Mathématiques CRM proceedings and lecture notes, vol 53. American Mathematical Society, pp 79–106
19. Hodges W (1993) Model theory. Cambridge University Press, Cambridge
20. Hoffman B (2013) A logical treatment of special relativity, with and without faster-than-light observers. BA Thesis, Lewis and Clark College, Oregon, 63pp. arXiv:1306.6004 [math.LO]
21. Konev B, Lutz C, Ponomaryov D, Wolter F (2010) Decomposing description logic ontologies. In: Proceedings of 12th conference on the principles of knowledge representation and reasoning, Association for the advancement of artificial intelligence, pp 236–246
22. Lutz C, Wolter F (2009) Mathematical logic for life science ontologies. In: Ono H, Kanazawa M, de Queiroz R (eds) Proceedings of WOLLIC-2009, LNAI 5514. Springer, pp 37–47
23. Madarász, JX (2002) Logic and relativity (in the light of definability theory). PhD Dissertation, Eötvös Loránd University. http://www.math-inst.hu/pub/algebraic-logic/diszi.pdf
24. Madarász JX, Székely G (2013) Special relativity over the field of rational numbers. Int J Theor Phys 52(5):1706–1718
25. Makkai M (1985) Ultraproducts and categorical logic, vol 1130. Methods in mathematical logic, Springer LNM, Berlin, pp 222–309

26. Makkai M (1993) Duality and definability in first order logic, vol 503. Memoirs of the American Mathematical Society, Providence
27. Makkai M, Reyes G (1977) First order categorical logic. Lecture Notes in Mathematics, vol 611. Springer, Berlin
28. Pambuccian V (2004/05) Elementary axiomatizations of projective space and of its associated Grassman space. Note de Matematica 24(1):129–141
29. Pambuccian V (2005) Groups and plane geometry. Studia Logica 81:387–398
30. Pambuccian V (2007) Alexandrov-Zeeman type theorems expressed in terms of definability. Aequationes Math 74:249–261
31. Previale F (1969) Rappresentabilità ed equipollenza di teorie assomatiche i. Ann Scuola Norm Sup Pisa 23(3):635–655
32. Schelb U (2000) Characterizability of free motion in special relativity. Found Phys 30(6):867–892
33. Schutz JW (1997) Independent axioms for Minkowski space-time. Longman, London
34. Suppes P (1959) Axioms for relativistic kinematics with or without parity. In: Henkin L, Tarski A, Suppes P (eds) Symposium on the axiomatic method with special reference to physics, North Holland, pp 297–307
35. Suppes P (1972) Some open problems in the philosophy of space and time. Synthese 24:298–316
36. Szabó LE (2009) Empirical foundation of space and time. In: Suárez M, Dorato MM, Rédei M (eds) EPSA07: launch of the European philosophy of science association, Springer. http://phil.elte.hu/leszabo/Preprints/LESzabo-madrid2007-preprint.pdf
37. Szczerba LW (1977) Interpretability of elementary theories. In: Butts RE, Hintikka J (eds) Logic, foundations of mathematics and computability theory (Proceeding of fifth international congress of logic, methodology and philosophical of science, University of Western Ontario, London), Part I. Reidel, Dordrecht, pp 129–145
38. Székely G (2009) First-order logic investigation of relativity theory with an emphasis on accelerated observers. PhD Dissertation, Eötvös Lorand University, Faculty of Sciences, Institute of Mathematics, Budapest, 150pp. ArXiv:1005.0973[gr-qc]
39. Székely G (2010) A geometrical characterization of the twin paradox and its variants. Studia Logica 95(1–2):161–182
40. Tarski A (1936) Der Wahrheitsbegriff in den formalisierten Sprachen. Studia Philosophica 1:152–278
41. Tarski A, Mostowski A, Robinson RM (1953) Undecidable theories. North-Holland, Amsterdam
42. Tarski A (1959) What is elementary geometry? In: Henkin L, Suppes P, Tarski A (eds) The axiomatic Method with Special Reference to Geometry and Phsics. North-Holland, Amsterdam, pp 16–29
43. Tarski A, Givant SR (1999) Tarski's system of geometry. Bull Symbolic Logic 5(2):175–214
44. Taylor EF, Wheeler JA (1963) Spacetime physics. Freeman, San Francisco
45. Weatherall JO (2011) Are Newtonian gravitation and geometrized Newtonian gravitation theoretically equivalent? (Unpublished manuscript)

Part II
Dynamics of Knowledge
and Belief Over Time

Chapter 7
Dynamic Epistemic Logics

Jan van Eijck

Abstract Dynamic epistemic logic, broadly conceived, is the study of rational social interaction in context, the study, that is, of how agents update their knowledge and change their beliefs on the basis of pieces of information they exchange in various ways. The information that gets exchanged can be about what is the case in the world, about what changes in the world, and about what agents know or believe about the world and about what others know or believe. This chapter gives an overview of dynamic epistemic logics, and traces some connections with propositional dynamic logic, with planning and with probabilistic updating.

7.1 Introduction

Logic is broadly conceived in Johan van Benthem's work as the science of information processing in evolving contexts. In most logic textbooks, with [18] as a notable exception, it is assumed that the reasoning processes that constitute the subject matter of the discipline take place in the head of an ideal reasoner. The validity of an argument, such as a step of Modus Ponens establishes a fact with the help of another fact plus an implication, and the agent performing these steps is kept out of the picture.

But in fact, the pieces of information that are put together by means of applications of Modus Ponens can have many different sources, involving different agents, and communication between them. Here is an early Chinese example that Johan is fond of quoting:

J. van Eijck (✉)
CWI, Science Park 123, 1098 XG Amsterdam, The Netherlands
e-mail: jve@cwi.nl

J. van Eijck
ILLC, Science Park 107, 1098 XG Amsterdam, The Netherlands

A. Baltag and S. Smets (eds.), *Johan van Benthem on Logic*
and Information Dynamics, Outstanding Contributions to Logic 5,
DOI: 10.1007/978-3-319-06025-5_7, © Springer International Publishing Switzerland 2014

Someone is standing next to a room and sees a white object outside. Now another person tells her that there is an object inside the room of the same colour as the one outside. After all this, the first person knows that there is a white object inside the room. This is based on three actions: an observation, then an act of communication, and finally an inference putting things together.

To give a full account of what goes on here, one has to represent the state of knowledge (or belief) of several agents, and model what goes on when they perceive facts in the world, when information is exchanged between them, and when they act on the world by making changes to it. This is what dynamic epistemic/doxastic logic is all about.

7.2 Knowledge, Belief and Change

The original account of belief and knowledge in terms of possible states of affairs or possible worlds is due to Hintikka [43], who proposed to analyze knowledge and belief with the tools of modal logic. Knowing about a fact p boils down to the ability to distinguish states of affairs where p is true from states of affairs where p is false, and the key notion of epistemic logic is that of an *indistinguishability relation* between possible worlds.

This analysis was taken up by cognitive scientists [34], computer scientists [29, 30, 41] and game theorists [4, 12, 58], and gradually extended to include interaction between different agents. It turned out that the notion of *common knowledge* plays a key part in the analysis of rational behaviour in games.

Dynamic epistemic/doxastic logic (see [23] for a textbook treatment) studies the evolution of knowledge and belief in the context of change. This change can be of various kinds:

- Changing beliefs about an unchanging world: in the Chinese room example the world does not change, but the first person learns something new about what is the case.
- Changing beliefs about a changing world: imagine a robot finding its way through a maze. The robot moves through the maze to a different spot and observes what it finds there. The change of location is a change in the states of affairs in the real world, the new observation causes the robot to change its belief state.
- Incorporating or failing to incorporate information about change in the world: a voter is taking part in an election process, but misses the communication about a change in the rules of the voting game.

In epistemic logic this is all expressed qualitatively, but there is an obvious relation to numerical ways of expressing rational belief. A change in belief could also be a change in the probability estimation that something is the case. Changes in the world may be thought of as being the result of indeterminate actions that occur with a certain probability. See below, Sect. 7.12 for probabilistic extensions of DEL that can cover such cases.

The distinction between qualitative ways and quantitative ways of expressing preference can also be found in game theory, where abstract strategic games are expressed in terms of individual preference relations on outcomes, while concrete strategic games use payoff functions that represent these preferences [55], and in probability theory, where the well-known Cox axioms list three conditions on 'degrees of belief' that are sufficient for a map to quantitative probabilities [21].

7.3 The Dynamic Turn in Epistemic/Doxastic Logic

A pioneer chapter shifting the focus from the study of information states *simpliciter* to information change is the work of Jan Plaza on the connection between public announcement and the generation of common knowledge [59]. This was followed up in a number of PhD theses under the supervision of Johan van Benthem: [35, 38, 45].

In an ILLC report from 2000, Johan van Benthem analyzes the kind of information update by means of world elimination that goes on in public announcement as model relativization [13], connecting truth in a full model with truth in a model relativized by some unary predicate. This explains why epistemic logic with public announcement can be axiomatized by means of reduction axioms that spell out the recursive definition of the relativization operation. The public announcement logic of epistemic logic with a common knowledge operator added admits no such axiomatization: the recursive definition of restriction on a common knowledge formula proceeds via relativized common knowledge.

A next key contribution is the proposal to view information updates themselves as Kripke models representing the agent's take on what happens when information is updated [6–8]. This generalizes information updating to the whole class of multi-agent Kripke models, not just multi-agent S5 models (an S5 model is a Kripke model where all accessibity relations are equivalence relations).

Action models are in fact epistemic/doxastic perspectives on communicative events. This epistemic/doxastic perspective on communication had already emerged in the AI literature, where epistemic reasoning was integrated into the situation calculus (essentially, a version of first order logic designed to describe changes in states of affairs) by Bob Moore [54]. The study of noisy sensors in [5] provides an epistemic perpective on communication in the situation calculus, by analyzing noisy observations as epistemic update events (S5 update models, in fact).

While action model update works for all Kripke models, multi-agent S5 updating of multi-agent S5 models is an important special case. The product of an S5 model with an S5 action model is again S5. The reason for this is that the S5 properties of reflexivity, transitivity and euclideaness are preserved under the update operation of restriction combined with product.

A sentence of first order logic is preserved under restriction and product if and only if the sentence is universal Horn, where a universal Horn sentence of FOL is the universal closure of a disjunction with at most one atom disjunct, and with the remaining disjuncts negations of atoms [52]. Reflexivity, transitivity and euclideaness are universal Horn, but seriality is not. This explains why the update of a KD45 epistemic model with a KD45 action model (a model where the accessibility relations are transitive, euclidean and serial) may yield a result that is not KD45. Information update of belief models with belief action models has a glitch.

This was one of the motivations to explore combinations of information update with information change as belief upgrade, in [2], followed by [9–11, 15]. The information update format was extended still further in [37].

Taking an extensional view on the update mechanisms involved in knowledge update, belief revision, belief change, and factual change, it turns out that these can all be described in terms of PDL style operations. Indeed, all of these update formats can be captured in a general logic of communication and change, and [16] proves a technical result that adding such update mechanisms to the logic of epistemic PDL (where the basic actions describe primitive epistemic/doxastic relations, and the PDL operators construct more complex epistemic/doxastic relations) does not increase the expressive power of the logic. This inspired further work on how PDL can be viewed (or reinterpreted) as a logic of belief revision, in [27], and as a multi-agent strategy logic, in [26].

However, this cannot be the whole story. This extensional view may illuminate the bare bones of DEL, but it disregards the flesh. Updating is a process where agents may follow specific update protocols, and it makes eminent sense to study possible formats of update rules. Such connections with protocol dynamics were explored in [64], and they led to interesting work on the use of DEL in planning [1].

Just as in matters of computation, it pays off to shift from an extensional to an intensional view. Extensionally, all we can say about the object of computation is that they are the recursive functions. Intensionally, we can say a lot more, by focussing on the *how* of computation. Similarly with communication: the extensional view disregards the inner workings of *how* information gets processed, and an intensional view on DEL brings this to light.

7.4 Announcements and Updating

The cartoon in Fig. 7.1 illustrates what goes on when public announcements are processed by perfectly rational agents, in this case logicians looking for something to drink. It is assumed that they all know what they want to drink themselves, but are uncertain about the wishes of the others. The cartoon tells the story of what happens then.

The question "Does everyone want beer?" triggers the following instruction, say for agent i (we use b_i for "i wants beer" and $\Box_i$ for "i knows", plus the usual boolean connectives):

- If $\Box_i (b_1 \wedge b_2 \wedge b_3)$ then i says "Yes".
- If $\Box_i \neg (b_1 \wedge b_2 \wedge b_3)$ then i says "No".
- Otherwise, i says "I don't know".

Fig. 7.1 Three logicians, from http://spikedmath.com. Reproduced with permission

These answers themselves serve as updates:

- i says "Yes": update with public announcement of $\Box_i (b_1 \wedge b_2 \wedge b_3)$
- i says "No": update with public announcement of $\Box_i \neg (b_1 \wedge b_2 \wedge b_3)$.
- i says "I don't know": update with public announcement of

$$\neg \Box_i (b_1 \wedge b_2 \wedge b_3) \wedge \neg \Box_i \neg (b_1 \wedge b_2 \wedge b_3).$$

The updates are instructions to *eliminate* worlds. The update with $\neg \Box_i (b_1 \wedge b_2 \wedge b_3) \wedge \neg \Box_i \neg (b_1 \wedge b_2 \wedge b_3)$ eliminates all worlds where $\Box_i (b_1 \wedge b_2 \wedge b_3)$ or $\Box_i \neg (b_1 \wedge b_2 \wedge b_3)$ holds.

To check a $\Box_i \phi$ formula, one has to check whether ϕ holds in all i-accessible worlds. If 1, 2, 3 are the three logicians standing or sitting next to each other in left-to-right order, then a situation where 1 wants beer and 2, 3 do not can be represented as ● ○ ○. The space of possibilities is given by:

$$\{\circ\,\circ\,\circ, \circ\,\circ\,\bullet, \circ\,\bullet\,\circ, \bullet\,\circ\,\circ, \bullet\,\circ\,\bullet, \bullet\,\bullet\,\circ, \bullet\,\bullet\,\circ, \bullet\,\bullet\,\bullet\}.$$

How about the epistemic accessibilities? These accessibilities should express that each agent initially only knows about her own state (thirsty for beer or not). If — represents the accessibility of the first logician, then then the state ○ ○ ○ should be connected by a — line to ○ ● ○, to ○ ○ ●, and to ○ ● ● (all states where the other

Fig. 7.2 Initial situation when three logicians enter a bar

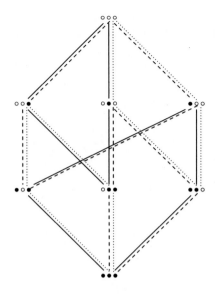

Fig. 7.3 After the announcement of the first logician

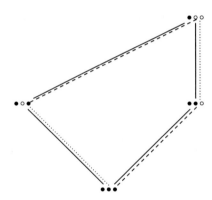

logicians have different wants). Thus, initially, the space of possibilities together with the epistemic accessibilities is given by the picture in Fig. 7.2, with • for "wants beer", solid lines for 1, dashed lines for 2 and dotted lines for 3, where 1, 2, 3 are the three logicians standing or sitting next to each other. Note that states s and t in the picture are linked by lines for agent i iff either in both of s, t agent i wants beer or in both of s, t agent i does not want beer.

After the first logician says "I don't know", the possibilities where

$$\Box_i (b_1 \wedge b_2 \wedge b_3) \text{ or } \Box_i \neg (b_1 \wedge b_2 \wedge b_3)$$

Fig. 7.4 After the
announcement of the
second logician

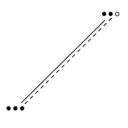

are true, drop out. Note that these are precisely the possibilities where she does not
want a beer herself (Fig. 7.3).

After the second logician says "I don't know", all remaining possibilities where
she does not want beer drop out: see Fig. 7.4. Now the third logician resolves the
case by saying either "yes" or "no", and if the bartender is also a perfect logician,
she knows in each case which of her customers to serve a beer.

The update process in the familiar muddy children example, where perfectly
rational children deduce from the assumption that at least one child has a muddy
forehead, and from the mud they see or fail to see on the foreheads of the other children
whether they themselves are muddy, is similar. Only the accessibility relations are
the converse of those in the thirsty logicians example: the thirsty logicians know what
they want to drink but do not know what the others want, while the muddy children
know about the muddiness of the others but not about their own muddiness.

7.5 Kripke Models and Action Model Update

The information update processes in the case of the three thirsty logicians, or of
the n muddy children, are special cases of a general procedure for updating epistemic models with action models due to Baltag, Moss, Solecki [7]. This handles the
communication between the drinking logicians, the muddy children, and much more
besides.

Let a finite set Ag of agents and a set $Prop$ of propositions be given. Then the
class of Kripke models over Ag and $Prop$ is given by:

Definition 7.1 A **Kripke Model** is a tuple (W, R, V) where

- W is a non-empty set of worlds.
- R is a function that assigns to every agent $a \in A$ a binary relation R_a on W.
- V is a valuation function that assigns to every $w \in W$ a subset of $Prop$.

An action model is like a Kripke model for Ag and $Prop$, with the difference that
the worlds are now called *actions* or *events*, and that the valuation has been replaced
by a map **pre** that assigns to each event e a formula of a suitable epistemic language
called the *precondition* of e. Let us fix the language first:

Definition 7.2 The multimodal language $\mathscr{L}$ over Ag and $Prop$ is given by the following BNF definition, where a ranges over Ag and p over $Prop$:

$$\phi :: = p \mid \neg\phi \mid \phi \wedge \phi \mid \Box_a\phi.$$

We assume the usual abbreviations for $\vee$, $\rightarrow$, $\leftrightarrow$, $\Diamond_a$.

The truth definition for this language is given by:

Definition 7.3 Let $\mathbf{M} = (W, R, V)$ and $w \in W$ in:

$$
\begin{array}{ll}
\mathbf{M} \models_w p & \text{iff } p \in V(w) \\
\mathbf{M} \models_w \neg\phi & \text{iff it is not the case that } \mathbf{M} \models_w \phi \\
\mathbf{M} \models_w \neg\phi_1 \wedge \phi_2 & \text{iff } \mathbf{M} \models_w \phi_1 \text{ and } \mathbf{M} \models_w \phi_2 \\
\mathbf{M} \models_w \Box_a\phi & \text{iff for all } v \text{ with } w R_a v : \mathbf{M} \models_v \phi.
\end{array}
$$

Action models over this language are defined by:

Definition 7.4 An **Action Model** is a tuple $(E, \mathbf{P}, \mathbf{pre})$ where

- E is a non-empty set of events.
- $\mathbf{P}$ is a function that assigns to every agent $a \in A$ a binary relation R_a on E.
- $\mathbf{pre}$ is a precondition function that assigns to every $e \in E$ a formula from $\mathscr{L}$.

From now on we call the regular epistemic models *static models*.

Updating a static model $\mathbf{M} = (W, R, V)$ with an action model $\mathbf{A} = (E, \mathbf{P}, \mathbf{pre})$ is defined as follows:

Definition 7.5 The update of static model $\mathbf{M} = (W, R, V)$ with an action model $\mathbf{A} = (E, \mathbf{P}, \mathbf{pre})$ succeeds if the set

$$\{(w, e) \mid w \in W, e \in E, \mathbf{M}, w \models \mathbf{pre}(e)\}$$

is non-empty. The update result is a new static model $\mathbf{M} \otimes \mathbf{A} = (W', R', V')$ with

- $W' = \{(w, e) \mid w \in W, e \in E, \mathbf{M}, w \models \mathbf{pre}(e)\}$,
- R'_a is given by $\{(w, e), (v, f)) \mid (w, v) \in R_a, (e, f) \in \mathbf{P}_a\}$,
- $V'(w, e) = V(w)$.

If the static model has a set of distinctive states W_0 and the action model a set of distinctive events E_0, then the distinctive worlds of $\mathbf{M} \otimes \mathbf{A}$ are the (w, e) with $w \in W_0$ and $e \in E_0$. The distinctive states are the states that can turn out to be the actual state of a static model. The distinctive events are the events that can turn out to be the actual event of an event model.

Below is an example pair of a static model with an update action model. The static model, on the left, pictures the result of a hidden coin toss, with three onlookers, Alice, Bob and Carol. The model has two distinctive worlds, marked in grey; h in a world means that the valuation makes h true, $\overline{h}$ in a world means that the valuation makes

h false in that world. The value h true means that the coin is facing heads up. The fact that both possibilities are distinctive means that both of these could turn out to be the actual world.

The R_a relations for the agents are assumed to be equivalences; reflexive loops for a, b, c at each world are omitted from the picture.

The static model on the left abbreviates two situations: the situation where the coin is facing heads up and it is common kwowledge among a, b, c that no-one knows this, and the situation where the coin is showing tails, and again it is common knowledge among a, b, c that no-one knows this. Imagine a situation where one of the agents tosses a coin under a cup and nobody has yet taken a look.

The action model on the right represents a test that reveals to a that the result of the toss is h, while b and c learn that a has learned the answer (without learning the answer themselves). Imagine the act of someone telling a the true value h, while b and c consider this possible. The distinctive event of the update is marked grey. The $\mathbf{P}_i$ relations are drawn, for two agents b, c. Reflexive loops are not drawn, so we do not see the $\mathbf{P}_i$ relation for a. The result of the update is shown here:

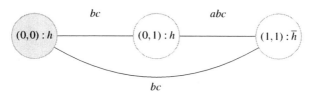

The result of the update is that the distinction mark on the $\overline{h}$ world has disappeared, that a now knows that the coin is showing heads, that b and c now know that a may know the face of the coin, but that b and c do not know, and all of this is common knowledge. In other words, the model makes each of the following formulas true in its actual world:

$$\Box_a h, \quad \neg\Box_b h, \quad \neg\Box_c h,$$
$$\Box_a(\Box_a h \vee \Box_a\neg h), \Box_b(\Box_a h \vee \Box_a\neg h), \Box_c(\Box_a h \vee \Box_a\neg h),$$
$$\Box_a(\neg\Box_b h \wedge \neg\Box_b\neg h), \Box_b(\neg\Box_b h \wedge \neg\Box_b\neg h), \Box_c(\neg\Box_b h \wedge \neg\Box_b\neg h),$$
$$\Box_a(\neg\Box_c h \wedge \neg\Box_c\neg h), \Box_b(\neg\Box_c h \wedge \neg\Box_c\neg h), \Box_c(\neg\Box_c h \wedge \neg\Box_c\neg h),$$
$$\Box_a\Box_b(\Box_a h \vee \Box_a\neg h), \ldots$$

The update operator, viewed abstractly, produces a restriction of a product of two models. It is folklore from model theory that a sentence of first order logic is perserved under restriction and product iff the sentence is universal Horn. A universal Horn sentence of FOL is the universal closure of a disjunction with at most one atom disjunct, while the remaining disjuncts are negations of atoms (see, e.g., [52]). The classes of S5 models or S5n models (multi-modal logics where all modalities are S5) are univeral Horn: the formulas for reflexivity, symmetry and transitivity can be written as Horn formulas. The classes of KD45 models or KD45n models are not universal Horn, for the seriality requirement cannot be expressed as a universal Horn sentence. Therefore, updating a static model where the accessibilities are KD45 with an update model where the accessibilities are KD45 does not guarantee a result where the accessibilities are again KD45.

7.6 Logics of Public Announcement

The language of public announcement logic is the extension of $\mathscr{L}$ with an operator $[\phi]\psi$ expressing that after public announcement of ϕ the formula ψ is true in the resulting model.

Definition 7.6 If $\mathbf{M} = (W, R, V)$ and $\phi \in \mathscr{L}$, then $\mathbf{M}^\phi = (W^\phi, R^\phi, V^\phi)$ is given by:

- $W^\phi = \{w \in W \mid \mathbf{M} \models_w \phi\}$.
- $R^\phi = \lambda a.\{(w, v) \mid w \in W^\phi, v \in W^\phi, wR_a v\}$.
- V^ϕ is the restriction of V to W^ϕ.

Then:

$$\mathbf{M} \models_w [\phi]\psi \quad \text{iff } \mathbf{M} \models_w \phi \text{ implies } \mathbf{M}^\phi \models_w \psi.$$

The logic of public announcements is now given by the reduction axioms from [59]:

Definition 7.7 The proof system for public announcement logic consists of the axioms and rules for multi-modal S5 epistemic logic (see [19]), plus the following reduction axioms:

$$\text{Atoms} \vdash [\phi]p \leftrightarrow (\phi \to p)$$
$$\text{Partial functionality} \vdash [\phi]\neg\psi \leftrightarrow (\phi \to \neg[\phi]\psi)$$
$$\text{Distribution} \vdash [\phi](\psi_1 \wedge \psi_2) \leftrightarrow ([\phi]\psi_1 \wedge [\phi]\psi_2)$$
$$\text{Knowledge announcement} \vdash [\phi]\Box_a\psi \leftrightarrow (\phi \to \Box_a[\phi]\psi)$$

plus the rules of inference for announcement generalization, given by:

$$\text{From } \vdash \psi \text{ infer } \vdash [\phi]\psi.$$

These axioms provide a recursive definition of the effect of public announcement, and they can be used to turn every formula from the enhanced language into an equivalent $\mathscr{L}$ formula. This allows us to prove completeness of the logic by means of a reduction argument.

Such an easy complete axiomatization is no longer available when we add an operator for common knowledge $C_B\phi$ to the language, where $C_B\phi$ expresses that ϕ is common knowledge for the agents in $B \subseteq A$. The semantics is given by:

Definition 7.8 Common knowledge among B:

$$\mathbf{M} \models_w C_B\phi \quad \text{iff } \mathbf{M} \models_w \phi \text{ for all } v \text{ with } (w, v) \in (R_B)^+,$$

where $R_B = \bigcup_{a \in B} R_a$, and $(R_B)^+$ denotes the transitive closure of R_B.

Still, the reduction method applies, once we are able to express the semantic intuitions for achieving common knowledge by announcement. Introduce an operator for relativized common knowledge, $C_B(\phi, \psi)$, with semantics given by:

Definition 7.9 Relativized common knowledge among B:

$$\mathbf{M} \models_w C_B(\phi, \psi) \quad \text{iff } \mathbf{M} \models_w \psi \text{ for all } v \text{ with } (w, v) \in (R_B^\phi)^+,$$

where $R_B^\phi = R_B \cap (W \times \{w \in W \mid \mathbf{M} \models_w \phi\})$.

Intuitively, $C_B(\phi, \psi)$ expresses that every ϕ-path through B accessibilities ends is a state where ψ holds. Let $E_B\phi$ abbreviate $\bigwedge_{a \in B} \Box_a\phi$. Then $E_B\phi$ expresses that ϕ is general knowledge among the agents in $B \subseteq A$.

Definition 7.10 The proof system for public announcement logic with relativized common knowledge consists of following axioms and rules:

Tautologies All instances of propositional tautologies

Knowledge Distribution $\vdash \Box_a(\phi \to \psi) \to (\Box_a\phi \to \Box_a\psi)$

Common Knowledge Distrib $\vdash C_B(\phi, \psi \to \chi) \to (C_B(\phi, \psi) \to C_B(\phi, \chi))$

Mix $\vdash C_B(\phi, \psi) \leftrightarrow E_B(\phi \to (\psi \wedge C_B(\phi, \psi)))$

Induction $\vdash (E_B(\phi \to \psi) \wedge C_B(\phi, \psi \to E_B(\phi \to \psi))) \to C_B(\phi, \psi)$

plus the following inference rules:

Modus Ponens From $\vdash \phi$ and $\vdash \phi \to \psi$ infer $\vdash \psi$.

$\Box$Necessitation From $\vdash \phi$ infer $\vdash \Box_a\phi$.

CNecessitation From $\vdash \phi$ infer $\vdash C_B(\psi, \phi)$.

The completeness of this system is proved in [16]. To better understand what goes on in the proof system, it is helpful to translate the statements of public announcement with relativized common knowledge into propositional dynamic logic (PDL), and to note that the above proof system essentially follows the usual PDL axioms. E.g., $C_B(\phi, \psi)$ gets the following PDL translation:

$$[(\bigcup_{a \in B} a; ?\phi)^+]\psi.$$

The connection with PDL will be worked out in the next section.

7.7 Connecting up with Epistemic PDL

PDL was designed as a general logic of (computational) action, as a generalization of Floyd-Hoare logic [32, 44]. In Floyd-Hoare logic, one studies correctness statements about programs, such as the following:

$$\{N = \gcd(x, y) \wedge x \neq y\}$$
$$\text{if } x > y \text{ then } x := x - y \text{ else } y := y - x$$
$$\{N = \gcd(x, y)\}.$$

Here the assertion $\{N = \gcd(x, y) \wedge x \neq y\}$ is called the precondition and the assertion $\{N = \gcd(x, y)\}$ the postcondition for the conditional program statement.

The Hoare specification asserts that the loop step in Euclid's GCD algorithm is correct: if x and y store integer numbers that are different, then their GCD does not change if you replace the largest number by the difference of the two numbers.

The general meaning of the Hoare triple $\{\phi\}$ π $\{\psi\}$ is: if a state satisfies the precondition ϕ, and program π is executed in that state, then any state that results from this execution will satisfy the postcondition ψ.

Vaughan Pratt saw that such Hoare triples can be viewed as implications in a logic where the program π appears as a modality [60]. The PDL guise of the Hoare correctness statement

$$\{\phi\} \pi \{\psi\}$$

is

$$\phi \rightarrow [\pi]\psi.$$

Hoare logic over programs for integer assignment is undecidable because it talks about variable assignment in the language of first order logic. It has rules like precondition strengthening and postcondition weakening:

$$\frac{\mathbb{N} \models \phi' \rightarrow \phi \quad \{\phi\} \pi \{\psi\}}{\{\phi'\} \pi \{\psi\}}$$

$$\frac{\{\phi\} \pi \{\psi\} \quad \mathbb{N} \models \psi \rightarrow \psi'}{\{\phi\} \pi \{\psi'\}}.$$

These use first order statements about natural numbers, which may be undecidable:

$$\mathbb{N} \models \phi' \rightarrow \phi.$$

But an extra abstraction step makes the logic decidable again. The basic building blocks for programs in Hoare logic are variable assignment statements $x := E$. Just replace these by arbitrary atomic actions a, and stipulate that the interpretation of a is some binary relation on an abstract set of states. Then PDL emerges as a general program logic, with assertions (formulas) and programs defined by mutual recursion, as follows (assume p ranges over a set of basic propositions $Prop$ and a over a set of basic actions Act):

Definition 7.11 PDL language:

$$\phi ::= \top \mid p \mid \neg\phi \mid \phi_1 \wedge \phi_2 \mid [\pi]\phi$$
$$\pi ::= a \mid ?\phi \mid \pi_1; \pi_2 \mid \pi_1 \cup \pi_2 \mid \pi^*.$$

This language is to be interpreted in multi-modal Kripke models $\mathbf{M} = (W, R, V)$, where W is a set of worlds or states, R is a function that assigns to every $a \in Act$ a binary relation $R_a \subseteq W^2$, and V is a valuation function that assigns to every $p \in Prop$ a subset of W.

Definition 7.12 Semantics of PDL. Let $\mathbf{M} = (W, R, V)$. The interpretations $[\![\phi]\!]^{\mathbf{M}}$ of formulas are subsets of W, and the interpretations $[\![\pi]\!]^{\mathbf{M}}$ are subsets of W^2. The clauses for the propositional atoms and the Boolean operators are as usual, the clause for $[\pi]\phi$ is

$$\{w \in W \mid \text{if for all } v \text{ with } w[\![\pi]\!]^{\mathbf{M}}v : v \in [\![\phi]\!]^{\mathbf{M}}\},$$

and the clauses for the programs are given by:

$$[\![a]\!]^{\mathbf{M}} = R_a$$
$$[\![?\phi]\!]^{\mathbf{M}} = \{(w, w) \in W^2 \mid w \in [\![\phi]\!]^{\mathbf{M}}\}$$
$$[\![\pi_1; \pi_2]\!]^{\mathbf{M}} = [\![\pi_1]\!]^{\mathbf{M}} \circ [\![\pi_2]\!]^{\mathbf{M}}$$
$$[\![\pi_1 \cup \pi_2]\!]^{\mathbf{M}} = [\![\pi_1]\!]^{\mathbf{M}} \cup [\![\pi_2]\!]^{\mathbf{M}}$$
$$[\![\pi^*]\!]^{\mathbf{M}} = ([\![\pi]\!]^{\mathbf{M}})^*.$$

Note the regular operations on relation on the righthand side: ∘ for relational composition, ∪ for union of relations, and * for Kleene star or reflexive transitive closure. Thus, the complex modalities are handled by the regular operations on relations.

We employ the usual abbreviations: $\bot$ is shorthand for $\neg\top$, $\phi_1 \vee \phi_2$ is shorthand for $\neg(\neg\phi_1 \wedge \neg\phi_2)$, $\phi_1 \rightarrow \phi_2$ is shorthand for $\neg(\phi_1 \wedge \phi_2)$, $\phi_1 \leftrightarrow \phi_2$ is shorthand for $(\phi_1 \rightarrow \phi_2) \wedge (\phi_2 \rightarrow \phi_1)$, $\langle\pi\rangle\phi$ is shorthand for $\neg[\pi]\neg\phi$, and $[\pi^+]\phi$ is shorthand for $[\pi; \pi^*]\phi$.

We now get that $[\pi]\phi$ is true in world w of $\mathbf{M}$ if it holds for all v with $(w, v) \in [\![\pi]\!]^{\mathbf{M}}$ that ϕ is true in v, and the Hoare assertion $\phi_1 \rightarrow [\pi]\phi_2$ is true in a world w if truth of ϕ_1 in w implies that it holds for all v with $(w, v) \in [\![\pi]\!]^{\mathbf{M}}$ that ϕ_2 is true in v.

Definition 7.13 The PDL language is completely axiomatized by the following PDL rules and axioms [48, 61]:

> Modus ponens and axioms for propositional logic
> Modal generalisation From $\vdash \phi$ infer $\vdash [\pi]\phi$

> Normality $\vdash [\pi](\phi \rightarrow \psi) \rightarrow ([\pi]\phi \rightarrow [\pi]\psi)$
> Test $\quad\ \vdash [?\phi]\psi \leftrightarrow (\phi \rightarrow \psi)$
> Sequence $\vdash [\pi_1; \pi_2]\phi \leftrightarrow [\pi_1][\pi_2]\phi$
> Choice $\quad \vdash [\pi_1 \cup \pi_2]\phi \leftrightarrow ([\pi_1]\phi \wedge [\pi_2]\phi)$
> Mix $\quad\ \ \vdash [\pi^*]\phi \leftrightarrow (\phi \wedge [\pi][\pi^*]\phi)$
> Induction $\vdash (\phi \wedge [\pi^*](\phi \rightarrow [\pi]\phi)) \rightarrow [\pi^*]\phi.$

In the previous Section, we already saw specific instances of the Mix and Induction axioms, in the proof system for public announcement logic with relativized common knowledge.

When the PDL language was designed, the basic actions a were thought of as abstract versions of basic programs (variable assignment statements, say). But nothing in the formal design forces this interpretation. The basic actions could be anything. PDL is a generic action logic for talking about actions as transitions from states of the world to other states of the world.

In PDL, no constraints are imposed on what the actions are. These could be changes in the world, but they could also be epistemic relations. *Epistemic PDL* is just PDL, but with the understanding that the accessibility relations express the *knowledge or belief* of agents.

Two extensions of the language are useful: an extension with a global modality G and an extension with a converse operator $\check{}$.

Definition 7.14 Epistemic PDL Action Expressions:

$$\pi ::= a \mid G \mid ?\phi \mid \pi_1; \pi_2 \mid \pi_1 \cup \pi_2 \mid \pi^* \mid \pi^{\check{}}.$$

Interpretation of the additions:

Definition 7.15 Semantics of G and $\check{}$:

$$[\![G]\!] = W^2$$
$$[\![\pi^{\check{}}]\!]^{\mathbf{M}} = ([\![\pi]\!]^{\mathbf{M}})^{\check{}}.$$

Thus, $[G]\phi$ expresses that everywhere in the model ϕ holds, and $\langle G \rangle \phi$ expresses that ϕ holds somewhere.

Definition 7.16 Proof system for epistemic PDL. Axioms and rules of PDL, plus the following. Axioms expressing that G is an S5-operator:

$$\text{Reflexivity } \vdash \phi \rightarrow \langle G \rangle \phi$$
$$\text{Symmetry } \vdash \phi \rightarrow [G]\langle G \rangle \phi$$
$$\text{Transitivity } \vdash \langle G \rangle \langle G \rangle \phi \rightarrow \langle G \rangle \phi$$
$$\text{Inclusion } \vdash \langle G \rangle \phi \rightarrow \langle \pi \rangle \phi.$$

Axioms for converse expressing the equivalences that reduce converse over a complex action to converse over atomic actions:

$$R_a \subseteq R_a^\smile \qquad \vdash \phi \leftrightarrow [a]\langle a^\smile \rangle \phi$$
$$R_a^\smile \subseteq R_a \qquad \vdash \phi \leftrightarrow [a^\smile]\langle a \rangle \phi$$
$$\text{Reduction for } G \vdash \langle G^\smile \rangle \phi \leftrightarrow \langle G \rangle \phi$$
$$\text{Reduction for } ?\phi \vdash \langle ?\phi^\smile \rangle \phi \leftrightarrow \langle ?\phi \rangle \phi$$
$$\text{Reduction for } ; \quad \vdash \langle (\pi_1 ; \pi_2)^\smile \rangle \phi \leftrightarrow \langle \pi_2^\smile ; \pi_1^\smile \rangle \phi$$
$$\text{Reduction for } \cup \vdash \langle (\pi_1 \cup \pi_2)^\smile \rangle \phi \leftrightarrow \langle \pi_1^\smile \cup \pi_2^\smile \rangle \phi$$
$$\text{Reduction for } * \quad \vdash \langle (\pi^*)^\smile \rangle \phi \leftrightarrow \langle (\pi^\smile)^* \rangle \phi.$$

It is well-known that adding global modality and converse to PDL does not change its computational properties: the logic remains decidable, satisfiability remains EXPTIME-complete [19, 42], and model checking is PTIME-complete [51]. Many further variations on the set-up of PDL are possible, and we will explore some of them below.

Note that it is not necessary to impose KD45 axioms for belief or S5 axioms for knowledge. Instead, we interpret the atoms as proto epistemic accessibility relations, and we construct the appropriate operators [27]. Let the interpretations of the atomic actions a be an arbitrary binary relation R_a.

Define $\sim_a$ as $(a \cup a^\smile)^*$. This operator is interpreted as the relation

$$(R_a \cup R_a^\smile)^*,$$

and this relation is symmetric, reflexive and transitive. Therefore, $\sim_a$ is an appropriate S5 operator for knowledge.

Some logicians (including Hintikka) have argued for dropping the symmetry requirement, and propose to use an S4 modality for knowledge. The corresponding epistemic PDL operator would be a^*.

To get a KD45 relation from an arbitrary binary relation R_a, consider: $(?[a]\bot; ?\top) \cup a; (a^\smile; a)^*$. Its interpretation S is the relation

$$\{(x, x) \mid \neg\exists z : (x, z) \in R_a\} \cup (R_a \circ (R_a^\smile \circ R_a)^*).$$

Then S is serial, for let x be an arbitrary member of the state set A. If there is no $y \in A$ with $(x, y) \in R_a$ we have $(x, x) \in S$. If there is such a y then $(x, y) \in S$. So in any case there is a $z \in A$ with $(x, z) \in S$.

It is also easy to see that S is transitive and euclidean. Therefore $(?[a]\bot; ?\top) \cup a; (a^{\smile}; a)^*$ can serve as a KD45 operator, and we have an appropriate way to interpret KD45 belief in epistemic PDL.

Abbreviate this operator as $\leqslant_a$. Observe that the interpretation of $\leqslant_a$ is included in that of $\sim_a$, so the following principle holds:

$$\langle\leqslant_a\rangle\phi \rightarrow \langle\sim_a\rangle\phi.$$

Contraposition gives:

$$[\sim_a]\phi \rightarrow [\leqslant_a]\phi.$$

This expresses that individual knowledge implies individual belief.

Also, the interpretation of $\leqslant_a; \sim_a$ is included in that of $\sim_a$, so the following holds as well:

$$[\sim_a]\phi \rightarrow [\leqslant_a][\sim_a]\phi.$$

Therefore, this interpretation of belief also gives reasonable connections between knowledge and belief in epistemic PDL.

Note that a relation R_a is S5 iff it holds that a and $\sim_a$ have the same interpretation. Similarly, a relation R_a is KD45 iff it holds that a and $\leqslant_a$ have the same interpretation.

The expressive power of epistemic PDL comes to the fore when we consider common knowledge and shared belief. An appropriate operator for common knowledge between agents a and b is readily defined as

$$(a \cup a^{\smile} \cup b \cup b^{\smile})^*,$$

or equivalently as $(\sim_a \cup \sim_b)^*$. This gives the equivalence relation with every state that is reachable by means of arbitrary numbers of forward or backward a or b steps in a single equivalence class, and this is how common knowledge was explained by philosophers [49], sociologists [33], economists [4] and computer scientists [40].

This is another good reason for using epistemic PDL rather than a logic with explicit modal operators for $\sim_a$ and $\leqslant_a$. Logics with explicit knowledge and belief modalities are naturally interpreted with respect to models where the corresponding relations have the appropriate properties. But for the case of belief, there is no guarantee that updating such models with 'reasonable' update models (where the accessibilities satisfy the same belief properties) results in a new model in the same class. If one uses epistemic PDL, there is no such problem. One can start with a model where the relation of an agent a is KD45 (meaning that a and $\leqslant_a$ have the same interpretation in that model), and after an update it may turn out that a and $\leqslant_a$ no longer have the same interpretation. In such a model $\leqslant_a$ still is a KD45 operator, so there is no problem in interpreting statements about belief.

7.8 Adding Factual Change

Factual change was already added to update models in LCC, the logic of communication and change of [16], which is basically a system that extends epistemic PDL with generic action update modalities, where the action update can also change the facts of the world. See also [22].

Consider again the model where a has found out the value of the coin, while b and c were onlookers.

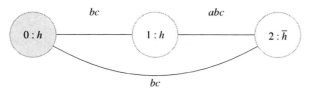

Here is a representation of the action of tossing the coin again, with a, b, c present, but without showing the result to any of a, b, c. Explanation: if the coin is tossed again, either the value of h does not change (expressed by $\top$), or it flips from True to False or vice versa (expressed by $h := \neg h$). Another way of representing this would by generating a history of coin flips h_1, h_2, and so on.

After an update with this fact changing action above, we get:

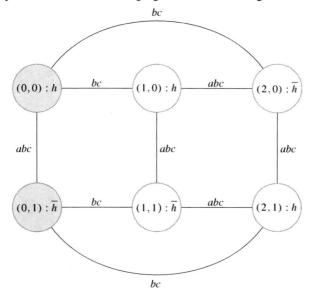

This model looks complicated, but it is bisimilar to the following model, where the coin may have fallen either way, and it is common knowledge that none of the agents knows which side is up. And that's intuitively right, for all players are aware that nobody knows anything about how the coin has fallen this time.

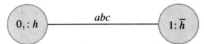

Another example of a simple fact-changing update is the flip of a coin to its other side, represented by the propositional substitution $h := \neg h$. For example, imagine a situation where b and c are aware of the fact that a coin is flipped to its other side (say by some trusted agent d), while a mistakenly believes that nothing has happened. This is modelled by the following action model:

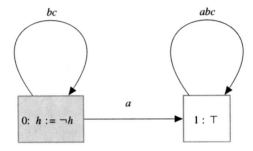

Note that we can no longer omit reflexive arrows, for this model is not reflexive for the a relation.

Consider again the model where a has found out the value of the coin, while b and c were onlookers. After an update with the fact changing action above, we get the result in Fig. 7.5.

Now $[<_a][\sim_a]h$ is true in the actual situation: a believes that she knows that the coin is showing heads. But $[(\sim_b \cup \sim_c)^*](\neg[\sim_a]h \wedge \neg[\sim_a]\neg h)$ is also true: b and c have common knowledge that a does *not* know what the coin is showing. Nobody knows what the coin is showing, but b and c know that a is mistaken about what she knows. This kind of situation occurs very often in everyday life: something happens, we mistake it for something else, and we end up with a false belief. This is one of the ways in which knowledge can decrade to mere belief.

7.9 Adding Belief Change

We will now also add belief change. This was not yet present in LCC, but it is studied extensively in Johan van Benthem's subsequent work [14]. See also [25, 27]. As a first example, consider the coin situation above, where a has been led astray by failing to observe some change in the world.

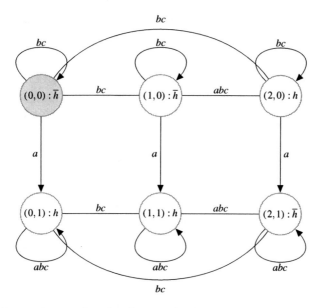

Fig. 7.5 A failure to observe a fact-changing event leading to false belief

Suppose a suddenly comes to believe that she was led astray. She (publicly) updates her belief by accepting that every conceivable state of affairs might be the true state of affairs. The action model for that is:

$$a := \sim_a$$

The result of updating the epistemic model of Fig. 7.5 with this is again an S5 model (so we can omit reflexive loops): see Fig. 7.6.

We see that relational change extends the expressive power of the updates, for it can *add* arrows, while action model update without relational change can only *delete* arrows.

We will now formally state how to modify the update process to accommodate both factual change and belief change. Let an action model with both kinds of changes be a quintuple

$$\mathbf{A} = (E, \mathbf{P}, \mathbf{pre}, \mathbf{Sub}, \mathbf{SUB})$$

where E, $\mathbf{P}$, **pre** are as before, **Sub** is a function that assigns a propositional binding (or propositional substitution) to each $e \in E$, and **SUB** is a function that assigns a relational binding to each $e \in E$. A propositional binding is a map from proposition letters to formulas, represented by a finite set of links

$$\{p_1 \mapsto \phi_1, \ldots, p_n \mapsto \phi_n\}$$

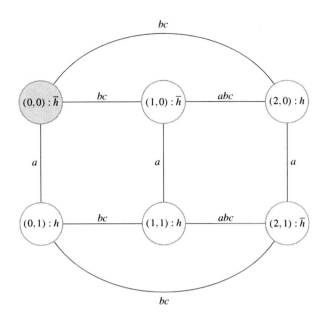

Fig. 7.6 A belief revision that creates an S5 model from a KD45 model

where the p_k are all different, and where no ϕ_k is equal to p_k. It is assumed that each p that does not occur in a left-hand side of a binding is mapped to itself.

Similarly, a relational binding is a map from agents to program expressions, represented by a finite set of links

$$\{a_1 \mapsto \pi_1, \ldots, a_n \mapsto \pi_n\}$$

where the a_j are agents, all different, and where the π_j are program expressions from the PDL language. It is assumed that each a that does not occur in the left-hand side of a binding is mapped to a. Use ϵ for the identity propositional or relational substitution.

Definition 7.17 The update execution of static model $\mathbf{M} = (W, P, V)$ with action model $\mathbf{A} = (E, \mathbf{P}, \mathbf{pre}, \mathbf{Sub}, \mathbf{SUB})$ is a tuple

$$\mathbf{M} \circledast \mathbf{A} = (W', P', V')$$

where

- $W' = \{(w, e) \mid \mathbf{M}, w \vDash \mathbf{pre}(e)\}$.
- P'_a is given by

$$\{((w_1, e_1), (w_2, e_2)) \mid$$
$$\text{there is a } \mathbf{SUB}(e_1)(a) \text{ path from } (w_1, e_1) \text{ to } (w_2, e_2) \text{ in } \mathbf{M} \otimes \mathbf{A}\}.$$

- $V'(p) = \{(w, e) \in W' \mid \mathbf{M}, w \vDash \mathbf{Sub}(e)(p)\}$.

The definition of P'_a refers to paths in the old style update product which is denoted with $\otimes$.

Consider the suggestive upgrade $\sharp_a\phi$ discussed in van Benthem and Liu [15] as a relation changer (*uniform* relational substitution):

$$\sharp_a\phi =_{\text{def}} ?\phi; a; ?\phi \cup ?\neg\phi; a; ?\neg\phi \cup ?\neg\phi; a; ?\phi.$$

This models a kind of belief change where preference links from ϕ worlds to $\neg\phi$ worlds for agent a get deleted. It can be modelled as the following example of public belief change.

Public Belief Change: Action model

$$G = (\{e\}, \mathbf{P}, \mathbf{pre}, \mathbf{Sub}, \mathbf{SUB})$$

where

- For all the $i \in Ag$, $\mathbf{P}_i = \{(e, e)\}$.
- $\mathbf{pre}(e) = \top$.
- $\mathbf{Sub}(e) = \varepsilon$.
- $\mathbf{SUB}(e) = \{a \mapsto \sharp_a\phi, b \mapsto \sharp_b\phi\}$.

In a picture (reflexive arrows omitted):

$$\boxed{a := \sharp_a\phi, b := \sharp_b\phi}$$

Note that our definition of $\circledast$ update implements *point-wise* relational substitutions, which is a more powerful mechanism than merely upgrading the relations *uniformly* everywhere in the model. This is illustrated by the following example.

Non-public Belief Change: Action model

$$G' = (\{e_0, e_1\}, \mathbf{P}, \mathbf{pre}, \mathbf{Sub}, \mathbf{SUB})$$

where

- For all $i \in Ag$, if $i \neq b$ then $\mathbf{P}_i = \{(e_0, e_0), (e_1, e_1)\}$,
  $\mathbf{P}_b = \{(e_0, e_0), (e_1, e_1), (e_0, e_1), (e_1, e_0)\}$
- $\mathbf{pre}(e_0) = \mathbf{pre}(e_1) = \top$.
- $\mathbf{Sub}(e_0) = \mathbf{Sub}(e_1) = \varepsilon$.
- $\mathbf{SUB}(e_0) = \{a \mapsto \sharp_a\phi\}$, $\mathbf{SUB}(e_1) = \varepsilon$.

Assume e_0 is the actual event.

This changes the belief of a while b remains unaware of the change. In a picture (reflexive arrows omitted, since this is an S5 model):

$$0 : a := \sharp_a \phi \quad\underset{\rule{0pt}{0pt}}{\overset{b}{\rule{3cm}{0.4pt}}}\quad 1 : \mathsf{T}$$

Let PDL^+ be the result of adding modalities of the form $[\mathbf{A}, e]\phi$ to PDL, with the following interpretation clause:

$$\mathbf{M}, w \models [\mathbf{A}, e]\phi \text{ iff } \mathbf{M}, w \models \mathbf{pre}(e) \text{ implies } \mathbf{M} \circledast \mathbf{A}, (w, e) \models \phi.$$

Then the completeness result for LCC extends to a completeness result for PDL^+. This can be proved by a patch of the LCC completeness proof in [16] where the action modalities are pushed through program modalities by program transformations.

7.10 Example: Navigation

Navigation problems provide a nice example of the interaction of information flow and change in the world. Consider the case of a robot in a maze. Assume a grid where the robot can move through a sequence of rooms, in some of the four directions North, East, South and West, but some of these directions may be blocked. Assume that the robot has a compass and a map, and that the robot can observe what its present location looks like (which of the four exits is blocked), but not what the next room looks like.

We can assume that the grid and its map look the same. Since the robot has a compass, it knows how to orient the map. Therefore, as soon as the robot uses its sensor it can distinguish the kind of room it is in. There are 15 possibilities:

$$\uparrow, \downarrow, \rightarrow, \leftarrow, \updownarrow, \leftrightarrow, \twoheadrightarrow, \twoheadleftarrow, \Rrightarrow, \Lleftarrow, \Leftrightarrow, \Rightarrow, \Leftarrow, \Uparrow, \Downarrow.$$

The possibility where the robot can get nowhere is ruled out: we assume the robot is not locked in a room. Initially the robot knows it could be anywhere in the grid, i.e., anywhere on the map. As soon as the robot senses it is in a room of (say) type $\Leftrightarrow$, the update that the robot makes is with the observation

$$\langle \uparrow \rangle \mathsf{T} \wedge \langle \rightarrow \rangle \mathsf{T} \wedge \langle \downarrow \rangle \mathsf{T} \wedge \langle \leftarrow \rangle \mathsf{T}.$$

This is to say: I am in a position where I can go North, East, South and West. This rules out all possible locations on the map except for those of type $\Leftrightarrow$.

So we assume that the rooms are states in a Kripke model, and that the modalities $\langle \uparrow \rangle, \langle \rightarrow \rangle, \langle \downarrow \rangle$ and $\langle \leftarrow \rangle$ are available, for moving one step in the indicated directions.

We can use modal formulas to express some obvious constraints on the model. The following formulas fix the relations between the directions:

$$[\uparrow]\langle\downarrow\rangle\top$$
$$[\downarrow]\langle\uparrow\rangle\top$$
$$[\rightarrow]\langle\leftarrow\rangle\top$$
$$[\leftarrow]\langle\rightarrow\rangle\top.$$

Here are two equivalent ways to represent the knowledge of the robot. Either assume that there is a basic proposition loc that is true in precisely one state, and represent the uncertainty of the robot as a set of identical maps, each with the proposition loc at a different place, or assume that there is just a single map, with a basic proposition loc pointing at the actual location of the robot and a basic proposition guess that is true at all places on the map that the robot has not yet ruled out as possibilities for where it might be.

Assume the second representation. Then initialize the value of guess to the set of all states on the map. The first update is when the robot uses its sensors to recognize the type of state. Let ϕ_{fit} be an abbreviation of the following formula:

$$\phi_{\text{fit}} := \bigwedge \{\langle x\rangle\top \mid x \in \{\uparrow, \rightarrow, \downarrow, \leftarrow\}, \langle x\rangle\top \text{ is true }\}$$
$$\wedge$$
$$\bigwedge \{[x]\bot \mid x \in \{\uparrow, \rightarrow, \downarrow, \leftarrow\}, \langle x\rangle\top \text{ is false }\}.$$

So if the robot finds itself in some type of location, and has not learned anything yet, but knows that the map is accurate and well oriented, then it will put guess equal to the set of locations that are of the same type as its current location. So the initial thing that the robot learns is:

$$\text{guess} := \phi_{\text{fit}}.$$

Now the robot can make a move, and learn from what it sees in the next location. Making a move is changing the location in the maze. So if the robot moves *North*, this is modelled as:

$$\text{loc} := \langle\downarrow\rangle\text{loc}.$$

Explanation: the old location is now *South* of the new location, so to define the new location in terms of the old, we must 'look back'. If the robot moves *South*, this is modelled as $\text{loc} := \langle\uparrow\rangle\text{loc}$, if the robot moves *East*, this is modelled as $\text{loc} := \langle\leftarrow\rangle\text{loc}$, if the robot moves *West*, this is modelled as $\text{loc} := \langle\rightarrow\rangle\text{loc}$. This was for modelling the change in the actual world. Now let's model what the robot learns about its location by observing the new location. After moving South, the robot updates its guess by means of:

$$\text{guess} := \langle\uparrow\rangle\text{guess} \wedge \phi_{\text{fit}}.$$

And so on for the other directions.

In the other representation, where the robot maintains a set of maps, updating consists of an update of loc as before, followed by a check loc $\rightarrow \phi_{\text{fit}}$ that eliminates all maps where the new location does not agree with the new observation given by ϕ_{fit}.

This was for the particular case of a maze, but it is clear that any kind of navigable world can be represented as a Kripke model. An interesting case is a model with non-functional accessibilities. If there are a-labelled arrows in different directions, it means that the robot cannot distinguish between two 'similar' actions that result in the robot ending up in different locations. In Kripke semantics, this just means that the following action can be executed in more than one way.

$$\text{loc} := \langle a^\vee \rangle \text{loc}.$$

For further information on the epistemics of navigation we refer to [56, 65].

7.11 Epistemic Planning and Protocol Languages

Navigation is a specific example of epistemic planning, for which the DEL framework is well-suited, because one can take atomic planning acts as event model updates. This generalizes the classical approach to planning in the presence of noisy sensors [36]. We give a summary, taking our cues from [1, 3, 20, 50].

A planning domain is a state transition system $\Sigma = (S, A, \gamma)$ where S is a finite or recursively enumerable set of states, A is a finite set of actions, and γ is a partial computable function $S \times A \hookrightarrow S$ (the state transition function). A planning task can now be viewed as a triple (Σ, s_0, G), where $\Sigma = (S, A, \gamma)$ is a planning domain, s_0 is a state in S, and G is a subset of S (the set of goal states). A solution to a planning task is a finite sequence of actions $a_1, a_2, \ldots, a_n$ (a plan), such that $\overline{\gamma}(a_1, a_2, \ldots, a_n)$ is defined and $\in G$, where $\overline{\gamma} : A^* \hookrightarrow S$ is defined by $\overline{\gamma}(a) = \gamma(s_0, a)$, $\overline{\gamma}(\overline{a}, a) = \gamma(\overline{\gamma}(\overline{a}), a)$ if $\overline{\gamma}(\overline{a})$ is defined, undefined otherwise. Informally, a plan succeeds if using the state transition function γ starting from (s_0, a_1) and following the plan, one can reach a goal state. An epistemic planning task is a special case of this where the s_0 is an epistemic state, the set A is a set of finite action models, and the set of goal states is represented by an epistemic formula ϕ_g.

An example of an epistemic plan for letting both a and b know that p without revealing to a that b knows p and without revealing to b that a knows p is first privately communicating p to a and then privately communicating p to b.

It is proved in [20] that the plan existence problem for three agent epistemic planning with factual change is undecidable. This result is strengthened in [3]: even without factual change, the plan existence problem for two agent epistemic planning in an S5 setting is undecidable.

For an intensional view of planning, we can use a version of PDL over action models, to define plan protocols (for still another guise of PDL, as a multi-agent strategy logic, see [26]).

Definition 7.18 The DEL protocol language for Ag and $Prop$ is given by:

$$\phi :: = p \mid \neg\phi \mid \phi \wedge \phi \mid \Box_a \phi \mid [\pi]\phi$$
$$\pi :: = (\mathbf{A}, e) \mid \pi \cup \pi \mid \pi; \pi \mid \pi^*$$

where p ranges over $Prop$, a ranges over Ag, and $(\mathbf{A}, e)$ is an action model (without factual change or belief change) for $\mathcal{L}$ with distinguished event e.

The truth conditions for the protocols π are given by:

Definition 7.19

$\mathbf{M} \models_w [\mathbf{A}, e]\phi$	iff $\mathbf{M} \models_w \mathbf{pre}(e)$ implies $\mathbf{M} \otimes \mathbf{A} \models_{(w,e)} \phi$
$\mathbf{M} \models_w [\pi_1 \cup \pi_2]\phi$	iff $\mathbf{M} \models_w [\pi_1]\phi$ and $\mathbf{M} \models_w [\pi_2]\phi$
$\mathbf{M} \models_w [\pi_1; \pi_2]\phi$	iff $\mathbf{M} \models_w [\pi_1][\pi_2]\phi$
$\mathbf{M} \models_w [\pi^*]\phi$	iff for any finite sequence $\pi; \ldots; \pi$ $\mathbf{M} \models_w [\pi; \ldots; \pi]\phi$

It follows from results in [53] that the satisfiability problem for this protocol language is undecidable. It is proved in [3] that the model checking problem for DEL protocols is also undecidable. This follows from a reduction of the plan existence problem to the model checking problem: an epistemic planning task $((\mathbf{M}, w), A, \phi_g)$ has a solution iff $\mathbf{M} \models_w \neg[A^*]\neg\phi_g$ holds.

7.12 Further Connections

An intriguing question about action model update that so far has only received a partial answer is: When are two action updates the same? More precisely, let us say that action models $\mathbf{A}$ and $\mathbf{B}$ are *equivalent* iff it holds for all static models $\mathbf{M}$ that $\mathbf{M} \otimes \mathbf{A} \Leftrightarrow \mathbf{M} \otimes \mathbf{B}$, where $\Leftrightarrow$ expresses the existence of a bisimulation that connects each distinctive point from $\mathbf{M} \otimes \mathbf{A}$ with a distinctive point from $\mathbf{M} \otimes \mathbf{B}$.

It turns out that action model bisimulation is not the appropriate structural notion to cover equivalence. In [28] a notion of *parametrized action emulation* is defined that characterizes action model equivalence. In [62, 63] this is replaced by a non-parametrized action emulation relation, and it is shown that this characterizes action model equivalence for canonical action models (action models with maximal consistent subsets of an appropriate closure language as preconditions). The question whether non-parametrized action emulation also characterizes action model equivalence for arbitrary action models is still open.

Another intriguing issue is the proper connection between epistemic/doxastic updating and probability theory. Useful overviews of logics of uncertainty are [57] and [39]. For the connection with DEL, see [17, 31, 46, 47] and the contribution of Kooi and Demey in the present volume. These proposals do not equate knowledge with certainty, but [24] does; this paper proposes a DEL logic (together with an epistemic model checking program) where the following principles hold ($P_a\phi$ is the probability that agent a assigns to ϕ):

$$\text{Certainty implies Truth} \quad P_a\phi = 1 \rightarrow \phi.$$
$$\text{Positive Introspection into Certainty} \quad P_a\phi = 1 \rightarrow P_a(P_a\phi = 1) = 1.$$
$$\text{Negative Introspection into Certainty} \quad P_a\phi < 1 \rightarrow P_a(P_a\phi < 1) = 1.$$

All these probabilistic versions of DEL incorporate Bayesian updating/learning; the difference is in whether Bayesian updates get analyzed as belief revision or as knowledge growth.

Acknowledgments Thanks to Guillaume Aucher, Alexandru Baltag, Johan van Benthem, Sonja Smets, for illuminating discussion, feedback and encouragement.

References

1. Andersen MB, Bolander T, Jensen MH (2012) Conditional epistemic planning. In: Proceedings of JELIA 2012, Lecture notes in artificial intelligence, vol 7519, pp 94–106
2. Aucher G (2003) A combined system for update logic and belief revision. Master's thesis, ILLC, Amsterdam
3. Aucher G, Bolander T (2013) Undecidability in epistemic planning. In: Proceedings of IJCAI 2013 (23rd international joint conference on artificial intelligence), August 2013
4. Aumann RJ (1976) Agreeing to disagree. Ann Stat 4(6):1236–1239
5. Bacchus F, Halpern J, Levesque H (1999) Reasoning about noisy sensors and effectors in the situation calculus. Artif Intell 111(1–2):171–208
6. Baltag A, Moss LS (2004) Logics for epistemic programs. Synthese 139(2):165–224
7. Baltag A, Moss LS, Solecki S (1998) The logic of public announcements, common knowledge, and private suspicions. In: Bilboa I (ed) Proceedings of TARK'98, pp 43–56
8. Baltag A, Moss LS, Solecki S (2003) The logic of public announcements, common knowledge, and private suspicions. Technical report, Dept of Cognitive Science, Indiana University and Dept of Computing, Oxford University, Bloomington and Oxford, 2003 (Updated version of [7])
9. Baltag A, Smets S (2006) Dynamic belief revision over multi-agent plausibility models. In: van der Hoek W, Wooldridge M (eds) Proceedings of LOFT'06, pp 11–24, Liverpool, 2006. University of Liverpool
10. Baltag A, Smets S (2008) The logic of conditional doxastic actions. In: Apt K, van Rooij R (eds) New perspectives on games and interaction. Texts in logic and games, vol 5. Amsterdam University Press, Amsterdam, pp 9–31
11. Baltag A, Smets S (2009) Group belief dynamics under iterated revision: fixed points and cycles of joint upgrades. In: Proceedings of the 12th conference on theoretical aspects of rationality and knowledge, TARK '09, pp 41–50, New York, NY, USA, 2009. ACM
12. Battigalli P, Bonanno G (1999) Recent results on belief, knowledge and the epistemic foundations of game theory. Res Econ 53:149–225
13. van Benthem J (2000) Information update as relativization. Technical report, ILLC, Amsterdam, 2000. http://staff.science.uva.nl/johan/Upd=Rel.pdf

14. van Benthem J (2007) Dynamic logic for belief revision. J Appl Non-Classical Logics 2:129–155
15. van Benthem J, Liu F (2007) Dynamic logic of preference upgrade. J Appl Non-Classical Logics 14(2):157–182
16. van Benthem J, van Eijck J, Kooi B (2006) Logics of communication and change. Inf Comput 204(11):1620–1662
17. van Benthem J, Gerbrandy J, Kooi B (2009) Dynamic update with probabilities. Stud Logica 93:67–96
18. van Benthem J, van Ditmarsch H, van Eijck J, Jaspars J (2012) Logic in action. Internet, 2012. http://www.logicinaction.org
19. Blackburn P, de Rijke M, Venema Y (2001) Modal logic. Cambridge tracts in theoretical computer science. Cambridge University Press, Cambridge
20. Bolander T, Andersen MB (2011) Epistemic planning for single- and multi-agent systems. J Appl Non-Classical Logics 21(1):9–34
21. Cox RT (1946) Probability, frequency, and reasonable expectation. Am J Phys 14(1):1–13
22. van Ditmarsch HP, van der Hoek W, Kooi BP (2005) Dynamic epistemic logic with assignment. In: Proceedings of the fourth international joint conference on autonomous agents and multi-agent systems (AAMAS 05), pp 141–148, New York, 2005. ACM Inc
23. van Ditmarsch HP, van der Hoek W, Kooi B (2006) Dynamic epistemic logic. Synthese library, vol 337. Springer, Dordrecht
24. van Eijck J Learning about probability. http://homepages.cwi.nl:/jve/software/prodemo
25. van Eijck J (2008) Yet more modal logics of preference change and belief revision. In: Apt KR, van Rooij R (eds) New perspectives on games and interaction. Texts in logic and games, vol 4, pp 81–104. Amsterdam University Press, Amsterdam
26. van Eijck J (2013) PDL as a multi-agent strategy logic. In: Schipper BC (ed) TARK 2013—Theoretical aspects of reasoning about knowledge, Proceedings of the 14th conference—Chennai, India, pp 206–215
27. van Eijck J, Wang Y (2008) Propositional dynamic logic as a logic of belief revision. In: Hodges W, de Queiros R (eds) Proceedings of Wollic'08, number 5110. Lecture notes in artificial intelligence, pp 136–148. Springer
28. van Eijck J, Ruan J, Sadzik T (2012) Action emulation. Synthese 185:131–151
29. Fagin R, Halpern JY, Moses Y, Vardi MY (1995) Reasoning about knowledge. MIT Press, Cambridge
30. Fagin R, Halpern JY, Moses Y, Vardi MY (1997) Knowledge-based programs. Distrib Comput 10(4):199–225
31. Fagin R, Halpern JY (1994) Reasoning about knowledge and probability. J ACM 41:340–367
32. Floyd RW (1967) Assigning meanings to programs. In: Proceedings AMS symposium applied mathematics, vol 19, pp 19–31, Providence, RI, 1967. American Mathematical Society
33. Friedell MF (1969) On the structure of shared awareness. Behav Sci 14(1):28–39
34. Gärdenfors P (1988) Knowledge in flux: modelling the dynamics of epistemic states. MIT Press, Cambridge
35. Gerbrandy J (1999) Bisimulations on planet Kripke. PhD thesis, ILLC, Amsterdam
36. Ghallab M, Nau DS, Traverso P (2004) Automated planning: theory and practice. Morgan Kaufmann, Amsterdam
37. Girard P, Seligman J (2012) General dynamic dynamic logic. In: Bolander T, Braüner T, Ghilardi S, Moss LS (eds) Advances in modal logic. College Publications, pp 239–260
38. Groeneveld W (1995) Logical investigations into dynamic semantics. PhD thesis, ILLC, Amsterdam
39. Halpern J (2003) Reasoning about uncertainty. MIT Press, Cambridge
40. Halpern JY, Moses YO (1984) Knowledge and common knowledge in a distributed environment. In: Proceedings 3rd ACVM symposium on distributed computing, pp 50–68
41. Halpern J (1987) Using reasoning about knowledge to analyse distributed systems. Annu Rev Comput Sci 2:37–68
42. Hemaspaandra E (1990) The price of universality. Notre Dame J Form Log 37(2):174–203

43. Hintikka J (1962) Knowledge and belief: an introduction to the logic of the two notions. Cornell University Press, Ithaca
44. Hoare CAR (1969) An axiomatic basis for computer programming. Commun ACM 12(10):567–580, 583
45. Jaspars J (1994) Calculi for constructive communication. PhD thesis, ITK, Tilburg and ILLC, Amsterdam
46. Kooi BP (2003) Knowledge, chance, and change. PhD thesis, Groningen University
47. Kooi Barteld P (2003) Probabilistic dynamic epistemic logic. J Logic Lang Inform 12(4):381–408
48. Kozen D, Parikh R (1981) An elementary proof of the completeness of PDL. Theoret Comput Sci 14:113–118
49. Lewis DK (1969) Convention: a philosophical study. Harvard University Press, Cambridge
50. Löwe B, Pacuit E, Witzel A (2011) Del planning and some tractable cases. In: Proceedings of the third international conference on logic, rationality, and interaction, LORI'11, pp 179–192, Berlin, Heidelberg, 2011. Springer-Verlag
51. Martin L (2006) Model checking propositional dynamic logic with all extras. J Appl Logic 4(1):39–49
52. McNulty G (1977) Fragments of first order logic, I: universal horn logic. J Symbolic Logic 42:221–237
53. Miller JS, Moss LS (2005) The undecidability of iterated modal relativization. Stud Logica 79(3):373–407
54. Moore RC (1985) A formal theory of knowledge and action. In: Hobbs JR, Moore RC (eds) Formal theories of the commonsense world. Ablex Publishing, Norwood, pp 319–358
55. Osborne MJ, Rubinstein A (1994) A course in game theory. MIT Press, Cambridge
56. Panangaden P, Sadrzadeh M (2011) Learning in a changing world, an algebraic modal logical approach. In: Proceedings of the 13th international conference on algebraic methodology and software technology, AMAST'10, pp 128–141, Berlin, Heidelberg, 2011. Springer-Verlag
57. Paris JB (1994) The uncertain reasoner's companion—a mathematical perspective. Cambridge tracts in theoretical computer science, vol 39. Cambridge University Press, Cambridge
58. Perea A (2012) Epistemic game theory: reasoning and choice. Cambridge University Press, Cambridge
59. Plaza JA (1989) Logics of public communications. In: Emrich ML, Pfeifer MS, Hadzikadic M, Ras ZW (eds) Proceedings of the 4th international symposium on methodologies for intelligent systems, pp 201–216
60. Pratt V (1976) Semantical considerations on Floyd-Hoare logic. In: Proceedings 17th IEEE symposium on foundations of computer, Science, pp 109–121
61. Segerberg K (1982) A completeness theorem in the modal logic of programs. In: Traczyck T (ed) Universal algebra and applications. Polish Science Publications, Warsaw, pp 36–46
62. Sietsma F (2012) Logics of communication and knowledge. PhD thesis, ILLC, Amsterdam
63. Sietsma F, van Eijck J (2013) Action emulation between canonical models. J Philos Logic (accepted for publication)
64. Wang Y (2010) Epistemic modelling and protocol dynamics. PhD thesis, ILLC, Amsterdam
65. Wang Y, Li Y (2012) Not all those who wander are lost: dynamic epistemic reasoning in navigation. In: Proceedings of AIML 2012, Advances in Modal Logic, vol 9, 559–580, College Publications

Chapter 8
Belief Revision and Dynamic Logic

Patrick Girard and Hans Rott

Abstract We explore belief change policies in a modal dynamic logic that explicitly delineates knowledge, belief, plausibility and the dynamics of these notions. Taking a Kripke semantics counterpart to Grove semantics for AGM as a starting point, we analyse belief in a basic modal language containing epistemic and doxastic modalities. We critically discuss some philosophical presuppositions underlying various modelling assumptions commonly made in the literature, such as the limit assumption and negative introspection for knowledge. Finally, we introduce in the language a general dynamic mechanism and define various policies of iterated belief expansion, revision, contraction and two-dimensional belief change operations.

8.1 Introduction

The history of belief revision is one in which researchers from various fields have tackled the same problem from different perspectives, but it originated from philosophy. After earlier work of Isaac Levi and William Harper, Carlos Alchourrón, Peter Gärdenfors and David Makinson (often referred to by the acronym "AGM") initiated the formal study of belief change operations in the 1980s. They analyzed belief change using three kinds of models: partial meet contractions and revisions (in terms of maximal non-implying sets [1]), safe contractions and revisions (in terms of minimal implying sets of sentences [2]), and entrenchment-based contractions and revisions (based on the comparative retractability of sentences [25]). Grove [29]

P. Girard (✉)
Department of Philosophy, The University of Auckland,
18 Symonds Street, Private Bag 92019, Auckland 1142, New Zealand
e-mail: p.girard@auckland.ac.nz

H. Rott
Department of Philosophy, University of Regensburg, 93040 Regensburg, Germany
e-mail: hans.rott@ur.de

A. Baltag and S. Smets (eds.), *Johan van Benthem on Logic*
and Information Dynamics, Outstanding Contributions to Logic 5,
DOI: 10.1007/978-3-319-06025-5_8, © Springer International Publishing Switzerland 2014

provided a possible worlds semantics for partial meet contraction. Iterated belief change was addressed in the 1990s, with important contributions by Boutilier [16], Darwiche and Pearl [17] and Nayak [37]. More recently, van Benthem [9], and Baltag and Smets [4, 5] modelled epistemic and doxastic states and their transformations on simple relational structures, introducing a new complexity due to the analysis of higher-order belief and knowledge. The relevant problems were attacked in a plethora of ways that may sometimes be differentiated by the degree of abstraction adopted.[1] Influenced by van Benthem, we choose our level of abstraction in line with a modular and minimalist attitude. This means that we only assume constraints on models when they become necessary, with philosophical awareness, and we use formal languages as simple as we can.

- Minimality: We try to keep the assumptions on models to be minimal, and we try to keep our basic object language as simple as possible.
- Modularity: We analyse the key notions of knowledge and belief in terms of two primitive relations representing indistinguishability and plausibility. We also break down complex dynamic notions into simpler parts using the programmatic PDL language.

8.2 Grove Systems of Spheres

In retrospect it looks as if AGM followed a purely syntactic approach to belief revision. We think that this is not quite correct, for both belief sets and inputs for belief change were essentially individuated only up to logical equivalence. Still it was a very important step for the program of belief revision when Adam Grove [29] provided a possible worlds semantics for AGM's partial meet contractions. He used systems of spheres to model belief change, closely following the seminal work of Lewis [18] on counterfactuals. We summarise Grove's semantics in some detail, and later show how to adapt the notation and interpretation to our needs.

Given a language $\mathscr{L}$, define a *theory* T over $\mathscr{L}$ as a set of $\mathscr{L}$-sentences that are logically closed, i.e., for which $\mathsf{cl}(T) \subseteq T$. Here $\mathsf{cl}(\cdot)$ is an operation that forms the logical closure of a set of sentences from $\mathscr{L}$. A possible world for $\mathscr{L}$ is an entity that assigns, for each atom of $\mathscr{L}$, a truth value 1 (for "true") or 0 (for "false").[2] We denote by W the set of all possible worlds, and by $[\![\varphi]\!]$ or $[\![T]\!]$ the set of all possible worlds that satisfy a sentence φ or a set of sentences T. For a set of worlds V, we denote by $\mathsf{Theory}(V)$ the set of sentences true at each world in V.

---

[1] This caused in turn a proliferation of acronyms such as DEL for 'dynamic epistemic logic' which limits our freedom in using those that would naturally arise with our terminology throughout the chapter. We thus use EDL to stand for 'epistemic doxastic logic'. The reader should try not to get confused by this choice.

[2] Grove [29] himself used maximal consistent sets of sentences rather than possible worlds, but this difference need not concern us here.

A *Grove system of spheres* centered around a set $\mathscr{B}$ of possible worlds for $\mathscr{L}$ is a set $\$_{\mathscr{B}} = \{S | S \subseteq W\}$ that satisfies the following conditions[3]:

1. $\$_{\mathscr{B}}$ is totally ordered by set-containment $\subseteq$,
2. $\mathscr{B} \in \$_{\mathscr{B}}$,
3. $\mathscr{B} \subseteq U$ for every $U \in \$_{\mathscr{B}}$,
4. *The limit assumption:* For any sentence φ, if there is a sphere $S \in \$_{\mathscr{B}}$ which intersects $[\![\varphi]\!]$, then there is a smallest sphere $S' \in \$_{\mathscr{B}}$ which intersects $[\![\varphi]\!]$.

A Grove system of spheres is called

5a. *universal* if in addition, $W = \bigcup \$_{\mathscr{B}}$; or
5b. *strongly universal* if in addition, $W \in \$_{\mathscr{B}}$.

The term "universal" is taken over from Lewis [18, p. 16]. Grove originally required his doxastic systems of spheres to be strongly universal. We choose to deviate from this by requiring not even universality. We do not want to rule out that an agent may consider some metaphysically possible worlds inconceivable.

Belief in a Grovean sphere system is identified with the innermost sphere $\mathscr{B}$ of $\$_{\mathscr{B}}$.[4] More precisely, a belief set in Grove spheres is identified with the set of all sentences Theory($\mathscr{B}$) that are true throughout the innermost sphere $\mathscr{B}$ in $\$_{\mathscr{B}}$. The belief set Theory($\mathscr{B}$) is the one to be revised with the AGM operators of expansion, contraction and revision, which we will discuss below. We represent a Grove system of spheres centered around a set $\mathscr{B}$ in the following way, shading the innermost sphere as characterizing the belief set $\mathscr{B}$:

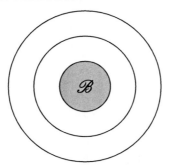

The sphere formulation is in no way necessary, as Grove notes: "a system of spheres is really an ordering on the set of worlds" [29, p. 160]. To see this, we can reformulate Grove spheres in terms of a relation $\leq$ on W, which we call a *Grove relation*. A Grove relation centered around a set $\mathscr{B} \subseteq W$ is a relation $\leq_{\mathscr{B}}$ on W that satisfies the following conditions[5]:

---

[3] We choose the notation $\mathscr{B}$ as a mnemonic device for 'belief'.

[4] Grove does not talk about beliefs, but about theories, and his semantics is about theory change broadly construed. However, his semantics is directly tailored to accommodate the AGM postulates, so we focus exclusively on a doxastic interpretation of his system.

[5] Unlike Grove, we read $x \leq_{\mathscr{B}} y$ as "y is more plausible than x according to $\leq$" and talk of *maximal* worlds instead of *minimal* worlds. Katsuno and Mendelzon [32] is a seminal reference for the use of ordering semantics within the belief revision community.

1. $\leq_{\mathscr{B}}$ is connected and transitive,
2. $x \in W$ is $\leq_{\mathscr{B}}$-maximal iff $x \in \mathscr{B}$.
3. *The limit assumption for* $\leq_{\mathscr{B}}$: For any sentence φ, if there are any worlds at which φ is true, there are $\leq_{\mathscr{B}}$-maximal worlds w at which φ is true, so $\{x \in [\![\varphi]\!] \mid y \leq_{\mathscr{B}} x$ for all $y \in [\![\varphi]\!]\}$ is non-empty.

In a Grove relation, the set $\{w \in W \mid v \leq_{\mathscr{B}} w$ for all $v \in W\}$ is the belief set, identified by the maximal elements in $\leq_{\mathscr{B}}$. We again refer to that set with $\mathscr{B}$. We represent Grove relations in the following way:

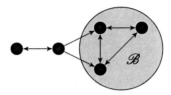

We have not drawn all arrows here, assuming that it is easy to see how the relations are transitively closed (Later we shall be even more economical in our use of arrows).

Formally, the difference between Grove spheres and relations is almost only one of taste, just requiring a Gestalt switch of the following kind:

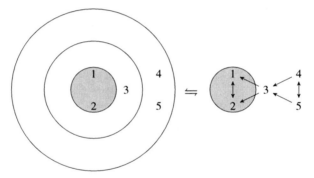

We can be more specific about the connection between Grove spheres and orderings: A Grove ordering $\leq_{\mathscr{B}}$ is obtained from a system of spheres $\$_{\mathscr{B}}$ by defining $v \leq_{\mathscr{B}} w$ (read: "v is at most as plausible as w") iff for all $S \in \$_{\mathscr{B}}$ such that $v \in S$ it also holds that $w \in S$; the field $W_{\mathscr{B}}$ of $\leq_{\mathscr{B}}$ is $\bigcup \$_{\mathscr{B}}$. Conversely, a system of Grove spheres $\$_{\mathscr{B}}$ is obtained from a Grove ordering $\leq_{\mathscr{B}}$ of W by collecting all sets S of the form $S_w = \{v \in W : w \leq_{\mathscr{B}} v\}$.

This modelling importantly allows for *ties* between the plausibilities of possible worlds, and this can of course be expressed in both ways of modelling. In the systems of spheres modelling, this means that for two neighbouring spheres S and S' in $\$_{\mathscr{B}}$ with $S \subseteq S'$, the set $S' - S$ has in general more than just one element. Expressed in orderings, this means that for many worlds w there may be any number of distinct worlds v such that both $v \leq_{\mathscr{B}} w$ and $w \leq_{\mathscr{B}} v$. The correspondence between systems of spheres and weak orderings is not perfect, however. The (strong) universality of a system of spheres $\$_{\mathscr{B}}$ cannot be expressed by a weak ordering alone. This is why we need to specify the field $W_{\mathscr{B}}$ of $\leq_{\mathscr{B}}$. We also take into account *inconsistent* belief states, represented by sphere systems $\$_{\mathscr{B}}$ that contain the empty sphere $\emptyset$ as an element. But if a sphere in $\$_{\mathscr{B}}$ is empty, it generates the same weak ordering $\leq_{\mathscr{B}}$, in the sense just explained, as the sphere system $\$_{\mathscr{B}} - \{\emptyset\}$. We take the case of empty innermost spheres seriously, because we want to address belief expansion in Sect. 8.5.2. This requires representing trivialisation under expansion with information inconsistent with current beliefs. As the equivalent of empty spheres is not available in ordering semantics, we employ domain restriction in order to achieve similar results.

We need to comment on the interpretation of Grove spheres, or Grove orderings, understood as semantics for the classical AGM style belief revision. Semantically, a whole Grovean system of spheres (or a Grove ordering, or any other model for AGM-style one-shot belief revision) *represents* or, loosely speaking, *is* a single agent's belief state. In some way (though it is hard to say in what way exactly), it encodes a first-person point of view: the set of worlds considered (doxastically) possible by the agent, and their comparative plausibilities as judged by the agent. Like the AGM approach, this modelling does not make room for *belief at a certain possible world*, it only encodes the beliefs and conditional beliefs as they are present for a single agent at the actual world ("here and now").

Another philosophical point: one should *not* identify knowledge with the "outer belief modality" in Grove spheres, i.e., with "irrevocable belief" in the sense of Segerberg [46]. Knowledge is not the same as irrevocable belief. While knowledge implies truth, irrevocable belief need not do so. Baltag and Smets's [5, p. 16] claim that the idea of identifying knowledge and irrevocable belief "can be traced back to Stalnaker [47]" is not quite correct; Stalnaker just defines $\Box\varphi$ as $\neg\varphi > \varphi$, without any epistemic interpretation.[6] Segerberg [46] uses this reading only "unofficially" and alternatively speaks of *doxastic commitment*. Leitgeb and Segerberg [34, p. 176] allow themselves some philosophical looseness, too, and use the slogan *K stands for "knowledge", not for 'knowledge'*. Our feeling is that this identification is made only for the sake of convenience, because it is then easy to define knowledge in terms of (conditional) belief—something that epistemologists were never able to achieve—, and it is easy to argue for positive and negative introspection concerning knowledge—something that epistemologists have never wanted to achieve. If agents were infallible, then irrevocable belief would come close to knowledge, because then irrevocable belief would be guaranteed to be true. But human agents are not infallible, even their most deeply rooted beliefs may turn out to be false. We will return to this topic in Sect. 8.3.

---

[6] Alexandru Baltag (p.c.) has reminded us that Stalnaker [47, p. 102] endorses a condition akin to universality (see Stalnaker's semantic condition (2)). Thus irrevocable belief must be true, and our argument against its identification with knowledge is no longer applicable. This is correct, but endorsing universality in this context means arguing that only metaphysical necessities can be known, since no metaphysically possible world gets epistemically excluded. For this reason, we strongly prefer not to require universality.

Finally, even though Grove spheres are inspired by Lewis systems of spheres [18], Lewis explicitly rejects the limit assumption. Lewis' argument is about similarity relations on worlds underlying counterfactual reasoning, but similar worries hold for plausibility orders: there are no most similar worlds in which Franck Ribéry is taller than 180 cm. Lewis' semantics for counterfactuals does not depend on the limit assumption. Likewise, we can define belief and conditional belief using sequences of increasingly plausible worlds that do not require the limit assumption. Furthermore, there are AGM-inspired doxastic transformations of expansion, revision and contraction that do not require the limit assumption. Hence, following our minimalist attitude, we refrain to assume the limit assumption until necessary.[7]

8.3 Epistemic Doxastic Logic

In contrast to the use of Grove models within the AGM paradigm, we will model knowledge and belief with Kripke structures. Thus what we call 'epistemic doxastic logic' in this chapter (EDL for short) stands in the tradition of epistemic logic going back at least as far as the seminal work of Hintikka [31]. More recently, epistemic logic has been dynamified in the Dutch tradition and is now referred to as 'dynamic epistemic logic' (DEL for short). The acronym DEL is sometimes used more generally to also include doxastic transformations (see for instance van Benthem [21]). We use DEL in this general sense, and we reserve the acronym EDL for our own version of DEL in this chapter. For us, EDL is the static core of our analysis of knowledge and belief. We will introduce dynamics in the next section and adapt our terminology accordingly.

In EDL, all theorizing begins with knowledge-at-a-possible-world-w or belief-at-a-possible-world-w. So belief and knowledge are "local" in this way. A whole EDL model does *not* represent some agent's or several agents' doxastic and epistemic states. The worlds in an EDL model are ways our actual world might be, metaphysically, *not* epistemically speaking.

One should not get confused by some "global" structures within EDL models. It might look strange that *within a cell of the epistemic partition of the worlds*, there is only one doxastic plausibility relation. In the local interpretation, which we believe is the correct one, it would be more precise to say that within such a cell, every world has the same plausibility relation assigned to it, and that this relation is assigned to the cell rather than to every world within the cell only for notational convenience.

---

[7] Our reservation to assume the limit assumption from the outset is not only driven by philosophical concerns. To find an adequate axiomatisation of the limit assumption in our framework is not easy, and presently we can only make an informed conjecture that, in the context of the other axioms we are using, the axiom known as the 'Löb's axiom' is exactly what we need. We will come back to this point below.

If possible worlds are considered as indices of evaluation, we might say that in our modelling, the doxastic relation is really just as *indexical* as the epistemic relation.[8]

That the plausibility relation is the same at each world within a cell, however has a very good justification. A plausibility relation is a formal encoding of the agent's doxastic state. The fundamental assumption is that agents fully know their own doxastic attitudes.[9] There has to be an equally fundamental EDL axiom that captures this assumption. If we had just beliefs, the axiom should be something like $B\varphi \rightarrow KB\varphi$. But since we work with plausibility relations that allow us to express something like degrees of beliefs (entrenchments), we capture that agents are fully aware of their degrees of belief.[10]

Such a fundamental axiom is an expression of full (epistemic) introspection of doxastic attitudes. By viewing a global relation *within a cell of the knowledge partition* just as an abbreviated way of specifying that this very relation is assigned to each world within the cell, we can justify the use of global relations (restricted to cells) by the above introspection principles, while still staying firmly on the ground of the local tradition of DEL.

It does not make sense to look at an EDL model and ask what an agent believes or knows *in that model*. One can only ask what an agent believes or knows *at a certain world* in that model. The situation changes in pointed models that come with a distinguished world. This is why Baltag, van Ditmarsch and Moss [6], for instance, define an 'epistemic state' (p. 387) or a 'doxastic state' (p. 397) as tuples (M, s), where M is a (relational/plausibility) model and s is a state or world.[11] Unfortunately, the authors do not comment at all on why they add a distinguished world. Summing up, a pointed model represents the actual world, with all the actual beliefs being in turn represented in terms of possible (conceivable) worlds. It thus makes sense to ask what an agent believes or knows *in a pointed model*.

If our interpretation is right, it hardly makes sense to drop worlds from the model as a result of some doxastic action. Worlds in EDL models are metaphysically possible worlds, not doxastically or epistemically possible worlds, as in Grove models. But as DEL theorists point out, a doxastic or epistemic action normally changes what is true in a world like any other action. Unlike the case of non-doxastic or non-epistemic actions, the effects of doxastic and epistemic actions can be represented in the model. The corresponding model transformation consists in manipulating the relevant relations between possible worlds, that is, in actions like cutting links, refining partitions, or shifting plausibilities.[12] This tension between world-elimination and link-cutting is a salient one for us. We will come back to it in Sect. 8.4.1.

---

[8] The indexical stance is represented in the models of Board [14, pp. 60–61], van Ditmarsch [20, p. 237], and van Benthem [9, p. 138], while Baltag and Smets [3, p. 12] and [4, pp. 17–23] prefer a presentation in terms of global plausibility relations.

[9] Similar ideas of extending positive and negative introspection assumptions about belief were advanced by Stalnaker [48, p. 145] [49, p. 189] and Board [14, pp. 60–61]. See Demey [19, p. 387] on "uniform" epistemic plausibility models.

[10] We will return to this point later, see the comment on definitions (8.2)–(8.4) in Sect. 8.3.2.

[11] Similarly for van Ditmarsch's [20, p. 237] 'doxastic-epistemic state'.

[12] Perhaps making some worlds "infinitely implausible".

The established tradition in modal logic as applied in computer science and game theory is to model knowledge by an equivalence relation that represents *indistinguishability for the agent*. The idea is that an agent knows that φ if and only if φ is true in all possible worlds that the agent cannot distinguish from the real world. We should like to emphasize that this is decidedly *not* the ordinary notion of knowledge. For knowledge in anything close to the ordinary, general sense, and the sense studied by epistemologists, negative introspection is a paradigmatic *non-theorem* rather than a theorem for epistemic logic.[13] More often than we like, we fail to know because we are wrong. In many such cases, we believe that we know p, but this belief is wrong. Thus, we don't know (because p is false), but don't know that we don't know (an exemplification of an important kind of unknown unknowns). It is abundantly clear that the Brouwerian principle $\neg\varphi \rightarrow K\neg K\varphi$ and the interaction principle $BK\varphi \rightarrow K\varphi$ are invalid for the ordinary, general notion of knowledge,[14] but they come out as valid according to the standard DEL semantics. The notion of knowledge that is modelled by S5 structures is the knowledge of agents that are infallible. But humans are not. We can justify taking such structures only by restricting ourselves to specific domains or contexts in which agents do not make any mistakes, i.e., in which their doxastic possibilities do not rule out the actual world. Such contexts are provided by certain fields of research in computer science and game theory. We pretend that we are working in some such context and just ignore this problem as a matter of idealisation.

We thus conceive of the combination of the epistemic and doxastic relations in the following way. The epistemic relation creates a partition of the domain W, and each equivalence class of the partition contains a Grove relation (or a system of spheres). That we only have a single Grove relation (or a single system of spheres) for each epistemic indistinguishability class reflects the idea that an agent has the same doxastic state in each possible world within a cell of the knowledge partition. So we assume that agents are fully knowledgeable about (have full introspection concerning) their own doxastic states. If they were not, we would need to assign Grove spheres to each world individually. We represent the doxastic epistemic structure of a single agent as follows:

---

[13] Even positive introspection for knowledge, the so-called KK thesis, has been much contested, especially in the light of the success of externalist and reliabilist accounts in epistemology. Very recently, some new defenders of KK have entered the scene, see Okasha [38] and Greco [28].

[14] Lenzen [35] and, following Lenzen, Stalnaker [49] offer very strong arguments in favour of defining $B\varphi$ as $\neg K\neg K\varphi$. In the light of this definition. Brouwer's principle, which corresponds to the symmetry of the accessibility relation, just *means* $B\varphi \rightarrow \varphi$, i.e., infallibility! Lenzen [35, p. 43], Lamarre and Shoham [33, pp. 415, 420] and Stalnaker [49, p. 179] advocate the principle of strong belief $B\varphi \rightarrow BK\varphi$ (the term "strong belief" is Stalnaker's, Lamarre and Shoham use the term "certainty"). Taken together with the strong belief principle, the interaction principle $BK\varphi \rightarrow K\varphi$ implies the undesirable $B\varphi \rightarrow K\varphi$. – In his attempt to maintain negative introspection for knowledge, Halpern [30] proposes to restrict the principle that knowledge implies belief to nonmodal sentences.

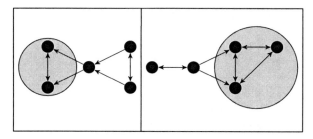

The domain is partitioned in two equivalence classes, and each "cell" contains a Grove relation, with maximal elements shaded in gray. The maximal elements of each cell are the states at which beliefs at a world in that very cell are to be evaluated. All worlds within a cell refer to the same belief set. So we picture Grovean sphere systems only within the cells of the knowledge partition (and each such cell contains exactly one such system). We impose this as a (the one standard) constraint on the interaction between the doxastic and epistemic relation, namely that the former is a subrelation of the latter: $\leq\, \subseteq\, \sim$.

The fact that we have Grove relations $\leq$ within each knowledge cell means that the domain of such Grove relations is not the whole of W (i.e., that the relevant Grove systems are not universal). The structural properties of Grove relations are all restricted to each individual indistinguishability cell. There are no plausibility comparison across cells. Hence, if we want to stick to the single relation modelling, we have to use more complicated structures than Grove orderings.

A *generalized Grove relation* is a reflexive and transitive relation $\leq$ over W such that the relation $\sim$, defined by

$$u \sim v \quad \text{if and only if} \quad \text{either } u \leq v \text{ or } v \leq u$$

is an equivalence relation.[15] Notice that we do not assume the limit assumption for the generalized Grove relation. This definition derives the epistemic relation from the doxastic relation: Indistinguishability means comparability in terms of plausibility. The definition guarantees that $\sim$ contains $\leq$. It is not required that generalized Grove relations are connected over the whole of W. They may have, and typically do have many belief sets (sets of doxastically possible worlds) on which systems of spheres are centered—one such structure in each cell of the partition. So $\leq$ never makes any plausibility comparisons across cells. But it is easily verified that each cell with respect to $\sim$ is a Grove relation—without the limit assumption.

---

[15] Given the transitivity of $\leq$, a condition equivalent to the transitivity of $\sim$ thus defined is the requirement of weak connectedness of $\leq$ both forwards and backwards, also known as "no branching of $\leq$ to the left or to the right":

If $w \leq u$ and $w \leq v$, then either $u \leq v$ or $v \leq u$ (NBR)

If $u \leq w$ and $v \leq w$, then either $u \leq v$ or $v \leq u$ (NBL)

This requirement is used by Baltag et al. [6, p. 396].

8.3.1 EDL *Language and Semantics*

We define *epistemic doxastic models* (for single agents), or EDL models for short, as structures $M = \langle W, \sim, \leq, <, V \rangle$, with W a non-empty set of worlds, an V a valuation function assigning sets of possible worlds to propositional variables. We take $\leq$ to be a generalised Grove relation. We define $\sim$ as above, and take $<$ as the strict subrelation of $\leq$ in the usual way: $w < v$ iff $w \leq v$ and $v \not\leq w$.[16] We read $w \leq v$ as 'v is at least as plausible as w', and $w < v$ as 'v is (strictly) more plausible than w'. To talk about EDL models, we use a basic EDL modal language with three modalities corresponding to the three accessibility relations:

$$\varphi ::= p \mid \neg\varphi \mid (\varphi \wedge \psi) \mid [\sim]\varphi \mid [\leq]\varphi \mid [<]\varphi$$

As usual, we define dual diamond operators as, for example, $\langle\sim\rangle\varphi := \neg[\sim]\neg\varphi$.

For the interpretation of the EDL language, we extend the valuation V to a valuation $[\![\cdot]\!]^M$ assigning semantic values, or sets of possible worlds, to the sentences of the EDL langauge. Hence, in each epistemic doxastic model $M = \langle W, \sim, \leq, <, V \rangle$, semantic values $[\![\varphi]\!]^M \subseteq W$ are given by:

$$[\![p]\!]^M = V(p)$$
$$[\![\neg\varphi]\!]^M = W \setminus [\![\varphi]\!]^M$$
$$[\![(\varphi \wedge \psi)]\!]^M = [\![\varphi]\!]^M \cap [\![\psi]\!]^M$$
$$[\![[\sim]\varphi]\!]^M = \{w \in W \mid \text{if } w \sim v, \text{ then } v \in [\![\varphi]\!]^M, \text{ for every } v \in W\}$$
$$[\![[\leq]\varphi]\!]^M = \{w \in W \mid \text{if } w \leq v, \text{ then } v \in [\![\varphi]\!]^M, \text{ for every } v \in W\}$$
$$[\![[<]\varphi]\!]^M = \{w \in W \mid \text{if } w < v, \text{ then } v \in [\![\varphi]\!]^M, \text{ for every } v \in W\}$$

For convenience, we sometimes use the more common notation $M, w \models \varphi$ instead of $w \in [\![\varphi]\!]^M$.

8.3.2 *Knowledge and Belief in* EDL

For reasons of simplicity and continuity with much of the literature, we assume that agents know that φ in a world w just in case φ is true in all worlds v that they cannot distinguish from w:

$$M, w \models K\varphi \quad \text{iff} \quad \text{for all } v \text{ such that } v \sim w, M, v \models \varphi.$$

---

[16] The presence of $\sim$ and $<$ is thus redundant, as they are definable in terms of $\leq$ in models, but we keep them for reasons that will become clear later.

In other words,

$$K\varphi := [\sim]\varphi \tag{8.1}$$

We thus have a semantics for $K\varphi$ that is widely accepted in computer science and game theory. As we explained above, we have philosophical scruples about it, but we wish to keep the focus of this chapter on belief.

So much for the epistemic part. For doxastic operators, we need to do more work. Our strategy is to derive the analysis of (various kinds of) belief by using the more primitive modalities $[\leq]$ and $[<]$. Following a strategy that can be traced back to at least Boutilier [15, p. 44], we have an intended semantics for belief that is in line with the AGM tradition:

$$M, w \models B\varphi \quad \text{iff} \quad \text{there is a } u \text{ such that } u \sim w \text{ and for all } v \text{ with } u \leq v, \, M, v \models \varphi.$$

Notice the epistemic constraint $v \sim w$ in the semantic definition of $B\varphi$, in order to guarantee that beliefs are independently evaluated in each class of the epistemic partition. Given our assumption of connectedness inside each epistemic class, this semantics says that at some point along the plausibility order, φ is true for every world at least as plausible. We can explicitly define this notion of belief in the EDL language:

$$B\varphi := \langle\sim\rangle[\leq]\varphi \tag{8.2}$$

The right-hand-side of Eq. (8.2) says that some worlds among the epistemically indistinguishable worlds have $[\leq]\varphi$, which is precisely the semantics of $B\varphi$. Assuming the limit assumption would guarantee that φ is true in all maximal worlds in the model, which is the more common definition of belief in the AGM tradition.

We can also express that a belief in φ is stronger or more entrenched than a belief in ψ, which we denote by $B(\psi \prec \varphi)$.

$$B(\psi \prec \varphi) := \langle\sim\rangle(\neg\psi \wedge [\leq]\varphi) \tag{8.3}$$

If we set $\psi = \bot$ in definition (8.3), we recover definition (8.2). Another doxastic notion that we can define is conditional belief of φ given ψ (cf., for instance, [4, 5, 12, 14]):

$$B(\varphi \mid \psi) := \langle\sim\rangle\psi \rightarrow \langle\sim\rangle(\psi \wedge [\leq](\psi \rightarrow \varphi)) \tag{8.4}$$

If we set $\psi = \top$ in definition (8.4), we again recover exactly definition (8.2). Notice that we didn't need to use the strict modality $[<]\varphi$ to define belief so far. But if we were to work over partial orders, we could use it to define belief, comparative entrenchment of belief and conditional belief in the following way:

$$B\varphi := [\sim](\neg\varphi \rightarrow \langle<\rangle(\varphi \wedge [<]\varphi)) \tag{8.5}$$

$$B(\psi \prec \varphi) := [\sim](\neg\varphi \rightarrow \langle<\rangle((\varphi \wedge \neg\psi) \wedge [<]\varphi)) \tag{8.6}$$

$$B(\varphi \mid \psi) := [\sim]((\psi \wedge \neg\varphi) \rightarrow \langle<\rangle((\psi \wedge \varphi) \wedge [<](\psi \rightarrow \varphi))) \tag{8.7}$$

Again, if we set $\psi = \bot$ in definition (8.6) or $\psi = \top$ in definition (8.7), we recover definition (8.5), as expected.

Finally, if we set $\psi = \neg\varphi$ in definition (8.4) or (8.7), we get the notion of irrevocable belief (Segerberg [46]), i.e., belief that is sustained even in the face of contradicting evidence. Since $\langle < \rangle(\varphi \wedge \neg\varphi)$ cannot be true anywhere, $B(\varphi \mid \neg\varphi)$ reduces to $[\sim]\varphi$. But this means that irrevocable belief reduces to knowledge-as-indistinguishability. In contrast to Baltag and Smets [4, 5, pp. 14–15, 28], we have argued against such an identification on Sect. 8.2, because even the strongest beliefs may be wrong. But with the means used in this chapter we cannot express the difference.[17]

We can now address the idea that agents have full introspection of their doxastic states. This is in fact built firmly into our notions of belief. Notice that the definitions (8.2)–(8.7) of belief have either $\langle \sim \rangle$ or $[\sim]$ as their main operator.[18] Given our semantics for the indistinguishability relations, it is clear that we get the characteristic S5 axioms of positive and negative introspection with respect to $[\sim]$. But given definition (8.1), this means that if the agent has certain beliefs at a world w that can be expressed by $B\varphi$, $B(\psi \prec \varphi)$ or $B(\varphi|\psi)$, then, by the very definition of these expressions, $KB\varphi$, $KB(\psi \prec \varphi)$ or $KB(\varphi|\psi)$, respectively, are also true at w.[19]

---

[17] Here are a few hints how this situation could get remedied. The picture is basically that within each indistinguishability cell ($\sim$-cell), there is a single system of spheres $ that need not exhaust this cell. In order to characterize $\bigcup \$_{\mathscr{B}}$, we suggest to extend epistemic doxastic models by a new relation $\rightsquigarrow$ that helps representing non-universal systems of spheres (see Sect. 8.2—but now everything happens within every single $\sim$-cell). $\rightsquigarrow$ should be a serial, transitive and Euclidean subrelation of the global indistinguishability relation $\sim$ that specifies a unique set of "conceivable" worlds within each set of indistinguishable worlds. Intuitively, $u \rightsquigarrow v$ for worlds u and v means that $u \sim v$ and v is within the relevant $\sim$-cell's system of spheres. Thus, if $u \sim v$ and $u \rightsquigarrow w$, then also $v \rightsquigarrow w$. We would also need to harmonise $\rightsquigarrow$ with $<$ (and $\leq$), by conditions like 'If $u \sim v$ and $w \rightsquigarrow u$ but not $w \rightsquigarrow v$, then $v < u$' and 'If $u \sim v$ and there is no w such that $w \rightsquigarrow u$ or $w \rightsquigarrow v$, then neither $u < v$ nor $v < u$.' We would then use $\langle \rightsquigarrow \rangle$ and $[\rightsquigarrow]$ rather than $\langle \sim \rangle$ and $[\sim]$ in the definitions (8.4) and (8.7) of conditional belief. Correspondingly, the notion of irrevocable belief would reduce to $[\rightsquigarrow]\varphi$ rather than $[\sim]\varphi$. Knowledge that φ (in the indistinguishability sense) would then imply irrevocable belief that φ, but not vice versa, as desired.

[18] Definition (8.4) is an exception. But even on this definition, we have $B(\varphi|\psi) \rightarrow KB(\varphi|\psi)$. It is easy to see this. Assume that $M, w \models B(\varphi|\psi)$ for some w. For $M, w \models KB(\varphi|\psi)$, we need to show that $M, v \models B(\varphi|\psi)$ for all v such that $w \sim v$. By definition, $M, w \models B(\varphi|\psi)$ means that either $M, w \models [\sim]\neg\psi$ or $M, w \models \langle \sim \rangle(\psi \wedge [\leq](\psi \rightarrow \varphi))$. But the truth value of both of these sentences are independent of the world v of evaluation, as long as $w \sim v$. So either $M, v \models [\sim]\neg\psi$ or $M, v \models \langle \sim \rangle(\psi \wedge [\leq](\psi \rightarrow \varphi))$, and thus $M, v \models B(\varphi|\psi)$, as desired.

[19] The modality '$[\leq]$' is referred to as "knowledge" by Lamarre and Shoham [33, p. 418], as "knowledge according to the defeasibility analysis" by Stalnaker [49, Sect. 6], and as "safe belief", "defeasible knowledge" and "Stalnaker knowledge" by Baltag and Smets [4, see in particular pp. 27–32]. In contrast to $K\varphi$ and $B\varphi$, the truth value of $[\leq]\varphi$ is in general not constant within a $\sim$-cell. The early chapter of Lamarre and Shoham is interesting: It disavows negative introspection for knowledge and finds strong belief ("certainty") that φ to be equivalent with $\neg K \neg K\varphi$—points we acclaim from a philosophical perspective. But it also finds knowledge that φ to be equivalent with the conditional belief $B(\varphi|\neg\varphi)$—a result we object to from a philosophical perspective. This unexpected conjunction is connected with the fact that Lamarre and Shoham let not only knowledge, but also conditional belief and conditional certainty vary from world to world, and thus disavow negative introspection for conditional belief and conditional certainty, too.

One advantage of our modular and minimalist approach is that we can express precisely what we mean when we talk of knowledge and (comparative firmness of) belief. We are very economical with the assumptions we impose on our models, and we relegate them to explicit definitions in our language. We refrain as much as we can to use background assumptions.

8.3.3 Axiomatisation of EDL

The axiomatisation of *static* EDL is based on standard propositional logic. Figure 8.1 shows the axioms for the static part of our modal logic in building blocks. We have the standard set of S5 axioms for the knowledge-as-indistinguishability modality [$\sim$]. For the non-strict plausibility relation [$\leq$] which takes $\leq$ as an accessibility relation, we have the S4.3 axioms that correspond to connected relations. For strict plausibility [$<$], we have K4 plus (Mod$<$). This latter axiom is interesting. As far as we know, it has not been used in epistemic or doxastic logics so far, but van Benthem [7, p. 200] has stated and analyzed it early on in the context of temporal logic. Being a Sahlqvist formula (cf. Blackburn, de Rijke and Venema [13]), (Mod$<$) enforces the modularity (or 'almost-connectedness' or 'virtual connectivity'[20]) to the right of the strict plausibility relation $<$: If $u < v$, $u < w$, $u < z$ and $v < w$, then either $v < z$ or $z < w$. Finally, we need interaction principles between [$\sim$], [$\leq$] and [$<$]. These principles are there to counteract the modal undefinability of $<$ in terms of $\leq$, as has been noted in van Benthem, Girard and Roy [10]. They guarantee that the relation $<$ is *adequate* under bulldozing (cf. Segerberg [45]) of the canonical model, so that $w < v$ iff $w \leq v$ and $v \not\leq w$. In a similar fashion, it is well-known that we cannot modally express that $\sim$ is the same as $\leq \cup \leq^{-1}$, but the canonical model can be adapted accordingly. Since the interaction principles are fairly strong, we are not claiming that our axiomatisation is free of redundancies. We prefer to have fully independent axiomatizations for each of our modal operators instead.

We have not assumed the limit assumption up until now, and we are still inclined against endorsing it. However, should one insist to include it, we suggest to add the so-called Löb axiom (Löb$<$), which corresponds to transitivity and converse well-foundedness (and thus irreflexivity), as an optional extra. It is known to exclude infinite chains and so is the natural counterpart to the limit assumption in ordering semantics.[21] While the limit assumption is not important in the static contexts of (conditional and unconditional) belief, it will turn out to be necessary for many important belief change operations on epistemic doxastic models.

---

[20] The most descriptive term 'modularity' was suggested by Ginsberg [26, p. 49]; 'almost-connectedness' is due to van Benthem [7, 8, pp. 194, 232], 'virtual connectivity' to Alchourón and Makinson [2, p. 415]. Notice that there is also a different sense of 'almost-connectedness' in the literature (see Doble et al. [22]).

[21] See Blackburn, de Rijke and Venema [13, pp. 130–132]. It would be nice to have a more compact axiomatisation of K4 plus (Mod$<$) and (Löb$<$). At this point, we can only conjecture that adding

Modal logic for ~, ≤ and <, and their interaction	
(Nec~)	$\vdash \varphi \;\;\Rightarrow\;\; \vdash [\sim]\varphi$
(K~)	$[\sim](\varphi \to \psi) \to ([\sim]\varphi \to [\sim]\psi)$
(T~)	$[\sim]\varphi \to \varphi$
(4~)	$[\sim]\varphi \to [\sim][\sim]\varphi$
(5~)	$\neg[\sim]\varphi \to [\sim]\neg[\sim]\varphi$
(Nec≤)	$\vdash \varphi \;\;\Rightarrow\;\; \vdash [\leq]\varphi$
(K≤)	$[\leq](\varphi \to \psi) \to ([\leq]\varphi \to [\leq]\psi)$
(T≤)	$[\leq]\varphi \to \varphi$
(4≤)	$[\leq]\varphi \to [\leq][\leq]\varphi$
(.3≤)	$[\leq]([\leq]\varphi \to \psi) \vee [\leq]([\leq]\psi \to \varphi)$
(.3≤)◊	$(\langle\leq\rangle\varphi \wedge \langle\leq\rangle\psi) \to (\langle\leq\rangle(\varphi \wedge \langle\leq\rangle\psi) \vee \langle\leq\rangle(\varphi \wedge \psi) \vee \langle\leq\rangle(\psi \wedge \langle\leq\rangle\varphi))$
(Nec<)	$\vdash \varphi \;\;\Rightarrow\;\; \vdash [<]\varphi$
(K<)	$[<](\varphi \to \psi) \to ([<]\varphi \to [<]\psi)$
(4<)	$[<]\varphi \to [<][<]\varphi$
(Mod<)	$([<](\varphi \vee [<]\chi) \wedge [<](\chi \vee [<]\psi) \wedge \neg[<]\chi) \to [<](\varphi \vee [<]\psi)$
(Mod<)◊	$(\langle<\rangle(\varphi \wedge \langle<\rangle\psi) \wedge \langle<\rangle\chi) \to (\langle<\rangle(\varphi \wedge \langle<\rangle\chi) \vee \langle<\rangle(\chi \wedge \langle<\rangle\psi))$
(Löb<)	$[<]([<]\varphi \to \varphi) \to [<]\varphi$
(Löb<)◊	$\langle<\rangle\varphi \to \langle<\rangle(\varphi \wedge \neg\langle<\rangle\varphi)$
(~≤)	$[\sim]\varphi \to [\leq]\varphi$
(≤<)	$[\leq]\varphi \to [<]\varphi$
(<≤ 1)	$[<]\varphi \to [\leq][<]\varphi \qquad$ (<≤ 1◊) $\langle\leq\rangle\langle<\rangle\varphi \to \langle<\rangle\varphi$
(<≤ 2)	$[<]\varphi \to [<][\leq]\varphi \qquad$ (<≤ 2◊) $\langle<\rangle\langle\leq\rangle\varphi \to \langle<\rangle\varphi$
(QuAdeq)	$([<]\psi \wedge [\leq](\psi \vee [\leq]\varphi)) \to ([\leq]\psi \vee \varphi)$
(QuAdeq)◊	$(\varphi \wedge \langle\leq\rangle\psi) \to (\langle<\rangle\psi \vee \langle\leq\rangle(\psi \wedge \langle\leq\rangle\varphi))$

Fig. 8.1 Axiomatisation of static EDL. For the reader's convenience, we have added a few variants using the possibility operator $\Diamond$, since these are sometimes more intuitive

8.4 Epistemic Doxastic PDL Logic

Epistemic doxastic PDL logic, EDPDL for short, is a variant of the now well-established PDL logic (propositional dynamic logic), whose first interpretation over relational structures can be found in Pratt [41], and further elaborated in Fischer and Ladner [24]. The original purpose of PDL was to provide a logic of programs in a modal framework, taking programs as modal operators or binary relations between states (transitions between states of a machine). The interpretation of PDL modalities $\langle\pi\rangle\varphi$ according to [24] is: 'π can terminate with φ holding on termination'. However,

---

(Footnote 21 continued)

Löb is sufficient to get a complete axiomatisation with the limit assumption, but we have to leave open the problem of showing the logic to be (weakly) complete with respect to the relevant class of frames.

PDL
13. $\vdash \varphi \;\Rightarrow\; \vdash [\pi]\varphi$
14. $\vdash [\pi](\varphi \rightarrow \psi) \rightarrow ([\pi]\varphi \rightarrow [\pi]\psi)$
15. $\vdash \langle \pi \rangle \varphi \leftrightarrow \neg[\pi]\neg\varphi$
16. $[\pi_1 ; \pi_2]\varphi \leftrightarrow [\pi_1][\pi_2]\varphi$
17. $[\pi_1 \cup \pi_2]\varphi \leftrightarrow ([\pi_1]\varphi \wedge [\pi_2]\varphi)$

Fig. 8.2 Axiomatization of general PDL part

this interpretation is not what we are after. The part of PDL that is relevant to us is the modal calculus of relation combination, which we exploit to formalise belief change. Formally, we extend the language of EDL by adding the PDL operations of test, choice and composition to the basic ingredients of our language:

$$\pi ::= \;\sim\; | \;\leq\; | \;<\; | \;\varphi? \;|\; \pi \cup \pi' \;|\; \pi \;;\; \pi'$$
$$\varphi ::= p \;|\; \neg\varphi \;|\; (\varphi \wedge \psi) \;|\; [\pi]\varphi$$

As usual, we define dual box operators $[\pi]\varphi := \neg\langle\pi\rangle\neg\varphi$ for each program π. In this notation, the special modalities $[\pi]\varphi$ with $\pi \in \{\sim, \leq, <\}$ are just the basic modalities $[\sim]\varphi$, $[\leq]\varphi$ and $[<]\varphi$ of the previous section. In each epistemic doxastic model $M = \langle W, \sim, \leq, <, V \rangle$, semantic values $[\![\varphi]\!]^M \subseteq W$ and $[\![\pi]\!]^M \subseteq W^2$ are given by:

$$[\![p]\!]^M = V(p)$$
$$[\![\neg\varphi]\!]^M = W \setminus [\![\varphi]\!]^M$$
$$[\![(\varphi \wedge \psi)]\!]^M = [\![\varphi]\!]^M \cap [\![\psi]\!]^M$$
$$[\![\langle\pi\rangle\varphi]\!]^M = \{u \in W \;|\; u[\![\pi]\!]^M v \text{ and } v \in [\![\varphi]\!]^M, \text{ for some } v \in W\}$$
$$[\![\sim]\!]^M = \;\sim$$
$$[\![\leq]\!]^M = \;\leq$$
$$[\![<]\!]^M = \;<$$
$$[\![\varphi?]\!]^M = \{\langle u, u \rangle \;|\; u \in [\![\varphi]\!]^M\}$$
$$[\![\pi_1; \pi_2]\!]^M = \{\langle u, v \rangle \;|\; u[\![\pi_1]\!]^M w \text{ and } w[\![\pi_2]\!]^M v, \text{ for some } w \in W\}$$
$$[\![\pi_1 \cup \pi_2]\!]^M = [\![\pi_1]\!]^M \cup [\![\pi_2]\!]^M$$

As usual, we also write $u[\![\pi]\!]^M v$ for $\langle u, v \rangle \in [\![\pi]\!]^M$ and $M, u \models \varphi$ for $u \in [\![\varphi]\!]^M$. Incorporating PDL in our axiomatisation is simple, especially since we are only appealing to the fragment of PDL without the Kleene star. In line with our modular approach, the PDL operators or test, choice and composition are recursively introduced (Fig. 8.2).

8.4.1 Doxastic **PDL** *Transformations*

To formalise belief change, we use a special case of PDL-transformations as defined in Girard, Seligman and Liu [27]. This latter chapter was directly motivated by Fact 19 from van Benthem and Liu [11]: every relation-changing operation that is definable in PDL without iteration has a complete set of reduction axioms in dynamic epistemic logic. We fully exploit this idea in the remainder of the chapter. For the details of the general case of PDL-transformations, the reader should consult Sect. 1 of [27]. We give here a self-contained specification of the special case of PDL-transformations required for our purposes. Basically, a *doxastic* **PDL***-transformation* Λ is a transformation on models that has two components: (1) a domain restriction provided by some sentence denoted $|\Lambda|$,[22] and (2) PDL-definable transformations of the relations $\sim$, $\leq$ and $<$. Even though we feel unconfortable about world-elimination, as we already explained, we will use domain restrictions to differentiate between expansion and revision. We are very much aware of philosophical difficulties that may ensue, and will treat them with care.

Given a model $M = \langle W, \sim, \leq, <, V \rangle$ and a PDL-transformation Λ, the result of transforming M with Λ is the model $\Lambda M = \langle \Lambda W, \Lambda(\sim), \Lambda(\leq), \Lambda(<), \Lambda V \rangle$. ΛM is computed by setting $\Lambda W = [\![|\Lambda|]\!]^M$ and taking all relations $\Lambda(\sim)$, $\Lambda(\leq)$ and $\Lambda(<)$ that are defined explicitly for each particular program to be restricted to $(\Lambda W)^2$. Likewise the valuation ΛV is simply the valuation V restricted to the domain ΛW.

We also define computable translations φ^Λ of sentences corresponding to PDL-transformations on models:

$$
\begin{aligned}
& & \sim^\Lambda & = \Lambda(\sim)\,;\,|\Lambda|? \\
p^\Lambda & = p & \leq^\Lambda & = \Lambda(\leq)\,;\,|\Lambda|? \\
(\neg\varphi)^\Lambda & = \neg\varphi^\Lambda & <^\Lambda & = \Lambda(<)\,;\,|\Lambda|? \\
(\varphi \wedge \psi)^\Lambda & = (\varphi^\Lambda \wedge \psi^\Lambda) & (\varphi?)^\Lambda & = (\varphi^\Lambda)? \\
(\langle\pi\rangle\varphi)^\Lambda & = \langle\pi^\Lambda\rangle\varphi^\Lambda & (\pi_1\,;\,\pi_2)^\Lambda & = \pi_1^\Lambda\,;\,\pi_2^\Lambda \\
& & (\pi_1 \cup \pi_2)^\Lambda & = \pi_1^\Lambda \cup \pi_2^\Lambda
\end{aligned}
$$

As demonstrated in [27], the translation φ^Λ guarantees that the following lemma holds:

Lemma 8.1 For each state u of ΛM and v of M, and for each sentence φ,

$$M, u \models \varphi^\Lambda \quad \textit{iff} \quad \Lambda M, u \models \varphi, \textit{ and}$$

$$u[\![\pi^\Lambda]\!]^M v \quad \textit{iff} \quad v \in \Lambda W \textit{ and } u[\![\pi]\!]^{\Lambda M} v.$$

We represent PDL-transformations in the following way:

---

[22] The accustomed reader will recognise this as something very much like a public announcement.

Name of a PDL-transformation	
Λ	$\|\Lambda\|$
	$\sim := \Lambda(\sim)$
	$\leq := \Lambda(\leq)$
	$< := \Lambda(<)$

A well-known special case of a PDL-transformation is the public announcement of a sentence φ, first studied by Plaza in 1989 and now republished in [40]:

Public Announcement	
$\varphi!$	φ
	$\sim := \sim$
	$\leq := \leq$
	$< := <$

In this representation, '$\sim := \sim$' means that the relation $\sim$ is assigned to its restriction to the new domain in the new model, i.e., $\varphi!(\sim) = \sim \cap (\varphi!W)^2 = \sim \cap (\llbracket\varphi\rrbracket^M)^2$. Thus, all a public announcement does is to restrict the domain by eliminating $\neg\varphi$-worlds. All relations are kept as they were, but restricted to the new domain. To ease notation, we omit writing identity assignment such as '$\sim := \sim$' in transformations. We also omit writing the domain restriction when $|\Lambda| = \top$. Thus the public announcement transformation can be succinctly written as:

Public Announcement	
$\varphi!$	φ

With this established, we expand our language of doxastic epistemic PDL logic with modalities $[\Lambda]\varphi$ for each PDL-transformation Λ with the following semantics:

$$M, w \models [\Lambda]\varphi \quad \text{iff} \quad \Lambda M, w \models \varphi.$$

We stipulate that $M, w \models [\Lambda]\varphi$ is vacuously true in case $M, w \not\models |\Lambda|$.

A technical difficulty with PDL-transformations is that they may not always transform models into new models of the right kind. A doxastic epistemic model M may be transformed by a PDL-transformation Λ in such a way that ΛM is not a doxastic epistemic model. To avoid this issue, we do not accept any possible doxastic PDL-transformation, but instead provide a class of doxastic PDL-transformations that are *proper*, in the sense that they always return doxastic epistemic model.

8.5 AGM Operations

In this section we provide a class of PDL-transformations which are candidates of unrestricted transformations in the style of Alchourrón, Gärdenfors and Makinson (compare [1, 2, 17, 25, 29, 32]); we refer to them as "AGM operations" or "doxastic transformations". We base our selection on those given in Rott [42].[23] We first study the operations of *belief expansion* and *belief revision*, which can be nicely treated in EDPDL logic. We then move to a study of the operation of *belief contraction*. Our selection is partly for the sake of exposition, but we include standard DEL doxastic change operations to be found in recent work such as van Benthem [9] or Baltag and Smets [4].

8.5.1 Expansion and Revision

The operations of expansion and revision are about adding beliefs to belief states. We start with three types of doxastic change, that we categorise as conservative, radical and moderate. A conservative doxastic transformation by φ is one that only shifts around maximal φ-worlds (or, in the case of contraction, $\neg\varphi$-worlds), and leaves the ordering between the other worlds intact. A moderate doxastic transformation by φ is one that shifts around all φ-worlds (or, in the case of contraction, $\neg\varphi$-worlds) in a uniform way. A radical doxastic transformation by φ is one that only preserves φ-worlds (or, in the case of contraction, almost only $\neg\varphi$-worlds).

We use the following abbreviations:

$$\sim^{\varphi} \quad :: = (\varphi?; \sim; \varphi?)$$
$$\leq^{\varphi} \quad :: = (\varphi?; \leq; \varphi?)$$
$$<^{\varphi} \quad :: = (\varphi?; <; \varphi?)$$
$$\max \varphi :: = (\varphi \wedge [<]\neg\varphi)$$

Notice that the sentence $\max \varphi$ is only true in the most plausible φ-worlds, i.e., φ-worlds such that all worlds more plausible, if there are any, satisfy $\neg\varphi$.[24]

8.5.2 Expansion

We start with expansion. Generally speaking, expanding one's beliefs with φ is to start believing φ without caring about consistency. If φ is not consistent with what the agent believes, then her beliefs trivialise and she now believes $\perp$. But it

---

[23] Any PDL transformation which outputs a Grove relation would be formally legitimate. To categorise this general class of transformations is still an open problem in GDDL, and we will not address it here, as our main concern is with AGM motivated transformations.

[24] Slightly abusing the term "maximality", one could also experiment with putting $\max(\varphi) ::= (\varphi \wedge ([<]\neg\varphi \vee [\leq]\varphi))$, but we will not pursue this idea in the present chapter.

is important to note that her belief *state* does *not* trivialise. In the semantics using Grovean systems of spheres, we can represent this nicely by adding an empty sphere to $\$_{\mathscr{B}}$. Unfortunately, no such picture is possible with the pure ordering approach. We have to stipulate here that if φ is inconsistent with the beliefs supported by $\leq$, which we write as $\neg[\sim]\langle\leq\rangle\varphi$, then the belief set that results from the expansion is the trivial one, $\mathsf{cl}(\bot)$. We achieve this by introducing the domain restriction $|[\sim]\langle\leq\rangle\varphi|$ in expansion transformations. Notice that this restriction doesn't really restrict the domain, because of our underlying assumption that the plausibility order is uniform inside epistemic classes. So either all worlds in a class satisfy $[\sim]\langle\leq\rangle\varphi$, or none do. In the latter case, the agent ends-up believing $\bot$.

We first look at conservative expansion. Conservative expansion by φ reorders maximal φ-worlds and leaves the rest of the model intact. That is, the order stays intact among the worlds that are not maximal φ-worlds (those that either do not satisfy φ or for which there are strictly more plausible worlds that satisfy φ), and it makes every maximal φ-world equally plausible to each other as well as strictly more plausible than any other world:

Conservative expansion	
$\mathsf{CE}\varphi$	$[\sim]\langle\leq\rangle\varphi$
	$\leq := \leq^{\neg\mathsf{max}(\varphi)} \cup (\sim ; \ \mathsf{max}\,\varphi?)$
	$< := <^{\neg\mathsf{max}(\varphi)} \cup (\neg\,\mathsf{max}\,\varphi? ; \sim ; \ \mathsf{max}\,\varphi?)$

To say that beliefs of agents trivialise under expansion with information φ that is inconsistent with their beliefs amounts to saying that $M, w \models [\mathsf{CE}\varphi]B\bot$ in case $M, w \not\models [\sim]\langle\leq\rangle\varphi$. Notice that conservative expansion is not successful if we do not assume the limit assumption. If there are no maximal φ-worlds, then nothing happens to the doxastic structure.

Second, consider the operation of moderate expansion, which differs from the conservative operation by reordering all φ-worlds instead of only the maximal ones. Hence, moderate expansion preserves the order among the φ-worlds and among the $\neg\varphi$-world, and makes every φ-world strictly more plausible than every $\neg\varphi$-world. In this case, the old maximal φ-worlds become the most plausible ones overall. Formally, moderate expansion $\mathsf{ME}\varphi$ is defined by:

Moderate expansion	
$\mathsf{ME}\varphi$	$[\sim]\langle\leq\rangle\varphi$
	$\leq := \leq^{\varphi} \cup \ \leq^{\neg\varphi} \cup (\neg\varphi? ; \sim ; \ \varphi?)$
	$< := <^{\varphi} \cup \ <^{\neg\varphi} \cup (\neg\varphi? ; \sim ; \ \varphi?)$

Finally, radical expansion is an action which reduces to a domain restriction. If φ is consistent with the agent's beliefs, then we only keep the φ-worlds. Thus, if there are maximal worlds that are φ-worlds, radical expansion deletes all $\neg\varphi$-worlds; otherwise beliefs trivialise. This is all that is required, so relations are simply

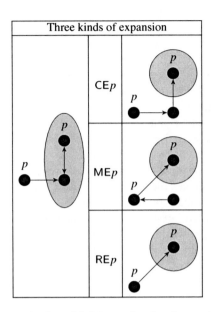

Fig. 8.3 Three kinds of expansion by p. Models are closed under transitivity and reflexivity

restricted as they were to the new domain. Formally, radical expansion $\mathsf{RE}\varphi$ is defined by:

Radical expansion	
$\mathsf{RE}\varphi$	$\varphi \wedge [\sim]\langle\leq\rangle\varphi$

Figure 8.3 displays the three kinds of expansion acting on the same model. Every expansion returns the same set of maximal states. The difference is in the ordering of the remaining worlds. We have chosen a model in which some p-worlds are among the maximal worlds. Radical expansion restricts the domain to p-worlds, exemplifying the way in which it is radical compared to the other ones.

Now that we have precise definitions of expansion as doxastic transformation, we can specify distinguished modalities for each of them: $[\mathsf{CE}\varphi]$, $[\mathsf{ME}\varphi]$, and $[\mathsf{RE}\varphi]$. So for instance, the sentence $[\mathsf{CE}\varphi]B\psi$ says that ψ is believed after moderately expanding with φ. As we have no restriction on iterations of doxastic actions, we can also express and analyse complex sentences such as the validities $[\mathsf{CE}p][\mathsf{RE}p]B\psi \leftrightarrow [\mathsf{RE}p]B\psi$ and $[\mathsf{RE}p][\mathsf{RE}q]\psi \leftrightarrow [\mathsf{RE}(p \wedge q)]\psi$. In a multi-agent language, we could also analyse higher-order beliefs about doxastic change. For instance, with s = "Robert is a spy" and l = "Robert is a liar", the sentence $B_r[\mathsf{RE}_b s]B_b l$ expresses that "Robert believes that Bernadette believes that Robert is a liar after radically expanding her belief by the fact that Robert is a spy".

8.5.3 Revision

Revision is exactly like expansion, except that agents do not get trivial beliefs when revising with information that was not consistent with their initial beliefs. Thus the only difference between revision and expansion is in the way the beliefs are retrieved from a plausibility ordering. For revisions, the standard rules apply, and thus an agent simply cannot have an inconsistent belief set! We can accommodate this nicely with conservative and moderate revision, but radical revision is problematic. We can get two interpretations, but neither works properly as a revision. We start with conservative and moderate revision:

Conservative revision	
$CR\varphi$	$\leq := \leq^{\neg max(\varphi)} \cup (\sim; \; max\,\varphi?)$
	$< := <^{\neg max(\varphi)} \cup (\neg\, max\,\varphi?; \sim; \; max\,\varphi?)$
Moderate revision	
$MR\varphi$	$\leq := \leq^{\varphi} \cup \; \leq^{\neg\varphi} \cup (\neg\varphi?; \sim; \; \varphi?)$
	$< := <^{\varphi} \cup \; <^{\neg\varphi} \cup (\neg\varphi?; \sim; \; \varphi?)$

Moderate revision precisely corresponds to the *lexicographic upgrade*, and conservative revision to the *elite change* of van Benthem [9, p. 141]. We will see in Sect. 8.5.4 that these operations can be regarded as the natural limiting cases of a common idea (viz., that of bounded revision). However, there is an important difference. While moderate revision has no need whatsoever for the limit assumption, conservative revision needs it badly. Like conservative expansion, conservative revision might not be successful without the limit assumption. If it is not met, then there may not be any maximal φ-worlds and conservative revision may not effect anything.

We can give two interpretations of radical revision as described in Rott [42]. An important aspect of radical revision by φ is that $\neg\varphi$-worlds can never be recovered. One way of incorporating this is by using a domain restriction $|\varphi|$ that removes $\neg\varphi$-worlds from the model altogether. Another way is to have no domain restriction, like in conservative and moderate revision, but cut every link between φ and $\neg\varphi$-worlds.

The first approach, in which we guarantee irrevocable revision with a domain restriction, is the following doxastic transformation:

Radical revision, version 1	
$RR\varphi$	φ

This version of radical revision by φ is the same as a public announcement of φ as we've analysed above. In the terminology of van Benthem [9], radical revision is a change under *hard information*, whereas conservative and moderate revisions are two alternatives of change under *soft information*.

The way in which this version of radical revision captures the irrevocability of φ is by deleting all the $\neg\varphi$-worlds. This is indeed a radical way of guaranteeing that $\neg\varphi$-worlds cannot be recovered. This approach also captures success of revision for Boolean sentences, but not without a price. The price to pay is that beliefs of agents trivialise in $\neg\varphi$-worlds when radically revising by φ. Indeed, assume that $M, w \models \neg\varphi$ in some model M. Because of the domain restriction $|\varphi|$, we get that $M, w \models [\mathsf{RR}\varphi]B\bot$. Success comes at the price of triviality, which is not what revision operators have been invented for.

One way to make sure that belief sets do not trivialise under revision is to avoid restricting the domain, as in conservative and moderate revision, with the following transformation:

Radical revision, version 2	
$\mathsf{RR}\varphi$	$\sim := \sim^\varphi \cup \sim^{\neg\varphi}$
	$\leq := \leq^\varphi \cup \leq^{\neg\varphi}$
	$< := <^\varphi \cup <^{\neg\varphi}$

The way in which this version of radical revision is irrevocable comes from the definition of our epistemic relation. We did not introduce a free transition as a basic program. We have been using the relation $\sim$ instead. For instance, in moderate revision, we can make sure that all φ-worlds become more plausible than $\neg\varphi$-worlds by re-defining $<$ as $<^\varphi \cup <^{\neg\varphi} \cup (\neg\varphi?\,;\ \sim;\ \varphi?)$, where $\sim$ is used to create new plausibility links in $(\neg\varphi?\,;\ \sim;\ \varphi?)$. Now, in our second interpretation of radical revision, once we cut links between φ and $\neg\varphi$-worlds, these links are no longer recoverable! This is nice, but also comes with its own cost, again when radical revision is evaluated in $\neg\varphi$-worlds.[25]

Let us highlight the problem about interpreting (possibly untruthful) public announcement and radical revision with the help of an example. Take a very simple model with two epistemically indistinguishable and equiplausible worlds w and v, with p and q true at w, but false at v (Fig. 8.4). Consider a radical revision by p. If radical revision goes by domain restriction, v simply vanishes, and we get that both $[\mathsf{RR}p]Bp$ and $K[\mathsf{RR}p]Bp$ are true at w. However, if radical revision goes by a "link-cutting" action then again $[\mathsf{RR}p]Bp$ is true at w, but false at v—surprisingly, even $[\mathsf{RR}p]B\neg p$ is true at v. Hence, since $v \sim w$, $K[\mathsf{RR}p]Bp$ is false at w. Suppose that in the initial situation the agent is *actually* located at v, but cannot distinguish v from w. So all the agent knows or believes at the beginning is $p \leftrightarrow q$. Now the agent receives a public (but not truthful!) announcement that p and as a result performs some kind of radical revision on p. What would happen? Metaphysically, world v would not

---

[25] For complex sentences φ that involve doxastic operators, it is possible that φ becomes true again at a $\neg\varphi$-world after other doxastic transformations. The old φ-worlds are irrevocable. It is the worlds that are irrevocable, not sentences. Only Boolean sentences (those without modalities) are truly irrevocable. This is related to the consideration of the AGM success postulate: Only Boolean sentences are guaranteed to be successful. If one revises by a sentence that says "p is true but you don't know it", then one does not get success.

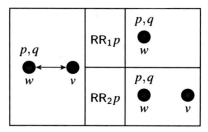

Fig. 8.4 Small example illustrating a problem with radical revision. Models are closed under reflexivity

cease to exist—this is what makes domain restriction strange.[26] The agent would (wrongly) believe that p, but of course she would not know that p. But she would not know or believe that $\neg p$ either – this is what makes link-cutting problematic. Intuitively, while the beliefs have changed—and this is why $[RRp]Bp$ should come out true at v—, the knowledge has not increased. The agent located at v is still not able to epistemically distinguish her world from w. None of our modellings have this option.

The versions of revision we have been investigating are illustrated as operating on the same initial model in Fig. 8.5.

8.5.4 Two-Dimensional Belief Change Operators

We continue our brief overview of revision operations in the framework of **EDL** with two-dimensional change operations in the sense of Rott [44]. These models are meant to increase the expressive power of purely qualitative, relational, thus non-numerical models for belief change. The extent to which an input sentence φ is accepted, is specified by a reference sentence ψ. The first two-dimensional belief change operation we consider is *bounded revision*. The idea of bounded revision is to accept φ *as long as* ψ holds along with φ—and just a little longer. Bounded revision satisfies (generalizations of) the semantically motivated postulates of Darwiche and Pearl [17], as well as a "Same beliefs condition" according to which the posterior *beliefs* of the agent should not depend on the reference sentence (although the posterior *belief state* does). For further motivation we refer to [44]. Bounded revision $\mathsf{BdR}_\psi \varphi$ is defined by:

It is not difficult to verify that bounded revision reduces to the unary operation of conservative revision if the reference sentence ψ is fixed to $\bot$, and that it reduces to the unary operation of moderate revision if the reference sentence ψ is fixed to $\top$.

---

[26] How can we evaluate a sentence at a world v which has vanished in the course of the evaluation? Above, we have stipulated that $M, w \models [\Lambda]\varphi$ to be vacuously true in case $M, w \not\models |\Lambda|$, in order to avoid facing the main clause $\Lambda M, w \models \varphi$ when w fails to be in ΛW of ΛM. But evidently, this is not a solution to the problem of untruthful public announcements or radical revisions.

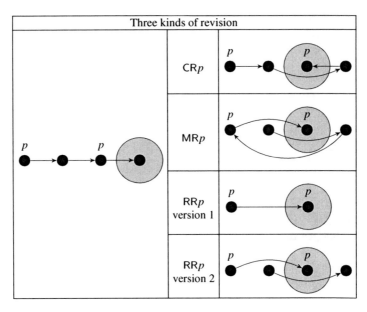

Fig. 8.5 Three kinds of revision by p. Models are closed under transitivity and reflexivity

Bounded revision	
$\mathsf{BdR}_\psi\varphi$	$\leq := \leq^{\varphi \wedge [<](\varphi \to \psi)} \cup \leq^{\neg(\varphi \wedge [<](\varphi \to \psi))} \cup$
	$(\neg(\varphi \wedge [<](\varphi \to \psi))?; \sim; (\varphi \wedge [<](\varphi \to \psi))?)$
	$< := <^{\varphi \wedge [<](\varphi \to \psi)} \cup <^{\neg(\varphi \wedge [<](\varphi \to \psi))} \cup$
	$(\neg(\varphi \wedge [<](\varphi \to \psi))?; \sim; (\varphi \wedge [<](\varphi \to \psi))?)$

In general, bounded revision requires the limit assumption, since for instance, if φ and ψ are inconsistent with each other, minimal φ-worlds are needed to ensure the success of the revision operation. By a deliberate choice of the reference sentence ψ, however, one may in many cases make sure that there is a broad enough range of worlds that satisfy $\varphi \wedge [<](\varphi \to \psi)$, and then the operation performs well even if the model does not satisfy the limit assumption. Figure 8.6 gives an illustration of bounded revision in a finite model.

Another interesting two-dimensional operation is *revision by comparison* (Fermé and Rott [23]). It is motivated by the same concerns as bounded revision. But while the idea of bounded revision is to accept φ *as long as* ψ *holds* along with it (and a little longer), revision by comparison accepts φ with a *strength* that at least equals that of the acceptance of ψ. In contrast to bounded revision, revision by comparison does not satisfy the Darwiche-Pearl postulates. In its intended cases of application, it is a revision operation, but it can also have the effects of a contraction operation (see Sect. 8.5.5): If the reference sentence is too weak (more precisely, if the input sentence is at least as surprising as the negation of the reference sentence), then the revision fails, and instead a severe contraction with respect to the reference sentence

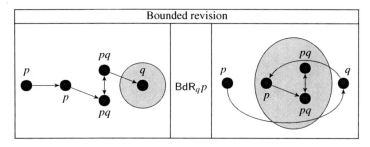

Fig. 8.6 Bounded revision by p with respect to reference sentence q. Models are closed under transitivity and reflexivity

is performed, provided that there are maximal nonmodels of the reference sentence.[27] In at least one way of presenting it (namely by manipulations of prioritized belief bases, cf. Rott [42]), revision by comparison is an extremely natural belief change operation.

The definition of revision by comparison given in Fermé and Rott [23, p. 14] can be represented in our framework as follows:

Revision by comparison	
$\mathsf{RbC}_\psi\varphi$	$\leq := \leq^\varphi \cup (\neg\lceil\leq\rceil\psi?; \leq) \cup (\neg\varphi?; \sim; \lceil<\rceil\psi?)$
	$< := <^\varphi \cup (\neg\lceil<\rceil\psi?; <) \cup (\neg\varphi?; \sim; (\varphi \wedge \lceil\leq\rceil\psi)?)$

Figure 8.7 gives an illustration of revision by comparison in a finite model. We are representing two cases, the successful one in which the input sentence gets accepted, and the unsuccessful one in which the reference sentence gets withdrawn.

If we set the reference sentence to $\top$, then revision by comparison reduces to a unary revision operation that is more radical than moderate revision but somewhat less radical than the revision operations we have called radical:

Radical revision, version 3	
$\mathsf{RR}_3\varphi$	$\leq := \leq^\varphi \cup (\neg\varphi?; \sim)$
	$< := <^\varphi \cup (\neg\varphi?; \sim; \varphi?)$

We deal with another interesting limiting case of revision by comparison obtained by setting the input sentence to $\bot$ in the next section.[28]

---

[27] The limit assumption guarantees this. If, however, the limit assumption is not satisfied, revision by comparison as defined below may fail to make the input sentence at least as firmly accepted as the reference sentence.

[28] Also compare Rott [43].

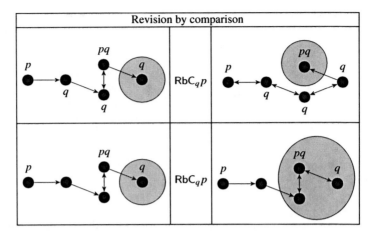

Fig. 8.7 Revision by comparison by p with respect to stronger and weaker reference sentences q: the successful case of a revision and the unsuccessful case reducing to a severe withdrawal of q. Models are closed under transitivity and reflexivity

8.5.5 Contraction

The operation of contraction is about withdrawing beliefs from belief states. The main idea is that a contraction by φ is effected by promoting the maximal $\neg\varphi$-worlds, and possibly some more worlds, to the ranks of the maximal worlds (i.e., the maximal $\top$-worlds).

To be successful, each of the following operations requires the use of max $\neg\varphi$, and most of them require the use of max $\top$ as well. Thus the difficulty with belief contraction is that we need to identify maximal states: the states where $[<]\varphi$ or, respectively, $[<]\bot$ hold. But without something like the limit assumption, there is no guarantee that maximal states exist in models. We can still define the operations with PDL-transformations, as we did for all other operations, but unless we have some means of ensuring the existence of maximal worlds, contraction even with atomic information might not be successful. Now, the way we have proposed to get the limit assumption is by introducing the Löb axiom. We know that our logic is sound over the appropriate class of frames in which there are maximal worlds, but we do not know how to prove completeness. Setting this technical question aside for future research, we proceed in this section assuming that models always have maximal worlds, which we identify as those worlds in which max $\neg\varphi$ or $[<]\bot$ is true.

Our first contraction operation is very simple. It has been studied by various authors and is perhaps best known under the names *severe withdrawal* (Pagnucco and Rott [39]) and *mild contraction* (Levi [36]). The incisions into sets of beliefs induced by severe withdrawal are substantially greater than those induced by (iterable generalisations of) AGM contraction functions.

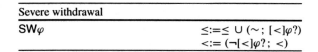

Severe withdrawal	
$SW\varphi$	$\leq := \leq \cup (\sim; [<]\varphi?)$ $< := (\neg[<]\varphi?; <)$

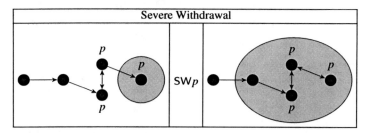

Fig. 8.8 Severe withdrawal of p. Models are closed under transitivity and reflexivity

Notice that severe withdrawal has no need of identifying maximal $\top$-worlds. But if the limit assumption is not met, then there may be no maximal $\neg\varphi$-worlds and severe withdrawal may weaken the beliefs of the agent without getting rid of φ. Figure 8.8 gives an illustration of a successful severe withdrawal:

It is easy to check that a revision by comparison $\mathsf{RbC}_\psi\varphi$ reduces to a severe withdrawal of the reference sentence, $\mathsf{SW}\psi$, if we substitute $\bot$ for φ.[29]

The following three kinds of contraction are modelled in analogy to conservative, moderate and radical revision. In line with the basic AGM theory, the way the corresponding contraction operations proceed is by putting the maximal $\neg\varphi$-worlds and the maximal $\top$-worlds on a par, in a maximal position. We now move to the investigation of conservative, moderate and radical contraction (Fig. 8.9).

Conservative contraction, like conservative revision above, keeps most of the structure intact and reorders maximal worlds. First, the order is preserved among the non-maximal $\neg\varphi$-worlds. Second, the maximal $\neg\varphi$-worlds are upgraded on top of non-maximal φ-worlds and made as plausible as the maximal $\top$-worlds. Formally, conservative contraction is the following doxastic transformation:

Conservative contraction	
$CC\varphi$	$\leq := \leq^{\neg\max(\neg\varphi)} \cup(\sim; \max \neg\varphi?) \cup (\sim; \max \top?)$ $< := <^{\neg\max(\neg\varphi)} \cup ((\neg\max\neg\varphi \wedge \neg\max\top)?; \sim; \max\neg\varphi?)$

Moderate contraction is defined in analogy to moderate revision, but it is hard to come up with a motivation for it. Why should the idea of being open-minded about φ result in a belief state that gives a lot of credit to $\neg\varphi$? We present it for reasons of uniformity [42].

---

[29] Notice that the transformation $(\neg[\leq]\psi?; \leq) \cup (\sim; [<]\psi?)$ is identical to the transformation $\leq \cup(\sim; [<]\psi?)$, because $M, w \models [\leq]\psi$ and $w \leq v$ taken together imply $M, v \models [<]\psi$.

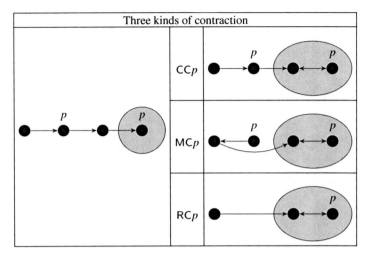

Fig. 8.9 Four kinds of contractions with respect to p. Models are closed under transitivity and reflexivity

Moderate contraction
$MC\varphi$ $\leq:=\leq^{\neg\varphi} \cup \leq^{\varphi} \cup((\varphi \wedge \neg \max \top)?; \sim; \neg\varphi?)$ $\cup(\sim; \max \top?) \cup (\sim; \max \neg\varphi?)$ $<:=<^{\varphi} \cup <^{\neg\varphi} \cup((\varphi \wedge \neg \max \top)?; \sim; \neg\varphi?)$ $\cup((\neg \max \neg\varphi \wedge \neg \max \top)?; \sim; \max \top?)$

We finally turn to radical contraction to which similar, and even stronger, cautionary remarks concerning its reasonableness apply.

Radical contraction
$RC\varphi$ $|\neg\varphi \vee \max \top|$ $\leq:=\leq^{\neg\varphi} \cup(\sim; \max \neg\varphi?;) \cup (\sim; \max \top?)$ $<:=<^{\neg\varphi} \cup((\neg\varphi \wedge \neg \max \neg\varphi)?; \sim; \max \top?)$

8.6 Conclusion

We have explored **AGM** belief change policies in a modal dynamic logic that explicitly delineates knowledge, belief, plausibility and the dynamics of these notions. Taking a Kripke semantics counterpart to Grove semantics for **AGM** as a starting point, we used a basic modal language containing one epistemic modality $[\sim]\varphi$ and two plausibility modalities $[\leq]\varphi$ and $[<]\varphi$, and defined several notions of belief. We critically discussed the philosophical presuppositions underlying various modelling assumptions commonly made in the literature, such as negative introspection for knowledge and the limit assumption. Then, we introduced PDL-transformations to define various policies of iterated belief expansion, revision, contraction and two-dimensional belief change operations. EDPDL thus formalises our minimalist and modular attitude.

Acknowledgments We would like to thank Alexandru Baltag, Johan van Benthem and Michael Hillas for very valuable comments on a previous version of this chapter. We dedicate this chapter to Johan van Benthem, whom we admire as a researcher and who has been, in many and various ways, a friend and mentor to both of us. Thank you so much, Johan!

References

1. Alchourrón CE, Gärdenfors P, Makinson D (1985) On the logic of theory change: partial meet contraction and revision functions. J Symbolic Logic 50:510–530
2. Alchourrón CE, Makinson D (1985) On the logic of theory change: safe contraction. Stud Logica 44:405–422
3. Baltag A, Smets S (2006) Conditional doxastic models: a qualitative approach to dynamic belief revision. Electron Notes Theoret Comput Sci 165:5–21 (2006). (Proceedings WoLLIC 2006)
4. Baltag A, Smets S (2008) A qualitative theory of dynamic interactive belief revision. In: Bonanno G, van der Hoek W, Wooldridge M (eds) Logic and the foundations of game and fdecision theory (LOFT 7). Texts in logic and games, vol 3. Amsterdam University Press, Amsterdam, pp 11–58
5. Baltag A, Smets S (2008) The logic of conditional doxastic actions. In: Apt KR, van Rooij R (eds) New perspectives on games and interaction. Texts in logic and games, vol 4. Amsterdam University Press, Amsterdam, pp 9–31
6. Baltag A, van Ditmarsch H, Moss LS (2008) Epistemic logic and information update. In: Adriaans P, van Benthem J (eds) Philosophy of information. Handbook of the philosophy of science. Elsevier, Amsterdam, pp 361–455
7. van Benthem J (1982) Later than late: on the logical origin of the temporal order. Pac Philos Q 63:193–203
8. van Benthem J (1983) The logic of time. A model-theoretic investigation into the varieties of temporal antology and temporal discourse. Reidel, Dordrecht
9. van Benthem J (2007) Dynamic logic for belief revision. J Appl Non-class Logic 17:129–155
10. van Benthem J, Girard P, Roy O (2009) Everything else being equal: a modal logic for ceteris paribus preferences. J Philos Logic 38:83–125
11. van Benthem J, Liu F (2007) Dynamic logic of preference upgrade. J Appl Non-Class Logics 17:157–182
12. van Benthem J, Roy O, van Otterloo S (2006) Preference logic, conditionals, and solution concepts in games. In: Lagerlund, L, Sliwinski S (eds) Modality matters: twenty-five essays in honour of Krister Segerberg, vol 53. Uppsala Philosophical Studie. Uppsala Universitet, Uppsala
13. Blackburn P, de Rijke M, Venema Y (2001) Modal logic. Cambridge University Press, New York
14. Board O (2004) Dynamic interactive epistemology. Games Econ Behav 49:49–80
15. Boutilier C (1994) Unifying default reasoning and belief revision in a modal framework. Artif Intell 68:33–85

16. Boutilier C (1996) Iterated revision and minimal change of conditional beliefs. J Philos Logic 25:263–305
17. Darwiche A, Pearl J (1997) On the logic of iterated belief revision. Artif Intell 89:1–29
18. David K (1973). Counterfactuals. Blackwell, Oxford
19. Demey L (2011) Some remarks on the model theory of epistemic plausibility models. J Appl Non-class Logics 21:375–395
20. van Ditmarsch H (2005) Prolegomena to dynamic logic for belief revision. Synthese 147:229–275
21. van Ditmarsch H, van der Hoek W, Kooi B (2007) Dynamic epistemic logic, vol 1. Springer, New York
22. Doble CW, Doignon J-P, Falmagne J-C, Fishburn PC (2001) Almost connected orders. Order 18:295–311
23. Fermé E, Rott H (2004) Revision by comparison. Artif Intell 157:5–47
24. Fischer MJ, Ladner RE (1979) Propositional dynamic logic of regular programs. J Comput Syst Sci 18:194–211
25. Gärdenfors P, Makinson D (1988) Revisions of knowledge systems using epistemic entrenchment. In: Vardi M (ed) Proceedings of the second conference on theoretical aspects of reasoning about knowledge (TARK'88), pp 83–95. Morgan Kaufmann, Los Altos, CA
26. Ginsberg ML (1986) Counterfactuals. Artif Intell 30:35–79
27. Girard P, Seligman J, Liu F (2012) General dynamic dynamic logic. In: Bolander T, Braüner T, Ghilardi S, Moss LS (eds) Advances in modal logics, vol 9. College, London, pp 239–260
28. Greco D. Could KK be OK? J Philos. (Forthcoming)
29. Grove A (1988) Two modellings for theory change. J Philos Logic 17:157–170
30. Halpern JY (1996) Should knowledge entail belief? J Philos Logic 25:483–494
31. Hintikka J (1962) Knowledge and belief, vol 414. Cornell University Press, Ithaca
32. Katsuno H, Mendelzon AO (1991) Propositional knowledge base revision and minimal change. Artif Intell 52:263–294
33. Lamarre P, Shoham Y (1994) Knowledge, certainty, belief, and conditionalisation (abbreviated version). In: Proceedings of KR'94, pp 415–424
34. Leitgeb H, Segerberg K (2007) Dynamic doxastic logic: why, how, and where to? Synthese 155:167–190
35. Lenzen W (1979) Epistemologische Betrachtungen zu [S4, S5]. Erkenntnis 14:33–56
36. Levi I (2004) Mild contraction: evaluating loss of information due to loss of belief. Clarendon Press, Oxford
37. Nayak A (1994) Iterated belief change based on epistemic entrenchment. Erkenntnis 41:353–390
38. Okasha S (2013) On a flawed argument against the KK principle. Analysis 73:80–86
39. Pagnucco M, Rott H (1999) Severe withdrawal—and recovery. J Philos Logic 28:501–547
40. Plaza J (2007) Logics of public communications. Synthese 158:165–179
41. Pratt VR (1976) Semantical considerations on Floyd-Hoare logic. In: Technical report, MIT, Cambridge
42. Rott H (2009) Shifting priorities: simple representations for twenty-seven iterated theory change operators. In: Makinson D, Malinowski J, Wansing H (eds) Towards mathematical philosophy. Trends in logic, vol 28. Springer, Dordrecht, pp 269–296
43. Rott H (2006) Revision by comparison as a unifying framework: severe withdrawal, irrevocable revision and irrefutable revision. Theoret Comput Sci 355:228–242
44. Rott H (2012) Bounded revision: Two-dimensional belief change between conservative and moderate revision. J Philos Logic 41:173–200
45. Segerberg K (1971) An essay in classical modal logic. Filosofiska studier utgivna av Filosofiska föreningen och Filosofiska institutionen vid Uppsala universitet, vol 1. Filosofiska föreningen och Filosofiska institutionen vid Uppsala universitet
46. Segerberg K (1998) Irrevocable belief revision in dynamic doxastic logic. Notre Dame J Formal Logic 39:287–306

47. Stalnaker R (1968) A theory of conditionals. In: Rescher N (ed) Studies in logical theory. APQ monograph series, vol 2. Blackwell, Oxford, pp 98–112
48. Stalnaker R (1996) Knowledge, belief and counterfactual reasoning in games. Econ Philos 12:133–163
49. Stalnaker R (2006) On logics of knowledge and belief. Philos Stud 128:169–199

Chapter 9
Temporal Aspects of the Dynamics of Knowledge

Valentin Goranko and Eric Pacuit

Abstract Knowledge and time are fundamental aspects of agency and their interaction is in the focus of a wide spectrum of philosophical and logical studies. This interaction is two-fold: on the one hand, knowledge evolves over time; on the other hand, in the subjective view of the agent, time only passes when her knowledge about the world changes. In this chapter we discuss models and logics reflecting the temporal aspects of the dynamics of knowledge and offer some speculations and ideas on how the interaction of temporality and knowledge can be systematically treated.

9.1 Introduction

Knowledge and time are fundamental aspects of agency and their interaction is in the focus of a wide spectrum of philosophical and logical studies. This interaction is a two-way street. On the one hand, knowledge evolves over time as the agent may learn new information, but also forgets previously known facts or their truth value may change over time. On the other hand, one can argue that, in the subjective view of the agent, time only passes when her knowledge about the world changes; and this is certainly the case when the agent has a watch and continuously keeps herself aware of the current time, but also when the agent has no other concept or measure of time except as a succession of events.

V. Goranko (✉)
Technical University of Denmark, 2800 Kongens Lyngby, Denmark
e-mail: vfgo@dtu.dk

V. Goranko
University of Johannesburg, Auckland Park, Johannesburg 2006, South Africa

E. Pacuit
Department of Philosophy, University of Maryland, College Park, MD, USA

A. Baltag and S. Smets (eds.), *Johan van Benthem on Logic*
and Information Dynamics, Outstanding Contributions to Logic 5,
DOI: 10.1007/978-3-319-06025-5_9, © Springer International Publishing Switzerland 2014

Each of these two concepts—knowledge and time—has been extensively formalized and studied in various logical frameworks, respectively forming the families of epistemic and temporal logics, since the 1960s, starting with the seminal works of Hintikka [45] and Prior [54]. Adding a multi-agent perspective makes the interaction between knowledge and time much more complex and versatile not only because of the intrinsic complexity of multi-agent epistemics, but also because of the partial information that individual agents have about the actual succession of events, and the problems arising with the synchronization of their communication.

Studies in formal logic gradually started reflecting on that interaction and a variety of logics combining knowledge and temporality started appearing in the 1980s. First, these were temporal-epistemic (aka, epistemic-temporal) logics [29, 30, 40, 41, 43, 53], coming mainly from the field of distributed computing and looking at the purely observable, explicit effect of the change of knowledge over time, but not at the reasons for such changes. Later, dynamic-epistemic logics emerged with the idea to focus on the causal aspects of the dynamics of knowledge, while leaving the temporality implicit, simply as succession of epistemic updates [3, 12, 26]. A strong impetus for new developments in logic reflecting the interaction between time, agents' knowledge and agents' abilities came from multi-agent systems where the so called "alternating-time temporal logic" [2] and numerous variations and extensions emerged in the early 2000s [36, 46, 60]. Another line of discussions related to issues that arise when developing logics that combine knowledge and temporality is in chapters on modal logics for reasoning about *strategies* and strategic analysis of multi-agent protocols in game situations [15, 49, 51]. A few approaches for combining the temporal, epistemic, and dynamic aspects have been recently proposed, too (see Sect. 9.6), but no commonly embraced Unification Theory has emerged yet. In this chapter we do not propose such theory, either, but rather offer some speculations on how it might look, and make some steps towards it.

This chapter is devoted to Johan van Benthem's influential contributions in this area. They go back at least to the 1990s and were first presented as a systematic research program in [9], later followed, inter alia, by [6, 10–13, 17] and culminated, so far, in [14] and [16]. While we do not purport here to survey Johan's work in this area, we certainly derive inspiration from it to chart some follow-up developments.

9.2 Preliminaries

We provide here only some basics of epistemic and temporal models and logics. For more details see the chapter [19] and the references at the end.

9.2.1 Models and Logics of (Static) Knowledge

9.2.1.1 Relational Models

The models that we discuss in this chapter are all instances of a *relational model*. Let At be a (finite) set of atomic sentences. A *relational model* (based on At) is a tuple $\langle W, \mathscr{R}, V \rangle$ where W is a nonempty set whose elements are called *possible worlds* or *states*; $\mathscr{R}$ is a set of relations on W, i.e., for each $R \in \mathscr{R}$, $R \subseteq W \times W$; and $V : \text{At} \to \wp(W)$ is a valuation function mapping atomic propositions to sets of states. Elements $p \in \text{At}$ are intended to describe ground facts about the situation being modeled, such as "the red card is on the table". The set W is intended to represent the different possible "scenarios" (elements of W are called possible worlds or states). The valuation function V associates with every ground fact the set of situations where that fact holds.

9.2.1.2 Epistemic Models

A basic epistemic model is a relational model with a single relation R, hereafter denoted as $\sim$, which represents the agent's knowledge, in terms of its epistemic uncertainties, as traditional in epistemic logic. The relation $\sim$ is the agent's *indistinguishability relation*, i.e., $q \sim q'$ means that the agent(s) is (are) not able to discern between the possible worlds q and q'; thus, both worlds appear identical from the agent's perspective. The indistinguishability relations representing the epistemic uncertainties are traditionally assumed to be *equivalence relations*. The knowledge of the agent is then determined as follows: the agent *knows* a property O in the world q if O is the case in all states indistinguishable from q for that agent.[1]

A multi-agent epistemic model involves an indistinguishability relation $\sim_i$ for every agent i. Formally, a *multi-agent epistemic structure* is a tuple $\mathscr{S} = \langle \mathbb{A}, \text{St}, \{\sim_i \mid i \in \mathbb{A}\}, V \rangle$ where $\mathbb{A}$ is the set of agents, St is the set of states (possible worlds) and $\sim_i$ is the indistinguishability relation over St associated with the agent i, for each $i \in \mathbb{A}$. Then, a *multi-agent epistemic model* (MAEM) is defined by adding a valuation to a multi-agent epistemic structure.

9.2.1.3 Epistemic Logics

Basic epistemic logic. A simple propositional modal language is often used to describe epistemic models. Let $\mathscr{L}_{EL}$ be the (smallest) set of sentences generated by the following grammar:

$$\varphi := p \mid \neg\varphi \mid \varphi \wedge \varphi \mid \mathrm{K}\varphi$$

---

[1] More generally, by varying the properties of the relation R, these models can also represent other informational attitudes of the agent, such as beliefs.

where $p \in \mathsf{At}$ (the set of atomic propositions). The additional propositional connectives $(\rightarrow, \leftrightarrow, \vee)$ are defined as usual and the dual of K, often denoted L, is defined as follows: $L\varphi := \neg K \neg \varphi$. The intended interpretation of $K\varphi$ is "according to the agent's current (hard) information, φ is true" (more standardly "the agent knows that φ is true"). Truth of the above language is defined as follows: Let $\mathcal{M} = \langle W, \sim, V \rangle$ be an epistemic model. For each $w \in W$, φ is true at state w, denoted $\mathcal{M}, w \models \varphi$, is defined by induction on the structure of φ:

- $\mathcal{M}, w \models p$ iff $w \in V(p)$
- $\mathcal{M}, w \models \neg\varphi$ iff $\mathcal{M}, w \not\models \varphi$
- $\mathcal{M}, w \models \varphi \wedge \psi$ iff $\mathcal{M}, w \models \varphi$ and $\mathcal{M}, w \models \psi$
- $\mathcal{M}, w \models K\varphi$ iff for all $v \in W$, if $w \sim v$ then $\mathcal{M}, v \models \varphi$ $\square$

Multi-agent epistemic logics. Besides the individual knowledge for each agent, the multi-agent epistemic framework involves several very natural and important notions of *multi-agent knowledge* and respective knowledge operators, for every non-empty set of agents A. These operators with their intended interpretations are:

- $K_A\varphi$, saying *'Every agent in the group A knows that φ'*. When $A = \{i\}$ we write K_i instead of $K_{\{i\}}$.
- $D_A\varphi$, saying *'It is a distributed knowledge amongst the agents in the group A that φ'*, intuitively meaning that the collective knowledge of all agents in the group A implies φ. For instance, if $K_a\varphi$ and $K_b(\varphi \rightarrow \psi)$ hold for some agents a and b, then $K_{a,b}(\varphi \wedge (\varphi \rightarrow \psi))$ holds, and therefore $K_{a,b}\psi$ holds, too, by the closure of knowledge under logical consequence.
- $C_A\varphi$, saying *'It is a common knowledge amongst the agents in the group A that φ'*, intuitively meaning that not only every agent in A knows φ, but also that every agent in A knows that every agent in A knows φ, and every agent in A knows that, etc., ad infinitum.

The language of the multi-agent epistemic logic builds on the basic epistemic logic by adding some or all of these operators and the formulae are defined by the following recursive definition:

$$\varphi := p \mid \neg\varphi \mid \varphi \wedge \varphi \mid D_A\varphi \mid C_A\varphi,$$

where p ranges over At and A ranges over the set $\mathscr{P}^+(\mathbb{A})$ of non-empty subsets of $\mathbb{A}$. The individual knowledge $K_i \varphi$ is definable as $D_{\{i\}}\varphi$, and then the group knowledge $K_A\varphi$ is definable as $\bigwedge_{i \in A} K_i \varphi$.

The formal semantics of the multi-agent epistemic operators at a state in a multi-agent epistemic model $\mathcal{M} = (\mathbb{A}, \mathsf{St}, \{\sim_i \mid i \in \mathbb{A}\}, V)$ is given by the clauses:

(K$_\mathbf{A}$) $\mathcal{M}, q \models K_A\varphi$ iff $\mathcal{M}, q' \models \varphi$ for all q' such that $q \sim_A^E q'$, where $\sim_A^E = \bigcup_{i \in A} \sim_i$.

(C$_\mathbf{A}$) $\mathcal{M}, q \models C_A\varphi$ iff $\mathcal{M}, q' \models \varphi$ for all q' such that $q \sim_A^C q'$, where $\sim_A^C$ is the transitive closure of $\sim_A^E$.

(D$_A$) $\mathcal{M}, q \models D_A \varphi$ iff $\mathcal{M}, q' \models \varphi$ for all q' such that $q \sim_A^D q'$, where $\sim_A^D = \bigcap_{i \in A} \sim_i$.

For more a more detailed discussion on epistemic models and the relevant modal logics see the chapter [19].

9.2.2 Temporal Models and Logics

Temporal reasoning stems from philosophical analysis of time and temporality, initiated in the Antiquity by Diodorus Chronos and Aristotle, but only formalized in precise logical terms first by Arthur Prior in his historical work culminating with his seminal book "Past, Present and Future" [54].

9.2.2.1 Temporal Models

There are various ontological assumptions for the nature of time, reflecting on the types of time flows and models used to formalize temporal reasoning: instant-based or interval-based, discrete or dense, continuous or not, endless or not, linear or branching, etc. The simplest formal model of time, aka *temporal frame*, is $\langle T, \preceq \rangle$ where T is a nonempty set of *time instants* or *moments* and $\preceq$ is a time precedence relation on T, which is generally a partial order, often assumed linear or tree-like, i.e., every time instant having a linearly ordered by $\preceq$ set of predecessors. More abstractly, one can adopt time intervals (periods) or entire time histories as primitive temporal entities and build models based on these, as demonstrated in [7].

In our context here, time is not an abstract flow of moments but rather a metaphor for the discrete succession of events—explicit or implicit—that determine the time instants and represents the passing of time. Such time flow can be linear, corresponding to a single time line (trace, history, etc.) or branching, corresponding to a non-deterministically evolving future of possible succession of events. Thus, the only observable effect of time passing is a discrete transition of one 'snapshot' of the world to another.

Depending on whether one is more interested in the sequence of events causing the passing of time or in the actual sequence of time states ('snapshots') of the world, this concept of time formalizes in either event-based temporal structures (sometimes called 'protocols') or in (temporal) transition systems. The former are more prominent in theories of events and agency as well as in distributed computing, whereas the latter are fundamental for the applications of temporal logics in computer science for model verification. Event-based protocols will be introduced later, and here we briefly present the basics of transition systems.

9.2.2.2 Transition Systems and Computations in Them

Formally, these are simply relational frames, consisting of a set of states and transition relations between them, possibly labelled by different types of actions. The states in a transition system can be thought of as program states, control states, configuration states, memory registers, etc. The actions can represent agents' actions, autonomous processes, or simply program instructions. Formally, a *labeled transition system* is a structure $\mathscr{T} = \langle \mathsf{St}, \{\overset{a}{\longmapsto}\}_{a \in \mathsf{Act}} \rangle$ consisting of a non-empty set St of *states*; a non-empty set Act of *actions* or *transitions*, and a binary *transition relation* $\overset{a}{\longmapsto} \subseteq \mathsf{St} \times \mathsf{St}$ associated with every action $a \in \mathsf{Act}$. The intuition is that each $a \in \mathsf{Act}$ acts, possibly non-deterministically, on states and produces *successor states*. We write $s \overset{a}{\longmapsto} t$ to indicate that the action a can transform the state s into the state t and say that s is an *a-predecessor* of t, while t is an *a-successor* of s. The successor relation between states generates a branching time discrete temporal structure. A labelled transition system that involves only one type of action is called a *simple transition system*, or just a *transition system* simpliciter. Then we omit the label and typically denote it by (St, R),

A state may have various properties. For instance, a state of a transition system modeling a computing process can be initial, accepting, safe or unsafe, critical or terminal for a given process, etc. Such properties of states can be indicated by special *atomic propositions*. The set of such propositions that are declared true at a given state is the *description* of that state. A transition system where every state is assigned such description is an *interpreted transition system*. Formally, this is a pair $\mathscr{M} = \langle \mathscr{T}, \mathsf{L} \rangle$ where $\mathscr{T}$ is a transition system and $\mathsf{L} : \mathsf{St} \to 2^{\mathsf{At}}$ is a *state description mapping* that assigns to every state s the set of atomic propositions from a fixed set At, that are true at s. Abstractly, interpreted transition systems are simply relational models, with valuation uniquely derived from the state description.

A *path (run, execution)* in a transition system $\mathscr{T}$ is a (finite or infinite) sequence of states and actions transforming every state into its successor: $s_0 \overset{a_0}{\longmapsto} s_1 \overset{a_1}{\longmapsto} s_2 \ldots$. Thus, a path is a linear time flow representing a possible time history.

A *computation*, or *trace*, in an interpreted transition system $(\mathscr{T}, \mathsf{L})$ is a (finite or infinite) sequence of state descriptions and respective actions along a path: $\mathsf{L}(s_0) \overset{a_0}{\longmapsto} \mathsf{L}(s_1) \overset{a_1}{\longmapsto} \mathsf{L}(s_2) \ldots$. Thus, a computation, intuitively, is the *observable effect* (the 'trace') of a path in a transition system. It can be regarded as a record of all successive intermediate results of the computing process. The idea is that the information encoded by the state descriptions includes all that is essential in the computation, including the values of all important variables. However, agents typically can only observe part of the state description, which represents their current information about the world. Usually, unless otherwise specified, we assume that the transition relation R is *serial*, or *total*, i.e., every state has at least one R-successor. When the actions are not important, one can represent paths and computations simply as sequences $s_0, s_1, s_2, \ldots$ and respectively $\mathsf{L}(s_0), \mathsf{L}(s_1), \mathsf{L}(s_2), \ldots$, or, more abstractly, as mappings $\sigma : \mathbb{N} \to \mathsf{St}$, respectively $\sigma : \mathbb{N} \to 2^{\mathsf{At}}$.

9.2.2.3 Basic Temporal Logic

The basic temporal language $\mathfrak{L}_t$, essentially due to Prior, is a propositional bimodal language, containing, besides a fixed set of atomic propositions and boolean connectives, the temporal operators H, G respectively referring to "always in the past" and "always in the future". The set of formulae is recursively defined by:

$$\varphi = p \mid \neg\varphi \mid (\varphi \wedge \varphi) \mid G\varphi \mid H\varphi$$

The dual temporal connectives, referring to "sometime in the past" and "sometime in the future", are defined as usual:

$$F\varphi := \neg G\neg\varphi, \, P\varphi := \neg H\neg\varphi.$$

Temporal model is a tuple $\langle T, \preceq, V \rangle$ where $\langle T, \preceq \rangle$ is a temporal frame and V is a valuation. Truth of a temporal formula at an instant t in a temporal model $M = \langle T, \preceq, V \rangle$ is defined in a traditional modal logic style, assuming that $\preceq$ is the accessibility relation associated with G and its converse $\succeq$ is associated with H:

- $M, t \models G\varphi$ if $M, s \models \varphi$ for every $s \in T$ such that $t \preceq s$.
- $M, t \models H\varphi$ if $M, s \models \varphi$ for every $s \in T$ such that $t \succeq s$.

Several additional temporal operators can be added, especially in a discrete setting, such as "Nexttime" N, "Since" S, "Until" U, etc. For further general references on temporality and temporal logics see [4, 7, 8].

9.3 From Static to Dynamic Reasoning about Knowledge: Temporal-Epistemic Frameworks

The traditional temporal models implicitly assume complete and fixed, unchangeable knowledge of all agents at all times. Furthermore, the epistemic models described above are *static*, or rather *timeless*. They describe what the agents know and believe in a fixed 'snapshot of the world'. Thus, none of these reflects the deficiencies, nor the dynamics, of knowledge over time.

There are various ways to relate models of time with knowledge and provide a framework for logical reasoning about their interaction. The syntactic merger seems easy: one can simply put together the desired repertoires of temporal and epistemic operators in a common logical language. However, the conceptual modeling of their interaction and the formal semantics capturing that interaction are the main challenges. In this chapter, we discuss further some known and some new ideas of how that can be done. The relevant literature is rich and diverse, and we only mention and briefly discuss some selected sources further in the text. A discussion on some of these approaches can also be found in the chapter [19].

- One of the general formal construction is *fusion* [33] of temporal and epistemic models into temporal-epistemic models, where possible worlds are regarded both as time instants and as epistemic alternatives. This construction is technically simple and elegant but conceptually deficient because it neither reflects nor explains the *temporal dynamics* of knowledge.
- Another generic formal approach is *temporalization* of a logical system [32], in this case of the epistemic logic. Semantically, it is based on temporal-epistemic models obtained by taking a temporal model and associating every time point in it with an epistemic model.
- *Protocol-based epistemic temporal models* are refinements of the temporalization construction, obtained as collections of epistemic models, related over time by protocols assigning a model to every time instant. The protocols are directed trees representing the possible sequences of events (or, actions) effecting the evolution of knowledge.
- A more refined approach alternates adding temporal and epistemic layers. It starts e.g., with a temporal model and then adds a 'cloud' of epistemic alternatives representing the uncertainties for each agent at every moment of time. Further, all epistemic alternatives are "temporalized" by adding time stamps to each of them and then extended with full time lines, thus creating a bundle of interleaved temporal models over epistemic states. Then the resulting models are endowed with clouds of epistemic alternatives for each time moment, etc. The alternation of adding epistemic alternatives and time lines until saturation or forever. The limit of that construction is the intended temporal-epistemic model.
- A similar, yet somewhat technically different idea is implemented in Halpern and Vardi's *interpreted systems* (generalizing and extending "interpreted transition systems" as defined earlier) [30, 31] built on sets of 'runs', each representing a possible evolution in time of a system consisting of several processors running in parallel, each having their own local state and all these local states composing into a 'global state'. Every agent can only observe its local component of the state and this partial information creates the agent's epistemic indistinguishability relation.
- Alternatively, one can start with an epistemic model and then add a temporal structure to each of the possible worlds: Following this approach, a "possible world" is no longer a primitive object in the model. The "possible worlds" of the above model are constructed from more basic objects, such as *events, local states* and *moments*. Thus, one can describe the different model transformations that are intended to represent different "epistemic actions". This is a "change-based" view of knowledge dynamics.
- A global epistemic approach takes a class of pointed temporal models and adds epistemic uncertainties between them (for each agent), producing an epistemic 'super-model'.
- An epistemic analogue of STIT models, where the choice relation of an agent is interpreted as his epistemic relation over possible alternative futures, which changes dynamically over time as the future is gradually revealed. See [18] for an initial discussion of this idea.

In the rest of this section we present in more details and discuss some of these approaches.

9.3.1 From Adding Epistemic Clouds to Fusion of Temporal and Epistemic Models

There is a whole spectrum of possible extensions of a temporal model with an epistemic dimension, reflecting the agents' awareness and knowledge of time and the degree of synchrony between agents. The conceptually simplest approach is to start with a temporal model—be it for linear or branching time—and to expand every time instant in it with a 'cloud' of epistemic alternatives for each agent. The resulting formal models can be defined as $\langle T, \preceq, \{W_t\}_{t \in T}\rangle$ where $\langle T, \preceq \rangle$ is a temporal model and for each $t \in T$ the set W_t is an epistemic model consisting of the alternative worlds that some of the agents consider possible at moment t.

In order to give semantics of a temporal epistemic language in such models we have to restrict it to formulae where temporal operators cannot be nested in epistemic operators. The semantics is then a straightforward combination of the temporal and purely epistemic semantic clauses. With such language one can reason about what one will know tomorrow, sometime, or always, but not what one knows now about what will be true tomorrow, sometime, or always. In order to interpret such statements, and further nesting of temporal and epistemic operators, the models defined above must be enriched with time stamps for every epistemically alternative world, or with alternative time lines passing through them, and then adding epistemic clouds for each instant on these alternative time lines, then arranging these in timelines, etc. Eventually, in the limit of that construction we obtain a full fusion of temporal and epistemic models: $\langle \mathbb{A}, \mathsf{St}, \preceq, \{\sim_i \mid i \in \mathbb{A}\}, V \rangle$, where the possible worlds incorporate the time instants and $\preceq$ represents time precedence over a possible timeline in the model. Depending on how the epistemic alternatives relate to the temporal knowledge of the agents, a variety of models can emerge here.

One extremity is a *fully synchronous system* where there is a global clock observable by all agents at all times. In this case, all epistemic alternatives of a possible world in the temporal model share the same time stamp. These alternatives can, however, appear or disappear in time, as the agent learns or forgets. However, these epistemic alternatives may, but need not, evolve over time which therefore renders it possibly meaningless to reason about what the agent may know in the future or has known in the past.

The other extremity is a *fully asynchronous system* where the agents have no knowledge, or possibly not even concept, of time. In this case, their epistemic alternatives may have different time stamps, or no time stamps at all. One can imagine that in this case agents have instantaneous knowledge at every time instant, but no memory at all, and the evolution of their knowledge is exogenous for them.

Finally, we note one particular construction of a temporal model with 'epistemic clouds' based on *partial observability*: every agent can only observe the truth value of some atomic propositions, which naturally creates the cloud of alternative possibilities for the actual world, of which it has a partial view. Note that this partial observability can vary over time for each agent, thus creating more elaborate scenarios about the dynamics of their knowledge.

9.3.2 Interpreted Systems as Temporal-Epistemic Models

Various proposals in the distributed systems literature of the 1980s [41, 43, 53] gradually crystalized in Halpern and Vardi's *interpreted systems* [40, 42], further developed in [30]; see further references in the latter. Interpreted systems model the evolution of the knowledge of one or several agents (processors) over time and are technically very similar to the fusion of temporal and epistemic models discussed above.

9.3.2.1 Interpreted Systems

Informally, an *interpreted system* is defined for a fixed set of agents $\mathbb{A}$ and builds on a *state space* St consisting of *(global) states*, where a global state can be viewed as a tuple of *local states*, one for each agent. More generally, a global state can be regarded as an abstract entity, of which every agent only has a partial *local view*. This allows for some parts of the state to be visible by several agents, and others—possibly by none.

A basic concept in an interpreted system is a *run*: an infinite sequence of global states from St; formally, a mapping $r{:}\mathbb{N} \to \mathsf{St}$. Generally, an interpreted system $\mathscr{I}$ may comprise any non-empty set $\mathscr{R}$ of runs on its state space. Then, a pair (r, n), where $r \in \mathscr{R}$ and $n \in \mathbb{N}$ is a *(time) point* on the run r. Thus, the set of points in $\mathscr{I}$ is $P(\mathscr{I}) = \mathscr{R} \times \mathbb{N}$. The point (r, n) corresponds to a unique state $r(n)$; however, different points may correspond to the same state.

The knowledge of every agent in an interpreted system is determined, as in pure epistemic logic, in terms of its uncertainty. The agent's uncertainty here is between different *time points*, however, Halpern and Vardi reduce it to uncertainty between states: two points t_1 and t_2 are *indistinguishable* for an agent i iff their corresponding states have the same local component for i, i.e., iff i has the same local view on them. This is denoted $t_1 \sim_i t_2$, where $\sim_i$ is the *indistinguishability relation* on $\mathscr{R} \times \mathbb{N}$ for i.

Finally, an interpreted system involves labeling of the points with sets of atomic propositions from a fixed set PROP. The label of a point is supposed to describe all essential features of that point. Thus, formally, an interpreted system is a tuple:

$$\mathscr{I} = \langle \mathbb{A}, \mathsf{St}, \mathscr{R}, \{\sim_i\}_{i \in \mathbb{A}}, \mathsf{L} \rangle$$

where $\mathbb{A}$ is the set of agents, St is the global state space, $\mathscr{R}$ is the set of runs, for each $i \in \mathbb{A}$ the relation $\sim_i$ is an equivalence relation of indistinguishability on $P(\mathscr{I}) = \mathscr{R} \times \mathbb{N}$ for the agent i, and $\mathsf{L} : P(\mathscr{I}) \to \mathrm{PROP}$ is the labeling function.

Note that for any given agent i, a run is a sequence of *local views*, that is, of point clusters with respect to $\sim_i$.

Given an interpreted system $\mathscr{I}$, we will refer to the relation $\{((r, n), (r, n+1)) \in P(\mathscr{I}) \times P(\mathscr{I}) \mid r \in \mathscr{R}, n \in \mathbb{N}\}$ as the *temporal relation* induced in $\mathscr{I}$.

A *computation* in the interpreted system $\mathscr{I}$ corresponding to the run $r : \mathsf{St} \to \mathbb{N}$ is the observable—by an external observer who has full view of all states— effect of that run: $\mathsf{L}(r(0)), \mathsf{L}(r(1)), \ldots$.

Now, the dynamics of the knowledge of an agent can be modeled in terms of the evolution of its local views in the course of a run or a computation. One can argue that the local view and, respectively, the knowledge of an agent should be based on the *label* of the current point, rather than on the point itself. This approach can be implemented by assigning to every agent i a subset PROP_i of *observable for i* atomic propositions [30, 43].

9.3.2.2 Some Important Properties of Interpreted Systems

Following [42] one can identify some key properties of interpreted systems that turn out to be crucial for the computational complexity of the problem of deciding satisfiability in them. An interpreted system $\mathscr{I} = \langle \mathbb{A}, \mathsf{St}, \mathscr{R}, \{\sim_i\}_{i \in \mathbb{A}}, \mathsf{L} \rangle$ has the property of:

- **Unique initial state** if $r(0) = r'(0)$ for all runs $r, r' \in \mathscr{R}$.
- **No forgetting** if for every $i \in \mathbb{A}$, if $((r, n) \sim_i (r', n'))$ then for all $k \leq n$ there exists a $k' \leq n'$ such that $((r, k) \sim_i (r', k'))$.
- **No learning** if for every $i \in \mathbb{A}$, if $((r, n) \sim_i (r', n'))$ then for all $k \geq n$ there exists a $k' \geq n'$ such that $((r, k) \sim_i (r', k'))$.
- **Synchrony** if for every $i \in \mathbb{A}$, if $((r, n) \sim_i (r', n'))$ then $n = n'$.

The property of **Synchrony** expresses the idea that the agents are able to perceive time and have a common clock. **No learning** expresses that agents do not learn over time, in the sense that if a coalition A of agents cannot distinguish two runs at a given time, it will not be able to do so later on. Likewise, **No forgetting** means that if A at a given time point can tell two different runs apart, it must have been able to do so at any previous point in time.[2]

9.3.2.3 Temporal-Epistemic Logics Over Interpreted Systems

Based on the choice of language: single-agent or multi-agent, linear time or branching time, including or not operators for common knowledge, as well as on the combi-

---

[2] This is somewhat more complicated in the asynchronous case, see [42, 44] for discussion and explanation.

nations of the semantic properties listed above, Halpern and Vardi identify in [40] and [42] a total of 96 temporal-epistemic logics and analyze the complexities of the satisfiability problems in them. The languages of these logics and their semantics are a fairly straightforward combination of the temporal and epistemic logics presented in Sect. 9.2. For instance, the formulae of the multi-agent linear time temporal epistemic logic with the temporal operators X ("next") and U ("until") of the logic LTL and individual and common knowledge (of all agents) operators are built as follows:

$$\varphi := p \mid \neg\varphi \mid (\varphi \wedge \varphi) \mid X\varphi \mid (\varphi\, U\varphi) \mid K_i\, \varphi \mid C\varphi$$

where $i \in \mathbb{A}$. The essential semantic clauses are:

$\mathcal{M}, (r, n) \models X\varphi$ iff $\mathcal{M}, (r, n + 1) \models \varphi$;

$\mathcal{M}, (r, n) \models \varphi\, U\psi$ iff $\mathcal{M}, (r, i) \models \psi$ for some $i \geq n$ such that $\mathcal{M}, (r, j) \models \varphi$ for every $n \leq j < i$;

$\mathcal{M}, (r, n) \models K_i\, \varphi$ iff $\mathcal{M}, (r', n') \models \varphi$ for every (r', n') such that $((r, n) \sim_i (r', n'))$.

For branching time logics, quantifiers over runs are added to the language, with semantics:

$\mathcal{M}, (r, n) \models \exists\varphi$ iff $\mathcal{M}, (r', n') \models \varphi$ for some r' such that $r(n) = r'(n')$.

Some of the properties listed above can be expressed by suitable axioms. For instance, **No learning** in linear-time interpreted systems corresponds to the axiom $XK_i\, \varphi \rightarrow K_i\, X\varphi$, whereas **No forgetting** corresponds to $K_i\, X\varphi \rightarrow XK_i\, \varphi$. For more, see [44] for the linear time logics and [48] for the branching time logics.

It turns out that when time and knowledge do not interact, or only weak forms of interaction are imposed (e.g., only synchrony) then these temporal-epistemic logics are computationally reasonably behaved, with EXPTIME complexity of their satisfiability/validity problems [40, 42] and allow relatively simple tableau-based decision procedures, even when the full epistemic repertoire, with common and distributed knowledge for each group of agents, is added to the language [38, 39]. However, most of these logics that involve more than one agent *whose knowledge interacts with time* (e.g., who do not learn or do not forget)—turn out undecidable (with common knowledge), or decidable but with non-elementary time lower bound (without common knowledge) [40, 42]. See [17] for a discussion of these issues. Even in the single-agent case, the interaction between knowledge and time proved to be quite costly (pushing the complexities of deciding satisfiability up to EXPSPACE and 2EXPTIME) (ibid.).

9.3.3 Protocol Based Epistemic-Temporal Models: Modeling Uncertainty About What Has Happened

This approach goes back to [53]; see also [52]. We fix a finite set of agents $\mathbb{A}$ and a (possibly infinite) set of events Σ. There is a large literature addressing the many subtleties surrounding the very notion of an *event* and when one event *causes* another event. However, for this chapter we take the notion of event as primitive. What is needed is that if an event takes place at some time t, then the fact that the event took place can be observed by a relevant set of agents at t. Compare this with the notion of an event from probability theory. If we assume that at each clock tick a coin is flipped exactly once, then "the coin landed heads" is a possible event. However, "the coin landed head more than tails" would not be an event, since it cannot be observed at any one moment. As we will see, the second statement will be considered a *property* of histories, or sequences of events. A Σ-*history* is a finite sequence of events from Σ. We write Σ^* for the set of Σ-histories. From now on we consider Σ fixed and call the elements of Σ^* just histories. For any history h, we denote by $\mathsf{len}(h)$ the length of h and we write he for the history h followed by the event e. Given $h, h' \in \Sigma^*$, we write $h \preceq h'$ if h is a prefix of h', and $h \prec_e h'$ if $h' = he$ for some event e.

There are several simplifying assumptions that we adopt. Since histories are sequences of (discrete) events, we assume the existence of a global discrete clock. The length of the history then represents the amount of time that has passed. Thus, this implies that we are assuming a finite past with a possibly infinite future. Furthermore, we assume that at each clock tick, or moment, *some* event—which need not be directly observable by any agent—takes place. Thus, we may include a special event $\mathsf{e_t}$ representing a "clock tick".

Definition 9.1 (*ETL Models*) Let Σ be a set of events and At a set of atomic propositions. An *epistemic temporal model (ETL model)* is a tuple $\langle \mathbb{A}, \Sigma, \mathsf{H}, \{\sim_i\}_{i \in \mathbb{A}}, V \rangle$ where $\mathbb{A}$ is a set of agents, H is a set of histories closed under prefixes, for each $i \in \mathbb{A}$, $\sim_i$ is an equivalence relation on H and V a valuation function ($V : \mathsf{At} \to \wp(\mathsf{H})$).

An ETL model describes how the agents' *knowledge* evolves over time. Formally, ETL models are very similar to interpreted systems, introduced in the previous section (consult [50] for an extended discussion). The domain of an ETL model (set of histories closed under non-empty prefixes) is called a *protocol*. Histories in a protocol are the analogues of the global states in an interpreted system. In addition, the protocol describes the temporal structure, with h' such that $h \prec_e h'$ representing the point in time after e has happened in h. The relations $\sim_i$ represent the uncertainty of the agents about how the current history has evolved. Thus, $h \sim_i h'$ means that from agent i's point of view, the history h' looks the same as the history h.

Assumptions about the domain of an ETL model corresponds to "fixing the playground" where the agents will interact. In other words, the protocol not only describes the temporal structure of the situation being modeled, but also any *causal* relationships between events (eg., sending a message must always preceed receiving that message) plus the motivations and dispositions of the participants (eg., liars send messages that they *know*—or believe—to be false). Thus the "knowledge" of agent i at a history h in an ETL model is derived from both i's observational powers (via the $\sim_i$ relation) and i's information about the "protocol" generating the histories in the model.

Analogously to properties of interpreted systems, we identify the following key properties of an ETL model: Let $\mathcal{M} = \langle \Sigma, \mathsf{H}, \{\sim_i\}_{i \in \mathbb{A}}, V \rangle$ be an ETL model. $\mathcal{M}$ satisfies:

- **Synchronicity** iff for all $h, h' \in \mathsf{H}$, if $h \sim_i h'$ then $\mathsf{len}(h) = \mathsf{len}(h')$
- **Perfect Recall** iff for all $h, h' \in \mathsf{H}$, $e, e' \in \Sigma$ with $he, h'e' \in \mathsf{H}$, if $he \sim_i h'e'$, then $h \sim_i h'$
- **Uniform No Miracles** iff for all $h, h' \in \mathsf{H}$, $e, e' \in \Sigma$ with $he, h'e' \in \mathsf{H}$, if there are $h'', h''' \in \mathsf{H}$ with $h''e, h'''e' \in \mathsf{H}$ such that $h''e \sim_i h'''e'$ and $h \sim_i h'$, then $he \sim_i h'e'$.

Note that the properties defined above only refer to the underlying *frames* of the ETL models.

Remark 9.1 (*Alternative Definition of Perfect Recall*) Johan van Benthem gives an alternative definition of Perfect Recall in [12]:

$$\text{if } he \sim_i h' \text{ then there is an event } f \text{ with } h' = h''f \text{ and } h \sim_i h''.$$

This property is equivalent over the class of ETL models to the above definition of Perfect Recall and synchronicity. The formulation of Perfect Recall given in the former definition above is closer to the one found in the computer science literature on verifying multiagent systems (cf. [30]) and the game theory literature (cf. [21]).

ETL models describe how the agents' knowledge changes during a given sequence of events. The example in Fig. 9.1 illustrates the type of knowledge flow that ETL models describe. Suppose that there is a deck of red and black cards. An agent is observing the cards being placed on a table. Suppose that a red card is placed face down on the table (the agent can see that there is a card on the table, but not the color of the card). The next two cards that will be chosen by the dealer are a black card followed by a red card. Furthermore, both cards will be placed faced up on the table. An ETL model describing this situation runs as follows: There are four events $\Sigma = \{\mathsf{R}_d, \mathsf{R}_u, \mathsf{B}_d, \mathsf{B}_u\}$ where R_d is the event[3] "a red card is placed on the table face down", R_u is the event "a red card is placed on the table face up" (similarly for B_d and B_u). We are interested in describing how the agent's knowledge changes during the history $h = \mathsf{R}_d\mathsf{B}_u\mathsf{R}_u$. The set H is the set of Σ-histories of length ≤ 3 depicted in Fig. 9.1. The dotted lines represent agent i's information cells. For a history h, we write $[h]_i$ for equivalence class of h under $\sim_i$. We make the following observations about this model, to be followed by some analysis in later sections. First

---

[3] To be more precise, R_d is an *event type* (similarly for the other events in Σ).

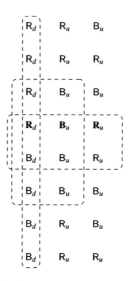

Fig. 9.1 Epistemic temporal model of the card dealing

some notation, if h is a history and $t \in \mathbb{N}$, we write h_t for the initial segment of h of length t and $h_{(t)}$ for the tth event in h. The actual history is $h = \mathsf{R}_d\mathsf{B}_u\mathsf{R}_u$.

- Restricting the set of "admissible" histories H allows the agent to incorporate knowledge of the "rules of the game" into his information. For example, the agent "knows" that the game proceeds by putting one card face down on the table followed by two cards placed face up. Furthermore, we assume that the agent cannot distinguish the two event R_d and B_d; however she can distinguish between the two events R_u and B_u (as well as between R_u and R_d, for example).
- After the first card is placed on the table face down, the agent does not know whether the card is black or red. This follows since the equivalence class of h is $[h]_i = \{h'_1 \mid h' \in \mathsf{H}\}$.
- On history h, after the first card is placed on the table (at moment 1), it is settled that the next card will be black, but the agent does not know this. This follows since $h_{(2)} = \mathsf{B}_u$ (the next event will be that a black card is placed face up on the table) and there is a $h' \in \mathsf{H}$ such that $h_1 \sim_i h'_1$ and $h'_{(2)} = \mathsf{R}_u$.
- After the second card is placed on the table, the agent learns that it is a black card. This follows since $[h_2]_i = \{h' \in \mathsf{H} \mid h'_{(2)} = \mathsf{B}_u\}$.

9.3.4 Adding Epistemics to Temporal Models: Modeling Uncertainty About What Will Happen

Let $\langle T, \preceq \rangle$ be a temporal model, e.g., T is a nonempty set of moments and $\preceq$ is the predecessor relation on T. A *full history* h is a maximal linearly ordered set of moments. Let $\mathscr{H}$ be the set of all full histories. For $h \in \mathscr{H}$, we write h/m for the pair (h, m) where $m \in h$. Each pair h/m is associated with a set of atomic propositions: the non-epistemic facts that are true at moment m given full history h. Let $\mathscr{H}_m = \{h \mid h \text{ is a full history with } m \in h\}$ be the set of full histories passing through moment m.

At each moment m there is a relation $\sim_i^m$ representing i's information uncertainties at moment m. This is intended to be a "forward-looking" notion of knowledge representing the information that agent i has about how the situation will evolve from moment m onwards. For simplicity, we assume that each $\sim_i^m$ is an equivalence relation, though this is not crucial for what follows. We further suppose that there is a distinguished moment $0 \in T$ representing the initial state of affairs. Then, $\sim_i^0$ represents i's initial uncertainty about all possible histories.

We have left open exactly which set of histories $\sim_i^m$ ranges over. There are two natural choices. The first choice is to let $\sim_i^m \subseteq \mathscr{H}_m \times \mathscr{H}_m$. So, if $h \sim_i^m h'$ then both h and h' are histories containing moment m. This builds in the assumption that the agent correctly observes all actions leading up to this moment. In addition, we may impose a stronger perfect recall condition:

- For all m and m', if $m \preceq m'$, then $\sim_i^{m'}$ is a *refinement* of $\sim_i^m$. I.e., if $h, h' \in \mathscr{H}_{m'}$ and $h \sim_i^{m'} h'$, then $h \sim_i^m h'$.

Recall the example discussed in the previous section. There is a deck of red and black cards, cards are being chosen one at a time and placed on a table in front of an agent i. A branching time model of this situation looks as follows:

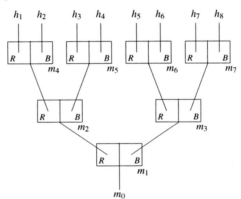

There are eight moments $T = \{m_0, \ldots, m_7\}$. If we assume that for each $m \in T$, $\sim_i^m = \mathscr{H}_m \times \mathscr{H}_m$, then we have the following observations:

- At moment m_1, the agent does not know whether the card chosen by the dealer is red or black.
- At the pair h_1/m_1, the card chosen by the dealer is red, but the agent does not know this. Indeed, the agent thinks that it may be black.
- At moment m_3, the agent knows that the card chosen at moment m_1 was black, but does not know the card the dealer is currently holding.

- At m_1, it is not settled yet which card the dealer will choose during the next round and the agent knows this.

Note that we assume that $\sim_i^m \subseteq \mathcal{H}_m \times \mathcal{H}_m$, and so, the agent's uncertainty at moment m ranges only over the set of histories running through moment m. Alternatively, we can assume that the agents uncertainty ranges over all histories: for each $m \in T$, $\sim_i^m \subseteq \mathcal{H} \times \mathcal{H}$. A natural constraint here is:

- For all h, h' if $h \sim_i^m h'$ then there is some $m' \preceq m$ such that $h, h' \in \mathcal{H}_{m'}$ and $h \sim_i^{m'} h'$

This means that if the agent cannot distinguish between h and h' at moment m, there must be some earlier moment m' such that both h and h' run through m and the agent could not distinguish h and h' at that moment. The flow of knowledge can be described as follows:

- At m_1, we have $\sim_i^{m_1} = \mathcal{H} \times \mathcal{H}$.
- The equivalence classes of $\sim_i^{m_j}$ with $j = 2, 3$ are
  $\{\{h_1, h_2, h_5, h_6\}, \{h_3, h_4, h_7, h_8\}\}$
- The equivalence classes of $\sim_i^{m_j}$ with $j = 4, 5, 6, 7$ are
  $\{\{h_1, h_5\}, \{h_2, h_6\}, \{h_3, h_7\}, \{h_4, h_8\}\}$

Given these definitions, we make the following observations:

- At moment m_1 on history h_1, the card chosen by the dealer is red, but the agent does not know this (this follows since $h_1 \sim_i^{m_1} h_5$).
- at moment m_2 on history h_1, the card chosen by the dealer is red and the agent knows this (this follows since it is true on all the histories that are $\sim_i^{m_2}$-equivalent to h_1 the red card is chosen at moment m_2). Furthermore, the agent still does not know that the card chosen at m_1 was red (this follows since $h_1 \sim_i^{m_2} h_5$).

9.3.5 Comparing Modeling Formalisms

We conclude this section with some brief comparisons between the various epistemic temporal constructions and models.

- Interpreted systems are a special kind of fusion of temporal and epistemic models, and protocol-based models can be regarded as a special kind of interpreted systems, where the runs are chains of histories along branches of the protocol-tree.
- The temporal models extended with uncertainties between histories are technically closely related to STIT models, see [18] for an initial discussion. Indeed, one can simply take a STIT model and treat the partition of all histories passing through a given point determined by the possible choices of an agent as arising from the epistemic indistinguishability relation between these histories for the agent. Note that the standard additional requirement in STIT models, that every selection of choices by all agents intersects in a single history, can now be interpreted as saying

that the agents have a complete distributed knowledge about the entire actual future (the 'thin red line').

We also note that there is an important distinction in the STIT literature between "moments" and "instants". The general idea is that instants represents the general flow of time while moments are specific "realizations" of the instances. Formally, an instant i is a partition of the moments such that every history intersects each instant at exactly one moment (i.e., each $i \in$ i, for all $h \in \mathscr{H}$, $|i \cap h| = 1$). For example, in the above model m_2 and m_3 both occur at the first instant. We may be interested in an agent's knowledge at a particular instant: after the second card flip, the agent knows the color of the card: at each moment in the second instant, it is true that agent knows the color of the card.

9.4 Looking Inside the Dynamics of Knowledge: Dynamic Epistemic Logic

The models introduced in the previous section each provide a "grand stage" where histories of some social interaction unfold constrained by some underlying *protocol*. Temporal-epistemic models present the observable effect of the dynamics of knowledge over time but do not reflect the causes for that dynamics. In this section, we introduce an alternative framework to reason about the dynamics of knowledge. The focus in this section is on "epistemic actions" that transform models describing the agents' current information. A number of elegant logical systems have been devised to reason about such epistemic actions (see [14] and the chapters [28] and [1] for overviews).

Similar to the way relational structures are used to capture the information the agents have about a *fixed* social situation, an *event model* describes the agents' information about which actual events are currently taking place. The temporal evolution of the situation is then computed from some initial epistemic model through a process of successive model updates, effected by a product construction between the epistemic model and the event model.

Definition 9.2 Suppose that $\mathscr{L}_{EL}$ is the epistemic modal language. An *event model* is a tuple $\langle S, \longrightarrow, \mathsf{pre} \rangle$, where S is a nonempty set of *primitive events*, for each $i \in \mathbb{A}$, $\longrightarrow \subseteq S \times S$ and $\mathsf{pre} : S \to \mathscr{L}_{EL}$ is the *precondition function*.

The only difference with a relational model is that the precondition function assigns a single formula to each primitive event. The intuition is that $\mathsf{pre}(e)$ describes what must be true in order for the event e to happen. Given two primitive events e and f, $e \longrightarrow f$ means "if event e takes place, then i *thinks* it is event f". The information provided by an event model can be incorporated into a relational structure using the following operation [3]:

Definition 9.3 (*Product Update*) The *product update* $\mathcal{M} \otimes \mathcal{E}$ of a relational model $\mathcal{M} = \langle W, R, V \rangle$ and event model $\mathcal{E} = \langle S, \longrightarrow, \mathsf{pre} \rangle$ is the relational model $\langle W', R', V' \rangle$ with:

1. $W' = \{(w, e) \mid w \in W, e \in S \text{ and } \mathcal{M}, w \models \mathsf{pre}(e)\}$;
2. $(w, e) R'(w', e')$ iff wRw' in $\mathcal{M}$ and $e \longrightarrow e'$ in $\mathcal{E}$; and
3. for all $p \in \mathsf{At}$, $(s, e) \in V'(p)$ iff $s \in V(p)$.   □

The following abstract example illustrates this operation. Suppose that, initially, the agent knows that p is the case, but thinks that both q and $\neg q$ are (epistemically) possible. This epistemic model $\mathcal{M}$ is represented on the left in the picture below, with an edge from state w to state v provided the agent cannot distinguish between w and v. Suppose that p describes the actual event, but the agent (mistakenly) thinks she observes q. This can be described by the following event model $\mathcal{E}$. The result of performing this action on the epistemic model $\mathcal{M}$ is *calculated* using Definition 9.3:

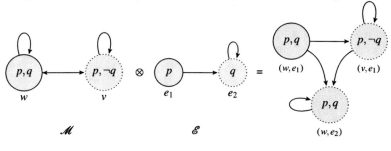

The first thing to notice is that the model $\mathcal{M} \otimes \mathcal{E}$ is not an epistemic model since the relation is not an equivalence relation. But this makes sense since the agent was misinformed or uncertain about precisely what she observed.

9.4.1 Comparing ETL and DEL

Both ETL and DEL are logical frameworks that are intended to describe the flow of information in a social interactive situation. Summarizing the results found in [6, Sect. 3], this section shows how these two "competing" logical frameworks can be rigorously compared. We will

(i) illustrate how DEL product update (Definition 9.3) may be used to generate interesting ETL frames, and
(ii) describe the observational powers of the agents presupposed in the DEL setting.

The key observation is that by repeatedly updating an epistemic model with event models, the machinery of DEL (i.e., Definition 9.3) in effect creates ETL models. Note that an ETL model contains not only a description of how the agents' information changes over time, but also "protocol information" describing *when* each event *can* be performed. Thus, in rigorously comparing DEL with ETL models, the protocol information must be made explicit.

Let $\mathbb{E} = \{(\mathscr{E}, e) \mid \mathscr{E}$ an event model and $e \in \mathscr{E}\}$ be the class of all pointed event models. A *DEL protocol* (called a *uniform protocol* in [6]) is a set $\mathsf{P} \subseteq \mathbb{E}^*$ closed under the (non-empty) initial segment relations.[4] Given a DEL protocol P, let σ denote an element of P, i.e., σ is a sequence of pointed event models. We write σ_n for the initial segment of σ of length $n \leq \mathsf{len}(\sigma)$ and write $\sigma_{(n)}$ for the nth component of σ. For example, if $\sigma = (\mathscr{E}_1, e_1)(\mathscr{E}_2, e_2)(\mathscr{E}_3, e_3) \cdots (\mathscr{E}_n, e_n)$, then $\sigma_3 = (\mathscr{E}_1, e_1)(\mathscr{E}_2, e_2)(\mathscr{E}_3, e_3)$ and $\sigma_{(3)} = (\mathscr{E}_3, e_3)$. Given a sequence $\sigma \in \mathbb{E}^*$, we abuse notation and write $\mathsf{pre}(\sigma_{(n)})$ for $\mathsf{pre}(e_n)$ where $\sigma_{(n)} = (\mathscr{E}_n, e_n)$. Furthermore, we write $\sigma_{(n)} \longrightarrow_i \sigma'_{(n)}$ provided $\sigma_{(n)} = (\mathscr{E}, e)$ and $\sigma'_{(n)} = (\mathscr{E}, e')$ and $e \longrightarrow_i e'$ is in $\mathscr{E}$. Finally, let $Ptcl(\mathbb{E})$ be the class of all DEL protocols, i.e., $Ptcl(\mathbb{E}) = \{\mathsf{P} \mid \mathsf{P} \subseteq \mathbb{E}^*$ is closed under initial segments$\}$.

The main idea is to start from an initial (pointed) epistemic model and construct an ETL model by repeatedly applying product updates.

Definition 9.4 Given a pointed epistemic model $\mathscr{M}, w$ and a finite sequence of pointed event models σ, we define the σ-*generated epistemic model*, $(\mathscr{M}, w)^\sigma$ as $(\mathscr{M}, w) \otimes \sigma_{(1)} \otimes \sigma_{(2)} \otimes \cdots \otimes \sigma_{(\mathsf{len}(\sigma))}$. We will write $\mathscr{M}^\sigma$ for $(\mathscr{M}, w)^\sigma$ when the state w is clear from context.

Definition 9.5 Let $\mathscr{M}, w$ be a pointed epistemic model, and P a DEL protocol. The ETL model generated by $\mathscr{M}$ and P, Forest$(\mathscr{M}, \mathsf{P})$, represents all possible evolutions of the system obtained by updating $\mathscr{M}$ with sequences from P. More precisely, Forest$(\mathscr{M}, \mathsf{P}) = \langle \Sigma, \mathsf{H}, \{\sim_i\}_{i \in \mathbb{A}}, V \rangle$, where $\langle \mathsf{H}, \{\sim_i\}_{i \in \mathbb{A}}, V \rangle$ is the union of all models of the form $\mathscr{M}^\sigma$ with $\sigma \in \mathsf{P}$.

Since any DEL protocol P is closed under prefixes, for any epistemic model $\mathscr{M}$, Forest$(\mathscr{M}, \mathsf{P})$ is indeed an ETL model. Now, given a class of DEL protocols **X**, let

$$\mathbb{F}(\mathbf{X}) = \{\mathsf{Forest}(\mathscr{M}, \mathsf{P}) \mid \mathscr{M} \text{ an epistemic model and } \mathsf{P} \in \mathbf{X}\}$$

If $\mathbf{X} = \{\mathsf{P}\}$ then we write $\mathbb{F}(\mathsf{P})$ instead of $\mathbb{F}(\{\mathsf{P}\})$.

Note that not all ETL models can be generated by a DEL protocol. Indeed, such generated ETL models satisfy synchronicity, perfect recall and uniform no miracles (see Sect. 9.3.3 for definitions). The main result (Theorem 9.2 in this section is a characterization of the ETL models that are generated by some (uniform) DEL protocol. This is an improvement of an earlier characterization result from [12] and provides a precise comparison between the DEL and ETL frameworks.

Suppose that $\mathscr{H}$ is an ETL frame, which satisfies synchronicity, perfect recall and uniform no miracles. We can easily read off an epistemic *frame* with a set of states W and relations R_i for each agent $i \in \mathbb{A}$ on W, to serve as the initial model, where the histories of length 1 are the states and the uncertainty relations are simply

---

[4] The *preconditions of DEL* also encode protocol information of a 'local' character, and hence they can do some of the work of global protocols, as has been pointed out by van Benthem [12].

copied. Furthermore, we can define a "DEL-like" protocol $\mathsf{P}_{\mathscr{H}}$ with the construction given below in the proof of Theorem 9.2, consisting of sequences of event models where the precondition function assigns to the primitive events *sets of finite histories*. Intuitively, if e is a primitive event, i.e., a state in an event model, then $\mathsf{pre}(e)$ is the set of histories where e can "be performed". Thus, we have a comparison of the two frameworks at the level of frames provided we work with a modified definition of an event model. However, the representation theorem is stated in terms of models, so we need additional properties. In particular, at each level of the ETL model we will need to specify a *formula* of $\mathscr{L}_{EL}$ as a pre-condition for each primitive event e (cf. Definition 9.2). As usual, this requires that the set of histories preceding an event e be *bisimulation-closed* (see [20] for a definition of bisimulations and [6] for the precise definition needed here). However, as is well-known, bisimulation-invariance alone is typically not enough to guarantee the existence of such a formula. More specifically, there are examples of *infinite* sets that are bisimulation closed but not definable by any formula of $\mathscr{L}_{EL}$ (However, it will be definable by a formula of epistemic logic with *infinitary* conjunctions—see [20] for a discussion). Thus, if the set of histories at some level in which an event e can be executed is infinite, there may not be a formula of $\mathscr{L}_{EL}$ that defines this set to be used as a pre-condition for e. Such a formula will exist under an appropriate *finiteness assumption*: at each level there are only finitely many histories in which e can be executed,[5] i.e., for each n, the set $\{h \mid he \in H \text{ and } \mathsf{len}(h) = n\}$ is finite.

One final assumption is needed since we are assuming that product update does not change the ground facts. An ETL model $\mathscr{H}$ satisfies *propositional stability* provided for all histories h in $\mathscr{H}$, events e with he in $\mathscr{H}$ and all propositional variables P, if P is true at h then P is true at he. We remark that this property is not crucial for the results in this section and can be dropped provided we allow product update to change the ground facts (cf. [5]).

Theorem 9.2 (Representation Theorem) *Let* $\mathbf{X}_{DEL}$ *be the class of uniform DEL protocols. An ETL model* $\mathscr{H}$ *is in* $\mathbb{F}(\mathbf{X}_{DEL})$ *if and only if* $\mathscr{H}$ *satisfies propositional stability, synchronicity, perfect recall, uniform no miracles, and bisimulation invariance.*

Consult [6, Theorem 1] for the proof. Note that the finiteness assumption can be dropped at the expense of allowing preconditions to come from a more expressive language (specifically, infinitary epistemic logic). Alternatively, as remarked above, we can define the preconditions to be *sets* of histories, instead of formulas of some logical language.

In [25] Dégremont, Löwe, and Witzel provide an alternative merging approach by mapping an epistemic model and a protocol of pointed models to an epistemic temporal structure. The resulting epistemic temporal structure need not be synchronous, so the authors argue that synchronicity is not an inherent property of DEL, but

---

[5] Note that this property may be violated even in an ETL model generated from only finitely many events.

rather of the translation used in [6]. They provide a different translation that produces asynchronous ETL models and discuss a minimal temporal extension of DEL that removes the ambiguities between the possible translations. In this context, they discuss the question of which epistemic-temporal properties are intrinsic to DEL and which ones are properties of the translation.

9.5 The Dynamics of Knowledge and Abilities in Multi-Player Games

In this section we discuss a particular aspect of the dynamics of knowledge in multi-agent systems, effected in the course of playing (abstract) multi-player games. Games offer a variety on perspectives on this topic, and we refer the reader to the chapter [23] for some of them. A most important specific issue arising in multi-player games on which we focus here is the *interaction between knowledge and abilities* of players to achieve their objectives in the play of the game. This interaction has two equally important directions. On the one hand, the abilities of the players to guarantee achievement of their objectives (i.e., to have a winning strategy) crucially depend on their information about the game as such, and about the particular play of the game. On the other hand, the players' knowledge changes dynamically in the course of playing the game. That dynamic interaction crucially affect the players' abilities in the play the game.

Building on [37], here we will outline a logical framework, capturing the dynamics of the interplay between knowledge and abilities of players in multi-player games. In order to avoid having to deal with fundamental issues of the concept of knowledge, we hereafter prefer to use the more neutral, and at the same time more general, notion of *information*, which refers not only to the knowledge or uncertainties, but also to beliefs and confusions of the agents/players.

9.5.1 A Priori vs Empirical Information of Players

First, some brief terminological remarks. Traditionally, in Game Theory the notion of 'incomplete information' refers to the knowledge or uncertainties of the player about the *structure and rules of the game*, while the notion of 'imperfect' information' refers to the knowledge or uncertainties of the player about *the course of the play* of the game, e.g. about the state in which the game currently is, or the history of the play, or the moves/actions taken by other players. Instead, we introduce the notions of 'a priori information' and 'empirical information'. Intuitively, a player's *a priori information* is the information (incl. knowledge, beliefs and uncertainties) that the player has about the game as such, *prior* to the actual play of the game about its *rules, protocol,* or *structure*. On the other hand, *empirical information* refers to the information that

Fig. 9.2 Learning from experience

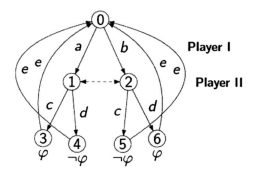

a player builds by way of observations, recollections, communication and reasoning made during the course of play. The a priori information plays its role only at the beginning of the game. Ultimately, it is the empirical information that determines the players' abilities in the game.

9.5.2 Some Examples

We now present several simple examples in order to illustrate the concepts of a priori information and the empirical information of a player. In all examples that follow we consider simple turn-based or concurrent games, played by two players, I and II, consisting in them making series of moves. States in the game are labelled with numbers, with 0 indicating the start state. Outcomes of the games are only qualitative, expressed in terms of the truth of certain propositions.

Example 9.1 (Learning from experience) In the game on Fig. 9.2 Player I moves first from state 0, then II moves, and then the game restarts from state 0. Player II has incomplete *a priori* information about the game: he is not able to distinguish a priori between states 1 and 2. For instance, that means that he cannot observe or recognize the actions *a* or *b* of Player I. For convenience we will usually refer to Player I as female and to Player II as male.

So, does Player II have the ability to eventually guarantee his desired outcome φ? If his uncertainty persists throughout the game, then clearly not. However, if Player II can observe the action of Player I and use some memory, he can learn from his experience: after choosing an action at random at the first round of the game if it does not lead to the desired outcome, then he can revise his strategy to achieve φ in the next round, by playing the other action if Player I repeats the same action and vice versa. Thus, the experience during the play can enrich the player's information and thus enhance his abilities.

Example 9.2 (Getting confused) In the game on Fig. 9.3 Player II has perfect a priori information about the game.

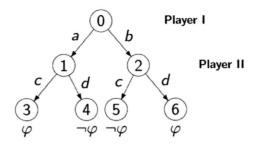

Fig. 9.3 Getting confused

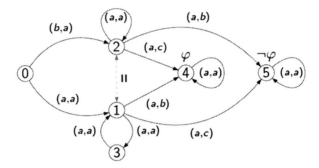

Fig. 9.4 Learning by experiments

So, does **II** have the a priori ability to guarantee the desired outcome φ? Yes, player **II** has a simple strategy for that. But, suppose that in the course of the actual play of the game, player **II** turns out unable to *observationally* distinguish states 1 and 2 after player **I** made her move, e.g., by failing to observe player **I**'s action, or due to malfunctioning sensors. If that happens, then Player **II** is no longer able to guarantee outcome φ. Thus, the experience in the play of the game can be negative, too and can lead to loss of information and abilities.

Example 9.3 (*Learning by experiments*) In the concurrent game on Fig. 9.4 essentially Player **I** determines the move from state 0, and thereafter **II** is in control. Player **II**'s objective is to reach a φ-state but she cannot distinguish a priori states 2 and 3 and cannot observe the action of Player **I** at state 0. However, Player **II** has all the information about the game which is provided on Fig. 9.4, except that he cannot see the labels 1 and 2.

So, does **II** can guarantee reaching the outcome φ? A priori, not. However, by performing a suitable experiment at the information set of states {2, 3} player **II** can generate sufficient *empirical information* to enable himself distinguish between these states, by following the strategy: play action a and observe the result. If that is state 2, then play c, and if it is state 3 (which player **II** can distinguish from state 2), then play a again, and then b.

Remark 9.2 Note that the reasoning described in the last example can be naturally regarded as 'a priori reasoning', as it may take place before any actual play of the game. However, one should distinguish the results of 'a priori reasoning' in this intuitive sense from the notion of 'a priori information' we use here. Since the results of the 'a priori reasoning' in this example applies to possible *plays* of the game, this information is 'empirical information' in the sense in which we use this term here.

In summary: players' information changes dynamically during the play of the game. The main problem arising here is: how to formally compute that dynamics?

9.5.3 Formalizing the a Priori and Empirical Information

Here we propose a formal modeling framework incorporating the a priori information and computing the empirical information of the players.

The models upon which the framework is built are variation of the concurrent game structures with incomplete information [2, 36, 46, 60]. The key extension to these structures consists in two 'information relations' per player. The first, called 'a priori information relation' relates one state to another if the player considers the second state a possible *structural alternative* for the first, i.e. a priori alternative in the game structure. This relation can also be used to represent structural uncertainties and beliefs. The second, called 'empirical information relation', relates a *run*, i.e., initial segment of a play of the game, to (possibly) another one which the player considers a possible 'observational' or 'empirical alternative' to the first one. This relation can be used to represent empirical uncertainties and beliefs arising in the course of the play.

9.5.3.1 Concurrent Game Structures

Definition 9.6 (*Concurrent game structure*) A *concurrent game structure* (CGS) is a tuple $\langle \mathbb{A}, Q, \text{Act}, d, \text{out} \rangle$, where:

- $\mathbb{A} = \{1, 2, \ldots, k\}$ is a finite set of players.
- Q is a non-empty set of *states*.
- $d : Q \times \mathbb{A} \longrightarrow 2^{\text{Act}}$ is a function that for every state q and a player i assigns the subset of actions available to player i at state q.
  The set of actions available to i at q will be denoted $\text{Act}_i(q)$.
  A *joint action* at a given state q, denoted by σ_q (or simply by σ when q is fixed by the context), is a tuple $(\alpha_1, \alpha_2, \ldots, \alpha_k)$, where $\alpha_i \in \text{Act}_i(q)$ for every $i \leq k$, consisting of a collection of actions, one for each player, that may be performed at state q. Given a joint action $\sigma = (\alpha_1, \alpha_2, \ldots, \alpha_k)$, we write σ^i to indicate α_i. We write $\text{Act}(Q)$ for the set of all joint actions from states in Q.
- $\text{out} : Q \times \text{Act}(Q) \to Q$ is the *transition function*, that maps a state q and a joint action at q to a unique *successor state* in Q. The set of all successor states of q

will be denoted by $succ(q)$. Thus, $q' \in succ(q)$ if there is a move vector σ such that out $(q, \sigma) = q'$.

One can think of a CGS as capturing the structure and the rules of a game.

Definition 9.7 (*Games, runs and plays*) A *game* is a pair $\langle \mathscr{S}, q \rangle$ consisting of a CGS $\mathscr{S}$ with a set of states Q and an *initial state* $q \in Q$. Given a game $\mathscr{G} = \langle \mathscr{S}, q \rangle$, a *play* of $\mathscr{G}$ is an infinite sequence $\lambda = q_0, \sigma_0; q_1, \sigma_1; q_2, \sigma_2 \ldots$ of alternating states and joint actions applied at them, such that $q_0 = q$ and $q_{i+1} = $ out (q_i, σ_i) for every $i \in \mathbb{N}$. A *(finite) run* in $\mathscr{G}$ is a finite initial segment of a play ending with a state: $q_0, \sigma_0; q_1, \sigma_1; \ldots; q_n$. One-state runs will be identified with the respective states.

A *q-play (resp. q-run)* is a play (resp. run) where $q_0 = q$. We denote the set of q-plays by $Play(q)$ and the set of all q-runs by $Run(q)$. We also denote the set of all plays in $\mathscr{S}$ by $Play(\mathscr{S})$ and the set of all runs in $\mathscr{S}$ by $Run(\mathscr{S})$.

Example 9.4 The game on Fig. 9.4 defines a CIGS $\mathscr{S} = \langle \mathbb{A}, Q, \mathsf{Act}, d, \mathsf{out} \rangle$, where:

- $\mathbb{A} = \{\mathbf{I}, \mathbf{II}\}$, $Q = \{0, \ldots, 7\}$, $\mathsf{Act} = \{a, b, c\}$.
- $d : Q \times \mathbb{A} \longrightarrow 2^{\mathsf{Act}}$ and out $: Q \times \mathsf{Act}(Q) \to Q$ are defined as on the figure, e.g.:
  $d(0, \mathbf{I}) = \{a, b\}, d(0, \mathbf{II}) = \{a\}$; out $(0; (a, a)) = 1$, out $(0; (b, a)) = 2$, etc.

Here are some plays in the game $(\mathscr{S}, 0)$:

- $0, (a, a); 1, (a, a); 3, (a, a); 1, (a, a); 3, (a, a) \ldots$
- $0, (a, a); 1, (a, c); 5, (a, a); 5, (a, a); 5, (a, a) \ldots$
- $0, (a, a); 1, (a, a); 3, (a, a); 1, (a, b); 4, (a, a); 4, (a, a) \ldots$
- $0, (b, a); 2, (a, a); 2, (a, a); 2, (a, c); 4, (a, a); 4, (a, a) \ldots$

9.5.3.2 Concurrent Informational Game Structures

Definition 9.8 (*Concurrent informational game structure*) A *concurrent informational game structure* (CIGS) is a tuple $\langle \mathscr{S}; \{\overset{a}{\leadsto}_i\}_{i \in \mathbb{A}}, \{\overset{e}{\leadsto}_i\}_{i \in \mathbb{A}} \rangle$, where $\mathscr{S} = \langle \mathbb{A}, Q, \mathsf{Act}, d, \mathsf{out} \rangle$ is a CGS and for every $\mathbf{i} \in \mathbb{A}$:

- $\overset{a}{\leadsto}_i \subseteq Q \times Q$ is a *a priori information relation* for the player $\mathbf{i}$;
- $\overset{e}{\leadsto}_i \subseteq Run(\mathscr{S}) \times Run(\mathscr{S})$ is an *empirical information relation* for the player $\mathbf{i}$, which coincides with $\overset{a}{\leadsto}_i$ when restricted to one-state runs.

The intuition: $q_1 \overset{a}{\leadsto}_i q_2$ holds if the player $\mathbf{i}$ considers the state q_2 as a possible *a priori* alternative of the state q_1 in the game structure $\mathscr{M}$. Likewise, $\rho_1 \overset{e}{\leadsto}_i \rho_2$ if the player $\mathbf{i}$ considers the run ρ_2 a possible alternative of the run ρ_1 in $\mathscr{M}$ at its last state. Initially, the empirical information of the player about a play is simply her a priori information about the game. As the play progresses, the player may on one hand gain some additional information about the structure of the game, and on the other hand may acquire some uncertainties or wrong beliefs about the history and the current state of the play. Here are some particular cases:

- If $\overset{a}{\leadsto}_i$ is the equality, then the player **i** has a complete (a priori) information about the game $\mathcal{M}$. Likewise, if $\overset{e}{\leadsto}_i$ restricted to the runs of a given play in $\mathcal{M}$ is the equality (i.e. does not associate any run of the play with any different run), then the player **i** maintains a perfect empirical information throughout that play.
- If the player **i** has no wrong beliefs, but only uncertainties about the game, then $\overset{a}{\leadsto}_i$ is an equivalence relation of *a priori indistinguishability*. Likewise, if the player **i** has no wrong beliefs, but only uncertainties about the states of a play, then $\overset{e}{\leadsto}_i$ is an equivalence relation of *empirical indistinguishability*.
- Furthermore, if the player can keep a count of the number of moves made in the play, or has a 'clock' showing how many time units have passed since the beginning of the play (where every time unit corresponds to one transition from a state to a successor state), then $\overset{e}{\leadsto}_i$ can only relate runs of the same length (recall the property of synchrony in temporal epistemic models). In general, however, it is conceivable that a player may 'forget' the length of the current run.
- In many cases it is reasonable to assume that the games considered are 'tree-like', where every state is associated with a unique run. Then the relations $\overset{e}{\leadsto}_i$ can be regarded as relations on states, rather than runs.

A more refined approach towards the different types of information and abilities would be to distinguish purely *observational* information acquired in the course of a play by only observing the current states through which the play goes from the accumulative empirical information which also involves recollections and reasoning.

Example 9.5 The game on Fig. 9.4 defines a CIGS $\mathcal{T} = \langle \mathcal{S}; \{\overset{a}{\leadsto}_i\}_{i\in A}, \{\overset{e}{\leadsto}_i\}_{i\in A} \rangle$, where $\mathcal{S}$ is the CGS defined in the previous example, and the a priori information relations are defined as follows:

- $\overset{a}{\leadsto}_I$ is the equality: $\{(q, q) \mid q \in Q\}$;
- $\overset{a}{\leadsto}_{II} = \{(q, q) \mid q \in Q\} \cup \{(1, 2), (2, 1)\}$.

The empirical relations:

- $\overset{e}{\leadsto}_I$ is the equality of runs.
- $\overset{e}{\leadsto}_{II}$ extends $\overset{a}{\leadsto}_{II}$ by relating every run to itself, but it also relates the run 0, (a, a); 1 with 0, (b, a); 2 and no other runs, because once the game is past any of these 2 runs, the a priori uncertainty of player **II** disappears (assuming that he has basic observational abilities and memory).

9.5.3.3 Computing the Empirical Information

Note that, while both sets of relations, $\overset{a}{\leadsto}$ and $\overset{e}{\leadsto}$ are part of the definition of a CIGS, only the former should be assumed to be given explicitly a priori, while the latter is to be computed in the course of the play. It is not possible to give a general rule of how the empirical relations of the players are computed, as that would depend on their observational abilities, memory, reasoning skills, etc. Computing the empirical information is one of the main problems in the development of this framework. Here we only outline a conceptual proposal for a mechanism computing the empirical information during the play of the game, as follows:

- Before the play begins, the empirical information of the players in the CIGS is their a priori information. It determines an "a priori" multi-agent epistemic model associated with the CIGS.
- Every transition in the CIGS generates an "information update model" à la DEL, which represents the epistemic updates for the payers generated by that transition.
- That update model is applied to the current epistemic model associated with the CIGS, to produce an updated epistemic model, which represents the empirical information relations between all runs of length being at most the length of the current history of the play.
- The players use the so obtained empirical information relations to determine their next actions, possibly following an 'empirical strategy' based on the empirical information represented by these relations.
- The collective action determines the next transition, and the cycle repeats.

The procedure of computing the information update models is the engine of the entire mechanism and depends on the abilities of the players to observe, memorize and recall, communicate, reason, etc. Some simple cases of that procedure are being developed in [35].

9.5.3.4 Logical Framework for Computing Empirical Strategic Abilities

Using the a priori and empirical relations, one can refine the notions of strategies and strategic abilities of players, underlying the semantics of the Alternating-time temporal logic ATL, [2] in order to distinguish between 'objective' abilities (what the player can achieve *if* they had perfect information), 'a priori' abilities (what the player can achieve based on her a priori information about the game), and 'empirical' abilities (what the player can achieve given that the player can take advantage of, or suffer disadvantage from, experience of actual play). Furthermore, these can be used to provide a formal semantics of an enrichment of ATL with incomplete information, with separate operators for stating objective, a priori, and empirical abilities of players and coalitions. For further detail on all these we refer the reader to [37] and the work in preparation [35].

9.6 Putting the Temporal, Dynamic and Epistemic Frameworks Together

Besides [17] and [6], a few other publications have appeared recently that propose combining temporal, dynamic and epistemic frameworks. We briefly survey the more popular of them here.

In his Ph. D. thesis [57] and in the subsequent chapter [58] Sack combines temporal logic with public announcement logic (PAL) and dynamic epistemic logic (DEL). Adding next-time and previous-time operators to PAL allows formalizing the muddy children and the 'sum and product' puzzles. He also discusses relationships between the announcements and the new knowledge that agents acquire. Adding a full past-time operator to DEL also helps obtaining a complete axiomatization. In [59] Sack proposes a new version of temporal DEL (TDEL) with (mostly) unparametrized past operators in a language with DEL-action signatures. This TDEL does not involve protocols and the update modality semantics explicitly changes the epistemic temporal structure.

In [47] Hoshi and Yap consider a version of temporal DEL (TDEL) with a parametrized past operator. In order to axiomatize that extension, they develop transformation a given model into a certain normal form. The authors suggest further applications of such extensions of DEL to the theories of agency and learning.

In [55] Renne, Sack, and Yap introduce a new type of arrow in the DEL action models in order to enable reasoning about epistemic temporal dynamics in multi-agent systems that need not be synchronous. Their framework provides a new perspective on the work in [6], in particular, while in each of the two approaches the epistemic temporal models generated by standard update frames necessarily satisfy certain structural properties such as synchronicity, [55] discusses which these structural properties are due to the inherent structure of the update models themselves. In the extended version [56] they relate DETL and TDEL and provide a completeness theorem for DETL with respect to well-behaved epistemic temporal models.

In [27] van Ditmarsch, van der Hoek and Ruan discuss a relation between DEL with the usual semantics on relational models, and a temporal epistemic logic with semantics in interpreted systems à la [30]. In particular, from a given 'epistemic state', i.e. pointed epistemic model and a DEL formula they construct an interpreted system that satisfies the translation of the formula in the respective temporal epistemic logic.

9.7 Concluding Remarks

The dynamics of agents' knowledge is a conceptually rich, deep and multi-faceted topic. Here we have discussed only some aspects of that dynamics, mainly related to its temporality rather than its causes and effects. Furthermore, we have focused on the dynamics of *knowledge* rather than other informational attitudes, such as *beliefs*. Consult [22, 24] and the chapter [34] for a discussion of the temporal aspects of beliefs.

In summary, while a number of models and logics have been proposed in the past. This chapter has shown that there is much more in common between these different logical systems than once thought. Yet, a "Unified Theory" of the dynamics of knowledge over time is still to be developed, if ever. However, our overall goal was not to argue that any one framework is the "right model", or even that there is a single such "Unified Theory", but rather that there is a coherent collection of logical systems each focused on modeling the dynamics of knowledge from a different perspective. Actually, we are more inclined to believe that there is no unique model, not even unique 'right' methodology for modeling that dynamics, but that the pluralism of relevant approaches is its inherent valuable feature.

This pluralistic viewpoint of ours, together with the natural limitations of time and space in which this chapter had to be placed, are our excuses for leaving untouched a number of relevant studies and approaches, including: learning theory, interactive epistemology, situation calculus, etc. We do, however, refer the unslaked reader to Johan van Benthem's recent collection of though-provoking essays on related topics in [14, 16].

Acknowledgments We thank Johan van Benthem, Joshua Sack and Dominik Klein for useful comments and references. Valentin Goranko's work on this chapter was completed during his sabbatical visit to the Centre International de Mathématiques et Informatique de Toulouse. Eric Pacuit's work was supported by an NWO Vidi Grant 016.094.345.

References

1. Ågotnes T, van Ditmarsch H (2014) Knowledge games and coalitional abilities. In: Baltag A, Smets S (eds) Johan van Benthem on Logic and Information Dynamics, Springer, Dordrecht, pp 451–485 (Chapter 16 in this volume)
2. Alur R, Henzinger TA, Kuperman O (2002) Alternating-time temporal logic. J ACM 49(5):672–713
3. Baltag A, Moss L, Solecki S (1998) The logic of common knowledge, public announcements and private suspicions. In: Gilboa I (ed) Proceedings of the 7th conference on theoretical aspects of rationality and knowledge (TARK 98), pp 43–56
4. van Benthem J (1984) Tense logic and time. Notre Dame J. Formal Logic 25:1–16
5. van Benthem J, van Eijck J, Kooi B (2006) Logics of communication and change. Inf Comput 204(11):1620–1662
6. van Benthem J, Gerbrandy J, Hoshi T, Pacuit E (2009) Merging frameworks of interaction. J Philos Logic 38(5):491–526
7. van Benthem J (1993) The logic of time, 2nd edn. Kluwer Academic Publishers, Dordrecht
8. van Benthem J (1995) Temporal logic, Oxford University Press, Oxford, pp 241–350
9. van Benthem J (1996) Exploring logical dynamics. Center for the study of language and information, Stanford
10. van Benthem J (2001) Games in dynamic-epistemic logic. Bull Econ Res 53(4):219–248
11. van Benthem J (2001) Logics for information update. In: van Benthem J (ed) Proceedings of TARK 2001, Morgan Kaufmann, San Francisco, pp 51–67
12. van Benthem J (2002) 'One is a lonely number': on the logic of communication. In: Chatzidakis Z, Koepke P, Pohlers W (eds) Logic colloquium '02, pp 96–129. ASL and A. K. Peters, Available at http://staff.science.uva.nl/~johan/Muenster.pdf
13. van Benthem J (2004) What one may come to know. Analysis 64(2):95–105
14. van Benthem J (2011) Logical dynamics of information and interaction. Cambridge University Press, Cambridge
15. van Benthem J (2013) Reasoning about strategies. In: Coecke B, Ong L, Panagaden P (eds) Abramsky festrschrift, LNCS, vol 7860, pp 336–347
16. van Benthem J (2013) Logic in games. The MIT Press, Cambridge
17. van Benthem J, Pacuit E (2006) The tree of knowledge in action: towards a common perspective. In: Advances in modal logic, pp 87–106

18. van Benthem J, Pacuit E (2014) Connecting logics of choice and change. In: Nuel Belnap on indeterminism and free action, outstanding contributions to logic, vol 2, pp 291–314
19. Bezhanishvili N, van der Hoek W (2014) Structures for epistemic logic. In: Baltag A, Smets S (eds) Johan van Benthem on Logic and Information Dynamics, Springer, Dordrecht, pp 339–380 (Chapter 12 in this volume)
20. Blackburn P, de Rijke M, Yde V (2002) Modal logic, Campbridge University Press, Campbridge
21. Bonanno G (2004) Memory and perfect recall in extensive games. Games Econ Behav 47:237–256
22. Bonanno G (2012) Belief change in branching time: AGM-consistency and iterated revision. J Philos Logic 41(1):201–236
23. Bonanno G, Dégremont C (2014) Logic and game theory. In: Baltag A, Smets S (eds) Johan van Benthem on Logic and Information Dynamics, Springer, Dordrecht, pp 421–449 (Chapter 15 in this volume)
24. Dégremont C (2010) The temporal mind. Observations on the logic of belief change in interactive systems. Ph. D. thesis, Institute for Logic, Language and Computation (DS-2010-03)
25. Dégremont C, Löwe B, Witzel A (2011) The synchronicity of dynamic epistemic logic. In: Proceedings of TARK'2011, pp 145–152
26. van Ditmarsch H, van der Hoek W, Kooi B (2008) Dynamic epistemic logic. Springer, Dordecht
27. van Ditmarsch H, van der Hoek W, Ruan J (2012) Connecting dynamic epistemic and temporal epistemic logics. J IGPL (to appear)
28. van Eijck J (2014) Dynamic epistemic logics. In: Baltag A, Smets S (eds) Johan van Benthem on Logic and Information Dynamics, Springer, Dordrecht, pp 175–202 (Chapter 7 in this volume)
29. Engelfriet J (1996) Minimal temporal epistemic logic. Notre Dame J Formal Logic 37(2):233–259
30. Fagin R, Halpern JY, Moses Y, Vardi MY (1995) Reasoning about knowledge. The MIT Press, Boston
31. Fagin R, Halpern JY, Vardi MY (1992) What can machines know? On the properties of knowledge in distributed systems. J ACM 39(2):328–376
32. Finger M, Gabbay DM (1992) Adding a temporal dimension to a logic system. J Logic, Lang Inf 1:203–233
33. Gabbay DM, Kurucz A, Wolter F, Zakharyaschev M (2003) Many-dimensional modal logics: theory and applications, Vol 148. Elsevier, Amsterdam
34. Girard P, Rott H (2014) Belief revision and dynamic logic. In: Baltag A, Smets S (eds) Johan van Benthem on Logic and Information Dynamics, Springer, Dordrecht, pp 203–233 (Chapter 8 in this volume)
35. Goranko V (2012) Modelling the dynamics of knowledge and strategic abilities of players in multi-player games. In: Work in progress, Technical University of Denmark, Kongens Lyngby
36. Goranko V, Jamroga W (2004) Comparing semantics of logics for multi-agent systems. Synthese 139(2):241–280
37. Goranko V, Hawke P (2010) On the dynamics of information and abilities of players in multi-player games. In: Proceedings of LOFT'2010
38. Goranko V, Shkatov D (2009) Tableau-based decision procedure for full coalitional multiagent temporal-epistemic logic of linear time. In: Decker K, Sichman J, Sierra C, Castelfranchi C (eds) Proceedings of AAMAS'2009
39. Goranko V, Shkatov D (2009) Tableau-based decision procedure for the full coalitional multi-agent logic of branching time. In: Baldoni et al M (eds) MALLOW'2009
40. Halpern JY, Vardi MY (1988) Reasoning about knowledge and time in asynchronous systems. In: Simon J (ed) Proceedings of STOC'1988, ACM, pp 53–65
41. Halpern JY (1987) Using reasoning about knowledge to analyze distributed systems. Ann Rev Comput Sci 2:37–68
42. Halpern JY, Vardi MY (1989) The complexity of reasoning about knowledge and time i: lower bounds. J Comput Syst Sci 38(1):195–237
43. Halpern JY, Moses Yoram (1990) Knowledge and common knowledge in a distributed environment. J ACM 37(3):549–587

44. Halpern JY, van der Meyden R, Vardi M (2004) Complete axiomatizations for reasoning about knowledge and time. SIAM J Comput 33(2):674–703
45. Hintikka J (1962) Knowledge and belief: an introduciton to the logic of the two notions. Cornell University Press, Ithaca
46. van der Hoek W, Wooldridge M (2003) Cooperation, knowledge, and time: alternating-time temporal epistemic logic and its applications. Stud Logica 75(1):125–157
47. Hoshi T, Yap A (2009) Dynamic epistemic logic with branching temporal structures. Synthese 169(2):259–281
48. van der Meyden R, Wong K-S (2003) Complete axiomatizations for reasoning about knowledge and branching time. Stud Logica 75(1):93–123
49. van Otterloo S (2005) A strategic analaysis of multi-agent protocols. Ph. D. thesis, Institute for Logic, Language and Computation
50. Pacuit E (2007) Some comments on history based structures. J Appl Logic 5(4):613–624
51. Pacuit E, Simon S (2012) Reasoning with protocols under imperfect information. Rev Symbolic Logic 4:412–444
52. Parikh R, Ramanujam R (2003) A knowledge based semantics of messages. J Logic, Lang Inf 12:453–467
53. Parikh R, Ramanujam R (1985) Distributed processes and the logic of knowledge. In: Logic of programs. Lecture notes in computer science, Vol 193. Springer, New York, pp 256–268
54. Prior AN (1967) Past, present and future. Oxford University Press, Oxford
55. Renne B, Sack J, Yap A (2009) Dynamic epistemic temporal logic. In: Proceedings of LORI, pp 263–277
56. Renne B, Sack J, Yap A (2010) Dynamic epistemic temporal logic
57. Sack J (2007) Adding temporal logic to dynamic epistemic logic. PhD thesis, Indiana University, Bloomington
58. Sack J (2008) Temporal languages for epistemic programs. J Logic, Lang Inf 17(2):183–216
59. Sack J (2010) Logic for update products and steps into the past. Ann Pure Appl Logic 161(12):1431–1461
60. Schobbens P-Y (2004) Alternating-time logic with imperfect recall. Electron Notes Theoret Comput Sci, 85(2):82–93

Chapter 10
Logic and Learning

Nina Gierasimczuk, Vincent F. Hendricks and Dick de Jongh

Abstract Learning and learnability have been long standing topics of interests within the linguistic, computational, and epistemological accounts of inductive inference. Johan van Benthem's vision of the "dynamic turn" has not only brought renewed life to research agendas in logic as the study of information processing, but likewise helped bring logic and learning in close proximity. This proximity relation is examined with respect to learning and belief revision, updating and efficiency, and with respect to how learnability fits in the greater scheme of dynamic epistemic logic and scientific method.

The research of Nina Gierasimczuk is funded by an Innovational Research Incentives Scheme Veni grant 275-20-043, The Netherlands Organisation for Scientific Research (NWO).

N. Gierasimczuk (✉)
Institute for Logic, Language and Computation,
University of Amsterdam, Science Park 107,
1098 XG Amsterdam, The Netherlands
e-mail: nina.gierasimczuk@gmail.com

D. de Jongh
Institute for Logic, Language and Computation,
University of Amsterdam, Science Park 107,
1098 XG Amsterdam, The Netherlands
e-mail: d.h.j.dejongh@uva.nl

V. F. Hendricks
Department of Media, Cognition and Communication,
University of Copenhagen, Njalsgade 80,
2300 Copenhagen S, Denmark
e-mail: vincent@hum.ku.dk

A. Baltag and S. Smets (eds.), *Johan van Benthem on Logic and Information Dynamics*, Outstanding Contributions to Logic 5, DOI: 10.1007/978-3-319-06025-5_10, © Springer International Publishing Switzerland 2014

10.1 Learning and the Dynamic Turn in Logic

For well over a decade, Johan van Benthem has been pushing the agenda of the *dynamic turn* in logic forward:

> ...over the past decades computer science has also begun to influence the research agenda of logic. Traditionally, logic is about propositions and inference. Its account of this is declarative, in terms of languages and semantic models that represent information. But inference is in the first place an information-generating *process*, and just one among many at that. [...]These days, in the same spirit, modern logic is undergoing a *Dynamic Turn*, putting activities of inference, evaluation, belief revision or argumentation at centre stage, not just their products like proofs or propositions. [6, p. 503]

The classical conception of logic as the study of propositions, valid arguments, and information representation may be extended to logic as the study of inference broadly conceived and of correct information processing. Once this step is made, logic may serve as the gateway for studying, modelling and optimizing belief revision processes, strategies in games, procedures for decision, deliberation and action, rational agent interaction, and . . . learning [10].

The dynamic turn from deduction and representation to active inference and information-generating processes comes from the influence computer science has exercised on logic. But from computer science come also the first ideas of formal learning theory. The concept of identification in the limit has been introduced as a computational counterpart of the process of language acquisition [25]. It inspired a group of mathematicians and computer scientists, and led to a number of results concerning (learning of) recursively enumerable sets. This culminated in the book *Systems that Learn* (Osherson et al. [44], later extended to Jain et al. [31]). From the perspective of linguistics, a promising line was given by Angluin [2] to Gold's scheme bending it to learning recursive languages generated by traditional types of grammars like context-free grammars.

In general, formal learning theory is about *reliable processes* for information acquisition as Kevin T. Kelly, explains:

> A learning problem specifies (1) what is to be learned, (2) a range of relevant possible environments in which the learner must succeed, (3) the kinds of inputs these environments provide to the learner, (4) what it means to learn over a range of relevantly possible environments, and (5) the sorts of learning strategies that will be entertained as solutions. A learning strategy solves a learning problem just in case it is admitted as a potential solution by the problem and succeeds in the specified sense over the relevant possibilities. A problem is solvable just in case some admissible strategy solves it. [34, p. 1]

"Learning problem", "possible environments", "learner", "success", "strategies" and "solvability" sound like computer science terms, but they feature prominently in the dynamic turn of logic as well. And for good reason too. Formal learning theory takes its point of departure with the problem of finding true or empirically adequate, general theories from an ongoing stream of particular, empirical data. The basic idea is to seek procedural justification in terms of reliable truth-tracking performance, rather than in philosophical intuition or other more or less unregimented prescriptions for scientific inquiry—reliable or not. For example, one of the first publications in formal learning theory involved Putnam's computational critique of the learning power of Carnap's confirmation theory [47].[1] Putnam's thread has been taken up by Glymour, Kelly, Schulte and again Osherson et al., and has been applied to more traditional epistemological issues. Such issues include explications of empirical underdetermination and simplicity, critiques of Bayesianism, Ockham's razor, justification of inductive inference, causal discovery, belief revision, and epistemic logic.

Many of these concerns and applications are congruent with the agenda of the dynamic turn in logic. By way of example, the axioms of belief revision [20] may be interpreted as prescriptions for methods of learning [35, 40] for which both their respective learning powers and the relative merits in terms of efficiency and speed may be assessed [32]. How belief revision fairs with respect to learning understood as conditioning and lexicographic revision (central to dynamic epistemic logic) has been investigated in [4, 22]. Similarly, the classical axioms of epistemic logic may be viewed as epistemic learning goals. The learning problem to be settled is then what sort of learners will be able to converge to, and in what sense, the validity of these axioms. Bridging logic and learning is about adding a long-run perspective to epistemic logic in which agents are taken to be mechanisms that learn over time. This is achieved by merging branching alethic-temporal logic with possible environments, learners, success, and strategies, all concepts from formal learning theory. On top of the model of all branching empirical data streams a formal language is introduced that includes epistemic modalities whose indices are learning mechanisms. The idea is then to look for reflections between epistemic axioms in the logic and the structural features of the learning mechanisms [27, 28]. Pursuing the line of formal languages one can formulate conditions for limiting learning in dynamic epistemic and doxastic logic [21].

In general, the connection between formal learning theory and dynamic epistemic logic benefits both paradigms. On the one hand, learning theory receives the fine-structure of well-motivated local learning actions and qualitative logical perspective, which in the long run offers a chance of generic reasoning calculi about inductive learning. On the other hand, dynamic epistemic logic gets a long-term 'horizon' which it missed, criteria for choosing appropriate update rules, and adequate learnability conditions.

Logic and learning are now being brought into close proximity. A decade ago these close encounters were already on the horizon of van Benthem's vision of the dynamic turn uniting logic, computation and learning:

---

[1] The terms "identifiability", "learnability", and "solvability" are often used interchangeably in formal learning theory. Preferring one over the others is usually determined by the wider, often philosophical, methodological, or technical context. "Identifiability" is used in technical contexts, concerned with choosing (identifying) one among many possibilities (e.g., Turing machines or grammars). "Learnability" is a broader quasi-psychological notion often assumed to be (accurately) modelled by identifiability. Finally, "solvability" occurs in more logic-oriented works, and denotes the possibility of deciding on an issue, e.g., whether a hypothesis is true or false. Obviously, the latter can also be viewed as a kind of identifiability.

Update, revision, and learning form a coherent family of issues, going upward from short term to long-term behaviour. [6, p. 510]

No better occasion than this to discuss how exactly these central concepts from logic and learning fit together; belief revision and reliability, updating and efficiency, epistemic logic in relation to expressibility and logic and scientific method.

10.2 Belief Revision and Learning

Consider the following scenario. An agent faces uncertainty about the actual state of affairs. She wants to come up with a conjecture that (if not completely, then at least substantially) describes the phenomenon she is confronted with. The progress of this inquiry is driven not only by internal deliberations, but also by observations, outcomes of experiments, those performed by herself and those communicated to her by others. The incoming information triggers occasional changes in her beliefs. It seems natural to assume that we, humans, are naturally equipped with cognitive mechanisms that make such changes possible. The way to mathematically model these mechanisms arises via adopting a high-level perspective of studying the long term belief evolution and its effects—studying not only learning, but also the possible *success* of the learning process. In this context several simple questions have been intensively researched. How to distinguish some policies of changing-ones-mind as more "desirable" than others? How *reliable* are possible belief-revision and knowledge-update policies?

Logical theories of belief revision construct models for belief states in ways that make the latter amenable to changes triggered by appropriately represented information. They propose ways in which the new information gets incorporated into and changes an agent's belief state. Several such theories have already been investigated in light of inductive inference. Here are three note-worthy attempts, of which the most recent one [4, 22] will be dealt with in more detail. Of the other two attempts the first one uses formal learning theory to evaluate belief revision policies and is due to [39, 40]. They rely on a first-order framework for inductive inquiry and within this setting a special class of learners that mimic a belief-revising agent is introduced. The belief revision procedure is that of the AGM paradigm [1], and thus contraction driven. It has been demonstrated, among other things, that a revision method that strongly resembles AGM revision is not universal, i.e., there are problems that are solvable (learnable) in the limit, but cannot be solved by any AGM-learner. The second approach [32, 33, 35] is concerned with the reliability of some belief revision policies, this time for the possible worlds interpretations of belief, given by a variety of authors [14, 15, 26, 43, 48]. The inductive inquiry framework adopted here is that of prediction: the successive data received by the agent are true reports of successive outcomes of some discrete, sequential experiment. The goal of learning is to arrive at a sufficiently informative belief state that allows predicting how the sequence will evolve in the unbounded future. The investigation of the learning power of the procedures listed above indicates that the simple conditioning-based revision may be found among the most powerful.

In the remainder of this section focus is on some examples of belief revision methods and their convergence properties—the truth-tracking power. Before going here attention is directed towards a specification of the basic setting. The remaining part of this section summarises the results of [4, 22], the reader is referred to those for more details and proofs.

10.2.1 Epistemic Spaces, Belief Revision, and Learning

An agent's uncertainty is, as usual, represented by an epistemic space (S, Φ) consisting of a set S of epistemic possibilities, or possible worlds, together with a family of propositions $\Phi \subseteq \mathscr{P}(S)$. As in epistemic logic these propositions represent facts or observables being true or false in any of the possible worlds under consideration. The agent will receive information about a possible world (the actual one), and this stream of data is modelled as an open-ended (infinite) sequence of propositions. For now, abstract away from any time or memory restrictions, so it is assumed that the information keeps on arriving indefinitely in a piecemeal fashion. Such an infinite stream $\varepsilon = (\varepsilon_1, \varepsilon_2 \ldots)$ of successive propositions from Φ will be called a stream for $s \in S$ just in case the set $\{\varepsilon_n : n \in \mathbb{N}\}$ of all propositions in the stream coincides with the set $\{P \in \Phi : s \in P\}$ of all propositions that are true in the given world.

Given such representation a learning method is a function L that on input of an epistemic space (S, Φ) and a finite sequence of observations $\sigma = (\sigma_0, \ldots, \sigma_n)$ outputs a hypothesis. The hypothesis is then a set of possible worlds, i.e., a proposition. In other words, $L((S, \Phi), \sigma) \subseteq S$. Now we can define a condition of learning that closely resembles identification in the limit.

Definition 10.1 Let us take an epistemic space (S, Φ).

A world $s \in S$ is *learnable in the limit by a method L* if, for every observational stream ε for s, there exists a finite stage n such that $L((S, \Phi), \varepsilon_0, \ldots, \varepsilon_k) = \{s\}$ for all $k \geq n$.

The epistemic space (S, Φ) is said to be *learnable in the limit by L* if all its worlds are learnable in the limit by L.

Finally, the epistemic space (S, Φ) is *learnable in the limit* just in case there is a learning method that can learn it in the limit.

The above notion of learning is additionally motivated by the fact that the epistemic state resulting from a *successful* learning process need not be as strong as irrevocable knowledge, i.e., the S5 type of knowledge. It rather matches the defeasible type of knowledge proposed by Lehrer [37, 38] and others, formalized by Stalnaker [49] and rediscovered in modal logics under the name of 'safe belief'. The strength of safety is in the guarantee that it provides: a safe belief is not endangered by new veritistic observations. In other words, defeasible knowledge emerges when stability is reached. The need for such a notion appeared in many different frameworks: from reaching an agreement in a conversational situation (see, e.g., [37, 38]) to considerations in philosophy of science pertaining to infallible scientific knowledge (see, e.g., [28]).

The above described learning method outputs conjectures aiming at one that would uniquely describe the actual world. Such a definition does not however give any insight into the details of the underlying deliberation process—what makes the learning method choose one conjecture over another? To address this question we will, in a manner of speaking, be 'plugging-in a belief-revision engine'.

In order to make an epistemic space account for beliefs one may enrich it with a plausibility order. The idea is that although remaining uncertain between several options, the agent holds some of them most entrenched—those seem to her simply more plausible or more probable, or more elegant than other options—and hence she *believes* in what is true in all those best states. We set a plausibility space to be $(S, \Phi, \leq)$, where $\leq$ is a total preorder on S, called plausibility order.

Three qualitative belief revision policies have received substantial attention in dynamic epistemic logic: conditioning, lexicographic revision, and minimal revision. In particular, the first may be related (via its eliminative nature) to public announcement logic [46], the remaining two have been given a logical treatment and a complete axiomatisation by van Benthem [8].

Definition 10.2 Take a plausibility space $(S, \Phi, \leq)$ and a proposition $p \in \Phi$. Below we will call any $s \in p$, a 'p-world'.

1. *Conditioning* of the plausibility space $(S, \Phi, \leq)$ with the proposition p results in removing all inconsistencies with p, i.e., the operation gives a new plausibility space $(S', \Phi', \leq')$, where S' includes only the p-worlds and Φ' as well as $\leq'$ are cut down to the new domain, S'.
2. *Lexicographic revision* of the plausibility space $(S, \Phi, \leq)$ with the proposition p results in keeping the same states in S but promoting all the p-worlds to be more plausible than all those that are not p-worlds, and within the two clusters the order remains unchanged.
3. *Minimal revision* of the plausibility space $(S, \Phi, \leq)$ with the proposition p results in promoting the most plausible p-worlds to be the most plausible overall, the rest of the order remaining the same. As in the case of lexicographic revision, S stays the same throughout the process.

A belief-revision method is then a function R that upgrades a plausibility state, i.e., it associates to any plausibility space $(S, \Phi, \leq)$ and any sequence $\sigma = (\sigma_0, \ldots, \sigma_n)$, some new plausibility space $R((S, \Phi, \leq), \sigma) := (S^\sigma, \Phi^\sigma, \leq^\sigma)$, with $S^\sigma \subseteq S$, $\Phi^\sigma = \{P \cap S^\sigma : P \in \Phi\}$, and $\leq^\sigma$ is $\leq$ revised by method R under sequence σ.

Now we are ready to merge the learning function and the revision function into a belief revision based learning method. First, take an epistemic space (S, Φ) together with a prior-plausibility assignment given by some $\leq_S$. From the resulting plausibility space and a belief revision function R obtain in a canonical way the learning method L^R, given by:

Table 10.1 Universality of belief revision policies under different kinds of data

	Conditioning	Lexicographic	Minimal
Positive	Yes	Yes	No
Positive and negative	Yes	Yes	No
Fair	No	Yes	No

$$L^R((S, \Phi), \sigma) := \min\ R((S, \Phi, \leq_S), \sigma),$$

where $\min(S, \Phi, \leq)$ is defined to be the set of all the least elements of S with respect to $\leq$ (if such least elements exist) or $\emptyset$, otherwise.

Definition 10.3 An epistemic space (S, Φ) is *learnable in the limit by a belief-revision method R* if there exists some prior plausibility order $\leq_S$ such that (S, Φ) is learnable in the limit by the canonical learning method $L^R(S, \Phi, \leq_S)$.

Learning methods differ in their learning power. One may investigate the issue of the learnability range by looking for the most powerful among them, those that are universal—those that can learn any epistemic state that is learnable by any other method.

10.2.2 Learning Power of Belief Revision

The results of universality of the aforementioned belief revision policies are summarised in Table 10.1. For the sake of completeness, here we report on the universality of learning by belief revision policies under three different conditions. The first one is learning from streams of positive data, which for any possible worlds s enumerate only propositions true in s. The second is learning from streams of positive and negative data, where data streams enumerate propositions and negations of propositions true in s. Finally, fair streams represent unfriendly conditions—when some observational errors may occur. For this, we give up soundness of data streams, i.e., the condition that in the data stream for a possible world s only the information that is true in s can occur, and replace it by a "fairness" assumption: errors occur only finitely often and are always eventually corrected. Unsurprisingly, this can be destructive for conditioning. If erroneous observations are possible, then eliminating worlds that do not fit the observations is risky business.

For the above-listed universality results a non-standard setting, allowing non-well-founded plausibility orders, is essential, i.e., neither of those methods is universal with respect to well-founded prior plausibility orders.[2]

---

[2] In general some of the proofs require a construction of an appropriate prior plausibility order. For this some classical learning-theoretic concepts and results are used, i.e., locking sequences introduced by Blum and Blum [13], as well as finite tell-tale sets and the simple non-computable version of Angluin's theorem [2], see also the next section.

The results summarized in this section and developed further in the original work provide additional insight with respect to the motivation for belief-revision operators in epistemic logic. In particular, the different capabilities conditioning and lexicographic have to deal with errors lends formal justification to the intuition that the intention of performing lexicographic revision means having less trust in the source of information. On the other hand, the results on minimal revision challenge its popular characteristics as the safest revision policy. Moreover, the most popular approach to modelling beliefs in possible worlds semantics, namely by guaranteeing the well-foundedness of the underlying preorders turn out to be restrictive for learnability.

The above setting and its results are learning-theoretic in spirit, but they also contribute to the study of truth-tracking and truth approximation within the dynamic epistemic logic tradition. The inductive inference perspective leads to studying new relevant features of iterated revision: data-retention, conservatism, history-independence and ways in which these influence the learning process (see [4, 22]). Note that some limit phenomena within iteration scenarios in doxastic-epistemic logic have been studied before in the context of game theory, involving plausibility changes in games in a learning process with active agents trying to both 'learn' and 'teach' (see [9]), and in the context of belief revision, where one can observe a trade-off between initial plausibility order and plausibility order built up from local cues during the learning process (see [5]).

10.3 Conclusive Update and Efficiency

In formal learning theory the particular way of learning is not prescribed but usually supposed to obey certain computability constraints whereas in dynamic epistemic logic there are intuitively clear, determinate manners in which models are updated disregarding computability. It is interesting to compare the two aspects of determinateness and computability. In the previous section we introduced the basic (non-effective) version of convergence. In formal learning theory, learning is commonly studied as an effective procedure and learners are taken to be recursive functions.

Let us see how this may work in the present setting. Again, consider the epistemic space (S, Φ). Firstly, assume that it can consist of at most countably many possibilities in S and countably many relevant propositions given in the set Φ. Moreover, we will introduce the condition of uniform decidability of the epistemic space. An epistemic space is uniformly decidable just in case there is a computable function f that for each pair consisting of a possible world and a proposition decides whether the proposition is true or false in the possible world.

Definition 10.4 An epistemic space (S, Φ) is *uniformly decidable* just in case there is a computable function $f : S \times \Phi \to \{0, 1\}$ such that:

$$f(w, p) = \begin{cases} 1 & \text{if } s \in p, \\ 0 & \text{if } s \notin p. \end{cases}$$

In epistemic logic it is common to assume that checking whether or not an atomic proposition holds within a possible world is treated as primitive and its complexity is left out. Hence, the assumption of uniform decidability does not seem to be restrictive with respect to the traditional setting. It does however seem non-trivial in the analysis of some epistemic situations, e.g., scientific scenarios, where performing such an atomic test may be hard. This simple and appealing condition is used to investigate the properties of convergence to knowledge.

Since the overall number of possibilities is at most countable we can name them with natural numbers. Similarly, the set of propositions Φ is countable. So, assume that for an epistemic space (S, Φ), $S = \{s_1, s_2, s_3, \ldots\}$ and $\Phi = \{p_1, p_2, p_3, \ldots\}$. In this context it is easy to see that the function f that gives the uniform decidability of an epistemic space can may thought of as a number theoretic function $f : \mathbb{N} \times \mathbb{N} \to \{0, 1\}$. Throughout this section unless specified otherwise, assume the epistemic spaces to be uniformly decidable. Moreover, for sake of simplicity assume that in S there are no multiple worlds that make exactly the same propositions true (this assumption is not essential, see [23]).

In this new setting one may easily define the effective version of learnability in the limit.

Definition 10.5 Take an epistemic space (S, Φ).

A world $s_m \in S$ is *effectively learnable in the limit by a function L* if L is recursive and for every observational stream ε for s, there exists a finite stage n such that $L((S, \Phi), \varepsilon_0, \ldots, \varepsilon_k) = \{m\}$ for all $k \geq n$.

The epistemic space (S, Φ) is said to be *effectively learnable in the limit by L* if L is recursive and all the worlds in S are learnable in the limit by L.

Finally, the epistemic space (S, Φ) is *effectively learnable in the limit* just in case there is a recursive learning function that can learn it in the limit.

The remaining part of this section summarizes the results of Gierasimczuk [22], Gierasimczuk and de Jongh [23], the reader should consult those for proofs and a more detailed and rigorous presentation.

10.3.1 Conclusive Update

The above notion of learnability in the limit guarantees the existence of a method that allows for *convergence* to a correct hypothesis. Observe, that the exact moment at which a correct hypothesis has been reached is not known and in general can be uncomputable. Things are different if we require learning to be conclusive, i.e., if the learner is supposed to definitely decide on one answer after a finite amount of information. We can think of this condition as of one in which the learning function is allowed to answer only once—the gameshow case. Clearly, such a conjecture has to be based on certainty. In other words, the learner must know that the answer she gives is true as there is no chance of a change of mind later. In order to define such convergence we will extend the range of learning function L by $\uparrow$, the answer corresponding to the output "I do not know". In the definition below by $\varepsilon \lceil n$ we mean the initial segment of ε of length n, i.e., the sequence $(\varepsilon_0, \varepsilon_1, \ldots, \varepsilon_{n-1})$.

Definition 10.6 Learning function L is (at most) *once defined* on (S, Φ) iff for any stream ε for any world in S and any $n, k \in \mathbb{N}$ such that $n \neq k$ it holds that $L(\varepsilon \lceil n) = \uparrow$ or $L(\varepsilon \lceil k) = \uparrow$.

Accordingly, define conclusive learnability[3] in the following way (note the important difference in the main condition when compared to Definition 10.5).

Definition 10.7 Take an epistemic space (S, Φ).

A world $s_m \in S$ is *conclusively learnable in an effective way by a function L* if L is recursive, once-defined, and for every observational stream ε for s, there exists a finite stage n such that $L((S, \Phi), \varepsilon_0, \ldots, \varepsilon_k) = \{m\}$.

The epistemic space (S, Φ) is said to be *conclusively learnable in an effective way by L* if L is recursive and all its worlds in S are conclusively learnable in an effective way by L.

Finally, the epistemic space (S, Φ) is *conclusively learnable in an effective way* just in case there is a recursive learning function that can conclusively learn it in an effective way.

The necessary and sufficient condition for conclusive learnability (finite identifiability) involves a modified, stronger notion of finite tell-tale [2], the *definite finite tell-tale set*, (DFTT, for short) [36, 42]. Here we give a version adapted to our needs.

Definition 10.8 Let (S, Φ) be an epistemic space. A set $D_i \subseteq \Phi$ is a *definite finite tell-tale set* (DFTT) for s_i in S if:

1. D_i is finite,
2. $s_i \in \bigcap D_i$, and
3. for any $s_j \in S$, if $s_j \in \bigcap D_i$ then $s_i = s_j$.

Theorem 10.1 *An epistemic space (S, Φ) is conclusively learnable in an effective way just in case there is a recursive function $f : \mathbb{N} \to \mathscr{P}^{<\omega}(\Phi)$ such that for each $n \in \mathbb{N}$, $f(n)$ is a finite definite tell-tale set for s_n.*

Hence, a possible world is conclusively learnable just in case it makes a finite conjunction of propositions true that is not true in any other possible world.

10.3.2 Eliminative Power and Complexity

Following the idea of propositional update, knowing that one hypothesis is true clearly means being able to exclude all other possibilities. This leads to the qualitative notion of *eliminative power* of a proposition, which stands for the set of possibilities that this proposition excludes.

---

[3] We will use the name "conclusive learnability" interchangeably with "finite identifiability" which is also sometimes referred to as "identification with certainty".

Definition 10.9 Consider a uniformly decidable epistemic space (S, Φ), and a proposition $x \in \Phi$. The *eliminative power* of x with respect to (S, Φ) is determined by a function $El_{(S,\Phi)} : \Phi \to \mathscr{P}(\mathbb{N})$, such that:

$$El_{(S,\Phi)}(x) = \{i \mid s_i \notin x \ \& \ s_i \text{ in } S\}.$$

Additionally, for $X \subset \Phi$ we write $El_{(S,\Phi)}(X)$ for $\bigcup_{x \in X} El_{(S,\Phi)}(x)$.

In other words, function $El_{(S,\Phi)}$ takes x and outputs the set of indices of all the possible worlds in (S, Φ) that are inconsistent with x, and therefore, in the light of x, can be "eliminated". If one were to link this notion to the epistemic logic terminology, eliminative power of a proposition is the complement of its extension in the epistemic space. This idea applies in a similar way to any formula of any (epistemic) modal language. We may now characterise finite identifiability in terms of eliminative power.

Proposition 10.1 *A set D_i is a definite tell-tale of s_i in S iff*

1. D_i *is finite, and*
2. $El_{(S,\Phi)}(D_i) = \mathbb{N} - \{i\}$.

We will now proceed to analyze the computational complexity of finding DFTTs. In order to do so attention is restricted to finite collections of finite sets. One may question the purpose of further reduction of sets that are already finite. As a matter of fact, if a finite collection of finite sets is finitely identifiable, then each set is already its own DFTT, but obviously finite sets can be much larger than their minimal DFTTs. A simple observation to start with:

Proposition 10.2 *Let (S, Φ) be such that S and Φ are both finite. For any s_i in S, $El_{(S,\Phi)}\{x \mid s_i \in x\}$ can be computed in polynomial time w.r.t. the size of epistemic space (i.e., $card(\Phi) \times card(S)$).*

Then the computational problem of conclusive learnability of an epistemic space is defined in the following way.

Definition 10.10 (FIN- ID *Problem*)
Instance: A finite epistemic space (S, Φ), a world s_i in S.
Question: Is s_i conclusively learnable within (S, Φ)?

Theorem 10.2 FIN- ID *Problem is in P.*

This result does not settle the issue of the efficiency of conclusive learning. In this context an interesting notion is the minimality of DFTTs. Finding the minimal, and even better, the minimal-size DFTTs may be viewed as the task of an efficient teacher, who looks for an optimal sample that allows conclusive learning. There are two nonequivalent ways in which DFTTs can be minimal. Call D_i a *minimal* DFTT of s_i in (S, Φ) just in case all the elements of the sets in D_i are essential for finite identification of s_i in (S, Φ), i.e., taking any element out of the set D_i will decrease the set's eliminative power with respect to (S, Φ), in such a way that it will no longer be a DFTT. In other words, minimal DFTTs of a state contain sufficient information to exclude other possibilities and involve no redundant data.

Definition 10.11 Take an epistemic space (S, Φ), and s_i in S. A *minimal DFTT of* s_i *in* (S, Φ) is a D_i, such that:

1. D_i is a DFTT for s_i in (S, Φ), and
2. for all $X \subset D_i$, X is not a DFTT for s_i in (S, Φ).

Proposition 10.3 *Let (S, Φ) be a finitely identifiable finite epistemic space. Finding a minimal DFTT of s_i in (S, Φ) can be done in polynomial time w.r.t. $card(\{x \mid s_i \in x\})$.*

Note that a possible world may well have many minimal DFTTs of different cardinalities. This is enough reason to introduce a second notion of minimality—*minimal-size* DFTT. Minimal-size DFTTs are the minimal DFTTs of smallest cardinality.

Definition 10.12 Consider an epistemic space (S, Φ), and s_i in S. A *minimal-size DFTT of s_i in (S, Φ)* is a D_i, such that

1. D_i is a DFTT for s_i in (S, Φ), and
2. there is no DFTT D_i' for s_i such that $card(D_i') < card(D_i)$.

How hard is it to find minimal-size DFTTs? In order to answer this question we will first specify the corresponding computational problem.

Definition 10.13 (MIN- SIZE *DFTT Problem*)
Instance: A finite epistemic space (S, Φ), a possible world $s_i \in S$ and a positive integer $k \leq card(\{p \mid s_i \in p\})$.
Question: Is there a DFTT $X_i \subseteq S_i$ of size $\leq k$?

Theorem 10.3 *The* MIN- SIZE DFTT *Problem is NP-complete.*

10.3.3 Preset Learning and Fastest Learning

Attention is now devoted to learners who can be seen as taking a more prescribed course of action by basing their conjectures on symptoms, i.e., on their knowledge of (some) DFTTs. Of course, if an s_i has a DFTT, it will have many, usually infinitely many DFTTs, e.g., each finite set of propositions true in s_i, which is a superset of a DFTT for s_i is a DFTT for s_i as well. Hence, it is more useful to express the learner's access to DFTTs by means of a so-called *dftt-function*. Such a function, let us call it f_{dftt}, is supposed to decide on an input of a finite set $X \subset \Phi$ and an $i \in \mathbb{N}$, whether it considers X to be a DFTT for s_i ($f_{dftt}(X, i) = 1$) or not ($f_{dftt}(X, i) = 0$). In case of a finitely identifiable epistemic space (S, Φ) there exists a dftt-function that recognizes for each $i \in \mathbb{N}$ at least one X as a DFTT for $s_i \in S$.

A learner that uses such a dftt-function in the process of identification is called a *preset learner*. Intuitively speaking, each time the learner receives a new input, and all the answers before have been ↑, the learner looks for the first world that accounts for the propositions listed in the sequence enumerated so far. Assume that this world's index is i. Then among the content of the sequence observed so far the learner looks for a subset X for which $f_{dftt}(X, i) = 1$. If the learner finds one, it answers with i, otherwise it answers with another ↑. It has been shown that if an epistemic space is finitely identifiable at all it is finitely identifiable by a preset learner; it is also the case that the preset learners are exactly those learners that react solely to the set-theoretic content of the information received, disregarding the order and multiplicity of the information.

If a preset learner L is based on a dftt-function that recognizes all DFTTs for all S_i, then the learner will make the proper conjecture always at the earliest possible stage of inquiry. Refer to such a learner as a *fastest learner*. It is clear that the procedure of the fastest learner is closely related to the DEL approach. There is an interesting question whether the fastest learner is always a recursive one. In [23] it has been shown that this is not the case.

Intuitively, *fastest learner* finitely identifies a world s_i as soon as objective 'ambiguity' between languages has been lifted. In other words, define the extreme case of a finite learner who settles on the right language as soon as *any* DFTT for it has been enumerated. We will characterize such a fastest learner as a preset learner based on the collection of all DFTTs.

Take again a finitely identifiable epistemic space (S, Φ), and s_i in S. Now, consider the collection $\mathbb{D}_i$ of all DFTTs of s_i in (S, Φ). For any sequence of data σ, $set(\sigma)$ stands for the set of propositions occurring in σ.

Definition 10.14 (S, Φ) is *finitely identifiable in the fastest way* if and only if there is a learning function L such that, for each ε and for each $i \in \mathbb{N}$,

$$L(\varepsilon\lceil n) = i \ \text{ iff } \ \exists D_i^j \in \mathbb{D}_i \ (D_i^j \subseteq set(\varepsilon\lceil n))\&$$
$$\neg\exists D_i^k \in \mathbb{D}_i \ (D_i^k \subseteq set(\varepsilon\lceil n - 1)).$$

Refer to such L as a *fastest learning function*.

Theorem 10.4 *There is a uniformly decidable epistemic space that is finitely identifiable, but for which no recursive function F exists such that for each i, $F(i)$ is the set of all minimal DFTTs for s_i.*

Proposition 10.4 *There is a uniformly decidable epistemic space that is finitely identifiable, but for which no recursive function F exists such that for each i, $F(i)$ is the set of all minimal-size DFTTs for S_i.*

Time to turn to the more general question whether every finitely identifiable class has a fastest learner. The answer is negative—there are finitely identifiable classes of languages which cannot be finitely identified in the fastest way.

Theorem 10.5 *There is a uniformly decidable epistemic space that is finitely identifiable, but is not finitely identifiable in the fastest way.*

Theorem 10.5 shows that fastest finite identifiability is properly included in finite identification and hence also in preset finite identification. Therefore, we have demonstrated the existence of yet another kind of learning, even more demanding than finite identification. Speaking in terms of conclusive update, our considerations show that in some cases, even if computable convergence to certainty is possible, it is not computable to reach that certainty the moment in which objective ambiguity disappears.

In the light of these discoveries about preset learning there is an additional computational justification for introducing multi-agency to this setting. It is interesting to switch the perspective from the single agent, learning-oriented view, to the two agent game of learner and *teacher* (see [3, 24]). The responsibility of effective learning, in the line with natural intuitions, is in the hands of the teacher, whose computational task is to find samples of information that guarantee optimal learning. Intuitively, it is not very surprising that the task of finding such minimal samples can be more difficult than the complexity of the actual learning. As such, computing the minimal(size) DFTTs seems to go beyond the abilities of the learner and is not necessary in order to be rational or successful. However, such a task is naturally performed by a teacher.

10.4 Epistemic Logic and Learning

This section devotes more attention to the syntactic counterparts of the logical approach to learnability. The previously chosen semantics may be reflected in an appropriate syntax for knowledge, belief, and their changes over time, both in dynamic and temporal settings.

The approach to inductive learning in light of dynamic epistemic and epistemic temporal logic is as follows: Take the initial class of sets to be possible worlds in an epistemic model, which mirrors the learner's initial uncertainty over the range of sets. The incoming pieces of information are taken to be events that modify the initial model. We will show that iterated update on epistemic models based on finitely identifiable classes of sets is bound to lead to the emergence of irrevocable knowledge. In a similar way identifiability in the limit leads to the emergence of safe (truthful and stable) belief. From here we consider a general temporal representation of learning in the limit. The relationship between dynamic epistemic logics and temporal epistemic logics has been studied (see [11, 12]). Given this correspondence, the study of convergence brings about new interesting problems.

10.4.1 Learning and Dynamic Epistemic Logic

The uncertainty range of the agent is revised as new pieces of data (in the form of propositions) are received. The information comes from a completely trusted source, and as such causes the agents to eliminate the worlds that do not satisfy it. In learning theory the truthfulness of incoming data is often assumed, and therefore, in principle, it is justified to use propositional and epistemic update as a way to conduct inquiry (for such interpretation of update see [7]). The following assumes basic knowledge of dynamic epistemic logic (for an overview see Chap. 6 of this book).

Epistemic states may be transformed into epistemic models, and plausibility states into doxastic models in order to deal with the epistemic languages. The initial learning model is a simple single-agent epistemic model whose structure corresponds to the initial epistemic space.

Definition 10.15 Let us take an epistemic space (S, Φ). For every proposition in $p_n \in \Phi$ we take a symbol $\mathrm{p_n} \in \mathrm{PROP}$. Moreover, we will use the set NOM, which contains a nominal symbol i for every $i \in \mathbb{N}$. *The initial learning model* $\mathcal{M}_{(S,\Phi)}$ is a triple:

$$\langle W, \sim, V \rangle,$$

where $W := S$, $\sim \; := W \times W$, $V : \mathrm{PROP} \cup \mathrm{NOM} \to \mathscr{P}(W)$, such that $s_i \in V(\mathrm{p_n})$ iff $s_i \in p_n$ in (S, Φ), and for any $\mathrm{i} \in \mathrm{NOM}$ we set $V(\mathrm{i}) = \{s_i\}$.

Similarly, every epistemic plausibility space $(S, \Phi, \leq)$ can be straightforwardly turned into an epistemic doxastic model $\mathcal{M}_{(S,\Phi,\leq)} = \langle W, \sim, \leq, V \rangle$.

On such models, as on other epistemic models one may interpret epistemic and doxastic logic languages in a standard way. Dynamic versions of such logics include some additional operators that allow describing changes taking place within a model. One particular logic of this type is public announcement logic (PAL, see [46]), where basic epistemic logic is extended to account for update with a specific 'φ-announcement' expression, written as $!\varphi$.

Definition 10.16 (*Syntax of* $\mathscr{L}_{\mathrm{PAL}}$) The syntax of epistemic language $\mathscr{L}_{\mathrm{PAL}}$ is defined as follows:

$$\varphi := p \mid \neg\varphi \mid \varphi \vee \varphi \mid K_a\varphi \mid [A]\varphi$$
$$A := !\varphi$$

where $p \in \mathrm{PROP}$, $a \in \mathscr{A}$, where $\mathscr{A}$ is a set of agents.

Definition 10.17 (*Semantics of* $\mathscr{L}_{\mathrm{PAL}}$) For the epistemic fragment $\mathscr{L}_{\mathrm{EL}}$ the interpretation is as usual (see Chap. 6). The remaining clause of $\mathscr{L}_{\mathrm{PAL}}$ is as follows.

$$\mathcal{M}, w \models [!\varphi]\psi \text{ iff if } \mathcal{M}, w \models \varphi \text{ then } \mathcal{M} \mid \varphi, w \models \psi$$

It has been shown that epistemic update performed on finitely identifiable class of sets leads to irrevocable knowledge.

Theorem 10.6 *The following are equivalent:*

1. *An epistemic space (S, Φ) is finitely identifiable.*
2. *For every $s_i \in S$ and every data stream ε for s_i there is an $n \in \mathbb{N}$ such that for all $m \geq n$, $\mathcal{M}_{(S,\Phi)}, s_i \models [!(\bigwedge set(\varepsilon \lceil m))] K \mathtt{i}$.*

A similar in spirit, but more complex result may be obtained for identifiability in the limit and doxastic version of public announcement logic whose language includes the expression $B_a\varphi$, interpreted on epistemic doxastic models in the following way.

Definition 10.18 (*Semantics of $\mathscr{L}_{\text{DOX-PAL}}$*)

$\mathcal{M}, w \models B_a\phi$ iff there is $v \in W$ such that $v \sim_a w$ and $v \leq_a w$
and for all $s \in W$ such that $s \leq_a v$ and $s \sim_a w$ it holds that
$\mathcal{M}, s \models \varphi$

Theorem 10.7 *The following are equivalent:*

1. *(S, Φ) is identifiable in the limit.*
2. *There is a plausibility preorder $\leq \subseteq S \times S$ such that for every $s_i \in S$ and every data stream ε for s_i there is $n \in \mathbb{N}$ such that for all $m \geq n$, $\mathcal{M}_{(S,\Phi,\leq)}, s_i \models [!(\bigwedge set(\varepsilon \lceil m))] B \mathtt{i}$.*

The results on the universality of lexicographic revision in [4, 22] allow drawing a corollary that the above theorem will also hold for the dynamic logic of lexicographic upgrade, in which case the update operator ! is replaced with ⇑. However, such results for the dynamic logic of minimal upgrade (also known as elite change, with the operator ↑) cannot be obtained.

On the grounds of different results from [4, 22], the plausibility preorder mentioned in Theorem 10.7 sometimes must be non well-founded, allowing models without minimum words according to $\leq$. Decision is thus necessary as to how to interpret the belief operator, $B_a\varphi$. Above, a more general, limiting interpretation of belief operator has been introduced—the agent believes that ϕ in a world w just in case she considers a more plausible world v such that all words that are more plausible than v satisfy φ [22, Chap. 6]).

The last remark concerns the meaning of $K \mathtt{i}$ ($B \mathtt{i}$) in the characterising formulas—they stand for the knowledge (belief) of what is the actual state. Going back to our original motivation, that of formal learning theory, it is contingent on what ones take to being the right, finite, generatively complete description of the world. In terms of propositional knowledge and belief this corresponds to the following: whatever is true in the actual world I know (believe) that it is true and vice versa. In other words, we may say: $K \mathtt{i}$ iff for any $p \in Prop$ such that $s_i \in p$ we have that $p \leftrightarrow Kp$, and similarly for $B \mathtt{i}$.

10.4.2 Learning and Temporal Logic

May one achieve a more complete description of learning in the limit with modal logic? To that end, a logic that allows quantifying over time and over possible histories is needed. Temporal logic offers such a view.[4]

First, consider what could serve as candidate for a temporal unfolding of an possible world s_i in an epistemic space (S, Φ). Instead of viewing a world as a set of atomic propositions, we can represent it as a set of (infinite) histories—possible streams of observations that could take place provided s_i is the actual state. What are the possible streams of information within s_i depends on, let us say, the "nature" of the possibility. Are the events that it generates sequential, is it possible that they permute, will some of them repeat, must everything that is true eventually occur? The learning-theoretic paradigm considered here offers a particular set of answers to those questions. Data streams of a given possibility s_i enumerate all and only the propositions true in s_i, the order of propositions does not matter, and repetitions can occur without restrictions.

Hence, thinking of the propositions true in s_i as events that might occur in s_i, build a temporal structure describing possible future evolutions at s_i. The learning theoretic paradigm requires that those are finite prefices of certain infinite data streams. The latter are infinite sequences of propositions true at s_i which enumerate all and only those propositions, possibly with repetitions. The finite sequences that may be observed are hence determined by a "protocol" that permits certain infinite streams at s_i. Each possible word in our initial uncertainty range can be assigned such a temporal representation. Note that for different possibilities we get mutually disjoint protocols. However, in each point of time the agent observes only a finite sequence of events and obviously such a finite prefix of an infinite data stream can be consistent with more than one possibility. Observing such sequences would not give the agent enough information to distinguish between the two worlds. In such case the agent is uncertain between the two finite sequences not only with respect to how the finite sequence will develop in the future, but also unsure as to the original possibility that generated the sequence. The temporal forest is transformed into an epistemic temporal forest, where the uncertainty relation of the agent will relate identical finite sequences, and the valuation is copied from the initial epistemic space. The resulting structure is an epistemic temporal model (see [18, 19, 45]) that represents all possible evolution of a learning scenario. Below we will call such a structure $\mathrm{For}((S, \Phi), P)$, a forest built from an epistemic space (S, Φ) and a protocol P.

A relevant epistemic temporal language $\mathscr{L}_{\mathrm{ETL}^*}$ contains the following expressions:

$$\varphi := p \mid \neg\varphi \mid \varphi \vee \varphi \mid K\varphi \mid F\varphi \mid A\varphi$$

where p ranges over a countable set of proposition letters PROP. $K\varphi$ reads: 'the agent knows that φ'. Symbol F stands for future, and we define G to mean $\neg F\neg$. $A\varphi$ means: 'in all infinite continuations conforming to the protocol, φ holds'.

---

[4] This section overviews the approach given in [17, 22].

$\mathscr{L}_{ETL^*}$ is interpreted over epistemic temporal frames, $\mathscr{H}$, and pairs of the form (ε, h), the former being a maximal, infinite history in our trees, and the latter a finite prefix of ε (see [41, 45]) in the usual way. The modality 'A' refers to the particular infinite sequences that belong to the chosen protocol. It may be viewed as an operator that performs a global update on the overall temporal structure, 'accepting' only those infinite histories that conform to the protocol.

To give a temporal characterization of finite identifiability the following idea must be expressed: In the epistemic temporal forest, for any starting, bottom node s_i it is the case that for all infinite data streams in the future there will be a point after which the agent will know that she started in s_i, which means that she will remain certain about the part of the forest she is in. The designated propositional letters from PROP_{NOM} correspond to the partitions, which can also be viewed as underlying theories that allow predicting further events.[5] Formally, with respect to finite identifiability of sets, the following theorem holds.

Theorem 10.8 *The following are equivalent:*

1. (S, Φ) *is finitely identifiable.*
2. $\text{For}((S, \Phi), P) \models \text{i} \rightarrow AFGK\text{i}.$

In order to give a temporal characterization of identifiability in the beliefs of the learner must be expressed. Therefore, the temporal forests should include a plausibility ordering. Conditioning (update) is a universal learning method from truthful data. In other words, in the case of identifiability in the limit, eliminating the worlds of an epistemic plausibility model is enough to reach stable and true belief. This allows for considering very specific temporal structures that result from updating a doxastic epistemic model with purely propositional information.[6] The epistemic temporal models need to be extended with a doxastic preorder on infinite data streams, and the epistemic temporal language needs to be extended with a doxastic counterpart, the belief operator B, that works in a way similar to dynamic doxastic epistemic logic.

As in the case of finite identifiability we will now provide a formula of doxastic epistemic temporal logic that characterises identifiability in the limit.

Theorem 10.9 *The following are equivalent:*

1. (S, Φ) *is identifiable in the limit.*
2. *There exists a plausibility preorder* $\leq \subseteq S \times S$ *s.t.* $\text{For}((S, \Phi, \leq), P) \models \text{i} \rightarrow AFGB\text{i}.$

The above results show that the two prominent approaches, learning theory and epistemic modal-temporal logics, may be joined together to describe the notions of belief and knowledge involved in inductive inference. Bridging the two approaches benefits both sides. For formal learning theory, to create a logic for it is to provide

---

[5] The characterisation involving designated propositional letters can be replaced with one that uses nominals as markers of bottom nodes. For such an approach see Dégremont and Gierasimczuk [16].

[6] For more complex actions performed on plausibility models in the context of the comparison between dynamic doxastic and doxastic temporal logic see van Benthem and Dégremont [11].

additional syntactic insight into the process of inductive learning. For logics of epistemic and doxastic change, it enriches their present scope with different learning scenarios, i.e., not only those based on the incorporation of new data but also on generalisation.

The temporal logic based approach to inductive inference gives a straightforward framework for analyzing various domains of learning on a common ground. In terms of protocols, sets may be seen as classes of specific histories—their permutation-closed complete enumerations. Functions, on the other hand, may be viewed as 'realities' that allow only one particular infinite sequence of events. We can think of many intermediate concepts that may be the object of learning. Interestingly, the identification of protocols, that seems to be a generalization of the set-learning paradigm provides what has been the original motivation for epistemic temporal logic from the start: identifying the current history that the agent is in, including its order of events, repetitions, and other constraints.

10.5 Logic, Learning, and Scientific Method

Logic and learning theory may also be bridged by considering the methodological merits of learners for identifying classical axioms of epistemic logic. This means treating epistemic axioms as learning goals and then considering the methodological constraints on learners for converging to the truth of such axioms (Hendricks [27, 28] and Kevin T. Kelly's chapter, Chap. 11). Axioms T, K, 4, 5 for instance, all present different learning problems for definitions of knowledge based on limiting convergence for both assessment and discovery methods. Assessment methods take as inputs finite evidence sequences and hypothesis, and map them to onto truth or falsity, while discovery methods conjecture hypotheses (sets of possible worlds) in response to incoming evidence. It turns out that the validity of canonical axioms of epistemic logic may be acutely sensitive to the methodological constraints enforced on the methods of scientific inquiry whether based on assessment or discovery. A method of scientific discovery may be consistent in the sense that it only conjectures something consistent with current evidence, consistently expectant insofar as it conjectures something consistent with the evidence and expects to see more of the same, or may be infallible in the sense that it only conjectures something which is entailed by the evidence observed so far. Now, these different methodological constraints come into play once one attempts to validate axioms of epistemic logic. No amount of methodology is going to help validating the axiom of negative introspection (5) and the reason is intuitively this: Not knowing means not having converged, which does not entail—not even for the infallible learner—knowledge of lack of convergence to the truth. Similarly, arguments since the times of American pragmatism have conveyed that positive introspection and limiting convergence are irreconcilable since positive introspection demands knowledge of the modulus of convergence to knowledge yet limiting definition entails exactly this very modulus of convergence may not be known—there just is a time, such that for each later time, the method has settled for the truth and will not oscillate again, but it may not necessarily be known when this time will be. However, there is a way to circumvent this conclusion—and thus eat the cake and have it. If the axiom of positive introspection is allowed a diachronic interpretation such that the consequent of knowing that one knows either happens later than the antecedent of knowing or would have obtained later even had things been otherwise, then it is possible to validate the axiom 4 assuming that the discovery method in question is consistently expectant [28, 29]. This result has two significant horns: It demonstrates that axioms of epistemic logic may have a temporal dimension of importance given the dichotomy between synchronic and diachronic interpretations of the axioms, and that their very validity is contingent upon what the method of inquiry decides to do. Additionally, methods of inquiry may work together—assessment methods are definable in terms of discovery methods and vice versa which turns out to being an important feature when considering the transmissibility of knowledge from one agent to another [28]. This is all as it should be. One of the important methodological benefits of treating agent indices of epistemic logic as learning functions is to activate agents in such a way that they play crucial roles in validating the epistemic axioms apparently describing the very rationality of epistemic agency for single agents and multiple agents interacting [30].[7]

The same goes for other methodological recommendations to be found in contemporary literature of formal epistemology. Axioms of belief revision may likewise be interpreted as recommendations for learners and the question then becomes how well these recommendations fare with respect to convergence to the truth—sometimes they do quite well, sometimes they create truth-tracking disasters and inductive amnesia [32, 35].

Combining logic and learning provides a stronghold for epistemological and methodological agent-interactive studies. Such studies, earlier on reserved for either epistemic logic or formal learning theory telling respectively their partial stories, are now given a chance to make for the full story. Computational epistemology is strictly speaking not about knowledge but about learning, but of course learning is about knowledge acquisition. And there you have it as Johan van Benthem would have it: learning as part of the dynamic turn in logic.

References

1. Alchourrón CE, Gärdenfors P, Makinson D (1985) On the logic of theory change: partial meet contraction and revision functions. J Symb Logic 50(2):510–530
2. Angluin D (1980) Inductive inference of formal languages from positive data. Inf Control 45(2):117–135
3. Balbach FJ, Zeugmann T (2009) Recent developments in algorithmic teaching. In: Dediu AH, Ionescu AM, Martín-Vide C (eds) LATA'09: Proceedings of 3rd International Conference on Language and Automata Theory and Applications, Tarragona, Spain, 2–8 April 2009. Lecture Notes in Computer Science, vol 5457. Springer, The Netherlands, pp 1–18

---

[7] Hintikka took, from the very beginning, the axioms of epistemic logic to describe a strong kind of agent rationality.

4. Baltag A, Gierasimczuk N, Smets S (2011) Belief revision as a truth-tracking process. In: Apt K (ed) TARK'11: Proceedings of the 13th Conference on Theoretical Aspects of Rationality and Knowledge, Groningen, The Netherlands, 12–14 July 2011. ACM, New York, pp 187–190

5. Baltag A, Smets S (2009) Learning by questions and answers: from belief-revision cycles to doxastic fixed points. In: Ono H, Kanazawa M, Queiroz R (eds) WoLLIC'09: Proceedings of 16th International Workshop on Logic, Language, Information and Computation, Tokyo, Japan, 21–24 June 2009. Lecture Notes in Computer Science, vol 5514. Springer, The Netherlands, pp 124–139

6. van Benthem J (2003) Logic and the dynamics of information. Minds Mach 13(4):503–519

7. van Benthem J (2006) One is a lonely number: on the logic of communication. In: Chatzidakis Z, Koepke P, Pohlers W (eds) LC'02: Proceedings of Logic Colloquium 2002. Lecture Notes in Logic, vol 27. ASL & A.K. Peters, Cergy-Pontoise, pp 96–129

8. van Benthem J (2007) Dynamic logic for belief revision. J Appl Non-Classical Logics 2: 129–155

9. van Benthem J (2007) Rational dynamics and epistemic logic in games. Int Game Theory Rev 9(1):13–45

10. van Benthem J (2011) Logical dynamics of information and interaction. Cambridge University Press, Cambridge

11. van Benthem J, Dégremont C (2010) Bridges between dynamic doxastic and doxastic temporal logics. In: Bonanno G, Löwe B, van der Hoek W (eds) LOFT'08: Revised selected papers of 8th Conference on Logic and the Foundations of Game and Decision Theory. Lecture Notes in Computer Science, vol 6006. Springer, New York, pp 151–173

12. van Benthem J, Gerbrandy J, Hoshi T, Pacuit E (2009) Merging frameworks for interaction: DEL and ETL. J Philos Logic 38(5):491–526

13. Blum L, Blum M (1975) Toward a mathematical theory of inductive inference. Inf Control 28:125–155

14. Boutilier C (1993) Revision sequences and nested conditionals. IJCAI'93: Proceedings of the 13th International Joint Conference on Artificial Intelligence. Chambery, France, pp 519–525

15. Darwiche A, Pearl J (1997) On the logic of iterated belief revision. Artif Intell 89:1–29

16. Dégremont C, Gierasimczuk N (2009) Can doxastic agents learn? On the temporal structure of learning. In: He X, Horty J, Pacuit E (eds) LORI 2009: Proceedings of Logic, Rationality, and Interaction, 2nd International Workshop, Chongqing, China, 8–11 Oct 2009. Lecture Notes in Computer Science, vol 5834. Springer, Berlin, pp 90–104

17. Dégremont C, Gierasimczuk N (2011) Finite identification from the viewpoint of epistemic update. Inf Comput 209(3):383–396

18. Emerson EA, Halpern JY (1986) "Sometimes" and "not never" revisited: on branching versus linear time temporal logic. J ACM 33(1):151–178

19. Fagin R, Halpern JY, Moses Y, Vardi MY (1995) Reasoning about knowledge. MIT Press, Cambridge

20. Gärdenfors P (1988) Knowledge in flux-modelling the dynamics of epistemic states. MIT Press, Cambridge

21. Gierasimczuk N (2009) Bridging learning theory and dynamic epistemic logic. Synthese 169(2):371–384

22. Gierasimczuk N (2010) Knowing one's limits. Logical analysis of inductive inference. PhD thesis, Universiteit van Amsterdam, The Netherlands

23. Gierasimczuk N, de Jongh D (2013) On the complexity of conclusive update. Comput J 56(3):365–377

24. Gierasimczuk N, Kurzen L, Velázquez-Quesada FR (2009) Learning and teaching as a game: a sabotage approach. In: He X et al. (eds) LORI 2009: Proceedings of Logic, Rationality, and Interaction, 2nd International Workshop, Chongqing, China, 8–11 Oct 2009. Lecture Notes in Computer Science, vol 5834. Springer, Berlin, pp 119–132

25. Gold EM (1967) Language identification in the limit. Inf Control 10:447–474

26. Goldszmidt M, Pearl J (1996) Qualitative probabilities for default reasoning, belief revision, and causal modeling. Artif Intell 84:57–112

27. Hendricks V (2003) Active agents. J Logic Lang Inf 12(4):469–495
28. Hendricks VF (2001) The convergence of scientific knowledge: a vew from the limit. Kluwer Academic Publishers, Dordrecht
29. Hendricks VF (2007) Mainstream and formal epistemology. Cambridge University Press, New York
30. Hintikka J (1962) Knowledge and belief: an introduction to the logic of the two notions. Cornell University Press, Cornell
31. Jain S, Osherson D, Royer JS, Sharma A (1999) Systems that learn. MIT Press, Chicago
32. Kelly KT (1998a) Iterated belief revision, reliability, and inductive amnesia. Erkenntnis 50: 11–58
33. Kelly KT (1998b) The learning power of belief revision. TARK'98: Proceedings of the 7th Conference on Theoretical Aspects of Rationality and Knowledge. Morgan Kaufmann Publishers, San Francisco, pp 111–124
34. Kelly KT (2004) Learning theory and epistemology. In: Niiniluoto I, Sintonen M, Smolenski J (eds) Handbook of epistemology. Kluwer, Dordrecht (Reprinted. In: Arolo-Costa H, Hendricks VF, van Benthem J (2013) A formal epistemology reader. Cambridge University Press, Cambridge)
35. Kelly KT, Schulte O, Hendricks V (1995) Reliable belief revision. Proceedings of the 10th International Congress of Logic, Methodology, and Philosophy of Science. Kluwer Academic Publishers, Dordrecht, pp 383–398
36. Lange S, Zeugmann T (1992) Types of monotonic language learning and their characterization. COLT'92: Proceedings of the 5th Annual ACM Conference on Computational Learning Theory, Pittsburgh, 27–29 July 1992. ACM, New York, pp 377–390
37. Lehrer K (1965) Knowledge, truth and evidence. Analysis 25(5):168–175
38. Lehrer K (1990) Theory of knowledge. Routledge, London
39. Martin E, Osherson D (1997) Scientific discovery based on belief revision. J Symb Logic 62(4):1352–1370
40. Martin E, Osherson D (1998) Elements of scientific inquiry. MIT Press, Cambridge
41. van der Meyden R, Wong K (2003) Complete axiomatizations for reasoning about knowledge and branching time. Studia Logica 75(1):93–123
42. Mukouchi Y (1992) Characterization of finite identification. In: Jantke K (ed) AII'92: Proceedings of the International Workshop on Analogical and Inductive Inference, Dagstuhl castle, Germany, 5–9 Oct 1992. Lecture Notes in Computer Science, vol 642. Springer, Berlin, pp 260–267
43. Nayak AC (1994) Iterated belief change based on epistemic entrenchment. Erkenntnis 41(3):353–390
44. Osherson D, Stob M, Weinstein S (1986) Systems that learn. MIT Press, Cambridge
45. Parikh R, Ramanujam R (2003) A knowledge based semantics of messages. J Logic Lang Inf 12(4):453–467
46. Plaza J (1989) Logics of public communications. In: Emrich M, Pfeifer M, Hadzikadic M, Ras Z (eds) Proceedings of the 4th International Symposium on Methodologies for Intelligent Systems. Springer, New York, pp 201–216
47. Putnam H (1975) 'Degree of Confirmation' and inductive logic, vol 1, chap 17. Cambridge University Press, Cambridge (Reprinted. In: Schilpp PA (ed) (1999) The philosophy of Rudolf Carnap. Library of living philosophers, vol 11)
48. Spohn W (1988) Ordinal conditional functions: a dynamic theory of epistemic states. In: Skyrms B, Harper WL (eds) Causation in decision, belief change, and statistics, vol II. Kluwer, Dordrecht
49. Stalnaker R (2009) Iterated belief revision. Erkenntnis 70(2):189–209

Chapter 11
A Computational Learning Semantics for Inductive Empirical Knowledge

Kevin T. Kelly

Abstract This chapter presents a new semantics for inductive empirical knowledge. The epistemic agent is represented concretely as a learner who processes new inputs through time and who forms new beliefs from those inputs by means of a concrete, computable learning program. The agent's belief state is represented hyper-intensionally as a set of time-indexed sentences. Knowledge is interpreted as avoidance of error in the limit and as having converged to true belief from the present time onward. Familiar topics are re-examined within the semantics, such as inductive skepticism, the logic of discovery, Duhem's problem, the articulation of theories by auxiliary hypotheses, the role of serendipity in scientific knowledge, Fitch's paradox, deductive closure of knowability, whether one can know inductively that one knows inductively, whether one can know inductively that one does not know inductively, and whether expert instruction can spread common inductive knowledge—as opposed to mere, true belief—through a community of gullible pupils.

11.1 Introduction

Science formulates general theories. Can such theories count as knowledge, or are they doomed to the status of *mere* theories, as the anti-scientific fringe perennially urges? The ancient argument for inductive skepticism urges the latter view: no finite sequence of observations can rule out the possibility of future surprises, so universal laws and theories are unknowable.

A familiar strategy for responding to skeptical arguments is to rule out skeptical possibilities as "irrelevant" [8]. One implementation of that strategy, motivated by the possible worlds semantics for subjunctive conditionals, is to ignore worlds "distant from" or "dissimilar to" the actual world. If you are really looking at a cat on a

K. T. Kelly (✉)
Department of Philosophy, Carnegie Mellon University, Pittsburgh, PA 15213, USA
e-mail: kk3n@andrew.cmu.edu

A. Baltag and S. Smets (eds.), *Johan van Benthem on Logic and Information Dynamics*, Outstanding Contributions to Logic 5, DOI: 10.1007/978-3-319-06025-5_11, © Springer International Publishing Switzerland 2014

mat under normal circumstances, you wouldn't be a brain in a vat hallucinating a non-existent cat if it weren't there, so your belief is *sensitive* to the truth [26, 35]. Or if you were to believe that there is a cat on the mat, most worlds in which your belief is false are remote worlds involving systematic hallucinations, so your belief is *safe* [31, 37, 42].

So much for "ultimate", brain-in-a-vat skepticism applied to particular perceptual beliefs. But what about inductive skepticism concerning general scientific laws and theories? Belief in such laws and theories does not seem "safe". For example, if the true law were not of the form $Y = bX + a$, would science have noticed that fact *already*? Are all worlds in which the true law has the form $Y = cX^2 + bX + a$ safely bounded away from $Y = bX + a$ worlds in terms of similarity—regardless how small c is?[1] One can formally ignore small values of c by the *ad hoc* assumption that they are farther from the $c = 0$ world than are worlds in which c is arbitrarily close to 0. But the resulting discontinuity in similarity is questionable and, in any event, the conditional "if there were a quadratic effect, it would have been so large that we would have noticed it already" is implausible, however one contrives to satisfy it. In fact, the history of science teaches that we have been wrong on fundamental matters in the past, due to pivotal but small effects (e.g., the relativistic corrections to classical mechanics), and that we cannot guard against more such surprises in the future [22]. So although subjunctive semantics appears to provide a plausible response to ultimate, brain-in-a-vat skepticism concerning ordinary perceptual knowledge, it is still overwhelmed by inductive skepticism, since, in that case, the nearby possibilities are exactly the *skeptical* ones.

The best that one can expect of even ideally diligent, ongoing scientific inquiry is that it detects and roots out error *eventually*. So if there is inductive knowledge, it must allow for a time lag between the onset of knowledge and the detection and elimination of error in other possible worlds. There is a venerable tradition, expounded by [5, 11, 16, 27, 32, 33] and subsequently developed by computer scientists and cognitive scientists into a body of work known as computational learning theory [15], that models the epistemic agent as a *learner* who processes information through time and who stabilizes, eventually, to true, inductive beliefs.

Inductive learning is a matter of finding the truth eventually. It is natural to think of inductive knowledge that ϕ as having learned that ϕ. Having learned that ϕ implies that one has actually stabilized to true belief that ϕ and that one would have *converged to true belief* whether ϕ otherwise. The proposed semantics is more lenient—one has knowledge that ϕ if and only if one has actually converged to true belief that ϕ (as in having learned) and one would have *avoided error* whether ϕ otherwise— one might simply suspend belief forever if the data are so unexpected that one no longer knows what is going on. Allowance for suspension of belief agrees better

---

[1] Nozick [26] and Roush [35] argue that we would have noticed the failure of known laws already because, if a given uniformity weren't true, some distinct uniformity would have been. But in the polynomial example, all the regularities are law-like. Nor can one object that all linear laws are closer to a linear law than any quadratic law is, since the knowledge claim in question is that the true law is linear, so sensitivity forces one to move to non-linear laws. Vogel [38] presents additional objections to tracking as an adequate account of inductive knowledge.

with scientific practice. Moreover, it turns out to be necessary if the consequences of known theories are to be knowable by the same standard.[2]

The semantics is not proposed as a true analysis of inductive knowledge in the traditional, exacting sense.[3] There may be no such thing, and it may not matter whether there is, since what matters in philosophy is not so much how we do talk, as how we *will* talk, after shopping in the marketplace of ideas. In that spirit, the semantics is proposed as a useful, unified, explanatory framework for framing problems and conceptual issues at the intersection of inductive knowledge, inductive learning, information, belief, and time. Such issues include: the relation of learnability to knowability, how deductive inference produces new inductive knowledge from old, how inductive knowledge can thrive in a morass of inconsistency, why scientific knowledge should allow for a certain kind of luck or "serendipity", how one can know that one knows, why one can't know that one doesn't know, how to know your own Moore sentence, and how expert instruction can spread common inductive knowledge through a population of passive pupils.

One common theme running through the development that follows is *epistemic parasitism*. Inference is not an argument or a mere, formal relation. It is the execution of a procedure for generating new beliefs from old. If inference produces new knowledge from old, it is because the inference procedure is guaranteed to produce new beliefs that satisfy the truth conditions for knowledge from beliefs that already do. Therefore, the semantics *explains how*, rather than merely *assumes that*, certain patterns of inference turn old knowledge into new. The basic idea is that the new knowledge is *parasitic* on the old because the inference pattern generates beliefs whose convergence tracks the convergence of the given beliefs. A related theme is the *hyper-intensionality* of belief. It is not assumed that the belief state of the agent is deductively closed or consistent, or that the learning method of the agent follows some popular conception of idealized rationality. Rather, rationality is something a computable agent can only approximate, and the desirability of doing so should be explained, rather than presupposed, by the semantics of learning and knowledge.

Inclusion of the entire learning process within models of epistemic logic is consonant with the current trend in epistemic logic [4] toward more dynamic and potentially explanatory modeling of the agent. Recently, there have been explicit studies of truth tracking and safety analyses of knowledge [14] and of inductive learning within a modal logical framework [10]. Earlier, Hendricks [13] proposed to develop learning models for inductive knowledge, itself, and a rather different proposal was sketched, informally, in [20].[4] This chapter places learning semantics on a firm, formal basis. For decades, Johan van Benthem has strongly encouraged the development of connections between learning theory and epistemic logic, both personally and in print, so it is a particular pleasure to contribute this study to his festschrift.

---

[2] Alternatively, one could simply stipulate that the deductive consequences of inductive knowledge are known [35], but then one would have no explanation why or how they are known, aside from the stipulation.

[3] An long list of improvements is provided just prior to the conclusion.

[4] The differences are described, in detail, below.

11.2 Syntax

Let $G = \{1, \ldots, N\}$ be indices for a group of N individuals. Let $\mathbf{L_{atom}} = \{\mathsf{p}_i : i \in \mathbb{N}\}$ be a countable collection of atomic sentences. Define the modal language $\mathbf{L_{BIT}}$ (belief, information, time) in the usual way, with the classical connectives, including $\perp$, and the modal operators presented in the following table, where Δ is understood to be a finite subset of $\mathbf{L_{BIT}}$. The unusually rich base language reflects Scotts's [36] advice to seek more interesting epistemic principles in interactions among modal operators. In the following glosses, let t^* be the time of the *epistemic context* at which "*i* knows that ϕ" is assessed and let $t \geq t^*$ be the *time of evaluation*, which may lie in the future, due to the evaluation of a future tense operator. The aim is to analyze convergent belief that ϕ *was* true at t^*, so one must keep a "clean copy" of t^* in the model in order to determine whether *i* believes at some later time $t > t^*$ that ϕ was true at t^* [17].[5]

Time

$@_t\, \phi$ *At:* it is true at t that ϕ.

$\mathsf{N}\, \phi$ *Now:* It is true at t^* that ϕ.

$\langle\mathsf{F}\rangle\, \phi$ *Future tense:* it is true at $t' \geq t$ that ϕ.

$\langle\dot{\mathsf{F}}\rangle\, \phi$ *Future context tense:* In epistemic context $t^{**} \geq t^*$, it is true that ϕ.

Information and Belief

$[\,\mathsf{I}\,]_i\, \phi$ *Information:* information has been made available to *i* by t^* that ϕ is true at t^*.

$[\mathsf{D}]_i\, \phi$ *Determination:* it is determined by information available to *i* at t^* and by the method of *i* at t^* that ϕ is true at t^*.

$[\mathsf{B}]_i\, \phi$ *Virtual belief:* the learning method of *i* at t^* directs *i* to believe that ϕ is true at t^*.

Methodology

$\langle\mathsf{M}\rangle_i\, \phi$ *Methodological feasibility:* it is feasible for *i* that ϕ is true.

$\psi\, \langle\mathsf{MD}\rangle_{i,\Delta}\, \phi$ *Conditional methodological feasibility:* given that ψ is true, it is feasible for *i* to ensure that ϕ is true without altering *i*'s learning disposition concerning the truth of the premises in Δ.

$\mathsf{S}_i\, \Delta$ *Inferential stability:* if *i* modifies her method in a way that holds her learning disposition with respect to statements in Δ fixed, then her future beliefs concerning the statements in Δ also remain unaltered—because *i* is insensitive to any changes in her sensory inputs that might result when other agents notice the changes to her method.

---

[5] Alternatively, one could introduce first-order quantifiers over temporal variables, but it is conceptually vivid to treat tense as a modality freely permutable with other modalities.

Let $\mathbf{L}_{@BIT}$ denote the set of all $\mathbf{L}_{BIT}$ sentences that are prefixed by an operator $@_t$ for some $t \in \mathbb{N}$. Extend $\mathbf{L}_{BIT}$ with definitions as follows. For primitive operators $[X]$, $\langle Y \rangle$, introduce the dual operators:

$$\langle X \rangle \, \phi := \neg [X] \neg \, \phi; \quad [Y] \, \phi := \neg \langle Y \rangle \neg \phi.$$

A tilde above a box operator $[X]_i$ indicates the "whether" form of the operator, which is defined as follows, unless noted otherwise:

$$[\tilde{X}] \, \phi := [X] \, \phi \lor [X] \neg \phi.$$

Introduce the standard notation:

$$\mathsf{B}_i := [\mathsf{B}]_i; \quad \mathsf{F} := \langle \mathsf{F} \rangle; \quad \mathsf{G} := [\mathsf{F}];$$

and similarly for $\dot{\mathsf{F}}$, $\dot{\mathsf{G}}$. Clean up notation in the following way:

$$\mathsf{S}_i \, \delta := \mathsf{S}_i \{\delta\};$$
$$\psi \langle \mathsf{MD} \rangle_{i,\delta} \, \phi := \psi \langle \mathsf{MD} \rangle_{i,\{\delta\}} \, \phi.$$

When Γ, Δ are finite subsets of $\mathbf{L}_{BIT}$ and X_i is an arbitrary modal operator, let:

$$\mathsf{X}_i \Gamma := \bigwedge_{\gamma \in \Gamma} \mathsf{X}_i \, \gamma;$$

$$\Delta \to \Gamma := \bigwedge_{\delta \in \Delta} \delta \to \bigwedge_{\gamma \in \Gamma} \gamma;$$

11.3 Computational Learning Models

Let E denote the set of possible *external worlds*. In a Kantian spirit, learning semantics imposes no structure or restrictions whatever on E. Let $T = \mathbb{N}$ be interpreted as discrete *stages of inquiry*. Let $G = \{1, \ldots, N\}$ be interpreted as a finite set of *agents*. Agent $i \in G$ is assumed to have some overall, discrete, physical sensory state at t that will be called the agent's current *input* at t. Think of $S = \mathbb{N}$ as code numbers for possible inputs. Inputs are not assumed to have propositional meanings (they are never assigned truth values), but their occurrence makes propositional information *available*. Let S^* be the set of all finite sequences of inputs, so each $\sigma \in S^*$ is a possible *input history*.

It is assumed that each agent's belief state is maintained by a *learning function L* that returns a verdict (1 for "believe" and 0 for "don't believe") for each sentence ϕ in $\mathbf{L}_{@BIT}$ in light of the current input history σ:

$$L : S^* \times \mathbf{L}_{@\mathsf{BIT}} \to \{0, 1\}.$$

Let $\phi_c(x, y)$ be the binary partial recursive function computed by the Turing machine with Gödel index c.[6] Learning function L is computable if and only if there exists $c \in \mathbb{N}$ such that:

$$L_c(\sigma, \phi) = \phi_c(\langle\sigma\rangle, \ulcorner\phi\urcorner),$$

for all $\sigma \in S^*$ and $\phi \in \mathbf{L}_{@\mathsf{BIT}}$, where $\langle.\rangle$ is an effective encoding of S^* into the natural numbers and $\ulcorner.\urcorner$ is an effective Gödel numbering of $\mathbf{L}_{@\mathsf{BIT}}$. Let C denote the set of all $c \in \mathbb{N}$ such that $\phi_c(\langle.\rangle, \ulcorner.\urcorner)$ is a learning function. Elements of C are called *learning methods*.

Each learning method covers all future contingencies, but i's learning method can change from time to time, through maturation, education, or mishap. A *joint method trajectory* is a function:

$$\mathbf{c} : (G \times T) \to C,$$

that assigns a learning method $c \in C$ to each agent $i \in G$ at each time $t \in T$. A *possible world* is an arbitrary pair $w = (e_w, \mathbf{c}_w)$, such that $e_w \in E$ and $\mathbf{c}_w$ is a joint method trajectory. Let $c_{i,w,t} = \mathbf{c}_w(i, t)$. Let W denote the set of all possible worlds.

A *preliminary computational learning model* (PCLM) for agents G is a quadruple $\mathfrak{M}_{t^*} = (E, \mathbf{s}, V, t^*)$ such that E is a non-empty set, $t^* \in T$ and:

$$\mathbf{s} : (G \times W \times T) \to S;$$
$$V : (\mathbf{L}_{\mathsf{atom}} \times T) \to \mathsf{Pow}(W).$$

The function V is the usual *valuation function*, according to which $V(\mathsf{p}, t)$ is the proposition expressed by atomic sentence p at arbitrary time t. The distinguished time t^* is the time of the epistemic context under discussion [17]. Think of $s_{i,w,t} = \mathbf{s}(i, w, t)$ as the input that w presents to i at t in w. Call $\mathbf{s}$ the *input assignment function* and s_i the input assignment function for agent i. Define the *input stream* of i in w at t and the *input history* of i in w up to, but not including t as follows:

$$s_{i,w} = (s_{i,w,0}, \dots, s_{i,w,t}, \dots);$$
$$s_{i,w}|t = (s_{i,w,0}, \dots, s_{i,w,t-1}).$$

One major aim of this study is to provide a precise semantics for learnability, knowability, and the feasibility of knowing some things given that you know other things. It is assumed that changing the method of learner i does not cause changes to the external world or to the methods of the other agents. Therefore, the nearest

---

[6] Lower-case ϕ is also standardly employed in logic as a sentential in logical axiom schemata. Context readily disambiguates the two uses.

world to w in which i uses method d at t is just the world $w[d/i, t]$ that results from substituting method d for agent i's method $c_{i,w,t}$ in w at t.[7]

Counterfactual shifts of method open the door to the medieval problem of information concerning future contingents, for since $s(i, w, t)$ depends on w, which specifies i's method trajectory $c_{i,w}$, a crystal ball can send signals to i about the methods employed by i or other agents in the future, so counterfactual changes of method in the future could cause changes to past inputs. Learning semantics assumes that past inputs are preserved under future method choices. A *computational learning model* (CLM) is, accordingly, a PCLM that satisfies:

$$s_{i,w}|t = s_{i,w[d/i,t]}|t, \tag{11.1}$$

for all $i \in G$, $d \in C$, $w \in W$, and $t \in T$.

11.4 Information, Belief, and Determination

The input history $s_{i,w}|t$ of i in w has no truth value—it is a temporal sequence of sensory states—but it *makes available* to i in w at t the following, propositional *information*[8]:

$$\mathbf{I}(i, w, t) = \{w' \in W : s_{i,w}|t = s_{i,w'}|t\}.$$

In Kripke semantics for modal epistemic logic, available information is represented in terms of the *accessibility* relation "w' is possible in light of all the information available to i in w at t":

$$\mathcal{I}_{i,t}(w, w') \Leftrightarrow w' \in I_{i,w,t}.$$

For fixed i and t, $(W, \mathcal{I}_{i,t}, V)$ is a standard Kripke model. Since $\mathcal{I}_{i,t}$ is an equivalence relation, the corresponding modal operator is S5, as is often assumed (e.g., [3]). Making propositional information available via physical signals is not the same thing as inserting that information directly into i's beliefs—it is still up to i's learning function $L_{c_{i,w,t}}$ to interpret the signals, to recover the information they afford, and to incorporate it smoothly into i's belief system.

---

[7] I.e.:

$$(\mathbf{c}[d/i, t])(i', t') = \begin{cases} d & \text{if } i' = i \ \wedge \ t' = t; \\ \mathbf{c}(i', t') & \text{otherwise.} \end{cases}$$

$$w[d/i, t] = (e_w, \mathbf{c}_w[d/i, t]).$$

[8] Cf. [23] for a similar proposal.

Possibilities of error that are incompatible with the information currently available will be deemed irrelevant to learning and knowledge. Furthermore, it does not seem that i needs to have been *informed* of her own learning method—the method merely has to *determine* success in light of available information. Accordingly, define the *determination assignment function*[9]:

$$\mathbf{D}(i, w, t) = \{w' \in I_{i,w,t} : c_{i,w,t^*} = c_{i,w',t^*}\}.$$

Then $D_{i,w,t} = \mathbf{D}(i, w, t)$ is the strongest proposition determined at t by the information and by the learning strategy possessed by i in w at t^*. The induced binary relation $\mathcal{D}_{i,t}(w, w')$ is again an equivalence relation that refines $\mathcal{I}_{i,t}(w, w')$.

Belief is handled very differently, as the concrete, hyper-intensional outcome of learning. The usual consistency, closure, or rationality assumptions are not imposed, because they are false. The actual belief state of i in w at t is produced by i's actual learning method at t:

$$\mathbf{B}_{\mathsf{act}}(i, w, t) = \{\phi \in \mathbf{L}_{@\mathsf{BIT}} : L_{c_{i,w,t}}(s_{i,w}|t, \phi) = 1\}.$$

However, in the long run, we are all dead and then we don't believe anything, so we never actually converge to the truth. An alternative account, in the spirit of Nozick [26], is that agent i would converge to true belief if *she* were to continue to use her current method forever:

$$\mathbf{B}_{\mathsf{ctr}}(i, w, t) = \{\phi \in \mathbf{L}_{@\mathsf{BIT}} : L_{c_{i,w,t^*}}(s_{i,w[c_{i,w,t^*}/i,t]}|t, \phi) = 1\}.$$

However, that would make it impossible for i to know inductively that all humans are mortal, since i would be immortal if she were literally to retain her current learning disposition forever. Alternatively, one can focus on what i's current learning method *directs* i to believe in the future, just as one can speak of the outputs of an algorithm on inputs larger than any concrete machine running the algorithm will ever receive or of linguistic competence concerning sentences that will never be uttered due to resource limitations:

$$\mathbf{B}(i, w, t) = \{\phi \in \mathbf{L}_{@\mathsf{BIT}} : L_{c_{i,w,t^*}}(s_{i,w}|t, \phi) = 1 \wedge \phi \in \mathbf{L}_{\mathsf{BIT}}\}.$$

Refer to $B_{i,w,t} = \mathbf{B}(i, w, t)$ as the *virtual* belief state of i in w at t.

---

[9] That idea is also sketched in [23]. It trivializes knowledge of one's own learning method. See the discussion in Sect. 11.17.4 below for a potential, contextualist remedy. Also, it fails to rule out brain-in-a-vat worlds. See Sect. 11.17.1 for a discussion of making determination safe or sensitive.

11.5 Learning Semantics

Let $\mathfrak{M}_{t^*} = (E, \mathbf{s}, V, t^*)$ be a CLM. Define the proposition $\|\phi\|^t_{\mathfrak{M}_{t^*}}$ expressed by ϕ in $\mathfrak{M}_{t^*}$ inductively as follows. In the base case:

$$\|\mathsf{p}\|^t_{\mathfrak{M}_{t^*}} = V(\mathsf{p}, t).$$

The connectives and $\bot$ have their standard, classical interpretations. For the temporal operators, define:

$$\|\langle \mathsf{F} \rangle\, \phi\|^t_{\mathfrak{M}_{t^*}} = \bigcup_{t' \geq t} \|\phi\|^{t'}_{\mathfrak{M}_{t^*}};$$

$$\|\langle \dot{\mathsf{F}} \rangle\, \phi\|^t_{\mathfrak{M}_{t^*}} = \bigcup_{t' \geq t} \|\phi\|^{t'}_{\mathfrak{M}_{t'}};$$

$$\|\mathsf{N}\, \phi\|^t_{\mathfrak{M}_{t^*}} = \|\phi\|^{t^*}_{\mathfrak{M}_{t^*}};$$

$$\|@_{t'}\, \phi\|^t_{\mathfrak{M}_{t^*}} = \|\phi\|^{t'}_{\mathfrak{M}_{t^*}}.$$

Operator $\mathsf{F} = \langle \mathsf{F} \rangle$ is a future tense operator that includes the present time. Its dual is the "henceforth" operator G. Operator $\dot{\mathsf{F}} = \langle \dot{\mathsf{F}} \rangle$ is similar, except that it moves the epistemic context forward. Operator N resets the time t of evaluation to the time t^* of the epistemic context. Operator $@_{t'}$ resets the time of evaluation to the specified time t'.

Information and determination are defined propositionally, in the standard way, and both are S5 operators, for reasons already discussed.

$$\|[\mathsf{I}]_i\, \phi\|^t_{\mathfrak{M}_{t^*}} = \{w \in W : I_{i,w,t} \subseteq \|\phi\|^{t^*}_{\mathfrak{M}_{t^*}}\};$$

$$\|[\mathsf{D}]_i\, \phi\|^t_{\mathfrak{M}_{t^*}} = \{w \in W : D_{i,w,t} \subseteq \|\phi\|^{t^*}_{\mathfrak{M}_{t^*}}\}.$$

Virtual belief, on the other hand, is entirely hyper-intensional, as it should be. Note that the time at which ϕ is believed to be true is always referred back to t^*, via the $@_{t^*}$ operator.

$$\|[\mathsf{B}]_i\, \phi\|^t_{\mathfrak{M}_{t^*}} = \{w \in W : @_{t^*}\phi \in B_{i,w,t}\}.$$

Methodological feasibility says that there is some method that i might have adopted that would achieve ϕ at t^* in w. It is used to express theses concerning learnability and knowability.

$$\|\langle \mathsf{M} \rangle_i\, \phi\|^t_{\mathfrak{M}_{t^*}} = \{w \in W : (\exists c \in C)\, w[c/i, t^*] \in \|\phi\|^t_{\mathfrak{M}_{t^*}}\}.$$

Methodological feasibility does not say that i can *guarantee* or *see to it* that ϕ is true. That stronger modality is expressed by $\langle \mathsf{M} \rangle_i [\mathsf{D}]_i$.

The remaining two operators are more subtle, and work together as a team. To motivate conditional feasibility, consider the familiar logical thesis that the knowledge of i is closed under known consequence:

$$(\mathsf{K}_i\,\phi \;\wedge\; \mathsf{K}_i(\phi \rightarrow \psi)) \rightarrow \mathsf{K}_i\,\psi.$$

Granted, modus ponens is an easy inference to perform, but nothing like that thesis is even remotely true. Perhaps it is intended as a regulative ideal or as an obligation, but ideals are approachable and ought implies can, so the more proximate and concrete question is whether satisfaction of the thesis is *feasible*, in the sense that there is an *inference procedure i* could adopt that would *guarantee* that i knows that ψ *given* that she knows both ϕ and $\phi \rightarrow \psi$.

From the viewpoint of learning, effectively performing inferences amounts to an effective modification $h(c)$ of one's learning program c. Think of $\Delta \subseteq \mathbf{L}_{@\mathsf{BIT}}$ as a finite set of premises. One examines the verdicts of c for sentences in Δ (including, perhaps, past verdicts), and then one possibly reverses the verdicts of c concerning some sentences (i.e., conclusions) outside of Δ. In that way, agent i can effectively modify her learning program c without having access either to c, itself, or to its raw, sub-cognitive, sensory inputs. Inference, therefore, makes sense as an evolutionary strategy—given some reptilian learning wet-ware that is hard to modify genetically without lethal effects, tack on some higher-level cognitive wet-ware that can intercept and modify the learning wet-ware's verdicts. That learning-theoretic conception of inference is made precise as follows. First, the *verdict* of learning method c concerning $\delta \in \mathbf{L}_{@\mathsf{BIT}}$ in response to σ is the ordered pair:

$$v_c(\sigma, \delta) = (L_d(\sigma, @_{t^*}\,\delta),\; L_d(\sigma, @_{t^*}\neg\delta)).$$

Define $c \equiv_\Delta d$ to hold if and only if learning programs c, d have identical verdicts for each $\delta \in \Delta$ and $\sigma \in S^*$. Define $c \equiv_{\Delta,\sigma} d$ to hold if and only if learning programs c, d have identical verdicts for each $\delta \in \Delta$ and for each initial segment τ of input history σ. Let h be a total recursive function that assumes values in C. Say that h *preserves* premises in Δ if and only if $h(c) \equiv_\Delta c$, for all $c \in C$. Say that h *depends only* on premises in Δ if and only if:

$$c \equiv_{\Delta,\sigma} d \;\Rightarrow\; v_{h(c)}(\sigma, \phi) = v_{h(d)}(\sigma, \phi).$$

for all $c, d \in C$, $\sigma \in S^*$, and $\phi \in \mathbf{L}_{@\mathsf{BIT}}$. Then h is an *inference procedure* with premises in Δ if and only if h is a total recursive function with range included in C that preserves premises in Δ and that depends only on premises in Δ.

Conditional feasibility expresses the existence of an inference procedure that *guarantees* the situation in the consequent, *given* the situation described in the antecedent. Accordingly, let $w \in \|\psi\langle\mathsf{MD}\rangle\!\!\mapsto_{i,\Delta}\phi\|^t_{\mathfrak{M}_{t^*}}$ if and only if there exists inference procedure h with premises in Δ such that, for all $u \in I_{i,w,t^*}$:

$$u \in \|\psi\|^t_{\mathfrak{M}_{t^*}} \Rightarrow u[h(c_{i,u,t})/i, t] \in \|\phi\|^t_{\mathfrak{M}_{t^*}}.$$

The notation $\psi\langle MD]\mapsto_{i,\Delta} \phi$ is mnemonic—existence of h is like $\langle M \rangle_i$, the guarantee is like $[D]_i$, and the assumption that the antecedent holds is like a conditional.

Inference—even deductive inference—can be subtly treacherous in learning semantics. Suppose that i contemplates changing her learning strategy c to d, which generates exactly the same verdict on δ that c does, in every possible input situation. Assumption (11.1) guarantees that d results in the same belief whether δ that c does *given the same inputs*, but the change from c to d could modify or even shut off the flow of *future* inputs to i because other agents detect the change in i (think of a poorly blinded social psychology experiment). Furthermore, the change from c to d could make δ false if the truth of δ depends on what some or all of the agents believe (e.g., i is a major player in the market). Either way, i's election to adopt inferential strategy d could be empirically or semantically *self-defeating*, in the sense that premise δ of the intended inference becomes untestable or false as a consequence of the inference being performed. Happily, good experimental design can prevent one's valid inferences from being self defeating, so it is useful to have vocabulary expressing that such preventive measures have successfully been carried out for some intended set of premises Δ. It is too strong to say that the inputs to i would be exactly the same whether i uses c or d, because i would presumably receive at least some information concerning her own beliefs. It suffices that neither the truth of the premises in Δ nor the verdicts of i concerning them is affected by the change. Define $w \in \|S_i \Delta\|^t_{\mathfrak{M}_{t^*}}$ to hold if and only if for all $d \in C$ such that $c_{i,w,t^*} \equiv_\Delta d$ and for all $u \in D_{i,w,t^*}$, $t \geq t^*$, and $\delta \in \Delta$, if we set $u' = u[d/i, t^*]$ and $c = c_{i,u,t^*}(= c_{i,w,t^*})$, then:

$$u \in \|\delta\|^{t^*}_{\mathfrak{M}_{t^*}} \Leftrightarrow u' \in \|\delta\|^{t^*}_{\mathfrak{M}_{t^*}};$$
$$v_c(s_{i,u}|t, \delta) = v_d(s_{i,u'}|t, \delta).$$

That concludes the truth conditions for $\mathbf{L_{BIT}}$. Let $\Gamma \subseteq \mathbf{L_{BIT}}$. Define validity in a model and logical validity as follows:

$$\mathfrak{M}_{t^*} \models \phi \Leftrightarrow W = \|\phi\|^{t^*}_{\mathfrak{M}_{t^*}};$$
$$\models \phi \Leftrightarrow \mathfrak{M}_{t^*} \models \phi, \text{ for each CLM } \mathfrak{M}_{t^*}.$$

Note that validity in a model initializes time to the model's current epistemic context time t^*. Finally, logical entailment and equivalence are defined as follows[10]:

$$\phi \models \psi \Leftrightarrow \models (\phi \to \psi);$$
$$\phi \equiv \psi \Leftrightarrow \models (\phi \leftrightarrow \psi).$$

---

[10] N.b. substitution of equivalents for equivalents under temporal operators does not preserve validity [17]. For example, $\models G(\phi \leftrightarrow \phi)$ and $\phi \equiv N \phi$, but $\not\models G(\phi \leftrightarrow N \phi)$.

11.6 Example: Outcomes of a Repeated Experiment

CLMs accommodate a boggling range of learning situations, but a collection of very elementary models suffices to illustrate many of the results that follow. Assume that each agent i passively observes the successive values of a repeated experiment whose outcomes are effectively coded as natural numbers. In the spirit of empiricism, identify possible external worlds with infinite outcome sequences $\varepsilon : \mathbb{N} \to \mathbb{N}$. Let E_0 denote the set of all such sequences. Define, for $k \in \mathbb{N}$:

$$s_0(i, w, t) = \varepsilon_w(t);$$
$$V_0(\mathsf{p}_k, t) = \{\varepsilon \in E_0 : \varepsilon(t) = k\} \times \mathsf{C}^N;$$
$$\mathfrak{N}_{t*} = (E_0, s_0, V_0, t^*).$$

Temporal operators allow for compact expression of a range of increasingly complex statements:

$$\mathsf{p}_k : \text{the current outcome is } k;$$
$$\mathsf{Gp}_k : \text{the outcome will be } k;$$
$$\mathsf{Fp}_k : \text{the outcome is } k \text{ from now on;}$$
$$\mathsf{FG}\,\mathsf{p}_k : \text{the outcome will stabilize to value } k;$$
$$\mathsf{GF}\,\mathsf{p}_k : \text{the outcome is } k \text{ infinitely often.}$$

A hypothesis ϕ is *objective* for i just in case i has the information available that i cannot alter the truth value of ϕ by changing her learning method. Objectivity simpliciter is objectivity for every agent.

$$\mathsf{O}_i\,\phi := [\mathsf{I}]_i (\phi \leftrightarrow [\mathsf{M}]_i\,\phi);$$
$$\mathsf{O}_G\,\phi := \bigwedge_{i \in G} \mathsf{O}_i\,\phi.$$

A special feature of model $\mathfrak{N}_{t*}$ is that objectivity implies inferential stability:

$$\mathfrak{N}_{t*} \models \mathsf{O}_i\,\phi \to \mathsf{S}_i\,\phi, \tag{11.2}$$

since inputs do not depend on methods at all, and neither does the truth of an objective statement.

11.7 Example: Agency, Games, and Experimentation

The agents in model $\mathfrak{N}_{t*}$ are isolated natural scientists who passively receive inputs from a fixed experiment. But even a solipsistic scientist can choose how to interact with nature, and communication among scientists can produce cascades of interactive, doxastic effects. Although $\mathbf{L_{BIT}}$ has no vocabulary describing acts other than belief, CLMs can represent arbitrarily complex social interactions involving such acts. The trick is to locate agents' diachronic strategies for non-doxastic actions within the "external world" $e \in E$. Then, all of the valid theses of learning semantics are valid for game-theoretic applications.

Here is one way to do it. Let $X \subseteq \mathbb{N}$ be a set of potential actions. Assuming that the actions are observable by all of the agents, let $S = X^N$. Then S^* contains all possible, finite play histories. Let A denote the set of all $a \in \mathbb{N}$ such that ϕ_a is a unary total recursive function with range included in X. The *disposition to act* computed by a looks at the current input history and chooses how to act:

$$A_a(\sigma) = \phi_a(\langle\sigma\rangle).$$

Since belief depends on inputs, one special way for actions to depend on inputs is for them to depend on beliefs.

Dispositions to act can change through time, just as dispositions to believe can. A *joint disposition trajectory* $\mathbf{a} : (G \times T) \to A$ assigns a profile of dispositions to the agents at each time. In purely social applications, the "external world" e can be identified with $\mathbf{a}$, so possible worlds are pairs $w = (\mathbf{a}, \mathbf{c})$. In experimental science, one agent can represent nature and the rest of the agents can be used to model socially distributed scientific inquiry. Each agent i receives as input the actions of every agent (including herself). Let $\sigma * s$ denote the concatenation of signal $s \in S$ to finite sequence $\sigma \in S^*$. The joint input assignment is then definable in stages as follows:

$$\dot{s}_{i,(\mathbf{a},\mathbf{c})}|0 \qquad = ();$$
$$\dot{s}_{i,(\mathbf{a},\mathbf{c})}|(t + 1) = \dot{s}_{i,(\mathbf{a},\mathbf{c})}|t * (A_{a_{i,t+1}}(\dot{s}_{i,(\mathbf{a},\mathbf{c})}|t) : i \leq N).$$

In the long run, all the players of an infinite game are dead, as are the dispositional properties of societies, economies, and terrestrial organisms. Hence, it may be more natural to think of the agents as *virtually* studying one another's and nature's *current* reactive dispositions, just as was done for belief[11]:

$$s_{i,(\mathbf{a},\mathbf{c})}|t^* \qquad = \dot{s}_{i,(\mathbf{a},\mathbf{c})}|t^*;$$
$$s_{i,(\mathbf{a},\mathbf{c})}|(t^* + t + 1) = s_{i,(\mathbf{a},\mathbf{c})}|t * (A_{a_{i,t^*}}(s_{i,(\mathbf{a},\mathbf{c})}|(t^* + t)) : i \leq N).$$

Either way, requirement (11.1) is satisfied.

In extensive form games, each agent receives some utility in each world at each time, as a result of what all the agents do. The utilities may also shift through time, if we interpret the agents as playing different games from time to time. Evolving utilities may be absorbed into the external world. The games just described assume perfect information. Of course, s can easily be made to censor some actions.

---

[11] The base case assumes that information gathered by means of earlier dispositions remains available.

11.8 Correctness and Error

Define "i is in *error* that ϕ" as follows:

$$\mathsf{E}_i\,\phi := \mathsf{B}_i\,\phi \wedge \mathsf{N}\neg\phi.$$

Error *whether* ϕ is defined according to the general definition of "whether" presented in Sect. 11.2 above.

$$\tilde{\mathsf{E}}_i\,\phi := \mathsf{E}_i\,\phi \vee \mathsf{E}_i\neg\phi.$$

It follows that i cannot be in error whether ϕ unless i believes that ϕ or believes that $\neg\phi$. That definition is straightforward, if belief is deductively closed, but it is very weak for hyper-intensional belief—e.g., belief that ϕ does not count as an error whether $\neg\phi$. However, in order to interpret successful learning whether ϕ, all that is required is some unambiguous convention for i "getting ϕ wrong", and the proposed convention suffices in a minimal way, by recording the learning function's verdict whether ϕ. Stronger, but finite, demands on deductive acumen would not alter the results that follow, except to complicate their proofs. In a similar spirit, *correctness that* ϕ is absence of error whether ϕ together with belief that ϕ and correctness *whether* ϕ is defined as absence of error whether ϕ together with the verdict for ϕ:

$$\mathsf{C}_i\,\phi := \neg\tilde{\mathsf{E}}_i\,\phi \wedge \mathsf{B}_i\,\phi;$$
$$\tilde{\mathsf{C}}_i\,\phi := \neg\tilde{\mathsf{E}}_i\,\phi \wedge \tilde{\mathsf{B}}_i\,\phi.$$

Correctness whether ϕ could have been defined in the usual way as correctness that ϕ or correctness that $\neg\phi$, but that concept depends on whether i believes that $\neg\neg\phi$. The proposed definition depends only on i's verdict whether ϕ.[12]

11.9 Inductive Learning

In computational learning theory, *inductive learning* whether ϕ is understood as guaranteed convergence of i's current learning method to correct belief whether ϕ. That is elegantly formalizable in $\mathbf{L_{BIT}}$ as follows:

$$\tilde{\mathsf{L}}_i\,\phi := [\mathsf{D}]_i \mathsf{FG}\tilde{\mathsf{C}}_i\,\phi.$$

The truth conditions for $\tilde{\mathsf{L}}_i\,\phi$ can be expressed entirely in terms of the proposition $\|\phi\|^{t^*}_{\mathfrak{M}_{t^*}}$ and the verdicts of i's learning method: $w \in \|\tilde{\mathsf{L}}_i\,\phi\|^{t}_{\mathfrak{M}_{t^*}}$ if and only if for all $u \in D_{i,w,t}$,

---

[12] Thanks to Ted Shear for this point.

$$u \in \|\phi\|_{\mathfrak{M}_{t^*}}^{t^*} \Rightarrow \left(\lim_{t \to \infty} L_{c_{i,w,t^*}}(s_{i,u}|t, @_{t^*}\phi) = 1 \wedge \right. \tag{11.3}$$

$$\left. \lim_{t \to \infty} L_{c_{i,w,t^*}}(s_{i,u}|t, @_{t^*}\neg\phi) = 0 \right);$$

$$u \notin \|\phi\|_{\mathfrak{M}_{t^*}}^{t^*} \Rightarrow \left(\lim_{t \to \infty} L_{c_{i,w,t^*}}(s_{i,u}|t, @_{t^*}\neg\phi) = 1 \wedge \right. \tag{11.4}$$

$$\left. \lim_{t \to \infty} L_{c_{i,w,t^*}}(s_{i,u}|t, @_{t^*}\phi) = 0 \right).$$

That is essentially equivalent to saying, in computational learning theory, that i's *current* method c_{i,w,t^*} *decides* ϕ in the limit [19], except that learning semantics allows the data to depend on the learning method.

11.10 Inductive Learnability

Just as the theory of computability concerns what can be computed, rather than how we actually compute, computational learning theory focuses on learnability—the feasibility of learning—rather than on the actual psychology of learning. Learning semantics affords at least four grades of feasibility:

$$\langle M \rangle_i [D]_i \phi \models [D]_i \langle M \rangle_i \phi \models \langle M \rangle_i \phi \models \langle M \rangle_i \langle D \rangle_i \phi. \tag{11.5}$$

In the case of learnability, those concepts collapse to $\langle M \rangle_i \tilde{L}_i \phi$—the last entails the first, since $\tilde{L}_i$ begins with $[D]_i$, which is an S5 operator:

$$\langle M \rangle_i [D]_i \tilde{L}_i \phi \equiv [D]_i \langle M \rangle_i \tilde{L}_i \phi \equiv \langle M \rangle_i \tilde{L}_i \phi \equiv \langle M \rangle_i \langle D \rangle_i \tilde{L}_i \phi. \tag{11.6}$$

Concretely, $w \in \|\langle M \rangle_i \tilde{L}_i \phi\|_{\mathfrak{M}_{t^*}}^t$ if and only if there exists $d \in C$ such that (11.3) and (11.4) hold with d substituted for c_{i,w,t^*} in u, for all $u \in I_{i,w[d/i,t^*],t}$. If ϕ satisfies $O_i \phi$ in $\mathfrak{M}_{t^*}$, one can also substitute $I_{i,w,t}$ for $I_{i,w[d/i,t^*],t}$, in which case the truth conditions for learnability are essentially the same as the conditions for decidability in the limit [19].[13]

Universal truths and existential truths about the future are inductively learnable in the empirical model $\mathfrak{M}_{t^*}$—just believe the universal hypothesis until it is refuted and believe its negation thereafter, and follow the dual strategy in the existential case:

$$\mathfrak{M}_{t^*} \models \langle M \rangle_i \tilde{L}_i G p_k; \tag{11.7}$$
$$\mathfrak{M}_{t^*} \models \langle M \rangle_i \tilde{L}_i F p_k. \tag{11.8}$$

But not every empirical hypothesis is inductively learnable. Kant [18] observed that hypotheses like the finite or infinite divisibity of matter or the existence of a first

---

[13] The differences concern mere conventions for coding the acceptance, rejection, or suspension of belief of i with respect to ϕ.

moment in time "outpace all possible experience". In terms of learnablity, he was right. Suppose that the laboratory returns a 1 whenever an allegedly fundamental particle is split and returns a 0 when an attempted split fails. Then finite divisibility of matter can be expressed as $FG\,p_0$ and infinite divisibility of matter can be expressed as $GF\,p_1$. Both hypotheses are evidently only remotely connected with current experience. In fact, neither is inductively learnable in $\mathfrak{N}_{t*}$:

$$\mathfrak{N}_{t*} \models \neg\tilde{L}_i FG\,p_k; \tag{11.9}$$

$$\mathfrak{N}_{t*} \models \neg\tilde{L}_i GF\,p_k. \tag{11.10}$$

It suffices to show, via a standard, learning theoretic diagonal argument, that no c satisfies convergence conditions (11.3) and (11.4).[14]

Learning semantics is a flexible framework for inductive learning and learnability that allows one, for the first time, to rigorously iterate the learning operator, in order to analyze such statements as that it is learnable whether someone else is learning whether ϕ. But in order to provide the sharpest possible contrast between learning semantics and traditional possible worlds semantics for epistemic logic, the focus of this study is on the semantics of inductive knowledge, to which we now turn.

11.11 Inductive Knowledge

Agent i *has learned* whether [that] ϕ if and only if i is learning whether ϕ and, henceforth, i correctly (virtually) believes whether [that] ϕ:

$$\tilde{Led}_i\,\phi := G\tilde{C}_i\,\phi \wedge \tilde{L}_i\,\phi;$$
$$Led_i\,\phi := GC_i\,\phi \wedge \tilde{L}_i\,\phi \equiv \tilde{Led}_i\,\phi \wedge \phi.$$

Having learned inductively whether ϕ may sound odd, since the culmination of inductive inquiry depends on what i's current learning method would do in the future. But such locutions are actually quite common: e.g., "I have quit smoking for good".

It is natural to suppose that inductive knowledge is having learned, but there is a powerful argument to the contrary: learnability is not preserved under logical consequence; for recall (11.7), (11.9), and (11.10) and note that $G\,\phi$ entails both $GF\,\phi$ and $FG\,\phi$. Since having learned entails learnability, it follows that knowability is not closed under logical consequence. And the examples sound bad: we would know that the laws of quantum mechanics apply invariably, but it would be unknowable that they apply infinitely often or all but finitely often. It sounds better to say that we know the latter two statements *because* we know the first.

---

[14] Proofs of selected theses are presented in the Appendix.

Pursuing that idea, suppose that i's only reason for believing that $GF\phi$ is that she believes $G\phi$ and suppose that her reason for believing $G\phi$ is that it has stood up to severe testing so far (a single counterexample would refute $G\phi$). It is a traditional theme in the philosophy of science that general theories are not testable until they are *articulated* with auxiliary assumptions [9]. Semantically speaking, "articulation" amounts to the substitution of a logically stronger, testable hypothesis for the untestable hypothesis, itself. Thus, one may think of $G\phi$ as a testable articulation of $GF\phi$, since it posits a particularly simple *way* in which $GF\phi$ *might* be true. Then i stabilizes to true belief that $GF\phi$ as soon as i stabilizes to true belief that $G\phi$, so the actual convergence requirement is met also for $GF\phi$. But what if $G\phi$ were to be refuted, say at time t? Maybe i has plausible ideas about how to re-articulate $GF\phi$ (e.g., as $@_{t+1}G\phi$). In order to learn by such a strategy, i would require a full contingency plan for re-articulating $GF\phi$ that somehow hits upon a true articulation eventually in *every* possible world in which $GF\phi$ is true. But it has already been shown that no such contingency plan exists for $GF\phi$, since $GF\phi$ is not learnable.

Another venerable theme in the philosophy of science is that there is "no logic of discovery" [12, 30], which means, roughly, that science need not have an explicit contingency plan for what to propose when old hypotheses are refuted, so far as scientific knowledge is concerned. The standard arguments for that conclusion are analogical and historical.[15] The argument from analogy is that a theorem is still a theorem no matter how one came to conjecture it, so scientific knowledge likewise does not depend on how one came to think up the hypothesis. The historical argument is that major scientific findings have been hit upon by luck. For a celebrated example, the chemist Kekulé claimed to discover the carbon ring structure of benzine by dreaming of a snake biting its tail [2, 12]. It does not seem to count against Kekulé's subsequent knowledge of that hypothesis that he possessed no systematic contingency plan for dreaming up alternative molecular structures, had the ring hypothesis failed. Scientists refer to luck that does not undermine scientific knowledge as *serendipity*. Kekule's dream was serendipitous in that sense, as is all luck in hitting upon a true hypothesis. Since untestable hypotheses like $GF\phi$ cannot be learned, they can be known only with serendipity. So allowance for serendipity, the practice of testing testable articulations of untestable hypotheses, and the slogan that there is "no logic of discovery" are both grounded in the closure of inductive knowability under logical consequence, a fundamental, epistemological consideration.

Suppose that i is commanded by her thesis advisor to investigate $GF\phi$ by severely testing $G\phi$. We know that i lacks a full logic of discovery for $GF\phi$, since $GF\phi$ is not learnable. Suppose, plausibly, that she has far less—if $G\phi$ is ever refuted, she has no idea what is going on, suspends belief forever whether $GF\phi$, and switches to a more lucrative career in finance. *If* her advisor was right (serendipity), then she has already converged to true belief that $GF\phi$ and, since her belief that $GF\phi$ is based *solely* on her belief that $G\phi$, she is also guaranteed to eliminate error with respect to $GF\phi$ eventually. Her (actual) convergence to true belief that the untestable hypothesis is

---

[15] A notable exception is [32].

true is serendipitous, but her eventual avoidance of error is not lucky at all—it is guaranteed by her commitment to suspend belief forever if $GF\phi$ is refuted.

In light of the preceding considerations, it is proposed that inductive knowledge that ϕ is actual convergence to true belief that ϕ along with guaranteed, eventual avoidance of error whether ϕ[16]:

$$\tilde{K}_i\,\phi := G\tilde{C}_i\,\phi \,\wedge\, [D]_i\,FG\neg\tilde{E}_i\,\phi;$$
$$K_i\,\phi := GC_i\,\phi \,\wedge\, [D]_i\,FG\neg\tilde{E}_i\,\phi \,\equiv\, \tilde{K}_i\,\phi \,\wedge\, \phi.$$

Thus, having learned whether ϕ is sufficient, but not necessary, for knowing whether ϕ:

$$\models \tilde{Led}_i\,\phi \to \tilde{K}_i\,\phi; \tag{11.11}$$
$$\models Led_i\,\phi \to K_i\,\phi. \tag{11.12}$$

In fact, learning is *equivalent* to guaranteed, eventual arrival at knowledge—a nice example of a plausible validity expressible in L_{BIT} but not in the traditional, pure K_i fragment.[17]

$$\tilde{L}_i\,\phi \equiv [D]_i\,F\tilde{K}_i\,\phi. \tag{11.13}$$

In terms of concrete learning methods, the first conjunct of $\tilde{K}\,\phi$ is true in w at t if and only if:

---

[16] Hendricks [13] presents several concepts of empirical knowledge, the closest of which to the following proposal is "realistic reliable true belief" or *RRT knowledge*. Hendricks' informal gloss of RRT knowledge (p. 181) amounts to the following idea in the present notation: $Krrt_i\,\phi := G\,\phi \wedge \tilde{L}_i\,\phi$ (the operator $[D]_i$ is dropped from the $\tilde{L}_i\,\phi$ condition in the accompanying formal statement—presumably unintentionally). RRT knowledge is very different from inductive knowledge as defined here. First of all, RRT knowledge requires that $G\,\phi$, which would make it impossible for i to know, for example, that she believes that ϕ, if that belief state is transient. Learning semantics sidesteps that difficulty by evaluating the proposition believed at the "now" of utterance. Second, RRT knowledge does not require $GB_i\,\phi$, so RRT knowledge does not even imply belief that ϕ, much less stable belief that ϕ—it may be years until the learning process succeeds. Finally, RRT knowledge does imply learning whether ϕ, which implies that RRT knowability cannot be closed under deductive consequence, as has just been explained. Hendricks' claim that RRT knowledge validates the axioms of modal system S4 (Proposition 11.3, p. 208) is therefore false. The discrepancy is explained by the fact that, just prior to the proof of Proposition 11.3, Hendricks inadvertently modifies the concept of RRT knowledge a second time (p. 194) to $G\,\phi$ conjoined with the existence of a future time t' such that it is determined now that i believes that ϕ forever after t'—whether or not ϕ is true.

[17] Thesis (11.13) is invalid with $\dot{F}$ in place of F. It is crucial that the doxastic future under consideration is virtual rather than actual.

$$w \in \|\phi\|_{\mathfrak{M}_{t^*}}^{t^*} \Rightarrow ((\forall t \geq t^*) \, L_{c_{i,w,t^*}}(s_{i,w}|t, @_{t^*}\phi) = 1 \, \wedge \qquad (11.14)$$
$$(\forall t \geq t^*) \, L_{c_{i,w,t^*}}(s_{i,w}|t, @_{t^*}\neg\phi) = 0);$$

$$w \notin \|\phi\|_{\mathfrak{M}_{t^*}}^{t^*} \Rightarrow ((\forall t \geq t^*) \, L_{c_{i,w,t^*}}(s_{i,w}|t, @_{t^*}\phi) = 0 \, \wedge \qquad (11.15)$$
$$(\forall t \geq t^*) \, L_{c_{i,w,t^*}}(s_{i,w}|t, @_{t^*}\neg\phi) = 1);$$

and the second conjunct is true in w at t if and only if for all $u \in I_{i,w,t}$:

$$u \in \|\phi\|_{\mathfrak{M}_{t^*}}^{t^*} \Rightarrow \lim_{t \to \infty} L_{c_{i,w,t^*}}(s_{i,u}|t, @_{t^*}\neg\phi) = 0; \qquad (11.16)$$

$$u \notin \|\phi\|_{\mathfrak{M}_{t^*}}^{t^*} \Rightarrow \lim_{t \to \infty} L_{c_{i,w,t^*}}(s_{i,u}|t, @_{t^*}\phi) = 0. \qquad (11.17)$$

Note that (11.16) and (11.17) weaken the corresponding conditions (11.3) and (11.4) for having learned.

11.12 Inductive Knowability

Learning semantics again affords at least four notions of inductive knowability, in descending strength:

$$\langle M \rangle_i [D]_i K_i \, \phi \;\models\; [D]_i \langle M \rangle_i K_i \, \phi \;\models\; \langle M \rangle_i K_i \, \phi \;\models\; \langle M \rangle_i \langle D \rangle_i K_i \, \phi. \qquad (11.18)$$

The four conditions of knowability are all logically distinct, due to the actual convergence condition for knowledge. But, due to that condition, the first three cannot be adopted as general theses concerning inductive knowability; for, as theses, they imply that i has the information that she has the power to make ϕ true immediately.

$$\models \langle M \rangle_i K_i \, \phi \Rightarrow \; \models [I]_i \langle M \rangle_i \, \phi. \qquad (11.19)$$

That leaves the weakest option, which requires only that it be feasible for i to make it *possible* that she knows now—an idea consonant with serendipity:

$$\langle MD \rangle_i \, \phi := \langle M \rangle_i \langle D \rangle_i \tilde{K}_i \, \phi \qquad (11.20)$$
$$\equiv \langle M \rangle_i \langle D \rangle_i (G\tilde{C}_i \, \phi \wedge [D]_i FG\neg\tilde{E}_i \, \phi) \qquad (11.21)$$
$$\equiv \langle M \rangle_i (\langle D \rangle_i G\tilde{C}_i \, \phi \wedge [D]_i FG\neg\tilde{E}_i \, \phi); \qquad (11.22)$$

where the last equivalence is again due to $[D]_i$ being S5. Condition (11.22) expands to the existence of $d \in C$ such that for some $u \in I_{i,w,t}$:

$$u[d/i, t] \in \|\phi\|^t_{\mathfrak{M}_{t*}} \Rightarrow ((\forall t \geq t^*) \, L_d(s_{u,i}|t, @_{t*} \phi) = 1 \, \wedge \qquad (11.23)$$
$$(\forall t \geq t^*) \, L_d(s_{u,i}|t, @_{t*} \neg\phi) = 0);$$
$$u[d/i, t] \notin \|\phi\|^t_{\mathfrak{M}_{t*}} \Rightarrow ((\forall t \geq t^*) \, L_d(s_{u,i}|t, @_{t*} \phi) = 0 \, \wedge \qquad (11.24)$$
$$(\forall t \geq t^*) \, L_d(s_{u,i}|t, @_{t*} \neg\phi) = 1);$$

and for all $u \in I_{i,w,t}$:

$$u[d/i, t] \in \|\phi\|^t_{\mathfrak{M}_{t*}} \Rightarrow \lim_{t \to \infty} L_d(s_{i,u}|t, @_{t*} \neg\phi) = 0; \qquad (11.25)$$
$$u[d/i, t] \notin \|\phi\|^t_{\mathfrak{M}_{t*}} \Rightarrow \lim_{t \to \infty} L_d(s_{i,u}|t, @_{t*} \phi) = 0. \qquad (11.26)$$

Conditions (11.23) and (11.24) are trivially satisfiable by dogmatically believing that ϕ and conditions (11.25) and (11.26) are trivially satisfiable by skeptically suspending belief whether ϕ. But the conditions are not jointly trivial—the possibility of having converged to the truth risks the possibility of error infinitely often, unless one has an appropriate plan in place for when to suspend judgment, as Popper [30] insisted. For example, weak knowability can fail when even the total input stream does not determine the truth of ϕ in any world. In that case, say that ϕ is *globally underdetermined*—venerable candidates include "the Absolute is lazy" and Poincare [29] perfect trade-off between shrinking forces and geometry. The logical positivists attempted to rule out globally underdetermined hypotheses by deeming them meaningless, on empiricist grounds, but that leaves open the question whether freedom from global underdetermination implies knowability. Learning semantics validates something close to that in the empiricist model $\mathfrak{M}_{t*}$, as long as the input stream is computable. Recall the strategy, discussed above, of guessing a testable articulation ψ of ϕ, believing ϕ until ψ is refuted, and suspending judgment thereafter. It witnesses the following, liberal knowability condition for objective hypotheses in $\mathfrak{M}_{t*}$:

Proposition 11.1 *Suppose that* $w \in \|O_G \phi\|^{t^*}_{\mathfrak{M}_{t*}}$ *and there exists* $u \in I_{i,w,t^*} \cap \|\phi\|^{t^*}_{\mathfrak{M}_{t*}}$ *with computable input stream* $s_{i,u}$. *Then* $w \in \|\langle MD \rangle_i \, K_i \, \phi \|^{t^*}_{\mathfrak{M}_{t*}}$.

As a corollary, we have the following knowability result, in contrast to the nonlearnability results (11.9) and (11.10) above[18]:

$$\mathfrak{M}_{t*} \models \langle MD \rangle_i (K_i G \, p_k \, \wedge \, K_i F \, p_k \, \wedge \, K_i FG \, p_k \, \wedge \, K_i GF \, p_k). \qquad (11.27)$$

The restriction to $\mathfrak{M}_{t*}$ and to objective ϕ rules out global underdetermination. The assumption that $s_{i,w}$ is computable is also crucial. For example, take the setting to be $\mathfrak{M}_{t*}$ restricted to worlds that present binary data. Add a new atomic sentence q with the valuation $V(q) = \{w \in W : s_{i,u} = g\}$, where g is a fixed, total, non-computable,

---

[18] Just let u satisfy $s_{i,u,t} = s_{i,w,t}$ for $t < t^*$ and $s_{i,u,t} = k$ for $t \geq t^*$.

binary-valued function. Call the resulting model $\mathfrak{B}_{t^*}$. Then we have[19]:

$$w \notin \| \langle MD \rangle_i K_i \, q \|_{\mathfrak{B}_{t^*}}^{t^*}. \tag{11.28}$$

This short foray into the logic of inductive knowability illustrates that the proposed semantics focuses attention precisely where it should—on concrete, methodological considerations like computability and global underdetermination. Furthermore, it is of interest that allowance for serendipitous knowledge allows not only for deductive closure of knowability, but also for a considerable broadening of the scope of inductive knowability beyond that of learnability.[20]

11.13 Fitch's Paradox

It has just been shown that, in learning semantics, the question of inductive knowability raises concrete, familiar, methodological issues. Since traditional epistemic logic makes no contact with learning, either in its syntax or in its models, it focuses attention on the more arcane problem of unknowability due to epistemic self-reference. Although self-referential paradoxes are remote from the concrete business of science, questions of genuine epistemological interest, such as whether it is possible for science to know inductively that it does not know inductively, open the logical floodgates to self-referential curiosities. Alas, one cannot simply ignore them. At the very least, one must construct a firewall against them that does not trivialize the principles of interest.

Consider, for example, the *Moore sentence* for ϕ, defined as follows:

$$Mo_i \, \phi := \phi \wedge \neg K_i \, \phi.$$

The Moore sentence is not knowable in standard epistemic logic, for suppose that i knows that $Mo_i \, \phi$. Then, since knowledge is true, $Mo_i \, \phi$ is also be true, so $\neg K_i \, \phi$ is true. But since $Mo_i \, \phi$ is known, so is conjunct ϕ of $Mo_i \, \phi$, so $K_i \, \phi$ is true. Contradiction. The proof requires only (i) that the conjuncts of a known conjunction are known and (ii) that knowledge is true, both of which are valid in standard, possible world semantics.

That is hardly surprising in itself, but it leads directly[21] to *Fitch's paradox*, the statement that any agent for whom every truth ϕ is knowable is *already omniscient*.[22]

---

[19] The restriction to binary sequences in (11.28) matters. If the range of inputs at each stage might be infinite, then one can add an atomic sentence to $\mathfrak{N}_{t^*}$ that is knowable but true only in worlds that are empirically *infinitely* uncomputable (cf. [19], 7.19).

[20] Cf. Sect. 11.14.1 below for a formal discussion of deductive closure of inductive knowledge.

[21] The ingenious step was taken by Church [7] in an anonymous referee report on Fitch's manuscript.

[22] It suffices that $\Diamond_i$ be the dual of an alethic necessity operator satisfying the rule of necessitation.

$$(\forall \phi)\, (\phi \rightarrow \Diamond_i K_i\, \phi) \rightarrow (\forall \phi)\, (\phi \rightarrow K_i\, \phi). \tag{11.29}$$

For suppose that the consequent of (11.29) is false. Then $(\exists \phi)$ $Mo_i\, \phi$ is true. But $Mo_i\, \phi$ is not knowable. So $Mo_i\, \phi$ is a counterexample to the antecedent of (11.29).

Fitch's paradox is not really paradoxical after the "gotcha" moment when one realizes that denying the consequent yields a true Moore sentence. If "every truth" is restricted to "every scientifically interesting, objective truth", the paradox evaporates. Nonetheless, there is a specialist literature devoted to refuting Fitch's paradox, some authors going so far as to blame proof by contraposition [41]. Therefore, it may be of interest to revisit the question whether the Moore sentence is knowable in learning semantics. The standard argument that $Mo_i\, \phi$ is not knowable assumes that i knows the conjuncts of any conjunction i knows. That step evidently fails in learning semantics, because even belief is not closed under deductive consequence. But inferring ϕ, ψ from $\phi \wedge \psi$ is the easiest of inferences—one need only erase the $\wedge$. It would, therefore, be more sporting to show that $Mo_i\, \phi$ is knowable by an agent whose beliefs are *conjunctively cogent* in the sense that:

$$Coco_i\,(\phi, \psi) := [\,|\,]_i (B_i(\phi \wedge \psi) \leftrightarrow (B_i\, \phi \wedge B_i\, \psi)).$$

Learning semantics yields a novel, positive verdict[23]:

$$\mathfrak{N}_{t*} \models (O_i\, \phi \wedge \neg[D]_i\, \phi \wedge \langle MD \rangle_i K_i\, \phi) \rightarrow \tag{11.30}$$
$$\rightarrow \langle MD \rangle_i (K_i Mo_i\, \phi \wedge Coco_i\,(\phi, \neg K_i\, \phi)).$$

Of course, some sort of aphasia is required to know one's own Moore sentence, but the aphasia now plausibly concerns learning, rather than a trivial, deductive inference.[24] Suppose that i is irrecoverably dogmatic that ϕ. When an acquaintance accuses i of not knowing that ϕ, even though ϕ is true (the evidence for ϕ is abundant), i takes a detached interest in the accusation. Since i's admitted dogmatism precludes her from knowing that ϕ, the knowability of $Mo_i\, \phi$ reduces, for i, to that of ϕ, so i can know $Mo_i\, \phi$ by basing her belief whether $Mo_i\, \phi$ on the evidence concerning ϕ. Since i is dogmatically attached to ϕ, she can maintain conjunctive cogency by believing that $\neg K_i\, \phi$ exactly when she believes that $Mo_i\, \phi$.

The preceding discussion notwithstanding, learning models still permit one to construct self-referential monstrosities by "brute force", using the valuation function: e.g., an atomic sentence can be interpreted to say "i does not believe that she knows me". Such models trivially invalidate the thesis that it is feasible to know that one knows what one knows. The real purpose of the doxastic stability operator S_i is to protect otherwise plausible theses of epistemic logic from that self-referential

---

[23] Note that there is no temporal equivocation here between the time at which $Mo_i\, \phi$ is known and the time at which ϕ is not known, as there is in solutions proposed in temporal dynamic epistemic logic (e.g., [43]).

[24] Alternative learning strategies within the same agent are a familiar theme in the epistemology literature—e.g., [26].

onslaught. Under the hypothesis that $S_i \phi$ obtains, knowledge, learning and having learned are preserved under counterfactual changes of method that do not modify the agent's current learning disposition with respect to ϕ.[25]

Proposition 11.2 *Suppose that* $u \in \|S_i \triangle\|_{\mathfrak{M}_{i*}}^{t*}$ *and* $d \equiv_{\triangle} c_{i,u,t*}$ *and* $\phi \in \triangle$. *Then:*

$$u \in \|K_i \phi\|_{\mathfrak{M}_{i*}}^{t*} \Rightarrow u[d/i, t^*] \in \|K_i \phi\|_{\mathfrak{M}_{i*}}^{t*}, \tag{11.31}$$

and similarly for $\tilde{K}_i$, $\tilde{L}_i$, $\tilde{Led}_i$ *and* Led_i.

11.14 Epistemic Logic Redux

The idea in traditional epistemic logic is to mine intuitions for principles stated entirely in terms of K_i and then to solve backwards for conditions on the accessibility relation that validate them. Modal semantics then serves as a silent bookkeeper that faithfully manages the iteration of K_i, subject to those assumptions.[26] Here is a standard menu of potential principles one might impose:

$$
\begin{aligned}
&\text{N} &&: K_i \phi, \text{ if } \models \phi; \\
&\text{K} &&: K_i(\phi \rightarrow \psi) \rightarrow (K_i \phi \rightarrow K_i \psi); \\
&\text{T} &&: K_i \phi \rightarrow \phi; \\
&\text{B} &&: \phi \rightarrow K_i \neg K_i \neg \phi; \\
&4 &&: K_i \phi \rightarrow K_i K_i \phi; \\
&0.2 &&: \neg K_i \neg K_i \phi \rightarrow K_i \neg K_i \neg \phi; \\
&0.3 &&: K_i(K_i \phi \rightarrow K_i \psi) \vee K_i(K_i \psi \rightarrow K_i \phi); \\
&0.4 &&: \phi \rightarrow (\neg K_i \neg K_i \phi \rightarrow \neg K_i \phi); \\
&5 &&: \neg K_i \phi \rightarrow K_i \neg K_i \phi.
\end{aligned}
$$

For example, principle T says that knowledge is true. In conventional possible worlds semantics, that corresponds to the imposition of reflexivity on the model's accessibility relation. Learning semantics also validates T in its standard form, with an explanation—knowledge requires that one has converged to correct belief:

$$\text{T:} \quad \models K_i \phi \rightarrow \phi. \tag{11.32}$$

The rest of the principles on the menu are plainly wrong for cognitively realistic agents. The standard response is to re-interpret K_i vaguely in terms of abilities, obligations, or ideals, but that changes the subject from knowledge to *je ne sais quoi*.

---

[25] In the author's opinion, finding a semantics for S_i such that $S_i \phi$ is both plausible and yet strong enough to yield the following invariance property proved to be the crux of the entire subject.

[26] In the preceding section, it was shown that this timid, non-explanatory strategy is still subject to error.

It is proposed, instead, to replace material implication $\rightarrow$ with conditional feasibility $\langle MD \vdash\!\!\!\rightarrow_{i,\phi}$. Then, thesis 4 says, plausibly, that there exists a computable inferential procedure that turns knowledge that ϕ into knowledge that one knows that ϕ. The question addressed in this section is which, if any, of the traditional candidate axioms is valid under that interpretation, and under what restrictions, when K_i is interpreted, without equivocation, as inductive knowledge.

11.14.1 Deductive Cogency

Let Δ be a finite set of premises and let Γ be a finite set of conclusions. Suppose that Δ implies Γ, in light of i's information. Maybe i knows neither Δ nor Γ. But is there any concrete, inferential disposition i could set up in herself to guarantee that if she knows the premises in Δ then she knows the conclusions in Γ as well? Yes, if the premises are inferentially stable, for learning semantics validates the following principle, for finite, disjoint Δ, $\Gamma \subseteq L_{BIT}$ and for arbitrary, finite superset Δ' of Δ that is disjoint from Γ:

$$FD: \quad \models \left(S_i\, \Delta' \land [\,|\,]_i\, (\Delta \rightarrow \Gamma) \land K_i\, \Delta \right) \langle MD \vdash\!\!\!\rightarrow_{i,\Delta'} K_i\, \Gamma; \quad (11.33)$$

When $\Delta = \varnothing$ and $\Gamma = \{\phi\}$, thesis (11.33) collapses to a feasible version of the rule N of necessitation:

$$FN: \quad \models [\,|\,]_i\, \phi\, \langle MD \vdash\!\!\!\rightarrow_{i,\Delta'} K_i\, \phi. \quad (11.34)$$

When $\Delta = \{\psi,\ \psi \rightarrow \phi\}$ and $\Gamma = \{\phi\}$, thesis (11.33) collapses to a feasible version of the standard axiom K:

$$FK: \quad \models \left(S_i\, \Delta' \land K_i\, \psi \land K_i(\psi \rightarrow \phi) \right) \langle MD \vdash\!\!\!\rightarrow_{i,\Delta'} K_i\, \phi. \quad (11.35)$$

One may not infer rashly from FN and FK, as one may from the corresponding, traditional axioms N and K, that the knowledge of i is closed under logical consequence, or even that it might be someday. The extension of knowledge by deductive inference must proceed, as it does in the real world, by dint of concrete, cognitive exertion. An inference method that witnesses thesis (11.33) is *pure deductive inference*—inferring elements of Γ from premises Δ, and for *no other reason*. Then convergence to correct belief that Δ in the actual world results in convergence to true belief that Γ in the actual world and guaranteed, eventual avoidance of error regarding the premises in Δ results in guaranteed, eventual avoidance of error regarding the conclusions in Γ. In that sense, pure deductive inference makes knowledge that Γ *epistemically parasitic* on knowledge that Δ. If the parasitic relationship is disrupted, because i has independent reasons for believing some conclusion $\gamma \in \Gamma$, then i might be disposed to fall into error with respect to γ infinitely often in some possible worlds compatible with current information. The validity of (11.33) is closely bound to allowance for serendipity. It has already been shown in terms of $G\, p_k$ and $GF\, p_k$ that (11.33) fails for learning:

$$\text{Thesis (33) is invalid with } \tilde{\mathsf{L}}_i, \widetilde{\mathsf{Led}}_i, \mathsf{Led}_i \text{ in place of } \mathsf{K}_i. \qquad (11.36)$$

Serendipity raises a cautionary moral about the role of deduction in natural science. The world of science is a "dappled" pastiche of mutually incompatible models and theories and missed connections [6]. Heisenberg and Schrödinger even battled over logically equivalent hypotheses, each of which was rigorously tested over distinct domains of phenomena.[27] When contradictions are found, scientists steer around them until some other experts resolve them, as long as the claims in question remain individually testable. When new logical connections are found between formerly disparate research programs, caution is exercised regarding the drawing of inferences from one program to the other until they are cross-checked by new data. Learning semantics explains that logical conservatism. For suppose that there are two independent research programs studying hypotheses ϕ and ψ, respectively, on the basis of disparate sets of phenomena and then it is discovered by a mathematician that ψ is a deductive consequence of ϕ. What to do? Inferring ψ from ϕ would generate new knowledge that ψ from knowledge that ϕ if inquiry whether ϕ has culminated. But if inquiry whether ψ has culminated in knowledge that $\neg\psi$, then inferring ψ from ϕ would *destroy* knowledge that $\neg\psi$. The contrapositive inference from $\neg\psi$ to $\neg\phi$ is fraught with a similar risk of destroying knowledge that ϕ. Hyper-intensional refusal to fire either inference is guaranteed to preserve knowledge of whichever hypothesis is known and leaves the door open to future empirical evidence to resolve the conflict. So far as inquiry after the truth is concerned, deductive consistency may be a hob-goblin, indeed.

11.14.2 Reflection

Suppose that i knows that ϕ. Evidently, she may fail to know that she knows that ϕ—she may not even conceive of the question whether she knows that ϕ unless she is challenged. Or ϕ may say "i does not believe that she knows me". But inattention and self-referential tricks aside, is i even *capable* of knowing that she knows, even though no bell rings [16] when inductive inquiry succeeds? The prospects seem grim:

> ...[Learning in the limit] does not entail that [the learner] knows he knows the answer, since [the learner] may lack any reason to believe that his hypotheses have begun to converge [24].

True, i cannot know *infallibly* that she knows some general truth infallibly, because she cannot even know the general truth infallibly. But there is an easy and natural

---

[27] For a version of the history, cf. [40]. Learning semantics allows for the possibility that each scientist knew his own formulation of quantum mechanics at the same time he disputed the competing formulation. Even neighborhood semantics [36], which models belief as a set of propositions, cannot model that situation.

inferential strategy i can adopt to know *inductively* that she knows inductively that ϕ, and so on, to arbitrary iterations. Define iterated knowledge by recursion:

$$\mathsf{K}_i{}^0\phi := \phi;$$
$$\mathsf{K}_i{}^{k+1}\phi := \mathsf{K}_i\mathsf{K}_i{}^k\phi.$$

Define the sets of sentences:

$$K_i^k(\phi) = \{\mathsf{K}_i{}^{k'}\phi : k' \leq k\};$$
$$K_i^\omega(\phi) = \bigcup_{k\in\mathbb{N}} K_i^k(\phi).$$

Then for each finite Δ containing ϕ and disjoint from $K_i^\omega(\phi)$, we have:[28]

$$\text{F4}^*: \quad \models (\mathsf{S}_i\,\Delta\,\wedge\,\mathsf{K}_i\,\phi)\,\langle\mathsf{MD}\rangle\!\!\mapsto_{i,\Delta}\,K_i^\omega(\phi). \tag{11.37}$$

As a consequence, we have the following, feasible version of the standard (infeasible) reflection principle 4, for each k:

$$\text{F4}: \quad \models (\mathsf{S}_i\,\phi\,\wedge\,\mathsf{K}_i\,\phi)\,\langle\mathsf{MD}\rangle\!\!\mapsto_{i,\phi}\,\mathsf{K}_i{}^k\,\phi. \tag{11.38}$$

A simple inferential strategy that witnesses (11.38) when $k = 2$ is for i to believe at t that she knew that ϕ at t^* if she never stopped believing that ϕ from t^* until t and to believe that she did not know that ϕ if the alternative case obtains. That inference is intuitive: if i remembers that she retracted ϕ between t^* and the current time t, then the retraction shakes her confidence that she knew that ϕ already at t^*. Otherwise, from i's viewpoint, she had persuasive evidence for ϕ at t^* and nothing in particular has dissuaded her since then, so of course she thinks she knew that ϕ at t^*.

In contrast to the situation for deductive closure, learning that one is learning is easy—learning implies that it is determined that one is learning and whatever is determined can be learned by believing it no matter what and never believing its negation. Having learned whether one has learned whether and having learned that one has learned that are both valid by the same inferential strategy invoked to validate (11.38). So we have:

Thesis (11.38) remains valid with $\tilde{\mathsf{K}}_i$, $\tilde{\mathsf{L}}_i$, $\tilde{\mathsf{Led}}_i$, Led_i in place of K_i. (11.39)

---

[28] Strictly speaking, one must restrict $K_i^\omega(\phi)$ to some finite $K_i^k(\phi)$ for the statement to be well-formed, but the proof of validity works for the unrestricted version.

11.14.3 The Unknowable Unknown

For Plato [28], the least flattering epistemic condition is *hubris*—failure to know that one does not know. The first step on the path of inquiry is to eliminate hubris. Thereafter, one comes to know and to know that one knows. But is the fateful, first step feasible? Learning semantics delivers a negative verdict for inductive knowledge, even in the empiricist model $\mathfrak{N}_{t^*}$.

$$\text{F5}': \quad \mathfrak{N}_{t^*} \not\models (\mathsf{S}_i \, \phi \, \wedge \, \neg \mathsf{K}_i \, \phi) \, \langle \mathsf{MD} \}\mapsto_{i,\phi} \, \mathsf{K}_i \neg \mathsf{K}_i \, \phi. \qquad (11.40)$$

The convergence required for knowing that one knows parasitically tracks the convergence of knowledge itself. But failure to know inductively may be witnessed only by ugly surprises in the distant future, and the requirement to have converged already to true belief that one will not be surprised in the future occasions the problem of induction, with which we began. For example, suppose that i has seen enough evidence to convince her that $\mathsf{G}\,\mathsf{p}_k$ until such time as some non-k input is received, causing her to drop her belief that $\mathsf{G}\,\mathsf{p}_k$. Call i's learning method c. Method c yields inductive knowledge that $\mathsf{G}\,\mathsf{p}_k$ in the constantly k world w in which $\mathsf{G}\,\mathsf{p}_k$ is true. Now, suppose that i possesses some magical inferential technique h that guarantees i knowledge now that she does not know that $\mathsf{G}\,\mathsf{p}_k$ if she does not know that $\mathsf{G}\,\mathsf{p}_k$ and that the inferential technique does not alter i's beliefs whether $\mathsf{G}\,\mathsf{p}_k$. Then learning method $h(c)$ must be guaranteed to yield knowledge immediately that c does not produce knowledge that $\mathsf{G}\,\mathsf{p}_k$. Let w_m be the "grue-like" world in which i receives input k until stage m and $k+1$ thereafter. Statement $\mathsf{G}\,\mathsf{p}_k$ is false in w_m, so $h(c)$ stabilizes to belief that $\neg \mathsf{K}_i \mathsf{G}\,\mathsf{p}_k$ immediately in w_m, for *every* m. So $h(c)$ converges to $\neg \mathsf{K}_i \mathsf{G}\,\mathsf{p}_k$ in world w, since w_m agrees empirically with w until m. But, ironically, i *knows* that $\mathsf{G}\,\mathsf{p}_k$ in w because $\mathsf{G}\,\mathsf{p}_k$ is objective in $\mathfrak{N}_{t^*}$ and h holds i's beliefs whether ϕ fixed. So $h(c)$ fails to avoid error in the limit whether $\neg \mathsf{K}_i \mathsf{G}\,\mathsf{p}_k$.

In fact, slight variants of the preceding argument suffice to invalidate the feasible versions of all of the proposed axioms between .4 and 5, so among the standard axioms, only T, FD, and F4 are valid in learning semantics:

$$
\begin{array}{llll}
\text{FB:} & \mathfrak{N}_{t^*} \not\models & (\mathsf{S}_i \, \phi \, \wedge & \neg\phi) \, \langle \mathsf{MD}\}\mapsto_{i,\phi} \quad \mathsf{K}_i \neg \mathsf{K}_i \, \phi; \\
\text{F.2:} & \mathfrak{N}_{t^*} \not\models & (\mathsf{S}_i \, \phi \, \wedge & \neg \mathsf{K}_i \neg \mathsf{K}_i \, \neg\phi) \, \langle \mathsf{MD}\}\mapsto_{i,\phi} \quad \mathsf{K}_i \neg \mathsf{K}_i \, \phi; \\
\text{F.3:} & \mathfrak{N}_{t^*} \not\models & ((\mathsf{S}_i \, \phi \wedge \mathsf{S}_i \, \psi \, \wedge & \mathsf{K}_i \neg \mathsf{K}_i \quad \phi) \, \langle \mathsf{MD}\}\mapsto_{i,\phi,\psi} \mathsf{K}_i \neg \mathsf{K}_i \, \psi) \, \vee \\
& \vee & ((\mathsf{S}_i \, \phi \wedge \mathsf{S}_i \, \phi \, \wedge & \mathsf{K}_i \neg \mathsf{K}_i \quad \psi) \, \langle \mathsf{MD}\}\mapsto_{i,\phi,\psi} \mathsf{K}_i \neg \mathsf{K}_i \, \phi); \\
\text{F.4:} & \mathfrak{N}_{t^*} \not\models & (\mathsf{S}_i \, \phi \wedge \quad \neg\phi \, \wedge & \neg \mathsf{K}_i \, \neg\phi) \, \langle \mathsf{MD}\}\mapsto_{i,\phi} \quad \mathsf{K}_i \neg \mathsf{K}_i \phi.
\end{array}
$$

It suffices to let $\phi = \mathsf{G}\,\mathsf{p}_k$ and $\psi = \mathsf{G}\,\mathsf{p}_{k'}$, for distinct k, k'.

The same examples refute the corresponding versions of the preceding theses, for knowing whether, having learned whether, and having learned that:

Theses (11.40), (FB–F.4) remain invalid with $\tilde{\mathsf{K}}_i$, $\mathsf{L\tilde{e}d}_i$, Led_i in place of K_i. (11.41)

However, it is trivially feasible for i to be learning whether i is not learning whether ϕ when i is not learning whether ϕ—it suffices for i to believe that she is not learning whether ϕ no matter what, since learning begins with operator $[D]_i$:

$$\text{F5L:} \quad \models (S_i \, \phi \, \wedge \, \neg \tilde{L}_i \, \phi) \, \langle MD \rangle \mapsto_{i,\phi} \tilde{L}_i \neg \tilde{L}_i \, \phi. \tag{11.42}$$

11.15 Joint Inductive Knowledge

Plato's original question in the *Meno* [28] was not what knowledge *is*, but whether virtue can be *taught*. Plato assumed that knowledge can be taught, but when knowledge is inductive, that assumption raises an *epistemological* question. Evidently, a knowledgable expert can *exhibit* her inductive knowledge to her pupils, and on a good day, she might even induce true belief in them, but can she really transfer her inductive *knowledge* to them? In a cooperative epistemic enterprise like education, it is natural to assume that knowledge supervenes *jointly* on the learning strategies of the pupils and of the instructor. In that spirit, this section presents an alternative, *joint* version of learning semantics that is friendlier to cooperative epistemic efforts. In the following section, it is shown how it is jointly feasible for the expert and a room full of pupils to acquire common knowledge of the expert's inductive knowledge.

Let $w \in W$, $\mathbf{c}_{w,t} = (c_{w,1,t}, \ldots c_{w,N,t})$ and $\mathbf{d} \in C^N$. Then let $u[\mathbf{d}/t]$ denote the result of substituting $\mathbf{d}$ for $\mathbf{c}_{w,t}$ in w at t. A *joint* CLM satisfies the following, joint invariance postulate, for each $i \in G$, $w \in W$, $\mathbf{d} \in C^N$, and $t \in T$:

$$s_{i,w}|t = s_{i,w[\mathbf{d}/t']}|t. \tag{11.43}$$

Joint information and determination are defined as follows:

$$I_{G,w,t} = \bigcup_{i \in G} I_{i,w,t};$$
$$D_{G,w,t} = \{u \in I_{G,w,t} : \mathbf{c}_{u,t} = \mathbf{c}_{w,t}\};$$

with corresponding operators:

$$\|[I]_G \, \phi\|^t_{\mathfrak{M}_{t*}} = \{w \in W : I_{G,w,t} \subseteq \|\phi\|^{t*}_{\mathfrak{M}_{t*}}\};$$
$$\|[D]_G \, \phi\|^t_{\mathfrak{M}_{t*}} = \{w \in W : D_{G,w,t} \subseteq \|\phi\|^{t*}_{\mathfrak{M}_{t*}}\}.$$

Joint information is weaker than individual information, but joint determination compensates, somewhat, by holding everyone's method fixed. Joint information and determination are no longer guaranteed to be S5 operators, but they *can* be—e.g., everyone gets the same information—so it is useful to have a concise notation for expressing that special case in the object language:

$$\|\mathsf{IS5}_G\|^{t^*}_{\mathfrak{M}_{t*}} = \{w \in W : (\forall u \in I_{G,w,t^*})\, I_{G,u,t^*} = I_{G,u,t^*}\}.$$

Define joint inductive knowledge for i as before, but with joint determination in place of personal determination:

$$\mathsf{K}_{G,i}\,\phi := \mathsf{GC}_i\,\phi \wedge [\mathsf{D}]_G \mathsf{FG} \neg \tilde{\mathsf{E}}_i\,\phi.$$

Joint methodological feasibility expresses the existence of a methodological coordination among the agents that brings about ϕ:

$$\|\langle \mathsf{M}\rangle_G\,\phi\|^t_{\mathfrak{M}_{t*}} = \{w \in W : (\exists \mathbf{d} \in C^N)\, w[\mathbf{d}/t^*] \in \|\phi\|^t_{\mathfrak{M}_{t*}}\}.$$

To define joint conditional feasibility, let $\mathbf{h} = (h_1, \ldots, h_N)$ be an N-sequence of total recursive functions taking values in C, let $\mathbf{h}(\mathbf{c}) = (h_1(c_1), \ldots, h_N(c_N))$, and let $\mathbf{\Delta}$ be an N-sequence of finite subsets of $\mathbf{L_{BIT}}$. Say that $\mathbf{h}$ *preserves* premises in $\mathbf{\Delta}$ if and only if h_i preserves premises in Δ_i, for each $i \in G$ and, similarly, say that $\mathbf{h}$ *depends only* on premises in $\mathbf{\Delta}$ if and only if h_i depends only on premises in Δ_i, for each $i \in G$. Then $\mathbf{h}$ is a *joint inference procedure* if and only if $\mathbf{h}$ is an N-sequence of total recursive functions taking values in C that preserves premises in $\mathbf{\Delta}$ and that depends only on premises in $\mathbf{\Delta}$. Finally, as before, let $\|\psi \langle \mathsf{MD}\rangle\!\!\rightarrow_{G,\mathbf{\Delta}}\,\phi\|^t_{\mathfrak{M}_{t*}}$ denote the set of all $w \in W$ for which there exists joint inference procedure $\mathbf{h}$ such that for all $u \in I_{G,w,t}$:

$$u \in \|\psi\|^t_{\mathfrak{M}_{t*}} \Rightarrow u[\mathbf{h}/t^*] \in \|\phi\|^t_{\mathfrak{M}_{t*}}.$$

It remains only to define a joint version of inferential stability. Define $\mathbf{c} \equiv_{\mathbf{\Delta}} \mathbf{d}$ to hold if and only if $c_i \equiv_{\Delta_i} d_i$, for all $i \in G$. Let $w \in \|\mathsf{S}_{G,i}\mathbf{\Delta}\|^t_{\mathfrak{M}_{t*}}$ hold if and only if for all $\mathbf{d} \in C^N$ such that $\mathbf{c}_{w,t^*} \equiv_{\mathbf{\Delta}} \mathbf{d}$ and for all $u \in D_{G,w,t^*}, t \geq t^*$, and $\delta \in \Delta_i$, if we set $u' = u[\mathbf{d}/t^*]$ and $c_i = c_{i,u,t^*}(= c_{i,w,t^*})$ then:

$$u \in \|\delta\|^{t^*}_{\mathfrak{M}_{t*}} \Leftrightarrow u' \in \|\delta\|^{t^*}_{\mathfrak{M}_{t*}};$$
$$v_{c_i}(s_{i,u}|t,\delta) = v_{d_i}(s_{i,u'}|t,\delta).$$

Crucially, a joint version of Proposition 11.2 holds:

Proposition 11.3 *Suppose that $\phi \in \Delta_i$ and $u \in \|\mathsf{S}_{G,i}\mathbf{\Delta}\|^{t^*}_{\mathfrak{M}_{t*}}$ and let $\mathbf{d} \in C^N$ satisfy $\mathbf{d} \equiv_{\mathbf{\Delta}} \mathbf{c}_{u,t^*}$. Then:*

$$u \in \|\mathsf{K}_{G,i}\,\phi\|^{t^*}_{\mathfrak{M}_{t*}} \Rightarrow u[\mathbf{d}/t^*] \in \|\mathsf{K}_{G,i}\,\phi\|^{t^*}_{\mathfrak{M}_{t*}}. \tag{11.44}$$

11.16 Common Inductive Knowledge

Given the joint perspective outlined in the preceding section and some basic assumptions about how the expert and pupils interact, it is jointly feasible for the expert and her pupils to jointly acquire the expert's inductive knowledge that ϕ. It suffices that the pupils believe that ϕ if the expert does and suspend belief that ϕ otherwise. Each pupil is then an epistemic parasite of the expert, just as the expert is an epistemic parasite of herself when she infers deductive consequences of what she knows.[29] Educated pupils and news media science reporters can serve, in turn, as experts, resulting in a cascade of joint scientific knowledge through the population—as long as, at the core, some expert has direct inductive knowledge based on experience.[30]

It is a further question whether the pupils and the expert can jointly know that they know, know that they know that they know, etc, all the way to joint, common inductive knowledge that ϕ. Define joint, *mutual, inductive knowledge* to level n as follows:

$$K_G{}^0 \phi := \phi;$$
$$K_G{}^{k+1} \phi := \bigwedge_{i \in G} K_{G,i} K_G{}^k \phi.$$

Define *common inductive knowledge* that ϕ as the set of sentences:

$$K_G^\omega(\phi) = \{K_G{}^k \phi : k \in \mathbb{N}\}.$$

It is plausible that a completely trusted, infallible, public announcement that ϕ can generate common knowledge that ϕ. It is less obvious that common *inductive* knowledge is feasible in a room full of computationally bounded pupils who trust their instructor. Learning semantics yields a positive verdict, based on epistemic parasitism and serendipity, in close analogy to the validity argument for F4.

The expert must communicate with the pupils in some way in order to instruct them. It suffices that the pupils receive information sufficient to correctly believe whether the expert believes that ϕ. Let $e \in G$ be the teacher and let $G_- = G \setminus \{e\}$ be the set of pupils. Define the operator "*e teaches* the pupils in G_- whether ϕ" as follows:

$$T_{G,e} \phi := \bigwedge_{j \in G_-} [\mathsf{I}]_G G\tilde{C}_j B_e \phi.$$

Now it is possible to state the *joint feasibility of common inductive knowledge* thesis, which is valid if Δ_e contains ϕ and Δ_i is disjoint from $K_G^\omega(\phi)$, for all $i \in G$:

---

[29] Indeed, the pupils can know consequences of what the expert knows by deriving them directly from what the expert believes, by the same sort of argument.

[30] More generally, the core expertise is grounded in a research group, but the story with respect to the rest of the population is the same.

$$\text{FC:} \quad \models (\text{IS5}_G \land \mathsf{T}_{G,e}\,\phi \land \mathsf{S}_{G,e}\Delta \land \mathsf{K}_{G,e}\,\phi)\,\langle\text{MD}\rangle_{\vdash G,\Delta}\,K_G^\omega(\phi). \quad (11.45)$$

Although the FC principle concerns common inductive knowledge generated and promulgated by a single expert, it sets the stage for a series of similar results that involve common inductive knowledge generated through the cooperation of a team of experts—a topic of current interest in social epistemology (e.g., [25]).

In dynamic epistemic logic, there are models in which public announcements generate common knowledge of what has been announced [3]. But how do public announcements result in anything more than common knowledge of the fact that the announcement was made? Plausibly, common knowledge of what has been announced is common inductive knowledge grounded in the community's joint strategy to disbelieve sources caught in inconsistencies or lies. One potential extension of FC is to validate the possibility of common inductive knowledge of what is reported in a public announcement in models that allow for false announcements.

A familiar assumption in game theory is that the agents have common knowledge of rationality [1]. But how is such knowledge possible and where does it come from? Standard possible worlds semantics has nothing to say, short of a veridical public announcement that all players are rational, but learning semantics provides a plausible, explanatory story. Recall the game-theoretic model described in Sect. 11.7 above. Violation of the kth level of mutual rationality is detectable by horizontal play in a centipede game of corresponding length. If all of the agents have the disposition to continue playing down at the first move in ever longer centipede games, learning semantics provides a determinate, explanatory, account of how common knowledge of rationality is jointly feasible in such a group. And if every agent is disposed to cooperate by playing sideways for a while, the group can just as easily develop inductive common knowledge of partial cooperation![31]

11.17 Conclusion and Future Directions

Learning semantics provides a rich, consistent, and workable conceptual framework for modeling interactions between, and iterations of, belief, information, and time, and inductive versions of learning, learnability, having learned, knowing, knowability, and common knowledge. The key feature of the semantics is an assignment of concrete, computational learning methods to each agent at each time. That makes it possible to define inductive learning and knowledge in terms of actual convergence to the truth and guaranteed, eventual avoidance of error, on the basis of increasing information through time.

Learning semantics has three important advantages over traditional possible worlds models, for applications involving inductive knowledge and learning. (1) It sidesteps inductive skepticism. (2) It imposes no logical or rational idealizations on the agent's belief states or learning procedures. (3) Its semantic arguments pro-

---

[31] This application is due to Jennifer Jhun, personal communication.

vide concrete, methodological explanations why some principles should be valid and others invalid.

It has been shown that learning semantics validates a cognitively plausible version of the familiar modal system S4 and plausibly refutes all of the standard axioms that have been proposed for epistemic logic beyond S4, when material implication is replaced with conditional feasibility. So the logical sky does not fall, after all, when belief and learning are modeled in a cognitively plausible way. The valid versions of the S4 axioms are explained by epistemic parasitism—the fact that an inferred statement can inherit the convergence conditions essential for knowledge from the convergence conditions possessed by known premises. The invalidity of the remaining axioms is explained by the fact that no inferential procedure can detect immediately that convergence might fail in the future, due to unforeseen surprises. Epistemic parasitism also explains how a knowledgable teacher can *convey* her inductive knowledge to her pupils, as opposed to merely instilling true belief in them, and how inductive common knowledge can spread through a community of passive scientific consumers. Generalization of that idea to inductive learning from the behavior of other learners provides a new understanding of the feasibility of common knowledge of rationality (or of irrational cooperation) in games. Learning semantics explains the scope of learnability in terms of concrete, non-learnability arguments of the sort that are familiar in computational learning theory. It also explains how allowance for serendipity in inductive knowledge both broadens the scope of knowability beyond that of learnability and guarantees that knowability (as opposed to knowledge, itself) is closed under logical consequence. Finally, learning semantics provides a surprising, but plausible, explanation of how one can know one's own Moore sentence $\phi \wedge \neg K_i \phi$ without ever failing to derive its conjuncts, and without equivocating on the times at which they come to be known. Traditional possible worlds semantics is irrevocably committed to the contrary conclusion.

The explanatory advantages of learning semantics come with a familiar, scientific cost—any formal model of a complex process must abstract, to some extent, from some potentially relevant details. But that is never an argument for giving up on explanation entirely. Instead, one checks whether improvements in the fidelity of one's model result in greater explanatory scope. In that spirit, the chapter closes with a tentative discussion of some potential refinements and extensions of the framework developed above. A repeated theme in the ensuing discussion is the importance of greater attention to the epistemic context.

11.17.1 Sensitivity and Safety

Learning semantics was designed to deal with inductive skepticism. It does nothing to avert brain-in-a-vat skepticism—the entire input stream could be the same, whether or not ϕ is true. Relevant alternatives semantics was designed to deal with brain-in-a-vat skepticism, but cannot handle inductive skepticism. Therefore, relevant alternatives semantics and learning semantics are not so much competitors as mutually essential *partners*: the former tosses out virulent but distant possibilities of error that would preclude even convergence to the truth, and the latter eventually weeds out the arbitrarily nearby possibilities of error we couldn't have noticed *yet*. Learning semantics would accommodate the full advantages of both approaches if the key modality D_i were re-interpreted in terms of sensitivity or safety. The change is not entirely trivial, since the "fact" that D_i is an S5 operator is appealed to repeatedly in the preceding development, and each such appeal must be re-examined.

11.17.2 Inductive Statistical Knowledge

Most scientific hypotheses are probabilistic—even the variables of deterministic equations are measured with random error. Such hypotheses can be tested, but a statistical test provides a guaranteed bound on chance of error only when the hypothesis is rejected. So if general statistical hypotheses are knowable, they are knowable only inductively.

A plausible semantics for inductive knowledge of statistical hypothesis ϕ is that i believes that ϕ with high chance that remains high in the actual world and the chance that i believes that ϕ goes to zero if ϕ is false. More ambitiously, one might require, in addition, that the chance that i believes that ϕ converges monotonically to 1 in the actual world. The interpretation of error probabilities requires some temporal gymnastics, as it does in frequentist statistics itself. Chance is a kind of disposition that governs future events. The fairness of a coin determines chances for sequences of future flips. But the coin might be bent later, after which *different* chances govern sequences of future events—the situation is much the same as it was for learning dispositions. For the chance disposition operative at t, every outcome prior to t has chance 0 or 1, depending on whether it actually occurred. Therefore, the chance that a belief at t^* based on a sample already taken by t^* is either 0 or 1 according to the chances operative at t^*. So non-trivial error probabilities must pertain to chances operative at some *reference time* t^{**} prior to sampling—e.g., when the experimental design was originally put into motion. Then the truth of ϕ should also be assessed with respect to the chances operative at t^{**} rather than those operative at t^*.

Since epistemic parasitism pertains to convergence in probability as well as to deterministic convergence, it is anticipated that all of the preceding arguments that depend on epistemic parasitism should generalize to the statistical setting. Also, assuming that successive samples are independent and identically distributed (i.i.d.), successive samples probably provide a better approximation to the fixed, underlying sampling distribution, so the positive results concerning learnability and knowability are also expected to carry over. However, if the sampling distribution may change from time to time, as in time series analysis, extra assumptions are required for convergence in probability to the truth—e.g., that the process under study is periodic, or is driven by hidden states that recur infinitely often [39]. Analyzing the connection between such assumptions and statistical, inductive knowability is a scientifically relevant, new direction for modal epistemic logic.

Aside from its intrinsic interest, the extension of learning semantics to probabilistic theories addresses a puzzle concerning the inductive knowability of future, random outcomes. It is plausible that stochastic theories and models can be known inductively, if inductive knowledge is possible at all. It is far less plausible that random outcomes like coin tosses can be known in advance, even inductively. But the non-statistical version of learning semantics underwrites such knowledge—just make a lucky guess at the outcome (serendipity) and believe the guess until the flip is observed, and drop it if it happens to be wrong [13]. The good news is that future coin flips are no longer knowable in statistical learning semantics—the chance of correct belief in the proposition ϕ that the toss will come up heads at future time t is the *joint probability* $p(\mathsf{B}_i \phi \wedge \phi) \leq p(\phi) = 1/2$. What about highly probable future events, such as that your ticket will lose the lottery? They are knowable inductively if their chances of occurring meet the standard for being "high" in the actual convergence condition, but no probabilistic outcome with chance less than one is knowable on the stricter version of the semantics that requires convergence to chance 1 of belief in the actual world.

11.17.3 Questions and Coherence

In light of the aim to model belief more realistically, the logical consistency requirements necessary for knowledge whether ϕ were pared down to the bare minimum required to recover an unambiguous verdict on ϕ for each agent. However, that goes too far. Recall that scientist i can know that the true input sequence is ε by guessing that it is ε until ε is refuted. Suppose that scientist i simultaneously believes *every* hypothesis of the form "the input stream is exactly primitive recursive sequence ε", and is disposed to drop each such hypothesis when it disagrees with the data. Suppose, by serendipity, that the true input stream ε is primitive recursive, so the hypothesis corresponding to ε is true. Then i knows that the future will conform to ε, even though i also believes every possible primitive recursive input stream compatible with current information. That makes inductive knowledge too easy. Furthermore, for someone as aphasic as i, the very concept of belief is called into question. What would i predict to happen at the next stage? Certainly not what she "knows" will happen, since she cannot pick her known theory out of the heap of her alternative, incompatible beliefs. Science may be incoherent overall, but each of its insular paradigms is coherent enough to generate consensus concerning determinate predictions. So normal science within a paradigm is not trivial in the sense under discussion, even though science may remain globally incoherent across paradigms forever. That idea could be modeled in learning semantics by adding a *question under discussion* (q.u.d.) to the epistemic context. The proposal is supported by the current trend in linguistics toward explaining diverse discourse phenomena in terms of such a question [34].

Knowledge of an answer to the q.u.d. requires that the beliefs of the scientist pick out a unique answer, which rules out the easy knowledge just described.

The advantages of hyper-intensionality are retained. Inconsistency across question contexts is permitted and even contradictions within a context that do not result in ambiguity concerning the answer selected are still permitted. The correct answer may even be rejected under some logically equivalent formulation, as long as no formulation of any alternative answer is accepted.

11.17.4 Feasibility Contextualism

Epistemic contextualists (e.g., [23]) hold that the standards for knowledge vary from one context to another—e.g., *raising* a skeptical doubt shifts the epistemic context to one in which the doubt becomes epistemically relevant, so one no longer knows what one knew before the doubt was raised. The idea is appealing, because it does justice both to the plausibility of ordinary knowledge claims and to the apparent force of skeptical doubts. It also addresses a puzzle concerning the psychology of learning. According to learning semantics, it is trivial to know one's own method because the modality D_i holds it fixed and, in the joint version of learning semantics, it is trivial to know what everyone else's method is, because $D_{G,i}$ holds them all fixed. But, according to epistemic contextualism, when the statement known concerns those very methods, possible worlds involving alternative methods become relevant.

Another plausible, but distinct way in which epistemic standards plausibly depend on context is the intrinsic feasibility of answering the question under discussion. For if "knowledge" is a social encomium whose function is to motivate the overall truth-conduciveness of socially distributed inquiry, then that encomium provides maximum guidance over the full range of epistemic contexts if it is bestowed only when the agent achieves the best standard of truth-conduciveness achievable with respect to the question in context. Call that natural idea *feasibility contextualism*. For example, concrete, cat-on-the-mat beliefs that can be decided by observation should be, so such knowledge must be safe or sensitive. General laws cannot be known safely or sensitively, but they are learnable, so knowledge should require that they have been learned. More general, untestable theories are unlearnable, but can be known with serendipity, so knowledge with serendipity suffices in that case.

Feasibility contextualism explains why scientists concerned with an inductive question ignore general, philosophical arguments for inductive skepticism, even though they remain fastidious concerning measurement and data analysis. When general theories are at issue, epistemic standards adjust to accommodate knowledge of them, so safety and sensitivity in the short run are no longer required, but error-detection in the limit can still be optimized by catching the errors as soon as possible. Feasibility contextualism also explains why scientists sometimes brand a hypothesis as "metaphysical" if it is difficult to find a plausible, testable articulation of it. In such cases, we simply run out of applicable senses of truth-conduciveness, so skepticism is back on the table.

Furthermore, feasibility contextualism helps to resolve a residual puzzle about prediction. It may seem that inductive knowledge, even of future, *deterministic* outcomes is too easy—just guess the outcome and wait to see what happens. But it seems fine—exemplary, even—to deduce the same prediction from an inductively known, universal law. There is a temptation to reach for dark, metaphysical explanations—the law endows the prediction with some ontological "oomph" that a bare prediction lacks. Here is a more concrete, linguistic explanation. When one *infers* a prediction from a law, the law remains in context along with the prediction, and when both the law and the prediction are in context, the operative standard for knowledge is naturally understood to be the strongest standard applicable to *both*. Thus, when the prediction is not inferred from a law, the standard of waiting for sensitivity or safety holds sway, but in light of inferring the prediction from a law, the operative standard is inductive. The idea also explains why the same jarring of intuitions does not accompany the inference of "infinitely often" from "always", for in that case the weaker standard already applies to the conclusion.

11.17.5 Justification and Truth-Conduciveness

Scientists prefer unified, cross-testable, explanatory theories over dis-unified, untestable, *ad hoc* theories, a preference popularly known as *Ockham's razor*. Learning semantics, as developed above, does not explain that preference, because a serendipitous guess at a complex law can count as knowledge just as much as a serendipitous guess at a simple one. But the addition of feasibility contextualism suggests such an explanation.[32]

Suppose that the question under discussion is "what is the true form of the polynomial law connecting X and Y?" More precisely, assuming that there exists finite set $S \subseteq \mathbb{N}$ such that the true law has form $Y = f_\theta(X) = \sum_{i \in S} \theta_i X^i$, with $\alpha_i \neq 0$ for each $i \in S$, what is S? Assume that the data are arbitrarily small open rectangles in the XY plane guaranteed to intersect the curve $Y = f_\theta(X)$.[33] Then there is an important structural relationship between the question and the potential information received by i: any information true of a simpler answer is also compatible with the truth of every more complex answer, whereas some information received if a complex answer is true rules out all simpler and incomparable answers. Instead of viewing those properties as merely symptomatic of the simplicity order, take them as definitive, relative to the question in context.[34] The resulting concept of empirical simplicity assumes alternative guises, depending on the question in context and on the space of possible, future, information states. If one is empirically hunting for new particles or other objects, extra particles make the theory more complex.

---

[32] For the details, cf. [21].

[33] In the statistical setting sketched above, the data can be understood, more realistically, as data points sampled independently from the joint distribution generated by the model $Y = f(X) + e$, where e is a normally distributed random variable independent from X and Y that has mean 0 that represents all stray sources of inaccuracy in measurement. Running up the sample size corresponds to narrowing the rectangles in the non-statistical semantics.

[34] That is an over-simplification, but it points in the right direction. Cf. [21] for a better proposal.

If one is selecting among theories with free parameters and the parameterization is well-behaved, additional parameters add extra complexity. If one compares theories that entail different symmetry groups, breaking symmetry adds complexity. If one compares theories with more or fewer causes, extra causes add complexity. And so on. It follows from the general definition of empirical simplicity that every learning method capable of inductively learning the true answer to the question can be forced to believe in each successively more complex answer before ultimately converging to the true one. That is an unavoidable, structural feature of the question's semantics, relative to the space of possible information states.

Truth conduciveness is efficient pursuit of the truth. Efficient pursuit entails that one close with the quarry as directly as possible—a random walk or gratuitous aerobatic loops or U-turns during the approach stretch the very concept of pursuit. Gratuitous doxastic loops and U-turns correspond to needless retractions and re-visitations of former beliefs. Thus, retraction minimization is not a mere, pragmatic afterthought—it is constitutive of the very concept of truth-conduciveness. Therefore, feasibility contextualism implies that retractions prior to convergence should be minimized, relative to the current question context. So parties to the question context should forgive methods that change their minds from simpler to more complex theories, since every learning method for the question can be forced to retract that often prior to convergence—but they should forgive no more retractions than those. It can also be shown that the *only* learning methods for the question that minimize worst-case retractions are those that follow Ockham's razor, by selecting the uniquely simplest theory compatible with available information. So Ockham's razor is explained by feasibility contextualism.

The preceding explanation assumes that a fairly rich question is in context, but what if only the known law is in context? Think of the belief $Y = f_\theta(X)$ as posing the default, binary question "yes or no" unless a more refined question is in context. There is a learning strategy that retracts at most once when the answer is $Y = f_\theta(X)$ (no, yes) and at most twice when the contrary answer is true (yes, no, yes). No tighter bounds are feasible, so that performance is also optimally truth-conducive. The *only* optimal methods are methods that wait for law forms simpler than $Y = f_\theta(X)$ to be refuted before yielding a positive verdict for $Y = f_\theta(X)$. Thus, feasibility contextualism still entails that $Y = f_\theta(X)$ cannot be known unless it is believed in accordance with Ockham's razor.

Acknowledgments This work was supported generously by the John Templeton Foundation, award number 24145. The author has had the basic ideas in mind since the early 1990s and has enjoyed discussions on the topic (in approximately temporal order, and keeping in mind the author's fading memory) with Oliver Schulte, Vincent Hendricks, Stig Andur Pedersen, Nuel Belnap, Johan van Benthem, Horacio Arlo-Costa, Hans Kamp, Fred Dretske, Peter van der Schraaf, Jonathan Vogel, Sherrilyn Roush, Ivan Verano, Hanti Lin, Timothy Williamson, Gregory Wheeler, Alexandru Baltag, Sonja Smets, Gerhard Schurz, Liam Bright, Wesley Holliday, Thomas Icard, Alexandru Radelescu, Clark Glymour, Kevin Zollman, Eric Martin, Konstantin Genin, and Ted Shear, none of whom should be blamed for the outcome. The author is also indebted to Ted Shear and Johan van Benthem for detailed comments on drafts. Preliminary versions of the system have been presented in talks at Carnegie Mellon University (2006), the Logic and Methodology Workshop at Stanford (2011), the workshop on Games Interactive Rationality, and Learning in Lund (2012, 2013), and the Workshop on Social Dynamics of Information Change in Amsterdam (2013).

Proofs of Propositions

Proof of Proposition 11.1 Just let $L_d(\sigma, \psi)$ return 1 if $\psi = @_{t^*} \phi$ and σ is an initial segment of $s_{i,u,t}$ and return 0 otherwise. $\square$

Proof of Proposition 11.2 Abbreviate:

$$c = c_{i,u,t^*};$$
$$x = u[d/i, t^*].$$

Assume that $\phi \in \Delta$ and that:

$$d \equiv_\Delta c; \tag{11.46}$$
$$u \in \|\mathsf{S}_i \Delta\|_{\mathfrak{M}_{t^*}}^{t^*}; \tag{11.47}$$
$$u \in \|\mathsf{K}_i \phi\|_{\mathfrak{M}_{t^*}}^{t^*}. \tag{11.48}$$

From (11.48) we have:

$$u \in \|\mathsf{GC}_i \phi\|_{\mathfrak{M}_{t^*}}^{t^*}; \tag{11.49}$$
$$y \in \|\mathsf{FG}\neg\tilde{\mathsf{E}}_i \phi\|_{\mathfrak{M}_{t^*}}^{t^*}, \text{ for all } y \in D_{i,u,t^*}. \tag{11.50}$$

It suffices to show that:

$$x \in \|\mathsf{GC}_i \phi\|_{\mathfrak{M}_{t^*}}^{t^*}; \tag{11.51}$$
$$y \in \|\mathsf{FG}\neg\tilde{\mathsf{E}}_i \phi\|_{\mathfrak{M}_{t^*}}^{t^*}, \text{ for all } y \in D_{i,x,t^*}. \tag{11.52}$$

From (11.46) to (11.47), we have that:

$$u \in \|\phi\|_{\mathfrak{M}_{t^*}}^{t^*} \Leftrightarrow x \in \|\phi\|_{\mathfrak{M}_{t^*}}^{t^*}; \tag{11.53}$$
$$u \in \|\mathsf{G}[\mathsf{B}]_i \phi\|_{\mathfrak{M}_{t^*}}^{t^*} \Leftrightarrow x \in \|\mathsf{G}[\mathsf{B}]_i \phi\|_{\mathfrak{M}_{t^*}}^{t^*}; \tag{11.54}$$
$$u \in \|\mathsf{G}\langle\mathsf{B}\rangle_i \phi\|_{\mathfrak{M}_{t^*}}^{t^*} \Leftrightarrow x \in \|\mathsf{G}\langle\mathsf{B}\rangle_i \phi\|_{\mathfrak{M}_{t^*}}^{t^*}. \tag{11.55}$$

So requirement (11.51) follows from (11.49).

For requirement (11.52), let $y \in D_{i,x,t^*}$. Then $s_{i,y}|t^* = s_{i,x}|t^* = s_{i,u[d/i,t^*]}|t^*$. So $s_{i,y}|t^* = s_{i,u}|t^*$, by (11.1). Let $z = y[c/i, t^*]$. So $s_{i,z}|t^* = s_{i,u}|t^*$, again by (11.1) and, hence, $z \in D_{i,u,t^*}$. So it follows from (11.50) that:

$$z \in \|\mathsf{FG}\neg\tilde{\mathsf{E}}_i\,\phi\|_{\mathfrak{M}_{t^*}}^{t^*};\tag{11.56}$$

and from (11.46) to (11.47) that:

$$y \in \|\phi\|_{\mathfrak{M}_{t^*}}^{t^*} \Leftrightarrow z \in \|\phi\|_{\mathfrak{M}_{t^*}}^{t^*};\tag{11.57}$$

$$y \in \|\mathsf{FG[B]}_i\,\phi\|_{\mathfrak{M}_{t^*}}^{t^*} \Leftrightarrow z \in \|\mathsf{FG[B]}_i\,\phi\|_{\mathfrak{M}_{t^*}}^{t^*};\tag{11.58}$$

$$y \in \|\mathsf{FG\langle B\rangle}_i\,\phi\|_{\mathfrak{M}_{t^*}}^{t^*} \Leftrightarrow z \in \|\mathsf{FG\langle B\rangle}_i\,\phi\|_{\mathfrak{M}_{t^*}}^{t^*}.\tag{11.59}$$

Requirement (11.52) follows directly from (11.56) to (11.59). □

Proof of Proposition 11.3 Let $\mathbf{d} \in C^N$ and let $u \in W$. Abbreviate:

$$\mathbf{c} = \mathbf{c}_{i,u,t^*};$$
$$x = u[\mathbf{d}/t^*].$$

Assume that $\phi \in \Delta_i$ and that:

$$d_i \equiv_\phi c_i;\tag{11.60}$$

$$u \in \|\mathsf{S}_{G,i}\Delta\|_{\mathfrak{M}_{t^*}}^{t^*};\tag{11.61}$$

$$u \in \|\mathsf{K}_{G,i}\,\phi\|_{\mathfrak{M}_{t^*}}^{t^*}.\tag{11.62}$$

Proceed as in the preceding proof, with D_{G,u,t^*}, D_{G,x,t^*} in place of D_{i,u,t^*}, D_{i,x,t^*}. The argument for requirement (11.51) is the same as before. For requirement (11.52), let $y \in D_{G,x,t^*}$. So $y \in D_{i,x,t^*}$, for some $i \in G$. Then $s_{i,y}|t^* = s_{i,x}|t^* = s_{i,u[\mathbf{d}/t^*]}|t^*$. So $s_{i,y}|t^* = s_{i,u}|t^*$, by (11.43). Let $z = y[\mathbf{c}/t^*]$. So $s_{i,z}|t^* = s_{i,u}|t^*$, again by (11.43) and, hence, $z \in D_{i,u,t^*} \subseteq D_{G,u,t^*}$. Continue as in the preceding proof. □

Proofs of Selected Statements

Proof of (11.7) *and* (11.8) Let $w \in W$ be given. To witness the first claim, define learning method c so that:

$$L_c(\sigma, \phi) = \begin{cases} 1 & \text{if } \phi = @_{t^*}\mathsf{G}\,p_k \text{ and } (\forall t : t^* \le t \le \mathsf{lh}(\sigma))\,\sigma(t) = k; \\ 1 & \text{if } \phi = @_{t^*}\neg\mathsf{G}\,p_k \text{ and } (\exists t : t^* \le t \le \mathsf{lh}(\sigma))\,\sigma(t) \neq k; \\ 0 & \text{otherwise.} \end{cases}$$

The method that witnesses the second claim is similar, except that $\neg$ and $\neq$ are moved from the second clause to the first. □

Proof of (11.9) *and* (11.10) The proof of the second statement is similar to that of the first. For the first statement, suppose for contradiction that c satisfies (11.3) and (11.4).

It suffices to construct $\varepsilon \in E_0$ such that (*) both (11.3) and (11.4) are false in arbitrary world w such that $e_w = \varepsilon$. A purely learning theoretic argument suffices. Construct ε by adding chunks in successive stages as follows, where $c = h(c_{w',i,t^*})$. At stage 0, present σ. Let $n > 0$. At stage $2n$, present k until L_c returns 1 for $@_{t^*}\mathsf{FG}\,\mathsf{p}_k$. Learning function L_c must return 1 for $@_{t^*}\mathsf{FG}\,\mathsf{p}_k$ eventually, because if L_c never takes the bait, you continue to present k and L_c fails to converge to belief that $@_{t^*}\mathsf{FG}\,\mathsf{p}_k$ even though it is true, contradicting the hypothesis. At that point, proceed to stage $2n + 1$. At stage $2n + 1$, the demon presents $k + 1$ until L_c returns 0 for $@_{t^*}\mathsf{FG}\,\mathsf{p}_k$. Learning function L_c must return 0 for $@_{t^*}\mathsf{FG}\,\mathsf{p}_k$ eventually, because if L_c never takes the bait, you continue to present $k + 1$ and L_c fails to converge to belief that $@_{t^*}\neg\mathsf{G}\,\mathsf{p}_k$ even though it is true, contradicting the hypothesis. At that point, proceed to stage $2n + 2$. You pass through each stage, producing ε that satisfies (*). □

Proof of (11.28) The proof follows [19, Proposition 7.15]. Suppose the contrary. Then we can use the witnessing L_d and $u \in I_{i,w,t^*}$ to compute $g(t)$, for $t \geq t^*$ (for $t < t^*$, use a lookup table). Say that finite input sequence σ of length t is t'-dead if and only if $L_d(\sigma', @_{t^*}\phi) = 0$, for each extension σ' of σ of length t'. By (11.23), $g|(t + 1)$ is never t'-dead, but by König's Lemma and (11.26), there exists $t' \geq t + 1$ such that every σ of length $t + 1$ that is distinct from $g|t$ is t'-dead. Then $g|(t + 1)$ is the unique sequence σ that is not t'-dead. Return the last entry of that sequence. □

Proof of (11.31) By hypothesis, ϕ is knowable in w at t^*. Since ϕ is knowable, let L_c and world $u \in I_{i,w,t^*}$ witness that fact. Let L_d believe that ϕ in all circumstances and believe, deny, or suspend belief for both $\neg\mathsf{K}_i\,\phi$ and $\mathsf{Mo}_i\,\phi$ whenever L_c does the same for ϕ. Since ϕ is assumed to be false in some world compatible with information, i does not know that ϕ. Recall that in $\mathfrak{N}_{t^*}$, (i) the inputs to i do not depend on i's learning method and (ii) the truth value of ϕ does not depend on i's learning method. Due to L_d's dogmatic belief that ϕ, the case hypothesis, and (i) and (ii), there is no world in I_{i,w,t^*} in which $\mathsf{K}_i\phi$ is true, so we have that $[\,\mathsf{I}\,]_i(\mathsf{Mo}_i\,\phi \leftrightarrow \phi)$ is true in w. So by (i) and (ii), agent i knows that $\mathsf{Mo}_i\,\phi$. By construction, i is conjunctively cogent with respect to $\mathsf{Mo}_i\,\phi$. □

Proof of (11.33) Let Δ, Γ be finite and mutually disjoint subsets of $\mathbf{L}_{\mathsf{BIT}}$. Let $\Delta \subseteq \Delta'$ and $\Delta' \cap \Gamma = \varnothing$. Define total recursive g such that:

$$g(c, \langle\sigma\rangle, \ulcorner\phi\urcorner)) = \begin{cases} 1 & \text{if } \phi = @_{t^*}\gamma \ \wedge \ \gamma \in \Gamma \ \wedge \ (\forall\delta \in \Delta)\, L_c(\sigma, @_{t^*}\delta) = 1; \\ 0 & \text{if } \phi = @_{t^*}\neg\gamma \ \wedge \ \gamma \in \Gamma; \\ L_c(\sigma, \phi) & \text{otherwise.} \end{cases}$$

The following lemma is a familiar consequence of the s-m-n theorem of recursive function theory:

$$(\forall \text{ t.r. } f)(\exists \text{ t.r. } h)(\forall c, x, y \in \mathbb{N}) \ \phi_{h(c)}(x, y) = f(c, x, y). \tag{11.63}$$

Apply (11.63) to obtain total recursive h such that $L_{h(c)}(\sigma, \phi) = g(c, \langle\sigma\rangle, \ulcorner\phi\urcorner)$. By the definition of h and the fact that Δ' is disjoint from Γ, we have that:

$$c \equiv_{\Delta'} h(c), \tag{11.64}$$

for each $c \in C$, and that for all $z \in W$, $t \in T$ and $\gamma \in \Gamma$:

$$L_{h(c)}(s_{i,z}|t, \, @_{t^*} \neg \mathsf{K}_i{}^k \, \gamma) = 0; \tag{11.65}$$
$$L_{h(c)}(s_{i,z}|t, \, @_{t^*} \mathsf{K}_i{}^k \, \gamma) = 1 \, \Leftrightarrow \, (\forall \delta \in \Delta) \, L_{h(c)}(s_{i,z}|t', \, @_{t^*} \delta) = 1. \tag{11.66}$$

Suppose that $u \in I_{i,w,t^*}$ satisfies:

$$u \in \|\mathsf{S}_i \Delta'\|_{\mathfrak{M}_{t^*}}^{t^*}; \tag{11.67}$$
$$u \in \|[\mathsf{I}]_i(\Delta \to \Gamma)\|_{\mathfrak{M}_{t^*}}^{t^*}; \tag{11.68}$$
$$u \in \|\mathsf{K}_i \Delta\|_{\mathfrak{M}_{t^*}}^{t^*}. \tag{11.69}$$

Abbreviate:

$$c = c_{i,u,t^*};$$
$$x = u[h(c)/i, t^*].$$

So from (11.64), (11.67) and (11.69), obtain via Proposition 11.2 that for each $\delta \in \Delta$:

$$x \in \|\mathsf{K}_i \, \delta\|_{\mathfrak{M}_{t^*}}^{t^*}. \tag{11.70}$$

So for each $\delta \in \Delta$:

$$x \in \|\mathsf{GC}_i \, \delta\|_{\mathfrak{M}_{t^*}}^{t^*}; \tag{11.71}$$
$$y \in \|\mathsf{FG}\neg\tilde{\mathsf{E}}_i \, \delta\|_{\mathfrak{M}_{t^*}}^{t^*}, \text{ for all } y \in D_{i,x,t^*}. \tag{11.72}$$

It suffices to show the following requirements, for each $\gamma \in \Gamma$:

$$x \in \|\mathsf{GC}_i \, \gamma\|_{\mathfrak{M}_{t^*}}^{t^*}; \tag{11.73}$$
$$y \in \|\mathsf{FG}\neg\tilde{\mathsf{E}}_i \, \gamma\|_{\mathfrak{M}_{t^*}}^{t^*}, \text{ for all } y \in D_{i,x,t^*}. \tag{11.74}$$

Let $\gamma \in \Gamma$. For requirement (11.73), we have by (11.1) that $x \in I_{i,u,t^*}$, so (11.68) and (11.71) yield that:

$$x \in \|\gamma\|_{\mathfrak{M}_{t^*}}^{t^*}. \tag{11.75}$$

So (11.71) and (11.75), together with properties (11.65)–(11.66), yield requirement (11.73). For requirement (11.74), suppose that $y \in D_{i,x,t^*}$. So by (11.1), $y \in I_{i,u,t^*}$. So (*) together with (11.68) and (11.72) yield requirement (11.74). □

Proof of statement (11.38) Define total recursive f as follows:

$$f(c, \langle \sigma \rangle, g(\psi)) = \begin{cases} 1 & \text{if } (\exists k)\, \psi = @_{t*}K_i{}^k\phi \wedge \\ & (\forall t' : t^* \le t' \le t)(\psi = @_{t*}K_i{}^k\phi \wedge L_c(\sigma|t', \phi) = 1); \\ 0 & \text{if } (\exists k)\, \psi = @_{t*}K_i{}^k\phi \wedge \\ & (\exists t' : t^* \le t' \le t)(\psi = @_{t*}K_i{}^k\phi \wedge L_c(\sigma|t', \phi) = 0); \\ 0 & \text{if } (\exists k)\, \psi = @_{t*}\neg K_i{}^k\phi \wedge \\ & (\forall t' : t^* \le t' \le t)(\psi = @_{t*}K_i{}^k\phi \wedge L_c(\sigma|t', \phi) = 1); \\ 1 & \text{if } (\exists k)\, \psi = @_{t*}\neg K_i{}^k\phi \wedge \\ & (\exists t' : t^* \le t' \le t)(\psi = @_{t*}K_i{}^k\phi \wedge L_c(\sigma|t', \phi) = 0); \\ L_c(\sigma, \phi) & \text{otherwise.} \end{cases}$$

Apply (11.63) to obtain h such that $L_{h(c)}(\sigma, \psi) = f(c, \langle \sigma \rangle, \ulcorner \phi \urcorner)$, for all $c \in \mathbb{N}$. Suppose that Δ includes ϕ and is disjoint from $K_i^\omega(\phi)$. By the definition of h, we have that for all $c \in C$:

$$c \equiv_\Delta h(c); \tag{11.76}$$

so h preserves Δ. Moreover, by construction, h depends only on Δ. Furthermore, for all $z \in W$, $t \in T$, and $k \in \mathbb{N}$:

$$L_{h(c)}(s_{i,z}|t, @_{t*}K_i{}^k\phi) = 1 \Leftrightarrow (\forall t' : t^* \le t' \le t)\, L_{h(c)}(s_{i,z}|t', @_{t*}\phi) = 1; \tag{11.77}$$

$$L_{h(c)}(s_{i,z}|t, @_{t*}\neg K_i{}^k\phi) = 1 \Leftrightarrow (\exists t' : t^* \le t' \le t)\, L_{h(c)}(s_{i,z}|t', @_{t*}\phi) = 0. \tag{11.78}$$

Suppose that $u \in I_{i,w,t^*}$ satisfies:

$$u \in \|S_i \Delta\|_{\mathfrak{M}_{t*}}^{t^*}; \tag{11.79}$$

$$u \in \|K_i \phi\|_{\mathfrak{M}_{t*}}^{t^*}. \tag{11.80}$$

Abbreviate:

$$c = c_{i,u,t^*};$$

$$x = u[h(c)/i, t^*].$$

From (11.76), (11.79) and (11.80), obtain via Proposition 11.2 that $x \in \|K_i \phi\|_{\mathfrak{M}_{t*}}^{t^*} = \|K_i{}^1\phi\|_{\mathfrak{M}_{t*}}^{t^*}$. Therefore, $x \in \|\phi\|_{\mathfrak{M}_{t*}}^{t^*} = \|K_i{}^0\phi\|_{\mathfrak{M}_{t*}}^{t^*}$. So we have the base case $x \in \|K^1(\phi)\|_{\mathfrak{M}_{t*}}^{t^*}$.

Next, assume for induction that $x \in \|K^{k+1}(\phi)\|_{\mathfrak{M}_{t*}}^{t^*}$. So:

$$x \in \|K_i K_i{}^k\phi\|_{\mathfrak{M}_{t*}}^{t^*}; \tag{11.81}$$

and, therefore:

$$x \in \|\mathsf{GC}_i\mathsf{K}_i{}^k\,\phi\|_{\mathfrak{M}_{t_*}}^{t^*};\tag{11.82}$$

$$y \in \|\mathsf{FG}\neg\tilde{\mathsf{E}}\mathsf{K}_i{}^k\,\phi\|_{\mathfrak{M}_{t_*}}^{t^*},\ \text{ for all } y \in D_{i,x,t^*}.\tag{11.83}$$

For $x \in \|K^{k+2}(\phi)\|_{\mathfrak{M}_{t_*}}^{t^*}$, it suffices to show that: $x \in \|\mathsf{K}_i\mathsf{K}_i\mathsf{K}_i{}^k\,\phi\|_{\mathfrak{M}_{t_*}}^{t^*}$. For that, it suffices, in turn, to show:

$$x \in \|\mathsf{GC}_i\mathsf{K}_i\mathsf{K}_i{}^k\,\phi\|_{\mathfrak{M}_{t_*}}^{t^*};\tag{11.84}$$

$$x \in \|\mathsf{FG}\neg\tilde{\mathsf{E}}\mathsf{K}_i\mathsf{K}_i{}^k\,\phi\|_{\mathfrak{M}_{t_*}}^{t^*},\ \text{ for all } y \in D_{i,x,t^*}.\tag{11.85}$$

Requirement (11.84) expands to the requirements:

$$x \in \|\mathsf{K}_i\mathsf{K}_i{}^k\,\phi\|_{\mathfrak{M}_{t_*}}^{t^*};\tag{11.86}$$

$$x \in \|\mathsf{G[B]}_i\mathsf{K}_i\mathsf{K}_i{}^k\,\phi\|_{\mathfrak{M}_{t_*}}^{t^*};\tag{11.87}$$

$$x \in \|\mathsf{G}\langle\mathsf{B}\rangle_i\mathsf{K}_i\mathsf{K}_i{}^k\,\phi\|_{\mathfrak{M}_{t_*}}^{t^*}.\tag{11.88}$$

Requirement (11.86) is just (11.81). Hence, (11.82) yields:

$$x \in \|\mathsf{G[B]}_i\mathsf{K}_i{}^k\,\phi\|_{\mathfrak{M}_{t_*}}^{t^*};\tag{11.89}$$

$$x \in \|\mathsf{G}\langle\mathsf{B}\rangle_i\mathsf{K}_i{}^k\,\phi\|_{\mathfrak{M}_{t_*}}^{t^*}.\tag{11.90}$$

Requirements (11.87)–(11.88) follow from (11.89) to (11.90) and properties (11.77)–(11.78) of h.

For requirement (11.85), suppose that $y \in D_{i,x,t^*}$. It suffices to show that for all $y \in D_{i,x,t^*}$:

$$y \in \|\mathsf{GF[B]}_i\neg\mathsf{K}_i\mathsf{K}_i{}^k\,\phi\|_{\mathfrak{M}_{t_*}}^{t^*} \Rightarrow y \notin \|\mathsf{K}_i\mathsf{K}_i{}^k\,\phi\|_{\mathfrak{M}_{t_*}}^{t^*};\tag{11.91}$$

$$y \in \|\mathsf{GF[B]}_i\mathsf{K}_i\mathsf{K}_i{}^k\,\phi\|_{\mathfrak{M}_{t_*}}^{t^*} \Rightarrow y \in \|\mathsf{K}_i\mathsf{K}_i{}^k\,\phi\|_{\mathfrak{M}_{t_*}}^{t^*}.\tag{11.92}$$

For requirement (11.91), suppose that:

$$y \in \|\mathsf{GF[B]}_i\neg\mathsf{K}_i\mathsf{K}_i{}^k\,\phi\|_{\mathfrak{M}_{t_*}}^{t^*}\tag{11.93}$$

Then by property (11.78) of h, there exists $t \geq t^*$ such that $y \notin \|\mathsf{B}_i\,\phi\|_{\mathfrak{M}_{t_*}}^{t}$, so by property (11.77), we have that $y \notin \|\mathsf{B}_i\mathsf{K}_i{}^k\,\phi\|_{\mathfrak{M}_{t_*}}^{t^*}$. So $y \notin \|\mathsf{K}_i\mathsf{K}_i{}^k\,\phi\|_{\mathfrak{M}_{t_*}}^{t^*}$.

For requirement (11.92), suppose that:

$$y \in \|\mathsf{GF[B]}_i\mathsf{K}_i\mathsf{K}_i{}^k\,\phi\|_{\mathfrak{M}_{t_*}}^{t^*}\tag{11.94}$$

For the consequent $y \in \|\mathsf{K}_i\mathsf{K}_i{}^k\,\phi\|_{\mathfrak{M}_{t_*}}^{t^*}$, it suffices, as usual, to show the requirements:

$$y \in \|\mathsf{GC}_i\mathsf{K}_i{}^k\,\phi\|_{\mathfrak{M}_{t*}}^{t^*};\tag{11.95}$$

$$z \in \|\mathsf{FG}\neg\tilde{\mathsf{E}}\mathsf{K}_i{}^k\,\phi\|_{\mathfrak{M}_{t*}}^{t^*}, \text{ for all } z \in D_{i,y,t^*}.\tag{11.96}$$

Requirement (11.96) is just (11.83), since $D_{i,y,t^*} = D_{i,u,t^*}$. Requirement (11.95) expands to:

$$y \in \|\mathsf{K}_i{}^k\,\phi\|_{\mathfrak{M}_{t*}}^{t^*};\tag{11.97}$$

$$y \in \|\mathsf{G[B]}_i\mathsf{K}_i{}^k\,\phi\|_{\mathfrak{M}_{t*}}^{t^*};\tag{11.98}$$

$$y \in \|\mathsf{G\langle B\rangle}_i\mathsf{K}_i{}^k\,\phi\|_{\mathfrak{M}_{t*}}^{t^*}.\tag{11.99}$$

For requirement (11.97), we have from (11.94) and property (11.77) of h that $y \in \|\mathsf{GF[B]}_i\mathsf{K}_i{}^k\,\phi\|_{\mathfrak{M}_{t*}}^{t^*}$. So $y \in \|\mathsf{K}_i{}^k\,\phi\|_{\mathfrak{M}_{t*}}^{t^*}$, by (11.83). For requirement (11.98), note that (11.94), along with property (11.77) of h implies that $y \in \|\mathsf{G[B]}_i\,\phi\|_{\mathfrak{M}_{t*}}^{t^*}$, which implies requirement (11.98) in light of property (11.77) and requirement (11.99) in light of property (11.78). □

Proof of statement (11.39) For the Led_i case, follow the proof of (11.38) with Led_i in place of K_i and $\tilde{\mathsf{C}}_i$ in place of $\tilde{\mathsf{E}}_i$. For the $\tilde{\mathsf{L}}_i$ case, make corresponding substitutions and ignore the actual convergence requirements. For the $\mathsf{L\tilde{e}d}_i$ case, add cases for actual convergence to true belief that $\neg\phi$. For the $\tilde{\mathsf{K}}_i$ case, do the same, but retain $\tilde{\mathsf{C}}_i$ in place of $\tilde{\mathsf{E}}_i$. □

Proof of statement (11.40) Let $w = (\varepsilon, \mathbf{c})$ be a world in $\mathfrak{N}_{t*}$. Let total recursive h preserve belief whether $\phi = \mathsf{G}\mathsf{p}_k$. Let c^* be as in the proof of statement (11.7). Let $\mathbf{c} \in C^N$ and let $w_{\varepsilon'} = (\varepsilon', \mathbf{c}[c^*/i]_{t*})$, for arbitrary $\varepsilon' \in E_0$. Let $\tau(t) = \varepsilon(t)$ for $t < t^*$ and let $\tau(t) = k$ for $t \geq t^*$. Let $\tau_t(t') = \tau(t')$ for $t' \geq t$ and let $\tau_t(t') = k+1$ for $t' \geq t$. It is easy to verify that for all $t \geq t^*$:

$$w_\tau \in \|\mathsf{K}_i\mathsf{G}\mathsf{p}_k\|_{\mathfrak{N}_{t*}}^{t^*};\tag{11.100}$$

$$w_{\tau_t} \in \|\neg\mathsf{K}_i\mathsf{G}\mathsf{p}_k\|_{\mathfrak{N}_{t*}}^{t^*}.\tag{11.101}$$

Since the truth of $\mathsf{G}\mathsf{p}_k$ does not depend on methods in $\mathfrak{N}_{t*}$, we have for all $t \geq t^*$ that:

$$w_{\tau_t} \in I_{i,w,t^*} \cap \|\mathsf{S}_i\,\mathsf{G}\mathsf{p}_k\|_{\mathfrak{N}_{t*}}^{t^*} \cap \|\neg\mathsf{K}_i\mathsf{G}\mathsf{p}_k\|_{\mathfrak{N}_{t*}}^{t^*}.\tag{11.102}$$

So it suffices to show that $w_{\tau_t}[h(c^*)/i, t^*] \notin \|\mathsf{K}_i\neg\mathsf{K}_i\mathsf{G}\mathsf{p}_k\|_{\mathfrak{N}_{t*}}^{t^*}$. For that it suffices to show that at least one of the following statements holds:

$$w_{\tau_t}[h(c^*)/i, t^*] \notin \|\mathsf{GC}_i\neg\mathsf{K}_i\,\phi\|_{\mathfrak{M}_{t*}}^{t^*};\tag{11.103}$$

$$w_{\tau_t}[h(c^*)/i, t^*] \notin \|[\mathsf{D}]_i\mathsf{FG}\neg\tilde{\mathsf{E}}_i\neg\mathsf{K}_i\,\phi\|_{\mathfrak{M}_{t*}}^{t^*}.\tag{11.104}$$

Case 1: $w_{\tau_t}[h(c^*)/i, t^*] \notin \|GB_i \neg K_i G p_k\|_{\mathfrak{N}_{t*}}^{t^*}$, for some $t \geq t^*$. So (11.103) holds, in light of (11.101).

Case 2: $w_{\tau_t}[h(c^*)/i, t^*] \in \|GB_i \neg K_i G p_k\|_{\mathfrak{N}_{t*}}^{t^*}$, for all $t \geq t^*$. Then since $\tau|t = \tau_t|t$, for each $t \geq t^*$, we have that $w_\tau[h(c^*)/i]_{\geq t^*} \in \|GB_i \neg K_i G p_k\|_{\mathfrak{N}_{t*}}^{t^*}$. Note that $w_\tau \in I_{w_{\tau_t}, i, t^*}$ by construction and (11.1). So (11.104) holds, in light of (11.100). □

Invalidity of statements (FB–F.4) One merely has to check that the respective antecedents of the various conditionals are satisfied by each world w_{τ_t} in the proof of (11.40). For (FB), observe that $\neg\phi$ is true in w_{τ_t}, by construction. For (F.2), observe that c^* suspends belief concerning $\neg K_i \neg\phi$. For (F.4), observe both that c^* suspends belief concerning $\neg\phi$ and that $\neg\phi$ is true in w_{τ_t}. For (F.3), let $w \in W$ and let total recursive h preserve both ϕ and ψ. To refute the second disjunct of (F.3) in w, let c^{**} follow the strategy of c^* with respect to ϕ, except that c^{**} believes that $\neg K_i \psi$ no matter what. Then, due to c^{**}'s suspension of belief whether ψ at t^*, we have that c^{**} witnesses the truth of $K_i \neg K_i \psi$ in every world, so the argument for (11.40) establishes the falsehood of the second disjunct of (F.3) in w. Reversing the roles of ϕ and ψ establishes that the first disjunct of (F.3) is also false in w. □

Proof of statement (11.45) Define total recursive f_e just as in the proof of (11.38), except that $K_i^k \phi$ is replaced with $K_G^k \phi$. For $j \in G_-$, define total recursive f_j just like f_e, but with the condition $L_c(\sigma|t', B_i \phi) = 1)$ in place of condition $L_c(\sigma|t', \phi) = 1)$. Apply (11.63) to each f_i to obtain respective, total recursive function h_i. Let $\mathbf{h} = (h_1, \ldots, h_N)$.

Suppose that $\phi \in \Delta_e$ and that $\Delta_i \cap K_G^\omega = \emptyset$, for each $i \in G$. By the definition of $\mathbf{h}$, we have that for all $\mathbf{c} \in C^N$:

$$\mathbf{c} \equiv_\Delta \mathbf{h}(\mathbf{c}); \tag{11.105}$$

so $\mathbf{h}$ preserves Δ. By construction, h depends only on Δ. Furthermore, for all $i \in G$, $z \in W, t \in T$, and $k \in \mathbb{N}$:

$$L_{h(c)}(s_{i,z}|t, @_{t^*} \neg K_G^k \phi) = 1 \Leftrightarrow L_{h(c)}(s_{i,z}|t, @_{t^*} K_G^k \phi) = 0; \tag{11.106}$$

Suppose that $u \in I_{i,w,t^*}$ satisfies:

$$u \in \|IS5_G\|_{\mathfrak{M}_{t*}}^{t^*}; \tag{11.107}$$

$$u \in \|T_{G,e} \phi\|_{\mathfrak{M}_{t*}}^{t^*}; \tag{11.108}$$

$$u \in \|S_{G,e} \Delta\|_{\mathfrak{M}_{t*}}^{t^*}; \tag{11.109}$$

$$u \in \|K_{G,e} \phi\|_{\mathfrak{M}_{t*}}^{t^*}. \tag{11.110}$$

Abbreviate:

$$\mathbf{c} = \mathbf{c}_{i,u,t^*};$$
$$x = u[\mathbf{h}(\mathbf{c})/i, t^*].$$

From (11.105), (11.109) and (11.110), obtain via Proposition 11.3 that:

$$x \in \|\mathsf{K}_{G,e}\,\phi\|_{\mathfrak{M}_{t^*}}^{t^*}. \qquad (11.111)$$

Note that for $j \in G_-$ and $z \in W$ we have by the definition of $\mathbf{h}$ that:

$$L_{h_e(c_e)}(s_{i,z}|t, @_{t^*}\mathsf{K}_G{}^k\,\phi) = 1 \Leftrightarrow (\forall t' : t^* \leq t' \leq t)\, L_{h_e(c_e)}(s_{i,z}|t', @_{t^*}\phi) = 1; \qquad (11.112)$$

$$L_{h_j(c_j)}(s_{i,z}|t, @_{t^*}\mathsf{K}_G{}^k\,\phi) = 1 \Leftrightarrow (\forall t' : t^* \leq t' \leq t)\, L_{h_j(c_j)}(s_{i,z}|t', @_{t^*}\mathsf{B}_{G,e}\,\phi) = 1; \qquad (11.113)$$

Let $y \in D_{G,x,t^*} \subseteq I_{G,x,t^*}$. So $y \in I_{G,u,t^*}$ by (11.43). Then by (11.108), we have for all $j \in G_-$ that $y \in \|\mathsf{G}\tilde{\mathsf{C}}_j\mathsf{B}_e\,\phi\|_{\mathfrak{M}_{t^*}}^{t^*}$. Hence, by (11.112)–(11.113), we have for all $i \in G$, $y \in D_{G,x,t^*}$, and $k \in \mathbb{N}$:

$$L_{h_i(c_i)}(s_{i,y}|t, @_{t^*}\mathsf{K}_G{}^k\,\phi) = 1 \Leftrightarrow (\forall t' : t^* \leq t' \leq t)$$
$$\times\, L_{h_e(c_e)}(s_{i,y}|t', @_{t^*}\phi) = 1; \qquad (11.114)$$

By (11.111), (11.106), and (11.114), we have that $x \in \|\mathsf{K}_{G,j}\,\phi\|_{\mathfrak{M}_{t^*}}^{t^*}$, for all $j \in G_-$, so again by (11.111) we have $x \in \|\mathsf{K}_G{}^1\,\phi\|_{\mathfrak{M}_{t^*}}^{t^*}$, and hence, that $x \in \|\phi\|_{\mathfrak{M}_{t^*}}^{t^*} = \|\mathsf{K}_G{}^0\,\phi\|_{\mathfrak{M}_{t^*}}^{t^*}$. Thus, we have the base case $\|K_G^1(\phi)\|_{\mathfrak{M}_{t^*}}^{t^*}$.

Next, assume for induction that $x \in \|K_G^{k+1}(\phi)\|_{\mathfrak{M}_{t^*}}^{t^*}$ and show that $x \in \|K_G^{k+2}(\phi)\|_{\mathfrak{M}_{t^*}}^{t^*}$. By the induction hypothesis, we have, for each $i \in G$ that:

$$x \in \|\mathsf{K}_{G,i}\mathsf{K}_G{}^k\,\phi\|_{\mathfrak{M}_{t^*}}^{t^*}; \qquad (11.115)$$

and, therefore:

$$x \in \|\mathsf{GC}_i\mathsf{K}_G{}^k\,\phi\|_{\mathfrak{M}_{t^*}}^{t^*}; \qquad (11.116)$$

$$y \in \|\mathsf{FG}\neg\tilde{\mathsf{E}}\mathsf{K}_G{}^k\,\phi\|_{\mathfrak{M}_{t^*}}^{t^*}, \text{ for all } y \in D_{G,x,t^*}. \qquad (11.117)$$

For $x \in \|K_G^{k+2}(\phi)\|_{\mathfrak{M}_{t^*}}^{t^*}$, it suffices to show, for each $i \in G$, that: $x \in \|\mathsf{K}_{G,i}\mathsf{K}_{G,i}\mathsf{K}_G{}^k\,\phi\|_{\mathfrak{M}_{t^*}}^{t^*}$. For that, it suffices, in turn, to show:

$$x \in \|\mathsf{GC}_i\mathsf{K}_{G,i}\mathsf{K}_G{}^k\,\phi\|_{\mathfrak{M}_{t^*}}^{t^*}; \qquad (11.118)$$

$$x \in \|\mathsf{FG}\neg\tilde{\mathsf{E}}\mathsf{K}_{G,i}\mathsf{K}_G{}^k\,\phi\|_{\mathfrak{M}_{t^*}}^{t^*}, \text{ for all } y \in D_{G,x,t^*}. \qquad (11.119)$$

Requirement (11.118) expands to the requirements:

$$x \in \|K_{G,i}K_G{}^k \phi\|^{t^*}_{\mathfrak{M}_{t^*}}; \tag{11.120}$$

$$x \in \|G[B]_i K_{G,i}K_G{}^k \phi\|^{t^*}_{\mathfrak{M}_{t^*}}; \tag{11.121}$$

$$x \in \|G\langle B\rangle_i K_{G,i}K_G{}^k \phi\|^{t^*}_{\mathfrak{M}_{t^*}}. \tag{11.122}$$

Requirement (11.120) is just (11.115). Hence, (11.116) yields:

$$x \in \|G[B]_i K_G{}^k \phi\|^{t^*}_{\mathfrak{M}_{t^*}}; \tag{11.123}$$

$$x \in \|G\langle B\rangle_i K_G{}^k \phi\|^{t^*}_{\mathfrak{M}_{t^*}}. \tag{11.124}$$

Requirements (11.121)–(11.122) follow from (11.123) to (11.124) and properties (11.106) and (11.114) of **h**.

For reuirement (11.119), suppose that $y \in D_{G,x,t^*}$. It suffices to show that for all $y \in D_{G,x,t^*}$:

$$y \in \|GF[B]_i \neg K_{G,i}K_G{}^k \phi\|^{t^*}_{\mathfrak{M}_{t^*}} \Rightarrow y \notin \|K_{G,i}K_G{}^k \phi\|^{t^*}_{\mathfrak{M}_{t^*}}; \tag{11.125}$$

$$y \in \|GF[B]_i K_{G,i}K_G{}^k \phi\|^{t^*}_{\mathfrak{M}_{t^*}} \Rightarrow y \in \|K_{G,i}K_G{}^k \phi\|^{t^*}_{\mathfrak{M}_{t^*}}. \tag{11.126}$$

For requirement (11.125), suppose that $y \in \|GF[B]_i \neg K_{G,i}K_G{}^k \phi\|^{t^*}_{\mathfrak{M}_{t^*}}$. Then by properties (11.114) and (11.106) of h, we have that $y \notin \|G[B]_i K_G{}^k \phi\|^{t^*}_{\mathfrak{M}_{t^*}}$. So $y \notin \|K_{G,i}K_G{}^k \phi\|^{t^*}_{\mathfrak{M}_{t^*}}$.

For requirement (11.126), suppose that:

$$y \in \|GF[B]_i K_{G,i}K_G{}^k \phi\|^{t^*}_{\mathfrak{M}_{t^*}} \tag{11.127}$$

For the consequent $y \in \|K_{G,i}K_G{}^k \phi\|^{t^*}_{\mathfrak{M}_{t^*}}$, it suffices, as usual, to show the requirements:

$$y \in \|GC_i K_G{}^k \phi\|^{t^*}_{\mathfrak{M}_{t^*}}; \tag{11.128}$$

$$z \in \|FG\neg \tilde{E}_i K_G{}^k \phi\|^{t^*}_{\mathfrak{M}_{t^*}}, \text{ for all } z \in D_{G,y,t^*}. \tag{11.129}$$

Requirement (11.129) is just (11.117), since $D_{G,y,t^*} = D_{G,u,t^*}$ by (11.107).[35] Requirement (11.128) expands to:

---

[35] This is the proof's only appeal to the S5 property for information.

$$y \in \|K_G{}^k \phi\|_{\mathfrak{M}_{t*}}^{t^*};$$ (11.130)

$$y \in \|G[B]_i K_G{}^k \phi\|_{\mathfrak{M}_{t*}}^{t^*};$$ (11.131)

$$y \in \|G\langle B\rangle_i K_G{}^k \phi\|_{\mathfrak{M}_{t*}}^{t^*}.$$ (11.132)

For requirement (11.130), we have from (11.127) and property (11.114) of h that $y \in \|GF[B]_i K_G{}^k \phi\|_{\mathfrak{M}_{t*}}^{t^*}$. So $y \in \|K_G{}^k \phi\|_{\mathfrak{M}_{t*}}^{t^*}$, by (11.117). For requirement (11.131), note that (11.127), along with property (11.114) of **h** implies that $y \in \|G[B]_i \phi\|_{\mathfrak{M}_{t*}}^{t^*}$, which again, in light of property (11.114) implies requirement (11.131). Requirement (11.132) is then immediate by property (11.106) of **h**.   □

References

1. Aumann R (1995) Backward induction and common knowledge of rationality. Games Econ Behav 8:6–19
2. Benfey OT (1958) August Kekulé and the birth of the structural theory of organic chemistry in 1848. J Chem Educ 35:21–23
3. van Benthem J (2010) Modal logic for open minds. CSLI Lecture Notes, Standord
4. van Benthem J (2011) Logical dynamics of information and interaction. Cambridge University Press, Cambridge
5. Carnap R (1945) On inductive logic. Philos Sci 12:72–97
6. Cartwright N (1999) The dappled world: a study of the boundaries of science. Cambridge University Press, Cambridge
7. Church A (2009) Referee reports on Fitch's 'A Definition of Value'. In: Salerno J (ed) New essays on the knowability paradox. Oxford University Press, Oxford, pp 13–20
8. Dretske F (1981) Knowledge and the flow of information. The MIT Press, Cambridge (Mass.)
9. Duhem P (1914) La théorie physique son objet et sa structure, 2nd edn. Chevalier et Rivi'ere, Paris
10. Gierasimczuk N (2010) Knowing one's limits: logical analysis of inductive inference. ILLC Dissertation Series DS-2010-11, Institute for Logic, Language, and Computation
11. Gold EM (1967) Language Identification in the limit. Inf Control 10:447–474
12. Hempel C (1945) Studies in the logic of confirmation. Mind 213:1–26
13. Hendricks V (2001) The convergence of scientific knowledge. Springer, Dordrecht
14. Holliday W (2012) Epistemic closure and epistemic logic I: Relevant alternatives and subjunctivism. J Philos Logic (Forthcoming)
15. Jain S, Osherson D, Royer J, Sharma A (1999) Systems that learn: an introduction to learning theory, 2nd edn. Bradford, New York
16. James W (1896) The will to believe. New World 5:327–347
17. Kamp H (1971) Formal properties of 'Now'. Theoria 27:227–274
18. Kant I (1782/1787) Critique of pure reason (trans: Guyer P, Wood A). Cambridge University Press, Cambridge (1997)
19. Kelly KT (1996) The logic of reliable inquiry. Oxford University Press, New York
20. Kelly KT (2001) The logic of success. Br J Philos Sci 51:639–666
21. Kelly KT (2010) Simplicity, truth, and probability. In: Bandyopadhyay PS, Forster M (eds) Handbook for the philosophy of statistics. Elsevier, Dordrecht
22. Laudan L (1981) A confutation of convergent realism. Philos Sci 48:19–49
23. Lewis D (1996) Elusive knowledge. Australas J Philos 74:549–567
24. Martin E, Osherson D (1998) Elements of scientific inquiry. The MIT Press, Cambridge (Mass.)

25. Mayo-Wilson C (2011) The problem of piece-meal induction. Philos Sci 78:864–874
26. Nozick R (1981) Philosophical explanations. Harvard University Press, Cambridge
27. Peirce CS (1878) How to make our ideas clear. Popular Sci Mon 12:286–302
28. Plato (1949) Meno (trans: Jowett B). Bobbs-Merrill, Indianapolis
29. Poincare H (1904) Science and hypothesis. The Walter Scott Publishing Co, New York
30. Popper KR (1935) Logik der Forschung. Springer, Vienna
31. Pritchard D (2007) Anti-luck epistemology. Synthese 158:277–298
32. Putnam H (1963) 'Degree of Confirmation' and inductive logic. In: Schilpp A (ed) The philosophy of Rudolf Carnap. Open Court, LaSalle
33. Reichenbach H (1949) The theory of probability. Cambridge University Press, London
34. Roberts C (2012) Information structure: towards an integrated formal theory of pragmatics. Semant Pragmat 5:1–69
35. Roush S (2007) Tracking truth: knowledge, evidence, and science. Oxford University Press, Oxford
36. Scott D (1970) Advice in modal logic. In: Lambert K (ed) Philosophicawl problems in logic, pp 143–173
37. Sosa E (1999) How to defeat opposition to Moore. Philos Perspect 13:141–154
38. Vogel J (1987) Tracking, closure, and inductive knowledge. In: Luper-Foy S (ed) The possibility of knowledge: Nozick and his critics. Rowman and Littlefield, Totowa
39. Wei W (1989) Time series analysis: univariate and multivariate methods. Addison-Wesley, New York
40. van der Werden A (1973) From matrix mechanics and wave mechanics to unified quantum mechanics. In: Mehra J (ed) The physicist's conception of nature. Reidel, Dordrecht, pp 276–293
41. Williamson T (1993) Verificationism and non-distributive knowledge. Australas J Philos 71:78–86
42. Williamson T (2000) Knowledge and its limits. Oxford University Press, Oxford
43. Yap A, Tomohiro H (2009) Dynamic epistemic logic and branching temporal structure. Synthese 169:259–281

Chapter 12
Structures for Epistemic Logic

Nick Bezhanishvili and Wiebe van der Hoek

Abstract In this chapter we overview the main structures of epistemic and doxastic logic. We start by discussing the most celebrated models for epistemic logic, i.e., epistemic Kripke structures. These structures provide a very intuitive interpretation of the accessibility relation, based on the notion of information. This also naturally extends to the multi-agent case. Based on Kripke models, we then look at systems that add a temporal or a computational component, and those that provide a 'grounded' semantics for knowledge. We also pay special attention to 'non-standard semantics' for knowledge and belief, i.e., semantics that are not based on an underlying relation on the sets of states. In particular, we discuss here neighbourhood semantics and topological semantics. In all of these approaches, we can clearly point at streams of results that are inspired by work by Johan van Benthem. We are extremely pleased and honoured to be part of this book dedicated to his work and influences.

12.1 Introduction

Epistemic modal logic in a narrow sense studies and formalises reasoning about *knowledge*. In a wider sense, it gives a formal account of the informational attitude that agents may have, and covers notions like knowledge, belief, uncertainty, and hence incomplete or partial information. As is so often the case in modal logic, such formalised notions become really interesting when studied in a broader context. When doing so, epistemic logic in a wider sense in fact relates to most of the other

N. Bezhanishvili (✉)
Institute for Logic, Language and Computation, University of Amsterdam, Amsterdam,
The Netherlands
e-mail: N.Bezhanishvili@uva.nl

W. van der Hoek
Department of Computer Science, University of Liverpool, Liverpool, UK
e-mail: wiebe@liv.ac.uk

A. Baltag and S. Smets (eds.), *Johan van Benthem on Logic*
and Information Dynamics, Outstanding Contributions to Logic 5,
DOI: 10.1007/978-3-319-06025-5_12, © Springer International Publishing Switzerland 2014

chapters in this book. What if we add a notion of time or action (Chap. 20): how does an agent revise its beliefs (cf. Chap. 7), or update its knowledge (Chap. 6)? And even if we fix one of the notions of interest, say knowledge, if there are many agents, how can we ascribe some level of knowledge to the group, and how do we represent knowledge of one agent about the knowledge (or ignorance, for that matter) of another (cf. Sect. 12.2.1)? What are reasonable requirements on the interaction between knowledge and strategic action (Chap. 14), and how is uncertainty dealt with in more general, qualitative models of agency (Chap. 11)?

Hintikka, notably through [69], is broadly acknowledged as the father of modern epistemic modal logic. Indeed, [69] gives an account of knowledge and belief based on Kripke models. In a nutshell, crucial for this semantics is the notion of a set of *states* or *worlds*, together with a binary relation for each agent, determining which worlds 'look the same', for the agent, or 'carry the same information'. Many disciplines realised the importance of the formalisation of knowledge, using Kripke semantics (or a close relative of it). Examples of such disciplines are Artificial Intelligence (notably Moore's [91] on actions and knowledge) philosophy [70], game theory (see Aumann's formalisation of common knowledge, [4]. Aumann's survey [5] on interactive epistemology can easily be recast using a Kripke semantics), and agents (the underlying semantics of the famous BDI approach by Rao and Georgeff for instance [96] is based on Kripke models). For more references to those disciplines, we refer to the chapters on the relevant topics in this book.

Another important aspect of this chapter is to review the neighbourhood and topological semantics of epistemic and doxastic logic. Topological semantics of modal logic originates from the ground-laying work of McKinsey and Tarski [87]. In recent years there has been a surge of interest in this semantics not least because of its connection to epistemic and doxastic logic. van Benthem (not surprisingly) has been in the centre of the recent developments in the area.

In short, the aim of this chapter is to explain some of the most popular semantic structures used to model informational attitudes, and at several places we have plenty of opportunity to point at van Benthem's contribution to the field. In fact, Johan's work spins over the different semantics of epistemic logic that we discuss here. It builds bridges between many different areas. Therefore, we cannot think of a better place for publishing this chapter than a volume dedicated to Johan's contributions.

The chapter is organised as follows. In Sect. 12.2, we briefly introduce a family of modal epistemic languages that are interpreted on the structures to be discussed. We also discuss the most popular axiom systems for multi-agent knowledge and belief. Then, in Sect. 12.3, we introduce probably the most celebrated structures for epistemic logic, i.e., epistemic Kripke structures. Based on Kripke models, we then add a temporal or a computational component, and also provide a 'grounded' semantics for knowledge. In Sect. 12.4 we consider 'non-standard', or 'generalised' semantics for knowledge and belief, i.e., semantics that are not based on an underlying relation on the sets of states. In particular, we discuss here neighbourhood semantics and topological semantics. In Sect. 12.5, we conclude.

12.2 Epistemic Logic: Language and Axiom Systems

Let us first agree on a formal language for reasoning about information of agents.

Definition 12.1 (*A Suite of Modal Epistemic Languages*) We assume a set At $=$ $\{p, q, p_1, \dots\}$ of atomic propositions, a set of agents Ag $= \{1, \dots, m\}$ and a set of modal operators Op. Then we define the language L(At, Op, Ag) by the following BNF:

$$\varphi := p \mid \neg\varphi \mid (\varphi \wedge \varphi) \mid \Box\varphi$$

where $p \in$ At and $\Box \in$ Op.

Abbreviations for the connective $\vee$ ('disjunction'), $\rightarrow$ ('implication') and $\leftrightarrow$ ('equivalence') are standard. Moreover the dual $\Diamond\varphi$ of an operator $\Box\varphi$ is defined as $\neg\Box\neg\varphi$. Typically, the set Op depends on Ag. For instance, the language for multi-agent epistemic logic is L(At, Op, Ag) with Op $= \{K_a \mid a \in$ Ag$\}$, that is, we have a knowledge operator for every agent. $K_a\varphi$ reads 'agent a knows that φ', so that $K_a\varphi \vee K_a\neg\varphi$ would indicate that agent a knows *whether* φ (which should be contrasted with the 'propositional' validity $(K_a\varphi \vee \neg K_a\varphi)$ and the 'modal' validity $K_a(\varphi \vee \neg\varphi)$). The dual of K_a is often written M_a. So for instance $M_a\varphi \wedge M_a\psi \wedge \neg M_a(\varphi \wedge \psi)$ says that agent a holds both φ and ψ to be possible, although he knows that φ and ψ do not both hold. For a language in which one wants to study interaction properties between knowledge and belief, we would have Op $= \{K_a, B_a \mid a \in$ Ag$\}$. A typical interaction property in such a language would be

$$K_a\varphi \rightarrow B_a\varphi \tag{12.1}$$

but of course not the other way around, since one would like the two notions of knowledge and belief not to collapse: [76] for instance assume (12.1) and $B_a\varphi \rightarrow K_aB_a\varphi$ as an axiom, but warn that 'the interesting formula $B_a\varphi \rightarrow B_aK_a\varphi$ is not included in our system', the reason for it being that knowledge and belief would become the same. This lead [73] to study 'how many' interaction between the two notions one can allow before they become the same: the latter study is in fact an application of *correspondence theory*, a notion developed by van Benthem in his PhD thesis [9], to which we will come back later (note also that Chap. 22 in this volume is dedicated to this topic).

So what then are the properties of knowledge and belief proper, and how do the two notions differ? To start with the latter question, in modal logic it is often assumed that knowledge is *veridical*, where belief is *not*. In other words, knowledge satisfies $K_a\varphi \rightarrow \varphi$ as a principle, while for belief, it is consistent to say that a believes certain φ, although φ is in fact false. Of course, agent a will not consider this a possibility: indeed, in the 'standard' logic for belief, we have that $B_a(B_a\varphi \rightarrow \varphi)$ is valid. The axioms **Taut** and **K**$_\Box$ and the inference rules **MP** and **Nec**$_\Box$ form the modal logic K (Table 12.1). For knowledge, one then often adds veridicality (**T**), and *positive*- (**4**)

Table 12.1 Basic modal and epistemic and doxastic axioms

Basic modal properties		Epistemic and Doxastic properties	
Taut	All instantiations of propositional tautologies	**D**	$\neg\Box\varphi$
$\mathbf{K}_\Box$	$\Box(\varphi \to \psi) \to (\Box\varphi \to \Box\psi)$	**T**	$\Box\varphi \to \varphi$
MP	From φ and $\varphi \to \psi$, infer ψ	**4**	$\Box\varphi \to \Box\Box\varphi$
$\mathbf{Nec}_\Box$	From φ, infer $\Box\varphi$	**5**	$\neg\Box\varphi \to \Box\neg\Box\varphi$

and *negative introspection* (**5**). For belief, veridicality is usually replaced by the weaker axiom *consistency* (**D**). If there are m agents (i.e., m knowledge operators $K_1, \ldots, K_m$), the axioms of $K + \{\mathbf{T}, \mathbf{4}\}$ are referred to as $S4_m$, the axioms of $K + \{\mathbf{T}, \mathbf{4}, \mathbf{5}\}$ are referred to as $S5_m$, and we call the agents in the latter case *epistemic agents*. The arguably most popular logic for belief $K + \{\mathbf{D}, \mathbf{4}, \mathbf{5}\}$ is usually denoted $KD45_m$. In fact, agents that are veridical and negatively introspective must already be positively introspective (and hence epistemic agents), i.e., $K + \{\mathbf{T}, \mathbf{5}\} \vdash \mathbf{4}$.

A *normal modal logic* is a set of formulas L containing all instances of axioms of K and closed under the rules **MP** and $\mathbf{Nec}_\Box$. We write $L \vdash \varphi$ if φ is a theorem of L.

12.2.1 Multi-agent Notions

To speak with van Benthem, *One is a lonely number* [12], and the notions of knowledge and belief become only more interesting in a *multi-agent* setting (and, as [12] also argues, in a *dynamic* setting, but for this, we refer to Chap. 6). Let $A \subseteq \mathsf{Ag}$ be a set of agents. One can then introduce an operator that says that everybody in A knows something: $E_A\varphi = \bigwedge_{a \in A} K_a\varphi$ (instead of E_{Ag}, write E). Obviously, this does not expand the logic's expressivity, but it *does* indeed decrease the descriptive complexity [50]: even in $S5_m$, having the operator E_A (if $\mid A \mid \geq 4$) makes the language more succinct.

One could in a similar way, using disjunctions, define a notion of 'somebody knows'. However, arguably a more interesting (and logically stronger) notion is that of *distributed knowledge* $D_A\varphi$ in a group A of φ. For instance, if a knows that every modal logician is interested in epistemic logic, and b knows that van Benthem is a modal logician, then there is distributed knowledge among a and b that van Benthem is interested in epistemic logic, even if none of the agents needs to know this.

Arguably the most interesting epistemic group notion is that of common knowledge of a group. Common knowledge of φ is supposed to mean that everybody knows φ, and moreover, everybody knows *that*, and everybody knows .... If our language would allow for infinite formulas, common knowledge would be captured by the infinite conjunction

$$E\varphi \wedge EE\varphi \wedge EEE\varphi \wedge \ldots \tag{12.2}$$

Table 12.2 Axioms and inference rules for group-, common- and distributed knowledge

Everybody's and common knowledge		Distributed knowledge	
E	$E\varphi \leftrightarrow \bigwedge_{a \in \mathsf{Ag}} K_a \varphi$	D_1	$\bigvee_{a \in \mathsf{Ag}} K_a \varphi \to D\varphi$
K_C	$C(\varphi \to \psi) \to (C\varphi \to C\psi)$	K_D	$D(\varphi \to \psi) \to (D\varphi \to D\psi)$
Mix	$C\varphi \to (\varphi \wedge EC\varphi)$	T	$D\varphi \to \varphi$
Ind	$C(\varphi \to E\varphi) \to (\varphi \to C\varphi)$	5	$\neg D\varphi \to D\neg D\varphi$
Nec_C	From φ, infer $C\varphi$	Nec_D	From φ, infer $D\varphi$

Phrased negatively, φ is *not* common knowledge as long as somebody considers it possible that somebody considers it possible that …somebody considers it possible that φ is false. Common knowledge explains why social laws (like a green traffic light) work: when approaching a green light, I not only know that I have right of way, but I also know that you know this, and that you know that I know it, etc. In games, common knowledge of rationality explains why certain strategies can be singled out as being in equilibrium (see Chap. 14). The axioms for common knowledge are K_C, **Mix, Ind** and inference rule Nec_C from Table 12.2. If L_m is a logic with m operators K_a, then adding the axioms **E**, K_C, **Mix, Ind** and rule Nec_C is denoted by L_m^C. Similarly for L_m^D for L with the axioms for distributed knowledge added. Sometimes, the axiom **Ind** is replaced by the inference rule

From $\varphi \to E(\psi \wedge C\varphi)$ infer $\varphi \to C\psi$.   **(RInd)**

Axioms and inference rules for the epistemic group notions discussed here are given in Table 12.2. They are usually added to $S5_m$. Notions of common belief and distributed belief also exist: for those, one usually adds slightly weaker axioms.

As for instance explained by van Benthem in [14], we can define common knowledge $C\varphi$ also as a fixed point of the following operator:

$$\varphi \wedge Ex \tag{12.3}$$

A fixed point ψ of this operator satisfies $\psi = \varphi \wedge E\psi = \varphi \wedge E(\varphi \wedge E\psi)\ldots$ in which one recognises the **Mix** axiom. Moreover, the **Ind** axiom states that we have a greatest fixed point, which can be obtained by iterated application of the operator to $\top$, giving $\varphi \wedge E\top, \varphi \wedge E(\varphi \wedge E\top), \varphi \wedge E(\varphi \wedge E(\varphi \wedge E\top))$, etc., see Sect. 12.3.1 for more details.

Common knowledge is obviously the strongest epistemic notion discussed here, while distributed knowledge is the weakest (see 12.4). As a consequence, common knowledge will be typically obtained for 'weak' formulas φ only (even if everybody in a group knows that Santa Claus does not exist, this does not have to be common knowledge), while distributed knowledge may pertain to 'strong' statements (no matter how large the group is, there is distributed knowledge about the fact whether there are two members sharing their birthday). In terms of Fagin et al. [47], common knowledge is what 'any fool' knows, while distributed knowledge characterises what

Table 12.3 Axioms and inference rules for linear temporal logic with next and until

Next		Next and until	
$\mathbf{K}_\bigcirc$	$\bigcirc(\varphi \to \psi) \to (\bigcirc\varphi \to \bigcirc\psi)$		
$\mathbf{T_2}$	$\bigcirc\neg\varphi \leftrightarrow \neg\bigcirc\varphi$	$\mathbf{T_3}$	$\varphi U \psi \leftrightarrow \psi \vee (\varphi \wedge \bigcirc(\varphi U \psi))$
$\mathbf{Nec}_\bigcirc$	From φ, infer $\bigcirc\varphi$	$\mathbf{RT}$	From $\varphi' \to \neg\psi \wedge \bigcirc\varphi$, infer $\varphi' \to \neg(\varphi U \psi)$

the 'wise man' knows. It is not difficult to see that when one adds the principles of Table 12.2 to $S5_m$, both the wise man and the fool are epistemic agents.

$$C\varphi \;\Rightarrow\; E\varphi \;\Rightarrow\; K_a\varphi \;\Rightarrow\; D\varphi \;\Rightarrow\; \varphi \tag{12.4}$$

12.2.2 Knowledge and Time

One of the most prominent themes in van Benthem's work in the last two decades is that of *dynamics*. There is a complete chapter (Chap. 6) in this volume dedicated to *Dynamic Epistemic Logic*. A simple setting to study dynamics of epistemics is obtained by combining temporal and epistemic logic (temporal logic is the subject of Chap. 20). Popular temporal models of agency are linear time models or else trees. For both, one can use Linear Time Logic (LTL) to reason about them. In the latter case, properties of the tree are those true on all of its branches (in CTL, one can quantify over branches as well). In LTL, one uses operators for $\bigcirc\varphi$ ('in the next state'), $\square$ ('always in the future'), $\Diamond$ ('some time in the future') and U (where $\varphi U \psi$ denotes 'φ holds until ψ is true'). When we want to refer to the memory of the agents, also past-time operators are used, allowing for $\bullet\varphi$ ('in the previous moment'), $\blacksquare$ ('always in the past') and $\blacklozenge$ ('some time in the past').

Some axioms for linear temporal time logic with future operators are given in Table 12.3. Let us call the logic consisting of them LTL. The future operators 'some time' and 'always' can be defined as $\Diamond\varphi = \neg\varphi U \varphi$ and $\square\varphi = \neg\Diamond\neg\varphi$, respectively. Axiom $\mathbf{T_2}$ says that $\bigcirc$ is functional (this is the $\leftarrow$-direction, saying there is at most one next state) and serial (the $\rightarrow$-direction, saying there is at least one next state). $\mathbf{T_3}$ defines until: 'φ until ψ' is equivalent to saying that 'either $\neg\psi$, or φ holds while in the next state, φ until ψ'. The rule $\mathbf{RT}$ explains how $\neg(\varphi U \psi)$ can be inferred, and this rule is reminiscent of the induction rule (**RInd**) for common knowledge (cf. [47, Theorem 8.1.1(e)]).

Similarly to common knowledge, the until operator also allows a fixed point definition as the least fixed point of $\psi \vee (\varphi \wedge \Diamond x)$. As always, things become more interesting when we look at properties that *relate* the modalities (for knowledge and time in this case) that we have. Typical mix properties for knowledge and time are then for instance

$$K_a\bigcirc\varphi \to \bigcirc K_a\varphi \;\;\&\;\; \bigcirc K_a\varphi \to K_a\bigcirc\varphi$$
$$\text{(perfect recall (PR) \& no surprise (NS))}$$

NS is sometimes called *no learning*: it expresses that everything that one will know in the next state, is currently already known to hold next. Readers interested in these notions should also consult Chap. 20 by Goranko and Pacuit in this volume.

12.3 Relational Epistemic Structures for Knowledge

We now present a semantics for our formal language, based on Kripke models.

12.3.1 Kripke Models

Definition 12.2 (*Kripke models and epistemic models*) A *Kripke model M* for $L(At, Op, Ag)$ is a tuple $\langle S, R, V \rangle$ where S is a set of states, or worlds, R associates each $\square \in Op$ with an accessibility relation $R(\square) \subseteq S \times S$. Rather than $(s, t) \in R(\square)$ we write $s R_\square t$. Finally, V assigns to each atom $p \in At$ a set of states $V(p) \subseteq S$: those are the states in M where p is true. A tuple $\langle S, R \rangle$ is called a *frame*. For $M = \langle S, R, V \rangle$, we will sloppily write $s \in M$ for $s \in S$. Truth of φ in a pair M, s (with $s \in M$) is then defined as follows:

$$
\begin{array}{lll}
M, s \models p & \text{iff} & s \in V(p) \\
M, s \models \varphi \wedge \psi & \text{iff} & M, s \models \varphi \text{ and } M, s \models \psi \\
M, s \models \neg\varphi & \text{iff} & \text{not } M, s \models \varphi \\
M, s \models \square\varphi & \text{iff} & \text{for all } t \text{ such that } s R_\square t, M, t \models \varphi.
\end{array}
$$

For $F = \langle S, R \rangle$, the notion $F \models \varphi$ is defined as $\forall V, \forall s, \langle S, R, V \rangle, s \models \varphi$. In that case, we say that φ is *valid in F*. We write $F \models L$, if $F \models \varphi$ for each $\varphi \in L$. If there exists $s \in S$ and a valuation V, such that $\langle S, R, V \rangle, s \models \varphi$, then we say that φ *is satisfiable in F*. If Γ is a set of formulas, we say Γ *is satisfiable in F* if there is $s \in S$ and a valuation V, such that $\langle S, R, V \rangle, s \models \varphi$ for each $\varphi \in \Gamma$. Validity of φ on a model M is defined as $M, s \models \varphi$ for all $s \in M$. The class of all Kripke models $\langle S, R, V \rangle$ with m accessibility relations $R(\square)$ is denoted $\mathcal{K}_m$.

Let $\mathcal{C}$ be some class of models. If $M \models \varphi$ for each $M \in \mathcal{C}$, then we say that φ is valid in $\mathcal{C}$, and write $\mathcal{C} \models \varphi$. Examples of classes of models are $\mathcal{K}_m$ (all Kripke models with m relations), $\mathcal{S}4_m$ (models with m relations, all being reflexive and transitive), $\mathcal{KD}\triangle\triangledown45_m$ (all relations being serial, transitive and Euclidean) and $\mathcal{S}5_m$ (all relations are equivalence relations). Also, $\mathcal{U}_m$ is the class of models where all m relations are the universal relation. If $\mathcal{C}_m$ is a class of models for m agents, $\mathcal{C}_m^C$ is the class obtained by adding a relation R_C, which is the transitive closure of the union of the m relations. Likewise, $\mathcal{C}_m^D$ has a relation R_D which is the intersection of the relations in the model.

If Op contains one modal operator for each agent, we often write R_a rather than for instance R_{K_a} or R_{B_a}. When all the operators are epistemic operators K_a, we

Fig. 12.1 A simple two-agent one-atom scenario

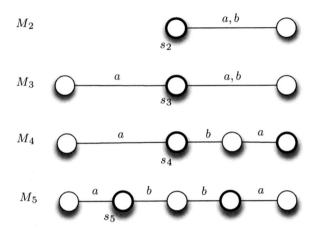

Fig. 12.2 Four 'different' models M_i, s_i for $p \wedge \neg K_a p \wedge \neg K_b p$. States where p is true have a *thick circle*

write $\sim_a$ for R_{K_a}, and we assume that $\sim_a$ is an *equivalence relation*. A model with such relations is called an *epistemic model*, and will be denoted $M = \langle S, \sim, V \rangle$. A pair M, s is also called a *pointed Kripke model* or *pointed epistemic model*. So $S5_m$ represents the class of all epistemic models.

In epistemic models, the interpretation of $s \sim_a t$ is that 'states s and t look similar for a', or 'in s and t, agent a has the same information', or, 'given state s, agent a considers it possible that the state is t'. These informal readings make it plausible that $\sim_a$ is an equivalence relation indeed.

An extremely simple multi-agent scenario involving two agents a and b and one atom p is given in Fig. 12.1. The pointed model M_1, s_1 models a situation where it is given that "p, but a and b don't know it". Let us denote this scenario by σ. Alternative models for the same scenario are given in Fig. 12.2. In our representation of such a model, states in which p is true are denoted with a thick circle, and a line between two states labeled with an agent means that the two states are similar for that agent— we omit reflexive arrows which are supposed to be present in all states.

We already mentioned van Benthem's pioneering work in *Correspondence Theory* [9, 10]. This theory establishes a formal connection between first-order properties of the accessibility relation on the one hand, and axioms or formula schemes, on the other. For instance, the axiom **T** corresponds to reflexivity, **4** to transitivity and **5** corresponds to the underlying accessibility relation being Euclidean. Since a relation that is reflexive, transitive and Euclidean is an equivalence relation, this then helps us establish that the logic $S5_m$ is sound and complete wrt epistemic models (the doxastic logic $KD45_m$ is sound and complete wrt models where the accessibility relations R_{B_a} are serial, transitive and Euclidean). See also Sect. 12.3.2, in particular Theorem 12.1.

By way of illustration of a proof of correspondence, let us follow [13] to show the correspondence between **4** and transitivity.

Fact 12.1 (Fact 1.1 [13]) $F, s \models \Box p \rightarrow \Box\Box p$ iff F's accessibility relation R is transitive at the point s: i.e., $F, s \models \forall yz((sRy \ \& \ yRz) \Rightarrow sRz)$.

Proof If the relation is transitive, $\Box p \rightarrow \Box\Box p$ clearly holds under every valuation. Conversely, let $F, s \models \Box p \rightarrow \Box\Box p$. It means that this axiom holds for every valuation V, so in particular when $V(p) = \{y \mid sRy\}$. For this V, the antecedent of the model formula holds at s, and hence so does $\Box\Box p$. By definition of V, this implies that R is transitive.

Given an epistemic model $\langle S, \sim, V \rangle$, it turns out that the group epistemic notions E, C and D can all be interpreted as modal operators with respect to some binary relation that is defined in terms of the individual relations $\sim_a$. More precisely, the operator E is the necessity operator for the relation $\sim_E = \bigcup_{a \in \mathsf{Ag}} \sim_a$: in order for $E\varphi$ to be true at M, s, the formula φ needs to be true in all successors of s, no matter which agent we choose.[1] In Fig. 12.1 for instance, we have $M_1, s_1 \models EM_a\neg p$ (both a and b know that a considers a $\neg p$-state possible) while $M_1, s_1 \models \neg EM_a p$ (since b considers it possible that a knows $\neg p$). One can also use correspondence theory to see that D can be interpreted as the modal operator for a relation $\sim_D$, with $\sim_D \subseteq \bigcap_{a \in \mathsf{Ag}} \sim_a$. At the end of Sect. 12.3.2, we will argue that for completeness, one can even replace the '$\subseteq$' by '$=$'. In terms of Fig. 12.1 again, we have $M_1, s_1 \models Dp$.

For common knowledge, the corresponding property is not first order definable, but van Benthem explains in [13] how it corresponds with a property in First-Order Logic with Least Fixed Points, see also [22].

We briefly recall the semantics of modal μ-calculus (e.g., [35]), skipping some well-known details. The formulas of modal μ-calculus are modal formulas extended with the formulas of type $\mu x\varphi$ and $\nu x\varphi$ for φ positive in x (i.e., if each occurrence of x is under the scope of an even number of negations). Let $\langle S, R \rangle$ be a Kripke frame. For each modal μ-formula φ and a valuation V, we define the semantics $[\![\varphi]\!]_V$ of φ by induction on the complexity of φ. If φ is a propositional variable, a constant, or is of the form $\psi \wedge \chi, \psi \vee \chi, \neg\psi, \Box\psi$ or $\Diamond\psi$, then the semantics of φ is defined as above. For each valuation V, we denote by V_x^U a new valuation such that $V_x^U(x) = U$ and $V_x^U(y) = V(y)$ for each propositional variable $y \neq x$ and $U \in \mathcal{P}(S)$.

Let φ be positive in x, then

$$[\![\mu x\varphi]\!]_V = \bigcap\{U \in \mathcal{P}(S) : [\![\varphi]\!]_{V_x^U} \subseteq U\}. \tag{12.5}$$

$$[\![\nu x\varphi]\!]_V = \bigcup\{U \in \mathcal{P}(S) : [\![\varphi]\!]_{V_x^U} \supseteq U\}. \tag{12.6}$$

---

[1] For easy of readability, we give the group notions with $A = \mathsf{Ag}$: cases for $A \subseteq \mathsf{Ag}$ are similar.

We will skip the index V if it is clear from the context. Note that $[\![\mu x \varphi]\!]_V$ and $[\![\nu x \varphi]\!]_V$ are, respectively, the least and greatest fixed points of the map $f_{\varphi,V}$: $\mathcal{P}(S) \to \mathcal{P}(S)$ defined by $f_{\varphi,V}(U) = [\![\varphi]\!]_{V_x^U}$. That φ is positive in x guarantees that $f_{\varphi,V}$ is monotone. Therefore, by the celebrated Knaster-Tarski theorem these fixed points exist and are computed as in (12.5) and (12.6). The least and greatest fixed points can also be reached by iterating the map $f_{\varphi,V}$. In particular, for an ordinal α we let $f_{\varphi,V}^0(\emptyset) = \emptyset$, $f_{\varphi,V}^\alpha(\emptyset) = f_{\varphi,V}(f_{\varphi,V}^\beta(\emptyset))$ if $\alpha = \beta + 1$, and $f_{\varphi,V}^\alpha(\emptyset) = \bigcup_{\beta < \alpha} f_{\varphi,V}^\beta(\emptyset)$, if α is a limit ordinal, and we let $f_{\varphi,V}^0(S) = S$, $f_{\varphi,V}^\alpha(S) = f_{\varphi,V}(f_{\varphi,V}^\beta(S))$ if $\alpha = \beta + 1$, and $f_{\varphi,V}^\alpha(S) = \bigcap_{\beta < \alpha} f^\beta(S)$, if α is a limit ordinal. Then $[\![\mu x \varphi]\!]_V = f_{\varphi,V}^\alpha(\emptyset)$, for some ordinal α such that $f_{\varphi,V}^{\alpha+1}(\emptyset) = f_{\varphi,V}^\alpha(\emptyset)$ and $[\![\nu x \varphi]\!]_V = f_{\varphi,V}^\alpha(S)$, for some ordinal α such that $f_{\varphi,V}^{\alpha+1}(S) = f_{\varphi,V}^\alpha(S)$.

Thus, we have two different ways of computing fixed point operators resulting in the same semantics. As we will see in the next section this is no longer the case in topological semantics. Now we have all the formal machinery for giving a fixed point definition of common knowledge. We let

$$C\varphi = \nu x(\varphi \wedge Ex). \tag{12.7}$$

A fixed point formula $\mu x \varphi$ ($\nu x \varphi$) is called *constructive* if the least (greatest) fixed point can be reached after countably many iterations of $f_{\varphi,V}$. Fontaine [49] gives a syntactic description of all *continuous* fixed point formulas that form a sub-fragment of all constructive formulas. Using this description it is easy to see that $C\varphi$ is the continuous and hence in the constructive fragment of all fixed point formulas. Therefore, in order to compute common knowledge we need only countably infinite iterations.

It is easy to see that $C\varphi$ expresses the reflexive transitive closure, i.e., 'some φ-world is reachable in finitely many $\sim_E$-steps' [13, Example 6]. Next we will compute common knowledge following our fixed point definition in some of the models shown in Fig. 12.2. In M_1 we have $V(p) = \{s_1\}$. So if $\varphi = p \wedge \Box_a x \wedge \Box_b x$, then

$$f_{\varphi,V}^0(S) = [\![\varphi]\!]_{V_x^S}$$
$$= V(p) \cap [\![\Box_a x]\!]_{V_x^S} \cap [\![\Box_b x]\!]_{V_x^S}$$
$$= \{s_1\} \cap S \cap S = \{s_1\}.$$

Then

$$f_{\varphi,V}^1(S) = \{s_1\} \cap [\![\Box_a x]\!]_{V_x^{f_{\varphi,V}^0(S)}} \cap [\![\Box_b x]\!]_{V_x^{f_{\varphi,V}^0(S)}}$$
$$= \{s_1\} \cap [\![\Box_a x]\!]_{V_x^{\{s_1\}}} \cap [\![\Box_a x]\!]_{V_x^{\{s_1\}}}$$
$$= \{s_1\} \cap \emptyset \cap \emptyset = \emptyset.$$

Finally, observe that $f_{\varphi,V}(\emptyset) = \emptyset$. So we reached the least fixed point and $[\![Cp]\!] = [\![vx\varphi]\!] = \emptyset$.

Now consider the second model and the formula $\sigma = \neg K_a p \wedge \neg K_b p$. It is easy to see that in M_2 we have $[\![\sigma]\!]_V = S$. Let $\varphi = \sigma \wedge \Box_a x \wedge \Box_b x$. Then $f^0_{\varphi,V}(S) = [\![\varphi]\!]_{V^S_x} = [\![\sigma]\!]_V \cap [\![\Box_a x]\!]_{V^S_x} \cap [\![\Box_b]\!]_{V^S_x} = S \cap S \cap S = S$. This means that S is the greatest fixed point of $f_{\varphi,V}$. So $[\![\check{C}\sigma]\!] = [\![vx\varphi]\!] = S$. We leave it up to the reader to compute common knowledge of various formulas in other models depicted in Fig. 12.2.

Note that $M_1, s_1 \models E\varphi \leftrightarrow C\varphi$, but also that $M_4, s_4 \models E\neg K_b p \wedge \neg C \neg K_b p$. The pointed epistemic model M_2, s_2 not only models the scenario $\sigma : p \wedge \neg K_a p \wedge \neg K_b p$ but also that this is common knowledge: $M_2, s_2 \models C\sigma$. It is the only pointed model M_i, s_i ($i \leq 5$) with this property.

Correspondence properties make modal logic a flexible tool to model epistemic and doxastic logics: once one has decided on the desired properties of the informational attitude, like negative introspection, the Kripke models obtained need just to satisfy an additional property, like Euclideaness. It also helps provide a neat analysis of informational group notions. There are also some drawbacks using Kripke models for knowledge and belief: we will come back to this in Sect. 12.4.1.

12.3.2 Completeness

In this section we briefly recall soundness and completeness of some important modal logics. Let L be a (normal) modal logic defined in Sect. 12.2. Recall that a (normal) modal logic L is called *sound* wrt a class K of Kripke frames if $F \models L$ for each $F \in K$. Logic L is called *complete* wrt K if for each formula φ, if φ is L-consistent (i.e., $L \cup \{\varphi\} \not\vdash \bot$), then there is $F \in K$ such that φ is satisfied in F. A frame F is called an L-*frame* if $F \models L$. It is easy to see that if L is sound and complete wrt some class K, then it is sound and complete wrt the class of all L-frames. L is called *strongly complete* wrt a class K of Kripke frames if for each set of formulas Γ, if Γ is L-consistent (i.e., $L \cup \Gamma \not\vdash \bot$), then there is $F \in K$ such that Γ is satisfied in F.

Recall also that a transitive frame $F = \langle S, R \rangle$, is called *rooted* if there exists $s \in S$, called a *root*, such that for each $s' \in S$ with $s' \neq s$ we have sRs'. It is well known that if a logic is sound a complete, then it is sound and complete wrt a class of rooted L-frames.

A standard method for proving completeness of modal logics is via the canonical model construction. We briefly review it here. In the next section we explain how this construction is generalised to the topological setting. All the details can be found in any modal logic textbook, e.g., [34] or [37].

Given a logic L, one considers the set S^C of all maximal L-consistent sets of formulas. A relation R^C on S^C is defined in the following way: for each $\Gamma, \Delta \in S^C$, $\Gamma R^C_\Box \Delta$ if for each formula φ we have $\Box\varphi \in \Gamma$ implies $\varphi \in \Delta$. Finally, the valuation V^C on S^C is defined by $\Gamma \in V^C(p)$ if $p \in \Gamma$. The model $M^C = \langle S^C, R^C, V^C \rangle$ is called the *canonical model* of L. Then one proves the Truth Lemma stating that for each formula φ and $\Gamma \in S^C$:

$$M^C, \Gamma \models \varphi \text{ iff } \varphi \in \Gamma.$$

Now suppose φ is L-consistent. Then by the Lindenbaum Lemma (see, e.g., [34, 37]), $\{\varphi\}$ can be extended to a maximal consistent set Γ. By the Truth Lemma, $M^C, \Gamma \models \varphi$. Thus, we found a frame $\langle S^C, R^C \rangle$ that satisfies φ. In order to finish the proof we need to show that $\langle S^C, R^C \rangle$ is an L-frame. If the latter is satisfied, then L is called *canonical*. Therefore, canonical modal logics are Kripke complete.

It is a classical result of modal logic that if a normal modal logic L is axiomatised by Sahlqvist formulas, then L is canonical, and hence Kripke complete, see e.g., [34] or [37]. Together with the Sahlqvist-van Benthem correspondence result discussed in the previous section, this theorem guarantees that every logic axiomatised by Sahlqvist formulas is sound and complete wrt a first-order definable class of Kripke frames. As a result we obtain that epistemic and doxastic logics $S4_m$, $S5_m$, $KD45_m$ are all sound and complete with respect to corresponding classes of Kripke frames discussed in the previous section.

We now summarise a number of completeness results for epistemic logics in the following theorem. Proofs and extensions of them can be found in [47, Chap. 3.1], and [88, Chap. 2] for epistemic logics, in [88, Chap. 1] and [33, Chap. 4] for normal modal logics in general and in [53] for LTL. The set of models $\mathcal{LIN}$ is the set of all linear orders: think of them as $M = \langle \mathbb{N}, Succ, V \rangle$, where $x\ Succ\ y$ iff $y = x + 1$.

Theorem 12.1 *In the following, $m \geq 1$. Item 6 presents a logic and a semantics to which it is sound and complete. All the other items present logics that are strongly sound and complete with respect to the mentioned semantics:*

1 K_m and $\mathcal{K}_m$	5 $S5_m$ and $\mathcal{S}5_m$
2 $S4_m$ and $\mathcal{S}4_m$	6 $S5_m^C$ and $\mathcal{S}5_m^C$
3 $KD45_m$ and $\mathcal{KD}\triangle\triangledown45_m$	7 $S5_m^D$ and $\mathcal{S}5_m^D$
4 $S5_1$ and $\mathcal{U}_1$	8 LTL and $\mathcal{LIN}$

Note that by (12.2), when only finite formulas are allowed, we will not be able to find a strong completeness result for common knowledge: the set $\{Ep, EEp, \ldots\} \cup \{\neg Cp\}$ is consistent, but not satisfiable. For logics with distributed knowledge, we saw that in the canonical model, we only have $R_D^C \subseteq \cap_{a \in \text{Ag}} R_a^C$. To also obtain the converse, for any two sets Γ and Δ for which we have $\Gamma(\cap_{a \in \text{Ag}} R_a^C) \Delta$, but not $\Gamma R_D^C \Delta$, one can replace Δ by n copies $\Delta_1, \ldots, \Delta_n$, with $\Gamma R_i^C \Delta_i$. Of course, in the context of for instance $S5$, one needs to take care that the relations remain an equivalence relation, but this can be done: for a discussion see for instance [51].

12.3.3 Expressivity and definability of Epistemic Models

Speaking with van Benthem's [15, p. 32], one can ask: 'When are two information models the same?' For instance, although all our five pointed models M_i, s_i verify the same scenario σ, do they differ in some other sense?

Definition 12.3 ((*Bi-*)*simulation*) Let $M = \langle S, \sim, V \rangle$ and $M' = \langle S' \sim', V' \rangle$ be two epistemic models. A *simulation* between M and M' is a relation $R \subseteq S \times S'$ such that

Harmony   If $s R s'$ then for all $p \in$ At, $s \in V(p)$ iff $s' \in V'(p)$.
Forth   For all $a \in$ Ag, if $s \sim_a t$ and $s R s'$, then for some $t' \in S'$, $t R t'$ and $s' \sim'_a t'$.

R is called a *bisimulation* if it moreover satisfies.

Back   For all $a \in$ Ag, if $s' \sim'_a t'$ and $s R s'$, then for some $t \in S$, $t R t'$ and $s \sim_a t$.

If $s R s'$ and R is a simulation, we say that M, s simulates M', s'; if R is a bisimulation, we say that M, s and M', s' are bisimilar.

As an example, note that M_1, s_1 simulates M_3, s_3, while M_1, s_1 and M_5, s_5 are bisimilar. Roughly speaking, if M, s simulates M', s', then ignorance (i.e., an M_a-formula) is preserved from M, s to M', s', and knowledge is preserved in the other direction. A bisimulation preserves both.

Lemma 12.1 [15, Invariance Lemma] *Let M and M' be finite models. Let* $\mathsf{L} = \mathsf{L}(\mathsf{At},$ $\mathsf{Op}, \mathsf{Ag})$, *with* $\mathsf{Op} = \{K_a \mid a \in \mathsf{Ag}\}$. *Then the following are equivalent:*

(a) M, s *and* M', s' *are bisimilar,*
(b) M, s *and* M', s' *satisfy the same formulas* $\varphi \in \mathsf{L}$.

Proof For (a) $\Rightarrow$ (b) one can follow a standard argument using induction on φ. For the converse, let (b) be given and define $x R x'$ as x and x' satisfy the same formulas from L. Clearly **Atoms** holds for R, and also, $s R s'$. To show **Forth**, suppose $x R x'$ while $x \sim_a y$ for some agent a. Suppose there is no state y' in S' with $y \sim_a y'$ for which $y R y'$ holds, i.e., for every y' with $y \sim_a y'$ there is a formula $\chi^{x+}_{y'-}$ true in x, but false in y'. Let χ be $\bigwedge_{\{y'|y \sim_a y'\}} \chi^{x+}_{y'-}$, then $M, x \models \chi$ while $M', x' \models \neg\chi$, contradicting $x R x'$. **Back** is proven similarly.

This lemma implies that our pointed models M_1, s_1 and M_2, s_2 are *not* bisimilar, since $M_a K_b \neg p$ is true in the first, but not in the second.

Where the invariance lemma says that 'bisimulation has exactly the expressive power of the modal language' [11, p. 56] the following State Definition Lemma says that every pointed epistemic model can be characterised by an epistemic formula in the language with common knowledge.

Lemma 12.2 [15, State Definition Lemma] *For each finite pointed epistemic model* M, s *there is a formula* $\varphi \in$ L(At, Op, Ag), *with* Op $= \{K_a \mid a \in$ Ag$\} \cup \{C\}$ *such that the following are equivalent (where M' is finite):*

(a) $M', s' \models \varphi$,
(b) M, s *is bisimilar to* M', s'.

The conditions in both lemmas are necessary: finite epistemic states are not definable up to *simulation* in the language with common knowledge, nor is bisimulation to finite epistemic models definable in the language without common knowledge [40].

For later reference, we conclude this section by stating van Benthem's characterisation theorem for modal logic. The standard translation ST_x takes a modal formula and returns a first-order formula using the clauses $ST_x(p) = Px$, it commutes with the Boolean connectives and stipulates that $ST_x(\Box\varphi) = \forall y(x R_\Box y \Rightarrow ST_y(\varphi))$.

Theorem 12.2 ([10]) *The following are equivalent for first-order formulas* $\Phi(x)$:

1. $\Phi(x)$ *is invariant under bisimulation,*
2. $\Phi(x)$ *is equivalent to $ST_x(\varphi)$ for some modal formula φ.*

12.3.4 Epistemic Temporal Frames

van Benthem and Pacuit [18] formalise a notion of time using so-called *epistemic temporal frames* $\mathcal{F} = \langle \Sigma, \mathcal{H}, \sim \rangle$, where Σ is a set of events (say, possible moves in a game) and $\mathcal{H}$ is a set of histories. For this chapter, $\mathcal{F}$ can be thought of as a finitely branching rooted tree labeled with events. The histories are then nothing else than strings of events. Figure 12.3 provides an example. Frame $\mathcal{F}$ in this figure

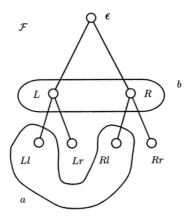

Fig. 12.3 An epistemic temporal frame $\mathcal{F}$

denotes a game where agent a can decide in node ϵ to move L or R, after which b can move either l or r (so $\Sigma = \{L, R, l, r\}$). For epistemic temporal frames, the indistinguishability relation is defined over *histories*, in $\mathcal{F}$ of Fig. 12.3 for instance, we have $L \sim_b R$ (agent b does not know which move a starts with) and $Ll \sim_a Rl$ (if b plays l, agent a forgets what his own initial move has been).

The semantic counterpart of no surprise (**NS**) would then say that for all finite histories $H, H' \in \mathcal{H}$ and all events $e \in \Sigma$ with $He, H'e \in \mathcal{H}$, if $H \sim_a H'$, then $He \sim_a H'e$. The converse of this would guarantee **PR**. One might be tempted to think that this converse ensures $K_a \bigcirc \varphi \rightarrow \bigcirc K_a \varphi$, but this is not the case for φ that refer to what is the case *now* (like, 'it is 3 am') or that refer to ignorance (like 'a does not know that ψ'), knowledge of such properties may be given up, even (or especially when) provided with more information (see [47, p. 130] for further discussion). A *bounded agent* does not have perfect recall, but instead has a finite bound on the number of preceding events which they can remember. van Benthem and Pacuit [18] call an agent *synchronised* if $H \sim_a H'$ can only occur for histories H and H' that have the same length (so the agent would know how many moves have been played, or, more generally, know the time of the global clock). In $\mathcal{F}$ of Fig. 12.3, both agents are synchronised, agent b does not satisfy no surprise (he cannot distinguish the histories L and R, but if in both the same action (say l) is performed, he can distinguish the result), while agent a does not satisfy perfect recall: he cannot distinguish Ll and Rl, although he knew the difference between L and R.

Following the pioneering [61] of Halpern and Vardi on the complexity of reasoning about knowledge and time, van Benthem and Pacuit highlight in [18] how several choices in the formalism can have quite dramatic consequences for the decidability and computational complexity (of the validity problem) of the underlying logic. Choices that heavily influence the complexity, regard for instance the language (does it include an operator for common knowledge, do we allow for temporal operators for the past and for the future?), structural conditions on the underlying event structure (what if we give up some conditions of an epistemic temporal frame, or look at forests rather than trees?) and conditions on the reasoning abilities of the agents (perfect recall, no surprise, synchronisation, bounded agents). Moreover, [18] marks the start of a research paradigm that compares and links existing approaches to epistemic logic (Kripke models, interpreted systems [47]), and 'Parikh style' logic [95], time (history based structures [95], runs [47]), and dynamics, including PDL-style logic [64] and dynamic epistemic logic (see Chap. 6). van Benthem further helped clarify the link between interpreted systems, epistemic temporal logic and dynamic epistemic logic in [25].

The chapter in this volume by Goranko and Pacuit presents a more comprehensive survey of temporal epistemic frameworks. For examples of completeness results regarding systems for knowledge and time, we refer to Theorem 12.3. For a general discussion on completeness and complexity issues for such logics, and further references, we refer to van Benthem and Pacuit's [18].

12.3.5 Interpreted Systems

In the 1980s, computer scientists became interested in epistemic logic. This line of research flourished in particular by a stream of publications around Fagin, Halpern, Moses and Vardi. Their important textbook [47] surveys their work on epistemic logic over a period of more than ten years. The emphasis in this work is on *interpreted systems* (IS) as an underlying model for their framework, a semantics that also facilitates reasoning about knowledge during *computation runs* in a natural way. The key idea behind IS is two-fold:

- It provides for a so-called *grounded semantics* of epistemic logic;
- It adds a dynamic and computational component to this through the notions of *run* and *protocol*.

Where in an epistemic model the equivalence relations $\sim_a$ are *given*, in an interpreted system they are *grounded* in the notion of *observational equivalence*. To be more precise (for formal definitions we refer to [47]), let L_a be a set of possible local states for agent a. For example, when modeling a distributed computation, such a local state could provide the value of the variables associated with processor a, or in a card game it could be the enumeration of cards held by player a. Moreover, let L_e be a set of possible states for the environment. This state could have information about a global clock, or keep track of whose turn it is in a card game. The set of global states of an interpreted system with m agents is then $\mathcal{G} = L_e \times L_1 \times \cdots \times L_m$. If $\mathbf{s} = \langle s_e, s_1, \ldots, s_m \rangle \in \mathcal{G}$ with $\mathbf{s}_a$ we mean s_a ($a \in \{e\} \cup \mathsf{Ag}$). An example of an interpreted system is $\mathcal{I}$ of Fig. 12.4, where the environment is not modelled (it is constant, say), $L_x = \{0, 1\}$, $L_y = \{0, 1, 2\}$ and $L_z = \mathbb{N}$.

Two global states $\mathbf{s} = \langle s_e, s_1, \ldots, s_m \rangle$ and $\mathbf{s}' = \langle s'_e, s'_1, \ldots, s'_m \rangle$ are now defined to be indistinguishable for agent i, written $\mathbf{s} \sim^I_i \mathbf{s}'$, if i's local state is the same in both, i.e., if $s_i = s'_i$. This is clearly an equivalence relation, and hence this notion of interpreted system gives rise to knowledge of veridical and introspective agents. In the most general case, we will not always consider the full cartesian product $\mathcal{G}$ but

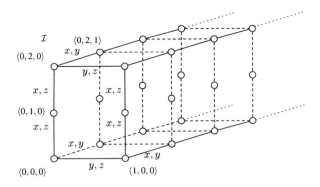

Fig. 12.4 An interpreted system $\mathcal{I}$

some subset $\mathcal{J} \subseteq \mathcal{G}$ of it. This represents situations where overall constraints of the system prevent some global states from being part of the model.

A *run over* $\mathcal{J} \subseteq \mathcal{G}$ is a function $r : \mathbb{N} \to \mathcal{J}$. Intuitively, this captures a computation, or a behaviour of the system. If $r(m) = \mathbf{s}$, then $r_i(m) = \mathbf{s}_i$ is the local state of agent i in run r at time m. A pair (r, m) is called a point. An interpreted system $\mathcal{I}$ is a pair $\langle \mathcal{R}, V^I \rangle$, where $\mathcal{R}$ is a set of runs and $V^I(p)$ denotes for each propositional variable p ('x is 3', or 'i holds card diamond 9') the set of global states in which it is true. In other words, we assume that the truth of atoms does not depend on 'where we are in the run', but only on the global state (in particular, if a run r visits the same global state twice, i.e., $r(m) = r(m + k)$, for some $m, k \in n$, then the truth of atoms is the same in both points). Moreover, to quote [47, p. 112], '*Quite often, in fact, the truth of a primitive proposition q of interest depends, not on the whole global state, but only on the component of some particular agent*'. In such cases, the valuation $V^I(q)$ respects the *locality* of q, which means that, if $\mathbf{s} \sim_i \mathbf{s}'$, then $\mathbf{s} \in V^I(q)$ iff $\mathbf{s}' \in V^I(q)$. In such a case, the fact that i knows the truth of such a property is common knowledge. To be more precise, suppose that there is a property $x_i = 0$, which is true exactly when in i's local state, the variable x_i is equal to 0. Then, we have $C(x_i = 0 \to K_i(x_i = 0))$.

It is easy to see that an interpreted system $\mathcal{I} = \langle \mathcal{R}, V^I \rangle$ gives rise to an epistemic model $M_I = \langle S, \sim, V \rangle$, by taking for S all the points generated by $\mathcal{R}$, and where $(r, m) \sim_i (r', m')$ iff $r(m) \sim^I_i r'(m')$ and $(r, m) \in V(p)$ iff $r(m) \in V^I(p)$ (so, $\sim$ and V defined over points (r, m) is determined by $\sim^I$ and V^I on global states $r(m)$).

If we now define $\mathcal{I}, r, m \models \varphi$ as $M_I, r(m) \models \varphi$, we have an interpretation for the individual and group epistemic notions discussed in Sect. 12.2.1. For *full* interpreted systems, where $\mathcal{G}$ is the full cartesian product $L_e \times L_1 \times \cdots \times L_m$, we have that common knowledge is constant over all runs. This is so since for every two global states $\mathbf{s} = \langle s_e, s_1, \ldots, s_m \rangle$ and $\mathbf{s}' = \langle s'_e, s'_1, \ldots, s'_m \rangle$ there is a third state $\mathbf{t} = \langle s_e, s_1, s'_2, \ldots \rangle$ 'epistemically connecting them'. The notion of a run in an interpreted system also directly facilitates the interpretation of temporal formulas: we define for instance $\mathcal{I}, r, m \models \bigcirc \varphi$ as $\mathcal{I}, r, m + 1 \models \varphi$.

For our example system $\mathcal{I}$ we assume to have propositional atoms like $x = 0$, $z = 9$. We also identify three runs, r_0, r_1 and r_2. In all of them, the variable z is increased by 1 in each step, where $z = 0$ in $(r_0, 0)$ and $(r_2, 0)$ and $z = 1$ in $(r_1, 0)$. In both r_0 and r_2, the values of $\langle x, y \rangle$ are a clockwise walk through the xy plane: $\langle x, y \rangle$ $= \langle 0, 0 \rangle, \langle 0, 1 \rangle, \langle 0, 2 \rangle, \langle 1, 2 \rangle, \langle 1, 1 \rangle, \langle 1, 0 \rangle, \langle 0, 0 \rangle, \ldots$. In r_2, the variables x and y are both 0 at even places, and both 1 at odd places.

$$r_0 : \langle 0, 0, 0 \rangle \ \langle 0, 1, 1 \rangle \ \langle 0, 2, 2 \rangle \ \langle 1, 2, 3 \rangle \ \langle 1, 1, 4 \rangle \ \langle 1, 0, 5 \rangle \ \langle 0, 0, 6 \rangle \ \ldots$$
$$r_1 : \langle 0, 0, 1 \rangle \ \langle 0, 1, 2 \rangle \ \langle 0, 2, 3 \rangle \ \langle 1, 2, 4 \rangle \ \langle 1, 1, 5 \rangle \ \langle 1, 0, 6 \rangle \ \langle 0, 0, 7 \rangle \ \ldots$$
$$r_2 : \langle 0, 0, 0 \rangle \ \langle 1, 1, 1 \rangle \ \langle 0, 0, 2 \rangle \ \langle 1, 1, 3 \rangle \ \langle 0, 0, 4 \rangle \ \langle 1, 1, 4 \rangle \ \langle 0, 0, 4 \rangle \ \ldots$$

Let $\mathcal{I}_{01}$ consist of the runs r_0 and r_1 whereas $\mathcal{I}_{12}$ has the runs r_1 and r_2. We then have, in $\mathcal{I}_{02}, r, \langle 0, 0, 0 \rangle$:

$$K_x x = 0 \wedge \neg K_x y = 0 \wedge E \,\square\, (x = 0 \leftrightarrow K_x x = 0) \wedge \neg K_x \bigcirc x = 1 \wedge K_z \bigcirc z = 1$$

In order to semantically characterise perfect recall in an interpreted system, let, for an agent i, his *local-state sequence at the point* (r, m) be the sequence of local states he has seen in run r up to time m, without consecutive repetitions. So, for the run r_0 above, the local state sequence for agent x at time 4 equals $\langle 01 \rangle$, for agent y it is $\langle 0, 1, 2, 1 \rangle$, and for z it is $\langle 0, 1, 2, 3, 4 \rangle$. We now say that i has perfect recall pr if whenever $(r, m) \sim_i (r', m')$, then i has the same local-state sequence at (r, m) and (r', m'). In the system $\mathcal{I}_{02}$, agent z has perfect recall, but in $\mathcal{I}_{01}$, he has *not*. To see the latter, we have $(r_0, 1) = \langle 0, 1, 1 \rangle \sim_z \langle 0, 0, 1 \rangle = (r_1, 0)$, whereas the state sequence for z in $(r_0, 1)$ is $\langle 01 \rangle$ while in $(r_1, 0)$ it is $\langle 1 \rangle$. Indeed, it is easy to see that we have $\mathcal{I}_{01}, r_0, \langle 0, 0, 0 \rangle \models K_z \bigcirc y = 1 \wedge \neg \bigcirc K_z y = 1$.

An interpreted system $\mathcal{I} = \langle \mathcal{R}, V^I \rangle$ satisfies *sync* if agents know what time it is, i.e., if for all agents i, we have that $(r, m) \sim_i (r', m')$ implies $m = m'$.

Theorem 12.3 *We have the following* (see [47, Chap. 8]).

1. *Both $S5_m + LTL$ and $S5_m^C$ are sound and complete with respect to the set of all interpreted systems $\mathcal{INT}_m$ for m agents.*
2. *Both $S5_m + LTL$ and $S5_m^C$ are sound and complete with respect to the set of synchronised interpreted systems $\mathcal{INT}_m^{sync}$.*
3. *$S5_m + LTL + \mathbf{PR}$ is sound and complete with respect to the set of synchronised interpreted systems with perfect recall $\mathcal{INT}_m^{sync, pr}$.*

The first item of Theorem 12.3 suggests that the static, non-temporal validities of interpreted systems are axiomatised by $S5_m$, and hence that interpreted systems are in some sense equivalent to Kripke models. This idea was taken up by Lomuscio and Ryan in e.g., [82], roughly (the analysis in [82] is appropriately done at the level of frames, we give here a summary on the level of models) as follows. In order to link interpreted systems with $S5_m$ structures, [82] restricts itself to structures (1) without dynamic component (i.e., systems without runs), (2) where the state space is the full cartesian product $\mathcal{G}$ and (3) where the environment is not modelled in a global state. This leads to a notion of *hypercube*, which is just $L_1 \times \cdots \times L_2$, where L_i is as before, as is the agents' accessibility relation. Call the set of hypercubes for m agents $\mathcal{H}_m$. From what we have said above, it follows that in a hypercube H, common knowledge is constant, i.e., for all states $\mathbf{s}$ and $\mathbf{t}$, we have $H, \mathbf{s} \models C\varphi$ iff $H, \mathbf{t} \models C\varphi$ (if $R_C\mathbf{s}\mathbf{u}$, then $R_1 \langle t_1, t_2, \ldots, t_m \rangle \langle t_1, u_2, \ldots u_m \rangle$ and $R_2 \langle t_1, u_2, \ldots u_m \rangle \langle u_1, u_2, \ldots u_m \rangle$, hence $R_C\mathbf{t}\mathbf{u}$). However, to show that this discriminates the validities in $\mathcal{H}_m$ from those in $S5_m$, one would need a universal modality. But we also have the following, which shows that hypercubes behave different form $S5_m$ models (recall that distributed knowledge $D\varphi$ is true in a state s if φ holds in all t for which $s R_D t$, where $R_D = \sim_1 \cap \cdots \cap \sim_m$):

Observation 12.1 *Let $H \in \mathcal{H}$. Let $i, j \in$ Ag. Recall that $M_i\varphi = \neg K_i\neg\varphi$.*

1 *In H, we have R_D is the identity, that is, $\mathbf{s}R_D\mathbf{t}$ iff $\mathbf{s} = \mathbf{t}$.*
2 *For all global sates $\mathbf{s}^1, \ldots, \mathbf{s}^m$, there is a global state $\mathbf{s}$ with $\mathbf{s} \sim_i \mathbf{s}^i$, for all $i \leq m$.*

From these semantic properties, we derive the following validities on hypercubes:

3 $\mathcal{H}_m \models \varphi \leftrightarrow D\varphi$.
4 $\mathcal{H}_m \models M_i K_j\varphi \rightarrow K_j M_i\varphi$.

However, those validities do not transfer to $\mathcal{S}5_m$:

5 $\mathcal{S}5_m \not\models \varphi \leftrightarrow D\varphi$.
6 $\mathcal{S}5_m \not\models M_i K_j\varphi \rightarrow K_j M_i\varphi$.

Proof Item 1 follows from the fact that $\mathbf{s} = \langle s_1, \ldots, s_m \rangle R_D \langle t_1, \ldots t_m \rangle = \mathbf{t}$ iff $s_1 = t_1 \& \ldots \& s_m = t_m$ iff $\mathbf{s} = \mathbf{t}$. This immediately implies item 3. For item 5, observe that in M_2, s_2 of Fig. 12.2 it holds that $p \wedge \neg Dp$. For item 2, take $\mathbf{s} = \langle s_1^1, s_2^2, \ldots, s_m^m \rangle$ (i.e., take agent 1's local state from $\mathbf{s}^1$, agent 2's local state from $\mathbf{s}^2$, etc). Obviously, $\mathbf{s} \sim_i \mathbf{s}^i$. One can use a correspondence theory argument to show that this implies item 4 (see e.g., [82, Lemma 9]). For item 6, consider the model M_1 in Fig. 12.1. We extend this model to M_1' as follows: it makes q true in the two right-most states. Then we have $M_1', s_1 \models M_a K_b \neg q \wedge M_b K_a q$, in other words, $M_1' \not\models \neg K_a \neg K_b \neg q \rightarrow K_b M_a \neg q$.

Observation 12.1 implies that hypercubes, the static part of interpreted systems, are a special kind of $\mathcal{S}5_m$ models, which verify some additional properties. In fact, the following theorem (for its proof we refer to that of [82, Theorem 20]) shows that Observation 12.1 in fact sums up everything that separates $\mathcal{H}_m$ from $\mathcal{S}5_m$:

Theorem 12.4 (Based on Theorem 20 of Lomuscio and Ryan [82]) *Let $\mathcal{H}\mathcal{S}5_m \subset \mathcal{S}5_m$ be the set of $\mathcal{S}5_m$ models $M = \langle S, \sim, V \rangle$ that satisfy:*

1. $\forall st \in S\ s(\sim_1 \cap \cdots \cap \sim_m)t$ *iff* $s = t$
2. $\forall s_1, \ldots, s_m \in S \exists s \in S$ *such that* $\forall i \in$ Ag $s \sim_i s_i$.

Then the validities (of the language with operators K_i, C and D) in $\mathcal{H}\mathcal{S}5_m$ and $\mathcal{H}_m$ are the same.

So, the grounded semantics for knowledge, where it is explained where the accessibility relations come from, when implemented through hypercubes, the static counterpart of full interpreted systems, has as a consequence that we get the two additional properties 3 and 4 of Observation 12.1 for knowledge, as compared to $\mathcal{S}5_m$. Of course, on can give up the condition of *full* interpreted systems (in which case property 4 would disappear), or think about different ways of groundedness in the first place.

Many theories of multi-agent systems, which try to model notions like knowledge, belief, intentions, commitments, obligations and actions of agents are embedded in the philosophical brand of modal logic, in a way that is similar to what we discuss here for the knowledge of agents. Computational groundedness was put forward (cf. [108])

to make such theories more relevant to practitioners in multi-agent systems and distributed artificial intelligence in general. It is therefore no surprise that attempts to make such intentional notions (see also Chap. 11 of this volume) grounded are not limited to the notion of knowledge only. For instance, Su and others [103] provided a grounded model for the notions of knowledge, belief and certainty. Roughly, a state in their models has an external and an internal part: the external part determines what of the system is visible, and what is not visible, while the internal part specifies for each agent his perception of the visible part of the environment state and the plausible invisible parts of the invisible part of the environment state that the agent thinks possible. Lomuscio and Sergot even use the notion of interpreted system to show 'how it can be trivially adapted to provide a basic grounded formalism for some deontic issues' [83, p. 3]. Their models are basically hypercubes, where each local state L_i is then partitioned in a set of green states (allowed states of computation) and red states (disallowed states). This enables them to define a notion $\mathcal{O}_i\varphi$, with the meaning that 'in all the possible correctly functioning alternatives of agent i, φ is the case'.

12.4 Generalised Structures for Knowledge

Kripke structures provide a very natural way to model uncertainty and (lack of) information, and they are conceptually relatively easy. Depending on the kind of uncertainty one wants to model, one can often employ correspondence theory and, in a modular way, add additional constraints on the agents' accessibility relations. But there is also a criticism using this semantics, going in the other direction even if we do not impose any additional constraints on those relations, do we not get properties (of knowledge or belief) that are in fact *too strong*? This problem is known as the *logical omniscience* problem, and neighbourhood semantics is developed partially with the aim to address this. Finally, there is a stream of *topological* models for epistemic languages, which have their own virtues.

12.4.1 Neighbourhood Semantics

So what are possible shortcomings of using relational models for knowledge and belief? First of all, although this is not implied by the semantics, it is almost always assumed that all agents are equal: their knowledge and beliefs all satisfy the same properties. Indeed, in $S5_m$ we have, for all φ, that $\vdash K_a\varphi$ iff $\vdash \varphi$ iff $\vdash K_b\varphi$. Van Benthem and Liu are among the first to take seriously that 'epistemic agents may have different powers of observation and reasoning' [17], and allow for a 'diversity of logical agents'. Secondly, if one wants to express that the beliefs of agent a are correct by adding the axiom $B_a\varphi \rightarrow \varphi$ to a logical system, this property becomes *globally valid*: every agent knows it, it would even become *common knowledge*, and in a temporal setting it will hold *forever*. A first step to address this was made in [41].

A more fundamental criticism against using normal modal logic to model information of agents is known as *logical omniscience*. No agent is a perfect reasoner, so no agent will know all tautologies (of $S5$, or even the weakest normal modal logic K). This observation questions the intuitive soundness of **Nec**. Indeed, security protocols for communication or authentication that use cryptographic keys are based on the assumption that agents are *not* able to oversee all the consequences of the underlying theory (like inferring whether a given number is prime).

A similar criticism is sometimes used against axiom **K**: whereas an agent applying **K** once seems rather innocent, having it as an axiom implies that the agent can apply it as often as he likes. As an example, suppose that an agent knows what day of the week is today, and that he also knows which day of the week it is on any given day, if he would know this about the previous day. This would imply that the agent knows which day of the week it is on 25 of August 6034! For a weaker notion like belief such criticisms are even more compelling. It is argued that humans for instance might well believe φ in 'one frame of mind' (e.g., 'I pursuit an academic career') and something that is incompatible with it, in another ('I aim to become rich'). Some formal manifestations of logical omniscience are the axiom **K**, the validity $\Box(\varphi \wedge \psi) \leftrightarrow (\Box\varphi \wedge \Box\psi)$, the inference rule **Nec** and, some argue, the derived rule **Eq**: from $\varphi \leftrightarrow \psi$, infer $\Box\varphi \leftrightarrow \Box\psi$.

The idea that it should be possible to believe φ in one frame of mind and $\neg\varphi$ in another is one of the motivating requirements that lead to *neighbourhood semantics*. Here, rather than states that are considered possible by the agent, we have *sets* of states: each such set represents a possible frame of mind the agent can be in.

Definition 12.4 A *neighbourhood model* $M = \langle S, N, V \rangle$ where S is a set of states and $N : Op \rightarrow W \rightarrow 2^{2^S}$ assigns a neighbourhood $N_\Box(s) \subseteq 2^S$ to every state s, for every operator $\Box \in Op$. As before, $V(p) \subseteq S$ is the valuation function of the model. The pair $F = \langle S, N \rangle$ is a *neighbourhood frame*. Given a model M, defining $[\![\varphi]\!]_M$ (or simply $[\![\varphi]\!]$ if M is clear) to be $[\![\varphi]\!] = \{s \in S \mid M, s \models \varphi\}$, the relevant truth condition for modal operators is

$$M, s \models \Box\varphi \text{ iff for some } T \in N_\Box(s), T = [\![\varphi]\!].$$

In terms of knowledge: φ is known at s if the denotation of φ in M is one of the neighbourhoods of s. Neighbourhood models are more general than relational Kripke models: given $M = \langle S, R, V \rangle$ one can define $M = \langle S, N, V \rangle$ by

$$N_\Box(s) = \{U \subseteq S : R(s) \subseteq U\}.$$

Then, for any $s \in S$, the models M, s and M, s satisfy the same formulas. The other direction does not hold: indeed, under neighbourhood semantics, the property $(\Box\varphi \wedge \Box\psi) \rightarrow \Box(\varphi \wedge \psi)$ is not valid. If a neighbourhood model $M = \langle S, N, V \rangle$ is *augmented*, there *does* exist an equivalent relational model for it, where M is augmented if for all s, (1) $\cap_{T \in N_\Box(s)} T \in N(s)$, (2) $T_1 \cap T_2 \in N_\Box(s)$ only if $T_1, T_2 \in N_\Box(s)$, and (3) If $T_1 \in N_\Box(s)$ and $T_1 \subseteq T_2$, then $T_2 \in N_\Box(s)$.

One can 'recover' epistemic properties like veridicality and introspection in neighbourhood semantics by putting further constraints on the neighbourhood function N. Moreover, it is possible to use this semantics for multi-agent logics: the notion of 'everybody knows' for instance is then the modal operator for the neighbourhood function $N_E = \bigcap_{a \in Ag} N_a$. For common knowledge this can be done as well: we here follow [81]. It is not difficult to see that

$$\mathsf{M}, s \models K_a K_b \varphi \text{ iff } \{t \in S \mid [\![\varphi]\!] \in N_b(t)\} \in N_a(s) \tag{12.8}$$

In order to manipulate such expressions, it is convenient to define an algebraic operator $\circ$ on neighbourhoods as follows. Let $T \subseteq S$.

$$T \in N_1 \circ N_2(s) \text{ iff } \{t \in S \mid T \in N_2(t)\} \in N_1(s) \tag{12.9}$$

Equation (12.8) then becomes: $\mathsf{M}, s \models K_a K_b \varphi$ iff $[\![\varphi]\!] \in N_a \circ N_b(s)$. In this context, it is best to interpret common knowledge as the infinite conjunction

$$E\varphi \wedge E(\varphi \wedge E\varphi) \wedge E(\varphi \wedge E(\varphi \wedge E\varphi)), \ldots \tag{12.10}$$

In normal modal logic (12.10) is equivalent to (12.2), but using a neighbourhood semantics it is not! Let the special neighbourhood system $\mathcal{E}$ be defined by $T \in \mathcal{E}(s)$ iff $s \in T$. We then have $N \circ \mathcal{E} = \mathcal{E} \circ N = N$ for every N. Keeping in mind (12.10) define now a sequence of neighbourhood systems as follows.

$$N_0 = N_E \text{ and for any ordinal } \eta, N_\eta = N_E \circ (\bigcap_{\zeta < \eta} N_k \cap \mathcal{E}) \tag{12.11}$$

We now assume that the systems N_a in a model M are closed under supersets, i.e., $T \in N_a(s)$ and $T \supseteq T'$ implies that $T' \in N_a(s)$. This notion is sometimes also called *monotony*, and 'makes for smoother theory', quoting van Benthem et al. [23]. On such models, we have

Lemma 12.3 [81, Lemma 5] *Let ξ and η be ordinals. If $\xi < \eta$, then $N_\eta \subseteq N_\xi$.*

By Lemma 12.3, for any $s \in S$ the sequence $N_\eta(s)$ is a decreasing sequence of sets. Hence, there is a smallest ordinal o_s such that for all $\eta \geq o_s$, $N_\eta(s) = N_{o_s}(s)$. Now take $\delta = sup\{o_s \mid s \in S\}$: we have $N_\eta = N_\delta$ for all $\eta \geq \delta$. So the neighbourhood system against which common knowledge is interpreted is $N_C = N_\delta$.

This semantics is characterised by an axiomatisation given by Lismont [80], summarised in Table 12.4.

It is also possible to generalise the notion of bisimulation (to *behavioural equivalence*) to neighbourhood models, as well as to have a suitable notion of standard translation to a two-sorted first-order language, where the crucial clause for the translation is $\mathsf{ST}_x(\Box\varphi) = \exists T(xNT \wedge \forall y(TEy \leftrightarrow \mathsf{ST}_y(\varphi)))$, where xNT iff $T \in N(x)$ and TEy iff $y \in T$. With such an apparatus in place, [63] has been able to prove

Table 12.4 The axioms and rules above are added to the propositional **Taut** and **MP**

Common Knowledge for neighbourhood models			
E	$E\varphi \leftrightarrow \bigwedge_{a \in \mathsf{Ag}} K_a \varphi$	**FP**	$C\varphi \rightarrow E(C\varphi \wedge \varphi)$
Ind	From $\phi \rightarrow E\varphi$, infer $E\varphi \rightarrow C\varphi$	**Mon**	From $\phi \rightarrow \psi$, infer $\Box\phi \rightarrow \Box\psi$ $\Box \neq E$

a 'van Benthem-style' characterisation theorem for modal logic using a neighbourhood semantics. For completeness of modal logics wrt neighbourhood semantics we refer to e.g., [38] and [62]. An example of a logic that is Kripke incomplete, but is complete wrt neighbourhood frames can be found in [58].

Neighbourhood semantics are a very powerful tool for reasoning about games as well, if a neighbourhood is interpreted as a set of states a player can enforce. van Benthem et al. use this semantics to define their *concurrent game logic* ([23], and Chap. 14 of this book). Interestingly, van Benthem and Pacuit [19] have given an interpretation reminiscent of the notion of groundedness (see Sect. 12.3.5) to that of neighbourhoods: rather than using neighbourhoods as a technical device to study weak modal logics, they 'concretely' interpret a neighbourhood as an 'evidence set' of an agent who then can reason about the evidence, beliefs and knowledge—and their dynamics—he entertains.

12.4.2 Topological Semantics

Next we will discuss topological semantics of epistemic and doxastic logic. Topological semantics is closely related to Kripke and neighbourhood semantics. As we will see below, the standard Kripke semantics of *S*4 corresponds to special (Alexandroff) topological spaces. So topological semantics generalises the Kripke semantics of epistemic logic. On the other hand, topological models coincide with the neighbourhood models of *S*4. Nevertheless, it is useful to think in topological terms as it gives us an elegant and, at the same time, powerful mathematical machinery to investigate non-standard models of epistemic logic. In topological models of intuitionistic logic, open sets are treated as 'observable properties'. In domain theory, Scott domains are special posets equipped with the so-called Scott topology, where points are interpreted as 'pieces of information' or 'results of a computation'. Modal (epistemic) logic also provides a useful formalism to reason about (topological) spaces connecting it to the area of *spatial logic*.

Topological semantics also brings concrete benefits to the semantics of epistemic logic as observed by van Benthem and Sarenac [20]. (1) Topological products provide a way of merging the knowledge of two agents with no new information arising. (2) More importantly, they address Barwise's criticism of the Kripke semantics as the two ways of computing common knowledge, discussed in previous sections, no longer coincide. (3) Topological products also address Barwise's other criticism of Kripke semantics about modelling shared epistemic situation. (4) Finally, in some important cases (i.e., for distributed knowledge) topological interpretations of epistemic notions nicely complement the relational interpretations.

12.4.2.1 Topological Spaces: Connection with Kripke and Neighbourhood Frames

A *topological space* is a pair (X, τ), where X is a non-empty set and $\tau \subseteq \mathcal{P}(X)$ contains X and $\emptyset$ and is closed under finite intersections and arbitrary unions. Elements of τ are called *open sets*. Complements of open sets are called *closed sets*. An open set containing $x \in X$ is called an *open neighbourhood* of x. The *interior* of a set $A \subseteq X$ is the largest open set contained in A and is denoted by $\mathrm{Int}(A)$. The *closure* of A is the least closed set containing A and is denoted by $\overline{A}$. In other words, $\mathrm{Int}(A) = \bigcup \{U \in \tau : U \subseteq A\}$ and $\overline{A} = \bigcap \{F : X \setminus F \in \tau, A \subseteq F\}$. It is easy to check that $\overline{A} = X \setminus \mathrm{Int}(X \setminus A)$.

A topological space (X, τ) is called an *Alexandroff space* if τ is closed under infinite intersections. It is easy to see that a topological space is Alexandroff iff every point has a least open neighbourhood (the intersection of all its open neighbourhoods). It is also well known that Alexandroff spaces correspond to reflexive and transitive Kripke frames. Indeed, given an Alexandroff space (X, τ) one can define a reflexive and transitive binary relation R_τ on X by putting $x R_\tau y$ iff $x \in \overline{\{y\}}$ (that is, every open set that contains x also contains y). Conversely, suppose X is a set with a reflexive and transitive relation R. We say that $U \subseteq X$ is an *upset* if for each $x, y \in X$, $x R y$ and $x \in U$ imply $y \in U$. We define τ_R as the set of all upsets of (X, R). Then (X, τ_R) is a topological space and $R(x) = \{y \in X : x R y\}$ is the least open neighbourhood containing the point x. Thus, (X, τ_R) is Alexandroff. It is easy to check that this correspondence is one-to-one. Therefore, reflexive and transitive Kripke frames can be seen as particular examples of topological spaces. This connection between reflexive and transitive orders and topologies is at the heart of the translation between the plausibility and evidence models of dynamic epistemic logic, see [19, Sect. 5] for details.

Now we will quickly review the connection between topological spaces and neighbourhood frames. Let (X, N) be a neighbourhood frame satisfying the following five conditions:

1. for each $x \in X$ we have $U \in N(x)$ and $U \subseteq V$ imply $V \in N(x)$.
2. for each $x \in X$ we have $U, V \in N(x)$ implies $U \cap V \in N(x)$.
3. for each $x \in X$ we have $N(x) \neq \emptyset$.
4. for each $x \in X$ we have $U \in N(x)$ implies $x \in U$.
5. for each $x \in X$ and $U \in N(x)$ there exists $V \in N(x)$ such that $V \subseteq U$ and for each $y \in V$ we have $V \in N(y)$.

Let (X, τ) be a topological space. Then a set A is called a *neighbourhood* of x if $x \in A$ and there is an open neighbourhood U of x (i.e., $U \in \tau$ with $x \in U$) such that $U \subseteq A$. Let $N_\tau(x) = \{A : A \text{ is a neighbourhood of } x\}$. Then it is easy to check that (X, N_τ) is a neighbourhood frame satisfying conditions (1)–(5).

Conversely, if (X, N) is such that it satisfies (1)–(5) we define a topology τ_N on X by $\tau_N = \{U : U \in N(x)$ for each $x \in U\}$. Then (X, τ_N) is a topological space. Moreover, it is not difficult to check that this correspondence is one-to-one. We refer to e.g., [74, Theorem 2.6] for all the details. We would like to mention that conditions (1)–(5) are exactly those that correspond to the axioms of the modal logic $S4$ (see, e.g., [38]). To be more precise the transitivity axiom ($\Box p \rightarrow \Box\Box p$) correspondence to condition $(5')$ below.

$5'$. for each $x \in X$ and $U \in N(x)$ there exists $V \in N(x)$ such that for each $y \in V$ we have $U \in N(y)$.

But it is easy to show that a neighbourhood frame (X, N) satisfies (1)–$(5')$ iff it satisfies (1)–(5). Thus, topological spaces correspond to neighbourhood frames of the modal logic $S4$.

12.4.2.2 Topological Models of Epistemic Logic

A triple $M = (X, \tau, v)$ is a *topological model* if (X, τ) is a topological space and v a map from the propositional variables to $\mathcal{P}(X)$. We assume that we work with the modal language introduced in Definition 12.1. Truth of a formula φ in the model M at a point x, written as $M, x \models \varphi$, is defined inductively as follows:

$$
\begin{aligned}
M, x &\models p && \text{iff} \quad x \in v(p) \\
M, x &\models \varphi \wedge \psi && \text{iff} \quad M, x \models \varphi \text{ and } M, x \models \psi \\
M, x &\models \neg\varphi && \text{iff} \quad \text{not } M, x \models \varphi \\
M, x &\models \Box\varphi && \text{iff} \quad \exists U \in \tau \text{ such that } x \in U \text{ and } \forall y \in U\, M, y \models \varphi.
\end{aligned}
$$

Let $[\![\varphi]\!]_v = \{x \in X : M, x \models \varphi\}$. We will skip the index if it is clear from the context. It is easy to see that the last item is equivalent to $[\![\Box\varphi]\!] = \mathrm{Int}([\![\varphi]\!])$. Moreover, as $\Diamond\varphi = \neg\Box\neg\varphi$, we have that $[\![\Diamond\varphi]\!] = \overline{[\![\varphi]\!]}$. A pointwise definition of the semantics of $\Diamond$ is as follows:

$$
M, x \models \Diamond\varphi \quad \text{iff} \quad \forall U \in \tau \text{ such that } x \in U, \exists y \in U \text{ with } M, y \models \varphi.
$$

Note that if (X, τ) is an Alexandroff space, then the above definition of the semantics of formulas coincides with the one defined in Sect. 12.3.1 for Kripke models. Also if we view topological models as particular examples of neighbourhood models, then the above semantics coincides with the semantics of formulas in neighbourhood models defined in Sect. 12.4.1. The notion of satisfiability and validity of formulas in topological models as well as topological soundness and completeness of logics is defined in the same way as in Sect. 12.4.1.

Let us look at an example of topological interpretations. Let $\mathbb{R}$ be the real line where a topology τ on $\mathbb{R}$ is given by open intervals and their unions. Let also $v(p) = [0, 1) = \{r \in \mathbb{R} : 0 \le r < 1\}$. We invite the reader to check that

- $\llbracket \Box p \rrbracket = (0, 1)$,
- $\llbracket \Diamond p \rrbracket = [0, 1]$,
- $\llbracket \Diamond p \wedge \Diamond \neg p \rrbracket = \{0, 1\}$,
- $\llbracket p \wedge \Diamond p \wedge \Diamond \neg p \rrbracket = \{0\}$,
- $\llbracket \neg p \wedge \Diamond p \wedge \Diamond \neg p \rrbracket = \{1\}$.

Now we briefly discuss why topological models are of interest from the epistemic logic point of view. Topological semantics of modal logic precedes Kripke semantics and dates back to the 1930s. Already back then topological models were used to model knowledge in the context of intuitionistic logic (see e.g., [104]). Open sets can be interpreted as 'pieces of evidence', e.g., about location of a point. This reflects on the Brouwer-Heyting-Kolmogorov semantics, which informally defines intuitionistic truth as provability and specifies the intuitionistic connectives via operations on proofs. One could extend this reading to modal logic and give an epistemic interpretation to $\Box_a p$ in a topological model as: there exists a piece of evidence for agent a (i.e., an open set in a's topology), which validates the proposition p. We point out again that in [19] neighbourhood models are used to model the evidence of agents. Thus, the topological/neighbourhood model setting does not just refine the analysis of deduction or static attitudes, but also allows for a richer repertoire of dynamic information-carrying events. As we will see below, topological models also give a (nice) way to 'naturally' merge the knowledge of different agents (see van Benthem and Sarenac [20] for more discussion on topologies as models of epistemic logic).

Finally, going a bit beyond epistemic logic, we remark that a related view of connecting topology to computer science proved to be very influential. In fact, many topological concepts provide natural interpretations to important notions of computability theory. For example, data type corresponds to a topological space, piece of data to a point, semi-decidable property (observable property, affirmable property) to an open set, computable function to a continuous map, etc. We refer to [1, 46, 100, 105] for a thorough investigation of this line of research.

12.4.2.3 Topo-bisimulations

Similarly to the relational semantics, in order to understand the expressive power of modal languages on topological models one needs to define the corresponding notion of a bisimulation. This has been done by van Benthem and Aiello in [2].

Definition 12.5 A *topological bisimulation* or simply a *topo-bisimulation* between two topological models $M = (X, \tau, \nu)$ and $M' = (X', \tau', \nu')$ is a non-empty relation $T \subseteq X \times X'$ such that if xTx' then:

Harmony $x \in \nu(p)$ iff $x' \in \nu'(p)$, for each $p \in P$,
Forth $(x \in U \in \tau) \Rightarrow (\exists U' \in \tau') (x' \in U' \& (\forall y' \in U')(\exists y \in U)(yTy'))$
Back $(x' \in U' \in \tau') \Rightarrow (\exists U \in \tau) (x \in U \& (\forall y \in U)(\exists y' \in U')(yTy'))$.

In other words T is a bisimulation if T-image and T-inverse image of an open set is open. Two topological models are *topo-bisimilar* if there is a topo-bisimulation between them.

Let us look at some examples. First note that if two topological models are based on Alexandroff spaces, then a topo-bisimulation is the same as the standard Kripke bisimulation (Definition 12.3) between the corresponding reflexive and transitive Kripke models. Recall that a topology on the real line $\mathbb{R}$ is given by open intervals and their unions. Let $W = \{s, s^-, s^+\}$ and let $\preceq$ be the reflexive closure of $\{(s, s^-), (s, s^+)\}$. Then $(W, \preceq)$ is the so-called *2-fork*. Obviously the 2-fork is reflexive and transitive. So it corresponds to an Alexandroff space. It is now an easy exercise to check that the relation T between $\mathbb{R}$ and W defined as: $T(0, s)$, $T(r, s^-)$ for each $r \in \mathbb{R}$ with $r < 0$ and $T(r, s^+)$ for each $r \in \mathbb{R}$ with $r > 0$, is a topo-bisimulation. In fact, there is a deeper connection between these two structures. We refer to [3, 21] for more details on the connection of the spatial logic of $\mathbb{R}$ and the logic of the 2-fork.

Let $\mathbb{Q}$ be the set of rational numbers equipped with the topology induced from the reals. That is, open sets of $\mathbb{Q}$ are intersections of $\mathbb{R}$-open sets with $\mathbb{Q}$. Now it is easy to check that $T \subseteq \mathbb{R} \times \mathbb{Q}$ defined as: $T(z, z)$ for each $z \in \mathbb{Z}$, $T(r, q)$ for each $r \in \mathbb{R}$ and $q \in \mathbb{Q}$ with $z < r, q < z + 1$, for each $z \in \mathbb{Z}$ is a topo-bisimulation.

In the two cases above we assumed that $v(p) = \emptyset$ for each p. Now consider an example where this is not the case. Let $\mathbb{R}$ and $\mathbb{R}'$ be two isomorphic copies of the reals, with $v(p) = [0, 1]$ and $v'(p) = (0', 1')$. Then there is no topo-bisimulation between $(\mathbb{R}, v)$ and $(\mathbb{R}', v')$ relating 0 to any point $r' \in \mathbb{R}'$. To see this, note that by the basic case of the topo-bisimulation, $r' \in (0', 1')$. But then, by the forth condition, for $U' = (0', 1')$ there exists a neighbourhood U of 0 such that every point in U is topo-bisimilar to some point in $(0', 1')$. But this is impossible as each neighbourhood of 0 contains a point $t < 0$. Then t cannot be topo-bisimilar to any point in $(0', 1')$ as t does not satisfy p. Thus, there is no topo-bisimulation between $(\mathbb{R}, v)$ and $(\mathbb{R}', v')$ relating 0 to some point in $\mathbb{R}'$.

Similarly to bisimilar Kripke models, topo-bisimilar models satisfy the same modal formulas. That is, if $x \in X$ and $x' \in X'$ and xTx', then $M, x \models \varphi$ iff $M', x' \models \varphi$, for each formula φ. The converse is true for finite models. The notion of bisimilarity can also be expressed using games (see e.g., [2]).

The celebrated van Benthem's characterisation theorem (Theorem 12.2), states that on Kripke models modal logic is the bisimulation invariant fragment of first-order logic. An analogue of this theorem for topo-bisimulations and the language L_t (an analogue of the first-order language for topological spaces) was proved in [36].

12.4.2.4 Topological Completeness

Now we turn to deductive systems and the issues of axiomatisation and topological completeness. Note that the interior and closure operators satisfy the following well-known Kuratowski axioms (see e.g., [42]):

1. $\mathrm{Int}(X) = X$, $\overline{\emptyset} = \emptyset$,
2. $\mathrm{Int}(A \cap B) = \mathrm{Int}(A) \cap \mathrm{Int}(B)$, $\overline{A \cup B} = \overline{A} \cup \overline{B}$.
3. $\mathrm{Int}(A) \subseteq A$ $A \subseteq \overline{A}$.
4. $\mathrm{Int}(A) \subseteq \mathrm{Int}(\mathrm{Int}(A))$, $\overline{\overline{A}} \subseteq \overline{A}$.

It is easy to see that the above implies that $S4$ is sound with respect to topological semantics. In fact, 1–4 above are the axioms of $S4$ translated into topological terms. For completeness, we need to show that if φ is $S4$-consistent, then there exists a topological model (X, τ, ν) satisfying φ. As we know (e.g., by the standard canonical Kripke model argument), if φ is $S4$-consistent, then φ is satisfiable in a Kripke model with a reflexive and transitive relation. As every reflexive and transitive Kripke frame corresponds to an Alexandroff space, the completeness follows. van Benthem and his collaborators, however, gave a different, elegant and more self-contained proof of this result by introducing a *topo-canonical model* (a topological analogue of a canonical Kripke model) [3, 16]. We will quickly sketch the basic idea of this construction.

The *topo-canonical model* of $S4$ (in fact, instead of $S4$ we can consider any logic L over $S4$) is a triple $M^{\mathcal{C}} = (X^{\mathcal{C}}, \tau^{\mathcal{C}}, \nu^{\mathcal{C}})$, where $X^{\mathcal{C}}$ is the set of all maximal $S4$-consistent sets. Elements of $\tau^{\mathcal{C}}$ are unions of the sets $U_\varphi = \{\Gamma \in X^{\mathcal{C}} : \Box\varphi \in \Gamma\}$. In other words, $\{U_\varphi : \varphi$ is any formula$\}$ forms a basis for $\tau^{\mathcal{C}}$. Finally, we put $\Gamma \in \nu^{\mathcal{C}}(p)$ if $p \in \Gamma$. Then $(X^{\mathcal{C}}, \tau^{\mathcal{C}}, \nu^{\mathcal{C}})$ is a topological model. Moreover, the Truth Lemma holds for this model. That is,

$$M^{\mathcal{C}}, \Gamma \models \varphi \text{ iff } \varphi \in \Gamma.$$

Now if φ is $S4$-consistent, then by the Lindenbaum Lemma (see, e.g., [34, 37]) $\{\varphi\}$ can be extended to a maximal $S4$-consistent set Γ. By the Truth Lemma, $M^{\mathcal{C}}, \Gamma \models \varphi$, which finishes the proof.

As mentioned above, the topo-canonical model construction can be defined for any normal modal logic L extending $S4$. In analogy with the relational case, in the topological setting too one can define the notion of canonicity. A logic $L \supseteq S4$ is called *topo-canonical* if the topo-canonical model of L is based on a topological space validating L. Topo-canonical logics have been thoroughly investigated in [30].

So $S4$ is sound and complete with respect to all topological spaces. However, next question is whether one can find 'good' topological spaces for which $S4$ is sound and complete. The classical result of McKinsey and Tarski [87] states that $S4$ is sound and complete with respect to any dense-in-itself metrizable separable space. This includes the real line $\mathbb{R}$, and in general any Euclidean space $\mathbb{R}^n$, the Cantor space $\mathbf{C}$, the space of rational numbers $\mathbb{Q}$, etc. There is a number of different proofs of completeness of $S4$ with respect to these structures, see e.g., [3, 16, 28, 71, 89, 90] for an overview. Here we only give a sketch of the basic idea underlying most of these proofs. Strong completeness of $S4$ with respect to any dense-in-itself metric separable space was recently shown in [77] and [79]. A full axiomatization of the space of rational numbers $\mathbb{Q}$ in the language with topological and temporal modalities F and P was given in [98]. Recently, [72] gave a full axiomatization of the real line $\mathbb{R}$ in the same language.

As $S4$ is sound with respect to all topological spaces, it is obviously sound with respect to each of the topological spaces mentioned above. For completeness, assume that φ is $S4$-consistent. Then as $S4$ has the finite model property (see e.g., [34, 37]) there exists a finite rooted, model $M = (W, R, V)$, with a root r, such that R is reflexive and transitive and $M, r \models \varphi$. Sometimes it is useful to exploit here the completeness of $S4$ with respect to other structures, say an infinite binary tree, etc. As R is reflexive and transitive, (W, R) could be viewed as an Alexandroff space and, thus, M is a topological model. Now let (X, τ) be the topological space for which we want to prove the completeness of $S4$ (e.g., $\mathbb{R}$, $\mathbb{Q}$, $\mathbf{C}$, etc.). If we manage to define a valuation v on X so that M and (X, τ, v) are topo-bisimilar, then as topo-bisimilar points satisfy the same formulas, (X, τ, v) will satisfy φ. In order to show that such a valuation and bisimulation exist, it is sufficient to prove that there exists a continuous and open map $f : X \to W$. Recall that f is *continuous* if the inverse f-image of an open set is open, and f is *open* if the direct f-image of an open set is open. Suppose such a map exists. Then we define $v(p) = f^{-1}(V(p))$. Moreover, the graph of this map will be a topo-bisimulation. Thus, the proof of completeness is reduced to defining a continuous and open map. This is not an easy task and there are many different constructions for different topological spaces. We refer to [16], and the references therein, for the details on this and on the topological completeness results obtained via this method.

So far we saw only one (epistemic) logical system associated with topological semantics - the modal logic $S4$. In the next section we will discuss few different ways of obtaining (epistemic) topological logics beyond $S4$. We want to reiterate that the ideas and insights of van Benthem were instrumental in advancing these research directions.

First we briefly discuss topological models with restricted valuations. When evaluating formulas in, for example, the real line $\mathbb{R}$, instead of the whole powerset, one could consider evaluating formulas as intervals and their finite (or countable) unions. In the real plane one could take (unions) of convex sets, polygons, or rectangles. Such evaluations, and corresponding logics have been studied in [3, 21, 107], see also [16]. In particular, [21] shows that such restricted valuation can capture the difference between dimensions of Euclidean spaces. This cannot be done with standard interpretations as the logic of any Euclidean space is $S4$. We refer to [75] for a thorough study of the computational aspects of the logics arising from Euclidean spaces.

12.4.2.5 Topological Products

Taking products is a natural way of combining two Kripke complete modal logics [52]. The problem with this is that the resulting logic might become much more complex than the original ones. For example, the logic $S4 \times S4$ is undecidable [55], whereas $S4$ is decidable. Moreover, when taking products of Kripke frames, the resulting frame always validates the extra axioms of *commutativity* ($\Box_1\Box_2 p \leftrightarrow \Box_2\Box_1 p$) and *Church-Rosser* ($\Diamond_1\Box_2 p \rightarrow \Box_2\Diamond_1 p$). In [24] van Benthem and his collaborators defined topological products of topological spaces. We briefly recall this construction.

Let (X, τ) and (X', τ') be topological spaces. Suppose $A \subseteq X \times X'$. We say that A is *horizontally open* (in short, *H-open*) if for any $(x, x') \in A$, there exists $U \in \tau$ such that $x \in U$ and $U \times \{x'\} \subseteq A$. *Vertically open sets* (in short, *V-open sets*) are defined similarly. We let τ_1 and τ_2 denote the collection of all horizontally and vertically open subsets of $X \times X'$, respectively. It is easy to see that τ_1 and τ_2 are topologies. Modal operators $\Box_1$ and $\Box_2$ in a product model $M = (X \times X', \tau_1, \tau_2, \nu)$ are interpreted as follows.

$$M, (x, x') \models \Box_1\varphi \quad \text{iff} \ \exists U \in \tau_1 \text{ such that } x \in U \text{ and } \forall y \in U, M, (y, x') \models \varphi.$$
$$M, (x, x') \models \Box_2\varphi \quad \text{iff} \ \exists U \in \tau_2 \text{ such that } x' \in U \text{ and } \forall z \in U, M, (x, z) \models \varphi.$$

Consider as an example $(\mathbb{R} \times \mathbb{R}, \tau_1, \tau_2)$, and let $\nu(p) = [0, 1) \times \{0\}$. Then it is easy to see that

- $[\![\Box_1 p]\!] = (0, 1) \times \{0\}$,
- $[\![\Diamond_1 p]\!] = [0, 1] \times \{0\}$,
- $[\![\Box_2 p]\!] = \emptyset$,
- $[\![\Diamond_2 p]\!] = [0, 1) \times \{0\}$.

This operation can be extended to a notion of a product of two topologically complete modal logics. Given topologically complete uni-modal logics L and L', their *topological product* is the bi-modal logic of the product frames $X \times X'$, where X is a topological frame for L and X' for L', respectively.

Surprisingly enough, topological products turned out to be very well behaved. [24] shows that the topological product of $S4$ with itself is the same as the logic $S4 \otimes S4$, $S4$ *fusion* $S4$, which is decidable. The fusion is just the smallest bi-modal logic that contains $S4$-axioms for both modalities. Thus, no extra axiom is valid on product topological spaces. The Church-Rosser and commutativity axioms can be refuted on the product space $\mathbb{R} \times \mathbb{R}$. Moreover, the logic $S4 \otimes S4$ is sound and complete with respect to the product space $\mathbb{Q} \times \mathbb{Q}$. It is still an open question to find an axiomatisation of the logic of $\mathbb{R} \times \mathbb{R}$ [78].

One could view topological products of epistemic logics as a new and interesting way of merging the knowledge of two agents [20]. The fact that unlike relational products, no new axiom is valid in the topological case, shows that when we 'topologically merge' (via taking topological products of uni-modal epistemic logics) the knowledge of two agents, no new information arises. Topological products have other nice properties. Recall that in relational semantics the *distributed knowledge of agents* (which describes what a group would know if its members decided to merge their information) is expressed by taking the D_A operator – the Box operator of the intersection of the relations. In topological products one expresses distributed knowledge by taking the interior of the join of the topologies. By the join of two topologies we mean the least topology that contains both topologies. We note that in the topological case the join is not always very interesting e.g., for $\mathbb{Q} \times \mathbb{Q}$, the join of horizontal and vertical topologies is just the discrete topology (all sets are open). We refer to [20] for all the details on this.

However, there is not always an analogy between relational and topological semantics of epistemic logic. In fact, (in some way) topological models provide a richer landscape for interpreting epistemic logic, than the relational semantics. In his well-known chapter [7] Barwise underlined that a proper analysis of common knowledge must distinguish the following three approaches:

1. countably infinite iteration of individual knowledge modalities,
2. the fixed point view of common knowledge as 'equilibrium',
3. agents having a shared epistemic situation.

The relational semantics of epistemic logic cannot properly distinguish these three approaches (see Sect. 12.3.1), whereas topological semantics (topological products) suits this purpose perfectly well. Recall that in the Kripke semantics the approximation of the common knowledge operator stabilises in $\kappa \leq \omega$ steps. It was noted by van Benthem and Sarenac [20] that this is no longer the case in topological semantics, thus addressing Barwise's criticism.

In topological models common knowledge as equilibrium is expressed by taking the intersection of topologies. We will quickly sketch this argument. Let τ_1 and τ_2 be two topologies. Here we do not need to assume that these topologies are the vertical and horizontal topologies of a product space. Recall from the previous section that $C\varphi = \nu x(\varphi \wedge \Box_1 x \wedge \Box_2 x)$. Let $\psi = \varphi \wedge \Box_1 x \wedge \Box_2 x$. Then $C\varphi = \nu x \psi$. As it was also discussed in the previous section, $[\![C\varphi]\!]_V = [\![\nu x \psi]\!]_V = \bigcup\{U \in \mathcal{P}(W) : U \subseteq [\![\psi]\!]_{V_x^U}\}$. Now observe that for each $i = 1, 2$ we have $[\![\Box_i x]\!]_{V_x^U} = \mathrm{Int}_{\tau_i}(U)$. Hence, for $i = 1, 2$ we have $U \subseteq [\![\Box_i x]\!]_{V_x^U}$ iff U is τ_i-open. Thus, $U \subseteq [\![\psi]\!]_{V_x^U}$ iff $U \subseteq [\![\varphi]\!]_V$ and U is τ_1 and τ_2-open. Therefore, $[\![C\varphi]\!]_V = \bigcup\{U \subseteq [\![\varphi]\!]_V : U \in \tau_1 \cap \tau_2\} = \mathrm{Int}_{\tau_1 \cap \tau_2}([\![\varphi]\!])$. So topologically the common knowledge corresponds to the interior of the intersection of the topologies.

It is proved in [20] that the (countably) infinite iterations of the individual knowledge modalities may not be (horizontally or vertically) open. Hence, computing $C\varphi$ as fixed equilibrium and as countable iterations of $\varphi \wedge E\top$ in topological models diverge. This fact captures the difference between (1) and (2) above. A similar observation was made in [81, Proposition 4] for neighbourhood frames.

In a topological setting one can also analyse a 'shared situation' when there is a new group concept τ that only accepts very strong collective evidence for any proposition. This corresponds to adding the standard product topology on top of the horizontal and vertical topologies. Thus, formally speaking, we have three operators, the horizontal and vertical $\Box_1$, $\Box_2$ and also $\Box$ of the standard product topology. This again addresses Barwise's critical comments on common knowledge in the topological setting by showing that (3) can also differ from (1) and (2). It was proved in [24] that the logic of such spaces is the ternary modal logic obtained by adding the axiom $\Box p \to (\Box_1 p \wedge \Box_2 p)$ to the three dimensional fusion logic $S4 \otimes S4 \otimes S4$.

Next we give an example that illustrates how to compute topologically common and shared knowledge. This example also shows that these two notions differ.

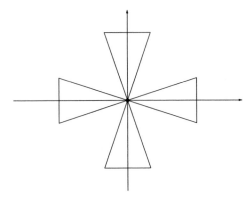

Fig. 12.5 The cross-valuation on the product space

Consider the topological product $\mathbb{R} \times \mathbb{R}$ with horizontal and vertical topologies. Let $\square$ be interpreted as the interior of the standard product topology. We let $v(p)$ be the 'cross' or the 'pair of orthogonal bow ties' depicted in Fig. 12.5 (we leave it up to the readers imagination to derive an appropriate name for this valuation). We assume that $v(p)$ has no boundaries, but the point $(0, 0)$ belongs to it. A very similar example can be found in [16, 20, 24]. Obviously one could give a formal mathematical description of $v(p)$, but we prefer to stick with a picture that provides sufficient intuition. Then it is easy to check that

- $[\![\square_1 p]\!] = [\![\square_2 p]\!] = [\![Cp]\!] = v(p)$,
- $[\![\square p]\!] = v(p) \setminus \{(0, 0)\}$.

Finally, we note that another epistemic logic $S5$ also admits topological semantics, but a much more exotic one than $S4$. We say that a topological space is a *clopen space* if every closed subset of it is also open. Examples of such spaces are discrete spaces (every subset is open), and also the spaces with the trivial topology (only the empty set and the whole space are open). We recall that $S5$ can be obtained from $S4$ by adding the axiom $p \rightarrow \square\lozenge p$ or alternatively by adding the axiom $\lozenge p \rightarrow \square\lozenge p$. Translating this into topological terms, a topological space (X, τ) validates $S5$ iff $\overline{A} \subseteq \text{Int}(\overline{A})$ for each $A \subseteq X$. This is equivalent to the fact that every closed subset of X is open. This means that $S5$ is sound with respect to clopen spaces. For the topological completeness of $S5$ one can apply Kripke completeness of $S5$ and the argument that was used in the topological completeness of $S4$, see [16] for the details.

For the topological completeness of other extensions of $S4$ such as the logic $S4.Grz$ we refer to [43] and [27]. Questions on modal definability of (classes) of topological spaces were studied in [54]. Topological completeness of $S4$ with the universal modality with respect to connected spaces was proved in [99] and further generalised in [28]. Measure-theoretic semantics of this logic can be found in [48].

Another approach (similar in essence, but different in technicalities) to topological interpretations of epistemic logic goes via the so-called *subset spaces* [92].

This interpretation uses a bimodal language. This led to introducing topologic—the axiomatic system (in this extended language) sound and complete wrt all topological spaces [56, 57]. Completeness of topologic via canonical model construction was proved in [39]. We will not discuss this approach here, but will instead refer to an overview article [94] for all the details and references. Further generalisations and stronger completeness results for topologic and multi-agent epistemic logics of subset spaces have been obtained by Heinemann in [65–68]. However, Heinemann achieves this by adding more than 'just' several knowledge operators K_i to subset space logic: they are definable from other operators that are added, and they do not have the $S5$ properties of knowledge. Only recently, Wang ad Ågotnes [106] gave a complete axiomatisation for multi-agent epistemic subset space logic as a generalisation of the single agent case.

12.4.2.6 Topological Models of Doxastic Logic

In this section we discuss a different topological semantics of modal logic via the derived set operator. As we will see below this semantics admits doxastic interpretations, nicely complementing the epistemic semantics of the closure and interior operators discussed in the previous section. We will again concentrate on the issues of expressivity (bisimulations) and topo-completeness for this semantics.

Let (X, τ) be a topological space. We recall that a point x is called a *limit point* (limit points are also called *accumulation points*) of a set $A \subseteq X$ if for each open neighbourhood U of x we have $(U \setminus \{x\}) \cap A \neq \emptyset$. A point $x \in A$ is called an *isolated point of A* if $x \notin d(A)$. Let $d(A)$ denote all limit points of A. This set is called the *derived set* and d is called the *derived set operator*. For each $A \subseteq X$ we let $t(A) = X \setminus d(X \setminus A)$. We call t the *co-derived set operator*. Also recall that there is a close connection between the derived set operator and the closure operator. In particular, for each $A \subseteq X$ we have $\overline{A} = A \cup d(A)$. Thus, the derived set operator is more expressive than the closure operator, see [29] for a discussion on this. Unlike the closure operator there may exist elements of A that are not its limit points. In other words, in general $A \not\subseteq d(A)$. To see this, consider the real line $\mathbb{R}$ and let $A = [0, 1] \cup \{2\}$. Then $2 \notin d(A)$, but $2 \in A$.

Let (X, τ) be an Alexandroff space. That is, τ is the set of all upsets for some reflexive and transitive relation R. By spelling out the definition of the derived set, we observe that $x \in d(A)$ for some $A \subseteq X$ iff there is $y \in A$ such that $x \neq y$ and xRy. So $d(A) = \{x \in X : \exists y \in A, \text{ such that } xRy \ \& \ x \neq y\}$ (see [44] for more details on the derived set operator in Alexandroff spaces).

Let $M = (X, \tau, \nu)$ be a topological model. We now define a new semantics for $\square$ and $\Diamond$ using the derived set operator. All the Boolean cases are the same as before, the only difference is in the way the modal operators are interpreted.

$M, x \models \square\varphi$ iff $\exists U \in \tau$ such that $x \in U$ and $\forall y \neq x$ with $y \in U$, $M, y \models \varphi$.
$M, x \models \Diamond\varphi$ iff $\forall U \in \tau$ such that $x \in U$, $\exists y \neq x$ with $y \in U$ and $M, y \models \varphi$.

We again assume that $[\![\varphi]\!]_v = \{x \in X : M, x \models \varphi\}$, and we skip the index if it is clear from the context. It is easy to see that $[\![\Box\varphi]\!] = t([\![\varphi]\!])$ and $[\![\Diamond\varphi]\!] = d([\![\varphi]\!])$.

In this context too the main questions to ask are what are bisimulations for topological models with this new interpretation and what kind of completeness results one can obtain. We first briefly address the bisimulation issue and then move to completeness and doxastic interpretations. In fact, the notion of a d-bisimulation of topological models is the same as a topo-bisimulation with the only difference that in Definition 5 in the forth and back conditions we add $y' \neq x'$ and $y \neq x$. We call these bisimulations d-bisimulations. The notion of d-bisimilarity is defined similarly to topo-bisimilarity. Then one can prove that d-bisimilar points satisfy the same modal formulas. Bezhanishvili et al. [29] defines d-morphisms and shows that d-morphisms are functional d-bisimulations. (d-morphism is a continuous and open map such that the inverse image of each point is a discrete space in the induced topology. This definition turned out to be very useful as checking whether a map between two topological spaces is a d-morphism is relatively easy.)

Fixed point operators have an interesting role to play in topological semantics of derived set operator as well. For example, consider a simple formula $\mu x \Box x$. We note again that we interpret fixed point formulas in topological structures in the same way as in Kripke structures (see Sect. 12.3.1). Then $\mu x \Box x$ is valid in a topological space X iff X is scattered. To see this, recall that X is *scattered* iff every non-empty subsets U of X has an isolated point. That is, for $U \subseteq X$,

$$U \neq \emptyset \Rightarrow U \setminus d(U) \neq \emptyset. \tag{12.12}$$

It was noted in Esakia [43] that (a logical formulation) of (12.12) is equivalent (over topological spaces and also over transitive Kripke frames) to the Gödel-Löb axiom $(\Box(\Box p \rightarrow p) \rightarrow \Box p)$. Now it is easy to see that (12.12) is equivalent to the following, for $U \subseteq X$,

$$U \neq X \Rightarrow t(U) \not\subseteq U. \tag{12.13}$$

Thus, the only subset U of X such that $t(U) \subseteq U$ is the whole space X. Therefore,

$$X = \bigcap\{U \subseteq X : t(U) \subseteq U\} = \bigcap\{U \subseteq X : [\![\Box x]\!]_{V_x^U} \subseteq U\} = [\![\mu x \Box x]\!]. \tag{12.14}$$

So X is scattered iff $\mu x \Box x$ is valid in X. As $\mu x \Box x$ has no free variables, validity and satisfiability for this formula are equivalent.

The fact that scatteredness of a space can be captured by fixed point formulas is not very surprising as scatteredness is a topological analogue of dual well-foundedness [43] and it was shown in [13] (see also [22]) that $\mu x \Box x$ together with the transitivity axiom expresses dual well-foundedness of a Kripke structure. A similar observation in algebraic terms (for the so-called diagonalisable algebras) has been made already in [60].

Now we turn to the issue of soundness and completeness. First recall (see e..g., [42]) that the derived and co-derived set operator satisfy the following axioms.

1. $t(X) = X$, $\quad\quad\quad\quad\quad\quad\quad\quad d(\emptyset) = \emptyset$,
2. $t(S \cap T) = t(S) \cap t(T)$, $\quad\quad d(S \cup T) = d(S) \cup d(T)$.
3. $A \cap t(A) \subseteq t(t(A))$ $\quad\quad\quad d(d(A)) \subseteq A \cup d(A)$.

Recall that $wK4$ is a modal logic obtained form the basic modal logic K by adding to it the following *weak transitivity axiom* $(p \wedge \Box p) \rightarrow \Box\Box p$. The logic $wK4$ is sound and complete with respect to Kripke frames with weakly transitive relations, where the relation is *weakly transitive* if xRy, yRz and $x \neq z$ imply xRz [44].

The three properties of derived and co-derived set operators listed above imply that $wK4$ is sound with respect to topological semantics. In fact, Esakia [44] proved that $wK4$ is also complete with respect to all topological spaces. A topological space (X, τ) is said to satisfy the T_D-*separation axiom* (is a T_D-*space*, for short) if every point of X is the intersection of a closed and an open set. In fact, this condition is equivalent to $d(d(A)) \subseteq d(A)$, for each $A \subseteq X$ [42]. This implies that $K4$ is sound with respect to T_D-spaces. Esakia [44] showed that $K4$ is also complete with respect to T_D-spaces. For topological d-semantics of the provability logic GL, polymodal provability logic GLP, the logics of the rationals, real line and Euclidean spaces, as well as the logics of all T_0-spaces, Stone and spectral spaces and many more, we refer to [8, 16, 29, 31, 32, 43–45, 71, 84–86, 97].

We close this section by reviewing topological completeness of the doxastic modal logic $KD45$, see [102] and [94]. The fact, mentioned above, that in general $A \not\subseteq d(A)$ yields that the reflexivity axiom $p \rightarrow \Diamond p$ (equivalently $\Box p \rightarrow p$) is not sound with respect to this semantics, which makes this semantics suitable for doxastic logic. Recall that Kripke frames of $KD45$ are serial, transitive and Euclidean. As we saw above, the topological reading of the transitivity axiom gives us T_D-spaces. It is well known that the seriality axiom $\Box p \rightarrow \Diamond p$ is equivalent to $\Diamond\top$. Translating this into the topological terms we obtain the condition $d(X) = X$. This means that every point of X is a limit point. Such spaces are called *dense-in-itself*. Finally, the topological reading of the Euclidean axiom $\Diamond p \rightarrow \Box\Diamond p$ results in the condition $d(A) \subseteq t(d(A))$. It is easy to see that a set A is closed iff $d(A) \subseteq A$. Dualising this, we obtain that a set A is open iff $A \subseteq t(A)$. Thus, $d(A) \subseteq t(d(A))$ is equivalent to $d(A)$ being open.

The above leads to the following definition. A topological space (X, τ) is called a *DSO-space*[2] if it is dense-in-itself T_D-space such that $d(A)$ is an open set for each $A \subseteq X$. The discussion above shows that $KD45$ is sound with respect to DSO-spaces. Next we give an example of a DSO-space. Let $(\mathbb{N}, \tau)$ be the set $\mathbb{N}$ of natural numbers equipped with the topology $\tau = \{\emptyset, \text{ all cofinite sets}\}$. Then it is not hard to check that for each $A \subseteq \mathbb{N}$ we have:

---

[2] DSO stands for *Derived Sets are Open*.

$$d(A) = \begin{cases} \emptyset, & \text{if } A \text{ is finite,} \\ \mathbb{N}, & \text{if } A \text{ is infinite.} \end{cases}$$

This implies that $(\mathbb{N}, \tau)$ is a DSO-space.

Now we turn to the issue of completeness of *KD*45 with respect to DSO-spaces. First note that with every weakly transitive frame (X, R) we can associate a topology τ_R of all R-upsets. Observe that this topology will be the same as the topology of all upsets of the reflexive closure of R. Indeed, $A \subseteq X$ is an upset iff it is an upset for the reflexive closure of R.

Let (X, R) be a *KD*45-frame (that is, serial, Euclidean, transitive). Note that (X, τ_R) defined above, in general, is not a DSO-space. In fact, it is a DSO-space iff there are no distinct points $x, y \in X$ such that $x R y$. To see this, note that if such points exist, then $d(\{y\})$ is not an upset: we have $x \in d(\{y\})$, and $x R y$, but $y \notin d(\{y\})$ (use the definition of d on Alexandroff spaces in the beginning of this section). So $\{y\}$ is such that $d(\{y\})$ is not open. Thus, (X, R) does not correspond to a DSO-space. The converse direction is similar. Therefore, we cannot use directly the Kripke completeness of *KD*45 for deriving its topological completeness, as we did for *S*4. Nevertheless, one could still use Kripke completeness of *KD*45 to obtain topological completeness. We sketch the proof. All the details can be found in [94, 102].

Assume that φ is a *KD*45-consistent formula. Then, by Kripke completeness of *KD*45, there exists a Kripke model (W, R, v), where R is serial, transitive and Euclidean relation such that (W, R, v) satisfies φ. Now let us take a product of this frame (seen as the Alexandroff space) with the DSO-space $(\mathbb{N}, \tau)$ discussed above. Then one can show that $\mathbb{N} \times W$ is a DSO-space (with the standard product topology). Moreover, the second projection is a d-morphism, and hence its graph is a d-bisimulation. Here we use the notion of a d-morphism between topological spaces and Kripke frames [29]. A map between a topological space and Kripke frame is called a d-morphism if it is continuous and open, the inverse image of an irreflexive point is a discrete subspace and the inverse image of a reflexive point is a dense-in-itself subspace. So $\mathbb{N} \times W$ satisfies φ, which proves the completeness of *KD*45 for DSO-spaces.

In fact, one can strengthen this result and give an alternative proof of completeness avoiding products. [26] gives a characterisation of rooted *KD*45 frames. Using this characterisation it is easy to see that there exists a d-morphism from $(\mathbb{N}, \tau)$ to any rooted *KD*45-frame. This implies that if a formula φ is *KD*45-consistent, then it is satisfied in a DSO-space $(\mathbb{N}, \tau)$. Thus, *KD*45 is sound and complete with respect to not only all DSO-spaces, but also with respect to just $(\mathbb{N}, \tau)$.

In spite of the elegance of the derived set semantics for belief, it is also vulnerable to some criticism. One of the main problems is the fact that $\text{Int}(A) = A \cap t(A)$. Therefore, the derived set semantics for belief leads to the identification of knowledge with true belief, which goes against the unanimous opinion of epistemologists, and the numerous 'Gettier-type' counter-arguments [59]. Baltag et al. [6] and Özgün [93] then propose an alternative topological semantics for belief, where the belief operator is interpreted as the *closure of the interior operator* (that is, $\llbracket \Box \varphi \rrbracket = \text{Cl}(\text{Int}\llbracket \varphi \rrbracket)$), and prove that $KD45$ is complete in this semantics with respect to extremally disconnected topological spaces. We recall that a topological space (X, τ) is *extremally disconnected* if the closure of every open set is again open. Also note that every DSO-space is extremally disconnected, but not vice versa. Therefore, the latter semantics is applicable to a wider class of models than Steinsvold's semantics. Moreover, the formalism of [6, 93] fits well with Stalnaker's conception of 'strong belief as subjective certainty' [101], embodied in his axiom $B\varphi \rightarrow BK\varphi$, which is satisfied in that setting. For more details on this new semantics of belief, topological belief revision etc., we refer to [6, 93].

We hope that all these results illustrate that topological spaces provide interesting and insightful semantics for both epistemic and doxastic modal logic.

12.5 Conclusion

In this chapter, we have focussed on modal logics for knowledge and belief, especially their semantics. Starting with epistemic Kripke structures, we showed how Johan's results on correspondence theory often makes it possible to build an epistemic logic to which one can add a number of appealing axioms. Correspondence theory then makes it possible to quickly come up with classes of Kripke models wrt which those logics are sound and complete. Also, using a Kripke model for knowledge, it is conceptually simple to add relations to such a model that model time, or some other kind of dynamics. Johan, with his collaborators, has contributed to this field by showing how several of such dynamic epistemic logics are related. Their epistemic temporal frames provide a broad class of structures to which one can related interpreted systems, and logics for updates and revision.

Having a class of structures at hand, natural questions are when two structures are different, and what can be expressed in that class. Johan's characterisation theorem gives an answer for the case of normal modal logics, we have shown in this chapter how this theorem has been adapted or generalised to other classes of structures.

An important class of structures for epistemic logic is obtained by moving to a so-called neighbourhood semantics, or the closely related semantics based on a topology. Those semantics give an alternative and independent view on epistemic logic. The latter for instance can discriminate three aspects of common knowledge, which seem to be intertwined under the Kripke semantics.

Johan's contribution to logics of knowledge and belief is to be found in the technical results he has provided in the field of modal logic in general and that of epistemic logic in particular, but equally important are the themes he has consistently pursued: knowledge and ignorance are mostly interesting in a multi-agent setting, they only come to live in a dynamic context, and, while there is a multitude of schools studying epistemic logic, a close analysis tells us that they have more in common than even those schools themselves often tend to think!

Acknowledgments The authors would like to thank Guram Bezhanishvili, David Gabelaia, Ian Hodkinson, Clemens Kupke and Levan Uridia for many interesting discussions and useful suggestions. Thanks also go to Aybüke Özgün for pointing out a few small errors. The first author would also like to acknowledge the support of the Netherlands Organization for Scientific Research grant 639.032.918 and the Rustaveli Science Foundation of Georgia grant FR/489/5-105/11.

References

1. Abramsky S (1991) Domain theory in logical form. Ann Pure Appl Logic 51(1–2):1–77
2. Aiello M, van Benthem J (2002) A modal walk through space. J Appl Non-Classical Logics 12(3–4):319–363
3. Aiello M, van Benthem J, Bezhanishvili G (2003) Reasoning about space: the modal way. J Logic Comput 13(6):889–920
4. Aumann RJ (1976) Agreeing to disagree. Ann Stat 4(6):1236–1239
5. Aumann RJ (1999) Interactive epistemology I: Knowledge. Int J Game Theory 28:263–300
6. Baltag A, Bezhanishvili N, Özgün A, Smets S (2013) The topology of belief, belief revision and defeasible knowledge. In: Proceedings of the 4th international workshop logic, rationality, interaction (LORI 2013), pp 27–41. Springer
7. Barwise J (1988) Three views of common knowledge. In: Proceedings of the second conference on theoretical aspects of reasoning about knowledge, Pacific Grove, CA. Morgan Kaufmann, Los Altos, pp 365–379
8. Beklemishev L, Gabelaia D (2013) Topological completeness of the provability logic GLP. Ann Pure Appl Logic 164(12):1201–1223
9. van Benthem J (1976) Modal correspondence theory. PhD thesis, University of Amsterdam
10. van Benthem J (1985) Modal logic and classical logic. Bibliopolis, Napoli
11. van Benthem J (2001) Logics for information update. In: Proceedings of the 8th conference on theoretical aspects of rationality and knowledge, TARK '01, pp 51–67, San Francisco, CA, USA. Morgan Kaufmann
12. van Benthem J (2002) One is a lonely number: on the logic of communication. Technical report, ILLC, University of Amsterdam, 2002. Report PP-2002-27 (material presented at the Logic Colloquium 2002)
13. van Benthem J (2006) Modal frame correspondences and fixed-points. Studia Logica 83(1/3):133–155
14. van Benthem J (2006) Open problems in logical dynamics. In: Gabbay DM, Goncharov SS, Zakharyaschev M (eds) Mathematical problems from applied logic I, vol 4. International mathematical series. Springer, New York, pp 137–192. doi:10.1007/0-387-31072-X_3
15. van Benthem J (2011) Logical dynamics of information and interaction. Cambridge Book University, Cambridge
16. van Benthem J, Bezhanishvili G (2007) Modal logics of space. In: Aiello M, Pratt-Hartmann IE, van Benthem J (eds) Handbook of spatial logics. Springer, Dordrecht, pp 217–298
17. van Benthem J, Liu F (2004) Diversity of logical agents in games. Philosophia Scientiae 8(2):165–181
18. van Benthem J, Pacuit E (2006) The tree of knowledge in action: towards a common perspective. In: Governatori G, Hodkinson I, Venema Y (eds) Proceedings of advances in modal logic, vol 6. King's College Press, London, pp 87–106
19. van Benthem J, Pacuit E (2011) Dynamic logics of evidence-based beliefs. Studia Logica 99:61–92
20. van Benthem J, Sarenac D (2004) The geometry of knowledge. In: Aspects of universal logic. Travaux Logic, vol 17. University of Neuchâtel, Neuchâtel, pp 1–31
21. van Benthem J, Bezhanishvili G, Gehrke M (2003) Euclidean hierarchy in modal logic. Studia Logica 75(3):327–344

22. van Benthem J, Bezhanishvili N, Hodkinson I (2012) Sahlqvist correspondence for modal mu-calculus. Studia Logica 100(1–2):31–60
23. van Benthem J, Ghosh S, Liu F (2008) Modelling simultaneous games in dynamic logic. Synthese 165:247–268
24. van Benthem J, Bezhanishvili G, ten Cate B, Sarenac D (2006) Multimodal logics of products of topologies. Studia Logica 84(3):369–392
25. van Benthem J, Hoshi T, Gerbrandy J, Pacuit E (2009) Merging frameworks for interaction. J Philos Logic 38(5):491–526
26. Bezhanishvili N (2002) Pseudomonadic algebras as algebraic models of doxastic modal logic. MLQ Math Logic Q 48(4):624–636
27. Bezhanishvili G, Mines R, Morandi P (2003) Scattered, hausdorff-reducible, and hereditarily irresolvable spaces. Topol Appl 132(3):291–306
28. Bezhanishvili G, Gehrke M (2005) Completeness of S4 with respect to the real line: revisited. Ann Pure Appl Logic 131(1–3):287–301
29. Bezhanishvili G, Esakia L, Gabelaia D (2005) Some results on modal axiomatization and definability for topological spaces. Studia Logica 81(3):325–355
30. Bezhanishvili G, Mines R, Morandi P (2008) Topo-canonical completions of closure algebras and Heyting algebras. Algebra Universalis 58(1):1–34
31. Bezhanishvili G, Esakia L, Gabelaia D (2010) The modal logic of Stone spaces: diamond as derivative. Rev Symb Logic 3(1):26–40
32. Bezhanishvili G, Esakia L, Gabelaia D (2011) Spectral and T_0-spaces in d-semantics. In: 8th International Tbilisi symposium on logic, language, and computation. Revised selected papers. LNAI, vol 6618, pp 16–29
33. Blackburn P, de Rijke M, Venema Y (2001) Modal logic. Cambridge tracts in theoretical computer science, vol 53. Cambridge University Press, Cambridge
34. Blackburn P, de Rijke M, Venema Y (2001) Modal logic. Cambridge University Press, New York
35. Bradfield J, Stirling C (2007) Modal mu-calculus. In: Blackburn P, van Benthem J, Wolter F (eds) Handbook of modal logic. Elsevier, New York, pp 721–756
36. ten Cate B, Gabelaia D, Sustretov D (2009) Modal languages for topology: expressivity and definability. Ann Pure Appl Logic 159(1–2):146–170
37. Chagrov A, Zakharyaschev M (1997) Modal logic. The Clarendon Press, Oxford
38. Chellas BF (1980) Modal logic. Cambridge University Press, Cambridge
39. Dabrowski A, Moss L, Parikh R (1996) Topological reasoning and the logic of knowledge. Ann Pure Appl Logic 78(1–3):73–110
40. van Ditmarsch H, Duque Fernández D, van der Hoek W (2011) On the definability of simulability and bisimilarity by finite epistemic models. In: Leite J, Torroni P, Ågotnes T, Boella G, van der Torre L (eds) Computational logic in multi-agent systems (CLIMA XII). LNCS, vol 6814. Springer, New York, pp 74–87
41. van Ditmarsch H, van der Hoek W, Kooi B (2009) Knowing more—from global to local correspondence. In: Boutillier C (ed) Proceedings of IJCAI-09, pp 955–960
42. Engelking R (1989) General topology, vol 6, 2nd edn. Heldermann Verlag, Berlin
43. Esakia L (1981) Diagonal constructions, the Löb formula and rarefied Cantor's scattered spaces. In: Studies in logic and semantics. Metsniereba, Tbilisi, pp 128–143
44. Esakia L (2001) Weak transitivity—a restitution. In: Karpenko AS (ed) Logical investigations, vol 8 (Russian) (Moscow, 2001). Nauka, Moscow, pp 244–255
45. Esakia L (2004) Intuitionistic logic and modality via topology. Ann Pure Appl Logic 127(1–3):155–170
46. Escardo M (2004) Domain theory in logical form. Electron Notes Theor Comput Sci 87:21–156
47. Fagin R, Halpern JY, Moses Y, Vardi MY (1995) Reasoning about Knowledge. The MIT Press, Cambridge, Massachusetts
48. Fernández-Duque D (2010) Absolute completeness of $S4_u$ for its measure-theoretic semantics. In: Beklemishev L, Goranko V, Shehtman V (eds) Advances in modal logic, vol. 8. College Publications, London, pp 100–119

49. Fontaine G (2010) Modal fixpoint logic: some model theoretic questions. PhD thesis, University of Amsterdam
50. French T, van der Hoek W, Iliev P, Kooi BP (2013) Succinctness of epistemic languages. Artif Intell 197:56–85
51. van der Hoek W, Meyer J-J Ch (1992) Making some issues of implicit knowledge explicit. Int J Found Comput Sci 3(2):193–224
52. Gabbay DM, Kurucz A, Wolter F, Zakharyaschev M (2003) Many-dimensional modal logics: theory and applications.Studies in logic and the foundations of mathematics, vol 148. North-Holland, Amsterdam
53. Gabbay D, Pnueli A, Shelah S, Stavi J (1980) On the temporal analysis of fairness. In: Proceedings of the 7th ACM SIGPLAN-SIGACT symposium on principles of programming languages, POPL '80. ACM, New York, NY, USA, pp 163–173
54. Gabelaia D (2001) Modal definability in topology. Master's Thesis, available as ILLC report: MoL-2001-10: 2001
55. Gabelaia D, Kurucz A, Wolter F, Zakharyaschev M (2005) Products of 'transitive' modal logics. J Symb Logic 70(3):993–1021
56. Georgatos K (1993) Modal logics for topological spaces. ProQuest LLC, Ann Arbor, MI. Thesis (PhD)-City University of New York
57. Georgatos K (1994) Reasoning about knowledge on computation trees. Logics in artificial intelligence. LNCS, vol 838. Springer, Berlin, pp 300–315
58. Gerson MS (1975) An extension of $S4$ complete for the neighbourhood semantics but incomplete for the relational semantics. Studia Logica 34(4):333–342
59. Gettier E (1963) Is justified true belief knowledge? Analysis 23:121–123
60. Goldblatt R (1985) An algebraic study of well-foundedness. Studia Logica 44(4):423–437
61. Halpern JY, Vardi MY (1989) The complexity of reasoning about knowledge and time. i. lower bounds. J Comput Syst Sci 38(1):195–237
62. Hansen HH (2003) Monotonic modal logics. Master's Thesis, available as: ILLC report: PP-2003-24
63. Hansen HH, Kupke C, Pacuit E (2007) Bisimulation for neighbourhood structures. In: Haveraanen M, Montanari U, Mossakoswki T (ed) CALCO. LNCS, vol 4624. Springer, New York, pp 279–293
64. Harel D, Kozen D, Tiuryn J (2000) Dynamic logic. Foundations of computing series. The MIT Press, Cambridge
65. Heinemann B (2008) Topology and knowledge of multiple agents. In: Geffner H, Prada R, Machado Alexandre I, David N (eds) Advances in artificial intelligence IBERAMIA 2008, vol. 5290 of Lecture Notes in Computer Science. Springer, Berlin Heidelberg, pp. 1–10
66. Heinemann B (2010) The cantor space as a generic model of topologically presented knowledge. In: Ablayev F, Mayr E (eds) Computer science theory and applications, vol. 6072 of Lecture Notes in Computer Science. Springer Berlin Heidelberg, pp. 169–180
67. Heinemann B (2010) Logics for multi-subset spaces. J Appl Non-Classical Logics 20(3):219–240
68. Heinemann B (2010) Using hybrid logic for coping with functions in subset spaces. Studia Logica 94(1):23–45
69. Hintikka J (1962) Knowledge and belief, an introduction to the logic of the two notions. Cornell University Press, Ithaca, New York (Republished in 2005 by King's College, London)
70. Hintikka J (1986) Reasoning about knowledge in philosophy. In: Halpern JY (ed) Proceedings of the 1986 conference on theoretical aspects of reasoning about knowledge. Morgan Kaufmann Publishers, San Francisco, pp 63–80
71. Hodkinson I (2011) Simple completeness proofs for some spatial logics of the real line. In: Proc. 12th Asian logic conference, Wellington (to appear)
72. Hodkinson I (2013) On the Priorean temporal logic with 'around now' over the real line. J Logic Comput
73. van der Hoek W (1993) Systems for knowledge and beliefs. J Logic Comput 3(2):173–195
74. Joshi KD (1983) Introduction to general topology. A halsted press book. Wiley, New York

75. Kontchakov R, Pratt-Hartmann I, Wolter F, Zakharyaschev M (2010) Spatial logics with connectedness predicates. Log Methods Comput Sci 6(3):3:5, 43
76. Kraus S, Lehmann D (1988) Knowledge, belief and time. Theor Comput Sci 58:155–174
77. Kremer P (2013) Strong completeness of S4 wrt the real line. Manuscript
78. Kremer P The incompleteness of $S4 \oplus S4$ for the product space $\mathbb{R} \times \mathbb{R}$. Studia Logica (to appear)
79. Kremer P (2013) Strong completeness of S4 wrt any dense-in-itself metric space. Rev Symb Logic 6:545–570
80. Lismont L (1993) La connaissance commune en logique modale. Math Logic Q 39(1):115–130
81. Lismont L (1994) Common knowledge: relating anti-founded situation semantics to modal logic neighbourhood semantics. J Logic Lang Inf 3(4):285–302
82. Lomuscio A, Ryan M (1997) On the relation between interpreted systems and Kripke models. In: Proceedings of the AI97 workshop on theoretical and practical foundations of intelligent agents and agent-oriented systems. LNAI, vol 1441, pp 46–59
83. Lomuscio A, Sergot M (2000) Investigations in grounded semantics for multi-agent systems specification via deontic logic. Technical Report, Imperial College, London
84. Lucero-Bryan J (2010) Modal logics of some subspaces of the real numbers: diamond as derivative. ProQuest LLC, Ann Arbor, MI. Thesis (PhD)-New Mexico State University
85. Lucero-Bryan J (2011) The d-logic of the rational numbers: a fruitful construction. Studia Logica 97(2):265–295
86. Lucero-Bryan J (2013) The d-logic of the real line. J Logic Comput 23(1):121–156
87. McKinsey JCC, Tarski A (1944) The algebra of topology. Ann Math 2(45):141–191
88. Meyer J-J Ch, van der Hoek W (1995) Epistemic logic for AI and computer science. Cambridge tracts in theoretical computer science, vol 41. Cambridge University Press, Cambridge
89. Mints G (1999) A completeness proof for propositional S4 in cantor space. In: Orlowska E (ed) Logic at work. Studies in fuzziness soft computing, vol 24. Physica, Heidelberg, pp 79–88
90. Mints G, Zhang T (2005) A proof of topological completeness for $S4$ in (0, 1). Ann Pure Appl Logic 133(1–3):231–245
91. Moore RC (1977) Reasoning about knowledge and action. In: Proceedings of the 5th international joint conference on artificial intelligence (IJCAI-77), Cambridge, Massachusetts
92. Moss L, Parikh R (1992) Topological reasoning and the logic of knowledge. In: Proceedings of the 4th conference on theoretical aspects of reasoning about knowledge, TARK, pp 95–105
93. Özgün A (2013) Topological models for belief and belief revision. Master's Thesis, available as ILLC report: MoL-2013-13: 2013
94. Parikh R, Moss L, Steinsvold C (2007) Topology and epistemic logic. In: Aiello M, Pratt-Hartmann I, van Benthem J (eds) Handbook of spatial logics. Springer, Dordrecht, pp 299–341
95. Parikh R, Ramanujam R (1985) Distributed processes and the logic of knowledge. In: Parikh R (ed) Logics of Programs. LNCS, vol 193. Springer, Berlin, pp 256–268
96. Rao AS, Georgeff MP (1991) Modeling rational agents within a BDI-architecture. In: Fikes R, Sandewall E (eds) Proceedings of knowledge representation and reasoning (KR&R-91). Morgan Kaufmann Publishers, San Francisco, pp 473–484
97. Shehtman V (1990) Derived sets in Euclidean spaces and modal logic. Report X-1990-05, University of Amsterdam
98. Shehtman V (1993) A logic with progressive tenses. In: de Rijke M (ed) Diamonds and defaults. Kluwer Academic Publishers, New York, pp 255–285
99. Shehtman V (1999) "Everywhere" and "here". J Appl Non-Classical Logics 9(2–3):369–379 (Issue in memory of George Gargov)
100. Smyth MB (1992) Topology. In: Abramsky S, Gabbay DM, Maibaum TSE (eds) Handbook of logic in computer science, vol 1. Oxford University Press, New York, pp 641–761
101. Stalnaker R (2006) On logics of knowledge and belief. Philosophical Studies 128(1):169–199
102. Steinsvold C (2007) Topological models of belief logics. ProQuest LLC, Ann Arbor, MI. Thesis (PhD)-City University of New York

103. Su K, Sattar A, Governatori G, Chen Q (2005) A computationally grounded logic of knowledge, belief and certainty. In: Proceedings of AAMAS, pp 149–156. ACM Press
104. Troelstra AS, van Dalen D (1988) Constructivism in mathematics. Studies in logic and the foundations of mathematics: an introduction, vol I, II. North-Holland Publishing, Amsterdam
105. Vickers S (1989) Topology via logic. Cambridge tracts in theoretical computer science, vol 5. Cambridge University Press, Cambridge
106. Wang Y, Ågotnes T (2013) Multi-agent subset space logic. Accepted for IJCAI 2013
107. Wolter F, Zakharyaschev M (2000) Spatial reasoning in RCC-8 with Boolean region terms. In: Proceedings of the 14th European conference on artificial intelligence (ECAI-2000), pp 244–248, Berlin. IOS Press
108. Wooldridge M (2000) Computationally grounded theories of agency. In: Durfee E (ed) Proceedings of ICMAS. IEEE Press, New Jersey, pp 13–20

Chapter 13
Logic and Probabilistic Update

Lorenz Demey and Barteld Kooi

Abstract This chapter surveys recent work on probabilistic extensions of epistemic and dynamic-epistemic logics (the latter include the basic system of public announcement logic as well as the full product update logic). It emphasizes the importance of higher-order information as a distinguishing feature of these logics. This becomes particularly clear in the dynamic setting: although there exists a clear relationship between usual Bayesian conditionalization and public announcement, the probabilistic effects of the latter are in general more difficult to describe, because of the subtleties involved in higher-order information. Finally, the chapter discusses some applications of probabilistic dynamic epistemic logic, such as the Lockean thesis in formal epistemology and Aumann's agreement theorem in game theory.

13.1 Introduction

Epistemic logic and probability theory both provide formal accounts of information. Epistemic logic takes a *qualitative* perspective on information, and works with a modal operator K. Formulas such as $K\varphi$ can be interpreted as 'the agent knows that φ', 'the agent believes that φ', or, more generally speaking, 'φ follows from the agent's current information'. Probability theory, on the other hand, takes a *quantitative* perspective on information, and works with numerical probability functions P. Formulas such as $P(\varphi) = k$ can be interpreted as 'the probability of φ is k'. In the

L. Demey (✉)
Center for Logic and Analytical Philosophy, KU Leuven—University of Leuven, Kardinaal Mercierplein 2, 3000 Leuven, Belgium
e-mail: Lorenz.Demey@hiw.kuleuven.be

B. Kooi
Faculty of Philosophy, University of Groningen, Oude Boteringestraat 52,
9712 GL Groningen, The Netherlands
e-mail: B.P.Kooi@rug.nl

A. Baltag and S. Smets (eds.), *Johan van Benthem on Logic and Information Dynamics*, Outstanding Contributions to Logic 5,
DOI: 10.1007/978-3-319-06025-5_13, © Springer International Publishing Switzerland 2014

present context, probabilities will usually be interpreted subjectively, and can thus be taken to represent the agent's degrees of belief or credences.

With respect to one and the same formula φ, epistemic logic is able to distinguish between three epistemic attitudes: knowing its truth ($K\varphi$), knowing its falsity ($K\neg\varphi$), and being ignorant about its truth value ($\neg K\varphi \wedge \neg K\neg\varphi$). Probability theory, however, distinguishes infinitely many epistemic attitudes with respect to φ, viz. assigning it probability k ($P(\varphi) = k$), for every $k \in [0, 1]$. In this sense probability theory can be said to provide a much more *fine-grained* perspective on information.

While epistemic logic thus is a coarser account of information, it certainly has a wider scope. From its very origins in Hintikka's [34], epistemic logic has not only been concerned with knowledge about 'the world', but also with knowledge about knowledge, i.e. with *higher-order information*. Typical discussions focus on principles such as positive introspection ($K\varphi \rightarrow KK\varphi$). On the other hand, probability theory rarely talks about principles involving higher-order probabilities, such as $P(\varphi) = 1 \rightarrow P(P(\varphi) = 1) = 1$.[1] This issue becomes even more pressing in multi-agent scenarios. Natural examples might involve an agent a not having any information about a proposition φ, while being certain that another agent, b, does have this information. In epistemic logic this is naturally formalized as

$$\neg K_a\varphi \wedge \neg K_a\neg\varphi \wedge K_a(K_b\varphi \vee K_b\neg\varphi).$$

A formalization in probability theory might look as follows:

$$P_a(\varphi) = 0.5 \wedge P_a(P_b(\varphi) = 1 \vee P_b(\varphi) = 0) = 1.$$

However, because this statement makes use of 'nested' probabilities, it is rarely used in standard treatments of probability theory.

An additional theme is that of dynamics, i.e. *information change*. The agents' information is not eternally the same; rather, it should be changed in the light of new incoming information. Probability theory typically uses Bayesian updating to represent information change (but other, more complicated update mechanisms are available as well). Dynamic epistemic logic interprets new information as changing the epistemic model, and uses the new, updated model to represent the agents' updated information states. Once again, the main difference is that dynamic epistemic logic takes (changes in) higher-order information into account, whereas probability theory does not.

For all these reasons, the project of *probabilistic epistemic logic* seems very interesting. Such systems inherit the fine-grained perspective on information from probability theory, and the representation of higher-order information from epistemic

---

[1] A notable exception is 'Miller's principle', which states that $P_1(\varphi \mid P_2(\varphi) = b) = b$. The probability functions P_1 and P_2 can have various interpretations, such as the probabilities of two agents, subjective probability (credence) and objective probability (chance), or the probabilities of one agent at different moments in time—in the last two cases, the principle is also called the 'principal principle' or the 'principle of reflection', respectively. This principle has been widely discussed in Bayesian epistemology and philosophy of science [29, 32, 38, 40, 41].

logic. Their *dynamic* versions provide a unified perspective on changes in first- and higher-order information. In other words, they can be thought of as incorporating the complementary perspectives of (dynamic) epistemic logic and probability theory, thus yielding richer and more detailed accounts of information and information flow.

The remainder of this chapter is organized as follows. Section 13.2 introduces the static framework of probabilistic epistemic logic, and discusses its intuitive interpretation and technical features. Section 13.3 focuses on a rather straightforward type of dynamics, namely public announcements. It describes a probabilistic version of the well-known system of public announcement logic, and compares public announcement and Bayesian conditionalization. In Sect. 13.4 a more general update mechanism is introduced. This is a probabilistic version of the 'product update' mechanism in dynamic epistemic logic. Section 13.5, finally, indicates some applications and potential avenues of further research for the systems discussed in this chapter.

13.2 Probabilistic Epistemic Logic

In this section we introduce the static framework of probabilistic epistemic logic, which will be 'dynamified' in Sects. 13.3 and 13.4. Section 13.2.1 discusses the models on which the logic is interpreted. Section 13.2.2 defines the formal language and its semantics. Finally, Sect. 13.2.3 provides a complete axiomatization.

13.2.1 Probabilistic Kripke Models

Consider a finite set I of agents, and a countably infinite set $Prop$ of proposition letters. Throughout this chapter, these sets will be kept fixed, so they will often be left implicit.

Definition 13.1 A *probabilistic Kripke frame* is a tuple $\mathbb{F} = \langle W, R_i, \mu_i \rangle_{i \in I}$, where W is a non-empty finite set of states, $R_i \subseteq W \times W$ is agent i's epistemic accessibility relation, and $\mu_i : W \to (W \rightharpoonup [0, 1])$ assigns to each state $w \in W$ a partial function $\mu_i(w) : W \rightharpoonup [0, 1]$, such that

$$\sum_{v \in \text{dom}(\mu_i(w))} \mu_i(w)(v) = 1.$$

Definition 13.2 A *probabilistic Kripke model* is a tuple $\mathbb{M} = \langle \mathbb{F}, V \rangle$, where $\mathbb{F}$ is a probabilistic Kripke frame (with set of states W), and $V : Prop \to \wp(W)$ is a valuation.

Note that in principle, no conditions are imposed on the agents' epistemic accessibility relations. However, as is usually done in the literature on (probabilistic) dynamic epistemic logic, we will henceforth assume these relations to be *equivalence relations* (so that the corresponding knowledge operators satisfy the principles of the modal logic S5).

The function $\mu_i(w)$ represents agent i's probabilities (i.e. degrees of belief) at state w. For example, $\mu_i(w)(v) = k$ means that at state w, agent i assigns probability k to state v being the actual state. From a mathematical perspective, this is not the most general approach: one can also define a *probability space* $\mathbb{P}_{i,w}$ for each agent i and state w, and let $\mu_i(w)$ assign probabilities to sets in a σ-algebra on $\mathbb{P}_{i,w}$, rather than to individual states. In this way one can easily drop the requirement that frames and models have finitely many states. This approach is taken in [28] for static probabilistic epistemic logic, and extended to dynamic settings in [47]. However, because all the characteristic features of probabilistic (dynamic) epistemic logic already arise in the simpler approach, in this chapter we will stick to this simpler approach, and take $\mu_i(w)$ to assign probabilities to individual states. These functions are additively extended from individual states to sets of states, by putting (for any set $X \subseteq \text{dom}(\mu_i(w))$):

$$\mu_i(w)(X) := \sum_{x \in X} \mu_i(w)(x).$$

A consequence of our simple approach is that all sets $X \subseteq \text{dom}(\mu_i(w))$ have a definite probability $\mu_i(w)(X)$, whereas in the more general approach, sets X not belonging to the σ-algebra on $\mathbb{P}_{i,w}$ are not assigned any definite probability at all. A similar distinction can be made at the level of individual states. Because $\mu_i(w)$ is a partial function, states $v \in W - \text{dom}(\mu_i(w))$ are not assigned any definite probability at all. An even simpler approach involves putting $\mu_i(w)(v) = 0$, rather than leaving it undefined. In this way, the function $\mu_i(w)$ is total after all. From a mathematical perspective, these two approaches are equivalent. From an informal perspective, however, there is a clear difference: $\mu_i(w)(v) = 0$ means that agent i is certain (at state w) that v is not the actual state, whereas $\mu_i(w)(v)$ being undefined means that agent i has no opinion whatsoever (at state w) about v being the actual state. Again, because all the characteristic features of probabilistic (dynamic) epistemic logic already arise without this intuitive distinction, we will opt for the even simpler approach, and henceforth assume that all probability functions are total.

To summarize: the approach adopted in this chapter is the simplest one possible, in the sense that definite probabilities are assigned to 'everything': (i) to all *sets* (there is no σ-algebra to rule out some sets from having a definite probability), and (ii) to all *states* (the probability functions $\mu_i(w)$ are total on their domain W, so no states are ruled out from having a definite probability).

We finish this subsection by mentioning two typical properties of probabilistic Kripke frames.[2] In the next subsection we will show that these properties correspond to natural principles about the interaction between knowledge and probability.

Definition 13.3 Consider a probabilistic Kripke frame $\mathbb{F}$ and an agent $i \in I$. Then $\mathbb{F}$ is said to be *i-consistent* iff for all states w, v: if $(w, v) \notin R_i$ then $\mu_i(w)(v) = 0$.

---

[2] See [33] for a further discussion of these and other properties, and their correspondence to knowledge/probability interaction principles.

Furthermore, $\mathbb{F}$ is said to be *i-uniform* iff for all states w, v: if $(w, v) \in R_i$ then $\mu_i(w) = \mu_i(v)$.

13.2.2 Language and Semantics

The language $\mathscr{L}$ of (static) probabilistic epistemic logic is defined by means of the following Backus-Naur form:

$$\varphi \ ::= \ p \ | \ \neg\varphi \ | \ \varphi_1 \wedge \varphi_2 \ | \ K_i\varphi \ | \ a_1 P_i(\varphi) + \cdots + a_n P_i(\varphi) \geq b$$

—where $p \in Prop, i \in I, 1 \leq n < \omega$, and $a_1, \ldots, a_n, b \in \mathbb{Q}$. We only allow rational numbers as values for $a_1, \ldots, a_n, b$ in order to keep the language countable. As usual, $K_i\varphi$ means that agent i knows that φ, or, more generally, that φ follows from agent i's information. Its dual is defined as $\hat{K}_i\varphi := \neg K_i\neg\varphi$, and means that φ is consistent with agent i's information.

Formulas of the form $a_1 P_i(\varphi_1) + \cdots + a_n P_i(\varphi_n) \geq b$ are called *i-probability formulas*.[3] Note that mixed agent indices are not allowed; for example, $P_a(p) + P_b(q) \geq b$ is *not* a well-formed formula. Intuitively, $P_i(\varphi) \geq b$ means that agent i assigns probability at least b to φ. We allow for linear combinations in i-probability formulas, because this additional expressivity is useful when looking for a complete axiomatization [28], and because it allows us to express comparative judgments such as 'agent i considers φ to be at least twice as probable as ψ': $P_i(\varphi) \geq 2P_i(\psi)$. This last formula is actually an abbreviation for $P_i(\varphi) - 2P_i(\psi) \geq 0$. In general, we introduce the following abbreviations:

$$
\begin{array}{lll}
\sum_{\ell=1}^{n} a_\ell P_i(\varphi_\ell) \geq b & \text{for} & a_1 P_i(\varphi_1) + \cdots + a_n P_i(\varphi_n) \geq b, \\
a_1 P_i(\varphi_1) \geq a_2 P_i(\varphi_2) & \text{for} & a_1 P_i(\varphi_1) + (-a_2)P_i(\varphi_2) \geq 0, \\
\sum_{\ell=1}^{n} a_\ell P_i(\varphi_\ell) \leq b & \text{for} & \sum_{\ell=1}^{n}(-a_\ell)P_i(\varphi_\ell) \geq -b, \\
\sum_{\ell=1}^{n} a_\ell P_i(\varphi_\ell) < b & \text{for} & \neg(\sum_{\ell=1}^{n} a_\ell P_i(\varphi_\ell) \geq b), \\
\sum_{\ell=1}^{n} a_\ell P_i(\varphi_\ell) > b & \text{for} & \neg(\sum_{\ell=1}^{n} a_\ell P_i(\varphi_\ell) \leq b), \\
\sum_{\ell=1}^{n} a_\ell P_i(\varphi_\ell) = b & \text{for} & \sum_{\ell=1}^{n} a_\ell P_i(\varphi_\ell) \geq b \wedge \sum_{\ell=1}^{n} a_\ell P_i(\varphi_\ell) \leq b.
\end{array}
$$

Note that because of its recursive definition, the language $\mathscr{L}$ can express the agents' higher-order information of any sort: higher-order knowledge (for example $K_a K_b\varphi$), but also higher-order probabilities (for example $P_a(P_b(\varphi) \geq 0.5) = 1$), and higher-order information that mixes knowledge and probabilities (for example, $K_a(P_b(\varphi) \geq 0.5)$ and $P_a(K_b\varphi) = 1$).

---

[3] The agents' probabilities are thus explicitly represented in the logic's object language $\mathscr{L}$. Other proposals provide a probabilistic semantics for an object language that is itself fully classical (i.e. that does not explicitly represent probabilities). See [26] for a recent overview of the various ways of combining logic and probability.

The formal semantics for $\mathscr{L}$ is defined as follows. Consider an arbitrary probabilistic Kripke model $\mathbb{M}$ (with set of states W) and a state $w \in W$. We will often abbreviate $[\![\varphi]\!]^{\mathbb{M}} := \{v \in W \mid \mathbb{M}, v \models \varphi\}$. Then:

$$
\begin{array}{lll}
\mathbb{M}, w \models p & \text{iff} & w \in V(p), \\
\mathbb{M}, w \models \neg\varphi & \text{iff} & \mathbb{M}, w \not\models \varphi, \\
\mathbb{M}, w \models \varphi \wedge \psi & \text{iff} & \mathbb{M}, w \models \varphi \text{ and } \mathbb{M}, w \models \psi, \\
\mathbb{M}, w \models K_i\varphi & \text{iff} & \text{for all } v \in W \colon \text{if } (w, v) \in R_i \text{ then } \mathbb{M}, v \models \varphi, \\
\mathbb{M}, w \models \sum_{\ell=1}^{n} a_\ell P_i(\varphi_\ell) \geq b & \text{iff} & \sum_{\ell=1}^{n} a_\ell \mu_i(w)([\![\varphi_\ell]\!]^{\mathbb{M}}) \geq b.
\end{array}
$$

Furthermore, we also define:

- $\mathbb{M} \models \varphi$ iff $\mathbb{M}, w \models \varphi$ for all $w \in W$,
- $\mathbb{F} \models \varphi$ iff $\langle \mathbb{F}, V \rangle \models \varphi$ for all valuations V on the frame $\mathbb{F}$,
- $\models \varphi$ iff $\mathbb{F} \models \varphi$ for all frames $\mathbb{F}$.

As promised, we will now provide correspondence results for the frame properties defined at the end of the previous subsection:

Lemma 13.1 *Consider a probabilistic Kripke frame $\mathbb{F}$. Then:*

1. *$\mathbb{F}$ is i-consistent iff $\mathbb{F} \models K_i p \to P_i(p) = 1$,*
2. *$\mathbb{F}$ is i-uniform iff $\mathbb{F} \models (\varphi \to K_i\varphi) \wedge (\neg\varphi \to K_i\neg\varphi)$ for all i-prob. formulas φ.*

From a technical perspective, this lemma indicates how the notion of *frame correspondence* from modal logic [8, 9, 20] can be extended into the probabilistic realm. From an intuitive perspective, this lemma sheds some new light on the various interactions between epistemic and probabilistic information. Probabilistic epistemic logic distinguishes between epistemic impossibility ($(w, v) \notin R_i$) and probabilistic impossibility ($\mu_i(w)(v) = 0$). For example, when a fair coin is tossed, an infinite series of tails is probabilistically impossible, but epistemically possible [37, p. 384]. Item 1 of Lemma 13.1 establishes a connection between the principle that knowledge implies certainty, and the property of consistency (epistemic impossibility entails probabilistic impossibility). Similarly, item 2 establishes a connection between the principle that agents know their own probabilistic setup, and the property of uniformity (the impossibility of epistemic uncertainty about probabilities).

13.2.3 Proof System

Probabilistic epistemic logic can be axiomatized in a highly modular fashion. An overview is given in Fig. 13.1. The propositional and epistemic components shouldn't need any further comments. The probabilistic component is a straightforward translation into the formal language $\mathscr{L}$ of the well-known Kolmogorov axioms of probability; it ensures that the formal symbol $P_i(\cdot)$ behaves like a real probability function. Finally, the linear inequalities component is mainly a technical tool to

1. propositional component
    - all propositional tautologies and the modus ponens rule
2. epistemic component
    - the S5 axioms and rules for the K_i-operators
3. probabilistic component
    - $P_i(\varphi) \geq 0$
    - $P_i(\top) = 1$
    - $P_i(\varphi \wedge \psi) + P_i(\varphi \wedge \neg \psi) = P_i(\varphi)$
    - if $\vdash \varphi \leftrightarrow \psi$ then $\vdash P_i(\varphi) = P_i(\psi)$
4. linear inequalities component
    - $\sum_{\ell=1}^{n} a_\ell P_i(\varphi_\ell) \geq b \leftrightarrow \sum_{\ell=1}^{n} a_\ell P_i(\varphi_\ell) + 0 P_i(\varphi_{n+1}) \geq b$
    - $\sum_{\ell=1}^{n} a_\ell P_i(\varphi_\ell) \geq b \leftrightarrow \sum_{\ell=1}^{n} a_{p(\ell)} P_i(\varphi_{p(\ell)}) \geq b$
      for any permutation p of $1, \ldots, n$
    - $\sum_{\ell=1}^{n} a_\ell P_i(\varphi_\ell) \geq b \wedge \sum_{\ell=1}^{n} a_\ell' P_i(\varphi_\ell) \geq b' \rightarrow$
      $\sum_{\ell=1}^{n} (a_\ell + a_\ell') P_i(\varphi_\ell) \geq b + b'$
    - $\sum_{\ell=1}^{n} a_\ell P_i(\varphi_\ell) \geq b \leftrightarrow \sum_{\ell=1}^{n} d a_\ell P_i(\varphi_\ell) \geq db$ (for any $d > 0$)
    - $\sum_{\ell=1}^{n} a_\ell P_i(\varphi_\ell) \geq b \vee \sum_{\ell=1}^{n} a_\ell P_i(\varphi_\ell) \leq b$
    - $\sum_{\ell=1}^{n} a_\ell P_i(\varphi_\ell) \geq b \rightarrow \sum_{\ell=1}^{n} a_\ell P_i(\varphi_\ell) > b'$ (for any $b' < b$)

Fig. 13.1 Componentwise axiomatization of probabilistic epistemic logic

ensure that the logic is strong enough to capture the behavior of linear inequalities of probabilities.

Using standard techniques the following theorem can be proved [28]:

Theorem 13.1 *Probabilistic epistemic logic, as axiomatized in Fig. 13.1, is sound and complete with respect to the class of probabilistic Kripke frames.*

The notion of completeness used in this theorem is *weak* completeness ($\vdash \varphi$ iff $\models \varphi$), rather than *strong* completeness ($\Gamma \vdash \varphi$ iff $\Gamma \models \varphi$). These two notions do not coincide in probabilistic epistemic logic, because this logic is not *compact*; for example, every finite subset of the set $\{P_i(p) > 0\} \cup \{P_i(p) \leq k \mid k > 0\}$ is satisfiable, but the entire set is not.

13.3 Probabilistic Public Announcement Logic

In this section we discuss a first 'dynamification' of probabilistic epistemic logic, by introducing public announcements into the logic. Section 13.3.1 discusses updated probabilistic Kripke models, and introduces a public announcement operator into the formal language to talk about these models. Section 13.3.2 provides a complete axiomatization, and Sect. 13.3.3 focuses on the role of higher-order information in public announcement dynamics.

13.3.1 Semantics

Public announcements form one of the simplest types of epistemic dynamics. They concern the truthful and public announcement of some piece of information φ by an external source. That the announcement is *truthful* means that the announced information φ has to be true; that it is *public* means that all agents $i \in I$ learn about it simultaneously and commonly. Finally, the announcement's source is called 'external' because it is not one of the agents $i \in I$ (and will thus not be explicitly represented in the formal language).

Public announcement logic [27, 31, 44] represents these announcements as updates that change Kripke models, and introduces a dynamic public announcement operator into the formal language to describe these updated models. This strategy can straightforwardly be extended into the probabilistic realm.

Syntactically, we add a dynamic operator $[!\cdot]\cdot$ to the static language $\mathcal{L}$, thus obtaining the new language $\mathcal{L}^!$. The formula $[!\varphi]\psi$ means that after any truthful public announcement of φ, it will be the case that ψ. Its dual is defined as $\langle !\varphi \rangle \psi :=$ $\neg[!\varphi]\neg\psi$, and means that φ can truthfully and publicly be announced, and afterwards ψ will be the case. These formulas thus allow us to express 'now' (i.e. *before* any dynamics has taken place) what will be the case 'later' (*after* the dynamics has taken place). These formulas are interpreted on a probabilistic Kripke model $\mathbb{M}$ and state w as follows:

$$\mathbb{M}, w \models [!\varphi]\psi \quad \text{iff} \quad \text{if } \mathbb{M}, w \models \varphi \text{ then } \mathbb{M}|\varphi, w \models \psi,$$
$$\mathbb{M}, w \models \langle !\varphi \rangle \psi \quad \text{iff} \quad \mathbb{M}, w \models \varphi \text{ and } \mathbb{M}|\varphi, w \models \psi.$$

Note that these clauses not only use the model $\mathbb{M}$, but also the updated model $\mathbb{M}|\varphi$. The model $\mathbb{M}$ represents the agents' information *before* the public announcement of φ; the model $\mathbb{M}|\varphi$ represents their information *after* the public announcement of φ; hence the public announcement of φ *itself* is represented by the update mechanism $\mathbb{M} \mapsto \mathbb{M}|\varphi$, which is formally defined as follows:

Definition 13.4 Consider a probabilistic Kripke model $\mathbb{M} = \langle W, R_i, \mu_i, V \rangle_{i \in I}$, a state $w \in W$, and a formula $\varphi \in \mathcal{L}^!$ such that $\mathbb{M}, w \models \varphi$. Then the *updated probabilistic Kripke model* $\mathbb{M}|\varphi := \langle W^\varphi, R_i^\varphi, \mu_i^\varphi, V^\varphi \rangle_{i \in I}$ is defined as follows:

- $W^\varphi := W$,
- $R_i^\varphi := R_i \cap (W \times [\![\varphi]\!]^{\mathbb{M}})$ (for every agent $i \in I$),
- $\mu_i^\varphi : W^\varphi \to (W^\varphi \to [0, 1])$ is defined (for every agent $i \in I$) by

$$\mu_i^\varphi(v)(u) := \begin{cases} \frac{\mu_i(v)(\{u\} \cap [\![\varphi]\!]^\mathbb{M})}{\mu_i(v)([\![\varphi]\!]^\mathbb{M})} & \text{if } \mu_i(v)([\![\varphi]\!]^\mathbb{M}) > 0 \\ \mu_i(v)(u) & \text{if } \mu_i(v)([\![\varphi]\!]^\mathbb{M}) = 0, \end{cases}$$

- $V^\varphi := V$.

The main effect of the public announcement of φ in a model $\mathbb{M}$ is that all links to $\neg\varphi$-states are deleted; hence these states are no longer accessible for any of the agents. This procedure is standard; we will therefore focus on the probabilistic components μ_i^φ.

First of all, it should be noted that the case distinction in the definition of $\mu_i^\varphi(v)(u)$ is made for strictly technical reasons, viz. to ensure that there are no 'dangerous' divisions by 0. In all examples and applications, we will be using the 'interesting' case $\mu_i(v)([\![\varphi]\!]^\mathbb{M}) > 0$. Still, for general theoretical reasons, *something* has to be said about the case $\mu_i(v)([\![\varphi]\!]^\mathbb{M}) = 0$. Leaving $\mu_i^\varphi(v)(u)$ undefined would lead to truth value gaps in the logic, and thus greatly increase the difficulty of finding a complete axiomatization. The approach taken in this chapter is to define $\mu_i^\varphi(v)(u)$ simply as $\mu_i(v)(u)$ in case $\mu_i(v)([\![\varphi]\!]^\mathbb{M}) = 0$—so the public announcement of φ has *no effect* whatsoever on $\mu_i(v)$. The intuitive idea behind this definition is that an agent i simply *ignores* new information if she previously assigned probability 0 to it. Technically speaking, this definition will yield a relatively simple axiomatization.

One can easily check that if $\mathbb{M}$ is a probabilistic Kripke model, then $\mathbb{M}|\varphi$ is a probabilistic Kripke model as well. We focus on $\mu^\varphi(v)$ (for some arbitrary state $v \in W^\varphi$). If $\mu_i(v)([\![\varphi]\!]^\mathbb{M}) = 0$, then $\mu_i^\varphi(v)$ is $\mu_i(v)$, which is a probability function on $W = W^\varphi$. If $\mu_i(v)([\![\varphi]\!]^\mathbb{M}) > 0$, then for any $u \in W^\varphi$,

$$\mu_i^\varphi(v)(u) = \frac{\mu_i(v)(\{u\} \cap [\![\varphi]\!]^\mathbb{M})}{\mu_i(v)([\![\varphi]\!]^\mathbb{M})},$$

which is positive because $\mu_i(v)(\{u\} \cap [\![\varphi]\!]^\mathbb{M})$ is positive, and at most 1, because $\mu_i(v)(\{u\} \cap [\![\varphi]\!]^\mathbb{M}) \leq \mu_i(v)([\![\varphi]\!]^\mathbb{M})$—and hence $\mu_i^\varphi(v)(u) \in [0, 1]$. Furthermore,

$$\sum_{u \in W^\varphi} \mu_i^\varphi(v)(u) = \sum_{u \in W} \frac{\mu_i(v)(\{u\} \cap [\![\varphi]\!]^\mathbb{M})}{\mu_i(v)([\![\varphi]\!]^\mathbb{M})} = \sum_{\mathbb{M}, u \models \varphi} \frac{\mu_i(v)(u)}{\mu_i(v)([\![\varphi]\!]^\mathbb{M})} = 1.$$

It should be noted that the definition of $\mu_i^\varphi(v)$—in the interesting case when $\mu_i(v)([\![\varphi]\!]^\mathbb{M}) > 0$—can also be expressed in terms of conditional probabilities:

$$\mu_i^\varphi(v)(u) = \frac{\mu_i(v)(\{u\} \cap [\![\varphi]\!]^\mathbb{M})}{\mu_i(v)([\![\varphi]\!]^\mathbb{M})} = \mu_i(v)(u \mid [\![\varphi]\!]^\mathbb{M}).$$

In general, for any $X \subseteq W^\varphi$ we have:

$$\mu_i^\varphi(v)(X) = \frac{\mu_i(v)(X \cap [\![\varphi]\!]^\mathbb{M})}{\mu_i(v)([\![\varphi]\!]^\mathbb{M})} = \mu_i(v)(X \mid [\![\varphi]\!]^\mathbb{M})$$

In other words, after the public announcement of a formula φ, the agents calculate their new, updated probabilities by means of *Bayesian conditionalization* on the information provided by the announced formula φ. This connection between public announcements and Bayesian conditionalization will be explored more thoroughly in Sect. 13.3.3.

Example 13.1 We finish this subsection by discussing a simple example. Consider the following scenario. An agent does not know whether p is the case, i.e. she cannot distinguish between p-states and $\neg p$-states. (In fact, p happens to be true.) Furthermore, the agent has no specific reason to think that one state is more probable than any other; therefore it is reasonable for her to assign equal probabilities to all states. This example can be formalized by the following model: $\mathbb{M} = \langle W, R, \mu, V \rangle$, $W = \{w, v\}$, $R = W \times W$, $\mu(w)(w) = \mu(w)(v) = \mu(v)(w) = \mu(v)(v) = 0.5$, and $V(p) = \{w\}$. (We work with only one agent in this example, so agent indices can be dropped.) This model is a faithful representation of the scenario described above; for example:

$$\mathbb{M}, w \models \neg Kp \wedge \neg K\neg p \wedge P(p) = 0.5 \wedge P(\neg p) = 0.5.$$

Now suppose that p is publicly announced (this is indeed possible, since p was assumed to be actually true). Applying Definition 13.4 we obtain the updated model $\mathbb{M}|p$, with $W^p = W$, $R = \{(w, w)\}$, and

$$\mu^p(w)(\llbracket p \rrbracket^{\mathbb{M}|p}) = \mu^p(w)(w) = \frac{\mu(w)(\{w\} \cap \llbracket p \rrbracket^{\mathbb{M}})}{\mu(w)(\llbracket p \rrbracket^{\mathbb{M}})} = \frac{\mu(w)(w)}{\mu(w)(w)} = 1.$$

Using this updated model $\mathbb{M}|p$, we find that

$$\mathbb{M}, w \models [!p]\big(Kp \wedge P(p) = 1 \wedge P(\neg p) = 0\big).$$

So after the public announcement of p, the agent has come to know that p is in fact the case. She has also adjusted her probabilities: she now assigns probability 1 to p being true, and probability 0 to p being false. These are the results that one would intuitively expect, so Definition 13.4 seems to yield an adequate representation of the epistemic and probabilistic effects of public announcements.

13.3.2 Proof System

Public announcement logic can be axiomatized by adding a set of *reduction axioms* to the static base logic [27]. These axioms allow us to recursively rewrite formulas containing dynamic public announcement operators as formulas without such operators; hence the dynamic language $\mathscr{L}^!$ is equally expressive as the static $\mathscr{L}$. Alternatively, reduction axioms can be seen as 'predicting' what will be the case after the public announcement has taken place in terms of what is the case before the public announcement has taken place.

1. static base logic

  • probabilistic epistemic logic, as axiomatized in Fig. 13.1

2. necessitation for public announcement

  • if $\vdash \psi$ then $\vdash [!\varphi]\psi$

3. reduction axioms for public announcement

$$[!\varphi]p \leftrightarrow \varphi \to p$$
$$[!\varphi]\neg\psi \leftrightarrow \varphi \to \neg[!\varphi]\psi$$
$$[!\varphi](\psi_1 \wedge \psi_2) \leftrightarrow [!\varphi]\psi_1 \wedge [!\varphi]\psi_2$$
$$[!\varphi]K_i\psi \leftrightarrow \varphi \to K_i[!\varphi]\psi$$
$$[!\varphi]\textstyle\sum_\ell a_\ell P_i(\psi_\ell) \geq b \leftrightarrow \varphi \to$$
$$\big(P_i(\varphi) = 0 \wedge \textstyle\sum_\ell a_\ell P_i(\langle!\varphi\rangle\psi_\ell) \geq b\big) \vee$$
$$\big(P_i(\varphi) > 0 \wedge \textstyle\sum_\ell a_\ell P_i(\langle!\varphi\rangle\psi_\ell) \geq bP_i(\varphi)\big)$$

Fig. 13.2 Axiomatization of probabilistic public announcement logic

This strategy can be extended into the probabilistic realm. For the static base logic, we do not simply take some system of epistemic logic (usually S5), but rather the system of probabilistic epistemic logic described in Sect. 13.2.3 (Fig. 13.1), and add the reduction axioms shown in Fig. 13.2. The first four reduction axioms are familiar from classical (non-probabilistic) public announcement logic. Note that the reduction axiom for i-probability formulas makes, just like Definition 13.4, a case distinction based on whether the agent assigns probability 0 to the announced formula φ. The significance of this reduction axiom, and its connection with Bayesian conditionalization, will be further explored in the next subsection.

Once again, standard techniques suffice to prove the following theorem [37]:

Theorem 13.2 *Probabilistic public announcement logic, as axiomatized in Fig. 13.2, is sound and complete with respect to the class of probabilistic Kripke frames.*

13.3.3 Higher-Order Information in Public Announcements

In this subsection we will discuss the role of higher-order information in probabilistic public announcement logic. This will further clarify the connection, but also the distinction, between (dynamic versions of) probabilistic epistemic logic and probability theory proper.

In the previous subsection we introduced a reduction axiom for i-probability formulas. This axiom allows us to derive the following principle as a special case:

$$(\varphi \wedge P(\varphi) > 0) \longrightarrow \big([!\varphi]P_i(\psi) \geq b \leftrightarrow P(\langle!\varphi\rangle\psi) \geq bP_i(\varphi)\big). \qquad (13.1)$$

The antecedent states that φ is true (because of the truthfulness of public announcements) and that agent i assigns it a strictly positive probability (so that we are in the 'interesting' case of the reduction axiom). To see the meaning of the consequent more clearly, note that $\vdash \langle !\varphi \rangle \psi \leftrightarrow (\varphi \wedge [!\varphi]\psi)$, and introduce the following abbreviation of conditional probability into the formal language:

$$P_i(\beta \mid \alpha) \geq b \quad := \quad P_i(\alpha \wedge \beta) \geq b P_i(\alpha).$$

Principle (13.1) can now be rewritten as follows:

$$(\varphi \wedge P(\varphi) > 0) \longrightarrow ([!\varphi]P_i(\psi) \geq b \leftrightarrow P([!\varphi]\psi \mid \varphi) \geq b). \qquad (13.2)$$

A similar version can be proved for $\leq$ instead of $\geq$; combining these two we get:

$$(\varphi \wedge P(\varphi) > 0) \longrightarrow ([!\varphi]P_i(\psi) = b \leftrightarrow P([!\varphi]\psi \mid \varphi) = b). \qquad (13.3)$$

The consequent thus states a connection between the agent's probability of ψ after the public announcement of φ, and her conditional probability of $[!\varphi]\psi$, given the truth of φ. In other words, after a public announcent of φ, the agent updates her probabilities by Bayesian conditionalization on φ. The subtlety of principle (13.3), however, is that the agent does not take the conditional probability (conditional on φ) of ψ *itself*, but rather of the *updated* formula $[!\varphi]\psi$.

The reason for this is that $[!\varphi]P_i(\psi) = b$ talks about the probability that the agent assigns to ψ after the public announcement of φ has *actually* happened. If we want to describe this probability as a conditional probability, we cannot simply make use of the conditional probability $P_i(\psi \mid \varphi)$, because this represents the probability that the agent *would* assign to ψ if a public announcement of φ *would* happen— hypothetically, not actually! Borrowing a slogan from van Benthem: "The former takes place once arrived at one's vacation destination, the latter is like reading a travel folder and musing about tropical islands." [11, p. 417]. Hence, if we want to describe the agent's probability of ψ after an actual public announcement of φ in terms of conditional probabilities, we need to represent the effects of the public announcement of φ on ψ explicitly, and thus take the conditional probability (conditional on φ) of $[!\varphi]\psi$, rather than ψ.

One might wonder about the relevance of this subtle distinction between actual and hypothetical public announcements. The point is that the public announcement of φ can have effects on the truth value of ψ. For large classes of formulas ψ, this will not occur: their truth value is not affected by the public announcement of φ. Formally, this means that $\vdash \psi \leftrightarrow [!\varphi]\psi$, and thus (the consequent of) principle (13.3) becomes:

$$[!\varphi]P_i(\psi) = b \leftrightarrow P_i(\psi \mid \varphi) = b$$

—thus wiping away all differences between the agent's probability of ψ after a public announcement of φ, and her conditional probability of ψ, given φ. A typical class of such formulas (whose truth value is unaffected by the public announcement of φ)

is formed by the Boolean combinations of proposition letters, i.e. those formulas which express *ontic* or *first-order information*. Since probability theory proper is usually only concerned with first-order information ('no nested probabilities'), the distinction between actual and hypothetical announcements—or in general, between actual and hypothetical learning of new information—thus vanishes completely, and Bayesian conditionalization can be used as a universal update rule to compute new probabilities after (actually) learning a new piece of information.

However, in probabilistic epistemic logic (and its dynamic versions, such as probabilistic PAL), higher-order information *is* taken into account, and hence the distinction between actual and hypothetical public announcements has to be taken seriously. Therefore, the consequent of principle (13.3) should really use the conditional probability $P_i([!\varphi]\psi \mid \varphi)$, rather than just $P_i(\psi \mid \varphi)$.[4]

Example 13.2 To illustrate this, consider again the model defined in Example 13.1, and put $\varphi := p \land P(\neg p) = 0.5$. It is easy to show that

$$\mathbb{M}, w \models P(\varphi \mid \varphi) = 1 \land P([!\varphi]\varphi \mid \varphi) = 0 \land [!\varphi]P(\varphi) = 0.$$

Hence the probability assigned to φ after the public announcement is the conditional probability $P([!\varphi]\varphi \mid \varphi)$, rather than just $P(\varphi \mid \varphi)$. Note that this example indeed involves higher-order information, since we are talking about the probability of φ, which itself contains the probability statement $P(\neg p) = 0.5$ as a conjunct. Finally, this example also shows that learning a new piece of information φ (via public announcement) does *not* automatically lead to the agents being certain about (i.e. assigning probability 1 to) that formula. This is to be contrasted with probability theory, where a new piece of information φ is processed via Bayesian conditionalization, and thus always leads to certainty: $P(\varphi \mid \varphi) = 1$. The explanation is, once again, that probability theory is only concerned with first-order information, whereas the phenomena described above can only occur at the level of higher-order information.[5,6]

---

[4] Romeijn [45] provides an analysis that stays closer in spirit to probability theory proper. He argues that the public announcement of φ induces a shift in the interpretation of ψ (in our terminology: from ψ to $[!\varphi]\psi$, i.e. from $[\![\psi]\!]^{\mathbb{M}}$ to $[\![\psi]\!]^{\mathbb{M}|\varphi}$), and shows that such meaning shifts can be modeled using Dempster-Shafer belief functions. Crucially, however, this proposal is able to deal with the case of ψ expressing *second-order* information (e.g. when it is of the form $P_i(p) = b$), but not with the case of higher-order information *in general* (e.g. when ψ is of the form $P_j(P_i(p) = b) = a$, or involves even more deeply nested probabilities) [45, p. 603].

[5] Similarly, the *success postulate* for belief expansion in the (traditional) AGM framework [1, 30] states that after expanding one's belief set with a new piece of information φ, the updated (expanded) belief set should always contain this new information. Also here the explanation is that AGM is only concerned with first-order information. (Note that we talk about the success postulate for belief *expansion*, rather than belief *revision*, because the former seems to be the best analogue of public announcement in the AGM framework.)

[6] The occurrence of higher-order information is a *necessary* condition for this phenomenon, but not a *sufficient* one: there exist formulas φ that involve higher-order information, but still $\models [!\varphi]P_i(\varphi) = 1$ (or epistemically: $\models [!\varphi]K_i\varphi$).

13.4 Probabilistic Dynamic Epistemic Logic

In this section we will move from a probabilistic version of public announcement
logic to a probabilistic version of 'full' dynamic epistemic logic. Section 13.4.1
introduces a probabilistic version of the *product update* mechanism that is behind
dynamic epistemic logic. Section 13.4.2 introduces dynamic operators into the formal
language to talk about these product updates, and discusses a detailed example.
Section 13.4.3, finally, shows how to obtain a complete axiomatization in a fully
standard (though non-trivial) fashion.

13.4.1 Probabilistic Product Update

Classical (non-probabilistic) dynamic epistemic logic models epistemic dynamics by
means of a product update mechanism [4, 5]. The agents' *static* information (what is
the current state?) is represented in a Kripke model $\mathbb{M}$, and their *dynamic* information
(what type of event is currently taking place?) is represented in an update model $\mathbb{E}$.
The agents' new information (after the dynamics has taken place) is represented
by means of a product construction $\mathbb{M} \otimes \mathbb{E}$. We will now show how to define a
probabilistic version of this construction.

Before stating the formal definitions, we show how they naturally arise as probabilistic generalizations of the classical (non-probabilistic) notions. The probabilistic
Kripke models introduced in Definition 13.2 represent the agents' static information,
in both its epistemic and its probabilistic aspects. This static probabilistic information is called the *prior probabilities of the states* in [17]. We can thus say that when
w is the actual state, agent i considers it epistemically possible that v is the actual
state $((w, v) \in R_i)$, and, more specifically, that she assigns probability b to v being
the actual state $(\mu_i(w)(v) = b)$.

Update models are essentially like Kripke models: they represent the agents' information about events, rather than states. Since probabilistic Kripke models represent
both epistemic and probabilistic information about *states*, by analogy probabilistic
update models should represent both epistemic and probabilistic information about
events. Hence, they should not only have epistemic accessibility relations R_i over
their set of events E, but also probability functions $\mu_i : E \to (E \to [0, 1])$. (Formal
details will be given in Definition 13.5.) We can then say that when e is the actually
occurring event, agent i considers it epistemically possible that f is the actually
occurring event $((e, f) \in R_i)$, and, more specifically, that she assigns probability b
to f being the actually occurring event $(\mu_i(e)(f) = b)$. This dynamic probabilistic
information is called the *observation probabilities* in van Benthem et al. [17].

Finally, how probable it is that an event e will occur, might vary from state to
state. We assume that this variation can be captured by means of a set Φ of (pairwise
inconsistent) sentences in the object language (so that the probability that an event e
will occur can only vary between states that satisfy *different* sentences of Φ). This will

be formalized by adding to the probabilistic update models a set of preconditions Φ, and probability functions $\mathsf{pre}\colon \Phi \to (E \to [0, 1])$. The meaning of $\mathsf{pre}(\varphi)(e) = b$ is that if φ holds, then event e occurs with probability b. In van Benthem et al. [17] these are called *occurrence probabilities*.[7]

We are now ready to formally introduce probabilistic update models:

Definition 13.5   A *probabilistic update model* is a tuple $\mathbb{E} = \langle E, R_i, \Phi, \mathsf{pre}, \mu_i \rangle_{i \in I}$, where E is a non-empty finite set of events, $R_i \subseteq E \times E$ is agent i's epistemic accessibility relation, $\Phi \subseteq \mathscr{L}^{\otimes}$ is a finite set of pairwise inconsistent sentences called *preconditions*, $\mu_i\colon E \to (E \to [0, 1])$ assigns to each event $e \in E$ a probability function $\mu_i(e)$ over E, and $\mathsf{pre}\colon \Phi \to (E \to [0, 1])$ assigns to each precondition $\varphi \in \Phi$ a probability function $\mathsf{pre}(\varphi)$ over E.

All components of a probabilistic update model have already been commented upon. Note that we use the same symbols R_i and μ_i to indicate agent i's epistemic and probabilistic information in a probabilistic Kripke model $\mathbb{M}$ and in a probabilistic update model $\mathbb{E}$—from the context it will always be clear which of the two is meant. The language $\mathscr{L}^{\otimes}$ that the preconditions are taken from will be formally defined in the next subsection. (As is usual in this area, there is a non-vicious simultaneous recursion going on here.)

We now introduce occurrence probabilities for events at states:

Definition 13.6   Consider a probabilistic Kripke model $\mathbb{M}$, a state w, a probabilistic update model $\mathbb{E}$, and an event e. Then the *occurrence probability of e at w* is defined as

$$\mathsf{pre}(w)(e) = \begin{cases} \mathsf{pre}(\varphi)(e) & \text{if } \varphi \in \Phi \text{ and } \mathbb{M}, w \models \varphi \\ 0 & \text{if there is no } \varphi \in \Phi \text{ such that } \mathbb{M}, w \models \varphi. \end{cases}$$

Since the preconditions are pairwise inconsistent, $\mathsf{pre}(w)(e)$ is always well-defined. The meaning of $\mathsf{pre}(w)(e) = b$ is that in state w, event e occurs with probability b. Note that if two states w and v satisfy the same precondition, then always $\mathsf{pre}(w)(e) = \mathsf{pre}(v)(e)$; in other words, the occurrence probabilities of an event e can only vary 'up to a precondition' (cf. supra).

The probabilistic product update mechanism can now be defined as follows:

Definition 13.7   Consider a probabilistic Kripke model $\mathbb{M} = \langle W, R_i, \mu_i, V \rangle_{i \in I}$ and a probabilistic update model $\mathbb{E} = \langle E, R_i, \Phi, \mathsf{pre}, \mu_i \rangle_{i \in I}$. Then the *updated model* $\mathbb{M} \otimes \mathbb{E} := \langle W', R'_i, \mu'_i, V' \rangle_{i \in I}$ is defined as follows:

- $W' := \{(w, e) \mid w \in W, e \in E, \mathsf{pre}(w)(e) > 0\}$,
- $R'_i := \{((w, e), (w', e')) \in W' \times W' \mid (w, w') \in R_i \text{ and } (e, e') \in R_i\}$ (for every agent $i \in I$),

---

[7] Occurrence probabilities are often assumed to be *objective frequencies*. This is reflected in the formal setup: the function pre is not agent-dependent.

- $\mu_i' : W' \to (W' \to [0, 1])$ is defined (for every agent $i \in I$) by

$$\mu_i'(w, e)(w', e') := \frac{\mu_i(w)(w') \cdot \mathsf{pre}(w')(e') \cdot \mu_i(e)(e')}{\sum_{\substack{w'' \in W \\ e'' \in E}} \mu_i(w)(w'') \cdot \mathsf{pre}(w'')(e'') \cdot \mu_i(e)(e'')}$$

if the denominator is strictly positive, and $\mu_i'(w, e)(w', e') := 0$ otherwise,
- $V'(p) := \{(w, e) \in W' \mid w \in V(p)\}$ (for every $p \in Prop$).

We will only comment on the probabilistic component of this definition (all other components are fully classical). After the dynamics has taken place, agent i calculates at state (w, e) her new probability for (w', e') by taking the arithmetical product of (i) her *prior probability* for w' at w, (ii) the *occurrence probability* of e' in w', and (iii) her *observation probability* for e' at e, and then normalizing this product. The factors in this product are not *weighted* (or equivalently, they all have weight 1)—van Benthem et al. [17] also discusses weighted versions of this update mechanism, and shows how one of these weighted versions corresponds to the rule of *Jeffrey conditioning* from probability theory [36]. Finally, note that $\mathbb{M} \otimes \mathbb{E}$ might fail to be a probabilistic Kripke model: if the denominator in the definition of $\mu_i'(w, e)$ is 0, then $\mu_i'(w, e)$ assigns 0 to all states in W'. We will not care here about the interpretation of this feature, but only remark that technically speaking it is harmless and, perhaps most importantly, still allows for a reduction axiom for i-probability formulas (cf. Sect. 13.4.3).

13.4.2 Language and Semantics

To talk about these updated models, we add dynamic operators $[\mathsf{E}, \mathsf{e}]$ to the static language $\mathscr{L}$, thus obtaining the new language $\mathscr{L}^\otimes$. Here, E, e are formal names for the probabilistic update model $\mathbb{E} = \langle E, R_i, \Phi, \mathsf{pre}, \mu_i \rangle_{i \in I}$ and event $e \in E$ (recall our remark about the mutual recursion of the dynamic language and the updated models). The formula $[\mathsf{E}, \mathsf{e}]\varphi$ means that after the event e has occurred, it will be the case that φ. It has the following semantics:

$$\mathbb{M}, w \models [\mathsf{E}, \mathsf{e}]\psi \quad \text{iff} \quad \text{if } \mathsf{pre}(w)(e) > 0, \text{ then } \mathbb{M} \otimes \mathbb{E}, (w, e) \models \psi.$$

Example 13.3 Consider the following scenario. While strolling through a flee market, you see a painting that you think might be a real Picasso. Of course, the chance that the painting is actually a real Picasso is very slim, say 1 in 100,000. You know from an art encyclopedia that Picasso signed almost all his paintings with a very characteristic signature. If the painting is a real Picasso, the chance that it bears the characteristic signature is 97 %, while if the painting is not a real Picasso, the chance that it bears the characteristic signature is 0 % (nobody is capable of imitating Picasso's signature). You immediately look at the painting's signature, but determining whether it is Picasso's characteristic signature is very hard, and—not being an expert art historian—you remain uncertain and think that the chance is 50 % that the painting's signature is Picasso's characteristic one.

Your initial information (before having looked at the painting's signature) can be represented as the following probabilistic Kripke model: $\mathbb{M} = \langle W, R, \mu, V \rangle$, where $W = \{w, v\}$, $R = W \times W$, $\mu(w)(w) = \mu(v)(w) = 0.00001$, $\mu(w)(v) = \mu(v)(v) = 0.99999$, and $V(\text{real}) = \{w\}$. (We work with only one agent in this example, so agent indices can be dropped.) Hence, initially you do not rule out the possibility that the painting in front of you is a real Picasso, but you consider it highly unlikely:

$$\mathbb{M}, w \models \hat{K}\text{real} \wedge P(\text{real}) = 0.00001.$$

The event of looking at the signature can be represented with the following update model: $\mathbb{E} = \langle E, R, \Phi, \text{pre}, \mu \rangle$, where $E = \{e, f\}$, $R = E \times E$, $\Phi = \{\text{real}, \neg\text{real}\}$, $\text{pre}(\text{real})(e) = 0.97$, $\text{pre}(\text{real})(f) = 0.03$, $\text{pre}(\neg\text{real})(e) = 0$, $\text{pre}(\neg\text{real})(f) = 1$, and $\mu(e)(e) = \mu(f)(e) = \mu(e)(f) = \mu(f)(f) = 0.5$. The event e represents 'looking at Picasso's characteristic signature'; the event f represents 'looking at a signature that is not Picasso's characteristic one'.

We now construct the updated model $\mathbb{M} \otimes \mathbb{E}$. Since $\mathbb{M}, v \not\models \text{real}$, it holds that $\text{pre}(v)(e) = \text{pre}(\neg\text{real})(e) = 0$, and hence (v, e) does not belong to the updated model. It is easy to see that the other states (w, e), (w, f) and (v, f) do belong to the updated model. Furthermore, one can easily calculate that $\mu'(w, e)(w, e) = 0.0000003$ and $\mu'(w, e)(w, f) = 0.0000097$, so $\mu'(w, e)([\![\text{real}]\!]^{\mathbb{M} \otimes \mathbb{E}}) = 0.0000003 + 0.0000097 = 0.00001$, and thus

$$\mathbb{M}, w \models [\mathbb{E}, e]P(\text{real}) = 0.00001.$$

Hence, even though the painting in front of you is a real Picasso (in state w), after looking at the signature (which is indeed Picasso's characteristic signature!—the event that actually happened was event e) you still assign a probability of 1 in 100,000 to it being a real Picasso.

Note that if you had been an expert art historian, with the same prior probabilities, but with the reliable capability of recognizing Picasso's characteristic signature— let's formalize this as $\mu(e)(e) = 0.99$ and $\mu(e)(f) = 0.01$—, then the same update mechanism would have implied that

$$\mathbb{M}, w \models [\mathbb{E}, e]P(\text{real}) = 0.00096.$$

In other words, if you had been an expert art historian, then looking at the painting's signature would have been highly informative: it would have led to a significant change in your probabilities.

13.4.3 Proof System

A complete axiomatization for probabilistic dynamic epistemic logic can be found using the standard strategy, viz. by adding a set of reduction axioms to static probabilistic epistemic logic. Implementing this strategy, however, is not entirely trivial. The reduction axioms for non-probabilistic formulas are familiar from classical (non-probabilistic) dynamic epistemic logic, but the reduction axiom for i-probability formulas is more complicated.

First of all, this reduction axiom makes a case distinction on whether a certain sum of probabilities is strictly positive or not. We will show that this corresponds to the case distinction made in the definition of the updated probability functions (Definition 13.7). In the definition of $\mu_i'(w, e)$, a case distinction is made on the value of the denominator of a fraction, i.e. on the value of the following expression:

$$\sum_{\substack{v \in W \\ f \in E}} \mu_i(w)(v) \cdot \text{pre}(v)(f) \cdot \mu_i(e)(f). \tag{13.4}$$

But this expression can be rewritten as

$$\sum_{\substack{v \in W \\ f \in E \\ \varphi \in \Phi \\ \mathbb{M}, v \models \varphi}} \mu_i(w)(v) \cdot \text{pre}(\varphi)(f) \cdot \mu_i(e)(f).$$

Using the definition of $k_{i,e,\varphi,f}$ (cf. Fig. 13.3), this can be rewritten as

$$\sum_{\substack{\varphi \in \Phi \\ f \in E}} \mu_i(w)(\llbracket \varphi \rrbracket^{\mathbb{M}}) \cdot k_{i,e,\varphi,f}.$$

Since E and Φ are finite, this sum is finite and corresponds to an expression in the formal language $\mathscr{L}^{\otimes}$, which we will abbreviate as σ:

$$\sigma := \sum_{\substack{\varphi \in \Phi \\ f \in E}} k_{i,e,\varphi,f} P_i(\varphi).$$

This expression can be turned into an i-probability formula by 'comparing' it with a rational number b; for example $\sigma \geq b$. Particularly important are the formulas $\sigma = 0$ and $\sigma > 0$: exactly these formulas are used to make the case distinction in the reduction axiom for i-probability formulas.[8]

Next, the reduction axiom for i-probability formulas provides a statement in each case of the case distinction: $0 \geq b$ in the case $\sigma = 0$, and χ (as defined in Fig. 13.3) in the case $\sigma > 0$. We will only explain the meaning of χ in the 'interesting' case

---

[8] Note that E and Φ are components of the probabilistic update model $\mathbb{E}$ named by E; furthermore, the values $k_{i,e,\varphi,f}$ are fully determined by the model $\mathbb{E}$ and event e named by E and e, respectively (consider their definition in Fig. 13.3). Hence any i-probability formula involving σ is fully determined by $\mathbb{E}$, e, and can be interpreted at any probabilistic Kripke model $\mathbb{M}$ and state w.

1. static base logic

   • probabilistic epistemic logic, as axiomatized in Fig. 13.1

2. necessitation for $[E,e]$

   • if $\vdash \psi$ then $\vdash [E,e]\psi$

3. reduction axioms

$$[E,e]p \leftrightarrow \text{pre}_{E,e} \to p$$

$$[E,e]\neg\psi \leftrightarrow \text{pre}_{E,e} \to \neg[E,e]\psi$$

$$[E,e](\psi_1 \wedge \psi_2) \leftrightarrow [E,e]\psi_1 \wedge [E,e]\psi_2$$

$$[E,e]K_i\psi \leftrightarrow \text{pre}_{E,e} \to \bigwedge_{(e,f)\in R_i} K_i[E,f]\psi$$

$$[E,e]\sum_\ell a_\ell P_i(\psi_\ell) \geq b \leftrightarrow \text{pre}_{E,e} \to$$

$$\left(\sum_{\substack{\varphi\in\Phi\\f\in E}} k_{i,e,\varphi,f} P_i(\varphi) = 0 \wedge 0 \geq b\right) \vee$$

$$\left(\sum_{\substack{\varphi\in\Phi\\f\in E}} k_{i,e,\varphi,f} P_i(\varphi) > 0 \wedge \chi\right)$$

using the following definitions:

• $\text{pre}_{E,e} := \bigvee_{\substack{\varphi\in\Phi\\\text{pre}(\varphi)(e)>0}} \varphi$

• $k_{i,e,\varphi,f} := \text{pre}(\varphi)(f) \cdot \mu_i(e)(f) \in \mathbb{R}$

• $\chi := \sum_\ell \sum_{\substack{\varphi\in\Phi\\f\in E}} a_\ell k_{i,e,\varphi,f} P_i(\varphi \wedge \langle E,f\rangle \psi_\ell) \geq \sum_{\substack{\varphi\in\Phi\\f\in E}} bk_{i,e,\varphi,f} P_i(\varphi)$

Fig. 13.3 Axiomatization of probabilistic dynamic epistemic logic

$\sigma > 0$. If $\mathbb{M}, w \models \sigma > 0$, then the value of (13.4) is strictly positive (cf. supra), and we can calculate:

$$
\begin{aligned}
\mu_i'(w,e)(\llbracket\psi\rrbracket^{\mathbb{M}\otimes\mathbb{E}}) &= \sum_{\mathbb{M}\otimes\mathbb{E}.(w',e')\models\psi} \mu_i'(w,e)(w',e') \\
&= \sum_{\substack{w'\in W, e'\in E \\ \mathbb{M},w'\models\langle E,e'\rangle\psi}} \frac{\mu_i(w)(w')\cdot\text{pre}(w')(e')\cdot\mu_i(e)(e')}{\sum_{\substack{v\in W\\f\in E}}\mu_i(w)(v)\cdot\text{pre}(v)(f)\cdot\mu_i(e)(f)} \\
&= \frac{\sum_{\substack{\varphi\in\Phi\\f\in E}}\mu_i(w)(\llbracket\varphi\wedge\langle E,f\rangle\psi\rrbracket^{\mathbb{M}})\cdot k_{i,e,\varphi,f}}{\sum_{\substack{\varphi\in\Phi\\f\in E}}\mu_i(w)(\llbracket\varphi\rrbracket^{\mathbb{M}})\cdot k_{i,e,\varphi,f}}.
\end{aligned}
$$

Hence, in this case ($\sigma > 0$) we can express that $\mu_i'(w,e)(\llbracket\psi\rrbracket^{\mathbb{M}\otimes\mathbb{E}}) \geq b$ in the formal language, by means of the following i-probability formula:

$$\sum_{\substack{\varphi\in\Phi\\f\in E}} k_{i,e,\varphi,f} P_i(\varphi \wedge \langle E,f\rangle\psi) \geq \sum_{\substack{\varphi\in\Phi\\f\in E}} bk_{i,e,\varphi,f} P_i(\varphi).$$

Moving to linear combinations, we can express that $\sum_\ell a_\ell\mu_i'(w,e)(\llbracket\psi_\ell\rrbracket^{\mathbb{M}\otimes\mathbb{E}}) \geq b$ in the formal language using an analogous i-probability formula, namely χ (cf. the definition of this formula in Fig. 13.3).

We thus obtain the following theorem [17]:

Theorem 13.3 *Probabilistic dynamic epistemic logic, as axiomatized in Fig. 13.3, is sound and complete with respect to the class of probabilistic Kripke frames.*

13.5 Further Developments and Applications

Probabilistic extensions of dynamic epistemic logic are a recent development, and there are various open questions and potential applications to be explored. In this section we discuss a selection of such topics for further research; more suggestions can be found in [17] and [15, ch. 8].

We distinguish between *technical* and *conceptual* open problems.[9] A typical technical problem that needs further research is the issue of *surprising information*. In the update mechanisms described in this chapter, the agents' new probabilities are calculated by means of a fraction whose denominator might take on the value 0. The focus has been on the 'interesting' (non-0) cases, and the 0-case has been treated as mere 'noise': a technical artefact that cannot be handled convincingly by the system. However, sometimes such 0-cases *do* represent very intuitive scenarios; for example, one can easily think of an agent being absolutely certain that a certain proposition φ is false ($P(\varphi) = 0$), while that proposition is actually true, and can thus be announced! In such cases, the system of probabilistic public announcement logic described in Sect. 13.3 predicts that the agent will simply *ignore* the announced information (rather than performing some sensible form of *belief revision*). More can, and should be said about such cases [2, 6, 46].

Another technical question is whether other representations of *soft information* can learn something from the probabilistic approach to dynamic epistemic logic. Probabilistic Kripke models represent the agents' soft information via the probability functions μ_i, and interpret formulas of the form $P_i(\varphi) \geq b$. Plausibility models, on the other hand, represent the agents' soft information via a (non-numerical) plausibility ordering $\leq_i$, and interpret more qualitative notions of belief [7, 12, 15, 22]. In particular, if we use $Min_{\leq_i}(X)$ to denote the set of $\leq_i$-minimal states in the set X, then the formula $B_i\varphi$ is interpreted in a plausibility model $\mathbb{M}$ as follows:[10]

$$\mathbb{M}, w \models B_i\varphi \quad \text{iff} \quad \text{for all } v \in Min_{\leq_i}(R_i[w]) : \mathbb{M}, v \models \varphi.$$

The product update for probabilistic Kripke models described in Definition 13.7 takes into account prior probabilities ($\mu_i(w)(v)$ for states w and v), observation probabilities ($\mu_i(e)(f)$ for events e and f), and occurrence probabilities ($\mathsf{pre}(w)(e)$ for a state w and event e). One can also define a product update for plausibility models; a widely used rule to define the updated plausibility ordering is the so-called 'priority update' [7, 15]:

---

[9] In practice, this distinction will not always be clear-cut, of course.

[10] As usual, $R_i[w]$ denotes the set $\{v \in W \mid (w, v) \in R_i\}$.

$$(w, e) \leq_i (v, f) \quad \text{iff} \quad e <_i f \text{ or } (e \cong_i f \text{ and } w \leq_i v).$$

The updated plausibility ordering thus gives priority to the plausibility ordering on events, and otherwise preserves the original plausibility ordering on states as much as possible. In analogy with the probabilistic setting, the plausibility orderings on states and events can be called the 'prior plausibility' and 'observation plausibility', respectively. However, the notion of occurrence probability does *not* seem to have a clear analogue in the framework of plausibility models. van Benthem [16] defines a notion of 'occurrence plausibility', which can be expressed as $e \leq_w f$: at state w, event e is at least as plausible f to occur (this ordering is not agent-dependent; recall Footnote 7). New product update rules thus have to merge *three* plausibility orderings: prior plausibility, observation plausibility, and occurrence plausibility. van Benthem [16] makes some proposals for such rules, but finding a fully satisfactory definition remains a major open problem in this area.

An important conceptual issue that is currently actively being investigated, is the exact relation between the quantitative (probabilistic) and qualitative perspectives on soft information. A widespread proposal is to connect *belief* with *high probability*, where 'high' means 'above some treshold $\tau \in (0.5, 1]$'; formally: $B_i \varphi \Leftrightarrow P_i(\varphi) \geq \tau$. An immediate problem of this proposal is that belief is standardly taken to be closed under conjunction, while 'high probability' is not closed under conjunction (unless $\tau = 1$). Despite this initial problem, there's also a lot in favor of this proposal. Plausibility models not only interpret a notion of belief, but also a notion of *conditional belief*: $B_i^{\alpha} \varphi$ means that agent i believes that φ, conditional on α. The connection between belief and high probability can perfectly be extended to conditional belief and high conditional probability:

$$B_i^{\alpha} \varphi \quad \Leftrightarrow \quad P_i(\varphi \mid \alpha) \geq \tau. \tag{13.5}$$

Furthermore, (conditional) belief and high (conditional) probability seem to display highly similar dynamic behavior under public announcements. We saw in Sect. 13.3.3 that $[!\varphi]P_i(\psi) \geq \tau$ can sometimes diverge in truth value from $P_i(\psi \mid \varphi) \geq \tau$, because of the presence of higher-order information. In exactly the same way (and for the same reason), $[!\varphi]B_i\psi$ and $B_i^{\varphi}\psi$ can diverge in truth value on plausibility models. Furthermore, (conditional) belief and high (conditional) probability have exactly the 'same' reduction axiom. This means that (13.6) (which is interpreted on probabilistic Kripke models) and (13.7) (which is interpreted on plausibility models) are intertranslatable, using principle (13.5) above:

$$[!\varphi]P_i(\psi \mid \alpha) \geq \tau \quad \leftrightarrow \quad \left(\varphi \to P_i(\langle !\varphi \rangle \psi \mid \langle !\varphi \rangle \alpha) \geq \tau \right), \tag{13.6}$$

$$[!\varphi]B_i^{\alpha}\psi \quad \leftrightarrow \quad \left(\varphi \to B_i^{\langle !\varphi \rangle \alpha} \langle !\varphi \rangle \psi \right). \tag{13.7}$$

The significance of these observations is further discussed in [24].

Several fruitful *applications* of probabilistic dynamic epistemic logic can be expected in the fields of *game theory* and *cognitive science*. In recent years, dynamic epistemic logic has been widely applied to explore the epistemic foundations of game theory [10, 13, 18]. However, given the importance of probability in game theory (for example, in the notion of mixed strategy), it is surprising that very few of these logical analyses have a probabilistic component.[11] Probabilistic dynamic epistemic logic provides the required tools to explore the epistemic *and* the probabilistic aspects of game theory.

For example, [21, 23] uses a version of probabilistic public announcement logic to analyze the role of common knowledge and communication in Aumann's well-known agreeing to disagree theorem. Classically, this theorem is stated as follows: "If two people have the same prior, and their posteriors for an event [φ] are common knowledge, then these posteriors are equal" [3, p. 1236]. If we represent the experiments (with respect to which the agents' probabilities are called 'prior' and 'posterior') with a dynamic operator [EXP], then this version can be formalized as (13.8), which is derivable in the underlying logical system:

$$[\text{EXP}]C\big(P_1(\varphi) = a \wedge P_2(\varphi) = b\big) \to a = b. \tag{13.8}$$

However, this version does not say *how* the agents are to obtain this common knowledge; it just assumes that they have been able to obtain it one way or another. The way to obtain common knowledge is via a certain communication protocol, which is described explicitly in the intuitive scenario that is used to motivate and explain this theorem. Once this communication dynamics is made explicitly part of the story, common knowledge of the posteriors need no longer be *assumed* in the formulation of the agreement theorem, since it will now simply *follow* from the communication protocol. If we represent the communication protocol with a dynamic operator [DIAL(φ)], this new version of the theorem can be formalized as (13.9):

$$[\text{EXP}][\text{DIAL}(\varphi)]\big(P_1(\varphi) = a \wedge P_2(\varphi) = b\big) \to a = b. \tag{13.9}$$

The notion of common knowledge is thus less central to the agreement theorem than is usually thought: if we compare (13.8) and (13.9), it is clear that common knowledge and communication are two sides of the same coin; the former is only needed to formulate the agreement theorem if the latter is not represented explicitly.

Another potential field of application is cognitive science. The usefulness of (epistemic) logic for cognitive science has been widely recognized [14, 35, 43]. Of course, as in any other empirical discipline, one quickly finds out that real-life human cognition is rarely a matter of all-or-nothing, but often involves degrees (probabilities). Furthermore, a recent development in cognitive science is toward probabilistic (Bayesian) models of cognition [42]. If epistemic logic is to remain a valuable tool here, it will thus have to be a thoroughly 'probabilized' version. For example, probabilistic dynamic epistemic logic has been used to model the cognitive phenomenon of surprise and its epistemic aspects [25, 39].

---

[11] The logic in [19] does have a probabilistic component, but this logic is fully *static*.

13.6 Conclusion

In this chapter we have introduced probabilistic epistemic logic, and several of its dynamic versions. These logics provide a standard epistemic (possible-worlds) analysis of the agents' hard information, and supplement it with a fine-grained probabilistic analysis of their soft information. Higher-order information of any kind (knowledge about probabilities, probabilities about knowledge, etc.) is represented explicitly. The importance of higher-order information in dynamics is clear from our discussion of the connection between public announcements and Bayesian conditionalization. The probabilistic versions of both public announcement logic and dynamic epistemic logic with product updates can be completely axiomatized in a standard way (via reduction axioms). The fertility of the research program of probabilistic dynamic epistemic logic is illustrated by the variety of technical and conceptual issues that are still open for further research, and its (potential) use in analyzing theorems and phenomena from game theory and cognitive science.

Acknowledgments The authors wish to thank Alexandru Baltag, Johan van Benthem and Sonja Smets for their valuable feedback on earlier versions of this chapter. Lorenz Demey is financially supported by a PhD fellowship of the Research Foundation–Flanders (FWO).

References

1. Alchourrón C, Gärdenfors P, Makinson D (1985) On the logic of theory change: partial meet contraction and revision functions. J Symb Logic 50:510–530
2. Aucher G (2003) A combined system for update logic and belief revision. Master's thesis, Institute for Logic, Language and Computation, Universiteit van Amsterdam
3. Aumann R (1976) Agreeing to disagree. Ann Stat 4:1236–1239
4. Baltag A, Moss LS (2004) Logics for epistemic programs. Synthese 139:1–60
5. Baltag A, Moss LS, Solecki S (1998) The logic of common knowledge, public announcements, and private suspicions. In: Gilboa I (ed) Proceedings of the 7th conference on theoretical aspects of rationality and knowledge (TARK '98), pp 43–56
6. Baltag A, Smets S (2008) Probabilistic dynamic belief revision. Synthese 165:179–202
7. Baltag A, Smets S (2008) A qualitative theory of dynamic interactive belief revision. In: Bonanno G, van der Hoek W, Woolridge M (eds) Texts in logic and games, vol 3. Amsterdam University Press, pp 9–58
8. van Benthem J (1983) Modal logic and classical logic. Bibliopolis, Napoli
9. van Benthem J (2001) Correspondence theory. In: Gabbay DM, Guenthner F (eds) Handbook of philosophical logic, revised 2nd edn, vol 3. Kluwer, Dordrecht, pp 325–408
10. van Benthem J (2001) Games in dynamic epistemic logic. Bull Econ Res 53:219–248
11. van Benthem J (2003) Conditional probability meets update logic. J Logic Lang Inf 12:409–421
12. van Benthem J (2007) Dynamic logic for belief revision. J Appl Nonclass Logics 17:129–155
13. van Benthem J (2007) Rational dynamics and epistemic logic in games. Int Game Theory Rev 9:13–45
14. van Benthem J (2008) Logic and reasoning: do the facts matter? Stud Logica 88:67–84
15. van Benthem J (2011) Logical dynamics of information and interaction. Cambridge University Press, Cambridge
16. van Benthem J (2012) A problem concerning qualitative probabilistic update. p 10. (Unpublished manuscript)

17. van Benthem J, Gerbrandy J, Kooi BP (2009) Dynamic update with probabilities. Stud Logica 93:67–96
18. Bonanno G, Dégremont C (2014) Logic and game theory. In: Baltag A, Smets S (eds) Johan van Benthem on logic and information dynamics. Springer, Dordrecht, pp 421–449 (Chapter 15 in this volume)
19. de Bruin B (2010) Explaining games: the epistemic programme in game theory. Springer, Dordrecht
20. Conradie W, Ghilardi S, Palmigiano A (2014) Unified correspondence. In: Baltag A, Smets S (eds) Johan van Benthem on logic and information dynamics. Springer, Dordrecht, pp 933–975 (Chapter 36 in this volume)
21. Demey L (2010) Agreeing to disagree in probabilistic dynamic epistemic logic. Master's thesis, Institute for Logic, Language and Computation, Universiteit van Amsterdam
22. Demey L (2011) Some remarks on the model theory of epistemic plausibility models. J Appl Nonclass Logics 21:375–395
23. Demey, L. (2013) Contemporary epistemic logic and the Lockean thesis. Found Sci 18:599–610
24. Demey, L. (2014) Agreeing to disagree in probabilistic dynamic epistemic logic. Synthese 191:409–438
25. Demey L (forthcoming) The dynamics of surprise. Logique Anal
26. Demey L, Kooi B, Sack J (2013) Logic and probability. In: Zalta EN (ed) The Stanford encyclopedia of philosophy, Stanford
27. van Ditmarsch H, van der Hoek W, Kooi B (2007) Dynamic epistemic logic. Springer, Dordrecht
28. Fagin R, Halpern J (1994) Reasoning about knowledge and probability. J ACM 41:340–367
29. van Fraassen B (1984) Belief and the will. J Philos 81:235–256
30. Gärdenfors P (1988) Knowledge in flux. MIT Press, Cambridge
31. Gerbrandy J, Groeneveld W (1997) Reasoning about information change. J Logic Lang Inf 6:147–169
32. Halpern JY (1991) The relationship between knowledge, belief, and certainty. Ann Math Artif Intell 4:301–322 (errata in the same journal, 26:59–61, 1999)
33. Halpern JY (2003) Reasoning about uncertainty. MIT Press, Cambridge
34. Hintikka J (1962) Knowledge and belief: an introduction to the logic of the two notions. Cornell University Press, Ithaca
35. Isaac A, Szymanik J, Verbrugge R (2014) Logic and complexity in cognitive science. In: Baltag A, Smets S (eds) Johan van Benthem on logic and information dynamics. Springer, Dordrecht, pp 787–824 (Chapter 30 in this volume)
36. Jeffrey R (1983) The logic of decision, 2nd edn. University of Chicago Press, Chicago
37. Kooi BP (2003) Probabilistic dynamic epistemic logic. J Logic Lang Inf 12:381–408
38. Lewis D (1980) A subjectivist's guide to objective chance. In: Jeffrey RC (ed) Studies in inductive logic and probability, vol 2. University of California Press, Berkeley, pp 263–293
39. Lorini E, Castelfranchi C (2007) The cognitive structure of surprise: looking for basic principles. Topoi 26:133–149
40. Meacham CJG (2010) Two mistakes regarding the principal principle. Br J Philos Sci 61:407–431
41. Miller D (1966) A paradox of information. Br J Philos Sci 17:59–61
42. Oaksford M, Chater N (2008) The probabilistic mind: prospects for Bayesian cognitive science. Oxford University Press, Oxford
43. Pietarinen AV (2003) What do epistemic logic and cognitive science have to do with each other? Cogn Syst Res 4:169–190
44. Plaza J (1989) Logics of public communications. In: Emrich ML, Pfeifer MS, Hadzikadic M, Ras ZW (eds) Proceedings of the 4th international symposium on methodologies for intelligent systems: poster session program, Oak Ridge National Laboratory, Oak Ridge, pp 201–216 (Reprinted in: Synthese 158:165–179, 2007)
45. Romeijn JW (2012) Conditioning and interpretation shifts. Stud Logica 100:583–606
46. Rott H, Girard P (2014) Belief revision and dynamic logic. In: Baltag A, Smets S (eds) Johan van Benthem on logic and information dynamics. Springer, Dordrecht, pp 203–233 (Chapter 8 in this volume)
47. Sack J (2009) Extending probabilistic dynamic epistemic logic. Synthese 169:241–257

Chapter 14
Belief as a Simplification of Probability, and What This Entails

Hannes Leitgeb

Abstract There are concepts of belief on different scales of measurement. In particular, it is common practice to ascribe beliefs to a person both in terms of a categorical (all-or-nothing) concept of belief and in terms of a numerical concept of degree of belief; the formal structure of categorical belief being the subject of doxastic logic, the formal structure of degrees of belief being the topic of subjective probability theory. How do these two kinds of belief relate to each other? We derive an answer to this question from one basic norm: rational categorical belief ought to be a simplified version of subjective probability, where the corresponding concept of simplification can be made mathematically precise in terms of minimizing sums of errors of the result of approximating the probability of a proposition by means of belief or disbelief in the proposition. As it turns out, essentially (glossing over a couple of details) the answer to our question is: a rational person's set of doxastically accessible worlds must have a stably high probability with respect the person's subjective probability measure.

14.1 Introduction: Belief as a Simplification of Subjective Probability

Rational belief can be expressed on different scales of measurement: in terms of numerical degrees of belief, or plausibility orders, or categorical "all-or-nothing" belief, or on a scale that lies somewhere in between the numerical and the ordinal scale, or between the ordinal and the categorical one.

Let us just focus on numerical rational belief now, which we are going to identify with subjective probability, and rational categorical ("all-or-nothing") belief, which we identify with what is studied in doxastic logic and its standard possible

H. Leitgeb (✉)
Ludwig Maximilian University, Munich, Germany
e-mail: hannes.leitgeb@lmu.de

A. Baltag and S. Smets (eds.), *Johan van Benthem on Logic
and Information Dynamics*, Outstanding Contributions to Logic 5,
DOI: 10.1007/978-3-319-06025-5_14, © Springer International Publishing Switzerland 2014

worlds semantics. Belief in both senses ultimately concerns an attitude towards propositions—*sets of possible worlds*—as these are the contents of belief, but at the same time both of these attitudes are actually derivable from attitudes towards the *possible worlds* themselves (at least if the underlying set of possible worlds is finite, as we are going to assume throughout the chapter): for the probability $P(X)$ of a proposition X is just the sum of the probabilities of (the singletons of) the members w of X, and a proposition X is believed if and only if it is a superset of a distinguished set (say, B_W) of doxastically possible or accessible worlds w. So belief in the sense of probability theory and belief in the sense of doxastic logic both incorporate ways of reducing the complexity of an attitude towards $2^{|W|}$ propositions to the complexity of taking a stand on the smaller number $|W|$ of members of the overall set W of possible worlds; in both cases the acceptance of certain rationality postulates—the principles of probability in the case of numerical belief, the principles of doxastic logic in the case of categorical belief—has an effect of simplification.

And whatever one's view on the exact relationship between degrees of belief and categorical belief might be, it is very plausible that the concept of simplification will play a role again when one aims to specify that very relationship—as the following norm seems to be valid:

- Rational categorical belief ought to be a simplified version of subjective probability.

When arguing for this norm, we should distinguish between two possible cases here, without having to commit ourselves to either of them. The cases concern the difference between externally ascribing beliefs and internally having them.

Either there is actually just one phenomenon "out there"—one kind of rational belief state—but we are able to talk about this phenomenon in terms of concepts of belief on different scales. Then the norm above amounts to:

- If one is rational, then one ought to ascribe rational categorical beliefs and subjective probabilities to a person so that the ascribed system of categorical beliefs is a simplified version of the ascribed probability measure.

This norm is plausible, if only to guarantee that one's rational ascription of a person's categorical belief system ends up in some kind of harmony with one's rational ascription of the same person's subjective probability measure; and this should be the case in order to minimize the threat of "schizophrenic" situations in which the categorical beliefs that one ascribes to a person would take her to be disposed to act in one way but where the ascription of the same person's subjective probability function would take her to be disposed to act differently. So a belief ascriber ought to obey the norm above in order for her different kinds of belief ascription to (more or less) cohere.

The other possible case is that there are two distinct phenomena "out there"—for instance, one and the same person having both a categorical belief system and a separate degree of belief system as parts of her overall cognitive system—and by 'categorical belief' and 'subjective probability' we refer to these distinct systems, respectively. Then the norm from before really means

- If a person is rational, then her system of rational categorical beliefs ought to be a simplified version of her subjective probability measure.

and again the norm sounds convincing: after all, a rational person's categorical belief states should end up in some kind of harmony with the same person's subjective probability measure in order to minimize the threat of "schizophrenic" situations in which the person's categorical beliefs would dispose her to act in one way but where the same person's subjective probability function would dispose her to act differently. A believer ought to obey the norm above in order for her different kinds of belief to (more or less) cohere. In either interpretation the norm seems valid.

Although the norms from above might sound like reductions of belief to probability, they need not be interpreted as such: for instance, categorical belief might well be more fundamental, in some sense, than degrees of belief, and if so, the norm above would entail that a person's degree of belief function ought be such that the given belief system happens to be a simplification thereof. Or maybe neither of categorical belief and degrees of belief is prior to the other, in which case the norm above merely expresses a join constraint on both of them. Independently of how one views this question of "priority", the norm above does sound right.

This said, in whatever way the norms above are interpreted, they are all vague and potentially ambiguous. The most urgent question being: what exactly does the term 'is a simplified version of' mean as used in the norms above?

Given a subjective probability measure P and a set B_W of doxastically accessible worlds (which determines rational categorical belief in the usual manner of possible worlds semantics), and accepting that the numerical belief scale is more fine-grained than the categorical belief scale and hence information is being lost by passing from P to B_W, *how can we measure the "error" of using B_W as one's set of doxastically possible worlds relative to the subjective probability measure P?* Or in other words: *How can we measure the error of "digitalizing" the "analogue" information that is contained in P in terms of the set B_W?* Any satisfying answer to this question will allow us to state a satisfying precisification of the norms above in the form

- It should be the case that the set B_W of doxastically possible worlds minimizes error relative to P.

Instead of 'minimizes error relative to P' we might just as well have said 'maximizes approximation of P' or the like; never mind the exact wording.

In what follows, we will determine an answer to the question from before, and we will present the precise norm on B_W and P that derives from it. Section 14.2 will clarify the notion of *minimizing error relative to P* in one possible way; and at the end of the section we will prove a theorem which will characterize the sets of doxastically possible worlds that minimize error in that sense. Section 14.3 will illustrate the theorem by means of a little example, and it will draw some conclusions from this.[1]

---

[1] In a different paper (H. Leitgeb, unpublished [5]), we approximate subjective probability in terms of plausibility orders of possible worlds, that is, by belief on an ordinal scale; in this way, also the dynamical aspects of belief can be taken into account. (The method of approximating probability

14.2 The Explication of Simplification

Let W be a given non-empty and finite set of possible worlds; for instance, one might think of W as the set of logically possible worlds for a given simple propositional language with finitely many propositional letters. We will use 'X', 'Y', 'Z' in order to denote propositions, that is, subsets of W. By 'P' we will always denote a probability measure over the power set of W. Whenever possible, we will suppress explicit universal quantifiers over propositions and probability measures.

As a first approximation, we suggest to measure the error of simplifying P in terms of choosing an "approximating" set X of doxastically possible worlds as follows:

$$Err_P(X) = \sum_{Z:X\subseteq Z} err_P(Bel(Z)) + \sum_{Z:X\cap Z=\varnothing} err_P(Dis(Z)).$$

In the first of the two sums, we consider all the propositions Z that one would end up believing if X were one's set of doxastically possible worlds—that is: all supersets Z of X—and we sum up the individual errors (relative to P) of believing any such Z; these individual errors are denoted by '$err_P(Bel(Z))$'. The second sum concerns all the propositions Z that one would end up disbelieving if X were one's set of doxastically possible worlds—all propositions disjoint from X—and we sum up the individual errors (relative to P) of disbelieving any such Z; these individual errors are denoted by '$err_P(Dis(Z))$'. Adding up the values of these two sums yields the overall error (relative to P) of taking X to be one's set of doxastically possible worlds; that overall error we denote by '$Err_P(X)$'.

We will say more about how to determine individual errors $err_P(Bel(Z))$ and $err_P(Dis(Z))$ later in the chapter. Note that it would be possible to simplify the sum above by identifying *disbelieving* Z with *believing* $W \setminus Z$, where $W \setminus Z$ is the set-theoretic complement of Z with respect to W: our only reason for not doing so is that we do not need to—none of our arguments will rely on this identification.

The two crucial assumptions that are built into our proposal of measuring error in terms of the sum above are: (i) being indifferent about a proposition Z—neither believing nor disbelieving it—does not count as an error at all (for such Zs are simply disregarded in the sum above); and (ii) individual errors are aggregated to an overall or total error by adding them up. Neither of these assumptions is unproblematic, of course. As far as (ii) is concerned, one might perhaps think of $Err_P(X)$ as a kind of expected epistemic (dis)utility of the act of choosing X to be one's set of doxastically possible worlds, in which case the summation of single (dis)utilities is a plausible thing to do. (Famously, Hempel [2] and Levi [6] compute such overall expected epistemic utilities, even though they do so differently.) As far as (i) is concerned, the idea might be that if one does not take a stand on a proposition, then one may well

---

(Footnote 1 continued)
by belief in Leitgeb, unpublished, is completely different from the one that will be employed in the next section.) Unfortunately, we will not be able to deal with this in the present chapter—sorry for remaining on the static side here, Johan!

be overly cautious, failing to take a risk, not doing anything, but one would not have made a proper *mistake*—at least not in the usual sense of doing something wrong. In any case, in what follows we will take these two assumptions for granted.

Now consider two sets X and Y, where $X \subseteq Y$: clearly then, by determining overall errors as sketched above, $Err_P(X) \geq Err_P(Y)$. In the extreme case, when Y is identical to W, then if $Y = W$ is a person's set of doxastically possible worlds, the person will end up believing only the tautological proposition W and disbelieving the contradictory proposition $\varnothing$, neither of which will be an error—so presumably $Err_P(W)$ will be just 0. Hence, if one's sole aim were to minimize total error, choosing W to be one's set of doxastically possible worlds would be the way to go.

But that does not sound right: one might well rationally intend to be less cautious than that. Therefore, our ultimate goal of finding the "best" approximation of P in terms of a set B_W of doxastically possible worlds cannot just consist in choosing the set that minimizes overall error. And this is a well-known pattern: a rational person should not just aim *not to believe any falsehoods*, she should also aim *to believe substantial truths*; and although we are not in the business of comparing belief to truth here, since our goal is really to compare belief to probability, the pattern remains the same: the norms from the last section should not dictate a person to be overly cautious; a brave believer should not be punished just for being brave.

This leads us to the following thought: whenever we intend to compare propositions in terms of the error (relative to a probability measure P) of choosing either of them as one's set of doxastically possible worlds, let us only consider propositions that are rivals of each other in the following obvious sense:

Definition 14.1 X is a rival of Y iff $X \not\subseteq Y, Y \not\subseteq X$.

If X is a rival of Y, and hence Y is a rival if X, it is not the case that the set of believed propositions according to X is a subset of the set of believed propositions according to Y, nor *vice versa*; and it is not the case that the set of disbelieved propositions according to X is a subset of the set of disbelieved propositions according to Y, nor *vice versa*. Therefore it will at least not be settled from the start which of the two will end up yielding a total error (relative to P) that is greater than or equal to the total error of the other.

But that is still not good enough. Let us assume that we want to compare two rivals X and Y in terms of the overall errors that they determine relative to some P; and suppose that the cardinality of X, $|X|$, is small, whereas the cardinality $|Y|$ of Y is large: when determining $Err_P(X)$ there will then be many more propositions Z for which individual errors will be summed up than this will be the case for $Err_P(Y)$— for there will be many more supersets of X than supersets of Y, and many more sets will be disjoint from X than there will be sets disjoint from Y. In such a case, unless compensated in some manner, X would seem to be disadvantaged to Y again, and minimizing overall error might again lead to a general bias towards choosing one's set of doxastically possible worlds to be large (cautious) rather than small (brave). And this would be so although X and Y are rivals in the sense introduced before. Once again this is not what we intend.

One way of addressing this problem would be to compare only propositions X and Y of the same cardinality. But we will follow a slightly different path here. Consider some rivals X and Y: if they are not of the same cardinality, we may still introduce a partition π of W—a coarse-graining of the underlying space of worlds, or formally: a set of pairwise disjoint subsets ("partition cells") of W the union of which is W again—so that the sets $X \setminus Y$, $Y \setminus X$, $X \cap Y$, $W \setminus (X \cup Y)$ remain "separated", that is, every partition cell of π is a subset of one of these four sets, but where the number of partition cells that are subsets of $X \setminus Y$ is identical to the number of partition cells that are subsets of $Y \setminus X$, and consequently the number of partition cells that are subsets of X is also identical to the number of partition cells that are subsets of Y. By partitioning W in this way we will thus "make" X and Y have the same cardinality. If we then also modify the definition of total error so that only propositions Z are taken into account that "respect" that partition π—so that Z does not "split" any of π's partition cells—then the same number of sets Z will be considered when determining the overall errors for X and Y, respectively, and neither of X and Y will be disadvantaged anymore. That is the proposal that we are going to put forward now in more formal terms.

Let us say that X respects a partition π (of W) if and only if X is a union of partition cells of π. Further down below we will apply a ternary notion of respecting: X and Y respect a partition π (of W) if and only if all of the sets $X \setminus Y$, $Y \setminus X$, $X \cap Y$, $W \setminus (X \cup Y)$ are unions of partition cells of π.

We can now adapt out original conception of overall error in the way that only propositions are taken into account that respect a given partition:

Definition 14.2 For all partitions π of W, for all X that respect π:

$$Err_{\pi, P}(X) = \sum_{\substack{Z:\ (i)\ Z\ \text{respects}\ \pi,\ (ii)\ X \subseteq Z}} err_P(Bel(Z)) + \sum_{\substack{Z:\ (i)\ Z\ \text{respects}\ \pi,\ (ii)\ X \cap Z = \varnothing}} err_P(Dis(Z))$$

Next, for rivals X and Y, we regard X as a better simplification or approximation of P than Y, when there is a partition of W according to which X and Y cover the same number of partition cells and X leads to less overall error than Y relative to P, but where there is no partition of W according to which X and Y cover the same number of partition cells and Y leads to less overall error than X relative to P—a kind of dominance conception of simplification.

Formally: When a proposition X respects a partition π, let us denote the number of partition cells of π that are subsets of X by '$|X|_\pi$'. We can then define:

Definition 14.3 If X is a rival of Y, then:
X produces fewer errors than Y (relative to P) iff

- there exists a partition π of W, such that X and Y respect π, $|X|_\pi = |Y|_\pi$, and $Err_{\pi, P}(X) < Err_{\pi, P}(Y)$;
- there does not exist a partition π of W, such that X and Y respect π, $|X|_\pi = |Y|_\pi$, and $Err_{\pi, P}(Y) < Err_{\pi, P}(X)$.

In this way the problem of comparing rivals of different cardinality can be circumvented.

Finally, in order to determine for a given proposition Y which of its rivals X, if any, produces fewer errors than Y in the sense just defined, we have to introduce some assumptions on individual errors. The two principles P1 and P2 that we are going to presuppose for that purpose are quite weak, and they make precise in what sense, and to what extent (on an ordinal scale), a believed or disbelieved proposition is affected by error relative to a probability measure:

P1: $err_P(Bel(Z)) < (\leq) err_P(Bel(Z'))$ iff $P(Z) > (\geq) P(Z')$.
P2: $err_P(Dis(Z)) < (\leq) err_P(Dis(Z'))$ iff $P(Z) < (\leq) P(Z')$.

The easiest way of thinking about these principles is this: If a proposition has probability 1, then believing it is not an error at all; there is no positive chance that one could be wrong about this. If a proposition has a probability of less than 1, then believing it does amount to an error of some extent, and the smaller the probability of that proposition is, the larger is the error. If the probability of a proposition is 0— and thus its probability has maximal distance from 1—then, presumably, believing it amounts to the maximal possible error. And the same holds, *mutatis mutandis*, for probability 0 and disbelief. Again none of this is sacrosanct, but at the same time this method of determining errors of belief or disbelief in single propositions is certainly not implausible.

Given these definitions of 'rival', '$Err_{\pi,P}$', 'produces fewer errors than (relative to P)', and the two principles P1 and P2, we are ready to determine at least a partial answer to our main question from the last section: *how can we measure the "error" of using B_W as one's set of doxastically possible worlds relative to the subjective probability measure P?*

In order to spell out this answer, we need one final concept:

Definition 14.4 X is P-stable iff for all $w \in X$: $P(\{w\}) > P(W \setminus X)$.

Non-empty P-stable sets X are such that even the smallest of their subsets have a probability greater than the complement of X. This condition is similar to Snow's [8] and Benferhat et al.'s [1] "big-stepped probabilities" condition on strict total orders of worlds to the effect that the probability of a singleton set of a world w is greater than the sum of probabilities of all singleton sets of worlds w' that are less preferred than w in the strict total order. We have applied the same concept of P-stability (or a slight variant of it) also in Leitgeb [3, 4],

Our usage of the term 'stability' is explained by the following fact: if X is P-stable in the sense above (and X is non-empty) then $P(X|Y) > \frac{1}{2}$ for every Y that has non-empty intersection with X and which, therefore, has also positive probability. Thus, when a non-empty set X is P-stable, it retains a probability greater than that of its complement as long as one conditionalizes P on a proposition Y that is consistent with X (and for which, therefore, conditionalization is well-defined, that is, where $P(Y) > 0$). In other words: it is not easy to "get rid of" the high probability of a P-stable set by means of probabilistic update. For that reason, this notion of probabilistic stability is also closely related to Skyrms' [7] notion of the *resiliency* of the (subjective) probability of a proposition, which he applied successfully in his theory of objective chance.

One can show that P-stable sets are never rivals of each other; for every two P-stable sets X and Y, X is a subset of Y or vice versa. And although P-stability might seem to be a pretty restrictive notion, it is easy to see that there are in fact lots of (non-empty) P-stable sets around: if P is defined on a finite sample space W, the least set of probability 1 is P-stable; much more importantly, for a given finite W again, almost all probability measures P on W allow for a non-empty P-stable proposition of probability *less than 1*: this qualification 'almost all' can be stated precisely in terms of taking the Lebesgue measure of the set of standard geometrical representations of such probability measures—the geometrical representations of all probability measures P *except for a set of Lebesgue measure 0* have the property that there is a non-empty P-stable set X with $P(X) < 1$ (see Leitgeb [3] for the formal details).

The following theorem makes it clear why we introduced this concept of P-stability into the present context. The upshot will be: in the sense specified by the theorem, *it is the P-stable sets which are the best approximations of P.*

Theorem 14.5

1. *If X is P-stable, and X is a rival of Y, then X produces fewer errors than Y relative to P.*
2. *If X is not P-stable, then there is a Y, such that X is a rival of Y, and it is not the case that X produces fewer errors than Y relative to P.*
3. *Let (*) P be such that for all $w \in W$ and for all Z with $w \notin Z$, $P(\{w\}) \neq P(Z)$: If X is not P-stable, then there is a Y, such that X is a rival of Y, and Y produces fewer errors than X relative to P.*

Proof We start with some general considerations:

(i) If X and Y are rivals, then clearly there is partition π of W, such that X and Y respect π, and $|X|_\pi = |Y|_\pi$: simply coarse-grain the sets $X \setminus Y$ and $Y \setminus X$ so that both of them contain the same number of partition cells as subsets; on $X \cap Y$ and $W \setminus (X \cup Y)$ let the partition cells be singleton sets.

(ii) Whenever X and Y respect π, and $|X|_\pi = |Y|_\pi$, there is a bijection f of π, such that f is the identity mapping on all partitions cell that are subsets of either of $X \cap Y$ and $W \setminus (X \cup Y)$, and where for all partition cells $p \subseteq X \setminus Y$ it holds that $f(p)$ is a partition cell that is a subset of $Y \setminus X$: this is because our assumptions entail that $|X|_\pi = |X \cap Y|_\pi + |X \setminus Y|_\pi$, $|Y|_\pi = |X \cap Y|_\pi + |Y \setminus X|_\pi$, $|X|_\pi = |Y|_\pi$, and thus $|X \setminus Y|_\pi = |Y \setminus X|_\pi$, which is why such a bijection exists.

(iii) If f is a bijection of π, define a mapping F from $\{Z : Z$ respects $\pi\}$ to itself by means of: $F(Z) = \cup_{p\in\pi:p\subseteq Z} f(p)$. It is easy to see then that F is a bijection.

(iv) Whenever X and Y respect π, f is a bijection of π that has the properties stated in (ii), and F is defined from f as explained in (iii), then for every Z that respects π the following is the case: $X \subseteq Z$ iff $Y \subseteq F(Z)$, and $X \cap Z = \varnothing$ iff $X \cap F(Z) = \varnothing$. This follows immediately from the definition of F and the properties of f.

(v) Whenever X and Y respect π, and f is a bijection of π that has the properties stated in (ii), then F (as defined in (iii)) maps the members of $\{Z: Z$ respects $\pi, X \subseteq Z\}$ bijectively to the members of $\{Z: Z$ respects $\pi, Y \subseteq Z\}$, and it maps the members of $\{Z: Z$ respects $\pi, X \cap Z = \varnothing\}$ bijectively to the members of $\{Z: Z$ respects $\pi, Y \cap Z = \varnothing\}$. This follows by applying (iii) and (iv).

(vi) Whenever X and Y respect π, and f is a bijection of π that has the properties stated in (ii), and F is defined as in (iii), then for every Z that respects π the following is the case: $F(Z) = F((Z \cap (X \setminus Y)) \cup (Z \cap (X \cap Y)) \cup (Z \cap (Y \setminus X)) \cup (Z \cap (W \setminus [X \cup Y]))) = F(Z \cap (X \setminus Y)) \cup F(Z \cap (X \cap Y)) \cup F(Z \cap (Y \setminus X)) \cup F(Z \cap (W \setminus [X \cup Y]))$, by X and Y respecting π and the properties of F and f, and $F(Z \cap (X \setminus Y)) \subseteq (Y \setminus X)$, $F(Z \cap (X \cap Y)) = Z \cap (X \cap Y)$, $F(Z \cap (Y \setminus X)) \subseteq (X \setminus Y)$, $F(Z \cap (W \setminus [X \cup Y])) = Z \cap (W \setminus [X \cup Y])$, by the properties of F and f again.

Now we turn to the proofs of 1–3 which follow from applying (i)–(vi).

For (1): First of all, we show that there exists a partition π of W, such that X and Y respect π, $|X|_\pi = |Y|_\pi$, and $Err_{\pi,P}(X) < Err_{\pi,P}(Y)$.

By assumption, X is a rival of Y. Hence by (i), there is partition π of W, such that X and Y respect π, and $|X|_\pi = |Y|_\pi$. (**) But for all such partitions π, with (ii), it follows that there is a bijection f of π, such that f is the identity mapping on all partitions cell that are subsets of either of $X \cap Y$ and $W \setminus (X \cup Y)$, and where for all partition cells $p \subseteq X \setminus Y$: $f(p)$ is a partition cell that is a subset of $Y \setminus X$. Furthermore, since $X \setminus Y$ is non-empty by the assumption that X and Y are rivals, there is a world $w \in X \setminus Y$; and because X is P-stable, by assumption again, for all $w \in X \setminus Y$ it holds that $P(\{w\}) > P(W \setminus X)$; since $W \setminus X \supseteq Y \setminus X$, it follows that $P(\{w\}) > P(Y \setminus X)$ and hence also that $P(X \setminus Y) > P(Y \setminus X)$. With (vi) this implies for all Z that respect π: If $X \subseteq Z$ (and so with (iv) $Y \subseteq F(Z)$) then $P(Z) \geq P(F(Z))$ and thus by P1, $err_P(Bel(Z)) \leq err_P(Bel(F(Z)))$. And if $X \cap Z = \varnothing$ (and so with (iv) $Y \cap F(Z) = \varnothing$), then $P(F(Z)) \geq P(Z)$ and thus by P2, $err_P(Dis(F(Z))) \geq err_P(Dis(Z))$. Moreover, at least for $Z = X \setminus Y$ even the strict inequality $err_P(Bel(Z)) < err_P(Bel(F(Z)))$ is satisfied. Taking these findings together with (v), by the definition of overall error in terms of sums of individual errors, it follows that $Err_{\pi,P}(X) < Err_{\pi,P}(Y)$.

Secondly, there does not exist a partition π of W, such that X and Y respect π, $|X|_\pi = |Y|_\pi$, and $Err_{\pi,P}(Y) < Err_{\pi,P}(X)$: This follows from the previous proof if considered from step (**) above, as what we showed there was that for all partitions π of W, such that X and Y respect π, and $|X|_\pi = |Y|_\pi$, it holds that $Err_{\pi,P}(X) < Err_{\pi,P}(Y)$.

So we have that X produces fewer errors than Y (relative to P), by the definition of 'produces fewer errors than'.

For (2): If X is not P-stable, then there must be a $w \in X$, such that $P(\{w\}) \leq P(W \setminus X)$. Let Y be defined as $[X \setminus \{w\}] \cup [W \setminus X] (= W \setminus \{w\})$: It follows that X is a rival to Y. Since $X \setminus Y = \{w\}$, the only partition π of W, such that X and Y respect π, and $|X|_\pi = |Y|_\pi$ is such that $|X \setminus Y|_\pi = |Y \setminus X|_\pi = 1$. Analogously to (1) above, it follows for this partition π, from the fact that $P(X \setminus Y) = P(\{w\}) \leq P(W \setminus X) = P(Y \setminus X)$, that $Err_{\pi,P}(X) \geq Err_{\pi,P}(Y)$. Hence, it is not the case that X produces fewer errors than Y (relative to P).

For (3): Let P be as described in (*) (in claim 3 in the statement of the theorem above), and assume X not to be P-stable: there must be a $w \in X$ again, such that $P(\{w\}) \leq P(W \setminus X)$, and by (*) this must actually be a strict inequality: $P(\{w\}) < P(W \setminus X)$. If Y is defined as in (2), then, analogously to what was shown before, there is just one partition π such that X and Y respect π and $|X|_\pi = |Y|_\pi$, but now the stronger statement $Err_{\pi,P}(X) > Err_{\pi,P}(Y)$ follows. Therefore, Y produces fewer errors than X (relative to P). □

Note that the main statement in (3) is a strengthening of (2) for the class of probability measures that satisfies condition (*) in (3). For a given finite W, this condition is actually satisfied by almost all probability measures P (what this means can be made precise again in terms of the Lebesgue measure of geometrical representatives of probability measures on W).

In the final section, we are going to discuss and illustrate the consequences of this theorem.

14.3 Conclusions: A Precise Norm on Belief and Probability

In the last section we answered our original question *how could we measure the "error" of using a set B_W as one's set of doxastically possible worlds relative to the subjective probability measure P?* The answer was: in terms of $Err_{\pi,P}$. Accordingly, we defined a notion of *produces fewer errors than ... relative to P* which enabled us to compare different choices of such sets B_W as far as the overall error was concerned by which they are affected relative to P (taking into account certain salient partitions π of the set of possible worlds).

Now what does the theorem from the last section tell us about how to choose one's set B_W of doxastically possible worlds relative to a degree of belief function P? Condition (*) in (3) holds for almost all probability measures on a given finite W, as mentioned at the end of the last section; if P is one of these measures, then the situation can be summarized as follows: If a set B_W is P-stable, then by (1) B_W does better than any rival Y in terms of overall error. But if a set B_W is not P-stable, then one can do better in terms of overall error: for by (3) there is a rival of B_W that produces less error than B_W if used as a simplification of P. Summing up: if condition (*) is satisfied, choosing a set B_W of doxastically possible worlds so that it is P-stable is the only way of minimizing overall error amongst the rivals of B_W.

Let us illustrate this in terms of an example:

Example 14.6 Let $W = \{w_1, \ldots, w_8\}$ be a set of eight possible worlds. Let P be the probability measure on the power set algebra on W that is given by: $P(\{w_1\}) = 0.54$, $P(\{w_2\}) = 0.342$, $P(\{w_3\}) = 0.058$, $P(\{w_4\}) = 0.03994$, $P(\{w_5\}) = 0.018$, $P(\{w_6\}) = 0.002$, $P(\{w_7\}) = 0.00006$, $P(\{w_8\}) = 0$. Figure 14.1 depicts this finite probability space.

The (non-empty) P-stable sets in this case are:

$$\{w_1\}, \{w_1, w_2\}, \{w_1, \ldots, w_4\}, \{w_1, \ldots, w_5\}, \{w_1, \ldots, w_6\}, \{w_1, \ldots, w_7\}$$

only the last of which has probability 1.

Fig. 14.1 An example
measure

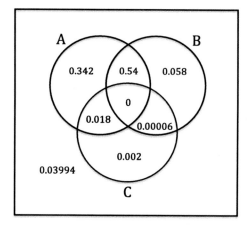

Let π be the trivial partition of W in terms of singleton sets. Now consider, e.g.,

$$Err_{\pi,P}(\{w_1, w_2\}) = \sum_{Z:\{w_1,w_2\}\subseteq Z} err_P(Bel(Z)) + \sum_{Z:\{w_1,w_2\}\cap Z=\varnothing} err_P(Dis(Z)).$$

It holds that $\{w_2, w_3\}$ is a rival of $\{w_1, w_2\}$, both sets trivially respect π, $|\{w_1, w_2\}|_\pi = |\{w_2, w_3\}|_\pi = 2$, and

$$Err_{\pi,P}(\{w_2, w_3\}) = \sum_{Z:\{w_2,w_3\}\subseteq Z} err_P(Bel(Z)) + \sum_{Z:\{w_2,w_3\}\cap Z=\varnothing} err_P(Dis(Z)).$$

$\{w_1, w_2\}$ is P-stable, while $\{w_2, w_3\}$ is not, and indeed

$$Err_{\pi,P}(\{w_1, w_2\}) < Err_{\pi,P}(\{w_2, w_3\}),$$

as can be seen from comparing the errors for single Zs along the function F that was defined (on the basis of an f) in the proof of the theorem from the last section. For instance: $F(\{w_1, w_2, w_5\}) = \{w_3, w_2, w_5\}$, $\{w_1, w_2\} \subseteq \{w_1, w_2, w_5\}$, $\{w_2, w_3\} \subseteq \{w_3, w_2, w_5\}$, and

$$err_P(Bel(\{w_1, w_2, w_5\})) < err_P(Bel(\{w_3, w_2, w_5\}))$$

by P1 from the last section, because $P(\{w_1, w_2, w_5\}) > P(\{w_3, w_2, w_5\})$. Accordingly, e.g., $F(\{w_3, w_5\}) = \{w_1, w_5\}$, $\{w_1, w_2\}\cap\{w_3, w_5\} = \varnothing$, $\{w_2, w_3\}\cap\{w_1, w_5\} = \varnothing$, and

$$err_P(Dis(\{w_3, w_5\})) < err_P(Dis(\{w_1, w_5\}))$$

by P2 from the last section, since $P(\{w_3, w_5\}) < P(\{w_1, w_5\})$. Overall: $\{w_1, w_2\}$ produces fewer errors than $\{w_2, w_3\}$ (relative to P).

Here is another set X to consider: $\{w_1, w_2, w_3\}$; this set is not P-stable. Now we construct a set Y as described in (2) and (3) of the theorem from the last section: e.g., since $P(\{w_3\}) < P(W \setminus \{w_1, w_2, w_3\})$, let $Y = [\{w_1, w_2, w_3\} \setminus \{w_3\}] \cup [W \setminus \{w_1, w_2, w_3\}] = W \setminus \{w_3\}$: X is a rival of Y, $P(X \setminus Y) < P(Y \setminus X)$, and if we use the trivial partition π into singleton sets again, it holds that $Err_{\pi, P}(\{w_1, w_2, w_3\}) > Err_{\pi, P}(W \setminus \{w_3\})$. Y produces fewer errors than X (relative to P).

Amongst all P-stable sets, of course, none can produce fewer errors than the largest P-stable set, that is, $\{w_1, \ldots, w_7\}$. But this should not be interpreted in the way that a rational person with degree of belief function P ought to have $\{w_1, \ldots, w_7\}$ as her set of doxastically accessible worlds; for none of the other P-stable sets is a rival of $\{w_1, \ldots, w_7\}$, and so it would be misleading to compare the errors that they produce with the error produced by $\{w_1, \ldots, w_7\}$; taking $\{w_1, \ldots, w_7\}$ to determine one's belief would simply be the most cautious option that is available to an agent who obeys the rationality requirements developed in this chapter, but that does not in itself necessitate $\{w_1, \ldots, w_7\}$ to be the "right" choice.

Let us take stock: we searched for a precisification of the final norm of section one,

- It should be the case that the set B_W of doxastically possible worlds minimizes error relative to P.

If this norm is made precise in terms of the concepts of the last section, then based on the theorem from the last section, this leads us to the following precise norm for a particular class of subjective probability measures:

- If P satisfies condition (*) in claim (3) from the theorem in the last section, it should be the case that the set B_W of doxastically possible worlds is P-stable.

And since condition (*) is almost always satisfied (for a given non-empty and finite set W of possible worlds), the norm really amounts to:

- Almost always it should be the case that the set B_W of doxastically possible worlds is P-stable.

As mentioned before, this is not a reduction of belief (B_W) to probability (P). For instance, if belief is, in some sense, prior to probability, then given any choice $B_W \subseteq W$ of doxastically possible worlds, the norm would entail that one ought to determine a probability measure P on W, such that B_W ends up being P-stable. The simplest such P would assign positive probability to the (singletons of) members of X and zero probability to the (singletons of) members of $W \setminus X$. But if B_W is a proper subset of W, one can show that infinitely many other measures P would do the job as well. However, by far not all measures P would do so: that is, the norm above does have bite even when belief is prior to probability.

Moreover, if neither belief nor subjective probability is prior to the other, then we have found, and justified, an interesting and formally precise bridge norm between belief and probability that imposes a rationality constraint on belief and degrees of belief *jointly*. And it is a norm that derives ultimately from a specific understanding of belief as a simplification of probability.

References

1. Benferhat S, Dubois D, Prade H (1997) Possibilistic and standard probabilistic semantics of conditional knowledge. J Logic Comput 9:873–895
2. Hempel Carl G (1962) Deductive-nomological versus statistical explanation. In: Feigl H, Maxwell G (eds) Minnesota studies in the philosophy of science III. University of Minnesota Press, Minneapolis
3. Leitgeb H (2013) Reducing belief simpliciter to degrees of belief. Ann Pure and Appl Logic 164:1338–1389
4. Leitgeb H (2014) The stability theory of belief. Philos Rev 123(2):131–171
5. Leitgeb H (Unpublished) On the best approximation of probability by belief
6. Levi I (1967) Gambling with the truth. An essay on induction and the aims of science. MIT, Cambridge
7. Skyrms B (1977) Resiliency, propensities, and causal necessity. J Philos 74(11):704–713
8. Snow P (1998) Is intelligent belief really beyond logic? In: Proceedings of the eleventh international Florida artificial intelligence research society conference. American Association for Artificial Intelligence, pp 430–434

Part III
Games

Chapter 15
Logic and Game Theory

Giacomo Bonanno and Cédric Dégremont

Abstract Johan van Benthem has highlighted in his work that many questions arising in the analysis of strategic interaction call for logical and computational analysis. These questions lead to both formal and conceptually illuminating answers, in that they contribute to clarifying some of the underlying assumptions behind certain aspects of game-theoretical reasoning. We focus on the insights of a part of the literature at the interface of game theory and mathematical logic that gravitates around van Benthem's work. We discuss the formal questions raised by the perspective consisting in taking games as models for formal languages, in particular modal languages, and how eliminative reasoning processes and solution algorithms can be analyzed logically as epistemic dynamics, and discuss the role played by beliefs in game-theoretical analysis and how they should be modeled from a logical point of view. We give many pointers to the literature throughout the chapter.

15.1 Introduction

In the past twenty years the interface between game theory and logic has grown at a fast pace. Johan van Benthem has been taking a very active part in developing this relationship both directly, in terms of his writings, and indirectly, by seeding ideas that have been inspiring developments at the interface between the two fields. While we refer to [32], parts IV–VI, for a discussion of the use of game-theoretical concepts to understand fundamental logical concepts, we are delighted to contribute

G. Bonanno (✉)
Department of Economics, University of California at Davis, Davis, USA
e-mail: gfbonanno@ucdavis.edu

C. Dégremont
IRIT, Université Paul Sabatier, Toulouse, France
e-mail: cedric.uva@gmail.com

A. Baltag and S. Smets (eds.), *Johan van Benthem on Logic
and Information Dynamics*, Outstanding Contributions to Logic 5,
DOI: 10.1007/978-3-319-06025-5_15, © Springer International Publishing Switzerland 2014

to this volume in honor of Johan van Benthem by focusing on some of his insights and contributions to game theory from a logical perspective.

These contributions have both a formal and conceptual aspect. They are formal in that they show how models of strategic interaction and game-theoretical concepts can be embedded in a broader formal context and analyzed from the perspective and with the tools of mathematical logic. They are also conceptual in that they contribute to clarifying some of the underlying assumptions behind certain aspects of game-theoretical reasoning. The late Michael Bacharach was one of the first to appreciate the usefulness of logic in reasoning about games:

> Game theory is full of deep puzzles, and there is often disagreement about proposed solutions to them. The puzzlement and disagreement are neither empirical nor mathematical but, rather, concern the meanings of fundamental concepts ('solution', 'rational', 'complete information') and the soundness of certain arguments (that solutions must be Nash equilibria, that rational players defect in Prisoner's Dilemmas, that players should consider what would happen in eventualities which they regard as impossible). Logic appears to be an appropriate tool for game theory both because these conceptual obscurities involve notions such as reasoning, knowledge and counterfactuality which are part of the stock-in-trade of logic, and because it is a prime function of logic to establish the validity or invalidity of disputed arguments ([6], p. 21).

For example, the tools of modal logic have made it possible to give an explicit formulation to concepts that were previously stated either informally or indirectly, such as the notion of rationalizability [33, 79] as an expression of the notion of common belief in rationality (see [85] and, for an overview of the epistemic foundations of game-theoretic solution concepts, [14]). The writings of Johan van Benthem have been equally useful in pointing out new insights that modal logic can contribute to the analysis of games.

This chapter is organized as follows. The next section has some background and preliminaries that the reader might consider skipping at a first reading. In Sect. 15.3, we discuss several ideas put forward by van Benthem: identifying modal languages to reason about extensive games, showing how they can be interpreted and how they can characterize important classes of games and strategically stable strategies. We also show briefly how the approach bridges game theory and computational analysis. Finally we point out how van Benthem's ideas shed new light on the question of under what conditions two games can be considered the same. In Sect. 15.4 we present the ideas developed by van Benthem [23], showing how eliminative reasoning processes and solution algorithms can be analyzed logically as principles of dynamic epistemic logic and under what conditions the convergence of their iteration can be analyzed in fixed-point modal languages. Section 15.5 discusses the role played by revisable beliefs in game-theoretical analysis and how they should be modeled from a logical point of view. Building on this, and following recent results by van Benthem and Gheerbrant [26] we discuss how backward induction can be given different interpretations and, especially, how all can be proven equivalent from the perspective of fixed-point first-order languages.

Johan van Benthem's own views on the relationship between logic and game theory are expressed in [22].

15.2 Preliminaries

15.2.1 Notation

Let S be a set and let A be a finite set and for each $a \in A$, let $R_a \subseteq S \times S$. We let $\wp(S)$ be the power set of S and we let $\#S$ be the cardinality of S. We let R_A^* be the reflexive transitive closure of $\bigcup_{a \in A} R_a$, so that $sR_A^* t$ if and only if either $s = t$ or there is a finite A-path from s to t. We also write R^* for R_A^* when A is clear from the context. ω is the set of natural numbers.

15.2.2 Game Theory

We assume some basic familiarity with the concept of a strategic game and of an extensive game with (im)perfect information. We will define concepts formally but a reader completely unfamiliar with game theory might like to consult an introduction to game theory such as [75].

15.2.3 Basic Modal Logic

Let τ be a non-empty countable set. Let PROP be a non-empty countable set (of propositional letters). The basic modal language $ML(\tau, PROP)$ is recursively defined as follows:

$$\varphi ::= p \mid \neg \varphi \mid \varphi \vee \varphi \mid \langle a \rangle \varphi$$

where p ranges over PROP and a over τ. A model for $ML(\tau, PROP)$ is a relational structure $\mathbb{M} = \langle W, (R_a)_{a \in \tau}, V \rangle$ where W is a non-empty set, $R_a \subseteq W \times W$ and $V : PROP \to \wp(W)$. $(W, (R_a)_{a \in \tau})$ is called a τ-frame. We also write $|\mathbb{M}| = W$.

Definition 15.1 (*Semantics of ML(τ, PROP)*) We interpret $ML(\tau, PROP)$ on pointed relational models as follows:

$$
\begin{array}{ll}
\mathbb{M}, w \models p & \text{iff } w \in V(p) \\
\mathbb{M}, w \models \neg \varphi & \text{iff it is not the case that } \mathbb{M}, w \models \varphi \\
\mathbb{M}, w \models \varphi \vee \psi & \text{iff } \mathbb{M}, w \models \varphi \text{ or } \mathbb{M}, w \models \psi \\
\mathbb{M}, w \models \langle a \rangle \varphi & \text{iff there is some } v \text{ with } wR_a v \text{ and } \mathbb{M}, v \models \varphi \\
\mathbb{M}, w \models [a]\varphi & \text{iff for all } v \text{ such that } wR_a v \text{ we have } \mathbb{M}, v \models \varphi
\end{array}
$$

where $p \in PROP$, $a \in \tau$. We will write $\top$ for $p \vee \neg p$ and $\bot$ for $\neg \top$. Other connectives $(\wedge, \to, \leftrightarrow)$ are defined in the usual way.

Given a model $\mathbb{M} = \langle W, (R_a)_{a \in \tau}, V \rangle$ and a formula $\varphi \in ML(\tau, PROP)$ we write $||\varphi||^{\mathbb{M}} := \{w \in W \mid \mathbb{M}, w \models \varphi\}$. Whenever $\mathbb{M}$ is clear from the context, we simply write $||\varphi||$ for $||\varphi||^{\mathbb{M}}$. Given a class C of relational models, we write $C \models \varphi$ whenever for every $\mathbb{M} \in C$ and $w \in |\mathbb{M}|$ we have $\mathbb{M}, w \models \varphi$ and we say that φ is valid on C. The same notion for classes of frames is defined by universally quantifying over the possible valuations of (the relevant) propositional letters. Satisfiability is the dual, existential counterpart to validity, that is, φ is satisfiable over C iff $\neg\varphi$ is not valid over C.

Definition 15.2 (*Bisimulation*) A local bisimulation between two pointed relational models, $(\mathbb{M}, w)$ and $(\mathbb{M}', w')$, with $\mathbb{M} = \langle W, (R_a)_{a \in \tau}, V \rangle$ and $\mathbb{M}' = \langle W', (R'_a)_{a \in \tau}, V' \rangle$ is a binary relation $Z \subseteq W \times W'$ such that wZw' and also for any pair of worlds $(x, x') \in W \times W'$, whenever xZx' then for all $a \in \tau$:

1. x, x' verify the same proposition letters.
2. if $xR_a u$ in $\mathbb{M}$ then there exists $u' \in W'$ with $x'R'_a u'$ and uZu'.
3. if $x'R'_a u'$ in $\mathbb{M}'$ then there exists $u \in W$ with $xR_a u$ and uZu'.

We say that $\mathbb{M}, w$ and $\mathbb{M}', w'$ are bisimilar $(\mathbb{M}, w \underline{\leftrightarrow} \mathbb{M}', w')$ if there exists a local bisimulation between $\mathbb{M}, w$ and $\mathbb{M}', w'$. We say that $\mathbb{M}$ and $\mathbb{M}'$ are bisimilar $(\mathbb{M} \underline{\leftrightarrow} \mathbb{M}')$ if there are $w \in W$ and $w' \in W'$ such that $(\mathbb{M}, w) \underline{\leftrightarrow} (\mathbb{M}', w')$.

15.2.4 Epistemic Logic

An interesting special case of a modal logic as defined above is epistemic logic. We only briefly recall the basic concepts of epistemic logic. For a more exhaustive introduction to epistemic logic, the reader can consult, e.g., [55, ch. 2]. Relational structures can compactly represent the information agents have about the world and about the information possessed by the other agents.

Definition 15.3 An *epistemic model* is a relational structure $(W, N, (\sim_i)_{i \in N}, V)$ where N is a finite set and for each $i \in N$, $\sim_i$ is a binary equivalence relation on W.

To explicitly talk about knowledge one may use the language of basic epistemic logic.

Definition 15.4 (*Syntax of $\mathscr{L}_{\mathrm{EL}}$*) The syntax of epistemic language $\mathscr{L}_{\mathrm{EL}}$ is recursively defined as follows:

$$\varphi := p \mid \neg\varphi \mid \varphi \vee \varphi \mid K_i\varphi$$

where $p \in \mathrm{PROP}$, $i \in N$. We write $\langle i \rangle \varphi$ for $\neg K_i \neg\varphi$.

We also write $\mathscr{L}_{\mathrm{EL}}(N, \mathrm{PROP})$, when we need to clarify the intended set N and the intended set PROP. The semantics of $\mathscr{L}_{\mathrm{EL}}$ is as expected and we only give the modal truth clause.

$$\mathbb{M}, w \models K_i\varphi \text{ iff for all } v \text{ such that } w \sim_i v \text{ we have } \mathbb{M}, v \models \varphi$$

Table 15.1 EL (also called $S5_N$) axiom system

PL	$\vdash \varphi$ if φ is a substitution instance of a tautology of propositional logic

For $i \in N$,

Nec	if $\vdash \varphi$, then $\vdash K_i\varphi$
K	$\vdash K_i(\varphi \to \psi) \to (K_i\varphi \to K_i\psi)$
T	$\vdash K_i\varphi \to \varphi$
4	$\vdash K_i\varphi \to K_iK_i\varphi$
5	$\vdash \neg K_i\varphi \to K_i\neg K_i\varphi$
MP	if $\vdash \varphi \to \psi$ and $\vdash \varphi$, then $\vdash \psi$

Table 15.2 Axiom system **MEL**

$\mathbf{C_GFP} \vdash C_G\varphi \to \bigwedge_{i \in G} K_i(\varphi \wedge C_G\varphi)$
$\mathbf{C_GIR}$ If $\vdash \varphi \to \bigwedge_{i \in G} K_i(\varphi \wedge \psi)$ then $\vdash \varphi \to C_G\psi$

Standard definitions such as truth sets, satisfiability and validity are of course a special case of the ones given in the previous section. Epistemic logic is fully axiomatized by the axiom system given in Table 15.1.

We write $K_i[w] := \{v \in W \mid w \sim_i v\}$. For any non-empty group of agents $G \subseteq N$ we write $R_G^*[w] := \{v \in W \mid w \sim_G^* v\}$. Let φ be a formula of epistemic logic. If $R_G^*[w] \subseteq ||\varphi||$ then for any $n \in \omega$ and sequence $i_0, \ldots, i_{n-1}$ with range G, $K_{i_0} \ldots K_{i_{n-1}} \varphi$ holds at w. If the conjunction of all such finite sequences is true at w, it intuitively means that φ is common knowledge at w. But this conjunction is not finitary. We can introduce a new formula $C_G\varphi$, for each $G \subseteq N$, with semantics

$$\mathbb{M}, w \models C_G\varphi \text{ iff for all } v \text{ such that } w \sim_G^* v \text{ we have } \mathbb{M}, v \models \varphi$$

We call the resulting logic $\mathscr{L}_{MEL}$ (for multi-agent epistemic logic). $\mathscr{L}_{MEL}$ is obviously no longer compact, but the logic is still invariant under basic bisimulations [18]. **Axiomatization.** The set of formulas of $\mathscr{L}_{MEL}$ valid over the class of all epistemic models can be axiomatized by extending EL with the axioms in Table 15.2.

Fagin et al. [55, ch. 2] has a completeness proof for **MEL**.

15.2.5 Model Theory

We assume some basic familiarity with first-order logic (**FO**). For an introduction see e.g., [52].

Given an operator: $F : \wp(U) \to \wp(U)$ we say that:

1. F is monotone, if for all $X, Y \subseteq U$ whenever $X \subseteq Y$ we have $F(X) \subseteq F(Y)$

2. F is inflationary, if $X \subseteq F(X)$ for all $X \subseteq U$
3. F is deflationary, if $F(X) \subseteq X$ for all $X \subseteq U$.

Given an arbitrary operator $F : \wp(U) \to \wp(U)$, let F_{infl} (F_{def}) be the inflationary (respectively deflationary) operator associated with F, defined as follows: $F_{infl}(X) = X \cup F(X)$ (respectively $F_{def}(X) = X \cap F(X)$). Note that for an inflationary (respectively a deflationary) operator F we have $F = F_{infl}$ (respectively $F = F_{def}$). If F has a least (greatest) fixed point, we denote it by $\mathbf{lfp}(F)$ (respectively $\mathbf{gfp}(F)$). Consider the sequence defined by:

$$X^0 = \emptyset; X^\lambda = F_{infl}\left(\bigcup_{\eta < \lambda} X^\eta\right)$$

It can be shown that this sequence is inductive and stabilizes at some ordinal $\kappa \leq \#U$ (see [73]). Call $\mathbf{ifp}(F) = X^\kappa$ the inflationary fixed point of F. The deflationary fixed point of F is defined analogously as the limit of the sequence

$$X^0 = U; X^\lambda = F_{def}\left(\bigcap_{\eta < \lambda} X^i\right)$$

Theorem 15.1 *(Knaster-Tarski Every monotone operator F has a least fixed point* $\mathbf{lfp}(F)$ *and a greatest fixed point* $\mathbf{gfp}(F)$. *Moreover,*

$$\mathbf{lfp}(F) = \bigcap\{X \subseteq U \mid F(X) \subseteq X\}$$

$$\mathbf{gfp}(F) = \bigcup\{X \subseteq U \mid X \subseteq F(X)\}$$

Let R be an n-ary relation symbol, let $\bar{x}$ be an n-tuple of variables and let $\bar{t}$ be a n-tuple of terms. We say that an occurrence of S is positive, if it is in the scope of an even number of negations. We say that a formula $\varphi(R, \bar{x})$ is positive in R if all occurrences of R are positive.

- FO(LFP) is the extension of FO with least fixed points. Formally it extends FO with the following formation rule: if $\varphi(R, \bar{x})$ is a formula *positive in R*, then $[\mathbf{lfp}_{R,\bar{x}}\varphi(R, \bar{x})](\bar{t})$ is a formula. Where $M \models [\mathbf{lfp}_{R,\bar{x}}\varphi(R, \bar{x})](\bar{a})$ iff $\bar{a} \in \mathbf{lfp}(F_\varphi)$.
- FO(IFP) is the extension of FO with inflationary fixed points. Formally it extends FO with the following formation rule: if $\varphi(R, \bar{x})$ is a formula, then $[\mathbf{ifp}_{R,\bar{x}}\varphi(R, \bar{x})](\bar{t})$ is a formula. Where $M \models [\mathbf{ifp}_{R,\bar{x}}\varphi(R, \bar{x})](\bar{a})$ iff $\bar{a} \in \mathbf{ifp}(F_\varphi)$.

Theorem 15.2 (Main Theorem of [62]) *For every FO(LFP) formula $\varphi(R, \bar{x})$, there is an FO(LFP) formula $\varphi^*(R, \bar{x})$ which is equivalent on all finite structures to* $[\mathbf{ifp}_{R,\bar{x}}\varphi(R, \bar{x})](\bar{t})$.

Corollary 15.3 [62] *FO(LFP) = FO(IFP) over finite structures.*

Theorem 15.4 [70] *For every FO(LFP) formula $\varphi(R, \bar{x})$, there is an FO(LFP) formula $\varphi^{\infty}(R, \bar{x})$ which is equivalent on all structures to $[\mathbf{ifp}_{R,\bar{x}}\varphi(R, \bar{x})](\bar{t})$.*

Corollary 15.5 [70] *FO(LFP) = FO(IFP) over all structures.*

15.2.6 Computability and Computational Complexity

We assume that the reader is familiar with the concept of a Turing machine. Our introduction will be somewhat informal and the reader is referred to [76] for a complete presentation of these topics. We refer to a set of (encodings of) inputs as a language. We say that a language L is *recursive* if there exists a Turing machine M that halts on an input w and accepts it whenever $w \in$ L, and halts and rejects the input otherwise. We say that a language L is *recursively enumerable* if there exists a Turing machine M that halts on an input w and accepts it whenever $w \in$ L, and either halts and rejects, or does not halt otherwise.

Besides computability, we are interested in those languages that can be recognized by Turing machines using limited number of computation steps or limited amount of working-tape cells. Somewhat informally speaking—for precise definitions, see [76]—given a function $f : \omega \to \omega$, let DTIME(f) (respectively NTIME(f)) be the class of languages which can be decided by a deterministic Turing machine in at most $f(n)$ steps (respectively by a non-deterministic Turing machine M such that all branches in the computation tree of M on x are bounded by $f(n)$) for any input x of size n with $n \geq n_0$ for some $n_0 \in \omega$. DSPACE(f) (respectively NSPACE(f)) is the class of languages which can be recognized by a deterministic Turing machine using (respectively by a non-deterministic Turing machine M such that, on all branches in the computation tree of M on x, it uses) at most $f(n)$ cells of the working-tape, for inputs of size $n \geq n_0$ for some constant $n_0 \in \omega$. We write

- PTIME $= \bigcup_{k \in \omega}$ DTIME(n^k)
- EXPTIME $= \bigcup_{k \in \omega}$ DTIME(2^{n^k})
- NP $= \bigcup_{k \in \omega}$ NTIME(n^k)
- PSPACE $= \bigcup_{k \in \omega}$ SPACE(n^k).

15.3 Games are Process Models

Games are process models: in two influential papers [19, 20] van Benthem proposes using modal languages to represent the internal structure of dynamic (or extensive-form) games. This starting point comes with important questions such as:

1. in what sense is a game a relational (epistemic, temporal) model for a modal language?

2. what classes of extensive forms can be modally characterized and in what language?
3. what is the computational complexity of the satisfiability problem for logics over epistemic temporal models?
4. what is the right notion of invariance for games?

15.3.1 Interpreting Epistemic-Temporal Languages Over Games

Modal languages can be naturally interpreted over dynamic games. In [19] van Benthem describes an extensive form as a tuple

$$\langle S, I, A, \{R_a\}_{a\in A}, \{\sim_i\}_{i\in I} \rangle$$

where S is a set of states (the nodes of the game tree), I is the set of players, A is the set of actions and, for every $a \in A$ and $i \in I$, R_a and $\sim_i$ are binary relations on S. If s is a decision node and $(s, t) \in R_a$ then there is a transition from node s to node t as a consequence of action a being taken by the player assigned to node s. Thus $\bigcup_{a\in A} R_a$ constitutes the game tree and the set of nodes $s \in S$ such that $(s, t) \in R_a$ for some $t \in S$ and $a \in A$ is the set of decision nodes. For every player $i \in I$, $\sim_i$ is an equivalence relation representing the state of information of the player at different stages of the game. As van Benthem notes (van Benthem [19], p. 229), this is an extension of the traditional definition of an extensive-form game where the uncertainty relation of player i is defined only on the set of decision nodes assigned to player i. This issue of specifying the information of a player also at decision nodes that belong to *other* players had earlier been studied in [13, 37, 82]. Finally, adding a valuation V that associates with every propositional letter $p \in$ PROP the set of nodes at which p is true, yields a *model* of the extensive-form game.

Among the atomic propositions van Benthem includes sentences such as $turn_i$, which is true precisely at the decision nodes assigned to player i (where it is player i's turn to move). Given a model, one can associate with every uncertainty relation $\sim_i$ a modal operator K_i with the interpretation of $K_i\varphi$ as "player i knows φ" and with the usual semantics (see Sect. 15.2.4). Similarly, with every transition relation R_a one can associate a modal operator $[a]$ with the interpretation of $[a]\varphi$ as "after action a it is the case that φ" with the expected semantics (cf. Sect. 15.2.3).

As usual, we can then try and determine which properties of our models can be characterized, at the level of models, but also—and this is naturally where modal logic's strength lies—at the level of frames. For the reader unfamiliar with modal logic, let us stress the difference: on the level of models, the modal language can surely not distinguish between a state that has at most one a-successor and a state that has many a-successors, unless these states satisfy different modal formulas. The following result explains this fact:

Theorem 15.6 (van Benthem [17]) *A formula of first-order logic is equivalent to the translation of a formula of modal logic iff it is invariant under bisimulations.*

However, determinacy of actions can be captured on the level of frames:

$$\langle a \rangle \varphi \;\rightarrow\; [a]\varphi$$

is valid on a class of frames iff these frames satisfy a-determinacy. In particular $\langle a \rangle \varphi \rightarrow [a]\varphi$ is valid a state w in some frame iff w has at most one a-successor in that frame.

15.3.2 Perfect Recall and von Neumann Extensive Forms

As van Benthem points out, game-theoretical assumptions such as: (1) 'all the nodes in the same information set have the same possible actions' and (2) 'a player knows when it his turn to move' can be characterized by the formula

$$turn_i \,\wedge\, \langle a \rangle \top \;\rightarrow\; K_i(turn_i \wedge \langle a \rangle \top).$$

Of particular interest is van Benthem's suggestion that the property of 'perfect recall' (defined below), which is traditionally incorporated in the definition of extensive form, can be expressed by the formula

$$turn_i \wedge K_i[a]\varphi \;\rightarrow\; [a]K_i\varphi \tag{vB}$$

which is very appealing, since it based on a simple commutation of the epistemic operator and the dynamic operator.

It turns out that van Benthem's two suggestions (to extend a player's uncertainty relation $\sim_i$ beyond player i's decision nodes and to characterize the property of perfect recall in terms of axiom (vB) are intimately connected and implicitly identify the subclass of extensive forms known as multi-stage or *von Neumann extensive forms*. Von Neumann extensive forms are defined (see [71], p. 52) by the property that any two decision nodes that belong to the same information set of a player have the same number of predecessors.

In order to make this more precise, we need a few definitions. Let S_i denote the nodes assigned to player i (player i's decision nodes) and let $R^* = R_A^*$. The property of perfect recall is defined as follows[1]:

For every player $i \in I$, for all nodes $t, y, y' \in S_i$ and $x \in S$ and for every action a, if $tR_a x$, xR^*y and $y \sim_i y'$ then there exist nodes $t' \in S_i$ and $x' \in S$ such that $t \sim_i t'$, $t'R_a x'$ and $x'R^*y'$. $\tag{PR}$

---

[1] The following definition is Selten's [84] reformulation of Kuhn's [71] original property which was stated in terms of pure strategies.

Kuhn [71] interpreted this property as "equivalent to the assertion that each player is allowed by the rules of the game to remember everything he knew at previous moves and all of his choices at those moves". Clearly, (PR) implies the following property, which captures the notion that at any of his decision nodes player i remembers what he knew at earlier decision nodes of his:

$$\text{If } t, y \in S_i \text{ and } tR^*y, \text{ then for every } y' \text{ such that } y \sim_i y' \qquad (KM)$$
$$\text{there exists a } t' \in S_i \text{ such that } t \sim_i t' \text{ and } t'R^*y'.$$

If, following van Benthem's suggestion, the uncertainty relation $\sim_i$ of player i is extended from S_i to the entire set S then it is natural to require that the memory property (KM) be preserved by the extension, that is, one would require the extended relation $\sim_i$ to satisfy the following property:

$$\text{If } tR^*y \text{ and } y \sim_i y', \text{ then there exists a } t' \text{ such that } t \sim_i t' \text{ and } t'R^*y'. \qquad (KM_{EXT})$$

Then we have the following result (see [38], inspired by [19]):

Proposition 15.1 *Fix an arbitrary extensive-form game G that satisfies property (KM). Then*
(a) there exists, for every player i, an extension of $\sim_i$ from S_i to S that satisfies (KM_{EXT}) if and only if G is a von Neumann extensive form,
(b) if G is a von Neumann extensive form then G satisfies (PR) if and only if axiom (vB) is valid in G relative to an extended relation $\sim_i$ that satisfies (KM_{EXT}).[2]

When the extensive form is not von Neumann, then a syntactic characterization of perfect recall is still possible, but it involves a slightly more complex axiom which contains an additional operator (corresponding to the relation R^*: see [38]).

Another line of analysis is concerned with identifying the epistemic-temporal properties characterizing certain types of epistemic updaters. For instance, product updaters ([10], see also Chap. 6 in this volume), are typically characterized by a form of perfect recall and a form of uniformity. We refer to [27, 30, 67] for more details on this line of research.

15.3.3 Backward Induction in Logic

The preceding modal languages could (indirectly) characterize classes of extensive forms of interest. Putting preferences and strategies into the picture, with corresponding modalities: with $sR_\sigma t$ meaning that t is the continuation of s given that players follow the strategy profile σ, van Benthem [29] shows how 'backward induction' as a property of a relation (induced by a profile of strategies) can be modally characterized by a simple PDL (see e.g., [54]) formula:

---

[2] A formula is valid in extensive form G if is true at every $s \in S$ in every model based on G.

Proposition 15.2 [29] *In generic extensive games, the relation σ is induced by the unique backward induction profile iff it satisfies the following axiom for all* $i \in N$:

$$(turn_i \wedge \langle \sigma^* \rangle (end \wedge p)) \rightarrow [move_i] \langle \sigma^* \rangle (end \wedge \langle \leq_i \rangle p)$$

The formula is essentially saying that a player cannot *unilaterally* deviate from σ at any stage in a way that can make her strictly better off. van der Hoek and Pauly [66] has more about similar definability results in the logic literature. In Sect. 15.5, we put this question in its broader mathematical perspective: that of fixed-point logics interpreted on trees [26, 59] and also discuss backward induction from the perspective of inductive reasoning and inductive belief update.

15.3.4 Existence of Extensive Games

If we reverse the perspective, instead of asking whether a given epistemic-temporal property holds of an extensive game with imperfect information, that is if the formula is true at a certain state in a certain epistemic-temporal model, we can ask whether we can construct a strategic situation respecting a collection of constraints. The problem is known in logic, and more generally theoretical computer science, as a satisfiability problem. Let L be a modal language and let C_P be the set of extensive temporal models that satisfy a property P. (Note that P could also be a collection of such properties.) Formally, the set of validities of L over C_P is the set $\{\varphi \in L : C_P \models \varphi\}$. The satisfiability problem for a modal language L over a class of models C_P, is to decide given any formula $\varphi \in L$ whether $\{M, w \mid M \in C_p, w \in |M|, M, w \models \varphi\} \neq \emptyset$, in which case the answer is positive. The following are important questions at the interface of logic and computer science:

1. Is the set of validities of L over C_P recursively enumerable?
2. If it is, is it recursive?
3. Is the satisfiability problem for L over C_P in EXPTIME? Is it in PSPACE?
4. Is it complete for these classes?

The answer to the first question would be positive, if we could identify a complete finite set of axioms. For a positive answer to the second question, on top of the previous axiomatization, we could show how to construct a model for any finite consistent set $S \subseteq L$, of size bounded by some $f(|S|)$. Negative answers to these questions can be proved by reduction from acceptance problems for Turing machines or from recurring tiling problems (see [64]). Interestingly, van Benthem and Pacuit [28] surveys how—depending on different assumptions we are making about epistemic-temporal agents (such as, e.g., perfect recall, no learning, synchronicity...)—the satisfiability problem of epistemic-temporal languages will lie on either side of the decidability border. One of the most important papers concerned with the assumptions that make the satisfiability problem of epistemic-temporal languages undecidable is Halpern and Vardi [63].

The third and fourth questions come, in a sense, second: when one is certain that a problem is algorithmically decidable, one can focus on exactly how many resources (time and working tape space) are required to decide it. In [28] van Benthem and Pacuit survey also such results and give pointers to the literature. Such results can ultimately be interpreted as describing the computability and the difficulty of deciding whether a list of game-theoretical assumptions are coherent with each other and whether it is possible to find a game satisfying a list of desirable constraints. In that sense the results discussed in the previous sections suggest that modal logics interpreted over game structures are a natural way to allow for the application of computational results to game theory.

Finally note that the satisfiability problem of the basic modal language is already PSPACE-complete, hence not considered tractable. By contrast, checking if a formula holds at some state in an epistemic-temporal model is tractable for the types of modal logics we have considered so far: for the most expressive of them, PDL, it can be done in a number of steps polynomial in the size of the formula and of the model. In Sect. 15.4 we will see that the logical analysis of strategic reasoning calls for more expressive fixed point logics, whose model-checking problem need not be tractable over arbitrary structures.

15.3.5 When are Two Extensive Forms the Same?

Besides asking the question "what are appropriate formal languages for games?", van Benthem [19] also raises the important question "when are two extensive forms the same?". A related question is: when is a transformation of an extensive form "inessential"? These are questions that could be explored further than they have been in the literature, and two immediate approaches come to mind.

For the logician, if a language has been fixed, the question is about finding the right notion of invariance. For modal languages, some adequate notion of bisimulation is usually the answer. For first-order languages, their fixed-point extensions and existential second-order languages, a winning strategy for Duplicator in some form of Ehrenfeucht-Fraïssé games is the answer [53, 56]. Hence we could have such a game to decide whether the difference between two games is essential or not. Johan van Benthem has mentioned this idea in talks (mentioning also the converse direction: interpreting languages over satisfiability games or over evaluation games, hence we could have a formula describing, indirectly, another formula).

For the game-theorist, the classical approach has been quite different: the issue being to define a different notion of game form, to show that every extensive form can be mapped into the proposed game form and to declare two extensive forms to be equivalent when they are mapped into the same "new" game form. This was done in the literature by mapping extensive forms into reduced normal forms [47, 68, 69, 86, 87] or into set-theoretic forms [36]. In both cases a corresponding set of "inessential" transformations of extensive forms were identified. As [36] puts it, these mappings offer a notion of *descriptive*, rather than strategic equivalence.

Fig. 15.1 Equivalent powers
for 1 and 2 in each game

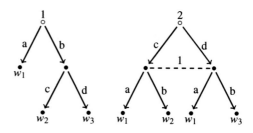

Interestingly, [19, 20] proposes also another notion of equivalence based on players' powers. For this purpose one needs to distinguish between terminal nodes and outcomes. Thus to the standard definition of extensive form one would add a set of outcomes W and a map f from the set of terminal nodes Z to W (if we take $Z = W$ and f to be the identity function, then we have the standard definition of extensive form). We can then say that the set $X \subseteq W$ of outcomes belongs to the *powers* of player i if player i has a strategy[3] that forces the play of the game to end at a terminal node associated with an outcome in X. For example, in Fig. 15.1 we have two different extensive forms (in particular, they have different sets of terminal nodes) which share the same set of outcomes $W = \{w_1, w_2, w_3\}$ and the same powers for each player (the powers of player 1 are $\{w_1\}$ and $\{w_2, w_3\}$ and the powers of player 2 are $\{w_1, w_2\}$ and $\{w_1, w_3\}$).[4]

Two extensive forms can then be defined to be "the same", or equivalent, if the powers of every player are the same in both. For example, the extensive forms of Fig. 15.1 are equivalent. As van Benthem notes ([20], p. 13) this is a notion of equivalence based on the associated "outcome-level normal form". It should also be noted that if instead of extensive forms one considers extensive-form *games* (obtained by associating with every terminal node a payoff for each player) and one identifies outcomes with payoff vectors, then the proposed equivalence coincides with the equivalence based on the reduced normal form [68, 87]. Indeed, as van Benthem notes ([19], p. 244, Proposition 6), the powers of the players remain the same under the Thompson [87] transformations.

So far the analysis has been restricted to the powers of the players at the root of the tree. However, one can similarly define the powers of a player at any node: the set $X \subseteq W$ of outcomes belongs to the powers of player i *at node* s if player i has a strategy that forces any play of the game that goes through node s to end at a terminal node associated with an outcome in X. van Benthem then defines a modal operator $\langle i \rangle$ for every player i with the intended interpretation of $\langle i \rangle \varphi$ as "player i has the power to bring about φ". Given a game G and a model $\mathbb{M}^G$ of G (obtained by adding a valuation that specifies which atomic propositions are true at every node) the validation rule for $\langle i \rangle \varphi$ at node s is thus: $\mathbb{M}^G, s \models \langle i \rangle \varphi$ if $\mathbb{M}^G, t \models \varphi$ for every $t \in X$ where X is one of the powers of player i at s. This approach thus uses the so-called neighborhood

---

[3] In the game theorist's sense of the word, namely a function that assigns to every information set of player i an action at that information set. van Benthem calls such objects 'uniform strategies'.

[4] The dotted line in the extensive-form on the right represents an information set of Player 1.

semantics for modal logic [46], whose universal validities are all the principles of the minimal modal logic except for distribution of $\langle i \rangle \varphi$ over disjunctions. One can then express interesting properties of games by means of this modal language. For example, van Benthem notes that the "consistency" property, according to which if X is a power of player i and Y is a power of player j at some node s then it must be that $X \cap Y \neq \varnothing$, can be characterized by the formula $\langle i \rangle \varphi \rightarrow \neg \langle j \rangle \neg \varphi$.

This is an interesting perspective on extensive-form games that deserves to be studied in more detail. In the logic literature, Coalition Logic [77], Game Logic [78], Alternating-time Temporal Logic [2], STIT (for "seeing to it that", [16])—and NCL [44] have semantics in that spirit. The reader can consult the above references for more information about them and consult [66] for a survey.

15.4 Reasoning in Games: Rational Dynamics

In order to determine reasonable and/or plausible outcomes for games under given epistemic assumptions, one needs an adequate view of how players will reason from their information to reach a decision. If reasoning is traditionally the object of logic, it is so in an external way: a finite set of valid principles are proven to be everything that agents need to draw all conclusions they need to draw about their environment. If one is interested in the consequences of an agent's current information, this is the relevant level of analysis. In the context of strategic interaction this is generally not enough, since we are also interested in the semantic processes corresponding to how agents update their beliefs when they receive new information, make additional assumptions and draw consequences, and even iterate such processes. In particular we are interested in the convergence of such reasoning processes. This is the subject matter of [23]: a logical approach to such reasoning processes, and the theoretical limits of any such approach. In this section, we follow closely the analysis in this chapter, factoring in our own, possibly different, way of looking at the topic.

Let us present the general program. As we have seen in the previous section, a game can be seen as a relational model for a modal language. More generally the epistemic aspects of a strategic situation can be described by a relational model that encodes players' preferences, players' information, and the actions they can take. A modal formula (of some sufficiently expressive language) can encode a notion of *rationality* based on the previous notions. Now assuming the rationality of the players, states in which the formula is not satisfied can be eliminated and the formula can be recursively interpreted in such submodels. To each formula corresponds a mapping on models, whose fixed-points we can hope to define in some fixed-point modal language. Moreover, for any formula and any game we might ask which (profiles of) strategies survives, given some assumptions about the epistemic model of the given game. Conversely, we might ask whether there exists some epistemic model satisfying certain properties such that a certain profile of strategies (or a certain strategy) survive the inductive elimination process. All these questions are both very natural from the point of view of mathematical logic and theoretical computer science, and the implementation of very natural questions in epistemic game theory. We illustrate this with the example of iterative solution algorithm for strategic games.

15.4.1 Epistemic Models of Games

A strategic game is a formal representation of a multi-agent decision situation, in which two or more agents have to make a decision, independently (rather than sequentially), 'that will influence one another's welfare' [74].

Definition 15.5 (*Strategic game*) A strategic game is a structure of the form

$$\Gamma = \langle N, (A_i)_{i \in N}, (\geq_i)_{i \in N} \rangle.$$

where N is a non-empty finite set of players, for each $i \in N$, A_i is a non-empty finite set of strategies and $\geq_i$ is a total preorder over $A = \times_{j \in N} A_j$.

Note that a strategic game is not by itself a model for some epistemic logic, but it can easily be made so. Let, for example, $\text{PROP}_i = A_i$ and let $\text{PROP}_\Gamma = \bigsqcup_{i \in N} A_i$. In [23] the result is called the full model over Γ and is defined as follows:

Definition 15.6 (*Full epistemic model over Γ*) The full epistemic model over

$$\Gamma = \langle N, (A_i)_{i \in N}, (\geq'_i)_{i \in N} \rangle$$

is the multi-agent S5(N) epistemic model

$$M(\Gamma) = \langle W, N, (\sim_i)_{i \in N}, (\geq_i)_{i \in N}, V \rangle$$

expanding Γ with

$$
\begin{array}{lll}
W & = & \times_{i \in N} A_i \\
((a_i)_{i \in N}, (b_i)_{i \in N}) \in \sim_i & \text{iff} & a(j) = b(j) \\
(a_i)_{i \in N} \in V(a_j^k) & \text{iff} & a(j) = a_j^k
\end{array}
$$

In words, the epistemic equivalence relation for j partitions the set of strategy profiles depending on the strategy used by j in that profile. In [31] van Benthem et al. also advocate the use of a modality for action freedom $[\approx_i]\varphi$ with the box semantics corresponding to the relation $\approx_i$ defined as follows:

$$a \approx_i b \text{ iff } a_{-i} = b_{-i}$$

—and very similar in spirit to (c)stit operators (see e.g., [16]) and NCL's [i] operator [44]—and coined by [31] 'action freedom modality' after a concept introduced in a talk by Jeremy Seligman.

The full epistemic model has notable non-epistemic properties, such as exactly one strategy is played at each state:

$$\bigwedge_{i \in N} \left(\bigvee_{p_i \in \mathrm{PROP}_i} p_i \wedge \bigwedge_{p_i, p_j \in \mathrm{PROP}_i, p_i \neq p_j} \neg (p_i \wedge p_j) \right) \qquad (\psi_\Gamma)$$

Fact 15.7 *Let $M(\Gamma)$ be the full epistemic model for some strategic game Γ, we have* $M(\Gamma) \models \psi_\Gamma$

Such models will also satisfy strategic introspection:

$$\bigwedge_{i \in N} \bigwedge_{p_i \in \mathrm{PROP}_i} (p_i \to K_i p_i) \qquad (\chi_\Gamma)$$

Fact 15.8 *Let $M(\Gamma)$ be the full epistemic model for some strategic game Γ. For every $p_i \in \mathrm{PROP}_i$, we have $M(\Gamma) \models \chi_\Gamma$*

But as far as higher-order knowledge is concerned, agents have very limited information in full epistemic models. To see that, we say that a formula $\varphi \in \mathscr{L}_{\mathrm{MEL}}(N, \mathrm{PROP}_\Gamma)$ is Γ-consistent if it is **MEL**$(C_G(\psi_\Gamma), C_G(\chi_\Gamma))$-consistent. The following is a variation on results in [18, 23].

Proposition 15.3 *Let Γ be a strategic game. Any Γ-consistent existential formula of the multi-agent epistemic language $\varphi \in \mathscr{L}_{\mathrm{MEL}}(N, \mathrm{PROP}_\Gamma)$ can be satisfied at some state in the full epistemic model $M(\Gamma)$.*

Proof (Sketch of the proof.) Existential formulas are equivalent to disjunctions of path formulas. Such a formula is satisfiable if such a path exists in $M(\Gamma)$ which can be ensured by two conditions: every state on the path should be propositionally satisfiable in $M(\Gamma)$ (this is what consistency with $C_G(\psi_\Gamma)$ ensures) and transitions should respect the epistemic grid structure of the game: i-transitions should preserve i-atoms (this is what consistency with $C_G(\chi_\Gamma)$ ensures). $\square$

Note that the preceding result comes in different flavors: if the language is richer—for example if it can express preferences or the intersection of basic relations—satisfiability will only be guaranteed for sets of formulas that are consistent with formulas of that language corresponding to the structural properties that any game model will satisfy. In other words valid properties on all game models that are invariant under the notion of bisimulation corresponding to a selected language, will need to be accounted for. But the core idea is the same: existential formulas will find a pointed full epistemic model in which they are satisfied.

However, in general, we might be interested in epistemic situations in which agents have non-trivial higher-order information. The looser the notion of an epistemic situation corresponding to a game—that is the less we would like to preserve of the epistemic structure of $M(\Gamma)$—the larger the collection of satisfiable sets of formulas. As an example, in submodels that are preserving the grid-like structure of the game, formulas that are inconsistent with the scheme

$$\langle i \rangle \langle j \rangle \varphi \leftrightarrow \langle j \rangle \langle i \rangle \varphi$$

will not be satisfiable. At the opposite end of the scale, we could work with the class of **KD45**-structures that preserve very minimal structural properties such as coherence of strategies and strategic introspection. Any $(\Gamma, \textbf{KD45})$-consistent set of formulas would then be satisfiable.

15.4.2 Assuming Rationality

Now that we have fixed our models, let us go back to our initial question. How can we model the reasoning processes of agents? Or, simply, to start with, how we can model a single reasoning step? A reasoning should essentially transform an epistemic model into another epistemic model. This is a very general statement. But there is also a very rich diversity of epistemic updates that can be modeled in logics of epistemic dynamics [7].

As far as the current analysis—following [23]—is concerned, we will for now restrict attention to what is arguably the simplest and most natural type of update: relativization. We restrict a model to the set of states that satisfy a certain formula. It is probably in the context of epistemic analysis that relativization is easiest to interpret: it is the result of a public announcement, whose reliability is not challenged by the agents. In this case, the information is "hard information" [31]. Softer types of information would in particular include information from only partially reliable sources, information that the agents consider as revisable (more on this issue in Sect. 15.5). But more than the result of a single step of eliminative reasoning, it is interesting to know what happens to game models if we recursively iterate such eliminative steps. Before we proceed, we will need a bit of notation. Given a relational model $M = \langle W, (R_a)_{a \in \tau}, V \rangle$ and a set $A \subseteq W$, let

$$M|_A = \langle A, (R'_a)_{a \in \tau}, V' \rangle$$

where $R'_a = R_a \cap (A \times A)$ and $V'(p) = V(p) \cap A$ for each $a \in \tau$ and $p \in$ PROP. We also write $M|_\varphi$ or M^φ for $M|_{\|\varphi\|^M}$.

Public announcement logic [7, 58, 81] is an extension of basic epistemic logic with public announcement operators $\langle \varphi \rangle \psi$ with semantics

$$M, w \models \langle \varphi \rangle \psi \text{ iff } M, w \models \varphi \text{ and } M|_\varphi, w \models \psi$$

Public announcement logic is actually exactly as expressive as basic epistemic logic. Now given a formula φ and a game Γ, inductively define a sequence of models

$\sigma(\Gamma, \varphi) = (M_\iota)_{\iota < \gamma}$ as follows:

$$M_0 = M(\Gamma); \quad M_\lambda = \left(\bigcap_{\eta < \lambda} M_\eta\right)^\varphi$$

We can ask two questions:

1. Is ψ true at some state in M_ι for some $\iota < \gamma$?
2. Is ψ true at some state in M_κ for the least ordinal κ such that $M_\kappa = (\bigcap_{\beta < \kappa} M_\beta)^\varphi$?

Here appears another direction of the program of applying logical analysis to games. In Sect. 15.3, modal logic was used to make explicit subphenomena of importance for strategic interaction, to identify the (model-theoretic, computational) properties of these logics that are relevant for game-theoretic analysis. Here the perspective is somewhat different, we abstract away from solution algorithms to analyze reasoning in games as a special case of reasoning about iterated relativization of relational structures in general.

The first question is concerned with iterated relativization [72]: is it the case that at any stage in the inductive sequence (that is at any step of the reasoning process) ψ holds? (We will address this question subsequently). The second question is concerned with the limit of iterated relativization. The first important observation to make is the following:

Proposition 15.4 [23] *Let φ be a modal formula. The limit of iterated φ-relativization is definable in modal iteration calculus, that is, in inflationary fixed-point modal logic.*

Proof (Sketch of the proof.) [23] The idea of the proof is to consider the relativization of φ to a fresh propositional variable X, $(\varphi)^X$. Now $M, w \models (\varphi)^X$ iff $M|_{V(X)}, w \models \varphi$. Hence the fixed-point of the deflationary induction for $X \leftarrow (\varphi)^X$ is the limit of iterated φ-relativization.   □

For arbitrary modal formulas, we cannot, in general, improve on this result and find an equivalent formula in the weaker modal μ-calculus. Consider the modal formula: $\varphi(a, b) := \langle a \rangle \top \vee (r \leftrightarrow [b] \bot)$ and consider labeled transition systems, with labels in $\{a, b\}$.

Proposition 15.5 [60] $(\mathbf{dfp}\, X \leftarrow (\varphi(a, b))^X)$ *is not equivalent to any MSO-formula.*

Proof (Sketch of the proof.) Grädel and Kreutzer [60] Define $T(n, m)$ to be a tree with two branches at the root: a branch consisting of n a-steps and a branch of m b-steps. Let r be true at the root. The idea of the proof is that the root survives in the deflationary fixed point of $X \leftarrow (\varphi(a, b))^X$ iff $n \geq m$ (see Fig. 15.2). But no finite tree automaton can accept $T(n, m)$ iff $n \geq m$. On trees, this is equivalent to the fact that there is no MSO-formula corresponding to $(\mathbf{dfp}\, X \leftarrow (\varphi(a, b))^X)$.   □

The undefinability of the limit of iterated relativization then follows from the fact that the modal μ-calculus is a fragment of MSO. However, van Benthem [23] shows the following:

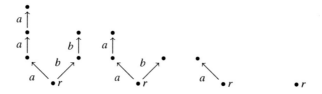

Fig. 15.2 $T_0 = T(3, 2)$, $T_1 = T(3, 2)|_{\varphi(a,b)}$, $T_2 = T_1|_{\varphi(a,b)}$, $T_3 = T_1|_{\varphi(a,b)} = T_3|_{\varphi(a,b)}$

Proposition 15.6 [23] *If φ is existential, the φ-relativization mapping is monotone.*

Hence, for an existential formula φ, the limit of iterated φ-relativization can be defined in the modal μ-calculus. What does this tell us about games? It tells us that reasoning about the limit of an assumption of rationality is equivalent to model checking a formula of the modal μ-calculus whenever the formula encoding this concept of rationality is existential, and that in general it is equivalent to model checking a formula of the modal iteration calculus (We refer to [48] for a presentation of modal iteration calculus and a comparison with the modal μ-calculus.). An important difference is that, while model-checking problem for the μ-calculus could be tractable Dawar et al. [48] show that the combined (and expressive) complexity of MIC is PSPACE-complete.

Let us now discuss the definability of iterated relativization. As suggested by van Benthem [21] iterated relativization is expressible in the modal iteration calculus. Miller and Moss [72] define a logic of iterated relativization extending public announcement with iterated public announcement operators $\langle \varphi^* \rangle \psi$ with semantics:

$$M, w \models \langle \varphi^* \rangle \psi \text{ iff } M, w \models \langle \varphi \rangle^n \psi \text{ for some } n \in \omega$$

and give a translation from the language of iterated relativization into the modal iteration calculus. Moreover, [72] shows that the satisifiability problem of the logic is highly undecidable (Σ_1^1-complete) by reduction from the tiling problem for recurring domino systems.

15.5 The Different Faces of Backward Induction

Backward induction (henceforth BI) in generic games of perfect information seems at first a very simple solution algorithm with limpid epistemic foundations. If it is common belief between Azazello and Behemoth that they will both play best-responses to their beliefs at every subgame, then in particular Azazello believes that Behemoth will play an action that maximizes his utility in subgames of length 1, hence will play according to the BI solution. Iterating the argument seems to provide us with the conclusion that BI play follows rationality and common belief of rationality. For a formal defense of this claim, the reader should consult [4]. Even if this was the

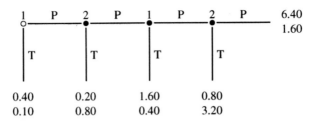

Fig. 15.3 A centipede game

end of the story, the logical analysis of this correspondence using fixed-point logics [26, 59] would still be insightful, and we will return to it. But there is more to this story. Consider the game in Fig. 15.3.

The BI solution (play T at every decision node) can be justified as follows: if 2 is rational, when the last node will be reached, 2 will play T. Since 1 expects 2 to play rationally, 1 should expect 2 to play T at the last node, hence if 1 is rational, 1 should play T at the penultimate node. The arguments iterates and 1 is argued to have a reason to play T at the first node, on the basis of common belief of rationality. Now, backward induction and any theory of rational behavior in general should be

immune to deviation from it. That is, it must never be to one's advantage to behave in a manner that the theory deems irrational. But in order to check this, one must be able to evaluate the effect of *not* conforming to the theory [83].

In particular a theory of rational behavior should have something to say about how rational agents should revise their beliefs when they observe decisions which are incompatible with the theory and how they should make decision after observing such a deviation. Now look back at the example, and assume that 1 deviates from the BI-path and plays P at the first node. What should 2 expect 1 would do in case 2 were to play P? We let the reader decide for herself or himself. Many results have shown sufficient or insufficient assumptions on the belief revision procedure used by the agents to guarantee BI compatible behavior. Our aim is not to survey them (we refer to [80] for such a discussion). We will also not cover related conceptual questions such as how beliefs should be modeled, what type of beliefs a player can have on her future and current decisions, and what types of counterfactual reasoning can be involved in strategic reasoning; and refer to [41, Sect. 4] for an overview in the context of an analysis of the epistemic foundations of BI. Rather, our aim, in this section, is to indicate how the question has triggered logical developments calling for logical analysis of concepts we did not discuss so far: (counterfactual) beliefs—and belief revision.

With this motivation in mind and drawing on both modal logic relational semantics and semantic models developed in the context of AGM [1] style belief revision theory (such as [61] spheres), Board [35] proposes a modal language interpreted over plausibility-based structures. Independently, several authors in the logic literature proposed similar models [8, 24, 51]. All these models have to do with the following idea: besides (or instead of) an epistemic relation giving the information of an agent at every state of a model or at every history in a tree (or an extensive form), introduce a plausibility pre-order $\leq_i$: where $x \leq_i y$ means x is at least as plausible as y from the perspective of i. Belief operators can then be interpreted in different ways. Typically, $B_i\varphi$ (read "i believes that φ") can be interpreted as meaning that all $\leq_i$-minimal elements in i's information set are φ-states. As the reader might suspect, some form of well-foundedness of the plausibility relation will then be a desirable feature for this belief operator to be well-defined. Some authors only require existence of minimal elements in cells of the information partition, which only calls for a local form of well-foundedness. Other authors prefer the $\leq_i$ to be a state-dependent relation and write $y \leq_{i,x} z$ giving them greater generality. This generally calls for additional assumptions if certain forms of positive or negative introspection are desired. But, throughout all these variations, the general idea remains the same.

We don't need to be more formal for now and we will proceed as follows. We start in Sect. 15.5.1 by illustrating the previous plausibility models with a foundational problem for interactive epistemology (in the sense of [5]): agreements and convergence to agreements. We also discuss a non-eliminative revision procedure: radical upgrade. We then return in Sect. 15.5.3 to backward induction from where we left it earlier and discuss the unifying analysis of BI in fixed-point logics developed in van Benthem and Gheerbrant [26], Gheerbrant [59]. Section 15.5.4 discusses a sequence of results by van Benthem and Gheerbrant [26] making explicit the link between strategic reasoning as a non-eliminative revision procedure and the backward induction solution.

15.5.1 Plausibility Models for the Interactive Epistemologist

Let us record the definition of an epistemic plausibility model as discussed above.

Definition 15.7 (*Epistemic Plausiblity Model*, [8]) An *epistemic plausibility model* $\mathbb{M} = \langle W, (\leq_i)_{i \in N}, (\sim_i)_{i \in N}, V \rangle$ has $W \neq \emptyset$, for each $i \in N$, $\leq_i$ is a pre-order on W and $\sim_i$ is a binary equivalence relation on W, and $V : \text{PROP} \to \wp(W)$.

Since we would like to define belief as truth in minimal states in an information cell, we need to make sure that such minimal elements do exist. We call an epistemic-plausibility model $\mathbb{M} = \langle W, (\leq_i)_{i \in N}, (\sim_i)_{i \in N}, V \rangle$ well-founded iff for every subset $X \subseteq W$, X has minimal elements. Clearly this condition is sufficient to guarantee that a belief operator $B_i\varphi$ with semantics:

$$M, w \models B_i\varphi \text{ iff } \min_{\leq_i} K_i[w] \subseteq ||\varphi||^M$$

is well-defined. In such models the plausibility ordering really encodes prior beliefs, while $\sim_i$ encodes the information of i, hence the above operator is really a posterior belief operator (in the sense of [3] posteriors). Similarly to the probabilistic case, for any $n \in \omega$, it is possible to construct a pointed epistemic-plausibility model M, w with common prior such that for all sequences σ of length $k \leq n$ over $\{1, 2\}$:

$$M, w \models K_{\sigma_0} \ldots K_{\sigma_k}(B_i p \wedge \neg B_j p)$$

However, again similarly to the probabilistic case, two agents cannot 'agree to disagree'.

Theorem 15.9 [50] *Common knowledge of disagreement is only possible in a model that is either not well-founded or for which the assumption of a common prior fails.*

Additionally it is possible to see that relativization via beliefs, that is, by public announcement of the beliefs of the agents about some formula φ, will converge to common knowledge of beliefs about φ, and hence, in well-founded models that satisfy common prior, to agreement. φ-relativization, as we mentioned earlier, is an eliminative type of update. It corresponds to the epistemic event in which all agents accept φ as true information, whose reliability cannot be put into question. Van Benthem [24] is concerned with softer types of updates, corresponding to information that can possibly turn out to be wrong. One procedure discussed in [24] is *radical plausibility upgrade*. It is more easily understood in the context of simple plausibility models (without epistemic relations). Radical plausibility upgrade with φ simply takes every φ-state in the plausibility ordering and puts them above all $\neg\varphi$-states. Within these two classes, the ordering of states remain unchanged. Call the resulting model $M \Uparrow \varphi$. It is possible to introduce a corresponding dynamic operator $[\Uparrow \varphi]\psi$ with semantics

$$M, w \models [\Uparrow \varphi]\psi \text{ iff } M \Uparrow \varphi, w \models \psi$$

The logic can be fully axiomatized by extending the axiomatization of some conditional doxastic logic interpreted over plausibility models with dynamic axioms. We refer to van Benthem [24] for details, but let us point out an important difference with the logic of public announcement (PAL), that the reader might expect: PAL validates the following axiom

$$[!p][!\neg p]B_i \bot$$

for propositional letters. This is no longer true for radical upgrade. The opposite is actually true

$$[\Uparrow p][\Uparrow \neg p]\neg B_i \bot$$

given any reasonable semantics of B_i, making radical plausibility upgrade a better candidate for iteration.

15.5.2 Belief Revision Over Time

As we have seen before, one update, one revision is usually not giving the full story. Much of our earlier analysis has a broader impact, now that we have semantics for beliefs and an approach to belief revision. It would take us out of the scope of this chapter to discuss such extensions in full details, but let us sketch some important questions that arise now that we are working with belief revision rather than knowledge update. First note that, in the same way that epistemic temporal models can be seen as generalizations (and, from a different perspective, as enrichments) of extensive forms, doxastic temporal models are similarly interestingly related to extensive games. Doxastic temporal logics interpreted over such models were introduced in Friedman and Halpern [57] and Bonanno [39], representing time globally as a bundle of possible histories where the beliefs of agents evolve as informational processes unfold.

As a detour, we should note that plausibility models and probability-based approaches are quite different in spirit. The plausibility structure is essentially concerned with different layers the agent can fall back on, in case her initial beliefs are defeated by new information, while probability approaches are concerned with the relative likelihood of different alternatives. The latter offer a rich basis for fine-grained decision-making rules, while the first one is robust to surprising information. There are of course systems at the interface of the two—such as lexicographic probability systems [34]—that have proven useful for the analysis of epistemic foundations of solution concepts (see e.g., [43]).

Now the dynamic approach discussed in the previous section can also be extended to deal with sequences and repetition of belief revision. Van Benthem and Dégremont [25], Dégremont [49] discuss the relation between the temporal and the dynamic approach to belief update, and logics at their interface. Iterated scenarios as discussed in Sect. 15.4 can be revisited for the more sophisticated type of updates we have just discussed. Baltag and Smets [9] has some important pioneering results and Baltag et al. [12], as well as Hendricks et al. [65], show their relevance for a logical approach to learning theory.

15.5.3 Unifying Perspectives on Backward Induction: Fixed-Point Logic on Trees

Our earlier analysis ended by mentioning the definability of backward induction over trees in PDL. Again think of backward induction (henceforth BI), as a subset of the successor relation on trees. Van Benthem and Gheerbrant [26], Gheerbrant [59] are interested in the unification of characterization of BI using extensions of FO with fixed points. The formula used in the previous result can be shown to correspond to a local concept of rationality expressible in FO with transitive closure for binary relations and the mentioned references have details. Van Benthem and Gheerbrant prefer however a different notion of rationality they call CF2 (for confluence):

$$\text{CF2} : \bigwedge_{i \in N} \forall x \forall y ((turn_i(x) \wedge \sigma(x, y)) \rightarrow (move(x, y) \wedge \forall z (move(x, z)$$
$$\rightarrow \exists u \exists v (end(u) \wedge end(v) \wedge \sigma^*(y, v) \wedge \sigma^*(z, u) \wedge u \leq_i v))))$$

and show that the relation BI—or a more permissive one on arbitrary games, yet equivalent on generic games—can be defined in FO(LFP) as a greatest-fixed point. But first they prove the corresponding semantic result:

Theorem 15.10 [26, 59] *BI is the largest subrelation of the move relation in a finite game tree satisfying the two properties that (a) the relation has a successor at each non-terminal node, and (b) CF2 holds.*

Let X be a relational symbol *not* in the above vocabulary. Syntactically, the definability of their brand of (the relation) BI in FO(LFP) is as follows:

$$BI(x, y) = [\mathbf{gfp}_{X,x,y}(move(x, y) \wedge \bigwedge_{i \in N} (turn_i(x) \rightarrow \forall z(move(x, z)$$
$$\rightarrow \exists u \exists v(end(u) \wedge end(v) \wedge X^*(y, v) \wedge X^*(z, u) \wedge u \leq_i v))))](x, y)$$

where $X^*(y, v)$ means that there exists an X-path from the interpretation of y to the interpretation of v, which is naturally definable in FO(LFP).

Interestingly inductively computing the interpretation of BI(x,y) [26, 59] in a given game tree is essentially equivalent to inductively computing a backward induction type solution algorithm. This illustrates that both the static and the dynamic perspective on games can ultimately be unified in the fixed point logic approach to both of them.

15.5.4 Backward Induction and Iterated Plausibility Upgrade

Let us go back to the example represented in Fig. 15.3. If you expect players to conform to BI at any stage of the game, you expect in particular that 1 will play T at the first node. In general, the left to right ordering really corresponds to an ordering in terms of plausibility given that you expect players to play according to BI. In van Benthem and Gheerbrant's [26] words "Backward Induction creates expectations for players". How it creates them, is something we have yet to see. It is very reasonable to expect the BI procedure would generate such a plausibility ordering inductively.

Before we proceed, we will need a bit of notation and terminology. Assume some finite extensive game. Let $Z(x)$ be the set of terminal nodes that can still be reached from x. Let Z_1, Z_2 be sets of terminal nodes of some finite tree. Given a total ordering $\leq$ over the terminal nodes and its complement $>$, we write $Z_1 >_{\forall\forall} Z_2$ iff for all $z_1 \in Z_1$ and $z_2 \in Z_2$ we have $z_1 > z_2$. Define $Z_1 \leq_{\forall\forall} Z_2$ similarly. Now call Z_1, Z_2 *ancestor-connected* iff is there is a node x with two-children y and z such that $Z(y) = Z_1$ and $Z(z) = Z_2$. Very roughly speaking, at x, the player who is to move decides between the set Z_1 and the set of Z_2.

Van Benthem and Gheerbrant define a relation of plausibility over leaves of a finite extensive game as a total ordering $\leq$ of the terminal nodes. Given a set of terminal nodes Z_1, let $B_i[Z_1] := \min_{\leq_i} Z_1$. Now consider the following notion of belief-based dominance.

Definition 15.8 ([26, 59]) Given a plausibility relation $\preceq_i$, we say that a move to a node x for player i, $\preceq_i$ *dominates a move to a sibling y of x in beliefs* if $B_i(Z(x)) >_{\forall\forall} B_i(Z(y))$. A move to x is said to be *rational*, if it is not dominated in beliefs by a move to a sibling.

Now inductively evaluate formulas in larger subgames starting from subgames of length 1. Assume the current subgame has a root x with at least two children z and y, if z dominates y in belief z, then upgrade $Z(z)$ over $Z(y)$. Call this procedure iterated radical upgrade of rationality in belief.

Theorem 15.11 [59] *On finite games, the BI strategy is encoded in the final plausibility ordering at the limit of iterated radical upgrade of rationality in belief.*

A different approach, building on a similar semantic framework, is taken in [11]. It uses the concept of 'stable true belief'. Essentially, a belief in φ is a true stable belief, if it is known to be robust to truthful announcements (robust to non-trivializing relativization). Their main result is that common stable true belief of rationality implies the BI outcome. They provide a logical characterization of that result in the sense that the previous theorem is a validity in some modal language. The reader should consult Baltag et al. [11] for details. In general, this line of research has the particularity of having a genuinely syntactic dimension to it. The idea of recasting game-theoretic arguments in proof-theoretic terms is something that we have omitted: the interested reader should consult [40, 45, 88].

15.6 Perspectives

We have seen that natural questions arising within game-theoretical analysis call for logical and computational analysis. Van Benthem et al. [31] argue that the meaning of this is certainly not only that some problems in game theory would require tools from logic, model theory and computational complexity to be solved. Rather, for van Benthem et al. [31], intelligent interaction, as constituted of informational processes—such as revising beliefs, adjusting strategy, changing goals or preferences —is the object of an emerging "more finely-structured theory of rational agency", that they think of as a "joint off-spring [...] of logic and game theory" and call "theory of play". In that sense, the results and analyses we have discussed in the previous sections, are elements of such a general theory of intelligent interaction, in which game-theoretical, logical and computational methods are simultaneously called for.

15.7 Conclusion

Showing that many problems in strategic interaction are ultimately logical and computational problems is one of the directions of Johan van Benthem's explorations in the past decade. Not only are games natural models for modal logics, allowing for enlightening characterization of classes of games or of relational structures that generalize games. But it is also possible to model the reasoning procedures of agents about games and, the convergence of these procedures. From a logical perspective, the analysis of these problems resides within the expressive power of fixed point logics. The logical analysis projects into computational analysis: from the computational perspective, the latter problems are much more demanding. Unlike properties definable in modal logic, checking if a given fixed-point logic definable property holds about a game is generally not tractable. And if we move to the satisfiability problem, we find that we cross the computability border (if we have not crossed it already by making dangerous assumptions about our models). Hence van Benthem's contribution to this direction of the interface between logic and games has really been two-fold: on the one hand, isolating subphenomena of importance for strategic interaction (such as belief revision) and making their principles explicit by logical analysis; and, on the other hand, putting games into a broader mathematical picture, giving a unifying logical point of view at which the correspondence between static and dynamic approaches to games and their solution naturally appears as two faces of the same mathematical object.

References

1. Alchourrón CE, Gärdenfors P, Makinson D (1985) On the logic of theory change: partial meet contraction and revision functions. J Symbolic Logic 50(2):510–530
2. Alur R, Henzinger TA, Kupferman O (1997) Alternating-time temporal logic. In: Proceedings of the 38th IEEE symposium on foundations of computer science
3. Aumann RJ (1976) Agreeing to disagree. Ann Stat 4(6):1236–1239
4. Aumann RJ (1995) Backward induction and common knowledge of rationality. Games Econ Behav 8(1):6–19
5. Aumann RJ (1999) Interactive epistemology I: knowledge. Int J Game Theory 28(3):263–300
6. Bacharach M (1994) The epistemic structure of a theory of a game. Theory Decis 37:7–48
7. Baltag A, Moss LS (2004) Logics for epistemic programs. Synthese 139(2):165–224
8. Baltag A, Smets S (2006) Dynamic belief revision over multi-agent plausibility models. In: Bonanno G, van der Hoek W, Wooldridge M (eds) Logic and the foundations of game and decision theory: proceedings of LOFT'06, TLG, AUP, pp 11–24
9. Baltag A, Smets S (2009) Group belief dynamics under iterated revision: fixed points and cycles of joint upgrades. In: Heifetz A (ed) TARK, pp 41–50
10. Baltag A, Moss LS, Solecki S (1998) The logic of public announcements, common knowledge, and private suspicions. In: TARK '98: proceedings of the 7th conference on theoretical aspects of rationality and knowledge, Morgan Kaufmann Publishers Inc., San Francisco, pp 43–56
11. Baltag A, Smets S, Zvesper JA (2009) Keep 'hoping' for rationality: a solution to the backward induction paradox. Synthese 169(2):301–333
12. Baltag A, Gierasimczuk N, Smets S (2011) Belief revision as a truth-tracking process. In: Apt KR (ed) TARK, ACM, pp 187–190
13. Battigalli P, Bonanno G (1997) Synchronic information, knowledge and common knowledge in extensive games. In: Bacharach M, Gérard-Varet LA, Mongin P, Shin H (ed) Epistemic logic and the theory of games and decisions, Theory and decision library, Kluwer Academic, pp 235–263. Reprinted in (Battigalli and Bonanno, 1999b)

14. Battigalli P, Bonanno G (1999a) Recent results on belief, knowledge and the epistemic foundations of game theory. Res Econ 53(2):149–225
15. Battigalli P, Bonanno G (1999b) Synchronic information, knowledge and common knowledge in extensive games. Res Econ 53(1):77–99
16. Belnap N, Perloff M (2001) Facing the future: agents and choices in our indeterminist world. Oxford University Press, Oxford
17. van Benthem J (1983) Modal logic and classical logic. Bibliopolis, Naples
18. van Benthem J (1997) Exploring logical dynamics. Center for the Study of Language and Information, Stanford
19. van Benthem J (2001) Games in dynamic-epistemic logic. Bull Econ Res 53(4):219–248
20. van Benthem J (2002) Extensive games as process models. J Logic Lang Inform 11(3):289–313
21. van Benthem J (2006) One is a lonely number. In: Chatzidakis Z, Koepke P, Pohlers W (eds) Logic Colloquium '02, ASL & A.K. Peters, Wellesley
22. van Benthem J (2007a) Five questions on game theory, Automatic Press, pp 9–19
23. van Benthem J (2007b) Rational dynamics and epistemic logic in games. Int Game Theory Rev 9(1): 377–409 (Erratum reprint, Volume 9(2), 2007, 377–409)
24. van Benthem J (2007c) Dynamic logic for belief change. J Appl Non-Classical Logics 17(2): 129–155
25. van Benthem J, Dégremont C (2010) Bridges between dynamic doxastic and doxastic temporal logics. In: Bonanno et al (2008) pp 151–173
26. van Benthem J, Gheerbrant A (2010) Game solution, epistemic dynamics and fixed-point logics. Fundamenta Informaticae 100(1–4):19–41
27. van Benthem J, Liu F (2004) Diversity of logical agents in games. Philosophia Scientiae 8:163–178
28. van Benthem J, Pacuit E (2006) The tree of knowledge in action: towards a common perspective. In: Hodkinson I, Governatori G, Venema Y (eds) Advances in modal logic, vol 6. College Publications
29. van Benthem J, van Otterlo S, Roy O (2005) Preference logic, conditionals and solution concepts in games. In: Lindström S, Lagerlund H, Sliwinski R (eds) Modality matters. Festschrift for Krister Segerberg, University of Uppsala, pp 61–76 (also available on the ILLC prepublication repository: PP-2005-28)
30. van Benthem J, Gerbrandy J, Hoshi T, Pacuit E (2009) Merging frameworks for interaction: DEL and ETL. J Phil Logic 38(5):491–526
31. van Benthem J, Pacuit E, Roy O (2011) Toward a theory of play: a logical perspective on games and interaction. Games 2(1):52–86
32. van Benthem J (2014) *Logic in Games*. Cambridge, MA: MIT Press
33. Bernheim BD (1984) Rationalizable strategic behavior. Econometrica 52(4):1007–1028
34. Blume L, Brandenburger A, Dekel E (1991) Lexicographic probabilities and choice under uncertainty. Econometrica 59(1):61–79
35. Board O (2004) Dynamic interactive epistemology. Games Econ Behav 49:49–80
36. Bonanno G (1992a) Set-theoretic equivalence of extensive-form games. Int J Game Theory 20(4):429–447
37. Bonanno G (1992b) Players' information in extensive games. Math Soc Sci 24(1):35–48
38. Bonanno G (2003) A syntactic characterization of perfect recall in extensive games. Res Econ 57(3):201–217
39. Bonanno G (2007) Axiomatic characterization of the AGM theory of belief revision in a temporal logic. Artif Intell 171(2–3):144–160
40. Bonanno G (2010) A syntactic approach to rationality in games with ordinal payoffs. In: Bonanno et al (2008) pp 59–86
41. Bonanno G (2013) A dynamic epistemic characterization of backward induction without counterfactuals. Games Econ Behav 78:31–43
42. Bonanno G, Löwe B, van der Hoek W (eds) (2008) Logic and the foundations of game and decision theory—LOFT 8, 8th International conference, Amsterdam, 3–5 July 2008 (Revised Selected Papers, volume 6006 of Lecture Notes in Computer Science, 2010. Springer.)

43. Brandenburger A, Friedenberg A, Keisler HJ (2008) Admissibility in games. Econometrica 76(2):307–352
44. Broersen J, Herzig A, Troquard N (2007) Normal coalition logic and its conformant extension. In: Samet D (ed) TARK'07, PUL, pp 91–101
45. de Bruin B (2008) Common knowledge of rationality in extensive games. Notre Dame J Formal Logic 49(3):261–280
46. Chellas BF (1980) Modal logic: an Introduction. Cambridge University Press, New York
47. Dalkey N (1953) Equivalence of information patterns and essentially determinate games. In: Kuhn HW, Tucker AW (eds) Contributions to the theory of games, vol II, Annals of mathematics studies, no. 28. Princeton University Press, NJ, pp 217–243
48. Dawar A, Grädel E, Kreutzer S (2004) Inflationary fixed points in modal logic. ACM Trans Comput Logic 5(2):282–315
49. Dégremont C (2010) The temporal mind. Observations on the logic of belief change in interactive systems. ILLC, ILLC Dissertation Series DS-2010-03, PhD thesis, Universiteit van Amsterdam
50. Dégremont C, Roy O (2012) Agreement theorems in dynamic-epistemic logic. J Philos Logic 41(4):735–764
51. van Ditmarsch H (2005) Prolegomena to dynamic logic for belief revision. Synthese 147:229–275
52. Ebbinghaus H-D, Flum J, Thomas W (1994) Mathematical logic, 2nd edn. Springer, Undergraduate texts in mathematics
53. Ehrenfeucht A (1961) An application of games to the completeness problem for formalized theories. Fundamenta Mathematicae 49:129–141
54. van Eijck J, Stokhof M (2006) The gamut of dynamic logics. In: Gabbay D, Woods J (eds) Handbook of the history of logic, vol 6—Logic and the modalities in the 20th century, Elsevier, pp 499–600
55. Fagin R, Halpern JY, Moses Y, Vardi MY (1995) Reasoning about knowledge. MIT Press, Cambridge
56. Fraïssé R (1954) Sur quelques classifications des systèmes de relations. Publ Sci Univ Alger Sér A. 1:35–182
57. Friedman N, Halpern JY (1997) Modeling belief in dynamic systems, part I: foundations. Artif Intell 95(2):257–316
58. Gerbrandy J (1999) Bisimulations on Planet Kripke. PhD thesis, ILLC, Universiteit van Amsterdam, 1999. ILLC Dissertation Series DS-1999-01
59. Gheerbrant A (2010) Fixed-Point logics on trees. PhD thesis, ILLC Dissertation Series DS-2010-08, University of Amsterdam
60. Grädel E, Kreutzer S (2003) Will deflation lead to depletion? on non-monotone fixed point inductions. In: LICS, IEEE Computer Society, ISBN 0-7695-1884-2, p 158
61. Grove A (1988) Two modellings for theory change. J Philos Logic 17:157–170
62. Gurevich Y, Shelah S (1986) Fixed-point extensions of first-order logic. Ann Pure Appl Logic 32:265–280
63. Halpern JY, Vardi MY (1989) The complexity of reasoning about knowledge and time. I. Lower bounds. J Comput Syst Sci 38(1):195–237
64. Harel D (1985) Recurring dominoes: making the highly undecidable highly understandable. In: Topics in the theory of computation, ISBN 0-444-87647-2, Elsevier, New York, pp 51–71
65. Hendricks V, de Jongh D, Gierasimczuk N (2014) Logic and Learning. In: Baltag A, Smets S (eds) Johan van Benthem on Logic and Information Dynamics. Springer, Dordrecht, pp 267–288 (Chapter 10 in this volume)
66. van der Hoek W, Pauly M (2006) Modal logic for games and information. In: Blackburn P, van Benthem J, Wolter F (eds) The handbook of modal logic, Elsevier, pp 1077–1148
67. Hoshi T (2009) Epistemic dynamics and protocol information. PhD thesis, Stanford
68. Kohlberg E, Mertens J-F (1986) On the strategic stability of equilibria. Econometrica 54(5):1003–1037

69. Krentel WD, McKinsey JCC, Quine WV (1951) A simplification of games in extensive form. Duke Math J 18:885–900
70. Kreutzer S (2004) Expressive equivalence of least and inflationary fixed-point logic. Ann Pure Appl Logic 130(1–3):61–78
71. Kuhn HW (1953) Extensive games and the problem of information. In: Kuhn HW, Tucker AW (eds) Contributions to the theory of games, vol II, Annals of mathematics studies, no 28, Princeton University Press, NJ, pp 193–216
72. Miller JS, Moss LS (2005) The undecidability of iterated modal relativization. Studia Logica 79(3):373–407
73. Moschovakis YM (1974) Elementary Induction on Abstract Structures. North-Holland, New York
74. Myerson RB (1991) Game theory: analysis of conflict. Harvard University Press, Cambridge
75. Osborne M (2004) An introduction to game theory. Oxford University Press, New York
76. Papadimitriou CH (1994) Computational Complexity. Addison Wesley, Reading
77. Pauly M (2002) A modal logic for coalitional power in games. J Logic Comput 12(1):149–166
78. Pauly M, Parikh R (2003) Game logic—an overview. Studia Logica 75(2):165–182
79. Pearce DG (1984) Rationalizable strategic behavior and the problem of perfection. Econometrica 52(4):1029–1050
80. Perea A (2007) Epistemic foundations for backward induction: an overview. In: van Benthem J, Gabbay D, Löwe B (eds) Interactive logic. Proceedings of the 7th Augustus de Morgan workshop, vol 1 of texts in logic and games, Amsterdam University Press, pp 159–193
81. Plaza JA (1989) Logics of public communications. In: Emrich ML, Pfeifer MS, Hadzikadic M, Ras ZW (eds) Proceedings of the 4th international symposium on methodologies for intelligent systems: poster session program, Oak Ridge National Laboratory, pp 201–216
82. Quesada A (2001) On expressing maximum information in extensive games. Math Soc Sci 42(2):161–167
83. Reny PJ (1993) Common belief and the theory of games with perfect information. J Econ Theory 59(2):257–274
84. Selten R (1975) Reexamination of the perfectness concept for equilibrium points in extensive games. Int J Game Theory 4:25–55
85. Stalnaker R (1994) On the evaluation of solution concepts. Theor Decis 37:49–73
86. Susan E, Reny PJ (1994) On the strategic equivalence of extensive form games. J Econ Theory 62(1):1–23
87. Thompson FB (1952) Equivalence of games in extensive form. RAND research memorandum 759, The RAND Corporation
88. Zvesper JA (2010) Playing with information. ILLC, ILLC Dissertation Series DS-2010-02, PhD thesis, Universiteit van Amsterdam

Chapter 16
Knowledge Games and Coalitional Abilities

Thomas Ågotnes and Hans van Ditmarsch

Abstract We present our recent work on the interaction of knowledge and coalitional abilities, on quantifying over information change, and on imperfect information games wherein the actions are public announcements or questions with informative answers. Such case studies should be seen as an implementation of the general programmatic ideas found in Johan van Benthem's recent books *Logical Dynamics of Information and Interaction* and *Logic in Games*.

16.1 Introduction

While the logic of knowledge, *epistemic logic*, has been studied for some time [51, 65, 80, 95], the focus on the logical principles of the *dynamics* of knowledge, i.e., of how knowledge changes in a multi-agent system, is more recent and is currently an active research topic. These principles can be studied by combining epistemic logic and temporal or dynamic logic, or by modelling information-changing actions and events and their epistemic preconditions and postconditions explicitly as in *dynamic epistemic logic* [44]. Another currently active research topic in the area of multi-agent systems is combining logic and *game theory* [66].

Already in [14] van Benthem pointed out various interesting research questions on the intersection of epistemic logic and game theory, and this has grown into a full-fledged research programme carried by an increasingly sizable community [17, 18]. In many real games *knowledge about the game* plays a role, as in imperfect

T. Ågotnes
Department of Information Science and Media Studies, University of Bergen, P.O. Box 7802,
5020 Bergen, Norway
e-mail: thomas.agotnes@infomedia.uib.no

H. van Ditmarsch (✉)
LORIA, CNRS—Université de Lorraine, BP 239, 54506 Vandoeuvre-lès-Nancy, France
e-mail: hans.van-ditmarsch@loria.fr

A. Baltag and S. Smets (eds.), *Johan van Benthem on Logic
and Information Dynamics*, Outstanding Contributions to Logic 5,
DOI: 10.1007/978-3-319-06025-5_16, © Springer International Publishing Switzerland 2014

information games such as Bridge, while other games, like Cluedo, are *games about knowledge*. In this paper, we survey some of our recent work in this area.

We first give a brief overview of epistemic logic including one of the most popular dynamic epistemic logics, namely public announcement logic, as well as a brief introduction to one of the most popular logical frameworks in the area of logics for games, namely logics of *coalitional ability*. We also briefly review some key game theoretical concepts. In Sect. 16.3 we discuss the combination of epistemic logic and coalitional ability logics, and how it can be used to express knowledge dynamics. Section 16.4 is on propositional quantification in epistemic logic, and in particular quantifications involving coalitions of agents. This should be seen as zooming in the general setting of Sect. 16.3 on more specific interaction between coalitional ability and knowledge by restricting the actions of agents to be a well-known type of information-changing actions: public announcements. Several variants of resulting logics are discussed. In Sect. 16.5 we turn to the question of which among her information-changing actions a rational agent would or should choose. This presupposes a preference over epistemic states. Given that and a Kripke structure we can induce different types of games: *public announcement games* (Sect. 16.5.1) and *question-answer games* (Sect. 16.5.2). Section 16.6 presents some "real" games involving dynamic knowledge. Such bottom-up input has fuelled the theoretical development of the area in the past and we expect that many as yet unanswered very concrete questions (what is an optimal strategy in Cluedo?) will continue to fuel the further development. In the concluding section we review the topics we have treated with respect to the extensive literature on logic and games and in particular the recent [16, 18].

16.2 Background

16.2.1 Logic

We will consider several variants of propositional modal logic. Each variant has a *language*, a class of *models* each with a *state space*, and a *satisfaction* relation $\models$ which is a binary relation between *pointed models* (consisting of a model and a state in the domain of that model) and formulae. Expression $M, s \models \varphi$ means that φ is true in the pointed model M, s, whereas $\models \varphi$ means that φ is *valid*, i.e., that $M, s \models \varphi$ for all pointed models M, s. The *model checking problem* is the problem of deciding whether $M, s \models \varphi$, when M, s, φ are given. The *satisfiability problem* is the problem of deciding whether there exists a pointed model such that $M, s \models \varphi$, when φ is given. We say that a logic is *decidable* if the satisfiability problem is decidable.

We henceforth assume that a set Θ of *atomic propositions* (or primitive propositions) and a set $N = \{1, \ldots, n\}$ of *agents* is given, and we will use the usual derived propositional connectives without explicit introduction.

instantiations of propositional tautologies

$K_i(\varphi \to \psi) \to (K_i\varphi \to K_i\psi)$ distribution

$K_i\varphi \to \varphi$ truth

$K_i\varphi \to K_iK_i\varphi$ positive introspection

$\neg K_i\varphi \to K_i\neg K_i\varphi$ negative introspection

From φ and $\varphi \to \psi$, infer ψ modus ponens

From φ, infer $K_i\varphi$ necessitation of knowledge

Fig. 16.1 Axiomatisation of EL

16.2.2 Epistemic Logic

Epistemic logic extends propositional logic with epistemic operators of the form K_i, where i is an agent. Formally, the language $\mathscr{L}_{EL}$ of EL is defined by the grammar:

$$\varphi ::= p \mid \neg\varphi \mid \varphi \wedge \varphi \mid K_i\varphi$$

where $i \in N$ and $p \in \Theta$. We write $\hat{K}_i\varphi$ for the dual $\neg K_i\neg\varphi$.

A *Kripke model* or *epistemic model* over N and Θ is a tuple $M = (S, \{\sim_1, \ldots, \sim_n\}, V)$, where S is a set of states, for each agent i $\sim_i \subseteq S \times S$ is an epistemic indistinguishability relation that is assumed to be an equivalence relation, and $V : \Theta \to 2^S$ maps primitive propositions to the states in which they are true. A *pointed Kripke structure* is a pair (M, s) where s is a state in M. The interpretation of $\mathscr{L}_{EL}$ formulae in a pointed Kripke structure is defined as follows.

$M, s \models p$ iff $p \in V(s)$

$M, s \models K_i\varphi$ iff for every t such that $s \sim_i t$, $M, t \models \varphi$

$M, s \models \neg\varphi$ iff not $M, s \models \varphi$

$M, s \models \varphi \wedge \psi$ iff $M, s \models \varphi$ and $M, s \models \psi$

We can add *epistemic group operators* to the language as well: $E_G\varphi$, $C_G\varphi$, and $D_G\varphi$ mean that everyone in the group $G \subseteq N$ knows φ, that φ is common knowledge in G, and that φ is distributed knowledge in G, respectively. Formally, $M, s \models X_G\varphi$ iff $M, t \models \varphi$ for all t such that $s \sim_G^X t$, where: $\sim_G^E = \bigcup_{i \in G} \sim_i$, $\sim_G^C$ is the transitive closure of $\sim_G^E$, and $\sim_G^D = \bigcap_{i \in G} \sim_i$.

A sound and complete axiomatisation of all valid $\mathscr{L}_{EL}$ formulae is shown in Fig. 16.1. (We will use epistemic group operators in various semantic settings but will not give axiomatizations involving those as well.)

The idea of modelling information change (as in belief revision) with dynamic modal operators goes back to van Benthem [22]. The perhaps simplest logic of this type, wherein epistemic operators are also explicit, is *public announcement logic* (PAL) [57, 86]. The language $\mathscr{L}_{PAL}$ of PAL extends the EL language with *public announcement* operators:

$$\varphi ::= p \mid \neg\varphi \mid \varphi \wedge \varphi \mid K_i\varphi \mid [\varphi]\varphi$$

axiom schemata and rules of EL

$[\varphi]p \leftrightarrow (\varphi \rightarrow p)$ atomic permanence
$[\varphi]\neg\psi \leftrightarrow (\varphi \rightarrow \neg[\varphi]\psi)$ announcement and negation
$[\varphi](\psi \wedge \chi) \leftrightarrow ([\varphi]\psi \wedge [\varphi]\chi)$ announcement and conjunction
$[\varphi]K_i\psi \leftrightarrow (\varphi \rightarrow K_i[\varphi]\psi)$ announcement and knowledge
$[\varphi][\psi]\chi \leftrightarrow [\varphi \wedge [\varphi]\psi]\chi$ announcement composition

Fig. 16.2 Axiomatisation of PAL

Expression $\langle\varphi\rangle\psi$ is shorthand for the dual $\neg[\varphi]\neg\psi$. The interpretation of the new clause $[\varphi]\psi$ is as follows:

$M, s \models [\varphi]\psi$ iff $M, s \models \varphi$ implies that $M|\varphi, s \models \psi$

where $M|\varphi = (S', \{\sim'_1, \ldots, \sim'_n\}, V)$ is such that

$$S' = \{s' \in S : M, s' \models \varphi\},$$
$$\sim'_i = \sim_i \cap (S' \times S') \text{ and}$$
$$V'(p) = V(p) \cap S'.$$

The update of M by φ, $M|\varphi$, is the submodel of M obtained by removing states where φ is false. Intuitively, $[\varphi]\psi$ means that if φ is truthfully publicly announced, then ψ will be true afterwards. A sound and complete axiomatisation of all valid $\mathscr{L}_{PAL}$ formulae is shown in Fig. 16.2.

We will sometimes make use of the *positive fragment* of $\mathscr{L}_{PAL}$ [46], which essentially consists of all formulae only containing negation immediately preceding atomic propositions:

$$\varphi ::= p \mid \neg p \mid \varphi \wedge \varphi \mid \varphi \vee \varphi \mid K_i\varphi \mid [\neg\varphi]\varphi$$

More sophisticated and general frameworks model more complex events than public announcements [11, 34, 56, 77]. This includes the incorporation of factual change into languages that express epistemic change [24, 43], preference-based modelling of belief revision with dynamic modal operators [7, 12, 20, 39], the integration of dynamic epistemic logics with temporal epistemic logics [25], and various forms of quantification over propositional variables [2, 4, 10]. See [51, 80] for further details about epistemic logic, and [44] for further details about dynamic epistemic logic.

16.2.3 Game Theory

A *strategic game* is a quintuple $\mathbf{G} = \langle N, \{A_i : i \in N\}, S, o, \{u_i : i \in N\}\rangle$ where

- N is the finite set of *players*;
- for each $i \in N$, A_i is the set of *strategies* (or *actions*) available to i. We note that $A = \times_{i \in N} A_i$ is the set of *strategy profiles*;

- S is a set of possible *outcomes*;
- $o : A \to S$ is the *outcome function* mapping strategy profiles to outcomes;
- for each $i \in N$, $u_i : S \to \mathbb{R}$ is the *payoff function* for i, mapping each outcome to a number.

When $(a_1, \ldots, a_n) \in A$, the notation $(a_1, \ldots, a_n)[a_i/a_i']$ stands for the profile wherein strategy a_i is replaced by a_i'. A (strategic) *game form* is a strategic game without utilities, i.e., of the form $\langle N, \{A_i : i \in N\}, S, o\rangle$.[1]

When we are not interested in the outcomes *per se*, we will sometimes write $u_i(a_1, \ldots, a_n)$ for $u_i(o(a_1, \ldots, a_n))$, and we will sometimes define a game as a tuple $\mathbf{G} = \langle N, \{A_i : i \in N\}, \{u_i : i \in N\}\rangle$ with the implicit convention that each strategy profile gives a distinct outcome.

A profile $(a_1, \ldots, a_n)$ is a (pure strategy) *Nash equilibrium* if and only if for all $i \in N$, for all $a_i' \neq a_i$, $u_i((a_1, \ldots, a_n)[a_i/a_i']) \leq u_i(a_1, \ldots, a_n)$. A strategy for an agent is *weakly dominant* if it is at least as good for that agent as any other strategy, no matter which strategies the other agents choose.[2] Formally, a strategy a_i for agent i is weakly dominant if and only if for all agents j, for all a_j', $u_i(a_1', \ldots, a_n') \leq u_i((a_1', \ldots, a_n')[a_i'/a_i])$. Clearly, a strategy profile where all the strategies are weakly dominant is a Nash equilibrium. A strategy for an agent is *strictly dominated* if there exists another strategy for that agent that always is better. Formally, a_i is strictly dominated if and only if there is a strategy $a_i' \neq a_i$ such that for all agents $j \neq i$, for all a_j', $u_i(a_1', \ldots, a_n') > u_i((a_1', \ldots, a_n')[a_i'/a_i])$.

See, e.g., [83] for further details.

16.2.4 Logics of Coalitional Ability

Logics of coalitional ability usually have a modal *coalition operator* $\langle G \rangle$ for each coalition $G \subseteq N$. A formula of the form $\langle G \rangle \varphi$ typically means that G can make φ true; that there is some joint action that the group G can do such that φ is guaranteed to be true afterwards. The dual $[G]\varphi$, defined as $\neg \langle G \rangle \neg \varphi$, means that φ will be true no matter what G does.

One of the most popular coalitional ability logics is Pauly's *Coalition Logic* (CL) [85]. The language of CL simply extends propositional logic with coalition operators.[3] (where $G \subseteq N$, and $p \in \Theta$):

---

[1] We remind the reader that $G \subseteq N$ typically names a group or coalition of agents, whereas $\mathbf{G}$, i.e. 'bold-G', names a game (C for coalition clashes with C for common knowledge).

[2] The literature differs in the definition of weakly dominant strategies. Another common definition in addition requires that the strategy is strictly better against at least one combination of actions by the other agents.

[3] The notation for coalition operators varies in the literature. In [85], Pauly in fact uses $[G]$ where we use $\langle G \rangle$. Pauly's interpretation of the operator uses a $\exists \forall$ pattern, and since we also will discuss other interpretations of the operator we choose to emphasise its existential nature and use the diamond notation $\langle G \rangle$.

$$\varphi ::= p \mid \neg\varphi \mid \varphi \wedge \varphi \mid \langle G \rangle \varphi$$

Coalition logic can be interpreted in *concurrent game structures* (CGSs) [6]. A CGS is a tuple

$$(S, V, ACT, d, \delta)$$

where

- S is a finite set of *states*.
- V is the *labeling function*, assigning a set $V(s) \subseteq \Theta$ to each $s \in S$.
- ACT is a finite set of *actions*.
- For each player $i \in N$ and state $s \in S$, $d_i(s) \subseteq ACT$ is the non-empty set of actions available to player i in s. $D(s) = d_1(s) \times \cdots \times d_n(s)$ is the set of *joint actions* in s. If $\mathbf{a} \in D(s)$, a_i denotes the ith component of $\mathbf{a}$.
- δ is the *transition function*, mapping each state $s \in S$ and joint action $\mathbf{a} \in D(s)$ to a state $\delta(s, \mathbf{a}) \in S$.

Note that a concurrent game structure implicitly associates a strategic game form $\mathbf{G}(s) = \langle N, \{A_i^s : i \in N\}, S, o^s \rangle$ with each state s: $A_i^s = d_i(s)$ and $o^s(a_1, \ldots, a_n) = \delta(s, (a_1, \ldots, a_n))$.

The CL interpretation of the coalition modality is as follows:

$M, s \models \langle G \rangle \psi$ iff there exists an action $a_i \in d_i(s)$ for each $i \in G$, such that for all possible actions $a_j \in d_j(s)$ for all $j \in N \setminus G$, $M, \delta(s, (a_1, \ldots, a_n)) \models \psi$.

The semantics of coalition logic can alternatively be given in terms of *effectivity functions* [85]. An effectivity function over a set of agents N and a set of states S is a function E that maps any coalition $G \subseteq N$ to a set of sets of states $E(G) \subseteq 2^S$. Intuitively, $X \in E(G)$ means that G can make some choice that will ensure that the outcome will be in X. Effectivity functions are implicit in strategic game forms. The α-*effectivity function* $E_\mathbf{G}$ of strategic game form $\mathbf{G}$ is defined as follows: $X \in E_\mathbf{G}(G)$ iff there exists an action $a_i \in A_i$ for each $i \in G$, such that for all possible actions $a_j \in A_j$ for all $j \in N \setminus G$, $o(a_1, \ldots, a_n) \in X$. An effectivity function E is called *playable* [85] iff: $X \in E(G) \ \& \ X \subseteq Y \Rightarrow Y \in E(G)$ (*outcome monotonicity*); $S \setminus X \notin E(\emptyset) \Rightarrow X \in E(N)$ (*N-maximality*); $\emptyset \notin E(G)$ (*Liveness*); $S \in E(G)$ (*Safety*); $G \cap G' = \emptyset \ \& \ X \in E(G) \ \& \ Y \in E(G') \Rightarrow X \cap Y \in E(G \cup G')$ (*superadditivity*). In [85] it is claimed that an effectivity function E is the α-effectivity function of a strategic game form iff E is playable; a property often referred to in the secondary literature. However, it has recently been shown [59] that this claim is in fact not correct: there are playable effectivity functions over infinite sets which are not α-effectivity functions of any strategic game forms. In [59], Goranko also shows that effectivity functions that in addition to being playable have the property $X \in E(N) \Rightarrow \exists x \in X, \{x\} \in E(N)$ (E is a *crown*), called *truly playable* effectivity functions, correspond exactly to α-effectivity functions.

instantiations of propositional tautologies

$$\neg\langle G\rangle\bot \qquad\qquad\qquad\qquad\qquad\qquad\qquad\qquad\qquad \bot$$
$$\langle G\rangle\top \qquad\qquad\qquad\qquad\qquad\qquad\qquad\qquad\qquad \top$$
$$\neg\langle\emptyset\rangle\neg\varphi \rightarrow \langle N\rangle\varphi \qquad\qquad\qquad\qquad\qquad\qquad\qquad N$$
$$\langle G\rangle(\varphi \wedge \psi) \rightarrow \langle G\rangle\psi \qquad\qquad\qquad\qquad\qquad\qquad M$$
$$(\langle G_1\rangle\varphi_1 \wedge \langle G_2\rangle\varphi_2) \rightarrow \langle G_1 \cup G_2\rangle(\varphi_1 \wedge \varphi_2) \quad \text{where } G_1 \cap G_2 = \emptyset \quad S$$
$$\dfrac{\varphi\leftrightarrow\psi}{\langle G\rangle\varphi\leftrightarrow\langle G\rangle\psi} \qquad\qquad\qquad\qquad\qquad\qquad\qquad\qquad E$$
$$\dfrac{\varphi, \varphi\rightarrow\psi}{\psi} \qquad\qquad\qquad\qquad\qquad\qquad\qquad\qquad\qquad MP$$

Fig. 16.3 Axiomatisation of CL

A *coalition model* [85] is a tuple $\mathcal{M} = \langle S, E, V\rangle$ where E gives a playable effectivity function $E(s)$ for each state $s \in S$, and V is a valuation function. The interpretation of the coalition modality in a coalition model is as follows:

$$\mathcal{M}, s \models \langle G\rangle\psi \text{ iff } \varphi^{\mathcal{M}} \in E(s)(G)$$

where $\varphi^{\mathcal{M}} = \{t \in S : \mathcal{M}, t \models \varphi\}$. It is easy to see that the two semantics coincide: $M, s \models \varphi$ iff $\mathcal{M}_\alpha, s \models \varphi$ for all φ, where $M = (S, \Theta, V, ACT, d, \delta)$ and $\mathcal{M}_\alpha = (S, E_\alpha, V)$ and $E_\alpha(s) = E_{\mathbf{G}(s)}$.

Figure 16.3 shows an axiomatisation of coalition logic [85]. It is sound and complete, both for concurrent game structures and for coalition models (over playable effectivity functions), and also for coalition models over truly playable effectivity functions [59].

Alternating-time Temporal Logic (ATL) [6] extends the coalition operators with a temporal dimension. $\langle G\rangle\bigcirc\varphi$, $\langle G\rangle\Diamond\varphi$, $\langle G\rangle\Box\varphi$ and $\langle G\rangle\varphi_1 \,\mathcal{U}\, \varphi_2$ means that G has a *strategy* to ensure that φ will be true in the *next state*, that φ will be true at *some state* in the future, that φ will be true at *all states* in the future, and that φ_1 will be true *until* φ_2 is true, respectively, no matter what the agents in $N \setminus G$ do. A strategy is a function that maps a sequence of states (a *history*) to an action for each agent in G. Strategies are sometimes assumed to be *memoryless*, represented by mapping single states to actions. Coalition logic can be seen as the next-time fragment of ATL. ATL can also be seen as an extension of the popular branching-time temporal logic *Compuation-Tree Logic* (CTL) [50].

In [14], van Benthem proposes an extension of propositional dynamic logic, interpreted over extensive form games, with *forcing modalities* of the form $\{G, i\}\varphi$. These are true at a game node if player i has a strategy for playing game G from there on, which forces a set of outcomes all which satisfy φ. These are called "game-internal" versions of the coalition logic modalities, and they are even closer to the "sometime-in-the-future" ATL modality $\langle G\rangle\Diamond\varphi$.

Another popular logical framework for coalitional ability is the logic of *seeing to it that*, or STIT logic [13]. CL and ATL are more directly related to games than STIT logic. However, although conceptually and technically quite different, STIT logic is nevertheless closely related to CL and ATL [31, 32, 94].

16.3 Epistemics and Coalitional Ability

As much of the work related to games in the general area of modal logic has focussed on logics of (coalitional) ability, extending such logics with an epistemic dimension is a natural perspective on dynamical epistemics and games. For example, such combinations allow us to express properties such as [67]:

- $K_i \varphi \rightarrow \langle i \rangle K_j \varphi$: if agent j knows φ, she can communicate that fact to agent j
- $K_i \varphi \wedge \neg K_j \varphi \wedge \neg K_k \varphi \wedge \langle i, j \rangle (K_i \varphi \wedge K_j \varphi \wedge \neg K_k \varphi)$: i can send private information to j without revealing it to k
- $\langle i \rangle \varphi \rightarrow K_i \psi$: knowledge of ψ is a *necessary epistemic precondition* for making φ true
- $\langle i \rangle \varphi \leftarrow K_i \psi$: knowledge of ψ is a *sufficient epistemic precondition* for making φ true.

Combinations of logics of coalitional ability and epistemic logic were first studied by van der Hoek and Wooldridge [67], who combine ATL with epistemic logic. We will here define a simple variant: coalition logic extended with epistemic operators, henceforth called CLK. The language $\mathscr{L}_{CLK}$ of CLK is defined by the following grammar:

$$\varphi ::= p \mid \neg \varphi \mid \varphi \wedge \varphi \mid K_i \varphi \mid \langle G \rangle \varphi$$

Semantically, CGSs extended with epistemic indistinguishability relations are called *concurrent epistemic game structures* (CEGSs). Formally, an CEGS is a tuple $(S, V, ACT, d, \delta, \{\sim_1, \ldots, \sim_n\})$ where (S, V, ACT, d, δ) is a CGS, each $\sim_i$ is an equivalence relation on S and for all i, s and t:

$$s \sim_i t \Rightarrow d_i(s) = d_i(t) \tag{16.1}$$

The condition that the same actions must be available in indiscernible states, i.e., that an agent knows which actions she has available, is commonly accepted [1, 70, 74]. The interpretation of the CLK language in a CGS state is as expected (combine the clauses for CL and EL).

16.3.1 Adding Temporal Operators and Strategies

ATEL [67] extends CLK to the full ATL language, i.e., it adds temporal operators, and also adds epistemic group operators C_G, D_G and E_G. The interpretation of the different operators in a state of a CEGS is as before. Strategies are usually assumed to be memoryless.[4] However, as first discussed in [70], strategies are usually restricted to be *uniform* in the sense that $f_i(s) = f_i(t)$ whenever $s \sim_i t$ for any i, s, t. In [14]

---

[4] One reason for this is that with perfect recall strategies and imperfect information the model checking problem is assumed to be undecidable [6, 90].

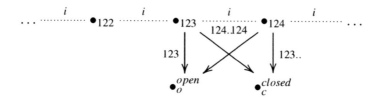

Fig. 16.4 Model M illustrating the difference between knowing *de re* and *de dicto*, for a single agent. Only a fragment of the model is shown. Dotted lines denote epistemic accessibility (the accessibility relation is assumed to be transitive), labelled arrows denote actions and transitions

epistemic forcing modalities are also considered, essentially modifying the definition of the forcing modality (see Sect. 16.2.4) to quantify only over uniform strategies. Many recent works relating strategies to knowledge have appeared [17, 42, 58]—we consider this beyond the scope of our contribution.

16.3.2 Knowing That Versus Knowing How

As pointed out by Jamroga [70], adding the dimension of imperfect information leads to several (often subtly) different "variants" of coalitional ability. First, there is a difference between having the ability to achieve something in the CL/ATL sense, i.e., between there existing an action that is guaranteed to be successful, and *knowing* that you have that ability. Even though you know which actions are available (ref. Eq. (16.1)), it might be that you consider it possible that the *consequences* of the actions are different from what they actually are. But there is also another important distinction.

Consider the following example (taken from [1]). An agent i is in front of a three-digit combination-lock safe. The agent does not know the combination. The correct combination is in fact 123. The example is modelled in Fig. 16.4. We have that $M, 123 \models \langle i \rangle open$—the agent has the ability to open the safe—again, in the sense that there exists an action (123) that will open the safe. But it is also in fact the case that the agent *knows* this, $M, 123 \models K_i \langle i \rangle open$, because $\langle i \rangle open$ holds also in every other state that agent i considers possible by the same argument. More precisely, the agent knows *that* she has the ability to open the safe. But, importantly, she does not know *how* to open the safe: there is no action that will open the safe in every state she considers possible. Following [74],[5] we can define the following three, increasingly stronger, variants of coalitional ability under incomplete information:

---

[5] The *de dicto/de re* distinction is well known [88] in logic/language in general and has been known in the area of reasoning about knowledge and action in AI for some time [81].

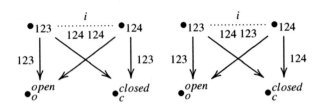

Fig. 16.5 A submodel of the model in Fig. 16.4 (*left*), and a modification (*right*)

1. Having the ability to make φ true, without necessarily knowing it.
2. Knowing *de dicto* that you have the ability to make φ true: in every state you consider possible you have the ability to make φ true (but not necessarily by using the same action in different states).
3. Knowing *de re* that you have the ability to make φ true: there is an action that will ensure that φ is true in every state you consider possible.

The first two are expressed in CLK by $\langle i \rangle \varphi$ and $K_i \langle i \rangle \varphi$, respectively. However, it is claimed [74] that knowledge *de re* is *not expressible* in CLK. While this is not formally shown in [74] or elsewhere, it is indeed easy to see that it is true, even in the single-agent case: in the model to the right in Fig. 16.5 the agent knows *de re* that she can open the safe; in the model to the left she does not, but it is easy to show that the two models satisfy exactly the same formulae—swapping the *names* of the two actions in state 124 does not change the interpretation of formulae in that state.

Extending the language in order to be able to express knowledge *de re* has turned out to be non-trivial, and many proposals have appeared and been studied [32, 64, 70–75, 84, 90]. A summary discussion of many of the proposed approaches can be found in [71].

One approach that is quite flexible is *Constructive Strategic Logic* (CSL) [71], which extends ATEL in two main ways. First, formulae are interpreted in *sets* of states rather than in single states. This makes it possible to define $M, S \models \langle i \rangle \varphi$, where S is a set of states, to be true whenever there is an action/strategy that will achieve φ in every state in S. Second, *constructive knowledge* operators $\mathbb{K}_i$ (and similar group operators) are added to the language, with the interpretation that $M, S \models \mathbb{K}_i \varphi$ iff $M, [S]_{\sim_i} \models \varphi$, where $[S]_{\sim_i} = \bigcup_{s \in S} [s]_{\sim_i}$. Thus, $M, \{s\} \models \mathbb{K}_i \langle i \rangle \varphi$ iff in s i knows *de re* that she can make φ true. For example, for the model M in Fig. 16.4 we have that $M, 123 \models \neg \mathbb{K}_i \langle i \rangle open$.

16.4 Quantification in Dynamic Epistemic Logic

On the one hand, logics of coalitional ability, like CL and ATL, formalise reasoning about joint actions by agents in a given coalition. On the other hand, dynamic epistemic logics like PAL formalise actions that can be seen as external events with epistemic preconditions and postconditions. It is natural, then, to combine these approaches by considering coalitional ability where actions are information changing actions of the type described in dynamic epistemic logic such as public announcements—after all, in many events agents take a part. Here, we consider two variants of the coalition logic language but where actions are public announcements (or other public events): the logics GAL and CAL. From the perspective of PAL, this corresponds to adding *quantification* over possible announcements to the logical language. We begin by considering a basic extension of PAL with quantification, a logic called APAL. Our targetted GAL and CAL can be seen as variations on the basic theme set in this APAL.

16.4.1 Arbitrary Public Announcement Logic

Interpreting the standard modal diamond so that $\Diamond\varphi$ means "there is an announcement after which φ" was suggested by van Benthem [23]. The result of extending public announcement logic extended with such a diamond is called *Arbitrary Public Announcement Logic* (APAL) by Balbiani et al. [9], who also study the logic in detail. One motivation for this type operator goes back to the Fitch paradox of knowability [54].

Formally, the language $\mathscr{L}_{APAL}$ of APAL is defined by adding a $\Box$ modality to $\mathscr{L}_{PAL}$, interpreted as follows (with the dual $\Diamond\varphi \equiv \neg\Box\neg\varphi$):

$$M, s \models \Box\varphi \text{ iff for all } \psi \in \mathscr{L}_{EL}, M, s \models [\psi]\varphi$$

The quantification ranges over the fragment without quantifiers rather than over the full $\mathscr{L}_{APAL}$ language, in order to make the semantics well defined. Here we should note that EL is equally expressive as PAL, so quantifying over PAL (the quantifier-free fragment) rather than EL indeed has the same meaning. This logic APAL is strictly more expressive than PAL, is axiomatisable and has various pleasing properties. See [9, 10] for further details. However, the reason to introduce the logic is now, subsequently, to introduce two versions of this logic with coalitional operators.

16.4.2 Group Announcement Logic

The interpretation of the APAL modality as quantifying over "all possible announcements" is somewhat inaccurate, given that some public events are not announcements (a typical example is the 'announcement' of no child stepping forward in the muddy children problem). In fact, the logic PAL is a logic of publicly *observed* events, but where these events should be seen as external to the modelled system. Now in PAL there is a trick to view the announcement made by an agent modelled in the system (i.e., for which we have an agent name i in the Kripke model, with an associated accessibility relation) as such an external event.

When an agent in the system announces that φ is true, and it is common knowledge that all announcements are truthful, it is not merely φ that is made public but also that fact that the announcer knew φ at the moment the announcement was made. The latter information is implicitly "announced" by the act of announcing φ. Thus, public announcements made by agents in the system are of the form $K_i\varphi$, where i is the agent making the announcement. This backdoor makes it possible to introduce a form of agency in dynamic epistemic logic, and thus coalitional operators.

Group Announcement Logic (GAL) [4] has modal operators $\langle i \rangle$, where i is an agent, quantifying over the announcements that can be made by i. A formula of the form $\langle i \rangle \varphi$ thus means that i can make some announcement such that φ becomes true.

This is naturally extended to *group announcements*. A group announcement for a group $G \subseteq N$ is a formula of the form

$$\bigwedge_{i \in G} K_i \varphi_i$$

where $\{\varphi_i : i \in G\}$ are formulae. A group announcement is an action that can be made collectively by the group, each member announcing (that she knows) some formula (although "can" here does not say anything about how they *agree* on which group announcement to make, it simply means that this is a possible outcome if everyone announces something). Quantification over group announcements gives an interpretation of coalition operators $\langle G \rangle$: $\langle G \rangle \varphi$ means that G can make a group announcement such that φ will become true. Note that there is a fundamental difference to the interpretation in coalition logic here, in addition to the fact that actions are announcements, namely that the interpretation here does not take into account that other agents can act at the same time. This point is further discussed in the next section.

Formally, the language $\mathscr{L}_{GAL}$ of GAL is defined by the following grammar, extending PAL with coalition operators:

$$\varphi ::= p \mid \neg\varphi \mid \varphi \wedge \varphi \mid K_i\varphi \mid [G]\varphi \mid [\varphi]\varphi$$

where $i \in N$, $G \subseteq N$. The dual $\langle G \rangle \varphi$ is defined by abbreviation as $\neg[G]\neg\varphi$. The interpretation in pointed Kripke models is as for PAL, extended with the following clause:

$\mathscr{M}, s \models [G]\varphi$ iff for every set $\{\psi_i : i \in G\} \subseteq \mathscr{L}_{EL}$, $\mathscr{M}, s \models [\bigwedge_{i \in G} K_i\psi_i]\varphi$

and the interpretation of the dual becomes:

$\mathscr{M}, s \models \langle G \rangle \varphi$ iff there exists a set $\{\psi_i : i \in G\} \subseteq \mathscr{L}_{EL}$ such that $\mathscr{M}, s \models \langle \bigwedge_{i \in G} K_i\psi_i \rangle \varphi$

As an example, the PAL formula

$$\langle K_i\varphi_i \rangle \langle K_j\varphi_j \rangle (K_i\psi \wedge K_j\psi \wedge \neg K_k\psi)$$

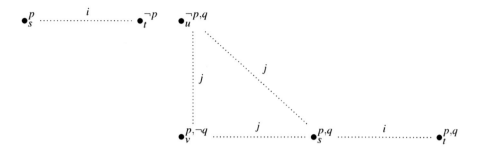

Fig. 16.6 Model M_1 (*left*) and M_2 (*right*)

expresses the fact that i can make the announcement φ_i after which j can make the
announcement φ_j after which both i and j will both know some formula ψ but agent
k will not. In GAL, however, this can be weakened to

$$\langle i \rangle \langle j \rangle (K_i \psi \wedge K_j \psi \wedge \neg K_k \psi)$$

expressing the fact that i and j can make *some* announcements (in the same sequence)
achieving the goal. The GAL formalism can potentially be useful to model check
security protocols, e.g., to answer questions of the type "does there exist a protocol
consisting of an announcement by i followed by an announcement by j achieving
the goal that i, j both know ψ but such that (eavesdropper) k remains ignorant."

In GAL the coalition operator has the following property [4]:

$$\models \langle G \rangle \langle G \rangle \varphi \leftrightarrow \langle G \rangle \varphi \tag{16.2}$$

This is significant, because it means that an alternative, equivalent, interpretation
of $\langle G \rangle \varphi$ is that "G can make a *sequence* of announcements after which φ will
become true". This seems like a type of property of interest, e.g., for model checking
communication and also for security protocols. However, it also might seem slightly
suspicious: doesn't it mean that all communication protocols can be reduced to
a single message in each direction? This is true in one sense, but here we must
be careful to discern between the different variants of "ability" under incomplete
information that we discussed in Sect. 16.3.

A natural question is whether, given the intimate connection between knowledge
and action in GAL, the mentioned variants of ability (in Sect. 16.3.2) actually are
different also in GAL or whether they coincide. That they indeed are different is
illustrated in Fig. 16.6. We have that:

- We have that $M_1, s \models \langle i \rangle p \wedge \neg K_i \langle i \rangle p$: agent i has the ability to make p come
  true in state s, but she does not know it. This is trivial, as public announcements
  do not change the value of factual propositions, and p was already true in s.

axiom schemata and rules of EL

$[G]\varphi \to [\bigwedge_{i \in G} K_i \psi_i]\varphi$ where $\psi_i \in \mathscr{L}_{EL}$ group and specific announcement
From φ, infer $[\psi]\varphi$ necessitation of announcement
From φ, infer $[G]\varphi$ necessitation of group announcement
From $\varphi \to [\theta][\bigwedge_{i \in G} K_i p_i]\psi$, infer $\varphi \to [\theta][G]\psi$ deriving group announcement / $R([G])$
        where $p_i \notin \Theta_\varphi \cup \Theta_\theta \cup \Theta_\psi$

Fig. 16.7 Axiomatisation of GAL. Θ_φ denotes the set of atomic propositions occurring in φ

- Let $\varphi = K_j q \wedge (\neg K_j p \vee \hat{K}_i (K_j p \wedge \neg K_j q))$. $M_2, s \models K_i \langle i \rangle \varphi$: in this case, agent i knows *de dicto* that she can make φ true. In particular, she can announce $K_i q$ in s, and $K_i p$ in t. However, neither announcing $K_i q$ in t nor $K_i p$ in s makes φ true, and there is in fact *no* announcement that i can make in s and t that will both make φ true in s and in t. Thus, i does not know *de re* that she can make φ true.

Like for CLK, the combination of coalition logic and epistemic logic, the proposition that i knows *de dicto* that she can make φ true can in GAL be expressed by $K_i \langle i \rangle \varphi$. As discussed in Sect. 16.3, knowledge *de re* is not expressible in CLK. There are two main differences between CLK and GAL: first, actions in GAL are of a particular type, namely truthful announcements, and hence action and knowledge are inter-related, and, second, the interpretation of coalition modalities in CL(K) uses a double $\exists\forall$ quantification, where the second quantifier ranges over all possible announcements by "the other" agents. However, note that if we restrict the set of agents to only a single agent, the $\exists\forall$ pattern for a coalition operator $\langle i \rangle$ becomes just $\exists$ like in GAL—and knowledge *de re* is still not expressible in single-agent CLK as shown in Sect. 16.3.

Consider:

$$\langle i \rangle K_i \varphi \tag{16.3}$$

In single-agent CLK it expresses the fact that agent i can choose some action after which it is true that she knows φ. The meaning in GAL is the same, with the restriction that that action is a public announcement. It is therefore perhaps surprising that (16.3) in fact expresses knowledge *de re* in the GAL semantics: it is true exactly iff agent i knows *de re* that she can make φ true ([4, Proposition 26]) (in both the single- and multi-agent case). Note that this does not hold for CLK, when general actions are allowed: in the model in Fig. 16.4, proposition (16.3) with $\varphi = open$ is true but as discussed in Sect. 16.3 agent i does not know *de re* that she can open the safe.

Thus, when actions are truthful public announcements, knowledge *de re* of ability can be expressed, unlike in the case of general actions. This is due to the intimate connection between actions as public announcements and knowledge, and it also depends on the S5 properties of knowledge.

Figure 16.7 shows a sound and complete [4, Theorem 15] axiomatisation of GAL. It is also known that the model checking problem for GAL (and APAL) is PSPACE-complete [4, Theorem 15].

When it comes to expressiveness, it is known that GAL is strictly more expressive than EL and PAL in the case of more than one agent but equally expressive in the single-agent case, and that GAL is not as expressive as APAL (i.e., there are APAL formulae that cannot be expressed in GAL). It is not known whether APAL is strictly more expressive than GAL. An open question is if the logics are incomparable.

16.4.3 Coalition Announcement Logic

As mentioned, a fundamental difference between GAL and CL is that the interpretation of the coalition modalities in the former does not take into account that other agents can act at the same time. In CL $\langle G \rangle$ is interpreted using a $\exists\forall$ pattern ("there exist actions for agents in G such that for all simultaneous actions for agents in $N \setminus G$ ..."), while in GAL $\langle G \rangle$ is interpreted like a normal modality, using the $\exists$ quantifier ("there exist actions for G.."). A consequence is that GAL *is* not a coalition logic; there are valid CL formulae that are not valid in GAL. For example, the superadditivity axiom S (Fig. 16.3) is not valid in GAL.

Coalition Announcement Logic (CAL) [2] is a variant of GAL where $\langle G \rangle \varphi$ means that G can make a group announcement such that no matter what the agents outside G announce at the same time, φ will be true. This brings us back full circle to coalition logic.

The language of CAL is the same as the language of GAL. The interpretation of the coalition operators is as follows:

> $M, s \models \langle G \rangle \varphi$ iff for every agent $i \in G$ there exists a formula $\psi_i \in \mathscr{L}_{EL}$ such that for every formula $\psi_j \in \mathscr{L}_{EL}$ for each of the agents $j \notin G$ we have that $M, s \models \bigwedge_{i \in G} K_i \psi_i \wedge [K_1 \psi_1 \wedge \cdots \wedge K_n \psi_n] \varphi$

This interpretation gives a $\exists\forall$ pattern of quantifiers. Note that in the definition the second quantifier is over all possible formulae for agents outside G, but the use of the "box" version of the public announcement operator ensures that only the formulae actually known by those agents play a role. This gives the following interpretation of the dual:

> $M, s \models [G] \varphi$ iff for all formulae $\psi_i \in \mathscr{L}_{EL}$ for every agent $i \in G$ there is a formula $\psi_j \in \mathscr{L}_{EL}$ for each of the agents $j \notin G$, such that $M, s \models \bigwedge_{i \in G} K_i \psi_i$ implies that $M, s \models \langle K_1 \psi_1 \wedge \cdots \wedge K_n \psi_n \rangle \varphi$

The logic CAL is a coalition logic, in the sense that all formulae that are valid in CL are valid also in CAL (the axioms in Fig. 16.3 are valid in CAL, and the rules are validity preserving). Of course, CAL and CL do not coincide; unlike the latter the former has both public announcement operators and epistemic operators. However, even the fragment of CL obtained by restricting the language to the CL language does not coincide with CL; the following is valid in CAL but not in CL (where p is an atomic proposition).

$$\langle G \rangle p \leftrightarrow p \qquad\qquad P$$

Valid properties of the interaction between coalition operators and public announcement operators include:

$$\langle K_1\psi_1 \wedge \cdots \wedge K_n\psi_n \rangle\varphi \rightarrow \langle N \rangle\varphi \qquad\qquad PAN$$
$$\langle \emptyset \rangle\varphi \rightarrow [K_1\psi_1 \wedge \cdots \wedge K_n\psi_n]\varphi \qquad\qquad PA\emptyset$$

Valid interaction properties of coalition operators and epistemic operators include (for any i and G):

$$\langle G \rangle \hat{K}_i\varphi \rightarrow \hat{K}_i \langle G \rangle\varphi \qquad\qquad KG$$

This formula is not valid in CLK, thus showing that CAL and CLK also are different.

The fact that CAL "is" a coalition logic, means that the coalition operators can be given a neighbourhood semantics. For simplicity, we consider the fragment without the public announcement operators, i.e., the fragment obtained by restricting the language to the CLK language $\mathscr{L}_{CLK}$:

$$\varphi ::= p \mid \neg\varphi \mid \varphi \wedge \varphi \mid K_i\varphi \mid [G]\varphi$$

In order to be able to interpret epistemic operators, we must extend coalition models with indistinguishability relations. An *epistemic coalition model (ECM)* is a tuple

$$\mathscr{M} = (S, E, V, \{\sim_1, \ldots, \sim_n\}),$$

where (S, E, V) is a coalition model and each $\sim_i$ is an epistemic indistinguishability relation. We can use ECMs to interpret coalition operators exactly as in coalition logic:

$$\mathscr{M}, s \models \langle G \rangle\varphi \Leftrightarrow \varphi^{\mathscr{M}} \in E(s)(G)$$

(and as usual for the other operators including epistemic operators). We can thus say that a pointed Kripke structure (M, s) and a pointed ECM $(\mathscr{M}, s')$ are $\mathscr{L}_{CLK}$-*equivalent* iff they satisfy the same formulae, i.e., if for all $\varphi \in \mathscr{L}_{CLK}$,

$$M, s \models \varphi \Leftrightarrow \mathscr{M}, s' \models \varphi \qquad\qquad (16.4)$$

In order to define an equivalent ECM from a Kripke structure, the state space must be extended in order to account for states corresponding to model updates. Any Kripke structure can be extended in a very simple way to a structure over which we can define an equivalent effectivity function, without changing satisfiability at any of the original states. Simply take the *power model* consisting of the union of all subsets of the original Kripke model. Formally, the power model $\hat{M}$ of a Kripke model M is defined as follows. A *submodel* of a Kripke model M is a model where the states are a subset of the states in M, and the valuation function and indistinguishability relations are restrictions of those in M to the state space of the submodel. The power model $\hat{M}$ of M is obtained by taking the disjoint union of M and every proper submodel of M, after renaming the states of the proper submodels such that the state spaces are disjoint. We say that two pointed Kripke models are $\mathscr{L}_{CLK}$-equivalent iff they satisfy the same $\mathscr{L}_{CLK}$ formulae. It is easy to see that for any model M and state s in M, (M, s) and $(\hat{M}, s)$ are $\mathscr{L}_{CLK}$-equivalent.

Given a Kripke model M, we can now define the induced ECM $\mathscr{M}^M$ as follows. $\mathscr{M}^M = (\hat{S}, E, \hat{V}, \{\hat{\sim}_1, \ldots, \hat{\sim}_n\})$ where $\hat{M} = (\hat{S}, \hat{V}, \{\hat{\sim}_1, \ldots, \hat{\sim}_n\})$ is the power model of M and

$$
X \in E(s)(G) \Leftrightarrow
\begin{cases}
\exists \varphi : \varphi^{\hat{M}} \subseteq X \text{ and } \hat{M}, s \models \langle G \rangle \varphi & G \neq N \\
\forall \varphi : \varphi^{\hat{M}} \subseteq \hat{S} \setminus X \text{ implies } \hat{M}, s \not\models \langle \emptyset \rangle \varphi & G = N
\end{cases}
$$

for any G and $s \in \hat{S}$. It can now be shown that for any s in M, (M, s) and $(\mathscr{M}^M, s)$ are $\mathscr{L}_{CLK}$-equivalent.

Interpretation of the full CAL language, with announcement operators, in induced power models is straightforwardly defined using submodels in the power model instead of model updates. Thus, for any Kripke structure there is an ECM which satisfies exactly the same CAL formulae.

16.4.4 Open Problems

There are several open theoretical problems related to meta-logical properties of GAL and CAL. First, let us look at GAL. Decidability of GAL is an open problem. It turns out that APAL is undecidable [55], but this result does not seem to be immediately be adaptable to GAL. While APAL is more expressive than GAL, it is not known whether it is *strictly* more expressive, i.e., whether APAL and GAL is expressively incomparable. Few meta-logical properties of CAL have been studied. The complete axiomatisation is an open problem. A complete characterisation of ECMs corresponding to Kripke models is not known. Finally, the relative expressivity between GAL and CAL is not completely understood. It may be that CAL is definable in GAL, but this has not been proved. If so, this would provide an interesting outlook on van Benthem's forcing operators [14] (an implementation so to speak), as they are intimately tied up to CAL—as for GAL, the operator here also quantifies over sequences of actions.

Many extensions of GAL and CAL are possible, e.g., with temporal modalities in the style of ATL, or with more sophisticated information-changing actions than public announcements.

16.5 Dynamic Epistemic Games

Dynamic epistemic logic describes actions available to agents; their epistemic preconditions and postconditions. Typically, an agent has several different actions available. Which action will she choose? That depends, of course, on her preferences over different resulting multi-agent epistemic states. Assuming such preferences, we have a game theoretic scenario, as epistemic states typically would depend on the actions of several agents [3]. The game-theoretic tool-chest can then be used to analyse which actions rational agents should or would choose. Here, we discuss two particular types of games. First, *public announcement games* where actions are public announcements, and, second, *question-answer games*, where actions are questions another agent is obliged to answer. Both types of games are strategic form games, and preferences are binary, represented as goal formulae.

16.5.1 Public Announcement Games

In [3] *public announcement games* (PAGs) are studied. These are strategic form games where actions are public announcements interpreted in a given Kripke structure. These games are implicit in a Kripke structure, under the additional assumption that each agent has a—typically epistemic—*goal formula* representing the preferences of the agent.

Formally, an *epistemic goal structure* (EGS) is a tuple $\langle M, \{\gamma_1, \ldots, \gamma_n\}\rangle$ where M is a finite Kripke model and where for each agent i the $\gamma_i \in \mathcal{L}_{PAL}$ is a goal formula. A *pointed epistemic goal structure* is a tuple (EGS, s) where $s \in M$. A strategic form game, *the state game* $\mathbf{G}(\text{EGS}, s)$, is associated with a state s of an epistemic goal structure EGS $= \langle M, \{\gamma_i, \ldots, \gamma_n\}\rangle$ as follows.

- $N = \{1, \ldots, n\}$,
- $A_i = \{\varphi_i \in \mathcal{L}_{PAL}\}$,
- $u_i(\varphi_1, \ldots, \varphi_n) = \begin{cases} 1 \text{ if } M, s \models \langle \underline{K}_1\varphi_1 \wedge \cdots \wedge \underline{K}_n\varphi_n\rangle\gamma_i \\ 0 \text{ otherwise} \end{cases}$

where $\underline{K}_i\varphi_i = K_i\varphi$ iff $M, s \models K_i\varphi_i$, and $\underline{K}_i\varphi_i = \neg K_i\varphi$, otherwise.

Strategies are possible announcements, and an agent gets a positive payoff iff her goal is satisfied after all agents make their announcement at the same time. An example is shown in Fig. 16.8, the EGS consists of the Kripke structure and the goal formulae γ_{Ann} and γ_{Bill}, and for each state in the Kripke structure the associated state game is shown.

The observant reader will have noted that state games by definition have infinitely many strategies, while the games in Fig. 16.8 shows only two strategies for each player in each state game. However, two strategies φ and ψ are *equivalent* for an agent i on a Kripke model M when $\{[\![K_i\varphi]\!], [\![\neg K_i\varphi]\!]\} = \{[\![K_i\psi]\!], [\![\neg K_i\psi]\!]\}$ (according to this simplification, Anne announcing p_A or $\neg p_A$ in Fig. 16.8 is in fact the same strategy). For a finite Kripke model there is always a finite number of equivalent strategies. If the model is bisimulation contracted there are exactly 2^{m-1} non-equivalent strategies for an agent with m equivalence classes.

State games are similar to Boolean games [61, 62] in that they have binary utilities represented by logical formulae. However, we cannot simply analyse a pointed public announcement game as a Boolean game, because (see Fig. 16.8) the agents typically

$$\gamma_{Ann} = (K_B p_A \vee K_B \neg p_A) \rightarrow (K_A p_B \vee K_A \neg p_B)$$
$$\gamma_{Bill} = (K_A p_B \vee K_A \neg p_B) \rightarrow (K_B p_A \vee K_B \neg p_A)$$

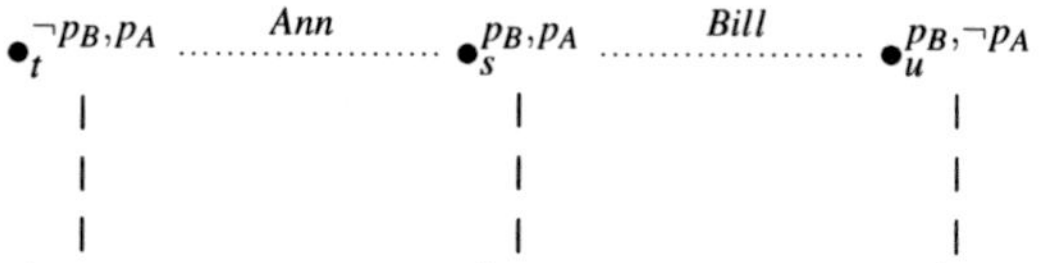

Fig. 16.8 Epistemic goal structure and associated state games

don't know the state game. In state s in the model in Fig. 16.8, *Ann* considers it possible that the game in state t is actually being played. And state games in different states might, e.g., have different Nash equilibria.

More sophisticated techniques must be used, and [3] explores two directions. First, in the tradition of epistemic logic, we can look at what agents *know*. For example, when do agents know that the state game has certain properties such as a Nash equilibrium? Second, in the tradition of game theory, we can use solution concepts developed particularly for games with incomplete information, such as Bayes-Nash equilibria.

When it comes to knowledge of game properties, let us first consider the existence of weakly dominant strategies. First, it might be the case that there is a weakly dominant strategy for one of the players, without that player knowing it. This is the case in state t in the model in Fig. 16.8, where p_A is weakly dominant for *Ann* without *Ann* knowing it (p_A is not weakly dominant in the other state she considers possible, s). But, second, the *de re/de dicto* disctinction (Sect. 16.3.2) again comes into play as well. It might be that an agent knows *that* there is weakly dominant strategy, without knowing *which strategy* is weakly dominant: different strategies may be weakly dominant in the different states she considers possible.

However, when an agent's goal is *positive*, i.e., the goal formula is in the positive fragment (see Sect. 16.2.2), it can be shown [3, Corollary 11] that an agent always has a weakly dominant strategy *de re* in any state game. Thus, *with a positive goal, the agent always knows what to do*.

What about knowledge of Nash equilibria? While for weakly dominant strategies it is enough that a single agent (the agent that can exectute the strategy) knows that it is weakly dominant, a Nash equilibrium should be common knowledge to be an realistic outcome. However, [3, Theorem 13] shows that *common knowledge of non-trivial Nash equilibria is impossible*, in the following sense: such equilibria must consist of the trivial announcement (the announcement $\top$ that does not change the model) for every agent.

	a_B^1	a_B^2	a_B^3	a_B^4
a_A^1	22	32	21	31
a_A^2	23	33	22	32
a_A^3	12	22	22	32
a_A^4	13	23	23	33

A_A

$a_A^1 : t,s \mapsto \top \quad u \mapsto \top$
$a_A^2 : t,s \mapsto \top \quad u \mapsto \neg p_A$
$a_A^3 : t,s \mapsto p_A \quad u \mapsto \top$
$a_A^4 : t,s \mapsto p_A \quad u \mapsto \neg p_A$

A_B

$a_B^1 : u,s \mapsto \top \quad t \mapsto \top$
$a_B^2 : u,s \mapsto \top \quad t \mapsto \neg p_B$
$a_B^3 : u,s \mapsto p_B \quad t \mapsto \top$
$a_B^4 : u,s \mapsto p_B \quad t \mapsto \neg p_B$

Fig. 16.9 Induced game for the EGS in Fig. 16.8. Strategies for *Ann* and *Bill*, resp., shown to the right. Payoffs are written without dividing by the number of states, for ease of presentation (the equilibria do not depend on this). The Nash equilibria are underlined

Turning to "standard" solution concepts, [3] attempts to view an epistemic goal structure as a single strategic form game. Given an epistemic goal structure $AG = \langle M, \{\gamma_1, \ldots, \gamma_n\}\rangle$ with $M = (S, \{\sim_1, \ldots, \sim_n\}, V)$, the *induced public announcement game* $\mathbf{G}(\text{EGS})$ is defined as follows:

- $N = \{1, \ldots, n\}$
- A_i is the set of functions $a_i : S \to \mathscr{L}_{PAL}$ with the following property:

  – Uniformity: $s \sim_i t \Rightarrow a_i(s) = a_i(t)$

- The payoffs are defined as follows. For any state s in AG, let $\mathbf{G}(AG, s) = (N, \{A_i^s : i \in N\}, \{u_i^s : i \in N\})$ be the state game associated with s (see above). Define, for any $(a_1, \ldots, a_n) \in A_1 \times \cdots \times A_n$:

$$u_i(a_1, \ldots, a_n) = \frac{\sum_{s \in S} u_i^s(a_1(s), \ldots, a_n(s))}{|S|}$$

These induced public announcement games are of course Bayesian games [63]. Nash equilibria of the induced game are exactly the Nash equilibria of the Bayesian game when defined from the EGS in a natural way.

The game induced from the EGS in Fig. 16.8 is shown in Fig. 16.9.

In the example in Fig. 16.8 the Nash equilibria of the induced game are "composed" of Nash equilibria of the induced games: (a, b) is a Nash equilibrium iff $(a(s'), b(s'))$ is a Nash equilbrium in the state game in s' for every state s'. But this is in general not the case: that (a, b) is a Nash equilibrium of the induced game is neither sufficient nor necessary for $(a(s'), b(s'))$ to be a Nash equilibrium in all state games. For positive goals, agents have weakly dominant strategies also in the induced game [3, Corollary 17].

16.5.2 Question-Answer Games

Instead of a game where the strategic moves are *announcements*, we can also conceive a deceptively similar (but in fact quite different) game where the strategies are *questions* and where those questions are answered. Such *question-answer games*

(QAGS) are studied in [5]. Instead of an announcement p ('p is true') one can be asked the question $?p$ ('Is p true?') and if your answer is 'yes', then the effect of that announcement and that question-answer combination, should be the same. Indeed, it is, of course the same. The requirement that announcements are truthful corresponds to the requirement that questions are answered truthfully. From here on, though, the picture diverges quickly. For example, to a typical question there are three answers: 'yes', 'no', and 'I don't know'. Nothing like that comes close to anything found in public announcement games. In public announcement games (over finite models), an agent can make a most informative announcement—declaring the actual equivalence class—and clearly *knows* what that most informative announcement is. But although there is a question with a most informative answer, the agent asking the question typically does not know which of all possible question elicits that answer. Here we present a bit of such question-answer games and emphasize in what respect they differ from public announcement games.

As an example, consider the Kripke model in Fig. 16.8. If the actual state is s, i.e., when p_A and p_B are both true, *Ann* is uncertain about p_B, and *Bill* is uncertain about p_A. In this model, given state s, *Ann* can make two different announcements: announce $K_a p_A$, which results in the model restriction to t and s, and announce $\top$, which does not result in a model restriction. Given the same model, and the same actual state s, *Ann* can also ask two different questions: she can ask for the truth about p_B, the strategy p_B?, or she can ask the trivial question, $\top$?. Now, the partition in fact is different: the answer to the first question is "yes" and the resulting restriction the states s and u. The answer to the second, trivial, question if of course also "yes". But that is a different partition, of course, than in the public announcement game.

In a game of questions, the strategic options for the player asking consist of the different informative answer she can receive in response from the other player. So we have a dual form of game. Well, if it's merely dual, we are quickly done. Instead of players making simultaneous announcements, we have players posing questions simultaneously and then answering each others' questions (in whatever order, simply assuming the obligation to answer the question truthfully). For the induced game, just take some mirror image from the public announcement game matrix, and there we have a question-answer game matrix. But it is not that simple. Because, in the induced game, the strategies are conditional to the equivalence classes of the questioning player, whereas the answers depend on the equivalence classes of the answering player. If a player x has 2 equivalence classes, so can always make 2 different announcements, whereas the player y has 3 equivalence classes, so can always make $2^2 = 4$ different announcements, then in the induced public announcement game a has $2 \cdot 2 = 4$ induced strategies, whereas b has $3 \cdot 4 = 12$ strategies. On the other hand, in the "similar" question-answer game, a has $2 \cdot 4 = 8$ induced strategies, whereas b has $3 \cdot 2 = 6$ strategies. A completely different game. In the three-state model example, *Ann* has four strategies in the induced game, namely, in words: 'If I know p_A then I ask p_B?, otherwise I ask p_B?', 'If I know p_A then I ask p_B?, otherwise I ask $\top$?', 'If I know p_A then I ask $\top$?, otherwise I ask p_B?', 'If I know p_A then I ask $\top$?, otherwise I ask $\top$?'.

Fig. 16.10 Example with
three answers to a question

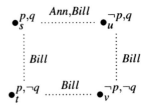

Let us look at an example wherein there are *three* answers to a question. Consider two players *Ann* and *Bill* who are uncertain about propositions p and q. In fact, b is utterly uncertain which of the four different valuations of p and q is the case, but a knows a bit more, either a knows that q but is uncertain about p, or a knows that p is true and q false, or a knows that p and q are both false. And these are the commonly known options to both players. The model of initial uncertainty is therefore as shown in Fig. 16.10.

What questions can *Bill* ask to *Ann*? Well, first, *Bill* can ask 'q?' to *Ann*. In that case, either *Ann* answers 'yes' (I know that q is true) or *Ann* answers 'no' (I know that q is false). Alternatively, *Bill* can ask 'p?' to *Ann*. Now, there are three possible answers, namely, 'yes', 'no', and 'don't know'. On information theoretic grounds, the second question should surely be preferred over the first one: as the answer partition due to the second question is a refinement of the answer partition due to the first question, an answer to the second question is always at least as informative as an answer to the first. Although, clearly, when modelling this as a strategic question-anwer game, both questions are simply different strategies in the game. Maybe *Bill*'s goal is to *prevent* becoming informed, so there is no immediate relation to payoff here.

The formal framework [5] for such games is as follows. In this case we assume merely two agents (instead of n—of course there are more-than-two-agent settings in which questions and answers make sense but the two-agent setting is the most natural), *Ann* and *Bill*, an epistemic model $M = (S, \{\sim_{Ann}, \sim_{Bill}\}, V)$ and two goal formulas γ_{Ann} and γ_{Bill}. In order to achieve their goals, agent *Ann* asks a question φ? to agent *Bill*, to which *Bill* is obliged to respond with 'yes' (I know that φ), 'no' (I know that $\neg\varphi$), or 'don't know' (I don't know whether φ). And similarly for *Bill* asking a question to *Ann*. We don't want to keep saying that all the time, so from now on we may refer to the two agents as i and j, where $i \neq j$, and i may be either *Ann* or *Bill*. We assume both agents ask their question simulaneously, and that subsequently both agents answer the question simultaneously.

Executing the strategy φ for agent i can be thought of as follows. Agent i asks φ? to j. If $M, s \models K_j\varphi$, then j answers (announces) "Yes, I know that φ". If $M, s \models K_j\neg\varphi$, then j answers "No, I know that $\neg\varphi$". Otherwise, j answers "I don't know whether φ". The resulting model restriction depends on both answers, e.g., if *Ann* asks φ? to which *Bill* responds $K_{Bill}\varphi$ and *Bill* asks ψ? to which

Ann responds $K_{Ann}\neg\psi$, the result is the restricted model $M|(K_{Bill}\varphi \wedge K_{Ann}\neg\psi)$. We can capture these alternatives with a construct $\overline{K}_i\varphi$, for 'agent i answers the question φ?', defined as follows. Given an epistemic model M and a state $s \in M$, if $M, s \models K_i\varphi$, then $\overline{K}_i\varphi \equiv K_i\varphi$; if $M, s \models K_i\neg\varphi$, then $\overline{K}_i\varphi \equiv K_i\neg\varphi$; and else $\overline{K}_i\varphi \equiv \neg(K_i\varphi \vee K_i\neg\varphi)$. This is reminiscent of the *resolution on φ* in [21].

The *state game* or *pointed question-answer game* $\mathbf{G}((M, s), \gamma_{Ann}, \gamma_{Bill})$ associated with state $s \in M$ of goals γ_{Ann} and γ_{Bill} for agents *Ann* and *Bill* respectively, is the strategic game defined by

- $N = \{Ann, Bill\}$;
- for $i = Ann, Bill$, $A_i = \{\varphi? \mid \varphi \in \mathscr{L}\}$;
- for $i = Ann, Bill$, $u_i^s(\varphi, \psi) = \begin{cases} 1 \text{ if } M, s \models \langle(\overline{K}_{Bill}\varphi \wedge \overline{K}_{Ann}\psi)\rangle\gamma_i \\ 0 \text{ otherwise} \end{cases}$

The state independent perspective on question-answer games, named the induced question-answer game, or simply question-answer game, is defined completely analogously to the induced public announcement game.

Given how we define strategies here, the notion for 'strategy equivalence' now becomes as follows. Let a model M be given. Two strategies $\varphi?$ and $\psi?$ for a pointed question-answer game for M are the same (equivalent) for agent i (Anne, or Bill) if

$$\{[\![K_j\varphi]\!]_M, [\![K_j\neg\varphi]\!]_M, [\![\neg(K_j\varphi \vee K_j\neg\varphi)]\!]_M\} =$$
$$\{[\![K_j\psi]\!]_M, [\![K_j\neg\psi]\!]_M, [\![\neg(K_j\psi \vee K_j\neg\psi)]\!]_M\}$$

where j is the other agent. Note that it is common knowledge to *Ann* and *Bill* if two strategies are the same, as we are comparing the denotations of formulas involving φ and ψ in the model, independent of the actual state. In the example in Fig. 16.10 above, b's question $?p$ is equivalent to the question $?\neg p$: they both result in the same partition into three a-classes.

We finish this small introduction to this variant of public announcement game by noting another difference between the two sorts of games. From a player's perspective, there is such a thing as a most informative announcement (tell them all you know). If all goals are positive, then the most informative announcement is a weakly dominant strategy in all points of the public announcement game. And because players know what their most informative announcement is, this is therefore an equilibrium strategy of the induced public announcement game. But the question that elicits the most informative answer from another player cannot be called the most informative question from the questioning player's point of view. In a different state in the same equivalence class for that player, the question to elicit the most informative answer may be a different question, as the responding player may be in a different equivalence class there. So even when all goals are positive, induced question-answer games may not have an equilibrium.

16.5.3 Open Problems

There are many interesting directions to go in, and various such directions are described in detail in the Chapter 'Knowledge Games' of [18]. On the logic side, we can consider different types of actions, apart from public announcements, we can also consider private announcements, and general action models. On the game theoretic side, different types of games and frameworks can be considered: strategic games, extensive games, even social choice [45] and mechanism design models [26]. The representation of preferences can also be generalized in different directions: instead of a single goal formula, we can consider sets of preferentially ordered goal formulae, or weighted formulae [49, 69, 76, 92], or CP-nets [28–30], and many other choices.

16.6 Knowledge in Real Games: From Chess to Sudoku

The playground where logical dynamics, agency, and game theory interact seems fairly large, as we have already seen, and there are many pieces of the puzzle we haven't mentioned yet. In multi-agent dynamics, the phenomena involving interaction of different subgroups are more interesting than those involving single agents and the set of all agents (the 'public') only. In the detailed dynamic scenarios in public announcement and question answer games we found a challenging way in which the moves in a game themselves are defined from a structure representing agents' uncertainty. However it may not have gone unnoticed that the dynamics were restricted to those of a public kind: public announcements. In this section we review some games that have a more challenging dynamics and report on the status quo of their analysis. For this, we resort to 'real' games, i.e., games like bridge and chess that are played for enjoyment. They are above the level of the 'small' (anything less than 20 states, say) Kripke models illustrating epistemic dynamics and agent interaction, but they are below the level of 'large' applied multi-agent systems with advanced knowledge representation involving ontologies etc.

16.6.1 Chess

Let us start with chess. For someone trying to join the interests of game theory and (multi-agent epistemic) logic, this is not the most interesting game. Two agents only, whereas the interesting phenomena begin with more than two. Perfect knowledge, whereas the interesting phenomena begin with imperfect knowledge. There are still interesting higher-order epistemic phenomena in opponent modelling in two-person perfect information games [48]. This is an interesting link between epistemic logic and games, relating to epistemically motivated heuristics in search (as opposed to information-theoretic heuristics), in view of bounded rationality.

16.6.2 Bridge

There are many chess players among computer scientists and mathematicians and there also seem to be many bridge players among them. Bidding in Bridge has clearly epistemic aspects. First, there are four players, and not two, as in chess. That seems more promising already. All moves are public. That's more of a pity, because the dynamics of public moves is that of public announcements—and for the communicative complexities of public announcements it does not matter whether the number of players is one, two, or twenty. (What matters are the static multi-agent epistemic features of the announced information.) But bidding in Bridge seems like subgroup dynamics. The messages in the bidding process are supposed to be maximally informative for the other team player and minimally informative for the opponent team. Are there ways to convey a general pattern of your knowledge or ignorance, or even to indicate in a next bid that you have understood the message sent by your team player in a prior bid? Bidding in bridge has been investigated in [52, 93]. This work later developed into information based security protocols in cryptography [53], including more-than-two-principal protocols (in this community, a principal is an agent participating in a protocol). But the thing in Bridge is that bids have to be public, even when their impact and intention is different for different agents. We pursue our search for knowledge games.

16.6.3 Cluedo

Cluedo (for Americans, Clue), is a murder-mystery board game wherein six partying guests are confronted with a dead body, and they are all suspected of the murder. The game board depicts the different rooms of the house wherein the murder is committed, and there are also a number of possible murder weapons. Six suspects (such as Professor Plum), nine rooms, and six possible murder weapons. These options constitute a deck of 21 cards, one of each kind is drawn and are considered the real murderer, murder weapon, and murder room. The other cards are shuffled (again), and distributed to the players. The game consists of moves that allow for the elimination of facts about card ownership, until the first player to guess the murder cards correctly has won.

The part of the game that interests us is when on the game board a room is reached by a player who may then voice a suspicion, such as 'I think Ms. Scarlett did it, with a knife, in the kitchen'. This question is addressed to another player and interpreted as a request to admit or deny ownership of these cards for that player. If the addressed player doesn't have any of the requested cards, she says so, but if a player holds at least one of the requested cards, she is obliged to show exactly one of those to the requesting player, and to that player only. The four other players cannot see which card has been shown, but of course know that it must have been one of the three.

Denying that you have Ms. Scarlett, or knife, or kitchen, is a public event of the public announcement kind that we have already seen. But the show action is different. One of the three requested cards must be in possession of that player, but the other players do not know which one. So no particular deal of cards can be eliminated by any player: it is *not* a straightforward model restriction as in the 'I don't have any of these cards' response. Let us illustrate what is going on by a simpler situation, wherein there are only three cards red, white, and blue (r, w, and b) and three players 1, 2, and 3 each holding one card (and only knowing their card). Now, player 1 shows player 2 her card. The informative transition is depicted below.

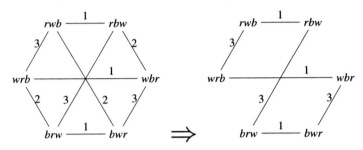

The name rwb at a node means that player 1 holds red, player 2 holds white, and player 3 holds blue, rbw that player 1 holds red, player 2 holds blue, and player 3 holds white, etc. Consider for a moment that rwb is the actual deal and that 1 shows red. As 2 holds white, 2 learns from that action what the card deal is. So 2's uncertainty, depicted as the 2-link between rwb and bwr, will disappear. But this, of course, will happen for any deal of cards. Also, 1 and 3 know that 2 will learn the card deal. The structure on the right incorporates all this information (see [35] for a detailed analysis of this phenomenon). Note that the number of distinct states in the structure has not changed, only the accessibility relation for one of the agents, 2.

In the red/white/blue action only a single card can be shown. In the Cluedo action, one of three cards is shown. This is a non-deterministic action, that may *in principle* increase the complexity of the epistemic state representing the uncertainty of players about the card deal and each other. An interesting observation about Cluedo is that this does not happen *in practice*, but only barely so. This sort of complexity of communication and uncertainty seems worthwhile to mention.

The complexity of a Kripke structure is determined by the number of its (non-bisimilar) worlds. (The standard measure is the number of worlds *and* the number of pairs in the acessibility relation, but for multi-S5 structures the latter is arguably a bad measure, as this would make the universal relation the most complex one; and anyway, to prove our point, the former is sufficient.) Suppose a player shows one of three requested cards, in Cluedo. If showing that card had been public, it would be a public announcement that one of your three cards is that card: a straightforward model reduction. There are 21 cards, and three cards per player. There are $\binom{21}{3}$ ways for a player to draw three cards from a pack of 21. If we assume a given card, then from the remaining 20 the player still has to draw two cards: $\binom{20}{2}$. Now $\binom{21}{3}$ divided by $\binom{20}{2}$ is $\frac{21 \cdot 20 \cdot 19}{3 \cdot 2 \cdot 1} \cdot \frac{2 \cdot 1}{20 \cdot 19} = \frac{3}{21}$. The uncertainty between which of three cards is shown means uncertainty (lots of links for the other players, but not for the showing and the requesting player) between three models of that size, so the epistemic complexity of the resulting model is $3 \cdot \frac{3}{21} = \frac{9}{21} < 1$: complexity *reduction*. Now suppose the player had not been asked to show one of three but one of eight cards: $8 \cdot \frac{3}{21} = \frac{24}{21} > 1$: a complexity *increase*. In other words, had the Cluedo rules been slightly different, the action of showing a card would have led to an increase and not to a decrease of complexity. Does the complexity decrease explain why Cluedo is 'playable'? Was this in the mind of the designer of that game (in the 1940s)? Can we design games based on complexity results for their dynamics?

There is yet another interesting aspect of the dynamic epistemics of Cluedo: you can win because someone else loses. In this way, it is not the strongest one-liner ever written, but let us explain. A move in Cluedo does not merely consist of asking other players questions about card ownership, but there is also the option to make an informed guess at what the murder cards are (and *only* the player whose turn it is may make that guess). This, you are allowed to do only once in the game. Let us now assume that all six Cluedo players are perfectly rational. Then, if you do *not* announce at the end of your move what the murder cards are, you do not know what they are yet—you have not eliminated a sufficient number of alternative card deals processing the answers to the question. This an informative public announcement. As a result of that, someone else may learn what the murder cards are. Let us give a concrete example wherein this 'ending your move'-aspect is an informative public announcement:

> Consider player 1, who starts the game, to reach the kitchen in her first move. She voices the suspicion 'I think that Scarlett has committed the murder in the kitchen with a knife'. None of the other players can show a card. Player 1 confidently writes down an accusation, checks the murder cards and announces that she has won.

Now, change the example a bit.

> Same as above, except that player 1 does not announce a win. The other players now deduce that player 1 holds at least one of the requested cards, because otherwise, she would have known what the murder cards are and would have made the accusation. But she didn't. So.

How about the game theory of Cluedo? The dynamic epistemic analysis of Cluedo game moves makes it possible to compute the precise informative effects of game actions: asking which of three cards another player has. If there is a measure for this complexity reduction in relation to the goal of the game (knowledge of the murder cards is an aspect of the valuation of a world in the Kripke model), and for each player, this would define the payoff for a move. Thus, we would have an extensive imperfect information game. It is not so clear what that measure is. This measure is about comparing sizes of different partitions of domains. Games like Mastermind face similar questions that are not satisfactorily answered, notwithstanding great efforts to solve them [82]. In Cluedo it is not even clear if 'asking for three cards that you do not have' is to be preferred over 'asking for three cards of which you hold one yourself'.

Other strategic considerations in Cluedo are even harder to model. It is your move, you know the murderer and the weapon but you still hesitate between the kitchen and the conservatory. Unless you guess the murder cards *now*, some other player may do so in the next round, and may win. Should you guess or not? The game of Pit, to be summarily discussed next, offers some more insight.

Logical dynamics of Cluedo and similar 'knowledge games' consisting of informative actions, are studied in [33, 36, 37, 47, 89]. The last interestingly uses temporal logics of knowledge for specification and verification of Cluedo, not explicit dynamic epistemics.

16.6.4 Pit

Pit is one of the few other examples of card games with logical dynamics involving subgroups of the public. In the Pit game (for *trading* pit—it's a market simulation card game) the players try to corner the market in coffee, wheat, oranges, or a number of other commodities, and it is like the 'Family Game' in that each of these commodities are distributed over the players in the form of cards, and the first player to gather a full suit of cards (i.e., nine cards) of any commodity, wins. The game moves consist of two players exchanging cards. This goes about as follows. A requirement to exchange cards is that they are of the same suit. Players shout the number of cards they wish to exchange, simultaneously, and two players shouting the same number may then make a change. For example, John has 2 apples, 3 oranges, and some other cards, Mary has 2 oranges and yet other cards. John could have shouted 1, 2, or 3 (changing some but not all of the cards of the same suit is also allowed), but goes for 2, and Mary goes for 2 as well. Shout, shout, ... And they make an exchange. John now has 5 orange cards! Still not 9, but better than before. The exchange action is somewhat similar to the move of showing a card in Cluedo: two players gain subgroup common knowledge, in this case, of the new ownership of the exchanged cards. The other players only learn that two players each have at least two cards of the same suit. This rules out some card distributions. This exchanging of cards continues until somebody gathers his suit of nine cards. This can, in principle, go on forever: it is an extensive game of imperfect information, with infinite (but highly repetitive, there is much symmetry) game tree branches. The game is investigated in [38, 41, 68, 87], where the first two publications focus on game theory, the third on asynchronous protocols, and the last on dynamic epistemics. In [38] some pure and mixed equilibria are given for a one-round strategic game simplification of a Pit game for three players and six cards (there is a clear direction how to generalize this to equilibria for the infinite version of the (extensive) game, using the symmetry in the infinite branches.

16.6.5 Sudoku

Unlike Cluedo and Pit, we assume that Sudoku does not need an introduction. Sudoku is a single-person game. It is merely combinatorics and elimination. Still, it has an interesting epistemic aspect that, we guess, surely the designers did not anticipate. This is rather the single player of the game, the Sudoku solver, playing against that designer. But again it reveals a truly dynamic epistemic aspect. Consider the following interim status of a Sudoku puzzle—we are interested in the remaining options for four relevant cells, the content of the other cells, blank in the figure, are irrelevant (they can be initially given, already determined, or still undetermined, i.e., 'blank' in the usual Sudoku sense):

67		67
67		267

So, in three of these four cells, 6 and 7 are the only remaining options and in the fourth cell 2, 6, and 7 are the remaining options. This is an argument demonstrating that that cell *must* contain the number 2. Suppose it is not 2. Then, there are two ways for a 6 and a 7 to be in the same column, row, and $3 \cdot 3$ block, that would lead towards solutions:

6		7		7		6
---	---	---		---	---	---
7		6		6		7

But a Sudoku puzzle always has a unique solution. So none of these two can be right. Therefore, the cell must contain a 2:

67		67
67		2

This phenomenon is treated in [40] and in [96], and possibly in numerous other (also) independent publications. It is a non-standard Sudoku move... Notice how *epistemic* this rule is, in the same sense as the Cluedo game move that makes you win because, on the assumption of perfect rationality, the other player did not declare a win. Assuming that the game is well-designed and only has one solution, a reasonable assumption, this elimination will eventually lead us to the solution. But if the designer intended some (6, 7, 6, 7) combination but overlooked the other solution, putting 2 in that place may result in no solution at all, instead of one of two. And this overlooking happens. The French newspaper Le Monde of 4 June 2008 had a Sudoku with two solutions. This did not result in indignant Letters to the Editor.

16.7 Conclusions: Logic in Games

We presented some results on the fine distinctions of 'de re' and 'de dicto' knowledge for individual agents and for coalitions; results on the power, knowledge, and ability of coalitions; how in dynamic epistemic logic we can simulate a coalitional notion of agency by quantifying over information change; and finally how various very concrete knowledge games present problems going beyond our presented solutions so far.

In a way, many of our proposals can be seen as implementations of the general program recently outlined in [16, 18], not surprisingly so, as this general program goes back a long way. In this contribution we only glanced at some other aspects involving knowledge and games, such as the characterization of game theoretic objects.

Noting the similarity between models of modal logic and extensive form games, van Benthem [14] suggests the use of propositional dynamic logic [60] for expressing properties of such games. He also proposes the use of PDL extended with epistemic operators to reason about extensive-form games extended with indistinguishability relations over the states (a slight generalisation of the standard information sets used in game theory to model imperfect information). This 2001 publication is also a source for topics as diverse as the distinction between 'knowledge de re' and the 'knowledge de dicto' that we discussed in Sect. 16.3.2.

In [15] van Benthem shows that solution concepts for games can be characterised using epistemic updates, such as *iterative elimination of strictly dominated strategies* (the IESDS algorithm). In [15], a Kripke model is obtained from a strategic form game by assuming that an agent knows her own strategy, but considers all possibilities of strategies chosen by other players to be open. This allows for an epistemic explanation of that IESDS algorithm. While the epistemic foundations for solution concepts, including the role of common knowledge of rationality in the IESDS algorithm, have been studied by many [8, 15, 27, 91], van Benthem [15] provides a new *dynamic* perspective.

van Benthem [15] also considers announcement of a stronger variant of rationality, *strong rationality*. Strong rationality is true if the current action of the player is a best response against *at least one* possible action of the opponent. It is argued in [15] that repeated announcements of strong rationality for both players correspond to Pearce's game-theoretic algorithm of *rationalizability*. Further in [15], it is noted that, in general, the use of epistemic actions to characterise game solutions may be *order dependent* in the sense that the effect of sequences of different epistemic assertions might depend on the order; [79] refines the approach in [15], in particular by studying conditions under which a stable outcome is determined independently of the order of the iterative information updates. The general approach in [15] is taken further in van Benthem's more recent [19]. This studies solutions for extensive form games via epistemic updates in detail.

In the chapter 'Knowledge Games' of [18], the public announcement games and question-answer games that we discussed in Sects. 16.5.1 and 16.5.2 are presented in detail. Somewhat beyond the scope of our logics with quantification over information change with a coalitional interpretation—the group announcement logic and coalition announcement logic—is the chapter on 'Sabotage games'. Removing an arrow from a model, *any* arrow, is yet another quantification on a different level from 'take any definable subset' (as in APAL). Concerning the chapter 'Forcing Powers' that presents the forcing modality in various incarnations, we mention again the possible definability of forcing in the coalition announcement logic (CAL—Sect. 16.4.3) or even already in group announcement logic, a challenging future case study. The *uniform forcing strategy* suggested in that 'Forcing Powers' chapter seems, as ever, to open up novel interactions between knowledge and games, that we have not addressed in this contribution.

A topic we have left aside is how various *strategy logics*: (i) contribute to the analysis of concrete knowledge games, (ii) as well as define (or indicate) versions of the coalitional logics presented in Sect. 16.3, (iii) or define versions of group announcement logic or coalition announcement logic within more restricted game rules (given strategies). Interesting is also how different agent roles ('types') can fix their strategies in the view of other agents, thus restricting search (and determining outcomes) [78], or the role of observation expectations on strategies [42]. Strategies, games, and logic are treated integrally in [17, 58, 78].

References

1. Ågotnes T (2006) Action and knowledge in alternating-time temporal logic. Synthese (Special Section on Knowledge, Rationality and Action) 149(2):377–409
2. Ågotnes T, van Ditmarsch H (2008) Coalitions and announcements. In: Padgham L, Parkes D, Muller J, Parsons S (eds) Proceedings of the 7th AAMAS. IFAMAAS/ACM DL, pp 673–680
3. Ågotnes T, van Ditmarsch H (2010) What will they say?—public announcement games. Synthese (Special Section on Knowledge, Rationality and Action) (to appear)
4. Ågotnes T, Balbiani P, van Ditmarsch H, Seban P (2010) Group announcement logic. J Appl Logic 8(1):62–81
5. Ågotnes T, van Benthem J, van Ditmarsch H, Minica S (2011) Question-answer games. J Appl Non-Class Logics 21(3–4):265–288
6. Alur R, Henzinger TA, Kupferman O (2002) Alternating-time temporal logic. J ACM 49:672–713
7. Aucher G (2003) A combined system for update logic and belief revision. Master's thesis, ILLC, University of Amsterdam, Amsterdam, the Netherlands, ILLC report MoL-2003-03
8. Aumann R, Brandenburger A (1995) Epistemic conditions for nash equilibrium. Econometrica 63(5):1161–1180
9. Balbiani P, Baltag A, van Ditmarsch H, Herzig A, Hoshi T, de Lima T (2007) What can we achieve by arbitrary announcements? A dynamic take on Fitch's knowability. In: Samet D (ed) Proceedings of TARK XI. Presses Universitaires de Louvain, Louvain-la-Neuve, Belgium, pp 42–51
10. Balbiani P, Baltag A, van Ditmarsch H, Herzig A, Hoshi T, de Lima T (2008) Knowable' as known after an announcement. Rev Symbol Logic 1(3):305–334
11. Baltag A, Moss LS, Solecki S (1998) The logic of public announcements, common knowledge, and private suspicions. In: Gilboa I (ed) Proceedings of the 7th conference on theoretical aspects of rationality and knowledge (TARK 98), pp 43–56
12. Baltag A, Smets S (2008) A qualitative theory of dynamic interactive belief revision. In: Proceedings of 7th LOFT, texts in logic and games 3. Amsterdam University Press, pp 13–60

13. Belnap N, Perloff M (1988) Seeing to it that: a canonical form for agentives. Theoria 54:175–199
14. van Benthem J (2001) Games in dynamic-epistemic logic. Bull Econ Res 53(4):219–248
15. van Benthem J (2007) Rational dynamics and epistemic logic in games. Int Game Theory Rev 9(1):13–45
16. van Benthem J (2011) Logical dynamics of information and interaction. Cambridge University Press, Cambridge
17. van Benthem J (2012) In praise of strategies. In: van Eijck J, Verbrugge R (eds) Games, actions and social software. LNCS vol 7010. Springer, Berlin pp 96–116
18. van Benthem J (2014) Logic in games. MIT Press, Cambridge
19. van Benthem J, Gheerbrant A (2010) Game solution, epistemic dynamics and fixed-point logics. Fundamenta Informaticae 100(1–4):19–41
20. van Benthem J, Liu F (2007) Dynamic logic of preference upgrade. J Appl Non-Class Logics 17(2):157–182
21. van Benthem J, Minica S (2009) Toward a dynamic logic of questions. In: He X, Horty JF, Pacuit E (eds) Logic, rationality, and interaction. Proceedings of LORI 2009. LNCS, vol 5834. Springer, Berlin, pp 27–41
22. van Benthem J (1989) Semantic parallels in natural language and computation. In Logic colloquium '87. North-Holland, Amsterdam
23. van Benthem J (2004) What one may come to know. Analysis 64(2):95–105
24. van Benthem J, van Eijck J, Kooi BP (2006) Logics of communication and change. Inf Comput 204(11):1620–1662
25. van Benthem J, Gerbrandy JD, Hoshi T, Pacuit E (2009) Merging frameworks for interaction. J Philos Logic 38:491–526
26. Bergemann D, Morris S (2005) Robust mechanism design. Econometrica 73(6):1771–1813
27. Binmore K (1992) Fun and games: a text on game theory. D. C. Heath and Company, Lexington, MA
28. Boutilier C, Brafman RI, Domschlak C, Hoos HH, Poole D (2003) CP-nets: a tool for representing and reasoning with conditional ceteris paribus preference statements. J Artif Intell Res 21:135–191
29. Boutilier C, Brafman RI, Hoos HH, Poole D (2004) Preference-based constrained optimization with CP-nets. Comput Intell 20(2):137–157
30. Boutilier C, Brafman RI, Hoos HH, Poole D (1999) Reasoning with conditional ceteris paribus preference statements. In: Proceedings of UAI99, pp 71–80
31. Broersen J, Herzig A, Troquard N (2006) Embedding Alternating-time Temporal Logic in Strategic STIT logic of agency. J Logic Comput 16(5):559–578
32. Broersen J, Herzig A, Troquard N (2007) A normal simulation of coalition logic and an epistemic extension. In: Proceedings of the 11th conference on theoretical aspects of rationality and knowledge, TARK '07. ACM, New York, pp 92–101
33. van Ditmarsch H (1999) The logic of knowledge games: showing a card. In: Postma E, Gyssens M (eds) Proceedings of the BNAIC 99. Maastricht University, pp 35–42
34. van Ditmarsch H (2000) Knowledge games. PhD thesis, University of Groningen, ILLC Dissertation Series DS-2000-06
35. van Ditmarsch H (2002) Descriptions of game actions. J Logic Lang Inf 11:349–365
36. van Ditmarsch H (2002) The description of game actions in Cluedo. In: Petrosian LA, Mazalov VV (eds) Game theory and applications, vol 8. Nova Science Publishers, Commack, NY, USA, pp 1–28
37. van Ditmarsch H (2002) Knowledge games. Bull Econ Res 53(4):249–273
38. van Ditmarsch H (2004) Some game theory of Pit. In: Zhang C, Guesgen HW, Yeap WK (eds) Proceedings of PRICAI 2004 (Eighth pacific rim international conference on artificial intelligence). LNAI, vol 3157. Springer, Berlin, pp 946–947
39. van Ditmarsch H (2005) Prolegomena to dynamic logic for belief revision. Synthese 147:229–275 (Knowledge, Rationality and Action)
40. van Ditmarsch H (2006) Serious play. Junctures 7:87–98 The Journal for Thematic Dialogue

41. van Ditmarsch H (2006) The logic of pit. Synthese 149(2):343–375 (Knowledge, Rationality and Action)
42. van Ditmarsch H, Ghosh S, Verbrugge R, Wang Y (2011) Hidden protocols. In: Apt KR (ed) Proceedings of 13th TARK, pp 65–74
43. van Ditmarsch H, van der Hoek W, Kooi BP (2005) Dynamic epistemic logic with assignment. In: Proceedings of the fourth international joint conference on autonomous agents and multi-agent systems (AAMAS 05). ACM Inc., New York, pp 141–148
44. van Ditmarsch H, van der Hoek W, Kooi BP (2007) Dynamic epistemic logic. Synthese library series, vol 337. Springer, Berlin
45. van Ditmarsch H, Lang J, Saffidine A (2013) Strategic voting and the logic of knowledge. In: Proceedings of 14th TARK—Chennai
46. van Ditmarsch H, Kooi B (2006) The secret of my success. Synthese 151:201–232
47. Dixon C (2006) Using temporal logics of knowledge for specification and verification-a case study. J Appl Logic 4(1):50–78
48. Donkers HHLM, Uiterwijk JWHM, van den Herik HJ (2003) Admissibility in opponent-model search. Inf Sci Inf Comput Sci 154(3–4):119–140
49. Elkind E, Goldberg LA, Goldberg P, Wooldridge M (2009) A tractable and expressive class of marginal contribution nets and its applications. Math Logic Q 55(4):362–376
50. Emerson EA (1990) Temporal and modal logic. In: van Leeuwen J (ed) Handbook of theoretical computer science volume B: formal models and semantics. Elsevier Science Publishers B.V, Amsterdam, The Netherlands, pp 996–1072
51. Fagin R, Halpern JY, Moses Y, Vardi MY (1995) Reasoning about knowledge. The MIT Press, Cambridge, MA
52. Fagin R, Naor M, Winkler P (1996) Comparing information without leaking it. Commun ACM 39(5):77–85
53. Fischer MJ, Wright RN (1996) Bounds on secret key exchange using a random deal of cards. J Cryptology 9(2):71–99
54. Fitch FB (1963) A logical analysis of some value concepts. J Symbol Logic 28(2):135–142
55. French T, van Ditmarsch HP (2008) Undecidability for arbitrary public announcement logic. In: Areces C, Goldblatt R (eds) Advances in modal logic. College Publications, pp 23–42
56. Gerbrandy JD (1999) Bisimulations on Planet Kripke. PhD thesis, University of Amsterdam, ILLC Dissertation Series DS-1999-01
57. Gerbrandy JD, Groeneveld W (1997) Reasoning about information change. J Logic Lang Inf 6:147–169
58. Ghosh S, van Benthem J, Verbrugge R (ed) (2013) Modeling strategic reasoning (to appear)
59. Goranko V, Jamroga W, Turrini P (2010) Strategic games and truly playable effectivity functions. In: Proceedings of the 8th European workshop on multi-agent systems, Paris, France
60. Harel D (1984) Dynamic logic. In: Gabbay D, Guenthner F (eds) Handbook of philosophical logic, vol II: Extensions of classical logic, volume 165 of Synthese Library, chapter II.10. D. Reidel Publishing Co., Dordrecht, pp 497–604
61. Harrenstein P (2004) Logic in conflict. PhD thesis, Utrecht University
62. Harrenstein P, van der Hoek W, Meyer J-JCh, Witteveen C (2001) Boolean games. In: van Benthem J (ed) Proceeding of the eighth conference on theoretical aspects of rationality and knowledge (TARK VIII). Siena, Italy, pp 287–298
63. Harsanyi JC (1967–1968) Games with incomplete information played by 'Bayesian' players, parts I, II, and III. Manag Sci 14:159–182, 320–334, 486–502
64. Herzig A, Troquard N (2006) Knowing how to play: uniform choices in logics of agency. In: Proceedings of AAMAS'06, pp 209–216
65. Hintikka J (1962) Knowledge and belief. Cornell University Press, Ithaca, NY
66. van der Hoek W, Pauly M (2006) Modal logic for games and information. In: van Benthem J, Blackburn P, Wolter F (eds) The handbook of modal logic. Elsevier, Amsterdam, The Netherlands, pp 1152–1180
67. van der Hoek W, Wooldridge M (2003) Cooperation, knowledge and time: Alternating-time temporal epistemic logic and its applications. Studia Logica 75:125–157

68. Holt CA (1996) Trading in a pit market. J Econ Perspect 10(1):193–203
69. Ieong S, Shoham Y (2005) Marginal contribution nets: a compact representation scheme for coalitional games. In: Proceedings of the sixth ACM conference on electronic commerce (EC'05), Vancouver, Canada
70. Jamroga W (2003) Some remarks on alternating temporal epistemic logic. In: Proceedings on FAMAS'03—Formal approaches to multi-agent systems. Warsaw, Poland, pp 133–140
71. Jamroga W, Ågotnes T (2007) Constructive knowledge: what agents can achieve under imperfect information. J Appl Non-Class Logics 17(4):423–475
72. Jamroga W, Ågotnes V (2006) What agents can achieve under incomplete information. In: Proceedings of AAMAS'06. ACM Press, pp 232–234
73. Jamroga W, Dix J (2006) Model checking ATL$_{ir}$ is indeed Δ_2^P-complete. In: Proceedings of EUMAS'06
74. Jamroga W, van der Hoek W (2004) Agents that know how to play. Fundamenta Informaticae 63:185–219
75. Jonker G (2003) Feasible strategies in alternating-time temporal epistemic logic. Master thesis, University of Utrecht
76. Lang J, Endriss U, Chevaleyre Y (2006) Expressive power of weighted propositional formulas for cardinal preference modelling. In: Proceedings of knowledge representation and reasoning (KR 2006), Lake District, England
77. van Linder B, van der Hoek W, Meyer J-JCh (1995) Actions that make you change your mind. In: Laux A, Wansing H (eds) Knowledge and belief in philosophy and artificial intelligence. Akademie, Berlin, pp 103–146
78. Liu F, Wang Y (2012) Reasoning about agent types and the hardest logic puzzle ever. Minds and Machines (Published online first)
79. Masuzawa T, Hasebe K (2011) Iterative information update and stability of strategies. Synthese 179(1):87–102. doi:10.1007/s11229-010-9835-y
80. Meyer J-J Ch, van der Hoek W (1995) Epistemic logic for AI and computer science. Cambridge University Press, Cambridge, England
81. Moore RC (1984) A formal theory of knowledge and action. In: Hobbs J, Moore RC (eds) Formal theories of the commonsense World. Ablex Publishing Corp., Norwood, NJ
82. Neuwirth E (1982) Some strategies for mastermind. Zeitschrift für Operat Res 26:257–278
83. Osborne MJ, Rubinstein A (1994) A course in game theory. The MIT Press, Cambridge, MA
84. van Otterloo S, Jonker G (2005) On epistemic temporal strategic logic. Electron Notes Theoret Comput Sci 126:77–92 (Proceedings of the 2nd International Workshop on Logic and Communication in Multi-Agent Systems)
85. Pauly M (2002) A modal logic for coalitional power in games. J Logic Comput 12(1):149–166
86. Plaza JA (1989) Logics of public communications. In: Emrich ML, Pfeifer MS, Hadzikadic M, Ras ZW (eds) Proceedings of the 4th international symposium on methodologies for intelligent systems: poster session program, pp 201–216. Oak Ridge National Laboratory, ORNL/DSRD-24
87. Purvis M, Nowostawski M, Cranefield S, Oliveira M (2004) Multi-agent interaction technology for peer-to-peer computing in electronic trading environments. In: Moro G, Sartori C, Singh M (eds) AP2PC. Lecture notes in computer science, vol 2872. Springer, Berlin, pp 150–161
88. Quine WV (1956) Quantifiers and propositional attitudes. J Philos 53(5):177–187
89. Renardel de Lavalette GR (1999) Memories and knowledge games. In: Gerbrandy J, Marx M, de Rijke M, Venema Y (eds) JFAK. Essays Dedicated to Johan van Benthem on the occasion of his 50th birthday. Amsterdam University Press, Amsterdam
90. Schobbens PY (2004) Alternating-time logic with imperfect recall. Electron Notes Theoret Comput Sci 85(2):82–93
91. Stalnaker R (1994) On the evaluation of solution concepts. Theory Decis 37:49–73. doi:10.1007/BF01079205
92. Uckelman J, Chevaleyre Y, Endriss U, Lang J (2009) Representing utility functions via weighted goals. Math Logic Q 55(4):341–361

93. Winkler P (1981) My night at the cryppie club. Bridge Magazine, August 1981:60–63. http://
    www.math.dartmouth.edu/pw/cryppie.htm
94. Wölfl S (2004) Qualitative action theory: a comparison of the semantics of alternating-time
    temporal logic and the Kutschera-Belnap approach to agency. In: Proceedings of JELIA'04,
    vol 3229. Lecture notes in artificial intelligence. Springer, Berlin, pp 70–81
95. von Wright GH (1951) An essay in modal logic. North Holland, Amsterdam
96. Zantema H (2007) De achterkant van SUDOKU. In: Gottmer H, NL (ed) Dutch. Translation
    of title: 'The Backside of Sudoku'

Chapter 17
On Definitive Solutions of Strategic Games

Sergei Artemov

Abstract In his dissertation of 1950, Nash based his concept of the solution to a game on the assumption that "a rational prediction should be unique, that the players should be able to deduce and make use of it". We study when such definitive solutions exist for strategic games with ordinal payoffs. We offer a new, syntactic approach: instead of reasoning about the specific model of a game, we deduce properties of interest directly from the description of the game itself. This captures Nash's deductive assumptions and helps to bridge a well-known gap between syntactic game descriptions and specific models which could require unwarranted additional epistemic assumptions, e.g., common knowledge of a model. We show that games without Nash equilibria do not have definitive solutions under any notion of rationality, but each Nash equilibrium can be a definitive solution for an appropriate refinement of Aumann rationality. With respect to Aumann rationality itself, games with multiple Nash equilibria cannot have definitive solutions. Some games with a unique Nash equilibrium have definitive solutions, others don't, and the criterion for a definitive solution is provided by the iterated deletion of strictly dominated strategies.

17.1 Introduction

Some classical strategic games have definitive solutions which follow logically from the game description and plausible principles of knowledge and rationality. Here is a quote from Nash's dissertation [19] which raises the issue of a deductive approach to solving games:

To Johan.

S. Artemov (✉)
The CUNY Graduate Center, 365 Fifth Avenue, rm. 4319, New York, NY 10016, USA
e-mail: sartemov@gc.cuny.edu

A. Baltag and S. Smets (eds.), *Johan van Benthem on Logic
and Information Dynamics*, Outstanding Contributions to Logic 5,
DOI: 10.1007/978-3-319-06025-5_17, © Springer International Publishing Switzerland 2014

> We proceed by investigating the question: what would be a rational prediction of the behavior to be expected of rational[ly] playing the game in question? By using the principles that a rational prediction should be unique, that the players should be able to deduce and make use of it, [...] one is led to the concept of a solution [...].

Another quote from [19] explains this issue even further:

> [...] we need to assume the players know the full structure of the game in order to be able to deduce the prediction for themselves.

Some game theorists consider the aforementioned assumption, that the player can deduce the strategies of other players, as rarely met. Perhaps the following quote from Pearce [21] represents this skepticism fairly:

> The rules of a game and its numerical data are seldom sufficient for logical deduction alone to single out a unique choice of strategy for each player. To do so one requires either richer information [...] or bolder assumptions about how players choose strategies.

Foundations for Nash's approach have been widely studied from probabilistic positions, and it is not feasible to provide a representative survey of the corresponding literature within the limits of this chapter; we mention [8, 10, 21], just to name few. There has also been a vast body of epistemic logic studies in the foundations of Game Theory, cf. [12, 14] for some recent surveys.

In this chapter, we offer a logical analysis of Nash's assumption that the player can *deduce* the strategies of others. This concept is of a deductive logical character which is represented by the notion of a *definitive solution* of a game as a strategy profile s such that it logically follows from the game description, including the epistemic and rationality assumptions, that each player plays s. Here "logically follows" can be understood twofold: as a logical deduction by certain rules from a set of formalized postulates, or as a logical entailment of a semantic nature on a class of models. Due to the basic soundness and completeness theorems of logic, these two approaches are theoretically equivalent and it is up to the user to choose which approach to follow.

Now dominant, the semantic approach can be traced to Aumann's seminal "Agreeing to Disagree" paper [7] and the notion of Aumann structures that model both structural and epistemic sides of games. Though flexible and convenient, this semantical approach was not quite foundationally satisfactory, first of all due to assumptions that

1. a given model, including possible epistemic states of players, adequately represents the game, which is normally described syntactically, and
2. the model itself is common knowledge for the players.

As emphatically stressed by Aumann himself (cf. [9]), this created tension between the syntactic character of the game description and model-theoretic tools of studying games.

It does not appear problematic to assume that the rules of the game, let us call them *GAME*, in plain English (or an appropriate logical-mathematical formalism), are commonly known, but this yields, generally speaking, neither (1) nor (2), since there can be many different models for a given game description.[1] Moreover, the standard method of producing a model of a syntactic set Γ, the so-called canonical model construction, generally speaking, offers an infinite model not necessarily equivalent to the desired model $\mathcal{M}$. So, in a general setting, when we assume *GAME* and then study a specific model $\mathcal{M}$ of *GAME*, we lose information: if a property F holds in $\mathcal{M}$, it does not necessary follow from the game description *GAME* and could be an artifact of the chosen model $\mathcal{M}$. One way out of this predicament could be assuming $\mathcal{M}$ to be the definition of the game. This, however, renders the game description implausible. Aumann in [9] writes about this alternative: *Your mother won't understand if you try to say it semantically.*

How serious is this threat of non-categoricity? Many simple epistemic scenarios are categorical and yield a specific model, e.g., for regular strategic games without additional semantical constraints,[2] the Muddy Children Puzzle, and others, the initial syntactic description is categorical and any model is equivalent (bisimilar) to the intended model. However, it is easy to offer a slight modification of the Muddy Children scenario which makes it non-categorical and the notion of the standard model meaningless. So, if we intend to consider games with general epistemic conditions, the categoricity consideration can become a serious matter.

The assumption about common knowledge of model $\mathcal{M}$ does not look plausible either: epistemic scenarios in plain English can be publicly announced to a group of players which makes *GAME* commonly known. However, publicly announcing even a simple Aumann structure does not seem realistic.[3]

17.2 New Format of Reasoning About Games

We suggest a modification of the format of reasoning about epistemic scenarios and games which retains the flexibility of the traditional model-theoretic approach and is free of its aforementioned deficiencies. The usual format of reasoning about games is

1. to assume the set of game rules *GAME*, including epistemic conditions given syntactically, in plain English or the usual logical-mathematical slang;
2. to pick a specific Aumann structure $\mathcal{A}$ that corresponds to *GAME*, often informally;
3. to tacitly assume 'common knowledge of $\mathcal{A}$' and not bother about its justification (whereas a straightforward rigorous formalization of common knowledge of $\mathcal{A}$ requires tools outside $\mathcal{A}$);
4. to reason about $\mathcal{A}$ using (3).

---

[1] This phenomenon is well known to logicians as 'non-categoricity'.

[2] cf. Sect. 17.7.4.

[3] This observation does not concern so-called computer knowledge when the programmer can program specific models to each computer.

The deficiency of this schema is obvious. Suppose you establish a property F using (4). Can you claim that F follows from the rules of the game? In general, no. Since $\mathcal{A}$ is a model of *GAME*, each fact that follows from *GAME* holds in $\mathcal{A}$, but not the other way around. It can easily happen that F holds in $\mathcal{A}$, but F does not follow from *GAME*. So this schema produces results about a specific model $\mathcal{A}$ of *GAME* with an additional and difficult-to-motivate assumption of the common knowledge of $\mathcal{A}$. Sometimes this schema works fine, e.g., when $\mathcal{A}$ is the only model of *GAME* (up to truth preserving bisimulations). However, such categoricity analysis is not normally performed and it is easy to provide examples in which categoricity does not hold. If step (1) in this schema is omitted and $\mathcal{A}$ itself is assumed to be the definition of the game, then the categoricity objection becomes void, but the problem with common knowledge of $\mathcal{A}$ persists.

We suggest the following format for studying games:

1. Assume the set of game rules *GAME* as above and common knowledge of *GAME*. This can be arranged in the language of *GAME* that contains the common knowledge modality **C**—we just assume $\mathbf{C}(P)$ for each postulate P of *GAME*.
2. Logically reason from *GAME* directly, e.g., to establish that a fact F follows from the rules of the game. If *GAME* is categorical, then reasoning in a specific model is equivalent to logical reasoning in *GAME*.
3. Use specific models of *GAME*, if needed, to check the consistency of *GAME* or to establish that a certain F does not follow from *GAME*.

In this schema, both previous objections, possible non-categoricity of *GAME*, and dubious assumptions about common knowledge of $\mathcal{A}$, are eliminated. The categoricity requirement is replaced by a much lighter condition: soundness of *GAME* in $\mathcal{A}$, i.e., that all postulates of *GAME* hold in $\mathcal{A}$. Common knowledge of $\mathcal{A}$ is no longer required and is replaced by the assumption of common knowledge of *GAME*. Additional bonuses of this approach include the possibility of utilizing logical intuition and reasoning from *GAME* informally in the logic of knowledge; informal reasoning using the rules of the game is often quite efficient and produces shorter proofs than rigorous model reasoning.[4] Another attractive feature of this approach is greater flexibility in using models $\mathcal{A}$ that can be chosen specifically to be counter-models for F without a commitment that $\mathcal{A}$ is a full description of the game.

All logical reasoning in this chapter can be made completely formal within a framework of multi-agent epistemic modal logic although we don't see sufficient incentive for doing so.[5] We adopt the view that the reader possesses the robust intuition of epistemic reasoning and that reference to the S5-based principles of knowledge[6] and to common knowledge is sufficient.

---

[4] By the same token, we use rigorous yet informal reasoning to establish the Pythagorean theorem, though such a proof could be completely formalized and derived in an axiomatic geometry.

[5] We continue our analogy with the Pythagorean theorem: one could try to formalize its proof completely only as a challenge or an exercise. A normal mathematically rigorous proof of it is not formal.

[6] Such as reflexivity and positive and negative introspection.

17.3 Content

As a case study, we consider the class of finite strategic games with ordinal payoffs. Since, for this class of games, the concept of mixed strategies and expected payoff is not well-founded, we have to consider pure strategies only.

We observe that games without Nash equilibria lack definitive solutions under any notion of rationality and that each Nash equilibrium can be a definitive solution for an appropriate refinement of Aumann rationality.

We show that with respect to basic Aumann rationality, all games with two or more Nash equilibria, and some games with a unique Nash equilibrium, do not have definitive solutions either.

Perhaps some of these impossibility results do not come as a surprise for an experienced game theorist. For example, Theorem 17.3 states that no game with more than one Nash equilibrium under normal epistemic assumptions and Aumann rationality can have a definitive solution. This corresponds to the intuition that choosing between several Nash equilibria, in addition to Aumann rationality, in Pearce's words, "requires either richer information . . . or bolder assumptions about how players choose strategies". However, there is a difference between empirically justified intuition and a rigorous proof. In computer programming, experts have no illusions that one could build a universal verifier which, for any given program P and input I, automatically decides whether P terminates on I. Turing's rigorous proof of this fact, known as the undecidability of the halting problem theorem, provided a basis for further fruitful studies of computability. If this chapter is successful, this could be a step towards logical studies of consistency and impossibility in Game Theory.

Furthermore, we show that the criterion for Nash's definitive solution in strategic games with ordinal payoffs and Aumann rationality is provided not by Nash equilibria, but rather by iterated deletion of strictly dominated strategies.

This chapter is an extended version of technical report [4] of 2010.

17.4 Logical Presentation of Strategic Games

We consider strategic games with n players $1, 2, \ldots, n$. A strategy profile

$$s = (s_1, s_2, \ldots, s_n)$$

is a collection of strategies s_i for players $i = 1, 2, \ldots, n$. Each strategy profile s uniquely determines the *outcome* in which each move is made according to s. We assume that everyone who knows the game can calculate i's payoff as determined by s. A strategy profile s is a **Nash equilibrium** if, given strategies of the other players, no player can profitably deviate (cf. [20] for rigorous definitions).

We assume that rules of the game are formally represented in an appropriate logical language as a set of formulas *GAME* as follows. Strategy j of player i is formally represented by a corresponding atomic sentence s_i^j stating that *player i has committed to strategy j.*

For profile $s = (s_1, s_2, \ldots, s_n)$, its formula representation will be $s_1 \wedge s_2 \wedge \ldots \wedge s_n$ which we will also call s. In this setting,

$$\bigwedge_{k \neq l} \neg(s_i^k \wedge s_i^l)$$

means that player i chooses only one strategy, and

$$s_i^1 \vee s_i^2 \vee \ldots \vee s_i^{m_i}$$

where $s_i^1, s_i^2, \ldots, s_i^{m_i}$ is the list of propositions for all strategies for i, reflects the assumption that one of these strategies has to be played. The fact that a profile $s = (s_1, s_2, \ldots, s_n)$ is at least as preferable as $s' = (s'_1, s'_2, \ldots, s'_n)$ for player i can be represented by a special preference formula

$$s \geq_i s'$$

and all these preferences are supposed to be common knowledge.

In principle, *GAME* may be a (possibly infinite) set of logical formulas with additional preference relations on strategy profiles, which contains a comprehensive game description including its epistemic conditions. Note that *GAME* is not necessarily consistent.

By "logically follows" we mean here a logical deduction denoted as "⊢". This notion is usually understood in logic as formal derivability. Our reasoning will not be completely formalized (cf. footnotes 4 and 5), but we assume could be if needed.

We use special logical symbols, **knowledge operators** (cf. [16]) $\mathbf{K}_1, \mathbf{K}_2, \ldots, \mathbf{K}_n$, to denote knowledge of players $1, 2, \ldots, n$ and assume the standard principles of knowledge, cf. [16, 20], a.k.a. S5 principles. For example, stating

$$\mathbf{K}_i(s_j^l)$$

says "i knows that j has chosen her strategy l".

We will use the "everybody knows" modality $\mathbf{E}$ as the abbreviation

$$\mathbf{E}F = \mathbf{K}_1 F \wedge \mathbf{K}_2 F \wedge \ldots \wedge \mathbf{K}_n F.$$

Common Knowledge of F, $\mathbf{C}F$, reflects a situation in which all propositions

$$\mathbf{K}_{i_1} \mathbf{K}_{i_2} \ldots \mathbf{K}_{i_m} F$$

hold for any $i_1, i_2, \ldots, i_m$.

An informal account of Aumann's notion of rationality can be found in [8]. Aumann states that for a rational player i,

> there is no strategy that i knows would have yielded him a conditional payoff …larger than that which in fact he gets.

Formal accounts of Aumann rationality based on epistemic models known as Aumann structures can be found in [8, 13] and other sources. Here we will adopt a syntactic formalization of Aumann rationality for strategic games.

A strategy profile $s = (s_1, s_2, \ldots, s_n)$ is **deemed possible** by player i if for each $j \neq i$,

$$\neg \mathbf{K}_i(\neg s_j).$$

This definition reflects the assumption that players consider all of their strategies possible whereas some of the other players' choices could be ruled out as impossible based on the rules and conditions of the game. Since we allow epistemic constraints in GAMES, the notion of possibility should be relativized: s is deemed possible by i at s' if

$$s' \rightarrow \text{ "}s\text{ is deemed possible by } i\text{ "}.$$

In particular, s is possible for each i at s. Indeed, by laws of logic, $\mathbf{K}_i(\neg s_j) \rightarrow \neg s_j$, hence $s_j \rightarrow \neg \mathbf{K}_i(\neg s_j)$.

Let (s_{-i}, x) denote the strategy profile obtained from s by replacing s_i by x. It follows from definitions that

s is deemed possible by i iff (s_{-i}, x) is deemed possible by i for every strategy x of i.

Definition 17.1 Let $s = (s_1, s_2, \ldots, s_n)$ be a strategy profile. A formula $R_i(s)$ stating that **player i is Aumann-rational at s** is, by definition, the natural formalization of the following: *for any strategy x of i, x, there is a profile s' deemed possible by i at s such that*

$$(s'_{-i}, s_i) \geq_i (s'_{-i}, x).$$

A formula

$$R(s) = \bigwedge_i R_i(s)$$

states that all players $i = 1, 2, \ldots, n$ are Aumann-rational at s.

Note that quantified sentences "for any x…" and "there is a profile s'…" are represented by propositional formulas since there are only finite sets of possible x's and s's and there is no need to invoke quantifiers over strategies or strategy profiles.

Definition 17.1 formalizes the most basic, Aumann rationality though there are also other, more elaborate notions of rationality. However, we assume that any notion of rationality should be at least as strong as Aumann rationality on each profile. Technically, if $r_i(s)$ is a formula representing some notion of rationality of i at s, then

$$r_i(s) \to R_i(s)$$

is assumed in the *GAME*.

In addition to the game rules we consider a **solution predicate**, a formula $Sol(s)$ that specifies the conditions under which strategy profile s is considered to be a solution of the game. Given rationality predicates $r_i(s)$, $i = 1, 2, \ldots, n$ and

$$r(s) = \bigwedge_i r_i(s), \tag{17.1}$$

the typical cases of $Sol(s)$ are

1. $r(s)$ informally stating that s is a profile at which all players are rational;
2. $\mathbf{E}r(s)$ claiming that s is a profile at which players' rationality is mutually known;
3. $\mathbf{C}r(s)$ stating common knowledge of rationality at s.

We assume that a solution predicate $Sol(s)$ contains condition 1 that all players are rational, i.e., for each profile s,

$$GAME \vdash Sol(s) \to r(s).$$

Solution predicate 3 corresponds to the familiar assumption of common knowledge of rationality.

The natural characteristic principle for the solution condition is *if a profile is played, then it ought to satisfy the solution constraints*:

$$s \to Sol(s) \tag{17.2}$$

which may be more recognizable in its contrapositive form

$$\neg Sol(s) \to \neg s,$$

meaning *if s does not satisfy the solution constraints, then s is not played.*

The familiar epistemic conditions: rationality, mutual knowledge of rationality and common knowledge of rationality, mean the corresponding conditions on the solution predicate.

Common knowledge of the game is formalized as common knowledge of each principle from *GAME*:

$$\textit{if } F \in GAME, \textit{ then } GAME \vdash \mathbf{C}F.$$

In particular, if *GAME* is commonly known, then

$$GAME \vdash \mathbf{E}(GAME).$$

We say that *GAME* has **common knowledge of the game and rationality** property, *CKGR*, if *GAME* is commonly known and the solution predicate $Sol(s)$ contains common knowledge of rationality

$$GAME \vdash Sol(s) \to \mathbf{Cr}(s)$$

for rationality predicates $r_1(s), r_2(s), \ldots, r_n(s)$ associated with the game and $r(s)$ being their conjunction as in (17.1).

Definition 17.2   A profile s is a **solution** of a game *GAME* if

$$GAME \vdash s.$$

For a consistent *GAME*, a solution, if it exists, is unique. The definition reflects the property that the rules of the game yield a unique strategy for each player.

   For example, Prisoner's Dilemma has a solution, and the Battle of Sexes does not have a solution in pure strategies on the basis of Aumann rationality only.[7]

Definition 17.3   A strategy profile s is a **definitive solution** of the game if

$$GAME \vdash \mathbf{E}(s).$$

The definition states that the rules of the game yield a unique strategy for each player and each player knows all these strategies. Each definitive solution is a solution. Hence, the definitive solution, if it exists, is unique.

Lemma 17.1   *For games with CKGR, each solution is a definitive solution.*

Proof   Indeed, since *CKGR*,

$$GAME \vdash \mathbf{E}(GAME).$$

By the rules of the (modal) logic of knowledge, from

$$GAME \vdash s,$$

one could conclude

$$\mathbf{E}(GAME) \vdash \mathbf{E}(s),$$

hence

$$GAME \vdash \mathbf{E}(s).$$

Informally, if the rules of the game yield a solution, and the rules are known to player i in full, then i knows this solution.   □

---

[7] This obvious observation, technically speaking, follows from Theorem 17.3.

If rationality of players is not known, it is easy to provide an example of a solution which is not definitive. Consider Prisoner's Dilemma in which players are rational but do not know about each other's rationality. By the domination argument, each player plays '*defect*' but none knows the choice of the other player.[8]

As another example of the notion of solution, consider the following **War and Peace dilemma**, **W&P**, introduced in [3].

> Imagine two neighboring countries: a big powerful B, and a small S. Each player can choose to wage war or keep the peace. The best outcome for both countries is peace. However, if both countries wage war, B wins easily and S loses everything, which is the second-best outcome for B and the worst for S. In situation (war_B, $peace_S$), B loses internationally, which is the second-best outcome for S. In ($peace_B$, war_S), B's government loses national support, which is the worst outcome for B and the second-worst for S.

The ordinal payoff matrix of this game is then

	war_S	$peace_S$
war_B	2,0	1,2
$peace_B$	0,1	3,3 .

There is one Nash equilibrium,

$$(peace_B, peace_S). \tag{17.3}$$

Let us assume Aumann rationality and *CKGR*. We claim that strategy profile (17.3) is the definitive solution to **W&P**. Indeed, S has a dominant strategy $peace_S$ and as a rational player, S has to commit to this strategy. This is known to B, since B knows the game and is aware of S's rationality. Therefore, as a rational player, B chooses $peace_B$. This reasoning can be carried out by any intelligent player. Hence it follows from the game description and *CKGR* that both players know solution (17.3) which is, therefore, the definitive solution of **W&P**.

Here is another game, **W&P2**, with the same payoff matrix in which players follow Aumann-Harsanyi rationality[9] that rules out Aumann-irrational strategies and then applies maximin[10] to make a choice. We assume that the payoff matrix is mutually known but players, though Aumann-Harsanyi-rational, are not aware of each other's rationality. In **W&P2**, S chooses $peace_S$ since it is S's dominant strategy. Since B considers both strategies for S, war_S and $peace_S$ possible,[11] both strategies for B, war_B and $peace_B$ are Aumann-rational. Then B should follow the maximin strategy, hence choosing war_B. The resultant strategy profile

---

[8] If player 1 knows that player 2 defects, then 1 knows that 2 is rational.

[9] Which is equivalent to the knowledge-based rationality studied in [3].

[10] Following Harsanyi's principle from [18] Sects. 6.2 and 6.3, Postulate A1.

[11] Otherwise, B would know that S is rational.

$$(war_B, peace_S) \tag{17.4}$$

is a unique solution of this game, but not a definitive solution. Indeed, B does not know S's choice $peace_S$.

17.5 Formalizing Nash Reasoning

Theorem 17.1 *A definitive solution of a game with rational players is a Nash equilibrium.*

Proof Consider a strategic game *GAME* with rationality predicates $r_i(x)$, $i = 1, 2, \ldots, n$, and solution predicate $Sol(x)$. Let $s = (s_1, s_2, \ldots, s_n)$ be a definitive solution of the game. We have to show that s is a Nash Equilibrium. Argue informally given *GAME*. We have

$$\mathbf{E}(s),$$

and hence for each i, j,

$$\mathbf{K}_i(s_j)$$

and hence for any strategy s'_j for j,

$$\mathbf{K}_i(\neg s'_j).$$

Therefore, s_j is the only strategy of j which is deemed possible by i.

Suppose s is not a Nash equilibrium, hence for some player i, the choice of strategy s_i is less preferable to some other choice x given the other players' strategies, so

$$s = (s_{-i}, s_i) <_i (s_{-i}, x)$$

for some strategy x of player i. Therefore, we found a player i and a strategy x of i such that for all profiles possible for i, $(s_{-i}, s_i) <_i (s_{-i}, x)$, i.e., i is not Aumann-rational at s, i.e., $\neg R_i(s)$.

Since s is the solution of *GAME*, s should satisfy the solution condition $Sol(s)$ which yields that all players are rational at s, i.e., $r(s)$. Since $r(s) \to R(s)$, all players should be Aumann-rational at s, $R(s)$; a contradiction.[12] $\qquad \square$

The proof of Theorem 17.1 demonstrates not only that players cannot derive a definitive solution from the rules of a game that does not have Nash equilibria, but that the mere existence of such a solution known to all players is incompatible with the rules of the game. In particular, this yields that no refinement of rationality, as

---

[12] Note that a default assumption that *GAME* is consistent is necessary since for inconsistent games, vacuously, each profile is a definitive solution.

long as it respects Aumann rationality, can possibly lead to a definitive solution in such a game.

Corollary 17.1 *No game with rational players and without Nash equilibria has a definitive solution under any notion of rationality.*

Note that this Theorem requires rationality but not any degree of *knowledge* of rationality. Roughly speaking, to spot that the solution is not a Nash equilibrium, we don't need players with knowledge of others' rationality: if the solution were not a Nash equilibrium in i's coordinate, it would directly contradict the i-th player's rationality.

An analogue of Theorem 17.1 has been obtained by Aumann and Brandenburger in [10]: *Suppose that each player is rational, knows his own payoff function, and knows the strategy choices of the others. Then the players' choices constitute a Nash equilibrium in the game being played.* Theorem 17.1 differs from that in [10] on several counts.

- The models of game and knowledge in [10] and this chapter are fundamentally different. For Aumann and Brandenburger, a game is a probability distribution on the set of strategy profiles (a belief system), and knowledge of F is probability 1 of the event F. In this chapter, we use a logic-based syntactic approach in which knowledge is represented symbolically by modal operators interpreted as strict, non-probabilistic knowledge. Since knowledge represented by modal operators is intrinsically linked to metareasoning (here a logical deduction from the game description, in accordance with Nash's aforementioned description of 1950), this logical model of game and knowledge allows us to draw impossibility conclusions that do not appear to be within the scope of probabilistic methods.
- The notions of rationality in [10] and in this chapter are of a quite different nature. In [10], rationality of i is maximization of i's expected payoff and is determined by the underlying belief system, whereas we allow as a rationality predicate any predicate which is at least as strong as Aumann rationality.

For the rest of the chapter, we consider two extreme notions of rationality: the most general Aumann rationality, and a highly specialized notion of bullet rationality.

17.6 On Stronger Notions of Rationality

In this section, we show that any Nash equilibrium can be a definitive solution for an appropriate notion of rationality.

Theorem 17.2 *Given a payoff matrix M and a Nash equilibrium e, not necessarily unique, there is a notion of rationality such that the corresponding game with CKGR has e as the definitive solution.*

Proof Let $s = (s_1, s_2, \ldots, s_n)$ be an arbitrary strategy profile and $e = (e_1, e_2, \ldots, e_n)$ a Nash equilibrium profile. A **bullet e-rationality**[13] is, by definition, a set of predicates $B_i^e(s)$ for $i = 1, 2, \ldots, n$ that hold at e and do not hold at any other strategy profile:

$$B_i^e(e) \in GAME \quad \text{and} \quad \neg B_i^e(s) \in GAME \text{ for any } s \neq e. \tag{17.5}$$

Informally, predicate $B_i^e(s)$ is used as a rationality of i predicate stating that

player i is rational only at profile e.

We define $B^e(s)$ as

$$B^e(s) = \bigwedge_i B_i^e(s).$$

Consider a strategic game *GAME* with

- payoff matrix M;
- bullet rationality $B_i^e(s)$ for $i = 1, 2, \ldots, n$;
- common knowledge of the game and rationality, *CKGR*;
- no other constraints except those explicitly mentioned above.

First, we show that *GAME* is consistent. For this, it suffices to find a model and a node at which all postulates of *GAME* hold. Consider the Aumann structure in which epistemic states are all strategy profiles,

$$\Omega = \{s \mid s \text{ is a strategy profile}\}, \quad \mathbf{s}(s) = s,$$

knowledge partitions are singletons

$$\mathcal{K}_i(s) = \{s\},$$

i.e., each profile is common knowledge in itself, standard truth relation '$\Vdash$'

$$s \Vdash s_i^j \text{ iff the } i\text{-th strategy in profile } s \text{ is } j,$$

and rationality predicates as defined in (17.5):

$$r_i(s) = B_i^e(s).$$

We claim that all assumptions of *GAME* hold at node e of the game model. All basic conditions on strategies and payoff preferences hold everywhere in the model. Rationality conditions (17.5) holds at e by definition and $B^e(e)$ is common knowledge

---

[13] The name is analogous to "bullet voting", in which the voter can vote for multiple candidates but votes for only one.

at e. The solution predicate is, by assumption, $\mathbf{CB}^e(s)$ and it holds at $s = e$. Therefore, the solution condition $s \to Sol(s)$ also holds at $s = e$.

It is easy to see that all players are Aumann-rational at e, since for each player, the possible strategies of others are those from e, and i cannot improve her payoff by changing her own strategy because e is a Nash equilibrium. Therefore, for each profile s,

$$B^e(s) \to R(s),$$

hence $B^e(s)$ is a legitimate rationality predicate.

Now we show that

$$GAME \vdash e.$$

Indeed, since for each $s \neq e$, $GAME$ proves $\neg B^e(s)$ and $GAME$ proves $\neg CB^e(s)$,

$$GAME \vdash \neg Sol(s)$$

and

$$GAME \vdash \neg s.$$

So, all profiles s different from e have been ruled out. However, $GAME$ assumes that each player has to choose a strategy: there should be at least one strategy profile chosen:

$$GAME \vdash \bigvee_s s$$

and e is the only remaining candidate,

$$GAME \vdash \bigwedge_{s \neq e} \neg s.$$

By propositional logic,

$$GAME \vdash (\bigwedge_{s \neq e} \neg s) \to e,$$

therefore

$$GAME \vdash e.$$

By Lemma 17.1,

$$GAME \vdash \mathbf{E}(e),$$

hence e is the definitive solution.   □

Such a "reverse engineered" bullet rationality is a technical notion which we do not offer as a viable practical notion of rationality. However, bullet rationality represents an epistemic condition which, when incorporated into the game description, can single out a given Nash equilibrium as a definitive solution. With a dash of good will, bullet rationality may be regarded as a theoretical prototype of Pearce's *bolder assumption about how players choose strategies* [21] that leads to a definitive solution in a multi-equilibrium situation.

Theorem 17.2 also has a nearby predecessor in [10] where it was stated that for any Nash equilibrium *s*, there is a belief system in which each player assigns probability 1 to *s*. Theorem 17.2 conveys basically the same message with "probability 1" replaced by the logical notion of strict knowledge which has helped to connect this result to Nash's notion of definitive solution.

17.7 Definitive Solutions for Aumann Rationality

Whereas Nash equilibria provide a general necessary condition for definitive solutions, a question of sufficient conditions, i.e., when definitive solutions actually exist, merits special analysis. Since tampering with the notion of rationality can render any Nash equilibrium a definitive solution, it makes sense to consider the definitive solution problem for a fixed notion of rationality. For the rest of the chapter, we consider games with basic Aumann rationality.

We first show that under Aumann rationality, a game with two or more Nash equilibria cannot have a definitive solution.

17.7.1 Regular Form of Strategic Games

A **regular strategic game** is a strategic game described by the following (finite) set of formulas *GAME*.

a. Conditions on strategy propositions s_i^j stating '*player i chooses strategy j.*' These conditions express that each player *i* chooses one and only one strategy:

$$(s_i^1 \vee \ldots \vee s_i^{m_i}) \text{ and } \neg(s_i^j \wedge s_i^l) \text{ for each } j \neq l.$$

b. A complete description of the preference relation for each player at each outcome.
c. Knowledge of one's own strategy $s_i \to \mathbf{K}_i(s_i)$.[14]
d. The solution condition $s \to Sol(s)$ for each *s* where solution predicate $Sol(s)$ is the formula stating common knowledge of Aumann rationality at *s*, $CR(s)$.
e. Common knowledge of **a–d** above.

For example, in the regular form of the War and Peace dilemma W&P, we can demonstrate that $(peace_B, peace_S)$ is a definitive solution. Indeed, it suffices to logically derive $peace_B \wedge peace_S$ from *GAME* of W&P and argue that this derivation can be performed by any player, hence

---

[14] This is the standard requirement of "measurability," cf. [8].

$$\mathbf{K}_i[peace_B \wedge peace_S] \quad \text{for each } i \in \{B, S\}.$$

Here is a derivation of $peace_B \wedge peace_S$ from *GAME* of W&P (an informal version of this derivation was presented in Sect. 17.4):

1. By (**b**) and (**e**), S knows that war_S is a strictly dominated strategy for S, hence S is not rational at all profiles with war_S;
2. by (**d**), none of these profiles can be a solution, hence $\neg war_S$;
3. by (**a**) and 2, $peace_S$;
4. by (**e**), B knows 1 and 2, which makes strategy war_S impossible for B;
5. from 4, B is not rational at $(war_B, peace_S)$, hence by (**d**), $\neg war_B$;
6. therefore $(peace_B \wedge peace_S)$, and, by (**e**), this conclusion is known to both players.

This example was intended to illustrate that the regular form of strategic games is sufficient for accommodating the usual epistemic reasoning in games.

17.7.2 Consistency Lemma

Lemma 17.2 *A regular strategic game GAME is consistent with the knowledge of any of its Nash equilibria: for each Nash equilibrium e,*

$$GAME + \mathbf{C}(e) \tag{17.6}$$

is consistent.

Proof It suffices to present an Aumann structure $\mathcal{M}$ in which, at some node, both *GAME* and $\mathbf{C}(e)$ hold. As in the proof of Theorem 17.2, we define $\mathcal{M}$ as

$$\Omega = \{s \mid s \text{ is a strategy profile}\}, \quad \mathbf{s}(s) = s,$$

knowledge partitions are singletons

$$\mathcal{K}_i(s) = \{s\},$$

the standard truth relation '$\Vdash$'

$$s \Vdash s_i^j \text{ iff the } i\text{-th strategy in profile } s \text{ is } j,$$

and Aumann rationality predicates $R_i(s)$.

By construction, $\mathcal{M}$ is omniscient, i.e., each fact which is true in a state is common knowledge in this state:

$$s \Vdash F \quad \text{yields} \quad s \Vdash \mathbf{C}F.$$

Let $e = (e_1, e_2, \ldots, e_n)$ be a Nash equilibrium of the game. We claim that

$$e \Vdash GAME \wedge \mathbf{C}(e).$$

Since model $\mathcal{M}$ is omniscient, it suffices to check that

$$e \Vdash GAME \wedge e.$$

Since $e \Vdash e$ holds by definition of '$\Vdash$,' it remains to show that

$$e \Vdash GAME.$$

We will check conditions (**a**–**e**) one by one.

- (**a**), (**b**), and (**c**) hold at each node by definition of '$\Vdash$ '.
- for (**d**) it suffices to check that each player is Aumann-rational at e. Player i knows all strategies of others, e_j with $j \neq i$, and deems any of i's own strategies x possible. However, by changing her strategy e_i at e, i cannot improve her payoff since e is a Nash equilibrium.
- (**e**) holds because model $\mathcal{M}$ is omniscient.

Alternatively, the consistency lemma (Lemma 17.2) can be also obtained by applying Proposition 5.4 (B) from [13] which pursues different goals.

The aforementioned result from [10] stating that for any Nash equilibrium e, there is a belief system in which each player assigns probability 1 to e, may be regarded as a natural probabilistic version of Lemma 17.2.

In this chapter, we take one more step and draw impossibility conclusions from the consistency lemma, thus connecting it with Nash's definitive solutions programme (cf. Sect. 17.7.3).

Corollary 17.2 *A regular strategic game is consistent with playing any of its Nash equilibria e: set $GAME + e$ is consistent.*

Proof Immediate from Lemma 17.2, since $\mathbf{C}(e) \rightarrow e$ in $GAME$.   □

17.7.3 No Definitive Solutions to Multi-equilibria Regular Games

Theorem 17.3 *No regular strategic game with more than one Nash equilibrium can have a definitive solution.*

Proof Suppose otherwise, i.e., that some Nash equilibrium e is a definitive solution of $GAME$

$$GAME \vdash e$$

for some regular game that has another Nash equilibrium e'. By (**a**), two different profiles are incompatible, hence

$$GAME \vdash \neg e',$$

which yields that

$$GAME + e'$$

is inconsistent. This contradicts Corollary 17.2.   □

Note that regular strategic games with some additional epistemic constraints can single out one of multiple Nash equilibria as a definitive solution. For example, take a regular game presented by *GAME* and let e be one of its multiple Nash equilibria. Consider a new game *GAME'* consisting of *GAME* with an additional condition that e is common knowledge:

$$GAME' = GAME + \mathbf{C}(e).$$

By Lemma 17.2, *GAME'* is consistent: it is easy to see that e is its definitive solution.

17.7.4 Definitive Solutions of Regular Games via IDSDS

In this Section, we observe that definitive solutions of regular strategic games with ordinal payoffs are completely described by the procedure of the Iterated Deletion of Strictly Dominated Strategies (*IDSDS*) rather then a unique Nash equilibrium. *IDSDS* iteratively deletes strategies which are strictly dominated by other pure strategies. Let S^∞ denote the set of strategy profiles which survive *IDSDS*. By construction, $S^\infty \neq \emptyset$.

The role of *IDSDS* has been well studied (cf. [1, 11, 13]) and Theorem 17.4 mainly connects these studies to the definitive solution framework.

Theorem 17.4 *A strategic regular game with Aumann rationality has a definitive solution s if and only if s is the only strategy profile that survives IDSDS.*

Proof (Sketch). We use the terminology of [13]. By Lemma 17.1, it suffices to prove the analogue of this theorem which speaks about "solution s" rather than "definitive solution s". Consider two cases.

Case 1 S^∞ contains states with different profiles, say s_1 and s_2. Then such a game does not have a definitive solution. Indeed, by Proposition 5.4 (B) from [13], there are states ω_1 and ω_2 corresponding to profiles s^1 and s^2 such that

$$\omega_i \Vdash GAME + s^i \text{ for } i = 1, 2.$$

Therefore, $GAME + s^i$ are consistent for $i = 1, 2$. Since s^1 and s^2 are incompatible,

$$GAME \nvdash s^i \text{ for } i = 1, 2.$$

Case 2 S^∞ is a singleton, e.g., $S^\infty = \{s\}$. We show that s is then the definitive solution. By the completeness theorem for the background modal logic of knowledge (normally, a multi-agent version of **S5**), it is sufficient to establish that

$$GAME \rightarrow s \tag{17.7}$$

holds at each node of each model (e.g., each Aumann structure). By Proposition 5.4 (A) from [13], for each state ω, if $\mathbf{s}(\omega) \notin S^\infty$, then the statement of common knowledge of rationality fails in ω, hence

$$\omega \not\Vdash \mathbf{s}(\omega) \rightarrow Sol(\mathbf{s}(\omega))$$

and

$$\omega \not\Vdash GAME.$$

Consider an arbitrary node ω. If $\mathbf{s}(\omega) \in S^\infty$, then $\mathbf{s}(\omega) = s$, hence $\omega \Vdash s$. If $\mathbf{s}(\omega) \notin S^\infty$, then, as above, $\omega \not\Vdash GAME$. In either case, (17.7) holds at ω.

Since (17.7) holds in each model,

$$GAME \vdash s,$$

and, by Lemma 17.1,

$$GAME \vdash \mathbf{E}(s),$$

hence s is the definitive solution of *GAME*.

Case 2 can also be derived from [1] which shows that players will choose only strategies that survive the iterated delition of strictly dominated strategies. □

17.8 Unique Nash Equilibrium Does not Yield a Definitive Solution

Consider the following game

$$\begin{bmatrix} 1,2 & 1,0 & 0,1 \\ 0,0 & 0,2 & 1,1 \end{bmatrix}$$

It has a unique Nash equilibrium $(1, 2)$, but no definitive solution within the scope of Aumann rationality, even if the game and rationality are commonly known. Indeed, each strategy in this game is Aumann-rational and hence cannot be deleted by *IDSDS*.

17.9 Discussion

We have seen that Nash's definitive solution paradigm is not at all universal: in many cases definitive solutions do not exist. For example, a direct count shows that though 75% of generic regular 2 × 2 games have definitive solutions, the proportion of solvable games quickly goes to 0 when the size of the game grows (the number of players or the number of strategies for each player). Even when definitive solutions exist, the notion of a Nash equilibrium does not provide sufficient criteria for them. In a way, the results of this chapter support Aumann's views [9]:

- Equilibrium is not the way to look at games. The most basic concept should be: *to maximize your utility given your information.*
- The starting point for realization of this concept should be *syntactic epistemic logic.*

In a game, one could expect epistemic and rationality conditions to be given, hence a methodologically correct way would be to consider whether a game has a definitive solution under given epistemic/rationality conditions.

For future work, one could apply similar methods for analyzing mixed strategies and settings with belief rather than knowledge conditions.

It makes sense to further explore the role of proof-theoretical methods in epistemic game theory. One possible avenue, along the lines of Johan van Benthem's 'rational dynamics' programme [11, 12], could be to add justifications—in particular, proofs as objects—to the logical analysis of games. The focus of such research could be to create a unified theory of reasoning and epistemic actions in the context of games. There is no action without reasoning for rational agents; reasoning is itself a kind of epistemic action, and takes other actions as inputs. A meaningful step in this direction was made by Renne in [22] in which he suggests interpreting proof terms t in the Logic of Proofs (cf. [2]) as strategies, so that $t{:}A$ may be read as

$$t \text{ is a winning strategy on } A.$$

In this light, the Logic of Proofs may thus be seen as a logic containing in-language descriptions of winning strategies on its own formulas.

Other major issues in epistemic game theory that the Logic of Proofs could help to address are awareness and the logical omniscience problem. The standard semantics for the logics of proofs and justifications, Fitting models [17], is a more expressive dynamic extension of Fagin-Halpern awareness models [15]; awareness models are Fitting models corresponding to one fixed proof term [23]. A coherent general treatment of the logical omniscience problem on the basis of proof complexity has been offered in [5, 6].

Acknowledgments The author is grateful to Adam Brandenburger for drawing attention to Nash's argument and supportive discussions. The author is indebted to Mel Fitting, Vladimir Krupski, Elena Nogina, Çağıl Taşdemir, and Johan van Benthem for many principal comments and suggestions. Special thanks to Karen Kletter for editing this text.

References

1. Apt K, Zvesper J (2010) Proof-theoretic analysis of rationality for strategic games with arbitrary strategy sets. In: Dix J, Leite J, Governatori G, Jamroga W (eds) Computational logic in multi-agent systems. Proceedings of the 11th international workshop, CLIMA XI, Lisbon, Portugal, August 16–17, 2010. Lecture Notes in Artificial Intelligence, vol 6245, pp 186–199
2. Artemov S (2001) Explicit provability and constructive semantics. Bull Symb Log 7(1):1–36
3. Artemov S (2009) Knowledge-based rational decisions. Technical Report TR-2009011, CUNY Ph.D. Program in Computer Science
4. Artemov S (2010) The impossibility of definitive solutions in some games. Technical Report TR-2010003, CUNY Ph.D. Program in Computer Science
5. Artemov S, Kuznets R (2006) Logical omniscience via proof complexity. In: Ésik Z (ed) Proceedings of the 20th international workshop on computer science logic, CSL 2006. Lecture notes in computer science, vol 4207, pp 135–149
6. Artemov S, Kuznets R (2009) Logical omniscience as a computational complexity problem. In: Heifetz A (ed) Proceedings of the twelfth conference theoretical aspects of rationality and knowledge, (TARK 2009), Stanford University, California, ACM, pp 14–23
7. Aumann R (1976) Agreeing to disagree. Ann Stat 4(6):1236–1239
8. Aumann R (1995) Backward induction and common knowledge of rationality. Games Econ Behav 8:6–19
9. Aumann R (2010) Interview. In: Hendricks VF, Roy O (eds) Epistemic logic: 5 questions. Automatic Press/VIP, New York, pp 21–33
10. Aumann R, Brandenburger A (1995) Epistemic conditions for Nash equilibrium. Econometrica: J Econom Soc 64(5):1161–1180
11. van Benthem J (2007) Rational dynamics and epistemic logic in games. Int Game Theory Rev 9(1):1345
12. van Benthem J, Pacuit E, Roy O (2011) Toward a theory of play: a logical perspective on games and interaction. Games 2:52–86
13. Bonanno G (2008) A syntactic approach to rationality in games with ordinal payoffs. In: Bonanno G, van der Hoek W, Wooldridge M (eds) Logic and the foundations of game and decision theory, texts in logic and games series. Amsterdam University Press, Amsterdam, pp 59–86
14. de Bruin B (2010) Explaining games: the epistemic programme in game theory. Springer, Berlin
15. Fagin R, Halpern JY (1988) Belief, awareness, and limited reasoning. Artif Intell 34:39–76
16. Fagin R, Halpern J, Moses Y, Vardi M (1995) Reasoning about knowledge. MIT Press, Cambridge
17. Fitting M (2005) The logic of proofs, semantically. Ann Pure Appl Log 132(1):1–25
18. Harsanyi JC (1986) Rational behaviour and bargaining equilibrium in games and social situations. Cambridge Books, Cambridge
19. Nash J (1950) Non-cooperative games. Ph.D. Thesis, Princeton University, Princeton
20. Osborne M, Rubinstein A (1994) A course in game theory. The MIT Press, Cambridge
21. Pearce D (1984) Rationalizable strategic behavior and the problem of perfection. Econometrica: J Econom Soc 52(4):1029–1050
22. Renne B (2009) Propositional games with explicit strategies. Inf Comput 207(10):1015–1043
23. Sedlár I (2013) Justifications, awareness and epistemic dynamics. In: Artemov S, Nerode A (eds) Proceedings of the logical foundations of computer science, LFCS 2013. Lecture notes in computer science, vol 7734, pp 307–318

Chapter 18
Logical Player Types for a Theory of Play

R. Ramanujam

Abstract In theory of play, a player needs to reason about other players' types that could conceivably explain how play has reached a particular node of the extensive form game tree. Notions of rationalizability are relevant for such reasoning. We present a logical description of such player types and show that the associated type space is constructible (by a Turing machine).

18.1 Reasoning About Games and in Games

> ... rationality is primarily a property of procedures of deliberation or other logical activities, and only secondarily a property of outcomes of such procedures [10].

Game theory tries to analyse situations where there are elements of conflict and cooperation among rational agents. The ultimate aim of the theory is to predict the behaviour of rational agents and prescribe a plan of action that needs to be adopted when faced with a strategic decision making situation. The theory therefore consists of the modelling part as well as the various solution concepts which try to predict the behaviour of such agents and prescribe what rational players should do. The effectiveness of a particular solution concept depends on how precise and effective the prescription turns out to be.

There are some fundamental problems in coming up with such a theory. Most real world interactions are extremely complex and modelling the complete process as a game may not be feasible. The usual approach to overcome this trouble is to consider an abstraction of the situation and to model this abstract setting as a game. Even though all the constituent elements of the situation cannot be preserved, the abstraction tries to retain the most relevant ones. The other challenge is due to the fact

R. Ramanujam (✉)
Institute of Mathematical Sciences, Chennai, India
e-mail: jam@imsc.res.in

A. Baltag and S. Smets (eds.), *Johan van Benthem on Logic and Information Dynamics*, Outstanding Contributions to Logic 5, DOI: 10.1007/978-3-319-06025-5_18, © Springer International Publishing Switzerland 2014

Fig. 18.1 An extensive form game

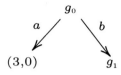

$$(3,0) \qquad\qquad g_1$$

Fig. 18.2 Normal form game g_1

	d_1	d_2	d_3
c_1	(2,2)	(2,1)	(0,0)
c_2	(1,1)	(1,2)	(4,0)

that game theory assumes that all players are *rational* whereas in many real world interactions, people often tend to act irrationally.

For finite extensive form games of perfect information, backward induction (BI) offers a solution that is simple and attractive as prediction of stable play. However, this critically depends on reasoning being backward, or bottom-up on the tree from the leaves to the root. In some games such as the famous example of the centipede game, this solution is somewhat counter-intuitive.

In general, an extensive form game can have several Nash equilibria apart from the one given by the backward induction solution. If this is the statement we make *about* the game, how does the player reason *in* the game?

18.1.1 Surprise Moves and Forward Induction

The following example is given by [24]: it is a two-player extensive form game in which the first player chooses a move a that ends the game or the move b that leads to a normal form game g_1, in which the players concurrently choose between $\{c_1, c_2\}$ and $\{d_1, d_2, d_3\}$, respectively (Figs. 18.1 and 18.2).

The backward induction solution advises player 1 to choose a, so player 2 does not expect to have any role. But suppose player 1 chooses b and the game does reach g_1. How should player 2 reason at this node? Should player 2 conclude that 1 is irrational and chose arbitrarily, or should 2 treat the subgame as a new game *ab initio* expecting rational play in the future?

Note that player 2 can ascribe a good reason for 1 to choose b: the expectation that 2 would choose d_3 in game g_1. (In this case, 1 can be expected to play c_2, and then player 2's best response would be d_2).

Such issues have been discussed extensively in the literature, and many resolutions have been suggested. Some of them go as follows:

- Players' actions are to be based on substantive stable common belief in future rationality [3, 16].
- Treat the first move of player 1 as a mistake, and either ignore past information or update beliefs accordingly [18].

- Players come in different *types*, and deviations from expected behaviour are interpreted according to players' knowledge of each others' type [5, 9].
- Players rationalize each other's behaviour [23].

Among these the last requires an explanation: according to this view, a player, at a game node, asks what rational strategy choices of the opponent could have led the history to this node. In such a situation, she must also ask whether the node could also have been reached by the opponent who does not only choose rationally herself, but who also believes that the other players choose rationally as well. This argument can be iterated, and leads to a form of **forward induction** [4, 24]. This leads to an interesting algorithm that can be seen as an alternative to backward induction [25].

A small point is worth noting here: the way forward induction (FI) is formalized as above, both BI and FI yield the same outcome [24] in extensive form games of perfect information. The strategies would in general be different, and this is in itself important for a theory of play. van Benthem [8] suggests an alternative viewpoint: rather than looking for a normal form subgame as above, he suggests that any sufficiently abstract representation of the subgame may result in FI yielding a different outcome. For instance, if the players were computationally limited, they would have only a limited view of a large subgame, and this is a very relevant consideration for a theory of play.

In general, how a player reasons *in* the game involves not only reasoning such as the above, but also computational abilities of the player. As the game unfolds, players have to record their observations, and a memory-restricted player needs to select what to record [6, 11].

Aumann and Dreze [2] make a strong case for the focus of game theory to shift from equilibrium computation to questions of how rational players should play. For zero sum games, the value of the game is unique and rational players will play to achieve this value. However, in the case of non zero sum games as mentioned above, multiple Nash equilibria can exist. This implies that players cannot extract an advice as to which strategy to employ from the equilibrium values. According to Aumann and Dreze, for a game to be well defined, it is also necessary that players have an expectation on what the other players will do. In estimating how the others will play, a rational player should take into account that others are estimating how he will play. The interactive element is crucial and a rational player should then play so as to maximize his utility, given how he thinks the others will play. The strategy specifications we introduce below are in the same spirit, since such a specification will be interactive in the sense of [2].

18.1.2 Players Matter

van Benthem [8] offers a masterly analysis of the many issues that distinguish reasoning about games and reasoning in games. Briefly, he points out that even if we consider BI as pre-game deliberation, there are aspects of dynamic belief revision to be considered; then there is the range of events that occur during play: players' observations, information received about other players, etc; then there is post-game reflection. As we move from deliberation to actual play, our interpretation of game theory requires considerable re-examination. We leave the reader to the pleasure of reading [8] for more on this, but pick up one slogan from there for discussion here: *the players matter*.

Briefly, reasoning inside games involves reasoning about actual play, and about the players involved. The standard game theoretic approach uses uniform algorithms (such as BI and FI) to talk of reasoning during play (including the actuality of surprise moves) and *type spaces* encode all hypotheses that players have about each other. However, the latter is again of the pre-game deliberative kind (as in BI), and abstracts all considerations of actual play into the type space. It is in this spirit that [24] talks of completeness of type spaces for FI, whereas the van Benthem analysis is a (clarion) call for dynamics in both aspects: dynamic decision making during play and a theory of player types that's dynamically constructed as well.

We suggest that this is a critical issue for logical foundations of game theory. A node of a game tree is a history of play, and unless all players have a logical explanation of how play got there, it is hard to see them making rational decisions at that point. The rationale that players employ then critically depends on perceived continuity in other players' behaviour, which needs to be construed during the course of play.

However, while this is easily said, it raises many questions that do not seem to have obvious answers. What would be a logic in which such reasoning as proceeds during play can be expressed? What would we ask of such a logic—that it provides formulas for every possible strategy that a player might employ in every possible game? That it be expressively complete to describe the (bewildering) diversity of player types? That we may derive stable strategy profiles using an inference engine underlying the logic? That we discover new strategies from the axioms and inference rules of the logic?

Several logics have been studied in the context of reasoning about strategic ability. [7] studies strategies in a dynamic logic, and in the context of alternating temporal logics, a variety of approaches have been studied [1, 17, 19, 29]. While these logics reason with the functional notion of strategy, a theory of play requires reasoning about the dynamics of player types as well.

18.1.3 Logic and Automata for Player Types

In this article, we take a very simple and minimalistic approach to these questions. We suggest that the logical language attempt to describe a universe of **constructible player types**. Therefore, players in this framework are of definable types and considerations of other players are also restricted to definable types. Rationalizability becomes relative to the expressiveness of the underlying formalism; we can perhaps call this notion 'extensive form *reasonability*'. We are less interested in completeness of the proposed language here, than in expressing interesting patterns of reasoning such as the ones alluded to above.

Our commitment is not only to simple modal logics to describe types, but also to realizing types by automata.[1] A number of reasons underlie this decision: for one, resource limitations of players critically affect course of play and selection of strategies. For another, automata present a nice tangible class of players that require rationale of the kind discussed above[2] and yet restrict the complexity that human players bring in. Further, automata theory highlights memory structure in players, and the selective process of observation and update.

Why is such an approach needed, or is indeed relevant, considering that an elegant topological construction of type spaces is already provided by Perea [4, 24] and others, with a completeness theorem as well? A crucial departure lies in the emphasis on constructivity and computability of types and strategies (rather than their existence). Moreover, if our attempt is not only to enrich the type space but also to provide *explanations* of types, logical means seem more attractive. The price to pay lies, of course, in the restrictive simplicity of the logics and automata employed, and is it is very likely that such reasoning is much less expressive than the topological type spaces.

This work continues the line of investigation initiated in [26]. A principal difference here is an explicit connection with the notion of extensive form rationalizability, which is achieved by the use of belief-like operators in the space of types.

18.2 Types as Formulas

Let N denote the set of players, we use i to range over this set. For technical convenience, we restrict our attention to two player games, i.e. we take $N = \{1, 2\}$. We often use the notation i and $\bar{\imath}$ to denote the players where $\bar{\imath} = 2$ when $i = 1$ and $\bar{\imath} = 1$ when $i = 2$. Let Act be a finite set of action symbols representing moves of players, we let a, b range over Act.

Strategizing during play involves making observations about moves, forming beliefs and revising them. Player types are constructed precisely in the same manner:

- Patterns of the form 'when condition p holds, player 2 chooses a' are observations by player 1 and help to assign a basic type to player 2.
- Such a process clearly involves nondeterminism to accommodate apparently contradictory behaviour, so a player needs to assign a disjunction of types to the other.
- The process of reasoning proceeds by case analysis: in situations such as x, the other player is seen to play conservatively whereas in other situations such as y, the type is apparently aggressive. Thus type construction is conjunctive as well.
- The planning of a player also includes how he responds to perceived opponent startegies that lie within this plan. Therefore type definition includes such responses.

---

[1] By automata, we refer only to finite state devices here, though probablistic polynomial time Turing machines are a natural class to consider as well [13].

[2] A surprise move by an opponent is perhaps much harder for an automaton to digest than for a human player.

- Rationalization: perceived behaviour can be explained by actual play being part of a strategy that involves the future as well, and this is articulated as a belief by the player about the opponent. Moreover, such beliefs include the opponent's beliefs about the player as well, and iterating the process builds a hierarchy of beliefs.

Above, we have spoken of the type of a player as it is ascribed by the opponent. Note that the same reasoning works for ascribing types 'from above' to a player. Such considerations lead us to the following syntax for player types.

18.2.1 Type Specifications

Let $P^i = \{p^i_0, p^i_1, \ldots\}$ be a (non-empty) countable set of observables for $i \in N$ and $P = \cup_{i \in N} P^i$. The syntax of type specifications is given by:

$$Type^i(P^i) := [\psi \mapsto a]^i \mid \sigma_1 + \sigma_2 \mid \sigma_1 \cdot \sigma_2$$

where $\psi \in Past(P^i)$ and $a \in Act$. ψ is intended to be a past play formula, to be defined below. Observe that since the atomic specifications are always indexed by the player identity, a type specification unambiguously points to a player, denoted by $pl(\sigma)$.

To complete the specification of types, we need to specify the syntax of $PF(P^i)$ which is given by a simple tense logic. For any countable set X, let $PF(X)$ be sets of **play formulas** given by the following syntax:

$$PF(X) := x \in X \mid \neg\psi \mid \psi_1 \vee \psi_2 \mid \langle a \rangle \psi \mid \Diamond\psi \mid \Diamondblack\psi \mid B_i\pi @ \bar{\imath}$$

where $a \in Act$ and $\pi \in Type^{\bar{\imath}}(P^{\bar{\imath}} \cap P^i)$.

Formulas in $PF(X)$ are interpreted at game positions. The formula $\langle a \rangle \psi$ talks about an a-edge in the game tree after which ψ holds. The formulas $\Diamondblack\psi$ and $\Diamond\psi$ assert ψ some time in the past or future, respectively. The boolean fragment of $PF(X)$ is denoted by $Bool(X)$. The "past tense" fragment of $PF(X)$ uses only boolean connectives and the $\Diamondblack$ modality, we denote this fragment by $Past(X)$.

Let $Act = \{a_1, \ldots, a_m\}$, we also make use of the following abbreviation.

- $null^i = [\top \mapsto a_1]^i + \cdots + [\top \mapsto a_m]^i$.

where $\top = p \vee \neg p$ for an observable $p \in P^i$. It will be clear from the semantics (which will be defined shortly) that any type of player i conforms to $null^i$, or in other words this is an empty specification. The empty specification is particularly useful for assertions of the form "there exists a type" where the property of the type is not of any relevance, and specifies a player type vacuously.

Before we proceed to the semantics, a remark on the two-layered syntax is in order. In principle, they can be combined into one, but there are good reasons to keep them separate: for one, note that negation is free in play formulas, but its scope in types is restricted to atomic types. This is deliberate, and has important implications for reasoning. Another reason is that types are intended as invariants on the game tree, though constructed locally, whereas play formulas are associated with game positions. The latter point will be clarified by the semantics presented below, but some preliminaries first.

18.2.2 Game Trees

Let $\mathbb{T}(Act) = (S, \Rightarrow, s_0)$ be a finite Act-labelled tree rooted at s_0 on the finite set of vertices S while $\Rightarrow : (S \times Act) \to S$ is a *partial* function specifying the edges of the tree. For a node $s \in S$, let $\vec{s} = \{s' \in S \mid s \overset{a}{\Rightarrow} s'$ for some $a \in Act\}$ and let $moves(s) = \{a \in Act \mid \exists s' \in S$ with $s \overset{a}{\Rightarrow} s'\}$. An action "$a$" is said to be enabled at a node s if $a \in moves(s)$. A node s is called a leaf node (or terminal node) if $\vec{s} = \emptyset$. For the rest of the paper we fix Act to be the finite set $Act = \{a_1, \ldots, a_m\}$, and refer to $\mathbb{T}$ to denote the tree $\mathbb{T}(Act)$.

An **extensive form game tree** is a tuple $T = (S, \Rightarrow, s_0, \lambda)$ where $\mathbb{T} = (S, \Rightarrow, s_0)$ is a tree. The set S denotes the set of game positions with s_0 being the initial game position. The edge function $\Rightarrow$ specifies the moves enabled at a game position and the turn function $\lambda : S \to N$ associates each game position with a player. Technically, we need player labelling only at the non-leaf nodes. However, for the sake of uniform presentation, we do not distinguish between leaf nodes and non-leaf nodes as far as player labelling is concerned. For $i \in N$, let $S^i = \{s \mid \lambda(s) = i\}$, and let $frontier(\mathbb{T})$ denote the set of all leaf nodes of T. When s is a game position in T, we denote the subtree rooted at s by T^s.

A **play** in the game T starts by placing a token on s_0 and proceeds as follows: at any stage, if the token is at a position s and $\lambda(s) = i$, then player i picks an action which is enabled for her at s, and the token is moved to s' where $s \overset{a}{\Rightarrow} s'$. Formally a play in T is a finite path $\rho : s_0 a_1 s_1 \cdots a_k s_k$ such that for all $0 \le j < k\ s_j \overset{a_j}{\Rightarrow} s_{j+1}$. A maximal path is one in which $s_k \in frontier(\mathbb{T})$. Note that each leaf node t denotes a play of the game which is the unique path from the root node s_0 to t. Let $Plays(T)$ denote the set of all plays in the game tree T.

18.2.3 Strategies and Plans

A **strategy** for player i is a function μ^i which specifies an enabled move at every game position of the player, i.e. $\mu^i : S^i \to Act$ which satisfies the condition: for all $s \in S^i$, $\mu^i(s) \in moves(s)$. For $i \in N$, we use the notation μ^i to denote strategies of player i and $\tau^{\bar{i}}$ to denote strategies of player $\bar{i}$. By abuse of notation, we will drop the superscripts when the context is clear and follow the convention that μ represents strategies of player i and τ represents strategies of $\bar{i}$. Let $\Omega^i(T)$ denote the set of all strategies for player i in the extensive form game tree T.

A strategy can be viewed as a collection of what Rubinstein [28] calls **plans of action**. The functional notion of strategy demands a choice for every player i node, whereas a plans specifes what a player would do only at nodes reachable from those the player had decided on earlier in the plan. This notion can be seen as an equivalence class of (classical) strategies that are outcome-equivalent. When we consider the set of plans for a player from *every* node of the game tree, we include all plans generated by strategies.

A plan is then a subtree of T in which for each player i node, there is a unique outgoing edge and for nodes belonging to player $\bar{\imath}$, every enabled move is included. Formally we define a **plan subtree** as follows: For $i \in N$, a plan tree at node s $T_\mu = (S_\mu, \Rightarrow_\mu, s, \lambda_\mu)$ associated with μ is the least subtree of T satisfying the following property:

- $s \in S_\mu$
- For any node $s \in S_\mu$,

  – if $\lambda(s) = i$ then there exists a unique $s' \in S_\mu$ and action a such that $s \overset{a}{\Rightarrow}_\mu s'$.
  – if $\lambda(s) \neq i$ then for all s' such that $s \overset{a}{\Rightarrow} s'$, we have $s \overset{a}{\Rightarrow}_\mu s'$.

A play $\rho : s_0 a_0 s_1 \cdots$ is said to be consistent with strategy μ if for all $j \geq 0$, $j : 0 \leq j < k$, we have $s_j \in S^i$ implies $\mu(s_j) = a_j$. A plan profile (μ, τ) consists of a pair of plans, one for each player. (We use μ, τ etc to denote both strategies and plans, the context making it clear which notion is in use.)

We say that a plan μ **potentially reaches** tree node s if s is in the subtree T_μ. Let $\Omega_i(s)$ denote the set of player i plans that potentially reach s. Given i, $\Omega_{\bar{\imath}}(s)$ then denotes the set of opponent plans that potentially reach s. Thus $T_\tau \in \Omega_{\bar{\imath}}(s)$ iff there exists $T_\mu \in \Omega_i(s)$ such that the play (μ, τ) visits s. This is important for rationalizability: a player conceives of what potential plans the opponent might be using that are consistent with the play reaching node s. The set $\Omega_{\bar{\imath}}(s)$ can be seen as a *belief set* for player i, in the sense that these are the hypothetical plans that i can attribute to $\bar{\imath}$ to explain the history of play leading to node s.[3]

The idea is to use the above constructs to specify properties of player types. For instance the interpretation of a player i specification $[p \mapsto a]^i$ where $p \in P^i$ is to choose move "a" at every player i game position where p holds. At positions where p does not hold, the player chooses any enabled move. $\sigma_1 + \sigma_2$ says that the type of player i conforms to the specification σ_1 or σ_2. The construct $\sigma_1 \cdot \sigma_2$ says that the type conforms to both specifications σ_1 and σ_2.

The construct $B_i \pi @\bar{\imath}$ describes (a kind of) *belief hierarchy*: player i believes that opponent behaviour corresponds to some complete plan π. Note that π, in turn, could be referring to some type σ' of player i, and so on. In this sense, a player holds a belief about opponent's strategy choices, about opponents' beliefs about other agents' choices, opponents' beliefs about others' beliefs etc. Since this is essentially how type spaces are defined, these specifications offer a compositional means for structuring type spaces.

---

[3] This observation can be formalized by defining an indistinguishability relation $\sim_i$ on player i nodes, but we do not pursue this here, since we do not attempt an axiomatization of the type space.

18.2.4 Semantics

Let $M = (T, V)$ where $T = (S, \Rightarrow, s_0, \lambda)$ is an extensive form game tree and $V : S \to 2^P$ a valuation function. The truth of a formula $\psi \in PF(P)$ at the state s, denoted $M, s \models \psi$ is defined as follows:

- $M, s \models p$ iff $p \in V(s)$.
- $M, s \models \neg\psi$ iff $M, s \not\models \psi$.
- $M, s \models \psi_1 \vee \psi_2$ iff $M, s \models \psi_1$ or $M, s \models \psi_2$.
- $M, s \models \langle a \rangle \psi$ iff there exists an s' such that $s \overset{a}{\Rightarrow} s'$ and $M, s' \models \psi$.
- $M, s \models \Diamond\psi$ iff there exists an s' such that $s \Rightarrow^* s'$ and $M, s' \models \psi$.
- $M, s \models \Diamond\!\!\!\!-\, \psi$ iff there exists an s' such that $s' \Rightarrow^* s$ and $M, s' \models \psi$.
- $M, s \models B_i \pi \,@\bar{\imath}$ iff there exists $\tau \in \Omega_{\bar{\imath}}(s)$ such that $(T_\tau, V_\tau), \models \pi$.

Type specifications are interpreted on plan subtrees of T. We assume the presence of two special propositions $\mathbf{turn}_1$ and $\mathbf{turn}_2$ that specify which player's turn it is to move. We also assume the existence of a special proposition *leaf* which holds at the terminal nodes. Formally, we assume that the valuation function satisfies the property:

- for all $i \in N$, $\mathbf{turn}_i \in V(s)$ iff $\lambda(s) = i$.
- $leaf \in V(s)$ iff $moves(s) = \emptyset$.

Recall that a plan μ of player i is a subtree $T_\mu = (S_\mu, \Rightarrow_\mu, s_0, \lambda_\mu)$ of T. Let V_μ denote the restriction of the valuation function V to S_μ. For a type specification $\sigma \in Type^i(P^i)$, we define the notion of μ **conforming** to σ (denoted $(T_\mu, V_\mu) \models \sigma$) as follows:

- $(T_\mu, V_\mu) \models [\psi \mapsto a]^i$ iff for every player i node $s \in S_\mu$, we have: $(T_\mu, V_\mu), s \models \psi$ implies $moves(s) = \{a\}$.
- $(T_\mu, V_\mu) \models \sigma_1 + \sigma_2$ iff $(T_\mu, V_\mu) \models \sigma_1$ or $(T_\mu, V_\mu) \models \sigma_2$.
- $(T_\mu, V_\mu) \models \sigma_1 \cdot \sigma_2$ iff $(T_\mu, V_\mu) \models \sigma_1$ and $(T_\mu, V_\mu) \models \sigma_2$.

Above, $\psi \in PF(P^i)$. Since s is an i node in a player i plan tree, it has a unique outgoing edge and therefore $moves(s)$ is a singleton set.

The belief formula is interpreted using potential rationalizable strategies. Here player i, at node s, considers how the history of play might have reached s, and postulates π as an explanation.[4] The logic is very similar to that in [20] but for the emphasis on rationalizability here.

We say that $\sigma \in Type^i(P^i)$ is **satisfiable** in T if there exists a plan μ such that $(T_\mu, V_\mu) \models \sigma$. Let $Sat^i(T)$ denote the set of player i's type specifications that are satisfiable in T. Further, given $\sigma \in Type^i(P^i)$ and a tree node s, let $\theta_i(\sigma, s) = \{\pi \mid$ for all plans μ such that $(T_\mu, V_\mu) \models \sigma$, if $\mu \in \Omega_i(s)$ there exists $\tau \in \Omega_{\bar{\imath}}(s)$ such that $(T_\tau, V_\tau) \models \pi\}$. This set can be thought of as the beliefs of player i implied by the type σ at the node s.

---

[4] Note that B_i is a "model changing" operator and thus 'dynamic', in the spirit of [8].

18.2.5 Types in Solution Concepts

The semantics of a player type is given as a set of the player's plan subtrees of the given game tree, based on observables. It is defined at every player i node, specifying player i's beliefs about opponents' strategies that could have resulted in play reaching that node. But since every opponent's type specifies the opponent's beliefs about others' strategy choices, this results in a recursive structure and we can build a hierarchy of types.

Note that the belief assertions can specify different strategic choices based on the past, and thus talk of how a player may, during the game, revise her beliefs, a form of dynamics.

This further suggests that we wish to *derive* types during play. Thus, rather than types as being fixed for the class of games, we consider types that start perhaps as heuristics, and *grow* during play. We take this up for discussion in Sect. 18.3.

Once we have the notion of types, it induces a notion of *local equilibrium* as follows. Consider player 1's response to player 2's strategy τ: here, 1's best response is not to τ, but to *every* type π that τ satisfies. Symmetrically, 2's best response is not to a straegy μ of player 1 but to every type σ that μ satisfies. Thus we can speak of the type pair (σ, π) being in equilibrium. We merely remark on this induced notion here, one well worth developing further on in future.

18.3 Growing Types

We have suggested that our definition of player types has been guided by concerns of constructibility and simplicity. Yet, we need to discuss how types as we have defined relate to the topological type spaces considered by game theorists, especially since forward induction is justified by the completeness of such spaces.

Let $\mathbb{T}(Act) = (S, \Rightarrow, s_0)$ be an extensive form game. A **type space** over $\mathbb{T}$ is a tuple $G = (U_i, \delta_i)_{i \in N}$ where each U_i is a compact topological space, representing the set of types for player i, and δ_i is a function that assigns to every type $u \in U_i$ and tree node s, a probability distribution $\delta_i(u, s) \in \Delta(\Omega_{\bar{\imath}}(s), U_{\bar{\imath}})$. Note that $\Omega_{\bar{\imath}}(s)$ represents the set of opponent strategies that potentially reach node s, $U_{\bar{\imath}} = \Pi_{j \neq i} U_j$ is the set of opponents' type combinations, and $\Delta(X)$ is the set of probability distributions on X with respect to the Borel σ-algebra.

In game theory, type spaces are typically defined for games of imperfect information, and the definition above coincides with the standard one when the information set for every player is a singleton. A natural question arises whether the concept makes sense for games of perfect information. In the discussions on forward induction, as for instance in [4, 24], the BI and FI solutions coincide, and the analysis differentiates games with nontrivial information sets. However, as van Benthem argues [8], there are other interpretations of forward induction that are relevant for a theory of play: when the game tree is large, a player at a tree node s may be able to reason only about a small inital fragment of the subtree issued at s, and the subsequent abstraction may be seen as imperfect information as well.[5] Moreover, rationalizing by the player i does induce an equivalence relation $\sim_i$ on the tree nodes in our analysis.

Note the similarity of our definition of types to the standard notion, without the use of probability distributions. The use of equivalence relations between nodes is an implicit form of qualitative expectations and we choose the simpler formalism as it is more amenable to modal logics. With these observations, consider the type space 'induced' in our framework.

Consider a model $M = (T, V)$ where $T = (S, \Rightarrow, s_0, \lambda)$ is an extensive form game tree and $V : S \to 2^P$ a valuation function. Then we define the **logical type space** over $\mathbb{T}$ to be a tuple $L = (Sat^i(T), \theta_i)_{i \in N}$, where $Sat^i(T)$ is the set of player i type specifications satisfiable in the game, and $\theta_i : (Sat^i(T) \times S) \to Sat^{\bar{\imath}}(T)$ was defined earlier. Recall that this set represents the beliefs that player i has about the opponent implied by the type σ at the node s.

Now one can see the close correspondence between the two definitions, as well as the differences. The type space G is globally defined, and can be seen as fixing an encoding of all possible beliefs of players about opponent behaviour *a priori*. In contrast, the type space L has more local structure, and is crucially determined by the expressiveness of the logic. The topological structure of the type space in G is replaced by logical structure in L. For instance, the types in L are downward closed: if $\sigma_1 \cdot \sigma_2$ is a type, then so are σ_1 and σ_2; it is closed under entailment: if σ_1 is a type and σ_1 entails σ_2, then σ_2 is a type, and so on. There are other symmetries such as: if $\pi \in \theta_i(\sigma, s)$ for some tree node s, then there exists a tree node t such that $\sigma \in \theta_{\bar{\imath}}(\pi, t)$. Characterizing this logical structure by a completeness theorem is an important question, but we do not proceed further on this here.

At this juncture, our claim to 'growing' types can be explained. Consider the root node s_0 in the tree T. Notice that the beliefs of the players about each other refer only to invariant properties in the game tree (as specified by the observables), and hence the only definite assertions are about the present, namely the root node itself. However, as play progresses, we have definite assertions about the past, as well as about the choices thus eliminated, and we have sharper type formulas. This process may be understood as a construction of the type space that proceeds top-down, starting from the root node and enriching players' beliefs based on observations as the game tree gets pruned by play.[6]

While this is a general picture, we focus on a specific question: does this 'construction' of a logical type require *unbounded* information? We now proceed to show that the required information is in fact *finite state*, and hence can be checked by an automaton. Further, we show that, in principle, a Turing machine can construct the type space.

---

[5] We refer the reader to [8] for a more detailed justification.

[6] A formal characterization of this process as a recursive function on the tree is in progress, but there are many technical challenges.

18.4 Types as Automata

Once we consider types to be logical, a natural question is whether a given type is consistent: we want a player to be a reasoner whose reasoning is coherent. It is this question we address here.

Note that a type corresponds to a set of plans and beliefs in our framework. We have spoken of a model in which a player records observations during play and rationalizes opponents' behaviour by considering what strategies might have led to opponents playing in a particular way. The meaning we offer for constructibility of such a type is a finite state automaton that 'plays out' such plans and rationalizes course of play.

The use of automata for finite memory strategies dates back to the work of Büchi and others [12] in the 1960's, and a rich theory of automata based strategies for regular games of infinite duration has been developed in the last few decades [15]. In [26] we apply this technique to develop a theory of structured strategies. Ramanujam and Simon [27] links this theory to game logic [21], by suggesting that game composition and strategy composition are interdependent, and develops a logic of composing game—strategy pairs. Ghosh [14] develops a similar framework as well. Paul et al. [22] extends the logic of strategies to consider *switching* whereby a player switches strategies during course of play, based on observations. This leads to a logical interpretation of mixed strategies, as switching between pure strategies over repeated play. We mention these lines of work here since the logical specification of types in this paper closely follows and builds on these papers, and the automaton construction below uses similar techniques. The crucial departure here is rationalization of course of play reflected in beliefs.

However, the essential elements of a logical structure of strategies and that of player types are the same, and these papers all emphasize top-down reasoning in games, thus contributing to a theory of play. Strategies are seen not as complete pre-selected plans that a player comes to a game with. A player is assumed to have access to a library of local heuristics as partial strategies, and composes them during course of play, selections depending on observations of events. In this sense, strategy specifications *are* player types, but without the belief structure addressed here.

18.4.1 Subformulas

For a type specification σ, let $SF(\sigma)$ denote the subformula closure of σ. For $\psi_0 \in PF(P)$, the subformula closure $SF(\psi_0)$ of ψ_0 and Σ_{ψ_0}, the *type vocabulary* of ψ_0, are defined by simultaenous induction as follows. $SF(\psi_0)$ is the least set containing ψ_0 and closed under the following conditions:

- $\neg\psi \in SF(\psi_0)$ iff $\psi \in SF(\psi_0)$ (where $\neg\neg\psi$ is considered the same as ψ).
- if $\psi_1 \vee \psi_2 \in SF(\psi_0)$ then $\{\psi_1, \psi_2\} \subseteq SF(\psi_0)$.
- if $\langle a \rangle \psi \in SF(\psi_0)$ then $\psi \in SF(\psi_0)$.

- if $\Diamond\psi \in SF(\psi_0)$ then $\psi \in SF(\psi_0)$ and for every $a \in Act$, $\langle a\rangle\Diamond\psi \in SF(\psi_0)$.
- if $\Diamond\!\!\!\!\!\Diamond\psi \in SF(\psi_0)$ then $\psi \in SF(\psi_0)$.
- If $B_i\pi\,@\bar{\imath} \in SF(\psi_0)$ then $\pi \in \Sigma_{\psi_0}$.
- If $\sigma \in \Sigma_{\psi_0}$ and $[\psi \mapsto a]^i \in SF(\sigma)$ then $\psi \in SF(\psi_0)$.

Fix a play formula ψ_0 and let CL denote $SF(\psi_0)$, Σ denote Σ_{ψ_0}.

Call $R \subseteq CL$ an *atom* if it is 'locally' consistent and complete: that is, for every $\psi \in CL$, $\neg\psi \in R$ iff $\psi \notin R$, for every $\psi_1 \vee \psi_2 \in CL$, $\psi_1 \vee \psi_2 \in R$ iff $\psi_1 \in R$ or $\psi_2 \in R$, and for every $\psi \in R$, if $\Diamond\psi \in CL$ then $\Diamond\psi \in R$, and if $\Diamond\!\!\!\!\!\Diamond\psi \in CL$ then $\Diamond\!\!\!\!\!\Diamond\psi \in R$. Let $\mathcal{AT}$ denote the set of atoms.

For a type specification σ, let $SF_\psi(\sigma)$ denote the past play subformulas in σ. Call $A \subseteq SF_\psi(\sigma)$ a *type atom* for σ if it is propositionally consistent and complete, as above. Let $\mathcal{T} - \mathcal{AT}$ denote the set of all type atoms (for all $\sigma \in \Sigma$).

An atom R is said to be *initial* if whenever $\Diamond\!\!\!\!\!\Diamond\psi \in R$ then $\psi \in R$ as well. Similarly, R is said to be *final* if whenever $\Diamond\psi \in R$ then $\psi \in R$ as well. Initial and final type atoms are defined similarly.

Let R and R' be atoms. Define $R \longrightarrow_a R'$ iff the following conditions hold:

- For every $\langle a\rangle\psi \in CL$, if $\psi \in R'$ then $\langle a\rangle\psi \in R$.
- If $\Diamond\!\!\!\!\!\Diamond\psi \in R$ then $\Diamond\!\!\!\!\!\Diamond\psi \in R'$.
- If $\Diamond\psi \in R'$ then $\Diamond\psi \in R$.

For type atoms C, C', $C \longrightarrow C'$ holds when $\Diamond\!\!\!\!\!\Diamond\psi \in C$ implies $\Diamond\!\!\!\!\!\Diamond\psi \in C'$.

Atoms and type atoms will be used to construct automata below.

18.4.2 Advice Automata

Clearly, every type specification defines a set of strategies. We now show that it is a *regular* set, recognizable by a finite state device, which we call an **advice** automaton.

For a game T, a nondeterministic advice automaton for player i is a tuple $\mathcal{A} = (Q, \delta, o, I)$ where Q is the set of states, $I \subseteq Q$ is the set of initial states, $\delta : (Q \times W \times Act) \to 2^Q$ is the transition relation, and $o : (Q \times W^i) \to Act$, is the output or advice function.

The language accepted by the automaton is a set of plans of player i. Given a plan subtree $\mu = (W_\mu, \longrightarrow_\mu, s_0)$ of player i, a run of $\mathcal{A}$ on μ is a Q-labelled tree $T = (W_\mu, \longrightarrow_\mu, \lambda)$, where λ maps each tree node to a state in Q as follows: $\lambda(s_0) \in I$, and for any s_k where $s_k \xrightarrow{a}_\mu s'_k$, we have $\lambda(s'_k) \in \delta(\lambda(s_k), s_k, a_k)$.

A Q-labelled tree T is accepted by the automaton $\mathcal{A}$ if for every tree node $s \in W_\mu^i$, if $s \xrightarrow{a}_T s'$ then $o(\lambda(s)) = a$. A plan μ is accepted by $\mathcal{A}$ if there exists an accepting run of $\mathcal{A}$ on μ.

The following lemma relates type specifications to advice automata.

Lemma 18.1 *Given a player $i \in N$ and a type specification σ, we can construct an advice automaton $\mathcal{A}_\sigma$ such that $(T_\mu, V_\mu) \in Lang(\mathcal{A}_\sigma)$ iff $(T_\mu, V_\mu) \models \sigma$.*

Proof The construction of automata is inductive, on the structure of specifications. Note that the plan satisfying the type is implemented principally by the output function of the advice automaton.

We proceed by induction on the structure of σ. We construct automata for atomic strategies and compose them for complex strategies.

$(\sigma \equiv [\psi \mapsto a])$: The automaton works as follows. Its states keep track of past formulas satisfied along a play as game positions are traversed, and that the valuation respects the constraints generated for satisfying ψ. The automaton also guesses a move at every step and checks that this is indeed a when ψ holds; in such a case this is the output of the automaton. Formally: $\mathcal{A}_\sigma = (Q_\sigma, \delta_\sigma, o_\sigma, I_\sigma)$, where

- $Q_\sigma = \mathcal{T} - \mathcal{AT} \times Act$.
- $I_\sigma = \{(C, x) | C \text{ is initial}, V(s_0) = (C \cap P_\sigma), x \in Act\}$.
- For a tree edge $s \xrightarrow{a} s'$ in T we have:
  $\delta_\sigma((C, x), s, a) = \{(C', y) | C \longrightarrow C', V(s') = (C' \cap P_\sigma), y \in Act\}$.
- $o((C, x), s) = \begin{cases} a \text{ if } \psi \in C \\ x \text{ otherwise} \end{cases}$

We now prove the assertion in the lemma that $\mu \in Lang(\mathcal{A}_\sigma)$ iff $\mu \models_i \sigma$. ($\Rightarrow$) Suppose $\mu \in Lang(\mathcal{A}_\sigma)$. Let $T = (W^1_\mu, W^2_\mu, \longrightarrow_T, \lambda)$ be the Q-labelled tree accepted by $\mathcal{A}_\sigma$. We need to show that for all $s \in W_\mu$, we have $\rho_s, s \models \psi$ implies $out(s) = a$.

The following claim, easily proved by structural induction on the structure of ψ, using the definition of $\longrightarrow$ on atoms, asserts that the states of the automaton check the play requirements correctly. Below we use the notation $\psi \in (C, x)$ to mean $\psi \in C$.

Claim For all $s \in W_\mu$, for all $\psi' \in SF_\psi(\sigma)$, $\psi' \in \lambda(s)$ iff $\rho_s, s \models \psi'$.

Assume the claim and consider any $s \in W_\mu$. From Claim 4, we have $\rho_s, s \models \psi$ implies $\psi \in \lambda(s)$. By the definition of o, we have $o(\lambda(s), s) = a$.

($\Leftarrow$) Suppose $\mu \models_1 [\psi \mapsto a]$. From the semantics, we have $\forall s \in W^1_\mu, \rho_s, s \models \psi$ implies $out(s) = a$. We need to show that there exists a Q-labelled tree accepted by $\mathcal{A}_\sigma$. For any s let the Q-labelling be defined as follows. Fix $x_0 \in Act$.

- For $s \in W^1_\mu$, let $\lambda(s) = (\{\psi' \in SF_\psi(\sigma) | \rho_s, s \models \psi'\}, out(s))$.
- For $s \in W^2_\mu$, let $\lambda(s) = (\{\psi' \in SF_\psi(\sigma) | \rho_s, s \models \psi'\}, x_0)$.

It is easy to check that $\lambda(s)$ constitutes an atom and the transition relation is respected. By the definition of o, we get that it is accepting.

$(\sigma \equiv \sigma_1 \cdot \sigma_2)$: By the induction hypothesis there exist $\mathcal{A}_{\sigma_1} = (Q_{\sigma_1}, \delta_{\sigma_1}, o_{\sigma_1}, I_{\sigma_1})$ and $\mathcal{A}_{\sigma_2} = (Q_{\sigma_2}, \delta_{\sigma_2}, o_{\sigma_2}, I_{\sigma_2})$ which accept all strategies satisfying σ_1 and σ_2 respectively. To obtain an automaton which accepts all strategies which satisfy $\sigma_1 \cdot \sigma_2$ we just need to take the product of $\mathcal{A}_{\sigma_1}$ and $\mathcal{A}_{\sigma_2}$.

$(\sigma \equiv \sigma_1 + \sigma_2)$: We take $\mathcal{A}_\sigma$ to be the disjoint union of $\mathcal{A}_{\sigma_1}$ and $\mathcal{A}_{\sigma_2}$. Since the automaton is nondeterministic with multiple initial states, we retain the initial states of both $\mathcal{A}_{\sigma_1}$ and $\mathcal{A}_{\sigma_2}$. If a run starts in an initial state of $\mathcal{A}_{\sigma_1}$ then it will never cross over into the state space of $\mathcal{A}_{\sigma_2}$ and vice versa.

Note that an advice automaton generates a set of plans for a player. On the other hand, a player's beliefs refer to other plans in the game tree and hence checking play formulas needs to work with the entire game tree. This is achieved by the following 'global' tree automaton construction.

18.4.3 Tree Automata

Fix $k > 0$. A game tree $T = (S, \Rightarrow, s_0, \lambda)$ is said to be k-ary, if for every node s, the out-degree of s is k, that is, $| \vec{s} | = k$.

A nondeterministic k-tree automaton over a k-ary tree $T = (S, \Rightarrow, s_0, \lambda)$ is a tuple $\mathcal{B} = (Q, \delta, I, F)$, where Q is the set of automaton states, $I, F \subseteq Q$ are the sets of initial and final states, and $\delta \subseteq (Q \times S \times Q^k)$ is the transition relation. A *run* of B is a map $\rho : S \to Q$ such that for every tree node s, if the set of children of s is $\{s_1, \ldots, s_k)\}$, then $(\rho(s), s, \rho(s_1) \ldots, \rho(s_k)) \in \delta$. We say that ρ is an *accepting* run if $\rho(s_0) \in I$ and for every leaf node s_ℓ, $\rho(s_\ell) \in F$. The emptiness problem for nondeterministic k-tree automata can be solved in polynomial time.

Given a $M = (T, V)$ and a play formula ψ_0 the objective is to construct a tree automaton $B(M, \psi_0)$ running over k-ary trees for some fixed $k > 0$. But the branching degree of nodes in M need not be uniform. However, since the automaton is parameterized by T, we can always 'normalize' the model in the following sense. Let $m = maxm_{s \in S}| \vec{s} |$. We can add 'dummy' edges from every node to leaf nodes labelled by a symbol not in Act so that the branching degree is made uniform. A tree automaton running on such a normalized tree would enter an accepting state on encountering a dummy node, so in effect any path ending in a dummy node is disregarded by the automaton and only plays that are present in M are checked for consistency requirements. (The advice automata corresponding to type specifications always output $a \in Act$, and hence the label on the dummy edges need not be checked either.) Therefore, without loss of generality we can assume that the model M is an m-ary tree.

Before we get to the tree automaton construction, we need one more step of preparation. We need to work with **deterministic** advice automata, rather than the nondeterministic ones we have associated with type specifications. For a game T, a deterministic advice automaton for player i is a tuple $\mathcal{A} = (Q, \delta, o, q_0)$ where Q is the set of states, $q_0 \in Q$ is the initial state, $\delta : (Q \times W \times Act) \to Q$ is the transition fuction, and $o : (Q \times W^i) \to 2^{Act}$ is the output or advice function. The notion of acceptance of a plan is modified so that the choice determined by the strategy is present in the automaton output (rather than being the same). The following proposition asserts that advice automata can be 'determinized'.

Proposition 18.1 *Give a game* $T = (S, \Rightarrow, s_0, \lambda)$ *and a nondeterministic advice automaton* $\mathcal{A}$ *on* T, *we can construct a deterministic advice automaton* $\mathcal{A}'$ *such that a plan* μ *is accepted by* $\mathcal{A}$ *iff it is accepted by* $\mathcal{A}'$.

We omit the proof since it is standard, achieved by a subset construction. For each $\sigma \in \Sigma_{\psi_0}$, let A_σ denote the deterministic advice automaton associated with it; let $A_1, \ldots, A_k$ be the set of these automata, with the product state space $(Q_1 \times \ldots Q_k)$.

18.4.4 Automaton Construction

Intuitively, the tree automaton works as follows. It keeps track of the atoms of ψ_0 and simulates the k advice automata in parallel, and when it encounters a player belief $B_i \pi @ \overline{\iota}$ it checks that the output of the advice automaton A_π is consistent. When it branches, it guesses for each branch which future requirements need to be satisfied along that particular branch. The acceptance condition checks that all final states are at final atoms.

A state of the tree automaton is a tuple $(s, R, (q_1, b_1) \ldots, (q_k, b_k))$ where s is a tree node, R is an atom, q_j is a state of the jth advice automaton, $b_j \in \{0, 1\}$ and the following conditions are satisfied:

- $V(s) = R \cap P$.
- For every $\Diamond \psi$ in SF_σ, for $\sigma \in \Sigma$, $\Diamond \psi \in q_\sigma$ iff $\Diamond \psi \in R$.
- If $B_i \pi_j @ \overline{\iota} \in R$ then $b_j = 1$.

Let U_{ψ_0} denote the set of states.

The boolean flag b_j marks the nodes reachable by following the advice output by the automata for type specifications. When play moves away from the plan constituted by the advice automaton, the flag is set to 0 and signals that the player belief is not justified.

Formally, given the model $M = (T, V)$ where $T = (S, \Rightarrow, s_0, \lambda)$ is an extensive form game tree, $V : S \to 2^P$ a valuation function, and a play formula ψ_0, the tree automaton $B(M, \psi_0)$ is defined to be the tuple $(U_{\psi_0}, \Delta, I, F)$ where U_{ψ_0} is as defined above, and:

- $I = \{(s_0, R, (q_1^0, 1) \ldots, (q_k^0, 1)) \mid R \text{ is an initial atom, and } q_j^0 \text{ is the initial state of } A_j\}$.
- $F = \{(s, R, (q_1, b_1) \ldots, (q_k, b_k)) \mid R \text{ is a final atom, and } s \text{ is a leaf node}\}$.
- $(x, s, y_1, \ldots y_m) \in \Delta$ if the following set of conditions are satisfied, where $x = (s, R, (q_1, b_1) \ldots, (q_k, b_k))$, and $y_i = (s_i, R^i, (q_1^i, b_1^i) \ldots, (q_k^i, b_k^i))$:

  – In the tree T, the moves at s are of the form $(a_1, s_1), \ldots, (a_m, s_m)$.
  – If $\langle a \rangle \psi \in R$, $a = a_i$ and $\psi \in R^i$ for some $i \in \{1, \ldots, m\}$.
  – If $\Diamond \psi \in R$, then $\Diamond \psi \in R^i$ for some $i \in \{1, \ldots, m\}$.
  – If $\lambda(s) = pl(\sigma_j)$, $1 \le j \le k$ and $b_j = 1$, then for all $i \in \{1, \ldots, m\}$, $b_j^i = 1$ iff $a_i \in o(q_j, s)$, and $q_j^i = \delta_j(q_j, s, a_i)$.

– If $\lambda(s) \neq pl(\sigma_j)$, $1 \leq j \leq k$ and $b_j = 1$, then for all $i \in \{1, \ldots, m\}$, $b_j^i = 1$ iff $q_j^i = \delta_j(q_j, s, a_i)$.

Lemma 18.2 *Given a model M and a play formula ψ_0, we have that $M, s_0 \models \psi_0$ iff $Lang(B(M, \psi_0)) \neq \emptyset$.*

Proof $(\Rightarrow)$: Suppose that $M, s_0 \models \psi_0$. We need to construct an accepting run of $B(M, \psi_0)$. For a node s, let $\nu(s) = \{\psi \mid M, s \models \psi\}$. Clearly $\nu(s)$ is an atom for every s, $\nu(s_0)$ is an initial atom, and $\nu(s_\ell)$ is a final atom for every leaf node s_ℓ.

We define the run inductively. For the root node s_0, we have: $\rho(s_0) = (s_0, \nu(s_0), (q_1^0, 1) \ldots, (q_k^0, 1))$, where q_j^0 is the initial state of the jth advice automaton. It is easy to see that $\rho(s_0) \in I$.

Inductively suppose that $\rho(s) \in U_{\psi_0}$ is defined and let the successors of s be given by: $s \overset{a_i}{\Rightarrow} s_i$, $1 \leq i \leq m$. Define $\rho(s_i) = (s_i, \nu(s_i), (q_1^i, b_1^i), \ldots, (q_k^i, b_k^i))$ where: if $\lambda(s) = pl(\sigma_j)$, $1 \leq j \leq k$ and $b_j = 1$, then for all $i \in \{1, \ldots, m\}$, $b_j^i = 1$ iff $a_i \in o(q_j, s)$, and $q_j^i = \delta_j(q_j, s)$. If $\lambda(s) \neq pl(\sigma_j)$, $1 \leq j \leq k$ and $b_j = 1$, then for all $i \in \{1, \ldots, m\}$, $b_j^i = 1$ iff $q_j^i = \delta_j(q_j, s, a_i)$.

We only need to verify that for each i, $\rho(s_i) \in U_{\psi_0}$ as well. The crucial condition to check is that, if $B_i \pi_j @ \bar{\iota} \in R$ then $b_j = 1$. This follows from the following claim:

Claim Let $\sigma_j \in \Sigma$, $1 \leq j \leq k$. The plan subtree μ is accepted by the advice automaton A_j iff for every node s present in the subtree μ, $b_j = 1$ in the tuple $\rho(s)$ defined above.

The proof of the claim follows by another inductive construction, this time of an accepting run of A_j. Note that for every player node, the flags are set to 1 along the moves output by A_j, and for every opponent node, flags are set for all successors. The flags continue to be 1 as long as A_j transitions are verified. Thus the 1-nodes form a connected plan subtree accepted by the automaton.

$(\Leftarrow)$: Suppose $Lang(B(M, \psi_0)) \neq \emptyset$. Let ρ be an accepting run of the tree automaton that labels tree nodes by automaton states: for tree node s, we have: $\rho(s) = (s, R, (q_1, b_1) \ldots, (q_k, b_k))$.

Claim For every $\psi \in CL$ and every $s \in S$, $M, s \models \psi$ iff $\psi \in R$.

The proof is by a routine induction on the structure of ψ using the definition of the automaton transition for verification of tense modalities and the advice automaton claim above for the belief modality.

Thus we have:

Theorem 18.1 *Checking consistency of given player type specifications on a game tree is decidable.*

An important problem for logical types is checking the consistency of type specifications. Recall that we defined the logical type space over T to be a tuple $L = (Sat^i(T), \theta_i)_{i \in N}$, where $Sat^i(T)$ is the set of player i type specifications satisfiable in the game, and $\theta_i : (Sat^i(T) \times S) \to Sat^{\bar{\iota}}(T)$ which associates with any player type specification σ and a tree node s, an opponent specification π such that the resulting play visits s.

An important consequence of the preceding lemma is that membership in $Sat^i(T)$ is decidable, and hence we have the following theorem.

Theorem 18.2 *For an extensive form game* $T = (S, \Rightarrow, s_0, \lambda)$*, the logical type space over* T *is recursively enumerable.*

18.4.5 Complexity

For a play formula ψ_0, let $|\psi_0|$ denote the size of the formula. The states of the tree automaton consist of the atoms of ψ_0, the states of the deterministic advice automata for types in the vocabulary of ψ_0, and the tree model is also encoded in the automaton. Since the number of type specifications in ψ_0 is bounded by $|\psi_0|$, the size of the tree automaton is doubly exponential in $|\psi_0|$ and linear in the size of M. Checking non-emptiness of the language of the automaton can be done in time polynomial in the size of M. Thus, we get a total complexity (of checking a play formula in a model, and hence checking consistency of types) that is doubly exponential in $|\psi_0|$ and polynomial in the size of M.

18.5 Discussion

We have tried to present a logical specification of types that can provide a framework for a theory of play. Specifically, in such a framework, one can specify patterns of reasoning displayed by forward induction and a form of rationalizability, for perhaps more general and partial solution concepts. The elements of such a logical language need to be more closely examined, this paper should be considered in the light of a preliminary attempt.

We need to delineate the expressiveness of the language precisely and thus explicate the epistemic universe of player types within which reasoning is carried out. For epistemic structures in the case of forward induction, complete type spaces can be given using compact topological spaces [25]. Axiomatizing type spaces using minimally structured logics such as the one presented here is an interesting question.

Acknowledgments A part of this article was written when I was visiting Peking University. I thank Yanjing Wang for stimulating discussions and warm hospitality at Beijing. I also thank Johan van Benthem for thoughtful and encouraging remarks that greatly helped in the formulation.

References

1. Ågotnes T, Walther D (2009) A logic of strategic ability under bounded memory. J Logic Lang Inf 18:55–77
2. Aumann R, Dreze JH (2005) When all is said and done, how should you play and what should you expect? http://www.ma.huji.ac.il/raumann/pdf/dp-387.pdf
3. Baltag A, Smets S, Zvesper J (2009) Keep hoping for rationality: a solution to the backward induction paradox. Synthese 169:301–333
4. Battigalli P, Sinisalchi M (1999) Hierarchies of conditional beliefs and interactive epistemology in dynamic games. J Econ Theory 88:188–230
5. van Benthem J (2009) Decisions, actions, and games: a logical perspective. In: Third Indian conference on logic and applications, pp 1–22
6. van Benthem J (2011) Exploring a theory of play. In: TARK, pp 12–16
7. van Benthem J (2012) In praise of strategies. In: Games, actions, and social software. LNCS, vol 7010. Springer, Berlin, pp 96–116
8. van Benthem J (2013) Logic in games. Forthcoming book
9. van Benthem J, Liu F (2004) Diversity of logical agents in games. Philosophia Scientiae 8:163–178
10. van Benthem J, Pacuit E, Roy O (2011) Toward a theory of play: a logical perspective on games and interaction. Games 2:52–86
11. de Bruin B (2010) Explaining games: the epistemic program in game theory. Synthese Library. Springer, New York
12. Büchi JR, Landweber LH (1969) Solving sequential conditions by finite-state strategies. Trans Am Math Soc 138:295–311
13. Fortnow L, Whang D (1994) Optimality and domination in repeated games with bounded players. In: Proceedings of the twenty-sixth annual ACM symposium on theory of computing. ACM, pp 741–749
14. Ghosh S (2008) Strategies made explicit in dynamic game logic. In: Proceedings of the workshop on logic and intelligent interaction, ESSLLI, pp 74–81
15. Grädel E, Thomas W, Wilke T (ed) (2002) Automata, logics and infinite games. Lecture notes in computer science, vol 2500. Springer, Berlin
16. Halpern J (2001) Substantive rationality and backward induction. Games Econ Behav 37:425–435
17. van der Hoek W, Jamroga W, Wooldridge M (2005) A logic for strategic reasoning. In: Proceedings of the fourth international joint conference on autonomous agents and multi-agent systems, pp 157–164
18. Hoshi T, Isaac A (2011) Taking mistakes seriously: equivalence notions for game scenarios with off equilibrium play. In: LORI-III, pp 111–124
19. Jamroga W, van der Hoek W (2004) Agents that know how to play. Fundamenta Informaticae 63(2–3):185–219
20. Liu F, Wang Y (2013) Reasoning about agent types and the hardest logic puzzle ever. Minds Mach 23(1):123–161
21. Parikh R (1985) The logic of games and its applications. Ann Discrete Math 24:111–140
22. Paul S, Ramanujam R, Simon S (2009) Stability under strategy switching. In: Proceedings of the 5th conference on computability in Europe, CiE 2009. LNCS, vol 5635. Springer, Berlin, pp 389–398
23. Pearce DG (1984) Rationalizable strategic behaviour and the problem of perfection. Econometrica 52:1029–1050
24. Perea A (2010) Backward induction versus forward induction reasoning. Games 1:168–188
25. Perea A (2012) Epistemic game theory. Reasoning and choice. Cambridge University Press, Cambridge
26. Ramanujam R, Simon S (2008) A logical structure for strategies. In: Logic and the foundations of game and decision theory (loft 7), volume 3 of texts in logic and games. Amsterdam University Press, pp. 183–208 (an initial version appeared in the proceedings of the 7th conference

on logic and the foundations of game and decision theory (LOFT06) under the title Axioms for composite, strategies, pp 189–198)

27. Ramanujam R, Simon S (2008) Dynamic logic on games with structured strategies. In: Proceedings of the 11th international conference on principles of knowledge representation and reasoning (KR-08). AAAI Press, pp 49–58
28. Rubinstein A (1991) Comments on the interpretation of game theory. Econometrica 59:909–924
29. Walther D, van der Hoek W, Wooldridge M (2007) Alternating-time temporal logic with explicit strategies. In: Proceedings of the 11th conference on theoretical aspects of rationality and knowledge (TARK-2007), pp 269–278

Chapter 19
An Alternative Analysis of Signaling Games

Gabriel Sandu

Abstract Evaluation games for first-order logic have arisen from the work of Hintikka in the 1970. They led to various applications to natural language phenomena (pronominal anaphora) in a single framework which is now known as Game-theoretical semantics (GTS). van Benthem [1, 2] observes that these games analyze the 'logical skeleton' of sentence construction: connectives, quantifiers, and anaphoric referential relationships, and that logic is still the driver of the analysis here. He emphasizes that GTS presupposes that the denotations of the basic lexical items such as predicates and object names have been settled but there is still the legitimate question of how to account for the linguistic conventions that settle the meanings of these basic items. To this purpose, different kind of games, known as signaling games have been developed from the 1960s stimulated by Lewis' work on conventions. This work has led to deeper connections with game-theory explored in the work of the Dutch school. In this chapter I give an alternative analyis of signaling games in terms of 2 player extensive game of imperfect information. The material presented here extends Sandu [7].

19.1 The Stag Hunt

The Stag Hunt is the prototype of a social contract. Here is its description in one of the standard textbooks on game theory:

> Each of a group of hunters has two options: she may remain attentive to the pursuit of a stag, or she may catch a hare. If all hunters pursue the stag, they catch it and share it equally; if any of the hunters devotes her energy to catching a hare, the stag escapes, and the hare

G. Sandu (✉)
Department of Philosophy, History, Culture and Art Studies, University of Helsinki, Helsinki, Finland
e-mail: gabriel.sandu@helsinki.fi

A. Baltag and S. Smets (eds.), *Johan van Benthem on Logic and Information Dynamics*, Outstanding Contributions to Logic 5, DOI: 10.1007/978-3-319-06025-5_19, © Springer International Publishing Switzerland 2014

belongs to the defecting hunter alone. Each hunter prefers a share of the stag to a hare
[6, p. 20].

As pointed out by Skyrms [8], the Stag Hunt appears in Hume's *Treatise* in the
scenario of two men who pull at the oars of a boat. They do it by an agreement or
convention though they have never given promises to each other. If both men row,
then the outcome is optimal for each of them. But if one decided not to row then they
don't get anywhere [3, p. 490]. Skyrms mentions also that the same game appears in
Hume's famous meadow draining problem:

> Two neighbours may agree to drain a meadow, which they possess in common, because
> it is easy for them to know each other's mind, and each may perceive that the immediate
> consequence of failing in his part is the abandonment of the whole project (Idem, p. 538).

The Stag Hunt is the topic of a whole book by Brian Skyrms: *The Stag Hunt and
the Evolution of the Social Structure* [8].

We notice that in Hume's first example, the worst thing for each man is for him
to row when the other does not row; and in Hume's second example the worst thing
is for a neighbour to drain when the other does not. Here is a representation of the
Stag Hunt as a 2-player cooperative strategic game:

	Stag	Hare
Stag	(2, 2)	(0, 1)
Hare	(1, 0)	(1, 1)

There are two equilibria: (*Stag*, *Stag*) and (*Hare*, *Hare*). But notice that, like in the
preceding examples, the worst thing for each hunter is for him to hunt the stag when
the other goes for the hare. On the other side, if one goes for the Hare, one does not
take any risk when the other hunter changes his option. For this reason the equilibrium
(*Stag*, *Stag*) is called a *risk dominant* equilibrium [8, p. 3].

Skyrms [8] sees Lewis signaling games as a variant of the Stag Hunt. I will follow
his analyis in the next section. Then I will reformulate Lewis signaling games in a
different way as win–lose games and draw some conclusions.

19.2 Lewis' Signaling Problems

The motivation for Lewis' work comes from Quine's well known attack against
the logical empiricists' conception according to which truths of logic are based
on conventions. Quine found this analyis circular, for he thought that conventions
presuppose already the use of language and logic. Lewis set himself the task to
provide an analysis of conventions which does not rely on such presuppositions. The
outcome (Lewis, [4]) is his well known notion of a *signaling problem:* a situation
which involves a communicator (sender) (C) and an audience (receiver) (A). C
observes one of several states m which he tries to communicate or "signal" to A,
who does not see m. After receiving the signal, A performs one of several alternative

actions, called responses. Every situation m has a corresponding response $b(m)$ that C and A agree is the best response to take when m holds. Lewis argues that a word acquires its meaning in virtue of its role in the solution of various signaling problems. Skyrms [8] makes an interesting comparison between Lewis' signaling problems and the Stag Hunt. Here are some of the details.

To model a Lewisian signaling problem, one fixes the following elements:

- A set S of situations or states of affairs, a set Σ of signals, and a set R of responses.
- A function $b : S \to R$ which maps each situation to its best response.
- An encoding function $f : S \to \Sigma$ employed by C to choose a signal for every situation.
- A decoding function $g : \Sigma \to R$ employed by A to decide which action to perform in response to the signal it receives.

A *signaling system* is a pair (f, g) of encoding and decoding functions which associates with each state of affairs the action which is optimal for the state, that is, $g \circ f = b$. When such a signaling system is settled, each signal acquires a "meaning": the action which is associated with it.

We consider for illustration a signaling system with two states, s_1 and s_2, two messages t_1 and t_2 and two actions a_1 and a_2. We know that a_1 is the best action for the state s_1 and a_2 is the best action for the state s_2. It is obvious that in these setting Sender has four strategies. We let f_1 denote the function which sends s_i to t_i, f_2 denote the (only) nontrivial permutation of f_1, f_3 and f_4 denote the constant function which always send the same message, t_1 and respectively t_2. As for the Receiver, he has also four functions: g_1 decodes t_i into a_i, g_2 is the (only) nontrivial permutation of g_1, and g_3 and g_4 are the constant functions which always decode every message to a_1 and respectively a_2. The following tables give the payoffs of the players for each state:

s_1	g_1	g_2	g_3	g_4	s_2	g_1	g_2	g_3	g_4
f_1	(1, 1)	(0, 0)	(0, 0)	(1, 1)	f_1	(1, 1)	(0, 0)	(1, 1)	(0, 0)
f_2	(0, 0)	(1, 1)	(0, 0)	(1, 1)	f_2	(0, 0)	(1, 1)	(1, 1)	(0, 0)
f_3	(1, 1)	(0, 0)	(0, 0)	(1, 1)	f_3	(0, 0)	(1, 1)	(1, 1)	(0, 0)
f_4	(0, 0)	(1, 1)	(0, 0)	(1, 1)	f_4	(1, 1)	(0, 0)	(1, 1)	(0, 0)

In order to compute the Nash equilibria, we need to assign a probability distribution ν over the states. If ν is a uniform probability distribution, that is, $\nu(s_1) = \nu(s_2) = 1/2$, then the resulting strategic game

	g_1	g_2	g_3	g_4
f_1	(1, 1)	(0, 0)	(1/2, 1/2)	(1/2, 1/2)
f_2	(0, 0)	(1, 1)	(1/2, 1/2)	(1/2, 1/2)
f_3	(1/2, 1/2)	(1/2, 1/2)	(1/2, 1/2)	(1/2, 1/2)
f_4	(1/2, 1/2)	(1/2, 1/2)	(1/2, 1/2)	(1/2, 1/2)

has 6 equilibria: (f_1, g_1), (f_2, g_2), (f_3, g_4), (f_4, g_3), (f_3, g_3), and (f_4, g_4).

However, only the first two equlibria are signaling systems: by deviating from each of them, the two players are worse off. Lewis call them *communicative equilibria*. The other 4 equilibria are such that by deviating from any of them the two players are neither better off not worse off. Lewis calls them *noncommunicative equilibria*. The analogy between communicative equilibria and noncommunicative equilibria, on one side and the risk dominant equilibria and equilibria which are not risk dominated in the Stag Hunt game should be obvious.

It seems that we have obtained at least a partial answer to the questions of what conventions are and how they are maintained: they are equilibria in Lewis' signaling systems which in addition have the special property of being risk dominant (i.e., they are strict). The others, the noncommunicative equilibria do not have this property. Skyrms points out (p. 54) that we do not yet have an answer to the question of how conventions are selected from the several risk dominant equilibria in the game. Lewis himself was aware of the fact that the property of equilibria being strict does not suffice for qualifying them as conventions. Strict equilibrium explains only why there is an incentive not to unilaterally deviate but if, for instance, you expect me to deviate, then you might believe to be better off by deviating as well. And if I believe you have such beliefs, I might deviate too. In other words, Lewis saw the need to back up the maintainance of conventions by a hierachy of interrelated beliefs which are common knowledge. Skyrms (p. 54) is somehow skeptical about the formation of such beliefs in the first place and offers interesting examples which show how equilibria are de facto selected without the assumption of common knowledge.

There is one feature of the signaling systems that is worth noting. The two communicative equilibria do not depend on the probability distribution over the set of states being uniform. In other words, any probability distribution over the set of states will give rise to the same set of equilibria in the combined game (although with different payoffs.) This is a problem to which we shall return. Meanwhile let us also take note of the fact that when there are more than three states, (together with the corresponding signals and optimal actions), then there are also partial signaling systems, that is, pairs (f, g) of strategies f of the Sender and strategies g of the Receiver which are equilibria but which transmit only partial information: there is only a partial match between states of affairs and the corresponding actions. Here is an example where the Sender's strategy f is

$$f : \begin{array}{ccc} s_1 & \longrightarrow & m_1 \\ s_2 & \longrightarrow & m_2 \\ s_3 & \longrightarrow & m_3 \end{array}$$

the Receiver's strategy g is

$$g : \begin{array}{ccc} m_1 & \longrightarrow & a_1 \\ m_2 & \longrightarrow & a_1 \\ m_3 & \longrightarrow & a_3 \end{array}$$

and the best actions are indicated in the matrix below:

	a_1	a_2	a_3
s_1	$(1, 1)$	$(0, 0)$	$(0, 0)$
s_2	$(0, 0)$	$(1, 1)$	$(0, 0)$
s_3	$(0, 0)$	$(0, 0)$	$(1, 1)$

That is, alternatively: $b(s_i) = a_i$.

19.3 Signaling Games as Win–Lose Extensive Game of Imperfect Information

We shall formulate Lewis' signaling systems as a 2-player win–lose extensive game of imperfect information played by the team of the Sender and Receiver (that we shall call the team of Eloise and denote it by ∃) against Nature (∀). We fix a finite set S of situations or states of affairs, a finite set Σ of signals, a finite set R of responses and a function $b : S \to R$ which maps each situation to its best response in the way indicated above. The game tree is straightforward. Each of its maximal branches has the form (x, z, y) where $x \in S$ is a choice by Nature, $z \in \Sigma$ is a choice by the Sender and $y \in R$ is a choice by the Receiver. For each such maximal branch the payoffs are determined by the following rule:

- If $y = b(x)$ then the team ∃ wins (payoff 1); otherwise ∀ wins.

The imperfect information comes from the following condition on the information set of the Receiver:

- (x, z) and (x', z') are equivalent (indistinguishable) for the Receiver whenever $z = z'$.

Strategies are defined as usual. A strategy for Nature is any member of S. We prefer to present the strategy of the team of Eloise as a pair of functions (f, g) such that $f : S \to \Sigma$ and $g : \Sigma \to R$. That is, f is a strategy function for the Sender and g is a strategy function for the Receiver. The fact that g is a unary function reflects the uniformity condition typical for games of imperfect information: given the condition above, the Receiver's strategy g depends only on z and not on both x and z.

Now when Nature follows a strategy $s \in S$ and the team of Eloise follows a strategy (f, g) a play of the game $(s, f(s), g(f(s)))$ is completed, which is a win for the team, if $g(f(s)) = b(s)$ and a win for Nature, otherwise.

When the number of states in S equals the number of signals in Σ, it is easy to see that there is a winning strategy for the team of Eloise, that is, a pair of functions (f, g) which guarantees a win no matter what is the strategy employed by Nature. In the particular case in which S consists of two states, s_1 and s_2, Σ consists of two signals, t_1 and t_2, R consists of two actions, a_1 and a_2 and the function b is as usual, the winning strategies are exactly the pairs (f_1, g_1) and (f_2, g_2) introduced in the preceding section.

19.4 Signaling Games as Win–Lose Strategic Games

We convert the extensive game from the previous section into a strategic game $\Gamma = (S_\exists, S_\forall, u_\exists, u_\forall)$. $S_\forall$, the set of strategies of Abelard is the set S; $S_\exists$, the set of strategies of Eloise is the set of pairs (f, g). The payoff functions $u_\exists$ and $u_\forall$ have been already specified: $u_\exists((f, g), s_i) = 1$ iff $g(f(s_i)) = b(s_i)$; and 0 otherwise. For Abelard: $u_\forall((f, g), s_i) = 1$ iff $g(f(s_i)) \neq b(s_i)$ and 0 otherwise.

When there are only two states s_1 and s_2, the strategic game looks like this:

	s_1	s_2
(f_1, g_1)	$(1, 0)$	$(1, 0)$
$\vdots$	$\vdots$	$\vdots$
(f_1, g_4)	$(1, 0)$	$(0, 1)$
$\vdots$	$\vdots$	$\vdots$
(f_2, g_1)	$(0, 1)$	$(0, 1)$
(f_2, g_2)	$(1, 0)$	$(1, 0)$
$\vdots$	$\vdots$	$\vdots$
(f_3, g_3)	$(0, 1)$	$(1, 0)$
$\vdots$	$\vdots$	$\vdots$
(f_4, g_4)	$(1, 0)$	$(0, 1)$

Notice how Lewis' initial signaling game has received a different twist: The Sender and the Receiver are now on the same side, and they try to coordinate no matter what the state (of Nature) turns out to be. The strategy pairs (f_1, g_1) and (f_2, g_2) weakly dominate all the other strategy pairs. Thus we can reduce the game to:

	s_1	s_2
(f_1, g_1)	$(1, 0)$	$(1, 0)$
(f_2, g_2)	$(1, 0)$	$(1, 0)$

In this smaller game there are 4 equilibria. All of them return the same expected utility to the players. The strategy pairs (f_3, g_4), (f_4, g_3), (f_3, g_3), and (f_4, g_4) which are the "noncommunicative equilibria" in the Lewis signaling systems, are not equilibria in the present game. This is as expected: each of them is weakly dominated by one of the strategies (f_1, g_1), (f_2, g_2).

Skyrms offers two main reasons why noncommunicative equilibria (i..e, those that are not signaling systems) in Lewis signaling games do not qualify as conventions. One of them, discussed earlier, is simply that such equilibria are not strict: a player is not worse off by unilateral deviation from them. He gives interesting arguments (p. 56) which support an additional reason: such equilibria never evolve because they are evolutionary unstable.

The account that emerged from our reformulation of signaling problems as win–lose games is different: it is irrational for the Sender–Receiver team to play such strategies. That is, in our account the strategies in question are not disregarded for empirical reasons, but for game-theoretical ones.

19.5 Mixed Strategy Equilibria

We fix finite sets S, Σ, R and a function $b : S \to R$ which maps each situation to its best response in the way indicated above. Let $\Gamma = (S_\exists, S_\forall, u_\exists, u_\forall)$ be the corresponding win–lose 2 player strategic game as specified in the previous section.

A mixed strategy ν for player $p \in \{\exists, \forall\}$ is a probability distribution over S_p, that is, a function $\nu : S_p \to [0, 1]$ such that $\sum_{\tau \in S_i} \nu(\tau) = 1$. ν is uniform over $S_i' \subseteq S_i$ if it assigns equal probability to all strategies in S_i' and zero probability to all the strategies in $S_i - S_i'$. Given a mixed strategy μ for $\exists$ and a mixed strategy ν for player $\forall$, the expected utility for player p is given by:

$$U_p(\mu, \nu) = \sum_{\sigma \in S_\exists} \sum_{\tau \in S_\forall} \mu(\sigma)\nu(\tau)u_p(\sigma, \tau).$$

We can simulate a pure strategy σ with a mixed strategy which assigns to σ probability 1. That is, when $\sigma \in S_\exists$ and ν is a mixed strategy for player $\forall$, we let

$$U_p(\sigma, \nu) = \sum_{\tau \in S_\forall} \nu(\tau)u_p(\sigma, \tau).$$

Similarly if $\tau \in S_\forall$ and μ is a mixed strategy for player $\exists$, we let

$$U_p(\mu, \tau) = \sum_{\sigma \in S_\exists} \mu(\sigma)u_p(\sigma, \tau).$$

Von Neuman's Minimax Theorem [10] tells us that every finite, two-person, constant-sum game has an equilibrium in mixed strategies. As an immediate corollary, we know that any two mixed strategy equilibria in a finite, two-person, constant-sum game return the same payoffs to the two players. These two results garantee that we can talk about *the value of a strategic game* Γ: it is the expected utility returned to player (team) $\exists$ by any equilibrium in the relevant strategic game.

The next results will help us to identify equilibria.

Proposition 19.1 *Let μ^* be a mixed strategy for player $\exists$ and ν^* a mixed strategy for player $\forall$ in a finite strategic, two player win–loss game Γ. The pair (μ^*, ν^*) is an equilibrium in Γ if and only if the following conditions hold:*

1. $U_\exists(\mu^*, \nu^*) = U_\exists(\sigma, \nu^*)$ *for every $\sigma \in S_\exists$ in the support of μ^**
2. $U_\forall(\mu^*, \nu^*) = U_\forall(\mu^*, \tau)$ *for every $\tau \in S_\forall$ in the support of ν^**

3. $U_\exists(\mu^*, \nu^*) \geq U_\exists(\sigma, \nu^*)$ for every $\sigma \in S_\exists$ outside the support of μ^*
4. $U_\forall(\mu^*, \nu^*) \geq U_\forall(\mu^*, \tau)$ for every $\tau \in S_\forall$ outside the support of ν^*.

The following result enables us to eliminate weakly dominated strategies.

Proposition 19.2 *Let* $\Gamma = (S_\exists, S_\forall, u_\exists, u_\forall)$ *be a strategic game as above. Then* Γ *has an equilibrium* $(\sigma_\exists, \sigma_\forall)$ *such that for each player p none of the strategies in the support of* σ_p *is weakly dominated in* Γ.

Definition 19.1 Let $\Gamma = (S_\exists, S_\forall, u_\exists, u_\forall)$ be a strategic game as above. For $\sigma, \sigma' \in S_\exists$, we say that σ' is payoff equivalent to σ if for every $\tau \in S_\forall : u_\exists(\sigma', \tau) = u_\exists(\sigma, \tau)$.

A similar notion is defined for Abelard.

The following result allows us to eliminate payoff equivalent strategies.

Proposition 19.3 *Let* $\Gamma = (S_\exists, S_\forall, u_\exists, u_\forall)$ *be a strategic game. Then* Γ *has an equilibrium* $(\sigma_\exists, \sigma_\forall)$ *such that for each player p there are no strategies in the support of* σ_p *which are payoff equivalent.*

The proofs of Propositions 19.2 and 19.3 may be found in [5]. We apply these results to Lewis signaling games.

We now go back to the win–lose strategic game in the previous section and consider also mixed strategy equilibria. The three propositions above ensure that the initial win–lose strategic game has the same value as the smaller game:

	s_1	s_2
(f_1, g_1)	$(1, 0)$	$(1, 0)$
(f_2, g_2)	$(1, 0)$	$(1, 0)$

The equilibrium $((f_1, g_1), s_1)$ corresponds to s_1 being "played" with probability 1 and (f_1, g_1) with probability 1; and analogously for the other equilibria $((f_1, g_1), s_2)$, $((f_2, g_2), s_1)$ and $((f_2, g_2), s_2)$. But there are many other mixed strategy equilibria, all returning the same expected utility to the players (this follows from the general properties of constant-sum games). In fact, for every probability distribution ν over the set of states, $((f_1, g_1), \nu)$ and $((f_2, g_2), \nu)$ are mixed strategy equilibria (we assimilate the pure strategy (f_1, g_1) to the mixed strategy which assigns 1 to (f_1, g_1) and 0 to all the others). In other words, the existence of the "communicative equilibria" in the game does not depend upon the probability distribution over the set of states. This reinforces our earlier conclusion.

19.6 Skyrms: Inventing the Code

19.6.1 The Emergence of Disjunction

Skyrms [9, p. 8] discusses an interesting example which is a variation of the signaling game with the number of signals being equal to the number of states. There are four states: $S = \{S_1, \ldots, S_4\}$, two Senders, and one Receiver. Both Senders see the state they are in, and try to signal them to the Receiver. The first Sender has only two signals, he can wave either a Red or Green flag, and so does the second who can wave either a Yellow or a Blue flag. We denote the first set of signals by $\Sigma_1 = \{R, G\}$ and the second one by $\Sigma_2 = \{Y, B\}$. There are now three functions to encode the two Senders and the Receiver's strategies:

- An encoding function $f_1 : S \to \Sigma_1$ employed by the first Sender to choose a signal for every situation he might observe
- An encoding function $f_2 : S \to \Sigma_2$ employed by the second Sender to choose a signal for every situation he might observe
- A decoding function $g : \Sigma_1 \times \Sigma_2 \to S$ employed by the Receiver to decode the pair of signals he receives
- Finally the best action function b which associates with each state s_i its best action $b(s_i) = a_i$. (Thus there are four actions.)

A signaling system will now be any triple (f_1, f_2, g) which is such that the Receiver's function g can decode the two messages $f_1(s)$ and $f_2(s)$ he receives in such a way that for each state s, $g(f_1(s), f_2(s)) = b(s)$.

It is straightforward to represent this game as a finite 2 player win–lose game played by Abelard (Nature) against the 3 player team of Eloise, consisting of two Senders and one Receiver. Here is an example of a signaling system:

$$f_1(s_1) = f_1(s_2) = R \quad f_1(s_3) = f_1(s_4) = G$$
$$f_2(s_1) = f_2(s_3) = Y \quad f_2(s_2) = f_2(s_4) = B$$

As for the Receiver's decoding function we have:

$$g_1(R, Y) = a_1 \; g_1(R, B) = a_2 \; g_1(G, Y) = a_3 \; g_1(G, B) = a_4$$

The strategic form of the game is now:

	s_1	s_2	s_3	s_4
(f_1, f_2, g_1)	$(1, 0)$	$(1, 0)$	$(1, 0)$	$(1, 0)$
$\vdots$	$\vdots$	$\vdots$	$\vdots$	$\vdots$

As in the earlier case, for every probability distribution μ over the set of states there is an equilibrium (σ_1, μ) in the game such that σ_1 assigns to (f_1, g_1) probability 1 (and 0 to all the others). Notice how the complex signals like (R, Y) arise out of the simpler signals R and Y. Here (R, Y) "means" s_1 but the simpler message R means either s_1 or s_2. Skyrms thinks that this signaling system provides an example of the emergence of the meaning of logical connectives (disjunction).

19.7 Less Signals Than States: The Indeterminacy of the Game

Lewis discusses only the case in which the number of signals equals the number of states. In such a case, that is, when the cardinality of the set S of situations or states of affairs is strictly less that the cardinality of the set Σ of signals, it is easy to see that there is no equilibrium in the game $\Gamma = (S_\exists, S_\forall, u_\exists, u_\forall)$ (we refer here to the win–lose game). For suppose, for a contradiction, that there is one, say $((f, g), s)$. The only outcomes of the game are 0 and 1. Obviously the outcome for the pair $((f, g), s)$ cannot be $(0, 1)$, for we can easily define a pair of strategies f_i, g_i such that $g_i(f_i(s)) = 1$. But this violates the condition:

$$u_\exists(((f, g), s)) \geq u_\exists((f_i, g_i), s)$$

So we must have $u_\exists(((f, g), s)) = 1$ and $u_\forall(((f, g), s)) = 0$. But given that $m < n$, the pair of functions (f, g) cannot pair all states with their best actions. In other words, there must exist distinct s_i and s_j such that $f(s_i) = f(s_j) = t$, for some signal t. Then $g(f(s_i)) = g(f(s_j))$. But given the constraints on the function b we know that $b(s_i) \neq b(s_j)$. Thereby $u_\exists((f, g), s_i) \neq u_\exists((f, g), s_j)$, that is, either $u_\exists((f, g), s_i) = 0$ or $u_\exists((f, g), s_j) = 0$, i.e., $u_\forall((f, g), s_i) = 1$ or $u_\forall((f, g), s_j) = 1$. The former violates the condition

$$u_\forall(((f, g), s)) \geq u_\forall((f, g), s_i)$$

and the latter violates the condition

$$u_\forall(((f, g), s)) \geq u_\forall((f, g), s_j)$$

Thus there is no equilibrium in the game $\Gamma = (S_\exists, S_\forall, u_\exists, u_\forall)$. To avoid the indeterminacy, we move to mixed strategies. Recall the strategic form of the game

	s_1	$\cdots$	s_n
(f_1, g_1)		$\cdots$	
(f_2, g_2)		$\cdots$	
$\vdots$		$\cdots$	
(f_p, g_p)		$\cdots$	

where $S_\forall = \{s_1, \ldots, s_n\}$, $S_\exists = \{(f_1, g_1), \ldots, (f_p, g_p)\}$ and, $f_i : S \to \Sigma$ and $g_i : \Sigma \to R$. The payoffs are determined as above.

In this game let $\mathbb{B} = \{B \subseteq S^M$ and $\mid B \mid = m\}$. Given that $m < n$, for every $B \subseteq \mathbb{B}$, there exists at least one pair (f, g) such that

A1 $f \upharpoonright B : B \to \Sigma^M$ is one-one and onto
A2 $g(f(s)) = b(s)$, for every $s \in B$

In other words, when (f, g) is played against s, the payoffs are $(1, 0)$ whenever $s \in B$, and $(0, 1)$ when $s \notin B$. That is, the Sender–Receiver team achieves coordination of states with their best actions for all the states s in B.

For any $B \subseteq \mathbb{B}$, let T_B be the collection of all pairs (f, g) which satisfy conditions (A1) and (A2). We notice that all the members of T_B are payoff equivalent. We choose one member from each T_B and collect them into a class $T_{\mathbb{B}}$. We apply the last Proposition and reduce the game Γ to a smaller game Γ_1 where the class of strategies of Eloise is restricted to $T_{\mathbb{B}}$.

Finally we notice that any strategy outside the set $T_{\mathbb{B}}$ is weakly dominated by some strategy in $T_{\mathbb{B}}$. By the second Proposition above we reduce the game Γ_1 to the smaller game Γ_2:

	s_1	$\cdots$	s_n
(f_1, g_1)		$\cdots$	
(f_2, g_2)		$\cdots$	
$\vdots$		$\cdots$	
(f_r, g_r)		$\cdots$	

where every (f_i, g_i) belongs to the set $T_{\mathbb{B}}$. It is enough to find an equilibrium in this game. We observe that

- Each pair $(f_i, g_i) \in T_{\mathbb{B}}$ gives m times the payoff $(1, 0)$ and $(n - m)$ times the payoff $(0, 1)$.

Let μ the uniform probability distribution over $T_{\mathbb{B}}$ and ν the uniform probability distribution over S. It can be shown that (μ, ν) is an equilibrium and the value of the game is m/n.

The conclusion is similar to the one for the case in which the number of signals equals the number of states: it is irrational for the Sender–Receiver team to play a strategy pair (including constant strategies) which coordinates on less than m states. The main argument above shows that any such strategy is weakly dominated by some strategy in $T_{\mathbb{B}}$.

There is a difference, however, between the two cases. In the first case, the existence of the equilibrium does not depend upon the probability distribution over the set of states: we saw that for every probability distribution ν over the set of states, $((f_1, g_1), \nu)$ and $((f_2, g_2), \nu)$ are mixed strategy equilibria which give the players the same expected utility. In the present case this is no longer so. We prefer to illustrate with an example where $n = 3$, and $m = 2$, but the point is perfectly general. By the argument sketched above, the game for $n = 3$, and $m = 2$ has the same value as the smaller game

	s_1	s_2	s_3
(f_1, g_1)	$(1, 0)$	$(1, 0)$	$(0, 1)$
(f_2, g_2)	$(1, 0)$	$(0, 1)$	$(1, 0)$
(f_3, g_3)	$(0, 1)$	$(1, 0)$	$(1, 0)$

We saw that the pair (μ, ν) is an equilibrium in this game, where μ is a uniform distribution over (f_1, g_1), (f_2, g_2) and (f_3, g_3); and ν is a uniform distribution over s_1, s_2 and s_3. Now it is true that the pair (μ, ν^*) where ν^* is any probability distribution over s_1, s_2 and s_3, gives the players the same expected utility as the equilibrium (μ, ν). To see this, suppose $\nu^*(s_1) = p_1$, $\nu^*(s_2) = p_2$, and $\nu^*(s_3) = p_3$. Then

$$U_{\exists}(\mu, \nu^*) = 1/3(p_1 + p_2) + 1/3(p_1 + p_3) + 1/3(p_2 + p_3) = 2/3$$

However, it is not any longer true in general that the pair (μ, ν^*) is an equilibrium in the game, due to the fact of the cardinality of signals being smaller than the cardinality of states. Here is a counter-example. Let ν^* be the probability distribution such that $\nu^*(s_1) = 1/2$, $\nu^*(s_2) = \nu^*(s_3) = 1/4$. Notice that

$$U_{\exists}((f_1, g_1), \nu^*) = (1 \times 1/2 \times 1) + (1 \times 1/4 \times 1) = 3/4.$$

Thus condition 1 of the first Proposition above is violated, and thereby (μ, ν^*) is not an equilibrium.

19.8 Expressing the Win–Lose Extensive Game in IF Logic

When introducing Lewis signaling problems in the third section, we first gave a description of a signaling problem in terms of a Sender sending a signal to the Receiver who undertakes an appropriate action after receiving it. This description has all the ingredients of a game of imperfect information. After that we modelled such a problem, following Lewis, in terms of a strategic, cooperative game.

In this section we replace the description with a sentence in a formal language interpreted by an appropriate extensive game of imperfect information. After that we will convert the extensive game into a strategic game. The analogy between the extensive game of imperfect information and the game in Sect. 19.4, as well as the analogy between the corresponding strategic game and the game in Sect. 19.5 will be obvious.

Consider the sentence φ_{sig}

$$\forall x \exists z (\exists y/\{x\})\{(S(x) \to (\Sigma(z) \land R(y) \land B(x, y))\}$$

intended to say: for every state x there is a signal z and an action y which is the best action for x. The slash indicates that the action y does not depend on x in a way to be made more precise below. The language with slashed quantifiers is known in the literature as Independence-Friendly language. Its properties are collected in Mann et al. [5].

The model $\mathbb{M}$

$$\mathbb{M} = (M, S^M, \Sigma^M, R^M, B^M)$$

interprets the nonlogical vocabulary of φ_{sig}. More exactly

$$M = \{s_1, \ldots, s_n, t_1, \ldots, t_m, a_1, \ldots, a_n\}$$
$$S^M = \{s_1, \ldots, s_n\}$$
$$\Sigma^M = \{t_1, \ldots, t_m\}$$
$$R^M = \{a_1, \ldots, a_n\}$$
$$B^M = \{(s_1, a_1), \ldots, (s_n, a_n)\}$$

The relation B^M represents the best function. We consider the case in which $m < n$.

The sentence φ_{sig} and the model $\mathbb{M}$ determine an extensive game of imperfect information $G(\varphi_{sig}, \mathbb{M})$ played by two players, Abelard (the universal quantifier) and Eloise (the existential quantifier). We can actually take Eloise to be a team consisting of two players corresponding to the two existential quantifiers. A play of the game $G(\varphi_{sig}, \mathbb{M})$ is a sequence consisting of the following moves:

- Abelard chooses $a \in M$ to be the value of the variable x;
- The first player in Eloise's team chooses $b \in M$ to be the value of z
- The second player in Eloise's team chooses $c \in M$ to be the value of y.

Let w be the assignment $w = \{(x, a), (z, b), (y, c)\}$. The team of Eloise wins the play if

$$M, w \vDash (S(x) \rightarrow (\Sigma(z) \wedge R(y) \wedge B(x, y))$$

Otherwise Abelard wins.

A strategy for Abelard is any $a \in M$. A strategy for the team of Eloise is a pair of functions (f, g) such that $f : M \rightarrow M$ and $g : M^2 \rightarrow M$. However, given the requirement that $(\exists y/\{x\})$("when choosing a value for y the second player of Eloise's team does not know the value for x chosen by Abelard earlier in the game") the function g must be uniform in $\{x\}$:

- For any (partial) assignments w and w', say $w = \{(x, a), (z, b)\}$ and $w' = \{(x, a'), (z, b')\}$, if $b = b'$ then $g(b) = g(b')$. In other words, we may take g to be also a unary function.

Game theoretical truth, $\mathbb{M}, s \vDash^+_{GTS} \varphi$, and game-theoretical falsity, $\mathbb{M}, s \vDash^-_{GTS} \varphi$, are defined by:

- $\mathbb{M}, s \vDash^+_{GTS} \varphi$ iff there is a winning strategy for Eloise in $G(\mathbb{M}, s, \varphi)$
- $\mathbb{M}, s \vDash^-_{GTS} \varphi$ iff there is a winning strategy for Abelard in $G(\mathbb{M}, s, \varphi)$.

When $m < n$ it can be shown that the game is non-determined: neither Abelard nor the team of Eloise has a winning strategy. Thus we can say that, for one thing, signaling problems with less signals than states provide a nice concrete example of an indeterminate IF sentence.[1]

19.9 Nash Equilibrium Semantics

We let $T_\exists$ be the set of pairs (h, k), say $T_\exists = \{(h_1, k_1), \ldots, (h_p, k_p)\}$ and $T_\forall$ be the set of strategies of Abelard, i.e. $T_\forall = M$.

When a strategy (h, k) of Eloise is played against a strategy t of Abelard, a play (history) of the extensive game of imperfect information $G(\varphi_{sig}, \mathbb{M})$ results. It results either in a win of Eloise or a win for Abelard, i.e. the payoffs of the players for each pair $(h, k), s$ are determined. We can now change the extensive game $G(\varphi_{sig}, \mathbb{M})$ from the previous section into a strategic game $\Gamma(\varphi_{sig}, \mathbb{M})$ which has the form:

	s_1	$\cdots$	s_n	t_1	$\cdots$	t_m	a_1	$\cdots$	a_n
(h_1, k_1)				$(1,0)$	$\cdots$	$(1,0)$	$(1,0)$	$\cdots$	$(1,0)$
(h_2, k_2)				$(1,0)$	$\cdots$	$(1,0)$	$(1,0)$	$\cdots$	$(1,0)$
$\vdots$				$\vdots$	$\vdots$	$\vdots$	$\vdots$	$\vdots$	$\vdots$
(h_p, k_p)				$(1,0)$	$\cdots$	$(1,0)$	$(1,0)$	$\cdots$	$(1,0)$

This strategic game is not the game Γ from Sect. 19.8. We will show how to reduce it to the strategic game Γ or rather to the game Γ_2 from the same section which is equivalent to it.

First notice that any two strategies $r, r' \in \Sigma^M \cup R^M$ are payoff equivalent for Abelard. We pick one of them, say t_i, ignore the rest, and reduce the game $\Gamma(\varphi_{sig}, \mathbb{M})$ to the smaller game Γ':

	s_1	$\cdots$	s_n	t_i
(h_1, k_1)				$(1,0)$
(h_2, k_2)				$(1,0)$
$\vdots$				$\vdots$
(h_p, k_p)				$(1,0)$

In virtue of the last Proposition in Sect. 19.7, we know that the values of Γ and Γ' are identical.

In Γ', let $\mathbb{B} = \{B \subseteq S^M$ and $\mid B \mid = m\}$. The next observation is that, given that $m < n$, for every $B \subseteq \mathbb{B}$, there exists at least one pair (h, k) such that

B1 $\quad h \restriction B : B \to \Sigma^M$ is one-one and onto

B2 $\quad k(h(s)) = s$, for every $s \in B$

---

[1] The significance of Lewis signaling games for the expressive power of IF logic is analyzed in detail in F. Barbero and G. Sandu, "Signalling in Independence-Friendly Logic", Logic Journal of the IGPL, 2014, Doi 10.1093/jigpal/jzu004.

In other words, when (h, k) is played against $s \in M$, the payoffs are $(1, 0)$ whenever $s \in B$, and $(0, 1)$ when $s \notin B$. Notice that conditions B1 and B2 are the same as our earlier conditions (A1), (A2), except for h and k having the whole universe as their domain.

For any $B \subseteq \mathbb{B}$, let T_B be the collection of all pairs (h, k) which satisfy conditions (B1) and (B2). By the same reasoning as above, we reduce the game Γ' to a smaller game Γ'' where the class of strategies of Eloise is restricted to $T_{\mathbb{B}}$.

We note that in the game Γ'' any strategy s_i weakly dominates the unique strategy t_i. To see this, fix s_i and pick $B \subseteq \mathbb{B}$, such that $s_i \notin B$. Then let (h, k) be the unique member of $T_{\mathbb{B}}$ which satisfies (B1) and (B2). By our remarks following (B2), we know that $u_\forall((h, k), s_i) = 1$.

We apply the second Proposition above, eliminate the strategy t_i and reduce the game Γ'' to the smaller game Γ''':

	s_1	$\cdots$	s_n
(h_1, k_1)			
(h_2, k_2)			
$\vdots$			
(h_p, k_p)			

Finally we notice that any strategy outside the set $T_{\mathbb{B}}$ is weakly dominated by some strategy in $T_{\mathbb{B}}$. Then we finally reduce the game Γ''' to the game

	s_1	$\cdots$	s_n
(h_1, k_1)		$\cdots$	
(h_2, k_2)		$\cdots$	
$\vdots$		$\cdots$	
(h_r, k_r)		$\cdots$	

where every (h_i, k_i) belongs to the set $T_{\mathbb{B}}$. This is the same as the game Γ_2 in Sect. 19.8.

19.10 Conclusions

We reformulated Lewis signaling games as win–lose extensive games of imperfect information in order to obtain an alternative account of conventions to Lewis' signaling problems. Lewis' signaling games made a distinction between strict (communicative) equilibria and nonstrict (noncommunicative) equilibria. Skyrms draws an interesting comparison between Lewis' account and the Stag Hunt game. He also gives supplementary reasons why noncommunicative equilibria do not evolve (e.g. they are not evolutionary stable). In our account noncommunicative strategies do not form an equilibrium: it is irrational for the Sender–Receiver team to play them. In addition, for the case in which the number of signals equals the number of states, the reformulation allows us to see that the communicative equilibria do not depend upon the probability distributions of the relevant states. We extended the analysis to the case in which there are less signals than states and realized that the strategies which form an equilibrium are the ones where the Sender and the Receiver coordinate over the maximal number of states. The value of the games in these cases can be nicely connected with a probabilistic interpretation of IF logic.

References

1. van Benthem J (2008) Games that make sense: logic, language and multi-agent interaction. In: Apt K, van Rooij R (eds) New perspectives on games and Interaction. Texts in logic and games 4. Amsterdam University Press, Amsterdam, pp 197–209
2. van Benthem J (2014) Logic in games. MIT Press. Cambridge
3. Hume D (1739) A treatise of human nature. In: Selby-Bigge LA (1949) Clarendon, Oxford
4. Lewis D (1969) Convention. Blackwell, Oxford
5. Mann AI, Sandu G, Sevenster M (2011) Independence-friendly logic: a game-theoretic approach. Cambridge University Press, Cambridge
6. Osborne MJ (2004) An introduction to game theory. Oxford University Press, Oxford
7. Sandu G (2014) Languages for imperfect information. In: van Benthem J, Ghosh S, Verbrugge R (eds) Modeling strategic reasoning, Institute for Logic, Language and Computation, Amsterdam and Department of Artificial Intelligence, University of Groningen
8. Skyrms B (2004) The stag hunt and the evolution of social structure. Cambridge University Press, Cambridge
9. Skyrms B (2006) "Signals". Presidential Address. Philosophy of Science Association
10. Von Neumann J (1928) Zur Theorie der Gesellschaftsspiele. Mathematische Annalen 100:295–320

Part IV
Agency

Chapter 20
Them and Us: Autonomous Agents In Vivo and In Silico

Peter Millican and Michael Wooldridge

Abstract The concept of *agency* is important in philosophy, cognitive science, and artificial intelligence. Our aim in this chapter is to highlight some of the issues that arise when considering the concept of agency across these disciplines. We discuss two different views of agency: agents as actors (the originators of purposeful deliberate action); and agents as intentional systems (systems to which we attribute mental states such as beliefs and desires). We focus in particular on the view of agents as intentional systems, and discuss Baron-Cohen's model of the human intentional system. We conclude by discussing what these different views tell us with respect to the goal of *constructing* artificial autonomous agents.

20.1 Introduction

As we look around our world and try to make sense of what we see, it seems that we naturally make a distinction between entities that in this chapter we will call "agents", and other objects. An agent in the sense of this chapter is something that seems to have a similar status to us as a self-determining actor. When a child deliberates over which chocolate to choose from a selection, and carefully picks one, we perceive agency: there is choice, and deliberate, purposeful, autonomous action. In contrast, when a plant grows from underneath a rock, and over time pushes the rock to one

---

With apologies to Will Hutton.

P. Millican (✉)
Hertford College, Oxford OX1 3BW, UK
e-mail: peter.millican@hertford.ox.ac.uk

M. Wooldridge
Department of Computer Science, University of Oxford, Oxford OX1 3QD, UK
e-mail: mjw@cs.ox.ac.uk

A. Baltag and S. Smets (eds.), *Johan van Benthem on Logic and Information Dynamics*, Outstanding Contributions to Logic 5, DOI: 10.1007/978-3-319-06025-5_20, © Springer International Publishing Switzerland 2014

side, we see no agency: there is action, of a kind, but we perceive neither deliberation nor purpose in the action.

The concept of agency is important for philosophers (who are interested in understanding what it means to be a self-determining being) and for cognitive scientists and psychologists (who seek to understand, for example, how some people can come to lack some of the attributes that we associate with fully realised autonomous agents, and how to prevent and treat such conditions). However, the concept of agency is also important for researchers in computer science and artificial intelligence, who wish to build computer systems that are capable of purposeful autonomous action (either individually or in coordination with each other). If such artificial agents are to interact with people, then it must be helpful also to understand how people make sense of agency.

The aim of this chapter is to survey and critically analyse various ways of conceptualising agents, and to propose what we consider to be a promising approach. Our discussion encompasses contributions from the literature on philosophy, cognitive science, and artificial intelligence. We start by examining two different views of agency:

- *First-personal view.* From this perspective, agents are purposeful originators of deliberate action, motivated by conscious purposes.
- *Third-personal view.* From this perspective, agents are entities whose behaviour can be predicted and explained through the attribution to them of beliefs, desires, and rational choice.

Cutting across these perspectives is the issue of *higher-order intentional reasoning*, by which an agent may adopt the third-personal view of other agents and adapt its behaviour accordingly, based in part on the intentional states that it attributes to those other agents. We shall see some evidence that such reasoning—a distinctive characteristic of human beings in social groups—provides a plausible evolutionary driver of our own brain size and conspicuous "intelligence". Following a discussion of the human intentional system and the condition of autism (drawing on work by Simon Baron-Cohen), we turn to the question of agency in silico, and ask what lessons can be learned with regard to the construction of artificial autonomous agents.

20.2 Agency from the First-Personal Perspective

We begin with the idea that *agents are the conscious originators of purposeful deliberate action*. As conscious beings ourselves, we naturally find this a compelling viewpoint, and it has understandably spawned many centuries of discussion about such thorny problems as free will, personal identity, and the relation between mind and body. Even if we leave these old chestnuts aside, however, the view raises other difficulties, which it will be useful to rehearse briefly.

First, there is the basic problem of how actions should be counted and individuated (which also arises, though perhaps less severely, from the third-personal perspective).

Consider the following classic example, due to Searle [31]. On 28 June 1914, the 19-year-old Yugoslav Nationalist Gavrilo Princip assassinated Archduke Franz Ferdinand of Austria, and thereby set in motion a chain of events that are generally accepted to have led to World War I, and hence the deaths of millions of people. This is, surely, one of the most famous deliberate actions in history. But if we try to isolate exactly *what action it was* that Princip carried out, we run into difficulties, with many different possibilities, including:

- Gavrilo squeezed his finger;
- Gavrilo pulled the trigger;
- Gavrilo fired a gun;
- Gavrilo assassinated Archduke Ferdinand;
- Gavrilo struck a blow against Austria;
- Gavrilo started World War I.

All six of these seem to be legitimate descriptions of what it was that Princip did, yet we are naturally reluctant to say that he simultaneously performed a host of actions through the simple squeezing of his finger. We would like to isolate some *privileged* description, but can be pulled in different directions when we attempt to do so. One tempting thought here is that the remote effects of what Princip did are surely no part of his action: allowing them to be so would mean that people are routinely completing actions long after they have died (as well as performing countless actions simultaneously, e.g., moving towards lots of different objects as we walk). This line of thought naturally leads us to identify the *genuine* action as the initiation of the entire causal process in Princip's own body—the part over which he exercised direct control in squeezing his finger. But if we go that far, should we not go further? Princip's finger movement was caused by his muscles contracting, which was in turn caused by some neurons firing, which was caused by some chemical reactions ... and so on. We seem to need some notion of *basic* or *primitive* action to halt this regress, but if such primitive actions are at the level of neuronal activity, then they are clearly not directly conscious or introspectible. This, however, makes them very doubtful paradigms of deliberate action, especially from the first-personal perspective whose focus is precisely on consciousness, and is therefore quite oblivious of the detailed activity of our muscles and neurons.

(As an aside, notice that when we consider the notion of agency in the context of computers, this threat of regress is, to some extent at least, mitigated. Computer processors are *designed* using an explicit notion of atomic action—in the form of an "atomic program instruction"—an indivisible instruction carried out by the processor.)

In reaction to these difficulties, a quite different tempting thought is precisely to appeal to our first-person experience, and to identify the *genuine* action with the effect that we consciously intend. But here we can face the problems of both too much, and too little, consciousness. For on the one hand, Princip plausibly *intended* at least four of the six action descriptions listed above, and again, this route will lead to posthumous action (since people routinely act with the conscious intention of bringing about effects after their death, such as providing for their children—see

[20, pp. 68–73] for the more general problem of trying to pin down the timing of extended actions). On the other hand, a great many of the actions that we perform intentionally are done without explicit consciousness of them, and the more expert we become at a skill (such as driving, riding a bike, typing, or playing the piano), the more likely we are to perform the actions that it involves with minimal consciousness of what we are doing (and indeed trying to concentrate on what we are doing is quite likely to disrupt our performance). Even when we do become fully conscious of acting in such a context—for example, when I suddenly swerve away on seeing a pedestrian fall into the road just ahead of my car—such activity is likely to *precede* our consciousness of it, and its emergency, "instinctive" nature anyway makes it an unlikely paradigm of conscious deliberate action.

In the face of these sorts of difficulties, many philosophers (notably Bratman [6]) have come to prefer an account of intentional action in terms of *plans*. Here, for example, is the first approximate formulation by Mele and Moser:

> A person, S, intentionally performs an action, A, at a time, t, only if at t, S has an action plan,
> P, that includes, or at least can suitably guide, her A-ing [25, p. 43].

They go on to add further conditions, requiring that S have an *intention* which includes action plan P, and also that S "suitably follows her intention-embedded plan P in A-ing" [25, p. 52] (for present purposes we can ignore here the additional conditions that Mele and Moser formulate to capture plausible constraints on evidence, skill, reliability, and luck). But importantly, intentionality is consistent with S's having "an intention that encompasses, ... *subconsciously*, a plan that guides her A-ing" [25, p. 45, emphasis added]. Seeing actions as falling into a pattern guided by a plan thus enables habitual or automatic actions to be brought into the account, whether they are conscious or not.

All this somewhat undermines the all-too-natural assumption that the first-personal point of view is specially privileged when it comes to the identification of, and understanding of, action. And as we shall see later, such theories of human action (e.g., Bratman's) have already borne fruit in work towards the design of practical reasoning computer agents. But in fact there is nothing here that precludes the idea that consciousness of what we are doing—and conscious reflection on it—plays a major role in *human* life and experience. A cognitive model that explains action in informational terms is perfectly compatible with the supposition that certain aspects of its operation may be available in some way to consciousness. For example, Goldman [19] sketches the model of action proposed by Norman and Shallice [28] and explains how conscious awareness "of the selection of an action schema, or a 'command' to the motor system" could play a role within it.

There might well, however, seem a threat here to our conception of human *free will*, if consciousness of what we are doing is seen as post-hoc monitoring of unconscious cognitive processes that have already taken place by the time we become aware of them. Such worries may be sharpened by recent research in neuropsychology, in which observations using MRI scanners indicated that the mental sensation of conscious decision can lag quite some time behind certain identifiable physiological conditions that are strongly correlated with the decision ultimately made.

Experiments carried out at the Max Planck Institute for Human Cognitive and Brain Sciences in Germany suggested that it was possible to detect that a person had already made a choice, and what that choice was, up to *10 seconds* before the person in question was consciously aware of it [34]. Interpretation of such results is highly controversial, and there is clearly more work to be done in this area. We have no space to explore the issues here, but would end with four brief comments. First, we see no significant conflict between the idea that our thought is determined by unconscious "subcognitive" processes and the claim that we are *genuinely* free. To confine ourselves to just one point from the familiar "compatibilist" arsenal of arguments, the term "free choice" is one that we learn in ordinary life, and it would be perverse to deny that paradigm cases of such choice (such as a child's choosing a chocolate, with which we started) are genuinely free—if *these* aren't cases of free choice, then we lose all purchase on the intended meaning of the term. Secondly, it is entirely unsurprising that our conscious thinking should be found to correlate strongly with certain events in the brain, and such correlation does not imply that "we" are not really in control. On the contrary, neural processes are apparently the mechanism *by which* we reason and make choices; that they determine our thoughts no more implies that "we" are not really thinking those thoughts than the transfer of visual signals along the optic nerve implies that "we" are not really seeing things (or, for that matter, that the electronic activity of its components implies that a computer is not really calculating things). Thirdly, we would resist any suggestion that the neurophysiological evidence points towards *epiphenomenalism*—the theory according to which mind and mental states are caused by physical (brain and body) processes, but are themselves causally inert (crudely but vividly, this takes the conscious mind to be a passenger in the body, under the illusion that it is a driver). If evolution has made us conscious of what we do, then it is overwhelmingly likely that this has some causal payoff, for otherwise it would be an outrageous fluke—utterly inexplicable in terms of evolutionary benefit or selection pressure—that our consciousness (e.g., of pains, or sweet tastes) should correlate so well with bodily events [27, Sect. 5]. Finally, there could indeed be some conflict between our intuitive view of action and the findings of neurophysiology if it turned out that even our most reflective decisions are typically physiologically "fixed" at a point in time when we feel ourselves to be consciously contemplating them. But given the implausibility of epiphenomenalism, and the evident utility of conscious reflection in our lives, we consider this scenario to be extremely unlikely (cf. [3, pp. 42–43]).

20.3 Agency from the Third-Personal Perspective

Returning to the motivation that introduced this chapter, suppose we are looking around us, trying to make sense of what we see in the world. We see a wide range of processes generating continual change, many of these closely associated with specific objects or systems. What standpoints can we adopt to try to understand these processes? One possibility is to understand the behaviour of a system with reference to what the philosopher Daniel Dennett calls the *physical stance* [12, p. 36]. Put simply, the idea of the physical stance is to start with some original configuration, and then use known laws of nature (physics, chemistry etc.) to predict how this system will behave:

> When I predict that a stone released from my hand will fall to the ground, I am using the physical stance. [...] I attribute mass, or weight, to the stone, and rely on the law of gravity to yield my prediction [12, p. 37].

While the physical stance works well for simple cases such as this, it is of course not practicable for understanding or predicting the behaviour of people, who are far too complex to be understood in this way.

Another possibility is the *design stance*, which involves prediction of behaviour based on our understanding of the purpose that a system is supposed to fulfil. Dennett gives the example of an alarm clock [12, pp. 37–39]. When someone presents us with an alarm clock, we do not need to make use of physical laws in order to understand its behaviour. If we know it to be a clock, then we can confidently interpret the numbers it displays as the time, because clocks are designed to display the time. Likewise, if the clock makes a loud and irritating noise, we can interpret this as an alarm that was set at a specific time, because making loud and irritating noises at specified times (but not otherwise) is again something that alarm clocks are designed to do. No understanding of the clock's internal mechanism is required for such an interpretation (at least in normal cases)—it is justified sufficiently by the fact that alarm clocks are designed to exhibit such behaviour.

Importantly, adopting the design stance towards some system does not require us to consider it as *actually* designed, especially in the light of evolutionary theory. Many aspects of biological systems are most easily understood from a design perspective, in terms of the adaptive functions that the various processes perform in the life and reproduction of the relevant organism, treating these processes (at least to a first approximation) *as though* they had been designed accordingly. The same can also apply to adaptive computer systems, whose behaviour is self-modifying through genetic algorithms or other broadly evolutionary methods. Understanding such systems involves the design stance at two distinct levels: at the first level, their overt behaviour—like that of biological systems—may be most easily predicted in terms of the appropriate functions; while at the second level, the fact that they exhibit such functional behaviour is explicable by their having been *designed* to incorporate the relevant evolutionary mechanisms.

A third possible explanatory stance, and the one that most interests us here, is what Dennett calls the *intentional stance* [11]. From this perspective, we attribute *mental states* to entities and then use a common-sense theory of these mental states to predict how the entity will behave, under the assumption that it makes choices in accordance with its attributed beliefs and desires. The most obvious rationale for this approach is that when explaining human activity, it is often useful to make statements such as the following:

> Janine *believes* it is going to rain.
> Peter *wants* to finish his marking.

These statements make use of a *folk psychology*, by which human behaviour is predicted and explained through the attribution of *attitudes*, such as believing and wanting, hoping, fearing, and so on (see, for example, [35] for a discussion of folk psychology). This style of explanation is entirely commonplace, and most people reading the above statements would consider their meaning to be entirely clear, without a second thought.

Notice that the attitudes employed in such folk psychological descriptions are *intentional* notions: they are directed towards some form of *propositional content*. In the above examples, the propositional content is respectively something like "it is going to rain" and "finish my marking", but although it is surprisingly hard to pin down how such content should be characterised or individuated (especially when it involves the identification or possibly misidentification of objects from different perspectives [26, Sect. 5]), we need not worry here about the precise details. Dennett coined the term *intentional system* to describe entities

> whose behaviour can be predicted by the method of attributing belief, desires and rational acumen [11, p. 49].

The intentional stance can be contrasted not only with the physical and design stances, but also with the *behavioural* view of agency. The behavioural view—most famously associated with B. F. Skinner—attempts to explain human action in terms of stimulus–response behaviours, which are produced via "conditioning" with positive and negative feedback. But as Pinker critically remarks,

> The stimulus-response theory turned out to be wrong. Why did Sally run out of the building? Because she believed it was on fire and did not want to die. [...] What [predicts] Sally's behaviour, and predicts it well, is whether she *believes* herself to be in danger. Sally's beliefs are, of course, related to the stimuli impinging on her, but only in a tortuous, circuitous way, mediated by all the rest of her beliefs about where she is and how the world works [29, pp. 62–63].

In practice, then, the intentional stance is indispensable for our understanding of other humans' behaviour. But it can also be applied, albeit often far less convincingly, to a wide range of other systems, many of which we certainly would not wish to admit as autonomous agents. For example, consider a conventional light switch, as described by Shoham:

> It is perfectly coherent to treat a light switch as a (very cooperative) agent with the capability of transmitting current at will, who invariably transmits current when it believes that we want it transmitted and not otherwise; flicking the switch is simply our way of communicating our desires [32, p. 6].

However, the intentional stance does not seem to be an *appropriate* way of understanding and predicting the behaviour of light switches: here it is far simpler to adopt the physical stance (especially if we are manufacturing light switches) or the design stance (if we are an ordinary user, needing to know only that the switch is designed to turn a light on or off). By contrast, notice that, at least as sketched by Shoham, an intentional explanation of the switch's behaviour requires the attribution to it of quite complex representational states, capable of representing not only the flowing or absence of current, but also our own desires (which, on this story, it acts to satisfy). So even if this intentional account provides accurate prediction of the switch's behaviour, it is wildly extravagant as an *explanation*: to attribute beliefs and desires to a switch is already implausible, but to attribute to it *higher-order* beliefs and desires is well beyond the pale.

20.4 Higher-Order Intentionality

Human beings are in the unusual position of being both intentional agents in the first-personal sense and also fertile ascribers of third-personal intentionality to other entities. Although above we have described the intentional stance as a third-person explanatory framework, that stance is not of course employed only by people of scientific inclination: indeed the intentional stance comes very naturally—and often far *too* naturally [27, Sect. 1]—to people in general.

This human predilection for the intentional stance seems to be intimately bound to our status as *social* animals. That is, the adaptive role of such intentional ascription seems to be to enable us to understand and predict the behaviour of other agents in society. In navigating our way through this complex social web, we become involved in *higher-order* intentional thinking, whereby the plans of individuals (whether ourselves or those we observe) are influenced by the anticipated intentional behaviour of other agents. The value of such thinking is clear from its ubiquity in human life and the extent to which we take it for granted in our communications. Take for example the following fragment of conversations between Alice and Bob (attributed by Baron-Cohen [2] to Pinker):

Alice: I'm leaving you.
Bob: Who is he?

The obvious intentional stance explanation of this scenario is simple, uncontrived, and compelling: Bob *believes* that Alice *prefers* someone else to him and that she is *planning* accordingly; Bob also *wants* to *know* who this is (perhaps in the *hope* of *dissuading* her), and he *believes* that asking Alice will *induce* her to tell him. It seems implausibly difficult to explain the exchange *without* appealing to concepts like belief and desire, not only as playing a role in the agents' behaviour, but also featuring explicitly in their own thinking and planning.

Adoption of the third- (or second-) person intentional stance is also a key ingredient in the way we *coordinate* our activities with each other on a day-by-day basis, as Pinker illustrates:

I call an old friend on the other coast and we agree to meet in Chicago at the entrance of a bar in a certain hotel on a particular day two months hence at 7:45pm, and everyone who knows us predicts that on that day at that time we will meet up. And we do meet up. [...] The calculus behind this forecasting is intuitive psychology: the knowledge that I *want* to meet my friend and vice versa, and that each of us *believes* the other will be at a certain place at a certain time and *knows* a sequence of rides, hikes, and flights that will take us there. No science of mind or brain is likely to do better [29, pp. 63–64].

All of this involves a mix of first- and higher-order intentional ascription, characterised by Dennett as follows:

> A *first-order* intentional system has beliefs and desires (etc.) but no beliefs and desires *about* beliefs and desires. [...] A *second-order* intentional system is more sophisticated; it has beliefs and desires (and no doubt other intentional states) about beliefs and desires (and other intentional states) — both those of others and its own [11, p. 243].

The following statements illustrate these different levels of intentionality:

> 1st order: Janine *believed* it was raining.
> 2nd order: Michael *wanted* Janine to *believe* it was raining.
> 3rd order: Peter *believed* Michael *wanted* Janine to *believe* it was raining.

In our everyday lives, it seems we probably do not use more than about three layers of the intentional stance hierarchy (unless we are engaged in an artificially constructed intellectual activity, such as solving a puzzle or complex game theory), and it seems that most of us would probably struggle to go beyond fifth-order reasoning.

Interestingly, there is some evidence suggesting that other animals are capable of and make use of at least *some* higher-order intentional reasoning. Consider the example of vervet monkeys, which in the wild make use of a warning cry indicating to other monkeys the presence of leopards (a threat to the monkey community):

> Seyfarth reports (in conversation) an incident in which one band of vervets was losing ground in a territorial skirmish with another band. One of the losing-side monkeys, temporarily out of the fray, seemed to get a bright idea: it suddenly issued a leopard alarm (in the absence of any leopards), leading *all* the vervets to take up the cry and head for the trees — creating a truce and regaining the ground his side had been losing. [...] If this act is not just a lucky coincidence, then the act is truly devious, for it is not simply a case of the vervet uttering an imperative "get into the trees" in the expectation that *all* the vervets will obey, since the vervet should not expect a rival band to honor *his* imperative. So either the leopard call is [...] a *warning* — and hence the utterer's credibility but not authority is enough to explain the effect, or our utterer is more devious still: he *wants* the rivals to *think* they are *overhearing* a command *intended* only for his own folk [10, p. 347].

One can, of course, put forward alternative explanations for the above scenario, which do not imply any higher-order intentional reasoning. But, nevertheless, this anecdote (amongst others) provides tentative support for the claim that some non-human animals engage in higher-order intentional reasoning. There are other examples: chimpanzees, for example, seem to demonstrate some understanding of how others see them, a behaviour that is indicative of such higher-order reasoning.

20.4.1 Higher-Order Intentionality and Species Intelligence

While there is evidence that some other animals are capable of higher-order intentional reasoning to a limited extent, there seems to be no evidence that they are capable of anything like the richness of intentional reasoning that humans routinely manage. Indeed, it is tempting to take the widespread ability to reason at higher orders of intentionality as a general indicator of species intelligence. This idea, as we shall see, can be given further support from Robin Dunbar's work on the analysis of social group size in primates [13].

Dunbar was interested in the following question: Why do primates have such large brains (specifically, neocortex size), compared with other animals? Ultimately, the brain is an (energetically expensive) information processing device, and so a large brain would presumably have evolved to deal with some important information processing requirement for the primate. But what requirement, exactly? Dunbar considered a number of primates, and possible factors that might imply the need for enhanced information processing capacity. For example, one possible explanation could be the need to keep track of food sources in the primate's environment. Another possible explanation could be the requirement for primates with a larger ranging or foraging area to keep track of larger spatial maps. However, Dunbar found that the factor that best predicted neocortex size was the primate's *mean group size*: the average number of animals in social groups. This suggests that the large brain size of primates is needed to keep track of, maintain, and exploit the social relationships in primate groups.

Dunbar's research suggests a tantalising question: given that we know the average human neocortex size, what does his analysis predict as being the average group size for humans? The value obtained by this analysis is now known as *Dunbar's number*, and it is usually quoted as 150. That is, given the average human neocortex size and Dunbar's analysis of other primates, we would expect the average size of human social groups to be around 150. Dunbar's number would remain a curiosity but for the fact that subsequent research found that this number has arisen repeatedly, across the planet, in connection with human social group sizes. For example, it seems that neolithic farming villages typically contained around 150 people. Of more recent interest is the fact that Dunbar's number has something to say about Internet-based social networking sites such as Facebook. We refer the reader to [14] for an informal discussion of this and other examples of how Dunbar's number manifests itself in human society.

If species neocortex size does indeed correlate strongly with social group size, then the most likely evolutionary explanation seems to be precisely the need for, and adaptive value of, higher-order intentional reasoning within a complex society. Whether hunting in groups, battling with conflicting tribes, pursuing a mate (perhaps against rivals), or gaining allies for influence and leadership (with plentiful potential rewards in evolutionary fitness), the value of being able to understand and anticipate the thinking of other individuals is obvious. We have already seen how higher-order intentional reasoning plays an important role in relationships between humans, to the extent that we routinely take such reasoning for granted in mutual communication. This being so, it is only to be expected that larger social groups would make more demands of such reasoning, providing an attractive explanation for the relationship with neocortex size that Dunbar identified (cf. his discussion in [14, p. 30]). This is further corroborated by evidence that higher-order intentional reasoning capabilities are approximately a linear function of the relative size of the frontal lobe of the brain [14, p. 181], which seems to be peculiar to primates, and is generally understood as that part of the brain that deals with conscious thought.

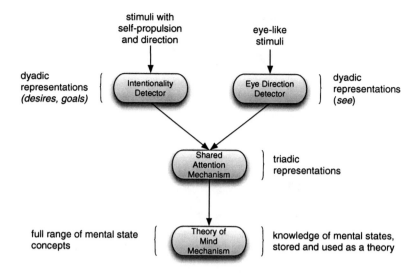

Fig. 20.1 Baron-Cohen's model of human intentional systems [2, p. 32]

20.5 The Human Intentional System

In this section, we briefly review a model of the *human* intentional system. The model was proposed by Baron-Cohen [2], an evolutionary psychologist interested in understanding of what he calls "mindreading"—the process by which humans understand and predict each other's mental states. A particular interest of Baron-Cohen's is the condition known as autism, which we discuss in more detail below.

Baron-Cohen's model of the human intentional system is composed of four main modules—see Fig. 20.1. Broadly speaking, the model attempts to define the key mechanisms involved in going from observations of processes and actions in the environment, through to predictions and explanations of agent behaviour. The four components of the model are as follows:

- the *Intentionality Detector (ID)*;
- the *Eye Direction Detector (EDD)*;
- the *Shared Attention Mechanism (SAM)*; and
- the *Theory of Mind Mechanism (ToMM)*.

The role of the Intentionality Detector (ID) is to:

[I]nterpet motion stimuli in terms of the primitive volitional mental states of goal and desire. [...] This device is activated whenever there is any perceptual input that might identify something as an agent. [...] This could be anything with self-propelled motion. Thus, a person, a butterfly, a billiard ball, a cat, a cloud, a hand, or a unicorn would do. Of course, when we discover that the object is not an agent — for example, when we discover that its motion is not self-caused, we can revise our initial reading. The claim, however, is that we readily interpret such data in terms of the object's goal and/or desire. [...] ID, then, is

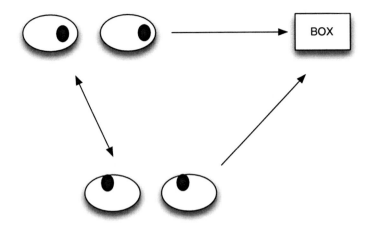

Fig. 20.2 The SAM builds triadic representations, such as "you and I see that we are looking at the same object" [2, p. 45]

very basic. It works through the senses (vision, touch, audition), [...and it will] interpret almost anything with self-propelled motion, or anything that makes a non-random sound, as an agent with goals and desires [2, pp. 32–33].

The output of the ID takes the form of primitive dyadic (two-place) intentional ascriptions, such as:

- She wants to stay dry.
- It wants to catch the wildebeest.

At broadly the same level as ID in Baron-Cohen's model is the Eye Direction Detector (EDD). In contrast to ID, which works on multiple types of perceptual input, the EDD is focussed around vision. Its basic role is as follows:

EDD has three basic functions: it detects the presence of eyes or eye-like stimuli, it computes whether eyes are directed towards it or toward something else, and it infers from its own case that if another organism's eyes are directed at something then that organism sees that thing. The last function is important because it [makes it possible to] attribute a perceptual state to another organism (such as "Mummy sees me") [2, pp. 38–39].

Dyadic representations such as those above provide a foundation upon which richer intentional acriptions might be developed, but they simply capture an attitude that an agent has to a proposition, and in this they are of limited value for understanding multi-agent interactions. The purpose of the Shared Attention Mechanism (SAM) is to build *nested*, triadic representations. Figure 20.2 illustrates a typical triadic representation: "You and I see that we are looking at the same object". Other examples of triadic representations include:

- Bob sees that Alice sees the gun.
- Alice sees that Bob sees the girl.

The Theory of Mind Mechanism (ToMM) is the final component of Baron-Cohen's model:

> ToMM is a system for inferring the full range of mental states from behaviour — that is, for employing a "theory of mind". So far, the other three mechanisms have got us to the point of being able to read behaviour in terms of *volitional mental states* (desire and goal), and to read eye direction in terms of *perceptual mental states* (e.g., see). They have also got us to the point of being able to verify that different people can be experiencing these particular mental states about the same object or event (shared attention). But a theory of mind, of course, includes much more [2, p. 51].

Thus, the ToMM goes from low-level intentional ascriptions to richer nested models. It is the ToMM to which we must appeal in order to understand Bob's question "Who is he?" when Alice says "I'm leaving you."

Baron-Cohen probably does not intend us to interpret the word "theory" in the ToMM to mean a theory in any formal sense (e.g., as a set of axioms within some logical system). However, this suggests an intriguing research agenda: To what extent can we come up with a *logical* ToMM, which can model the same role as the ToMM that we all have? While progress has been made on studying *idealised* aspects of *isolated* components of agency, such as knowledge [15, 21], attempting to construct an *integrated theory of agency* is altogether more challenging. We refer the reader to [9, 37, 38] for discussion and detailed references.

20.5.1 Autism

It is illuminating to consider what the consequences would be if some of the mechanisms of a fully-fledged intentional system were damaged or malfunctioning. Baron-Cohen hypothesises that the condition known as *autism* is a consequence of impairments in the higher-order mechanisms of the human intentional system: the SAM and/or ToMM. Autism is a serious, widespread psychiatric condition that manifests itself in childhood:

> The key symptoms [of autism] are that social and communication development are clearly abnormal in the first few years of life, and the child's play is characterized by a lack of the usual flexibility, imagination, and pretense. [...] The key features of the social abnormalities in autism [...] include lack of eye contact, lack of normal social awareness or appropriate social behaviour, "aloneness", one-sidedness in interaction, and inability to join a social group [2, pp. 62–63].

Baron-Cohen argues that autism is the result of failures in the higher-order components of the human intentional system described above, i.e., those mechanisms that deal with triadic representations and more complex social reasoning: the SAM and ToMM. He presents experimental evidence to support the claim that the ID and EDD mechanisms are typically functioning normally in children with autism [2]. For example, they use explanations such as "she wants an ice cream" and "he is going to go swimming" to explain stories and pictures, suggesting that the ID mechanism is functioning (recall that the role of the ID mechanism is to interpret apparently purposeful actions in terms of goals and desires). Moreover, they are able to interpret pictures of faces and make judgements such as "he is looking at me", suggesting that the EDD mechanism is functioning. However, autistic children seem unable to engage in shared activities, such as pointing to direct the gaze of another individual, suggesting that the SAM is not functioning properly. Finally, experiments indicate that autistic children have difficulty reasoning about the mental states of others, for example, trying to understand what others believe and why. Baron-Cohen takes this as a failure of the ToMM.

To evaluate Baron-Cohen's theory, consider how individuals with an impaired higher-order intentional system would behave. We might expect them to have difficulty in complex social settings and in predicting how others will react to their actions, to struggle when attempting to engage in group activities; and so on. And indeed, it seems these behaviours correlate well with the observed behaviours of autistic children.

20.6 Agency and Artificial Intelligence

Our discussion thus far has been divorced from the question of how we might actually *build* computer systems that can act as autonomous agents, and how far consideration of the nature of human agency can yield insights into how we might go about doing so. This question is of course central to the discipline of artificial intelligence—indeed one plausible way of defining the aim of the artificial intelligence field is to say that it is concerned with building artificial autonomous agents [30].

We start by considering the *logicist* tradition within artificial intelligence, which was historically very influential. It dates from the earliest days of artificial intelligence research, and is perhaps most closely associated with John McCarthy (the man who named the discipline of artificial intelligence—see, e.g., [24] for an overview of McCarthy's programme). As the name suggests, logic and logical reasoning take centre stage in the logicist tradition, whose guiding theme is that the fundamental problem faced by an agent—that of deciding what action to perform at any given moment—is reducible to a problem of purely logical reasoning. Figure 20.3 illustrates a possible architecture for a (highly stylized!) logical reasoning agent (cf. [16, pp. 307–328]):

- The agent has sensors, the purpose of which is to obtain information about the agent's environment. In contemporary robots, such sensors might be laser range finders, cameras, and radars, and GPS positioning systems [36].
- The agent has effectors, through which it can act upon its environment (e.g., robot arms for manipulating objects, wheels for locomotion).
- The two key data structures within an agent are a set Δ of logical formulae, which represent the *state* of the agent, and a set of rules, R, which represent the *theory* of the agent. The set Δ will typically include information about the agent's

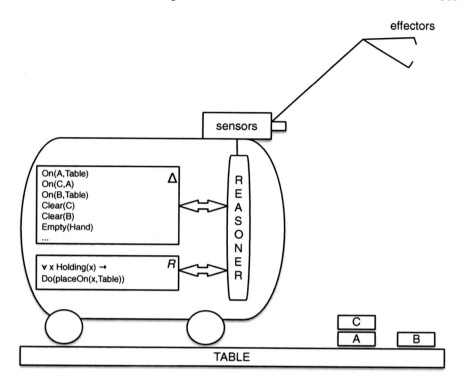

Fig. 20.3 An artificial agent that decides what to do via logical reasoning

environment, and any other information recorded by the agent as it executes. The rule set *R* will typically include both a *background theory* (e.g., information such as "if an object is on top of a block, then that block is not clear") and a *theory of rational choice* for the agent.

- *Transducers* transform raw sensor data into the symbolic logical form of Δ. Similarly, they map software instructions issued by the robot to commands for the actuators and effectors of the robot.
- A general-purpose *logical reasoning* component enables the agent to apply rules *R* to the agent's database Δ to derive logical conclusions; we also assume this component handles updating of Δ in the face of new sensor data, etc.

The agent continually executes a *sense-reason-act* loop, as follows:

- *Sense*: The agent observes its environment through its sensors, and after appropriate processing by transducers, this provides potentially new information in logical form; this new information is then incorporated into the agent's representation Δ.
- *Reason*: The reasoning component of the agent then tries to prove a sequent of the form $\Delta \vdash_R Do(\alpha)$, where α is a term that will correspond to an action available to the agent (e.g., an action with the robot arm). The idea is that, if the agent is able to prove such a sequent, then assuming the agent's representation Δ is correct, and

the rules R have been constructed appropriately, then α will be the appropriate ("optimal") action for the agent to take.

- *Act*: At this point, the action α selected during the previous phase stage is executed.

Thus, the "program" of the agent is encoded within its rules R. If these rules are designed appropriately, and if the various subsystems of the agent are operating correctly, then the agent will autonomously select an appropriate action to perform every time it cycles round the sense-reason-act loop.

The idea of building an agent in this way is seductive. The great attraction is that the rules R explicitly encode a theory of rational action for our agent. If the theory is good, then the decisions our agent makes will also be good. However, there are manifold difficulties with the scheme, chief among them being the following (see, e.g., [4] for a detailed discussion):

- The problem of representing information about complex, dynamic, multi-agent environments in a declarative logical form.
- The problem of translating raw sensor data into the appropriate declarative logical form, in time for this information to be of use in decision-making.
- The problem of automating the reasoning process (i.e., checking whether $\Delta \vdash_R Do(\alpha)$), particularly when decisions are required promptly.

Despite great efforts invested into researching these problems over the past half century, they remain essentially unsolved in general, and the picture we paint above of autonomous decision-making via logical reasoning does not represent a mainstream position in contemporary artificial intelligence research. Indeed, in the late 1980s and early 1990s, many researchers in artificial intelligence began to reject the logicist tradition, and to look to alternative methods for building agents. (See [8] for a detailed discussion of alternative approaches to artificial agency by Rodney Brooks, one of the most prominent and outspoken researchers against the logicist tradition and behind alternative proposals for building agents, and see [38] for a discussion and detailed references.)

Before we leave the logicist tradition of artificial intelligence, it is interesting to comment on the status of the logical representation Δ within an agent. Intuitively understood, the database Δ contains all the information that the agent has gathered and retained from its environment. For example, referring back to Fig. 20.3, we see that the agent has within its representation Δ the predicate $On(A, Table)$; and we can also see that indeed the block labelled "A" is in fact on top of the table. It is therefore very tempting to interpret Δ as being the *beliefs* of the agent, and thus assert that *the agent believes block "A" is on the table*. Under this interpretation, the presence of a predicate $P(a, b)$ in Δ would mean "the agent believes $P(a, b)$", and we would be inclined to say the agent's belief was correct if, when we examined the agent's environment, we found that the object a stood in relation P to object b (this assumes, of course, that we know what objects/relations a, b, and P are supposed to denote in the environment: the agent designer can presumably give us this mapping). See Konolige [22] for a detailed discussion of this subject.

20.6.1 A Refinement: Practical Reasoning Agents

The practical difficulties in attempting to realise the logicist vision of autonomous agents have led researchers to explore alternatives, and such exploration can also be motivated by the consideration that *we* don't seem to make decisions in that way! While there are surely occasions when many of us use abstract reasoning and problem solving techniques in deciding what to do, it is hard to imagine many realistic situations in which our decision-making is realised via logical proof. An alternative is to view decision-making in autonomous agents as a process of *practical reasoning*: reasoning directed towards action, rather than beliefs. That is, practical reasoning changes our actions, while theoretical reasoning changes our beliefs [6]:

> Practical reasoning is a matter of weighing conflicting considerations for and against competing options, where the relevant considerations are provided by what the agent desires/values/cares about and what the agent believes [7, p. 17].

Bratman [7] distinguishes two processes that take place in practical reasoning: *deliberation* and *means-ends reasoning*. Deliberation is the process of deciding *what we want to achieve*. As a result of deliberating, we fix upon some *intentions*: commitments to bring about specific states of affairs. Typically, deliberation involves considering multiple possible candidate states of affairs, and choosing between them. The second process in practical reasoning involves determining how to achieve the chosen states of affairs, given the means available to the agent; this process is hence called *means-ends* reasoning. The output of means-ends reasoning is a *plan*: a recipe that can be carried out by the agent, such that after the plan is carried out, the intended end state will be achieved.

Thus, after practical reasoning is completed, the agent will have chosen some intentions, and will have a plan that is appropriate for fulfilling these intentions. Under normal circumstances, an agent can proceed to execute its chosen plans, and the desired ends will result. The following *practical syllogism* provides a link between beliefs, intentions, plans, and action:

If I intend to achieve ϕ *and*
   I believe plan π will accomplish ϕ
Then I will do π.

This practical reasoning model has been hugely influential within the artificial intelligence community (see, e.g., [1, 18]). A typical architecture for a practical reasoning agent is illustrated in Fig. 20.4. The agent has three key data structures, which, as in the logicist tradition, are symbolic/logical representations. The agent's beliefs are a representation of the agent's environment; the agent's goal represents a state of affairs that the agent is currently committed to bringing about, and the agent's plan is a sequence of actions that the agent is currently executing. If the agent's beliefs are correct, and the plan is sound, then the execution of the plan will result in the accomplishment of the goal [23].

Architectures of the type shown in Fig. 20.4 are often referred to as *belief-desire-intention* (BDI) architectures. In this context, "desire" is usually considered as an

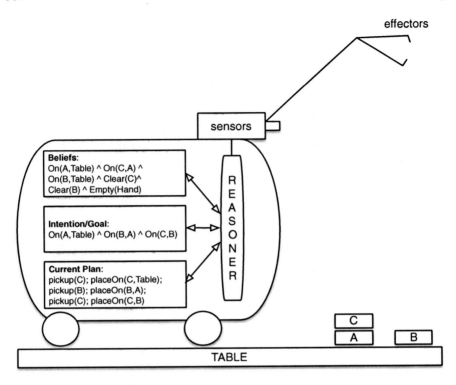

Fig. 20.4 A practical reasoning agent

intermediate state: the agent has potentially many conflicting desires, but chooses between them to determine the goal or intention that it will then fix on. In the BDI model, the sense-reason-act decision-making loop is modified as follows [37]:

- *Sense*: Observe the environment, and update beliefs on the basis of observations.
- *Option generation*: Given the current beliefs and intentions of the agent, determine what *options* are available, i.e., those states of affairs that the agent *could* usefully commit to bringing about.
- *Filtering*: Given the current beliefs, desires, and intentions of the agent, choose between competing options and commit to one. The chosen option becomes the agent's current *intention*.
- *Means-Ends Reasoning*: Given the current beliefs and intentions of the agent, find a plan such that, when executed in an environment where the agent's beliefs are correct, the plan will result in the achievement of the agent's intentions.
- *Action*: Execute the plan.

Various refinements can be made to this loop (e.g., so that an agent is not assumed to execute the entire plan before observing its environment again) [37], and of course the picture can be complicated considerably to take account of uncertainties and interaction with complex and changing situations (including game-theoretical consideration of the planning and behaviour of other agents).

The practical reasoning/BDI paradigm suffers from many of the same difficulties that beset the logicist paradigm. For example, the assumption of a logical representation of the agent's beliefs implies the need for transducers that can obtain logical representations from raw sensor data. In addition, the means-ends reasoning problem is computationally complex for logical representations of even modest richness [18]. However, various refinements permit efficient implementations of the architecture; perhaps the best known is the *reactive planning* class of architectures [5, 17]. The basic idea in such architectures is that the agent is equipped with a collection of plans, generated by the agent designer, which are labelled with the goals that they can be used to achieve. The means-ends reasoning problem then reduces to the comparatively tractable problem of searching through the plan library to try to find a plan that is suitable for the current intention.

Of course, one could now ask to what extent such an agent is genuinely *autonomous*, when in a sense "all it is doing" is assembling and executing plans made up of pre-compiled plan fragments. Such questions raise deep philosophical issues that we cannot address fully now, but here is a sketch of a response. First, there is much to be said for the idea that autonomy is a matter of degree: everything that we do is subject to constraints of various sorts (of ability, cost, law, physical possibility etc.), and all of these can vary in countless ways that extend or limit how far things are "under our control". Secondly, the autonomy attributable to a human—and by extension a computer system—depends in part on how far the reasoning employed in deciding on a course of action is "internal" to the agent: if the agent is performing a complex calculation, taking into account the various constraints and aims within a range of flexible possibilities, this demonstrates far more autonomy than an agent that is simply following orders without any internal reasoning or selection of choices. Thirdly, it follows that autonomy correlates quite strongly with the extent to which application of the third-person intentional stance assists in deep understanding and prediction of the system's behaviour. Some researchers, inspired by the utility of this view of such systems, have proposed the idea of *agent-oriented programming*, in which intentional stance notions such as belief, desire, and intention are first-class entities in a programming language for autonomous agents [33].

20.7 Conclusions

Our discussion has revealed many ways in which research on agency has led to a convergence in our understanding of human and of artificial agents. In both, the folk-psychological intentional stance—in terms of attributed beliefs and desires—has clear predictive value, and in both, our attempts at a deeper understanding that goes beyond folk psychology have led to plan-based models that shed light on our own behaviour, and also point the way towards practical development of artificial agents. Whether we should describe such an artificial system as a genuine "agent" and as having *literal* "beliefs" and "desires" is, of course, a matter for debate. But when a system is sufficiently sophisticated that its behaviour can only feasibly be understood or predicted in terms of belief-like, desire-like, and plan-like states and the interplay between those (rather than purely in terms of simple execution of pre-packaged instructions), we think there is a lot to be said for extending the boundaries of our concepts accordingly. As Millican [27, Sect. 3–4] has argued in the case of artificial intelligence, the development of such systems has faced us with a new problem which is not anticipated by the established boundaries of our traditional concepts. The concept of "agency", like that of "intelligence" is *open textured*, and how we mould it within this new context is largely a matter of decision, rather than mere analysis of our pre-existing conceptual repertoire.

There are several possible avenues for future research. One interesting open problem remains the extent to which we can develop *formal* theories that can predict and explain the behaviour of *human* agents; another is the extent to which we can link such formal theories with computer programs, in order to provide an account of their behaviour in terms of agentive concepts such as beliefs, desires, and rational choice.

Acknowledgments Wooldridge was supported by the European Research Council under Advanced Grant 291528 ("RACE").

References

1. Allen JF, Hendler J, Tate A (eds) (1990) Readings in planning. Morgan Kaufmann, San Mateo
2. Baron-Cohen S (1995) Mindblindness: an essay on autism and theory of mind. The MIT Press, Cambridge
3. Bayne T (2011) Libet and the case for free will scepticism. In: Swinburne R (ed) Free will and modern science. British Academy, Oxford, pp 25–46
4. Birnbaum L (1992) Rigor mortis. In: Kirsh D (ed) Foundations of artificial intelligence. The MIT Press, Cambridge, pp 57–78
5. Bordini R, Hübner JF, Wooldridge M (2007) Programming multi-agent systems in AgentSpeak using Jason. Wiley, Chichester
6. Bratman ME (1987) Intention, plans, and practical reason. Harvard University Press, Cambridge
7. Bratman ME (1990) What is intention? In: Cohen PR, Morgan JL, Pollack ME (eds) Intentions in communication. The MIT Press, Cambridge, pp 15–32
8. Brooks RA (1999) Cambrian intelligence. The MIT Press, Cambridge
9. Cohen PR, Levesque HJ (1990) Intention is choice with commitment. Artif Intell 42:213–261
10. Dennett DC (1983) Intentional systems in cognitive ethology. Behav Brain Sci 6:343–390
11. Dennett DC (1987) The intentional stance. The MIT Press, Cambridge
12. Dennett DC (1996) Kinds of minds. Phoenix, London
13. Dunbar RIM (1992) Neocortex size as a constraint on group size in primates. J Hum Evol 22:469–493
14. Dunbar RIM (2011) How many friends does one person need?: Dunbar's number and other evolutionary quirks. Faber and Faber, London
15. Fagin R, Halpern JY, Moses Y, Vardi MY (1995) Reasoning about knowledge. The MIT Press, Cambridge
16. Genesereth MR, Nilsson N (1987) Logical foundations of artificial intelligence. Morgan Kaufmann, San Mateo

17. Georgeff MP, Lansky AL (1987) Reactive reasoning and planning. In: Proceedings of the sixth national conference on artificial intelligence (AAAI-87), Seattle, pp 677–682
18. Ghallab M, Nau D, Traverso P (2004) Automated planning: theory and practice. Morgan Kaufmann, San Mateo
19. Goldman AI (1995) Action. In: Guttenplan S (ed) A companion to the philosophy of mind. Blackwell, Oxford, pp 117–121
20. Guttenplan S (ed) (1995) A companion to the philosophy of mind. Blackwell, Oxford
21. Hintikka J (1962) Knowledge and belief. Cornell University Press, Ithaca
22. Konolige K (1986) A deduction model of belief. Pitman Publishing, London, Morgan Kaufmann, San Mateo
23. Lifschitz V (1986) On the semantics of STRIPS. In: Georgeff MP, Lansky AL (eds) Reasoning about actions and plans. In: Proceedings of the 1986 workshop. Morgan Kaufmann, San Mateo, pp 1–10
24. McCarthy J (1990) Formalization of common sense: papers by John McCarthy. Ablex Publishing Corporation, New Jersey
25. Mele AR, Moser PK (1994) Intentional action. Nous 28(1):39–68
26. Millican P (1990) Content, thoughts, and definite descriptions. Proc Aristotelian Soc supplementary, 64:167–203
27. Millican P (2013) The philosophical significance of the Turing machine and the Turing test. In: Cooper SB, van Leeuwen J (eds) Alan Turing: his work and impact. Elsevier, Amsterdam, pp 587–601
28. Norman DA, Shallice T (1986) Attention to action; willed and automatic control of behaviour. In: Davidon RJ, Schwartz GE, Shapiro D (eds) Consciousness and self-regulation: advances in research and theory, vol 4. Plenum Press, New York, pp 1–18
29. Pinker S (1997) How the mind works. W. W. Norton & Co. Inc., New York
30. Russell S, Norvig P (1995) Artificial intelligence: a modern approach. Prentice-Hall, Englewood Cliffs
31. Searle JR (1983) Intentionality: an essay in the philosophy of mind. Cambridge University Press, Cambridge
32. Shoham Y (1990) Agent-oriented programming. Technical report STAN-CS-1335-90, Computer science department, Stanford University, Stanford, p 94305
33. Shoham Y (1993) Agent-oriented programming. Artif Intell 60(1):51–92
34. Soon CS, Brass M, Heinze H-J, Haynes J-D (2008) Unconscious determinants of free decisions in the human brain. Nat Neurosci 11(5):543–545
35. Stich SP (1983) From folk psychology to cognitive science. The MIT Press, Cambridge
36. Thrun S, Montemerlo M, Dahlkamp H, Stavens D, Aron A, Diebel J, Fong P, Gale J, Halpenny M, Hoffmann G, Lau K, Oakley C, Palatucci M, Pratt V, Stang P, Strohband S, Dupont C, Jendrossek L-E, Koelen C, Markey C, Rummel C, van Niekerk J, Jensen E, Alessandrini P, Bradski G, Davies B, Ettinger S, Kaehler A, Nefian A, Mahoney P (2007) Stanley: the robot that won the DARPA grand challenge. In: Buehler M, Iagnemma K, Singh S (eds) The 2005 DARPA grand challenge. Springer, Berlin, pp 1–43
37. Wooldridge M (2000) Reasoning about rational agents. The MIT Press, Cambridge
38. Wooldridge M, Jennings NR (1995) Intelligent agents: theory and practice. Knowl Eng Rev 10(2):115–152

Chapter 21
Incorporating Action Models into the Situation Calculus

Yongmei Liu and Hector J. Levesque

Abstract While both situation calculus and dynamic epistemic logics (DELs) are concerned with reasoning about actions and their effects, historically, the emphasis of situation calculus was on physical actions in the single-agent case, in contrast, DELs focused on epistemic actions in the multi-agent case. In recent years, cross-fertilization between the two areas has begun to attract attention. In this paper, we incorporate the idea of action models from DELs into the situation calculus to develop a general multi-agent extension of it. We analyze properties of beliefs in this extension, and prove that action model logic can be embedded into the extended situation calculus. Examples are given to illustrate the modeling of multi-agent scenarios in the situation calculus.

21.1 Introduction

While both situation calculus [19] and dynamic epistemic logics (DELs) [10] are concerned with reasoning about actions and their effects, historically, the emphasis of situation calculus was on physical actions in the single-agent case, in contrast, DELs focused on epistemic actions in the multi-agent case. In recent years, cross-fertilization between the two areas has begun to attract attention. In particular, van Benthem [7] proposed the idea that situation calculus and modal logic meet and merge. van Ditmarsch et al. [11] embedded a propositional fragment of the situation

Y. Liu (✉)
Department of Computer Science, Sun Yat-sen University,
Guangzhou 510006, China
e-mail: ymliu@mail.sysu.edu.cn

H. J. Levesque
Department of Computer Science, University of Toronto,
Toronto, ON M5S 3G4, Canada
e-mail: hector@cs.toronto.edu

A. Baltag and S. Smets (eds.), *Johan van Benthem on Logic
and Information Dynamics*, Outstanding Contributions to Logic 5,
DOI: 10.1007/978-3-319-06025-5_21, © Springer International Publishing Switzerland 2014

calculus into a DEL. Kelly and Pearce [12] incorporated ideas from DELs to handle regression for common knowledge in the situation calculus. Baral [5] proposed to combine results from reasoning about actions and DELs.

In a multi-agent setting, the agents in the domain may have different perspectives of the actions. Baltag et al. [2, 3] introduced a construct called an action model to represent these differences of perspectives. An action model consists of a set of actions, a precondition for each action, and a binary relation on the set of actions for each agent, which represents the agent's ability to distinguish between the actions. Moreover, they defined an operation by which an action model may be used to update a Kripke world to obtain a successor world modeling the effects of the action execution. They proposed a logic, called action model logic, to reason about action models and their effects on agents' epistemic states. van Benthem et al. [8] generalized the concept of action model to that of update model where each action is also associated with a postcondition. So action models can model events which bring about epistemic change, but update models can model events which can not only change agents' epistemic states but also the world state.

The situation calculus was first introduced by McCarthy and Hayes [16], and historically, one of its major concerns was how to solve the frame problem, that is, how to represent the effects of a world-changing action without explicitly specifying which conditions are not affected by the action. Reiter [18] gave a solution to the frame problem under some conditions in the form of successor state axioms. This solution to the frame problem has proven useful as the foundation for the high-level robot programming language Golog [15]. Scherl and Levesque [20, 21] extended Reiter's solution to cover epistemic actions in the single-agent case. Later, Shapiro et al. [22] extended their work to the multi-agent case, but they only considered public actions whose occurrence is common knowledge. In the last decade, Lakemeyer and Levesque [13, 14] proposed a logic called ES, which is a fragment of the situation calculus with knowledge. Recently, Belle and Lakemeyer [6] gave a multi-agent extension of ES, but as [22], they only considered public actions. So up to now, although there have been extensions of the situation calculus into the multi-agent case, they are not able to account for arbitrary multi-agent scenarios.

In this chapter, we incorporate action models into the situation calculus to develop a general multi-agent extension of it. We analyze properties of beliefs in this extension, and prove that action model logic can be embedded into the extended situation calculus. Examples are given to illustrate the modeling of multi-agent scenarios in the situation calculus.

The rest of the chapter is organized as follows. In the next section, we introduce the situation calculus and action model logic. In Sect. 21.3, we present a multi-agent extension of the situation calculus by incorporating action models. Section 21.4 analyzes properties of beliefs in the extended situation calculus, and Sect. 21.5 shows that action model logic can be embedded into the multi-agent situation calculus. In Sect. 21.6, we present two extended examples of modeling multi-agent scenarios in the situation calculus. Finally, we conclude and describe some future work.

21.2 Preliminaries

In this section, we introduce the situation calculus, and action model logic.

21.2.1 The Situation Calculus and Golog

The situation calculus [19] is a many-sorted first-order language suitable for describing dynamic worlds. There are three disjoint sorts: *action* for actions, *situation* for situations, and *object* for everything else. A situation calculus language $\mathscr{L}_{sc}$ has the following components: a constant S_0 denoting the initial situation; a binary function $do(a, s)$ denoting the successor situation to s resulting from performing action a; a binary predicate $s \sqsubset s'$ meaning that situation s is a proper subhistory of situation s'; a binary predicate $Poss(a, s)$ meaning that action a is possible in situation s; action functions; a finite number of relational and functional fluents, i.e., predicates and functions taking a situation term as their last argument; and a finite number of situation-independent predicates and functions.

The situation calculus has been extended to accommodate sensing and knowledge. Assume that in addition to ordinary actions that change the world, there are sensing actions which do not change the world but tell the agent information about the world. A special binary function $SR(a, s)$ is used to characterize what the sensing action tells the agent about the world. Knowledge is modeled in the possible-world style by introducing a special fluent $K(s', s)$, meaning that situation s' is accessible from situation s. Note that the order of the arguments is reversed from the usual convention in modal logic. Then knowing ϕ at situation s is represented as follows:

$$\mathbf{Knows}(\phi(now), s) \stackrel{def}{=} \forall s'.K(s', s) \supset \phi(s'),$$

where *now* is used as a placeholder for a situation argument. For example,

$$\mathbf{Knows}(\exists s^*.now = do(open, s^*), s)$$

means knowing that the *open* action has just been executed. When "*now*" only appears as a situation argument to fluents, it is often omitted.

Scherl and Levesque [20] proposed the following successor state axiom for the K fluent: (Throughout this paper, free variables are assumed to be universally quantified from outside.)

$$K(s', do(a, s)) \equiv \exists s^*.K(s^*, s) \wedge s' = do(a, s^*) \wedge SR(a, s^*) = SR(a, s).$$

Intuitively, situation s' is accessible after action a is done in situation s iff it is the result of doing a in some s^* which is accessible from s and agrees with s on the sensing result.

Based on the situation calculus, a logic programming language Golog [15] has been designed for high-level robotic control. The formal semantics of Golog is specified by an abbreviation $Do(\delta, s, s')$, which intuitively means that executing δ brings us from situation s to s'. It is inductively defined on δ as follows, where we omit the definition of procedures:

1. Primitive actions:
$$Do(\alpha, s, s') \stackrel{def}{=} Poss(\alpha, s) \wedge s' = do(\alpha, s).$$

2. Test actions:
$$Do(\phi?, s, s') \stackrel{def}{=} \phi[s] \wedge s = s'.$$
   Here, ϕ is a situation-suppressed formula, i.e., a situation calculus formula with all situation arguments suppressed, and $\phi[s]$ denotes the formula obtained from ϕ by taking s as the situation arguments of all fluents mentioned in ϕ.

3. Sequence:
$$Do(\delta_1; \delta_2, s, s') \stackrel{def}{=} (\exists s'').Do(\delta_1, s, s'') \wedge Do(\delta_2, s'', s').$$

4. Nondeterministic choice of two actions:
$$Do(\delta_1 \mid \delta_2, s, s') \stackrel{def}{=} Do(\delta_1, s, s') \vee Do(\delta_2, s, s').$$

5. Nondeterministic choice of action arguments: Execute $\delta(x)$ with a nondeterministically chosen argument x.
$$Do((\pi \, x)\delta(x), s, s') \stackrel{def}{=} (\exists x)Do(\delta(x), s, s').$$

6. Nondeterministic iteration: Execute δ zero or more times.
$$Do(\delta^*, s, s') \stackrel{def}{=} (\forall P).\{(\forall s_1)P(s_1, s_1) \wedge$$
$$(\forall s_1, s_2, s_3)[P(s_1, s_2) \wedge Do(\delta, s_2, s_3) \supset P(s_1, s_3)]\} \supset P(s, s').$$

Conditionals and loops are defined as abbreviations:

if ϕ **then** δ_1 **else** δ_2 **fi** $\stackrel{def}{=} [\phi?; \delta_1] \mid [\neg\phi?; \delta_2]$,

while ϕ **do** δ **od** $\stackrel{def}{=} [\phi?; \delta]^*; \neg\phi?$.

For example, the following is a Golog program which nondeterministically moves a block onto another, so long as there are at least two blocks on the table:

$$\textbf{while} \quad (\exists x, y)[ontable(x) \wedge ontable(y) \wedge x \neq y] \quad \textbf{do}$$
$$(\pi \, u, v)move(u, v) \quad \textbf{od}$$

21.2.2 Action Model Logic (AML)

In a nutshell, action model logic (AML) extends epistemic logic with reasoning about epistemic actions which bring about epistemic change. We now present the syntax and semantics of action model logic. We fix a finite set of agents $\mathscr{A}$ and a countable set of propositional atoms $\mathscr{P}$. We first define Kripke models.

Definition 21.1 A Kripke model M is a triple (S, R, V) where

- S is a set of states;
- For each agent i, R_i is a binary relation on S;
- For each $t \in S$, $V(t)$ is a subset of the atoms.

A pointed Kripke model is a pair (M, s_0) where M is a Kripke model and s_0 is a state of M.

Intuitively, a Kripke model represents the agents' uncertainty about the current world state. Here, S is the set of all possible world states; $p \in V(t)$ means that proposition p is true in state t; and $t R_i t'$ means that in state t, agent i thinks that t' might be the actual state.

An action model is a Kripke model of "actions", which represents the agents' uncertainty about the current action. The definition of an action model is similar to that of a Kripke model except: a truth assignment is associated to each state in a Kripke model, but a precondition is associated to each action in an action model.

Definition 21.2 An action model over a language $\mathscr{L}$ is a triple $(A, \rightarrow, pre)$ where

- A is a set of action points;
- For each agent i, $\rightarrow_i$ is a binary relation on A;
- For each action point e, $pre(e) \in \mathscr{L}$ is its precondition.

A pointed action model is a pair (N, e_0) where N is an action model and e_0 is an action point of N.

Given a Kripke model and an action model, by the product update operation, defined as follows, we obtain the new Kripke model resulting from executing the action model in the given Kripke model.

Definition 21.3 Let $M = (S, R, V)$ be a Kripke model, and $t_0 \in S$. Let $N = (A, \rightarrow, pre)$ be an action model, and $e_0 \in A$ such that $M, t_0 \models pre(e_0)$. The product of (M, t_0) and (N, e_0), denoted by $(M, t_0) \otimes (N, e_0)$, is a pointed Kripke model (M', t_0') where $M' = (S', R', V')$, and

- $S' = \{(t, e) \mid t \in S, e \in A, \text{ and } M, t \models pre(e)\}$;
- $t_0' = (t_0, e_0)$;
- $(t, e) R_i' (t', e')$ iff $t R_i t'$ and $e \rightarrow_i e'$;
- For each $(t, e) \in S'$, $V'((t, e)) = V(t)$.

Intuitively, (t, e) is the world state resulting from executing action e in state t. Note that e is an epistemic action: it does not change the world state, thus the truth assignment associated to (t, e) is the same as that associated to t. In state (t, e), agent i considers (t', e') as a possible state if she considers t' as a possible alternative of t and e' as a possible alternative of e.

The language of action model logic extends the language of epistemic logic with a construct $[N, e_0]\phi$, which intuitively means that formula ϕ holds after the execution of the pointed action model (N, e_0).

Definition 21.4 The language $\mathcal{L}_{am}$ of action model logic is recursively defined as follows:

1. Any propositional atom $p \in \mathcal{P}$ is an AML formula.
2. If ϕ and ψ are AML formulas, so are $\neg \phi$ and $(\phi \wedge \psi)$.
3. If ϕ is an AML formula, so are $B_i \phi$ and $C_{\mathcal{E}} \phi$, where $i \in \mathcal{A}$, $\mathcal{E} \subseteq \mathcal{A}$.
4. If ϕ is an AML formula and (N, e_0) is a pointed action model with a finite domain and such that for all action points e, $pre(e)$ is an AML formula, so is $[N, e_0]\phi$.

The following is a complexity measure of AML formulas as presented in [10]:

Definition 21.5 The complexity measure $c : \mathcal{L}_{am} \rightarrow \mathbb{N}$ is inductively defined as follows:

1. $c(p) = 1$;
2. $c(\neg \phi) = 1 + c(\phi)$;
3. $c(\phi \wedge \psi) = 1 + \max\{c(\phi), c(\psi)\}$;
4. $c(B_i \phi) = 1 + c(\phi)$;
5. $c(C_{\mathcal{E}} \phi) = 1 + c(\phi)$;
6. $c([N, e_0]\phi) = (4 + \max\{c(pre(e)) \mid e \text{ is an action point of } N\}) \cdot c(\phi)$.

We now present the semantics of action model logic:

Definition 21.6 Let $M = (S, R, V)$ be a Kripke model and t_0 a state of M. We interpret the formulas by induction on their complexity as follows:

1. $M, t_0 \models p$ iff $p \in V(t_0)$;
2. $M, t_0 \models \neg \phi$ iff $M, t_0 \not\models \phi$;
3. $M, t_0 \models \phi \wedge \psi$ iff $M, t_0 \models \phi$ and $M, t_0 \models \psi$;
4. $M, t_0 \models B_i \phi$ iff for all t such that $t_0 R_i t$, $M, t \models \phi$;
5. $M, t_0 \models C_{\mathcal{E}} \phi$ iff for all t such that $t_0 R_{\mathcal{E}} t$, $M, t \models \phi$, where $R_{\mathcal{E}}$ is the reflexive transitive closure of the union of R_i for $i \in \mathcal{E}$;
6. $M, t_0 \models [N, e_0]\phi$ iff if $M, t_0 \models pre(e_0)$, then $(M, t_0) \otimes (N, e_0) \models \phi$.

A formula ϕ is valid if it is true in any pointed Kripke model.

We end this section with an example:

Example 21.1 [10] Two stockbrokers Ann and Bob are having a little break in a Wall Street bar, sitting at a table. A messenger comes in and delivers a letter to Ann. On the envelope is written "urgently requested data on United Agents". Let atom p mean that "United Agents is doing well". Consider the following scenarios:

1. Bob sees that Ann reads the letter. From Bob's point of view, Ann could learn p or she could learn $\neg p$, and he cannot distinguish between these two actions. But Ann can certainly distinguish between them. Thus we get the following action model: $read = (A, \rightarrow, pre)$, where $A = \{0, 1\}$, $pre(0) = \neg p$, $pre(1) = p$, $\rightarrow_a$ is the identity relation, and $\rightarrow_b$ is the total relation.

2. Bob leaves the table; Ann may have read the letter while Bob is away. From Bob's point of view, there are 3 possibilities: Ann learns p, Ann learns $\neg p$, and Ann learns nothing, and he cannot distinguish between these actions. Thus the action model is: $mayread = (A, \rightarrow, pre)$, where $A = \{0, 1, t\}$, $pre(0) = \neg p$, $pre(1) = p$, $pre(t) = true$, $\rightarrow_a$ is the identity relation, and $\rightarrow_b$ is the total relation.

21.3 A Multi-agent Extension of the Situation Calculus

In this section, we present a multi-agent extension of the situation calculus by incorporating action models. Instead of Scherl and Levesque's K fluent, we now use a fluent $B(i, s', s)$, which means that agent i considers situation s' accessible from situation s. We introduce a special predicate $A(i, a', a, s)$, meaning that in situation s, agent i considers action a' as a possible alternative of action a.

We assume that there are two types of primitive actions: ordinary actions which change the world, and epistemic actions which do not change the world but informs the agent. We use the action precondition axiom to specify what the epistemic action tells the agent about the current situation. For example, we may have an epistemic action $ison(i, x)$ which tells agent i that switch x is on. This is axiomatized as:

$$Poss(ison(i, x), s) \equiv on(x, s).$$

In particular, there is a special epistemic action nil, meaning that nothing happens, with the axiom $Poss(nil, s) \equiv true$. Note that a sensing action which tells the agent whether ϕ holds can be treated as the nondeterministic choice of two epistemic actions: one is possible iff ϕ holds, and the other is possible iff $\neg\phi$ holds.

We propose the following successor state axiom for the B fluent:

$$B(i, s', do(a, s)) \equiv \exists s^*\exists a^*.B(i, s^*, s) \wedge A(i, a^*, a, s)\wedge$$
$$(Poss(a, s) \supset Poss(a^*, s^*)) \wedge s' = do(a^*, s^*).$$

Intuitively, for agent i, situation s' is accessible after action a is performed in situation s iff it is the result of doing some alternative a^* of a in some s^* accessible from s, and executability of a in s implies that of a^* in s^*. Note that when a is not possible in s, we do not care whether a^* is possible in s^*. In the multi-agent case, a domain of application is specified by a basic action theory of the form:

$$\mathcal{D} = \Sigma \cup \mathcal{D}_{ap} \cup \mathcal{D}_{ss} \cup \mathcal{D}_{aa} \cup \mathcal{D}_{una} \cup \mathcal{D}_{S_0}, \text{ where}$$

1. Σ are the foundational axioms:

   (F1) $do(a_1, s_1) = do(a_2, s_2) \supset a_1 = a_2 \wedge s_1 = s_2$
   (F2) $(\neg s \sqsubset S_0) \wedge (s \sqsubset do(a, s') \equiv s \sqsubseteq s')$

(F3) $\forall P.\forall s[Init(s) \supset P(s)] \wedge \forall a, s[P(s) \supset P(do(a, s))] \supset (\forall s)P(s)$, where
$Init(s) \stackrel{def}{=} \neg(\exists a, s')s = do(a, s')$.

(F4) $B(i, s, s') \supset [Init(s) \equiv Init(s')]$.

Intuitively, $Init(s)$ means s is an initial situation. A model of $\{F1, F2, F3\}$ consists of a forest of isomorphic trees rooted at the initial situations. F4 specifies that initial situations can be B-related to only initial situations.

2. $\mathscr{D}_{ap}$ is a set of action precondition axioms, one for each action function C, of the form $Poss(C(\mathbf{x}), s) \equiv \Pi_C(\mathbf{x}, s)$. This includes the precondition axioms for epistemic actions.

3. $\mathscr{D}_{ss}$ is a set of successor state axioms (SSAs) for fluents, one for each fluent F, of the form $F(\mathbf{x}, do(a, s)) \equiv \Phi_F(\mathbf{x}, a, s)$. This includes the SSA for the B fluent. The SSAs for ordinary fluents must satisfy the no-side-effect conditions, i.e., they are not affected by epistemic actions.

4. $\mathscr{D}_{aa}$ is a set of action alternative axioms, one for each action function C, of the form $A(i, a, C(\mathbf{x}), s) \equiv \psi_C(i, a, \mathbf{x}, s)$.

5. $\mathscr{D}_{una}$ is the set of unique names axioms for actions:

$$C(\mathbf{x}) \neq C'(\mathbf{y}), \quad \text{and} \quad C(\mathbf{x}) = C(\mathbf{y}) \supset \mathbf{x} = \mathbf{y},$$

where C and C' are distinct action functions.

6. $\mathscr{D}_{S_0}$ is a set of sentences about S_0.

In the rest of the paper, when we present a basic action theory, we will only present relevant axioms from $\mathscr{D}_{ap} \cup \mathscr{D}_{ss} \cup \mathscr{D}_{aa} \cup \mathscr{D}_{S_0}$.

Example 21.2 Consider a simple blocks world. There is a single physical action: $move(x, y)$, moving block x onto block y. There are two fluents: $clear(x, s)$, block x has no blocks on top of it; $on(x, y, s)$, block x is on block y. The following are the action precondition and successor state axioms:

$$Poss(move(x, y), s) \equiv clear(x, s) \wedge clear(y, s) \wedge x \neq y$$
$$on(x, y, do(a, s)) \equiv a = move(x, y) \vee$$
$$on(x, y, s) \wedge \neg(\exists z)a = move(x, z),$$
$$clear(x, do(a, s)) \equiv (\exists y)(\exists z)a = move(y, z) \wedge on(y, x, s) \vee$$
$$clear(x, s) \wedge \neg(\exists y)a = move(y, x).$$

We now axiomatize in the situation calculus the letter example of Sect. 21.2.2.

Example 21.3

1. Bob sees that Ann reads the letter. We introduce an epistemic action $read(e)$, which means that Ann senses the truth value of p with result e while Bob is observing her. The axioms are:

$$Poss(read(e), s) \equiv (e = 1 \equiv p(s)),$$
$$A(i, a, read(e), s) \equiv (i = ann \supset a = read(e)) \wedge (i = bob \supset \exists e'.a = read(e')).$$

So Ann can distinguish between $read(1)$ and $read(0)$, but Bob can't.

2. Bob thinks Ann may have read the letter. We introduce an epistemic action $mread(e)$, which means that Ann senses the truth value of p with result e while Bob is not sure about whether this happens. The axioms are:

$$Poss(mread(e), s) \equiv (e = 1 \equiv p(s)),$$
$$A(i, a, mread(e), s) \equiv (i = ann \supset a = mread(e)) \wedge$$
$$(i = bob \supset a = nil \vee \exists e'.a = mread(e')),$$
$$A(i, a, nil, s) \equiv (i = ann \supset a = nil) \wedge$$
$$(i = bob \supset a = nil \vee \exists e'.a = mread(e'))$$

So Bob can't distinguish between the three actions $mread(1)$, $mread(0)$, and nil.

Finally, we introduce some notation which will be used in the rest of the paper. Let $\phi(s)$ be a formula with a single situation variable s.

1. Agent i believes ϕ:

$$\mathbf{Bel}(i, \phi(now), s) \overset{def}{=} \forall s'.B(i, s', s) \supset \phi(s').$$

2. Agent i truly believes ϕ:

$$\mathbf{TBel}(i, \phi(now), s) \overset{def}{=} \phi(s) \wedge \mathbf{Bel}(i, \phi(now), s).$$

3. Agent i believes whether ϕ holds:

$$\mathbf{BW}(i, \phi(now), s) \overset{def}{=} \mathbf{Bel}(i, \phi(now), s) \vee \mathbf{Bel}(i, \neg\phi(now), s).$$

4. Let $\mathscr{E}$ be a subset of the agents. We let $C(\mathscr{E}, s', s)$ denote the reflexive transitive closure of $\exists i \in \mathscr{E}.B(i, s', s)$, which can be defined with a second-order formula:

$$C(\mathscr{E}, s', s) \overset{def}{=}$$
$$\forall P.\forall u\, P(u, u) \wedge \forall i \in \mathscr{E}, u, v, w[P(u, v) \wedge B(i, v, w) \supset P(u, w)] \supset P(s', s).$$

5. The agents commonly know ϕ:

$$\mathbf{CKnows}(\phi(now), s) \overset{def}{=} \forall s'.C(\mathscr{A}, s', s) \supset \phi(s'),$$

where $\mathscr{A}$ is the set of all agents.

21.4 Properties of Beliefs

In this section, we analyze properties of beliefs in our formalism. We begin with the main property of beliefs. We use $\psi_0(a, s)$ to denote the following formula:

$$\forall i.\mathbf{Bel}(i, \exists s^*\exists a^*.A(i, a^*, a, s) \wedge now = do(a^*, s^*) \wedge Poss(a^*, s^*) \wedge \mathscr{D}_{ss}[a^*, s^*], do(a, s)),$$

where $\mathscr{D}_{ss}[a^*, s^*]$ denotes the instantiation of the SSAs for ordinary fluents wrt a^* and s^*. This says that in the situation resulting from doing action a, each agent i believes that some alternative of a was possible and has happened. We use $\psi_{n+1}(a, s)$ to denote the following formula:

$$\forall i.\mathbf{Bel}(i, \exists s^*\exists a^*.A(i, a^*, a, s) \wedge now = do(a^*, s^*) \wedge$$
$$Poss(a^*, s^*) \wedge \mathscr{D}_{ss}[a^*, s^*] \wedge \Psi_n(a^*, s^*), do(a, s)).$$

Thus $\Psi_1(a, s)$ says that in the situation resulting from doing action a, each agent i believes that some alternative a^* of a was possible, has happened, and in the resulting situation, each agent believes that some alternative of a^* was possible and has happened. By the SSA for the B fluent, it is straightforward to prove:

Theorem 21.1 *For all n, $\mathscr{D} \models \forall a\forall s.Poss(a, s) \supset \Psi_n(a, s)$.*

Proof We prove by induction on n.
  Basis: $n = 0$. This directly follows from the SSA for the B fluent.
  Induction step: Assume that $\mathscr{D} \models \forall a\forall s.Poss(a, s) \supset \Psi_n(a, s)$. By the SSA for the B fluent, we have

$$\forall a\forall s.Poss(a, s) \supset \forall i.\mathbf{Bel}(i, \exists s^*\exists a^*.A(i, a^*, a, s) \wedge now = do(a^*, s^*) \wedge$$
$$Poss(a^*, s^*) \wedge \mathscr{D}_{ss}[a^*, s^*], do(a, s)).$$

By the induction hypothesis, we have

$$\forall a\forall s.Poss(a, s) \supset \forall i.\mathbf{Bel}(i, \exists s^*\exists a^*.A(i, a^*, a, s) \wedge now = do(a^*, s^*) \wedge$$
$$Poss(a^*, s^*) \wedge \mathscr{D}_{ss}[a^*, s^*] \wedge \Psi_n(a^*, s^*), do(a, s)),$$

which is $\forall a\forall s.Poss(a, s) \supset \Psi_{n+1}(a, s)$. $\square$

  Let $\phi(s)$ be a formula with a single situation variable s. We introduce an epistemic action $observe_\phi$, which tells the agent that ϕ holds in the current situation. The axiom is: $Poss(observe_\phi, s) \equiv \phi(s)$. It is easy to prove the following propositions. By an objective formula, we mean one which does not use the B fluent or the A predicate.

Proposition 21.1 *Let ϕ be an objective formula. Suppose that agent i is an observer of action $observe_\phi$ in situation σ, i.e., $\mathscr{D} \models \forall a.A(i, a, observe_\phi, \sigma) \equiv a = observe_\phi$. Then $\mathscr{D} \models \phi(\sigma) \supset \mathbf{Bel}(i, \phi, do(observe_\phi, \sigma))$.*

Proposition 21.2 *Let ϕ be an objective formula. Suppose that agent i is a* partial observer *of action $observe_\phi$ in situation σ, i.e.,*
$$\mathscr{D} \models \forall a.A(i, a, observe_\phi, \sigma) \equiv a = observe_\phi \vee a = observe_{\neg\phi}.$$
Then $\mathscr{D} \models \phi(\sigma) \wedge \neg\textbf{BW}(i, \phi, \sigma) \supset \neg\textbf{BW}(i, \phi, do(observe_\phi, \sigma))$.

Proposition 21.3 *Let ϕ be an objective formula. Suppose that agent i is* oblivious *of action α in situation σ, i.e., $\mathscr{D} \models \forall a.A(i, a, \alpha, \sigma) \equiv a = nil$.*
Then $\mathscr{D} \models Poss(\alpha, \sigma) \supset [\textbf{Bel}(i, \phi, \sigma) \equiv \textbf{Bel}(i, \phi, do(\alpha, \sigma))]$.

In the following, we show how we model some special types of actions and prove the desired properties. We first consider public sensing and reading actions: we say a sensing or reading action is public if its occurrence is common knowledge but only the performer of the action gets to know the result. The axioms are as follows:

- $sense_\phi(i, \textbf{x}, e)$ means agent i senses the truth value of $\phi(\textbf{x})$ and gets result e;
- $read_f(i, \textbf{x}, y)$ means agent i reads the value of $f(\textbf{x})$ and gets result y.

1. $Poss(sense_\phi(i, \textbf{x}, e), s) \equiv (e = 1 \equiv \phi(\textbf{x}, s))$
2. $A(j, a, sense_\phi(i, \textbf{x}, e), s) \equiv \exists e'(a = sense_\phi(i, \textbf{x}, e')) \wedge (j = i \supset a = sense_\phi(i, \textbf{x}, e))$
3. $Poss(read_f(i, \textbf{x}, y), s) \equiv f(\textbf{x}, s) = y$
4. $A(j, a, read_f(i, \textbf{x}, y), s) \equiv \exists y'(a = read_f(i, \textbf{x}, y')) \wedge (j = i \supset a = read_f(i, \textbf{x}, y))$

We let $sense_\phi(i, \textbf{x})$ denote $sense_\phi(i, \textbf{x}, 1) \mid sense_\phi(i, \textbf{x}, 0)$, and $read_f(i, \textbf{x})$ denote $(\pi y)read_f(i, \textbf{x}, y)$. It is easy to prove:

Proposition 21.4 *$\mathscr{D}$ entails the following:*

1. $Do(sense_\phi(i, \textbf{x}), s, s_1) \supset [B(j, s', s_1) \equiv$
$\exists s^*.B(j, s^*, s) \wedge Do(sense_\phi(i, \textbf{x}), s^*, s') \wedge (j = i \supset \phi(\textbf{x}, s) \equiv \phi(\textbf{x}, s^*))]$
2. $Do(read_f(i, \textbf{x}), s, s_1) \supset [B(j, s', s_1) \equiv$
$\exists s^*.B(j, s^*, s) \wedge Do(read_f(i, \textbf{x}), s^*, s') \wedge (j = i \supset f(\textbf{x}, s) = f(\textbf{x}, s^*))]$

This is the same as Shapiro et al.'s extension of Scherl and Levesque's SSA for the K fluent to public sensing and reading actions in the multi-agent case [22]. So our account of beliefs and actions subsumes theirs. As an easy corollary, we get

Proposition 21.5 *Let ϕ be an objective formula. Then $\mathscr{D}$ entails the following:*

1. $Do(sense_\phi(i, \textbf{x}), s, s_1) \supset \textbf{BW}(i, \phi(\textbf{x}), s_1) \wedge$
$(j \neq i \supset \textbf{Bel}(j, \textbf{BW}(i, \phi(\textbf{x})), s_1))$
2. $Do(read_f(i, \textbf{x}), s, s_1) \supset \exists y\textbf{Bel}(i, f(\textbf{x}) = y, s_1) \wedge$
$(j \neq i \supset \textbf{Bel}(j, \exists y\textbf{Bel}(i, f(\textbf{x}) = y), s_1))$

Bacchus et al. [1] considered noisy sensors: when an agent reads the value of $f(\textbf{x})$, she may get a value y such that $|f(\textbf{x}, s) - y| \leq b$ for some bound b. We introduce an epistemic action $nread_f(i, \textbf{x}, y)$ for this purpose, and let $nread_f(i, \textbf{x})$ denote $(\pi y)nread_f(i, \textbf{x}, y)$. The axioms are:

1. $Poss(nread_f(i, \mathbf{x}, y), s) \equiv |f(\mathbf{x}, s) - y| \leq b$
1. $A(j, a, nread_f(i, \mathbf{x}, y), s) \equiv (\exists y')a = nread_f(i, \mathbf{x}, y') \wedge$
$\quad \{j = i \supset (\exists y').a = nread_f(i, \mathbf{x}, y') \wedge |y - y'| \leq b\}$

As desired, we have

Proposition 21.6 $\mathscr{D} \models Do(nread_f(i, \mathbf{x}), s, s') \supset \exists y.\mathbf{Bel}(i, |f(\mathbf{x}) - y| \leq b, s')$.

Delgrande and Levesque [9] considered unintended actions: an agent wants to push button m, but she may push button n such that $|m - n| \leq b$. We introduce a physical action $npush(i, m, n)$, meaning that agent i wants to push button m but ends up pushing button n. We let $npush(i, m)$ denote $(\pi n)npush(i, m, n)$. The axioms are:

1. $Poss(npush(i, m, n), s) \equiv |m - n| \leq b$
2. $on(n, do(a, s)) \equiv \exists i, m.a = npush(i, m, n)$
3. $A(j, a, npush(i, m, n), s) \equiv (\exists m', n')a = npush(i, m', n') \wedge$
$\quad \{j = i \supset (\exists n').a = push(i, m, n')\}$

Proposition 21.7 $\mathscr{D} \models Do(npush(i, m), s, s') \supset \mathbf{Bel}(i, \exists n.|n - m| \leq b \wedge on(n), s')$.

Dynamic epistemic logics originated with public announcement logic, which reasons about the epistemic change brought about by public communications [17]. We have the following description for the action of publicly truthfully announcing ϕ:

1. $Poss(pub_\phi, s) \equiv \phi(s)$
2. $A(i, a, pub_\phi, s) \equiv a = pub_\phi$

Proposition 21.8 *Let ϕ be an objective formula. Then*
$\mathscr{D} \models \phi(s) \supset \mathbf{CKnows}(\phi, do(pub_\phi, s))$.

21.5 The Embedding Theorem

In this section, we prove that action model logic can be embedded into the extended situation calculus. We first define two functions: $\mathscr{B}$, which maps a formula ϕ in AML into a basic action theory encoding the action models involved in ϕ, and $\mathscr{F}$, which maps a formula in AML and a situation term into a formula in the situation calculus. We then prove that for any formula ϕ in AML, ϕ is valid in AML iff $\mathscr{B}(\phi) \models \mathscr{F}(\phi, S_0)$.

Definition 21.7 Let ϕ be a formula in AML. We define two sets $AM(\phi)$ and $Prop(\phi)$ recursively as follows:

1. $AM(\phi)$ is the set of action models N such that N appears in ϕ or there exist an action model N' which occurs in ϕ and an action point e of N' such that $N \in AM(pre(e))$;

2. $Prop(\phi)$ is the set of propositions p such that p appears in ϕ or there exist an action model N' which occurs in ϕ and an action point e of N' such that $p \in Prop(pre(e))$.

We define the vocabulary of our situation calculus language associated to ϕ as follows. For each $N \in AM(\phi)$, we introduce an action $c_N(x)$, where x ranges over the action points of N. For each $p \in Prop(\phi)$, we introduce a unary fluent $p(s)$.

Definition 21.8 Let ϕ be a formula in AML, and s a situation term. We define a situation calculus formula $\mathcal{F}(\phi, s)$ by induction on the complexity of ϕ as follows:

1. $\mathcal{F}(p, s) = p(s)$;
2. $\mathcal{F}(\neg\psi, s) = \neg\mathcal{F}(\psi, s)$;
3. $\mathcal{F}(\psi \wedge \eta, s) = \mathcal{F}(\psi, s) \wedge \mathcal{F}(\eta, s)$;
4. $\mathcal{F}(B_i\psi, s) = \forall s'.B(i, s', s) \supset \mathcal{F}(\psi, s')$;
5. $\mathcal{F}(C_{\mathcal{E}}\psi, s) = \forall s'.C(\mathcal{E}, s', s) \supset \mathcal{F}(\psi, s')$;
6. $\mathcal{F}([N, e_0]\psi, s) = \mathcal{F}(pre(e_0), s) \supset \mathcal{F}(\psi, do(c_N(e_0), s))$.

Note that we model the execution of a pointed action model (N, e_0) with the execution of the action $c_N(e_0)$ in the situation calculus.

Definition 21.9 Let ϕ be a formula in AML. Let $AM(\phi) = \{N_1, \dots, N_m\}$, where $N_k = (A_k, \rightarrow, pre)$, $k = 1, \dots, m$. Without loss of generality, we assume that the A_k's are pairwise disjoint. We construct a basic action theory $\mathcal{B}(\phi)$ as follows.

(A0) $e \neq e'$, where e and e' are distinct action points;
(A1) $Poss(c_{N_k}(x), s) \equiv \bigvee_{e \in A_k}[x = e \wedge \mathcal{F}(pre(e), s)]$;
(A2) $p(do(a, s)) \equiv p(s)$;
(A3) $A(j, a, c_{N_k}(x), s) \equiv \exists y.a = c_{N_k}(y) \wedge \bigvee_{(e,e') \in \rightarrow_j} x = e \wedge y = e'$;

Note that the reason we have A2 is that in action models, actions do not change the world. A3 specifies that the value of the A predicate is set according to the accessibility relations of the action models.

The following is the embedding theorem:

Theorem 21.2 *For any formula ϕ in AML, ϕ is valid in AML iff $\mathcal{B}(\phi) \models \mathcal{F}(\phi, S_0)$.*

To prove the embedding theorem, we first introduce a method to induce a Kripke model from a structure of the situation calculus. For a structure L and a syntactic object o, we let o^L stand for the denotation of o in L. We say that a situation is at level k if it results from performing a sequence of k actions in an initial situation. Let L be a structure of the situation calculus, τ a situation of L, and $\phi(s)$ a situation calculus formula with a free situation variable s. We use $L, \tau \models \phi(s)$ to denote that when s is interpreted as τ, L satisfies $\phi(s)$.

Definition 21.10 Let L be a structure of the situation calculus. Let τ be a level k situation of L. We define a Kripke model $M_L^\tau = (S, R, V)$ as follows:

- S consists of all level k situations of L;
- For any $t_1, t_2 \in S$, $t_1 R_i t_2$ iff $(i, t_2, t_1) \in B^L$;
- For each $t \in S$, $V(t)$ is the set of fluents p which are true at t in L.

We call M_L^τ the Kripke projection of L onto τ. Note that if τ_1 and τ_2 are at the same level, then $M_L^{\tau_1} = M_L^{\tau_2}$.

The proof of the embedding theorem is involved. We now explain the general idea of the proof. First, suppose $\mathscr{B}(\phi) \not\models \mathscr{F}(\phi, S_0)$. Let L be a model of $\mathscr{B}(\phi) \cup \{\neg \mathscr{F}(\phi, S_0)\}$. We show that the pointed Kripke model $(M_L^{\tau_0}, \tau_0)$ satisfies $\neg\phi$, where $\tau_0 = S_0^L$, and $M_L^{\tau_0}$ is the Kripke projection of L onto τ_0. The other direction is more complicated, and the reason is that for an action point e of an action model N, $pre(e)$ may involve action models. The precondition axiom for action C_N is defined with $\mathscr{F}(pre(e), s)$. When $pre(e)$ involves action models, $\mathscr{F}(pre(e), s)$ refers to future situations of s.

Now suppose ϕ is not valid. Let (M, t_0) be a model of $\neg\phi$. We construct a model L of $\mathscr{D} = \mathscr{B}(\phi)$ as follows. First, let $L \models \mathscr{D}_{una} \cup \{A0\}$. The initial situations of L are the states of M, and L interprets S_0 as t_0. The situations of L form a forest of isomorphic trees rooted at the initial situations, where the children of each situation one-to-one correspond to the actions. We interpret the A predicate according to A3. We interpret $Poss$ and the fluents by induction on the level of situations. For initial situations, we interpret the B and the p fluents according to M. Let τ be an initial situation. For each action model N and action point e of it, we let $L, \tau \models Poss(c_N(e), s)$ iff $M, \tau \models pre(e)$. Assume we have interpreted $Poss$ and all the fluents at level k situations. We interpret the fluents at level $k + 1$ situations according to $\mathscr{D}_{ss}$. Let τ be a level $k + 1$ situation. For each action model N and action point e of it, we let $L, \tau \models Poss(c_N(e), s)$ iff $M_L^\tau, \tau \models pre(e)$. We show that L is a model of $\mathscr{B}(\phi) \cup \{\neg \mathscr{F}(\phi, S_0)\}$.

In the above outline of the proof, for the $\mathscr{L}_{sc}$-structure L that we construct, we have that the Kripke projection of L onto the initial situations—$M_L^{t_0}$, is isomorphic to M. The isomorphism of Kripke models will play an important role in our proof. However, instead of requiring that two pointed Kripke models (M_1, t_1) and (M_2, t_2) be isomorphic, we only require that their reductions be isomorphic, i.e., the resulting pointed Kripke models are isomorphic after we remove the states of M_i not reachable from t_i for $i = 1, 2$. In the following, we define the concepts of isomorphism and reduction of Kripke models and study their basic properties.

Let (M_1, t_1) and (M_2, t_2) be two pointed Kripke models. We let $h : (M_1, t_1) \cong (M_2, t_2)$ denote that h is an isomorphism from (M_1, t_1) to (M_2, t_2), i.e.,

- h is a bijection from the states of M_1 to those of M_2, and $h(t_1) = t_2$;
- h preserves the accessibility relation, i.e., for any agent i and any two states t and t' of M_1, $t R_i t'$ iff $h(t) R_i h(t')$;
- h preserves the atoms, that is, for any atom p and state t of M_1, p holds at t iff p holds at $h(t)$.

Let (M, t_0) be a pointed Kripke model, we use $\mathscr{R}(M, t_0)$ to denote the pointed Kripke model (M', t_0), where M' is obtained from M by removing those states not reachable from t_0. It is easy to prove the following properties:

Proposition 21.9 *Let $h : \mathscr{R}(M_1, t_1) \cong \mathscr{R}(M_2, t_2)$, and t a state of M_1 reachable from t_1. Then $\mathscr{R}(M_1, t) \cong \mathscr{R}(M_2, h(t))$.*

Proof Since t is reachable from t_1, the states of $\mathscr{R}(M_1, t)$ are contained in those of $\mathscr{R}(M_1, t_1)$. It suffices to prove that for any state t' of M_1 reachable from t, $h(t')$ is reachable from $h(t)$, which holds because h preserves the accessibility relation. □

Proposition 21.10 *Let $h : \mathscr{R}(M_1, t_1) \cong \mathscr{R}(M_2, t_2)$. Then for any AML formula ϕ, $M_1, t_1 \models \phi$ iff $M_2, t_2 \models \phi$.*

Proof See Appendix.   □

In the above outline of proof, the $\mathscr{L}_{sc}$-structure L we construct is a model of $\mathscr{D} - \mathscr{D}_{ap}$ and it has a property defined as follows:

Definition 21.11 We say that an $\mathscr{L}_{sc}$-structure L has property C1 if for any situation τ, action model N and action point e of N, $L, \tau \models Poss(c_N(e), s)$ iff $M_L^\tau, \tau \models pre(e)$.

In the following, we study properties of models of $\mathscr{D} - \mathscr{D}_{ap}$ with C1. We first prove a proposition which shows that the execution of a pointed action model (N, e_0) can be modeled in the situation calculus with the execution of the action $c_N(e_0)$.

Proposition 21.11 *Let L be a model of $\mathscr{D} - \mathscr{D}_{ap}$ with C1, and τ_0 a situation of L. Let (M, t_0) be a pointed Kripke model, and (N, e_0) a pointed action model. Suppose that $h : \mathscr{R}(M, t_0) \cong \mathscr{R}(M_L^{\tau_0}, \tau_0)$ and $M, t_0 \models pre(e_0)$. Then $\mathscr{R}((M, t_0) \otimes (N, e_0)) \cong \mathscr{R}(M_L^{\tau_1}, \tau_1)$ where τ_1 is $do^L(c_N(e_0)^L, \tau_0)$.*

Proof See Appendix.   □

Next, we prove a lemma which shows that an AML formula ϕ can be modeled by the situation calculus formula $\mathscr{F}(\phi, s)$:

Lemma 21.1 *Let L be a model of $\mathscr{D} - \mathscr{D}_{ap}$ with C1, and τ_0 a situation of L. Suppose $h : \mathscr{R}(M, t_0) \cong \mathscr{R}(M_L^{\tau_0}, \tau_0)$. Then $M, t_0 \models \phi$ iff $L, \tau_0 \models \mathscr{F}(\phi, s)$.*

Proof We prove by induction on the complexity of ϕ.

1. ϕ is p. Then $M, t_0 \models p$ iff p is true at t_0 in M iff p is true at τ_0 in L (since $\mathscr{R}(M, t_0) \cong \mathscr{R}(M_L^{\tau_0}, \tau_0)$) iff $L, \tau_0 \models p(s)$.
2. ϕ is $\neg\psi$. Then $M, t_0 \models \neg\psi$ iff $M, t_0 \not\models \psi$ iff $L, \tau_0 \not\models \mathscr{F}(\psi, s)$ (by induction hypothesis) iff $L, \tau_0 \models \neg\mathscr{F}(\psi, s)$, i.e., $L, \tau_0 \models \mathscr{F}(\neg\psi, s)$.
3. ϕ is $\psi \wedge \psi'$. Similar to the above case.

4. ϕ is $B_i\psi$. Let t be a state of M such that t_0R_it. By Proposition 21.9, $\mathscr{R}(M, t) \cong$ $\mathscr{R}(M_L^{\tau_0}, h(t))$. By induction hypothesis, $M, t \models \psi$ iff $L, h(t) \models \mathscr{F}(\psi, s')$. So $M, t_0 \models B_i\psi$ iff for every t such that t_0R_it, $M, t \models \psi$ iff for every τ such that $(i, \tau, \tau_0) \in B^L$, $L, \tau \models \mathscr{F}(\psi, s')$ iff $L, \tau_0 \models \forall s'.B(i, s', s) \supset \mathscr{F}(\psi, s')$, i.e., $L, \tau_0 \models \mathscr{F}(B_i\psi, s)$.

5. ϕ is $C_{\mathscr{E}}\psi$. Similar to the above case.

6. ϕ is $[N, e_0]\psi$. By induction hypothesis, $M, t_0 \models pre(e_0)$ iff $L, \tau_0 \models \mathscr{F}(pre(e_0), s)$. By Proposition 21.11, if $M, t_0 \models pre(e_0)$, then $(M, t_0) \otimes (N, e_0) \cong M_L^{\tau_1}, \tau_1$, where $\tau_1 = do(c_N(e_0)^L, \tau_0)$. By induction hypothesis, $(M, t_0) \otimes (N, e_0) \models \psi$ iff $L, \tau_1 \models \mathscr{F}(\psi, s)$. Thus $M, t_0 \models [N, e_0]\psi$ iff if $M, t_0 \models pre(e_0)$ then $(M, t_0) \otimes (N, e_0) \models \psi$ iff if $L, \tau_0 \models \mathscr{F}(pre(e_0), s)$ then $L, \tau_1 \models \mathscr{F}(\psi, s)$ iff $L, \tau_0 \models \mathscr{F}(pre(e_0), s) \supset \mathscr{F}(\psi, do(c_N(e_0), s))$, which is $\mathscr{F}([N, e_0]\psi, s)$. $\square$

The above lemma requires that L be a model of $\mathscr{D} - \mathscr{D}_{ap}$ with C1. The lemma below shows that we can replace this requirement with the one that L is a model of $\mathscr{D}$.

Lemma 21.2 *Let L be a model of $\mathscr{D}$, and τ_0 a situation of L. Suppose h : $\mathscr{R}(M, t_0) \cong \mathscr{R}(M_L^{\tau_0}, \tau_0)$. Then $M, t_0 \models \phi$ iff $L, \tau_0 \models \mathscr{F}(\phi, s)$.*

Proof We prove by induction on the complexity of ϕ. Assume that the statement holds for all formulas less complex than ϕ. Then for any situation τ and action point e of action model N, since $pre(e)$ is less complex than ϕ and $\mathscr{R}(M_L^{\tau}, \tau) \cong \mathscr{R}(M_L^{\tau}, \tau)$, we have that $M_L^{\tau}, \tau \models pre(e)$ iff $L, \tau \models \mathscr{F}(pre(e), s)$. By $\mathscr{D}_{ap}$, $L, \tau \models Poss(c_N(e), s) \equiv \mathscr{F}(pre(e), s)$. Thus we have $L, \tau \models Poss(c_N(e), s)$ iff $M_L^{\tau}, \tau \models pre(e)$. So L satisfies C1. By applying Lemma 21.1, we have $M, t_0 \models \phi$ iff $L, \tau_0 \models \mathscr{F}(\phi, s)$. $\square$

Finally, we are ready to prove the embedding theorem:

Proof First, suppose $\mathscr{B}(\phi) \not\models \mathscr{F}(\phi, S_0)$. Let L be a model of $\mathscr{B}(\phi) \cup \{\neg\mathscr{F}(\phi, S_0)\}$. Let $\tau_0 = S_0^L$. Since $\mathscr{R}(M_L^{\tau_0}, \tau_0) \cong \mathscr{R}(M_L^{\tau_0}, \tau_0)$, by Lemma 21.2, $M_L^{\tau_0}, \tau_0 \models \phi$ iff $L \models \mathscr{F}(\phi, S_0)$. Thus $M_L^{\tau_0}, \tau_0 \models \neg\phi$. So ϕ is not valid.

Now suppose ϕ is not valid. Let (M, t_0) be a model of $\neg\phi$. We construct a model L of $\mathscr{D} = \mathscr{B}(\phi)$ as follows. First, let $L \models \mathscr{D}_{una} \cup \{A0\}$. The initial situations of L are the states of M, and L interprets S_0 as t_0. The situations of L form a forest of isomorphic trees rooted at the initial situations, where the children of each situation one-to-one correspond to the actions. Thus L satisfies the foundational axioms F1, F2, and F3. We interpret the A predicate according to A3. We now interpret $Poss$ and the fluents by induction on the level of situations:

1. τ is an initial situation. The B fluent restricted to the initial situations is exactly the same as the accessibility relation of M. For each unary fluent p, L interprets p at τ as M does. For each action model N and action point e of it, we let $L, \tau \models Poss(c_N(e), s)$ iff $M, \tau \models pre(e)$.

2. Assume we have interpreted $Poss$ and all the fluents at level k situations. We interpret the fluents at level $k + 1$ situations according to $\mathscr{D}_{ss}$. Let τ be a level $k + 1$ situation. For each action model N and action point e of it, we let $L, \tau \models Poss(c_N(e), s)$ iff $M_L^\tau, \tau \models pre(e)$. Recall that the Kripke model $M_L^\tau = (S, R, V)$ is defined as follows:

- S consists of all level $k + 1$ situations of L;
- For any $t_1, t_2 \in S$, $t_1 R_i t_2$ iff $(i, t_2, t_1) \in B^L$;
- For each $t \in S$, $V(t)$ is the set of fluents p which are true at t in L.

By the SSA for B, it is easy to see that L satisfies F4: $B(i, s, s') \supset [Init(s) \equiv Init(s')]$. Also, L has property C1: for any situation τ, action model N and action point e of N, $L, \tau \models Poss(c_N(e), s)$ iff $M_L^\tau, \tau \models pre(e)$. So L is a model of $\mathscr{D} - \mathscr{D}_{ap}$ with C1. We now prove that $L \models \mathscr{D}_{ap}$. Since $\mathscr{R}(M_L^\tau, \tau) \cong \mathscr{R}(M_L^\tau, \tau)$, by Lemma 21.1, $M_L^\tau, \tau \models pre(e)$ iff $L, \tau \models \mathscr{F}(pre(e), s)$. Thus $L, \tau \models Poss(c_N(e), s)$ iff $L, \tau \models \mathscr{F}(pre(e), s)$. So $L, \tau \models Poss(c_N(e), s) \equiv \mathscr{F}(pre(e), s)$.

So we have proved that L is a model of $\mathscr{B}(\phi)$. Obviously, we have $\mathscr{R}(M, t_0) \cong \mathscr{R}(M_L^{t_0}, t_0)$. By Lemma 21.2, $M, t_0 \models \phi$ iff $L \models \mathscr{F}(\phi, S_0)$. Recall that (M, t_0) is a model of $\neg\phi$. So L is a model of $\mathscr{B}(\phi) \cup \{\neg\mathscr{F}(\phi, S_0)\}$. Thus $\mathscr{B}(\phi) \not\models \mathscr{F}(\phi, S_0)$. □

21.6 Extended Examples

In this section, we present two extended examples of modeling multi-agent scenarios in the situation calculus. In the first example, the role of each agent is not common knowledge. The second one involves both physical and sensing actions.

Example 21.4 Ann senses the truth value of p. Bob and Carol are observing Ann. But Ann doesn't know the role of Bob or Carol. Bob and Carol do not know the role of each other. We introduce an epistemic action $obs(e, b, c)$, which means that Ann senses the truth value of p with result e, and $b = 1$ (resp. $c=1$) iff Bob (resp. Carol) is observing Ann. The axioms are as follows:

1. $Poss(obs(e, b, c), s) \equiv (e = 1 \equiv p(s))$
2. $A(i, a, obs(e, b, c), s) \equiv$
   $[i = ann \supset (\exists b', c')a = obs(e, b', c')] \wedge$
   $[i = bob \supset (b = 0 \supset a = nil) \wedge (b = 1 \supset (\exists e', c')a = obs(e', b, c'))] \wedge$
   $[i = carol \supset (c = 0 \supset a = nil) \wedge (c = 1 \supset (\exists e', b')a = obs(e', b', c))]$
3. $A(i, a, nil, s) \equiv a = nil$

The reason we have $[i = ann \supset (\exists b', c')a = obs(e, b', c')]$ is that Ann knows the sensing result but she doesn't know the role of Bob or Carol. The reason we have $[i = bob \wedge b = 1 \supset (\exists e', c')a = obs(e', b, c')]$ is that Bob is observing Ann but he does not know the role of Carol.

Assume that $\mathscr{D}_{S_0}$ contains $p(S_0) \wedge \mathbf{CKnows}(\forall i \neg \mathbf{BW}(i, p), S_0)$. Then $\mathscr{D}$ entails the following, where $S_1 = do(obs(1, 1, 1), S_0)$.

1. **BW**(ann, p, S_1);
2. $\neg$**BW**(bob, p, S_1);
3. **Bel**$(bob, \mathbf{BW}(ann, p), S_1)$;
4. $\neg$**Bel**$(ann, \mathbf{Bel}(bob, \mathbf{BW}(ann, p)), S_1)$;
5. $\neg$**Bel**$(carol, \mathbf{Bel}(bob, \mathbf{BW}(ann, p)), S_1)$.

Example 21.5 We use a simplified and adapted version of Levesque's Squirrel World. Squirrels and acorns live in a one-dimensional world unbounded on both sides. Each acorn and squirrel is located at some point, and each point can contain any number of squirrels and acorns. Acorns are completely passive. Squirrels can do the following actions:

1. $left(i)$: Squirrel i moves left a unit;
2. $right(i)$: Squirrel i moves right a unit;
3. $pick(i)$: Squirrel i picks up an acorn, which is possible when he is not holding an acorn and there is at least one acorn at his location;
4. $drop(i)$: Squirrel i drops the acorn he is holding;
5. $learn(i, n)$: Squirrel i learns that there are n acorns at his location. We use $smell(i)$ to denote $(\pi n)learn(i, n)$.

A squirrel can observe the action of another squirrel within a distance of 4, but if the action is a sensing action, the result is not observable. Initially, there are two acorns at each point. There are three squirrels: Nutty, Edgy, and Wally. Initially, they are all at point 0, holding no acorns, and have no knowledge of the number of acorns at each point, and the above is common knowledge. There are three ordinary fluents:

1. $hold(i, s)$: Squirrel i is holding an acorn in situation s;
2. $loc(i, p, s)$: Squirrel i is at location p in situation s;
3. $acorn(p, n, s)$: There are n acorns at location p in situation s.

For illustration, we only present some axioms of $\mathscr{D}$:

1. $Poss(pick(i), s) \equiv \neg hold(i, s) \wedge \exists p, n(loc(i, p, s) \wedge acorn(p, n, s) \wedge n > 0)$
2. $loc(i, p, do(a, s)) \equiv a = left(i) \wedge loc(i, p + 1, s) \vee$
   $\quad a = right(i) \wedge loc(i, p - 1, s) \vee loc(i, p, s) \wedge a \neq left(i) \wedge a \neq right(i)$
3. $A(j, a, pick(i), s) \equiv \exists p, p'[loc(i, p, s) \wedge loc(j, p', s) \wedge$
   $\quad (|p - p'| > 4 \supset a = nil) \wedge (|p - p'| \leq 4 \supset a = pick(i))]$
4. $A(j, a, learn(i, n), s) \equiv \exists p, p'[loc(i, p, s) \wedge loc(j, p', s) \wedge$
   $\quad (|p - p'| > 4 \supset a = nil) \wedge (j = i \supset a = learn(i, n))$
   $\quad (|p - p'| \leq 4 \wedge j \neq i \supset (\exists n')a = learn(i, n'))]$
5. **CKnows**$(\forall i.loc(i, 0) \wedge \neg hold(i) \wedge \forall p, n\neg\mathbf{Bel}(i, \neg acorn(p, n)), S_0)$
6. $\forall p.acorn(p, 2, S_0)$

Let $\phi(s, s')$ be a formula. We introduce the following abbreviation:

$$\mathbf{Bel}(i, \phi(now, prev), s) \stackrel{def}{=} \forall s'.B(i, s', s) \supset \exists s^*\exists a^*.s' = do(a^*, s^*) \wedge \phi(s', s^*).$$

We abbreviate Nutty, Edgy, and Wally with N, E, and W, respectively. Let $\delta_1 = smell(N); pick(N)$, $\delta_2 = right(N); drop(N)$, $\delta_3 = left(W)^2; right(E)^3$, and $\delta_4 = smell(W); pick(W); left(W); left(E)$. Then $\mathscr{D}$ entails the following:

1. $Do(\delta_1, S_0, s) \supset \mathbf{TBel}(N, acorn(0, 1), s) \wedge$
   $\mathbf{CKnows}(hold(N) \wedge \exists n \mathbf{TBel}(N, acorn(0, n)), s)$.
2. $Do(\delta_1; \delta_2, S_0, s) \supset \mathbf{CKnows}(\exists n(acorn(1, n, prev) \wedge acorn(1, n+1, now)), s)$.
   This says that the squirrels commonly know that there is one more acorn at point 1 now than previously.
3. $Do(\delta_1; \delta_2; \delta_3, S_0, s) \supset \mathbf{CKnows}(loc(W, -2) \wedge loc(N, 1) \wedge loc(E, 3), s)$.
4. $Do(\delta_1; \delta_2; \delta_3; \delta_4, S_0, s) \supset \mathbf{TBel}(N, hold(W) \wedge loc(W, -3) \wedge loc(E, 2), s) \wedge$
   $\mathbf{Bel}(E, \neg hold(W) \wedge loc(W, -2), s) \wedge \mathbf{Bel}(W, loc(E, 3), s)$.
   Note that now Edgy and Wally have incorrect beliefs about each other.

21.7 Conclusions

In this paper, by incorporating the idea of action models from DELs, we have developed a general multi-agent extension of the situation calculus. We analyzed properties of multi-agent beliefs in the situation calculus, and showed that we can provide a uniform treatment of special types of actions, such as public sensing and reading actions, noisy sensors and unintended actions, and public announcements. We showed that action model logic can be embedded into the situation calculus, and hence any multi-agent scenario which can be modeled in action model logic can be modeled in the situation calculus. Since DELs are propositional, an advantage of our work is the gain of more expressiveness and compactness in representation. We gave two extended examples to illustrate the modeling of multi-agent scenarios in the situation calculus.

There are a number of topics for future research. First of all, as mentioned in the introduction, van Benthem et al. [8] generalized the concept of action model to that of update model which can be used to model both epistemic and physical actions. They proposed a logic, called the logic of communication and change (LCC), to reason about update models. It would be interesting to explore if we can embed LCC into the situation calculus. Secondly, as shown in the Squirrel World example, because of unreliable sources of information, at certain points, agents may have incorrect beliefs about the world and other agents. When incorrect beliefs lead to inconsistent beliefs, belief revision is necessary for the agents to keep functioning in the world. The DEL community has done extensive work on multi-agent belief revision, and a good reference is [4]. The general idea is this: The semantic model is a plausibility model, where for each agent, there is a plausibility order on the set of states or actions. An agent believes ϕ if ϕ holds in the most plausible states. When we update a plausibility model by an action plausibility model, give priority to the action plausibility order. In the future, we would like to incorporate this line of work into the situation calculus. Thirdly, while the focus of the current paper is on the representation side, in the future, we would like to investigate reasoning in the multi-agent situation calculus. Finally, we would like to explore multi-agent high-level program execution and develop interesting applications of it.

Acknowledgments We thank Johan van Benthem, Hans van Ditmarsch, Alexandru Baltag and Sonja Smets for helpful discussions on the topic of this paper. The first author was supported by the Natural Science Foundation of China under Grant No. 61073053.

Appendix

Proposition 21.10 *Let* $h : \mathcal{R}(M_1, t_1) \cong \mathcal{R}(M_2, t_2)$. *Then for any AML formula* ϕ, $M_1, t_1 \models \phi$ *iff* $M_2, t_2 \models \phi$.

Proof We prove by induction on the complexity of ϕ. The cases that ϕ is p, $\neg\psi$, or $\psi \wedge \psi'$ are easy. We prove the remaining cases:

1. ϕ is $B_i\psi$. Let t be a state of M such that $t_1 R_i t$. By Proposition 21.9, $\mathcal{R}(M_1, t) \cong \mathcal{R}(M_2, h(t))$. By induction hypothesis, $M_1, t \models \psi$ iff $M_2, h(t) \models \psi$. So $M_1, t_1 \models B_i\psi$ iff for every t such that $t_1 R_i t$, $M_1, t \models \psi$ iff for every t' such that $t_2 R_i t'$, $M_2, t' \models \psi$ iff $M_2, t_2 \models B_i\psi$.
2. ϕ is $C_{\mathcal{E}}\psi$. Similar to the above case.
3. ϕ is $[N, e_0]\psi$. Let t be a state of M reachable from t_1. By Proposition 21.9, $\mathcal{R}(M_1, t) \cong \mathcal{R}(M_2, h(t))$. Let e be an action point of N. By induction hypothesis, $M_1, t \models pre(e)$ iff $M_2, h(t) \models pre(e)$. Now suppose $M_1, t_1 \models pre(e_0)$. We show that $\mathcal{R}((M_1, t_1) \otimes (N, e_0)) \cong \mathcal{R}((M_2, t_2) \otimes (N, e_0))$. Let (t, e) be a state of $\mathcal{R}((M_1, t_1) \otimes (N, e_0))$. Then $M_1, t \models pre(e)$ and (t, e) is reachable from (t_1, e_0). So $M_2, h(t) \models pre(e)$, and $(h(t), e)$ is reachable from (t_2, e_0) (since $t_2 = h(t_1)$ and h preserves the accessibility relation). Hence $(h(t), e)$ is a state of $\mathcal{R}((M_2, t_2) \otimes (N, e_0))$. Let $g((t, e)) = (h(t), e)$. It is easy to show that g is a bijection from the states of $\mathcal{R}((M_1, t_1) \otimes (N, e_0))$ to those of $\mathcal{R}((M_2, t_2) \otimes (N, e_0))$, g preserves the accessibility relation and the atoms. So if $M_1, t_1 \models pre(e_0)$, then $\mathcal{R}((M_1, t_1) \otimes (N, e_0)) \cong \mathcal{R}((M_2, t_2) \otimes (N, e_0))$. By induction hypothesis, $(M_1, t_1) \otimes (N, e_0) \models \psi$ iff $(M_2, t_2) \otimes (N, e_0) \models \psi$. Thus $M_1, t_1 \models [N, e_0]\psi$ iff if $M_1, t_1 \models pre(e_0)$ then $(M_1, t_1) \otimes (N, e_0) \models \psi$ iff if $M_2, t_2 \models pre(e_0)$ then $(M_2, t_2) \otimes (N, e_0) \models \psi$ iff $M_2, t_2 \models [N, e_0]\psi$. □

Proposition 21.11 *Let* L *be a model of* $\mathcal{D} - \mathcal{D}_{ap}$ *with C1, and* τ_0 *a situation of* L. *Let* (M, t_0) *be a pointed Kripke model, and* (N, e_0) *a pointed action model. Suppose that* $h : \mathcal{R}(M, t_0) \cong \mathcal{R}(M_L^{\tau_0}, \tau_0)$ *and* $M, t_0 \models pre(e_0)$. *Then* $\mathcal{R}((M, t_0) \otimes (N, e_0)) \cong \mathcal{R}(M_L^{\tau_1}, \tau_1)$ *where* τ_1 *is* $do^L(c_N(e_0)^L, \tau_0)$.

Proof To begin with, we show that for any state t of M reachable from t_0 and for any action point e of N, $M, t \models pre(e)$ iff $L, h(t) \models Poss(c_N(e), s)$. Since $h : \mathcal{R}(M, t_0) \cong \mathcal{R}(M_L^{\tau_0}, \tau_0)$, by Proposition 21.9, $\mathcal{R}(M, t) \cong \mathcal{R}(M_L^{\tau_0}, h(t))$. By Proposition 21.10, $M, t \models pre(e)$ iff $M_L^{\tau_0}, h(t) \models pre(e)$. By C1, $L, h(t) \models$

$Poss(c_N(e), s)$ iff $M_L^{\tau_0}, h(t) \models pre(e)$. Thus $M, t \models pre(e)$ iff $L, h(t) \models Poss(c_N(e), s)$.

We define a function g from the states of $\mathscr{R}((M, t_0) \otimes (N, e_0))$ to the situations of L as follows: $g((t, e)) = do^L(c_N(e)^L, h(t))$. We first show that g is an injection. Suppose that $do^L(c_N(e_1)^L, h(t_1)) = do^L(c_N(e_2)^L, h(t_2))$. Then by F1, $c_N(e_1)^L = c_N(e_2)^L$ and $h(t_1) = h(t_2)$. By $\mathscr{D}_{una}$, $e_1^L = e_2^L$. By A0, $e_1 = e_2$. Since h is an injection, $t_1 = t_2$.

We now show that g preserves the accessibility relation. Let (t_1, e_1) and (t_2, e_2) be two states of $\mathscr{R}((M, t_0) \otimes (N, e_0))$. Then $M, t_i \models pre(e_i)$, $i = 1, 2$. So $L, h(t_i) \models Poss(c_N(e_i), s)$. Thus $(t_1, e_1) R_i' (t_2, e_2)$ iff $t_1 R_i t_2$ and $e_1 \rightarrow_i e_2$ iff $(i, h(t_2), h(t_1)) \in B^L$ and $L, h(t_1) \models A(i, c_N(e_2), c_N(e_1), s)$ (by A3) iff $(i, g((t_2, e_2)), g((t_1, e_1))) \in B^L$ (by the SSA for B). Next, we show that g preserves the fluents. For any state (t, e) of $\mathscr{R}((M, t_0) \otimes (N, e_0))$, for any fluent p, p is true at (t, e) iff p is true at t iff p is true at $h(t)$ iff p is true at $do^L(c_N(e)^L, h(t))$, which is $g((t, e))$, by the SSA for p.

We now show that g is a function from the states of $\mathscr{R}((M, t_0) \otimes (N, e_0))$ to those of $\mathscr{R}(M_L^{\tau_1}, \tau_1)$. Let (t, e) be a state of $\mathscr{R}((M, t_0) \otimes (N, e_0))$. Then (t, e) is reachable from (t_0, e_0). Since g preserves the accessibility relation, $g(t, e)$ is reachable from $g(t_0, e_0)$, which is τ_1. Thus $g(t, e)$ is a state of $\mathscr{R}(M_L^{\tau_1}, \tau_1)$.

Finally, we show that g is a surjection. Let ω_1 be a situation of L that is reachable from τ_1 by a B-path. Since $M, t_0 \models pre(e_0)$, $L, \tau_0 \models Poss(c_N(e_0), s)$. By A3 and the SSA for the B fluent, there exist an action point e of N reachable from e_0 and a situation ω_0 of L reachable from τ_0, such that $\omega_1 = do^L(c_N(e)^L, \omega_0)$ and $L, \omega_0 \models Poss(c_N(e), s)$. Since $h : \mathscr{R}(M, t_0) \cong \mathscr{R}(M_L^{\tau_0}, \tau_0)$, there exists a state t of M reachable from t_0 such that $h(t) = \omega_0$. Thus $M, t \models pre(e)$, (t, e) is reachable from (t_0, e_0), and $\omega_1 = g((t, e))$. $\qquad\square$

References

1. Bacchus F, Halpern J, Levesque H (1999) Reasoning about noisy sensors in the situation calculus. Artif Intell 111:171–208
2. Baltag A, Moss L (2004) Logics for epistemic programs. Synthese 139(2):165–224
3. Baltag A, Moss LS, Solecki S (1998) The logic of public announcements, common knowledge, and private suspicions. In: Proceedings of the conference on theoretical aspects of rationality and knowledge (TARK-98)
4. Baltag A, Smets S (2008) A qualitative theory of dynamic interactive belief revision. In: Bonanno G, van der Hoek W, Wooldridge M (eds) Texts in logic and games, vol 3. Amsterdam University Press, Amsterdam
5. Baral C (2010) Reasoning about actions and change: from single agent actions to multi-agent actions. In: Proceedings of the international conference on principles of knowledge representation and reasoning (KR-10)
6. Belle V, Lakemeyer G (2010) Reasoning about imperfect information games in the epistemic situation calculus. In: Proceedings of the AAAI conference on artificial intelligence (AAAI-10).
7. van Benthem J (2011) McCarthy variations in a modal key. Artif Intell 175(1):428–439
8. van Benthem J, van Eijck J, Kooi B (2006) Newblock logics of communication and change. Inf Comput 204(11):1620–1662

9. Delgrande JP, Levesque HJ (2012) Belief revision with sensing and fallibe actions. In: Proceedings of the international conference on principles of knowledge representation and reasoning (KR-12)
10. van Ditmarsch H, van der Hoek W, Kooi B (2007) Dynamic epistemic logic. Springer, New York
11. van Ditmarsch H, Herzig A, de Lima T (2011) From situation calculus to dynamic epistemic logic. J Logic Comput 21(2):179–204
12. Kelly RF, Pearce AR (2008) Complex epistemic modalities in the situation calculus. In: Proceedings of the international conference on principles of knowledge representation and reasoning (KR-08)
13. Lakemeyer G, Levesque HJ (2004) Situations, si! situation terms, no! In: Proceedings of the international conference on principles of knowledge representation and reasoning (KR-04).
14. Lakemeyer G, Levesque HJ (2005) Semantics for a useful fragment of the situation calculus. In: Proceedings of the international joint conference on artificial intelligence (IJCAI-05).
15. Levesque HJ, Reiter R, Lespérance Y, Lin F, Scherl RB (1997) GOLOG: a logic programming language for dynamic domains. J Logic Progr 31(1–3):59–84
16. McCarthy J, Hayes PJ (1969) Some philosophical problems from the standpoint of artificial intelligence. Mach Intell 4:463–502
17. Plaza J (1989) Logics of public communications. In: Proceedings of the 4th international symposium on methodologies for intelligent systems (ISMIS-89)
18. Reiter R (1991) The frame problem in the situation calculus: a simple solution (sometimes) and a completeness result for goal regression. In: Lifschitz V (ed) Artificial intelligence and mathematical theory of computation: papers in honor of John McCarthy. Academic Press, San Diego, pp 359–380
19. Reiter R (2001) Knowledge in action: logical foundations for specifying and implementing dynamical systems. MIT Press, Cambridge
20. Scherl RB, Levesque HJ (1993) The frame problem and knowledge-producing actions. In: Proceedings of the AAAI conference on artificial intelligence (AAAI-93).
21. Scherl RB, Levesque HJ (2003) Knowledge, action, and the frame problem. Artif Intell 144(1–2):1–39
22. Shapiro S, Lespérance Y, Levesque HJ (1998) Specifying communicative multi-agent systems. In: Wobcke W et al. (eds) Agents and multi-agent systems–formalisms, methodologies, and applications. Lecture notes in computer science, vol 1441. Springer, New York, pp 1–14

Chapter 22
Roles, Rigidity, and Quantification in Epistemic Logic

Wesley H. Holliday and John Perry

Abstract Epistemic modal predicate logic raises conceptual problems not faced in the case of alethic modal predicate logic: Frege's "Hesperus-Phosphorus" problem—how to make sense of ascribing to agents ignorance of necessarily true identity statements—and the related "Hintikka-Kripke" problem—how to set up a logical system combining epistemic and alethic modalities, as well as others problems, such as Quine's "Double Vision" problem and problems of self-knowledge. In this paper, we lay out a philosophical approach to epistemic predicate logic, implemented formally in Melvin Fitting's First-Order Intensional Logic, that we argue solves these and other conceptual problems. Topics covered include: Quine on the "collapse" of modal distinctions; the rigidity of names; belief reports and unarticulated constituents; epistemic roles; counterfactual attitudes; representational versus interpretational semantics; ignorance of co-reference versus ignorance of identity; two-dimensional epistemic models; quantification into epistemic contexts; and an approach to multi-agent epistemic logic based on centered worlds and hybrid logic.

22.1 Introduction

In *Modal Logic for Open Minds*, van Benthem [10] remarks on the step from modal propositional logic to modal predicate logic: "it is important to perform this extension also from a practical point of view. Knowing objects like persons, telephone num-

W. H. Holliday (✉)
Department of Philosophy, University of California, 314 Moses Hall #2390,
Berkeley, CA 94720, USA
e-mail: wesholliday@berkeley.edu

J. Perry
Department of Philosophy, University of California, HMNSS Building, Room 1604,
900 University Avenue, Riverside, CA 92521, USA
e-mail: john.perry@ucr.edu

A. Baltag and S. Smets (eds.), *Johan van Benthem on Logic
and Information Dynamics*, Outstanding Contributions to Logic 5,
DOI: 10.1007/978-3-319-06025-5_22, © Springer International Publishing Switzerland 2014

bers, or even rules and methods is crucial to natural language and human agency"
(124). Indeed, talking about such knowledge, as well as beliefs and other cognitive
attitudes, is also crucial. In ordinary discourse, we freely combine talk about attitudes with predication, quantification, modals, tense, and other constructions. Here
are some examples, not all of which are exactly pieces of ordinary discourse, but all
of which seem readily intelligible, together with straightforward formalizations—
formalizations that we argue below won't quite do:

(1) Elwood believes that JvB wrote *Modal Logic for Open Minds*.

(1_f) $B_e\, W(j, m)$

(2) There is someone who Elwood believes wrote *MLOM*.

(2_f) $\exists x\, B_e\, W(x, m)$

(3) There is a book that Elwood believes JvB wrote.

(3_f) $\exists y (Bk(y) \wedge B_e\, W(j, y))$

(4) Elwood believes that JvB is J.F.A.K. van Benthem.

(4_f) $B_e\, j = j'$

(5) Elwood believes that JvB couldn't have been a computer, but could have been a
    computer scientist.

(5_f) $B_e(\neg \Diamond C(j) \wedge \Diamond CS(j))$

(6) There is someone who Elwood believes couldn't have been a computer, but could
    have been a computer scientist.

(6_f) $\exists x\, B_e(\neg \Diamond C(x) \wedge \Diamond CS(x))$

Syntactically, these formalizations combine predicates, variables, names, identity,
quantifiers, alethic modal operators, and epistemic modal operators. The progress of
logic has involved seeing how to build on the semantics for earlier items on the list to
treat items later on the list. In this paper, we will consider to what extent the addition
of *epistemic* operators requires departing from the semantics that works well for the
previous items on the list—or whether we can get away with the following.

> **Conservative Approach**: add epistemic modal operators with minimal departures
> from a base semantics for alethic modal predicate logic, e.g., without changing
> the semantics of *singular terms* or the nature of *possible worlds*.[1]

---

[1] We do not mean to suggest that there is a consensus on the proper semantics for alethic modal
predicate logic. What we have in mind here is the standard development of Kripke-style semantics
for modal predicate logic (see, e.g., [12]). To the extent that we support the Conservative Approach
so understood, one might expect that epistemic operators could be smoothly introduced into alethic
modal predicate logic developed in other ways as well. As another point of qualification, we are
not arguing for conservativeness with respect to the question of relational versus neighborhood
semantics for epistemic logic (see [3]).

There is a difficulty with the Conservative Approach, however, which we will call *the problem of the cognitive fix*. This difficulty has led a number of writers to abandon the Conservative Approach. Some have concluded that the "possible worlds" needed for epistemic logic differ from those needed for alethic modal logic. Some have concluded that names cannot be treated as "rigid designators" in epistemic logic, as they are in alethic modal logic. Some have supposed that the individuals we talk about in epistemic logic are not quite the same as the ones we talk about in alethic modal logic. In this essay, however, we defend the Conservative Approach.

In Sect. 22.2 we discuss the problem of the cognitive fix, its history, and how we propose to handle it. After introducing the formal framework in Sects. 22.3 and 22.4 we show how our approach resolves a related problem, the so-called Hintikka-Kripke Problem [4, 42, 46] for combined alethic-epistemic modal logic. In Sect. 22.5 we show how we deal with quantification into epistemic contexts, and in Sect. 22.6 we extend the framework to multi-agent epistemic logic. Finally, in Sect. 22.7 we conclude with a speculation about how our semantics for static epistemic predicate logic may lead to new directions in dynamic epistemic predicate logic.

In recent years, there has been a wealth of applications of epistemic logic, mostly using propositional languages. While many applications do not need the bells and whistles of variables and quantifiers, others do. This is the point in the quote above from van Benthem, who in addition to doing pioneering work in modal propositional logic has made notable contributions to modal predicate logic [7–9]. As he explains at the end of the chapter on Epistemic Logic in *Modal Logic for Open Minds*, "extending our whole framework to the predicate-logical case is a task that mostly still needs to be done—and who knows, it may be done by you!" (144). This is the kind of encouragement from Johan that has launched so many research projects. While we don't pretend to have completed the task, we hope that something here will spark ideas in Johan that will lead to further progress.

22.2 The Problem of the Cognitive Fix

Although the problems to be dealt with in this paper fall under the province of both linguists and philosophers, our project is one of philosophical logic, not natural language semantics or pragmatics. One of the projects of the philosophical logician is to design formal languages that are unambiguous, clear, and explicit, but that can also capture important kinds of claims about the world expressible in natural language—and that do so in philosophically illuminating ways. Our question in this paper is whether the philosophical logician can design such a language to formalize a class of claims of special interest to philosophers and epistemologists: claims about what agents believe and know. One of the chief obstacles to such a formal analysis, the problem of the cognitive fix, dates back to at least Frege and Russell.

22.2.1 Frege, Russell and the Problem of the Cognitive Fix

In their development of ideas and options for the foundations of first-order logic, Frege and Russell were not much concerned about necessity and possibility, what we will call the *alethic modalities*. But they were motivated by puzzles about the way names, variables, and identity work in the context of discourse about knowledge and belief, what we will call the *epistemic modalities*. How does the "cognitive significance" of '$a = a$' differ from that of '$a = b$'? How can George IV know that the author of *Waverley* wrote *Waverley*, yet be ignorant of the fact that Scott wrote *Waverley*, given that Scott is the author of *Waverley*? It seems that an agent can know or believe something about an individual, thought about in one way, or, as we shall say, relative to one "cognitive fix", while not believing the same thing of the same individual—and perhaps even believing the opposite—relative to another cognitive fix.[2] These differing cognitive fixes seem to enter into the truth-conditions of reports about what people know and believe. For example, consider:

(7) Elwood believes that the author of *De Natura Deorum* was a Roman.
(8) Elwood believes that the author of *De Fato* was a Roman.

It seems that (7) might be true, while (8) is false, in spite of the fact that Cicero authored both of the works mentioned. In "On Sense and Reference", Frege [28] treated the problem with his theory of indirect reference: in epistemic contexts such as these, the embedded sentence and its parts do not have their customary reference, but rather refer to their customary *sense*. Hence (7) and (8) do not really report a relation between Elwood and Cicero, the person designated by the descriptions.

In "On Denoting" Russell [59] used his theory of descriptions to reach a similar conclusion. For example, (7) is rendered as follows, using intuitive abbreviations:

(9) $B_e \exists x (Au(x, DeN) \wedge \forall y(Au(y, DeN) \to x = y) \wedge Rom(x))$.

With Russell's treatment, as with Frege's, (7) and (8) contain no reference to Cicero when properly understood. Since there is no reference to Cicero, we cannot substitute the apparently co-referring descriptions; for they do not actually co-refer. On Russell's account, the descriptions both *denote* Cicero. However, sameness of denotation does not support substitution of terms *salva veritate* in general, but only in "extensional" contexts, unlike belief contexts. Although their philosophical tools were quite different, it is generally agreed that Frege and Russell were recognizing a real phenomenon. At least on one permissible reading, the *de dicto* reading, statements like (7) and (8) do not report relations between Elwood and Cicero, and substitution of the singular terms that designate Cicero may fail to preserve truth.

Russell, however, also allowed a second reading of these sentences, where the quantifier takes wide scope:

(10) $\exists x (Au(x, DeN) \wedge \forall y(Au(y, DeN) \to x = y) \wedge B_e Rom(x))$.

---

[2] Wettstein [65] uses 'cognitive fix' in a more demanding sense, requiring not merely a way of thinking about an object, but also accurate beliefs about what distinguishes the object from others.

On this reading, (7) and (8) assert a relation between Elwood and Cicero, and the substitution of the descriptions will preserve truth. The reading Russell is getting at here, the *de re* reading, has also been widely, but not universally, recognized. In a de re belief report, a relation is asserted between the believer and the object about which she has a belief, and substitution of co-designative terms preserves truth.

The de re reading seems natural when names are used rather than descriptions:

(11) Elwood believes that Cicero was a Roman.
(12) Elwood believes that Tully was a Roman.

However, it is not at all obvious that this is correct. Names like 'Cicero' and 'Tully' give rise to the problem with which Frege begins "On Sense and Reference" [28]. The statement 'Cicero = Cicero' seems trivial, while the statement 'Cicero = Tully' contains valuable information; they have different cognitive significance. If Elwood is a rather desultory student, who has heard of Cicero in Philosophy class, where he was not identified as a Roman, and has heard of Tully in Classics, where he was clearly identified as a Roman, then Elwood might not know that Cicero was Tully. For him, 'Cicero = Tully' might contain just the information he will need on his midterm. Before he gets it, it seems that (11) might be false, while (12) is true.

Frege is usually understood to have treated names as having senses that pick out their referents by conditions they uniquely fulfill, what Carnap [15] was to call "individual concepts". So for Frege names behave like descriptions in epistemic contexts, referring to their customary senses. In a perfect language, the sense of a proper name would be established by the rules of the language. But Frege realized that for ordinary proper names this may not be so; the relevant sense may be clear from context, perhaps constructed from the speaker's beliefs or generally accepted beliefs about the object. It is the sense of the proper name, not the individual picked out, that individuates the proposition referred to by the that-clause in a belief report; thus, Frege does not allow for de re belief ascriptions in any straightforward way.

Russell thought that statements like (11) and (12) *would* report de re beliefs, so long as 'Cicero' and 'Tully' were taken to be "logically proper names". As his views developed, however, he came to doubt that any ordinary names were logically proper names. To assign a logically proper name, he thought, we have to be acquainted with the thing named, and he came to think that we are acquainted only with our own sense data, the properties and relations that obtain among our sense data, and perhaps our selves. All our beliefs about ordinary objects are de dicto; such objects are not known by acquaintance but only by description [60]. For Russell, there are de re beliefs, but not about tables and chairs and other people.[3]

Thus, neither Frege nor Russell would countenance the formalizations $(1f)$–$(6f)$ of (1)–(6).

---

[3] It seems that this leaves Russell without a solution to the problem of the cognitive fix, relative to logically proper names. For a discussion of this issue, see [66].

22.2.2 Quine on the Collapse of Modal Distinctions

Perhaps 1953 was the nadir of modal logic in philosophy. Quine argued in his "Three Grades of Modal Involvement" [55] that the whole enterprise was based on a mistake. His argument basically transfers the problem of the cognitive fix from the epistemic realm to the alethic, concluding that if we must have alethic modalities, they should be limited to the *de dicto*, without quantification into alethic contexts.

Granting—only for the sake of argument—that analyticity made sense, Quine could understand necessity and possibility as properties of sentences, *being analytic* and *being synthetic*. If we think it is analytic, and hence necessary, that philosophers are thoughtful, and we are thoughtful philosophers, we should express this as

(13) 'Philosophers are thoughtful' is necessary.

That's the "first grade" of modal involvement: modality expressed as predicates of sentences. The second grade of modal involvement, to which less thoughtful philosophers sink, blurs the use-mention distinction; they say things like:

(14) Necessarily, the heaviest philosopher in the room is thoughtful;
(15) Possibly, the heaviest object in the room is thoughtful;

or, in logical symbols:

(16) $\Box$(the heaviest philosopher in the room is thoughtful);
(17) $\Diamond$(the heaviest object in the room is thoughtful).

If we restrict the aspiring modal logician to statements like this, we will have allowed her what we might think of as *de dicto* alethic modality. The modal operator blocks the ordinary interpretation of the material inside, much like quotation would.

Nothing in the theory of necessity as analyticity allows us to make sense of the third grade of modal involvement, that is, *de re* alethic modality, as would be required to make sense of "quantifying in":

(18) $\exists x \,\Box(x$ is thoughtful).

And in fact, Quine thinks that this makes no sense, at least if we retain an orthodox interpretation of predication, identity, names, variables, and quantifiers.

The basic problem with the move to (18), Quine thought, is that it requires the aspiring modal logician to give unequal treatment to singular and general terms. Surely she must insist on *extensional opacity*: "Russell was a human" can be necessary, while "Russell was a featherless biped" is contingent, even if actually all and only humans are featherless bipeds. How we designate a *class* is crucial for the modal status of a sentence. But to make sense of quantification into modal contexts, the modal logician must maintain that it makes sense for an open sentence like

(19) $\Box(x$ is thoughtful)

to be true of an object absolutely, not relative to some description or other. For the alethic modalities at least, she must insist that how we designate *individuals* doesn't matter. She should, then, accept *referential transparency* and *substitutivity*:

If t and t' are co-designative singular terms, substitution of t for t' preserves truth, even in alethic modal contexts.

But combining extensional opacity and referential transparency won't work, or so Quine maintained. He used a version of what is now called "the slingshot" to argue that if we accept both doctrines, then the modalities "collapse": all sentences with the same truth value have the same modal status—all necessary or all contingent.
   Assume that

(A) van Benthem is a human.
(D) van Benthem is a logician.

are both true, but (A) is necessary while (D) is contingent. So (A) and (D) have different modal statuses. Then it seems we should grant that (B) has the same modal status as (A), and (C) has the same modal status as (D):

(B) $\{x \mid x = \emptyset \,\&\,$ van Benthem is a human$\} = \{\emptyset\}$.
(C) $\{x \mid x = \emptyset \,\&\,$ van Benthem is a logician$\} = \{\emptyset\}$.

The only way (B) could be false is if van Benthem were not a human. Assuming (A) is necessary, that's impossible, so (B) must also be necessary. It's sufficient for (C) to be false that van Benthem not be a logician. Assuming (D) is contingent, that's indeed possible, so (C) must also be contingent. But note that (B) and (C) differ only in that co-referential expressions are substituted one for the other on the left side of the identity sign. So by *substitutivity*, the sentences that result from prefixing 'Necessarily' to (B) and (C) must have the same truth value, i.e., (B) and (C) must have the same modal status, and hence so must (A) and (D). The modalities collapse.
   Føllesdal [27] identified a flaw in Quine's argument.[4] The aspiring modal logician need not claim that *all* co-referential singular terms are intersubstitutable in alethic modal contexts, only that some are; and the step from (B) to (C) will only work if the class abstracts on the left sides of the identity signs are among them.
   What would make a class of singular terms such that co-referential terms in the class are always substitutable for each other in alethic modal contexts? It would be so if each term is a *rigid designator* in Kripke's [41] sense, designating the same object in every possible world (where the object exists). But we can also explain why it would be so in a more general way. Consider the following:

(20) $H(\text{Cicero})$;
(21) $\square\, H(\text{Cicero})$.

The standard semantics for (20) tells us that it is true if and only if the open sentence '$H(x)$' is true of *the individual designated by* 'Cicero', whether 'Cicero' is treated

---

[4] See [53] for a fuller account of the slingshot and Føllesdal's treatment of it.

as a hidden description or a logically proper name. Let us say that if 'Cicero' is a *modally loyal* term, then the semantics for (21) is such that (21) is true *iff* the open sentence '$\Box H(x)$' is true of that same individual. The same goes for $\Box H$(Tully), assuming 'Tully' is also a modally loyal term designating the same individual. And an analogous point holds for any other sequence of $\Box$'s and $\Diamond$'s applied to H(Cicero). To have the desired loyalty, the individual supplied for each sentence needs to be the same; and no further conditions that can vary with co-referential terms can be brought into the truth-conditions for sentences that are formed by adding operators.

There are at least two ways terms can be modally loyal. Føllesdal's [27] "genuine names" are loyal because they are not hidden descriptions or some other complex names built up from general terms with descriptive content. Kripke [41] argued that ordinary names, like 'Cicero' and 'Tully', are genuine in this sense.

A second source of loyalty, which could be called *sequestering*, can be illustrated by Kaplan's [40] operator 'Dthat'. 'Dthat' converts descriptions into what Kaplan calls "terms of direct reference". On a standard theory of descriptions, 'the ϕ' *denotes* the unique object, if there is one, that is ϕ. According to Kaplan, the singular term 'Dthat(the ϕ)' *directly refers* to the object denoted by 'the ϕ'. Whatever may be suggested by 'directly', the cash value in alethic modal contexts is this: it is the object denoted by 'the ϕ' that needs to satisfy the open sentence that results from replacing all ocurences of 'Dthat(the ϕ)' in the original sentence by the same variable; and this is *all* that 'Dthat(the ϕ)' makes available; in particular, the relevance of the condition of being a ϕ is exhausted in determining the reference and plays no further role in the semantics of modal sentences in which 'Dthat(the ϕ)' occurs.

Genuine names and sequestered descriptions suggest a solution to Quine's argument. Substitutivity holds for some singular terms, but not all. The class abstracts at the beginning of (B) and (C) are not among the singular terms for which substitutivity holds. The class abstracts are not genuine names or sequestered descriptions, but rather complex names whose semantics is based on that of general terms.[5]

---

[5] What if we amend (B) and (C) so that they begin with sequestered terms? Extending 'Dthat' to class abstracts, we have:

(A) van Benthem is a human.
(B') Dthat($\{x \mid x = \emptyset \,\& \,$ van Benthem is a human $\}$) = $\{\emptyset\}$.
(C') Dthat($\{x \mid x = \emptyset \,\& \,$ van Benthem is a logician $\}$) = $\{\emptyset\}$.
(D) van Benthem is a logician.

Could Quine still argue that (A) and (B'), and (C') and (D), have the same modal status? (D) is clearly contingent; the multi-talented van Benthem could have been a computer scientist. Is (C') contingent? The result of applying Dthat to a description or class abstract is supposed to be a rigid designator, or more generally, a modally loyal term in the sense defined above. Thus, evaluating

$\Box$Dthat($\{x \mid x = \emptyset \,\&\,$ van Benthem is a logician $\}$) = $\{\emptyset\}$

amounts to checking for every possible world whether '$y = \{\emptyset\}$' is true there, where y is assigned the object that is designated by 'Dthat($\{x \mid x = \emptyset \,\& \,$ van Benthem is a logician $\}$)' *in the actual world* (or the world of the context of utterance). Since the object designated by 'Dthat($\{x \mid x = \emptyset \,\& \,$ van Benthem is a logician $\}$)' in the actual world is $\{\emptyset\}$, the check succeeds, so (C') is necessary. Hence (C') and (D) do not have the same modal status; so the modalities do not collapse.

22.2.3 *Names in Epistemic Logic*

While it no doubt remains for alethic modal logic to reach its zenith, Quine's attempts to undermine it are but a dim memory of older philosophers. Modally loyal names are ubiquitous, both syntactically simple genuine names, functioning basically like variables except with reference fixed in the Lexicon,[6] and complex but sequestered singular terms, as in Kaplan's [40] *Demonstratives*.

It is widely held, however, that modally loyal names, understood as they are in alethic logic, do not work for epistemic logic. Alethic modal logic can perhaps ignore the problem of the cognitive fix. But cognition is the stuff that reports of belief and knowledge seem to be about, and the problem won't go away.

The problem is perhaps clearer in the case of sequestered terms than in that of genuine names. The beliefs that might motivate a competent speaker to say or write

(22) Dthat(the author of *Satan in the Suburbs*) resigned from UCLA.

might differ from the beliefs that would motivate him to say or write

(23) Dthat(the author of *Marriage and Morals*) resigned from UCLA.

Someone might be in a position to sincerely and confidently utter the second, without having any idea about the first. The features of sequestering that allow the aspiring alethic modal logician to evade Quine's argument do not seem to automatically help the aspiring epistemic logician with the problem of the cognitive fix.

Even with genuine names, the problem of the cognitive fix doesn't go away. When we are talking about Cicero's modal properties, it doesn't seem to matter whether we call him 'Cicero' or 'Tully'. But as we saw above, it seems to matter a lot when we are talking about what people know and believe.

The apparent mismatch between the alethic and epistemic modalities comes out in the contrast between the following:

(24) If it is necessary that Tully was a Roman, then it is necessary that Cicero was a Roman.

(24_f) $\Box Rom(Tully) \rightarrow \Box Rom(Cicero)$

(25) If Elwood believes that Tully was a Roman, then Elwood believes that Cicero was a Roman.

(25_f) $B_e Rom(Tully) \rightarrow B_e Rom(Cicero)$

Under the standard interpretation of alethic modality, $(24)/(24_f)$ is valid. If Tully was a Roman in every possible world, then so was Cicero. But intuitively, (25) is not valid. Perhaps Tully is a Roman in every world compatible with what Elwood believes, for Elwood realizes that Tully was a Roman. But if he doesn't believe

---

[6] In what follows we consider approaches that assimilate proper names in natural language to *constants* in a formal modal language. We do not have room to discuss alternative approaches, e.g., that assimilate names to predicates [13] or variables [20].

that Tully is Cicero, and doesn't believe that Cicero was a Roman, it seems there are worlds, compatible with what he believes, in which Cicero isn't a Roman. But Cicero *is* Tully. So how can (25) be false? There are two main schools of thought.

One approach is to deny the disanalogy between (24) and (25) and claim that (25) is also valid. If Elwood believes that Tully was a Roman, then he believes that Cicero was too. Epistemic contexts are transparent for names. According to a view in the philosophy of language that we will call the *the Heroic Pragmatic theory*,[7] if one says, "Elwood believes that Tully was a Roman, but Elwood doesn't believe that Cicero was a Roman", then one says something false. But one may nevertheless convey something true; the choice of 'Tully' for the first clause and 'Cicero' for the second may convey a Gricean implicature that is true, namely that Elwood wouldn't use the name 'Cicero' to express his belief. By denying the disanalogy between (24) and (25), the Heroic Pragmatist supports the Conservative Approach of Sect. 22.1 for extending alethic to epistemic modal logic, but we think at too high an intuitive cost.

The other approach, more common among philosophical logicians, accepts the intuitive disanalogy between (24) and (25), and rejects the Conservative Approach. While the syntax of epistemic modal logic may parallel that of alethic modal logic, as (25_f) parallels (24_f), the semantics cannot. According to the second approach, although 'Cicero = Tully' is true in all possible worlds, in order to invalidate (25_f) we must allow 'Cicero = Tully' to be false in some *doxastic possibility*. We will call this *the Special Semantics theory*, since it claims that we must modify our semantics when we move from alethic to epistemic modal logic. Some explain this by claiming that in epistemic contexts, names are not rigid designators (see Sect. 22.4.1). Others explain it by a difference between doxastic/epistemic possibilities and possible worlds.

Both approaches accept an assumption that we reject about how the philosophical logician should translate natural language reports into the modal language.

Complement = Operand Hypothesis: in epistemic logic, as in alethic modal logic, the formula to which the modal operator is applied—the *operand*—is the formalization of the sentence embedded in the 'that'-clause—the *complement sentence*—in the natural language belief/knowledge ascription.

Consider the relation between (24) and (24_f). To form the antecedent of the latter, we attach the alethic operator '$\Box$' to the formalization of the sentence embedded in the that-clause of the former, 'Tully was a Roman': $\Box Rom(Tully)$; similarly with the consequent. This is a natural step, and goes with thinking of '$\Box$' as being basically a translation of the natural language phrase 'it is necessary that' (or 'necessarily').

It is then natural to follow the same procedure with epistemic operators, as we did in going from (25) to (25_f). But in doing so, we build in an assumption that is not part of the formal framework of modal logic. In doing epistemic modal logic, we want to use the apparatus of modal logic to analyze belief reports. It does not follow that the sentence operated upon, in the analysis, needs to be the same as the sentence embedded in the 'that'-clause in the natural language ascription. This assumption, though natural, is on our view incorrect. This assumption, which one can think of as

---

[7] See, for example, [5, 62].

a form of *syntactic conservatism*, has to be rejected in order to maintain, in a way that takes account of intuitions, our approach of *semantic* conservatism.

The reason is that natural language reports of belief and knowledge are not fully explicit. On our view, they are not only about objects, but also about the cognitive fixes via which those objects are cognized. One can represent such cognitive fixes with bits of language (names or definite descriptions), mental particulars (ideas or notions), or entities to which these somehow correspond. Frege's *Sinne* and Carnap's individual concepts are examples of the latter. This is the course we follow. But the entities we choose are what we call *agent-relative roles* (cf. [51]).

We assume only that in every case of cognition about an object, there is some relation that holds between the cognizer and the object—the object plays some role vis-a-vis the cognizer—in virtue of which this is the object cognized. Carnap's individual concepts are supposed to provide the same way of thinking about the object for any agent. But in the general case, this requirement is too strong.

Here is a variation of an example from Quine [56] that makes this clear.

Suppose Ralph knows that the shortest spy is a spy. It doesn't seem that this would make the FBI interested in him. But there is a stronger condition that would make the FBI interested, which we could express by saying: there is someone Ralph knows to be a spy, or, to make the structure a bit more explicit, there is someone such that Ralph knows that *he* is a spy. But surely we can't say that just because Ralph knows that the shortest spy is a spy, because anyone who knows English knows that.

Suppose that Ralph sees his neighbor Ortcutt on the beach, but doesn't recognize him (so he wouldn't say "that man is Ortcutt"). Ralph sees Ortcutt take papers from a bag marked 'CIA' and hand them to an obvious Bolshevik. Here the relevant agent-relative role is *being the man seen*.[8] Ortcutt plays this role relative to Ralph. It is in virtue of his playing this role that Ralph's thought—the one he might express with "that man is a spy"—is *about* Ortcutt. Roles correspond to cognitive fixes.

The police arrive, and they take Ortcutt into custody, but Ralph still doesn't recognize him. After interviewing Ralph, one policeman reports to the other: "Ralph believes that Ortcutt is a spy". The name 'Ortcutt' is associated by the police with a certain role, *being the person identified on the mug sheets as 'Ortcutt'*. But that role is not the one that the police are claiming to play a role in *Ralph*'s cognitive economy. Clearly, their interest in his testimony is due to the fact that Ortcutt was the man Ralph was *seeing*. As we construe the policeman's remark, he says that Ralph had a belief about a certain man, Ortcutt, via the role, *being the man seen*.

As this suggests, on our view roles are "unarticulated constituents" of the reports. That is, there is no morpheme in the report that refers to them, but they figure in the truth-conditions of what is said. This means, in effect, that to carry out a formalization you need a story (context) and not just a sentence. The classic example motivating the idea of unarticulated constituents is an utterance of "It is raining", made in Palo Alto, by way of calling off a tennis match scheduled there [50]. No morpheme refers to Palo Alto, but clearly it is the fact that it is raining in Palo Alto that makes the utterance true. While the classic example has proven quite controversial, no doubt

---

[8] Cf. Lewis [44, p. 543] on *watching* as a "relation of acquaintance".

because of the complexity of weather phenomena of all sorts, the same idea was applied to belief reports in [19].[9]

Thus, on our view, singular terms employed in complement sentences of belief-reports have two functions. Their semantic function is to identify the objects the belief is about. Typically, they also have the pragmatic function of providing evidence about which cognitive fixes, or roles, are relevant. On this much, we agree with the Heroic Pragmatists. We disagree, however, in the way we think cognitive fixes are relevant. The Heroic Pragmatist sees them as triggers for Gricean implicatures. We see them as semantic parameters of the whole belief report. In our formalization of belief-reports, two kinds of information will be fully explicit: the objects the beliefs are about, and the cognitive fixes on such objects that are involved in the beliefs.

Thus we see virtue in each of the approaches we don't take. On the one hand, the Heroic Pragmatic approach sees correctly that the semantic job of the names in the complement sentence is to refer to the objects believed about. On the other hand, the Special Semantics approach sees correctly that the terms in doxastic operand sentences are not modally loyal (or rigid) in the way that names are in alethic operand sentences. However, each approach is wrong about something too. The Heroic Pragmatic approach does not give cognitive fixes their rightful semantic place. The Special Semantics approach assumes that non-transparency requires non-rigid *names*, whereas in our formalization below, the names will occur in transparent positions, but not within the scope of a doxastic operator; the opacity will be due to thinking of objects via the roles they play, which can differ from world to world.

What, then, should be said about (25) and (25$_f$)? Let's first look at the sort of case that leads one to think they are invalid. Remember our desultory student Elwood, who knows from his Classics class that Tully was a Roman, and who has heard of Cicero in his Philosophy class, but hasn't learned that he was a Roman there.

On the picture of proper names we have in mind, when people use names they exploit causal/historical networks that support conventions for using the name to refer to objects—typically objects that are the sources of the networks [52]. Let r_1 be the role of *being the person at the source of the 'Tully' network exploited by Elwood*, and let r_2 be the role of *being the person at the source of the 'Cicero' network exploited by Elwood*.[10] Although in the actual world, Cicero (= Tully) plays both of these roles, in other worlds compatible with what Elwood believes, different individuals may play these roles—the networks may have different sources.

---

[9] While our strategy is based on the approach of Crimmins and Perry [19], those authors took cognitive fixes to be *notions* and took notions to be unarticulated constituents of belief-reports. Subsequently Perry has developed an account that is basically similar, but takes cognitive fixes to be *notion-networks*, basically intersubjective routes through notions. Of course, traditionally cognitive fixes have been taken to be individual concepts. We believe that the concept of a *role* provides a general framework into which all of these candidates can be fit.

[10] Cf. Lewis [44, p. 542]: "If I have a belief that I might express by saying "Hume was noble", I probably ascribe nobility to Hume under the description "the one I have heard of under the name of 'Hume' ". That description is a relation of acquaintance that I bear to Hume. This is the real reason why I believe *de re* of Hume that he was noble".

Let z_1 be what we call a *role-based* (*object*) *variable*, whose interpretation (relative to a current assignment) in any world w is the object that plays role $\mathbf{r}_1$ in w, and let z_2 be another role-based variable, whose interpretation in any w is the object that plays $\mathbf{r}_2$ in w. Suppose someone utters "Elwood believes that Tully was a Roman, but Elwood doesn't believe that Cicero was a Roman", where the unarticulated role associated with the use of 'Tully' is $\mathbf{r}_1$ and the unarticulated role associated with the use of 'Cicero' is $\mathbf{r}_2$. Then we claim that the truth of the report imposes the following conditions on the actual world and the space of Elwood's doxastic alternatives (where $\boxdot \varphi$ means that φ is true in all of Elwood's doxastic alternatives):

(26) $Tully = z_1 \wedge \boxdot Rom(z_1) \wedge Cicero = z_2 \wedge \neg \boxdot Rom(z_2).$

In other words, Tully is the person who plays role $\mathbf{r}_1$ in the actual world, and in all worlds v compatible with what Elwood believes, the person who plays $\mathbf{r}_1$ in v is a Roman; but while Cicero is the person who plays role $\mathbf{r}_2$ in the actual world, there is some world u compatible with what Elwood believes such that the person who plays $\mathbf{r}_2$ in u is not a Roman. Since this can happen, (26) can be true. This does not require any possible world compatible with what Elwood believes in which Cicero (= Tully) is not Tully (= Cicero). It simply requires that there are worlds, compatible with what he believes, in which the object that plays role $\mathbf{r}_1$ does not play role $\mathbf{r}_2$.

The examples above involve two kinds of roles exploited in thinking about things: *perceptual* roles and *name-network* roles. Both kinds of roles afford ways of finding out more about an object; they are what Perry calls *epistemic roles* [52].[11] The idea that an agent's believing of an object (*de re*) that it has a property requires, as with (26), that the object be the unique player of some epistemic role for the agent in the actual world, such that in all worlds compatible with the agent's beliefs, the player of the role in that world has the property in question, is closely related to the accounts of *de re* belief proposed by Kaplan [39] and Lewis [44, Sect. 8], which have been influential in linguistics (cf. [17]). When we add the idea that belief reports involving names have such roles as unarticulated constituents, we can make sense of the difference between (24) and (25) (see Sect. 22.4).

Accounts of *de re* belief in the style of Kaplan and Lewis are not without challenges. Ninan [47] has argued that such accounts do not generalize to handle other attitudes such as *imagination*; and Yalcin [67] has argued that it is difficult to make such accounts compatible with semantic *compositionality*. We will briefly discuss the first problem in Sect. 22.3.1. For a critical discussion of compositionality taking into account unarticulated constituents, we refer to Crimmins [18, Chap. 1].

On our view, belief reports get at important cognitive aspects of agents, and keeping track of them helps us understand agents' behavior. First and foremost, in this paper we are interested in using a formal language to describe these aspects of agents, not in doing the formal semantics of natural language belief reports. This is in line with how epistemic logic is used in theoretic computer science and AI, as a tool to model the information that agents have and how they update it, not the meanings of English sentences containing the words 'knows' and 'believes'.

---

[11] Cf. Lewis [45, 10f] on "relations of epistemic rapport" or "relations of acquaintance".

The same can be said of quantified epistemic logic. Although our examples will be drawn from philosophy rather than computer science or AI, notions of de re belief and "double vision" can be applied to computers as well as humans (cf. [6]); once again, one may be interested in the phenomena themselves, rather than in English talk about them. With this distinction in mind, even the Heroic Pragmatist about belief talk could agree with much of the formal modeling to follow, if he does not disagree with the basic picture of agents acquiring information about objects via the roles these objects play for the agents. For the purpose of modeling agents themselves, whether we take facts about roles to be involved in the semantics or the pragmatics of belief talk is not crucial.

In the next section we introduce the formal framework with which we will implement these ideas, Fitting's [24, 25, Sect. 5] First-Order Intensional Logic (FOIL).

22.3 Formal Framework

Our formal language is a slight variant of that of Fitting's [25, Sect. 5] FOIL.[12]

Definition 22.1 (*Language*) Fix a set $\{n_1, n_2, \dots\}$ of constant symbols (or *names*); two disjoint sets $\{x_1, x_2, \dots\}$ and $\{r_1, r_2, \dots\}$ of *object variables* and *role variables*; and for every $m \in \mathbb{N}$, a set $\{Q_1^m, Q_2^m, \dots\}$ of *m-place relation symbols*. The *terms t* and *formulas* φ of our language are generated by the following grammar ($i, k \in \mathbb{N}$):

$$t ::= n_i \mid x_i$$
$$\varphi ::= Q_i^k(t_1, \dots, t_k) \mid t = t' \mid \mathsf{P}(t, r_i) \mid \neg\varphi \mid (\varphi \wedge \varphi) \mid \boxed{a}\varphi \mid \boxed{d}\varphi \mid \forall x_i \varphi \mid \forall r_i \varphi.$$

We will often replace the subscripts on variables with suggestive letters or words (or drop them altogether) and write out italicized English names in place of $n_1, n_2, \dots$. We define $\exists y\varphi$ as $\neg\forall\neg y\varphi$ for y an object or role variable, and sometimes we use $\square$ to represent both $\boxed{a}$ and $\boxed{d}$. For the new symbols, we give the following readings:

> $\mathsf{P}(t, r_i)$ "t plays role r_i for the agent";
> $\boxed{a}\varphi$ "it is alethically necessary that φ";
> $\boxed{d}\varphi$ "it is doxastically necessary for the agent that φ".

In Sect. 22.6, we extend the language to describe the beliefs of multiple agents. Of course, one could also introduce an epistemic necessity operator $\boxed{e}$ for knowledge (see [34, 35] for discussion of additional issues raised by knowledge).

---

[12] There are a few small differences. First, Fitting allows relation symbols Q (but not =) to apply to what we call *role variables*—his *intensional variables*—whereas to simplify the definition of the language (to avoid typing relations), we do not; this is why we do not count role variables among the terms t in Definition 22.1. Second, we include constant symbols in the language, whereas for convenience Fitting does not. Third, we have a bimodal language with $\boxed{a}$ and $\boxed{d}$, whereas Fitting has a monomodal language with $\square$. Finally, where we write $\mathsf{P}(t, r_i)$, Fitting would write $\mathsf{D}(r_i, t)$.

Our models are a standard kind of modal models with constant domains.[13]

Definition 22.2 (*Constant Domain Models*) A (constant domain) model for the language of Definition 22.1 is a tuple $\mathcal{M} = \langle W, R_a, R_d, D, F, V \rangle$ such that:

(1) W is a non-empty set of *worlds*;
(2) R_a is a binary *alethic accessibility* relation on W;
(3) R_d is a binary *doxastic accessibility* relation on W;
(4) D is a non-empty set of *objects*;
(5) F is a non-empty set of *roles*, which are partial functions from W to D;
(6) V is a *valuation* function such that for a relation symbol Q_i^k and world w, $V(Q_i^k, w)$ is a set of k-tuples of objects; and for a name n_i, $V(n_i, w)$ is an object, for which we assume the following:

    (a) R_a-*rigidity of names* – for all $w, v \in W$, if $w R_a v$, then $V(n_i, w) = V(n_i, v)$;
    (b) R_d-*rigidity of names* – for all $w, v \in W$, if $w R_d v$, then $V(n_i, w) = V(n_i, v)$.

For $w \in W$, the pair $\mathcal{M}, w$ is a *pointed model*.

Instead of stating (a) and (b), we could simply require that for all names n_i and worlds $w, v \in W$, $V(n_i, w) = V(n_i, v)$. However, for the purposes of our discussion in Sect. 22.4, it helps to distinguish (a) and (b); for according to a common view, we should assume only the R_a-rigidity of names, not the R_d-rigidity of names.

Functions from W to D are traditionally thought of as *individual concepts*, whereas we think of the partial functions from W to D in F as representing *agent-relative roles*.[14] The distinction is not just terminological. The role view leads us to reject constraints on F that have been proposed assuming the individual concept view (in Sect. 22.5), and it leads to important differences in the multi-agent case (in Sect. 22.6).

It would be natural to assume that R_a is a *reflexive* relation—perhaps even an equivalence relation—while R_d is a *serial* relation. But nothing here will turn on properties of the relations, so we prefer to define the most general model classes.

The next step in introducing the semantics is to define variable assignments.

Definition 22.3 (*Variable Assignment*) Given a model $\mathcal{M}$, a *variable assignment* μ maps each object variable x_i to some $\mu(x_i) \in D$ and each role variable r_i to some $\mu(r_i) \in F$. For $d \in D$, let $\mu[x_i/d]$ be the assignment such that $\mu[x_i/d](x_i) = d$ and for all other x_j, $\mu[x_i/d](x_j) = \mu(x_j)$. For $f \in F$, $\mu[r_i/f]$ is defined analogously.

We can now define the interpretations of the two types of terms in our language.

---

[13] These models are almost the same as those for "contingent identity systems" in [48] (cf. [37]) and [54], but for a few differences: we follow Fitting in allowing F to contain *partial* functions; Parks does not deal with constants; and while Priest does deal with constants, he treats them as *non-rigid*. The differences between our models and Fitting's [24, 25] are that we deal with constants, and Fitting defines V so that predicates can apply not only to elements of D, but also to elements of F (cf. [37]).

[14] We are not suggesting that all there is to a role is a partial frunction from W to D; but such a function captures an important aspect of a role, namely the players of the role across worlds.

Definition 22.4 (*Interpretation of Terms*) The interpretation $[t]_{\mathcal{M}, w, \mu}$ of a term t in model $\mathcal{M}$ at world w with respect to assignment μ is an object given by:

$$[n_i]_{\mathcal{M}, w, \mu} = V(n_i, w);$$
$$[x_i]_{\mathcal{M}, w, \mu} = \mu(x_i).$$

It follows from Definition 22.4 and parts (a) and (b) of Definition 22.2 that

for all n_i and $w, v \in W$, if $wR_a v$, then $[n_i]_{\mathcal{M}, w, \mu} = [n_i]_{\mathcal{M}, v, \mu}$;

for all n_i and $w, v \in W$, if $wR_d v$, then $[n_i]_{\mathcal{M}, w, \mu} = [n_i]_{\mathcal{M}, v, \mu}$.

This is the sense in which a name n_i is a *rigid designator*. Given the first point, we can call n_i an "alethically rigid designator", and given the second point, we can call n_i a "doxastically rigid designator". We will return to these points in Sect. 22.4.

Finally, we are ready to state the truth definition for our version of FOIL.

Definition 22.5 (*Truth*) Given a pointed model $\mathcal{M}, w$, an assignment μ, and a formula φ, we define $\mathcal{M}, w \vDash_\mu \varphi$ (φ is true in $\mathcal{M}$ at w with respect to μ) as follows:

$\mathcal{M}, w \vDash_\mu Q_i^k(t_1, \ldots, t_k)$ iff $\langle [t_1]_{\mathcal{M}, w, \mu}, \ldots, [t_k]_{\mathcal{M}, w, \mu} \rangle \in V(Q_i^k, w)$;

$\mathcal{M}, w \vDash_\mu t = t'$ iff $[t]_{\mathcal{M}, w, \mu} = [t']_{\mathcal{M}, w, \mu}$;

$\mathcal{M}, w \vDash_\mu P(t, r_i)$ iff $\mu(r_i)$ is defined at w and $[t]_{\mathcal{M}, w, \mu} = \mu(r_i)(w)$;

$\mathcal{M}, w \vDash_\mu \neg\varphi$ iff $\mathcal{M}, w \nvDash_\mu \varphi$;

$\mathcal{M}, w \vDash_\mu (\varphi \wedge \psi)$ iff $\mathcal{M}, w \vDash_\mu \varphi$ and $\mathcal{M}, w \vDash_\mu \psi$;

$\mathcal{M}, w \vDash_\mu \forall x_i \varphi$ iff for all $d \in D$, $\mathcal{M}, w \vDash_{\mu[x_i/d]} \varphi$;

$\mathcal{M}, w \vDash_\mu \forall r_i \varphi$ iff for all $f \in F$, $\mathcal{M}, w \vDash_{\mu[r_i/f]} \varphi$;

$\mathcal{M}, w \vDash_\mu \boxed{a}\varphi$ iff for all $v \in W$, if $wR_a v$ then $\mathcal{M}, v \vDash_\mu \varphi$;

$\mathcal{M}, w \vDash_\mu \boxed{d}\varphi$ iff for all $v \in W$, if $wR_d v$ then $\mathcal{M}, v \vDash_\mu \varphi$.

For a complete axiomatization of FOIL, we refer the reader to [25, Sect. 5].[15]

Despite having only a single domain D of objects, we can capture the idea of world-relative varying domains with the standard device of a one-place *existence predicate* E, thinking of $V(E, w)$ as the non-empty domain of objects that exist in world w. We can then define the *actualist quantifier* by $\forall_a x \varphi := \forall x(Ex \to \varphi)$.[16]

Instead of using *role variables* $r_1, r_2, \ldots$, a number of authors use what we call *role-based object variables* $z_1, z_2, \ldots$[17] These are treated in the same way by an assignment μ, so that $\mu(z_i) = \mu(r_i) \in F$, but their interpretations differ:[18]

---

[15] Some minor changes must be made, e.g., since we include constants in the language (recall note 12), but we will not go into the details here.

[16] One may then wish to add the assumption that for all $f \in F$, if $f(w) = d$, then $d \in V(E, w)$, i.e., if an object d plays a role for the agent in w, then d exists in w, validating $P(t, r_i) \to \exists_a x\, t = x$.

[17] See, for example, [1, 14, 48] and [54, Sect. 8].

[18] We did not define the interpretation of role variables in Definition 22.4, since we do not officially count them as terms (recall note 12), and they only appear in the $P(t, r_i)$ clause in Definition 22.5, where the assignment μ takes care of them directly; but the definition would be $[r_i]_{\mathcal{M}, w, \mu} = \mu(r_i)$.

$$[r_i]_{\mathscr{M}, w, \mu} = \mu(r_i);$$
$$[z_i]_{\mathscr{M}, w, \mu} = \mu(z_i)(w).$$

Hence r_i picks out a role, whereas z_i picks out the object that plays that role, which may vary between worlds. The truth clause for quantification with z variables is:

$$\mathscr{M}, w \vDash_\mu \forall z_i \varphi \text{ iff for all } f \in F, \mathscr{M}, w \vDash_{\mu[z_i/f]} \varphi.$$

One complication with this approach is that if we allow F to contain *partial* functions, then $\mu(z_i)$ may be undefined at w, in which case $[z_i]_{\mathscr{M}, w, \mu}$ is undefined. Hence we have to deal with evaluating atomic formulas containing undefined terms. Typically authors who use such variables assume that F is a set of total functions.[19]

Let us see how we can translate a language with z variables into our language.

Definition 22.6 (*Z-translation and abbreviation*) For an atomic formula $At(t_1, \ldots, t_n)$ (including identity formulas), possibly containing x variables and z variables (instead of r variables), let $\mathbf{z}$ be the sequence of z variables contained therein (in order of first occurrence); let I be the set of indices of these variables; and let $\mathbf{x}$ be the sequence of x variables obtained from $\mathbf{z}$ by replacing each z_i with x_{i^*}, where i^* is the least $j \geq i$ such that x_j is not in $At(t_1, \ldots, t_n)$ and $j \neq k^*$ for any z_k preceding z_i in $\mathbf{z}$. Define the *Z-translation*:

$$Z(At(t_1, \ldots, t_n)) = \exists \mathbf{x}(\bigwedge_{i \in I} P(x_{i^*}, r_i) \wedge At(t_1, \ldots, t_n)_{\mathbf{x}}^{\mathbf{z}}),$$

where $\exists \mathbf{x}$ abbreviates a string of quantifiers and $\varphi_{\mathbf{x}}^{\mathbf{z}}$ indicates simultaneous substitution of each element of $\mathbf{x}$ for the corresponding element of $\mathbf{z}$, and

$$Z(\neg\varphi) \quad = \neg Z(\varphi);$$
$$Z((\varphi \wedge \psi)) = (Z(\varphi) \wedge Z(\psi));$$
$$Z(\forall x_i \varphi) \quad = \forall x_i Z(\varphi);$$
$$Z(\forall z_i \varphi) \quad = \forall r_i Z(\varphi);$$
$$Z(\Box\varphi) \quad = \Box Z(\varphi).$$

If $\alpha = Z(\beta)$, then we call β a *Z-abbreviation* of α.

For example, our formula from Sect. 22.2.3,

(26) $Tully = z_1 \wedge \boxdot Rom(z_1) \wedge Cicero = z_2 \wedge \neg\boxdot Rom(z_2),$

is the Z-abbreviation of

(27) $\exists x_1(P(x_1, r_1) \wedge Tully = x_1) \wedge \boxdot\exists x_1(P(x_1, r_1) \wedge Rom(x_1)) \wedge$
    $\exists x_2(P(x_2, r_2) \wedge Cicero = x_2) \wedge \neg\boxdot\exists x_2(P(x_2, r_2) \wedge Rom(x_2)),$

which is equivalent to

---

[19] The exception among the authors referenced in note 17 is Carlson, who allows F to contain partial functions and uses a three-valued semantics to deal with undefined terms.

(28) $P(Tully, r_1) \wedge \boxtimes \exists x_1(P(x_1, r_1) \wedge Rom(x_1)) \wedge$
       $P(Cicero, r_2) \wedge \neg \boxtimes \exists x_2(P(x_2, r_2) \wedge Rom(x_2)).$

We will use Z-abbreviations repeatedly in order to compare our formalizations of belief ascriptions with those of other authors whose systems use z variables.[20]

There has been much discussion in the literature on quantified modal logic about whether quantifiers should take regular object variables x_i or what we called role-based object variables z_i (see [29, 30, Sect. 13]). We think both types of quantification are useful.[21] In Sect. 22.5, we will suggest that trying to make quantification with z variables also do the normal work of quantification with x_i variables forces one to postulate otherwise unnatural constraints on the functions assigned to z_i variables.

There is another important point about variables and quantifiers that will apply throughout. For simplicity, the belief reports we consider are cases of *role provision*,[22] where the relevant roles are clear from context, so we formalize the report with an *open* formula with *free* role variables to which the relevant roles are assigned by a context-sensitive assignment μ. In other cases of *role constraint*, there may not be any unique roles clear from context, although context supplies a set of relevant roles, so we formalize the report by prefixing role quantifiers; e.g., (28) goes to

(29) $\exists r_1[P(Tully, r_1) \wedge \boxtimes \exists x_1(P(x_1, r_1) \wedge Rom(x_1))] \wedge$
       $\exists r_2[P(Cicero, r_2) \wedge \neg \boxtimes \exists x_2(P(x_2, r_2) \wedge Rom(x_2))].$

Then we can think of the set F of roles over which the quantifiers range as context sensitive, so that a change of context can be represented by a change of models from $\mathcal{M} = \langle W, R_a, R_d, D, F, V \rangle$ to $\mathcal{M} = \langle W, R_a, R_d, D, F', V \rangle$ as in dynamic epistemic logic. Or we could put many sets of roles in one model and superscript the role variables so that, e.g., r_k^i and r_k^j range over sets $\mu(i)$ and $\mu(j)$ of roles, where the context-sensitive assignment μ also maps numbers (superscripts) to sets of roles, as in Aloni's approach with z-variables at the end of Sect. 22.5. Then the quantifiers in a sentence could range over distinct role sets. Role provision could be seen as the special case of role constraint where the cardinality of the relevant set of roles is 1.

22.3.1 Extension for Counterfactual Attitudes

The framework presented so far is designed to handle doxastic/epistemic attitudes. As noted in Sect. 22.2.3, Ninan [47] has raised a challenge for accounts of these

---

[20] Remember that in (27) and (28), we read $\boxtimes \varphi$ as "it is doxastically necessary that φ" or "in all worlds compatible with the agent's beliefs, φ". The whole of (27) gives the condition that the truth of the belief report imposes on the actual world and the space of the agent's doxastic alternatives, so we would not read the second conjunct as "the agent believes that there exists …".

[21] Belardinelli and Lomuscio [6] include both x and z variables in their multi-agent quantified epistemic logic. Instead of distinguishing two types of variables, we could instead distinguish two types of quantifiers, in the tradition of Hintikka's [33] distinction between $\exists y$ and Ey. By understanding $\exists z$ quantification in terms of agent-relative roles, we are following Perry [51].

[22] This point is inspired by Crimmins and Perry [19] on notion provision vs. notion constraint.

attitudes in the style of Kaplan and Lewis, on the grounds that they do not generalize to handle "counterfactual attitudes" like *imagining*, *wishing*, and *dreaming*.

To see the problem, let us return to the example of Ralph and Ortcutt from Sect. 22.2.3. Now Ralph looks at Ortcutt on the beach and imagines that *no one ever saw that man*. We might report this as

(30) Ralph imagines that no one ever saw Ortcutt.

Ninan suggests that the truth-conditions of (30) should be stated in terms of what is true in the *worlds compatible with what Ralph is imagining*. By analogy with our treatment of the Tully-Cicero case in (28), we might try to formalize (30) as (31), relative to an assignment μ such that $\mu(r)$ is the role of *being the man seen*:

(31) $P(Ortcutt, r) \wedge \boxed{i}\exists x(P(x, r) \wedge NeverSeen(x))$,

where $\boxed{i}\varphi$ means that φ holds in all worlds compatible with Ralph's imagining. But (31) won't work, because it should be false throughout our intended model; since $\mu(r)$ is supposed to be the role of *being the man seen*, no one who plays that role in a world can be in the extension of the *NeverSeen* predicate in that world.

The reason that the schema for belief, represented in (28), does not work for imagination is that the schema for belief commits us to the following.

Doxastic Match Hypothesis: for an agent to believe of an object o via a role **r** that it has a property, the agent must also believe of o via **r** *that it plays* **r**.

As a corollary, if the agent does not have *any* true beliefs about the roles that o plays in his life, then the agent cannot have any *de re* beliefs about o. We think this hypothesis is plausible for a suitably general notion of "belief" (positive doxastic attitude), but Ninan is right to reject the analogous hypothesis for imagination.

Imaginative Match Hypothesis: for an agent to "imagine of" object o via role **r** that it has a property, the agent must also imagine of o via **r** *that it plays* **r**.

The example above, where Ralph sees Ortcutt and imagines that no one ever saw that man, is a counterexample to the Imaginative Match Hypothesis. It is not clear, however, whether Ralph can see Ortcutt and *believe* that no one ever saw that man.

Inspired by Lewis [45], Ninan's solution to the problem of counterfactual attitudes is (roughly) the following: we can *stipulate* that an object in a given "imagination world" v is supposed to represent Ortcutt, rather than finding the object as the player in v of a role that Ortcutt plays in the actual world. On this view, to check at a world whether Ralph is imagining that Ortcutt has a given property, we simply check whether the *stipulated representatives* of Ortcutt across Ralph's imagination worlds have that property. But there is an additional subtlety, arising due to "double vision" cases. Suppose that for *Ralph*, Ortcutt plays both the roles of *being the man seen* and *being the man called 'Bernard Ortcutt'*. Now Ralph looks at Ortcutt on the beach and says: "I'm imagining a situation in which that guy (Ralph points at the man on the waterfront) is distinct from Bernard Ortcutt, and in which I never saw that guy, and in which Ortcutt never goes by the name 'Bernard Ortcutt' " [47]. For handling this kind of case, Ninan generalizes the proposal stated above: we can stipulate that one object in an imagination world v represents Ortcutt relative to one role (*being the man seen*), while a different object in v can represent Ortcutt relative to another role (*being the man called 'Bernard Ortcutt'*).

Let us sketch how Ninan's treatment of counterfactual attitudes can be accommodated in the framework of FOIL. For simplicity, in Sect. 22.3 we did not define the full language of FOIL as presented by Fitting. In the full language, there are *intensional predicates* that we can apply to function variables. Previously we called function variables 'role variables', but let us now consider adding partial functions to F in our models that are not thought of as roles. We introduce a two-place intensional predicate Stip such that for function variables r and s,[23] Stip(r, s) is a formula. Intuitively, think of the truth of Stip(r, s) at world w relative to an assignment μ as telling us the following about the function $\mu(s)$: in every world v, $\mu(s)(v)$ is the individual in v that "represents" $\mu(r)(w)$, relative to role $\mu(r)$, "by stipulation".[24] In essence, Ninan proposes truth-conditions for (30) that are equivalent to that of

(32) $\mathsf{P}(Ortcutt, r) \wedge \exists s (\mathsf{Stip}(r, s) \wedge \Box \exists x (\mathsf{P}(x, s) \wedge Never\,Seen(x)))$.

In other words, Ortcutt plays the role of *being the man seen* for Ralph, and in every world compatible with what Ralph imagines, the thing that *by stipulation* represents Ortcutt *relative to that role* has the property of never being seen. It is easy to see how this approach will also handle Ralph's double vision case above.

In what follows, we will return to the simpler treatment of belief that assumes the Doxastic Match Hypothesis. However, it is noteworthy that the FOIL framework has the flexibility to model attitudes for which such a hypothesis is not reasonable.

22.4 Names in Alethic and Epistemic Logic

In this section, we discuss the treatment of names in alethic and epistemic/doxastic logic. In Sect. 22.4.1, we begin by discussing a problem about names that has lead some to reject standard semantics for epistemic/doxastic predicate logic as incoherent. In Sect. 22.4.4, we will propose a solution to this problem based on the ideas of Sect. 22.2.3.

22.4.1 The "Hintikka-Kripke Problem"

Consider the difference between the following:

---

[23] We could have two sorts of function variables, $r_1, r_2, \ldots$ for roles and $s_1, s_2, \ldots$ for non-role functions. Or we could indicate the difference between role functions and non-role functions by a one-place predicate Role whose extension contains only functions to be thought of as roles.

[24] According to this intuitive understanding, the extension of Stip should be a *functional* relation: if Stip(r, s) holds, then Stip(r, s') should not hold for any $s' \neq s$.

(33) Hesperus = Phosphorus, but it's not necessary that Hesperus = Phosphorus.
(34) Hesperus = Phosphorus, but Elwood doesn't believe that Hesperus = Phosphorus.

Following Kripke [41], (33) should be formalized by a sentence in our formal language that is *unsatisfiable*. By contrast, following Hintikka [32], (34) should be formalized by a sentence that is *satisfiable*. Translating (33) as

(35) $Hesperus = Phosphorus \land \neg\boxed{a}\,Hesperus = Phosphorus,$

if we assume the R_a-rigidity of names from Definition 22.2, then (35) is unsatisfiable, as desired. Similarly, if we translate (34) as

(36) $Hesperus = Phosphorus \land \neg\boxed{d}Hesperus = Phosphorus$

and assume the R_d-rigidity of names, then (36) is also unsatisfiable—but we want our formalization of (34) to be *satisfiable*. This has been called the "Hintikka-Kripke Problem" [42]. To solve it, we have two choices: give a different formalization of (34) or give up R_d-rigidity. The second seems to have become common among logicians working on epistemic predicate logic, following Hintikka's [34] view that "in the context of propositional attitudes even grammatical proper names do not behave like 'logically proper names' ". For example, in their textbook on first-order modal logic, Fitting and Mendelsohn [26] write that "the problem, quite clearly, lies with the understanding that these names are rigid designators. We see that, although they are rigid within the context of an *alethic* reading of $\Box$, they cannot be rigid under an *epistemic* reading of $\Box$". This suggests that we should not assume the R-rigidity of names when R is an epistemic/doxastic accessibility relation (also see [54]). The question is how we are supposed to understand the failure of R_d-rigidity. As Linksy [46] writes:

> Hence, these names do not denote the same thing in all doxastically possible worlds; that is, they are not doxastically rigid designators. How are we to make this situation intelligible to ourselves? Hesperus (= Phosphorus) is (are?) *two* objects in the world described by the sentences [true at some world]. It is not just that 'Hesperus' and 'Phosphorus' are names of different objects, for that is easily enough understood. The problem is that in this world Hesperus (= Phosphorus) is not Phosphorus (= Hesperus). That cannot be understood at all.

Let us call this the *impossible worlds* explanation of the failure of R_d-rigidity, which leads Linsky to believe that there are problems with analyzing belief ascriptions in terms of the semantics of $\Box$ that do not arise for alethic necessity (also see [4], which is even more critical). By contrast, Aloni [1, pp. 9–10] interprets the failure of R_d-rigidity in the weaker way that Linsky mentions:

> [I]n different doxastic alternatives a proper name can denote different individuals. The failure of substitutivity of co-referential terms (in particular proper names) in belief contexts does not depend on the ways in which terms actually refer to objects (so this analysis is not in opposition to Kripke's (1972) theory of proper names), *it is simply due to the possibility that two terms that actually refer to one and the same individual are not believed by someone to do so*.... Many authors...have distinguished semantically rigid designators from epistemically rigid designators—the former refer to specific individuals in counterfactual situations, the

latter identify objects across possibilities in belief states—and concluded that proper names are rigid only in the first sense.[25] [emphasis added]

The emphasized claim is in the tradition of Stalnaker [63]: "If a person is ignorant of the fact that Hesperus is Phosphorus, it is because his knowledge fails to exclude a possible situation in which, because causal connections between names and objects are different, one of those names refers to a different planet, and so the statement, 'Hesperus is Phosphorus' says something different than it actually says".

However, there are two problems with appealing to Stalnaker's idea in support of making (36) satisfiable by giving up R_d-rigidity: first, the appeal leads to serious problems about how to understand the valuation function V in modal models and about how R_d and R_a are related, as discussed in Sect. 22.4.2; second, Stalnaker's idea is inadequate to explain other attributions of ignorance, as discussed in Sect. 22.4.3. After explaining these problems, we will solve the Hintikka-Kripke problem in the way suggested by our discussion in Sect. 22.2.3: give a different formalization of (34).

22.4.2 Representational Versus Interpretational Semantics

Those who give up R_d-rigidity allow models in which, e.g., wR_dv, $V(Hesperus, w) = V(Phosphorus, w)$, but $V(Hesperus, v) \neq V(Phosphorus, v)$. One might think that this simply reflects Stalnaker's idea that 'Hesperus' and 'Phosphorus' refer to different things in v but the same thing in w. The problem is that this blurs an important distinction about how to understand the valuation V. This distinction is closely related to Etchemendy's [22] distinction between *representational* and *interpretational* semantics for classical logic, but here applied to models for modal logic.

When we switch from one model $\mathcal{M} = \langle W, R, D, F, V \rangle$ to another model $\mathcal{M}' = \langle W, R, D, F, V' \rangle$ that differs only with respect to its valuation function, one way to think of this switch is as a change in how we interpret the language. Intuitively, V may interpret the predicate *white* so that its extension $V(white, w)$ in any world w is the set of things in w that are white, while V' may interpret *white* so that its extension $V'(white, w)$ in any world w is the set of things in w that are green.

When we switch from one world w to another world v within the same model $\mathcal{M} = \langle W, R, D, F, V \rangle$ (so $w, v \in W$), the extension $V(white, w)$ of the predicate *white* in w may differ from the extension $V(white, v)$ of *white* in v. If this is so, then the set of white objects in w differs from the set of white objects in v. Similarly, if $V'(white, w)$ differs from $V'(white, v)$, then the set of green objects in w differs from the set of green objects in v. We could also have $V(white, w) \neq V(white, v)$ but $V'(white, w) = V'(white, v)$, in which case w and v differ with respect to which objects are white, but they are the same with respect to which objects are green.

The point is that we think of the difference between $V(white, w)$ and $V(white, v)$ as reflecting a difference in how the worlds w and v are, not a difference in how we

---

[25] The last of the quoted sentences occurs in footnote 7 of [1].

interpret the language—whereas we can think of the difference between $V(white, w)$ and $V'(white, w)$ as reflecting a difference in how we interpret the language.

Exactly the same points apply to names. If we think of the difference between $V(white, w)$ and $V(white, v)$ as reflecting a difference in how the worlds w and v are, not a difference in how we interpret the language, then we should also think of any difference between $V(Hesperus, w)$ and $V(Hesperus, v)$ as reflecting a difference in how the worlds w and v are, not a difference in how we interpret the language. (Though of course we can think of $V(Hesperus, w) \neq V'(Hesperus, w)$ as reflecting a difference in how we interpret the language.) This brings us back to Linsky's question of how we are to make intelligible that $V(Hesperus, w) = V(Phosphorus, w)$ but $V(Hesperus, v) \neq V(Phosphorus, v)$. For in this case v is not merely a world in which words refer to different things; it is a metaphysically impossible world.

Here is a crucial point: assuming that $V(Hesperus, w) = V(Phosphorus, w)$ but $V(Hesperus, v) \neq V(Phosphorus, v)$, we cannot have $wR_a v$, for otherwise we violate the R_a-rigidity of names and (35) becomes satisfiable, which no one wants. Hence it is not the case that $wR_a v$, so v is an impossible world (relative to w). This shows that v cannot simply be understood as a world largely like w but in which 'Hesperus' refers to something different than in w—for such a world should be metaphysically *possible* (relative to w). The upshot is that one cannot explain the failure of R_d-rigidity in the weaker way proposed at the end of Sect. 22.4.1.

A defender of that proposal might reply that the problem is instead with trying to combine alethic and doxastic operators in one system, as we have. But why should such a combination be problematic? We take it to be a virtue of our approach below that it allows such a combination for a full solution of the Hintikka-Kripke problem.

Although Stalnaker's idea from Sect. 22.4.1 cannot be implemented in standard modal semantics simply by giving up R_d-rigidity, it may be possible to implement with a more complicated two-dimensional framework. However, we will not investigate that possibility until Sect. 22.4.5, since in Sect. 22.4.3 we will raise a problem for the idea itself.

22.4.3 Ignorance of Co-reference Versus Ignorance of Identity

The problem is that ignorance of co-reference and ignorance of identity can come apart. Before explaining this idea, let us formally represent ignorance of co-reference. Suppose that our language contains not only names $n_1, n_2, \ldots$ of objects, but also names 'n_1', 'n_2', $\ldots$ of those names. Moreover, suppose that the object names $n_1, n_2, \ldots$ are also elements of the domain D of our models—they are also objects— and let us require that $V('n_i', w) = n_i$ for all names n_i and worlds w. Finally, suppose we have a two place naming predicate N such that $V(N, w)$ is the set of pairs $\langle n_i, d \rangle$ such that speakers in world w use n_i to *name* object d. With this setup, we can state the point from Sect. 22.4.2 about how to understand the V function as follows:

$\langle n_i, d \rangle \in V(N, w)$ need not imply $V(n_i, w) = d$ or vice versa.

In other words, how speakers in different worlds in our model use names in the language is one thing, reflected by $V(N, w)$; and how *we* who build the model are interpreting the names of the language is another, reflected by $V(n_i, w)$.

Now we can easily represent a world w in which an agent does not believe that 'Hesperus' and 'Phosphorus' co-refer by including a world v such that $w R_d v$ and:

- $\{o \in D \mid \langle Hesperus, o \rangle \in V(N, w)\} = \{o \in D \mid \langle Phosphorus, o \rangle \in V(N, w)\}$;
- $\{o \in D \mid \langle Hesperus, o \rangle \in V(N, v)\} \neq \{o \in D \mid \langle Phosphorus, o \rangle \in V(N, v)\}$.

In such a case, even given that $V(Hesperus, u) = V(Phosphorus, u)$ for all $u \in W$, the following are both satisfied at w:

(37) $\forall x (N('Hesperus', x) \leftrightarrow N('Phosphorus', x))$;
(38) $\neg \boxed{d} \forall x (N('Hesperus', x) \leftrightarrow N('Phosphorus', x))$,

so we have a case of ignorance of co-reference.

Before explaining how to formalize ignorance of identity, we will show how ignorance of co-reference and ignorance of identity can come apart.

Example 22.1 (The Registrar) The county registrar goes fishing regularly with his old friend Elwood. Unknown to the registrar, however, Elwood's identical twin, Egbert, occasionally substitutes for him on these fishing trips. Since the trips are mostly silent, Egbert has no problem keeping the deception from the registrar. The registrar regularly says things like "Hey Elwood, pass me a beer", while talking to Egbert.

It seems fair to say,

(39) Although Egbert is not Elwood, the registrar believes that Egbert is Elwood.

Traditionally, (39) would be rendered as

(40) $Egbert \neq Elwood \wedge \boxed{d} \, Egbert = Elwood$.

The truth of (40) requires that $Egbert = Elwood$ be true in all worlds compatible with what the Registrar believes. Its truth therefore requires that names are not "doxastically rigid" designators. At first blush, this seems to mean that only impossible worlds are compatible with what the registrar believes. In the spirit of Aloni, we might seek to avoid this unfortunate result by adopting the following analysis.

Co-Reference Mistake Analysis: the satisfiability of (40) is simply due to the possibility that two terms that actually refer to different individuals are believed by the registrar to co-refer, that is, to refer to the same individual.

The problem with this analysis becomes clear when we consider the rest of the story.

Example 22.1 (The Registrar Continued) The registrar, being the registrar, knows that Elwood has a brother, whom he thinks he has never met, although he sends him an invoice for his taxes each year, and often follow-up reminders. Based on his records, the registrar *knows* that 'Elwood' and 'Egbert' refer to different people.

Hearing the registrar say to Egbert, "Hey Elwood, your brother better pay his taxes", someone in on the deception might explain to a third party:

(41) Although the registrar knows that 'Elwood' and 'Egbert' refer to different people, he believes that Egbert is Elwood.

What (41) shows is that the Co-Reference Mistake Analysis doesn't work. The truth of (39) cannot always be understood in terms of false belief about co-reference, because the truth of (41) cannot be so understood. In Sect. 22.4.4, we will explain how, on our account, (41) can nonetheless be a reasonable and true thing to say.

22.4.4 The "Hintikka-Kripke Problem" Resolved

In Sect. 22.4.1, we noted two ways to respond to the Hintikka-Kripke Problem: give up R_d-rigidity or give a different formalization of (34). Having seen the problems with the first, let us consider the second. When someone says "Elwood does not believe that Hesperus is Phosphorus", our formalization of the claim is not

(42) $\neg \boxdot Hesperus = Phosphorus.$

In fact, it is not clear that (42) is the correct formalization of any natural language belief ascription. As Lewis [43, p. 360] said in another context, "why must every logical form find an expression in ordinary language?" Relatedly, consider:

(43) $Hesperus = Hesperus \rightarrow \boxdot Hesperus = Hesperus.$
(44) $Hesperus = Phosphorus \rightarrow \boxdot Hesperus = Phosphorus.$

In our framework, (43) *and* (44) are valid, but this does not mean that we would claim in natural language that "If Hesperus is Phosphorus, then Elwood believes that Hesperus is Phosphorus". For the function of belief ascriptions involving names in natural language is not just to say something about how the *objects named* show up across doxastic alternatives, which is all that (44) manages to capture.

Before handling the Hesperus and Phosphorus case, let us treat the story of Elwood and Egbert from Example 22.1, using ideas from Sect. 22.2.3. Recall:

(39) Although Egbert is not Elwood, the registrar believes that Egbert is Elwood.

In the context of Example 22.1, we propose to analyze (39) as follows. Elwood is playing a number of epistemic roles in the registrar's life, but the role that is contextually salient in Example 22.1 is *being the source of a cluster of memories—about his old friend and fishing buddy* ($\mathbf{r}_f$). Egbert is also playing a number of epistemic roles in the registrar's life, but the role that is contextually salient in Example 22.1 is *being the person seen and talked to* ($\mathbf{r}_s$). Consider a model $\mathcal{M}$, assignment μ, and role variables r_f and r_s such that for all worlds w in $\mathcal{M}$, $\mu(r_f)(w)$ is the player of role $\mathbf{r}_f$ in w, and $\mu(r_s)(w)$ is the player of role $\mathbf{r}_s$ in w. Then using Z-abbreviation (Definition 22.6), we formalize (39) in the style of (26) in Sect. 22.2.3, instead of (40):

(45) $Egbert \neq Elwood \land Elwood = z_f \land Egbert = z_s \land \boxed{d} z_f = z_s.$

If we translate (39) as (45), we can maintain the Conservative Approach of Sect. 22.1: there is no need to claim that names in doxastic logic are not "doxastically rigid" or that "doxastically possible worlds" are not plain possible worlds. Moreover, (45) can be true even if the registrar believes that 'Elwood' and 'Egbert' do not co-refer:

(46) $\boxed{d} \neg \exists x_1 (N(\text{'Elwood'}, x_1) \land N(\text{'Egbert'}, x_1)).$

Hence we can also handle the second part of the registrar story in (41).

Let us now return to the classic case of Hesperus and Phosphorus:

(47) Elwood does not believe that Hesperus is Phosphorus.

One might utter (47) in a variety of contexts. Suppose Elwood has never heard the words 'Hersperus' and 'Phosphorus', but he likes to look at the planet Venus early in the evening and at the same planet early in the morning, not realizing it is the same one. In this case, it would be natural to utter (47). Take a model $\mathscr{M}$, assignment μ, and role variables r_e and r_m such that for any world w in $\mathscr{M}$, $\mu(r_e)(w)$ is the star that Elwood likes to look at in the evening in w, and $\mu(r_m)(w)$ is the star that Elwood likes to look at in the morning in w. Using Z-abbreviation, we translate (47) as:

(48) $Hesperus = z_e \land Phosphorus = z_m \land \neg \boxed{d} z_e = z_m.$

Another context in which (47) makes sense is one where Elwood has heard of Hesperus and Phosphorus in Astronomy class, but since he wasn't paying attention, he has no idea what they are. Here the roles that Hesperus/Phosphorus plays in his life are simply *being the source of the 'Hesperus' network exploited by Elwood* (recall Sect. 22.2.3) and *being the source of the 'Phosphorus' network exploited by Elwood*. In this case, we translate (47) with a sentence of the same form as (48), only using role variables associated with these different roles in our model. What this shows is that the role-based analysis subsumes Stalnaker's [63, 85f] analysis (recall Sect. 22.4) in those contexts where Stalnaker's analysis works. But the role-based analysis also works in cases like the Elwood-Egbert story, where Stalnaker's does not.

Finally, since (48) can be true at a world where

(49) $\boxed{d} Hesperus = Phosphorus$

is true, belief attributions as in (47) do not pose a problem for a combined epistemic-alethic modal logic. The Hintikka-Kripke Problem is no longer a problem.

22.4.5 Two-Dimensional Epistemic Models

Recall Aloni's [1, p. 9] idea, similar to Stalnaker's [63], that Hesperus-Phosphorus style cases arise because of the "possibility that two terms that actually refer to one and the same individual are not believed by someone to do so". In Sect. 22.4.2, we argued that this idea cannnot be correctly implemented in standard modal semantics.

Let us now return to the suggestion that it may be implementable in a two-dimensional epistemic framework (cf. [61] and references therein).

Definition 22.7 (*2D Constant Domain Models*) A 2D (constant domain) model for the language of Definition 22.1 is a tuple $\mathcal{M} = \langle W, R_a, \mathbf{R}_d, D, F, \mathbf{V} \rangle$ where W, R_a, D, and F are defined as in Definition 22.2; $\mathbf{R}_d$ is a binary relation on $W \times W$; for any relation symbol Q_i^k and worlds $w, v \in W$, $\mathbf{V}(Q_i^k, w, v)$ is a set of k-tuples of objects; for any name n_i and worlds $w, v \in W$, $\mathbf{V}(n_i, w, v)$ is an object. We assume the general rigidity condition for names that for all $w, v, u \in W$, $\mathbf{V}(n_i, w, v) = \mathbf{V}(n_i, w, u)$.

For the sake of generality, we have defined $\mathbf{R}_d$ as a kind of 2D accessibility relation, following Israel and Perry [38] and Rabinowicz and Segerberg [57]. This raises the question of what $\langle w, v \rangle \mathbf{R}_d \langle w', v' \rangle$ is supposed to mean intuitively. However, here we will consider the class of models such that if $\langle w, v \rangle \mathbf{R}_d \langle w', v' \rangle$, then $w' = v'$ and $\langle x, v \rangle \mathbf{R}_d \langle w', v' \rangle$ for all $x \in W$. Hence all we need to know is whether the second coordinates are related, written as $v \mathbf{R}_d v'$. Take $v \mathbf{R}_d v'$ to mean that everything the agent believes in v is compatible with the hypothesis that v' is his actual world.

For a name n_i, take $\mathbf{V}(n_i, w, v) = d$ to mean that if we consider w as the actual world, then n_i (as used by speakers in w) names d in world v. Suppose w is a world in which the heavenly body that people see in the evening, that they call 'Hesperus', etc., is the same as the heavenly body that they see in the morning, that they call 'Phosphorus', etc. Hence if we consider w as actual, then we will have $\mathbf{V}(Hesperus, w, v) = \mathbf{V}(Phosphorus, w, v)$ for all worlds $v \in W$, regardless of how language is used (or whether there are any language users) in v. However, if we consider as actual a world w' in which the heavenly body that people see in the morning, that they call 'Hesperus', etc., is not the same as the heavenly body that they see in the morning, that they call 'Phosphorus', etc., then we will have $\mathbf{V}(Hesperus, w', v) \neq \mathbf{V}(Phosphorus, w', v)$ for all worlds $v \in W$.[26]

Variable assignments are defined as in Definition 22.3, but the interpretation of a name is now given relative to a pair of worlds instead of a single world.

Definition 22.8 (*Interpretation of Terms*) The interpretation $[t]_{\mathcal{M}, w, v, \mu}$ of a term t in a 2D model $\mathcal{M}$ at world v, *with world w considered as actual*, is an object given by:

$$[n_i]_{\mathcal{M}, w, v, \mu} = V(n_i, w, v);$$
$$[x_i]_{\mathcal{M}, w, v, \mu} = \mu(x_i).$$

Definition 22.9 (*2D Truth*) Given a 2D model $\mathcal{M}$ with $w, v \in W$, an assignment μ, and a formula φ, we define $\mathcal{M}, w, v \models_\mu \varphi$ as follows:

---

[26] One may try to apply a similar strategy to *predicate symbols* in order to model agents who do not believe/know that two (necessarily) co-extensive predicates are co-extensive.

$\mathcal{M}, w, v \vDash_\mu Q_i^k(t_1, \ldots, t_k)$ iff $\langle [t_1]_{\mathcal{M}, w, v, \mu}, \ldots, [t_k]_{\mathcal{M}, w, v, \mu} \rangle \in \mathbf{V}(Q_i^k, w, v)$;

$\mathcal{M}, w, v \vDash_\mu t = t'$ iff $[t]_{\mathcal{M}, w, v, \mu} = [t']_{\mathcal{M}, w, v, \mu}$;

$\mathcal{M}, w, v \vDash_\mu \mathsf{P}(t, r_i)$ iff $\mu(r_i)$ is defined at v and $[t]_{\mathcal{M}, w, v, \mu} = \mu(r_i)(v)$;

$\mathcal{M}, w, v \vDash_\mu \neg\varphi$ iff $\mathcal{M}, w, v \nvDash_\mu \varphi$;

$\mathcal{M}, w, v \vDash_\mu (\varphi \wedge \psi)$ iff $\mathcal{M}, w, v \vDash_\mu \varphi$ and $\mathcal{M}, w, v \vDash_\mu \psi$;

$\mathcal{M}, w, v \vDash_\mu \forall x_i \varphi$ iff for all $d \in D$, $\mathcal{M}, w, v \vDash_{\mu[x_i/d]} \varphi$;

$\mathcal{M}, w, v \vDash_\mu \forall r_i \varphi$ iff for all $f \in F$, $\mathcal{M}, w, v \vDash_{\mu[r_i/f]} \varphi$;

$\mathcal{M}, w, v \vDash_\mu \boxed{a}\varphi$ iff for all $v' \in W$, if $vR_a v'$ then $\mathcal{M}, w, v' \vDash_\mu \varphi$;

$\mathcal{M}, w, v \vDash_\mu \boxed{d}\varphi$ iff for all $w', v' \in W$, if $\langle w, v \rangle \mathbf{R}_d \langle w', v' \rangle$

then $\mathcal{M}, w', v' \vDash_\mu \varphi$.

Given our assumed constraints on $\mathbf{R}_d$, we can re-write the last clause as[27]

$$\mathcal{M}, w, v \vDash_\mu \boxed{d}\varphi \text{ iff for all } v' \in W, \text{ if } v\mathbf{R}_d v' \text{ then } \mathcal{M}, v', v' \vDash_\mu \varphi.$$

To see how this is supposed to solve the Hintikka-Kripke problem, consider

(50) $\boxed{a} n_1 = n_2 \wedge \boxed{d} n_1 \neq n_2$.

We have $\mathcal{M}, w, w \vDash_\mu \boxed{a} n_1 = n_2 \wedge \boxed{d} n_1 \neq n_2$ iff both of the following hold:

- for all $x \in W$, if $wR_a x$ then $\mathcal{M}, w, x \vDash_\mu n_1 = n_2$;
- for all $y \in W$, if $wR_d y$ then $\mathcal{M}, y, y \vDash_\mu n_1 \neq n_2$.

Clearly we can construct a model satisfying these conditions, so (50) is satisfiable.

There is much to be said about the 2D approach, but we will limit ourselves to two points. First, there is a way of understanding the 2D treatment of names in epistemic contexts as a *special case* of the role-based treatment. In particular, with every name n_i we can associate a role f_i such that

$$f_i(v) = \mathbf{V}(n_i, v, v).$$

Then if we map role variables r_1 and r_2 to f_1 and f_2, respectively, (50) will have the same truth value at any pair of worlds as its role-based translation:

(51) $\boxed{a} n_1 = n_2 \wedge \mathsf{P}(n_1, r_1) \wedge \mathsf{P}(n_2, r_2) \wedge \boxed{d} \exists x_1 \exists x_2 (\mathsf{P}(x_1, r_1) \wedge \mathsf{P}(x_2, r_2) \wedge x_1 \neq x_2)$.

Note, however, that there may be many other roles that the objects named by n_1 and n_2 play, besides f_1 and f_2. In effect, the 2D framework restricts us to just those roles.

This leads to the second point: it is not clear how the 2D framework can handle the case of the registrar in Example 22.1 in Sect. 22.4.3 without bringing in roles. In our world w, the registrar believes that 'Egbert' and 'Elwood' refer to different people, so for any world v compatible with his beliefs, *Egbert $\neq$ Elwood* should be true at v *considered as actual*. But then $\boxed{d} Egbert \neq Elwood$ will be true at w, so if this is the two-dimensionalists' formalization of 'the registrar believes that Egbert is not Elwood', then they face the problem raised in Sect. 22.4.3: in the context of

---

[27] Compare this to the "fixedly actually" operator of Davis and Humberstone [21].

Example 22.1 it seems true to say instead that 'the registrar believes that Egbert is Elwood'.

In what follows we return to the 1D framework. There are good reasons for multi-dimensionality to deal with terms like 'now' and 'actually', but it seems that roles are still needed to deal with belief attributions involving names.[28] In Sect. 22.5, we shall see how roles are also useful in formalizing belief attributions involving quantification.

22.5 Quantification into Epistemic Contexts

Having shown how a role-based analysis of attitude ascriptions resolves the Hintikka-Kripke Problem, we will now apply the analysis to quantification into epistemic contexts, comparing it to the analyses of Carlson [14] and Aloni [1].

First, consider the following sentence:

(52) "The police do not know who a certain person is" [14, p. 232].

How should we translate (52) into our formal language? As suggested in Sect. 22.2.3, we cannot answer this question simply by looking at the sentence out of context.

In our view, there seem to be two readings of (52), which are natural in different contexts. First, suppose the police pride themselves on keeping track of everyone in the area. However, someone has slipped through the cracks: Jones, whom the police do not know anything about. In this case, it makes sense to utter (52), understood as

(53) $\exists x_1 \neg \exists z_1 \boxed{e}\, x_1 = z_1,$

the Z-abbreviation of

(54) $\exists x_1 \neg \exists r_1 \boxed{e}\, \exists x_2 (P(x_2, r_1) \wedge x_1 = x_2),$

where we write $\boxed{e}$ in place of $\boxed{\square}$ for *epistemic* necessity.

Second, suppose Jones now plays the role for the police of *being the suspect chased*. However, for all the police know, they could be chasing Smith instead of Jones. In this case, it makes sense to utter (52), now understood as

(55) $\exists z_1 \neg \exists x_1 \boxed{e}\, x_1 = z_1,$

the Z-abbreviation of

(56) $\exists r_1 \neg \exists x_1 \boxed{e}\, \exists x_2 (P(x_2, r_1) \wedge x_1 = x_2).$[29]

Which translation is better depends on the context in which (52) is uttered.

Can we translate (52) with only z variables? Carlson [14] tries to do so with

---

[28] To deal in the 1D framework with an agent who does not believe, e.g., that something contains water iff it contains H20, we would need to generalize the notion of role so that *properties* (understood extensionally, intensionally, or hyper-intensionally) can play roles for an agent.

[29] Note that (55)/(56) does not require the existence of anyone who actually plays r_1. We can express a reading that requires the existence of a role-player with: $\exists z_1 (\exists x_1 z_1 = x_1 \wedge \neg \exists x_2 \boxed{e}\, x_2 = z_1).$

(57) $\exists z_1 \neg \exists z_2 \boxed{e} \; z_1 = z_2$,

where $\exists z_1$ and $\exists z_2$ both quantify over the same set of functions. He allows z variables to be mapped to partial functions and uses a three-valued semantics such that (57) is satisfiable in his framework. However, with only one kind of variable in that framework, there is no way to make a distinction like the one we have made between (53) and (55). Moreover, the intuitive meaning of (57) is not at all clear.

An advantage of having quantification over both objects and functions is that when one tries to do all the work with just quantification over functions, one is tempted to impose otherwise unnatural constraints on the set of functions. For example, Carlson [14] and Aloni [1] propose two conditions on the set F of functions, which they interpret as *individuating functions* and *individual concepts*, respectively: an *existence* condition and a *uniqueness* condition.

Definition 22.10 (*Existence Condition*) In $\mathscr{M} = \langle W, R_a, R_d, D, F, V \rangle$, F satisfies the *existence condition* iff for all $w \in W$, $d \in D$, there is some $f \in F$ with $f(w) = d$.

When we think of the functions in F as agent-relative roles, the existence condition is not plausible. For it is built in to the idea of agent-relativity that there may be an object that does not play any role in the cognitive life of our agent in a world.

Let us consider Aloni's argument for the existence condition. First, consider:

(58) If the president of Russia is a spy, then there is someone who is a spy.

Translating (58) as

(59) $S(p) \rightarrow \exists z S(z)$,

Aloni [1] notes that if we do not assume the existence condition on F, then (59) is not valid, whereas the translation of (58) should be valid. Hence we should assume the existence condition. However, this is too quick, because we have two options: assume the existence condition or give a different translation of (58). In the framework of Sect. 22.3, there is a clear candidate for the latter:

(60) $S(p) \rightarrow \exists x S(x)$,

where $\exists x$ quantifies over D. Unlike (59), (60) is valid in our framework. We take (60) to be the appropriate translation of (58). The existence assumption seems to be an artifact of trying to make $\exists z$ do all the work of two types of quantification.

Definition 22.11 (*Uniqueness Condition*) In $\mathscr{M} = \langle W, R_a, R_d, D, F, V \rangle$, F satisfies the *uniqueness condition* iff for all $w \in W$, $f, f' \in F$, if $f \neq f'$, then $f(w) \neq f'(w)$.

When we think of functions in F as agent-relative roles, the uniqueness condition is not plausible. For it is built in to the idea of roles that there can be an object in a world that plays multiple roles in the cognitive life of our agent.

Without uniqueness, we can easily handle ascriptions of ignorance such as:

(61) "There is someone who might be two different people as far as the police know" [14, p. 237].

Imagine, for example, that although the same person was both the thief and the getaway driver, for all the police know, different people played these roles—it was a two man job. Corresponding to (61), we have the satisfiable sentence

(62) $\exists z_1 \exists z_2 (z_1 = z_2 \wedge \neg \boxed{e} \, z_1 = z_2)$.

As one can easily check with Z-translation, (62) is true in the case where the same person plays the role of the thief and the role of the driver in the actual world, but in some world compatible with what the police know, there are two people involved.

Although the most natural way of making (62) satisfiable violates uniqueness, Carlson manages to make (62) satisfiable while requiring uniqueness. He does so by mapping z_1 and z_2 to the same partial function, which is defined at the world of evaluation but undefined at some epistemically accessible world. As Carlson [14, p. 238] puts it, "[(62)] in our interpretation does not imply that ...[z_1 and z_2] ...pick out two different properly cross-identified individuals in some alternative, only that they fail to refer to one and the same individual somewhere".

The problem with Carlson's analysis is that it does not generalize to capture other cases, such as the following elaboration of the heist example above:

(63) Someone who was at the crime scene might be two people as far as the police know, but the police know that whoever was there was a gangster.

It seems that any good formalization of (63) should imply the following:

(64) $\exists z_1 \exists z_2 (z_1 = z_2 \wedge \neg \boxed{e} \, z_1 = z_2 \wedge \boxed{e} (G z_1 \wedge G z_2))$.

However, (64) is unsatisfiable in Carlson's system. For if z_1 and z_2 are mapped to functions that are undefined at some epistemically accessible world, as required for $z_1 = z_2 \wedge \neg \boxed{e} \, z_1 = z_2$ to be true for Carlson, then $\boxed{e} (G z_1 \wedge G z_2)$ is not true. By contrast, since we reject uniqueness, (64) is satisfiable in our framework.

Aloni observes that without uniqueness, the following are not equivalent:

(65) $\neg \exists z_1 (z_1 = Ortcutt \wedge \boxed{A} Spy(z_1))$.
(66) $\exists z_1 (z_1 = Ortcutt \wedge \neg \boxed{A} Spy(z_1))$.

Indeed, without uniqueness, the truth of (66) is compatible with the falsity of (65). And without existence, the truth of (65) is compatible with the falsity of (66).

Given the non-equivalence of (65) and (66), Aloni [1] concludes that formal systems without the uniqueness condition "predict a structural ambiguity for sentences like 'Ralph does not believe Ortcutt to be a spy', with a wide scope reading asserting that Ralph does not ascribe espionage to Ortcutt under any (suitable) representation, and a narrow scope reading asserting that there is a (suitable) representation under which Ralph does not ascribe espionage to Ortcutt. This ambiguity is automatically generated by any system" that does not satisfy uniqueness, but Aloni doubts that there is any such ambiguity in natural language.

But the fact that (65) and (66) are not equivalent in a given formal system does not mean that the system "predicts a structural ambiguity" for the English sentence

(67) Ralph does not believe Ortcutt to be a spy.

Linguists predict ambiguities in English. Logical systems do not. (Repeating Lewis [43], "why must every logical form find an expression in ordinary language?")

On our view, the correct formalization of (67) depends on the context. Suppose we just had a long conversation about Ralph's only next door neighbor, Ortcutt, when you utter (67). Although Ortcutt plays several roles in Ralph's life, in this case the contextually salient role is *being the next door neighbor*. Consider a model $\mathcal{M}$, assignment μ, and role variable r_n such that for any world w in $\mathcal{M}$, $\mu(r_n)(w)$ is Ralph's next door neighbor in w. Using Z-abbreviation, we formalize (67) as

(68) $Ortcutt = z_n \wedge \neg\boxed{d}Spy(z_n)$.

Of course, (66) follows from (68). Now suppose the conversation turns to Ralph's beliefs about *the man he sees on the beach*, who unbeknownst to Ralph is Ortcutt. At this point, we might wonder whether in uttering (67) you had in mind the full strength of (65), from which it follows that Ralph does not believe of the man he sees on the beach—via the role r_b, say—that he is a spy. The coherence of wondering this suggests that it is not a problematic result that (65) and (66) are not equivalent.

While Aloni's framework requires both the existence and uniqueness conditions for any fixed set F of functions, it also allows that different sets of functions may serve as the domain of quantification in different contexts. To formalize this idea in the style of Aloni [1] with a language of z variables, expand the set of terms to include for all $i \in \mathbb{N}$ a set $\{z_1^i, z_2^i, \dots\}$ of variables associated with context i; second, redefine a model to be a tuple $\mathcal{M} = \langle W, R_a, R_d, D, \mathbf{F}, V \rangle$ such that $\mathbf{F} \subseteq \mathscr{P}(D^W)$ and for all $F \in \mathbf{F}$, $\mathcal{M} = \langle W, R_a, R_d, D, F, V \rangle$ satisfies Definition 22.2; third, redefine an assignment to be a function π like μ in Definition 22.3 but extended so that for all $i \in \mathbb{N}$, $\pi(i) \in \mathbf{F}$; finally, redefine the clause for quantification with z variables:

$$\mathcal{M}, w \vDash_\pi \forall z_j^i \varphi \text{ iff for all } f \in \pi(i), \mathcal{M}, w \vDash_{\pi[z_j^i/f]} \varphi.$$

Hence for each context i, z^i variables are associated with their own domain of quantification $\pi(i) \in \mathbf{F}$.[30] We can recover the semantics of Sect. 22.3 by requiring that $|\mathbf{F}| = 1$.

As suggested at the end of Sect. 22.3, one can easily generalize the FOIL semantics to allow many sets of roles, only we would use superscripts on role variables.

---

[30] Aloni considers it an advantage of this more general semantics that we can have

(69) $\mathcal{M}, w \vDash_\pi \exists z^i \varphi(z^i) \wedge \neg\exists z^j \varphi(z^j)$,

as if there is a shift in context mid-formula. Instead of doing this with one of Aloni's models, we could consider two regular models $\mathcal{M} = \langle W, R_a, R_d, D, \pi(i), V \rangle$ and $\mathcal{M}' = \langle W, R_a, R_d, D, \pi(j), V \rangle$, each associated with a different context, such that

(70) $\mathcal{M}, w \vDash_\mu \exists z \varphi(z)$ and $\mathcal{M}', w \vDash_{\mu'} \neg\exists z \varphi(z)$.

22.6 Multiple Agents and Points of View

In distinguishing roles from individual concepts, we emphasized the *agent-relativity* of roles. We will now make this relativity more explicit by extending our framework to a multi-agent language and multi-agent models. Typically the move from single to multi-agent epistemic logic is a matter of subscripting operators and relations by agent labels. We will take a different approach in two ways. First, instead of introducing many doxastic operators, we will introduce many "point of view" operators. Second, instead of subscripting these operators with agent labels, we will subscript them by *terms* of our language, so that agents will be individuals in our domain.[31]

Definition 22.12 (*Multi-Agent Language*) Given the same sets of basic symbols as in Definition 22.1, the multi-agent language is generated by the grammar ($i, k \in \mathbb{N}$):

$$t ::= n_i \mid x_i$$
$$\varphi ::= Q_i^k(t_1, \ldots, t_k) \mid t = t' \mid \mathsf{P}(t, r_i) \mid \neg\varphi \mid (\varphi \wedge \varphi) \mid \boxed{\varnothing}\varphi \mid \boxed{\varnothing}\varphi \mid \forall x_i \varphi \mid \forall r_i \varphi \mid \mathsf{pov}_t \varphi.$$

The only addition are the "point of view" operators pov_t, with the intended reading:

$$\mathsf{pov}_t \varphi \quad \text{"from the point of view of } t, \varphi\text{"}$$
$$\text{(or more technically, "centering on } t, \varphi\text{").}$$

The unsubscripted P predicate and $\boxed{\varnothing}$ operator can now be read as:

(Footnote 30 continued)

Aloni's motivation for considering (69) is the following kind of reasoning:

(I) Ralph believes that the man with the brown hat is a spy.
(II) The man with the brown hat is Ortcutt.
(III) So Ralph believes of Ortcutt that he is a spy.
(IV) Ralph believes that the man seen on the beach is not a spy.
(V) The man seen on the beach is Ortcutt.
(VI) So Ralph does *not* believe of Ortcutt that he is a spy.

Aloni concludes that

(71) $\exists z^1(z^1 = o \wedge \boxed{\varnothing}S(z^1)) \wedge \neg\exists z^2(z^2 = o \wedge \boxed{\varnothing}S(z^2))$

should be satisfiable, which it is in her semantics. However, it seems to us to be a mistake to conclude (VI) on the basis of (IV) and (V). Instead, by analogy with (I)-(III), one should conclude

(VI′) So Ralph believes of Ortcutt that he is not a spy.

Then we can express the compatibility of (III) and (VI′) by the satisfiable sentence

(72) $\exists z_1(z_1 = o \wedge \boxed{\varnothing}S(z_1)) \wedge \exists z_2(z_2 = o \wedge \boxed{\varnothing}\neg S(z_2))$.

This is not to say, however, that there are not other good motivations for the more general semantics.

[31] Thus, we have a *term-modal logic* in the sense of [23].

$P(t, r_i)$ "t plays role r_i (for the agent at the center)";
$\boxdot\varphi$ "it is doxastically necessary (for the agent at the center) that φ".

Thus, intuitively, $\text{pov}_t\boxdot\varphi$ indicates that *for individual t, it is doxastically necessary that φ*; and $\text{pov}_t P(t', r_i)$ indicates that *for individual t, t' plays role r_i*.[32] Note that allowing *variables* to occur as subscripts of the point of view operators significantly increases the language's expressive power, since these variables may then be bound by quantifiers. As a simple example, in this framework (as in those of [31] and [23]) one can express the likes of "someone believes φ", using $\exists x \, \text{pov}_x \boxdot\varphi$.

Definition 22.13 (*Multi-Agent Model*) A *multi-agent* constant domain model for the language of Definition 22.12 is a tuple $\mathcal{M} = \langle W, R_a, R_d, D, F, V \rangle$ such that:

- W, R_a, D, and V are as in Definition 22.2;
- R_d is a binary relation on $W \times D$;
- F is a set of partial functions from $W \times D$ to D.

The only difference between these models and those of Definition 22.2 is that the doxastic relation R_d and role functions in F now apply to Lewis's [44] "centered worlds", which are pairs of a possible world and an individual—the center.[33] For the doxastic relation, given any $w, w' \in W$ and $c, c' \in D$, we take $\langle w, c \rangle R_d \langle w', c' \rangle$ to mean that *it is compatible with individual c's beliefs in w that she is individual c' in w'* (as in [64, p. 70]), allowing an agent to have uncertainty both about the world and about *her own identity*. Since the relevant agent is given by the center c of the first centered world $\langle w, c \rangle$, we only need one R_d relation to represent the beliefs of multiple agents. Finally, for the roles, given any $f \in F$, $w \in W$, and $c \in D$, we take $f(w, c)$ to be the object that plays role f in world w *for the individual c*.

A variable assignment μ now maps role variables r_i to elements of F.

Definition 22.14 (*Truth*) Given a multi-agent model $\mathcal{M}$, $w \in W$, $c \in D$, assignment μ, and formula φ, define $\mathcal{M}, w, c \vDash_\mu \varphi$ as follows (with other clauses as in Definition 22.5):

$\mathcal{M}, w, c \vDash_\mu P(t, r_i)$ iff $\mu(r_i)$ is defined at $\langle w, c \rangle$ and $[t]_{\mathcal{M}, w, \mu} = \mu(r_i)(w, c)$;

$\mathcal{M}, w, c \vDash_\mu \boxdot\varphi$ iff for all $w' \in W$, if wR_aw' then $\mathcal{M}, w', c \vDash_\mu \varphi$;

$\mathcal{M}, w, c \vDash_\mu \boxdot\varphi$ iff for all $w' \in W$, $c' \in D$, if $\langle w, c \rangle R_d \langle w', c' \rangle$
$\qquad\qquad\qquad\qquad\qquad\qquad$ then $\mathcal{M}, w', c' \vDash_\mu \varphi$;

$\mathcal{M}, w, c \vDash_\mu \text{pov}_t\varphi$ iff $\mathcal{M}, w, [t]_{\mathcal{M}, w, \mu} \vDash_\mu \varphi$.

---

[32] In English, to say "from the point of view of t, φ", might suggest that t *believes* φ, but this it not the intended reading of $\text{pov}_t\varphi$, as its formal truth definition below makes clear.

[33] While we take a centered world to be any element of $W \times D$, one may wish to only admit pairs $\langle w, c \rangle$ such that c is an agent (in some distinguished set $Agt \subseteq D$) and c exists in w (using the existence predicate of Sect. 22.3, $c \in V(\mathsf{E}, w)$), but for simplicity we do not make these assumptions here. Also for simplicity, we are not putting explicit *times* into the centered worlds. Adding a temporal dimension to our framework would expand its range of application to further interesting issues.

Hence pov_t can be thought of as a "center shifting" operator. Like the similar $@_s$ operators of hybrid logic (see [2]), the pov_t operators are normal modal operators validating the K axiom ($\mathsf{pov}_t(\varphi \to \psi) \to (\mathsf{pov}_t\varphi \to \mathsf{pov}_t\psi)$) and necessitation rule (if φ is valid, so is $\mathsf{pov}_t\varphi$), plus self-duality ($\mathsf{pov}_t\varphi \leftrightarrow \neg\mathsf{pov}_t\neg\varphi$). One can define the agent-indexed $\boxdot_t\varphi := \mathsf{pov}_t\boxdot\varphi$ and $\mathsf{P}_t(t', r_i) := \mathsf{pov}_t\mathsf{P}(t', r_i)$, but there are logical reasons not to take these defined operators as primitive.[34]

Let us begin by analyzing Perry's [49] case of the man Heimson who believes he is Hume, with an added twist of "double vision". Suppose that Hume plays two roles in Heimson's life: first, Hume is the author of the books labeled 'David Hume' on Heimson's shelf, a role we will assign to the role variable r_a (for *author*); second, Hume is a pen pal of Heimson's (here taken to be a contemporary of Hume) who signs his letters with the name 'D.H.', a role we will assign to r_{pp} (for *pen pal*). Finally, Heimson is the person who Heimson finds out about by introspection, proprioception, etc., playing the *self-role* that we will assign to r_{self}.[35] The catch is that Heimson is confused about his own identity in such a way that we can say:

(73) Heimson believes that he is Hume and that D.H. lives far away.

We can formalize (73) as follows:[36]

(77) $\mathsf{pov}_{Heimson}\big[\mathsf{P}(Heimson, r_{self}) \wedge \mathsf{P}(Hume, r_a) \wedge \mathsf{P}(D.H., r_{pp}) \wedge$
$\quad \boxdot\exists x\exists y\exists z(\mathsf{P}(x, r_{self}) \wedge \mathsf{P}(y, r_a) \wedge \mathsf{P}(z, r_{pp}) \wedge x = y \wedge FA(z))\big].$

---

[34] By the truth definition, we have $\mathscr{M}, w, c \vDash_\mu \boxdot_t\varphi$ iff for all $w' \in W$, $c' \in D$, if $\langle w, [t]_{\mathscr{M}, w, \mu}\rangle R_d\langle w', c'\rangle$, then $\mathscr{M}, w', c' \vDash_\mu \varphi$. The problem with taking the $\boxdot_t$ operators as primitive instead of $\boxdot$ is that we would then lose important results of modal correspondence theory. For example, requiring that R_d be *reflexive* (thinking of it now as an *epistemic* accessibility relation) would not guarantee the validity of $\boxdot_t\varphi \to \varphi$, since the reflexivity of R_d would not guarantee that $\langle w, [t]_{\mathscr{M}, w, \mu}\rangle R_d\langle w, c\rangle$. But reflexivity would guarantee the validity of $\boxdot\varphi \to \varphi$, as desired.

[35] One may want to define the self-role such that for all worlds w and agents c, $f_{self}(w, c) = c$.

[36] A similar analysis applies to other well-known problems in the theory of reference, such as Castañeda's [16] puzzle about the first person. Surely through most of his life after 1884, Samuel Clemens believed that he wrote *Huckelberry Finn*. But one can imagine that in his dotage, Clemens held a copy of the book in his hand, saw that it was written by Mark Twain, but couldn't remember that 'Mark Twain' had been his pseudonym and had no inclination to say "I wrote this". Castañeda made the point, with many similar examples, that even in the latter case, we *could* say

(74) Samuel Clemens believes that he wrote *Huckelberry Finn*.

since he is Mark Twain, and he believes that Mark Twain wrote *Huckelberry Finn*. However, in the sense in which it was true through much of his life that he believed he wrote *Huckelberry Finn*, at this moment late in his life, it is not. There is a reading of (74) on which it is false.

In the case we are imagining, Samuel Clemens plays (at least) two roles in Samuel Clemens' life, the self-role $\mathbf{r}_{self}$ and the role $\mathbf{r}_{MT}$ of being the source of the 'Mark Twain' name-network that is exploited by the use of that name on the book he holds in his hands. Given this, we can distinguish between the two readings of (74), the first false and the second true, as follows:

(75) $\mathsf{pov}_{Samuel}[\mathsf{P}(Samuel, r_{self}) \wedge \boxdot\exists x(\mathsf{P}(x, r_{self}) \wedge Wrote(x, HF))];$
(76) $\mathsf{pov}_{Samuel}[\mathsf{P}(Samuel, r_{MT}) \wedge \boxdot\exists x(\mathsf{P}(x, r_{MT}) \wedge Wrote(x, HF))].$

Let us now see how this framework allows us to analyze multi-agent belief ascriptions. Building on Example 22.1 from Sect. 22.4.3, suppose that the registrar is looking at Egbert and gathers from Egbert's grin that he thinks there is a fish on the hook:

(78) The registrar believes that Egbert believes there is a fish on the hook.

Where r_s is assigned the role of *being the person seen*, we can translate (78) as:

(79) $\mathsf{pov}_{registrar}\big[\mathsf{P}(Egbert, r_s) \wedge \boxed{A}\exists x_2(\mathsf{P}(x_2, r_s) \wedge \mathsf{pov}_{x_2}\boxed{A}\exists x_3 FOH(x_3))\big]$

We take it that (78) entails:

(80) The registrar believes that someone believes there is a fish on the hook.

With a de re reading, we formalize (80) as

(81) $\mathsf{pov}_{registrar}\exists x_1 \exists r\big[\mathsf{P}(x_1, r) \wedge \boxed{A}\exists x_2(\mathsf{P}(x_2, r) \wedge \mathsf{pov}_{x_2}\boxed{A}\exists x_3 FOH(x_3))\big].$

Many other interesting multi-agent belief ascriptions can be handled in this way. Let us look at one more famous example, due to Richard [58]. A man m sees a woman w in a phone booth. As he watches her from his office window, he sees that an out-of-control steamroller is headed toward the phone booth. He waves wildly to warn her. At the same time, he is talking on the phone to a friend. She tells him of a strange man who is waving wildly to her, apparently believing she is in danger. Of course, unknown to m, he is talking to the very woman he his seeing, without realizing it. In this case, m might tell w over the phone, "I believe you are not in danger", while at the same time agreeing with her that "The man waving at you believes you are in danger". How is this coherent? The answer from Crimmins and Perry [19] is that the choice of words in the *subject* position of the belief reports ('I' or 'the man waving at you') can affect what is the relevant *role* via which the subject is said to believe something of the object: the first belief attribution is true iff m believes of w via the role $\mathbf{r}_{phoned}$ that she is not in danger, while the second is true iff m believes of w via the role $\mathbf{r}_{seen}$ that she is in danger. Mapping variables r_{phoned} and r_{seen} to these roles, we can describe m's doxastic state as follows:

(82) $\mathsf{pov}_m\big[\mathsf{P}(w, r_{phoned}) \wedge \boxed{A}\exists x_1(\mathsf{P}(x_1, r_{phoned}) \wedge \neg InDanger(x_1))\big];$

(83) $\mathsf{pov}_m\big[\mathsf{P}(w, r_{seen}) \wedge \boxed{A}\exists x_1(\mathsf{P}(x_1, r_{seen}) \wedge InDanger(x_1))\big];$

(84) $\mathsf{pov}_m\big[\mathsf{P}(w, r_{phoned}) \wedge \boxed{A}\exists x_1\big(\mathsf{P}(x_1, r_{phoned}) \wedge \neg InDanger(x_1) \wedge \exists! x_2$
      $\big(WavingAt(x_2, x_1) \wedge \mathsf{pov}_{x_2}\big[\mathsf{P}(x_1, r_{seen}) \wedge \boxed{A}\exists x_3(\mathsf{P}(x_3, r_{seen}) \wedge$
      $InDanger(x_3))\big]\big)\big)\big].$

We leave it to the reader to further explore the possibilities for formalizing multi-agent belief ascriptions in this framework. We also leave it to future work to investigate the differences between FOIL and the logic over our multi-agent models.

22.7 Conclusion

We began this paper with the idea that epistemic predicate logic faces a problem that alethic predicate logic does not: the problem of the cognitive fix. As we saw, this problem requires a different solution than the solution, based on modally loyal names, to the problems that Quine raised for alethic predicate logic. However, we have argued that the solution to the problem of the cognitive fix is not to treat names, worlds, or individuals differently in epistemic logic than in alethic logic. The solution does not require giving up what we called the Conservative Approach.

Instead, the solution requires giving up the idea that translating belief ascriptions into modal logic follows the simple pattern of translating necessity claims, what we called the Complement = Operand Hypothesis. We argued for an alternative approach to formalizing belief reports, based on making explicit the *unarticulated constituents* of such reports. Taking these unarticulated constituents to be the *roles* that the objects of belief play in the cognitive life of the believer, we carried out the formalizations in a version of Fitting's First-Order Intensional Logic. We applied the idea of agent-relative roles to the Hintikka-Kripke Problem for alethic-epistemic logic, to quantification into epistemic contexts, and to multi-agent belief ascriptions.

The move from individual concepts to agent-relative roles also opens up new ways of thinking about the dynamics of knowledge and belief. One can easily add to our framework the basic machinery of dynamic epistemic logic [11]: when an agent learns φ, update the model by cutting doxastic/epistemic accessibility links to $\neg\varphi$-worlds. But now we can consider not only the dynamics of the relations, but also the *dynamics of roles*. When an agent makes an observation of the world, it is not just that she receives information that rules out epistemic possibilities; in addition, the objects that she observes come to play various roles in her cognitive life, making it possible for her to have new thoughts about those objects. A dynamic epistemic predicate logic that allows the accessibility relations R_d to change should not freeze the set F of functions in place. While Fregean senses may be static, agent-relative roles are not. Perhaps it is not only the dynamics of *ruling out*, but also the dynamics of *roles* that belongs on the agenda of logical dynamics.

Acknowledgments For helpful discussion or comments on this paper, we thank Johan van Benthem, Russell Buehler, Thomas Icard, David Israel, Ethan Jerzak, Alex Kocurek, Daniel Lassiter, John MacFarlane, Michael Rieppel, Shane Steinert-Threlkeld, Justin Vlasits, and Seth Yalcin.

References

1. Aloni M (2005) Individual concepts in modal predicate logic. J Philos Log 34(1):1–64
2. Areces C, ten Cate B (2006) Hybrid logic. In: van Benthem J, Wolter F, Blackburn P (eds) Handbook of modal logic. Elsevier, Amsterdam, pp 821–868
3. Arló-Costa H, Pacuit E (2006) First-order classical modal logic. Studia Logica 84:171–210
4. Barnes KT (1976) Proper names, possible worlds, and propositional attitudes. Philosophia 6(1):29–38

5. Barwise J, Perry J (1983) Situations and attitudes. MIT Press, Cambridge
6. Belardinelli F, Lomuscio A (2009) Quantified epistemic logics for reasoning about knowledge in multi-agent systems. Artif Intell 173:982–1013
7. van Benthem J (1985) Modal logic and classical logic. Bibliopolis, Napoli
8. van Benthem J (1993) Beyond accessibility: functional models for modal logic. In: de Rijke M (ed) Diamonds and defaults. Kluwer Academic, Dordrecht, pp 1–18
9. van Benthem J (2010a) Frame correspondences in modal predicate logic. In: Feferman S, Sieg W (eds) Proofs, categories and computations: essays in honor of Grigori Mints. College Publications, London
10. van Benthem J (2010b) Modal logic for open minds. CSLI Publications, Stanford
11. van Benthem J (2011) Logical dynamics of information and interaction. Cambridge University Press, Cambridge
12. Braüner T, Ghilardi S (2006) First-order modal logic. In: van Benthem J, Wolter F, Blackburn P (eds) Handbook of modal logic. Elsevier, Amsterdam, pp 549–620
13. Burge T (1973) Reference and proper names. J Philos 70(14):425–439
14. Carlson L (1988) Quantified Hintikka-style epistemic logic. Synthese 74(2):223–262
15. Carnap R (1947) Meaning and necessity. University of Chicago Press, Chicago
16. Castañeda HN (1966) 'He': a study in the logic of self-consciousness. Ratio 7:130–157
17. Cresswell MJ, von Stechow A (1982) De Re belief generalized. Linguist Philos 5:503–535
18. Crimmins M (1992) Talk about belief. MIT Press, Cambridge
19. Crimmins M, Perry J (1989) The prince and the phone booth: reporting puzzling beliefs. J Philos 86(12):685–711
20. Cumming S (2008) Variabilism. Philos Rev 117(4):525–554
21. Davies M, Humberstone L (1980) Two notions of necessity. Philos Stud 38:1–30
22. Etchemendy J (1999) The concept of logical consequence. CSLI Publications, Stanford
23. Fitting M, Thalmann L, Voronkov A(2001) Term-modal logics. Studia Logica 69(1): 133–169
24. Fitting M (2004) First-order intensional logic. Ann Pure Appl Log 127(1–3):171–193
25. Fitting M (2006) FOIL axiomatized. Studia Logica 84(1):1–22
26. Fitting M, Mendelsohn RL (1998) First-order modal logic. Kluwer, Dordrecht
27. Føllesdal D (1961) Referential opacity and modal logic. Routledge, London (published 2004)
28. Frege G (1892) Über Sinn und Bedeutung. Zeitschrift für Philosophie und philosophische Kritik 100:25–50
29. Garson JW (2001) Quantification in modal logic. In: Gabbay D, Guenthner E (eds) Handbook of philosophical logic, 2nd edn, vol 3. Kluwer, Dordrecht, pp 267–323
30. Garson JW (2006) Modal logic for philosophers. Cambridge Universtiy Press, Cambridge
31. Grove AJ (1995) Naming and identity in epistemic logic part II: a first-order logic for naming. Artif Intell 74:311–350
32. Hintikka J (1962) Knowledge and belief: an introduction to the logic of the two notions. College Publications, London (Republished 2005)
33. Hintikka J (1969) On the logic of perception. Models for modalities. Reidel, Dordrecht, pp 151–183
34. Hintikka J (1970) Objects of knowledge and belief: acquaintances and public figures. J Philos 67:869–883
35. Holliday WH (2014) Epistemic closure and epistemic logic I: relevant alternatives and sub-junctivism. J Philos Log (forthcoming). doi:10.1007/s10992-013-9306-2
36. Holliday WH (2014) Epistemic logic and epistemology. In: Hansson SO, Hendricks VF (eds) Handbook of formal philosophy. Springer (forthcoming)
37. Hughes GE, Cresswell MJ (1996) A new introduction to modal logic. Routledge, London
38. Israel D, Perry J (1996) Where monsters dwell. In: Seligman J, Westerståhl D (eds) Proceedings of the logic, language and computation: the 1994 Moraga, CSLI Publications, pp 303–316
39. Kaplan D (1968) Quantifying In. Synthese 19:178–214
40. Kaplan D (1989) Demonstratives. In: Perry J, Wettstein H, Almog J (eds) Themes from Kaplan. MIT Press, Cambridge, pp 481–563
41. Kripke S (1980) Naming and necessity. Harvard University Press, Cambridge

42. Lehmann S (1978) The Hintikka-Kripke problem. Philosophia 8:59–70
43. Lewis D (1977) Possible-world semantics for counterfactual logics: a rejoinder. J Philos Log 6:359–363
44. Lewis D (1979) Attitudes De Dicto and De Se. Philos Rev 88(4):513–543
45. Lewis D (1983) Individuation by acquaintance and by stipulation. Philos Rev 92(1):3–32
46. Linsky L (1979) Believing and necessity. Theory Decis 11:81–94
47. Ninan D (2012) Counterfactual attitudes and multi-centered worlds. Semant Pragmat 5(5):1–57
48. Parks Z (1974) Semantics for contingent identity systems. Notre Dame J Formal Log 15(2): 333–334
49. Perry J (1977) Frege on demonstratives. Philos Rev 86:474–497
50. Perry J (1986) Thought without representation. Proc Aristotelian Soc 60:137–151
51. Perry J (2009) Hintikka on demonstratives. Revue internationale de philosophie 4(250): 369–382
52. Perry J (2012) Reference and reflexivity, 2nd edn. CSLI Publications, Stanford
53. Perry J (2013) Føllesdal and Quine's slingshot. In: Frauchiger M (ed) Reference, rationality, and phenomenology: themes from Føllesdal. Ontos Verlag, Frankfurt, pp 237–258
54. Priest G (2002) The hooded man. J Philos Log 31(5):445–467
55. Quine W (1953) Three grades of modal involvement. In: Proceedings of the 11th international congress of philosophy, vol 14. North-Holland Publishing, Amsterdam, pp 158–176
56. Quine W (1956) Quantifiers and propositional attitudes. J Philos 53(5):177–187
57. Rabinowicz W, Segerberg K (1994) Actual truth, possible knowledge. Topoi 13:101–105
58. Richard M (1983) Direct reference and ascriptions of belief. J Philos Log 12:425–452
59. Russell B (1905) On denoting. Mind 14(4):479–493
60. Russell B (1911) Knowledge by acquaintance and knowledge by description. Proc Aristotelian Soc 11:108–128
61. Schroeter L (2012) Two-dimensional semantics. In: Zalta EN (ed) The stanford encyclopedia of philosophy (winter 2012 edn)
62. Soames S (1989) Direct reference and propositional attitudes. In: Perry J, Wettstein H, Almog J (eds) Themes from kaplan. MIT Press, Cambridge, pp 393–419
63. Stalnaker RC (1984) Inquiry. MIT Press, Cambridge
64. Stalnaker RC (2010) Our knowledge of the internal world. Oxford University Press, Boston
65. Wettstein H (1988) Cognitive significance without cognitive content. Mind 97(385):1–28
66. Wishon D (2012) Russellian acquaintance and Frege's puzzle (manuscript)
67. Yalcin S (2012) Quantifying in from a Fregean perspective (manuscript)

Chapter 23
Stit Logics, Games, Knowledge, and Freedom

Roberto Ciuni and John Horty

Abstract This chapter has two main goals: highlighting the connections between Stit logics and game theory and comparing Stit logics with Matrix Game Logic, a Dynamic Logic introduced by van Benthem in order to model some interesting epistemic notions from game theory. Achieving the first goal will prove the flexibility of Stit logics and their applicability in the logical foundations of game theory, and will lay the groundwork for accomplishing the second. A comparison between Stit logics and Matrix Game Logic is already offered in recent work by van Benthem and Pacuit. Here, we push the comparison further by embedding Matrix Game Logic into a fragment of group Stit logic, and using the embedding to derive some properties of Matrix Game Logic—in particular, undecidability and the lack of finite axiomatizability. In addition, the embedding sheds light on some open issues about the so-called "freedom operator" of Matrix Game Logic.

23.1 Introduction

Johan van Benthem's career has been about research—often ground breaking, transformative research—but also, and especially in recent years, about building bridges and establishing conversations: across disciplines, between research communities, and among researchers from different nations and cultures. At a stage when so many others of his stature would be content with focusing inward, solidifying results, and protecting their territory, Johan has been looking in fresh directions,

R. Ciuni (✉)
Department of Philosophy II, Ruhr University Bochum, Universitätsstraße 150, 44780 Bochum, Germany
e-mail: ciuniroberto@yahoo.it

J. Horty
Philosophy Department, University of Maryland, College Park, MD 20742, USA
e-mail: horty@umiacs.umd.edu

A. Baltag and S. Smets (eds.), *Johan van Benthem on Logic and Information Dynamics*, Outstanding Contributions to Logic 5, DOI: 10.1007/978-3-319-06025-5_23, © Springer International Publishing Switzerland 2014

breaking down barriers, and seeking to involve others in a common enterprise. In twenty years time, logic will be a stronger, more integrated discipline because of his ambassadorial work; in fifty years, it will be stronger still.

This chapter contributes to one of the many conversations that Johan has begun—in this case, between those working in the tradition of Stit Logics, and those working with the different picture of agency underlying Dynamic Logic and Dynamic Epistemic Logic, a theory to which Johan himself has made seminal contributions.

Stit logics—which we here characterize, collectively, as STIT—were originated by Nuel Belnap and his many collaborators in a series of papers culminating with the monograph [3]; the framework was then connected to issues in decision theory, deontic logic, and cooperative game theory in [24]. STIT takes its name from the phrase "seeing to it that," which the theory interprets as a modal operator, known as the "stit operator," capturing the idea that an agent i sees to it that ϕ just in case ϕ is true at all states, or courses of events, compatible with a particular choice made by i. The main semantic ingredient of the theory is, accordingly, the notion of the *choices* available to an agent, which STIT characterizes—in a purely extensional way—as sets of states, or courses of events. Acting is then interpreted as selecting some such sets and excluding others. Beside this, STIT has two eye-catching features: it does not include labels for actions, which in turn find no expression in the language, and it assumes an independence condition according to which any choice of any agent is compatible with any choice of any other agent.

STIT has its roots in the field of formal philosophy, and has been applied to clarify some crucial conceptual issues in the theory of action and ethics—for example, connections between moral responsibility and the principle of "could have done otherwise" [3], the rigorous formulation of criteria for consequentialist theories of action [24], the judicial notion of *mens rea* [12], and the attribution of individual responsibility in cases of group agency [18]. However, in the twenty some years since its introduction,[1] the applications of STIT have slowly shifted to theoretical computer science and related areas, particularly the logical foundations of multi-agent systems, artificial intelligence, and game theory.

This shift has opened interesting issues. STIT and game theory talk different jargons and have been directed toward different targets. Though there are game-theoretical roots in STIT, a clear display of the connections between STIT and games has not been undertaken.[2] Also, the arena of formalisms for the logical foundations of games and multi-agent systems is very rich, and an analysis of the relations with prominent formalisms in this family constitutes a fascinating area of applications for STIT.

In the present chapter, we touch on both of these issues. First, we try to clarify the potential of STIT as a logical foundation for game theory by describing its adequacy for modeling certain game-theoretic notions. And second, we compare STIT with

---

[1] Belnap and Perloff [4] and von Kutschera [27] are usually regarded as the two papers that, independently, lay the foundations of STIT.

[2] See, however, the important earlier work by Kooi and Tamminga [25], Tamminga [29], and Turrini [5].

Matrix Game Logic, that is a Dynamic Logic first introduced by van Benthem in [6] and developed in his later [7] and [8]; in this comparitive project, we continue, and hope to advance, the conversation initiated by van Benthem and Pacuit [11].[3]

The chapter proceeds as follows. Section 23.2 introduces structures and semantics for a particular Stit logic together with a Hilbert style axiomatization, and reviews some interesting validities and formal properties. Section 23.3 focuses on a comparison between STIT and strategic games, and tries to fill the gap between these two areas; some interesting points of comparison are the possibility of reading STIT game-theoretically, and of isolating a "STIT component" within games. Sections 23.4 and 23.5 compare STIT with Matrix Game Logic. Section 23.5 contains the most novel result of the chapter: a mutual embedding between Matrix Game Logic and a particular STIT for group agency, with a consequent property transfer. Section 23.6 presents some conclusions.

23.2 STIT

STIT logics and stit operators abound, and the choice among them largely depends on one's purposes. We will rely here on the so-called "Chellas stit"[4] and we interpret our logic on *Choice Kripke frames*, with no temporal ordering on states of evaluation: more complex operators and the temporal dimension of agency are not needed for the comparisons we draw in later sections of this chapter.[5]

Choice Kripke Frames. Formally, a *Choice Kripke frame*—a CKF, for short—is a triple $\langle W, Ags, \{\sim_i^C \mid i \in Ags\} \rangle$ where:

- W is a non-empty set $\{w, w', w'' \dots\}$ of *states*
- Ags is a finite, non-empty set $\{1, \dots, n\}$ of individual *agents*
- For each agent $i \in Ags$, the relation $\sim_i^C \subseteq W \times W$ is a *choice-equivalence relation*

---

[3] Others have also tried to developed unified perspectives encompassing STIT and dynamic logics. See for instance Herzig and Lorini [20], which presents a dynamic logic of agency in the tradition of Propositional Dynamic Logics. In this framework, a basic stit operator can be defined as an existential quantifier over the actions of a given agent.

[4] This particular operator was first isolated, and given this name, by Horty and Belnap [22]; the name reflects the fact that the operator captures, in the different framework of stit semantics, ideas introduced much earlier by Chellas [15]. A comparison between Chellas's early work on agency and the later STIT can be found in Chellas [16], and also in Horty and Belnap [22].

[5] STIT is traditionally interpreted on branching-time structures (see [3] and [24]) where moments are linearly ordered toward the past but are not linearly ordered toward the future. In such structures, *choices* are sets of histories, and *histories* are in turn sets of *moments* which are (1) maximal with respect to inclusion and (2) linearly ordered toward the future. However, the most widespread stit operators do not express any temporal dimension, and thus the indeterministic framework can be replaced by Kripke frames where no temporal order is imposed. Such frames are used by Balbiani, Herzig, and Troquard et al. [1], and by Herzig and Schwarzentruber [21]; and we follow them in the present chapter.

satisfying the following conditions:

(R1) $\sim_i^C$ is an equivalence relation

(R2) for all $(w_1, \ldots, w_n) \in W^n$, $\bigcap_{i \in Ags}\{w \mid w \sim_i^C w_i\} \neq \emptyset$.

If $w \sim_i^C w'$, we say that w is choice-equivalent with w' for agent i, and we let $[w]_i$ be the equivalence class including the states which are choice-equivalent with w—that is, $[w]_i = \{w' \mid w \sim_i^C w'\}$. Condition R2 guarantees the *independence condition* mentioned earlier, that all the choices of all the different agents are compatible. This is a very demanding condition, but it turns to be a key element for the game-theoretical reading of STIT, which in turn forms the bridge between STIT and Matrix Game Logic; we return to this issue in Sect. 23.3.2. Also, R2 implies that $W \times W \subseteq \sim_i^C \circ \sim_j^C$, for all distinct pairs of indices i and j between 1 and n, and thus it corresponds to a *strong* form of *confluence*—this is why we will refer to R2 as to *strong confluence* property.

We define a *restricted Choice Kripke frame*—a CKF$^+$, for short—as a CKF satisfying the further condition:

(R3) For every state w, $[w]_1 \cap \cdots \cap [w]_n = \{w\}$.

R3 states that the combination of the choices of all the agents at w consists in w itself: the combined choices of all the agents are enough to determine a unique state of the world.[6] Also, we define

(D1) For every state w, $[w]_{\bar{i}} = \bigcap_{j \in \bar{i}}[w]_j$, where $\bar{i} = Ags/\{i\}$.

We refer to $\bar{i}$ as the *anti-group* of i—the group including all agents except i. D1 defines the choice of i's anti-group at w as the intersection of the choices of its members at w.[7]

In what follows, we will confine ourselves to CKF$^+$'s, which will make the comparison with strategic games easier.[8] To illustrate, Fig. 23.1 on the next page exemplifies a CKF$^+$: The two columns represent the choices $[w']_1$ and $[w''']_1$ of agent 1, while the rows represent the choices $[w']_2$ and $[w]_2$ of agent 2. A moment's consideration is enough to see that R1–R3 are satisfied by the structure represented by the figure; also, in the frame represented by Fig. 23.1, $\bar{1} = \{2\}$ and $\bar{2} = \{1\}$, the construction of the anti-groups trivially follows D1.

---

[6] In case this condition seems too strong, it is helpful to think of one of the agents as "nature," which removes any remaining indeterminacy once all the more ordinary agents have made their choices; this tactic was mentioned in [24, p. 91].

[7] D1 just encodes a special case of the game-theoretical principle of *additivity*, which characterizes the construction of all groups in group STIT; see condition R4 in Sect. 23.5.

[8] Actually, the correspondence between games and consequentialist CKF$^+$'s (see below for a definition) can also be established without imposing condition R3; see, for example, van Benthem and Pacuit [11] and Tamminga [29]. However, the condition makes the proof of such a correspondence much more straightforward and general. In addition the proofs which do not use R3 essentially rely on consideration about language, while the correspondence result which uses R3, established by Turrini in [5], relies only on the structures in question.

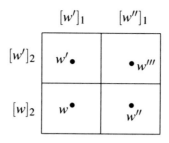

Fig. 23.1 A CKF$^+$ with two agents, 1 and 2 and two choices per agent

Language and Semantics. In addition to the set *Ags* of agents, assume a denumerable set of atomic formulas. Our language $\mathscr{L}_{\text{CSTIT}}$ then has the Backus-Naur form

$$\phi ::= p \mid \neg\phi \mid \phi_1 \wedge \phi_2 \mid [i]\phi \mid \Box\phi,$$

where *p* is atomic, where ϕ_1 and ϕ_2 are arbitrary formulas, and where $i \in Ags$; the other Boolean connectives are defined as usual on the basis of $\neg$ and $\wedge$. The symbol $[i]$ is the Chellas stit operator, so that $[i]\phi$ should be taken to mean that the agent *i* sees to it that ϕ—that is, that the truth of ϕ is guaranteed by a choice due to *i*. Its dual is the symbol $\langle i \rangle$, so that $\langle i \rangle\phi$ should be understood as meaning that ϕ is consistent with the choice exerted by *i*. Finally, the symbol $\Box$ is the usual universal modality, so that $\Box\phi$ should be taken to mean that ϕ holds in all the possible states. Its dual is the symbol $\Diamond$, so that $\Diamond\phi$ should be taken to mean that ϕ holds in some of the possible states.

   The formulas in $\mathscr{L}_{\text{CSTIT}}$ are evaluated on Choice Kripke Models—CKM$^+$, for short—where a CKM$^+$ $\mathscr{M}$ is a pair $\langle \mathscr{K}, V \rangle$, with $\mathscr{K}$ a CKF$^+$ and V a function from atomic formulas into sets of states at which they are true. The satisfaction relation $\models$ for formulas in $\mathscr{L}_{\text{CSTIT}}$ can then be defined as follows:

$$\begin{aligned}
&\mathscr{M}, w \models p && \text{iff } w \in V(p) \\
&\mathscr{M}, w \models \neg\phi && \text{iff } \mathscr{M}, w \not\models \phi \\
&\mathscr{M}, w \models \phi \vee \psi && \text{iff } \mathscr{M}, w \models \phi \text{ or } \mathscr{M}, w \models \psi \\
&\mathscr{M}, w \models [i]\phi && \text{iff for all } w', \text{ if } w' \in [w]_i \text{ then } \mathscr{M}, w' \models \phi \\
&\mathscr{M}, w \models \Box\phi && \text{iff for all } w' \in W, \mathscr{M}, w' \models \phi.
\end{aligned}$$

Truth in a CKM$^+$, in a CKF$^+$, and in all CKF$^+$'s is defined by the usual universal quantifications; and as usual, we will take $\mid \phi \mid^{\mathscr{M}} = \{w \mid \mathscr{M}, w \models \phi\}$ as the set of states from the model $\mathscr{M}$ satisfying ϕ. Figure 23.2 represents a CKM$^+$ built from the CKF$^+$ depicted in Fig. 23.1. It is easy to see that $[1]\phi$ is true at w' and w but false at w''' and w'', while $[2]\phi$ is true at w and w'' but false at w' and w'''.

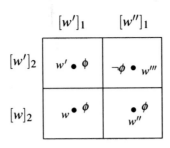

Fig. 23.2 A CKM based on Fig. 23.1

23.2.1 Axiomatics and Interesting Validities

Axiomatics. The axiomatization of CSTIT consisting of

(S5) S5's axioms for $\Box$ and each $[i]$

(MA) $\Box\phi \rightarrow [i]\phi$

(IA) $\bigwedge_{i \in Ags} \Diamond[i]\phi_i \rightarrow \Diamond \bigwedge_{i \in Ags}[i]\phi_i$

together with Modus Ponens and the Rule of Necessitation for $\Box$ and every stit operator $[i]$ is sound and complete relative to CKF$^+$'s.[9] The S5 properties follow from that fact the fact that $[w]_i$ is an equivalence class, for every agent i; this tells us also that CSTIT is a multi-modal S5—a very nice one, in fact, as we shall see below. IA is the well-known axiom of independence of agents, which states that any combination of independently possible actions, one for each agent, is jointly possible. Notice that its validity for all the agents $1, \ldots, n$ in Ags implies its validity for any smaller interaction of agents.[10] This is, as we mentioned earlier, a very strong principle and a key feature of STIT; we discuss it further in Sect. 23.3.2, after highlighting the connections between STIT and game theory.

Interesting Valid Formulas. Balbiani, Herzig, and Troquard [1] show that IA can be replaced by the axiom $\Diamond\phi \rightarrow \langle i \rangle \bigwedge_{j \in \bar{i}} \langle j \rangle \phi$ (more precisely, IA is provable from the new axiom, S5 and MA). If we contrapose the two-agent version of the axiom in [1], we get a principle of *Triviality of Coercion*,

(TC) $[i][j]\phi \rightarrow \Box\phi$ (for $i \neq j$),

which states that one agent can guarantee that another agent guarantees a certain proposition only if that proposition is itself trivial.[11] Figure 23.3 helps check the

---

[9] See Balbiani, Herzig, and Troquard et al. [1] for discussion. This axiomatization is due to Xu [31], where, however, the Chellas stit was replaced by the deliberative stit, and completeness is proved relative to trees endowed with choices and agents.

[10] Thus, for instance, the validity of IA implies the validity of $\Diamond[i]\phi \wedge \Diamond[j]\phi \rightarrow \Diamond([i]\phi \wedge [j]\phi)$—see [1, p. 391].

[11] TC also plays a role in replacing IA with the new axiom in [1]: the two-agent version of IA ($\Diamond[i]\phi \wedge \Diamond[j]\phi \rightarrow \Diamond([i]\phi \wedge [j]\phi)$) is derivable by TC, S5 and MA by Modus Ponens and

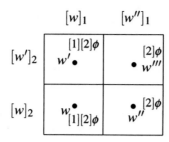

Fig. 23.3 $[1][2]\phi \to \Box\phi$. $[1][2]\phi$ holds at w, from which it follows by our semantics that $[2]\phi$ holds at both w and w', and then, by applying again the semantic clauses, that ϕ holds both at w and w'' and at w' and w'''. The fact that $[1][2]\phi$ holds somewhere thus implies that ϕ is true at all states in the model

validity of the principle. Of course, the equivalence $\Box\phi \leftrightarrow [1][2]\phi$ is provable from TC, MA and axiom 4 (included in S5); this allows us to define $\Box$ as an abbreviation of $[1][2]$. Also worth mentioning is the *Permutation Principle*

(PP) $[i][j]\phi \leftrightarrow [j][i]\phi$

which follows by TC, MA and axiom 4.[12]

Formal Properties of STIT and group CSTIT. CSTIT has very convenient formal properties, some of which are surprising in a multi-modal S5. First, the axiomatization above is complete with respect to CKF. Since R3 is not expressible in $\mathscr{L}_{\text{CSTIT}}$, the system does not distinguish the frames satisfying the condition, and thus it proves complete also relative to CKF$^+$'s.

Also, CSTIT is decidable and finitely axiomatizable.[13] This is trivial if we confine our attention to a two-agent CSTIT; in this case, the STIT property of strong confluence reduces to the confluence property of S5^2 ("squared S5"), which is decidable and finitely axiomatizable.[14] The interesting virtue of STIT is that it keeps these properties even when more than two operators are at stake. The key issue is indeed strong confluence, which in turn reduces *all* the confluences of arbitrary length to the confluence $\sim_i^C \circ \sim_j^C$, with $[i]$ and $[j]$ arbitrary. This has two interesting consequences: (1) It implies that CSTIT with $n > 2$

(Footnote 11 continued)
Necessitation of $\Box$, while the k-agent version can be proved by induction on the $k-1$ case and $\Diamond\phi \to \langle k \rangle \bigwedge_{1 \le i < k} \langle i \rangle \phi$, for $k > 1$ ([1, pp. 392–393]).

[12] Balbiani et al. [1] also introduces a third (and more compact) axiomatic system for CSTIT, consisting of S5 for $[i]$ agency operators, the definition $\Box\phi \leftrightarrow [1][2]\phi$ and the formula $\langle i \rangle\langle j \rangle\phi \to \langle k \rangle \bigwedge_{m \in \bar{k}} \langle m \rangle \phi$. S5 for $\Box$, MA, the axiom $\Diamond\phi \to \langle i \rangle \bigwedge_{j \in \bar{i}} \langle j \rangle \phi$—and hence IA—are provable from these new axioms by Modus Ponens and Necessitation. See [1, pp. 394 and 397].

[13] These results were first established by Xu [30]; see also Balbiani et al. [1].

[14] Balbiani et al. [1, p. 395] prove that the logic of two-agent Chellas stit is nothing but S5^2.

agents does not yield the product logic $S5^n$, so that, for example, CSTIT with 3 agents is not $S5^3$ ("cube S5"); the implicit virtue here is clear, since any product $S5^n$ with $n > 2$ is undecidable and not finitely axiomatizable.[15] (2) It implies that CSTIT does not encode grid structure enough to become undecidable.

STIT's view on agency and the many fashions of STIT. Our presentation has emphasized the distinctive marks of STIT mentioned in Sect. 23.1—the extensional view on choices, acting as state selection, the absence of action types in PDL or DEL style, and the independence of agents. CSTIT is not the only STIT logic available, but most STIT logics share these features.[16]

None of this tells us exactly how the stit operator is to be read. Here we just briefly mention two prominent readings. The first is the original reading due to Belnap and Perloff, and is today the most widespread among philosophers; according to this reading, "seeing to it that" captures the contribution of the agents to a change in the causal structure of the world.[17] The second points at a game-theoretical interpretation of STIT, and is suggested by [24] and other research in the computer science side of STIT; here the idea is to focus on what an agent guarantees, or "sees to," by following a winning strategy.

Comparing such readings here would go beyond the scope of the chapter; we note only that the former focuses on agency as a factor of change in spatial regions of the universe, while the second stresses the choice-making dimension of agency. The two readings thus suggest different applications and questions: Belnap's reading naturally calls for a specification of what "causal contribution" to a change is, and consequently requires also a picture of what the causal structure of the world is; by contrast, the game-theoretical reading highlights the logical dynamics of choice-making and multi-agent interaction, and can be adapted to the different structures provided by game theory, such as, for example, extended game forms or strategic games.

An interesting point of Belnap's reading is that it naturally calls for representing agency in time, since causation presupposes a temporal dimension.[18] The CKF's are not enough to this purpose, and thus it is no surprise that Belnap and colleagues use *branching-time* structures endowed with a set of agents and a choice function (the

---

[15] See Hirsch et al. [23].

[16] A noticeable exception is the combinations of STIT and actions explored by Xu [32].

[17] If we follow Belnap's reading, deliberative stit may prove more suitable than the Chellas stit, since the former does not allow for trivial truths to be seen to by any agent; the Chellas stit allows for this and does not seem to fit equally well the idea of a causal contribution to a change in the world.

[18] Such a component proves relevant also if we wish to represent the sequential aspect of choice-making in extended game forms, since a sequence of choice-making acts presuppose a temporal dimension.

so-called "BT + AC structures"),[19] which are indeed intended to capture the notion of an indeterministic causal change in the world.[20]

Finally, STIT traditionally does not define choices in relational terms, but rather in functional ones: the standard option is to introduce a choice function $Ch(i) : Ags \longmapsto \wp(W)$ which partitions W and such that for every state $w \in W$, $\bigcap \{Ch(i, w) \mid i \in Ags\} \neq \emptyset$. It is the easy to see that our relation $\sim_i^C$ can be defined in terms of $Ch(i)$ and vice versa. We use this interchangeability in Sect. 23.3, and we also define for every state $w \in W$, $\bigcap \{Ch(i, w) \mid i \in Ags\} = \{w\}$, in analogy with R3 and $Ch(\bar{i}, w) = \bigcap_{j \in \bar{i}} Ch(j, w)$ for every $w \in W$—in analogy with D1.

23.3 STIT and Strategic Games

Although STIT was first presented as a theory of the contribution of agents to changes in the causal structure of the world, it was soon applied to problems concerning choice-making, and much of the current research is in keeping this direction; this applies especially the application of STIT in the logical foundations of multi-agent systems. The reason for such a shift is that STIT presents some interesting connections with game theory, which is the most widespread framework of reference in studies on multi-agent systems.[21] Here, we consider some of these connections, focusing on *strategic games*. This is an indispensable move for the formal comparison between STIT and Matrix Game Logic in Sect. 23.4, and also helps clarify some features of STIT which have been debated, such as the independence of agents.

23.3.1 Bridging Two Worlds

A strategic game $\mathscr{G}$ is a five-tuple $\langle W, Ags, \{a_i \mid i \in Ags\}, o, \{\succeq_i \mid i \in Ags\} \rangle$, where W and Ags are as in CKF's; for each $i \in Ags$, a_i is an action available to

---

[19] To be more precise, the temporal component of BT + AC structures are *trees*. Along the years, other temporal components of indeterministic time for choices and agents have been introduced; see, for instance, the XSTIT *frames* due to Broersen [12], the *bundled choice trees* of Ciuni and Zanardo [19], and the *Temporal Kripke STIT models* of Lorini [28].

[20] Display of a temporal order is necessary to define the so-called 'stit operators for *non-instantaneous* agency', that is operators that express a temporal hiatus between choice and result. Examples of such operators are the fused xstit in Broersen [12], and the operators introduced in Ciuni and Mastop [18] and Ciuni and Zanardo [19]. A hiatus between choices and results can be also expressed by *combining* autonomous operators for agency and for temporal distinctions. Broersen follows this line in [13] and [14], as does Lorini [28]. A very complex stit operator is the original "achievement stit" due to Belnap and Perloff [4] which captures the cross-temporal dimension of agency by expressing the notion that a result holds at m due to a previous choice of i; variants of the achievement stit are proposed by Zanardo [33] and Ciuni [17].

[21] One should not forget that game-theoretical ideas were very important in STIT since its very beginning. This is clear from [3, pp. 283, 343–344], where the matrix representation of games is mentioned and independence of agents is explained with it, and where a comparison between extended game forms and BT + AC is briefly drawn.

agent i. We call A_i the set of actions $\{a_i, a_i', \ldots\}$ available to i.[22] Each sequence $\{a_1, \ldots, a_n\} \in \Pi_{i \in Ags} A_i$ is a *action profile*; such an action profile can also be denoted as $(a_i, a_{\bar{i}})$, which separates the action performed by the particular agent i from the actions performed by all the other agents from $\bar{i}$.

The function $o : \Pi_{i \in Ags} A_i \longmapsto W$ maps each action profile into a resulting state in W, according to the standards in the definition of a strategic game. Notice that each set in $\Pi_{i \in Ags} A_i$ represents a possible combination of actions in the game; as a consequence, all actions of any agent i are compatible with all the actions of any other agent j from $\bar{i}$; also, o is total, and thus this compatibility is represented at the level of the outcomes. We generalize the signature of o in two different ways. First, we let it take two arguments, so that $o(a_i, a_{\bar{i}})$ defines the outcome as resulting from a pair consisting of the action of the particular agent i, in the first place, taken together with the actions of all the other agents, from $\bar{i}$, in the second. Second, we denote the outcomes of a_i as $o(a_i)$, where $o(a_i) = \{w \mid w = o(a_i, a_{\bar{i}})$ for some $a_{\bar{i}} \in A_{\bar{i}}\}$ where $A_{\bar{i}} = \Pi_{j \in \bar{i}} A_j$. It is then clear that while the outcomes of action profiles are single states, the outcomes of the actions of individual agents are sets of states.

Finally, for each $i \in Ags$, the relation $\succeq_i$ is a reflexive preference ordering between outcomes of action profiles. The reading of $o(a_i, a_{\bar{i}}) \succeq_i o(a_i', a_{\bar{i}})$ is the standard one: agent i weakly prefers the state resulting from action profile $(a_i, a_{\bar{i}})$ to that resulting from action profile $o(a_i', a_{\bar{i}})$. The preference relation is easily extended to (outcomes of) actions of a given agent.

In order to draw our comparisons, we first extend CKF$^+$'s with preference relations $\{\succeq_i \mid i \in Ags\}$, thus obtaining *consequentialist* CKF$^+$'s, or CCKF$^+$'s for short. More exactly, a CCKF$^+$ $\mathscr{C}$ is a pair $\langle \mathscr{K}, \{\succeq_i \mid i \in Ags\}\rangle$, where $\mathscr{K}$ is a CKF$^+$. A model built from a CCKF$^+$—that is, a consequentialist CKM$^+$, or a CCKM$^+$, for short—is obtained by supplementing the CCKF$^+$ with an evaluation function V in the standard way.

Turrini [5] has proved an interesting correspondence between strategic games and CCKF$^+$'s in their functional version (see end of Sect. 23.2). Relying of the definition of the choice function $Ch(i)$, he first introduces the notion of *a choice structure* $Ch^{\mathscr{G}}$ *in a game* $\mathscr{G}$ as follows: $X \in Ch^{\mathscr{G}}(i)$ if and only if there is an action a_i such that $\{o(a_i, a_{\bar{i}}) \mid a_{\bar{i}} \in \Pi_{j \in \bar{i}} A_j\} = X$. He then proves:

Proposition 23.1 (Representation Theorem) *For every (functional)* CCKF$^+$ $\mathscr{C} = \langle W, Ags, Ch, \{\succeq_i \mid i \in Ags\}\rangle$, *there is a strategic game* $\mathscr{G}$ *such that* $Ch(i) = Ch^{\mathscr{G}}(i)$, *and vice versa.*[23]

---

[22] Since we do not deal with the sequential aspect of choice-making here, we prefer to use the term 'action' rather than 'strategy'.

[23] This result, established as Theorem 1 in [5], is actually stated there for *full* groups of agents ("coalitions" in the standard game-theoretical terminology) and their anti-groups. Notice that the result in [5] naturally extends to CKF$^+$'s without preference relation and strategic game forms, which obtain from games by dropping the preference relation.

The result can be adapted to our *relational* CCKF$^+$'s with no risk of loss, due to the definition of the function of choice at the end of Sect. 23.2.[24]

Turrini's observation has three interesting consequences. First, it implies that for every CCKF$^+$ $\mathscr{C} = \langle W, Ags, \{\sim_i^C | i \in Ags\}, \{\succeq_i | i \in Ags\} \rangle$ we can construct the corresponding *choice structure* of a game (call it $\mathscr{G}^\mathscr{C}$ for short); the converse also holds: the CCKF$^+$ $\mathscr{C}^{\mathscr{G}^\mathscr{C}}$ built on the choice structure $\mathscr{G}^\mathscr{C}$ is in turn nothing but $\mathscr{C}$. There is then a correspondence between CCKF$^+$'s and choice structures of games. Second, the choices of i in a CCKF$^+$ $\mathscr{C}$ are actually the outcomes of some action of i in the game with the corresponding choice structure $\mathscr{G}^\mathscr{C}$, or more exactly:

For every CCKF$^+$ $\mathscr{C}$, $w \sim_i^C w'$ iff $w, w' \in o(a_i)$ for some action $a_i \in A_i$ in the strategic game whose choice structure corresponds to CCKF$^+$.

We can express this also by saying that $Ch(i) = Ch^\mathscr{G}(i) = \{o(a_i) \mid a_i \in A_i\}$ in $\mathscr{G}^\mathscr{C}$. By R3 and D1, this allows to express (outcomes of) action profiles as intersections $[w]_i \cap [w']_{\bar{i}}$ of a choice of i and one of her anti-group: $w = o(a_i, a_{\bar{i}})$ iff $w = [w]_i \cap [w']_{\bar{i}}$ (for some $w' \in W$ and every CCKF$^+$ $\mathscr{C}$). Finally, proposition 1 also guarantees that CCKF$^+$ can represent a number of game-theoretical notions; the most paradigmatic examples is that of *weak dominance*:

Definition 23.1 (*Weak Dominance in a CCKF$^+$*).
$[w]_i \succeq_i [w']_i$ iff $[w]_i \cap [w'']_{\bar{i}} \succeq_i [w']_i \cap [w'']_{\bar{i}}$ for each $w'' \in W$, and $[w]_i \cap [w''']_{\bar{i}} \succ_i [w']_i \cap [w''']_{\bar{i}}$ for some $w''' \in W$—with $\succ$ being defined as usual as strict preference.

This idea, which corresponds to the standard game-theoretical definition of weak dominance,[25] was first introduced into STIT by [24], and has been the main focus of consequentialist work in the STIT tradition.[26] However, CCKF$^+$'s can also model other interesting notions of action preference, such as:

Definition 23.2 (*Best Choices in a CCKF$^+$*).
$[w]_i$ is a best choice for i given $[w']_{\bar{i}}$ iff $[w]_i \cap [w']_{\bar{i}} \succeq_i [w'']_i \cap [w']_{\bar{i}}$ for each $w'' \in W$.

which displays a clear correspondence with the game-theoretical notion of a *best action*.[27]

---

[24] Thus, from every game $\mathscr{G}$ we can construct the corresponding CCKF$^+$ $\mathscr{C}^\mathscr{G} = \langle W^\mathscr{G}, Ags^\mathscr{G}, \{\sim_i^{C\mathscr{G}} | i \in Ags\}, \{\succeq_i^{C\mathscr{G}} | i \in Ags\} \rangle$, where $w \sim_i^{C\mathscr{G}} w'$ iff $w' \in [w']_i$ for $[w]_i \in Ch^\mathscr{G}$.

[25] $o(a_i)$ is a *weakly dominant action* iff $o(a_i, a_{\bar{i}}) \succeq_i o(a_i', a_{\bar{i}})$ for all $a_i' \in A_i$ and all $a_{\bar{i}} \in A_{\bar{i}}$.

[26] See, for example, Kooi and Tamminga [25], Turrini [5], and Tamminga [29].

[27] $o(a_i)$ is a *best action* of i iff $o(a_i, a_{\bar{i}}) \succeq_i o(a_i', a_{\bar{i}})$ for all $a_i' \in A_i$.

23.3.2 Some Conceptual Insights

Proposition 1 and the related facts are revealing in many different respects. Though philosophers know STIT mainly under the causative reading, the theory may be naturally used for modeling game-theoretical notions: its strong ties with game theory allow us to trade notions defined in STIT with notions defined in games, and vice versa. In a nutshell, we can give STIT a game-theoretical reading without loosing any relevant feature of the framework, leading to some interesting consequences.

Game-theoretical Reading of stit. First, if we trade the notion of 'choice' with that of 'outcome of some action', it is clear that 'seeing to it that' equates with 'displaying a winning action'. For take a CCKF$^+$ $\mathscr{C}^{\mathscr{G}^{\mathscr{C}}}$ built on a game structure $\mathscr{G}^{\mathscr{C}}$. Due to proposition 1, for every choice $[w]_i$ defined in $\mathscr{C}^{\mathscr{G}^{\mathscr{C}}}$, there is some action $a_i \in A_i$ such that for every $w' \in W$, $w' \in o(a_i)$ iff $w' \in [w]_i$. Thus, $[i]\phi$ is true at w iff ϕ is true at all the states w' which are in the same outcome $o(a_i)$ as w. But any such states will also be in (the outcome of) some action of the rest of the agents; as a consequence, the fact that $[i]\phi$ holds at w implies that, for every $a_{\bar{i}} \in A_{\bar{i}}$, there is some state $w'' \in o(a_{\bar{i}})$ where ϕ holds. As a consequence, $\bar{i}$ cannot see to it that $\neg\phi$. In other words, if i sees to it that ϕ, i is performing a winning action to the effect that ϕ.

STIT, Games and Independence. The independence condition—R2 from Sect. 23.2—may sound strong and even surprising if we cast STIT against the background of the physical world and our role in its changes: from an intuitive standpoint, it is very infrequent that we are beyond any possibility of being deprived of our choices by others. However, the principle makes good sense if we read STIT game-theoretically. If we trade, once again, choices for outcomes of actions, R2 will amount to the assumption highlighted at the beginning of Sect. 23.3.1: each action profile (1) includes *one* action per agent,[28] and (2) correspond to *one* state. Independence is nothing but this, and thus proves a very game-theoretically oriented feature of STIT. The logical principles IA and TC, likewise in Sect. 23.2.1, follow from R2 and the truth-clause of $[i]$.

Independence and non-winning actions. There is an interesting feature of STIT which is not usually highlighted: in principle, you can have condition R2 *without* having an operator for agency which coincides with 'displaying a *winning action*'. This may become clear by analogy with *strictly competitive games*. In such games, the outcome function o may be defined according to the points (1) and (2) above, exactly as we just did for strategic games in general. At the same time, these games are characterized by the fact that no agent has a winning action: all actions of all the different agents are compatible, but no action of a single agent can ensure a given result: any relevant result in such games depend on the interaction of the different agents. One can have quite the same in a STIT setting, if the truth-clause of the stit operator is weakened; for instance, the probabilistic STIT presented by Broersen in

---

[28] This also explains why the function $Ch(i)$ is defined as a *partition* (see end of Sect. 23.2).

[10] and [9] retains the independence of agents at the level of the frames, but defines an operator which equates with displaying the choice that comes with the highest chance of success in getting the given result (and the latter is clearly compatible with failure in ensuring the result).

Independence, continued. Game theory goes much further than giving formal expression to the notion of "winning action," as strictly competitive games prove. Recent work in STIT logic shows that its potential is not confined to that notion. At the same time, it suggests that adapting STIT to a broader set of game-theoretical notions is compatible with retaining the independence of agents. Approaching some phenomena of game theory without imposing independence is clearly possible and equally sound. For instance, van Benthem and Pacuit [11] suggests dropping the totality of the outcome function o in order to model the game-theoretical notion of a *correlation*, where there is some form of dependence between some agent's choices. This suggestion is in line with a general tradition of Dynamic Logics in modeling game-theoretical notions. The suggestion is very reasonable, but if the temporal aspect of choice-making is acknowledged, STIT provides a natural alternative: agents are independent when it comes to their simultaneous choices, but the present choice of one agent may limit the choices available to others at subsequent moments.[29]

Preferences and Ought. The *consequentialist* CKF$^+$'s are a further proof of the entwinement of STIT and game theory. Indeed, CCKF$^+$'s have their origins in the application of STIT to a consequentialist perspective on action, which was first carried by Horty in [26] and [24]. Roughly speaking, a consequentialist perspective evaluates what an agent *ought to do* on the ground of the value that can be attached to the consequences of the agent's choices; if we read such consequences as the sets of states extending each choices, we will define our framework by assigning values to states and—indirectly—to choices, exactly as we did with the preference relations.

There is one point, however, where the analysis of [24] significantly differs from the present CCKF$^+$: in [24] the values (or preferences) are *agent-independent*, that is the values it imposes does not vary with the agent in question.[30] Here, we relaxed this condition and allowed for preferences to be agent-relative. This relaxation shows the match between the consequentialist perspective implicit in game theory and the perspective encoded in the STIT analyses based on [24]. Also, we need to go to agent-relative preferences if we wish to model other game-theoretical notions which are "intrinsically multi-agent". Think of the notion of a Nash Equilibrium: it implies a consideration of the preferences of *all* different agents, and keeping preferences agent-independent would make such a consideration trivial. The same applies to other phenomena, like the removal of strictly dominated strategies.

---

[29] This is no proof that STIT can deal with correlation as intended by [11], but is a general sign of the adaptability of STIT relative to the issue of independence.

[30] Also, notice that the "utilitarian STIT frames" introduced by [24] are grounded on branching-time structures. In such frames, the value attached to a history is not only agent-independent, but also *moment-independent*, that is it does not vary with time. This reminds the definition of preferences and priorities in standard rational-choice theory.

The main modal operator in [24] is the so-called "dominance ought operator" $\odot[i]$—where $\odot[i]\phi$ should be taken to mean that, if i exerts any of her weakly dominant choices, then ϕ is the case. The operator clearly models a notion of weak dominance. An interesting point is that allowing for agent-relative preferences, as we do here, also allows for the definition of alternative operators meaning, for example, that, if i exerts one of her *best choices*, then ϕ is the case, or if i exerts one choice of her in the *Nash Equilibrium*, then ϕ is the case. Finally, as in the work of Kooi and Tamminga [25] and Tamminga [29], we can consider the choices of i relative to the utility they have for j, and define an operator meaning, intuitively, that, if i exerts any choice of her *that is weakly dominant for j*, then ϕ is the case. This work, which opens the interesting issue of modeling the notion of "acting in the interest of someone else," was elaborated by Turrini [5] to apply also to notions of dominance taking into account the interests of other groups, or coalitions.

23.4 STIT and Matrix Game Logic: Ex Interim Knowledge

Matrix Game Logic—or MGL, for short—made its first appearance in van Benthem [6], in order to model the notion of iterated removal of strictly dominated strategies. The logic was later enriched with a notion of "freedom" that deserves attention, since it is thought to capture the margin of action that a combined choice of all the other agents leaves to a particular agent i. van Benthem and Pacuit [11] contains a very interesting comparison between STIT and MGL, and proves that there is an embedding of the former into the latter. Here, we continue the comparison with a mutual embedding between MGL's operator for *ex interim knowledge*—e.i.-knowledge, for short—and the Chellas stit. More important, we push the comparison further and show that a mutual embedding holds between MGL's operator for freedom and a Chellas stit for the agency of anti-groups. This determines interesting property transfers, sheds some light on the freedom operator, and allows some interesting considerations on the applications of STIT.

23.4.1 Matrix Game Logic for Epistemic Notions and STIT: A Formal Comparison

A matrix game frame, or MGF, is a structure $\mathcal{MG} = \langle \mathcal{G}, \{\sim_i | i \in Ags\}, \{\approx_i | i \in Ags\} \rangle$, specified as follows. $\mathcal{G}$ is a game, as defined in Sect. 23.3.1.[31] For each agent i, $\sim_i$ is an *equivalence relation* which represents the *ex interim (e.i.-) uncertainty* of i: if $w \sim_i w'$, then i is uncertain whether w or w' is the actual state *after i performs her action* (see more details below). Finally, $\approx_i$ is the "freedom relation for i": $w \approx_i w'$

---

[31] In its original version, MGL sees *action profiles* themselves as *states*. Thus, W and o are not included in the original definition of a MGF. Here we consider situations where there is no action-profile gap, which can be in turn seen as situations where the function o is total and no restriction is imposed on the construction of action profiles. In this case, MGF's can be defined as in the text.

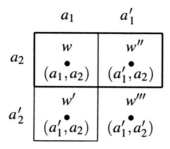

Fig. 23.4   A MGF with two agents

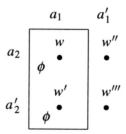

Fig. 23.5 $K_1\phi$ is true at w (and w') (we omit the freedom relation and do not specify action profiles here)

just in case w' is in the outcome $o(a_{\bar{i}})$ which also includes w; in other words, the relation models the "margin of freedom" of i—what i is free to achieve by changing her action, while the given action of its anti-group $\bar{i}$ is kept fixed. A matrix game model, or MGM, obtains by extending a MGF with an evaluation function V in the standard way.

Figure 23.4 exemplifies a MGF with two agents. Here, a_1 is the action 1 performs at w or w'. The thin box represents the class of states which are e.i.-equivalent with w; i is uncertain where state w or w' are the actual state of the world, since the latter crucially depends on what action 2 performs, and—coherently with the characterization of e.i.-knowledge in games, see below—i has only *e post* knowledge of this. The thick box represents the states which are "freedom equivalent" with w: w itself, which results from (a_1, a_2) and w'', which results from switching from a_1 to a_1' while keeping the same action of 2—that is, a_2. Let us set aside the freedom relation for now and focus on the e.i.-uncertainty relation $\sim_i$.

Games, Ex interim Knowledge, and Perfect Information. Typically, the game-theoretical literature distinguishes three kinds of knowledge: *ex ante* knowledge is knowledge of the rules of the game, *ex interim knowledge* is the knowledge one has after performing an action, *e post* knowledge is knowledge of what actions others play and what state results from the game. The games we consider here presuppose *perfect information*. In such games, agents are not uncertain about the state of the world, they know their own preferences and those of other agents, and they have common knowledge of rationality. Since what results from i's action is determined by the rules of the game, and since agents have no uncertainty about the states they are in before the game-round, it is easy to see that in games with perfect information e.i.-knowledge is essentially knowledge of what one does.

The epistemic fragment EGML. For the time being, let us drop the freedom relations $\approx_i$ from MGM's, thus obtaining *epistemic* MGM's, or EMGM's. We will likewise refer to the epistemic fragment of MGL, or EMGL, as that logic whose only operators are the agent-relative e.i.-knowledge operators K_i, with $i \in Ags$, and defined as follows:

$$\mathscr{MG}^{\mathcal{E}}, w \models K_i\phi \quad \text{iff for all } w' \sim_i w, \, \mathscr{MG}^{\mathcal{E}}, w' \models \phi$$

where $\mathscr{MG}^{\mathcal{E}}$ is an EMGM. Figure 23.5 gives an intuitive grasp of the truth-clause of $K_i\phi$. As is clear from van Benthem [6] and [8], and van Benthem and Pacuit [11], the $\sim_i$ relation satisfies strong confluence—we have, that is, $W \times W \subseteq \sim_i \circ \sim_j$ for all distinct pairs of indices i and j (between 1 and n). This is a plausible principle for e.i.-knowledge. In game theory, for each agent i e.i.-knowledge is basically knowledge of what is due to the action i performs. Suppose a_i is the action in question and consider any states $w, w' \in o(a_i)$. Of course, we have $w \sim_i w'$: i knows that, due to her action, some state in $o(a_i)$ results from the game round, but she does not know which does. This crucially depends on the action $\overline{i}$ plays, which is something i knows only *e post*, that is, when the result is settled. Thus, if $w, w' \in o(a_i)$ then $w \sim_i w'$. But also the converse holds. Suppose $w \in o(a_i)$ and $w' \notin o(a_i)$. In performing a_i, i also gets e.i.-knowledge of what states she is selecting away. Hence, $w \not\sim_i w'$. As a consequence, we have $w \sim_i w'$ iff $w, w' \in o(a_i)$. Strong confluence and equivalence are the only conditions defining $\sim_i$, and thus, by Proposition 1, we know that

For every EMGM $\mathscr{MG}^{\mathcal{E}}$, $w \sim_i w'$ iff $w \sim_i^C w'$ in $\mathscr{C}.\mathscr{MG}^{\mathcal{E}}$

where $\mathscr{C}.\mathscr{MG}^{\mathcal{E}}$ is the CCKF^+ corresponding to the given EMGM $\mathscr{MG}^{\mathcal{E}}$ (with $Ch(i) = \{o(a_i) \mid a_i \in A_i\}$ for every i). We can therefore define the following truth-preserving translation τ:

$$\tau([i]\phi) = K_i\phi$$

which in turn guarantees a mutual embedding between CSTIT and EMGL.[32] This comes with very convenient properties: indeed, it now follows from the similar properties of CSTIT that EMGL is decidable and finitely axiomatizable, no matter the number of agents.

Strong confluence also allows us to introduce $\square$ as short for $K_j K_i$, exactly as we did with $[j][i]$, and guarantees an interesting principle: $K_i K_j \phi \rightarrow \square\phi$.

---

[32] The translation τ above is already defined in and [8] and [11]. However, it does not define a mutual embedding there, since the full MGL is considered, and as we shall see, CSTIT is a proper fragment of it.

	a_1	a_1'	a_1''
a_2	w_1 • 2,3	w_2 • 2,2	w_3 • 1,1
a_2'	w_4 • 0,2	w_5 • 4,0	w_6 • 1,0
a_2''	w_7 • 0,1	w_8 • 1,4	w_9 • 2,0

Fig. 23.6 A game model with two agents, each with three actions available

	a_1	a_1'
a_2	w_1 • 2,3	w_2 • 2,2
a_2'	w_4 • 0,2	w_5 • 4,0
a_2''	w_7 • 0,1	w_8 • 1,4

Fig. 23.7 The sub-game obtaining by removing a_1''

The analysis of e.i.-knowledge above suffices to explain why: if i has e.i.-knowledge of what follows from the action of another agent j, this means that what results from j's action was trivial.

Ex Interim Knowledge and Best Actions between Matrix Game Logic and STIT.
A strategy, or action, a_i of agent i is *strictly dominated* if there is another strategy—or action—a_i' such that $o(a_i) \prec_i o(a_i')$—that is, a_i' is strictly preferred to a_i by i, no matter what action $a_{\bar{i}}$ is performed by $\bar{i}$. Since no agent would play a strictly dominated strategy, in foreseeing the moves of the other players, we may remove their strictly dominated strategies. This will create a sub-game and change the range of the agents' preferences; new strictly dominated strategies will emerge and will once again be removed, and so on, step-by-step. This is the procedure of iterated removal of strictly dominated strategies.

An example of iterated removal of strictly dominated strategies is given by Figs. 23.6 and 23.7. The pairs of numbers denotes utilities of the two agents. It is clear that agent 1 will not play a_2'', since it is a strictly dominated strategy for her. This action will then be removed, thus generating the sub-game in Fig. 23.7.

The iterated removal of strictly dominated strategies would then continue by removing a_2'', then a_1', and then a_2', thus generating three further sub-games. The last one is constituted by w_1 alone, which is the Nash Equilibrium of the initial game. Following a tradition in game theory, MGL explains iterated removal of strictly dominated strategies in *epistemic terms*: considerations about rationality, interests and strategies of others lead us to remove some strategies from an initial EMGM, and thus transform it in a sub-EMGM where other strategies become dominated. Becoming, model transformation, rationality: all this naturally calls for Dynamic Logic, and EMGL has been an answer to the call. An interesting consequence of the embedding above is that STIT can also capture these dynamics.

We cannot describe this in detail here, but we give the basic ingredients. First, we confine ourselves to finite EMGM's and CCKF$^+$'s—that is frames where each agent has a finite number of actions available. By the definition of action profiles and the outcome function, this suffices to guarantee a finite number of states. The notions involved are those of *best action* and the *epistemic notions* of *e.i.-knowledge*, *strong rationality* and *weak rationality*. Let us consider these in order.

Best actions are represented by van Benthem [6] as atoms b_1, b_2, b_3, …. They are agent-indexed and defined by the truth-clause: $\mathscr{C}^{\mathscr{MG}^{\mathscr{E}}}$, $w \models b_i$ iff there is an action $a_i \in A_i$ such that $w \in o(a_i, a_{\bar{i}})$ and $o(a_i, a_{\bar{i}}) \succeq_i o(a_i', a_{\bar{i}})$ for all $a_i' \in A_i$. In other words, the atom b_i is to be interpreted as meaning that i is performing her best action, and it is true at all and only those states which are in the outcome of some best action of i.[33] An extension of $\mathscr{L}_{\text{CSTIT}}$ with the same propositional constants is straightforward, and thus also this notion can be expressed by STIT. The notion of a best action is in turn indispensable to define the notions of *strong* and *weak* rationality.

Strong Rationality is expressed by the sentence $\neg K_i \neg b_i$ ('i does not know that she is not performing one of her best actions'). Basically, then, i is strongly rational if she knows she has at least one best action over the whole game; and note that K_i satisfies *negative introspection*, and thus we have $\neg K_i \neg b_i \leftrightarrow K_i \neg K_i \neg b_i$. Van Benthem proves that for every agent i, sentences of strong rationality hold in at least some state of a *finite* EMGM, though the same may fail for infinite EMGM or even for sub-EMGM.

Given our translation τ and the extensions with atoms, STIT can express strong rationality by $\neg[i]\neg b_i$, now taken to mean that i does not prevent herself from performing a best action. This reading points out at the purely agentive side of strong rationality: in a game-theoretical context, rational agents do not play actions different from their best ones.[34]

---

[33] It may seem that introducing linguistic atoms b_1, b_2, b_3, … to express the notion of a best action is a kind of trick. However, the move makes sense if the goal is not providing an analysis of the notion, but simply to give us linguistic means to express the fact that such an action is being performed.

[34] Theorem 6 in [6] is easily adapted to finite CCKF$^+$'s.

Weak rationality actually leads to the "dynamic" part of EMGL. While strong rationality consists in not choosing a strategy that is strictly dominated in the whole given EMGM, weak rationality consists in not choosing a strategy that is strictly dominated in the sub-EMGM in question. The difference can be appreciated by considering those cases where agents do not know that their current action is best relative to the whole game, but where they do know that such an action has no better alternative, where alternatives are now limited to the sub-EMGM considered. This also requires "relative best actions"—that is atoms b_1^*, b_2^*, b_3^*, ... Where W^* is the set of states of the sub-EMGM into account, the truth-clause for these new atoms is: $w^* \models b_i^*$ iff there is an action $a_i \in A_i$ with $w^* \in o(a_i, a_{\bar{i}})$ and such that $o(a_i, a_{\bar{i}}) \succeq_i o(a_i', a_{\bar{i}})$ for all $a_i' \in A_i$ and $o(a_i), o(a_i'), o(a_{\bar{i}}) \subseteq W^*$. In other words, b_i states that i is performing her best action relative to the actions which have not been removed.

Weak rationality is expressed by $\neg K_i \neg b_i^*$. The sentence in turn proves interesting since it is false at any state which extends the outcome of a strictly dominated strategy of i. Remarkably, given a state w in the solution zone for strictly dominated strategy removal, repeating assertions of weak rationality stabilizes at a sub-game which include w and whose domain is in that solution zone, an observation due to van Benthem [6, Theorem 7].

If we extend STIT with atoms for relative best actions, we get that weak rationality of i is expressed by $\neg [i] \neg b_i^*$: if i is rational, then she does not prevent herself from performing a relative best action. Also, it is easy to see that van Benthem's result, mentioned just above, can be straightforwardly adapted to STIT. The interesting point is that, where w is in the solution zone for strictly dominated strategies in a given game, the iterated removal of strictly dominated strategies stabilizes at a sub-game which includes w and solves the game if we iterate the choice of not preventing ourselves from playing our best action—where "iterating the choice" here means that we apply it at any sub-game resulting in the removal process.

A very brief conceptual insight. The mutual embedding has shown a surprising virtue of STIT: though designed to express *purely* agentive notions, in some situations it can also express interesting epistemic notions, such as *e.i.-knowledge, strong* and *weak rationality*. Thus, STIT also shows potential relative to certain notions which are crucial in the epistemic foundations of game theory. In particular, STIT can capture some crucial notions in iterated removal of strictly dominated strategies. The crucial issue here is what exact notions can be expressed by STIT. The ability to express strong rationality does not prove a striking result, once the mutual embedding between CSTIT and EMGL is considered. The interesting point is rather the expression of weak rationality. Indeed, such a notion seems to witness the dynamic character of the iterated removal process, and the surprising point is that STIT can frame some of this character, though it has not been designed to capture those dynamics of model-transformation which are captured by Dynamic Logics. At the same time, we need to be aware of the limits of such connections. The possibility of connecting e.i.-knowledge and seeing to it is confined to strategic games with perfect information: if an agent i could be wrong about the current state of the world, she could also be

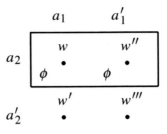

Fig. 23.8 $[\approx_1]\phi$ is true at w (and w'')

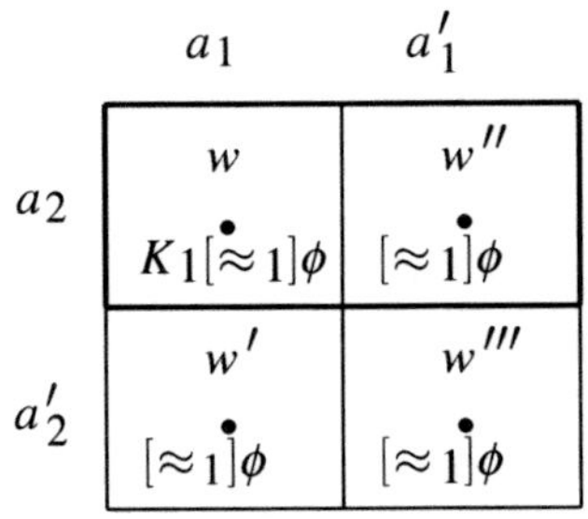

$$
\begin{array}{c|c|c|}
 & a_1 & a'_1 \\
\hline
a_2 & \begin{array}{c} w \\ \bullet \\ K_1[\approx_1]\phi \end{array} & \begin{array}{c} w'' \\ \bullet \\ [\approx_1]\phi \end{array} \\
\hline
a'_2 & \begin{array}{c} w' \\ \bullet \\ [\approx_1]\phi \end{array} & \begin{array}{c} w''' \\ \bullet \\ [\approx_1]\phi \end{array} \\
\hline
\end{array}
$$

Fig. 23.9 What follows from having $K_1[\approx_1]\phi$ true at w (we include the e.i.-uncertainty relation here)

confused about the results of her action, and thus she could see to it that ϕ without having e.i.-knowledge of ϕ. In this situation, the gap between the agentive dimension encoded by STIT and the epistemic dimension encoded by K_i resurfaces, and the latter must be explicitly introduced in the logic. Also, the ability to capture some dynamic features of the iterated removal process does not guarantee the possibility of capturing *any* dynamic aspect of action and game-theoretical interaction.

23.5 Matrix Game Logic, Freedom and STIT

Let us now return to 'full' MGL and the 'freedom relations', which, as we recall, are defined as follows: $w \approx_i w'$ iff $w, w' \in o(a_{\bar{i}})$ for some action $a_{\bar{i}}$ of $\bar{i}$. The definition of the "freedom operator" is where the new relation comes in:

$$\mathcal{MG}, w \models [\approx_i]\phi \text{ iff for all } w' \approx_i w, \text{ then } \mathcal{MG}, w' \models \phi.$$

Here, $[\approx_i]\phi$ should be taken to mean that i is left free to achieve ϕ, by $\bar{i}$'s current action. Figure 23.8 above provides an example.

Figure 23.8 is based on Fig. 23.4, but it omits labels for action profiles and the column representing the e.i.-knowledge of 1 at w and w'. What is left from Fig. 23.4 is then the margin of freedom of i at w and w''. $[\approx_1]\phi$ is true at w since it is true at all the states which are freedom-equivalent to w—namely, w itself and w''. The same

holds if we evaluate the formula at w''. Figure 23.9 on the previous page extends
Fig. 23.8 with the e.i.-knowledge of i at w (represented by the left column) and w''
(represented by the right column). The figure provides an easy way to check the
validity of $K_i[\approx_i]\phi \rightarrow \Box\phi$ below.

The new freedom operator carries a lot of information: it calls upon the action
of $\bar{i}$, as is clear from the definition of $\approx_i$, and tells us what options the current
action of $\bar{i}$'s leaves to i. This involves much more than talking about the individual
agent i and what action she performs, of course: it also implies reference to a more
complex concept, which hints at the way some of i's actions interact with those of
$\bar{i}$. As a consequence $[\approx_i]$ is a complex operator, and a systematic investigation of its
properties may prove difficult.

Two clear features are that $[\approx_i]$ is an S5 operator and that it satisfies a strong
confluence axiom in combination with K_i—that is, $W \times W \subseteq \sim_i \circ \approx_i$. These
observations are established by van Benthem and Pacuit in [8] and [11], and illustrated
in Fig. 23.9. As a result, we have the principle:

$$K_i[\approx_i]\phi \rightarrow \Box\phi.$$

This is an interesting principle: it states that only in case ϕ is trivial can the agent i
e.i.-know whether she is free to achieve ϕ; knowledge of what is not excluded by the
choice of an agent's own anti-group is trivial, and so not interesting. The validity of
the formula is easily checked with the help of Fig. 23.9: here, $K_i[\approx_i]\phi$ is true at w.
By this and the semantics of K_i, we have that $[\approx_i]\phi$ is true at w and w'. Since $[\approx_i]\phi$
is true at w, we have that $[\approx_i]\phi$ is true at w'', and since $[\approx_i]\phi$ is true at w', we have
that $[\approx_i]\phi$ is true at w'''. $[\approx_i]\phi$ is then true at all the states of the model represented
by Fig. 23.9. By this and the semantics of $[\approx_i]$, we have that ϕ is true at every state
of the model. As a consequence, we have $\Box\phi$ true at all of them.

Other questions remain open. Can $[\approx_i]$, $[\approx_j]$ be turned into stit operators? What
about a confluence property involving freedom operators only? Is MGL decidable
and finitely axiomatizable? These questions have not yet been settled in the literature.
We do this below, but in order for us to show the result, we need to extend CSTIT to
group CSTIT first.

A STIT logic for Groups. MGL is a proper extension of EMGL. Since there is a
mutual embedding between the latter and CSTIT, we have that CSTIT is a proper
fragment of MGL. However, the relation may change (and indeed changes) if we
consider *group* CSTIT, which allows us to talk, among the other things, of groups,
singletons, and anti-groups.[35] Let us start from a full group STIT and then isolate
the anti-group fragment. In STIT, a group is a set of agents; then, the set of all the
groups definable in a frame for group STIT is nothing but the power set $\wp(Ags)$ of
the set Ags of (individual agents).

---

[35] *Group* STIT was already present at the beginning of STIT; see Belnap [3] and Horty [24].
Both Belnap and Horty assume additivity; see [3, Definition 10–3], and [24, Definition 2.10]. Neither

A (relational) *group* CKF$^+$—henceforth, a GCKF$^+$—is a triple $\langle W, Ags, \{\sim_I^C \mid I \subseteq Ags\}\rangle$, where W and Ags are as in CCKF$^+$, and $\sim_I^C$ is a group choice-equivalence relation between states in W, such that:

(R1$'$) $\sim_I^C$ is an equivalence relation
(R2$'$) for all $(w_1, \ldots, w_k) \in W^k$ and $J_1, \ldots, J_k \in \wp(Ags)$, if $l \neq m$ implies $J_l \cap J_m = \emptyset$, then $\bigcap_{i \in \{1, \ldots, k\}} \{w \mid w \sim_{J_i}^C w_i\} \neq \emptyset$
(R3$'$) $\sim_{Ags}^C = Id_W$
(R4) $\sim_I = \sim_J \cap \sim_{I/J}$ where $J \subseteq I$
(R5) $\sim_I^C \subseteq \sim_J^C$ if $J \subseteq I$.

Here, R2$'$ is the group version of the strong confluence (notice the restriction to *disjoint* groups), while R3$'$ is grand group determinism: the grand group can determine a unique state of the world (Id_W being the identity relation on W, standardly defined: $Id_W = \{\{w, w'\} \mid w, w' \in W$ and $w = w'\}$)[36] and R4 is additivity the choice of a group is the intersection of the choices of its disjoint subgroups.[37] R3$'$ is clearly the group version of R3, while R4 allows to understand D1 as a special case of additivity. R5 is the so-called coalition monotonicity which holds that, if agents (and groups) join their efforts, they improve their result; the condition mirrors the standard assumptions about coalition effectivity functions in game theory. An individual agent i is here taken as a special case of groups, namely the singleton $\{i\}$, we keep the individual notation for the sake of readability. An analog to Proposition 1, above, also applies to GCKF$^+$'s—indeed the actual statement of Theorem 1 in Turrini [5] is about coalitions. Finally, the new operator $[I]$ is just the group version of $[i]$, so that: $[I]\phi$ is true at w in a model for CGKF$^+$ iff ϕ is true at every state $w' \sim_I^C w$.

The principles R1$'$, R2$'$, R3$'$ and R5 correspond to the following principles:

(P1) S5 axioms for $[I]$;
(P2) $[I][J]\phi \rightarrow \Box\phi$, where $I \cap J = \emptyset$
(P3) $\phi \leftrightarrow [Ags]\phi$
(P5) $[J]\phi \rightarrow [I]\phi$ if $J \subseteq I$.

And, as the reader will notice, many distinctive features of individual STIT are transferred to the group level, though with restrictions: for instance, we have

---

(Footnote 35 continued)
Belnap nor Horty assume condition R3$'$ below—that the joint agency of all the agents may determine a unique outcome—although, as mentioned earlier, Horty [24, p. 91] considers models that satisfy this condition. Basically, the conditions we present here build on those presented in Horty [24] by adding the standard game-theoretical condition of coalitional monotonicity. The latter can be actually derived by R2$'$, but we present it here as a basic condition in order to conform with the standard presentation of group STIT.

[36] The principle is also called "Rectangularity" in Turrini [5].

[37] The standard condition in game theory is actually *superadditivity*, which allows for the choice of I to be a subset of the choices of I's members; the condition is actually a consequence of coalition monotonicity. However, we prefer to include R4 in order to comply with the standard choice in group STIT, and also because it makes the construction of groups conceptually easier.

$[I][J]\phi \rightarrow \Box\phi$ if $I \cap J = \emptyset$, but otherwise the principle may fail. Also, it is easy to show that $\Diamond[I]\phi \wedge \Diamond[J]\psi \rightarrow \Diamond([I]\phi \wedge [J]\psi)$ holds *if* $I \neq J$. Herzig and Schwarzentruber have proved that group CSTIT with strictly more than two agents is undecidable and is not finitely axiomatizable; see [21, Theorems 22 and 23]. This is primarily due to the fact that strong confluence fails if $I \cap J \neq \emptyset$; if— additionally—the groups I and J overlap without being subsets one of another, then the logic can be mapped into $S5^n$, with n the number of agents in the group STIT in question. Since all extensions of $S5^3$ are undecidable and not finitely axiomatizable so is group CSTIT with three or more agents.[38]

STIT, Anti-groups and Freedom. Let us call *anti-group* CSTIT that fragment of *group* CSTIT where only agents in Ags and their anti-groups are included. Some interesting connections between individual agents and anti-groups are easily captured:

(R2″) $W \times W \subseteq \sim_i^C \circ \sim_{\bar{i}}^C$ for every $i \in Ags$
(R5′) $\sim_{\bar{j}}^C \subseteq \sim_{\bar{i}}^C$ for all $j \neq i$

R2″ holds since an agent and her anti-group satisfy by definition the disjointness proviso in R2′; R5′ holds because, by definition, any agent will be a subgroup of the anti-group of any other agent. As a consequence,

(P2′) $[i][\bar{i}]\phi \rightarrow \Box\phi$
(P5′) $[i]\phi \rightarrow \bigwedge_{j \neq i}[\bar{j}]\phi$

hold. The most interesting connection, however, is with MGL's $\approx_i$. Indeed, since the correspondence result in Sect. 23.3 extends to group STIT and coalitional games (see above), we have that: $w \sim_{\bar{i}}^C w'$ iff $w, w' \in o(a_{\bar{i}})$ for some action $a_{\bar{i}}$ of $\bar{i}$. From this, it follows that the relation $\sim_{\bar{i}}^C$ in GCKF$^+$ is nothing but the relation $\approx_i$ in MGM. We thus have:

For every MGM $\mathcal{M}\mathcal{G}$, $w \approx_i w'$ iff $w \sim_{\bar{i}}^C w'$ in $\mathscr{C}^{\mathcal{M}\mathcal{G}}$

and we can therefore define a truth-preserving translation τ' such that

$$\tau'([i]\phi) = K_i\phi$$
$$\tau'([\bar{i}]\phi) = [\approx_i]\phi$$

where $\mathscr{C}^{\mathcal{M}\mathcal{G}}$ is the GCKF$^+$ corresponding to the given MGM $\mathcal{M}\mathcal{G}$ (with $Ch(i) = \{o(a_i) \mid a_i \in A_i\}$ for every i). There is therefore a mutual embedding between *anti-group* CSTIT and MGL. This answers one question we asked earlier: $[\approx_i]$ can indeed be interpreted as a stit operator. And with this answer comes both bad news and good news.

---

[38] Again, see Hirsch et al. [23] for these results concerning $S5^n$.

Bad news. The bad news is that we can now conclude that the "full" MGL[39] is undecidable and not finitely axiomatizable. This follows from the mutual embedding, together with the fact that anti-group CSTIT with strictly more than two agents has these properties, which transmit to MGL.[40]

Good news. The good news is that we can gain insight into the properties of $[\approx_i]$ via established results about group stit operators—particularly those concerning $\bar{i}$. For instance, we can now see that $K_i\phi \rightarrow \bigwedge_{j\in\bar{i}}[\approx_j]\phi$ holds, from P5″ and τ'. This is an interesting principle: it states that, if i has e.i.-knowledge that ϕ, then she is not excluding that any other agent j achieves ϕ. If we dig into the conditions that define the e.i.-knowledge and freedom relations, it is evident that this principle is sensible: the agent i can have e.i.-knowledge that ϕ because her current action removes the possibility of achieving "non-ϕ" states. Thus, j also has a margin to achieve ϕ with her current action, while it is excluded that she achieves $\neg\phi$ with any of her available actions.

The mutual embedding also helps us understand the issue of strong confluence. Contrary to what happens with K_i, the freedom operator $[\approx_i]$ does not satisfy the strong confluence property: $[\approx_i][\approx_j]\phi \rightarrow \Box\phi$ does not hold, since $[\bar{i}][\bar{j}]\phi \rightarrow \Box\phi$ does not hold in group CSTIT with more than two agents, since, in that case, $\bar{i} \cap \bar{j} \neq \emptyset$. This failure implies that there are cases where ϕ is not trivial and yet i has a margin of freedom to let j have a margin of freedom to achieve ϕ—or equivalently: the current choice of i's anti-group does not imply that j's anti-group achieves $\neg\phi$. Transmission of freedom, it turns out, is not trivial, after all!

For analogous reasons, $\bigwedge_{i\in Ags}\Diamond[\approx_i]\phi_i \rightarrow \Diamond\bigwedge_{i\in Ags}[\approx_i]\phi_i$ also fails in MGL: even though two different agents 1 and 2 are left free to achieve ϕ_1 and ϕ_2 respectively, their results may be incompatible. Thus, agents are not independent in their margins of freedom. This sounds plausible: after all, the margins of freedom that one agent has *depend* on what the current choice of the other agents is.

23.6 Conclusions

In this chapter we have accomplished two main tasks. First, we have highlighted the ties between STIT and the basic settings of game theory. This has involved demonstrating the possibility of reading STIT game-theoretically and expressing game-theoretical notions in STIT's terms. The connection thus established proves a very good hint at the flexibility and richness of STIT theory.

---

[39] Here we mean MGL as defined in this chapter, not the full logic defined in van Benthem [8], which also includes an operator for preferences.

[40] See our observation above on the conditions for undecidability and failure of finite axiomatizability in group CSTIT. Of course, decidability and finite axiomatizability are restored if we confine to MGL with only two agents, so that $Ags = \{1, 2\}$. In that case, $\bar{1} = 2$ and $\bar{2} = 1$. The anti-groups thus collapse into different agents, and MGL with two agents actually collapse into EMGL with two agents—which is indeed decidable and finitely axiomatizable, since EMGL is, no matter the cardinality of Ags. Thus, the case with two agents does not hold much interest.

Second, we have considered the MGL logic of games, a form of Dynamic Logic, and have furthered the comparison begun by van Benthem and Pacuit [11] between STIT and MGL. This comparison has led, we believe, to some interesting results. First, as noted in that earlier work, the "epistemic fragment" of MGL has a mutual embedding with the logic CSTIT for individual agency; thus, STIT has the potential to capture the notions of ex interim knowledge and the assertions of weak and strong rationality. Also, decidability and finite axiomatizability transmit from CSTIT and the fragment of MGL.

It was established here, however, that full MGL, including the "freedom operator," has a mutual embedding with a group version of CSTIT which includes arbitrarily many individual agents and their anti-groups. This suffices to secure that full MGL is undecidable and not finitely axiomatizable.

However, the embedding also allows us to explore issues about the freedom operator—which is conceptually very rich—"through the mirror" of STIT. This helps us to notice that the freedom operator does not obey independence (for reasons which are explained by the very setting of group STIT), and shows an interesting relation between the ex interim knowledge of an agent and the margin of freedom left to all other agents.

This proves STIT to be illuminating, not only for its own sake, but also as a tool for developing formal and conceptual perspectives on other frameworks for agency. A thorough comparison with Dynamic Epistemic Logic in the style of Baltag, Moss, and Solecki [2] could be a further interesting step in bridging STIT and the dynamic framework. The ground for this has been provided in [11]. The merging of the two methodologies could prove extremely fruitful in the modeling of multi-agent situations where it is crucial to express whether the information update in the doxastic state of agent i has been brought about by i herself or passively received by other agents.

Acknowledgments This work was carried out while Roberto Ciuni was a Humboldt Postdoctoral Fellow working on the project 'A Tempo-Modal Logic for Responsibility Attribution in Many-Step Actions' (2011–2013). We wish to thank Alexandru Baltag and Yan Zhang for very helpful comments on earlier versions of this work.

References

1. Balbiani P, Herzig A, Troquard N (2008) Alternative axiomatics and complexity of deliberative STIT theories. J Philos Logic 37(4):387–406
2. Baltag A, Moss LS, Solecki S (1998) The logic of public announcements, common knowledge and private suspicions. In: Proceedings TARK, Morgan Kaufmann Publishers, Los Altos, pp 43–56. (updated versions through 2004)
3. Belnap N, Michael P, Ming X (2001) Facing the future: agents and choices in our indeterminist world. Oxford University Press, Oxford
4. Belnap N, Perloff M (1988) Seeing to it that: a canonical form for agentives. Theoria 54:175–199

5. van Benthem J (2007) Rational dynamics and epistemic logic in games. Int Game Theory Rev 9(1):13–45
6. van Benthem J (2011) Logical dynamics of information and interaction. Cambridge University Press, Cambridge
7. van Benthem J (forthcoming) Logic in games. The MIT Press, Cambridge (MA)
8. van Benthem J, Pacuit E (forthcoming) Connecting logics for choice and change. In: Müller Thomas (ed) Volume in honour of Nuel Belnap. Springer, Berlin (outstanding logicians series)
9. Broersen J (2011) A deontic epistemic stit logic distinguishing modes of Mens Rea. J Appl Logic 9(2):137–152
10. Broersen J (2011) Modeling attempt and action failure in probabilistic stit logic. In: Proceedings of twenty-second international joint conference on artificial intelligence (IJCAI 2011), pp 792–797
11. Broersen J (2011) Probabilistic stit logic. In: Proceedings 11th european conference on symbolic and quantitative approaches to reasoning with uncertainty (ECSQARU 2011). Lecture notes in artificial intelligence, vol 6717. Springer, Berlin, pp 521–531
12. Broersen J, Herzig A, Troquard N (2006) Embedding ATL in strategic STIT logic of agency. J Comput 16(5):559–578
13. Broersen J, Herzig A, Troquard N (2006) From coalition logic to STIT. Electron Notes Theoret Comput Sci 157:23–35
14. Chellas B (1969) The logical form of imperatives. PhD thesis, Philosophy Department, Stanford University
15. Chellas B (1992) Time and modality in the logic of agency. Studia Logica 51:485–517
16. Ciuni R (2010) From achievement stit to metric possible choices, logica 2009 yearbook. College Publications, London, pp 33–46
17. Ciuni R, Mastop R (2009) Attributing distributed responsibility in stit logic. In: Xiandong H, Horty J, Pacuit E (eds) Logic rationality, interaction (lecture notes in computer science, vol 5834. Springer, Berlin, pp 66–75
18. Ciuni R, Zanardo A (2010) Completeness of a branching-time logic with possible choices. Studia Logica 96(3):393–420
19. Herzig A, Lorini E (2010) A dynamic logic of agency I: STIT, abilities and powers. J Logic Lang Inform19(1):89–121
20. Herzig A, Schwarzentruber F (2008) Properties of logics for individual and group agency. In: Areces C, Goldblatt R (eds) Advances in modal logic, vol VII. College Publications, London, pp 133–149
21. Hirsch R, Hodkinson I, Kurucz A (2002) On modal logics between $K \times K \times K$ and $S5 \times S5 \times S5$. J Symbol Logic 67(1):221–234
22. Horty J (1996) Agency and obligation. Synthese 108(2):269–307
23. Horty J (2001) Agency and deontic logic. Oxford University Press, Oxford
24. Horty J, Belnap N (1995) The deliberative stit: a study of action, omission, and obligation. J Philos Logic 24(6):583–644
25. Kooi B, Tamminga A (2008) Moral conflicts between groups of agents. J Philos Logic 37:1–21
26. von Kutschera F (1986) Bewirken. Erkenntnis 24(3):253–281
27. Lorini E (2013) Temporal STIT logic and its application to normative reasoning. J Appl Non-Class Logics 23(4):372–399
28. Tamminga A (2013) Deontic logic for strategic games. Erkenntnis 78(1):183–200
29. Turrini P (2012) Agreements as norms. In: Ågotnes T, Broersen J, Elgesem D (eds) DEON 2012, LNAI 7393. Springer, Berlin, pp 31–45
30. Xu M (1994) Decidability of deliberative stit theories with multiple agents. In: Gabbay D, Ohlbach H (eds) Proceedings of the first international conference in temporal logic. Springer, Berlin, pp 332–348
31. Xu M (1998) Axioms for deliberative stit. J Philos Logic 27:505–552
32. Xu M (2010) Combinations of STIT and actions. J Logic Lang Inform 19(4):485–503
33. Zanardo A (2013) Indistinguishability, choices and logics of agency. Studia Logica 101(6): 1215–1236

Chapter 24
The Logic of Best Actions from a Deontic Perspective

Olivier Roy, Albert J. J. Anglberger and Norbert Gratzl

Abstract This chapter re-visits Johan van Benthem's proposal to study the logic of "best actions" in games. After introducing the main ideas behind this proposal, this chapter makes three general arguments. First, we argue that the logic of best action has a natural deontic rider. Second, that this deontic perspective on the logic of best action opens the door to fruitful contributions from deontic logic to the normative foundation of solution concepts in game theory. Third, we argue that the deontic logic of solution concepts in games takes a specific form, which we call "obligation as weakest permission". We present some salient features of that logic, and conclude with remarks about how to apply it to specific understandings of best actions in games.

24.1 Logic of Best Actions in Games

The term "logic of best action" was coined by Johan van Benthem in a number of recent papers at the intersection of modal logic and game theory [7–10]. Roughly, the goal is to study the logical properties of actions and strategies that are deemed "best" by some solution concepts for games, for instance backward induction or equilibrium play.

O. Roy (✉)
Universität Bayreuth, Bayreuth, Germany
e-mail: Olivier.Roy@uni-bayreuth.de

A. J. J. Anglberger
Ludwig-Maximilians-Universität, München, Germany
e-mail: Albert.Anglberger@lrz.uni-muenchen.de

N. Gratzl
Ludwig-Maximilians-Universität, München, Germany
e-mail: Norbert.Gratzl@lrz.uni-muenchen.de

A. Baltag and S. Smets (eds.), *Johan van Benthem on Logic*   657
and Information Dynamics, Outstanding Contributions to Logic 5,
DOI: 10.1007/978-3-319-06025-5_24, © Springer International Publishing Switzerland 2014

This idea is part of a broader endeavor of looking at "intelligent interaction" from a logical point of view. Indeed, the classical components of a game, *actions*, *preferences* and *beliefs*, have been the subject of extensive logical investigations. Preference logic has a long history. Its connection to game-theoretic concepts has gained considerable momentum in recent years [11, 21]. The logical, combinatoric and computational properties of actions and strategies have also attracted the attention of logicians [5, 12, 20, 34]. And last but not least, the so-called epistemic programme in game theory has brought to the fore the importance of *knowledge* and *beliefs* in games, and has a long-standing relationship with epistemic logic [9, 18].

Inasmuch as these classical components are in themselves interesting for logicians, put together they articulate *solution concepts*, which raise logical questions of their own. A solution concept is a proposal as to which action the player *will* or *should* choose in specific game-playing situations. Nash equilibrium is probably the most well-known, and also the most widely used solution concept in game theory. This is surely not the only solution concept on the market, though, and may not be the most interesting for logicians. Recursive solutions, involving the *iteration* of a given procedure, also raise many interesting logical questions, for instance regarding the existence, nature and complexity of fixed-points of such procedures. We will look at the example of backward induction shortly, but the recent interest in the dynamics of iterated strict and weak dominance is also a case at hand [1, 6, 31].

There is by now a large literature on logical characterizations of solution concepts. We discuss a number of recent contributions in what follows. Our running example will be the logical characterization of backward induction. We use that example because it brings out all the paradigmatic elements of recent logical approaches to game-theoretic solution concepts. With this in place we then briefly survey some other logical characterizations of other solution concepts. The goal here is not to survey the field thoroughly. This would be beyond the scope of this chapter. Rather, we want to highlight the distinction between these logical characterizations and what we call the more minimalistic approach of the logic of best action. The reader interested in a thorough survey of logical characterizations of solution concepts can look at van der Hoek and Pauly [22] for an overview of the first wave of such characterizations, and the work and references in [12, 28, 38] for a good idea of subsequent takes.

24.1.1 Solution Concepts in Rich Logical Languages: The Case of Backward Induction

In this section we present van Benthem and Gheerbrant's [10] logical characterization of backward induction. This will allow us to highlight the key elements of what we call a characterization in rich logical languages, and to contrast it later on with the more minimalistic approach of the logic of best action. Since we use backward induction as an illustrative example, we will not give precise definitions of this algorithm, nor

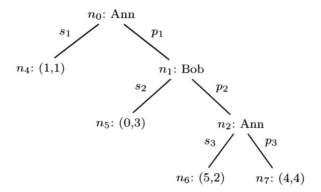

Fig. 24.1 An extensive game. Each non-terminal node represents a choice point for one of the players. Ann moves first (n_0). If she passes then Bob is to play (n_1), followed by Ann if he passes as well (n_2). The actions available at each node are the edges branching to other nodes. Each terminal node is associated with a payoff (x, y) for each player, with x and y respectively Ann's and Bob's

of the underlying game-theoretical notions like game trees, strategies and reachable nodes. The reader interested in these can consult [32] for the details.

Backward induction is an algorithm that computes a solution concept for games in extensive forms called *sub-game perfect equilibrium*. Games in extensive form are often represented as trees, for instance in Fig. 24.1. They explicitly encode the temporal structure of the play. For simplicity we only consider here games where no two terminal nodes give equal payoff for a player. This can be lifted, at the cost of somewhat complicating the backward induction algorithm. See e.g. [10].

A sub-game perfect equilibrium is a combination of strategies, one for each player, in which at any node in the tree, following that strategy never leads to choosing a strategy that is strictly dominated on the assumption that the others are following their sub-game perfect equilibrium strategy subsequently. In Fig. 24.1, (s_1, s_3) and s_2 are, respectively, Ann's and Bob's unique sub-game perfect equilibrium strategies. By contrast, the profile (s_1, p_3) and s_2 is also an equilibrium of this game, but it is not sub-game perfect. At n_2 Ann chooses p_3 even though she could have done better by going for s_3 instead.

The backward induction algorithm computes this solution by inductively excluding actions from the set of potentially available ones. It starts by considering the set S_0 of all actions as available. This is our basic case. In our example this is simply the set of all s_i and p_i for $i = 1, 2$ or 3. The inductive step proceeds as follows. Suppose S_k is defined. Then S_{k+1} is S_k minus all actions that are *strictly dominated* by another action in S_k. An action in S_k, i.e. a move from node n to n', is strictly dominated if there is another action, from n to n'', also in S_k such that all terminal nodes that are reachable from n'' by actions still available give a higher payoff for the player who is to play at n than all terminal nodes that are reachable from n'. In our example, as is often the case, the move from S_0 to S_1, i.e. the first round of elimination, only removes actions that lead directly to terminal nodes. Here only p_3, because it gives only 4 to Ann, while she gets 5 by playing s_3. No other action is strictly dominated. The key point is that this makes n_7 no more reachable at subsequent steps. In particular, p_2 becomes then strictly dominated in S_1, because n_6 is the only terminal node reachable from n_2 in that restricted set of actions, and this gives Bob 2, as opposed to the 3 he would get by playing s_2. By the same reasoning p_1 is eliminated at the third round, reaching $\{s_1, s_2, s_3\}$, where no action is strictly dominated, and which is also the sub-game perfect equilibrium of that game. There is of course an epistemic story to tell about this procedure, in terms of rationality and common belief in rationality at future nodes [2, 3]. We leave that aside for now.

In extensive games of perfect information, backward induction and sub-game perfect equilibrium are two sides of the same coin. The solution concept singles out a subset of the set of possible actions. In relational terms, closer to a modal logician's perspective on the matter, it identifies a *sub-relation of the move/edge relation in the game tree*. But this sub-relation is precisely the one reached when the backward induction algorithm terminates or, in other words, it is the *fixed point of recursively ruling out sub-game dominated moves*.

This is the starting point of van Benthem and Gheerbrant's [10] logical analysis of backward induction. Logically speaking, a transition from node n to move m is in part of a player's subgame-perfect equilibrium strategy if and only if it satisfies the formula (BI-LFP(FO)) below.

$$\nu S.(move(n, m) \wedge \bigwedge_i (turn_i(n) \rightarrow \forall z(move(n, z) \tag{BI-LFP(FO)}$$
$$\rightarrow \exists u, v(end(u) \wedge end(v) \wedge S^*(y, v) \wedge S^*(y, v) \wedge \pi_i(u) \leq \pi_i(v)))))$$

This formula belongs to LFP(FO), i.e. First-Order Logic augmented with Fixed-Point operators, here ν, for a largest fixed point.[1] Its essential feature is that the computation of this fixed point follows precisely (as a theorem) the backward induction algorithm just presented. Starting with the full set of actions it moves to more and more precise approximations until it stops at the set of actions that precisely corresponds to the sub-game perfect equilibrium strategy.

This characterization of backward induction in first-order logic with fixed-points takes many forms when one moves to propositional modal languages. For instance, in one containing operators $[move]$ and $[\leq_i]$ for the move and player i's preference relations, the fixed-point or algorithmic counterpart of sub-game perfect equilibrium becomes a confluence property. It is the unique strategy profile/sub-relation of the move relation σ that satisfies (BI-ML) below. The original result is from [13]. The correspondance with BI-LFP(FO) is proved in [10].

$$(turn_i \wedge \langle \sigma^* \rangle (end \wedge p)) \rightarrow [move]\langle \sigma^* \rangle (end \wedge \langle \leq_i \rangle p) \tag{BI-ML}$$

---

[1] The relation S in this formula is the sub-relation of *move* that will correspond to sub-game perfect equilibrium. It is a one-step relation in the tree, with its usual reflexive-transitive closure S^*. The other predicates used in this formula are meant to identify which player is to choose at a given node ($turn_i$), whether a node is terminal (*end*) and to make payoff comparisons.

Alternatively, one can capture sub-game perfect equilibrium in propositional languages with dynamic operators, modeling the backward induction process in terms of absolute deletion of strategies or of forming more and more precise, but nonetheless revisable expectations about the player's choices. Details on both point of views, and more, can be found in [10, 19].

24.1.2 Solution Concepts in Rich Logical Languages: More Examples and General Perspective

The crucial point about the characterizations of backward induction that we have just presented is that they *embed the logic of that solution concept in a language expressive enough to talk about its constituents*, namely actions, outcomes or terminal nodes, and preferences. By "the logic of that solution concept" we mean the valid formulas containing explicit operators for the sub-game perfect equilibrium actions. Such validities are but a small fragment of the logic within which the characterizations above are developed. For instance, the first-order fixed point logic in which BI-LFP(FO) is defined, also contains validities regarding the move relation (S), preference or utility comparisons ($\leq$), and the status of the nodes ($turn_i(n)$, $end(u)$). For instance, the logic contains the following valid formula, expressing the fact that no player is to play at terminal nodes:

$$\forall x \left(end(x) \leftrightarrow \neg (\bigvee_{i \in A} turn_i(x)) \right)$$

Most contemporary logical characterizations of solution concepts for games follow this methodology, i.e. they embed the logic of the solution concept at hand in rich logical languages. A good example of that is the deontic logic for strategic games developed by Kooi and Tamminga [27, 38]. This logic falls in the broad family of *stit*-based deontic systems, where the notion of *agency* occupies a central place [4, 23]. In Horty's work [23, 24], for instance, the non-deontic fragment of the logical language crucially contains *modalities for actions and/or agency*. These are the well-known *stit* operators, describing what the agents can achieve at each moment. The deontic operators themselves are interpreted on so-called consequentialist models, where the possible outcomes of one's actions are *comparable* along a given scale, ordinal or quantitative. In this framework deontic operators are thus much akin to modalities taking about preferences or relative goodness, or a specific pattern thereof, for instance strict dominance. Kooi and Tamminga [27] have shown that this combination of agentive and preferential modalities can be fruitfully brought to the analysis of game-theoretic scenarios, and especially to understand better the structure of conflicting obligations that arise in cases like the Prisonners' Dilemma. In subsequent work Tamminga [38] has shown that the resulting logic is expressive enough to define Nash equilibria in strategic games. Analogous results, although not cast in the framework of *stit* theory, can be found in [11, 22]. See also [30] for quite an original extension of the *stit* approach to obligations in games, and [28] for a characterization of solution concepts also using epistemic tools.

For the present chapter the important point is the following: these constitute successful instances of logical characterizations of game-theoretic solution concepts—for instance Nash equilibrium—within a rich modal language—for instance one capable of expressing notions of preferences, dominance, and agency.

24.1.3 Solution Concepts in Minimalistic Logical Languages: The Logic of Best Action

These characterizations in richer languages describe *indirectly* the core logical and computational properties of the solution concepts themselves.[2] Once one has shown that, say, sub-game perfect equilibrium is definable in a language that talks about actions and preferences, at the axiomatic level one can forget about this solution concept and focus entirely on the core principles for the action and preference fragments. The resulting proof system does contain validities about sub-game perfect equilibrium, but to recover them one needs to identify instances of usually quite complex formulas provided by the characterization results. This is an exercise that, to our knowledge, has rarely been engaged in. The situation is similar on the computational side. The complexity of the satisfiability and the model checking problems for these rich logics provide upper and lower bounds for the specific cases involving solution concepts. To be sure, these can be informative. But they do not answer the question of how difficult these specific cases are.

There are thus good logical and computational reasons to look at "best actions in games" in a minimalistic logical environment, that is to study the logic of what is deemed "best" according to given solution concepts, in isolation from the fine-grained apparatus used to talk about preferences, actions and beliefs. At the logical level, such logic would highlight the core validities involving these solution concepts. We indeed saw that, in extensive games, solution concepts can be seen as sub-relations of the move relation.[3] From that point of view, the question of the "core logical validities" is one of identifying, if possible, the complete logic of that specific sub-relation. To go back to the example of sub-game perfect equilibrium, the question is whether one can completely axiomatize the set of validities pertaining to operators [*BI*] interpreted on that specific sub-relation of the total move relation, as opposed to embedding them in richer logical systems. Doing this also promises computational insights on the difficulty of checking for consistency of statements about, say, sub-

---

[2] This is not to say that they are not illuminating, quite the contrary. A good example is the characterizations in [11, 21], that highlight the *ceteris paribus* character of equilibrium solutions.

[3] The same holds *mutatis mutandis* for games in strategic form.

game perfect equilibria (satisfiability), or of computing equilibria in given games (model checking).

But beyond that, there is also a general, conceptual motivation for investigating the logic of best actions in games coming from the *normative interpretation* of solution concepts, and more generally of game theory. Viewed this way, the "logic of best action" programme is a natural addition to *deontic logic*. This is what we argue now.

24.2 Logics of Best Action as Deontic Logics

We have already mentioned that solution concepts can be given either a descriptive or a normative interpretation. According to the first, to say that action *a* of a given game is compatible with a given solution concept is to say that this is what players *will* or *might* play in that game. It is well-known that solution concepts fare rather badly under this interpretation [17]. According to the normative interpretation, solution concepts are rather recommendations about what players *should* or *may* choose in a given game. Alternatively, they can be seen as descriptions of what *ideal* agents would/might choose.

Under the normative interpretation, solution concepts thus make recommendations to the players. They circumscribe the set of actions that *should* or *may* be played. In the game in Fig. 24.1, for instance, sub-game perfect equilibrium, in its normative interpretation, recommends Ann to choose s_1 at the beginning of the game, Bob to choose s_2 if he is to play, and Ann again to choose s_3 if n_3 is reached. Of course, it is controversial whether these are normatively valid recommendations, especially given the fact that in this game they lead to a non-pareto optimal outcome. We come back to that point in a moment.

Once one makes this move to the normative interpretation of solution concepts, operators for "best action" get a natural deontic reading. Under their usual interpretation, "best action" operators are normal modal operators interpreted on the sub-relation of the move relation that corresponds to a given solution concept:

(Best)	$[best]\varphi$ is true at a node/state *s* whenever all actions available at *s* compatible with the solution concept at hand lead to φ-state.

Under the normative interpretation, actions singled out by a given solution concept are, roughly, actions that the players should choose. Substituting this in the formulation above we get:

(Best-Should)	$[best]\varphi$ is true at a node/state *s* in a given tree whenever all actions that *should be taken* at *s* lead to φ-nodes.

Best action operators, from that point of view, are thus describing what the players should do at a given node in the tree. They are deontic operators. This congeniality with deontic formulas is even more salient if we think of solution concepts as singling out the set of actions A^* that would be played in ideal scenarios. Our clause above then becomes essentially the classical Kripke semantics for deontic operators [29].

(Best-DEON)	$[best]\varphi$ is true at a node/state s in a given tree whenever only φ-nodes/states are reachable by actions in the ideal set A^*

Studying the logic of best action generated by a given solution concept, from that perspective, is entering the deontic arena, and this raises interesting logical questions. How do such logics compare with existing deontic systems? Are these logics of best action prone to the known deontic paradoxes, by now the canonical rite of passage for deontic systems? These are questions that we come back to later on.

24.2.1 Arguments for a Deontic Perspective on the Logic of Best Actions

At this point the reader might legitimately wonder why one should look at the logic of best action from a deontic perspective. In the previous section we have argued that this perspective is natural once one takes the normative interpretation of solution concepts. But beyond naturalness, what kind of concrete contributions can one expect from this move to the deontic arena? We argue now that it can contribute to the philosophical debate concerning the normative status of solution concepts for games in two ways. The deontic logic of best action can be used to analyze normative conflicts that bear on agents in games, and it can be used to check for internal consistency and plausibility of the system of recommendations that stem from given solution concepts.

Let us first emphasize the classical difficulty faced by the normative interpretation of solution concepts. Look again at Fig. 24.1. The sub-game perfect equilibrium profile is not Pareto optimal. There are other combinations of strategies, for instance the one where both Ann and Bob always pass, that yield a strictly better payoff for both of them. In view of this, one may ask on what ground should the players, or what kind of reasons are there for them to play their sub-game perfect equilibrium.

The classical debate on that question usually opposes those who think that solution concepts like sub-game equilibrium yield the intuitively wrong recommendations in such cases to those who resort to classical decision-theoretic foundations, possibly supplemented by some epistemic conditions. The former think that games like the one in Fig. 24.1 constitute a decisive counter-example. It seems clear to them that players *should not*, or at the very least that it is not the case that they should play according to sub-game perfect equilibrium in that game. What they should do is to reach a Pareto optimal outcome. So, they conclude, sub-game perfect equilibrium is not normative [16].

Recent work in the so-called epistemic programme in game theory has pointed out, however, that in extensive games with perfect information sub-game perfect equilibrium is what utility maximizers would play under the common belief of future rationality [2, 3]. But maximization of expected utility, they argue, has an independent normative appeal. We know from classical representation theorems from Ramsey, de Finetti and Savage[4] that players who do not maximize expected utility will make incoherent decisions. If one agrees that such incoherence should be avoided then, at least under the common belief that the players will maximize expected utility at future nodes, playing according to this solution concept is what players should do [36].

In our view, this debate is a symptom of a deeper normative conflict, which can be fruitfully analyzed in deontic logic. The epistemic or decision-theoretic view on games emphasizes the normative appeal of maximization of expected utility. From an individual point of view, there are good reasons, namely avoiding incoherence, to maximize expected payoffs. Pareto optimality has a more collective character. It enjoins promoting *everyone*'s goodness when this can be done without requiring any one individual to sacrifice her own. Games like the one in Fig. 24.1 are ones where these two kinds of reason, individual and general, speak in favor of incompatible actions. This is a normative conflict.

Deontic logicians have proposed a great number of approaches to deal with such conflicts. To take a recent example, consider Horty's "conflict account" [25]. This view tolerates conflicting obligations. In certain cases, perhaps like Ann's in the game of Fig. 24.1, it states that the conflict is the correct description of the normative situation in which the agent is in. At the first node of the game in Fig. 24.1, Ann ought to skip (not pass) at n_0, *and* she ought to pass. It does not follow, still under Horty's account, that Ann has the contradictory obligation to pass and not to pass. Obligations do not "agglomerate" when the result would be inconsistent. Horty argues in fact that the account he proposes gives just the right balance between tolerance of conflicts and preservation of consistency, and that it furthermore meets many of the traditional objections that have been raised against the possibility of conflict between obligations.

We will not present the details of this view here, nor do we argue that it gives the correct answer to the question of how to balance individual and collective considerations in games. For now it is enough to observe that there are sophisticated accounts of how to deal with normative conflicts in the deontic logic literature. These accounts, given the bridge provided by the logic of best action viewed from a deontic perspective, can yield fresh insights into a classical debate at the foundation of the theory of games. Instead of viewing Ann's cases in Fig. 24.1 as epitomizing the debate between individualist and collectivist views on rational play in games, one should rather acknowledge that Ann is facing a normative conflict, and try to propose solutions that take that fact into account.

---

[4] Cf. [26] for a general presentation.

The deontic perspective on the logic of best action can also be used to study the *internal* consistency and plausibility of specific systems of recommendations that given solution concepts give rise to. Deontic conflicts arise when *different* normative sources generate opposing reasons for action. But even granting that a single solution concept carries genuine normative force, i.e. that the recommendations it makes are plausible, it can still happen that, as a whole, the normative *system* they give rise to is counter-intuitive. The classical deontic paradoxes [29] are a case in point. Are the recommendations stemming from, say, sub-game perfect equilibrium immune to such counter-intuitive consequences? We will argue below that there are very general reasons to think that this is not the case. If this argument is correct, deontic systems of rational recommendations in games meet an important standard on the deontic logic side.

Systems of norms can also give rise to inconsistent sets of recommendations in specific situations. Admissibility[5] and common knowledge thereof is a good illustration. There are games where it makes recommendations to the players that are individually perfectly intuitive, yet mutually inconsistent [37]. Such cases, we think, can cast doubts on the normative plausibility of this solution concept. These inconsistencies, clearly highlighted by a deontic analysis, are as important for the normative foundation of solution concepts as those that arise from alleged counter-intuitive examples like Fig. 24.1 for sub-game perfect equilibrium.

We are thus advocating here a *systemic* view of the normative interpretation of solution concepts, one in which the logic of best action, under a deontic interpretation, plays a key role. Of course to do this one needs to be more precise about how to extract deontic operators from best action formulas. Note that Best-DEON leaves us full flexibility in that respect. Best actions might be seen as obligatory, or just as well as permitted. We will argue now for a specific way to extract deontic operators from best actions, one that moves us away from the "normal" setting suggested by Best-DEON above.

24.3 The Specific Structure of Obligations and Permissions from Best Actions

Taking a deontic perspective on the logic of best action in games can thus yield important contributions in the debate on the normative status of solution concepts. In this section we turn to the deontic operators that would stem from this logic, and argue that they should be non-normal.

In many cases there will be more than one strategy profile that is deemed best according to a given solution concept. In pure coordination games (Fig. 24.2), for instance, neither strict nor weak dominance rule out any pure strategies. All actions

---

[5] That is not choosing weakly dominated strategies or, equivalently, maximizing expected utility under cautious beliefs. Cf. [33] for this equivalence.

Bob

		L	R
	T	1, 1	0, 0
Ann	B	0, 0	1, 1

Fig. 24.2 A coordination game

for either player are "best". In technical terms, the "best" sub-relation of the move relation is simply the whole move relation.

Two conclusions about the structure of obligations and permissions should be drawn from cases where there is not a unique best solution. First, the players *should*, or *ought to* pick a strategy *among* those that are deemed best according to the solution concept at hand. For one thing, they ought not to play strategies that fall outside the set of best ones. But strategies are mutually exclusive alternatives. Each player can only play one of them. So it makes no sense to think of the players as being required to choose all or even many "best" strategies. This would be a blatant violation of the ought-can principle. All they ought to choose is *a* best action.

Second, there is no further ground, as far as "best action" is concerned, to restrict further the set of actions among which the players are rationally required to choose. To be sure, other considerations, for instance morality or prudence, could be used as additional "filters" for the set of actions that are deemed best [15]. But purely in terms of the solution concept at hand, no best action can be excluded from the set of those among which the players should choose. In particular, *it is not the case that the players ought to play one specific action among the best ones.*

So as far as best actions are concerned, it is plausible to think that all the players ought to do is play one of them. They ought not to choose outside the set of what is best, but they are not required to choose any specific best action either.

What about permissions from best actions? The case of multiple best actions yields the dual picture, although not the dual operator, as we shall see presently. By the dual picture we mean the following. When there are multiple best actions, the players may play any of them, but no other. These and only these are individually permitted, although in the general case none is obligatory. Putting all this together we get:

(O/P-Best)	A player ought to play *a* best action. All specific best actions are permitted, although none is obligatory.

This understanding of obligation and permissions stemming from best actions in games gives rise to what we called elsewhere "obligations as weakest permissions" [35]. Under (O/P-Best) there is a unique set of obligated actions, namely the set of best actions itself. Viewing sets of actions as action types, "best" is the *unique* type of action an agent ought to perform. No logically weaker action type, i.e. one that is implied by being a "best" action, is obligatory, nor is any action type inconsistent with and/or independent of being a best action. Agents ought not, after all, to perform actions that are not best. But no logically stronger action type is obligatory either, at least as far as the criterion for best action alone is concerned. These are all permitted, and only those are permitted. And it seems plausible that every action type the extension of which is a combination/union of best actions should also be permitted, with the whole set of best actions as upper bound.

So we are thus advocating a very specific reading of obligations and permissions stemming from best action recommendations in games. Under this reading the unique action type/set of actions that is obligatory is the one singled out by the solution concept one is working with. To go back to the example of backward induction, what the agents ought to do, given that solution concept, is to play a strategy that is part of a sub-game perfect equilibrium. Any such strategy is permitted. But in cases where there are many of them, none is obligatory. The only thing that the agent ought to do is to pick one.

Obligations, in this reading, become the *weakest permitted action type* available to the agents. "Weakest" here means logically weakest, in the sense that any other permitted action type entails it. If the trivial or tautological action type is permitted, then this will always automatically be the only obligated one. So the logic of such obligations and permission, which we now turn to, cannot be normal, in the technical sense, on pain of trivialization.

24.4 The Logic of Obligation as Weakest Permission: An Overview

In this section we present the logical system that results from taking obligations as weakest permission. We argued in the previous section that this reading of deontic operators is particularly well-suited for rational recommendations in games, i.e. for recommendations stemming from solution concepts.

Before we look at the technical details it should be emphasized that the argument we gave in the last section applies to any system of obligations and permissions stemming from solution concepts. It is not tied to a specific one, for instance sub-game perfect equilibrium in the tree or Nash equilibrium in the matrix. It applies just as well to recommendations to each individual player—for instance, "Ann should maximize her expected utility"—as to recommendations to groups of players—for instance "Ann and Bob should achieve a Pareto-optimal outcome". In both cases, if the argument above is correct, obligations stemming from solution concepts will be weakest permissions.

For that reason we present the deontic logic of obligation as weakest permission in an abstract way, without mention of the agent(s) to whom these normative notions apply, or the specific source from which they stem. Our deontic operators are not "indexed" by specific agents and/or solution concepts. This is not to say that we think such specific cases are not interesting. Quite the contrary. In the next section and in the conclusion of this chapter we do give some more specific examples of how this abstract logic can be instantiated. But we think that already at this level some interesting logical principles arise.

The language $\mathcal{L}_D$ we work with is a usual propositional deontic logic extended with an alethic modality.

Definition 24.1 Let p be an element of a given, countable set of atomic propositions. Then $\mathcal{L}_D$ is defined as follows:

$$\varphi := p \mid \neg\varphi \mid \varphi \wedge \varphi \mid \Diamond\varphi \mid P\varphi \mid O\varphi$$

Formulas of the form $O\varphi$ and $P\varphi$ read, respectively, "φ is obligatory" and "φ is permitted". $\Diamond\varphi$ is the alethic modality, to be read "it is possible for the agent to choose a φ-action", with $\Box\varphi$ its usual dual operator.

In this language both O and P are primitive operators. The reason for this is that under the interpretation of obligation and permissions we want to give, the two notions turn out not to be duals.

When obligations are weakest permissions, on pain of trivialization the P operator cannot be a normal modality.[6] Recall that we argued that, from the perspective of best action, an action (type) is obligatory just in cases where it is permitted and no logically weaker proposition is permitted. But if P was normal then by necessitation $P\top$ would be a theorem, with $\top$ being any tautology.[7] Since there is no logically weaker proposition than $\top$, the tautology would be the only proposition that is ever obligatory. Hardly an interesting deontic system.

We are interpreting the deontic modalities using neighborhood semantics, by now the normal way to abnormality.

Definition 24.2 A *frame* $\mathcal{F}$ is a quadruple $\langle H, Alt, n_P, n_O \rangle$ where:

- H is a set of actions;
- Alt is an equivalence relation on H.
- n_P and n_O are neighborhood functions assigning a set of sets of actions to each element of H.

H is the set of all possible actions the agents could perform. We will see a concrete example of that in the next section. Alt partitions the set of all possible actions into different decision problems. The deontic modalities are interpreted using two neighbourhood functions n_P and n_O.

Definition 24.3 Let V be a valuation function for the atomic propositions in $\mathcal{L}_D$. Then for the propositions and boolean connectives the truth conditions are the usual. For modal formulas truth is defined as follows:

---

[6] A normal modal operator is one that satisfies the K axiom, i.e. that distributes over material implication, and the rule of necessitation. See [14] for details.

[7] We are here taking P as a "box", i.e. not as dual of O, which is in line with the interpretation we want to give to the two notions. But if one insists on them being dual, then $P\top$ comes out as a theorem whenever one assumes that obligations are always consistent.

- $\mathcal{M}, h \models \Diamond\varphi$ iff there is an h' such that $Alt(h, h')$ and $\mathcal{M}, h' \models \varphi$.
- $\mathcal{M}, h \models P\varphi$ iff $||\varphi|| \in n_P(h)$.
- $\mathcal{M}, h \models O\varphi$ iff $||\varphi|| \in n_O(h)$.

We thus evaluate formulas of $\mathcal{L}_D$ at specific actions (h). This might require the reader to suspend some well-engrained habits regarding truth as states/possible worlds in modal logic. To help fix intuitions one should keep in mind the standard translation of modal formulas into first-order ones [14]. Under this translation atomic propositions become (unary) predicates of objects. Here the objects are actions, and the predicates ($p, q,...$) are "true at an action" whenever that action has that property (according to the valuation). This idea extends easily to the conjunctions and negations. But this is not to say that the notion of a state or a situation the agent is in has completely disappeared from this interpretation. As mentioned earlier, one can view equivalence classes under Alt as a description of the decision problem the agent is facing, or the situation she is in.

The intended interpretation of obligations as weakest permissions translates into constraints on the neighborhood function. For P we remain very liberal. We only impose the following, which ensures that obligations are well-defined, at least in the finite case.

$$\text{If } X, Y \in n_P(h) \text{ then } X \cup Y \in n_P(h) \qquad \text{(Consistency)}$$

The real action happens at the level of obligations, which are required to be weakest permissions. To be precise, we require that obligations be the weakest *feasible* permissions. That is, obligations are the logically weakest types of action that can be performed in a given decision problem, which in our case corresponds to a cell of the partition induced by Alt. At the level of frames this translates into the following two conditions.

$$\text{If } X \in n_O(h) \text{ then } X \in n_P(h) \qquad \text{(Ought-Perm)}$$

$$\text{If } X \in n_O(h) \text{ then for all } Y \in n_P(h), Y \cap Alt[h] \subseteq X \cap Alt[h] \quad \text{(Weakest-Perm)}$$

The first condition simply states that obligated actions are permitted. The second is the key constraint. It makes an action type obligatory *only if* no weaker action type is both permitted and feasible. This is the reading of obligation we argued earlier makes most sense when looking at the recommendations stemming from best actions in games. We finally impose the classical "ought implies can" requirement,

$$\text{If } X \in n_O(h) \text{ then there is an } h' \in X \text{ such that } Alt(h, h') \qquad \text{(Ought-Can)}$$

In what follows we work with what we call "uniform" frames. These are frames where obligations and permissions remain constant within situations or decision problems, i.e. each equivalence class according to Alt. This is certainly the case from obligations stemming from best actions. In each decision problem there is a single set of permitted actions, and a single action type that is obligatory, namely the "best" ones.

$$\text{For all } h, h' \in H \text{ if } Alt(h, h') \text{ then } n_P(h) = n_P(h') \text{ and } n_O(h) = n_O(h')$$
$$\text{(Uniform)}$$

Table 24.1 The axiom system for deontic frames. Here D is either O or P

All propositional tautologies and $S5$ for $\Diamond$
For P:
(Union-closure) $(P\varphi \wedge P\psi) \to P(\psi \vee \varphi)$
Interaction axioms
(Ought-Perm) $O\varphi \to P\varphi$
(Ought-Can) $O\varphi \to \Diamond\varphi$
(Weakest-Perm) $O\varphi \to (P\psi \to \Box(\psi \to \varphi))$
(Uniformity$_D$) $D\varphi \to \Box(D\varphi)$
Modus Ponens, (Nec) for $\Diamond$ and the following: (RD) From $\varphi \leftrightarrow \psi$ infer $D\varphi \leftrightarrow D\psi$

Definition 24.4 A *deontic frame* is a frame that satisfies (Consistency), (Ought-Perm), (Weakest-Perm), (Ought-Can) and (Uniform).

From now on when we talk about frames we always mean deontic frames.

Validities in deontic frames already unveil interesting principles relating deontic notions to each other and with alethic modalities.

Theorem 24.1 *The axioms and rules in Table 24.1 are sound and complete with respect to the class of deontic frames.*

The proof of this theorem is a standard completeness proof for neighborhood system. All the axioms are canonical for their frame condition. The key one is of course Weekest-Perm, which precisely captures the homonymous frame condition. The reader can consult [35] for details. It should be emphasized again here that this is a general logic for obligation as weakest permission. We argued above that recommendations from solution concepts should take this form. But of course the specific logic of such solution concepts will be much richer, including additional principles capturing the behavior of the understanding of "best" at hand. The system just presented is simple because it is abstract.

The logical system shows its specificity in the formulas that it does not validate. For one thing, the usual duality between O and P, i.e. $O\varphi \leftrightarrow \neg P\neg\varphi$ is not valid on that class of frames. The right-to-left direction obviously fails. That $\neg\varphi$ is not permitted does not imply at all that φ is permitted, which is a pre-requisite for it to be obligatory. Failures of the converse direction are slightly more subtle, as they always involve permitted actions that are not available in the current decision problem. Indeed, one can show that the left-to-right direction of the dual is valid *if and only if* all permitted actions in a given decision problem are feasible alternatives. There is a similar interplay between alethic and deontic notions when one looks at the classical K axiom.

$$O(\varphi \to \psi) \to (O\varphi \to O\psi) \tag{K}$$

This formula is not valid in general, but it characterizes the class of deontic frames where *Alt* is the universal relation, i.e. where all possible actions are feasible alternatives, or the "root" of the tree in extensive games.

Finally, it should be noted that this logic allows for counter-examples to three of the most well-known deontic paradoxes: the contrary to duty paradox (first formula below), Ross's paradox (second formula below) and the good samaritan paradox (third formula below).

Fact 24.1 *None of the following formulae are valid:*

- $O \neg \varphi \rightarrow O(\varphi \rightarrow \psi)$
- $O\varphi \rightarrow O(\varphi \lor \psi)$
- $(O\varphi \land \Box(\varphi \rightarrow \psi)) \rightarrow P\psi$

We provide a counter-example to the first formula below. The other can be found in [35]. For now is it important to remark the following. Avoiding such paradoxes is usually the main motivation for moving to non-normal deontic logic. But this is not the case here. Our motivation is more basic: A normal deontic logic for obligations and permissions, under the present interpretation, would trivialize the logic, making only tautologies obligatory. The understanding of obligations that we propose here does avoid some of the well-known paradoxes, but this is a welcome consequence of the fact that we are forced into non-normality, rather than an explanation for it.

This avoidance of deontic paradoxes is also good news from the perspective of the normative interpretation of solution concepts given above. We argued above that all recommendations from best actions in games are best seen as giving rise to a system of norms where obligations are weakest permissions. Inasmuch as this argument is correct, none of these systems of recommendations are likely to fall within the classical deontic paradoxes.

24.4.1 Best Actions in Games as Weakest Permissions

We now apply the idea of reading obligations from best actions as weakest permission to the specific case of recommendations from sub-game perfect equilibrium in the game pictured in Fig. 24.1. The driving idea of this implementation is to take this solution concept to issue *negative* prescriptions. In this case a dominated action, and more generally a strategy that is not a sub-game perfect equilibrium, is one that one *ought not* to play.

Let H_{Ann} be the set of all actions for *Ann* in Fig. 24.1, i.e. $H_{Ann} = \{s_i, p_i\}$ for $i \in \{1, 3\}$. Suppose we are considering the obligations and permissions that Ann has at the root of the tree (n_0), i.e. where all of her actions are still available, so that *Alt* is the universal relation. At that point we say that only s_1 and s_3 are compatible with sub-game perfect equilibrium. So Ann ought not to play anything else. We thus set, for all $a_i \in H_{Ann}$, $n_O^{Ann}(a_i) = \{\{s_1, s_3\}\}$. The set of permitted action types $n_P^{Ann}(a_i)$ will then be defined as the downward closure of $n_O^{Ann}(a_i)$, excluding the impossible (empty) action: $n_P^{Ann}(a_i) = \{\{s_1\}, \{s_3\}, \{s_1, s_3\}\}$. This makes, in particular, every individual "best" action permitted, but none of them obligatory. This is a deontic frame. It is uniform and it satisfies the ought-can principle as well consistency and Weakest-Perm.

General properties of the logic of obligations as weakest permission can be seen to hold in this example too. The right-to-left direction of the dual, for instance, is not valid in this deontic frame. Suppose our language contains atomic propositions for the specific actions of Ann (s_i, p_i for $i = 1, 3$), and take a model based on the frame above, with the natural valuation for these atomic propositions. In this model $\neg p_3$, the proposition expressing the fact that Ann is not passing at the third node, is true at p_1, s_1 and s_3. The set of these actions is not in $n_p^{Ann}(a_i)$ for any a_i. So $\neg P_{Ann} \neg p_3$ is valid in that model. But since $\{p_3\} \notin n_O^{Ann}(a_i) = \{\{s_1, s_3\}\}$, we don't get $O_{Ann} p_3$ either.

The contrary to duty paradox, $O\neg\varphi \rightarrow O(\varphi \rightarrow \psi)$, also fails here. Take the same model as in the previous paragraph. Ann ought not to pass: $O_{Ann}\neg(p_1 \vee p_3)$ is valid in that model. But it is not the case that Ann ought to make following implication true: $(p_1 \vee p_3) \rightarrow \top$. This implication is true everywhere in our model, but H_{Ann} itself is not in $n_O^{Ann}(a_i)$.

One can abstract from that particular example. For any finite game G one can define the deontic frame $\mathcal{F}_G^i$ based on a solution concept S as follows. $H = \Pi_i S_i$, the set of all strategy profiles, $Alt[h] = H \times H$. $n_O(h) = \{\{h' \in S\}\}$ and $n_P(h) = \{X \subseteq \{h' \in S\} : X \neq \emptyset\}$. In this definition permitted action types are all those that cannot be executed outside the given solution concept. The only thing an agent *ought* to do in this construction is to play a best strategy.

24.5 Conclusions

In this chapter we have made three conceptual arguments regarding the logic of best action in games. We have argued first that it should be seen as a deontic logic. We have then argued that viewed this way, the logic of best action in games promises an interesting interplay between deontic logic and the foundation of the normative interpretations of solution concepts. Finally, we have argued that the deontic logic of best action in games should be developed in what we called "obligations as weakest permissions", which is, we have shown, an interesting, non-normal logical system.

It is worth emphasizing again that the logic of obligation as weakest permission provides, in our view, the *generic* structure of recommendations from solution concepts. The validities just mentioned reflect the way we constructed deontic frames from games and solution concepts, rather than intrinsic properties of, say, sub-game perfect equilibrium. Using another choice rule, for instance iterated weak dominance, would have resulted in the same deontic principles. The core of the proposal here is conceptual. Once one views rational recommendations in games in terms of weakest permissions, a view which we argued is natural, we are in the realm of the logic sketched above.

To study the deontic logic of best actions for concrete solution concepts one would, we think, have to generalize the language to *binary* deontic operators. These are operators of the form $O(\varphi, \psi)$ interpreted as "given φ, she ought to play a ψ-action". Such operators are well-known from the deontic logic literature. They seem particularly natural for game solutions, which are very often dependent on the *context* where the game is played. This context, in turn, is usually described in terms of what the players believe about and expect of each other. So $O(\varphi, \psi)$ would be read as "given that i believes that φ, she ought to play a ψ-strategy".

Once this epistemic parameter is made explicit, divergences in the recommendations from concrete solution concepts start to appear [36]. Let us mention one example. Both iterated weak and strict dominance are solution concepts that are not "upward monotonic". A strategy that is strictly dominated in small games might no longer be strictly dominated when one moves to a larger game. The same story can be told epistemically. A strategy that is strictly dominated given one's belief that ones opponent will play, say, A, might no longer be dominated upon suspending that belief. Formally, this means that the following would not be valid in the specific deontic logic of strict or weak dominance.

$$\not\models O(\varphi, \psi) \rightarrow O(\chi, \psi) \quad \text{when } \varphi \rightarrow \chi$$

Strict dominance, on the other hand, is downward monotonic. It is preserved under taking sub-games or refinement of the agents' information. So we would have the following for O_{SD}, the deontic logic of conditional best actions, given strict dominance.

$$\models O_{SD}(\chi, \psi) \rightarrow O_{SD}(\varphi, \psi) \quad \text{when } \varphi \rightarrow \chi$$

This principle would not be valid for weak dominance, which is notorious for also failing downward monotonicity.

$$\not\models O_{WD}(\chi, \psi) \rightarrow O_{WD}(\varphi, \psi) \quad \text{when } \varphi \rightarrow \chi$$

We leave the development of these specific systems for future work. For now we hope to have convinced the reader that the logic of best action, in itself and viewed from a deontic perspective, not only raises interesting logical questions, but can also make fundamental contributions to our the understanding of the normative character of solution concepts in games, and more generally to the norms and obligations that are inherent in any intelligent interaction.

Acknowledgments This research has been supported by the Alexander von Humboldt Foundation and by a LMU Research Fellowship as part of the LMU Academic Career Program and the Excellence Initiative.

References

1. Apt KR, Zvesper JA (2010) The role of monotonicity in the epistemic analysis of strategic games. Games 1(4):381–394
2. Baltag A, Smets S, Zvesper JA (2009) Keep hoping for rationality: a solution to the backward induction paradox. Synthese 169(2):301–333
3. Battigalli P, Siniscalchi M (2002) Strong belief and forward induction reasoning. J Econ Theory 106(2):356–391
4. Belnap ND, Perloff M, Xu M (2001) Facing the future: agents and choices in our indeterminist world. Oxford University Press, Oxford
5. van Benthem J (2002) Extensive games as process models. J Log Lang Inf 11(3):289–313
6. van Benthem J (2007) Rational dynamics and epistemic logic in games. Int Game Theory Rev, vol. 9
7. van Benthem J (2011) Exploring a theory of play. In: Proceedings of the 13th conference on theoretical aspects of rationality and knowledge. ACM, pp 12–16.
8. van Benthem J (2011) Logic in a social setting. Episteme 8(03):227–247
9. van Benthem J (2011) Logical dynamics of information and interaction. Cambridge University Press, Cambridge
10. van Benthem J, Gheerbrant A (2010) Game solution, epistemic dynamics and fixed-point logics. Fundamenta Informaticae 100(1):19–41
11. van Benthem J, Girard P, Roy O (2009) Everything else being equal: a modal logic for ceteris paribus preferences. J Philos Log 38(1):83–125
12. van Benthem J, Pacuit E, Roy O (2011) Toward a theory of play: a logical perspective on games and interaction. Games 2(1):52–86
13. van Benthem J, Roy O, van Otterloo S (2006) Preference logic, conditionals, and solution concepts in games. In Lagerlund H, Lindstrom S, Sliwinski R (eds) Modality matters: twenty-five essays in honour of Krister Segerberg. Uppsala Philosophical Studies, vol 53. Uppsala University Press, Uppsala.
14. Blackburn P, De Rijke M, Venema Y (2002) Modal logic, vol 53. Cambridge University Press, Cambridge
15. Braham M, van Hees M (2012) From wall street to guantánamo bay: the right and the good, practically speaking. http://www.rug.nl/staff/martin.van.hees/Ordering.pdf
16. de Bruin B (2010) Explaining games: the epistemic programme in game theory. Synthese library, Springer, Dordrecht
17. Camerer CF (2003) Behavioral game theory: experiments in strategic interaction. Princeton University Press, Princeton
18. Fagin R, Halpern JY, Moses Y, Vardi MY (1995) Reasoning about knowledge, vol 4. MIT Press Cambridge, Massachusetts
19. Gheerbrant AP (2010) Fixed-point logics on trees. Institute for Logic, Language and Computation, Amsterdam
20. Ghosh S (2008) Strategies made explicit in dynamic game logic. Logic and the foundations of game and decision theory.
21. Grossi D, Lorini E, Schwarzentruber F (2013) Ceteris paribus structure in logics of game forms. In: Proceedings of the 14th conference on theoretical aspects of rationality and knowledge. ACM.
22. van der Hoek W, Pauly M (2006) Modal logic for games and information. Handbook of modal logic, vol 3. Elsevier, Amsterdam, pp 1077–1148.
23. Horty JF (2001) Agency and deontic logic. Oxford University Press, Oxford
24. Horty JF (2011) Perspectival act utilitarianism. In: Girard P, Roy O, Marion M (eds) Dynamic formal epistemology. Springer, Berlin, pp 197–221
25. Horty JF (2012) Reasons as defaults. OUP, Oxford
26. Joyce JM (2004) Bayesianism. In: Mele AR, Rawling P (eds) The Oxford handbook of rationality. Oxford University Press, Oxford

27. Kooi B, Tamminga A (2008) Moral conflicts between groups of agents. J Philos Log 37(1):1–21
28. Lorini E, Schwarzentruber F (2010) A modal logic of epistemic games. Games 1(4):478–526
29. McNamara P (2010) Deontic logic. In: Zalta EN (ed) The Stanford encyclopedia of philosophy. Fall 2010 edition.
30. Olde Loohuis L (2009) Obligations in a responsible world. Logic, rationality, and interaction. Springer, Heidelberg, pp 251–262
31. Pacuit E, Roy O (2011) A dynamic analysis of interactive rationality. Logic, rationality, and interaction, pp 244–257.
32. Pacuit E, Roy O (forthcoming) Epistemic foundations of game theory. In: Zalta E (ed) Stanford encyclopedia of philosophy
33. Pearce DG (1984) Rationalizable strategic behavior and the problem of perfection. Econometrica: J Econometric Soc 52:1029–1050.
34. Ramanujam R, Simon S (2008) Dynamic logic on games with structured strategies. In: Proceedings of the conference on principles of knowledge representation and reasoning, pp 49–58.
35. Roy O, Anglberger A, Gratzl N (2012) The logic of obligation as weakest permission. In: Broersen J, Agotnes T, Elgensem D (eds) Deontic logic in computer science. Springer, Heidelberg, pp 139–150
36. Roy O, Pacuit E (2011) Interactive rationality and the dynamics of reasons. Manuscript.
37. Samuelson L (1992) Dominated strategies and common knowledge. Games Econ Behav 4(2):284–313
38. Tamminga A (2013) Deontic logic for strategic games. Erkenntnis 78:183–200

Chapter 25
When are Two Arguments the Same?
Equivalence in Abstract Argumentation

Dov Gabbay and Davide Grossi

> *[...] actual reasoning may be more like weaving a piece of cloth from many threads than forging a chain with links in linear mathematical proof style [...]*
>
> van Benthem [7, p. 83]

Abstract In abstract argumentation arguments are just points in a network of attacks: they do not hold premises, conclusions or internal structure. So is there a meaningful way in which two arguments, belonging possibly to different attack graphs, can be said to be equivalent? The paper argues for a positive answer and, interfacing methods from modal logic, the theory of argument games and the equational approach to argumentation, puts forth and explores a formal theory of equivalence for abstract argumentation.

25.1 Introduction

Abstract argumentation [14] is the theory of structures $\langle A, \rightarrow \rangle$—called *attack graphs* —as models of the sort of conflict that occurs in argumentation, where arguments (set A) interact by attacking one another (through the binary 'attack' relation $\rightarrow$).

D. Gabbay
King's College London, London, UK
e-mail: dov.gabbay@kcl.ac.uk

D. Gabbay
Bar Ilan University, Ramat Gan, Israel

D. Gabbay
University of Luxembourg, Luxembourg, Luxembourg

D. Grossi (✉)
University of Liverpool, Liverpool, UK
e-mail: d.grossi@liverpool.ac.uk

A. Baltag and S. Smets (eds.), *Johan van Benthem on Logic and Information Dynamics*, Outstanding Contributions to Logic 5, DOI: 10.1007/978-3-319-06025-5_25, © Springer International Publishing Switzerland 2014

On the one hand, this has proven to be a prolific abstraction from which to study structural properties of sets of arguments that form 'justified' or 'rational' positions in an argumentation (cf. [2, 3] for recent overviews). On the other hand, this perspective leaves the internal structure of arguments unspecified and arguments are nothing but points in a network of attacks. When looking at similarities between arguments from this point of view, issues such as having the same premises and conclusions, or exhibiting the same logical structure, become immaterial.

However even at this level of abstraction there is a telling sense in which two arguments a and a' belonging to two (possibly different) graphs $\langle A, \rightarrow \rangle$ and $\langle A', \rightarrow' \rangle$ can be said to be 'the same', or to be equivalent, namely if they 'behave' in the same way in the two graphs. Put otherwise, a and a' can be said to be equivalent if they interact in similar ways with the other arguments in their respective graphs. This point of view suggests a way of comparing arguments which is independent of their content, and which instead stresses the role they play in an argumentation through their interaction with other arguments.

Suggestively, this 'behavioral' view of the notion of equivalence of arguments ties in well with Toulmin's view of a theory of argumentation as something that is "field-invariant":

> What features of our arguments should we expect to be field-invariant: which features will be field-dependent? We can get some hints, if we consider the parallel between the judicial process, by which the questions raised in a law court are settled, and the rational process, by which arguments are set out and produced in support of an initial assertion. [...] One broad distinction is fairly clear. The sorts of evidence relevant in cases of different kinds will naturally be very variable. [...] On the other hand there will be, within limits, certain broad similarities between the orders of proceedings adopted in the actual trial of different cases, even when these are concerned with issues of very different kinds. [...] When we turn from the judicial to the rational process, the same broad distinction can be drawn. Certain basic *similarities of pattern and procedure* [our emphasis] can be recognized, not only among legal arguments but among justificatory arguments in general, however widely different the fields of the arguments, the sort of evidence relevant, and the weight of the evidence may be. [30, pp.15–17]

The paper aims at developing a theory of equivalence of arguments based on structural *similarities of pattern and procedure*. To this aim, the paper pushes further the application of modal logic techniques to abstract argumentation already argued for in a number of recent works (cf. [11, 16, 20–22]). It builds on the view of attack graphs $\langle A, \rightarrow \rangle$ as Kripke frames and presents a systematic exploration of the idea that argument equivalence can be expressed as equality of (fragments) of the modal theory of each argument. This idea naturally relates to the modal invariance notion of bisimulation [4][1] and with the theory of argument games, that is, 'argumentation procedures' modeled as two-player zero-sum games played on attack graphs.[2] Inspired by insights from [5, 8], we will look at a power-based notion of argument equivalence: two arguments can be said to be equivalent when the powers of the proponent and opponent in the argument games for the two arguments are, in some precise sense,

---

[1] The relevance of bisimulation in abstract argumentation was first emphasized in [20, 21].

[2] Cf. [24] for a recent overview of argument games.

the 'same'. Finally, we will see how this game-theoretic view of argumentation and argument equivalence ties in with the equational view of argumentation put forth in [15, 17, 18].

Structure of the paper. In Sect. 25.2 we concisely introduce the key concepts of abstract argumentation which will be used in the paper. Section 25.3 provides some modal logic preliminaries and Sect. 25.4 applies modal equivalence to define a notion of equivalence for arguments, with respect to Dung's grounded extension. Section 25.5 elaborates on that definition proposing a strategic variant of it based on the powers that a proponent and an opponent have in an argument game for the grounded extension. Section 25.6 relates the construction of winning strategies in such argument games to the equational approach to argumentation, and brings the three strands of the paper—the modal, the game-theoretic and the equational— together. Finally, conclusions follow in Sect. 25.7.

25.2 Preliminaries on Abstract Argumentation

The present section introduces the necessary preliminaries on abstract argumentation which set the stage of our investigations.

25.2.1 Attack Graphs

We start by the key notion of [14]:

Definition 25.1 (*Attack graph*) An attack graph—or Dung framework—is a tuple $\mathscr{A} = \langle A, \rightarrow \rangle$ where:

- A is a non-empty set—the set of arguments;
- $\rightarrow \subseteq A^2$ is a binary relation—the attack relation.

The set of all attack graphs on a given set A is denoted $\mathfrak{A}(A)$. The set of all attack graphs is denoted $\mathfrak{A}$. With $a \rightarrow b$ we indicate that a attacks b, and with $X \rightarrow a$ we indicate that $\exists b \in X$ s.t. $b \rightarrow a$. Similarly, $a \rightarrow X$ indicates that $\exists b \in X$ s.t. $b \leftarrow a$. An attack graph such that, for each $a \in A$ the cardinality $|\{b \mid a \leftarrow b\}|$ of the set of the attackers of a is finite, is called *finitary*.[3] Given an argument a, we denote by $R_{\mathscr{A}}(a)$ the set of arguments attacking a: $\{b \in A \mid b \rightarrow a\}$.

These relational structures (see Fig. 25.1 for an example) are the building blocks of abstract argumentation theory. Once A is taken to represent a set of arguments (or 'pieces of evidence' or 'information sources'), and $\rightarrow$ an 'attack' relation between arguments (so that $a \rightarrow b$ means "a attacks b"), the study of these structures provides

---

[3] This property is known in modal logic as *image-finiteness* of the accessibility relation of a Kripke frame [9, Chap. 2].

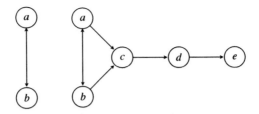

Fig. 25.1 Two attack graphs. The one on the *left* represents a full opposition between, for instance, two contradictory arguments. The one on the *right* represents an argumentation where two opposite arguments (*a* and *b*) both attack a same argument (*c*) which in turn defends a final argument (*e*) by attacking its attacker (*d*)

very general insights on how competing arguments interact and structural properties of subsets of A can be taken to formalize how collections of arguments form 'justifiable' positions in an argumentation.

25.2.2 Characteristic Functions of Attack Graphs

The formulation of all main argumentation theoretic properties makes use of two functions that can be naturally associated to each attack graph.

25.2.2.1 Characteristic Functions

The first one is a function called in [14] characteristic function, which we will call here the defense function.

Definition 25.2 (*Defense function*) Let $\mathscr{A} = \langle A, \rightarrow \rangle$ be an attack graph. The defense function $\mathsf{d}_{\mathscr{A}} : \wp(A) \longrightarrow \wp(A)$ for $\mathscr{A}$ is so defined:

$$\mathsf{d}_{\mathscr{A}}(X) = \left\{ x \in A \mid \forall y \in A : \text{IF } y \rightarrow x \text{ THEN } X \rightarrow y \right\}.$$

Given a set of arguments X, the n-fold iteration of $\mathsf{d}_{\mathscr{A}}$ is denoted $\mathsf{d}_{\mathscr{A}}^n$ for $0 \leq n < \omega$ and its infinite iteration is denoted $\mathsf{d}_{\mathscr{A}}^\omega$. For a given X, an infinite iteration generates an infinite sequence, or stream, $\mathsf{d}_{\mathscr{A}}^0(X), \mathsf{d}_{\mathscr{A}}^1(X), \mathsf{d}_{\mathscr{A}}^2(X), \dots$. A stream is said to stabilize if and only if there exists $0 \leq n < \omega$ such that $\mathsf{d}_{\mathscr{A}}^n(X) = \mathsf{d}_{\mathscr{A}}^{n+1}(X)$. Such set $\mathsf{d}_{\mathscr{A}}^n(X)$ is then called the limit of the stream. When clear from the context we will drop the reference to $\mathscr{A}$ in $\mathsf{d}_{\mathscr{A}}$.

Intuitively, for a given $\mathscr{A}$, function $\mathsf{d}_{\mathscr{A}}$ encodes for each set of arguments X, which other arguments the set X is able to defend within $\mathscr{A}$.

The second function was first introduced in [27, 28] and further studied in [14]. It is not known with a specific name in the literature. We call it here the neutrality function.

Definition 25.3 (*Neutrality function*) Let $\mathscr{A} = \langle A, \rightarrow \rangle$ be an attack graph. The neutrality function $n_{\mathscr{A}} : \wp(A) \longrightarrow \wp(A)$ for $\mathscr{A}$ is so defined:

$$n_{\mathscr{A}}(X) = \{x \in A \mid \text{NOT } X \rightarrow x\}$$

The definitions of n-fold iteration, stream, and stabilization are like in Definition 25.2.

Intuitively, given $\mathscr{A}$, function $n_{\mathscr{A}}$ encodes for each set X of arguments in $\mathscr{A}$, the arguments about which X is neutral in the sense of not attacking any of those arguments.

Example 25.1 (Defense and neutrality in Fig. 25.1) The functions applied to the symmetric graph on the left of Fig. 25.1 yield the following equations:

$$
\begin{array}{llll}
d(\emptyset) & = & \emptyset & n(\emptyset) & = \{a, b\} \\
d(\{a\}) & = & \{a\} & n(\{a\}) & = \{a\} \\
d(\{b\}) & = & \{b\} & n(\{b\}) & = \{b\} \\
d(\{a, b\}) & = & \{a, b\} & n(\{a, b\}) & = \emptyset.
\end{array}
$$

25.2.2.2 Properties of the Defense Function

We list here two properties of the defense function which will be used in the development of the paper.

The first one, monotonicity, expresses the property that larger sets of arguments are able to defend larger sets of arguments. This is enough to guarantee the existence of least and greatest fixpoints of the defense function, by the Knaster-Tarski theorem.[4]

The second one, continuity, expresses the property that in finitary graphs (i.e., graphs where arguments have at most a finite number of attackers, recall Definition 25.1), what is defended by a series of larger and larger sets of arguments is equivalent to the union of what each of those sets defends. As we will see later, continuity enables the possibility of studying processes of computation of argumentation-theoretic notions as iterated applications of the defense function.

Fact 25.1 (Monotonicity) *Let $\mathscr{A} = \langle A, \rightarrow \rangle$ be an attack graph. Function $n_{\mathscr{A}}$ is monotone, i.e., for any $X, Y \subseteq A$:*

$$X \subseteq Y \Longrightarrow d_{\mathscr{A}}(X) \subseteq d_{\mathscr{A}}(Y).$$

---

[4] The reader is referred to [12] for a detailed presentation of this result.

Table 25.1 Some of the key notions of abstract argumentation theory from Dung [14]

X is conflict-free in $\mathscr{A}$	iff	$X \subseteq n_{\mathscr{A}}(X)$
X is self-defended in $\mathscr{A}$	iff	$X \subseteq d_{\mathscr{A}}(X)$
X is admissible in $\mathscr{A}$	iff	$X \subseteq n_{\mathscr{A}}(X)$ and $X \subseteq d_{\mathscr{A}}(X)$
X is a complete set in $\mathscr{A}$	iff	$X \subseteq n_{\mathscr{A}}(X)$ and $X = d_{\mathscr{A}}(X)$
X is the grounded set in $\mathscr{A}$	iff	$X = \mathsf{lfp}.d_{\mathscr{A}}$

Fact 25.2 (Continuity [14]) *Let $\mathscr{A}$ be a finitary attack graph. If $\mathscr{A}$ is finitary, then $d_{\mathscr{A}}$ is continuous for any $X \subseteq A$, i.e., for any directed set $D \in \wp(\wp(A))$:*
$$d_{\mathscr{A}}\left(\bigcup_{X \in D} X\right) = \bigcup_{X \in D} d_{\mathscr{A}}(X).$$

Proof [RIGHT TO LEFT] Trivial. [LEFT TO RIGHT] Assume $a \in d_{\mathscr{A}}(\bigcup_{X \in D} X)$. By image-finiteness there exists $X \in D$ s.t. it contains all arguments that attack some of a's attackers. Hence $a \in \bigcup_{X \in D} d_{\mathscr{A}}(X)$. $\square$

25.2.3 Solving Attack Graphs

By 'solving' an attack graph we mean selecting a subset of arguments that enjoy some characteristic structural property. The idea behind Dung's semantics for argumentation is precisely that some structural properties of attack graphs can capture intuitive notions of justifiability of arguments or, if you wish, of standard of proof—what in argumentation are usually called *extensions*. Therefore, the study of structural properties of attack graphs delivers very general insights on how competing arguments interact and how collections of them form 'tenable' or 'justifiable' argumentative positions.

Table 25.1 recapitulates the basic notions of abstract argumentation which we will be touching upon in the paper. They are all formulated either as fixpoints ($X = f(X)$) or post-fixpoints ($X \subseteq f(X)$) of the defense and neutrality functions, or as combinations of the two.

Intuitively, conflict-freeness demands that the set of arguments at issue is not able to attack itself—it is neutral with respect to itself. Self-defense requires that the set of arguments is able to defend itself. An admissible set is then a set of arguments which is conflict-free and is able to defend all its attackers. So, as the name suggests, admissible sets can be thought of as 'admissible' positions within an attack graph. By considering those admissible sets which also contain all the arguments they are able to defend—viz., the admissible sets that are fixpoints of the defense function—we obtain the notion of complete set. It formalizes the idea of a fully exploited admissible position, that is, a position which has no conflicts, and which consists exactly of all the arguments that can be successfully defended. The grounded set represents what all complete extensions have in common. In a way, it formalizes what at least must be accepted as 'reasonable' within the graph.

Example 25.2 (Extensions in Fig. 25.1) Consider the graph on the right of Fig. 25.1. The grounded extension is $\emptyset$. There are two complete extensions: $\{a, d\}$ and $\{b, d\}$. An example of a conflict-free set which is not admissible is $\{c, e\}$.

25.2.4 Computing the Grounded Set

We now look at a process of computation of the grounded set. This will be related later to the notion of argument equivalence to be developed, and the availability of winning strategies for the proponent in argument games.

We will focus on finitary graphs (recall Definition 25.1). The case of non-finitary graphs is briefly discussed in Remark 25.1.

Theorem 25.1 (Computation of grounded extensions [14]) *Let $\mathscr{A}$ be a finitary attack graph:*

$$\mathsf{lfp}.\mathsf{d}_{\mathscr{A}} = \bigcup_{0 \leq n < \omega} \mathsf{d}^n_{\mathscr{A}}(\emptyset) \tag{25.1}$$

Proof First, we prove that $\bigcup_{0 \leq n < \omega} \mathsf{d}^n_{\mathscr{A}}(\emptyset)$ is a fixpoint by the following equations:

$$\mathsf{d}_{\mathscr{A}}\left(\bigcup_{0 \leq n < \omega} \mathsf{d}^n_{\mathscr{A}}(\emptyset)\right) = \bigcup_{0 \leq n < \omega} \mathsf{d}_{\mathscr{A}}(\mathsf{d}^n_{\mathscr{A}}(\emptyset))$$
$$= \bigcup_{0 \leq n < \omega} \mathsf{d}^n_{\mathscr{A}}(\emptyset)$$

where the first equation holds by the continuity of $\mathsf{d}_{\mathscr{A}}$, and the second since, by monotonicity, $\mathsf{d}^0_{\mathscr{A}}(\emptyset), \mathsf{d}^1_{\mathscr{A}}(\emptyset), \ldots$ is non-descending. Second, we proceed to prove that $\bigcup_{0 \leq n < \omega} \mathsf{d}^n_{\mathscr{A}}(\emptyset)$ is indeed the least fixpoint. Suppose, towards a contradiction that there exists Y s.t.: $\emptyset \subset Y = \mathsf{d}_{\mathscr{A}}(Y) \subset \bigcup_{0 \leq n < \omega} \mathsf{d}^n_{\mathscr{A}}(\emptyset)$. It follows that $\emptyset \subset Y = \mathsf{d}_{\mathscr{A}}(Y) \subset \mathsf{d}^n_{\mathscr{A}}(\emptyset)$ for some $0 \leq n < \omega$. But, by Fact 25.1, we have that $\mathsf{d}^n_{\mathscr{A}}(\emptyset) \subseteq \mathsf{d}^n_{\mathscr{A}}(Y)$. Contradiction. $\qquad \square$

Remark 25.1 (Non-finitary graphs) For infinite graphs which are not finitary, Theorem 25.1 could be generalized by ordinal induction:

$$\mathsf{d}^0_{\mathscr{A}}(\emptyset) = \emptyset$$
$$\mathsf{d}^{\alpha+1}_{\mathscr{A}}(\emptyset) = \mathsf{d}_{\mathscr{A}}(\mathsf{d}^\alpha_{\mathscr{A}}(\emptyset))$$
$$\mathsf{d}^\lambda_{\mathscr{A}} = \bigcup_{\alpha < \lambda} \mathsf{d}^\alpha_{\mathscr{A}}(\emptyset) \text{ (for } \lambda \text{ arbitrary limit ordinal).}$$

Fig. 25.2 A linear well-founded attack graph. The greatest and smallest fixpoint of the defense function coincide here: $\{a, c, e\}$. The set of arguments not belonging to the greatest fixpoint is $\{d, b\}$. Note, in particular, that while b is defended by set $\{a, b, c, d, e\}$ (namely by d), it is not defended by the set of arguments that is defended by $\{a, b, c, d, e\}$. So it does not belong to the greatest fixpoint of the defense function

By the monotonicity of $d_{\mathscr{A}}$ it can then be shown that there exists an ordinal α of cardinality at most $|A|$ such that: $\mathsf{lfp}.d_{\mathscr{A}} = d_{\mathscr{A}}^{\alpha}(\emptyset)$.[5]

25.2.4.1 Smallest and Greatest Fixpoints of the Defense Function

We have seen that the smallest fixpoint of the defense function $d_{\mathscr{A}}$ defines the so-called grounded extension of an attack graph. What about the largest: $\mathsf{gfp}.d_{\mathscr{A}}$? We will confine our discussion to finitary graphs.

The arguments that belong to $\mathsf{gfp}.d_{\mathscr{A}}$ are those which can always be defended by some other argument that can also in turn be defended. The dual of Theorem 25.1 offers a good perspective from which to appreciate the notion:

$$\mathsf{gfp}.d_{\mathscr{A}} = \bigcap_{0 \le n < \omega} d_{\mathscr{A}}^{n}(A)$$

i.e., the set consisting of arguments that are defended by the set of all arguments, and by the set that is defended by the set of all arguments and so on: $d_{\mathscr{A}}(A) \cap d_{\mathscr{A}}(d_{\mathscr{A}}(A)) \cap \ldots$ (see Figs. 25.2 and 25.3 for examples).

25.3 Attack Graphs and Modal Logic

The section recapitulates and slightly extends (in particular w.r.t. frame languages) the modal logic approach to abstract argumentation put forth in [20, 21].

25.3.1 Attack Graphs and Kripke Models

Once an attack graph is viewed as a Kripke frame, the addition of a function assigning names to sets of arguments—a labeling or valuation function—yields a Kripke model (or a state transition system).

---

[5] A proof of this statement in the general setting of complete partial orders can be found in [31, Corollary 3.7].

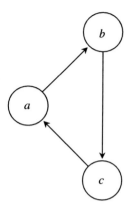

Fig. 25.3 A 3-cycle attack graph. Here the smallest fixpoint of the defense function is $\emptyset$ and the greatest fixpoint is $\{a, b, c\}$

Definition 25.4 (*Attack models*) Let **P** be a set of atoms. An attack model is a tuple $\mathcal{M} = \langle \mathcal{A}, \mathcal{V} \rangle$ where $\mathcal{A} = \langle A, \rightarrow \rangle$ is an attack graph and $\mathcal{V} : \mathbf{P} \longrightarrow \wp(A)$ is a valuation function. A pointed attack model is a pair $\langle \mathcal{M}, a \rangle$ with $a \in A$. The set of attack models is $\mathfrak{M}$.

Attack models are nothing but attack graphs together with a way of 'naming' sets of arguments or, to put it otherwise, of 'labeling' arguments.[6] So, the fact that an argument a belongs to the set $\mathcal{V}(p)$ in a given model $\mathcal{M}$ reads in logical notation as $(\mathcal{A}, \mathcal{V}), a \models p$. By using the language of propositional logic we can then form 'complex' labels φ for sets of arguments stating, for instance, that "a belongs to both the sets called p and q": $(\mathcal{A}, \mathcal{V}), a \models p \wedge q$.

In order to formalize argumentation-theoretic statements more than just propositional expressivity is needed. Let us mention a couple of examples: "*there exists* an argument in a set named φ attacking argument a" or "*for all* attackers of argument a there exist some attackers in a set named φ". These are statements involving a bounded quantification and they can be naturally formalized by a modal operator $\Diamond$ whose reading is: "there exists an attacking argument such that ...". This takes us to modal languages.

25.3.2 The 'Being Attacked' Modality

Interpret now the basic modal language on argumentation models as follows:

$$\mathcal{M}, a \models \Diamond\varphi \iff \exists b \in A : a \leftarrow b \text{ AND } \mathcal{M}, b \models \varphi$$

---

[6] It might be worth noticing that this is a generalization of the sort of labeling functions studied in argumentation theory (cf. [2, 10]).

An argument a belongs to the set called $\Diamond\varphi$ iff some argument b is accessible via the inverse of the attack relation and b belongs to φ or, more simply, iff a is attacked by some argument in φ.

This is enough expressivity to express the defense and neutrality functions in modal logic K. The two functions $\mathsf{d}_{\mathscr{A}}$ and $\mathsf{n}_{\mathscr{A}}$ correspond to the functions denoted in $\mathscr{L}$ by the modal expressions $\Box\Diamond$ and, respectively, $\neg\Diamond$ on a given graph $\mathscr{A}$.

Lemma 25.1 (*Defense and neutrality in modal logic*) *Let $\mathscr{A}$ be an attack graph and $\mathscr{V}$ a valuation function.*

$$\langle\mathscr{A},\mathscr{V}\rangle, a \models \Box\Diamond\varphi \iff a \in \mathsf{d}_{\mathscr{A}}([\varphi]_{\langle\mathscr{A},\mathscr{V}\rangle})$$
$$\langle\mathscr{A},\mathscr{V}\rangle, a \models \neg\Diamond\varphi \iff a \in \mathsf{n}_{\mathscr{A}}([\varphi]_{\langle\mathscr{A},\mathscr{V}\rangle}).$$

Proof (Sketch of proof) For $\Box\Diamond$ we have these equivalences:

$$[\Box\Diamond\varphi]_{\langle\mathscr{A},\mathscr{V}\rangle} = \{a \mid \forall b : \text{IF } a \leftarrow b \text{ THEN } b \leftarrow [\varphi]_{\langle\mathscr{A},\mathscr{V}\rangle}\}$$
$$= \mathsf{d}_{\mathscr{A}}([\varphi]_{\langle\mathscr{A},\mathscr{V}\rangle}).$$

The first equation holds by construction, the second and third are application of the the semantics of $\Box\Diamond$ and Definition 25.2. The reasoning for $\neg\Diamond\varphi$ is analogous.[7] $\quad\square$

In general, emphasizing the modal nature of $\mathsf{d}_{\mathscr{A}}$ and $\mathsf{n}_{\mathscr{A}}$ has the advantage of allowing us to use available modal principles in reasoning about argumentation-theoretic notions. All the theorems of logic K concerning $\Box\Diamond$- and $\neg\Diamond$-formulae can legitimately be seen as theorems of abstract argumentation. Here we list a few very simple theorems of K which carry interesting readings in terms of abstract argumentation theory.

Fact 25.3 *The following are theorems of* K:

$$\Box\Diamond\bot \leftrightarrow \Box\bot \tag{25.2}$$
$$\Box\Diamond\varphi \leftrightarrow \neg\Diamond\neg\Diamond\varphi \tag{25.3}$$
$$\Box\Diamond\Box\Diamond\bot \leftrightarrow \Box\Diamond\bot \vee \Box\Diamond\Box\Diamond\bot. \tag{25.4}$$

Formula 25.2 uses the trivial modal fact that $\Diamond\bot \leftrightarrow \bot$ to express that the set of arguments defended by the empty set corresponds to the set of arguments that have no attackers (dead ends). This equivalence will be constantly used in the remainder of the paper. Formula (25.3) is the modal counterpart of the equivalence of the defense function and the 2-fold iteration of the neutrality function, i.e., for any X and graph $\mathscr{A}: \mathsf{d}_{\mathscr{A}}(X) = \mathsf{n}_{\mathscr{A}}(\mathsf{n}_{\mathscr{A}}(X))$. Formula (25.4) states that, for any $\mathscr{A}$, the finite union of subsequent iterations of $\mathsf{d}_{\mathscr{A}}$ over $\emptyset$ is equivalent to the longest iteration.

---

[7] More generally, the claim is a direct consequence of the existence of a homomorphism from the term algebra $\mathsf{Term} = \langle\mathscr{L}, \wedge, \neg, \bot, \Diamond\rangle$ of language $\mathscr{L}$ (without universal modality) to the complex algebra $\mathsf{Set}_{\mathscr{A}} = \langle 2^A, \cap, -, \emptyset, f\rangle$ where $f : \wp(A) \longrightarrow \wp(A)$ such that $f(A) = \{a \in A \mid \exists b \in A : a \leftarrow b\}$ [9, Chap. 5].

In the remainder of the paper, in order to concisely express the nth iteration of $\Box\Diamond$ (resp., $\neg\Diamond$) we will write $(\Box\Diamond)^n$ (resp., $(\neg\Diamond)^n$).

Remark 25.2 (Frame language) When interested in the application of the characteristic functions solely to the set of all arguments, or to the empty set of arguments, all is needed to express d and n is a limited fragment of the language $\mathscr{L}$ introduced above. The fragment is defined by the following BNF:

$$\varphi ::= \bot \mid \neg\varphi \mid \varphi \wedge \varphi \mid \Diamond\varphi$$

This is a so-called *frame language*,[8] which does not use propositional atoms. In fact, this language does not need models to be interpreted, but simply attack graphs (Definition 25.1). It therefore expresses properties of pointed attack graphs: $\langle \mathscr{A}, a \rangle$. This will be the language we will be working with when defining a notion of argument equivalence with respect to the grounded set.

25.3.2.1 The Grounded Set in Modal Logic

As a consequence of Theorem 25.1 and Lemma 25.1—showing that d can be represented as $\Box\Diamond$—the grounded extension can, in any finitary graph $\mathscr{A}$, be expressed by the following infinite but countable disjunction (cf. Eq. (25.1)):

$$\bigvee_{0 \leq n < \omega} (\Box\Diamond)^n \bot \tag{25.5}$$

Clearly, in a finite $\mathscr{A}$ we will have a finite integer n where the stream $d_{\mathscr{A}}^\omega(\emptyset)$ reaches its limit, and we could then express the grounded extension by a finite disjunction $\bigvee_{0 \leq i \leq n} (\Box\Diamond)^i \bot$ or simply as $(\Box\Diamond)^i \bot$.

Similarly, it is worth observing that the greatest fixpoint of $d_{\mathscr{A}}$ for a given $\mathscr{A}$ is expressed by the following infinite conjunction:

$$\bigwedge_{0 \leq n < \omega} \neg(\Diamond\Box)^n \bot \tag{25.6}$$

i.e., it is neither the case that the current argument is attacked by a dead end, nor that it is attacked by an argument whose attackers are attacked by a dead end, and so on.

Remark 25.3 (Being attacked by the grounded set) Notice that arguments not belonging to the greatest fixpoint of d, i.e., satisfying $\neg \bigwedge_{0 \leq n < \omega} \neg(\Diamond\Box)^n \Diamond\bot$, are arguments attacked by the grounded set, i.e., arguments satisfying $\bigvee_{0 \leq n < \omega} \Diamond(\Box\Diamond)^n \bot$.

Remark 25.4 (Infinite attack graphs and the mu-calculus) In the general case, in order to express the grounded extension modally it is necessary to resort to the

---

[8] See Blackburn et al. [9, Chap. 3.1].

expressivity of the mu-calculus, where the grounded extension can be expressed by the following formula:

$$\mu p.\Box\Diamond p \tag{25.7}$$

denoting precisely the smallest fixpoint of function $\Box\Diamond$, i.e., in a given $\mathscr{A}$, the modal rendering of $\mathsf{d}_{\mathscr{A}}$ (Lemma 25.1). Similarly, $\nu p.\Box\Diamond p$ denotes the largest fixpoint. We refer the reader to [19, 21] for more information on the application of the modal mu-calculus to abstract argumentation.

25.3.2.2 Other Argumentation-Theoretic Notions in Modal Logic

We have shown how to express the grounded extension by a formula of the basic frame language. It must be clear that, from a modal point of view, the grounded extension is therefore a property of a pointed frame $\langle\mathscr{A}, a\rangle$, that is, the property of an argument in a graph.

How are the other notions of Table 25.1 to be formalized? In [21] it has been shown that logic K with the universal modality $\langle U\rangle$ suffices to express conflict-freeness, self-defense, admissibility and complete extensions. But in this case, the full modal language (with at least one atom p) is required:

$$\mathscr{V}(p) \text{ is conflict-free} \iff \langle\mathscr{A}, \mathscr{V}\rangle, a \models [U](p \to \neg\Diamond p)$$
$$\mathscr{V}(p) \text{ is self-defended} \iff \langle\mathscr{A}, \mathscr{V}\rangle, a \models [U](p \to \Box\Diamond p)$$
$$\mathscr{V}(p) \text{ is admissible} \iff \langle\mathscr{A}, \mathscr{V}\rangle, a \models [U](p \to \neg\Diamond p) \wedge [U](p \to \Box\Diamond p)$$
$$\mathscr{V}(p) \text{ is a complete set} \iff \langle\mathscr{A}, \mathscr{V}\rangle, a \models [U](p \to \neg\Diamond p) \wedge [U](p \leftrightarrow \Box\Diamond p).$$

These notions are therefore properties of pointed models $\langle\mathscr{M}, a\rangle$, that is, properties of arguments in a graph where a set of arguments has been labeled. In the remainder of the paper we will be concerned only with frame properties and will therefore be working with the frame language.

25.4 A Modal Notion of Argument Equivalence

The section develops a modal notion of argument equivalence characterizing the status of an argument in terms of a special family of modal formulae it satisfies.

25.4.1 When are Two Arguments Equivalent w.r.t. the Grounded Set?

Let us start by recalling a few observations from Sect. 25.3. For any graph $\mathscr{A}$, the set of arguments is partitioned in the set of arguments belonging to $\mathsf{lfp.d}_{\mathscr{A}}$ (the grounded set), those not belonging to $\mathsf{gfp.d}_{\mathscr{A}}$ (i.e., the arguments attacked by the grounded set, recall Remark 25.3), and the arguments belonging to $\mathsf{gfp.d}_{\mathscr{A}}\text{'}-\mathsf{lfp.d}_{\mathscr{A}}$ (i.e., the arguments neither belonging to the grounded set nor being attacked by it).[9] Figures 25.2 and 25.3 offer good examples for the identification of this tripartition.

So, from the point of view of the grounded set, what matters in a graph $\mathscr{A}$ is the status of an argument with respect to the three above sets, and hence with respect to membership to $\mathsf{lfp.d}_{\mathscr{A}}$ and $\mathsf{gfp.d}_{\mathscr{A}}$. A natural refinement of this idea in finitary graphs is to understand the status of an argument not only in terms of its membership to $\mathsf{lfp.d}_{\mathscr{A}}$ and $\mathsf{gfp.d}_{\mathscr{A}}$, but also in terms of 'when' it enters those sets, in the sense of which are the stages in the fixpoint computation to which the argument belongs,[10] i.e., at which n the argument comes to belong to $\mathsf{d}^n_{\mathscr{A}}(\emptyset)$ and at which it ceases to belong to $\mathsf{d}^n_{\mathscr{A}}(A)$. This suggests the following definition of status of an argument:

Definition 25.5 (*Status*) Let $\mathscr{A} = \langle A, \to \rangle$ be an attach graph. The status of $a \in A$ is defined as, for $1 \leq n < \omega$:

$$\mathbf{T}(a) = \left\{ (\Box\Diamond)^n \bot \mid \mathscr{A}, a \models (\Box\Diamond)^n \bot \right\} \cup \left\{ (\Box\Diamond)^n \top \mid \mathscr{A}, a \models (\Box\Diamond)^n \top \right\} (25.8)$$

Recall the modal principle: $(\Box\Diamond)^n \top \leftrightarrow \neg\Diamond(\Box\Diamond)^n \bot$. So, the status of an argument is the subset of its modal theory in the frame language which consists of formulae corresponding to iterations of the defense function over $\emptyset$ (i.e., $\bot$) and over the set of all arguments (i.e., $\top$).

To familiarize ourselves with the notion of argument status, let us mention this simple fact following from Theorem 25.1:

Fact 25.4 Let $\mathscr{A}$ be a finitary graph:

$$a \in \mathsf{lfp.d}_{\mathscr{A}} \iff \mathbf{T}(a) = \left\{ (\Box\Diamond)^m \bot \mid \exists n : n \leq m < \omega \right\} \cup \left\{ (\Box\Diamond)^n \top \mid 1 \leq n < \omega \right\}$$
$$a \in -\mathsf{gfp.d}_{\mathscr{A}} \iff \mathbf{T}(a) = \left\{ (\Box\Diamond)^m \top \mid \exists n : 1 \leq m \leq n \right\}$$
$$a \in \mathsf{gfp.d}_{\mathscr{A}} - \mathsf{lfp.d}_{\mathscr{A}} \iff \mathbf{T}(a) = \left\{ (\Box\Diamond)^n \top \mid 1 \leq n < \omega \right\}.$$

We can then say that two arguments are equivalent w.r.t. the grounded set (notation, $\mathscr{A}, a \equiv_{\mathrm{d}} \mathscr{A}', a'$) if and only if they have the same status:

$$\mathscr{A}, a \equiv_{\mathrm{d}} \mathscr{A}', a' \iff \mathbf{T}(a) = \mathbf{T}(a') \tag{25.9}$$

---

[9] It is worth observing that this three-set partition corresponds to the labeling of arguments as "in" (i.e., belonging to the extension at issue), "out" (i.e, being attacked by the extension at issue), and "undecided" (i.e., neither of the above) of the labeling-based semantics of argumentation [2, 10].

[10] It is worth stressing that this is a refinement of the common understanding of 'status of an argument' in the literature on argumentation theory.

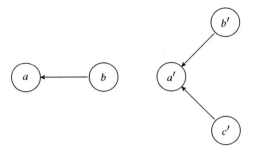

Fig. 25.4 Arguments a and a' have the same status: $\mathbf{T}(a) = \emptyset = \mathbf{T}(a')$

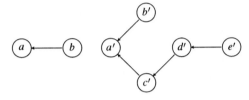

Fig. 25.5 Arguments b and c' have different statuses: $\mathbf{T}(b) = \{(\Box\Diamond)^n \bot \mid 1 \leq n < \omega\} \cup \{(\Box\Diamond)^n \top \mid 1 \leq n < \omega\}$, while $\mathbf{T}(c') = \{(\Box\Diamond)^n \bot \mid 2 \leq n < \omega\} \cup \{(\Box\Diamond)^n \top \mid 1 \leq n < \omega\}$. Both belong to the grounded sets of the respective graphs

Intuitively, two arguments are equivalent if and only if they belong to exactly the same stages of iteration of the defense function applied to the empty set, and to the same stages of iteration of the defense function applied to the set of all arguments. Figures 25.4 and 25.5 give an illustration of the definitions of status and status equivalence.

25.4.2 Status Equivalence and Frame Bisimulation

We recall the standard definition of the notion of frame bisimulation[11]:

Definition 25.6 (*Frame bisimulation* [4]) Let $\mathscr{A} = \langle A, \rightarrow \rangle$ and $\mathscr{A}' = \langle A', \rightarrow' \rangle$ be two attack graphs. A (frame-)bisimulation between $\mathscr{A}$ and $\mathscr{A}'$ is a non-empty relation $Z \subseteq A \times A'$ such that:

Zig: if aZa' and $a \leftarrow b$ for some $b \in A$, then $a' \leftarrow b'$ for some $b' \in A'$ and bZb';
Zag: if aZa' and $a' \leftarrow b'$ for some $b' \in A$ then $a \leftarrow b$ for some $b \in A$ and bZb'.

When a frame bisimulation exists linking $a \in A$ and $a' \in A'$ we write $\mathscr{A}, a \leftrightarrow \mathscr{A}', a'$.

Intuitively, a bisimulation is a process-like view of equivalence between attack graphs that links the walks along the attack relation—one might say the dialogues

---

[11] See [9, Chap. 2].

(cf. Sect. 25.5)—that one can do on one graphs to corresponding walks that one can do on the other.

By applying standard results from modal logic we can show that frame bisimulation implies status equivalence: two bisimilar arguments are also equivalent with respect to their status.

Fact 25.5 ($\leftrightarrow\ \subseteq\ \equiv_d$) *Let* $\langle \mathscr{A}, a \rangle$ *and* $\langle \mathscr{A}', a' \rangle$ *be two pointed attack graphs:*

$$\mathscr{A}, a \leftrightarrow \mathscr{A}', a' \implies \mathscr{A}, a \equiv_d \mathscr{A}', a'$$

Proof The claim is a direct consequence of Formula (25.9) and the fact that the basic modal language is bisimulation invariant (cf. [9]).

25.5 Status Equivalence and Argument Games

The picture of argumentation we have given so far is of a static kind, but argumentation calls intuitively for a process of interaction between arguers. In fact, although notions like the grounded extension formalize different static views of what makes a set of arguments a 'justifiable' or good position in an argumentation, these views can be made dynamic through two-player zero-sum games. Many researchers in the last two decades have focused on 'dialogue games' for argumentation, i.e., games able to adequately establish whether a given argument belongs or not to a given extension.[12]

The sort of results that drive this literature are called *adequacy* theorems and have, roughly, the following form: argument a has property S (e.g., belongs to the grounded extension) if and only if the proponent has a winning strategy in the dialogue game for property S (e.g., the dialogue game for the grounded extension) starting with argument a.

In this section we will see how the notion of bisimulation between arguments ties in with the theory of argument games.

25.5.1 Argument Games

The section recapitulates key definitions and results pertaining to an adequate game for the grounded extension.

25.5.1.1 Game for the Grounded Extension

Let us fix some further auxiliary notation before starting. Let $\mathbf{a} \in A^{<\omega} \cup A^{\omega}$ be a finite or infinite sequence of arguments in A, which we will call a *dialogue*. To

---

[12] The contributions that started this line of research is [13]. Cf. [24] for a recent overview.

Table 25.2 Winning conditions for the game for grounded given a terminal dialogue **a**

Length of a :	$\mathscr{P}$ wins if :	$\mathscr{O}$ wins if :
$\ell(\mathbf{a}) < \omega$	$\mathsf{t}(\mathbf{a}) = \mathscr{O}$	$\mathsf{t}(\mathbf{a}) = \mathscr{P}$
$\ell(\mathbf{a}) = \omega$	*Never*	*Always*

denote the nth element, for $1 \le n < \omega$, of a dialogue **a** we write $\mathbf{a}_n$, and to denote the dialogue consisting of the first n elements of **a** we write $\mathbf{a}|_n$. The last argument of a finite dialogue **a** is denoted $h(\mathbf{a})$. Finally, the length $\ell(\mathbf{a})$ of **a** is $n - 1$ if $\mathbf{a}|_n = \mathbf{a}$, and ω otherwise. We start with the formal definition:

Definition 25.7 (*Argument game for grounded* [13]) The game for the grounded extension is a function $\mathscr{D}$ which for each attack graph $\mathscr{A}$ yields a structure $\mathscr{D}(\mathscr{A}) = \langle N, A, \mathsf{t}, \mathsf{m}, \mathsf{p} \rangle$ where:

- $N = \{\mathscr{P}, \mathscr{O}\}$—the set of players consists of proponent $\mathscr{P}$ and opponent $\mathscr{O}$.
- A is the set of arguments in $\mathscr{A}$.
- $\mathsf{t} : A^{<\omega} \longrightarrow N$ is the *turn function*. It is a (partial[13]) function assigning one player to each finite dialogue in such a way that, for any $0 \le m < \omega$ and $\mathbf{a} \in A^{<\omega}$, if $\ell(\mathbf{a}) = 2m$ then $\mathsf{t}(\mathbf{a}) = \mathscr{O}$, and if $\ell(\mathbf{a}) = 2m + 1$ then $\mathsf{t}(\mathbf{a}) = \mathscr{P}$. i.e., even positions are assigned to $\mathscr{O}$ and odd positions to $\mathscr{P}$.
- $\mathsf{m} : A^{<\omega} \longrightarrow \wp(A)$ is a (partial) function from dialogues to sets of arguments defined as: $\mathsf{m}(\mathbf{a}) = R_{\mathscr{A}}(h(\mathbf{a}))$. I.e., the available moves at **a** are the arguments attacking the last argument of **a**. The set of all dialogues compatible with m—the legal dialogues of the game—is denoted D. Dialogues **a** for which $\mathsf{m}(\mathbf{a}) = \emptyset$ or such that $\ell(\mathbf{a}) = \omega$ are called terminal, and the set of all terminal dialogues of the game is denoted T.
- $\mathsf{p} : T \longrightarrow N$ is the payoff function given in Table 25.2, which associates a player—the winner—to each terminal dialogue.

The game is played starting from a given argument a. When a is explicitly given we talk about an instantiated game (notation, $\mathscr{D}(\mathscr{A})@a$).

The two players play the game by alternating each other ($\mathscr{O}$ starts) and navigating the attack graph along the 'being attacked' relation. The winning conditions state that $\mathscr{P}$ wins whenever she manages to state an argument to which $\mathscr{O}$ cannot reply, i.e., an argument with no attackers. Notice the asymmetry in the winning conditions of the payoff function for $\mathscr{P}$ and $\mathscr{O}$.

25.5.1.2 Adequacy

The different ways in which proponent and opponent can play an argument game are called strategies:

---

[13] The function is partial because only sequences compatible with the move function m below need to be considered.

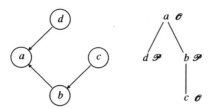

Fig. 25.6 An attack graph (*left*) and its dialogue game for grounded (*right*). Positions are labeled by the player whose turn it is to play. $\mathscr{P}$ wins the terminal dialogue *abc* but loses the terminal dialogue *ad*. $\mathscr{O}$ has a winning strategy that makes him win in one move

Definition 25.8 (*Strategies*) Let $\mathscr{D}(\mathscr{A}) = \langle N, A, \mathsf{t}, \mathsf{m}, \mathsf{p} \rangle$, $a \in A$ and $i \in N$. A strategy for i in the instantiated game $\mathscr{D}@a$ is a function: $\sigma_i : \{\mathbf{a} \in D - T \mid \mathbf{a}_0 = a \text{ AND } \mathsf{t}(\mathbf{a}) = i\} \longrightarrow A$ s.t. $\sigma_i(\mathbf{a}) \in \mathsf{m}(\mathbf{a})$. The set of terminal dialogues compatible with σ_i is defined as follows: $T_{\sigma_i} = \{\mathbf{a} \in T \mid \mathbf{a}_0 = a \text{ AND } \forall n \leq \ell(\mathbf{a}) \text{ IF } \mathsf{t}(\mathbf{a}|_n) = i \text{ THEN } \mathbf{a}_{n+1} = \sigma_i(\mathbf{a}|_n)\}$.

A strategy tells i which argument to choose, among the available ones, at each non-terminal dialogue $\mathbf{a}$ in $\mathscr{D}@a$. So, in the game for grounded, a strategy $\sigma_{\mathscr{P}}$ will encode the proponent's choices in dialogues of odd length, while $\sigma_{\mathscr{O}}$ will encode the opponent's choices in dialogues of even length. Observe that, in a game for grounded, a strategy $\sigma_{\mathscr{P}}$ and a strategy $\sigma_{\mathscr{O}}$—i.e., a strategy profile in the game-theory terminology—together determine one terminal dialogue or, in other words, $T_{\sigma_{\mathscr{P}}} \cap T_{\sigma_{\mathscr{O}}}$ is a singleton.

What matters of a strategy is whether it guarantees the player that plays according to it to win the game. This brings us to the notion of winning strategy:

Definition 25.9 (*Winning strategies and arguments*) Let $\mathscr{D}(\mathscr{A}) = \langle N, S, \mathsf{t}, \mathsf{m}, \mathsf{p} \rangle$, $a \in A$ and $i \in N$. A strategy σ is *winning for i* in $\mathscr{D}(\mathscr{A})@a$ if and only if for all $\mathbf{a} \in T_\sigma$ it is the case that $\mathsf{p}(\mathbf{a}) = i$. An argument a is *winning for i* iff there exists a winning strategy for i in $\mathscr{D}(\mathscr{A})@a$. The set of winning positions of $\mathscr{D}$ for i is denoted $Win_i(\mathscr{D}(\mathscr{A}))$. An argument a is *winning for i in k rounds* ($k \geq 0$) iff there exists a winning strategy σ_i in $\mathscr{D}@a$ such that for all $\mathbf{a} \in T_{\sigma_i}$, $\ell(\mathbf{a}) + 1 \leq k$, that is, i can always win in at most k rounds using σ_i. The set of winning positions in k rounds is denoted $Win_i^k(\mathscr{D})$.

Dialogue games are two-player zero-sum games with perfect information (and possibly infinite horizon).[14] See Fig. 25.6 for the illustration of a finite dialogue game.

Now all ingredients are in place to study the property we are interested in, viz. the adequacy of the game of Definition 25.7 with respect to the grounded extension. We first prove a slightly stronger result: an argument a belongs to the kth iteration of the defense function on the empty set of arguments, if and only if $\mathscr{P}$ has a winning strategy in the game initiated at a, which she can carry out in at most $2(k-1)$ rounds.

---

[14] These games are determined by the Gale-Stewart theorem since it can always be decided whether a dialogue is winning for $\mathscr{P}$, i.e., the winning positions for $\mathscr{P}$ are an open set (cf. [23, Chap. 6]).

Lemma 25.2 (*Strong adequacy of the game for grounded* [13]) *Let $\mathscr{D}(\mathscr{A})$ be the dialogue game for grounded on graph $\mathscr{A}$ and $a \in A$, for $1 \leq k < \omega$:*

$$a \in d_{\mathscr{A}}^{k}(\emptyset) \Longleftrightarrow a \in Win_{\mathscr{P}}^{2(k-1)}(\mathscr{D}(\mathscr{A})).$$

Proof We proceed by induction:

[**B**] The following equivalences prove the induction base:

$$a \in d_{\mathscr{A}}(\emptyset) \Longleftrightarrow \nexists b : b \to a \qquad \text{[Definition 25.2]}$$
$$\Longleftrightarrow a \in Win_{\mathscr{P}}^{0}(\mathscr{D}(\mathscr{A})) \text{ [Definition 25.9]}$$

[**S**] If $a \in d_{\mathscr{A}}^{n}(\emptyset) \Longleftrightarrow a \in Win_{\mathscr{P}}^{2(n-1)}(\mathscr{D}(\mathscr{A}))$ (IH) then we claim: $a \in d_{\mathscr{A}}^{n+1}(\emptyset) \Longleftrightarrow a \in Win_{\mathscr{P}}^{2n}(\mathscr{D}(\mathscr{A}))$. [LEFT TO RIGHT] Assume $a \in d_{\mathscr{A}}^{n+1}(\emptyset)$. This means that $\forall b : b \to a, \exists c : c \to b$ and such that $c \in d_{\mathscr{A}}^{n}(\emptyset)$ which, by IH, is equivalent to $c \in Win_{\mathscr{P}}^{2(n-1)}(\mathscr{D}(\mathscr{A}))$. So, by Definition 25.7, for any $\mathcal{O}$'s move b at position a, $\mathscr{P}$ has a counter-argument c from which she has a winning strategy forcing a win in at most $2(n-1)$ rounds. Hence, by Definition 25.9, $\mathscr{P}$ can win the game at a in $2(n-1)+2$ rounds, i.e., $a \in Win_{\mathscr{P}}^{2n}(\mathscr{D}(\mathscr{A}))$. [RIGHT TO LEFT] Assume $a \in Win_{\mathscr{P}}^{2n}(\mathscr{D}(\mathscr{A}))$. This means that, for any $\mathcal{O}$'s move b at a, $\mathscr{P}$ has a counter-argument c from which she has a winning strategy forcing a win in at most $2n - 2$ rounds. By IH, this is equivalent with $c \in d_{\mathscr{A}}^{n}(\emptyset)$ and by Definition 25.2 we conclude that $a \in d_{\mathscr{A}}^{n+1}(\emptyset)$. This completes the proof.

As a consequence, an argument belongs to the grounded extension of an argumentation framework if and only if the proponent has a winning strategy for the dialogue game for grounded (in that argumentation framework) instantiated at that argument.

Theorem 25.2 (*Adequacy of the game for grounded*) *Let $\mathscr{D}(\mathscr{A}) = \langle N, S, \mathsf{t}, \mathsf{m}, \mathsf{p} \rangle$ be the dialogue game for grounded on a finitary graph $\mathscr{A}$ and $a \in A$:*

$$a \in \mathsf{lfp}.d_{\mathscr{A}} \Longleftrightarrow a \in Win_{\mathscr{P}}(\mathscr{D}(\mathscr{A})).$$

Proof The claim is proven by the following series of equivalences:

$$a \in Win_{\mathscr{P}}(\mathscr{D}(\mathscr{A})) \Longleftrightarrow a \in \bigcup_{1 \leq k < \omega} Win_{\mathscr{P}}^{2(k-1)}(\mathscr{D}(\mathscr{A}))$$
$$\Longleftrightarrow a \in \bigcup_{1 \leq k < \omega} d_{\mathscr{A}}^{k}(\emptyset)$$
$$\Longleftrightarrow a \in \mathsf{lfp}.d_{\mathscr{A}}$$

The first equivalence holds by the winning conditions of Definition 25.7 and Definition 25.9: $\mathscr{P}$ wins if and only if she can force the game to reach an unattacked argument in an even number of steps. The second equivalence holds by Lemma 25.2 and the third one by Theorem 25.1.

Theorems like Theorem 25.2 play a significant role in the development of a formal theory of argumentation. Firstly, they guarantee that the argument game at issue is a sound (if the proponent has a winning strategy then the the argument is grounded) and complete (if the argument is grounded, then the proponent has a winning strategy) proof procedure with respect to the corresponding semantics. Secondly, literature in argumentation (e.g., [1]) has pointed out—convincingly in our view—that Dung's extensions can be soundly viewed as abstract models of standards of proof in debates, and that argument games are a viable abstraction of procedural rules, or protocols, for debates. (cf. [29]). Viewed in this light, adequacy is then the property of debate protocols successfully *implementing* a given standard of proof, like the grounded extension.[15]

25.5.2 Strategic Equivalence of Arguments and Status Equivalence

When can two arguments in two attack graphs be considered equivalent from the point of view of the above game? Intuitively, we might say that the two arguments are equivalent if the proponent (respectively, the opponent) has a winning strategy that allows her (respectively, him) to win the game in at most the same number of rounds. More precisely:

Definition 25.10 (*Strategic equivalence of arguments*) Two pointed attack graphs $\langle \mathscr{A}, a \rangle$ and $\langle \mathscr{A}', a' \rangle$ are *strategically equivalent* if and only if the two following conditions are met:

(i) For $0 \leq n < \omega$, if $\mathscr{P}$ can always win $\mathscr{D}(\mathscr{A})@a$ in at most $2n$ rounds, then she can always win $\mathscr{D}(\mathscr{A}')@a'$ in at most the same number of rounds, and vice versa;

(ii) For $0 \leq n < \omega$, if $\mathscr{O}$ can always win $\mathscr{D}(\mathscr{A})@a$ in at most $2n + 1$ rounds, then he can always win $\mathscr{D}(\mathscr{A}')@a'$ in at most the same number of rounds, and vice versa.

In other words, two arguments are strategically equivalent whenever they support the same 'powers' for the proponent and the opponent, that is, whenever they support winning strategies (for the proponent or the opponent) that can force a win in the game for grounded in at most the same number of rounds.

Example 25.3 Let us get back to Fig. 25.5. Consider arguments a and a'. These are strategically equivalent: $\mathscr{O}$ has a winning strategy for the arguments, on both graphs,

---

[15] We use the word "implement" here in the technical sense in which it is typically used in game theory [25, Chap. 10] or social software [26].

guaranteeing him a win in at most 1 round. Consider now arguments b and c'. $\mathscr{P}$ has a winning strategy on both games. But while she always wins in 0 rounds from b, she always wins in 2 rounds playing from c'. So, b and c' are not strategically equivalent.

Now, capitalizing on Lemma 25.1 and 25.2, this notion of strategic equivalence can be shown to be just a game-theoretic variant of the notion of status equivalence:

Theorem 25.3 *Let $\langle \mathscr{A}, a \rangle$ and $\langle \mathscr{A}', a' \rangle$ be two pointed attack graphs: $\mathscr{A}, a \equiv_d \mathscr{A}', a'$ if and only if $\langle \mathscr{A}, a \rangle$ and $\langle \mathscr{A}', a' \rangle$ are strategically equivalent.*

Proof Define the following set:

$$
W(a) = \begin{cases} \left\{ (\Box \Diamond)^n \bot \mid a \in \mathit{Win}^{2(n-1)}_{\mathscr{P}}(\mathscr{D}(\mathscr{A})), \text{ FOR } 1 \le n < \omega \right\} \\ \cup \\ \left\{ (\Box \Diamond)^n \top \mid \nexists b : a \leftarrow b \text{ AND } b \in \mathit{Win}^{2(n-1)}_{\mathscr{P}}(\mathscr{D}(\mathscr{A})), \text{ FOR } 1 \le n < \omega \right\} \end{cases}
$$

First of all, observe that, for any n, if $\exists b : a \leftarrow b$ AND $b \in \mathit{Win}^{2(n-1)}_{\mathscr{P}}(\mathscr{D}(\mathscr{A})$ then $\mathscr{O}$ has a winning strategy in a that forces a win in $2(n-1)+1$ rounds (in symbols, $a \in \mathit{Win}^{2n-1}_{\mathscr{O}}(\mathscr{D}(\mathscr{A}))$), and vice versa. So, by Lemma 25.1 and 25.2, it is not difficult to see that $\langle \mathscr{A}, a \rangle$ and $\langle \mathscr{A}', a' \rangle$ are strategically equivalent if and only if $W(a) = W(a')$ (recall that $(\Box \Diamond)^n \top \leftrightarrow \neg \Diamond (\Box \Diamond)^n \bot$). By the definition of W, Definition 25.5 and Lemma 25.1 it then follows directly that $\mathbf{T}(a) = \mathbf{T}(a')$.

We have thus shown that the modally defined notion of status equivalence for the grounded extension has a natural strategic variant based on the argument game for that extension. As a direct consequence of Fact 25.5 we also obtain that if two arguments are frame bisimilar, then they are strategically equivalent.

Getting back to the Toulmin's quote by which we opened the paper, Theorem 25.3 establishes an equivalence of arguments in terms of a procedural equivalence relating the ways proponent and opponent are able to argue with respect to the argument at issue. Two arguments in two different argumentations can be said to be equivalent whenever the powers—intended as the availability of a strategy to force a win in a fixed number of rounds—of the proponent and the opponent in the two graphs are the same. This ties in well with power-based notions of game equivalence as put forth, for instance, in [5, 8].

25.6 Games and Equations

In this final section we look at one more perspective on argument equivalence, based on the equational semantics of abstract argumentation [15].

25.6.1 The Equational Approach to Abstract Argumentation

Let us start with a few preliminaries. The equational approach to—or equational semantics of—argumentation consists in extracting from a given finite attack graph $\mathscr{A} = \langle A, \rightarrow \rangle$ a system of equations:

$$f(a_1) = 1 - \max(\{f(b) \mid a_1 \leftarrow b\})$$
$$f(a_2) = 1 - \max(\{f(b) \mid a_2 \leftarrow b\})$$
$$\cdots \quad \cdots$$
$$f(a_n) = 1 - \max(\{f(b) \mid a_n \leftarrow b\})$$

where $a_1, \ldots, a_n$ is an enumeration of the arguments in A, and $f : A \longrightarrow [0, 1]$ is a function from the sets of arguments to the real values between 0 and 1.[16] Intuitively, 0 represents a form of rejection of the argument, 1 a form of acceptance, and intermediate values a form of undecidedness.

As shown in [15], each solution f to one such system of equations defines a set of arguments $\{a \in A \mid f(a) = 1\}$ corresponding to a complete extension (see Table 25.1) of the underlying attack graph. The solution f_g such that $\{a \mid f_g(a) = 1\}$ is minimal corresponds therefore to the grounded extension, i.e., to $\mathsf{lfp.d}_{\mathscr{A}}$. So, the equational perspective looks at how values of acceptance or rejection propagate within the attack graph stabilizing into steady states—the solutions—that have a nice correspondence with Dung's theory.

Example 25.4 Consider the graph on the left of Fig. 25.1. The corresponding system of equations is:

$$f(a) = 1 - \max(\{f(b)\})$$
$$f(b) = 1 - \max(\{f(a)\})$$

This gives three solutions: $f'(a) = 1$ and $f'(b) = 0$, $f''(a) = 0$ and $f''(b) = 1$, $f'''(a) = 0.5$ and $f'''(b) = 0.5$. The latter minimizes the set $\{a \mid f_g(a) = 1\}$ and corresponds therefore to the grounded extension.

25.6.2 Playing Argument Games Through Equations

We now look at how to build winning strategies for the proponent in an argument game using solutions to the system of equation of a given attack graph.

Let $\mathscr{A}$ be an attack graph. Consider its argument game $\mathscr{D}(\mathscr{A})@a$ for grounded at argument a and the equational theory for $\mathscr{A}$ corresponding to its grounded extension. Consider a strategy for $\mathscr{P}$ with the following property:

---

[16] Other systems making use of different mathematical functions instead of $1 - \max(.)$ are discussed in [15]. See also [17] for an extensive exposition of the equational approach to argumentation.

$$\sigma^*_{\mathscr{P}}(\mathbf{a}) \in \left\{ a \mid f_g(a) = \max(\{b \mid b \in R(h(\mathbf{a}))\}) \right\} \quad \text{FOR } t(\mathbf{a}) = \mathscr{P} \quad (25.10)$$

Intuitively, the strategy consist in $\mathscr{P}$ maximizing at each of her choice nodes the value f_g among the arguments attacking the last argument in the dialogue. In other words, $\mathscr{P}$ uses the information encoded by f_g to pick her arguments.

We can show that if the set of dialogues generated by $\sigma^*_{\mathscr{P}}$ are all of even length smaller than $2n$ then $\mathscr{P}$ can force a win in at most $2n$ rounds and vice versa:

Theorem 25.4 (Equationally defined winning strategies) *Let $\mathscr{D}(\mathscr{A})@a$ be the argument game for grounded on $\mathscr{A}$ instantiated at a, for $0 \le n < \omega$:*

$$\forall \mathbf{a} \in T_{\sigma^*_{\mathscr{P}}} : \ell(\mathbf{a}) = 2m \le 2n \iff a \in Win^{2n}_{\mathscr{P}}$$

Proof (Sketch) [RIGHT TO LEFT] If $a \in Win^{2n}_{\mathscr{P}}$ then $\mathscr{P}$ can force a win in at most $2n$ rounds. By its definition (Formula (25.10)), $\sigma^*_{\mathscr{P}}$ must be a winning strategy. So, for any response $\sigma_{\mathscr{O}}$, $\mathrm{p}(\sigma^*_{\mathscr{P}}, \sigma_{\mathscr{O}}) = \mathscr{P}$ and hence the length $\ell(\sigma^*_{\mathscr{P}}, \sigma_{\mathscr{O}})$ must be even. Suppose now towards a contradiction that $\ell(\sigma^*_{\mathscr{P}}, \sigma_{\mathscr{O}}) > 2n$. $\mathscr{P}$ would then need in one case more than $2n$ rounds to win the game, against the assumption. [LEFT TO RIGHT] Similar. ∎

One might say that $\sigma^*_{\mathscr{P}}$ is some kind of 'canonical' strategy for $\mathscr{P}$. As a direct corollary we obtain: $\sigma^*_{\mathscr{P}}$ is a winning strategy if and only if $f_g(a) = 1$. That is, a strategy that maximizes f_g at each choice node is winning for $\mathscr{P}$ if and only if the f_g value of the first argument is 1, i.e., if and only if a belongs to the grounded set. Similarly, it directly follows that if two arguments are strategically equivalent, then $\sigma^*_{\mathscr{P}}$ is winning (in a given number of rounds) for the first argument if and only if it is winning (in the same number of rounds) for the second.

25.6.3 Bisimulation, Status Equivalence, Strategic Equivalence and Equational Semantics

The equational semantics of abstract argumentation helps us in bringing together all the results handled in the paper, highlighting a wealth of interconnections between the modal, the strategic and the equational views of abstract argumentation theory.

Concretely, we have seen that frame bisimulation implies the status equivalence of two arguments in two attack graphs, which is in turn equivalent to their strategic equivalence in argument games seen as equivalence of 'powers' of strategies of the proponent and the opponent. All these different types of equivalences force arguments to obtain the same values in terms of Dung's semantics (i.e., one belongs to the grounded set if and only if the other also does) and Gabbay's equational variants (i.e., the value of f_g is the same for both arguments), as well as guaranteeing that equationally defined strategies for the proponent are winning on the first graph only if they are winning on the second, and vice versa. Figure 25.7 depicts these relations diagrammatically.

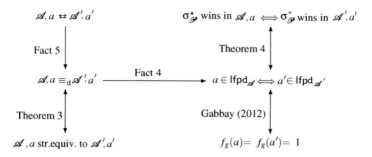

Fig. 25.7 A diagram relating the notions of frame bisimulation, status equivalence, strategic equivalence, sameness of values according to Dung's grounded semantics, sameness of value according to Gabbay's equational semantics for the grounded set, and equivalence of 'powers' of equationally defined winning strategies

25.7 Conclusions

The paper has touched upon several strands of research at the interface of Dung-style abstract argumentation, modal logic, games and equational systems. From this interdisciplinary vantage point the paper has advocated a notion of equivalence of arguments abstracting from their content and based on the way they 'behave' with respect to the other arguments in the attack graph with respect to some external criterion of 'justifiability', which in this paper has been assumed to be the grounded extension.

First of all, the paper has shown how modal logic puts at disposal a number of notions and tools that can be readily used to provide an analysis of this sort of equivalence of arguments based on their abstract patterns of interaction. Argument equivalence has been based on the notion of modal equivalence, and thereby related to the notion of (frame) bisimulation. This strengthens the many links between abstract argumentation and modal logic that have been object of several recent studies (e.g., [11, 16, 19–22]).

Second, the paper has shown how this static view of equivalence has a natural dynamic and strategic counterpart in argument games. In this view equivalent arguments are such that they support strategies for the proponent and opponent having the 'same powers' where power is intended as the possibility to guarantee a win in at most a given number of rounds of the game. This, together with the previous modal perspective, brings argumentation close to the thriving body of research into games and logical dynamics [6, 8], and offers the picture of a theory that goes well beyond its more 'traditional' boundaries of the static study of justification criteria for arguments

Finally, Gabbay's equational approach [15, 17] to abstract argumentation has been used in relation to argument games as a method for constructing winning strategies for the proponent, thereby providing a sort of 'canonical' characterization of strategies viewed as local maximizers of the values provided by solutions to the equational systems of the graphs. This lays an interesting bridge between the modal and game-theoretic view of abstract argumentation and the rich body of techniques made available by the equational view.

Acknowledgments We would like to thank Johan van Benthem for the many useful suggestions that helped us shape this last version of the paper.

References

1. Atkinson K, Bench-Capon T (2007) Argumentation and standards of proof. In: Proceedings of the 11th international conference on artificial intelligence and law (ICAIL'07), ACM, pp 107–116
2. Baroni P, Giacomin M (2009) Semantics of abstract argument systems. In: Rahwan I, Simari GR (eds) Argumentation in artifical intelligence, Springer, Dordrecht
3. Baroni P, Caminada M, Giacomin M (2011) An introduction to argumentation semantics. Knowl Eng Rev 26(4):365–410
4. van Benthem J (1983) Modal logic and classical logic. Monographs in philosophical logic and formal Linguistics, Bibliopolis, Berkeley
5. van Benthem J (2002) Extensive games as process models. J Logic Lang Inform 11:289–313
6. van Benthem J (2011) Logical dynamics of information and interaction. Cambridge University Press, Cambridge
7. van Benthem J (2012) The nets of reason. Argument Comput 3(2–3):83–86
8. van Benthem J (2014) Logic in games. MIT Press, Cambridge, MA
9. Blackburn P, de Rijke M, Venema Y (2001) Modal logic. Cambridge University Press, Cambridge
10. Caminada M (2006) On the issue of reinstatement in argumentation. In: Fischer M, van der Hoek W, Konev B, Lisitsa A (eds) Logics in artificial intelligence. Proceedings of JELIA 2006, pp 111–123
11. Caminada M, Gabbay D (2009) A logical account of formal argumentation. Studia Logica 93(2):109–145
12. Davey BA, Priestley HA (1990) Introduction to lattices and order. Cambridge University Press, Cambridge
13. Dung PM (1994) Logic programming as dialogue games. Technical report. Division of computer science, Asian Institute of Technology
14. Dung PM (1995) On the acceptability of arguments and its fundamental role in nonmonotonic reasoning, logic programming and n-person games. Artif Intell 77(2):321–358
15. Gabbay D (2011a) Introducing equational semantics for argumentation networks. In: Liu W (ed) Proceedings of ECSQARU 2011, no. 6717 in LNAI, pp 19–35
16. Gabbay D (2011b) Sampling logic and argumentation networks: a manifesto (vol 2). In: Gupta A, van Benthem J (eds) Logic and philosophy today, Studies in logic, vol 30, College Publications, pp 231–250
17. Gabbay D (2012) An equational approach to argumentation networks. Argument Comput 3 (2–3):87–142
18. Gabbay D (2013) Meta-Logical investigations in argumentation networks. College Publications, London
19. Gratie C, Florea AM, Meyer J (2012) Full hybrid mu-calculus, its bisimulation invariance and application to argumentation. Proc COMMA 2012:181–194
20. Grossi D (2009) Doing argumentation theory in modal logic. ILLC Prepublication Series PP-2009–24, Institute for Logic, Language and Computation

21. Grossi D (2010) On the logic of argumentation theory. In: van der Hoek W, Kaminka G, Lespérance Y, Sen S (eds) Proceedings of the 9th international conference on autonomous agents and multiagent systems (AAMAS 2010), IFAAMAS, pp 409–416
22. Grossi D (2011) Argumentation theory in the view of modal logic. In: McBurney P, Rahwan I (eds) Post-proceedings of the 7th international workshop on argumentation in Multi-Agent systems, no. 6614 in LNAI, pp 190–208
23. Kanamori A (1994) The Higher Infinite. Springer, Dordrecht
24. Modgil S, Caminada M (2009) Proof theories and algorithms for abstract argumentation frameworks. In: Rahwan I, Simari G (ed) Argumentation in AI, Springer, pp 105–132
25. Osborne MJ, Rubinstein A (1994) A course in game theory. MIT Press, Cambridge
26. Parikh R (2002) Social software. Synthese 132(3):187–211
27. Pollock JL (1987) Defeasible reasoning. Cogn Sci 11:481–518
28. Pollock JL (1991) A theory of defeasible reasoning. Int J Intell Syst 6(1):33–54
29. Prakken H (2009) Models of persuasion dialogue. In: Rahwan I, Simari G (eds) Argumentation in Artificial Intelligence, chap 14, Springer, Dordrecht
30. Toulmin S (1958) The uses of argument. Cambridge University Press, Cambridge
31. Venema Y (2008) Lectures on the modal μ-calculus. Renmin University, Beijing

Part V
Language and Cognition

Chapter 26
Three Etudes on Logical Dynamics and the Program of Natural Logic

Lawrence S. Moss

Abstract This chapter has three discussions related to one of Johan van Benthem's longstanding interests, the areas of interaction of logic and linguistics. We review much of what is known on the landscape of syllogistic logics. These are logics which correspond to fragments of language. The idea in this area is to have complete and decidable systems. Next we present a very simple form of dynamic logic, essentially a logic of two worlds with a back-and-forth arrow between them. This is then related to an issue in dynamic semantics, the logic of "and then". Our last discussion is related to an area which van Benthem again did so much to stimulate, the area of monotonicity reasoning in language. We connect the topic to reasoning in elementary mathematics. We formalize a monotonicity calculus following van Benthem and Sánchez Valencia, and we interpret this on hierarchies of preorders rather than sets.

26.1 Introduction

26.1.1 Overview of Natural Logic

Johan van Benthem has been a leading figure in work at the border of logic and linguistics which aims to craft logical systems which capture aspects of inference in natural language. His work on this topic overlaps with the main line of this volume, logical dynamics: he was the most forceful proponent of using dynamic logics in connection to model cognitive transitions, and of using dynamic versions of predicate logic and modal logic in connection with linguistic phenomena, and most recently of the uses of dynamic epistemic and doxastic logics in connection with strategic reasoning. Although many people work on dynamic logic in connection with

L. S. Moss (✉)
Department of Mathematics, Indiana University, Bloomington, IN47405, USA
e-mail: lsm@cs.indiana.edu

A. Baltag and S. Smets (eds.), *Johan van Benthem on Logic and Information Dynamics*, Outstanding Contributions to Logic 5,
DOI: 10.1007/978-3-319-06025-5_26, © Springer International Publishing Switzerland 2014

programming language semantics, I think it is fair to say that Johan van Benthem is the key connection between dynamic logic and fields like linguistics, cognitive science, and economics. His proposals, insights, and experiences have always been an inspiration to me.

The topic of this chapter in our tribute book concerns a strand of his work connected to language that is not exactly the main line of work mentioned above, but yet is fairly close. I refer to his work on *natural logic*. The ultimate goal is a logical system (or many of them) which is strong enough to represent the most common inferential phenomena in language and yet is weak enough to be decidable. The logical system should be tuned to the phenomena the way the Kripke semantics of modal logic is tuned to relations.

This chapter is not the place to go into all of van Benthem's contributions to logic and language, but in addition to mentioning the main lines of work and what I think is of lasting importance about them, I want to offer some *études*: simple versions of more complex phenomena that could be interesting. They come mainly from my teaching, and I would like to think that they could be useful to others.

The subject of natural logic might be defined as "logic for natural language, logic in natural language." As mentioned in van Benthem [3], his work on logic and semantics led to the realization that the "proposed ingredients" of a logical system for linguistic reasoning would consist of several "modules":

(a) Monotonicity Reasoning, i.e., Predicate Replacement,
(b) Conservativity, i.e., Predicate Restriction, and also
(c) Algebraic Laws for inferential features of specific lexical items.

All of these topics are prominent in his work starting perhaps with generalized quantifiers and conservativity (b) in the 1980s. We shall have something to say about all of them.

In a sense, point (c) on "algebraic" laws of specific lexical items might be covered by *syllogistic reasoning*; turning things around, the principles of syllogistic reasoning in an enriched setting can indeed start to look like algebraic laws. Syllogistic reasoning deals with very simple sentences; to a linguist its subject matter would not count as syntax in any serious sense. However, it is possible to go beyond the very limited fragment of the classical syllogistic and add transitive verbs, relative clauses, and adjectives. One still has a very limited syntax, but at least the resulting fragment has recursion (and so it is infinite). We commence with a discussion of what must be the smallest logical system in existence, the system where all of the sentences are of the form All X are Y; no other sentences are considered. Focusing on such a small fragment allows us to raise the general issues of logic very clearly and succinctly.

Our second étude concerns dynamic logic itself. I would like to offer a particularly easy way into the subject and into modal logic itself. Just as the syllogistic logic of All is arguably the simplest logical system, the logics in Sect. 26.3 are arguably the simplest modal and dynamic logics.

The last, and most sustained topic of the chapter has to do with the *natural logic* program initiated by [1, 2], and elaborated in [10]. (So this use of *natural logic* is somewhat narrower than what I have in mind in this contribution.) It develops a formal approach to *monotonicity* on top of categorial grammar (CG). This approach makes sense because the semantics of CG is given in terms of functions, and monotonicity in the semantic sense is all about functions. The success story of the natural logic program is given by the *monotonicity calculus*, a way of determining the *polarity* of words in a given sentence. Here is what this means, based on an example. Consider a very simple sentence:

$$\text{Every dog barks} \qquad (26.1)$$

Also, consider the following order statements:

$$\begin{array}{c} \text{old dog} \leq \text{dog} \leq \text{animal} \\ \text{barks loudly} \leq \text{barks} \leq \text{vociferates} \end{array} \qquad (26.2)$$

Think of these as implications: every old dog is a dog, every dog is an animal, etc. Suppose one is reasoning about a situation where (26.1) holds, and also has the inequalities in (26.2) as background information. Then it follows that every old dog barks and also that every dog vociferates. It would not necessarily follow that every animal barks and also that every dog barks loudly. That is, the inferences from (26.2) go "down" in the second argument and "up" in the second. We indicate this by writing every dog$^{\downarrow}$ barks$^{\uparrow}$. We also have other similar findings:

no dog$^{\downarrow}$ barks$^{\downarrow}$
not every dog$^{\uparrow}$ barks$^{\downarrow}$
some dog$^{\uparrow}$ barks$^{\uparrow}$
most dogs$^{\times}$ bark$^{\uparrow}$

The $\times$ in the last line indicates that there is no inference either way in the first argument of most. It is clear from these examples that the "direction of inference" is not determined by the words involved, but rather by aspects of the syntactic structure; however, something having to do with the particular determiners must in addition be involved. By the *polarity* of a word occurrence we mean the up and down arrows. Johan van Benthem's seminal idea in this area was to propose a systematic account of polarity, an account which works on something closer to real sentences than to logical representations. We present a version of it in Sect. 26.4.

26.2 The Simplest Fragment "of All"

We begin our contribution with the simplest logical fragment whatsoever. The sentences are all of the form All p are q. So we have a very impoverished language, with only one kind of sentence. But we'll have a precise semantics, a proof system, and a soundness/completeness theorem which relates the two.

Syntax and semantics

Our syntax is extremely small. We start with a collection **P** of *unary atoms* (for nouns). We write these as $p, q, \ldots$, Then the *sentences* of this first fragment $\mathcal{A}$ are the expressions

$$\text{All } p \text{ are } q$$

where p and q are any atoms in **P**.

The semantics is based on *models*. A model $\mathcal{M}$ for this fragment $\mathcal{A}$ is a structure

$$\mathcal{M} = (M, [\![\ ]\!])$$

consisting of a set M, together with an *interpretation* $[\![p]\!] \subseteq M$ for each noun $p \in \mathbf{P}$. The main semantic definition is *truth in a model*:

$$\mathcal{M} \models \text{All } p \text{ are } q \quad \text{iff} \quad [\![p]\!] \subseteq [\![q]\!]$$

We read this in various ways, such as $\mathcal{M}$ *satisfies* All p are q, or All p are q is true in $\mathcal{M}$.

From this definition, we get two further notions: If Γ is a set of sentences, we say that $\mathcal{M} \models \Gamma$ iff $\mathcal{M} \models \varphi$ for every $\varphi \in \Gamma$. Finally, we say that $\Gamma \models \varphi$ iff whenever $\mathcal{M} \models \Gamma$, also $\mathcal{M} \models \varphi$. We read this as Γ *logically implies* φ, or Γ *semantically implies* φ, or that φ *is a semantic consequence of* Γ.

With the syntax and semantics in place, we turn to the proof system for $\mathcal{A}$. A *proof tree over a set* Γ is a finite tree $\mathcal{T}$ whose nodes are labeled with sentences, and each node is either an element of Γ, or comes from its parent(s) by an application of one of the two rules below:

$$\frac{}{\text{All } p \text{ are } p} \qquad \frac{\text{All } p \text{ are } n \quad \text{All } n \text{ are } q}{\text{All } p \text{ are } q}$$

$\Gamma \vdash \varphi$ means that there is a proof tree $\mathcal{T}$ over Γ whose root is labeled φ.

The soundness of the system is trivial, and so we shall prove the completeness. Let Γ be a set of sentences in the fragment. Define $u \leq_\Gamma v$ to mean that

$$\Gamma \vdash \textit{All u are v}. \tag{26.3}$$

The rules of the logic imply that $\leq_\Gamma$ is always a preorder: reflexive and transitive.

Suppose that $\Gamma \models \text{All } p \text{ are } q$. Consider the model $\mathcal{M}$ whose universe is the set **P** of nouns in the fragment, and with

$$[\![u]\!] = \{v : v \leq u\}.$$

It is easy to check that with this definition, all sentences in Γ hold in the model. Returning to our overall task, we now see that $[\![p]\!] \subseteq [\![q]\!]$ in our model. But $p \in [\![p]\!]$,

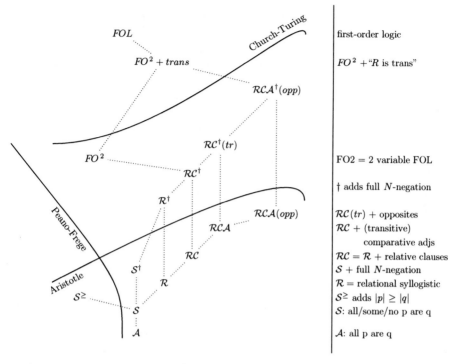

Fig. 26.1 Some systems of natural logic

via a one-point tree. So $p \in [\![q]\!]$ as well. This means that $p \leq q$, and that means that $\Gamma \vdash$ **All** p **are** q, as desired.

26.2.1 More on Syllogistic Logics

Quite a bit more is now known about syllogistic logics, and at this point there is the makings of a *landscape* of such logics.

This section presents logical systems modeled on syllogistic inference that attempt to model aspects of natural language inference. Most of these are listed in the chart in Fig. 26.1, but to keep the chart readable I have left off several of them.

The chart mostly consists of logical languages, each with an intended semantics. For example, *FOL* at the top is first-order logic. (Very short explanations of the systems appear to the right of the vertical line.) In the chart, the lines going down mean "subsystem," and sometimes the intention is that we allow a translation preserving the intended semantics. So all of the systems in the chart may be regarded as subsystems of *FOL* with the exception of $S^{\geq}$.

The smallest system in the chart is $\mathcal{A}$, a system even smaller than the classical syllogistic. Its only sentences are All p are q where p and q are variables. It is our intent that these variables be interpreted as plural common nouns of English, and that a model $\mathcal{M}$ be an arbitrary set M together with interpretations of the variables by subsets of M.

The next smallest system in the chart is $\mathcal{S}$, a system still even smaller than the classical syllogistic. It adds sentences Some p are q to $\mathcal{A}$.

Moving up, $\mathcal{S}^{\geq}$ adds additional sentences of the form there are at least as many p as q. The additions are not expressible in FOL, and we indicate this by setting $\mathcal{S}^{\geq}$ on the "outside" of the "Peano–Frege Boundary."

The language $\mathcal{S}^{\dagger}$ adds full negation on nouns to $\mathcal{S}$. For example, one can say All p are $\overline{q}$ with the intended reading "no p are q." One can also say All $\overline{p}$ are $\overline{q}$, and this goes beyond what is usually done in the syllogistic logic. The use of the $\dagger$ notation will be maintained throughout these notes as a mark of systems which contain complete noun-level negation.

Moving up the chart we find the systems $\mathcal{R}$, $\mathcal{RC}$, $\mathcal{R}^{\dagger}$, and $\mathcal{RC}^{\dagger}$. The system $\mathcal{R}$ extends $\mathcal{S}$ by adding *verbs*, interpreted as arbitrary relations. (The '$\mathcal{R}$' stands for 'relation.') This system and the others in this paragraph originate in [9]. So $\mathcal{R}$ would contain Some dogs chase no cats, and the yet larger system $\mathcal{RC}$ would contain *relative clauses* as exemplified in All who love all animals love all cats. The languages with the dagger such as $\mathcal{S}^{\dagger}$ and $\mathcal{R}^{\dagger}$ are further enrichments which allow subject nouns to be negated. This is rather un-natural in standard speech, but it would be exemplified in sentences like Every non-dog runs. The point: the dagger fragments are beyond the Aristotle boundary in the sense that they cannot be treated by the relatively simpler syllogistic logics. The only known logical systems for them use *variables* in a key way. The line marked "Aristotle" separates the logical systems above the line, systems which can be profitably studied on their own terms without devices like variables over individuals, from those which cannot. The chart continues with the addition of *comparative adjective phrases* with the systems $\mathcal{RC}(tr)$ and $\mathcal{RC}^{\dagger}(tr)$.

As a methodological point, we greatly prefer logical systems which are decidable, and in fact we would like to clarify the relationship of deduction in language and computational complexity theory. There are many decidable fragments of first-order logic, and even of second-order logic, but sadly these do not seem to be of very great relevance for representing linguistic inference. However, as the chart shows, there are many decidable logics for natural language inference, both of the purely syllogistic variety and those which also use variables.

26.3 Prolegomena to Dynamic Semantics

Dynamic semantics is the central thrust of this book, so surely it will be introduced in many of the individual articles. Nevertheless, I would like to offer an introduction that I think is (a) one of the simplest possible, and thus may be useful for didactic purposes; and (b) connected to an issue in natural language semantics, the difference between *and* and *and then*.

Ronnie is a rat[1] who is very interested in what he sees around him. His owners own a company that manufactures two-chambered rat houses. Each day, they put him in a different house. He runs back and forth all day long. At any given time, Ronnie is very interested in what he finds in the room he currently is in, and also what he sees in the room he is *not* in. For example, here is one of the houses.

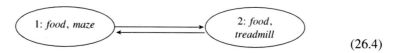

$$(26.4)$$

In room 1, there is food and a maze (and no treadmill); in room 2, there is food and a treadmill (and no maze).

He's a good reasoner, and so he needs a logical language in which to make assertions. The atomic sentences of this language should correspond to the presence of objects familiar to Ronnie. We'll write these atomic sentences as *water*, *food*, *maze*, etc.

Ronnie himself does not have access to the room numbers in the various houses. They are written on the outside, and he is stuck indoors. All that he can do is talk (in a particular room) about what is in the current room, what is in the other room, and more complicated variations on these assertions. By "more complicated variations" we mean combinations using the standard sentential connectives $\neg, \wedge, \vee, \rightarrow$, and $\leftrightarrow$, and (crucially) a modal operator $*$ that can be read as "in the other room." We want the language to work as illustrated below:

English	$\mathcal{L}(*)$
there is water in the room	*water*
there is food in the other room, and a maze in this room	$(*food) \wedge maze$
after switching rooms twice, the room has a maze	$**maze$

Here is how we can formalize this sort of reasoning. We add a new operator $*$ to basic propositional logic, and we read this operator as "switch". The resulting *static switching language* $\mathcal{L}(*)$ has complex sentences in it like

$$*(treadmill \rightarrow *\neg maze).$$

This would say that in the other room from where he currently is, if there is a treadmill, then in the other room (from there), there is no maze.

Models

A model for $\mathcal{L}(*)$ is an ordered pair $M = (R_1, R_2)$ of sets of atomic sentences. We'll draw them as graphs the way we have been doing. So what we drew as (26.4) above

---

[1] This material is adapted from [7].

is technically the ordered pair

$$(\{food, maze\}, \{food, treadmill\}).$$

The picture and the formal representation have the same content: in room 1, there is food and a maze (and no treadmill); in room 2, there is food and a treadmill (and no maze).

For the most part, the semantics is classical: for atomic p, $i \Vdash p$ iff $p \in i$ (remember that technically the rooms of houses are sets of atomic propositions). The exception is the $*$ operator. It works as follows:

$$1 \Vdash *\varphi \quad \text{iff } 2 \Vdash \varphi \qquad\qquad\qquad (26.5)$$
$$2 \Vdash *\varphi \quad \text{iff } 1 \Vdash \varphi.$$

Now the main semantic notions for this logic are as follows:

φ is satisfiable means φ is true in some room of some house

$\Gamma \models_G \varphi$ means for all houses M,
                       if all sentences in Γ are true in both rooms of M,
                       then φ is also true in both rooms

$\Gamma \models_L \varphi$ means for all houses M,
                       if all sentences in Γ are true in a given room i of M,
                       then φ is also true in this same room i

In modal terms, we are studying Kripke models whose accessibility relation is a transposition.

Remark 26.1 Here is an important point about our semantics of the binary connectives $\wedge$, $\vee$, $\rightarrow$, and $\leftrightarrow$ in in this logic. We consider some particular model, call it $\mathcal{M}$. Let us consider an example of the semantics, say $1 \Vdash (*p) \wedge q$. To evaluate this, we check whether this sentence is true in $\mathcal{M}$ or not, we check to see if $1 \Vdash *p$ and also if $1 \Vdash q$. It is tempting to think that after we see whether $1 \Vdash *p$, we are in room 2 rather than room 1. So on this view, we should ask about $1 \Vdash *p$ and $2 \Vdash q$. Indeed, this reflects the intuitions about moving between the rooms of a rat house with which we began this section. However natural this is, this is nevertheless not how the semantics actually works. The point of our later discussion is to reformulate the semantics to reflect this more "dynamic" process of evaluation.

To get an idea for what is going on with the two different semantic notions, note that $p \models_G *p$ but $p \not\models_L *p$. That is if we assume that p is true in both rooms of a given house, then $*p$ will also be true in both rooms. This verifies that $p \models_G *p$. To see that $p \not\models_L *p$, consider a house where p is true in exactly one room.

It is not hard to check the following facts. First, the notions of $\models_L \varphi$ and $\models_G \varphi$ coincide. Second, consider the axiom system presented below. It defines a

Hilbert-style proof system.

> **Axioms** All tautologies
> >   (Involution) $\varphi \rightarrow {**}\varphi$
> >   (Determinacy) ${*}\neg\varphi \leftrightarrow \neg{*}\varphi$
> >   (Normality) ${*}(\varphi \rightarrow \psi) \rightarrow ({*}\varphi \rightarrow {*}\psi)$

> **Rules** (Modus Ponens) From φ and $\varphi \rightarrow \psi$, infer ψ
> >   (Switch Rule)   From φ, infer ${*}\varphi$.

When hypotheses are used, we have

$$\varphi_1, \ldots, \varphi_n \models_G \psi \quad \text{iff} \quad \varphi_1, \ldots, \varphi_n \vdash \psi.$$

The basic idea is that in the proof system, we can convert every sentence φ into a disjunction of *state descriptions*: conjunctions of the form

$$\pm p_1 \wedge \cdots \wedge \pm p_n \wedge {*}{\pm} p_1 \wedge \cdots \wedge {*}{\pm} p_n$$

where each $\pm$ is either nothing or a negation sign, and $p_1, \ldots, p_n$ are the atomic propositions in φ. It is also possible to formulate a *natural deduction* proof system $\vdash_{ND}$ for this logic, using specialized introduction and elimination rules for reasoning about the other room. When one does this, we get a correspondence with the local notion of validity:

$$\varphi_1, \ldots, \varphi_n \models_L \psi \quad \text{iff} \quad \varphi_1, \ldots, \varphi_n \vdash_{ND} \psi.$$

26.3.1 Dynamic Semantics of "and Then"

So far, the point of this first part of our étude is that it gives what appears to be the simplest modal setting, so simple that the complete analysis of the logic is a minor variation on propositional logic, even simpler perhaps than what goes on with modal logics like S5. But we are not so interested in this chapter with the further study of $\mathcal{L}({*})$. Instead, we wish to reconsider our semantics of the conjunction, adding a dynamic reading. So this would be appropriate for symbolizing the English "and then" operation. We'll write this using the semicolon ;. We don't want φ ; ψ to be equivalent to φ ; ψ, since the former says "from where we start, φ is true, and from wherever we end up in verifying that assertion, ψ will then be true." However, we have an important architectural change to the semantics of our logic. Instead of defining a relation $i \models \varphi$, we need a relation of beginning and ending,

$$i[\varphi]j$$

meaning "if we start in room i, we can verify φ and we would end up in room j." The main clauses here are

$$
\begin{array}{ll}
i[p]j & \text{if } i = j \text{ and } p \in i \\
i[*\varphi]j & \text{if } (3 - i)[\varphi]j \\
i[\varphi \; ; \psi]j & \text{if for some } k, i[\varphi]k \text{ and then } k[\psi]j \\
i[\varphi \wedge \psi]j & \text{if } i[\varphi]j \text{ and } i[\psi]j.
\end{array}
\tag{26.6}
$$

For example, in our model from (26.4)

$$
1[*treadmill \; ; \neg maze]2.
$$

Indeed, we get a few equivalences:

$$
\begin{array}{ll}
\varphi \; ; (\psi \; ; \chi) & \leftrightarrow (\varphi \; ; \psi) \; ; \chi \\
*(\varphi \; ; \psi) & \leftrightarrow *\varphi \; ; \psi \\
p \; ; \varphi & \leftrightarrow p \wedge \varphi \\
*p \; ; \varphi & \leftrightarrow *(p \wedge \varphi) \\
(\varphi \vee \psi) \; ; \chi & \leftrightarrow (\varphi \; ; \chi) \vee (\psi \; ; \chi) \\
\varphi \; ; (\psi \vee \chi) & \leftrightarrow (\varphi \; ; \psi) \vee (\varphi \; ; \chi).
\end{array}
$$

The third and fourth properties are just for atomic p, or more generally for sentences with an even $*$-height. (It should also be noted that this semantics is *deterministic*: if $i[\varphi]j$ and $i[\varphi]k$, then $j = k$.)

With a little more work, one can see that the laws above are complete: all sentences may be written without ; using the laws, and so these laws together with our earlier ones are a complete set for the fragment.

One could and should go further, all the way to dynamic logic itself. In that setting, we would have not only sentences but also *programs*. The programs in this fragment would be $*$, the test programs $?\varphi$ for sentences φ, and programs obtained from these by composition. The sentences are then of the form $[\pi]\varphi$, where π is a program. It is possible to motivate dynamic logic from this simple example. The point for us is that modeling the expression *and then* is a way into all of this material.

26.4 Monotonicity and Polarity Explained by an Example from Algebra

Phenomena related to monotonicity and polarity in language have their subtleties to be sure, but they are also easy to explain via examples. Consider the following sentences.

1. Some bears[↑] dance[↑].
2. No bears[↓] dance[↓].

3. Most bears dance$^\uparrow$.
4. Not every bear$^\uparrow$ dances$^\downarrow$.
5. John sees every bear$^\downarrow$.
6. Mary sees no bear$^\downarrow$.
7. Any bear$^\downarrow$ in Hawaii would prefer to move to Alaska.
8. If any bear$^\downarrow$ dances, Harry will be happy.
9. If you play loud enough music, any bear$^\downarrow$ will start to dance.
10. Doreen didn't see any bears$^\downarrow$ in her dorm room.

Here is an explanation of the arrow notations. Suppose we are in a situation in which one of these sentences is true, and the arrow points up. If we replace that word by a "bigger" word, the sentence again is true (in the same situation). Similarly, replacing a word marked with a downward arrow with a "smaller" word again preserves truth. What we mean by "bigger" and "smaller" for nouns can be explained by subsets; for example brown bear ≤ bear ≤ animal. We shall call these $^\uparrow$ and $^\downarrow$ notations *polarity indicators*. One might think from many examples that every noun occurrence in a sentence has one of the two polarities ↑ or ↓, sentence (3) above shows that this is not to be.

It is clear from our examples that the words in a sentence have something to do with these polarity indicators. But the polarity indicators are not properties of words alone; they're properties of *words in a context*.

Furthermore, this phenomenon is not particular to nouns in the first place. One can do the same thing with other categories: from every dog$^\downarrow$ barks$^\uparrow$ and barks ≤ vociferates, we conclude that every dog$^\downarrow$ vociferate$^\uparrow$. From two ≤ three, we conclude that at most two ≤ at most three, and also that at least three ≤ at least two (notice the difference). And then we want to mark polarities on determiners, as in:

> (at least three)$^\uparrow$ people in town have (at most four)$^\uparrow$ cousins
> if (at least two)$^\downarrow$ people read this example, I will be happy.

The last issue we wish to bring up has to do with the word any. Looking back at examples (7–10) above, we can see that as a rule, any means every when both have the polarity ↑, and any means some when both have the polarity ↓. (Sadly, there is more to say on any, but for this chapter, we shall operate under these assumptions.)

The basic issue in van Benthem's program of natural logic was to give an account of how all of this works: what is the relation of monotonicity and polarity? How can we determine the polarity of words and sub-expressions inside a sentence? How can we devise a logical system that takes monotonicity and polarity into account?

We shall return to a general discussion on these matters after we make a digression into algebra, where things are simpler. But as a transition, let me quote Bernardi's explanation of the matter:

The differences between monotonicity and polarity could be summarized in a few words by saying that monotonicity is a property of functions .... On the other hand, polarity is a static syntactic notion which can be computed for all positions in a given formula. This connection between the semantic notion of monotonicity and the syntactic one of polarity is what one needs to reach a proof theoretical account of natural reasoning and build a natural logic [4].

26.4.1 Monotonicity in Elementary Mathematics

This issue of monotonicity in language is simplified somewhat when we move to a simpler setting, *monotonicity in elementary mathematics*. Consider a function on the reals such as

$$f(v, w, x, y, z) = \frac{x - y}{2^{z-(v+w)}}. \tag{26.7}$$

Suppose we fix numerical values for the variables v, w, x, y, z, and then suppose we also increase x a bit to some number $x' \geq x$. Does the value of f go up or down? That is, which is true:

$$f(v, w, x, y, z) \leq f(v, w, x', y, z) \quad \text{or} \quad f(v, w, x', y, z) \leq f(v, w, x, y, z)$$

A moment's thought shows that the first option is the correct one.

Next, suppose we fix values but this time move y up to some number y'. This time, the value goes down. We also can study z, v, and w the same way. We could summarize all of the observations by writing

$$f(v^\uparrow, w^\uparrow, x^\uparrow, y^\downarrow, z^\downarrow).$$

The arrows are indicators of *polarity*.

The example in (26.7) oversimplifies things because each variable occurs only once in the expression on the right. Really we would like to think about functions like $g(x, y) = (x - y)/2^x$. And then we would want to decorate the different occurrences of variables with polarity indicators: $g(x, y) = (x^\uparrow - y^\downarrow)/2^{x^\downarrow}$.

At this point, we turn the expressions of algebra into a formal language, and we use a syntax with *higher-type* operations. That is, we start with a single base type, r (for *reals*). We take our variables v, w, x, y, z to have type r. And then we have *typed constants*

$$\begin{array}{ll} plus : r \to (r \to r) & minus : r \to (r \to r) \\ times : r \to (r \to r) & div2 : r \to (r \to r). \end{array} \tag{26.8}$$

Please note that we have curried the syntax: instead of two-place functions of type $r \times r \to r$, we have functions from type r to functions of type $r \to r$. Also, *div2* is *not* supposed to be the usual division. The idea is that $div2(x)(y)$ should be $x \div 2^y$, not x/y. This complication is to make everything monotone, and I hope it's not too confusing.

We get terms in *Polish notation*:

$$\frac{\dfrac{minus : r \to (r \to r) \quad x : r}{minus\ x : r \to r} \quad y : r}{minus\ x\ y : r} \ .$$

To fit trees like this on a page, let's drop the types. The function f involved in our basic example of (26.7) is then typed as

$$\frac{\dfrac{\dfrac{minus\quad x}{minus\ x \quad y}}{div2 \quad minus\ x\ y} \quad \dfrac{minus\ z \quad \dfrac{plus\quad v}{plus\ v \quad w}}{minus\ z \quad plus\ v\ w}}{\dfrac{div2\ minus\ x\ y \qquad minus\ z\ plus\ v\ w}{div2\ minus\ x\ y\ minus\ z\ plus\ v\ w}} \ .$$

Can we determine the polarities of the variables from the tree? The first algorithm to do this is due to van Benthem; it was then explored in Sanchez.

We would like to present this using two colors. Instead, we do it using explicit down-arrows ↓ for the negative nodes. We won't show up-arrows ↑ in order to simplify the notation. So we go from the root to the leaves, either leaving nodes unmarked, or marking them ↓. The overall root is unmarked. (However, to understand the overall algorithm, we'll need to understand what would happen if we started with the root marked ↓.) The rule for propagating the ↓ notation is that we maintain arrows up the tree, except that the right children of nodes marked *minus* and *div2* flip arrows. The end result would be

$$\frac{\dfrac{\dfrac{minus\quad x}{minus\ x \quad y^{\downarrow}}}{div2 \quad minus\ x\ y} \quad \dfrac{(minus)^{\downarrow} \quad z^{\downarrow} \quad \dfrac{plus\quad v}{plus\ v \quad w}}{(minus\ z)^{\downarrow} \quad plus\ v\ w}}{\dfrac{div2\ minus\ x\ y \qquad (minus\ z\ plus\ v\ w)^{\downarrow}}{div2\ minus\ x\ y\ minus\ z\ plus\ v\ w}} \ . \qquad (26.9)$$

This agrees with what we saw before:

$$f(v^{\uparrow}, w^{\uparrow}, x^{\uparrow}, y^{\downarrow}, z^{\downarrow}).$$

Let us try to understand why this is the case. For each of our types σ, we have a *semantic space* $[\![\sigma]\!]$. The definition is by recursion on σ: $[\![r]\!] = \mathbb{R}$ (the real numbers), and

$$[\![\sigma \to \tau]\!] = [\![\sigma]\!] \to [\![\tau]\!] \qquad (26.10)$$

(the set of all functions from $[\![\sigma]\!]$ to $[\![\tau]\!]$). Each of our *constants* gets a semantics, and we have

$$[\![plus]\!] \in [\![r \to (r \to r)]\!],$$

and similarly for the other operators. Every term with variables gets a semantics

$$[\![t(v_1, \ldots, v_n)]\!] \in \overbrace{([\![r]\!] \times [\![r]\!] \times \cdots [\![r]\!])}^{n} \to [\![r]\!].$$

Lemma 26.1 (van Benthem) *Let* $t(v_1, \ldots, v_n)$ *be a term in which each variable occurs exactly once. Suppose we run the algorithm starting with the root colored with one of the two colors. Then* $[\![t(v_1, \ldots, v_n)]\!]$ *is a monotone function of the leaves with the same color, and an antitone function of the leaves with the opposite color.*

The proof is by induction on the term $t(v_1, \ldots, v_n)$.

However, this makes the determination of polarities an *external* feature of the syntax tree, something determined by an algorithm. We might like to have it work "in situ." Instead of using an algorithm, we'll take polarity-marked function symbols to be the lexical items ("letters") in a categorial grammar; later, we'll do the same thing with polarity-marked words of English.

To see what this means, we'll complicate our type system a little. Instead of having sets as our semantic spaces $[\![\tau]\!]$, we add some structure and take *preorders*: sets together with a reflexive and transitive order. We use letters like $\mathbb{P}$ and $\mathbb{Q}$ for preorders. As opposed to (26.10), we'll take $[\![\sigma \to \tau]\!]$ to be the set of *monotone* (order-preserving) functions from $[\![\sigma]\!]$ to $[\![\tau]\!]$. This set is itself a preorder, using the pointwise order:

$$f \le g \text{ in } [\![\sigma \to \tau]\!] \quad \text{iff} \quad \text{for all } x \in [\![\sigma]\!], f(x) \le g(x) \text{ in } [\![\tau]\!]. \qquad (26.11)$$

Finally, in addition to forming new types by using function spaces, we'll also use *opposite*. For any preorder $\mathbb{P}$, $-\mathbb{P}$ is the preorder with the same points as $\mathbb{P}$ but with $x \le y$ in $-\mathbb{P}$ iff $y \le x$ in $\mathbb{P}$.

This means that we now have $[\![r]\!] = (\mathbb{R}, \le)$, $[\![-r]\!] = -[\![r]\!] = (\mathbb{R}, \ge)$, $[\![r \to r]\!]$ = the monotone functions from $(\mathbb{R}, \le)$ to itself; $[\![r \to -r]\!]$ = the antitone functions from $(\mathbb{R}, \le)$ to itself; $[\![-r \to -r]\!]$ = the antitone functions from $(\mathbb{R}, \le)$ to itself (again, but with the opposite order). And now we revise our declarations in (26.8), as follows:

$$
\begin{array}{ll}
v^+ : r & v^- : -r \\
\vdots & \vdots \\
z^+ : r & z^- : -r \\
plus^+ : r \to (r \to r) & plus^- : -r \to (-r \to -r) \\
minus^+ : r \to (-r \to r) & minus^- : -r \to (r \to -r) \\
times^+ : r \to (r \to r) & times^- : -r \to (-r \to -r) \\
div2^+ : r \to (-r \to r) & div2^- : -r \to (r \to -r).
\end{array}
$$

We interpret them in the obvious way: for example,

$$[\![minus^+]\!](x)(y) = x - y = [\![minus^-]\!](x)(y).$$

Probably the important ones to look at are the typings of $minus^+$ and $minus^-$; our remarks hold also for $div2^+$ and $div2^-$, mutatis mutandis. For $minus^+$, note that for fixed x, $[\![minus^+]\!](x)$ is an antitone function from $\mathbb{R}$ to itself: if $y \leq y'$, then $x - y' \leq x - y$. Thus each $[\![minus^+]\!](x)$ is an element of the preorder $[\![-r \to r]\!]$. And if $x \leq x'$, then $[\![minus^+]\!](x) \leq [\![minus^+]\!](x')$ in $[\![-r \to r]\!]$. To see this, note that for $y \in [\![-r]\!]$, $x - y \leq x' - y$. The same considerations apply to $minus^-$.

At this point, we can go back and build terms out of our new syntax. For example:

$$\cfrac{div2^+ \quad \cfrac{\cfrac{minus^+ \quad x^+}{\cfrac{minus^+ x^+ \quad y^-}{minus^+ x^+ y^-}}}{div2^+ \; minus^+ \, x^+ \, y^-} \quad \cfrac{minus^- \quad z^-}{minus^- z^-} \quad \cfrac{\cfrac{plus^+ \quad v^+}{\cfrac{plus^+ v^+ \quad w^+}{plus^+ v^+ w^+}}}{minus^- z^- \, plus^+ v^+ w^+}}{div2^+ \; minus^+ x^+ y^- \, minus^- z^- \, plus^+ v^+ w^+} \tag{26.12}$$

Compare this with (26.9). I have omitted the types, but the important one is the type at the root, which is r. In fact, the only way to derive

$$div2 \; minus \; x \; y \; minus \; z \; plus \; v \; w : r$$

with any polarity indicators is the one in (26.12). The difference between our work in (26.9) and (26.12) is that the second presentation didn't depend on an algorithm, the polarities appear on the lexical items (the function symbols and variables).

What is missing at this point is a soundness result like Lemma 26.1. We'll get this after we revisit the system in the linguistic direction.

26.4.2 Higher-Order Terms Over Preorders and the Context Lemma

After our detour into monotonicity in algebra, we reconsider the project of grafting monotonicity and polarity information into the sentences derived from categorial grammars. The work below is self-contained, but it would be best appreciated by someone who is familiar with the basic architecture of categorial grammar; see van Benthem [2], for example.

Fix a set T_0 of *basic types*. Let T_1 be the smallest superset of T_0 closed in the following way:

1. If $\sigma, \tau \in T_1$, then also $(\sigma, \tau) \in T_1$.
2. If $\sigma \in T_1$, then also $-\sigma \in T_1$.

Let $\equiv$ be the smallest equivalence relation on T_1 such that the following hold:

1. $-(-\sigma) \equiv \sigma$.
2. $-(\sigma, \tau) \equiv (-\sigma, -\tau)$.

3. If $\sigma \equiv \sigma'$, then also $-\sigma \equiv -\sigma'$.
4. If $\sigma \equiv \sigma'$ and $\tau \equiv \tau'$, then $(\sigma, \tau) \equiv (\sigma', \tau')$.

Definition 26.1 $\mathcal{T} = \mathcal{T}_1/\equiv$. This is the set of *types over* $\mathcal{T}_0$.

The operations $\sigma \mapsto -\sigma$ and $\sigma, \tau \mapsto (\sigma, \tau)$ are well-defined on $\mathcal{T}$. We always use letters like σ and τ to denote elements of $\mathcal{T}$, as opposed to writing $[\sigma]$ and $[\tau]$. That is, we simply work with the elements of $\mathcal{T}_1$, but identify equivalent types.

Definition 26.2 Let $\mathcal{T}_0$ be a set of basic types. A *typed language over* $\mathcal{T}_0$ is a collection of *typed variables* $v : \sigma$ and *typed constants* $c : \sigma$, where σ in each of these is an element of $\mathcal{T}$. We generally assume that the set of typed variables includes infinitely many of each type. But there might be no constants whatsoever. We use $\mathcal{L}$ to denote a typed language in this sense.

Let $\mathcal{L}$ be a typed language. We form *typed terms* $t : \sigma$ as follows:

1. If $v : \sigma$ (as a typed variable), then $v : \sigma$ (as a typed term).
2. If $c : \sigma$ (as a typed constant), then $c : \sigma$ (as a typed term).
3. If $t : (\sigma, \tau)$ and $u : \sigma$, then $t(u) : \tau$.

Frequently we do not display the types of our terms.

Monotone functions on preorders, again

If $\mathbb{P}$ and $\mathbb{Q}$ are preorders, we let $[\mathbb{P}, \mathbb{Q}]$ be the preorder of all monotone functions from $\mathbb{P}$ to $\mathbb{Q}$. For a set X, we write $\mathbb{X}$ for the *flat preorder* on X: $x \leq y$ iff $x = y$. We also write $\mathbb{P} \times \mathbb{Q}$ for the obvious product.

Proposition 26.1 *For all preorders $\mathbb{P}$, $\mathbb{Q}$, and $\mathbb{R}$, and all sets X:*

1. *For each $p \in P$, the function $app_p : [\mathbb{P}, \mathbb{Q}] \to \mathbb{Q}$ given by $app_p(f) = f(p)$ is an element of $[[\mathbb{P}, \mathbb{Q}], \mathbb{Q}]$.*
2. $[\mathbb{P} \times \mathbb{Q}, \mathbb{R}] \cong [\mathbb{P}, [\mathbb{Q}, \mathbb{R}]]$.
3. $[\mathbb{X}, 2] \cong \mathcal{P}(X)$.
4. $-(-\mathbb{P}) = \mathbb{P}$.
5. $[\mathbb{P}, -\mathbb{Q}] = -[-\mathbb{P}, \mathbb{Q}]$.
6. $[-\mathbb{P}, -\mathbb{Q}] = -[\mathbb{P}, \mathbb{Q}]$.
7. *If $f : \mathbb{P} \to \mathbb{Q}$ and $g : \mathbb{Q} \to \mathbb{R}$, then $g \circ f : \mathbb{P} \to \mathbb{R}$.*
8. *If $f : \mathbb{P} \to -\mathbb{Q}$ and $g : \mathbb{Q} \to -\mathbb{R}$, then $g \circ f : \mathbb{P} \to \mathbb{R}$.*
9. $-(\mathbb{P} \times \mathbb{Q}) \cong -\mathbb{P} \times -\mathbb{Q}$.
10. $\mathbb{X} \cong -\mathbb{X}$.
11. $-[\mathbb{X}, \mathbb{P}] \cong [\mathbb{X}, -\mathbb{P}]$.

Semantics

For the semantics of our higher-order language $\mathcal{L}$ we use *models* $\mathcal{M}$ of the following form. $\mathcal{M}$ consists of an assignment of preorders $\sigma \mapsto [\![\sigma]\!]$ on $\mathcal{T}_0$, together with some data which we shall mention shortly. Before this, extend the assignment

$\sigma \mapsto [\![\sigma]\!]$ to $\mathcal{T}_1$ by

$$[\![(\sigma, \tau)]\!] = [\![\,[\![\sigma]\!], [\![\tau]\!]\,]\!]$$
$$[\![-\sigma]\!] = -[\![\sigma]\!].$$

An easy induction shows that if $\sigma \equiv \tau$, then $[\![\sigma]\!] = [\![\tau]\!]$. So we have $[\![\mathbb{P}]\!]_\sigma$ for $\sigma \in \mathcal{T}$. We use $\mathbb{P}_\sigma$ to denote the set underlying the preorder $[\![\sigma]\!]$.

The rest of the structure of a model $\mathcal{M}$ consists of an assignment $[\![c]\!] \in P_\sigma$ for each constant $c : \sigma$, and also a *typed map* f; this is just a map which to any typed variable $v : \sigma$ gives some $f(v) \in P_\sigma$.

Ground Terms and Contexts

A *ground term* is a term with no free variables. Each ground term $t : \sigma$ has a *denotation* $[\![t]\!] \in P_\sigma$ defined in the obvious way:

$$[\![c]\!] = \text{is given at the outset for constants } c : \sigma$$
$$[\![t(u)]\!] = [\![t]\!]([\![u]\!]).$$

A *context* is a typed term with exactly one variable, x. (This variable may be of any type.) We write t for a context. We'll be interested in contexts of the form $t(u)$. Note that if $t(u)$ is a context and if x appears in u, then t is a ground term; and vice-versa.

In the definition below, we remind you that subterms are not necessarily proper. That, is a variable x is a subterm of itself.

Definition 26.3 Fix a model $\mathcal{M}$ for $\mathcal{L}$. Let $x : \rho$, and let $t : \sigma$ be a context. We associate to t a set function

$$f_t : P_\rho \to P_\sigma$$

in the following way:

1. If $t = x$, so that $\sigma = \rho$, then $f_x : P_\sigma \to P_\sigma$ is the identity.
2. If t is $u(v)$ with $u : (\tau, \sigma)$ and $v : \tau$, and if x is a subterm of u, then f_t is $app_{[\![v]\!]} \circ f_u$. That is, $f_{t(u)}$ is

$$a \in P_\rho \mapsto f_u(a)([\![v]\!]).$$

3. If t is $u(v)$ with $u : (\tau, \sigma)$ and $v : \tau$, and if x is a subterm of v, then f_t is $[\![u]\!] \circ f_v$. That is, f_t is

$$a \in P_\rho \mapsto [\![u]\!](f_v(a)).$$

The idea of f_t is that as a ranges over its interpretation space P_ρ, $f_t(a)$ would be the result of substituting various values of this space in for the variable, and then evaluating the result.

Notice that we defined f_t as a set function and wrote $f_t : P_\rho \to P_\sigma$; it is not immediately clear that f_t is monotone. This is the content of the following result.

Lemma 26.2 (Context Lemma) *Let t be a context, where $t : \sigma$ and $x : \rho$. Then f_t is an element of $[\![\,[\![\rho]\!],\,[\![\sigma]\!]\,]\!]$.*

This Context Lemma is the main technical result on this system. It is a reformulation of van Benthem's result, Lemma 1. We'll see some examples in the next section.

The Context Lemma also allows one to generalize our work from the applicative (AB) grammars to the setting of categorial grammar (CG) using the Lambek Calculus. In detail, one generalizes the notion of a context to one which allows more than one free variable but requires that all variables which occur free have only one free occurrence. Suppose that $x_1 : \rho_1, \ldots, x_n : \rho_n$ are the free variables in such a generalized context $t(x_1, \ldots, x_n) : \sigma$. Then t defines a set function $f_t : \prod_i P_{\rho_i} \to P_\sigma$. A generalization of the Context Lemma then shows that f_t is monotone as a function from the product preorder $\prod_i [\![\rho_i]\!]$. So functions given by lambda abstraction on each of the variables are also monotone, and this amounts to the soundness of the introduction rules of the Lambek Calculus in the internalized setting.

26.4.3 An Example Grammar

We present a small example to illustrate the ideas. First, we describe a language $\mathcal{L}$ corresponding to this vocabulary. We take our set $\mathcal{T}_0$ of basic types to be $\{t, pr\}$. (These stand for *truth value* and *property*. In more traditional presentations in the Montague grammar/categorial grammar tradition, the type pr might be (e, t), where e is a type of *entities*.)

Here are the constants of the language $\mathcal{L}$ and their types:

1. We have typed constants

$$
\begin{array}{ll}
\text{every}^\uparrow : (-pr, (pr, t)) & \text{every}^\downarrow : (pr, (-pr, -t)) \\
\text{some}^\uparrow : (pr, (pr, t)) & \text{some}^\downarrow : (-pr, (-pr, -t)) \\
\text{no}^\uparrow \quad : (-pr, (-pr, t)) & \text{no}^\downarrow \quad : (pr, (pr, -t)) \\
\text{any}^\uparrow \quad : (-pr, (pr, t)) & \text{any}^\downarrow \quad : (-pr, (-pr, -t)). \\
\text{if} \qquad : (-t, (t, t)) &
\end{array}
$$

2. We fix a set of *unary atoms* corresponding to some plural nouns and lexical verb phrases in English. For definiteness, we take cat, dog, animal, runs, and walks. Each unary atom p gives two typed constants: $\text{p}^\uparrow : pr$ and $\text{p}^\downarrow : -pr$.
3. We also fix a set of *binary atoms* corresponding to some transitive verbs in English. To be definite, we take chase, see. Every binary atom r gives four type constants:

$$r_1^\uparrow : ((pr, t), pr) \qquad\qquad r_2^\uparrow : ((-pr, t), pr)$$

$$r_1^\downarrow : ((-pr, -t), -pr) \qquad\qquad r_2^\downarrow : ((pr, -t), -pr).$$

This completes the definition of our typed language $\mathcal{L}$.

These notations $\uparrow$ and $\downarrow$ are mnemonic; we could do without them.

As always with categorial grammars, a *lexicon* is a set of pairs consisting of words in a natural language together with terms. We have been writing the words in the target language in *italics*, and then terms for them are written in sans serif. It is very important that the lexicon allows a given word to appear with many terms. As we have seen, we need every to appear with every$^\uparrow$ and every$^\downarrow$, for example. We still are only concerned with the syntax at this point, and the semantics will enter once we have seen some examples.

Examples of Typed Terms and Contexts

Here are a few examples of typed terms along with their derivations.

$$\frac{\dfrac{\text{every}^\uparrow : (-pr, (pr, t)) \quad \text{man}^\downarrow : -pr}{\text{every}^\uparrow\,\text{man}^\downarrow : (pr, t)} \qquad \text{walks}^\uparrow : pr}{\text{every}^\uparrow\,\text{man}^\downarrow\,\text{walks}^\uparrow : t}$$

$$\frac{\dfrac{\text{some}^\uparrow : (pr, (pr, t)) \quad \text{man}^\uparrow : pr}{\text{some}^\uparrow\,\text{man}^\uparrow : (pr, t)} \qquad \text{walks}^\uparrow : pr}{\text{some}^\uparrow\,\text{man}^\uparrow\,\text{walks}^\uparrow : t}$$

$$\frac{\dfrac{\text{no}^\uparrow : (-pr, (-pr, t)) \quad \text{man}^\downarrow : -pr}{\text{no}^\uparrow\,\text{man}^\downarrow : (-pr, t)} \qquad \text{walks}^\downarrow : -pr}{\text{no}^\uparrow\,\text{man}^\downarrow\,\text{walks}^\uparrow : t} \qquad .$$

We similarly have the following terms:

$$\text{some}^\uparrow(\text{dog}^\uparrow)(\text{chase}_1^\uparrow(\text{every}^\uparrow(\text{cat}^\downarrow))) : t$$

$$\text{some}^\uparrow(\text{dog}^\uparrow)(\text{chase}_2^\uparrow(\text{no}^\uparrow(\text{cat}^\downarrow))) : t$$

$$\text{no}^\uparrow(\text{dog}^\downarrow)(\text{chase}_2^\downarrow(\text{no}^\uparrow(\text{cat}^\uparrow))) : t$$

$$\text{no}^\uparrow(\text{dog}^\downarrow)(\text{chase}_1^\downarrow(\text{every}^\downarrow(\text{cat}^\uparrow))) : t.$$

All four different typings of the transitive verbs are needed to represent sentences of English.

Here is an example of a context: $\text{no}^\uparrow(x : -pr)(\text{chase}_1^\downarrow(\text{every}^\downarrow(\text{cat}^\uparrow))) : t$. So x is a variable of type $-pr$. In any model, this context gives a function from interpretations of type $-pr$ to those of type $-t$. The Context Lemma would tell us that this function is a monotone function. Turning things around, it would be an *antitone* function from interpretations of type $-pr$ to those of type $-t$.

If.

We have taken if to be of type $(-t, (t, t))$. We also have

$$\frac{\text{every}^{\downarrow} : (pr, (-pr, -t)) \quad \text{dog}^{\uparrow} : pr}{\dfrac{\text{every}^{\downarrow} \text{dog}^{\uparrow} : (-pr, -t) \quad \text{walks}^{\downarrow} : -pr}{\text{every}^{\downarrow} \text{dog}^{\uparrow} \text{walks}^{\downarrow} : -t}}$$

and then we get

$$\frac{\text{if} : (-t, (t, t)) \quad \text{every}^{\downarrow} \text{dog}^{\uparrow} \text{walks}^{\downarrow} : -t \quad \vdots}{\dfrac{\text{if} : (-t, (t, t)) \text{every}^{\downarrow} \text{dog}^{\uparrow} \text{walks}^{\downarrow} : (t, t) \quad \text{some}^{\uparrow} \text{cat}^{\uparrow} \text{runs}^{\uparrow} : t}{\text{if every}^{\downarrow} \text{dog}^{\uparrow} \text{walks}^{\downarrow}, \text{some}^{\uparrow} \text{cat}^{\uparrow} \text{runs}^{\uparrow} : t}} \tag{26.13}$$

We do not need a negative version if$^{\downarrow}$.

Any.

The current approach enables a treatment of any that has any mean the same thing as every when it has positive polarity, and the same thing as some when it has negative polarity. For example, here is a sentence intended to mean everything which sees any cat runs:

$$\frac{\text{every}^{\uparrow} : (-pr, (pr, t)) \quad \text{see}_2^{\downarrow}(\text{any}^{\downarrow}(\text{cat}^{\downarrow})) : -pr}{\dfrac{\text{every}^{\uparrow}(\text{see}_2^{\downarrow}(\text{any}^{\downarrow}(\text{cat}^{\downarrow}))) : (pr, t) \quad \text{runs}^{\uparrow} : pr}{\text{every}^{\uparrow}(\text{see}_2^{\downarrow}(\text{any}^{\downarrow}(\text{cat}^{\downarrow}))(\text{runs}^{\uparrow}) : t}} .$$

The natural reading is for any to have an existential reading. Another context for existential readings of *any* is in the antecedent of a conditional. In the other direction, consider

$$\text{any}^{\uparrow}(\text{cat}^{\downarrow})(\text{see}_1^{\downarrow}(\text{any}^{\uparrow}(\text{dog}^{\downarrow}))) : t.$$

The natural reading of both occurrences is universal. By the same token, in more involved syntactic contexts, our grammar is inevitably going to make the wrong predictions, since the actual facts about the meaning of any are not built in to this framework.

Standard models

Up until now, we have only given one example of a typed language $\mathcal{L}$. Now we describe a family of models for this language. The family is based on sets with interpretations of the unary and binary atoms. To make this precise, let us call a *pre-model* a structure $\mathcal{M}_0$ consisting of a set M together with subsets $[\![p]\!] \subseteq M$ for unary atoms p and relations $[\![r]\!] \subseteq M \times M$ for binary atoms r.

Every pre-model $\mathcal{M}_0$ now gives a bona fide model $\mathcal{M}$ of $\mathcal{L}$ in the following way. The underlying universe M gives a flat preorder $\mathbb{M}$. We take $[\![pr]\!] = [\mathbb{M}, 2] \cong \mathcal{P}(M)$. We also take $[\![t]\!] = 2$.

We interpret the typed constants $p^\uparrow : pr$ corresponding to unary atoms by

$$[\![p^\uparrow]\!](m) = \mathsf{T} \quad \text{iff} \quad m \in [\![p]\!].$$

(On the right we use the interpretation of p in the model $\mathcal{M}$.) Usually we write $p^\uparrow$ instead of $[\![p]\!]$.

The constants $p^\downarrow : -pr$ are interpreted by the same functions.

The interpretations of $\mathsf{every}^\uparrow$, $\mathsf{some}^\uparrow$, and $\mathsf{no}^\uparrow$ are given as follows (taking the set X to be the universe M of $\mathcal{M}$): Define

$$\mathsf{every} \in [-[X, 2], [[X, 2], 2]]$$
$$\mathsf{some} \in [[X, 2], [[X, 2], 2]]$$
$$\mathsf{no} \in [-[X, 2], [-[X, 2], 2]]$$

in the standard way:

$$\mathsf{every}(p)(q) = \begin{cases} \mathsf{T} \text{ if } p \leq q \\ \mathsf{F} \text{ otherwise} \end{cases}$$
$$\mathsf{some}(p)(q) = \neg\mathsf{every}(p)(\neg \circ q)$$
$$\mathsf{no}(p)(q) = \neg\mathsf{some}(p)(q).$$

It is routine to verify that these functions really belong to the sets mentioned above. Each of these functions belongs to the opposite preorder as well, and we therefore also have

$$\mathsf{every} \in [[X, 2], [-[X, 2], -2]]$$
$$\mathsf{some} \in [-[X, 2], [-[X, 2], -2]]$$
$$\mathsf{no} \in [[X, 2], [[X, 2], -2]].$$

And the interpretations of $\mathsf{every}^\downarrow$, $\mathsf{some}^\downarrow$, and $\mathsf{no}^\downarrow$ are the same functions.

We interpret $\mathsf{any}^\uparrow$ the same way as $\mathsf{every}^\uparrow$, and $\mathsf{any}^\downarrow$ the same way as $\mathsf{some}^\uparrow$.

Recall that the binary atom r gives four typed constants $r_1^\uparrow$, $r_1^\downarrow$, $r_2^\uparrow$, and $r_2^\downarrow$. These are all interpreted in models in the same way, by

$$\mathsf{r}_*(q)(m) = q(\{m' \in M : [\![r]\!](m, m')\}).$$

It is clear that for all $q \in [\![(pr, t)]\!] = [[\mathbb{M}, \mathbb{2}], \mathbb{2}]$, $\mathsf{r}_*(q)$ is a function from M to $\{\mathsf{T}, \mathsf{F}\}$. The monotonicity of this function is trivial, since $\mathbb{M}$ is flat.

Using the Logic

We now discuss how the system works. Suppose we take as background assertions like

$$\text{amble} \leq \text{walk} \leq \text{moves}$$
$$\text{poodle} \leq \text{dog} \leq \text{animal}.$$

These kinds of assertions are something one might glean from a source such as WordNet. In addition, we might appeal to real-world knowledge in some form and assert that

$$\text{at least three} \leq \text{at least two} \leq \text{some}.$$

Suppose we would like to carry out an inference like

$$\frac{\text{at least three dogs amble}}{\text{at least two animals move}}$$

We parse the first sentence in our grammar. The system is small enough that there would be only one parse, and it is

$$\frac{\dfrac{\text{at least three}^\uparrow : (pr, (pr, t)) \quad \text{dogs}^\uparrow : pr}{\text{at least three}^\uparrow \text{ dogs}^\uparrow : (pr, t)} \quad \text{amble}^\uparrow : pr}{\text{at least three}^\uparrow \text{ dogs}^\uparrow \text{ amble}^\uparrow : t}$$

So using our parse and the Context Lemma (three times), we infer

$$\text{at least two}^\uparrow \text{ animals}^\uparrow \text{ move}^\uparrow : t.$$

Finally, we drop the $\uparrow$ signs to get the desired conclusion.

For another example, suppose we wish to assume that if **every dog walks, some cat runs** and infer if **every dog ambles, some cat moves**. We have already seen a parse of the assumption in (26.13). We parse to infer polarities:

$$\text{if every}^\downarrow \text{ dog}^\uparrow \text{ walks}^\downarrow, \text{ some}^\uparrow \text{ cat}^\uparrow \text{ runs}^\uparrow : t.$$

Our desired inference would then follow from the Context Lemma.

Incidentally, formal reasoning about monotonicity is finding its way into computational systems for linguistic processing (cf. [6] and [8]).

For more on this topic, see Icard and Moss [5]. That chapter is the beginning of what we hope will be a sustained development, mixing the monotonicity calculus with the typed lambda calculus. There are many interesting results and questions which are (at this early time) open, but the topic deserves a different kind of treatment from what we are aiming for in this chapter.

26.5 Conclusion

My aim in these études is to show that logical dynamics involves a variety of techniques, that it really has something to say about language, and finally that much of it can be presented in very elementary terms.

I am grateful to Johan for his continuing inspiration on all of these matters, and for stimulating discussion on these études.

References

1. van Benthem J (1986) Essays in Logical Semantics. In: Studies in linguistics and philosophy, vol 29. D. Reidel Publishing Co., Dordrecht.
2. van Benthem J (1991), Language in Action. In: Studies in logic and the foundations of mathematics, vol 130. North-Holland, Amsterdam.
3. van Benthem J (2008) A brief history of natural logic. In: Mitra MN, Chakraborty M, Löwe B, Sarukkai S (eds) Logic, Navya-Nyaya and applications, homage to Bimal Krishna Matilal. College Publications, London
4. Bernardi R (2002) Reasoning with polarity in categorial type logic. PhD thesis, University of Utrecht.
5. Icard III TF, Moss LS (2014) Recent progress on monotonicity, *Linguistic Issues in Language Technology* 9(7).
6. MacCartney B, Manning CD (2009) An extended model of natural logic. In: Proceedings of the eighth international conference on computational semantics (IWCS-8), Tilburg, Netherlands.
7. Moss LS (2013) Invitation to possible worlds. Indiana University, Bloomington (Unpublished text for a modal logic course)
8. Nairn R, Condoravdi C, Karttunen L (2006) Computing relative polarity for textual inference. Proceedings of ICoS-5 (inference in computational semantics). Buxton, UK, pp 20–21
9. Pratt-Hartmann I, Moss LS (2009) Logics for the relational syllogistic. Review of Symbolic Logic 2(4):647–683
10. Sánchez-Valencia VM (1991) Studies on natural logic and categorial grammar. PhD thesis, University of Amsterdam.

Chapter 27
From Good to Better: Using Contextual Shifts to Define Preference in Terms of Monadic Value

Sven Ove Hansson and Fenrong Liu

Abstract It has usually been assumed that monadic value notions can be defined in terms of dyadic value notions, whereas definitions in the opposite direction are not possible. In this paper, inspired by van Benthem's work, it is shown that the latter direction is feasible with a method in which shifts in context have a crucial role. But although dyadic preference orderings can be defined from context-indexed monadic notions, the monadic notions cannot be regained from the preference relation that they gave rise to. Two formal languages are proposed in which reasoning about context can be represented in a fairly general way. One of these is a modal language much inspired by van Benthem's work. Throughout the paper the focus is on relationships among the value notions "good", "bad", and "better". Other interpretations like "tall" and "taller" are equally natural. It is hoped that the results of this paper can be relevant for the analysis of natural language comparatives and of vague predicates in general.

27.1 Introduction

The Chinese character "好 *hào*" (like, prefer) appears frequently in many ancient texts. For instance, in *The Analects of Confucius*, when attitudes and behavioural manners were being discussed, we see "好学 *hàoxué*" (fond of learning), "好德 *hàodé*" (love virtues), "好色 *hàosè*" (love beauty) and "好乐 *hàoyuè*" (love music), etc. Interestingly, if the character "好" is read as *hǎo*, with the third tone in Chinese *pinyin*, it simply means "good". These two parallel usages of the same character

S. O. Hansson (✉)
Division of Philosophy, Royal Institute of Technology, Stockholm, Sweden
e-mail: soh@kth.se

F. Liu
Department of Philosophy, Tsinghua University, Beijing, China
e-mail: fenrong@tsinghua.edu.cn

A. Baltag and S. Smets (eds.), *Johan van Benthem on Logic and Information Dynamics*, Outstanding Contributions to Logic 5,
DOI: 10.1007/978-3-319-06025-5_27, © Springer International Publishing Switzerland 2014

"好" are common in the modern Chinese language, too. Nevertheless, it seems that the Mohists (approx. 479–221 B.C.) were the first to make general observations concerning "好 *hào*" :

且夫读书，非书也；好读书，好书也。且斗鸡，非鸡也；好斗鸡，好鸡也。...... 此
乃不是而然者也。

This can be translated into English as follows:

> Moreover reading a book is not a book, but to like reading books is to like books. A cockfight is not a cock, but to like cockfights is to like cocks. ... These are cases in which 'something is so though the instanced is not this thing.' (NO 16. Cf. [9, pp. 489–490].[1])

Here "something is so though the instanced is not this thing" is one of the four cases of making inferences discussed in the *Canons*. The above two sentences involving "好 *hào*" are used as examples to illustrate it. However no investigation of the logical properties of such cases seems to be available in the *Canons*.

In the West, studies of the logical aspects of value can be traced back to Aristotle's discussion in Book III of the *Topics* of the general properties of desirability, betterness, and other value concepts. Most of his discussion is devoted to issues that fall outside of the domain covered by modern preference logic, but some of the principles that Aristotle proposed are directly translatable into modern logical language, such as the following:

> Also, *A* is more desirable if *A* is desirable without *B*, but not *B* without *A* ([1, p. 165]; *Topics* III:2).

This is directly related to the modern discussion of the postulate

$$p > q \leftrightarrow p \,\&\, \neg q > q \,\&\, \neg p$$

and its variants.[2] However, the systematic study of the logic of value expressions began with the seminal books by Halldén [11] and by von Wright [22]. Just like Aristotle's text, both of these were devoted primarily to comparative concepts. (Aristotle's treatise begins with the words: "The question which is the more desirable, or the better, of two or more things, should be examined upon the following lines.")

Natural languages contain two major classes of non-numerical value predicates. The *monadic* (one-place) predicates express classificatory notions, i.e. they assign a value property to a single object of evaluation. Examples are "good", "bad", "best", "very bad", "almost worst", "fairly good", etc. The *dyadic* (two-place) predicates express comparative notions such as "better" ($>$), "equally good as" ($\approx$), and "at least as good as" ($\geq$).

---

[1] Here we follow Graham's numbering of the *Canons*. He made a hybrid text from *Xiaoqu* and part of *Daqu* under the title "Names and Objects" (abbreviated "NO") We also use his translations.

[2] Halldén [11, p. 28], von Wright [22, pp. 24–25, 40, 60]. For further references see Hansson [13, pp. 89–90].

   Preference logic, the logic of the dyadic predicates, is the most well-developed part of the logic of value concepts. It can be divided into two branches. One branch treats the objects of preferences as primitive and mutually exclusive entities. A preference for x over y is then equal to a preference for x without y over y without x, for the simple reason that all non-identical objects of evaluation are taken to be mutually exclusive.[3] These are *exclusionary* preferences [13]. Most of the discussion on the logic of exclusionary preferences has been devoted to the following two properties:

$x \geq y \ \lor \ y \geq x$ (completeness)

$x \geq y \geq z \ \rightarrow \ x \geq z$ (transitivity)[4]

and to various weakened versions of the latter such as:

$x > y > z \ \rightarrow \ x > z$ (quasi-transitivity)

$x_1 > x_2 > \cdots > x_n \ \rightarrow \ \neg(x_n > x_1)$ (acyclicity).

In the other branch of preference logic, the objects of preferences are potentially compatible states of affairs that are represented by sentences and therefore combinable through truth-functional operations. These are *combinative* preferences. In addition to the properties relevant for exclusionary preferences, a large number of additional properties can be applied to combinative preferences, such as:

$(p \lor q) \geq r \ \rightarrow \ p \geq r$

$p \geq q \ \rightarrow \ (p \lor q) \geq q$

$p \geq q \rightarrow \neg q \geq \neg p$.

The logic of monadic value predicates, taken by themselves, has not attracted much attention since there seem to be relatively few general principles to discover. For the two most studied of these predicates, namely good (G) and bad (B), the following principles appear to be credible:

$\neg(Gp \ \& \ Bp)$ (mutual exclusiveness)

$\neg(Gp \ \& \ G\neg p)$ (non-duplicity of "good")

$\neg(Bp \ \& \ B\neg p)$ (non-duplicity of "bad")

$Gp \rightarrow B\neg p$

$Bp \rightarrow G\neg p$.

Whereas the logic of "good" and "bad", as such, does not seem to offer many interesting developments, a much more interesting logic can be developed if we introduce the dyadic and the monadic predicates into one and the same framework. This is intuitively plausible, since our classificatory and comparative concepts appear to be closely connected to each other. This was implicitly recognized already by Aristotle, when he said that "if one thing exceeds while the other falls short of the same

---

[3] In the logic of exclusionary preferences we use the letters $x, y, z \ldots$ to denote the objects of evaluation. When the objects of evaluation have sentential structure we use the letters $p, q, r \ldots$.

[4] To simplify the notation, we contract series of two-place predicate expressions, thus writing $x \geq y \geq z$ for $x \geq y \ \& \ y \geq z$, and similarly $x > y \approx z$ for $x > y \ \& \ y \approx z$, etc.

standard of good, the one which exceeds is the more desirable" (*Topics*, III:3), which can be interpreted as a statement that:

$$Gp \,\&\, \neg Gq \rightarrow p > q$$

Other plausible connections between the monadic and dyadic predicates include:

$$Gp \,\&\, Bq \rightarrow p > q$$
$$p > q \rightarrow Gp \,\lor\, Bq \text{ (closeness)}$$
$$Gp \,\&\, q \geq p \rightarrow Gq \text{ (positivity of "good")}$$
$$Bp \,\&\, p \geq q \rightarrow Bq \text{ (negativity of "bad").}$$

Are the connections between the monadic and the dyadic predicates so close that one of the two classes of predicates can be defined in terms of the other? Several proposals have been put forward to define "good" and "bad" in terms of the dyadic predicates. The oldest of these proposals is Brogan's [5] definition of "good" as "better than its negation" and "bad" as "worse than its negation", i.e.:

$$G_N p \leftrightarrow p > \neg p \text{ (negation-related good)}$$
$$B_N p \leftrightarrow \neg p > p \text{ (negation-related bad).}$$

This proposal has been taken up by many other writers on the logic of value concepts.[5] An alternative approach was developed by Chisholm and Sosa [7] who identified a set of indifferent sentences, namely those that are neither better nor worse than their negations. Then "a state of affairs is good provided it is better than some state of affairs that is indifferent, and…a state of affairs is bad provided some state of affairs that is indifferent is better than it" [7, p. 246]. In formal terms:

$$G_I p \leftrightarrow (\exists q)(p > q \approx \neg q) \text{ (indifference-related good)}$$
$$B_I p \leftrightarrow (\exists q)(\neg q \approx q > p) \text{ (indifference-related bad)}[6]$$

Unfortunately, neither the negation-related nor the indifference-related definitions have plausible properties for all types of preference relations. There are important types of preference relations for which G_N does not satisfy the above-mentioned property of positivity, and neither does B_N satisfy the corresponding property of negativity. The same applies to G_I and B_I, and in addition there are preference

---

[5] See Mitchell [18, pp. 103–105], Halldén [11, p. 109], von Wright [22, p. 34] and [23, p. 162], and Åqvist [3].

[6] Other variants of the same basic approach replace Chisholm and Sosa's indifferent sentences by tautologies [8, p. 37], or contradictions [23, p. 164]. This approach gives rise to the following formal definitions:

$$G_\top p \leftrightarrow p > \top \text{ (tautology-related good)}$$
$$B_\top p \leftrightarrow \top > p \text{ (tautology-related bad)}$$
$$G_\perp p \leftrightarrow p > \perp \text{ (contradiction-related good)}$$
$$B_\perp p \leftrightarrow \perp > p \text{ (contradiction-related bad)}$$

However, the interpretation of what it means for something to be better or worse than a tautology or (in particular) a contradiction is quite problematic.

relations for which G_I and B_I do not satisfy the even more elementary postulates mutual exclusiveness and non-duplicity of "good" and "bad" [13, pp. 123–124]. The following definition was introduced in order to make the monadic predicates definable in a wider category of preference relations:

$G_C p \leftrightarrow (\forall q)(q \geq^* p \to q > \neg q)$ (canonical good)
$B_C p \leftrightarrow (\forall q)(p \geq^* q \to \neg q > q)$ (canonical bad)

($\geq^*$ is the ancestral of $\geq$, i.e. $p \geq^* q$ holds if and only if either $p \geq q$ or there is a series $r_1, \ldots, r_n$ of sentences such that $p \geq r_1 \geq \cdots \geq r_n \geq q$.)

It has been shown that whenever $\geq$ satisfies reflexivity, then G_C and B_C satisfy the required postulates for a plausible interpretation of "good" and "bad" (including mutual exclusivity, closeness, non-duplicity of both postulates, positivity of "good", and negativity of "bad").[7] Furthermore, this pair of predicates is a generalization of the negation-related pair G_N and B_N in the following sense: If the preference relation is such that G_N satisfies positivity and B_N satisfies negativity, then G_N and B_N coincide with G_C and B_C. In the other cases G_C and B_C still satisfy positivity and negativity. They are therefore a useful replacement for G_N and B_N when the latter are not plausible [13, p. 123].

These attempts to logically reduce the two categories, monadic and dyadic value predicates, to one basic category, all go in the same direction: Monadic predicates are defined in terms of the dyadic ones, not the other way around. This is somewhat problematic from a philosophical point of view since the monadic predicates seem to be more suitable as primitive concepts than the dyadic ones. (Leibniz' failed attempts to eliminate relational expressions from a regimented language are among the clearest expressions of that intuition, see Hill [15].) Also, moral judgements are often expressed in terms of absolute predicates, e.g. good or bad.

Linguistic evidence also points in the direction of reducing dyadic terms to monadic ones, rather than the other way around. The English "better" seems to originate from a comparative form of a Proto-Indo-European adjective meaning "good", and the French "meilleur" from a comparative form of a Proto-Indo-European word meaning "strong". The Chinese "更好 gènghǎo" (better) reads directly as "more or further good", which is a combination of a comparative "more" and an absolute "good". As these examples illustrate, in a wide range of languages the comparative form of adjectives is derived from the absolute form, not the other way around. In a survey of 123 languages, no instance of the opposite relationship between the two forms was found [21]. This discrepancy between logical primacy and apparent conceptual primacy makes it particularly interesting to investigate whether the opposite definitional direction is possible, i.e. whether the dyadic value concepts can be defined in terms of the monadic ones.

---

[7] An even weaker property than reflexivity, namely ancestral reflexivity ($p \geq^* p$), is sufficient for this result.

27.2 Context-Indexed Value Predicates

In an important article, van Benthem [4] developed an account of value terms in which he reversed the usual relationship between monadic and dyadic value predicates. Instead of defining "good" in terms of "better", he defined "better" in terms of "good". In order to do this he had to cut a Gordian knot—instead of the format "x is good" he used the format "x is good among Z", for some set Z of objects, to express the monadic value concept. This format is based on what he calls "the vital observation concerning adjectives such as 'big' (or 'hot' or 'good')", namely "that they are context-dependent". In order to express this conception of the monadic value predicates, we will write $G_Z x$ for "x is good among Z" and $B_Z x$ for "x is bad among Z". G_Z and B_Z will be called *context-indexed* value predicates, and Z is their context index (index of comparative context).

Contexts, in a wider sense of the word, can differ both extensionally and intensionally. When comparing the three horses Amulet, Blackie, and Cosmo we can think of them as riding horses, and then consider Blackie to be a bad horse among the three. Alternatively, we can think of them as draught horses and then consider him to be a good horse among the three. However, this type of difference is not treated by van Benthem in this framework.[8] The contexts for comparisons that he refers to are purely extensional, i.e. they refer to the reference group (set of objects under evaluation) and nothing else. "We are not concerned with verdicts like 'a great philosopher, not a great husband', which involve a shift in meaning" (p. 195). Using such extensional contexts, he defines dyadic predicates in terms of monadic predicates referring to the smallest context that includes both comparanda, thus:

x is α-er than y if and only if: In the context $\{x, y\}$, x is α while y is not α [4, p. 195].

In preference logic, it is universally taken for granted that worseness is nothing else than converse betterness. Therefore, van Benthem's recipe can be interpreted in two ways depending on whether we read $x > y$ as "x is better than y" or as "y is worse than x"[9]:

$$x > y \text{ if and only if } G_{\{x,y\}}x \text{ and } \neg G_{\{x,y\}}y \qquad (27.1)$$

$$x > y \text{ if and only if } B_{\{x,y\}}y \text{ and } \neg B_{\{x,y\}}x \qquad (27.2)$$

These two definitions are not equivalent. To see that, let x be good and not bad in the context $\{x, y\}$, and let y be neither good nor bad in the same context. Then $x > y$ holds according to definition (27.1) but not according to definition (27.2). This seems to be to speak in favour of definition (27.1), since we would expect $x > y$ to hold in this case. But next, let x be neither good nor bad in the context $\{x, y\}$, and let y be bad but not good in the same context. Then $x > y$ holds according to definition (27.2) but not according to definition (27.1), which seems to speak in favour of definition (27.2).

---

[8] For an attempt to deal with some such differences, see Hansson [14]. See also Stalnaker [20].

[9] This also applies to many other adjectives, hence we would say that x is longer than y if and only if y is shorter than x.

The following observation specifies conditions under which the two definitions are equivalent:

Observation 27.1 *Let $G_{\{x,y\}}$ and $B_{\{x,y\}}$ satisfy Mutual exclusiveness. Then definitions (27.1) and (27.2) of betterness are equivalent if and only if:*

$$\text{If } G_{\{x,y\}}x \vee B_{\{x,y\}}x \text{ then } G_{\{x,y\}}y \vee B_{\{x,y\}}y. \qquad \text{(Polarization}^{10}\text{)}$$

Proof For one direction, let $G_{\{x,y\}}$ and $B_{\{x,y\}}$ satisfy Mutual exclusiveness and Polarization. Then only the following five options are possible:

(a) $G_{\{x,y\}}x, \neg B_{\{x,y\}}x, G_{\{x,y\}}y,$ and $\neg B_{\{x,y\}}y$
(b) $\neg G_{\{x,y\}}x, B_{\{x,y\}}x, \neg G_{\{x,y\}}y,$ and $B_{\{x,y\}}y$
(c) $G_{\{x,y\}}x, \neg B_{\{x,y\}}x, \neg G_{\{x,y\}}y,$ and $B_{\{x,y\}}y$
(d) $\neg G_{\{x,y\}}x, B_{\{x,y\}}x, G_{\{x,y\}}y,$ and $\neg B_{\{x,y\}}y$
(e) $\neg G_{\{x,y\}}x, \neg B_{\{x,y\}}x, \neg G_{\{x,y\}}y,$ and $\neg B_{\{x,y\}}y$.

It can easily be verified that the definitions (27.1) and (27.2) yield the same result in all these cases.

For the other direction, let it be the case that $G_{\{x,y\}}$ and $B_{\{x,y\}}$ satisfy Mutual exclusiveness, but Polarization does not hold. Then one of the following four situations is the case:

(f) $G_{\{x,y\}}x, \neg B_{\{x,y\}}x, \neg G_{\{x,y\}}y,$ and $\neg B_{\{x,y\}}y$
(g) $\neg G_{\{x,y\}}x, B_{\{x,y\}}x, \neg G_{\{x,y\}}y,$ and $\neg B_{\{x,y\}}y$
(h) $\neg G_{\{x,y\}}x, \neg B_{\{x,y\}}x, G_{\{x,y\}}y,$ and $\neg B_{\{x,y\}}y$
(i) $\neg G_{\{x,y\}}x, \neg B_{\{x,y\}}x, \neg G_{\{x,y\}}y,$ and $B_{\{x,y\}}y$

It can easily be verified that the definitions (27.1) and (27.2) yield different results in all these cases. □

As we will now show with two examples, Polarization is far from uncontroversial as a condition on goodness and badness. First, suppose that three measures have been proposed as means to become a better piano player:

take piano lessons (x),
drink half a glass of water every day at exactly 3 a.m. (y), and
have three fingers amputated on each hand (z).

In the context $\{x, z\}$ it is quite clear that x is good and z is bad. It should also be clear that x is good in the context $\{x, y\}$. But what about y in the context $\{x, y\}$? According to Polarization it has to be either good or bad. But since it is quite inefficient for the purpose it would be strange to regard it as good in a context that contains x. It would also seem strange to treat it as bad, i.e. to assign to it the same position in $\{x, y\}$ that z has in $\{x, z\}$. A neutral position may seem more appropriate.

---

[10] The term "polarization" indicates that if x and y differ in terms of goodness and badness in the context $\{x, y\}$, then they are at opposite sides of the value scale, i.e. one of them is good and the other is bad.

In the other example we will assume that Ann's recommended daily calorie intake is 2,000 kcal. For each n, let k_n denote that she eats n kcal each day. We can then assume that k_{1999} is good in the context $\{k_{2000}, k_{1999}\}$ (and so is of course k_{2000}). Equally obviously, k_0 is bad in the context $\{k_{2000}, k_0\}$. For sufficiently high m (below 2,000) we expect k_m to be good in the context $\{k_{2000}, k_m\}$, but for sufficiently low values of m it will be bad. It would then be reasonable to think that somewhere on this scale we can find a value or range of values that are neither good nor bad. This however, is prohibited by Polarization. Generally speaking, if there is an option x and a continuum of options from y_0 to y_n, such that y_0 is bad in the context $\{x, y_0\}$ and y_n is good in the context $\{x, y_n\}$, then according to Polarization it holds for each y_m on the scale between y_0 and y_n that y_m is either good or bad in the context $\{x, y_m\}$, i.e. there is no neutral position on the value scale. Due to Observation 27.1, this always holds if the value scale corresponds to a preference relation $>$ that represents both betterness according to (27.1) and worseness according to (27.2).

This problem can be solved, however, if we replace (27.1) and (27.2) by a unified definition that takes both goodness and badness into account. As explained in Hansson [12, 13], "good" and "bad" should preferably not be seen as independent concepts. (It would be no good idea, for instance, to combine G_I, as defined above, with B_N.) Instead, the appropriate formal entity is a pair of a predicate for "good" and a predicate for "bad", such as $\langle G_C, B_C \rangle$, $\langle G_N, B_N \rangle$, or $\langle G_I, B_I \rangle$. Given such a pair of (context-indexed) predicates for "good" and "bad" we can then replace (27.1) and (27.2) by the following combined criterion:

$$x > y \text{ if and only if either } G_{\{x,y\}}x \ \& \ \neg G_{\{x,y\}}y \text{ or } B_{\{x,y\}}y \ \& \ \neg B_{\{x,y\}}x \quad (27.3)$$

Indifference and weak preference can be defined in the same vein:

$$x \approx y \text{ if and only if } G_{\{x,y\}}x \ \leftrightarrow \ G_{\{x,y\}}y \text{ and } B_{\{x,y\}}x \ \leftrightarrow \ B_{\{x,y\}}y \quad (27.4)$$

$$x \geq y \text{ if and only if either :}$$

$$\text{(i) } G_{\{x,y\}}x, \text{ (ii) } B_{\{x,y\}}y, \text{ or (iii) } \neg G_{\{x,y\}}x \ \& \ \neg G_{\{x,y\}}y \ \& \ \neg B_{\{x,y\}}x \ \& \ \neg B_{\{x,y\}}y \quad (27.5)$$

These definitions give rise to the standard relationships between strict preference, weak preference, and indifference:

Observation 27.2 *Let $\langle G, B \rangle$ be a pair of monadic predicates that satisfies mutual exclusiveness. Let $>$, $\approx$, and $\geq$ be the relations that are defined from $\langle G, B \rangle$ through definitions (27.3), (27.4), and (27.5). Then:*

1. $x \geq y \leftrightarrow x > y \ \lor \ x \approx y$
2. $x \approx y \leftrightarrow x \geq y \ \& \ y \geq x$
3. $x > y \leftrightarrow x \geq y \ \& \ \neg(y \geq x)$

Proof In this proof we suppress the context index of $G_{\{x,y\}}$ and $B_{\{x,y\}}$. For *Part 1* we have:

$x > y \ \lor \ x \approx y$

iff $(Gx \ \& \ \neg Gy) \lor (By \ \& \ \neg Bx) \lor ((Gx \leftrightarrow Gy) \ \& \ (Bx \leftrightarrow By))$ ((27.3) and (27.4))

iff $(Gx \ \& \ \neg Gy) \lor (By \ \& \ \neg Bx) \lor (Gx \ \& \ Gy) \lor (Bx \ \& \ By) \lor$

$\quad \lor (\neg Gx \ \& \ \neg Gy \ \& \ \neg Bx \ \& \ \neg By)$ (mutual exclusiveness)

iff $Gx \ \lor \ By \ \lor \ (\neg Gx \ \& \ \neg Gy \ \& \ \neg Bx \ \& \ \neg By)$

iff $x \geq y$ (27.5)

For *Part 2* we have:

$x \geq y \ \& \ y \geq x$

iff $(Gx \ \lor \ By \ \lor \ (\neg Gx \ \& \ \neg Gy \ \& \ \neg Bx \ \& \ \neg By)) \&$

$\quad \& \ (Gy \lor Bx \lor (\neg Gx \ \& \ \neg Gy \ \& \ \neg Bx \ \& \ \neg By))$ (27.5)

iff $(Gx \ \& \ Gy) \lor (Bx \ \& \ By) \ \lor \ (\neg Gx \ \& \ \neg Gy \ \& \ \neg Bx \ \& \ \neg By)$

(mutual exclusiveness)

iff $(Gx \ \& \ Gy \ \& \ \neg Bx \ \& \ \neg By) \lor (Bx \ \& \ By \ \& \ \neg Gx \ \& \ \neg Gy) \ \lor$

$\quad \lor (\neg Gx \ \& \ \neg Gy \ \& \ \neg Bx \ \& \ \neg By)$ (mutual exclusiveness)

iff $(Gx \leftrightarrow Gy) \ \& \ (Bx \leftrightarrow By)$

iff $x \approx y$ (27.4)

For *Part 3* we have:

$x \geq y \ \& \ \neg(y \geq x)$

iff $(Gx \ \lor \ By \ \lor \ (\neg Gx \ \& \ \neg Gy \ \& \ \neg Bx \ \& \ \neg By)) \ \& \ \neg Gy \ \& \ \neg Bx \ \& \ (Gx \lor Gy \lor$

$\quad Bx \lor By))$ (27.5)

iff $(Gx \ \& \ \neg Gy) \ \lor \ (By \ \& \ \neg Bx)$ (mutual exclusiveness)

iff $x > y$ (27.3)

$\qquad\qquad\qquad\qquad\qquad\qquad\qquad\qquad\qquad\qquad\qquad\qquad\qquad\qquad\qquad$ □

27.3 Properties of the Derived Preference Relations

In his 1982 article, van Benthem introduced a series of properties for monadic predicates, and derived from them the standard properties of strict preference. In what follows the monadic properties are expressed for an arbitrary predicate H, that may be equal to a context-indexed G or B. Following van Benthem, by a *difference pair* for H in a particular context Z we mean a pair of two objects x and y such that $H_Z x \ \& \ \neg H_Z y$. van Benthem's properties are as follows:

No reversal:
If $H_Z x \ \& \ \neg H_Z y$ then there is no context V such that $H_V y \ \& \ \neg H_V x$.

Upward difference:
If $Z \subseteq V$ and H has a difference pair in Z, then H has a difference pair in V.

Downward difference:
If x and y form a difference pair for H in V, and $\{x, y\} \subseteq Z \subseteq V$, then H has a difference pair in Z.

Observation 27.3 [4] *Let H be a predicate that satisfies No reversal, Upward difference, and Downward difference. Let > be the relation derived from H through definition* (27.1). *Then it satisfies:*

$\neg(x > x)$ *(irreflexivity)*

$x > y > z \rightarrow x > z$ *(quasi-transitivity)*

$x > z \rightarrow x > y \lor y > z$ *(virtual connectivity)*

Proof See van Benthem [4, pp. 196–197].   □

The following observation shows that van Benthem's three axioms for monadic predicates are also sufficient to ensure that a relation that is obtained from "good" and "bad" through definition (27.5) satisfies the standard properties for a weak preference relation.

Observation 27.4 *Let $\langle G, B \rangle$ be a pair of context-indexed monadic predicates and let $\geq$ be the dyadic predicate obtained from them through definition* (27.5). *Then:*

1. $\geq$ *satisfies completeness*
2. *If $\langle G, B \rangle$ satisfies mutual exclusiveness and both G and B satisfy No reversal, Upward difference, and Downward difference, then $\geq$ satisfies transitivity.*

Proof Part 1 follows truth-functionally from definition (27.5).

Part 2: Let $>$ be the strict part of $\geq$. We are first going to show that $x \geq y \leftrightarrow \neg(y > x)$. First let $y > x$. Then part 3 of Observation 27.2 yields $\neg(x \geq y)$. For the other direction, let $\neg(x \geq y)$. Then part 1 of the present observation yields $y \geq x$, and part 3 of Observation 27.2 yields $y > x$. Combining the two directions, we obtain $x \geq y \leftrightarrow \neg(y > x)$ as desired.

From this it follows that transitivity is equivalent with $x > z \rightarrow x > y \lor y > z$ (virtual connectivity). In order to show that virtual connectivity holds, let $x > z$. It follows from definition (27.3) that either $G_{\{x,z\}}x$ & $\neg G_{\{x,z\}}z$ or $B_{\{x,z\}}z$ & $\neg B_{\{x,z\}}x$.

Case i, $G_{\{x,z\}}x$ & $\neg G_{\{x,z\}}z$: We can distinguish between eight subcases:

(a) $G_{\{x,y,z\}}x$ & $G_{\{x,y,z\}}y$ & $G_{\{x,y,z\}}z$

(b) $G_{\{x,y,z\}}x$ & $G_{\{x,y,z\}}y$ & $\neg G_{\{x,y,z\}}z$

(c) $G_{\{x,y,z\}}x$ & $\neg G_{\{x,y,z\}}y$ & $G_{\{x,y,z\}}z$

(d) $G_{\{x,y,z\}}x$ & $\neg G_{\{x,y,z\}}y$ & $\neg G_{\{x,y,z\}}z$

(e) $\neg G_{\{x,y,z\}}x$ & $G_{\{x,y,z\}}y$ & $G_{\{x,y,z\}}z$

(f) $\neg G_{\{x,y,z\}}x$ & $G_{\{x,y,z\}}y$ & $\neg G_{\{x,y,z\}}z$

(g) $\neg G_{\{x,y,z\}}x$ & $\neg G_{\{x,y,z\}}y$ & $G_{\{x,y,z\}}z$

(h) $\neg G_{\{x,y,z\}}x$ & $\neg G_{\{x,y,z\}}y$ & $\neg G_{\{x,y,z\}}z$

Subcases (a) and (h) are impossible due to Upward difference. Subcases (e) and (g) are impossible due to No reversal. In subcases (c) and (d) it follows from Downward difference that either $G_{\{x,y\}}x$ & $\neg G_{\{x,y\}}y$ or $\neg G_{\{x,y\}}x$ & $G_{\{x,y\}}y$, and then from No reversal that $G_{\{x,y\}}x$ & $\neg G_{\{x,y\}}y$. It then follows from definition (27.3) that $x > y$. In subcases (b) and (f), $y > z$ follows in the same way.

Case ii, $B_{\{x,z\}}z$ & $\neg B_{\{x,z\}}x$: We can distinguish between the following eight subcases to be inspected:

(a) $B_{\{x,y,z\}}x$ & $B_{\{x,y,z\}}y$ & $B_{\{x,y,z\}}z$
(b) $B_{\{x,y,z\}}x$ & $B_{\{x,y,z\}}y$ & $\neg B_{\{x,y,z\}}z$
(c) $B_{\{x,y,z\}}x$ & $\neg B_{\{x,y,z\}}y$ & $B_{\{x,y,z\}}z$
(d) $B_{\{x,y,z\}}x$ & $\neg B_{\{x,y,z\}}y$ & $\neg B_{\{x,y,z\}}z$
(e) $\neg B_{\{x,y,z\}}x$ & $B_{\{x,y,z\}}y$ & $B_{\{x,y,z\}}z$
(f) $\neg B_{\{x,y,z\}}x$ & $B_{\{x,y,z\}}y$ & $\neg B_{\{x,y,z\}}z$
(g) $\neg B_{\{x,y,z\}}x$ & $\neg B_{\{x,y,z\}}y$ & $B_{\{x,y,z\}}z$
(h) $\neg B_{\{x,y,z\}}x$ & $\neg B_{\{x,y,z\}}y$ & $\neg B_{\{x,y,z\}}z$

Subcases (a) and (h) are impossible due to Upward difference. Subcases (b) and (d) are impossible due to No reversal. In subcases (e) and (f) it follows from Downward difference that either $B_{\{x,y\}}x$ & $\neg B_{\{x,y\}}y$ or $\neg B_{\{x,y\}}x$ & $B_{\{x,y\}}y$, and from No reversal that $\neg B_{\{x,y\}}x$ & $B_{\{x,y\}}y$. It then follows from definition (27.3) that $x > y$. In subcases (c) and (g), $y > z$ follows in the same way. □

According to Observation 27.4, we can derive a standard preference relation $\geq$ from a pair $\langle G, B\rangle$ of context-indexed monadic value predicates. (We will leave it to a later investigation whether weaker properties for $\langle G, B\rangle$ correspond to weaker properties for $\geq$.) An interesting question is whether the pair $\langle G, B\rangle$ can be regained from $\geq$, thus making the monadic and the dyadic representations of value interchangeable. There are useful such "back-and-forth" definitions in some other areas. One example is that in belief change, certain properties of belief contraction give rise (through the Levi identity) to certain properties of belief revision, which in their turn (through the Harper identity) give rise to the same properties of belief contraction that we began with, cf. Alchourrón et al. [2] Perhaps more to the point, the same type of connection holds between choice functions and preference relations [19]. However, the following example shows that a return path to the original monadic pair is not in general available.

Let the context-indexed pairs $\langle G, B\rangle$, $\langle G', B'\rangle$, and $\langle G'', B''\rangle$ have the following properties:

$G_{\{x,y\}}x$, $G_{\{x,y\}}y$, $\neg B_{\{x,y\}}x$, and $\neg B_{\{x,y\}}y$.
$\neg G'_{\{x,y\}}x$, $\neg G'_{\{x,y\}}y$, $\neg B'_{\{x,y\}}x$, and $\neg B'_{\{x,y\}}y$.
$\neg G''_{\{x,y\}}x$, $\neg G''_{\{x,y\}}y$, $B''_{\{x,y\}}x$, and $B''_{\{x,y\}}y$.

Then $\langle G, B\rangle$, $\langle G', B'\rangle$, and $\langle G'', B''\rangle$ all give rise through (27.5) to the same preference relation $\geq$ (such that $x \approx y$). Therefore, the monadic pair cannot be reconstructed from the preference relation that it gives rise to.

The basic underlying reason for this is of course that in order to determine what is "good" we do not only need to be able to compare different alternatives in terms of their goodness but also to determine how much goodness is required for something to be good (and similarly for "bad"). The second part of this information is not in general encoded in the preference relation itself. In order to encode it we need to introduce one or several reference points in the alternative set.

One possibility is to require that each index set contains an identifiable baseline, a "neutral element" with reference to which goodness can be determined. Let $n_V \in V$ be the neutral element assigned to the index set V. Then we can follow the model from indifference-related good and bad as explained in Sect. 27.1, and let any element $x \in V$ be good if and only if it is good in the smaller context $\{x, n_V\}$, i.e:

$$G_V x \text{ if and only if } G_{\{x,n_V\}} x$$

With the help of this construction, the goodness predicate can be reconstructed from the preference relation. The badness predicate can be reconstructed in the same way, most plausibly with the same baseline element, so that we have:

$$B_V x \text{ if and only if } B_{\{x,n_V\}} x$$

In this way, the monadic pair and the preference relation become interchangeable, with the help of the baseline element of the index set. The details of this approach remain to be developed.

27.4 Reasoning About Contexts

In the preceding sections, our focus has been on the relationship between the monadic and dyadic value notions, e.g. "good" and "bad" respectively "better". Following van Benthem's proposal, we have studied the formal properties of the dyadic preference relations that can be derived from a monadic value predicate with a context index. However, in van Benthem's seminal paper [4], the research agenda was more general. He was concerned not only with the value notions, but also with other notions expressible with monadic predicates in natural languages, for instance "tall", "small", "hot", etc. His general topic is the interplay between comparative judgements and the context-dependence of monadic judgements. This raises a natural and more general question: can a formal language be constructed that represents this interplay also for other types of judgements than value judgements?

Consider the following two examples in natural language:

(i) Alice is tall.
(ii) Alice was born on the 2nd of June in 1980.

One difference between these sentences is that (i) may not hold when our context shifts. Alice may be tall in one comparative context (such as that of her sisters) but not in another (such as that of her basketball team). The same applies to many other predicates such as "big", "hot", and "good". We will call them *context-dependent predicates*, and the properties that they represent are *context-dependent properties*. In contrast, (ii) exemplifies *context-independent properties*. Predicates expressing them are *context-independent predicates*.

As we have mentioned earlier, van Benthem's recipe for defining comparative concepts in terms of monadic ones is as follows:

x is α-er than y if and only if: In the context $\{x, y\}$, x is α while y is not α [4, p. 195].

Can this recipe be applied in any way to context-independent predicates? In order to answer that question we need to divide the context-independent predicates into two subcategories. Some of them, such as "dead", "ordained", and "pregnant", represent all-or-nothing properties. In English, such predicates are typically expressed by adjectives that do not have any comparative form. To the extent that it is at all meaningful to say that Anne is more pregnant than Beatrice it can only mean that Anne is pregnant while Beatrice is not. This applies irrespective of context. It applies for instance even in the context {Anne, Beatrice, Connie} where Connie is pregnant with triplets (but nevertheless not "more pregnant" than Anne) and in the context {Anne, Beatrice, David} where David lacks the physiological capacity to become pregnant (but is nevertheless not "less pregnant" than Beatrice who has that capacity). With this interpretation, van Benthem's recipe is applicable to this subcategory of context-independent predicates.

The other subcategory consists of those context-independent predicates, such as "anemic", asthmatic", and "snow-covered", that represent properties coming in degrees. van Benthem's definition strategy does not work for them. Peter who has 90 g Hb/l and Robert who has 120 are both anemic. In the context {Peter, Robert} as well as in any other context we would say that both are anemic. Therefore, the monadic predicate cannot be used to derive the comparative predicate "more anemic than", although there could be no doubt that Peter is more anemic than Robert. The reason for this is that even though the property expressed by the predicate comes in degrees there is a context-independent and reasonably well-defined limit specifying when an object does at all have the property in question. When a person with mild asthma is discussed in a context consisting of herself and ten other patients who all have more severe asthma, we do not stop calling her "asthmatic" (whereas a person who is called "tall" in other contexts will not be called "tall" in the context of her basketball team in which she is the shortest member). The failure of van Benthem's definition strategy for these predicates is problematic since it seems to indicate that his recipe cannot be used for all comparatives.

But let us focus on the cases in which the construction can be used. We will propose in outline two formal approaches into which it can be incorporated. The first of these is a simple extension of predicate logic, *context-indexed predicate logic*. It treats the context as inherent in the property and therefore assumes that predicates have an implicit index of context that can be written out in formal representations. Instead of writing Tx for "x is tall" we write $T_A x$ for "x is tall in the context A" (with the assumption that $x \in A$ and that A is a subset of the set of terms). A predicate T is context-independent if and only if $T_A x \leftrightarrow T_B x$ for all sets of terms (contexts) A and B that contain x. For each monadic predicate T there is a dyadic predicate $\overset{>}{T}$ defined according to van Benthem's definition:

$$\overset{>}{T}(a, b) \text{ if and only if } T_{\{a,b\}}a \And \neg T_{\{a,b\}}b$$

The other formal approach is a van Benthem-style modal logic that treats the context as inherent in the evaluation rather than in the property itself. Contexts are represented by sets of worlds. (See Stalnaker [20] for a justification of this representation of contexts.) Thus, instead of evaluating a proposition in relation to a possible world (the actual one) it is evaluated in relation to the combination of that world and a set of possible worlds surrounding it. For an example, consider Singh who finished conservatory a few years ago and now has a fairly well-paid job as a tutti player in a regional orchestra. How should we evaluate the statement "Singh is a successful musician"? If we think of alternative possible worlds in which he has great difficulties in earning a living as a musician, we would hold this statement to be true, but if we include comparisons with worlds in which he is a famous soloist we would probably not. This can be expressed as a difference between evaluations relative to the pairs $\langle s, X_1 \rangle$ and $\langle s, X_2 \rangle$, where s is the actual world and X_1 and X_2 are different sets of worlds.

To develop such a model it is useful to have three non-empty and disjoint sets of proposition letters, P, Q and Ω. The elements of P are context-dependent proposition letters, those of Q context-independent proposition letters, and those of Ω nominals. Nominals are sentences, introduced for technical reasons, that hold in exactly one world. (Some of the propositions in P and Q can be expressed with predicates, but that option will not be developed here.) A *modal context language* with basic resources for context-dependent expressions is given by the following Backus–Naur Grammar clause:

$$\phi ::= p_i \mid q_i \mid j \mid \neg\phi \mid \phi \,\&\, \psi \mid E\phi \mid <in>\phi \mid <up>\phi \mid <dn>\phi.$$

Hence all proposition letters $p_i \in P$, $q_i \in Q$, and $j \in \Omega$ are well-formed formulas. The set of well-formed formulas is closed under negation and conjunction. The language contains four modalities E, $<in>$, $<up>$ and $<dn>$; adding one of these symbols in front of a well-formed formula gives a well-formed formula. We use the diamond versions of the modalities. Their box versions $U\phi$, $[in]\phi$, $[up]\phi$, and $[dn]\phi$ can be introduced in the usual way, for instance, $U\phi$ is an abbreviation of $\neg E\neg\phi$.

The modality E refers to possibility, widely conceived. $E\phi$ holds at a state s in the context X if and only if ϕ is true at some state t in some context Y. This extends the standard meaning of the operator E as an existential modality in modal logic. $<in>$ refers to possibility within the present context, i.e. $<in>\phi$ holds relative to s and X if and only if there is some state $t \in X$ such that ϕ holds relative to t and X, thus $<in>$ acts as a (context-limited) existential modality.

Next, $<up>$ and $<dn>$ represent changes in context. In an evaluation relative to the actual world s and the context X, $<up>\phi$ holds if and only if there is some superset X' of X such that ϕ holds relative to s and X'. Similarly, $<dn>\phi$ holds if and only if there is some subset X' of X such that ϕ holds relative to s and X'. Hence the modality $<in>$ represents changes in the actual state but not in the context, whereas $<up>$ and $<dn>$ represent changes in context but not in the actual state, and E represents changes both in context and in the actual state.

In order to interpret the above language we introduce a modal context model

$$\mathcal{M} = \langle W, \Phi, \mathcal{V} \rangle$$

W is a set of possible states (possible worlds). Capital letters $X, Y, Z \ldots$ denote subsets of W. The elements of W are denoted $s, t \ldots$, and one of these elements is the actual state of the world (the actual world). Φ is the set of (possibly context-dependent) propositions. $\mathcal{V}$ is an evaluation function that, depending on the context X and the actual state s assigns a subset of W to each proposition. To any nominal j, $\mathcal{V}$ assigns a singleton subset of W, independently of the state and the context. For any context-independent proposition letter q_i, $\mathcal{V}$ assigns a subset of W that depends on the state of the world but is the same independently of the context. More precisely, the following holds for all $s, s', t, t' \in W$ and $X, X' \subseteq W$ and all $p_i \in P$, $q_i \in Q$, and $j \in \Omega$:

(i) $\mathcal{V}(\langle X, s, p_i \rangle) \subseteq \mathcal{P}(W)$
(ii) $\mathcal{V}(\langle X, s, q_i \rangle) = \mathcal{V}(\langle X', s, q_i \rangle) \subseteq \mathcal{P}(W)$
(iii) $\mathcal{V}(\langle X, s, j \rangle) = \mathcal{V}(\langle X', s', j \rangle) \subseteq \mathcal{P}(W)$
(iv) If $\{t, t'\} \subseteq \mathcal{V}(\langle X, s, j \rangle)$ then $t = t'$.

Given a modal context model $\mathcal{M}$, a context X, and a state s in X (representing the actual state of the world), we can evaluate any formula to see whether it is true at that state in that context. The truth-conditions for formulas in this language are defined recursively as follows:

(1) $\mathcal{M}, X, s \models p_i$ iff $s \in \mathcal{V}(\langle X, s, p_i \rangle)$.
(2) $\mathcal{M}, X, s \models q_i$ iff $s \in \mathcal{V}(\langle X, s, q_i \rangle)$.
(3) $\mathcal{M}, X, s \models j$ iff $s \in \mathcal{V}(\langle X, s, j \rangle)$.
(4) $\mathcal{M}, X, s \models \neg\phi$ iff   it is not the case that $\mathcal{M}, X, s \models \phi$.
(5) $\mathcal{M}, X, s \models \phi \,\&\, \psi$ iff $\mathcal{M}, X, s \models \phi$ and $\mathcal{M}, X, s \models \psi$.
(6) $\mathcal{M}, X, s \models E\phi$ iff   there are some t and Y s.t. $\mathcal{M}, Y, t \models \phi$.

(7) $\mathcal{M}, X, s \models {<}in{>}\phi$ iff   there is some $t \in X$ s.t. $\mathcal{M}, X, t \models \phi$.
(8) $\mathcal{M}, X, s \models {<}up{>}\phi$ iff   there is some Y s.t. $X \subseteq Y$ and $\mathcal{M}, Y, s \models \phi$.
(9) $\mathcal{M}, X, s \models {<}dn{>}\phi$ iff   there is some Y s.t. $s \in Y \subseteq X$ and $\mathcal{M}, Y, s \models \phi$.

This version of the modal context language does not contain predicates. Therefore it cannot be used to represent the relationship between sentences such as "i is tall" and "j is tall", and consequently it cannot represent properties such as van Benthems's No reversal, Upward difference and Downward difference that all refer to pairs of such sentences. However, closely analogous properties can be expressed without adding predicates to the language. Instead we can use sentences representing indexical propositions. The modal context language turns out to be quite suitable to represent such propositions. Consider the (indexical) sentence "It is hot here today" that was recently uttered by one of the authors when he was in Stockholm. Such a sentence can be true in a comparison of the state in which it was uttered to other states in which the utterer is in Sweden, but nevertheless false in a comparison involving a larger context including states of the world in which the utterer is in a country close to the Equator. Let this sentence be expressed by a context-dependent proposition ϕ. Let s be the present state of the world in which it is 30 °C in Stockholm when the sentence is uttered, and let t be a state in which it is 25 °C. There can then be a context X_1 (e.g. representing states in which he experiences Swedish summer days) and another context X_2 (e.g. representing states in which he experiences summer days anywhere in the world) such that (1) relative to X_1, ϕ is true in both s and t and (2) relative to X_2, ϕ is true in s but false in t. However, we can plausibly assume that there is no context X_3 relative to which ϕ is false in s but true in t. Our claim that there is no such context is closely related to the No reversal property proposed by van Benthem. In formal terms it can be expressed as follows[11]:

NR:
$$p \,\&\, <in>(\neg p \,\&\, j) \rightarrow \neg<up><dn>(\neg p \,\&\, <in>(p \,\&\, j))$$

For Upward Difference, again consider the same proposition ϕ ("It is hot here today") and again let X_1 represent the states in which the utterer experiences Swedish summer days. Furthermore we assume that X_1 contains the actual state s in which it is 30 °C and ϕ holds relative to the current context, and it also contains some other state u in which it is 22 °C and ϕ does not hold relative to the current context. Thus X_1 contains a difference pair. Let X_2 be any context that includes X_1. We then have strong intuitive reasons to assume that X_2 contains a difference pair. To see this, first suppose that X_2 contains some state v that is hotter than the hottest state in X_1 and that v is also unsurpassed in hotness within X_2. Then ϕ should hold in v relative to X_2. Let w be the state in X_2 that has the lowest temperature. Then w is at least as cold as the coldest state in X_1. The presence of v in X_2 should if anything strengthen our unwillingness to call w hot. We can therefore conclude that X_2 has a difference pair consisting of v and w. Next suppose instead that X_2 contains some state that is colder than the coldest state in X_1. We can then show with an analogous argument that X_2 should have a difference pair in this case as well. Such examples confirm the plausibility of Upward difference. It can be expressed as follows in the formal language:

UD:
$$<in> p \,\&\, <in> \neg p \rightarrow [up](<in> p \,\&\, <in\, \neg p)$$

---

[11] The subformula $<up><dn>$ is used as a means to reach (any) other reachable context while remaining in the same actual state. Its plausibility for this purpose depends on conditions that we will not discuss further here, namely (i) $<up><up>\phi \leftrightarrow <up>\phi$, (ii) $<dn><dn>\phi \leftrightarrow <dn>\phi$, and (the somewhat more questionable) $<up><dn>\phi \leftrightarrow <dn><up>\phi$.

For Downward difference, consider again the same sentence ϕ and the same contexts X_1 (in which the utterer experiences Swedish summer days) and X_2 (in which he experiences summer days anywhere in the world). Then clearly X_2 includes X_1. Furthermore let the two states s (in which he experiences 30 °C in Stockholm) and u (in which he experiences 22 °C in Stockholm) be included in X_1 and consequently also in X_2. Suppose that ϕ holds in s relative to X_2 but does not hold in u relative to X_2. It is a highly intuitive conclusion that X_1 contains both some state in which ϕ holds and some other state in which it does not hold. (More precisely: X_1 contains both some state in which ϕ holds relative to X_1 and some other state in which ϕ does not hold relative to X_1.) In formal terms this can be expressed as follows:

DD:

$$<in>(p \,\&\, j_1) \,\&\, <in>(\neg p \,\&\, j_2) \rightarrow [dn](<in>j_1 \,\&\, <in>j_2 \rightarrow <in>p \,\&\, <in>\neg p))$$

What we have presented in this section is just a first step towards a systematic study of modal context logic. The connections between intuitions for comparative predicates and indexical propositions that we have started to investigate need more detailed consideration. Another important topic is the full characterization in this framework, syntactically and semantically, of plausible properties for context-dependent propositions. The ideas on shifts in context presented by van Benthem in his seminal 1982 paper have opened up several highly interesting areas of investigation.

27.5 Conclusion and Future Work

In the value-theoretical literature it has usually been assumed that monadic value notions can be defined in terms of dyadic value notions, whereas definitions in the opposite direction are not possible. Building on crucial insights in van Benthem [4], we have shown that the latter direction is feasible with a method in which shifts in context have a crucial role. But these definitions are not reversible, i.e. we can define dyadic preference orderings from context-indexed monadic predicates, but these monadic predicates cannot be regained from the preference relation that they gave rise to. Although we have interpreted these results as referring to the value notions "good", "bad", and "better", other interpretations like "tall" and "taller" are equally natural. Therefore these results can be relevant for studies of natural language comparatives and of vague predicates in general.

Several issues have emerged in our analysis that we would like to return to in the future. Firstly, concerning interdefinability, a similar picture appeared in Liu [16] where one can derive preference relations from priority graphs, but in the other direction one can only obtain a "representation" from the preference relation, which need not be the original priority graph. This shows a clear analogy. Connections may exist between contexts and priority graphs, and this is a topic for future investigations. Secondly, on a more technical side, other (modal) logics have been proposed to characterize the notion of context. For instance, Buvac et al. [6] extended propositional logic with a new modality $ist(\kappa, \phi)$, which expresses that the sentence ϕ holds in the context κ. In their work, each context has its own vocabulary, i.e. a set of propositional atoms which are meaningful in that context. They also provided a complete and decidable proof system. We would like to further develop our system and compare it to theirs and possibly to others. Finally, studies in preference change have made some progress in recent years (e.g. [10] and [17]). The results we have obtained here suggest that preference change can be explored in a richer setting including context and monadic notions of good and bad. Moreover, context shifts might be a good locus for studying triggers or reasons for preference change. We leave these issues for future occasions.

Acknowledgments We would like to thank Alexandru Baltag and Sonja Smets for their efforts in putting together this volume. We would like to thank Johan van Benthem for his useful comments. Fenrong Liu is supported by Project 13AZX01B of the National Social Science Foundation of China and the Tsinghua University Initiative, Scientific Research Program 20131089292.

References

1. Aristotle (1952) Topics, translated by WA Pickard-Cambridge. In: WD Ross, JA Smith (eds) Works, vol 1. Encyclopædia Britannica, Chicago, pp 139–223
2. Alchourrón C, Gärdenfors P, Makinson D (1985) On the logic of theory change: partial meet functions for contraction and revision. J Symb Logic 50:510–530
3. Åqvist L (1967) Good samaritans, contrary-to-duty imperatives, and epistemic obligations. Noûs 1:361–379
4. van Benthem J (1982) Later than late: on the logical origin of the temporal order. Pac Philos Q 63:193–203
5. Brogan AP (1919) The fundamental value universal. J Philos Psychol Sci Methods 16:96–104
6. Buvac S, Buvac V, Mason IA (1995) Metamathematics of contexts. Fundam Informaticae 23(2–3):263–301
7. Chisholm RM, Sosa E (1966) On the logic of 'intrinsically better'. Am Philos Q 3:244–249
8. Danielsson S (1968) Preference and obligation. Filosofiska Föreningen, Uppsala
9. Graham AC (1978) Later Mohist logic, ethics and science. Chinese University Press, Hong Kong
10. Grüne-Yanoff T, Hansson SO (2009) Preference change: approaches from philosophy, economics and psychology. Theory and Decision Library. Springer, Berlin
11. Halldén S (1957) On the logic of 'Better'. Gleerup, Lund
12. Hansson SO (1990) Defining 'Good' and 'Bad' in terms of 'Better'. Notre Dame J Formal Logic 31:136–149
13. Hansson SO (2001) The structure of values and norms. Cambridge University Press, Cambridge
14. Hansson SO (2006) Category-specified value statements. Synthese 148:425–432
15. Hill J (2008) Leibniz, relations, and rewriting projects. Hist Philos Q 25:115–135
16. Liu F (2011) A two-level perspective on preference. J Philos Logic 40(3):421–439
17. Liu F (2011) Reasoning about preference dynamics, Series synthese library. Springer, Heidelberg, p 354
18. Mitchell ET (1950) A system of ethics. Charles Scribner's Sons, New York
19. Sen A (1970) Collective choice and social welfare. Holden-Day, San Francisco
20. Stalnaker R (1996) On the representation of context. In: Galloway T, Spence J (eds) SALT VI. Cornell University, Ithaca, New York, pp 279–294
21. Ultan R (1972) Some features of basic comparative constructions. In: Stanford working papers on language universals, vol 9. Stanford University Press, Stanford, pp 117–162

22. von Wright GH (1963) The logic of preference. Edinburgh University Press, Edinburgh
23. von Wright GH (1972) The logic of preference reconsidered. Theor Decis 3:140–169

Chapter 28
Arguing About Dynamic Meaning

Martin Stokhof

Abstract Whether, and if so in what sense, dynamic semantics establishes the need to move away from standard truth-conditional semantics, is a question that has been discussed in the literature on and off. This paper does not attempt to answer it, it merely wants to draw attention to an aspect that has hitherto received little attention in the discussion, viz., the question what role we assign to the use of formal systems in doing natural language semantics.

28.1 What Dynamics?

The question whether dynamic semantics constitutes a move away from static semantics, and if so, what that move involves and how it should be justified, has been discussed on and off since the first dynamic approaches appeared in the early 1980s.

Some have argued that the introduction of dynamic notions in semantics is superfluous and that whatever is treated by dynamic theories using their characteristic conceptual apparatus can be treated with equal descriptive adequacy by static semantics, and with greater explanatory success because it draws on more standard conceptual resources.

But ever since the advent of theories of dynamic interpretation and dynamic meaning, their proponents have tried to make the case that this development does constitute a legitimate, even necessary move beyond the truth conditional conception of meaning of the logical and philosophical traditions that had been one of the sources of inspiration of formal semantics in the late sixties, early seventies of the twentieth

Martin Stokhof would like to thank Johan van Benthem and the editors for their comments and their patience.

M. Stokhof (✉)
ILLC/Department of Philosophy, Universiteit van Amsterdam, Amsterdam, The Netherlands
e-mail: M.J.B.Stokhof@uva.nl

A. Baltag and S. Smets (eds.), *Johan van Benthem on Logic*
and Information Dynamics, Outstanding Contributions to Logic 5,
DOI: 10.1007/978-3-319-06025-5_28, © Springer International Publishing Switzerland 2014

century. Let us briefly review a couple of examples of the arguments that were employed.

Hans Kamp, in his seminal paper 'A Theory of Truth and Semantic Representation' [20] which introduced the framework of discourse representation theory (DRT), claimed that in addition to the truth-conditional concept of meaning, which focusses on the reference relation between expressions and external entities, there is a second conception, stemming from psychology, linguistics, and artificial intelligence, that is concerned with the articulation of 'the structure of the representations which speakers construct in response to verbal inputs.' Noting that these two conceptions, stemming as they do from quite diverse disciplinary traditions, had been separated for quite some time, Kamp states that:

> This separation has become an obstacle to the development of semantic theory, impeding progress on either side of the line of division it has created. The theory presented here is an attempt to remove this obstacle.

So the aim of DRT is to combine referential and representational notions of meaning. In DRT this is done basically by making semantic interpretation a two-step process. When processing a sequence of utterances or a text, the listener/reader creates a formal representation of its content by building a so-called 'discourse representation structure' (DRS), in which score is kept of what is being talked about, what is being said about the various entities, and how various pieces of information are related to each other. This complex representation then is assigned a truth conditional interpretation by defining an embedding of the DRS in a model theoretic structure, i.e., into the kind of model that is familiar from standard truth conditional semantics.

In this way both the representational and the referential aspects of meaning are being accounted for within DRT. What is important to note in the context of the questions that are central to this paper is that there is not one, unified concept of meaning that accounts for both aspects. Rather there are two different processes that are integrated, not conceptually, but at the level of the overall theory. One is the incremental build-up of a representation, and the other process is the non-incremental, 'in one fell swoop' embedding of the resulting representation in a model. The former is, sensu strictu, not a process of semantic interpretation. The latter is, but it is quite a standard one, at least in terms of the semantic concepts it employs. So, it seems appropriate to call DRT 'a dynamic theory of interpretation', rather than a semantic theory that incorporates a dynamic concept of meaning.

In a similar way, one of the main sources of inspiration and motivation of Heim's file change semantics (FCS; cf. [18, 19]), which was developed to give a non-quantificational account of definite and indefinite expressions, is a procedural take on how such expressions function: not by referring to something in the world, but by making available so-called 'discourse referents' (a notion that goes back to work by Karttunen in the late sixties of the last century). And Heim, like Kamp, ends up with a combination of a dynamic component that builds representations, so-called 'files', and a static, truth conditional semantics that interprets them:

> Roughly, the model of semantics that I am going to present will embody the following assumptions. The grammar of a language generates sentences with representations on various

levels of analysis, among them a level of 'logical form'. Each logical form is assigned a 'file change potential', i.e., a function from files into files. [...] The system moreover includes an assignment of truth conditions to files. Note that logical forms themselves are not assigned truth conditions, only files are. Only in an indirect way, i.e., via the files they affect, will logical forms be associated with truth conditions.

Thus FCS, like DRT, postulates a division of labour in semantics: semantic interpretation is (minimally) a two-step process. Assuming that the assignment of logical forms is part of the syntactic component of the grammar, the semantic component consists of two parts. First, it associates with each logical form an operation on a specific type of representation, the files. Thus a logical form itself is associated with a dynamic construct, viz., a file change potential, which, however, is not directly semantic in nature: it constructs a new file from an input file, but these files are themselves not semantic objects. They need to be interpreted and that is what the second component does: it assigns static truth conditions to files. So the dynamic aspects are accounted for in an indirect manner. It is not meaning as such that is dynamic, but rather the process of interpretation.

This indirect approach sets such theories as DRT and FCS apart from approaches that attempt to deal with dynamic aspects directly, by building them right into the concept of meaning itself. An example of a theory that introduces a concept of meaning that is different from the traditional, truth-conditional one, is update semantics, as developed by Veltman [31–33]. In the opening paragraph of 'Defaults in Update Semantics', Veltman refers to the standard definition of validity in terms of truth preservation, and then goes on to characterise his own approach, that of update semantics, as follows:

The slogan 'You know the meaning of a sentence if you know the conditions under which it is true' is replaced by this one: 'You know the meaning of a sentence if you know the change it brings about in the information state of anyone who accepts the news conveyed by it.' Thus, meaning becomes a dynamic notion: the meaning of a sentence is an operation on information states.

Similar quotes can be culled from other papers, e.g., [13, 15, 17]. In the latter paper, it is emphasised that the dynamics of meaning may affect various aspects of a situation, not just the information states of the speech participants, thus proposing that meaning be analysed in term of 'context change potentials'. This is of some importance as it signals that these later systems progress beyond the initial conception in crucial respects. And it also (partly) explains why in a footnote that occurs in the passage just quoted Veltman relates the dynamic conception of meaning, not just to the work of Kamp and Heim, but also to earlier work by Stalnaker, which focusses very much on information update.

As for Heim and Kamp's work, as we just saw there is a subtle, yet principled distinction between the concept of dynamic interpretation of DRT and FCS, and the dynamic semantics of which Veltman's update semantics is a specimen. The difference, as we indicated, resides in the concept of meaning itself. As for Stalnaker's work of the 1970s [25, 26], that is motivated in yet other ways, it seems. Stalnaker is primarily concerned with the analysis of assertion and presupposition, and focusses on the role these play in conversations of a particular kind, viz., information exchanges. Cf. the following quote from 'Assertion':

> [...] acts of assertion affect, and are intended to affect, the context, in particular the attitudes of the participants in the situation

Thus, assertion as a speech act is a dynamic entity. But the concept of meaning, in particular, the concept of the content of an assertion, remains a static entity.[1] For Stalnaker, what an assertion contributes in terms of content is a proposition, i.e., a set of possible worlds, and the context changing effect is accounted for, not at the level of meaning, but at the level of speech acts.

Thus what we have here is a third kind of dynamic theory. The work of Kamp and Heim can be characterised as concerned with dynamic interpretation of linguistic structures. The approach of Veltman c.s. is involved with the development of a dynamic conception of meaning. And Stalnaker employs a static notion of meaning in what is basically a speech act level account of dynamics. Thus what we have here are three different notions of what 'dynamics' in the context of natural language meaning might mean: dynamic assignment of static meanings, in the Heim and Kamp case; dynamic meaning as such, as in Veltman's update semantics; and dynamic employment of static meanings, as exemplified by Stalnaker's approach. If one would look closer at the literature, one would presumably find even more variations than these three, but in order to set the stage for the main question, this should suffice.

That main question is whether, and if so in what sense, these are really rival theories. There are empirical issues involved here, of course, and conceptual ones, but first let's look at the issue from a theoretical perspective.

It would appear that the differences between theories of dynamic interpretation and theories of dynamic semantics are centred around methodological questions concerning the internal organisation of grammar. The issue of representationalism and compositionality is a good example of such a methodological consideration. DRT and FCS adopt a level of representation in grammar that is different from both syntactical structure and meaning proper, and that mediates between form and meaning. Having such an intermediate level of representation in the grammar implies that a strong form of compositionality (often referred to as 'surface compositionality') no longer applies: it is not (structured) expressions that are interpreted directly, i.e., 'as is', but representations that are built from them in an incremental way.

The differences between dynamic semantics and the Stalnakerian approach are not concerned with the organisation of grammar, but rather seem to focus on the concept of meaning as such. Consider the difference between a dynamic approach, such as Veltman's, and Stalnaker's account of information change. The difference is subtle, but real, nevertheless. It basically comes down to this: is information change something that is brought about using an expression that has a static meaning, or does it reside in the meaning of the expression itself? Dynamic semantics takes the latter route, and it does so unequivocally. Cf. the following quote from [15]:

---

[1] Of course, the content of an assertion itself is a context-dependent entity, in many cases, but that does not turn it into a *dynamic* one.

> The general starting point of the kind of semantics that dynamic predicate logic is an instance
> of, is that the meaning of a sentence does not lie in its truth conditions, but rather in the way it
> changes (the representation of) the information of the interpreter. The utterance of a sentence
> brings us from a certain state of information to another one. The meaning of a sentence lies
> in the way it brings about such a transition.

The difference with Stalnaker's view is that the latter regards information change as an external effect of the use of expressions that have meanings that themselves are perfectly static. Information change is a pragmatic effect brought about by static semantic means. In that respect the Stalnakerian view builds on a standard hierarchical view on the relation between semantics and pragmatics that goes straight back to their traditional semiotic characterisations. The dynamic view departs from that view in two ways. It no longer subscribes to semantics as the study of the relation between language and the world, with the associated referential and truth-conditional conception of meaning. And consequently, it draws the line between semantics and pragmatics differently.

From this perspective DRT, FCS, and their kin are something of a mixed bag. They 'side', so to speak, with dynamic theories in regarding dynamic aspects as part of semantics, but they locate them in the process of building representations. These are then interpreted in a standard, truth conditional way, and in that respect these approaches are more Stalnakerian than Veltmanian. Dynamic semantics can be regarded as a kind of straightening out of these issues: by redefining semantics (and implicitly redefining pragmatics) it eliminates the need for the kind of representations that are characteristic of DRT and sundry systems. One of the original motivations for the development of dynamic predicate logic was exactly this: the elimination of what was regarded as an unnecessary and insufficiently motivated complication in the grammar.[2] This centred essentially around the wish to maintain a particular, strong form of compositionality. But of course one might argue that there are independent reasons for having those kinds of representations as part of the semantics (and thus for turning strong compositionality from a methodological principle into an empirical issue).

Be that as it may, one core issue now appears to be whether natural language meaning is better modelled in the standard way, i.e., in terms of a static truth-conditional concept of meaning combined with a pragmatic theory that accounts for dynamic effects such as information change, or in the dynamic way, by constructing meaning in terms of context change potential.

One may try to answer that question in two ways: by an appeal to empirical considerations, and by conceptual arguments. From the empirical perspective, one might reason that the standard picture is standard for good reasons, and one would need to change sides only if there are there empirical phenomena concerning natural language that really can only be accounted by embracing a dynamic picture. From the conceptual perspective, things appear to be less constrained: when one tries to determine what counts as a convincing conceptual consideration, empirical adequacy is an obvious necessary condition, but it does leave room for other considerations.

---

[2] Cf. Groenendijk and Stokhof [15, Sect. 5.2].

Perhaps it leaves too much room, as it is not obvious that if from the empirical perspective all things are equal when it comes to the static–dynamic choice, there are indeed any decisive conceptual considerations to base a choice on.

28.2 A Step Back

But before we enter into considerations concerning a choice, we'd better ask a preliminary question: does it really matter? Is there a difference between these two conceptions that is worth investigating to begin with?

One of the decisive steps forward in the development of dynamic semantics for natural language was made from a logical perspective. In a number of papers,[3] van Benthem established dynamic semantics as a subject matter in its own right by identifying a proper meta-theoretical framework for studying its properties. Where Boolean algebra provides the general mathematical framework in which standard semantics can be formulated and important meta-properties can be studied, van Benthem showed that relational algebra plays a similar role for dynamic theories. It provides a general framework in which concrete dynamic theories can be studied and compared. As a general meta-theoretical framework it is not confined to dynamic systems used in natural language semantics but also provides the tools to study similar approaches in logic itself, in cognition, artificial intelligence and the like. This kind of inquiry into the formal, meta-logical properties that are characteristic of various dynamic systems was taken up by a number of authors.[4]

One example to illustrate the kind of concerns that are at stake here. A central question is what exactly distinguishes static and dynamic systems from each other. One way to go about answering that question is by providing a formal characterisation of what makes a system static. Usually this is done in terms of formal properties of the updates, i.e., the operations that take states into states, that the system makes available. It turns out that there are several ways to do so, with slightly different consequences for what counts as static and what not. In a recent study [23] Rothschild and Yalcin have traced the history of these attempts in great detail, so what follows is just a very brief illustration, and the reader is urged to consult the Rothschild and Yalcin paper for further details.

van Benthem provided a first definition of staticness.[5] According to this characterisation a system is static if its updates are eliminative and (finitely) distributive. If states are sets of some kind, these properties come down to the following: an eliminative update results in a state that is a subset of the state to which it is applied, and a distributive one is an update that works 'point-wise', i.e., its effect can be defined

---

[3] Among others, [2, 3], and the papers collected in [4].

[4] For natural language semantics we should mention, among others, Vermeulen [34], Visser [35]. Cf. also [10] for a computational perspective on deduction in dynamic semantics, and [7] for a more recent overview.

[5] In van Benthem [1].

in terms of its effects on the singleton elements of the state on which it operates. Both properties are needed to maintain staticness. An illustration is provided in [14] where it is shown that Veltman's update semantics is dynamic because its updates are not distributive, while maintaining eliminativity, where for DPL it is the other way around: the updates of that system are distributive, but lack eliminativity.

A generalisation was provided by Veltman [33]: where the van Benthem characterisation has Boolean algebra as its backdrop, Veltman uses the more general concept of an information lattice. Staticness is the defined in terms of the properties of the lattice: if a system's information lattice satisfies idempotence, persistence, strengthening, and monotonicity, it is static.

Rothschild and Yalcin take this meta-theoretical approach another step further. They focus on what they call 'conversation systems', which abstract away as much as possible from particular features of the language under consideration and its associated semantics, and talk only about states and the operations on states that the semantics induces. A general characterisation of staticness is then given as follows: a conversation system is static if and only if the associated state system satisfies idempotence and commutativity. They show that this characterisation encompasses both that of van Benthem and that of Veltman, and that it can be used to prove the non-staticness of various systems, such as FCS, DPL, and update semantics.

That different characterisations can be found in the literature is explained by how close one stays to a specific system or set of systems, with less general characterisations allowing for more fine-grained analyses. A case in point: as we just saw, the original van Benthem characterisation of staticness in terms of eliminativity and distributivity allows us to not only classify both DPL and update semantics as non-static, but also to differentiate them in an informative way. Using the more general approach of Rothschild and Yalcin that possibility disappears: from their perspective both systems are non-static because they are both neither idempotent nor commutative. But this is, of course, the usual trade-off between generality and specificity.

Be that as it may, for our present purposes what is important is that as far as formal systems and their properties are concerned there is substance to the distinction between static and dynamic systems. However, that still leaves the question open whether from the point of view of the semantics of natural language the distinction makes sense as well. That we *can* describe natural language meaning both in a static as well as in a dynamic manner, using the appropriate formal systems, and that such descriptions differ in the meta-logical properties of the systems, employed, does not imply that we *need* to do so. So the question remains, but we can be certain that it concerns a substantial distinction.

28.3 Fact or Fiction?

As should be clear from the way in which various theories in the broad realm of 'dynamics' are introduced and motivated, there is general agreement that when we observe language in its actual use it is abundantly clear that this has dynamic effects.

This is not something that anybody would want to deny. The 'dynamic wave' that started in the 1980s then marks, minimally, an increased attention for such dynamic aspects. In that sense all the various approaches discussed above are part of the same development. But what is under discussion is, first of all, to what extent these dynamic effects need to be accounted for in a linguistic theory, and, second, if so, in what manner. The first question is answered positively by many,[6] on the basis of quite comparable argumentation. It is to the second one that answers starts to diverge.

So it is particularly the latter question, i.e., the question where and how in an overall account of language it is that we should account for the dynamic effects of language use, that is central to many of the discussions in the literature. Obviously, this is closely connected to the question where we are to draw the line between semantics and pragmatics. Now suppose we start with the traditional semiotic characterisation of semantics and pragmatics, with semantics being concerned with the relation between language and the world, and pragmatics with the use of language. One reason that this is a good starting point is that it is a relatively theory-independent description, one that is stated in terms that are neutral and descriptive. Now, by phrasing the phenomena in question as 'effects of language use' we might seem to have settled already on an answer. For if we go by the semiotic characterisation, it would seem obvious that dynamic effects, described as effects of language use, should be accounted for in pragmatics. This seems indeed to be one way of settling the matter.

Of course, it's not always that straightforward, if only because there are many alternative ways of describing what semantics and pragmatics are concerned with. By way of illustration, let us look at the following passage from a recent paper [22] by Karen Lewis:

> Dynamic contents encode (some of) the effects of an utterance on an arbitrary input context. By contrast, static contents do not encode any updates to the context. On a static view, the effect(s) of content on the context has to be explained pragmatically. *These are fundamentally different sorts of explanations. Semantics describes facts about natural language.* Pragmatics, on the other hand, describes facts about rational agents who engage in co-operative activities. [emphasis in original, ms]

Clearly, Lewis has in mind a particular conception of semantics and pragmatics, and of the associated the division of labour between the two, that is different from the traditional one. First pragmatics. On the one hand the description given here is much broader than the traditional one, as it does not mention language or language use and language users, but talks about agents in general. On the other hand, if we narrow down 'agents' to language users, we get a conception that is much narrower as these language users are now restricted to rational ones that engage in cooperative activities. Of course, language users do, at least sometimes, act as rational agents, and they do, again at least in some situations, engage in cooperative activities. But it will hardly do to try to force any aspect of language use into that restricted mould.

---

[6] It is a testimony to the impact of the generative tradition, though, that even today many authors would seem to work with a more or less principled distinction between 'language-as-a-system' and 'language-as-use', which echoes the competence—performance distinction that Chomsky used to define the proper domain of linguistics as a scientific endeavour. Cf. [30] for further discussion.

Rather, the conception of pragmatics that is at stake here is one that goes back to Grice, who used it in the execution of a quite specific, philosophically motivated program. Thus this conception of pragmatics is not one that we can appeal to in order to classify a certain set of phenomena, in this case the effects of language use that dynamic theories are concerned with, without an independent motivation of why pragmatics should be (only) this.

Then semantics. The crucial question here is whether the characterisation of semantics that Lewis uses, viz., 'semantics describes facts about natural language', is sufficiently specific to rule out a dynamic conception of natural language meaning. Taken literally, it does not seem to do so. If we analyse the meaning of a certain expression in natural language as consisting in a context-change potential, in what way do we go beyond describing 'facts about natural language'? As was already noted, it is not that there is some theory-independent way of identifying what those facts are, that we can appeal to in order to answer this question.

It appears that we need additional considerations if we are to conclude, as Lewis intends to, that the dynamic effects that we are concerned with here can not be part of semantics, but must be accounted for in pragmatics. Adopting a classical, truth-conditional and static semantics as that which 'describes facts about natural language' will do the job. And it is a way of looking at things that has a venerable ancestry, given that it is the most prominent account of the 'language–world' relationship that the semiotic conception claims semantics is concerned with. But, and this is crucial, it can hardly be appealed to as an argument. It is a stipulation, one that may be justified in a number of ways to be sure, but it is not in and by itself a move that has argumentative force.[7]

So, it seems we're stuck: an appeal to prior characterisations of what semantics and pragmatics are is unlikely to be both sufficiently restrictive and theory-neutral to allow us to reach a decision as to whether the relevant effects of language use should be accounted in one or the other. And per implication that means that along these lines we will not be able to adjudicate the question whether dynamic semantics is a bona fide theory of natural language semantics. So what are we to do?

One obvious suggestion would be that we have been barking up the wrong tree all along in looking for conceptual-methodological arguments to decide the issue, and that we rather should go back 'zu den Sachen selbst'. After all, natural language is an empirical phenomenon, and so is its semantics. Therefore, shouldn't we be able to conclude on empirical grounds that dynamic semantics is on the right track, or that static semantics is the empirically adequate characterisation of what natural language meaning is?

To be sure, in the conceptual-methodological considerations that one can find in the literature, empirical arguments are deemed relevant as well. And that appears only natural if only because the theoretically motivated preference for dealing with

---

[7] And we would do well to note that it is not that even if we accept the semiotic characterisation of semantics as a neutral starting point: what 'the world' is, is left underspecified in that characterisation, and there seems to be no a priori way of ruling out that information states of language users are part of 'the world'.

certain phenomena in semantics, or rather in pragmatics, comes with the obligation to show that it actually can be done in that way. And that needs to be shown, then. But do note that there is a matter of fact here only if we have fixed the format of a semantic and a pragmatic theory. Which means that we turn around in circles: we can appeal to empirical considerations only if we assume we have reached consensus on the conceptual issues. And the latter, it seems, can not be obtained solely by conceptual considerations but time and again steers us towards the empirical.

The question of empirical adequacy has been discussed in the literature in some detail.[8] But what is important to note is that apparently it is not easy to come up with 'hard evidence', i.e., with a phenomenon of which we can show that it can, or can not, be accounted for in a particular way. In fact, if we look back at, e.g., the discussion between Kamp and Groenendijk and Stokhof in the late 1980s, we note that the argument was never about empirical coverage *per se*, but about accounting for a set of empirical phenomena in a particular way.

Another illustration of the fact that empirical and theoretical motivations come as a mixture, is provided by Cresswell in [9]. He takes DPL as his point of departure and then develops an alternative, in the sense of a theory that has the same empirical coverage as DPL, that is static, i.e., truth-conditional. The details need not concern us here, what is relevant for our discussion is the way in which Cresswell characterises his own enterprise:

> My purpose in these last two sections has not been to adjudicate between the use of double assignments, as in Groenendijk and Stokhof, and 'namely' variables [which are the new logical tool that Cresswell introduces, MS] , but simply to point out that the translation scheme shows that there is no empirical difference between the two approaches.

This is quite representative of a lot of work that has been done in this area. One takes a fixed set of phenomena, a given account of them (static or dynamic), and then develops an alternative account (dynamic or static) that has the same empirical coverage. That is interesting and revealing, but the key question in the present context is: what does it tell us about the choice between dynamic and static accounts of meaning? Does it tell us anything at all?

According to Cresswell it does. In the passage just quoted, he continues to draw the following conclusion:

> And since the use of free variables does not constitute a departure from the standard truth-conditional account of meanings, then neither do the empirically equivalent dynamic theories of semantics.

But that seems a non sequitur. As was mentioned above, one of the main elements in the development of dynamic predicate logic was to show that DPL and DRT account for exactly the same facts. Yet, neither proponents of DRT nor those of DPL subsequently claimed that therefore the two theories are somehow the same. On the contrary, the very fact that both theories were able to account for the same phenomena focussed the discussion on their conceptual differences, and on the justification of

---

[8] Beside the older literature that has been referred to above, cf. e.g. the more recent [8, 11, 21, 24]; and [22] (already mentioned).

their respective conceptual and methodological assumptions.[9] In other words, when comparing theories we can not look *just* at their empirical coverage, we also need to take into account their conceptual apparatus.

But then it seems we are indeed back at square one. Conceptual considerations seem to be unable adjudicate because the relevant concepts—of semantics, meaning, pragmatics, and the like—are intrinsically theory-dependent. And empirical considerations don't really help us out because obviously empirical equivalence (or empirical difference for that matter) simply does not say enough. So we are in a conundrum.

In such a case, it is best to take a step back, and to take a closer look at what kind of enterprise we are dealing with here. That involves a lot of issues and a great number of considerations, The issues are complex and we can not hope to do justice to all of them. So in what follows we just want to point to a particular aspect that hitherto has received little or no attention in the discussion: the role of formal systems.

28.4 Another Take

What appears to be a central, though not too often explicitly thematised, factor in how one adjudicates the issue, concerns how one views the role of formal systems in natural language semantics.[10] By 'formal system' we mean here minimally a formal language plus an explicit model-theoretic and/or a proof-theoretic account of its logic. The 'logic' part can be more or less extensive, depending on what the system is meant to capture.[11] It should be noted that especially in linguistic applications, the specification of the logic is often subdued, which explains the tendency to phrase these issues in terms of the role of 'formal languages', rather than 'formal systems'.

Let us start with natural language semantics. As we have argued elsewhere [29], there are (minimally) two main perspectives on the role that formal systems have to play in natural language semantics. One perspective is that a formal system is primarily a tool, something that is used in formulating and evaluating a theory. But there is an earlier, and arguably still dominant perspective on which a formal system is viewed as a model for a natural language with its semantics. Here the central features of the formal language employed are supposed to model similar features of the natural language that is being described. On this view, the task of the semanticist is twofold: first to find a formal language which has the required properties; and second,

---

[9] As was already mentioned earlier, compositionality played a key role in that discussion, and it is interesting to note that there are indeed good arguments that compositionality is not an empirical issue, but a methodological principle. Cf. [16] for more discussion.

[10] Similar considerations apply to formal systems in other domains, e.g., in the kind of naturalistic philosophical analysis that is exemplified in dynamic epistemic logic. We can not go into these matters here, but cf. e.g. [6] for discussion.

[11] Another explanation is that in the early days of generative grammar, natural languages were primarily studied from a syntactic point of view, often in terms of structural properties familiar from the theory of formal languages.

to devise a systematic relation between natural language and formal language so that, in the relevant respects, the latter can be seen as going proxy for the former.

It is interesting to note that Stalnaker, in [27], construes a major difference between DRT and DPL precisely in these terms: the latter, but not the former, models natural language. Now, contrary to the suggestion in [27, p. 4], it is obvious that the proponents of dynamic semantics do intend these to be part of a semantic theory of natural language, i.e., to be combined with a systematic account of how syntactic structures are mapped onto representations in a formal language. To give just one example, Dynamic Montague Grammar [13] closely follows the lead of Montague's original set-up and gives an explicit definition of the mapping that takes natural language expressions (or rather, their derivation trees) into expressions of intensional logic, which then is interpreted in a dynamic way. From that perspective, Stalnaker's claim (*loc. cit*) that:

> Dynamic predicate logic, on the other hand [i.e., in contrast to DRT, MS], is only indirectly relevant to any natural language. It defines an artificial language with new kinds of dynamic variable binding operations, obviously different from anything in natural language, but presumably intended to model, approximately, some of the devices used in natural language.

is decidedly odd. First of all, as the references just given show, dynamic semantics is explicitly intended to be part of a systematic theory. And secondly, the use of an artificial language with elements that lack direct counterparts in natural language applies to *all* formal semantic theories that implement indirect interpretation, be they static or dynamic. It definitely applies to DRT as much as it does to DPL, but it also applies to static theories, including those that implement Stalnakerian ideas.

But the quoted passage is also interesting for another reason, viz., because of the view on the role of formal systems it appears to assume, or, rather, the ambivalence with regard to this role. On the one hand, Stalnaker maintains that the extended variable binding of the existential quantifier in DPL is 'obviously different from anything in natural language'. This is true, but in a rather trivial sense. On the other hand, the introduction of these devices is 'presumably intended to model, approximately, some of the devices used in natural language'. Forget about the 'approximately' for the moment, what is interesting is that apparently, for Stalnaker there is a tension in using a concept or structural property of a formal system that has no direct counterpart in natural language to model some aspect of that very same natural language. But this is confusing, as it appears to put unrealistic constraints on the relationship between a formal system and what it models, in this case a natural language.

Different views on to the role of formal systems are connected with different views on what a semantic theory should do, on what the relation is between a semantic theory and what it is a theory of, viz., the semantics of natural language. The modelling approach constructs the explanatory force of a semantic theory in terms of the successful modelling of (part of) natural language semantics by means of a formal system. The other main perspective, which views formal systems as tools, defines the task of a semantic theory as *describing* relevant aspects of the semantics of a natural language, and on the basis of such descriptions providing explanations of various regularities. On this view, the use of formal systems is akin to their role in other sciences, such as biology, or physics.[12] Here the primary criteria are expedience, in addition to economy, elegance and simplicity, but *not* similarity, in terms of structure and concepts. If we view semantic theory as a descriptive-explanatory enterprise, we would allow the theorist in principle the use of any formal system that gets the job done. Descriptive adequacy is the first and most important criterion. Of course, when two or more equivalent descriptions are available, other considerations become relevant, but similarity, it seems, is never a pre-condition, and hence not a principled consideration to judge semantic theories.

This appears uncontroversial. But there is a danger lurking here as well, for it would seem to follow, on this view of what the task of a semantic theory is, that we should be able to identify the facts and features that we intend to describe independently of the means that we bring to the task at hand. As we have seen above, that is a dangerous assumption: there may be facts in the sense of their being systematic patterns in judgements about entailment, or synonymy, or analyticity, but there certainly are no facts that can be classified as 'semantic' or 'pragmatic' independent from conceptual and methodological assumptions. And that implies that the key issue is whether the choice of a formal system can be made in terms that are not informed, one way or another, by such assumptions.

Of course, the alternative modelling approach is not better on this score, on the contrary. If we assume that the formal system that we use in our semantic or pragmatic analyses somehow models the relevant aspects of natural language because they share core features, we risk to loose any explanatory force. For this type of modelling can be considered adequate only if we have independent access to the relevant features: we can only judge whether some formal system accurately models aspects of natural language if we can compare them in the relevant respects. And that means that we have to be able to access the features of the formal system and those of natural language independently from each other.

With a formal system that is, of course, not a problem. We can investigate such a system and ascertain its properties. And we can, of course, simply design a system in such a way that it has the properties we want. In the case of a natural language, however, the situation is completely different. If a natural language is an empirical phenomenon then it is what it is, and has the properties that it has, quite independent from our access of them, and even quite independent of their accessibility.

Given this asymmetry, the use of formal systems in an descriptive–explanatory role in natural language analysis is puzzling. To put it bluntly: what is the point? If we need to have independent access to the relevant features of natural language that we want to model with a formal system in order to be able to decide whether the formal system is an adequate model, then what do we stand to gain? The answer here is not 'Nothing', since we can use formal systems as models in useful ways: to provide concise overviews of features that we are interested in, to come up with 'perspicuous presentations' of them. But one thing such models can not be, and that is descriptions with explanatory power.

---

[12] Of course, there are many differences as well, but these are not relevant for the main point that is at stake here.

To summarise: it seems that a formal system is regarded either as a tool or as a model, i.e., it is used either to describe or to express[13] features of natural language. In the first case the internal conceptual structure of the tool is of secondary importance (which is not to say that it is of no importance at all): what counts is whether it gets the job done, and it is first and foremost on those grounds that particular conceptions are justified (with other considerations coming in only as secondary). That is a pragmatic justification, one that comes without much ontological implications: that we can successfully describe a certain range of phenomena using a certain concept does not commit us to the existence of anything corresponding to the concept, not even after dutiful application of considerations of economy and simplicity. On the second view, however, the modelling one, this is essentially different: there the conceptual structure of the formal system is supposed to align with that of the natural language. But, and this is the crucial point, it can do so only by design, so to speak. We already need to have a good grip on the nature of what we want to model in order to be able to find a system that 'fits'. But explanatory value such a fit does not have, at least not as long as there are no independently, empirically motivated constraints on the formal systems that we can regard as candidate models.

So, if natural language semantics is an empirical discipline, one which has descriptive-explanatory goals, the proper perspective on the role that formal systems might play is a thoroughly pragmatic one. Formal systems are tools, selected first and foremost for their ability to get the job done, i.e., for the descriptive power that they provide. The tools in and of themselves don't need to have any essential characteristics in common with what they are applied to.

28.5 So What?

The consequence of these considerations for the issue that is at stake in this paper, viz., how to decide whether dynamic semantics is correct in claiming that natural language meaning is a dynamic concept, is straightforward: we can't decide the issue in a remotely theory-independent way. This is illustrated by the simple, but significant observation that the description of dynamic effects can be done in static terms, but the modelling of something dynamic has to be done in terms of something that is itself dynamic. These two perspectives are mutually exclusive since, as we argued above, they do not depend on any 'fact-of-the-matter', but represent choices to do things one way rather than another.

This does not make all discussion pointless, of course. Not all formal systems have the same descriptive power, so if we want to describe certain dynamic effects in static terms we still need to find the right system to do that. Analogously, if we want to model these effects we need a dynamic system that has the right dynamic features. Within each of these two settings there is ample room for discussion, as there can

---

[13] Or 'show'; cf. [28] for an extensive analysis of how the universalism of Wittgenstein's early work, with its associated distinction between 'saying' and 'showing', is connected with the two conceptions of the role of formal systems in natural language analysis outlined here.

be better and worse tools, and we need to go into the empirical details in order to be able to decide which is which. It is between these settings that discussion loses its point. There is no answer to the question whether natural language meaning 'is' dynamic, or not: we simply lack a theory-independent concept of natural language meaning that we can refer to in our attempts to decide the issue one way or another. If we ignore this, conceptual muddles and misguided discussions will be the result.

Such considerations as the above apply in a wider context that the question of dynamic meaning. They are not confined to natural language semantics, but also apply in other contexts, such as the application of logic in the analysis of philosophical concepts, in cognitive science, and so on. A classical example is provided by the study of modal concepts. Here we can use modal logics to model the properties of modal concepts,, but many of their features can also be described using a non-modal tool, such as first order logic. In this case too, basically the same considerations as outlined above apply. This is not an empirical issue, or one that can be decided on conceptual grounds. Ultimately is a a matter of choice, If we fail to see that, we end up in fruitless debates, or, to use van Benthem's poignant phrase, 'system imprisonment'[14]:

> Nevertheless, I am worried by what I call the 'system imprisonment' of modern logic. It clutters up the philosophy of logic and mathematics, replacing real issues by system-generated ones, and it isolates us from the surrounding world. I do think that formal languages and formal systems are important, and at some extreme level, they are also useful, e.g., in using computers for theorem proving or natural language processing. But I think there is a whole further area that we need to understand, viz., the interaction between formal systems and natural practice.

I read van Benthem here as arguing for a pragmatic stance that is akin in spirit to the one outlined in this paper. Indeed, there is a lot that we still need to understand about the role that formal systems can, and cannot, play in an adequate account of our 'natural practices', i.e., in coming to an understanding of the ways in which we reason, use language, and so on. But if we can agree that a pragmatic attitude provides a better starting point than the essentialistic perspective that has informed too much of our discussions thus far, we have gained something.

References

1. van Benthem J (1986) Essays in logical semantics. Springer, New York
2. van Benthem J (1989) Semantic parallels in natural language and computation. In: Ebbinghaus H-D et al (eds) Logic colloquium '87, studies in logic and the philosophy of mathematics, vol 129. North-Holland, Amsterdam, pp 331–375
3. van Benthem J (1991) General dynamics. Theor Linguist 17:159–201
4. van Benthem J (1996) Exploring logical dynamics. CSLI, Stanford
5. van Benthem J (1999) Wider still and wider: resetting the bounds of logic. In: The nature of logic, European review of philosophy, vol 4. The University of Chicago Press, Chicago

---

[14] In van Benthem [5].

6. van Benthem J (2013) Implicit and explicit stances in logic
7. van Benthem J, Muskens R, Visser A (1997) Dynamics. In: van Benthem JFAK, ter Meulen, Alice GB (eds) Handbook of logic and linguistics. Elsevier, Amsterdam, pp 587–648
8. Breheny R (2003) On the dynamic turn in the study of meaning and interpretation. In: Peregrin J (ed) Meaning: the dynamic turn. Elsevier, Dordrecht, pp 69–89
9. Cresswell MJ (2002) Static semantics for dynamic discourse. Linguist Philos 25:545–571
10. van Eijck J, de Vries F-J (1992) Dynamic interpretation and Hoare deduction. J Logic Lang Inf 1(1):1–44
11. Gauker C (2007) Comments on dynamic semantics. APA Central Division, Chicago
12. Groenendijk J, Janssen T, Stokhof M (eds) (1984) Truth, interpretation and information, Grassseries, vol 2. Foris, Dordrecht
13. Groenendijk J, Stokhof M (1990a) Dynamic Montague grammar. In: Kálmán L, Pólos L (eds) Papers from the second symposium on logic and language. Akadémiai Kiadó, Budapest, pp 3–48
14. Groenendijk J, Stokhof M (1990b) Two theories of dynamic semantics. In: van Eijck J (ed) Logics in AI. Springer, Berlin, pp 55–64
15. Groenendijk J, Stokhof M (1991) Dynamic predicate logic. Linguist Philos 14(1):39–100
16. Groenendijk J, Stokhof M (2005) Why compositionality? In: Carlson G, Pelletier J (eds) Reference and quantification: the Partee effect. CSLI, Stanford, pp 83–106
17. Groenendijk J, Stokhof M, Veltman F (1996) Coreference and modality. In: Lappin S (ed) Handbook of contemporary semantic theory. Blackwell, Oxford, pp 179–213
18. Heim I (1982) The semantics of definite and indefinite noun phrases. Ph.D. thesis, University of Massachusetts, Amherst (Published in 1989 by Garland, New York)
19. Heim I (1983) File change semantics and the familiarity theory of definiteness. In: Bäuerle R, Schwarze C, von Stechow A (eds) Meaning, use, and interpretation of language. De Gruyter, Berlin
20. Kamp H (1981) A theory of truth and semantic representation. In: Groenendijk J, Janssen T, Stokhof M (eds) Formal methods in the study of language, MC tracts, vol 135. Mathematical Centre, Amsterdam (Reprinted in Groenendijk et al. 1984, p 1–41)
21. Lewis KS (2011) Understanding dynamic discourse. Ph.D. thesis, Rutgers, New Brunswick, New Jersey
22. Lewis KS (2012) Discourse dynamics, pragmatics, and indefinites. Philos Stud 158(2):313–342
23. Rothschild D, Yalcin S (2012) On the dynamics of conversation
24. Schlenker P (2007) Anti-dynamics: presupposition projection without dynamic semantics. J Logic Lang Inf 16:325–356
25. Stalnaker R (1974) Pragmatic presuppositions. In: Munitz M, Unger P (eds) Semantics and philosophy. New York University Press, New York
26. Stalnaker R (1979) Assertion. In: Cole P (ed) Syntax and semantics 9—pragmatics. Academic Press, New York
27. Stalnaker R (1998) On the representation of context. J Logic Lang Inf 7:3–19
28. Stokhof M (2008) The architecture of meaning: Wittgenstein's Tractatus and formal semantics. In: Levy D, Zamuner E (eds) Wittgenstein's enduring arguments. Routledge, London, pp 211–244
29. Stokhof M (2012) The role of artificial languages. In: Russell G, Fara DG (eds) The Routledge companion to the philosophy of language. Routledge, London/New York, pp 544–553
30. Stokhof M, van Lambalgen M (2011) Abstraction and idealisation: the construction of modern linguistics. Theor Linguist 37(1–2):1–26
31. Veltman F (1984) Data semantics. In: Groenendijk J, Janssen TMV, Stokhof M (eds) Truth, interpretation and information. Foris, Dordrecht, pp 43–62
32. Veltman F (1986) Data semantics and the pragmatics of indicative conditionals. In: Traugott E et al (eds) On conditionals. Cambridge University Press, Cambridge
33. Veltman F (1996) Defaults in update semantics. J Philos Logic 25:221–261
34. Vermeulen CFM (1994) Incremental semantics for propositional texts. Notre Dame J Formal Logic 35(2):243–271
35. Visser A (1998) Contexts in dynamic predicate logic. J Logic Lang Inf 7:21–52

Chapter 29
Logic of and for Language, and Logic of and for Mind

Hans Kamp

Abstract This is a largely informal and increasingly speculative reflection on the implications and the importance of formal semantics and formal logic, not just because they throw light on the subject matters with which they deal, but also because of their power to make those who become acquainted with them better speakers and better agents. The earlier parts of the paper rehearse some familiar formal points about model-theoretic accounts of the semantics of natural languages and then ask in what ways treatments can be tested against the intuitions of competent speakers. It is argued that model-theoretic accounts are testable even when speakers have firm judgements only about some of the sentences of the language fragment treated by the theory; and, further, that once a speaker who has verified the treatment to his satisfaction against those judgements of which he feels certain, the theory may help him to deepen his understanding of his language and thus also to improve as a speaker. I then go on to argue that a similar dynamics governs our interaction with axiomatic theories of non-linguistic aspects of human behaviour, such as knowledge or action: We may convince ourselves that such a theory is right by observing that it agrees with the judgements about aspects of the formalized concepts about which we are confident and then let the theory guide us to a better understanding of other aspects of those concepts; and that may change us as agents. These considerations suggest that there is more in common between formal treatments of the semantics of natural language and formal logics of aspects of cognition and action than is usually assumed.

H. Kamp (✉)
IMS, Stuttgart University, Stuttgart, Germany
e-mail: hans@ims.uni-stuttgart.de

H. Kamp
Department of Linguistics, University of Texas at Austin, Austin, USA

H. Kamp
Department of Philosophy, University of Texas at Austin, Austin, USA

A. Baltag and S. Smets (eds.), *Johan van Benthem on Logic
and Information Dynamics*, Outstanding Contributions to Logic 5,
DOI: 10.1007/978-3-319-06025-5_29, © Springer International Publishing Switzerland 2014

29.1 Formal Languages as Tools for Sharpening our Understanding of Your Mother's Tongue

To really understand the meaning of a sentence, it has often been suggested, is to know what follows from it. That sounds rather circular. For the suggestion makes sense only when knowing what follows from a sentence isn't just knowing what other sentences follow from it, but also knowing what those sentences mean. But then . . .?

Nevertheless the suggestion has been immensely useful, definitely for more than a century, and probably for much longer. Given the apparent threat of circularity, it is worth asking oneself why.

The way logic was used in the analysis of language throughout the first two thirds of the 20-th century shows us an answer. Some of the statements we encounter in philosophy and mathematics are hard to process. In some sense we understand them when we hear or read them, but in another sense, which in mathematics, science or philosophy matters crucially, we don't, or it dawns on us after a while that we don't, or didn't. In such situations it is often useful to have a logical formalism at hand that we can use to highlight and profile those aspects of the sentence meaning that thus far we didn't see or appreciate enough: By 'translating', or 'symbolizing', the sentence into the formalism we bring features of its meaning into focus which we did not recognize clearly enough, but which are essential if we are to recognize those inferences (and non-inferences) that matter in the given philosophical, mathematical or scientific context. A classical example is the power of quantification theory to reveal the import and importance of quantifier scope. I still quite vividly remember the functional analysis class I attended as an undergraduate at Leiden University, in which several weeks were spent on explaining the difference between continuity and uniform continuity. We weren't taught any symbolic logic then and my first serious encounter with the predicate calculus did not occur until 3 years later, when I already was a graduate student in Amsterdam. In retrospect, it was quite clear that the difference between continuity and uniform continuity was just a matter of scope—the difference between 'for all x and for all ϵ there is a δ such that . . .' and 'for all ϵ there is a δ such that for all x . . .'. If we had been attuned to this distinction in those terms, I am sure that we could have proceeded much faster: A familiarity with a formalism such as the Predicate Calculus *helps us to understand better* the language which we use in any case to express the things we want to say.

That the Predicate Calculus is a formalism that can help us in this way is at least in part explained by the fact that it was the result of a careful, deep and astute analysis of how we can express, with great accuracy, mathematical and philosophical propositions in the language we have. Learning the Predicate Calculus is therefore to a large extent learning how to say things in it that we do know how to say in our mother's tongue—learning how our ways of saying those things in our mother's tongue can be rendered in the formalism. But even so, learning how to use such a formalism in this way—finding the correct representations in it for new sentences of the natural language we speak, which had not crossed our path and which resist immediate understanding—can be a true challenge. But—and this is what is remarkable—the challenge can often be met: we can find a formalization for a new sentence, and once we have found it we can first convince ourselves that it *is* the right formalization for this sentence and then exploit our knowledge of the formal properties of the formalism to get a better grip on the meaning of the sentence than we had to start with.

How this kind of two-way interaction between natural language and formal logic is possible is, it seems to me, one of the great mysteries of human cognition. (If ever there were true cases of the Paradox of Analysis,[1] these are prime examples; each application of a logical formalism to clarify or sharpen our understanding of a sentence we somehow understand well enough to find a symbolization for it, and to persuade ourselves that it is correct, is an instance of this paradox.)

This is the way that formal logic was used during the first and a good part of the second half of the last century—often with remarkable success—and in which it is still used today—and still often beneficially—within philosophy. But the results aren't fruitful invariably. In fact, the risk of things going wrong is real and prominent enough to make us want to find better ways of controlling our use of formal logics in the exploration of meaning—ways of demystifying the successes and guarding against the failures.

29.2 Logics as Specification Formalisms in Model-theoretic Semantics

The decisive break-through—this point is overly familiar but it needs an explicit statement here—was Montague's seminal work on the semantics of natural languages, in which he found a way of applying to the analysis of natural language the model-theoretic method that Tarski and his school had developed for formal languages such as Predicate Logic. When talking about this work today, it is important to give visibility to the magnitude of its achievement by stressing the strength of the then prevailing prejudice according to which such an enterprise must be doomed to failure. We have come to take it for granted that natural language meaning can be analysed in this way and that such analyses can advance our understanding of the languages to which we apply them. After all this is what Montague has shown us. But if anything was taken for granted at the time when he did his work, it was that it couldn't be done.

It is also important, however, to stress how *natural* the idea was of treating the semantics of human languages in the same way that one already knew how to do in

---

[1] The Paradox of Analysis was originally formulated in connection with the notion of conceptual analysis as propagated in the work of G. E. Moore. The 'paradox' is this: suppose I want to clarify a concept C and propose an analysis (or 'explanatory definition') A for it. Then my proposal meets the following predicament: either (i) C is understood well enough to make it possible to see that A is a correct analysis of it; but in that case the analysis cannot tell anything about C that wasn't already known; or (ii) A does tell us something about C that wasn't yet known; but then it isn't possible to verify that A is a correct analysis of C. See e.g. [1, 10, 22].

dealing with the semantics of the languages of formal logic. I remember Montague commenting more than once in private conversation in 1965 on what he saw as an oddity of formal logic as it was taught then [and in particular as it was presented in his own text book (written jointly with Donald Kalish, see [19])]: 'Everything in that book', he said, 'is spelled out in painstaking formal detail—the syntax of propositional and predicate logic and the method of proof. But we say nothing explicit about how you go from English sentences and English arguments to their symbolisations. That is taught by example; somehow the student must catch on.' It was in a way that gap which Montague then, shortly afterwards, proceeded to fill when he produced his path-breaking papers on natural language semantics.

I say 'in a way' because Montague saw his work on natural language semantics emphatically *not* as showing how the sentences of the fragments of English that are analysed in his papers can be translated into some formal language. In his papers on natural language, the natural language fragment is related directly to the models they specify, by a function that assigns to syntactically well-formed expressions semantic objects that are connected with those models—the 'semantic values' that are 'denoted' by those expressions in those models. Translation into some other formal language is a by-product of the way in which this function is defined; but this by-product shouldn't, he insisted, be confused with what he took to be the central task of a theory of meaning: showing how natural language expressions project onto their semantic values, by virtue of the meanings of their words and their own intrinsic grammatical structure.

And yet you cannot really avoid the by-product. For—first point—you need a precise way of talking about the models that are part of a model-theoretic account of meaning and about the semantic values associated with those models. And that talk is possible only in a highly regimented language, a formalism that is designed for the task. In fact, specifying such a formalism and specifying the models for the natural language fragment that is being treated are two sides of a single coin. If the formalism is to be truly up to its task, then it must describe those models and the objects associated with them *transparently*: the syntactic structure of its expressions must reflect the structure of the models and associated semantic objects as close and directly as possible. The second point has to do with the use that Montagovian accounts make of these formalisms. The formalism is used to specify 'canonical' descriptions for the semantic values of the well-formed expressions from the language or language fragment dealt with in the account, and in particular for the semantic values of its sentences. And the semantic values of sentences are like propositions or truth conditions in that for any given sentence A, the semantic value of A determines for any model M from the model class proffered by the account whether A is true in M or false. (From now on I will refer to the formalism that a given model-theoretic account uses for the purposes just indicated as its *specification formalism* and to the language or language fragment for which it provides a semantics as its *object language*.)

What has been said in the last paragraph about the part that specification formalisms play in model-theoretic treatments of meaning has the following implication. Suppose we want to find out what the treatment has to say about entailment relations between sentences of its object language. Suppose, more specifically, that we want to know the treatment's verdict on whether sentence A entails sentence B. Given the model-theoretic definition of entailment as preservation of truth, A will entail B according to the treatment iff the semantic values of A and B are such that if in any model M, A is true according to the semantic value assigned to A, then B is true in M according to the semantic value assigned to B. But this is nothing other than that the proposition-like object that is the semantic value of A entails the proposition-like object that is the semantic value of B. Or, to put the same point in slightly different terminology: Let us call a term τ of the specification formalism that the account uses to specify the semantic value of an expression α of the object language the *logical form* of α and denote this value as '$\tau(\alpha)$'. Then the account has it that A entails B iff $\tau(A)$ entails $\tau(B)$. In other words, entailment between object language sentences is reduced to entailment between their logical forms. So the logical form assignment that is part of the proposed model-theoretic account delivers an entailment-preserving translation from the object language into the specification formalism.

I have been going over this thoroughly familiar ground at such length because I wanted to bring out the inevitability with which model-theoretic accounts of semantics generate assignments of logical forms to expressions of their object languages. Insisting, as Montague does, that the point of such accounts is to articulate semantic relations between object language expressions and entities 'in the world' and that the specification formalism is just a tool that is needed to do that, makes no difference as far as that is concerned.

But nevertheless it is important to keep in mind *why* Montague insisted that the point of his account of natural language semantics was that of describing semantic relations between expressions and models, and not that of providing a semantics for the object language through translation into some other language for which the semantics has been cleared in advance. His reason for insisting was his conviction that there is no fundamental difference between natural languages and the artificial languages of symbolic logic; both are amenable to a semantic analysis that relates grammatically well-formed expressions to models. What specification languages are being used as part of what is needed to state these relations is immaterial to that general point.

29.3 Using Natural Languages as Specification Formalisms

In fact, there is no a priori reason why there couldn't be an account in which the role of specification formalism is assigned to some natural language or suitable fragment thereof. And as a matter of fact, proposals to this effect are well-documented. First, there is the long tradition of Davidsonian truth theory, in which a natural language is used at the meta-level to specify the truth-conditional semantics of the same or some other natural language [16, 17]. (Davidsonian Truth Theory shuns models, so it cannot be described as a *model-theoretic* treatment of natural language in which a natural language is used as specification formalism; but for the issue at hand this distinction between model theory and truth theory does not matter.) A second proposal to this effect is the one, going back to Boolos,[2] to use English (or some other natural language) as specification formalism when stating the semantics of second order logic. Essential to the natural language fragment used for this purpose is that it include plurals, which, it is contended, are optimally suited to capture the semantics of the second order quantification that distinguishes second from first order logic.[3]

The arguments given in support of these two proposals are different. The principal motivation for Davidsonian Truth Theory is that it captures what goes on when a speaker S with natural language L' interprets the language L of some other speaker S' (or the language of some group of people or speech community). An important special case of this is supposed to be that in which L' is superficially the same as L, and where there is a presumption that they are the same language in more than superficial appearance. A Davidsonian Truth Theory of this type, it is suggested, captures what is involved when S accepts S' as a speaker of the same language that he speaks himself.

The justification that is given for a Boolos-like approach to second order logic is that the student of second order logic, in his capacity of a competent speaker of his mother tongue—let us assume for the sake of argument that that language is English—has a full command of it, including its various constructions involving plurals, and that because of this, English is the perfect tool for analysing the semantics of formal systems involving second order quantification (and I suppose for analysing the semantics of natural languages as well).

These arguments in favour of the proposals are often accompanied by objections to approaches of the kind exemplified by Montague's work, which employ logical formalisms as specification languages. The charge of these criticisms is either that these formalisms lack the foundation that their use as specification languages pre-supposes, or else that they distort the semantics of the object language.[4] The best, it is implied, that can be expected of accounts that use such specification formalisms is that they don't get things wrong. But any hope that we could learn something from such an account about a language L we already have must be idle. For either we know L well enough to be able to verify that the account is correct; but then, what could the account teach us that we do not already know. Or else we do not know L

---

[2] See e.g. [11–13].

[3] More recently there have been concerted efforts to turn Boolos' insights into explicit semantic accounts of second order logic in which the meta-language, in which the truth definition of Second Order Logic is stated, is inspired by the way plurals enable us to say things that would otherwise have to be said by referring to sets or mereological complexes. This branch of logic and semantics is usually referred to as 'Plural Logic'. For a recent contribution see e.g. [24].

[4] Specific charges of these sorts have been levelled at semantic accounts of modalities in terms of possible worlds. On the one hand the notion of a 'possible world' is said to be ill-defined or incoherent, which disqualifies specification formalisms that refer to possible worlds. On the other hand there is the objection that specification formalisms that analyse modalities in terms of possible worlds try to capture the meaning of essentially non-extensional notions (the modalities) in extensional terms (i.e. in terms of possible worlds). Such accounts therefore cannot fail to distort the semantics of the object language.

this well. But then we have no basis for trusting the account and so have no reason to put any trust in the lessons it might seem to offer us.

For what I want to say here, it is the critical part of these proposals that matters most—the charge that model-theoretic accounts which make use of formal languages as specification formalisms must be either unverifiable or uninformative. These criticisms seem to me to manifest a pessimism that I do not share and that I think isn't shared by most of the 'language and logic' community. (This must be true in particular of those who apply Montague's methods to the analysis of natural languages. If they did share the pessimism, they wouldn't be doing what they are engaged in much of the time.)[5]

29.4 Formal Semantics of Natural Language and the Paradox of Analysis

One way to describe this clash of convictions and intuitions is in terms of the Paradox of Analysis. Both the criticisms of the model-theoretic method fielded by the advocates of Truth Theory and the confidence in our own language mastery that forms the basis of Boolos' proposal to use English in a formal explication of second order logic imply that where linguistic meaning is at issue, there is no room for instances of the Paradox. Those at whom the criticisms are targeted, on the other hand, and among them in particular those who believe that the semantics of plurals can be clarified through the use of logical systems as specification formalisms, demonstrate their conviction that progress is possible in ways that are paradoxical in the sense the Paradox is designed to bring out into the open.

To articulate the controversies in these terms is to acknowledge that the advocates of the methods of formal semantics have a predicament: How *is* it possible to find out things about your own language through the application of such methods? There is no simple answer to this query. (Explanations of how there can be instances of the Paradox of Analysis can never be simple. That's after all why it is called a paradox.) But let me try. The reason why we often seem to be in a position to ascertain that a certain logical analysis of a certain construction (or group of interacting constructions) of our language is correct is that we have a sufficiently firm grasp of the semantics of

---

[5] In the case of Plural Logic the skepticism towards formal specification formalisms comes hand in hand with a curious optimism about our competence as native speakers: that our untutored grasp of the use of plurals in the language or languages we speak is good enough to justify the use of such a natural language to explicate the semantic foundations of higher order logic. This optimism I share even less. Here is one reason (I know it is not conclusive): Boolos [11] presents a translation algorithm from second order logic into English in which second order quantification translates into constructions involving plurals. My own experience with this algorithm is that when I try to apply it to any but the simplest formulas of second order logic, my head invariably spins out of control; I cannot get a grasp of the English sentence obtained as a translation. What I feel I desperately need in order to make sense of it is a further, formal analysis of it. But if I understand Boolos correctly, that is just the wrong thing to ask for.

simple instances of those constructions: We understand the truth conditions of simple sentences containing them. If the analysis makes verifiably correct predictions about these simple cases, then that may be all that is needed to justify our confidence that the analysis is correct. And once we have reached that point, we are in a position to use the analysis to improve our understanding of more complicated instances of the constructions, and the more complicated interactions between them.

This is no different, I think, from the way in which formal arithmetic helps us to understand what the numbers are. By the time children learn decimal notation and the routines for adding and multiplying which exploit that notation in school, they already have a solid grasp of (at least) a range of the smaller natural numbers— starting with 1 and extending, say, and with gaps perhaps, to 1,000. Catching on to the decimal notation means on the one hand being able to use it to describe numbers that are familiar already, and perhaps also to see how the notation can be used to capture what is already known about arithmetical operations, such as counting things, and addition. But at some point the understanding of how decimal notation works takes over, as it were. It now conceptually functions as a kind of foundation to understanding the number system and not just as a formal device that can be used to navigate within a structure that is known independently.

In the case of natural language these limitations can take at least two different forms: (i) that of an incomplete grasp of the rules of syntax which permit in principle the formation of grammatical sentences of arbitrary length and complexity; and, more importantly for present purposes, (ii) that of an incomplete grasp of how complex syntactic structures are mapped onto meanings, so that even for certain strings that we do recognise as grammatical sentences we cannot get their meaning properly into focus. The revolution in the theory of language that was brought about in the fifties and sixties of the last century, largely through the work of Chomsky[6] and that of Montague, has encouraged the conviction that first language acquisition is a matter of zeroing in on its general syntactic generation or admission principles and on the related semantic rules once and for all. On this conception there is no room for a partial mastery of those rules, which enables the speaker to apply them reliably to fairly simple strings, but not to strings of greater complexity.[7]

But formal semantics and logic *can* improve our understanding of what complex expressions of our language mean. That is why teaching introductory logic—when it is done in the right way and falls into the right minds—can be so useful: it can be a stepping stone for the student to better understanding his own language. After all, being good at logic *is* to a large extent a matter of being clear about the semantics of the

---

[6] See in particular [14, 15].

[7] Chomsky has always been well aware of the fact that natural languages admit strings as grammatical that are too complex for human processing and makes an emphatic distinction between competence and performance to account for this fact. But the limitations that a speaker may demonstrate because of performance constraints—e.g. to fail to recognise a certain string as grammatical although it is grammatical by the very principles she has internalised, or to fail to assign the correct interpretation to a string that she has recognised as a grammatical sentence—must be sharply distinguished from the partial mastery spoken of here. Limited performance and partial mastery of a language are very different things.

language you speak. My own (admittedly subjective) experience with work in formal semantics tells me that the same is true there. Here too, exposure to contributions to semantic theory can improve our capabilities as speakers. (Unfortunately, if I am right, then quite large doses of the medicine are needed before any effects will show in either case. But perhaps I am a little too pessimistic on this point.)

29.5 Model-theoretic Semantics as a Theory of Natural Language Entailment

The reason, I suggested above, why formal semantic analyses can teach us something about our own language is that they permit a kind of bootstrapping: we can convince ourselves that the given analysis is right for the comparatively simple cases for which our understanding is good enough, and then use the analysis to improve our understanding of more complex cases. But what form can 'verifying the analysis of the simple cases' take? Here Montague may be cited once more. For Montague the empirical import of the model-theoretic treatment of a fragment Fr_L of a natural language L consists in the predictions it makes about entailment relations between the sentences of Fr_L. It is assumed that speakers of L can judge which L-sentences follow from which, and thus can test the treatment's entailment predictions against this evidence. (It is perhaps not all that clear that this is the only way in which empirical adequacy of such treatments can be tested. But that is a question that doesn't need an answer here.)

If it is true, as suggested above, that even competent speakers of a language may have a limited understanding of the meaning of its expressions and that this understanding is restricted to its simpler expressions, then their native ability to judge which sentences follow from which will of necessity be limited as well. This does not necessarily compromise Montague's criterion. For the range of reliable entailment judgements may nevertheless be extensive enough to allow for adequate testing. And when a model-theoretic treatment has passed those tests, it can then take on its role as instrument for sharpening and extending the speaker's intuitions about entailment relations involving sentences which so far had been out of reach.

In what follows I will adopt Montague's criterion and focus on this inferential aspect of linguistic meaning—i.e. on the entailment relations between the sentences of a language L. The question to what extent such a narrowing of focus is justified brings us back to the one implicit in the opening paragraph of this essay: What is the relation between entailment and linguistic meaning? So far I have made no real effort to answer that question. But in a way the model-theoretic method, which has been our central topic so far, contains at least a partial answer to it: The semantic values that applications of the method assign to expressions of the object language can on the one hand be thought of as capturing their meanings, while on the other hand they determine which sentences of the object language are entailed by which: entailment is defined as logical consequence; A entails B iff the class of models in which A is true according to the semantic value assigned to A is included in the class of models in which B is true, according to the semantic value assigned to B.[8]

N. B. The thesis that judgements about entailments are our only means for verifying whether a model-theoretic account of some natural language fragment is empirically correct points towards a question that is of considerable conceptual importance but that appears to be technically hard: Is it possible to recapture the semantic values that such an account assigns to the expressions of its object language just from the entailment relation it defines? An answer to this question will be important in that it will tell you something about the extent to which semantic values are theoretical constructs, which could be chosen this way or that without making a difference to what follows from what. If I am right, then there is a strong sense on the part of most working semanticists that there just isn't much flexibility on this point—that you 'can't get the entailments right if you do not get the semantic values right as well'. But as things stand, I do not know of any formal results that bear on the question.

Let me summarize and repeat: The model-theoretic method comes with the assumption that there is a clear connection between meaning and entailment in one direction: When all meanings are fixed, in the form of semantic values, then so is the entailment relation, which is identified with the semantic relation of logical consequence. About the converse relation—can semantic values be recovered once entailment relations are known?—we are still very much in the dark.

29.6 Entailment in Natural Language: Model Theory or Proof Theory?

So far much of the discussion has hovered around the question whether you can perceive a relation of entailment between A and B if you don't know what A and B mean. The tenor has been that that isn't possible, either because knowing what a sentence means just *is* being able to recognise entailments involving it, or because

---

[8] In a reaction of Johan van Benthem to an earlier draft of this contribution he expressed his doubts that intuitions about entailments are our only means of checking the empirical adequacy of model-theoretic treatments. I would like, especially in reaction to that remark, to state my own conviction here that, yes, I also believe that there are other ways of verifying such treatments. One rests on our ability to understand the descriptions that are typically provided by such treatments for the models of their model classes. Those descriptions usually enable us to tell, when we are confronted with an actual or imaginary situation and a sentence that is presented as a description of it, what model or models from the formally specified model class corresponds to this situation. In such cases we can check correctness of the treatment by comparing the truth value that the treatment assigns to the sentence in the relevant model with our own speaker's judgement whether or not the sentence is true in the given real or imagined situation.

There may be further ways in which model-theoretic treatments can be put to the test, but for me this is the most prominent one; and also, I suspect, the one most closely connected with the semanticists' hunch that empirically adequate model-theoretic treatments of natural language fragments have little room for manoeuvre in assigning semantic values to the expressions belonging to these fragments.

if you do not know the meaning of A and B, there just isn't anything for you to apply your ability to perceive entailments to. But is there really such a tight connection between entailment and meaning? The history of logic in the Western world, going back to Aristotle's theory of the syllogism, suggests otherwise. It suggests that entailment relations are a matter of form—that you can verify that A entails B by observing that their syntactic forms are related in the right way. And making such an observation is possible even when you lack a complete grasp of the meaning of A and B, in that sense in which we do not 'grasp the meaning of' a formula of predicate logic with uninterpreted predicates—such as, say, $(\exists x)P(x)$ or $(\exists x)(P(x)\&Q(x))$, but the choice is arbitrary—because we have no way of telling for any real situation in which we might find ourselves whether the formula is true of it; the question isn't even well-defined. Of course, to see that the forms of $(\exists x)P(x)$ or $(\exists x)(P(x)\&Q(x))$ are related in such a way that the second cannot but entail the first must presumably involve some kind of awareness that the formal relation between them is such that would the second be true the first would be true as well. But a full semantic understanding, of the kind that is possible only when it is known what particular properties are denoted by P and Q, is neither available nor needed.

One of the driving motivations of the episode of modern logic which started with Frege and Peano and found its apotheosis in the Unified Science program that issued from the Vienna Circle was the conviction that all cases of entailment could in last analysis be explained in terms of relations of form. That such a reduction should be possible in general is far from obvious. Just consider two simple examples. The premise 'x is red' entails the conclusion 'x is not green' and the premise 'x is a cube and y is a face of x' entails 'y is a square'; see (1a, b). But are these entailments a matter of form? Not obviously. At least not if we take the form of 'x is red' to be '$P(x)$' and the form of 'x is not green' '$\neg Q(x)$', or take that of 'x is a cube and y is a face of x' to be '$P(x)\& R(x,y)$' and that of 'y is a square' to be '$Q(y)$'.

(1)   a. x is red $\models x$ is not green
      b. x is a cube and y is a face of $x \models y$ is a square
      c. $P(x) \models \neg Q(x)$
      d. $P(x) \& R(x,y) \models Q(y)$
      e. $P(x), \neg(P(x)\& Q(x)) \models \neg Q(x)$.

To maintain that on closer analysis these entailments are a matter of form too we have to dig deeper. For instance, we can define 'cube', 'face' and 'square' in terms of fundamental concepts of geometry and substitute the definientia for P, R and Q in (1d). That still won't turn premise and conclusion of (1d) into expressions whose forms are related by logical entailment. But we get such a relation when we extend the premise with a conjunction of axioms that capture the essential properties of the fundamental geometrical notions in terms of which 'cube', 'face' and 'square' have been defined. Note well, however, that this reduction is acceptable only because we can argue that the added premise has a special status—that it can be established as true once and for all, and independently of the question whether there is an entailment relation between the particular premise and conclusion that make up the argument in (1b).

The case of (1a) is similar but also different. In order to account for this case as an entailment by form alone it is necessary to appeal to the fact that red and green are mutually exclusive predicates and add the formal expression of this fact as an additional premise, as in (1e). Premises and conclusion are evidently related by a purely formal relation of entailment and the extra premise $\neg(P(x) \& Q(x))$ can be justified on account of having a special status. (It can be seen as a 'Meaning Postulate', which is true analytically.)

This way of reducing entailments to relations of syntactic form involving additional premises—the so-called 'Axiomatic Method'—has been immensely fruitful in mathematics, philosophy and the mathematical sciences. But for all its success, it isn't obvious that all cases of entailment can be reduced to purely formal relations in this way.[9] But there is another point that matters more for what is at issue here: Even if all cases of entailment can be reduced to relations of form, this need not mean that speakers of a language L can recognise entailments between sentences of L only by seeing them as standing in such a formal relation.

29.7 Recognition of Entailment and Mental Representation of Content

Is it possible for speakers to recognise entailment relations without recognising them as relations of form? I do not know. The question is a hard one, and quite possibly one of the hardest of all challenges for Cognitive Science. The reason why it is such a hard question, and why there is little possibility of answering it at the present time, is that it is tied to fundamental issues of mental representation and those are issues about which we are still very much in the dark—far, far more than we should. We aren't much closer to an answer to the question how information is represented in the human mind now than we were 30 years ago, and until we are very much closer than we are today, hypotheses about how speakers can have intuitions about what is entailed by what must remain mere guesswork.

But of course this is not to deny that what information we have must be mentally represented in some form or other. This must be so in particular for information at the level of conscious awareness. For how else would it be possible for us to make the systematic and sophisticated use of the different bits of information at this level that we can and do make of it, in deliberating, planning and so on, or just in putting our thoughts into words or grasping the meaning of the words of others? So it seems safe to assume this much at least: That conscious information is registered by the mind in the form of some kind of representational format (or formats). Moreover, let us assume that this is so in particular for the information carried by sentences of the

---

[9] For some this question may have become a matter of terminology: two sentences simply won't count as standing in a relation of entailment unless their relation *can* be explicated in terms of form along the lines indicated. My own intuitions go against this. The truth conditions of many sentences of natural languages are fixed with enough definiteness, I believe, to make the question which of them are related by entailment a meaningful one independently of any formal reduction.

languages that people speak. And let us also make the further assumption that it is such mental representations that are the vehicles of semantic understanding: that an agent understands the meaning of an expression from a language she speaks if and only if she associates a mental content representation with it, and that, finally, it is the content representation that she associates with a given expression, or utterance of that expression, that identifies the meaning she attaches to it.

If we make these assumptions, there is another one which more or less imposes itself as well: When a speaker S judges that a sentence A entails some other sentence B, then that judgement will typically involve S's mental representations of those sentences: A is judged to entail B because a relation of entailment is perceived between the mental representations of A and B. But nothing more can be said about what it is like to perceive an entailment relation between mental representations until more is known about those representations themselves—whether they are like sentences of some Language of Thought or what the Language of Thought might exactly be like.[10] In this essay I will not make any assumptions about the form of mental representation. So there is nothing specific that I will be in a position to say about exactly what perceiving entailments between sentences of the languages we speak consists in (over and above the bare assumption that it is part of being a competent speaker of your language that there is a fair number of such judgements you can make). In particular, there is no basis for assuming that perceiving an entailment relation between mental representations is to perceive a formal relation in the sense of formal logic.

Let us sum up to where we have got:

1. Perception of entailments between sentences of some public language is often based on perceiving entailment relations between mental representations for those sentences.[11] This latter relation may, for all we have said, be a formal one, but as things stand we have no basis for making this assumption.

2. A model-theoretic treatment T à la Montague of some fragment Fr_L of a natural language L will assign, for any model M, M-related semantic values to the expressions of Fr_L, and in particular truth-conditions to the sentences of Fr_L. But it will also, inevitably, determine 'logical forms' for the sentences of Fr_L, expressed in the formalism that T uses to specify semantic values.

3. It may be possible for a speaker S of Fr_L to persuade herself of the empirical adequacy of T by verifying that the predictions T makes about entailment relations between certain sentences of Fr_L match the judgements she is able to independently make herself. This doesn't require her to be capable of making independent judgements about entailments between all sentences of Fr_L, and

---

[10] The standard reference is [18]. For views more directly responsible for these remarks see [20, 21].

[11] I say 'often' because I do not want to exclude cases where we compute entailments by applying some formal account of meaning that assigns logical forms to premises and conclusion, and thereby provides ways of formally deriving the conclusion from the premises, without any direct appeal to the content representations of those sentences that may be in the mind of the agent who carries out the formal deduction. More on this point later on.

not even between all sentences she can understand (i.e. for which she has or can readily determine mental representations). All that is needed is that she can come up with a reasonable number of such judgements.

4. When S has verified T in this way to her satisfaction and on that basis has come to accept it as true, then T may help S to improve her understanding of L there where so far it had been lacking. In particular, it can help her to perceive entailment relations in cases where thus far she could not.

29.8 Improving Language Skills Through Model-theoretic Semantics

So far so good. But how exactly can T help S to determine whether a sentence A from L entails a sentence B from L? Here are three possibilities. Two of these can be regarded as *external*, and the third as *internal*. We start with the two external possibilities.

(*External*)
  T can help S to determine whether A entails B:

  i. by enabling S to deduce from the truth conditions which T assigns to A and those which it assigns to B that whenever the former are satisfied the latter are satisfied as well.
  ii. by enabling S to deduce the logical form that T assigns to B from the logical form it assigns to A. (This presupposes of course that T's specification formalism comes equipped with some kind of proof theory.)

In addition to these external ways in which T could assist S—in which it serves as a kind of toolbox, but without necessarily changing S's linguistic competence as such—there is, I believe, also a third possibility. As I have been hinting more than once, I believe that it is possible for a speaker of L to learn something from a model-theoretic treatment of L or part of it *qua speaker of L* (and not just in her role of theoretician interested in the semantic properties of L as an object of linguistic study). In other words, it is possible that T will make S into a better speaker of L. In particular:

(*Internal*)
  T can help S to determine whether A entails B:

  iii. because of her exposure to T, S can now associate mental representations with A and B and recognise that these representations stand in a relation of entailment, but her recognition of this relation requires no further direct appeal to T.

There is more to be said about each of these three possibilities. And we will have more to say later on about the second one. But at this point I want to stress a much simpler and more obvious point: The situation of a speaker S of L who is confronted with a model-theoretic treatment T of some part of L involves three distinct modes of content representation: (i) the language L; (ii) S's mental representations of content (and in particular her content representations of well-formed expressions of L); and (iii) the specification formalism of T. This triad is reminiscent of the 'Semiotic Triangle'. Two of the vertices, the public language and the mode of mental representation, are the same for the two triangles. But there is a difference in their third vertices. In the Semiotic Triangle the third vertex consists of the referents of the representations belonging to language or thought; in our triad it consists of canonical descriptions that the theory T makes available for such referents.[12]

The reason why I mention this analogy is that, just as the referents of the third vertex of the Semiotic Triangle are related *both* to the expressions of the public language and to the mental representations of its speakers, so the canonical descriptions of the third vertex of our triad are related both to the expressions of L and to the mental representations of S. In the light of what has been said so far, however, there would seem to be one important difference between the two cases: In discussions of the Semiotic Triangle it is usually assumed that there are one or more direct channels that connect the mind with reality, so that there are two reference relations that commute with the correspondence relation between linguistic expressions and mental representations.[13] In contrast, the things that have thus far been said about our triad suggest that here the only way in which mental representations are connected with the canonical descriptions of the specification formalism is via the expressions of L: T specifies canonical descriptions of the semantics of expressions of L, S associates mental representations with expressions of L, and when two and two are put together we get canonical specifications for the semantics of mental representations. (So we only get the first of the two equations of the last footnote, and this equation functions as a definition of ref_{men}, rather than as a contingent claim about an independently grounded mapping function.)

But how clear is this difference between our triad and the Semiotic Triangle? Suppose S has accepted T and adopted it as a reliable guide to the semantics of L. Then we should expect that she will connect the logical forms that T provides for certain sentences from L not only with those sentences themselves but also with her own mental representations for those sentences. In this way she will be able to relate T not only to her understanding and use of L, but also as bearing on her deployment of those mental representations in cognitive processes that do not involve an overt use of language. And when this point is reached, there will be a direct connection between T and S's mental representations, one that completes the diagram of our triad in the same way that the Semiotic Triangle 'diagram' is complete, by filling in that part of it that by the looks of it, our triad was missing.

---

[12] A classical reference is [23].

[13] 'Commute' in the following sense: if ref_{lin} is the reference relation for expressions from the public language and ref_{men} the reference relation for mental representations, and mrep is a 1–1 correspondence relation between well-formed expressions of the public language and mental representations of their content, then $\text{ref}_{men} = \text{ref}_{lin} \circ \text{mrep}^{-1}$ and $\text{ref}_{lin} = \text{ref}_{men} \circ \text{mrep}$.

If this is right, then the practice of formal semantics can not only teach us something about linguistic meaning and, sometimes, thereby turn us into better users of our own language; it can also teach us things about our non-linguistic cognitive competence, enable us to improve it and to demonstrate that by better and surer performances in which we display this competence.

29.9 Natural Language and Johan van Benthem

It is at this point that we can at last make contact with the aims of the present book. This book is a volume in honour of Johan van Benthem. But its aim is not only to make visible to all and sundry the breadth and depth of his contributions to science and philosophy, but also to sketch out, to the best of our ability, the various directions in which those contributions are pointing. But what, the readers of this particular essay may have been asking themselves with growing bemusement, does this essay have to contribute to that general goal? I'll try to explain.[14]

In a recent paper, van Benthem observes that he 'left natural language a long time ago' [9]. There may be some sense in which that is sort of true. In the early parts of his career, going back to the 1970s and 1980s, many of his contributions were unequivocal contributions to natural language semantics and syntax. (I am thinking here in particular of his work on quantifiers, on the language of time and on Categorial Grammar and the Lambek Calculus as formalisations of the logic of the syntax–semantics interface. See e.g. [2, 3, 5].) The influence of that work is still very much present among those working on the syntax and semantics of natural languages today. Since those early days van Benthem has—this much is true—turned increasingly to issues that one might be inclined to classify as belonging to formal and philosophical logic or as belonging to the theory of representation, manipulation and communication of information, rather than as issues in natural language semantics. But is it really true that these more recent 'non-linguistic' contributions aren't contributions to natural language? The reflections above suggest that we should think twice before agreeing.

The point of those reflections was that once we accept that the knowledge and use of a natural language must involve mental representations of linguistic meaning or content, we must acknowledge that formal accounts of natural language meaning— of the kind that formal semanticists following Montague's method have been at pains to articulate—may qualify not just as theories of and guidelines to the meaning and use of public languages, but also as theories of and guidelines to aspects of cognition that, inasmuch as they are connected to language at all, are that only in the indirect sense that they make use of the same mental representation formats. But once we have arrived at this view of the import of theories that are developed as, and unambiguously present themselves as, analyses of natural language phenomena, we can't help asking whether, conversely, contributions to our understanding of apparently non-linguistic

---

[14] Here is a small selection of van Benthem's seminal contributions to logic that—I suggest—are relevant to various aspects of cognition: [4, 6–8].

cognitive behaviour could not, in their turn, be seen as contributions to the theory of natural language.

Our earlier considerations in this paper have already primed us to two ways in which this might be so. These two ways correspond to a distinction that has been in the background of much that I have been saying, but the time has come to make it explicit. The distinction is perhaps easiest to explain in connection with propositional attitudes such as beliefs and desires. People (a) entertain propositional attitudes, (b) they often express those attitudes in words and (c) they also use language to describe the attitudes they attribute to others (and sometimes to their former—and, in a suitably reflective mood, even their current—selves). This should, I think, be treated as a genuinely three-fold distinction and not as a binary one, as it often seems to be made out to be in the literature. Propositional attitudes are thoughts with propositional content. Entertaining some propositional attitude—holding a certain belief, say, or harbouring a certain desire—must therefore involve a representation of the content of that attitude, for the same reasons that mental content representations must be involved in all 'meaningful' use that we make of language. A person can express the contents of such representations overtly, in a language in which she communicates with others. This is what we do when we 'speak our minds' and it is something that we often do just for our own benefit, as a way of making sure that we really know what it is we believe or desire. Such 'speaking one's mind' is quite different, however, from what we do when we use our public language to describe the propositional attitudes we attribute to others, i.e. when we produce an 'attitude report'. (There is of course a sense in which we 'speak our mind' also when we make attitude reports, but then the thought we express is not the one (or ones) our report describes. What we express is *our* belief that the attributee has the attitude or attitudes we are ascribing to him.) On the whole the linguistic forms we typically use to describe the attitudes of others (and sometimes of ourselves) are notably different from those we use when we evince our own thoughts. But there are obviously also close connections between these two ways in which language can be used in relation to mental content, and it is one of the challenges for the semantics of attitude reports to articulate exactly what these connections are. This is not the place to go into the details of this, however. I mention the distinction (between the linguistic forms used in attitude reporting and those used in the plain linguistic expression of attitudes) only because it underscores the difference between the corresponding ways in which we correlate language with thought when we express our own opinions or wants and when we try to find the right words for what we take to be the thoughts of others.

This difference, moreover, is not restricted to expressions and descriptions of attitudinal *states*, it applies equally to *changes* of mind—that is, changes from one attitudinal state to another. On the one hand people will often articulate their thought processes while they are changing their minds; they will, when in the course of revising a belief—or of abandoning a desire or forming a new one, or when they conceive of a plan in order to realise a desire already in place—use words to express where their mind is going, and how it is getting there. (Such talk is dynamic in that it comes in step with the changing attitudes of which it is the running expression.) And on the other hand our attitude reports, too, are as often as not descriptions of how attributees change their mind, and not just of the attitudinal states that they are in at one particular time. (Much of the literature on attitude reports oddly ignores reports of attitude changes, but in real life these are perfectly natural and also quite common: it is often that and how people *change* their minds that we find particularly interesting and worth commenting on.)

29.10 Logic as the Study of Human Cognition and Human Behavior

There are all manner of reasons why people change their minds. Among them are internal processes of deliberation, such as: the kinds of practical reasoning that are involved in the making of plans; or the reasoning that is responsible for our reactions to the public announcement of various kinds of information (including information of which we already had private knowledge) and other kinds of reasoning about the knowledge and ignorance of other agents; or reasoning about the possible moves by other players in various socio-economic situations, in which we see ourselves as competitors with others, or, alternatively, as members of a collaborating collective that is working towards a common goal. This is evidently just a selection from a much larger range of possible internal causes of mind changes. And in addition there are of course all sorts of external causes of such changes, some of which will set in motion the internal cognitive processes that are then the more direct triggers of the attitudinal change in question. In fact, it is a selection in two quite different respects. On the one hand it selects from the various mental processes that produce mental state changes. On the other it is a way of pointing to just some of the areas in which van Benthem's work has been especially influential for our understanding of the mind and its ways of interacting with the outside world and where it has opened up new vistas.

In all these instances, the logics developed and investigated by van Benthem and others can be seen as models of certain aspects of human behaviour, and of human cognitive behaviour as a crucial part of it. And as models of cognitive behaviour they can, I venture to say, have an impact at two different levels: an impact on us as theorists of human reasoning and other forms of cognitive processing, and an impact on us as 'private citizens', as cognitive agents in our own right. As theorists of reasoning and other forms of cognitive information processing we can learn from those logical models how, at a carefully chosen level of abstraction, certain cognitive processes function, or ought to function if they are to function optimally. As cognitive agents we may, under the influence of the lessons we can learn from those logical models, become more conscious, more conscientious and more proficient in executing such processes.

This last suggestion parallels what I said earlier about the potential dual impact of model-theoretic treatments of the languages we speak—the impact on us as linguists and the impact on us as speakers. And as in the case of model-theoretic semantics I venture even one further step. About model-theoretic accounts of natural language meaning I suggested that their impact at the 'private' level need not be restricted to our competence as speakers, but may extend beyond that to our competence at non-verbal cognitive tasks. Likewise, I am now suggesting with regard to the logical models we are discussing at this point that their 'private' impact need not be restricted to our competence at those non-verbal tasks which they are meant to model in the first instance, but that it may extend also to our verbal competence. Moreover, the impact of such models on our verbal competence may be of two different kinds, corresponding to the two ways we can use language to relate to mental representations of content—to verbally express those representations, and to describe them as part of our verbal reports of mental states and events.[15]

All this must have come across as highly speculative, and increasingly so. The thought that we can become better speakers and better cognitive agents by immersing ourselves into the achievements of formal semantics cannot but have struck many as a display of naive optimism about the reformability of man by science that harks back to the 19th century, and which even beats that earlier incarnation by a good many lengths. The thought that logical models of cognition can have a similar impact on us as cognitive agents may have provoked a heightened incredulity. And to top it all, I ended with the suggestion that by studying logical analyses of non-verbal information processing we can become better *speakers*. There is little that I can offer in reply to the sceptic who wants to dismiss this as the wishful thinking of someone with a deep need to see our scientific engagement with language and cognition as a road to self-improvement and not just to scientific enlightenment. True, these are the reflections of my own impressions and hunches; and it would be disingenuous to try and sell them as anything else. But I'd like to add that as time has gone by these impressions have grown firmer and the hunches stronger. That surely won't convince anyone who isn't of the same persuasion to start with. But this is all I can say. I realise that I have been sticking out my neck quite a long way. But I will just leave it where it is.

---

[15] There is of course also a third way in which exposure to logical models of cognition can affect our command of language. New scientific theories come almost inevitably with new terms, or new uses of existing words, and it is part of learning what the theory has to say about its subject matter that its students must assimilate those new terms and new meanings; for only in this way will they be able to talk, and think, about what the theory has to say. In this way the theorist's vocabulary is extended with new elements to articulate new facts and hypotheses about this particular subject matter. But in that respect theories about human behaviour and cognition don't differ from theories on any other subject matter—astrophysics, nano-chemistry, population genetics, what have you.

It should have been clear at this point, but let it be stressed once more, that the impact of logical models that is spoken of in the text is *not* of this sort. It is an impact on our linguistic *core* competence—on our command of that part of our language which we share with all other competent adult speakers, and not just with some select handful, with whom we share a special profession or hobby (be it entomology, chemical engineering, trading in financial products, a passion for cricket or, for that matter, cognitive science).

29.11 On the Borderline of Linguistics and Cognition: Sub-lexical Semantics and Ontology

Let us come down from these heights of vertiginous speculation to a less subjective level. Whether the study of language or cognition can improve us as human beings is one thing. What the study of language can teach us about non-verbal cognition or the study of non-verbal cognition can teach us about language is another. Answers to these last two questions are not easy to come by either. And that is so in particular for the question what work like van Benthem's, which does reveal to us the structural and logical properties of apparently non-verbal aspects of human cognition, can tell us about human language: What can a semanticist learn from work of this kind?

I am in no doubt that there are answers to this question of which I am not aware. But let me mention one that I am aware of. It has to do with two perspective shifts that have occurred within formal semantics over the past decade or two: (a) a shift from the compositional to the lexical part of what constitutes linguistic meaning and (b) the shift from a lexicalist perspective, in which words are the smallest units of linguistic meaning, to a 'root-based' perspective, according to which words have internal structure, involving 'roots' and sub-lexical functional constituents. Both of these shifts have redirected our focus to the lower echelons of the meaning-constituting process, and more particularly to the meanings of the smallest meaningful units, on which the whole compositional edifice rests. With this shift towards lexical and sub-lexical semantics has come an increased awareness of the importance of ontology—in that broad sense of the term in which ontology isn't just about what sorts of things there are, but also about the distinctive properties of the different sorts of entities, and about the fundamental relations that those entities stand in to each other, and also those in which they stand to entities of other sorts. There has been a growing awareness that in order to understand the meanings of many words we need to be able to say much about the sorts of things they refer to and about the properties and relations that are constitutive of those sorts. The work on sub-lexical structure has reinforced this conviction. At the sub-lexical level ontological information is not only needed for the semantic specification of the smallest meaningful parts. Some of the operations which combine sub-lexical meaning constituents into the meanings of words also involve detailed ontological assumptions.

Much of recent ontology research has focussed on topics such as time, space, or mereology (primarily part-whole relations and relations of material constitution). That these parts of ontology are indispensable to natural language semantics has become a lieu commun. But the same is equally true of those parts of ontology that have to do with human cognition and its afferent and efferent interactions with the world (that is, with perception and action). For the vocabulary that we use to express what we think, or describe what we take to be on the minds of others, and how we see the world as acting on ourselves and on others, and how we act, or they act, on the world—all that vocabulary depends on sortal distinctions in the cognitive domain (and on the properties and relations involving the different sorts from that domain) just as so much of our other vocabulary depends on sortal distinctions, properties and relations within the realms of space, time, motion or mereology. It is about this cognition-related part of ontology that there is much we can learn from the logical analyses of which I have been speaking. And that is so in particular when the analysis includes a model-theoretic component, which is the case for nearly all of the analyses I have in mind. For the definition of the models will have to make all ontological commitments of the analysis explicit.

One conclusion from these last considerations is that it isn't as easy to 'get out of language' as van Benthem may have thought it is. But of course, if that is what he thinks, then that is just a case of underestimating the outer reaches of one's own work. A second conclusion, which echoes some of what I have already said, is that semanticists should, no less than logicians or philosophers, expose themselves to current work in logic, and that to a far greater extent than appears to be the case at present. And to that purpose they should acquire, and should make sure that the generations of semanticists after them acquire, the formal knowledge and sophistication without which such work is inaccessible.[16]

References

1. Ackerman F (1990) Analysis, language, and concepts: the second paradox of analysis. Nat Lang Semant 4:535–543
2. van Benthem J (1983) The logic of time. Reidel, Dordrecht
3. van Benthem J (1984) Questions about quantifiers. J Symbolic Logic 49:443–466
4. van Benthem J (1986) Essays in logical semantics. Reidel, Dordrecht
5. van Benthem J (1990) Categorial grammar and type theory. J Philos Logic 19:115–168
6. van Benthem J (1991) Language in action: categories, lambdas and dynamic logic. In: Studies in logic, vol 130. Elsevier, Amsterdam
7. van Benthem J (1996) Exploring logical dynamics. CSLI, Stanford
8. van Benthem J (2011) Logical dynamics of information and action. CUP, Cambridge
9. van Benthem J (2012) Dynamic logic in natural language. In: Graf T, Paprno D, Szabolcsi A, Tellings J (eds) Theories of everything: in honor of Ed Keenan. ICLA working papers in linguistics 17
10. Black M (1944) The "paradox of analysi". Mind 53:263–267
11. Boolos G (1984) To be is to be a value of a variable (or to be some values of some variables). J Philos 81:430–449
12. Boolos G (1985) Nominalist platonism. Phil Rev 94:327–344
13. Boolos G (1998) Logic, logic and logic. Harvard University Press, Cambridge

---

[16] The history of interactions between the fields of logic and linguistics is a topic that deserves an essay in its own right. What formal semantics of natural language owes to logic is obvious and known to all: It owes its existence to a logician who saw how methods from symbolic logic could be applied to the analysis of natural languages, and his insights are with us to this very day. That semantics can be a source of inspiration to formal logicians has been demonstrated, again and again, by outstanding contributions—to the logic of quantification by van Benthem, Keenan, Westerstahl, to name some of its most important contributors, to the logic and semantics of situations in the work of Barwise and others, and in countless contributions to the logic of conditionals and the logic of vagueness. What I am arguing here is that once again the pendulum is swinging back: that there is much that semanticists can learn from results in formal logic; and that that is so because of what those results have to tell us about various aspects of ontological structure.

14. Chomsky N (1965) Aspects of the theory of syntax. MIT Press, Cambridge
15. Chomsky N (1966) Cartesian linguistics: a chapter in the history of rationalist thought. Harper and Row, Cambridge
16. Davidson D (1967a) Rasical interprtation. Dialectica 27:314–328
17. Davidson D (1967b) Truth and meaning. Synthese 7:304–323
18. Fodor J (1975) The language of thought. Harvard University Press, Cambridge
19. Kalish D, Montague R (1964) Logic: techniques of formal reasoning. Harcourt, Brace & World, New York
20. Kamp H (1990) Prolegomena to a structural account of belief and other attitudes. In: Anderson CA, Owens J (eds) Propositional attitudes—the role of content in logic, language, and mind. University of Chicago Press and CSLI, Stanford, pp 27–90 (Chapter 2)
21. Kamp H, van Genabith J, Reyle U (2011) Discourse representation theory: an updated survey. In: Handbook of philosophical logic, vol XV. Elsevier, Amsterdam, pp 125–394
22. Moore G (1963) Analysis: a reply to my critics. In: Alston W, Nakhnikian G (eds) Readings in twentieth century philosophy. Free Press of Glencoe, Glencoe, pp 221–242
23. Ogden C, Richards I (1923) The meaning of meaning: a study of the influence of language upon thought and the science of symbolism. Routledge and Kegan Paul, London
24. Oliver A, Smiley T (2011) Plural logic. OUP, Oxford

Chapter 30
Logic and Complexity in Cognitive Science

Alistair M. C. Isaac, Jakub Szymanik and Rineke Verbrugge

Abstract This chapter surveys the use of logic and computational complexity theory in cognitive science. We emphasize in particular the role played by logic in bridging the gaps between Marr's three levels: representation theorems for non-monotonic logics resolve algorithmic/implementation debates, while complexity theory probes the relationship between computational task analysis and algorithms. We argue that the computational perspective allows feedback from empirical results to guide the development of increasingly subtle computational models. We defend this perspective via a survey of the role of logic in several classic problems in cognitive science (the Wason selection task, the frame problem, the connectionism/symbolic systems debate) before looking in more detail at case studies involving quantifier processing and social cognition. In these examples, models developed by Johan van Benthem have been supplemented with complexity analysis to drive successful programs of empirical research.

30.1 Introduction

How can logic help us to understand cognition? One answer is provided by the computational perspective, which treats cognition as information flow in a computational system. This perspective draws an analogy between intelligent behavior as we

A. M. C. Isaac (✉)
Department of Philosophy, University of Edinburgh, Edinburgh, Scotland
e-mail: a.m.c.isaac@ed.ac.uk

J. Szymanik (✉)
Institute for Logic, Language, and Computation, University of Amsterdam, Amsterdam, The Netherlands
e-mail: J.K.Szymanik@uva.nl

R. Verbrugge
Institute of Artificial Intelligence, University of Groningen, Groningen, The Netherlands
e-mail: L.C.Verbrugge@rug.nl

A. Baltag and S. Smets (eds.), *Johan van Benthem on Logic and Information Dynamics*, Outstanding Contributions to Logic 5, DOI: 10.1007/978-3-319-06025-5_30, © Springer International Publishing Switzerland 2014

observe it in human beings and the complex behavior of human-made computational devices, such as the digital computer. If we accept this analogy, then the behavior of cognitive systems in general can be investigated formally through logical analysis. From this perspective, logical methods can analyze:

1. the boundary between possible and impossible tasks;
2. the efficiency with which any possible task can be solved; and
3. the low-level information flow which implements a solution.

This chapter will survey some examples of the application of logical techniques in each of these areas.

In general, we will see a back-and-forth between logical analysis and empirical findings. This back-and-forth helps to bridge the gap between the normative and descriptive roles of logic. For example, one may believe that people should perform a certain way on a given task because that is the "logical" or "rational" thing to do. We observe that they do not, in fact, perform as predicted. This does not change our assessment that human behavior can be described in terms of logical operations, but it changes our analysis of which task exactly humans perform in response to a particular experimental setup. The Wason selection task provides an example of this sort (Sect. 30.2.1).

Likewise, suppose we analyze the complexity of a task and determine that it has no efficient solution. If we observe humans apparently solving this task, we use this analysis as evidence that they are in fact solving a different, simpler problem. More generally, complexity analysis can make predictions about the relationship between, for example, input size and solution speed for particular tasks. These predictions can be compared with empirical evidence to determine when subjects switch from one algorithm to another, as in the counting of objects in the visual field or quantifier processing (Sect. 30.5.3).

Given the ubiquity of logic and its flexibility as a tool for analyzing complex systems, we do not presume to cover all possible roles of logic in cognitive science.[1] However, focusing on the *computational perspective* highlights two properties of logical analysis essential for cognitive science: it can clarify conceptual debates by making them precise, and it can drive empirical research by providing specific predictions. After some additional background on the computational perspective and Marr's levels of analysis, we conclude this section with a brief discussion of the influence of Johan van Benthem and an outline of the remainder of the chapter.

---

[1] For more references on the interface between logic and cognition, see also the 2007 special issue of *Topoi* on "Logic and Cognition", ed. J. van Benthem, H. Hodges, and W. Hodges; the 2008 special issue of *Journal of Logic, Language and Information* on "Formal Models for Real People", ed. M. Counihan; the 2008 special issue of *Studia Logica* on "Psychologism in Logic?", ed. H. Leitgeb, including [8]; and the 2013 special issue of *Journal of Logic, Language and Information* on "Logic and Cognition" ed. J. Szymanik and R. Verbrugge.

30.1.1 The Computational Perspective

The Church-Turing Thesis states that all computation, in the intuitive sense of a mechanical procedure for solving problems, is formally equivalent to computation by a Turing machine. This is a conceptual claim which cannot be formally proved. However, all attempts so far to explicate intuitive computability, many of them independently motivated, have turned out to define exactly the same class of problems. For instance, definitions via abstract machines (random access machines, quantum computers, cellular automata, genetic algorithms), formal systems (the lambda calculus, Post rewriting systems), and particular classes of function (recursive functions) are all formally equivalent to the definition of Turing machine computability. These results provide compelling support for the claim that all computation is equivalent to Turing computation (see, for example, [27]).

If we accept that the human mind is a physical system, and we accept the Church-Turing Thesis, then we should also accept its psychological counterpart:

> The human mind can only solve computable problems.

In other words, cognitive tasks comprise computable functions. From an abstract perspective, a cognitive task is an information-processing task:

> Given some input (e.g. a visual stimulus, a state of the world, a sensation of pain), produce an appropriate output (e.g. perform an action, draw a conclusion, utter a response).

Generally, then, cognitive tasks can be understood as functions from inputs to outputs, and the psychological version of the Church-Turing Thesis states that the only realistic candidates for information-processing tasks performed by the human mind are computable functions.

Not everyone accepts the psychological version of the Church-Turing Thesis. In particular, some critics have argued that cognitive systems can do more than Turing machines. For example, learning understood as identifiability in the limit [59] is not computable (see [72] for an extensive discussion). Another strand of argumentation is motivated by Gödel's theorems. The claim is that Gödel's incompleteness results somehow demonstrate that the human mind cannot have an algorithmic nature. For example, Lucas [78] claimed: "Gödel's theorem seems to me to prove that Mechanism is false, that is, that minds cannot be explained as machines". He gives the following argument: A computer behaves according to a program, hence we can view it as a formal system. Applying Gödel's theorem to this system we get a true sentence which is unprovable in the system. Thus, the machine does not know that the sentence is true while we can see that it is true. Hence, we cannot be a machine. Lucas' argument was revived by Penrose [93] who supplemented it with the claim that quantum properties of the brain allow it to solve uncomputable problems. Lucas' argument has been strongly criticized by logicians and philosophers (e.g. [6, 96]), as has Penrose's (e.g. [39]).

If identifiability in the limit is the correct analysis of learning, or if the arguments from Gödel's theorems are correct, then we must accept the possibility of "hyper-" or "super-Turing" computation, i.e. the physical realization of machines strictly more powerful than the Turing machine. Examples of such powerful machines have been explored theoretically, for instance Zeno machines (accelerated Turing machines), which allow a countably infinite number of algorithmic steps to be performed in finite time (see [119] for a survey), or analog neural networks, which allow computation over arbitrarily precise real values (e.g. [108, 109]). However, no plausible account of how such devices could be physically realized has been offered so far. Both Penrose's appeal to quantum properties of the brain and Siegelmann's arbitrarily precise neural networks fail to take into account the noise inherent in any real-world analog system. [2] In general, any physical system, including the brain, is susceptible to thermal noise, which defeats the possibility of the arbitrarily precise information transfer required for hyper-computation.

However, there are two more interesting reasons to endorse the psychological version of the Church-Turing Thesis than simple physical plausibility. The first is its fruitfulness as a theoretical assumption. If we assume that neural computability is equivalent to Turing computability, we can generate precise hypotheses about which tasks the human mind can and cannot perform. The second is the close concordance between the computational perspective and psychological practice. Experimental psychology is naturally task oriented, because subjects are typically studied in the context of specific experimental tasks. Furthermore, the dominant approach in cognitive psychology is to view human cognition as a form of information processing (see e.g. [118]). The natural extension of this information processing perspective is the attempt to reproduce human behavior using computational models. Although much of this work employs Bayesian or stochastic methods[3] (rather than logic-based formalisms), it is predicated on the assumption of the psychological version of the Church-Turing Thesis.

---

[2] Siegelmann repeatedly appeals to a result in Siegelmann and Sontag [107] when arguing in later papers that analog neural networks do not require arbitrary precision (and are thus physically realizable). In particular, their Lemma 4.1 shows that for every neural network which computes over real numbers, there exists a neural network which computes over truncated reals (i.e. reals precise only to a finite number of digits). However, the length of truncation required is a function of the length of the computation—longer computations require longer truncated strings. Consequently, if length of computation is allowed to grow arbitrarily, so must the length of the strings of digits over which the computation is performed in a truncated network. Thus, one still must allow computation over arbitrarily precise reals if one is considering the computational properties of analog neural networks *in general*, i.e. over arbitrarily long computation times.

[3] For an overview, see the 2006 special issue of *Trends in Cognitive Sciences* (vol. 10, no. 7) on probabilistic models of cognition, or [129].

30.1.2 Marr's Levels

If we assume the psychological version of the Church-Turing Thesis, what does it tell us about how to analyze cognition? Marr [79] proposed a general framework for explanation in cognitive science based on the computational perspective. He argued that any particular task computed by a cognitive system must ultimately be analyzed at three levels (in order of decreasing abstraction):

1. the computational level (the problem solved or function computed);
2. the algorithmic level (the algorithm used to achieve a solution); and
3. the implementation level (how the algorithm is actually implemented in neural activity).

Considerations at each of these levels may constrain answers at the others, although Marr argues that analysis at the computational level is the most critical for achieving progress in cognitive science. Combining Marr's arguments with those of Newell [90], Anderson [1] defends the *Principle of Rationality*, which asserts that the most powerful explanation of human cognition is to be found via analysis at the computational level under the assumption that task performance has been optimized on an evolutionary timescale, and not via analysis of underlying mechanisms.

However, Marr's three-level system can only be applied *relative to* a particular computational question. For instance, a particular pattern of neural wiring may implement an algorithm which performs the computational function of detecting edges at a particular orientation. But of each neuron in that pattern, we can ask what is its computational function (usually, to integrate over inputs from other neurons) and how is this function implemented (electrochemical changes in the cell). Likewise, we may take a detected edge as an informational primitive when analyzing a more high-level visual function, such as object identification. Nevertheless, the most obvious examples of computational-level analysis concern human performance on behavioral tasks, and the obvious target for implementation-level analysis is neural wiring. Algorithmic analysis via complexity theory can then play the crucial role of bridging the gap between these two domains.

30.1.3 The Contributions of Johan van Benthem

Johan van Benthem's research intersects with cognitive science at many places, several of which are discussed elsewhere in this volume. The present chapter focuses on two specific contributions of Johan van Benthem's which, when supplemented with complexity analysis, have generated fruitful programs of empirical research: his analysis of quantifiers as automata (Sect. 30.5.2) and his models of interactive social reasoning (Sects. 30.6.1, 30.6.2).

The remainder of this chapter surveys applications of logic in cognitive science which are not directly connected to the specifics of Johan van Benthem's research.

Nevertheless, our aim in crafting this survey has been to illustrate a theme which can be found throughout Johan van Benthem oeuvre, namely an ecumenical approach to formal systems, emphasizing *commonality* and *fine structure* rather than conflict and divergence. Where some would seek to defend the advantages of one system over others, Johan van Benthem's seeks to find ways in which apparent competitors are at essence the same (commonality) or, if they do differ, exactly how they differ, and whether or not intermediary systems might lie in between (fine structure). The theme of commonality emerges in our discussions of the Wason selection task (Sect. 30.2.1) and the symbolic/connectionist debate (Sect. 30.3.2). We emphasize the fine structure approach in our discussions of non-monotonic logics and hierarchies of complexity.

Our discussion is organized around Marr's three levels. After arguing for the importance of logic at the computational level through the famous examples of the Wason selection task and the frame problem (Sect. 30.2), we survey the role logic can play in bridging the gap between the algorithmic and implementation levels (Sect. 30.3). We then pause to introduce the basics of computational complexity theory and the P-Cognition Thesis (Sect. 30.4), before investigating the role that complexity analysis can play in bridging the gap between the computational and algorithmic levels through the examples of quantifier processing (Sect. 30.5) and social reasoning (Sect. 30.6).

30.2 The Computational Level: Human Behavior

A number of results from experimental psychology seem to indicate that humans do not behave in accordance with the recommendations of classical logic. Yet logic forms the foundation for norms of ideal rationality. Even fallacies of probabilistic or decision-theoretic reasoning often rest on violations of basic logical principles. Consider, for example, the *conjunction fallacy*: after reading a short passage about Linda which describes her as a social activist in college, 85% of subjects rated the proposition that "Linda is a bank teller and is active in the feminist movement" as more probable than the proposition that "Linda is a bank teller" [133]. This is a fallacy because the axioms of probability ensure that $P(A\&B) \leq P(A)$ for all A and B, where this constraint itself follows from the basic axiom $A \wedge B \rightarrow A$ of propositional logic. Do results such as these demonstrate that humans are fundamentally irrational?

We argue that the apparent irrationality of human behavior does not undermine the use of logic in cognitive science, rather it provides evidence for the correct computational analysis of the task being performed. We examine the example of the Wason selection task, where apparently irrational behavior drove the development of increasingly sophisticated computational models. After discussing this specific example, we'll look at the frame problem, a more general challenge to the computational perspective. This problem motivated the development of non-monotonic logic as a means of providing a formal analysis of human reasoning. This tool will also prove useful when we examine the relationship between algorithm and implementation in Sect. 30.3.

30.2.1 *Is Human Behavior Logical?*

The Wason selection task [137, 138] is very simple. Subjects are shown four cards and told that all cards have a number on one side and a letter on the other. The faces visible to the subject read *D*, *K*, 3, and 7. The subject is then told "Every card which has a *D* on one side has a 3 on the other" and asked which cards they need to turn over to verify this rule. From a classical standpoint, the claim has the basic structure of the material conditional "*D* *is on one side* → 3 *is on the other side*", and the correct answer is to turn over cards *D* and 7. However, the most common answers (in order of decreasing frequency) are (1) *D* and 3; (2) *D*; (3) *D*, 3, and 7; and (4) *D* and 7. The classically correct answer ranks fourth, while the first-ranking answer includes an instance of the logical fallacy of *affirming the consequent*: judging that 3 is relevant for determining whether the conditional is true. Wason's robust and frequently reproduced result seems to show that most people are poor at modus tollens and engage in fallacious reasoning on even very simple tasks. Are we really this bad at logic? Even worse, Cheng and colleagues [20] suggest people may continue to do poorly at the task even when they have taken an introductory logic class! Does this imply that human behavior does not decompose into logical steps? Or that our neural wiring is somehow qualitatively different from the logical structure that can be found in typical computational devices?

As it turns out, there are complexities in the data. The original selection task involved an abstract domain of numbers and letters. When the problem is rephrased in terms of certain types of domain with which subjects are familiar, reasoning suddenly improves. For example, Griggs and Cox [60] demonstrate that if cards have ages on one side and types of drink on the other, subjects perform nearly perfectly when the task is to determine which cards to turn over to ensure that the rule "if a person is drinking beer, then that person is over 19 years old" is satisfied. This study builds upon earlier work by Johnson-Laird et al. [70], demonstrating a similar phenomenon when the task involves postal regulations.

What exactly is different between Wason's original setup and those involving underage drinking and postal regulations, and how should this difference affect our computational model? Johnson-Laird et al. [70] and Griggs and Cox [60] concluded that humans are better at logical reasoning in domains with which they are familiar: since the original Wason task involves an abstract domain of letters and numbers, subjects are confused and fail to reason correctly. Cosmides [28] and Cosmides and Tooby [29] argue that the results tell us something about cognitive architecture. In particular, they conjecture that questions about postal regulations and drinking laws trigger a "cheater detection module." The proposed module is said to be hard-wired to reason effectively in contexts where free-riders might undermine social structure, but provides no logical support for domain-general reasoning.

Stenning and van Lambalgen [117] propose an illuminating new logical analysis of the Wason selection task. They point out that Wason's assertion that there is only one correct answer is too quick, as it assumes a single interpretation of an ambiguous task. Subjects who interpret the described rule as stating some other kind of dependency between D's and 3's than that captured by the material conditional are not necessarily making an error. The key here is in figuring out the relevant difference between versions of the task on which subjects perform in accordance with classical rules and versions (such as the original) on which they do not. Is it because the latter are abstract and the former concrete? Because the latter are unfamiliar and the former familiar? Because the latter are domain-general while the former involve cheater detection? Stenning and van Lambalgen's novel suggestion here is that the crucial difference is in whether the subject interprets the task as merely *checking satisfaction of instances* or as actually *determining the truth of a rule*. In the case of familiar deontic rules, their truth is not at issue, only whether or not they are being satisfied. The deontic nature of these rules means that turning cards over cannot falsify the rules: underage drinking is still wrong, even if one discovers that it occurs. This strictly limits interpretation of the task to checking whether the rule has been satisfied. In contrast, the original version of the task may be interpreted as involving either a descriptive or a prescriptive rule, greatly increasing the cognitive burden on the subject.

None of these analyses of the Wason selection task abandons logic. Johnson Laird et al. [70] and Griggs and Cox [60] shift logical reasoning from innate ability to learned behavior in familiar domains. Cosmides and Tooby [29] locate logical reasoning within the hard wiring of domain-specific modules. Stenning and van Lambalgen [117] identify logical structure with neural structure and argue that apparent violations of logical principles are a side effect of task ambiguity.

30.2.2 The Frame Problem and Non-monotonic Logics

The Wason selection task was designed to probe classical deductive reasoning, but deduction does not exhaust logical inference. In a complex and changing world, cognitive agents must draw conclusions about their circumstances on the basis of incomplete evidence. Crucially, this evidence is *defeasible*, which means that conclusions drawn from it may be defeated by later evidence. For example, suppose I wake up in a strange place and hear voices around me speaking in Chinese; I might conclude that I am in a Chinese restaurant. When I feel the surface on which I lie gently undulating, however, I might revise my conclusion, deciding instead that I have been shanghaied and am currently a passenger on a Chinese junk. Although my evidence has increased, my conclusions have changed. Modeling this type of reasoning requires a *non-monotonic* framework.

Typically, a non-monotonic logic supplements an underlying classical logic with a new, non-monotonic connective and a set of inference rules which govern it. The rules describe a logic of *defeasible* inference, inferences which are usually safe, but which may be defeated by additional information. For example, from the fact that *this is a bird*, I can usually conclude that *this can fly*. This inference can be defeated, however, if I learn that *this is a penguin*. Symbolically, we want our system to ensure that $Bird(x) \Rightarrow Fly(x)$, but $Bird(x) \wedge Penguin(x) \nRightarrow Fly(x)$. Concepts which formalize this defeasibility include circumscription [80], negation as failure [24], and default reasoning [98].

The original motivation for non-monotonic logic came from consideration of a particular type of defeasible reasoning, namely reasoning about a changing world. When an event occurs, humans are able to reason swiftly and effectively about both those features of the world which change *and those which do not*. The problem of how to keep track of those features of the world which do not change is called the "frame problem" [81]. The frame problem comes in both a narrow and a broad version (see the discussion in [33]). The broad version concerns the potential relevance of any piece of information in memory for effective inference, and has troubled philosophers of cognitive science since at least [42]. The original, narrow problem, however, has been effectively solved by non-monotonic logic (see [105] for a complete history and detailed treatment).

The basic idea is easy to see. If we allow ourselves default assumptions about the state of the world, we can easily reason about how it changes. For example, we might assume as a default that facts about the world do not change unless they are explicitly addressed by incoming evidence. Learning that you ate eggs for breakfast does not change my belief that my tie is blue. Without the basic assumption that features of the world not mentioned by my incoming evidence do not change, I would waste all my computational resources checking irrelevant facts about the world whenever I received new information (such as checking the color of my tie after learning what you had for breakfast). This consideration inspired McCarthy's assertion that, not only do "humans use ... 'non-monotonic' reasoning," but also "it is required for intelligent behavior" [80].

A more sophisticated form of default reasoning is found in systems which employ "negation as failure". Such a system may derive $\neg A$ provided it cannot derive A. Negation as failure is frequently implemented in systems using Horn clauses, such as logic programming. Horn clauses state conditional relations such that the antecedent is a (possibly empty) conjunction of positive literals and the consequent is a single positive literal or *falsum*. In general, the semantics for systems involving negation as failure involve fixed points, e.g. finding the minimal model which satisfies all clauses (for a survey, see [40]).

Kraus et al. [71] provide a unified approach to a hierarchy of non-monotonic logics of varying strengths. Their insight was to generalize the semantics of [106], which used a preference ("plausibility") ordering over worlds as a model for non-monotonic inference. Kraus et al. [71] realized that increasingly strict constraints on this semantic ordering correspond to increasingly powerful sets of syntactic rules, and used this insight to define the systems $\mathbf{C} \subseteq \mathbf{CL} \subseteq \mathbf{P} \subseteq \mathbf{M}$, where $\mathbf{C}$ ("cumulative reasoning") is the weakest non-monotonic system they consider and $\mathbf{M}$ ("monotonic") is equivalent to standard propositional logic. Intermediary systems are characterized semantically by added constraints on the plausibility ordering over worlds and syntactically by the addition of stronger inference rules. For example, models for $\mathbf{C}$ are sets of worlds ordered by a relation $\prec$ which is asymmetric and well-founded. $\mathbf{C}$ is strengthened to the system $\mathbf{CL}$ by adding the inference rule *Loop*:

$$\frac{\alpha_0 \Rightarrow \alpha_1, \alpha_1 \Rightarrow \alpha_2, \ldots, \alpha_{k-1} \Rightarrow \alpha_k, \alpha_k \Rightarrow \alpha_0}{\alpha_0 \Rightarrow \alpha_k} \quad (Loop)$$

Semantically, models for **CL** add the constraint that $\prec$ be transitive, i.e. form a strict partial order. Kraus et al. [71] show that systems which use Horn clauses collapse the distinction between **CL** and **P**.

In solving the frame problem, non-monotonic logics have proved their power for modeling defeasible inference at the computational level. As we shall see in the next section, they are also powerful tools for analyzing the structure of the implementation level.

30.3 Between Algorithm and Implementation

How are computations performed in the brain? The answer which has dominated neuroscience since the late nineteenth century is the neuron hypothesis of Ramón y Cajal. He was the first to observe and report the division of brain tissue into distinct cells: neurons. More importantly, he posited a flow of information from axon to dendrite through this web of neural connections, which he denoted by drawing arrows on his illustrations of neural tissue. From the computational perspective, it is natural to identify this flow of information from neuron to neuron as the locus of the computation for solving cognitive tasks.

It is worth noting that this is not the only game in town. A plausible alternative to the neuron hypothesis comes from the dynamical systems perspective, which asserts that the behavior of a family of neurons cannot be reduced to signals communicated between them. Instead, this perspective asserts that computations should be modeled in terms of a dynamical system seeking basins of attraction. Neuroscientists such as Freeman [44, 45] find support for this view in observed neural dynamics, while philosophers such as van Gelder [51, 52] have argued that the dynamical systems perspective constitutes a substantive alternative to the computational perspective of Sect. 30.1.1. With the recent interest in embodied and extended theories of mind, the dynamical systems perspective has become ubiquitous (e.g. Gärdenfors, this volume; see [23] for a survey). In the context of the present discussion, however, we treat it not as an alternative to computationalism, but as a substantive hypothesis within the computational paradigm.

Logic provides an abstract symbolic perspective on neural computation. As such, it can never be the whole story of the implementation level (which by definition involves the physical instantiation of an algorithm). Nevertheless, logic can help bridge the gap between the implementation and algorithmic levels by analyzing structural similarities across different proposed instantiations. For example, if we subscribe to the neuron hypothesis, it is natural to look for logic gates in the wiring between neurons. But we may also look for logic gates in the wiring between families of neurons, or equivalent structure in the relations between basins of attraction in a dynamical system. Logical analysis can distinguish the commonalities across implementation-level hypotheses from their true disagreements.

30.3.1 Logical Neurons

The classic work of McCullogh and Pitts [82] proved the first representation theorem for a logic in an artificial neural network. In general, a representation theorem demonstrates that for every model of a theory, there exists an equivalent model within a distinguished subset. In this case, the "theory" is just a time-stamped set of propositional formulas representing a logical derivation, and the distinguished subset in question is the set of neural networks satisfying a particular set of assumptions, for example, neural firing is "all or none", the only delay is synaptic delay, and the network does not change over time. McCulloch and Pitts show the opposite direction as well: the behavior of any network of the specified type can be represented by a sequence of time-stamped propositional formulas. The propositions need to be time-stamped to represent the evolution of the network through time: the activations of neurons at time t are interpreted as a logical consequence of the activations of neurons at time $t - 1$.

McCulloch and Pitts had shown how neurons could be interpreted as performing logical calculations, and thus, how their behavior could be described and analyzed by logical tools. Furthermore, their approach was modular, as they demonstrated how different patterns of neural wiring could be interpreted as logic gates: signal junctions which compute the truth value of the conjunction, disjunction, or negation of incoming signals. The applications of this result are limited by its idealizing assumptions, however. As neurophysiology has enriched our understanding of neural behavior, the hypothesis of synchronized computations cascading through a structurally unchanging network has become too distant from neural plausibility to resolve debates about implementation in the brain.

Nevertheless, logical methods continue to provide insight into the structure of neural computation. In the face of an increasingly complex theory of neurophysiology, two obvious research projects present themselves. The first focuses on realistic models of individual neurons. Sandler and Tsitolovsky [103], for example, begin with a detailed examination of the biological structure of the neuron, then develop a model of its behavior using fuzzy logic. A second project focuses on artificial neural networks designed to mimic brain dynamics as closely as possible. For example, Vogels and Abbott [136] ran a number of simulations on large networks of integrate-and-fire neurons. These artificial neurons include many realistic features, such as a resting potential and a reset time after each action potential is generated. After randomly generating such networks, Vogels and Abbott [136] investigated their behavior to see if patterns of neurons exhibited the characteristics of logic gates. They successfully identified patterns of activation corresponding to NOT, XOR, and other types of logic gate within their networks.

The idealizing assumptions of these models continue to temper the conclusions which can be drawn from them. Nevertheless, there is a trend of increasing fit between mathematical models of neural behavior and the richness of neurophysiology, and logic continues to guide our understanding of neurons *as computational units*. But from the standpoint of cognitive science, an explanatory question remains: are these computational units the right primitives for analyzing cognition? More generally, is there some *in principle* difference between an analysis offered in terms of neural networks and one offered in terms of logical rules?

30.3.2 The Symbolic Versus Connectionist Debate

In an influential paper, Fodor and Pylyshyn [43] argued that (i) mental representations exhibit systematicity; (ii) representations in neural networks do not exhibit systematicity; therefore (iii) the appropriate formalism for modeling cognition is symbolic (not connectionist). Systematicity here is just the claim that changes in the meaning of a representation correspond systematically to changes in its internal structure (e.g. from my ability to represent "John loves Mary," it follows that I can also represent "Mary loves John"). Fodor and Pylyshyn claim that the only case in which representations in a neural network do exhibit systematicity is when the network is a "mere" implementation of a symbolic system.[4]

It is important to notice what is at stake here: if cognitive tasks manipulate representations, then the appropriate analysis of a cognitive task must respect the properties of those representations. The claim that explanations in cognitive science must be in terms of symbolic systems does not, however, restrict attention to the computational level. Paradigmatic examples of the symbolic approach in cognitive science such as [21] investigate the role of particular algorithms for solving information processing tasks (such as extracting syntactic structure from a string of words). Nevertheless, the claim is that somewhere between abstract task specification and physical implementation, explanatory power breaks down, and neural networks fall on the implementation side of this barrier.

The response from connectionist modelers was violent and univocal: Fodor and Pylyshyn [43] had simply misunderstood the representational properties of neural networks. Responses elucidated how representations in neural networks are "distributed" or "subsymbolic." Smolensky [113, 114], van Gelder [49, 50], Clark [22], and many others all emphasized the importance of acknowledging the distinctive properties of distributed representations in understanding the difference between neural networks and symbolic systems. Yet it is difficult to put one's finger on just what the essential feature of a distributed representation is which makes it qualitatively different from a symbolic representation. Since the late 1990s, the supposed distinction has largely been ignored as hybrid models have risen to prominence, such as the ACT-R architecture of Anderson and Lebiere [2], or the analysis of concepts in Gärdenfors [47]. These hybrid models combine neural networks (for learning) and symbolic manipulation (for high-level problem solving). Although pragmati-

---

[4] However, they do not indicate how such implementational networks avoid their general critique, see [18].

Fig. 30.1 Neural network for non-monotonic reasoning about birds

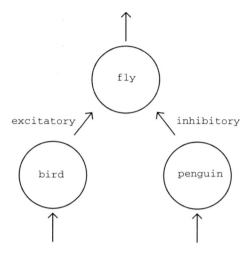

cally satisfying, the hybrid approach avoids rather than resolves questions about the essential difference between symbolic and distributed representations.

Is there some *in principle* difference between subsymbolic computation by neural networks over distributed representations and symbolic computation by Turing (or equivalent) machines? The representation theorem of McCullogh and Pitts [82] discussed above suggests differently, namely that logical theories and neural networks are essentially the same, i.e. their computational and representational properties are equivalent. Can this result be extended to a larger class of neural networks? The trick, it turns out, is to treat neural computation as non-monotonic.

It should be easy to see that some particular non-monotonic theories may be represented by neural networks. Consider the system discussed above for reasoning about birds: two input nodes (one for *Bird(x)* and one for *Penguin(x)*) and an output node (for *Fly(x)*) are all we need to model this system with a simple neural network (Fig. 30.1). So long as there is an excitatory connection from *Bird(x)* to *Fly(x)* and at least as strong an inhibitory connection from *Penguin(x)* to *Fly(x)*, this network will produce the same conclusions from the same premises as our non-monotonic theory. But this is just a specific case; a representation theorem for non-monotonic logics in neural networks would show us that for *every* non-monotonic theory, there is some neural network which computes the same conclusions. Such a theorem would demonstrate a deep computational equivalence between non-monotonic logics and neural networks.

As it turns out, representation theorems of this form have been given by several logicians coming from a variety of backgrounds and motivations. Balkenius and Gärdenfors [5] consider the inferential relationship between a fixed input to a neural network and its "resonant state," i.e. the stable activation state it reaches given that input. Partitioning the state space of these networks into *schemata*, informational components closed under conjunction, disjunction, and complementation, they show that the relation between input schemata and the corresponding resonant state satisfies the axioms of a non-monotonic logic. Hölldobler and collaborators [64, 65] prove a representation theorem for logic programs, demonstrating that for any logic program P, a three-layer, feed-forward network can be found which computes P. Pinkas [94] provides similar results for a wider class of neural networks and penalty logic. Penalty logic is a non-monotonic logic which weights conditionals with integers representing the "penalty" if that conditional is violated. Reasoning in penalty logic involves identifying the set of propositions which minimizes the overall penalty for a set of these weighted conditionals. Blutner [14] generalizes the work of Balkenius and Gärdenfors [5] using a strategy similar to that of Pinkas. He proves a representation theorem for weight-annotated Poole systems, which differ from penalty logic in that they weight consistent sets of sentences with a positive value, the number of possible "hypotheses" (sentences) which agree with them minus the number which disagree.

These results shed some light on the formal relationship between neural networks and symbolic systems, but they also serve a practical purpose. In practical applications, an algorithm for constructing a neural network from a set of non-monotonic inference rules has computational value because it can efficiently find the fixed point which maximizes satisfaction of those rules. Unfortunately, this computational efficiency can only be achieved on a case by case basis (we discuss this point in more detail in Sect. 30.4.2). Furthermore, the practical motivations behind this research directed attention towards networks with very simple nodes and favored proofs by construction, which identify individual nodes with atomic propositions. As a resolution to the connectionism / symbolic systems debate, then, these results need to be supplemented in two ways: first, by extension to more realistic neural networks; second, by extension to the case of truly distributed representations. We conclude this section by discussing results which address these two worries.

Stenning and van Lambalgen [117] see themselves as following the tradition of Hölldobler, but focusing on neural networks which plausibly represent actual structure in the brain. They identify the work of d'Avila-Garcez and colleagues [30] as a crucial step in this direction, extending results of the kind discussed in the previous paragraphs to networks made of nodes with sigmoid activation functions, which more realistically model the behavior of actual neurons. Building on this work, and with the goal of providing a neurally plausible algorithm for their analysis of the Wason selection task, Stenning and van Lambalgen prove a representation theorem for three-valued logic programs in coupled neural networks. Three-valued logic programs can assign three distinct truth values to proposition letters: true, false, or "undecided." Here, undecided plays a role similar to negation as failure, though using three truth values allows for greater flexibility than two. Coupled neural networks are sheets of linked isomorphic networks, such that each node in the first network has a link to a corresponding node in the second network.

Theorem 30.1 ([117]). *If P is a 3-valued logic program and* CN(P) *is the associated coupled neural network, then the least fixed-point model of P corresponds to the activation of the output layer of* CN(P).

Stenning and van Lambalgen's [117] goal is neural plausibility, and they motivate their coupled neural networks by appealing to the extensive evidence for isomorphic mappings between layers of neurons in many parts of the human brain. Nevertheless, it is not clear that any of these mappings exhibit the logical structure postulated by Stenning and van Lambalgen, nor whether this is the appropriate level of neural detail at which to model behavior on the Wason selection task.

Leitgeb identifies himself with the tradition originating with Balkenius and Gärdenfors [5], yet aims to establish a broader class of results. The theorems discussed so far typically address the relationship between a particular non-monotonic logic and a particular type of neural network. In contrast, Leitgeb [73, 74] proves a sequence of representation theorems for each system introduced by Kraus et al. [71] in distinguished classes of neural networks. These results involve inhibition nets with different constraints on internal structure, where an inhibition net is a spreading activation neural network with binary (i.e. firing or non-firing) nodes and both excitatory and inhibitory connections. Leitgeb [75] extends these results, proving representation theorems for the same logics into interpreted dynamical systems.

At the most abstract level, a dynamical system is a set of states with a transition function defined over them. An interpretation $\mathfrak{J}$ of a dynamical system is a mapping from formulas in a propositional language to regions of its state space. Leitgeb gets closure under logical connectives via the same strategy as [5], by assuming an ordering $\leq$ over informational states. If $S_{\mathfrak{J}}$ is an interpreted dynamical system, then $S_{\mathfrak{J}} \models \phi \Rightarrow \psi$ iff s is the resonant state of $S_{\mathfrak{J}}$ on fixed input $\mathfrak{J}(\phi)$ and $\mathfrak{J}(\psi) \leq s$. Call the set of all such conditionals $T S_{\mathfrak{J}}$, then

Theorem 30.2 ([75]). *If $S_{\mathfrak{J}}$ is an interpreted dynamical system, then the theory $T S_{\mathfrak{J}}$ is closed under the rules of the* [71] *system* **C**.

Theorem 30.3 ([75]). *If $T_{\Rightarrow}$ is a consistent theory closed under the rules of* **C**, *then there exists an interpreted dynamical system $S_{\mathfrak{J}}$ such that $T S_{\mathfrak{J}} = T_{\Rightarrow}$.*

Unlike the other results discussed here, Leitgeb takes pains to ensure that his representation theorems subsume the distributed case. In particular, the interpretation function may map a propositional formula to a set of nodes, i.e. distributing its representation throughout the network. From a philosophical standpoint, this result should raise questions for the debate between symbolic and connectionist approaches. Leitgeb has shown that any dynamical system performing calculations over distributed representations may be interpreted as a symbolic system performing non-monotonic inference. This result appears to show that there is no substantive difference in the representational power of symbolic systems and that of neural networks. If there is such a difference, the key to articulating it may be embedded somewhere in Leitgeb's assumptions. The key step here is the ordering over informational states of the network; it is an open question whether the states of actual networks to which connectionists attribute representational properties satisfy such an ordering. Consequently, there is work yet to be done in providing a full resolution to the symbolic systems / connectionism debate.

Even if there is no substantive difference between the representational capacities of symbolic systems and those of neural networks, there may be other principled differences between their computational powers, for instance their *computational efficiency*. We turn next to this question, and to other applications of complexity analysis in cognitive science.

30.4 Computational Complexity and the Tractable Cognition Thesis

The Church-Turing Thesis provides a distinction between those problems which are computable and those which are not. Complexity theory supplements the computational perspective with more fine-grained distinctions, analyzing the *efficiency* of algorithms and distinguishing those problems which have efficient solutions from those which do not. Efficiency considerations can help bridge the gap between computational and algorithmic levels of analysis by turning computational hypotheses into quantitative empirical predictions. Before looking at some specific examples in Sects. 30.5 and 30.6, we first introduce the basic complexity classes and defend the Tractable Cognition Thesis, which grounds the use of complexity analysis in cognitive science.

30.4.1 Tractable Problems

Some problems, although computable, nevertheless require too much time or memory to be feasibly solved by a realistic computational device. Computational complexity theory investigates the resources (time, memory, etc.) required for the execution of algorithms and the inherent difficulty of computational problems [3, 92]. Its particular strength is that it can identify features which hold for all possible solutions to a query, thereby precisely distinguishing those problems which have efficient solutions from those which do not.

This method for analyzing queries allows us to sort them into complexity classes. The class of problems that can be computed relatively quickly, namely in polynomial time with respect to the size of the input, is called PTIME (P for short). A problem is shown to belong to this class if one can show that it can be computed by a deterministic Turing machine in polynomial time. NPTIME (NP) is the class of problems that can be computed by nondeterministic Turing machines in polynomial time. NP-hard problems are problems that are at least as difficult as any problem belonging to NP. Finally, NP-complete problems are NP-hard problems that belong to NP, hence they are the most difficult NPTIME problems.

Of course, this categorization is helpful only under the assumption that the complexity classes defined in the theory are essentially different. These inequalities are usually extremely difficult to prove. In fact, the question whether $P \neq NP$ is considered one of the seven most important open mathematical problems by the Clay Institute of Mathematics, who have offered a \$1,000,000 prize for its solution. If we could show for any NP-complete problem that it is PTIME-computable, we would have demonstrated that P = NP. Computer science generally, and computational complexity theory in particular, operate under the assumption that these two classes are different, an assumption that has proved enormously fruitful in practice.

Intuitively, a problem is NP-hard if there is no *efficient* algorithm for solving it.[5] The only way to deal with it is by using brute-force methods: searching through all possible combinations of elements over a universe. Importantly, contrary to common suggestions in the cognitive science literature (e.g. [19]), computational complexity theory has shown that many NP-hard functions cannot be efficiently approximated [4]. In other words, NP-hard problems generally lead to combinatorial explosion.

If we identify efficiency with *tractability*, computational complexity theory provides a principled method for distinguishing those problems which can reasonably be solved from those which cannot. The following thesis was formulated independently by Cobham [25] and Edmonds [35]:

> The class of practically computable problems is identical to the PTIME class, that is, the class of problems which can be computed by a deterministic Turing machine in a number of steps bounded by a polynomial function of the length of a query.

This thesis is accepted by most computer scientists. For example, Garey and Johnson [48] identify discovery of a PTIME algorithm with producing a real solution to a problem:

> Most exponential time algorithms are merely variations on exhaustive search, whereas polynomial time algorithms generally are made possible only through the gain of some deeper insight into the nature of the problem. There is wide agreement that a problem has not been "well-solved" until a polynomial time algorithm is known for it. Hence, we shall refer to a problem as intractable, if it is so hard that no polynomial time algorithm can possibly solve it.

The common belief in the Cobham-Edmonds Thesis stems from the practice of programmers. NP-hard problems often lead to algorithms which are not practically implementable even for inputs of not very large size. Assuming the Church-Turing Thesis and P $\neq$ NP, we come to the conclusion that this has to be due to intrinsic features of these problems and not to the details of current computing technology.

The substantive content of the Cobham-Edmonds thesis will become more clear if we consider some examples. Many natural problems are computable in polynomial time in terms of the length of the input, for instance calculating the greatest common divisor of two numbers or looking something up in a dictionary. However, the task of deciding whether a given classical propositional formula is satisfiable (SAT) is NP-complete [26].

Thus, even very simple logics can give rise to extremely difficult computational problems. Descriptive complexity theory deals with the relationship between logical

---

[5] The intuitive connection between efficiency and PTIME-computability depends crucially on considering efficiency over arbitrarily large input size n. For example, an algorithm bounded by $n^{\frac{1}{3}\log\log n}$ could be used practically even though it is not polynomial (since $n^{\frac{1}{3}\log\log n} > n^2$ only when $n > e^{e^{10}}$, [61]). Conversely, an algorithm bounded by $n^{98466506514687}$ is PTIME-computable, but even for small n it is not practical to implement. We return to these considerations in the following sections.

definability and computational complexity. The main idea is to treat classes of finite models over a fixed vocabulary as computational problems: what strength of logical language is needed to define a given class of models? The seminal result in descriptive complexity theory is Fagin's theorem, establishing a correspondence between existential second-order logic and NP:

Theorem 30.4 ([37]). Σ_1^1 *captures* NP.

This means that for every property φ, it is definable in the existential fragment of second-order logic, Σ_1^1, if and only if it can be recognized in polynomial time by a non-deterministic Turing machine [68].

What do we know about the computational complexity of reasoning with non-monotonic logics? It turns out that typically the computational complexity of non-monotonic inferences is higher than the complexity of the underlying monotonic logic. As an example, restricting the expressiveness of the language to Horn clauses allows for polynomial inference as far as classical propositional logic is concerned. However, this inference task becomes NP-hard when propositional default logic or circumscription is employed. This increase in complexity comes from the fixed-point constructions needed to provide the semantics for negation as failure and other non-monotonic reasoning rules. In general, determining minimality is NP-hard (see [16], for a survey).

But now we face a puzzle. We were attracted to non-monotonic logics because they appeared to bridge the gap between neurally plausible computational devices and formal languages. Yet the Cobham-Edmonds thesis appears to show that computing properties defined in a non-monotonic language is an intractable task. In order to resolve this puzzle, we will need to examine the relationship between complexity considerations and computational devices.

30.4.2 The Invariance Thesis

The most common model of computation used in complexity analysis is the Turing machine, yet Turing machines have a radically different structure from that of modern digital computers, and even more so from that of neural networks. In order to justify the application of results from complexity theory (e.g. that a particular problem is intractable) in cognitive science, we need to demonstrate that they hold independent of any particular implementation.

The Invariance Thesis (see e.g. [36]) states that:

> Given a "reasonable encoding" of the input and two "reasonable machines," the complexity of the computation performed by these machines on that input will differ by at most a polynomial amount.

Here, "reasonable machine" means any computing device that may be realistically implemented in the physical world. The situation here is very similar to that of the Church-Turing Thesis: although we cannot prove the Invariance Thesis, the fact that it holds for all known physically implementable computational devices provides powerful support for it. Of course, there are well-known machines which are ruled out by the physical realizability criterion; for example, non-deterministic Turing machines and arbitrarily precise analog neural networks are not realistic in this sense. Assuming the Invariance Thesis, a task is difficult if and only if it corresponds to a function of high computational complexity, independent of the computational device under consideration, at least as long as it is reasonable.

It is worth discussing in a little more detail why neural networks fall within the scope of the Invariance Thesis. Neural networks can provide a speed-up over traditional computers because they can perform computations in parallel. However, from the standpoint of complexity theory, the difference between serial and parallel computation is irrelevant for tractability considerations. The essential point is this: any realistic parallel computing device only has a *finite* number of parallel channels for simultaneous computation. This will only provide a polynomial speed-up over a similar serial device. In particular, if the parallel device has n channels, then it should speed up computation by a factor of n (providing it can use its parallel channels with maximum efficiency). As the size of the input grows significantly larger than the number of parallel channels, the advantage in computational power for the parallel machine becomes less and less significant. In particular, the polynomial speed-up of parallel computation provides a vanishingly small advantage on NP-hard problems where the solution time grows exponentially. Therefore, the difference between symbolic and connectionist computations is negligible from the tractability perspective (see [102], particularly Sect. 6.6, for extended discussion).

These considerations should clarify and mitigate the significance of the representation theorems discussed in Sect. 30.3.2. How can we reconcile these results with the observation in Sect. 30.4.1 that fixed-point constructions are NP-hard? We provide a possible answer to this in the next section when discussing, for instance, the Fixed Parameter Tractability Thesis. Simply put, it may be the case that even though the general problem of reasoning with non-monotonic logics is intractable, in our everyday experience we only deal with a specific instance of that problem which, due to properties such as bounds on input size or statistical properties of the environment, yields tractable reasoning. Approaches such as those discussed in the next section allow us to rigorously describe properties of intractable problems that can reduce their general complexity. Then we may study whether the instances of the general problem that people routinely solve indeed constitute an easy subset of the more general problem.

30.4.3 The P-Cognition Thesis and Beyond

How can complexity considerations inform our theory of cognition? The general worry that realistic agents in a complex world must compute effective action despite limited computational resources is the problem of bounded rationality [110].

Simon [105] argued that bounded agents solve difficult problems with rough heuristics rather than exhaustive analyses of the problem space. In essence, rather than solve the hard problem presented to her by the environment, the agent solves an easier, more tractable problem which nevertheless generates an effective action. In order to apply this insight in cognitive science, it would be helpful to have a precise characterization of which class of problems can plausibly be computed by a realistic agent. The answer suggested by complexity theory is to adapt the Cobham-Edmonds Thesis:

> The class of problems which can be computed by a cognitive agent is approximated by the PTIME class, i.e. bounded agents can only solve problems with polynomial time solutions.

As far as we are aware, a version of the Cobham-Edmonds Thesis for cognitive science was first formulated explicitly in print by Frixione [46][6] and later dubbed the P-Cognition Thesis by van Rooij [102]. The P-Cognition Thesis states that a cognitive task is easy if it corresponds to a tractable problem, and hard if it corresponds to an intractable one.

The P-Cognition Thesis can be used to analyze which task an agent is plausibly solving when the world presents her with an (apparently) intractable problem. For example, Levesque [76] argues that the computational complexity of general logic problems motivates the use of Horn clauses and other tractable formalisms to obtain psychologically realistic models of human reasoning. Similarly, Tsatsos [132] emphasizes that visual search in its general bottom-up form is NP-complete. As a consequence, only visual models in which top-down information constrains visual search space are computationally plausible. In the study of categorization and subset choice, computational complexity serves as a good evaluation of psychological models [100].

Nevertheless, one might worry that the "worst-case scenario" attitude of computational complexity theory is inappropriate for analyzing the pragmatic problem-solving skills of real-world cognitive agents. Computational complexity is defined in terms of limit behavior. It answers the question: as the size of the input increases *indefinitely*, how do the running time and memory requirements of the algorithm change? The results of this analysis do not necessarily apply to computations with fixed or bounded input size.[7]

Our response to this worry is twofold. First, complexity analysis proves its value through the role it plays in fruitful ongoing programs of empirical research, such as those we discuss in Sects. 30.5 and 30.6. Second, there are natural extensions to computational complexity theory which avoid this criticism, yet rest on the same fundamental principles elucidated here. In the remainder of this section, we briefly survey some of these extensions.

Some problems are NP-hard on only a small proportion of possible inputs. Average-case complexity analysis studies the complexity of problems over randomly generated inputs, thereby allowing algorithms to be optimized for average inputs on

---

[6] See also [89].

[7] However, see [102].

problems for which only extraordinary inputs require exhaustive search (see e.g. [77] Chap. 4). But average-case complexity theory extends and supplements worst-case analysis; it does not, in general, replace it. For example, when deciding between competing tractable algorithmic hypotheses, average-case analysis can be used to compare their respective time-complexities with accuracy and reaction time data obtained via experimentation [57, 102]. This research does not undermine the use of complexity analysis in cognitive science, it supports and refines it.

Another extension of complexity theory which adds fine structure to the analysis of realistic agents divides a problem into parameters which can be independently analyzed for their contributions to its overall complexity. Such an analysis is useful, for example, if the intractability of a problem comes from a parameter which is usually very small, no matter how large the input (see [101] for examples). This way of thinking leads to parametrized complexity theory as a measure for the complexity of computational cognitive models. van Rooij [102] investigates this subject, proposing the Fixed-Parameter Tractability Thesis as a refinement of the P-Cognition Thesis. The FPT Thesis posits that cognitive agents can only solve problems which are tractable modulo the fixing of a parameter which is usually small in practice, and thereby subsumes many apparently "intractable" task analyses under the general P-Cognition perspective.

The proof of any formal analysis in empirical research is its success in driving predictions and increasing theoretical power. In the remainder of this chapter, we demonstrate the power of complexity analysis for driving research on quantifier processing and social cognition.

30.5 Between Computation and Algorithm: Quantifier Efficiency

Computational complexity theory has been successfully employed to probe the interaction between Marr's computational and algorithmic levels. Given a formal task analysis, complexity theory can make predictions about reaction time; conversely, abrupt changes in reaction time can provide evidence for changes in the algorithm employed on a task. For example, it is known that reaction time increases linearly when subjects are asked to count between 4 and 15 objects. Up to 3 or 4 objects the answer is immediate, so-called subitizing. For judgments involving more than 15 objects, subjects start to approximate: reaction time is constant and the number of incorrect answers increases dramatically [32]. Results such as this allow a fine-grained analysis of the algorithmic dynamics underlying a computational task.

This section illustrates how complexity theory can guide empirical research on algorithmic dynamics through the example of natural language quantifiers. Johan van Benthem's analysis of quantifiers in terms of finite state automata, when combined with complexity analysis, has produced a lively research program on quantifier processing, confirmed by empirical data on reaction times and neuroimaging.

30.5.1 Complexity and Natural Language

In Sect. 30.4.1 we saw how descriptive complexity can be used to analyze formal languages, but what about natural language? Some of the earliest research combining computational complexity with semantics can be found in Ristad's [99] *The Language Complexity Game*. Ristad carefully analyzes the comprehension of anaphoric dependencies in discourse. He considers a few approaches to describing the meaning of anaphora and proves their complexity. Finally, he concludes that the problem is inherently NP-complete and that all good formalisms accounting for it should be NP-complete as well.

More recently, Pratt-Harmann [95] shows that different fragments of natural language capture various complexity classes. More precisely, he studies the computational complexity of satisfiability problems for various fragments of natural language. Pratt-Hartmann [95] proves that the satisfiability problem for the syllogistic fragment is in PTIME, as opposed to the fragment containing relative clauses, which is NP-complete. He also describes fragments of language of even higher computational complexity, with non-copula verbs or restricted anaphora. Finally, he identifies an undecidable fragment containing unrestricted anaphora. Thorne [130] observes that the computational complexity of various fragments of English is inversely proportional to their frequency.

This work appears to challenge the P-Cognition Thesis. For, suppose that satisfiability is NP-complete, or even uncomputable, for fragments of natural language. Then it appears that we are forced not only to abandon the P-Cognition Thesis, but even to abandon the Church-Turing Thesis. This would imply that cognition involves super-Turing computation. As discussed above, this conclusion contradicts all available evidence on physical systems. While it does not defeat the possibility of formal analysis, since there is an extensive mathematics of super-Turing computation, it does complicate, and maybe defeat, the experimental investigation of the mind, which depends upon bounded and finite methods developed in continuity with the rest of empirical science. Complexity analysis provides a constructive way to move forward here. The negative results of Ristad [99] and Pratt-Hartmann [95] suggest that the brain is not solving the complete satisfiability problem when interpreting sentences of natural language. This motivates the search for polynomial time heuristics that might plausibly compute interpretations of sentences of natural language. For example, Pagin [91] tries to explain compositionality in terms of computational complexity, cognitive burden during real-time communication, and language learnability. He argues that compositionality simplifies the complexity of language communication.

30.5.2 *Johan van Benthem's Semantic Automata*

Johan van Benthem [7] was one of the first to emphasize and explore the tight connection between computation and meaning in natural language (see also [8]). He proposed treating "linguistic expressions as denoting certain 'procedures' performed within models for the language" [7].

Johan van Benthem formalized this proposal by identifying generalized quantifiers with automata. These *semantic automata* operate over sequences of elements from a universe, which they test for the quantified property. If they accept the sequence, the quantificational claim is true, if they reject it, the claim is false.

Before looking at some examples, let us recall the definition of a generalized quantifier. Intuitively, a generalized quantifier characterizes relations between properties over a universe; for example, "every A is B" makes a claim about the relationship between objects with property A and objects with property B. We can organize quantifiers by the number and arity of the properties required to define them. More formally:

Definition 30.1 Let $t = (n_1, \ldots, n_k)$ be a k-tuple of positive integers. A *generalized quantifier of type t* is a class $\mathbf{Q}$ of models of a vocabulary $\tau_t = \{R_1, \ldots, R_k\}$, such that R_i is n_i-ary for $1 \leq i \leq k$, and $\mathbf{Q}$ is closed under isomorphisms, i.e. if $\mathbb{M}$ and $\mathbb{M}'$ are isomorphic, then $\mathbb{M} \in \mathbf{Q}$ *iff* $\mathbb{M}' \in \mathbf{Q}$.

Therefore, formally speaking:

$$\forall = \{(M, A) \mid A = M\}.$$
$$\exists = \{(M, A) \mid A \subseteq M \text{ and } A \neq \emptyset\}.$$
$$\text{every} = \{(M, A, B) \mid A, B \subseteq M \text{ and } A \subseteq B\}.$$
$$\text{most} = \{(M, A, B) \mid A, B \subseteq M \text{ and } card(A \cap B) > card(A - B)\}.$$
$$\mathsf{D_n} = \{(M, A) \mid A \subseteq M \text{ and } card(A) \text{ is divisible by } n\}.$$

The first two examples are the standard first-order universal and existential quantifiers, both of type (1). They are classes of models with one unary predicate such that the extension of the predicate is equal to the whole universe in the case of the universal quantifier and is nonempty in the case of the existential quantifier. Quantifiers every and most of type (1, 1) are familiar from natural language semantics. Their aim is to capture the truth-conditions of sentences of the form: "Every A is B" and "Most A's are B." In other words, they are classes of models in which these statements are true. The divisibility quantifier of type (1) is a familiar mathematical quantifier.

Johan van Benthem's insight was to identify quantifier complexity with constraints on the type of automata required to verify quantifier claims over an arbitrary universe. Figures 30.2, 30.3, and 30.4 illustrate examples of these automata for some familiar quantifiers. Johan van Benthem was able to prove the following:

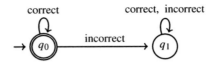

Fig. 30.2 This finite automaton checks whether every sentence in the text is grammatically correct. It inspects the text sentence by sentence, starting in the accepting state (*double circled*), q_0. As long as it does not find an incorrect sentence, it stays in the accepting state. If it finds such a sentence, then it already "knows" that the sentence is false and moves to the rejecting state, q_1, where it stays no matter what sentences come next

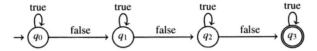

Fig. 30.3 This finite automaton recognizes whether at least 3 sentences in the text are false. This automaton needs 4 states. It starts in the rejecting state, q_0, and eventually, if the condition is satisfied, moves to the accepting state, q_3. Furthermore, notice that to recognize "at least 8" we would need 9 states, and so on

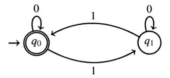

Fig. 30.4 Finite automaton checking whether the number of "1"s is even

Theorem 30.5 ([7]). *A monadic quantifier* Q *is first-order definable if and only if it can be recognized by a finite automaton without cycles.*

But this does not exhaust the class of finite automata. Finite automata with cycles can identify properties which depend on divisibility of the cardinality of the universe, such as whether the universe has an even or odd number of elements (see Fig. 30.4). Consequently, they are definable in first-order logic supplemented with special divisibility quantifiers, D_n, for each natural number n.

Theorem 30.6 ([88]). *A monadic quantifier* Q *is definable in the divisibility logic (i.e.* $FO(D_n)_{n<\omega}$*) iff it can be recognized by a finite automaton.*

However, for recognizing some higher-order quantifiers such as "less than half" or "most," we need computability models that make use of internal memory. Intuitively, to check whether sentence (1) is true we must identify the number of correct sentences and hold it in working memory to compare with the number of incorrect sentences.

(1) Most of the sentences in this chapter are grammatically correct.

Mathematically speaking, such an algorithm can be realized by an automaton supplemented with a push-down stack, a last-in / first-out form of memory.

Theorem 30.7 ([7]). *A quantifier Q of type (1) is definable in the arithmetic of addition iff it can be recognized by a push-down automaton.*

30.5.3 From Automata to Psycholinguistics

Johan van Benthem's formal identification of generalized quantifiers with automata can be used to generate empirical predictions about natural language processing. For example, Szymanik [120, 121] investigate whether the cognitive difficulty of quantifier processing can be assessed on the basis of the complexity of the corresponding minimal automata. He demonstrates that the computational distinction between quantifiers recognized by finite automata and those recognized by push-down automata is psychologically relevant: the more complex the automaton, the longer the reaction time and the greater the recruitment of working memory in subjects asked to solve the verification task. Szymanik and Zajenkowski [123] show that sentences with the Aristotelian quantifiers "some" and "every," corresponding to two-state finite automata, were verified in the least amount of time, while the proportional quantifiers "more than half" and "less than half" triggered the longest reaction times. When it comes to the numerical quantifiers "more than k" and "fewer than k," corresponding to finite automata with $k + 2$ states, the corresponding latencies were positively correlated with the number k. Szymanik and Zajenkowski [124, 128] explore this complexity hierarchy in concurrent processing experiments, demonstrating that during verification, the subject's working memory is qualitatively more engaged while processing proportional quantifiers than while processing numerical or Aristotelian quantifiers.

This work builds on recent neuroimaging research aimed at distinguishing the quantifier classes identified by van Benthem in terms of the neural resources they exploit. For example, McMillan and colleagues [83], in an fMRI study, show that during verification all quantified sentences recruit the right inferior parietal cortex associated with numerosity, but only proportional quantifiers recruit the prefrontal cortex, which is associated with executive resources such as working memory. These findings were later strengthened by evidence on quantifier comprehension in patients with focal neurodegenerative disease ([84]; see [67] for a survey of related results). Moreover, Zajenkowski and colleagues [140] compares the processing of natural language quantifiers in a group of patients with schizophrenia and a healthy control group. In both groups, the difficulty of the quantifiers was consistent with computational predictions. In general, patients with schizophrenia had longer reaction times. Their performance differed in accuracy only on proportional quantifiers, however, confirming the predicted qualitative increase in difficulty for quantifiers which require memory, and explainable in terms of the diminished executive control in schizophrenics.

In the next step, Szymanik [122] studied the computational complexity of multi-quantifier sentences. It turns out that there is a computational dichotomy between different readings of reciprocal sentences, for example between the following:

(2) Most of the parliament members refer indirectly to each other.
(3) Boston pitchers were sitting alongside each other.

While the first sentence is usually given an intractable NP-hard reading in which all pairs of parliament members need to be checked, the second one is interpreted by a PTIME-computable formula. This motivates the conjecture that listeners are more likely to assign readings that are simpler to compute. The psychological plausibility of this conjecture is still awaiting further investigation; however, there are already some interesting early results [104].

Szymanik [122] also asked whether multi-quantifier constructions could be computationally analyzed by extending van Benthem's framework. Threlkeld and Icard [116] give a positive answer by showing that if two quantifiers are recognizable by finite automata (push-down automata) then their iteration must also be recognizable by a finite automaton (push-down automaton). For instance, there are finite automata which recognize whether the following sentences are true:

(4) Some dots and every circle are connected.
(5) Every dot and some circles are connected.

This opens the road for further investigations into the delicate interplay between computability, expressibility, and cognitive load (see e.g. [126]).

The above results demonstrate how the P-Cognition Thesis can drive experimental practice. Differences in performance, such as in reaction times, can support a computational analysis of the task being performed [123]. Furthermore, one can track the changes in heuristics a single agent employs as the problem space changes [131].

30.6 The Complexity of Intelligent Interaction

Johan van Benthem begins his "Cognition as Interaction" [10] with a plea to cognitive scientists to move away from their myopic focus on single agents:

> It is intelligent social life which often shows truly human cognitive abilities at their best and most admirable. But textbook chapters in cognitive science mostly emphasise the apparatus that is used by single agents: reasoning, perception, memory or learning. And this emphasis becomes even stronger under the influence of neuroscience, as the only *obvious* thing that can be studied in a hard scientific manner are the brain processes inside individual bodies. Protagoras famously said that "Man is the measure of all things", and many neuroscientists would even say that it's just her brain. By contrast, this very brief chapter makes a plea for the irreducibly social side of cognition, as evidenced in the ways in which people communicate and interact. Even in physics, many bodies in interaction can form one new object, such as a Solar system. This is true all the more when we have a meeting of many minds!

We could not agree more. In this section we discuss how computational constraints can be taken seriously in the study of multi-agent social interactions. After examining a case study in detail, which illustrates how shifting from a global to a local perspective in a multi-agent reasoning scenario can reduce the complexity of representations, we'll briefly survey the role that games can play as an empirical testing ground for logical models of social reasoning.

30.6.1 Plausible Epistemic Representations

Other chapters in this volume extensively discuss epistemic logics, and in particular Dynamic Epistemic Logic, DEL (see [12] for a recent survey). We argue that although these logics can describe the epistemic reasoning in social interactions in an elegant way, they postulate cognitively implausible representations. This leads to discussion of a recent proposal for more computationally plausible models for epistemic reasoning which repurposes models developed by van Benthem for representing quantifiers.

Let us start with a classic example: the Muddy Children puzzle. Three kids are playing outside. When they come back home, their father says: (1) "At least one of you has mud on your forehead." Then, he asks the children: (I) "Can you tell for sure whether you have mud on your forehead? If yes, announce your status." The children commonly know that their father never lies and that they are all sincere and perfect logical reasoners. Each child can see the mud on others but cannot see his or her own forehead. After the father asks (I) once, the children are silent. When he asks the question a second time, however, suddenly all, in this case two, muddy children respond that they know they have mud on their foreheads. How is that possible?

DEL models the underlying reasoning with Kripke structures characterizing the agents' uncertainty. Let us give the three children names: a, b, and c, and use propositional letters m_a, m_b, and m_c to express that the corresponding child is muddy. Possible worlds correspond to distributions of mud on children's foreheads, for example, $w_5 : m_a$ stands for a being muddy and b and c being clean in world w_5. Two worlds are joined with an edge labelled with i, for agent $i \in \{a, b, c\}$, if i cannot distinguish between the two worlds on the basis of his information; for clarity we drop the reflexive arrows for each state. The standard epistemic modeling (see e.g. [34, 38]) is depicted in Fig. 30.5; the boxed state stands for the actual world, in this case, all three children are muddy.

Now, let us recall how the reasoning process is modeled in this setting. The first public announcement has the form: (1) $m_a \lor m_b \lor m_c$, and after its announcement, (1) becomes common knowledge among the children. As a result, the children perform an update, that is, they eliminate world w_8 in which (1) is false. Then the father asks for the first time: (I) who among them knows his status. The children reason as follows. In world w_6, c knows that he is dirty (there is no uncertainty for c between this world and any other world in which he is clean). Therefore, if the actual world were w_6, agent c would know his state and announce it. The situation is similar for a and b in w_5 and w_7, respectively. The silence of the children after (I) is equivalent to the announcement that none of them know whether they are muddy. Hence, all agents eliminate those worlds that do not make such an announcement true: w_5, w_6, and w_7.

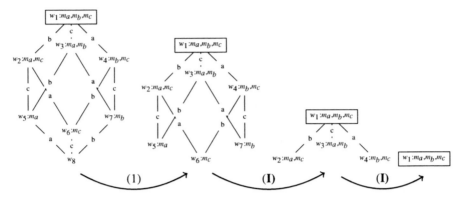

Fig. 30.5 The classical modeling of the Muddy Children puzzle. *Arrows* indicate the dynamic update to the model after the corresponding announcement

When (**I**) is announced a second time, it is again clear that if one of w_2, w_3, or w_4 were the actual world, the respective agents would have announced their knowledge. The children still do not respond, so at the start of the next round everyone knows – and in fact it is common knowledge – that the actual situation cannot be any of w_2, w_3, or w_4. Hence, they all eliminate these worlds leaving just the possibility w_1. All uncertainty disappears and they all know at the same time that they are dirty.

In spite of its logical elegance, the proposed solution is problematic from the standpoint of the Tractable Cognition Thesis. DEL flexibly models epistemic scenarios from a global perspective, but this power comes at a price: the relevant logics turn out to be very complex. First of all, DEL postulates exponential representations. Not only is it intuitively implausible that actual agents generate such exponential models of all possible scenarios, it is computationally intractable. The core of the problem is DEL's global perspective, shared with other modal logics, which assesses complexity from the modeler's point of view [115]. In fact, what we need is a new *local perspective*, a study of epistemic reasoning from the perspective of the agents involved. Such a shift in perspective can lead to logics and representations that are much more cognitively plausible [31, 55, 56]. Let's see how this applies to the Muddy Children puzzle.

In *Essays in Logical Semantics*, van Benthem [7] proposes not only computational semantics but also a geometrical representation for generalized quantifiers in the form of *number triangles*. Gierasimczuk and Szymanik [56] use number triangles to develop a new, concise logical modeling strategy for multi-agent scenarios, which focuses on the role of *quantitative* information in allowing agents to successfully converge on *qualitative* knowledge about their situation. As an example, let us consider a generalization of the Muddy Children puzzle by allowing public announcements based on an arbitrary generalized quantifier Q, for example, "Most children are muddy."

As we restrict ourselves to finite universes, we can represent all that is relevant for type (1) generalized quantifiers in a number triangle, which simply enumerates

Fig. 30.6 Number triangle representing the quantifier "at least 1"

all finite models of type (1). Let U be the universe of discourse, here all the father's children, and let A be the subset of muddy children. The node labeled (k, n) stands for a model in which $|U - A| = k$ and $|A| = n$. Now, every generalized quantifier of type (1) can be represented by putting "+" at those (k, n) that belong to **Q** and "–" at the rest. For example, the number triangle representation of the quantifier "at least 1" is shown in Fig. 30.6. Number triangles play a crucial role in generalized quantifier theory. Gierasimczuk and Szymanik [56] interpret the pairs (k, n) as possible worlds.

Gierasimczuk and Szymanik illustrate how the number triangle may be used to derive a more concise solution for the Muddy Children puzzle. As before, we consider three agents a, b, and c. All possibilities with respect to the *size* of the set of muddy children are enumerated in the fourth level of the number triangle. Let us also assume at this point that the actual situation is that agents a and b are muddy and c is clean. Therefore, the real world is $(1, 2)$, one child is clean and two are muddy:

$$\text{(3,0)} \qquad \text{(2,1)} \qquad \boxed{\text{(1,2)}} \qquad \text{(0,3)}$$

Now, let us focus on what the agents observe. Agent a sees one muddy child and one clean child. The same holds symmetrically for agent b. Their observational state can be encoded as $(1, 1)$. Accordingly, the observational state of c is $(0, 2)$. In general, if the number of agents is n, each agent can observe $n - 1$ agents. As a result, what agents observe is encoded in the third level of the number triangle:

$$\text{(2,0)} \qquad \text{(1,1)} \qquad \text{(0,2)}$$

$$\text{(3,0)} \qquad \text{(2,1)} \qquad \boxed{\text{(1,2)}} \qquad \text{(0,3)}$$

Each of the agents faces the question whether he is muddy. For example, agent a has to decide whether he should *extend* his observation state $(1, 1)$ to the left state $(2, 1)$ (a decides that he is clean) or to the right state $(1, 2)$ (a decides that he is muddy). The same holds for agent b. The situation of agent c is similar: his observational state is $(0, 2)$, which has two potential extensions, namely $(1, 2)$ and $(0, 3)$. In general, note that every observational state has two possible successors.

Given this representation, we can now analyze what happens in the Muddy Children scenario. Figure 30.7 represents the process, with the initial model at the top. First, the announcement is made: "At least one of you is muddy." This allows elimination of those possible states, i.e. those on the bottom row, that are not in

Fig. 30.7 Modeling of the
Muddy Children puzzle based
on number triangles

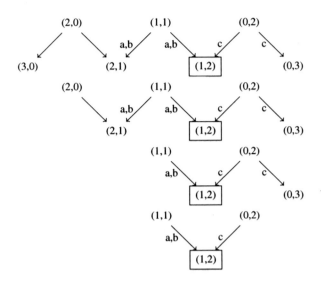

the quantifier (i.e. intersecting with the "−"s in Fig. 30.6). In this case, $(3, 0)$ is eliminated. The resulting model is the second from the top. Then the father asks: "Can you tell for sure whether or not you have mud on your forehead?" In our graph, this question means: "Does any of you have only one successor?" All agents commonly know that $(3, 0)$ has just been eliminated. Agent a considers it possible that the actual state is $(2, 1)$, i.e. that two agents are clean and one is muddy, so that he himself would have to be clean. But then he knows that there would have to be an agent whose observational state is $(2, 0)$—there has to be a muddy agent that observes two clean ones. For this hypothetical agent, the uncertainty disappeared just after the quantifier announcement (for $(2, 0)$ there is only one successor left). So, when it becomes clear that no one knows and the father asks the question again, the world $(2, 1)$ gets eliminated and the only possibility for agent a is now $(1, 2)$ via the right successor of $(1, 1)$, and this indicates that he has to be muddy. Agent b is in exactly the same situation. They both can announce that they know. Since c witnessed the whole process, he now knows that the only way for them to know was to be in $(1, 1)$, and therefore he concludes $(1, 2)$ as well.

In general, if there are n agents, we need levels n and $n + 1$ of the triangle to enumerate all possible scenarios (up to isomorphism). Therefore, the total size of the initial model is $2n + 1$. As a result, the proposed representation is exponentially more succinct than the standard DEL approach, which requires on the order of 2^n states. Strategies such as this are attractive when modeling the representations employed by realistic cognitive agents. Computational concerns constitute a plausibility test on such proposed representations.

Classic epistemic logic usually assumes an exponential representation including all possibilities (see e.g. [34]). The representation proposed here exploits a crucial feature of generalized quantifiers, namely closure under isomorphism, in order to increase the informational power of a message relative to the observational powers of the agents. Such informational "shortcuts" can be extremely powerful, yet have so far rarely been taken into account in the epistemic literature.

30.6.2 Games and Social Cognition

There is a close relationship between epistemic logics and extensive form games. Sequences of DEL models such as those depicted in Fig. 30.5 can be strung together to form a single model for Epistemic Temporal Logic [13, 66]. These models can be pictured as branching tree structures, and a distinguished subset of them, if supplemented with a preference ordering over end states, are equivalent to extensive form games [69]. These close formal relations reflect the fact that DEL is well-suited for modeling the higher-order reasoning (e.g. involving my belief about your belief about my belief, etc.) required to explain game-theoretical arguments [135]. An important concept common to both epistemic logic and game theory is backward induction, the process of reasoning backwards from the end of the game to determine a sequence of optimal actions [9]. Backward induction can be understood as an inductive algorithm defined on a game tree. The backward induction algorithm tells us which sequence of actions will be chosen by agents that want to maximize their own payoffs, assuming common knowledge of rationality. In game-theoretical terms, backward induction calculates subgame perfect equilibria for finite extensive form games.

Games have proved a powerful tool for studying the evolution of cooperation and of social cognition [15, 111, 112, 139]. Games also play a pivotal role in designing experiments for studying social cognition in actual use [17], recently with a particular focus on higher-order social cognition, e.g. the matrix game [63], the race game [58, 62], the road game [41, 97], and the Marble Drop game [85–87].

But how well do games capture the actual dynamics of mental computation during social reasoning? There are natural ways to view games as models of cognitive processes [9], yet we might wonder how well the normative analysis of these processes in traditional game theory captures the flexible dynamics of actual human reasoning. Johan van Benthem's [8], for instance, points out that it is not the realization of perfect solutions, but rather the human ability to use interaction to dynamically recover from mistakes, which is most impressive:

> As Joerg Siekmann once said, the most admirable moments in a mathematics seminar are not when someone presents a well-oiled proof, but when he discovers a mistake and recovers on the spot. Logic should understand this dynamic behaviour, which surely involves many more mechanisms than inference .... And on that view, logic is not the static guardian of correctness, as we still find it defined in most textbooks, but rather the much more dynamic, and much more inspiring *immune system of the mind*!

These considerations motivate a closer look at the fit between game-theoretic strategies and actual human reasoning.

In this spirit, many studies have proven that the application of higher-order social reasoning among adults is far from optimal (see, for example, [63, 134]). However,

Meijering and colleagues [85, 87] report an almost perfect performance by subjects whose reasoning processes have been facilitated by step-wise training, which progresses from simple decisions without any opponent, through games that require first-order reasoning ("he plans to go right at the next trapdoor"), to games that require second-order reasoning ("he thinks that I plan to go left at the last trapdoor"). Surprisingly, an eye-tracking study of the subjects solving the game suggests that backward induction is not necessarily the strategy that participants used; they seemed instead to favour a form of "forward reasoning plus backtracking" [86].

Another application of formal methods in this spirit has been implemented by Ghosh et al. [53] and Gosh and Meijering [54]. They formulate all reasoning strategies on an experimental task in a logical language and compare ACT-R models based on each strategy with the subject's actual performance in a sequence of games on the basis of reaction times, accuracy, and eye-tracking data. These comparisons allowed them to develop a "cognitive computational model" with similar task performance to that of the human subjects. As a final example, consider the recent work of [125]. The authors successfully used structural properties of the game, namely types of turn alternation and pay-off distributions, to predict the cognitive complexity of various trials. This complexity analysis accurately predicted changes in subjects' reaction times during game play.

These examples confirm Johan van Benthem's point: it is not static solution concepts, but dynamic algorithm switching, which characterizes realistic game play. Yet, far from being immune to logical analysis, such dynamic strategic reasoning can be formalized and explored through logical methods.

30.7 Conclusion

We have examined a number of applications of logical methods in cognitive science. These methods assume the computational perspective, which treats cognitive agents as realistic machines solving information processing tasks. The value of the computational perspective is in its fruitfulness as a research program: formal analysis of an information processing task generates empirical predictions, and breakdowns in these predictions motivate revisions in the formal theory. We have illustrated this back-and-forth between logical analyses and empirical results through a number of specific examples, including quantifier processing and higher-order social reasoning.

We have also emphasized the role that logical methods can play in clarifying concepts, with a particular focus on the role of non-monotonic logics in bridging Marr's algorithmic and implementation levels and the role of complexity theory in bridging his computational and algorithmic levels. Typically, formal methods generate "in principle" results, such as the boundary between logical and illogical solutions, tractable and intractable problems, or monotonic and non-monotonic reasoning. However, these in principle results do not constitute brittle hypotheses to be confirmed or disconfirmed: it is not the case that humans are simply "logical" or "illogical." Rather, such results form a bedrock on which a nuanced analysis of fine structure can be built. An apparent strict boundary between logical and illogical becomes an array of types of logic, an apparent strict boundary between tractable and intractable becomes a hierarchy of algorithmic types which differentially recruit processing resources. It is this fine structure which allows the subtle and fruitful interplay between logical and empirical methods.

There are morals here for both sides of the logical/empirical coin. Logicians can strengthen the relevance of their analyses for science by embracing complexity analysis. Not just formal semantics and logics of agency, but all logical models of cognitive behavior (temporal reasoning, learning, mathematical problem solving, etc.) can strengthen their relevance for empirical methods by embracing complexity analysis and the fine structure of the complexity hierarchy which rests upon it. Likewise, empirical scientists should recognize the degree of nuance a formal analysis can bring to empirical predictions: not just which task can be solved, but how quickly, and using what resources may all be predicted by an algorithmic analysis. The theme which unites these two facets of cognitive science is that *representations matter*. The content and structure of representations constrain performance on cognitive tasks. Both experiments and logical models probe the nature of these representations, and it is in converging on a single analysis through the back-and-forth of theoretical/empirical interaction that cognitive science progresses.

Acknowledgments We would like to thank Alexandru Baltag, Johan van Benthem, Peter Gärdenfors, Iris van Rooij, and Keith Stenning for many comments and suggestions. The first author would also like to thank Douwe Kiela and Thomas Icard for helpful discussions of this material; he was supported by NSF grant 1028130. The second author was supported by NWO Vici grant 277-80-001 and NWO Veni grant 639-021-232. The third author was supported by NWO Vici grant 277-80-001.

References

1. Anderson JR (1990) The adaptive character of thought: Studies in cognition. Lawrence Erlbaum, Mahwah
2. Anderson JR, Lebiere C (1998) The atomic components of thought. Lawrence Erlbaum, Mahwah
3. Arora S, Barak B (2009) Computational complexity: A modern approach (1st edn). Cambridge University Press, Cambridge
4. Ausiello G, Crescenzi P, Kann V, Gambosi G, Marchetti-Spaccamela A, Protasi M (2000) Complexity and approximation: Combinatorial optimization problems and their approximability properties. Springer, Berlin
5. Balkenius C, Gärdenfors P (1991) Nonmonotonic inference in neural networks. In: Allen JA, Fikes R, Sandewall E (eds) Proceedings of the second international conference on knowledge representation and reasoning. Morgan Kaufmann, San Mateo pp 32–39
6. Benacerraf P (1967) God, the devil, and Gödel. Monist 51:9–32
7. van Benthem J (1986) Essays in logical semantics. Reidel, Dordrecht
8. van Benthem J (1991) Language in action: Categories, lambdas and dynamic logic. North-Holland, Amsterdam (paperback edition 1995, MIT Press, Cambridge)
9. van Benthem J (2002) Extensive games as process models. J Logic Lang Inform 11(3):289–313

10. van Benthem J (2007) Cognition as interaction. In: Proceedings of symposium on cognitive foundations of interpretation. KNAW, Amsterdam, pp 27–38
11. van Benthem J (2008) Logic and reasoning: Do the facts matter? Stud Logica 88(1):67–84
12. van Benthem J (2011) Logical dynamics of information and interaction. Cambridge University Press, Cambridge
13. van Benthem J, Gerbrandy J, Hoshi T, Pacuit E (2009) Merging frameworks for interaction. J Philos Logic 38:491–526
14. Blutner R (2004) Nonmonotonic inferences and neural networks. Synthese 142:143–174
15. Bowles S, Gintis H (2011) A cooperative species: Human reciprocity and its evolution. Princeton University Press, Princeton
16. Cadoli M, Schaerf M (1993) A survey of complexity results for nonmonotonic logics. J Log Program 17(2/3&4):127–160
17. Camerer C (2003) Behavioral game theory: Experiments in strategic interaction. Princeton University Press, New Jersey
18. Chalmers D (1990) Why Fodor and Pylyshyn were wrong: The simplest refutation. In: Proceedings of the twelfth annual conference of the cognitive science society. pp 340–347
19. Chater N, Tenenbaum J, Yuille A (2006) Probabilistic models of cognition: Where next? Trends Cogn Sci 10(7):292–293
20. Cheng PW, Holyoak KJ, Nisbett RE, Oliver LM (1986) Pragmatic versus syntactic approaches to training deductive reasoning. Cogn Psychol 18:293–328
21. Chomsky N (1957) Syntactic structures. Mouton de Gruyter, The Hague
22. Clark A (1993) Associative engines. Bradford Books, Cambridge
23. Clark A (2011) Supersizing the mind: Embodiment, action, and cognitive extension. Oxford University Press, Oxford
24. Clark KL (1978) Negation as failure. In: Gallaire H, Minker J (eds) Logic and databases. Plenum Press, New York
25. Cobham J (1965) The intrinsic computational difficulty of functions. In: Proceedings of the 1964 international congress for logic, methodology, and philosophy of science. North-Holland, Amsterdam, pp 24–30
26. Cook SA (1971) The complexity of theorem-proving procedures. In STOC '71: Proceedings of the third annual ACM symposium on theory of computing. ACM Press, New York pp 151–158
27. Cooper BS (2003) Computability theory: CRC mathematics series. Chapman and Hall/CRC, London
28. Cosmides L (1989) The logic of social exchange: Has natural selection shaped how humans reason? studies with the Wason selection task. Cognition 31:187–276
29. Cosmides L, Tooby J (1992) Cognitive adaptations for social exchange. In: Barkow J, Cosmides L, Tooby J (eds) The adapted mind: Evolutionary psychology and the generation of culture. Oxford University Press, Oxford pp 163–228
30. d'Avila Garcez AS, Lamb LC, Gabbay DM (2002) Neural-symbolic learning systems: foundations and applications. Springer, London
31. Dégremont C, Kurzen L, Szymanik J (2014) Exploring the tractability border in epistemic tasks. Synthese 191(3):371–408
32. Dehaene S (1999) The number sense: How the mind creates mathematics. Oxford University Press, New York
33. Dennett DC (1984) Cognitive wheels: The frame problem of AI. In: Hookway C (ed) Minds, machines and evolution: Philosophical studies. Cambridge University Press, Cambridge
34. van Ditmarsch H, van der Hoek W, Kooi B (2007) Dynamic epistemic logic. Springer, Dordrecht
35. Edmonds J (1965) Paths, trees, and flowers. Can J Math 17:449–467
36. van Emde Boas P (1990) Machine models and simulations. In: van Leeuwen J (ed) Handbook of theoretical computer science. MIT Press, Cambridge pp 1–66
37. Fagin R (1974) Generalized first-order spectra and polynomial time recognizable sets. In: Karp R (ed) Proceedings of SIAM—AMS on complexity of computation, vol 7. American Mathematical Society pp 43–73

38. Fagin R, Halpern JY, Moses Y, Vardi MY (1995) Reasoning about knowledge. MIT Press, Cambridge
39. Feferman S (1995) Penrose's Gödelian argument. Psyche 2, online publication, available at http://www.calculemus.org/MathUniversalis/NS/10/04feferman.html
40. Fitting M (2002) Fixpoint semantics for logic programming: A survey. Theoret Comput Sci 278(1–2):25–51
41. Flobbe L, Verbrugge R, Hendriks P, Krämer I (2008) Children's application of theory of mind in reasoning and language. J Logic Lang Inform 17(4):417–442
42. Fodor JA (1983) The modularity of mind: An essay on faculty psychology. MIT Press, Cambridge
43. Fodor JA, Pylyshyn ZW (1988) Connectionism and cognitive architecture: A critical analysis. Cognition 28:3–71
44. Freeman WJ (2000) How brains make up their minds. Columbia University Press, New York
45. Freeman WJ (1972) Waves, pulses and the theory of neural masses. Prog Theor Biol 2:87–165
46. Frixione M (2001) Tractable competence. Mind Mach 11(3):379–397
47. Gärdenfors P (2000) Conceptual spaces: The geometry of thought. MIT Press, Cambridge
48. Garey MR, Johnson DS (1979) Computers and intractability. W. H. Freeman and Co., San Francisco
49. van Gelder T (1990) Compositionality: A connectionist variation on a classical theme. Cogn Sci 14:355–364
50. van Gelder T (1991) Classical questions, radical answers: Connectionism and the structure of mental representations. In: Horgan T, Tienson J (eds) connectionism and the philosophy of mind. Kluwer Academic Publishers, Dordrecht pp 355–381
51. van Gelder T (1995) What might cognition be, if not computation? J Philos 92(7):345–381
52. van Gelder T (1998) The dynamical hypothesis in cognitive science. Behav Brain Sci 21(5):615–628
53. Ghosh S, Meijering B, Verbrugge R (2010) Logic meets cognition: Empirical reasoning in games. In: Boissier O, others (eds) MALLOW, CEUR Workshop Proceedings, vol 627
54. Ghosh S, Meijering B, Verbrugge R (2014), Strategic reasoning: Building cognitive models from logical formulas. J Logic Lang Inform 23:1–29
55. Gierasimczuk N, Szymanik J (2011a) Invariance properties of quantifiers and multiagent information exchange. In: Kanazawa M, Kornai A, Kracht M, Seki H (eds) Proceedings of 12th biennial conference on the mathematics of language (MOL 12, Nara, Japan, September 6–8, 2011), vol 6878. Springer, Berlin pp 72–89
56. Gierasimczuk N, Szymanik J (2011b) A note on a generalization of the muddy children puzzle. In: Apt KR (ed) Proceedings of the 13th conference on theoretical aspects of rationality and knowledge. ACM Digital Library pp 257–264
57. Gierasimczuk N, van der Maas H, Raijmakers M (2013) Logical and psychological analysis of deductive Mastermind. J Logic Lang Inform 22(3):297–314
58. Gneezy U, Rustichini A, Vostroknutov A (2010) Experience and insight in the race game. J Econ Behav Organ 75(2):144–155
59. Gold EM (1967) Language identification in the limit. Inf Control 10:447–474
60. Griggs RA, Cox JR (1982) The elusive thematic-materials effect in Wason's selection task. Br J Psychol 73:407–420
61. Gurevich Y (1993) Feasible functions. London Math Soc Newsl 10(206):6–7
62. Hawes DR, Vostroknutov A, Rustichini A (2012) Experience and abstract reasoning in learning backward induction. Front Neurosci 6(23)
63. Hedden T, Zhang J (2002) What do you think I think you think? Strategic reasoning in matrix games. Cognition 85(1):1–36
64. Hitzler P, Hölldobler S, Seda AK (2004) Logic programs and connectionist networks. J Appl Logic 2(3):245–272
65. Hölldobler S, Kalinke Y (1994) Toward a new massively parallel computational model for logic programming. In: Proceedings of workshop on combining symbolic and connectionist processing. ECAI-94, Amsterdam

66. Hoshi T (2010) Merging DEL and ETL. J Logic Lang Inform 19:413–430
67. Hubbard E, Diester I, Cantlon J, Ansari D, van Opstal F, Troiani V (2008) The evolution of numerical cognition: From number neurons to linguistic quantifiers. J Neurosci 28(46):11,819–11,824
68. Immerman N (1999) Descriptive complexity. Springer Verlag, Berlin
69. Isaac A, Hoshi T (2011) Synchronizing diachronic uncertainty. J Logic Lang Info 20:137–159
70. Johnson-Laird PN, Legrenzi P, Legrenzi MS (1972) Reasoning and a sense of reality. Br J Psychol 63:395–400
71. Kraus S, Lehmann DJ, Magidor M (1990) Nonmonotonic reasoning, preferential models and cumulative logics. Artif Intell 44(1–2):167–207
72. Kugel P (1986) Thinking may be more than computing. Cognition 22(2):137–198
73. Leitgeb H (2003) Nonmonotonic reasoning by inhibition nets II. Int J Uncertainty, Fuzziness and Knowl Based Syst 11(2 (Supplement)):105–135
74. Leitgeb H (2001) Nonmonotonic reasoning by inhibition nets. Artif Intell 128(1–2):161–201
75. Leitgeb H (2005) Interpreted dynamical systems and qualitative laws: From neural networks to evolutionary systems. Synthese 146:189–202
76. Levesque HJ (1988) Logic and the complexity of reasoning. J Philos Logic 17(4):355–389
77. Li M, Vitányi P (2008) An introduction to Kolmogorov complexity and its applications (3rd edn). Springer, Berlin
78. Lucas JR (1961) Minds, machines and Gödel. Philosophy 36(137):112–127
79. Marr D (1983) Vision: A computational investigation into the human representation and processing visual information. W.H. Freeman, San Francisco
80. McCarthy J (1980) Circumscription–a form of non-monotonic reasoning. Artif Intell 13(1–2):27–39
81. McCarthy J, Hayes PJ (1969) Some philosophical problems from the standpoint of artificial intelligence. In: Michie D, Meltzer B (eds) Machine Intelligence, vol 4. Edinburgh University Press, Edinburgh pp 463–502
82. McCulloch WS, Pitts WH (1943) A logical calculus immanent in nervous activity. Bull Math Biophys 5:115–133
83. McMillan CT, Clark R, Moore P, Devita C, Grossman M (2005) Neural basis for generalized quantifier comprehension. Neuropsychologia 43:1729–1737
84. McMillan CT, Clark R, Moore P, Grossman M (2006) Quantifier comprehension in corticobasal degeneration. Brain Cogn 65:250–260
85. Meijering B, van Maanen L, Verbrugge R (2010) The facilitative effect of context on second-order social reasoning. In: Catrambone R, Ohlsson S (eds) Proceedings of the 32nd annual conference of the cognitive science society. Cognitive Science Society, Austin pp 1423–1428
86. Meijering B, van Rijn H, Taatgen NA, Verbrugge R (2012) What eye movements can tell about theory of mind in a strategic game. PLoS ONE 7(9):e45,961
87. Meijering B, van Rijn H, Verbrugge R (2011) I do know what you think I think: Second-order theory of mind in strategic games is not that difficult. In: Proceedings of the 33rd annual meeting of the cognitive science society, Cognitive Science Society, Boston
88. Mostowski M (1998) Computational semantics for monadic quantifiers. J Appl Non-Class Log 8:107–121
89. Mostowski M, Wojtyniak D (2004) Computational complexity of the semantics of some natural language constructions. Ann Pure Appl Logic 127(1–3):219–227
90. Newell A (1981) The knowledge level: Presidential address. AI Mag 2(2):1–20
91. Pagin P (2012) Communication and the complexity of semantics. In: Werning M, Hinzen W, Machery E (eds) Oxford handbook of compositionality. Oxford University Press, Oxford
92. Papadimitriou CH (1993) Computational complexity. Addison Wesley, Boston
93. Penrose R (1994) Shadows of the mind: A search for the missing science of consciousness. Oxford University Press, New York
94. Pinkas G (1995) Reasoning, nonmonotonicity and learning in connectionist networks that capture propositional knowledge. Artif Intell 77:203–247
95. Pratt-Hartmann I (2004) Fragments of language. J Logic Lang Inform 13(2):207–223

96. Pudlák P (1999) A note on applicability of the incompleteness theorem to human mind. Ann Pure Appl Logic 96(1–3):335–342
97. Raijmakers ME, Mandell DJ, van Es SE, Counihan M (2014) Children's strategy use when playing strategic games. Synthese 191:355–370
98. Reiter R (1980) A logic for default reasoning. Artif Intell 13:81–132
99. Ristad ES (1993) The Language complexity game (Artificial intelligence). MIT Press, Cambridge, MA,
100. van Rooij I, Stege U, Kadlec H (2005) Sources of complexity in subset choice. J Math Psychol 49(2):160–187
101. van Rooij I, Wareham T (2007) Parameterized complexity in cognitive modeling: Foundations, applications and opportunities. Comput J 51(3):385–404
102. van Rooij I (2008) The tractable cognition thesis. Cognitive Sci 32(6):939–984
103. Sandler U, Tsitolovsky L (2008) Neural cell behavior and fuzzy logic. Springer, New York
104. Schlotterbeck F, Bott O (2013) Easy solutions for a hard problem? The computational complexity of reciprocals with quantificational antecedents. J Logic Lang Inform 22(4):363–390
105. Shanahan M (1997) Solving the frame problem: A mathematical investigation of the common sense law of inertia. MIT Press, Cambridge
106. Shoham Y (1987) A semantical approach to nonmonotonic logics. In: Proceedings of the 2nd symposium on logic in computer science. CEUR Workshop Proceedings pp 275–279
107. Siegelmann HT, Sontag ED (1994) Analog computation via neural networks. Theoret Comput Sci 131:331–360
108. Siegelmann HT (1995) Computation beyond the Turing limit. Science 268:545–548
109. Siegelmann HT (2003) Neural and super-Turing computing. Mind Mach 13:103–114
110. Simon HA (1957) Models of man: Social and rational. Wiley, New York
111. Skyrms B (1996) The evolution of the social contract. Cambridge University Press, New York
112. Skyrms B (2004) The stag hunt and the evolution of social structure. Cambridge University Press, New York
113. Smolensky P (1987) Connectionism and cognitive architecture: A critical analysis. Southern J Philos 26(supplement):137–163
114. Smolensky P (1988) On the proper treatment of connectionism. Behav Brain Sci 11:1–74
115. Spaan E (1993) Complexity of modal logics. Ph.D. thesis, Universiteit van Amsterdam, Amsterdam
116. Steinert-Threlkeld S, Icard TF (2013) Iterating semantic automata. Linguist Philos 36:151–173
117. Stenning K, van Lambalgen M (2008) Human reasoning and cognitive science. MIT Press, Cambridge
118. Sternberg RJ (2002) Cognitive psychology. Wadsworth Publishing, Stamford
119. Syropoulos A (2008) Hypercomputation: Computing beyond the Church-Turing barrier (Monographs in computer science). Springer, Berlin
120. Szymanik J (2009) Quantifiers in time and space. Computational complexity of generalized quantifiers in natural language. Ph.D. thesis, University of Amsterdam, Amsterdam
121. Szymanik J (2007) A comment on a neuroimaging study of natural language quantifier comprehension. Neuropsychologia 45:2158–2160
122. Szymanik J (2010) Computational complexity of polyadic lifts of generalized quantifiers in natural language. Linguist Philos 33(3):215–250
123. Szymanik J, Zajenkowski M (2010a) Comprehension of simple quantifiers: Empirical evaluation of a computational model. Cognitive Sci 34(3):521–532
124. Szymanik J, Zajenkowski M (2011) Contribution of working memory in parity and proportional judgments. Belgian J Linguist 25(1):176–194
125. Szymanik J, Meijering B, Verbrugge R (2013a) Using intrinsic complexity of turn-taking games to predict participants' reaction times. In: Proceedings of the 35th annual conference of the cognitive science society. Cognitive Science Society, Austin

126. Szymanik J, Steinert-Threlkeld S, Zajenkowski M, Icard TF (2013b) Automata and complexity in multiple-quantifier sentence verification. In: Proceedings of the 12th International Conference on Cognitive Modeling, R. West, T. Stewart (eds.), Ottawa, Carleton University, 2013

127. Szymanik J, Verbrugge R (eds) (2013) Special issue of Journal of Logic, Language, and Information on Logic and Cognition, vol 22, Issue 3, 4

128. Szymanik J, Zajenkowski M (2010b) Quantifiers and working memory. In: Aloni M, Schulz K (eds) Amsterdam colloquium 2009, LNAI, vol 6042. Springer, Berlin, pp 456–464

129. Tenenbaum J, Kemp C, Griffiths T, Goodman N (2011) How to grow a mind: Statistics, structure, and abstraction. Science 331(6022):1279–1285

130. Thorne C (2012) Studying the distribution of fragments of English using deep semantic annotation. In: Proceedings of the ISA8 workshop

131. Tomaszewicz B (2012) Quantifiers and visual cognition. J Logic Lang Inform 22(3):335–356

132. Tsotsos J (1990) Analyzing vision at the complexity level. Behav Brain Sci 13(3):423–469

133. Tversky A, Kahneman D (1983) Extensional versus intuitive reasoning: The conjunction fallacy in probability judgment. Psychol Rev 90(4):293–315

134. Verbrugge R, Mol L (2008) Learning to apply theory of mind. J Logic Lang Inform 17(4):489–511

135. Verbrugge R (2009) Logic and social cognition: The facts matter, and so do computational models. J Philos Logic 38(6):649–680

136. Vogels TP, Abbott LF (2005) Signal propagation and logic gating in networks of integrate-and-fire neurons. J Neurosci 25(46):10786–10795

137. Wason PC (1968) Reasoning about a rule. Q J Exp Psychol 20:273–281

138. Wason PC, Shapiro D (1971) Natural and contrived experience in a reasoning problem. Q J Exp Psychol 23:63–71

139. de Weerd H, Verbrugge R, Verheij B (2013) How much does it help to know what she knows you know? An agent-based simulation study. Artif Intell 199–200: 67–92

140. Zajenkowski M, Styła R, Szymanik J (2011) A computational approach to quantifiers as an explanation for some language impairments in schizophrenia. J Commun Disord 44(6):595–600

Chapter 31
Computational Complexity and Cognitive Science: How the Body and the World Help the Mind be Efficient

Peter Gärdenfors

Abstract Computational complexity has been developed under the assumption that thinking can be modelled by a Turing machine. This view of cognition has more recently been complemented with situated and embodied cognition where the key idea is that cognition consists of an interaction between the brain, the body and the surrounding world. This chapter deals with the meaning of complexity from a situated and embodied perspective. The main claim is that if the structure of the world is taken into account in problem solving, the complexity of certain problems will be reduced in relation to Turing machine complexity. For example, search algorithms can be simplified if the visual structure of the world is exploited. Another case is the logical problem of language acquisition, claiming that children cannot learn language simply by considering the input. It is argued that this problem will not arise if it is taken into account that children's learning of grammatical features often exploits world knowledge.

31.1 The Notion of Complexity in Cognitive Science

Cognitive science comes in three flavours [6, pp. 83–84], [11]. The historically first is *classical computationalism*. The basic tenets are that the brain is a computer (Turing machine) and that all thinking is manipulation of symbols (e.g. [8, 9]). The second is *connectionism* (associationism). Here the central tenets are that the brain can be seen as a neural network and that thinking can be described as parallel distributed processing in such a network [25]. The third is *situated and embodied cognition* where the key idea is that cognition consists of an interaction between the brain, the body and the surrounding world. Thinking is not encapsulated in the brain but it leaks out into the world [6].

P. Gärdenfors (✉)
Cognitive Science, Lund University, Lund, Sweden
e-mail: Peter.Gardenfors@lucs.lu.se

A. Baltag and S. Smets (eds.), *Johan van Benthem on Logic and Information Dynamics*, Outstanding Contributions to Logic 5, DOI: 10.1007/978-3-319-06025-5_31, © Springer International Publishing Switzerland 2014

Classical computationalism entails that all cognitive science can be reduced to the study of computers and the algorithms that are run on them. Isaac et al. (this volume) formulate this idea crisply: "The human mind can only solve computable problems". Since the 1950's there has been a rapid development of computer science and its relation to logical formalisms. One problem area concerns the *complexity* of different kinds of computation. A difference between the analysis of algorithms and computational complexity theory is that an analysis of an algorithm for a particular problem can determine which amount of resources is used to solve the problem, whereas complexity theory asks a more general question about the minimal resources required among all possible algorithms that could be used to solve the same problem. The paper by Isaac et al. (this volume) is an overview of the consequences for cognitive science of the results concerning complexity and logical formalisms.

However, if one takes a different perspective on cognition, considerations concerning complexity will be of a different nature. The focus of this article will be the relation between complexity theory and situated and embodied cognition.

In order to bring out the contrast between the different kinds of cognitive science in relation to complexity, I want to highlight two assumptions of classical computationalism:

(1) All computation is (sequential) manipulation of symbols.
(2) The algorithms are run in a system (a computer or a brain) that is separated from the world—once the inputs are given to an algorithm it runs independently of what happens outside the system.

31.2 Complexity in Neural Networks

The second flavour of cognitive science is connectionism. In this tradition, Assumption (1), that all computation is manipulation of symbols, is abandoned. The neurons in a neural network are seen as processing information on the "subsymbolic" [27] or "subconceptual" [12] level. In general, connectionism kept Assumption (2), that computation is performed in a system that is separated from the world. In most applications, the neural network is given an input—in the form of a vector of values to its input layer—that is then processed by the system resulting in an output—a vector of values in its output layer.

However there are exceptions: In robotics, the *reactive systems* studied by Brooks [2] and others consist of comparatively simple processors, not necessarily parallel, that are in a constant interaction with the world. The research on reactive systems can be seen as precursors of the movement towards situated cognition. In these systems, it is no longer meaningful to separate input and output since they function as feedback loops, directly involving the surrounding world in its computations. Brooks [2] denies that a reactive system needs any internal representations at all. He takes the stance that robots do not need a model of the world to determine what to do next because they can simply sense it directly. He says that the world is its own best representation and that an efficient system should exploit this. However, his position has met with criticism (e.g. [18, 32]), even within the situated cognition camp.

As a part of the debate between classical computationalists and connectionists, it has been shown that all neural networks can be simulated by traditional computers (Turing machines) and vice versa. Hence many of the classical computationalists have claimed that the debate is a red herring. However, in these results complexity issues are eschewed.[1] Even though a Turing machine can simulate any neural network, it does not follow that the complexity of the algorithm for the Turing machine is of the same order as the one followed by the neural network.

Nowadays the area of complexity results concerning computation with neural networks is flourishing. A comprehensive survey is presented by Sima and Orponen [26]. They summarize the situation as that "a complexity theoretic taxonomy of neural networks has evolved, enriching the traditional repertoire of formal computational modes and even pointing out new sources of efficient computation" (p. 2728). However, one conspicuous lacuna in their survey is that the results they consider do not at all account for the *learning dynamics* of neural networks. This is, in my opinion, a serious limitation, since one of the main computational advantages of neural networks is that they can learn, albeit slowly, from the input they are presented with. Modelling such learning becomes much more difficult with classical symbolic computing.[2]

Isaac et al. (this volume) also discuss computation in neural networks, although their focus is on how systems for non-monotonic reasoning may be implemented. In particular they relate results in [20, 21] showing that any system performing computations over distributed representations may be interpreted as a classical computational system performing non-monotonic reasoning. These results support the view that anything that can be computed with a neural network can also be computed in a classical system.

31.3 Complexity in Situated Cognition

Next I turn to complexity issues in relation to situated cognition. The proponents of this position would claim that the brain is not made for checking the logical consistency of sentences or for handling any other NP-complete problem, but for surviving and reproducing in an environment that is partly predictable and partly unpredictable. The primary duty of the brain is to serve the body (the brain is a butler, not a boss). It does not function in solitude, but is largely dependent on the body it is employed by and the environment it is interacting with. In contrast, when

---

[1] For example, it is surprising that Marr [22] did not at all mention computational complexity in his description of the three levels of computation.

[2] There are attempts, however, in the work on adaptive Turing machines.

the brain is seen as a Turing machine, it has become customary to view it as an isolated entity, in accordance with Assumption (2) above.[3]

In addition to the two assumptions, the traditional complexity argument presumes that the problem is expressed in a representation where the primitive elements (the predicates) are independent of each other. This goes back to the ideals of logical positivism, in particular Carnap's [3] attempt to "reconstruct" the world in terms of atomic predicates. The position of situated cognition is, in contrast, that cognitive processes exploit (and mimic) the structures of the world itself, in particular the spatial layout of information.

Furthermore, situated cognition, at least partly, accepts the position that the world is its own best representation. As we saw, this is a central tenet of reactive systems [2]. Consequently, the brain does not need to construct detached representations of everything it interacts with.[4] Hence, situated cognition gives up both Assumptions (1) and (2) of classical computationalism. The position is succinctly formulated by Clark [6, p. 148]: "Structured, symbolic, representational, and computational views of cognition are mistaken. Embodied cognition is best studied by means of noncomputational and nonrepresentational ideas and explanatory schemes involving, e.g. the tools of Dynamical Systems theory".

In situated cognition, the visual system is not merely seen as an input device to the brain and the hand as enacting the will of the brain, but the eye-hand-brain is a coordinated system that functions as a feedback loop. For many tasks, it turns out that we solve problems more efficiently with our hands than with our brains. A simple example is the computer game Tetris where you are supposed to quickly turn, with the aid of the keys on the keyboard, geometric objects that come falling over a computer screen in order to fit them with the pattern at the bottom of the screen. When a new object appears, one can mentally rotate it to determine how it should be turned before actually touching the keyboard. However, expert players turn the object faster with the aid of the keyboard than they turn an image of the object in their brains [19]. This is an example of what has been called *interactive thinking*. The upshot is that a human who is manipulating representations in the head is sometimes a cognitively less efficient system than somebody interacting directly with the represented objects.

Clark [6, pp. 219–220] presents a fascinating example of a situated interaction between an organism and the world. It has been suggested that some aquatic animals, such as tunas and dolphins are simply not strong enough to propel themselves at the speeds they are observed to reach. Triantafullou and Triantafullou [29, p. 69] paradoxically claim that "it is even possible for a fish's swimming efficiency to exceed 100 %". The reason tunas and dolphins can be so efficient is that they in their swimming create and exploit swirls and vortices in the water that improve their propulsion and ability to manoeuver. In brief, the tunas and dolphins swim *with* the water, not in the water. The analogy I want to bring out is that our brains can be very

---

[3] This assumption is the basis for all sci-fi novels about a brain in the vat.

[4] In contrast to [2], the position does not deny, however, that the brain employs *some* detached representations, for example, when it is planning [13].

efficient, even with their limited resources, since they think *with* the world, not in the world.

It should be pointed out that ideas related to those of embodied and situated cognition that have become popular in the last decades have several predecessors. One example is the "ecological" psychology of Gibson [15] who rejected the idea that cognition is information processing and instead claimed that organisms could "pick up" all the necessary visual information directly from the environment and so that no computation was needed. Another tradition is the cybernetic movement in the middle of the 20th century (e.g. [31]) that studied feedback loops between an agent and the environment, again without exploiting any symbolic representations.

As far as I know, no strict account of the complexity of cognitive processes has been developed within the tradition of situated cognition. One reason for this is that it is difficult to develop formal models of how a situated approach influences complexity issues since we often do no know enough about what in the world the brain exploits directly and what it represents for itself.

One toy example, dear to researchers in classical AI, is how to determine whether a block x is above a block y in a tower of blocks (a typical robotics problem in the early days). In classical computation, this problem would be represented by a set of atomic statements of the type on(a, b), on(b, c), on(c, d)... and formulas expressing that the relation "above" is the transitive closure of "on". All this would be put into an inference engine that can determine the truth or falsity of above(x, y). The computational complexity of this problem is of the order n^2, where n is the number of blocks in the tower.

In contrast to the classical internal computation, a model within situated cognition would take into account that in the real world the blocks are *spatially organized* along the vertical dimension. The transitivity of the relation "above" is *built into* this spatial organization and need not be expressed in axioms, let alone be computed. A robot can simply visually scan the blocks from the bottom and see whether it encounters x or y first to determine the truth or falsity of above(x, y). The complexity of this procedure is of the order n, where n is the number of blocks, that is, it is linear in the number of blocks. The upshot is that *the geometric structure of the external word reduces the complexity of the problem.* This toy (sic) problem, illustrates in what sense the structure of the world helps offloading a cognitive system.[5]

More generally, one can consider the complexity of visual search problems. Tsotsos [30, p. 428] distinguishes between two variants: *bounded search* in which the visual target is explicitly provided in advance and *unbounded search* in which the target is defined only implicitly, for example, by specifying relationships it must have with other visual stimuli. He proves that unbounded visual search is NP-complete, while bounded visual search has linear complexity.

These theoretical results can be compared with the empirical results from Treisman [28] and her colleagues. In the experiments, two types of stimuli were used:

---

[5] In the terminology of Barwise and Shimojima's [1] "surrogate reasoning", this example is a "free ride" provided by the geometric constraints. However, the authors do not consider the reduction in complexity provided by "free rides".

disjunctive where the target can be identified by only one feature, such as color or orientation, and *conjunctive* where the target requires that more than one feature is identified. Both types are cases of bounded search in Tsotso's [30] terminology. Treisman [28] finds that conjunctive displays are identified in a response time that is linear with the number of items in the scene, just as predicted by Tsotsos' complexity result. However, for disjunctive stimuli, the target is found immediately—it simply "pops out"—independently of the number of items present. In this case, the human visual system somehow finds a solution that is more efficient in terms of complexity than what is predicted by Tsotsos' theoretical results.

31.4 Other Problems Relating to Complexity and Situated Cognition

In this section I will discuss complexity issues related to two well-known enigmas for classical computationalism in terms of situated cognition.

The first is the *frame problem* [7, 23]. Within the early AI community, it was hoped that if we could represent the knowledge necessary to describe the world and the possible actions in a suitable symbolic formalism, then by coupling this world description with a powerful inference machine one could construct an artificial agent capable of planning and problem solving. It soon turned out, however, that describing actions and their consequences in a symbolic form leads to a combinatorial explosion of the logical inferences that are needed. In other words, the complexity of the problem became insurmountable.

The crux is that symbolic representations are not well suited for representing causal connections or dynamic interactions in the world. Various escape routes were tried, but the frame problem persisted in one form or another. As a consequence, the entire program of building planning agents based on purely symbolic representations more or less came to a stall.

At the other extreme one finds the reactive systems that were presented earlier. Such systems are able to solve problems in the immediate environment without any symbolic representations simply by being directly situated in the world. On the other hand, reactive systems cannot form any plans that go beyond what is given in the environment.

Nowadays, many robotic systems take a middle road. They build up representations from their experience of the world, for example by constructing maps of their environment. Often, the representations are of a non-symbolic form. Some robots build on hybrid forms of representations, mixing symbols with maps and other non-symbolic forms (e.g. [4]). However, there exists no principled theory of how the computationally most efficient mixture between inner representations and immediate reactions to the environment is to be determined for a planning system. The problem is still in the hands of the engineers. Again, a suitable theory of the complexity of the problem is lacking.

A second enigma in the classical tradition is Chomsky's [5] *poverty of stimulus* argument, which claims that the grammar of a natural language cannot be learned by children because of the limited data available to them. In a more general form, this has become known as the *logical problem of language acquisition*, claiming that children cannot learn language simply by considering the input.[6] The argument can be structured as follows:

- All languages contain grammatical patterns that cannot be learned by children using *positive evidence* alone.
- Children are only presented with positive evidence for these patterns.
- Children learn the correct grammars for their native languages.

As a consequence, Chomsky argues, learning the grammar of a language must depend on some sort of *innate* linguistic capacity that provides additional knowledge to the children. In brief, language is too complex to be 100 % learned. Note that the logical problem of language acquisition presumes analogues of the assumptions (1) and (2), in particular that language processing is done independently of the world.[7]

From the perspective of situated cognition, a similar argument to the one presented in the previous section can be applied here. The key idea is that the child does not learn a language in the world, it learns it *with* the world, in particular together with other humans.

First of all, note that the problem of language acquisition, at least in Chomsky's version, does not concern how a language is learned, but how the *grammar* of a language is acquired. Formulating the problem in this fashion builds on the additional assumption that the grammar of a language is *independent* of its semantics (let alone, its pragmatics). However, outside the Chomskian congregation, this assumption is denied. Cognitive linguistics, for example, builds on the idea that the syntax of language is constrained, if not determined, by its semantics. And as soon as one then allows some connection between the semantics of a language and the world the language user is situated in, learning a grammar will, at least to some extent, be dependent on one's knowledge about the world.

Several experiments about language learning have shown how the learning of grammatical features exploits world knowledge (e.g. [10, 24]). For example Ramscar and Yarlett [24] show that children's world knowledge generates *expectations* about grammatical patterns. When such expectations are violated, for instance by an irregular plural form, the input can indeed function as negative evidence. In this way the argument from the poverty of stimulus is blocked.

Furthermore, a sentence is not just taken as input to the grammar crank in the child's brain and then determined to be grammatical or not—a sentence is *used* in a

---

[6] Several researchers have used Gold's [16] theorem to support this argument, but, as Johnson [17] shows, this result has little bearing on how people actually learn languages.

[7] Chomsky's early work concerned the relations between different kinds of formal automata and the (formal) languages they could identify. This is a typical problem of computationalism that builds on Assumptions (1) and (2). Chomsky seems, more or less, to have stuck to these assumptions throughout his career.

particular context. And the use of a sentence may provide constraints for its structure. Here, I do not wish to speculate on how the constraints can be specified. Suffice it to notice that such constraints will block the poverty of stimulus argument, at least in its current form.

31.5 Conclusion

In conclusion, Assumptions (1) and (2) of classical computationalism have been taken over implicitly in many other areas. Once they are brought out into the open, however, they are seen to be invalid for many kinds of cognitive problems. The main argument of this paper is that once we give up these assumptions, many problems that have seemed hopelessly complex for the classical computationalist may become more manageable, if a connectionist or situated perspective on cognition is adopted instead. And evolution is a tinkerer with limited resources: rest assured that if one solution to a problem is cheaper than another, evolution will, in the long run, select the cheap one.

Still, humans have evolved symbolic language. In my opinion [13, 14], the main reason for this is that it has improved our *planning capacities*. There are situations involving reasoning with numbers, reasoning with cases or reasoning with conditional assumptions where symbolic structures are required. My point in this paper is simply that there are cases of problem solving where less complex methods than those offered by symbolic thinking are sufficient and therefore more efficient.

Humans have also speeded up the evolutionary selection processes: We have created cultures and artefacts that greatly improve our problems solving capacities. We have invented pencil and paper, libraries and smartphones that offload our memories, allow us to share knowledge, and amplify our calculations. Tunas and dolphins create structures in the water that improve their swimming. Humans create structures in the world that improve their thinking. As Clark [6, p. 180] puts it: "Our brains make the world smart so that we can be dumb in peace! Or to look at it another way, it is the human brain *plus* these chunks of external scaffolding that finally constitutes the smart, rational inference engine we call mind".

It must be pointed out, though, that the theory of situated cognition still lacks a rigor that would make it possible to develop a parallel to the theory of complexity that exists for classical computationalism and to some extent also for connectionism. Barwise and Shimojima's [1] ideas about *constraint projection* is perhaps a first step in that direction. I can only hope that a more precise theory will be formulated that will allow comparisons with the results concerning the complexity of situated processes.

Acknowledgments I gratefully acknowledge support from the Swedish Research Council for the Linnaeus environment *Thinking in Time: Cognition, Communication and Learning*. Thanks to Holger Andreas, Johan van Benthem, Alistair Isaac, Giovanni Pezzulo and Jakub Szymanik for helpful comments on an earlier version of the paper.

References

1. Barwise J, Shimojima A (1995) Surrogate reasoning. Cogn Stud Bull Japan Cogn Sci Soc 2(4):7–27
2. Brooks R (1991) Intelligence without representation. Artif Intell 47:139–159
3. Carnap R (1928) Der logische Aubau der Welt. Felix Meiner Verlag, Hamburg
4. Chella A, Frixione M, Gaglio S (1997) A cognitive architecture for artificial vision. Artif Intell 89:73–111
5. Chomsky N (1980) Rules and representations. Basil Blackwell, Oxford
6. Clark A (1997) Being there: putting brain body and world together again. MIT Press, Cambridge
7. Dennett DC (1984) Cognitive wheels: the frame problem of AI. In: Hookway C (ed) Minds, machines and evolution: philosophical studies. Cambridge University Press, Cambridge, pp 129–151
8. Fodor JA (1975) The language of thought. Harvard University Press, Cambridge
9. Fodor JA, Pylyshyn ZW (1988) Connectionism and cognitive architecture: a critical analysis. Cognition 28:3–71
10. Foraker S, Regier T, Khetarpal N, Perfors A, Tenenbaum J (2009) Indirect evidence and the poverty of the stimulus: the case of anaphoric "one". Cogn Sci 33:287–300
11. Gärdenfors P (1999) Cognitive science: from computers to anthills as models of human thought. World Social Science Report, UNESCO Publishing, Paris, pp 316–327
12. Gärdenfors P (2000) Conceptual spaces: the geometry of thought. MIT Press, Cambridge
13. Gärdenfors P (2004) Cooperation and the evolution of symbolic communication. In: Oller K, Griebel U (eds) The evolution of communication systems. MIT Press, Cambridge, pp 237–256
14. Gärdenfors P (2012) The cognitive and communicative demands of cooperation. In: van Eijck J, Verbrugge R (eds) Games, actions and social software, LNCS 7010. Springer, Berlin, pp 164–183
15. Gibson JJ (1979) The ecological approach to visual perception. Houghton Mifflin, Boston
16. Gold E (1967) Language identification in the limit. Inf Control 10:447–474
17. Johnson K (2004) Gold's theorem and cognitive science. Philos Sci 71:571–592
18. Kirsh D (1991) Today the earwig, tomorrow man? Artif Intell 47:161–184
19. Kirsh D, Maglio P (1994) On distinguishing epistemic from pragmatic action. Cogn Sci 18:513–549
20. Leitgeb H (2001) Nonmonotonic reasoning by inhibition nets. Artif Intell 128:161–201
21. Leitgeb H (2003) Nonmonotonic reasoning by inhibition nets II. Int J Uncertainty Fuzziness Knowl Based Syst 11:105–135
22. Marr D (1982) Vision: a computational investigation into the human representation and processing visual information. Freeman, San Francisco
23. McCarthy J, Hayes PJ (1969) Some philosophical problems from the standpoint of artificial intelligence. In: Michie D, Meltzer B (eds) Machine intelligence, vol 4. Edinburgh University Press, Edinburgh, pp 164–183
24. Ramscar M, Yarlett D (2007) Linguistic self-correction in the absence of feedback: a new approach to the logical problem of language acquisition. Cogn Sci 31:927–960
25. Rumelhart DE, McClelland JL (1986) Parallel distributed processing: explorations in the microstructure of cognition. MIT Press, Cambridge
26. Sima J, Orponen P (2003) General purpose computation with neural networks: a survey of complexity theoretic results. Neural Comput 15:2727–2778
27. Smolensky P (1988) On the proper treatment of connectionism. Behav Brain Sci 11:1–23
28. Treisman A (1988) Features and objects: the fourteenth Bartlett memorial lecture. Q J Exp Psychol Sect A: Hum Exp Psychol 40:201–237
29. Triantafyllou MS, Triantafyllou GS (1995) An efficient swimming machine. Sci Am 272:64–70
30. Tsotsos J (1990) Analyzing vision at the complexity level. Behav Brain Sci 13:423–469
31. Wiener N (1948) Cybernetics: or control and communication in the animal and the machine. Hermann and Cie, Paris
32. Williams MA (2008) Representation = grounded information. In: Proceedings of the 10th Pacific Rim international conference on artificial intelligence, Hanoi, pp 473–484

Part VI
Styles of Reasoning

Chapter 32
Dynamic Versus Classical Consequence

Denis Bonnay and Dag Westerståhl

Abstract The shift of interest in logic from just reasoning to all forms of information flow has considerably widened the scope of the discipline, as amply illustrated in Johan van Benthem's recent book *Logical Dynamics of Information and Interaction*. But how much does this change when it comes to the study of traditional logical notions such as logical consequence? We propose a systematic comparison between classical consequence, explicated in terms of truth preservation, and a dynamic notion of consequence, explicated in terms of information flow. After a brief overview of logical consequence relations and the distinctive features of classical consequence, we define classical and dynamic consequence over abstract information frames. We study the properties of information under which the two notions prove to be equivalent, both in the abstract setting of information frames and in the concrete setting of Public Announcement Logic. The main lesson is that dynamic consequence diverges from classical consequence when information is not persistent, which is in particular the case of epistemic information about what we do not yet know. We end by comparing our results with recent work by Rothschild and Yalcin on the conditions under which the dynamics of information updates can be classically represented. We show that classicality for consequence is strictly less demanding than classicality for updates.

Johan van Benthem's recent book *Logical Dynamics of Information and Interaction* [8] can be seen as a passionate plea for a radically new view of logic. To be sure, the book is not a philosophical discussion of what logic is but rather an impressive

D. Bonnay (✉)
Département de Philosophie, Université Paris Ouest, 200 avenue de la République, 92000 Nanterre, France
e-mail: denis.bonnay@u-paris10.fr

D. Westerståhl
Department of Philosophy, Stockholm University, SE-106 91 Stockholm, Sweden
e-mail: dag.westerstahl@philosophy.su.se

A. Baltag and S. Smets (eds.), *Johan van Benthem on Logic and Information Dynamics*, Outstanding Contributions to Logic 5, DOI: 10.1007/978-3-319-06025-5_32, © Springer International Publishing Switzerland 2014

series of illustrations of what logic *can be*, with presentations of numerous logical languages and a wealth of meta-logical results about them. The view is called simply Logical Dynamics, and contrasted with more traditional views of logic, and also with the earlier view from e.g. [5], now called Pluralism, in which logic was seen as the study of consequence relations.

According to Logical Dynamics, logic is not only about reasoning, about what follows from what, but about all aspects of *information flow among rational agents*. Not just proof and inference, but observations, questions, announcements, communication, plans, strategies, etc. are first-class citizens in the land of Logic. And not only the output of these activities belong to logic, but also the processes leading up to it.

This is a fascinating and inspiring view of logic. But how different is it from a more standard view? In particular, what does it change for the analysis of logical consequence, which had been the focus of traditional logical enquiry? This paper attempts some answers to the latter question, with a view to get clearer about the former.

32.1 Introduction

… in line with the thrust of this book, I see a discipline as a dynamic activity, not as any of its static products: proofs, formal systems, or languages. Logic is a stance, a modus operandi, and perhaps a way of life. That is wonderful enough. [8, p. 302]

Logicians will surely recognize this: doing logic is approaching your subject from a certain stance, a certain view of what the interesting questions are, what tools to use, what kind of abstractions are called for. The stance itself may be hard to put into words but is recognizable to the practitioners. Logical Dynamics is not really recommending a new stance towards logic, it seems to us. The novelty lies in what is taken to be its subject matter.

A rough object versus meta level distinction is helpful: According to Logical Dynamics, many more object languages can and should be studied with logical methods than has traditionally been the case. But the form that this study takes is still of the familiar kind: (a) you introduce a formal language in which a particular variety of information flow can be expressed; (b) you provide a formal semantics and a deductive apparatus and see what can be proved from a choice of axioms; (c) you establish facts *about* these things: expressive power, definability, completeness, decidability, the structure of proofs, computational complexity, relations to other languages, etc. Indeed, this is precisely what a large part of *Logical Dynamics of Information and Interaction* is devoted to.

Reflection on logic from this perspective raises intriguing questions. What is it about a certain form of information flow that makes it apt for investigation by logical methods? What characterizes the syntactic constructs—the logical constants—used in the various object languages? Here we focus on just one aspect: the variety of consequence relations that emerge.

In fact, this variety is so great that it may in the end be just confusing to use the label "consequence relation" for all of them. Some have rather little to do with 'what follows from what' in any usual sense. But a distinction between two important kinds can be made: *dynamic* and *classical* relations. Very roughly: in order to draw the conclusion in the dynamic case, it may matter *how* the information from the premises was processed. For example, during that process some information may get lost. For classical consequence, on the other hand, only the actual information contained in the premises matters.

In particular, we are interested in how this distinction applies at the meta level. In contrast with the object level, where there seems to be no end to the variety of information-related activities that can be explored, the goals of meta level logical study seem rather fixed. One wants to know facts about the object level phenomena, that is, one wants the *truth* about them. Moreover, these truths are *mathematical* in a wide sense, and the only way you are allowed to assert a mathematical truth is to *prove* it. So the consequence relations operating at this level concern *reasoning towards truth*. This already separates them from a host of consequence relations resulting from object level phenomena. Does it in fact narrow the options down to just classical ones?[1]

To get a feeling for the issues involved here, and the variety of consequence relations on the market, let us look at a few examples.

- **Non-monotonicity 1**

  One way in which non-monotone consequence relations arise is when trying to model various kinds of reasoning under uncertainty, default reasoning, abductive reasoning, etc., where a conclusion is drawn tentatively, in awareness that it may have to be abandoned in the light of further evidence. There is no claim that the conclusion really *follows* from the premises, and hence no real clash with classical consequence. Such reasoning is what everyone—even the logician— often has to resort to in daily life. But obviously it is never accepted in mathematics as conclusive grounds for a claim, and likewise not in metalogical reasoning. A number theorist may perhaps say that Goldbach's Conjecture is *likely* to be true, with reference to the so far observed even numbers greater than 2, but never that it is true (unless she has a proof).

- **Non-monotonicity 2**

  A different motive for rejecting monotonicity is proposed by *relevant* (or relevance) logicians. The idea is that adding an irrelevant premiss is not allowed. One question here concerns whether relevant logic is thought to be, in the terminology of John Burgess, *descriptive* (of the practice of mathematicians) or *prescriptive* (wanting to reform that practice; see [3] and [21]). In any case, relevantists would seem to claim that relevance is or should be practiced at the meta level too.

But relevantists do aim at reasoning towards truth. And *no one* could argue that adding 'unnecessary' premises, say in the form of a Weakening rule, threatens to

---

[1] That is, classical in the sense just introduced. An intuitionistic consequence relation may well be classical in this sense.

lead from true premisses to false conclusions. To understand a relevantist stand on monotonicity, we need to be precise about which consequence relation is under discussion. Say that some relevant logic *Rel* is presented as a natural deduction system for deriving sequents of the form $\Gamma \succ \phi$. Then, if we define monotonicity

$$\Gamma \vdash_{Rel} \phi \quad \text{iff there is a finite } \Gamma_0 \subseteq \Gamma \text{ s.t. } \Gamma_0 \succ \phi \text{ is derivable in } Rel,$$

will hold. Thus, for example, $\{p, q\} \vdash_{Rel} p$ (since $\{p\} \succ p$ and $\{p\}$ is a finite subset of $\{p, q\}$), even though $\{p, q\} \succ p$ may not be derivable. So a natural classical consequence relation extends the non-monotone relevantist one, in a way that can never be harmful, by anybody's lights, for reasoning towards truth.

- **Substructural logic**
  Weakening is a structural rule, and much recent work in logic rejects or modifies various such rules, e.g. in Linear Logic. But most of this work is not about truth at all. For a clear example, consider the Lambek Calculus (see e.g. [4]). Here Weakening, Contraction, and Permutation fail, for the obvious reason that the calculus aims to describe natural language syntax. Adding words, permuting words, or contracting two occurrences of the same word, may destroy well-formedness. This has nothing to do with truth, and indeed nicely illustrates the difference between consequence relations that logicians *study*, and the ones they may *use* in their own reasoning.

- **Contraction-free reasoning about truth**
  Consider a language for talking about truth: it contains a truth predicate, all instances of Tarski's T-schema, and some means for self-reference. With classical logic this leads to inconsistency. It has long been noted that proof-theoretic derivations of the Liar or Curry's Paradox rely on the *Contraction* rule,[2] and it has been suggested that dropping Contraction is a natural way to avoid paradoxes; for a recent proposal, see [22].

This is a case where the difference between dynamic and classical consequence seems to matter. From a classical point of view, no one could seriously think that Contraction (as described in note 2) is an *invalid* rule. That would be like someone blaming your proof of a theorem B for using an assumption A *twice* (say, an earlier proved lemma), without explicitly saying so. First, you could retort that since A has been proved, you could easily repeat that proof twice in your proof of B. But really, the right answer is that the complaint makes no sense. When you claim that B *follows from* A (and possibly other assumptions), A and B are not *tokens*, although you need to use tokens of them when writing up the proof. They are *types*, or *propositions*. And with these abstract objects it makes no sense to ask how many copies of them you are using.

---

[2] See e.g. [9]. We here intend a rule of the type "If $\Gamma, \phi, \phi \vdash \psi$ then $\Gamma, \phi \vdash \psi$". The validity of this rule need not entail the validity of, say, $\phi \to (\phi \to \psi) \vdash \phi \to \psi$.

But these considerations do not apply to a dynamic notion of consequence. An additional *order* is introduced: you process one thing *before* another thing, and then of course there is no guarantee that the usual structural rules will hold. Even if propositions are still types (not tokens), their processing happens in time, as it were.[3]

• Public announcement

Our concern here, however, is not reasoning *about* truth, but reasoning with the aim of arriving at true propositions. Is there a role for dynamics here? Dynamic phenomena can themselves be expressed in richer logical languages. The clearest example is perhaps Public Announcement Logic (PAL) (introduced in [16]; see [8], ch. 3, for an overview). Here an announcement, when it can be made, changes the information state: situations where the announcement is false are discarded ('hard update'). Of course you cannot go around announcing anything: the claim has to be true in the current situation. What the announcement changes is agents' *knowledge*. In other words, this kind of reasoning only makes a difference when it is (also) about knowledge itself.

Although many non-classical consequence relations studied by logicians are irrelevant to our present concern with reasoning towards truth, the dynamic idea of consequence seems very different from the classical one. But note that rendering dynamic phenomena in a richer language like PAL is a *reduction* of dynamic consequence to classical consequence. That is, various dynamic consequence relations are expressible in PAL and similar logics, but PAL validity itself is classical. This is in fact part of the program in *Logical Dynamics of Information and Interaction*. The following quote is illustrative:

> Non-monotonicity is like a fever: it does not tell you which disease causes it. [8, p. 297]

Explicitly accounting for update phenomena in a richer language reveals the *causes* of non-monotonicity, but in a classical framework. The question is raised (but not answered) in the book whether this sort of reduction is always possible. That would be a very strong vindication of classical logic.

Our project in the rest of this paper is more modest. First, we specify precisely in what sense classical consequence is related to truth *preservation* (Sect. 32.2). Then we compare classical and dynamic consequence in an update-friendly framework, both abstractly (Sect. 32.3) and in the concrete setting of PAL (Sect. 32.4), and show when the two notions of consequence coincide. Finally, we consider (Sect. 32.5) a recent result in [18] on the conditions under which updates themselves are classical,

---

[3] Zardini presents a formal system without Contraction containing a 'naive theory of truth' and shows that his consequence relation satisfies truth preservation in the following form (simplified): if $\Gamma, \phi \vdash \psi$, then $\Gamma \vdash Tr(\langle \phi \rangle) \rightarrow Tr(\langle \psi \rangle)$, where $\langle \phi \rangle$ names ϕ in the theory. By contrast, Field in [12] seems to agree with our point about Contraction and Permutation [12, pp. 10–11], but argues for an approach that rejects Excluded Middle (is 'paracomplete') as well as truth preservation in Zardini's sense. But note that this is a special version of truth preservation, tied to the occurrence of a truth predicate, and to the meaning of the conditional. As we will see in the next section, since Field's preferred consequence relation appears to be reflexive and transitive, there is a clear sense in which it necessarily preserves truth.

and show that classicality of updates is a strictly stronger requirement than classicality of consequence.

32.2 Classical Consequence and Truth Preservation

Following the remarks on Contraction and Permutation above, we take, in this section, consequence relations to hold between sets of sentences (the premisses) and sentences (the conclusion). The classical idea of logical consequence is as necessary (in some sense) truth preservation. There is no question that such reasoning is safe if your goal is arriving at the truth. But there are many different consequence relations that enforce truth preservation, for example, intuitionist as well as classical (in the sense of accepting excluded middle etc.) ones. So let us point out what they all have in common.

To begin, there may seem to be two ways to think about truth and truth preservation. One is in terms of *truth at a point* (think: possible world). The other is about *truth in an interpretation*. But at the current abstract level, they are really equivalent.

To see this, fix a language L with its set $Sent_L$ of sentences. For truth at a point, suppose $\pi : Sent_L \to \wp(X)$ maps sentences to subsets of a set X of points, i.e. $\pi(\phi)$ is the set of points at which ϕ is *true*. Now consequence as necessary truth preservation (relative to π) is defined by

(1) $\Gamma \vdash^\pi \phi$ iff $\bigcap_{\psi \in \Gamma} \pi(\psi) \subseteq \pi(\phi)$

For the other approach to truth, let a *valuation* v assign truth values 1 or 0 to sentences, so we can take v to be a *subset* of $Sent_L$. Say that a sequent (Γ, ϕ) is *true in v* iff whenever $\Gamma \subseteq v$, we have $\phi \in v$. Note that sentence truth is a special case: ϕ is true in v iff $\phi \in v$ iff $(\emptyset, \phi)$ is true in v. For any set K of valuations we have a corresponding notion of truth-preserving consequence:

(2) $\Gamma \vdash_K \phi$ iff for all $v \in K$, (Γ, ϕ) is true in v.

To see that these approaches are equivalent, let π be given as above, and for each $a \in X$, let $v_a = \{\phi : \phi$ is true at $a\} = \{\phi : a \in \pi(\phi)\}$, and $K = \{v_a : a \in X\}$. Then:

(3) $\vdash^\pi = \vdash_K$

Conversely, if K is any set of valuations, let $\pi(\phi) = \{v \in K : \phi \in v\}$. Then again we obtain (3).

Since the two formats are equivalent, let us choose one: $\vdash_K$. Now stipulate that an arbitrary relation $\vdash$ between sets of sentences and sentences is *classical* iff it is reflexive and transitive in the following sense:

(R) If $\phi \in \Gamma$, then $\Gamma \vdash \phi$.

(T) If $\Delta \vdash \phi$, and for all $\psi \in \Delta$, $\Gamma \vdash \psi$, then $\Gamma \vdash \phi$.

Note that (R) and (T) entail monotonicity:[4]

---

[4] Another version of transitivity is Cut, in one of these two formulations:

(M) If $\Gamma \vdash \phi$ and $\Gamma \subseteq \Delta$, and $\Delta \vdash \phi$.

Now our point is this: reflexivity plus transitivity is *exactly* what constitutes necessary truth preservation in the above sense. This is a well-known fact, but let us spell it out.

Proposition 32.1 $\vdash$ *is classical if and only if* $\vdash = \vdash_K$, *for some K. In fact, we can take* $K = Val(\vdash)$, *so* $\vdash$ *is classical if and only if* $\vdash = \vdash_{Val}$ $(\vdash)$.

Proof It is obvious that each $\vdash_K$ satisfies (R) and (T). In the other direction, suppose $\vdash$ is classical. It is easy to verify from the definitions that $\vdash \subseteq \vdash_{Val}$ $(\vdash)$. To prove the converse inclusion, suppose $\Gamma \nvdash \phi$. Define the valuation v as follows: for any sentence ψ in L, $v(\psi) = 1$ iff $\Gamma \vdash \psi$. By (R), $v(\Gamma) = 1$, and by assumption, $v(\phi) = 0$. So it is enough to show that $v \in Val(\vdash)$. Suppose $\Delta \vdash \psi$ and $v(\Delta) = 1$; we must show that $v(\psi) = 1$. But it follows from (T) and the definition of v that $\Gamma \vdash \psi$, and we are done. □

To repeat, this is well-known. For example, in view of (3), Proposition 32.1 is a reformulation of a representation theorem in [8, p. 297].[5] But the result is relevant to our discussion. Any instance of, say, non-monotonicity guarantees *with respect to any class of interpretations*, that you will deduce a false conclusion from true premises. That is just the trivial direction of Proposition 32.1. The slightly less trivial direction says that if you have Reflexivity and Transitivity, there is always at least one class of interpretations with respect to which you can construe your consequence relation as necessary truth preservation.[6]

Finally, let us emphasize again that adhering to truth-preservational consequence relations is perfectly compatible with intuitionistic or (most forms of) relevant consequence. That choice depends on whether you regard particular rules, such as $\neg\neg\phi \vdash \phi$ or $\phi, \neg\phi \vdash \psi$, as valid.[7]

(Footnote 4 continued)
(C1) If $\Delta \vdash \psi$ and $\Gamma, \psi \vdash \phi$, then $\Delta, \Gamma \vdash \phi$.
(C2) If $\Gamma \vdash \psi$ and $\Gamma, \psi \vdash \phi$, then $\Gamma \vdash \phi$.

If all sets are finite, we have (cf. [20, pp. 17–18]): (R)+(T) $\Leftrightarrow$ (C1)+(R)+(M) $\Leftrightarrow$ (C2)+(R)+(M).

[5] Also in [4, p. 247], and Prop. 7.4 in [5]. We are not sure who first made observation contained in Proposition 32.1. It appears in [19], presented without proof as a familiar fact, but apparently it goes back at least to [15]. The above proof (as well as a generalization of the result to multiple-conclusion logics) is given in [20] and in [14]. Indeed, Proposition 32.1 is a cornerstone in the abstract theory of consequence relations and propositional connectives expounded and elaborated in [14], especially chs. 1.1 and 3.

[6] We may note that the thesis of *Logical Pluralism* in [2] is in fact that logic studies classical consequence relations defined as in (1) (they call the points *cases*). In particular, they observe (though with some hesitation) that relevant consequence relations should be monotone.

[7] That, for example, intuitionistic propositional logic consequence $\vdash^{IL}$ is classical in our sense means that Proposition 32.1 holds for it, i.e. $\vdash^{IL}$ is determined by $Val(\vdash^{IL})$. Since $\vdash^{IL} \subseteq \vdash^{CL}$, $Val(\vdash^{IL})$ extends $Val(\vdash^{CL})$ by allowing valuations that are not Boolean. More precisely, if $\mathcal{M}$ is any Kripke model for IL and $w \in |\mathcal{M}|$, the corresponding valuation $v_w^{\mathcal{M}}(\phi)$, consisting of the true sentences in $\mathcal{M}, w$, is equal to $Val(\vdash^{IL})$ (by completeness), and $v_w^{\mathcal{M}}$ is Boolean for $\wedge$ and $\vee$, but not necessarily for $\neg$ or $\rightarrow$, since one may have e.g. $\phi \notin v_w^{\mathcal{M}}$ and $\neg\phi \notin v_w^{\mathcal{M}}$.

32.3 Two Views About Consequence

Judging from the previous section, analyzing consequence in terms of truth preservation for propositions does not seem to leave many options open. However, taking
the intuitions of the dynamic perspective seriously, the picture becomes more complex. When asking whether something follows from some piece of information, one
may ask whether it follows from information already secured and fully available, or
whether it follows from information received along the way, which may not have been
preserved. The first question corresponds to the classical notion of consequence: we
look at where we are but not how we got there. In other words, we are not interested
in how information has been received, but only in what follows from the information
that we have. The second question corresponds to a thoroughly dynamic approach to
consequence: rather than looking at where we are, we look at how we got there. In
other words, we are interested in the information we received, which may end up not
being what we have kept. Indeed, these two notions are likely to come apart because
even hard information may not be preserved. Upon learning something which I did
not know, I cease not to know it; I did receive new information, which allowed
me to make some new inferences (e.g. that this something is true), but not all that
information is being preserved.

We will now try to capture these two intuitions by means of formal definitions in
an abstract dynamic setting.[8] Given a set of formulas L, an *abstract frame* for L is
a structure

$$\mathscr{F} = (\Sigma, \{[\phi]\}_{\phi \in L})$$

where Σ is a set and each $[\phi]: \Sigma \to \Sigma$ is a partial function. Elements of Σ are to be
construed as information states and each $[\phi]$ represents the effect of updating a given
information state with the information that ϕ. Functionality expresses the fact that
information updates are deterministic: the future state of information is completely
determined by the past information state and the extra information that has been
received. Partiality expresses the fact that not every piece of information is compatible
with every information state: the (true) information that $\neg p$ is not compatible with
the (truthful) announcement that p. Let the language L be fixed, together with a class
of abstract frames $\mathfrak{F}$ representing the relevant possible informational scenarios.[9] The
thoroughly dynamic notion of consequence may be captured by what is known as
Update to Test consequence:

---

[8] Technically speaking, in such a setting, there would be more possible definitions of dynamic
consequence than the two we are going to discuss—see [5] for more on this abstract stage setting,
and an investigation into some more possibilities. But we take these two to be representative of the
alternative between an essentially classical approach to consequence and an essentially dynamic
one.

[9] Our definitions are relativized to a class $\mathfrak{F}$. Alternatively, we could have defined absolute notions by
considering the greatest frame encompassing all possible informational scenarios. The relativized
notions will help us make precise the role played by some technical assumptions.

Definition 32.1 (*Update to Test consequence*) $\phi_1, \ldots, \phi_n \vDash_{UT}^{\mathfrak{F}} \psi$ iff for every $\mathscr{F} \in \mathfrak{F}$, $range([\phi_n] \circ \ldots \circ [\phi_1]) \subseteq fix([\psi])$.

(Here *range* assigns to every function in $\mathscr{F}$ its range and *fix* its set of fixed points.) This captures the dynamic notion because we ask whether we are in a fixed point for $[\psi]$ after shifting our information state along $[\phi_1], \ldots, [\phi_n]$. By contrast, the classical notion will be stated by considering an information state in which $[\phi_1], \ldots, [\phi_n]$ steadily hold; this is known in the dynamic literature as Test to Test consequence:

Definition 32.2 (*Test to Test consequence*) $\phi_1, \ldots, \phi_n \vDash_{TT}^{\mathfrak{F}} \psi$ iff for every $\mathscr{F} \in \mathfrak{F}$, $fix([\phi_1]) \cap \ldots \cap fix([\phi_n])\} \subseteq fix([\psi])$.

Indeed, this notion of consequence is classical in the sense of Sect. 32.2: it satisfies (R) and (T).

Comparing these two notions, $\vDash_{UT}^{\mathfrak{F}} \subseteq \vDash_{TT}^{\mathfrak{F}}$ but the converse inclusion does not hold in general. As suggested earlier, it is bound to fail whenever information is not preserved. So when do classical and dynamic consequence come together? The intuitive answer is that they do so when the information represented behaves classically: received information is persistent, it gets into the current informational state and stays there.

To express this formally, a few definitions are needed. An abstract frame $\mathscr{F} = (\Sigma, \{[\phi]\}_{\phi \in L})$ is *idempotent* iff for all $\sigma \in \Sigma$, for all $\phi \in L$, $\sigma[\phi] = \sigma[\phi][\phi]$.[10] It is *commutative* iff for all $\sigma \in \Sigma$ and all $\phi, \psi \in L$, $\sigma[\phi][\psi] = \sigma[\psi][\phi]$. It is *f-commutative* iff for all $\sigma \in \Sigma$ and all $\phi, \psi \in L$, if $\sigma[\phi] = \sigma$, then $\sigma[\phi][\psi] = \sigma[\psi][\phi]$. So *f*-commutativity is a restricted version of commutativity, which only works for *f*ixed points and in the *f*orward direction.[11]

Idempotence and f-commutativity are conveniently made into a package deal:

Proposition 32.2 *An abstract frame* $\mathscr{F} = (\Sigma, \{[\phi]\}_{\phi \in L})$ *is idempotent and f-commutative iff for all* $\phi \in L$, *for any (possibly empty) sequence* $\phi_1, \ldots, \phi_n$ *of formulas in L,*

$$[\phi][\phi_1]\ldots[\phi_n] = [\phi][\phi_1]\ldots[\phi_n][\phi]$$

Proof From Left to Right: The proof is by induction on the length of the sequence of formulas. For the empty sequence, this is idempotence. Consider a sequence $\phi_1, \ldots, \phi_n, \phi_{n+1}$ and let $\sigma \in \Sigma$ be such that $\sigma[\phi][\phi_1]\ldots[\phi_n][\phi_{n+1}]$ exists. By

---

[10] Here and below we use the arrow-like notation $\sigma[\phi] = \sigma'$, rather than the functional $[\phi](\sigma) = \sigma'$, to indicate that σ' is the result of updating σ with (the information that) ϕ:

$$\sigma \xrightarrow{\phi} \sigma'$$

Note that $[\phi][\psi]$ is the same as $[\psi] \circ [\phi]$. Note also that $[\phi]$ may be undefined for some σ. Throughout the paper, we take equalities $\sigma[\phi] = \sigma'[\psi]$ to mean that $\sigma[\phi]$ is defined iff $\sigma'[\psi]$ is, and that, when they are defined, they are equal.

[11] As pointed out to us by Seth Yalcin, f-commutativity can also be understood as a property of persistence of truth at a state.

induction hypothesis, $[\phi][\phi_1]\ldots[\phi_n] = [\phi][\phi_1]\ldots[\phi_n][\phi]$. Hence we may apply f-commutativity at $\sigma[\phi][\phi_1]\ldots[\phi_n] = \sigma'$ for ϕ and ϕ_{n+1}:

$$\sigma'[\phi][\phi_{n+1}] = \sigma'[\phi_{n+1}][\phi]$$

Replacing $\sigma'[\phi]$ by σ' in the left-hand side yields

$$\sigma[\phi][\phi_1]\ldots[\phi_n][\phi_{n+1}] = \sigma[\phi][\phi_1]\ldots[\phi_n][\phi_{n+1}][\phi]$$

as required.

From Right to Left: Idempotence is the case when the sequence is empty. As to f-commutativity, let $\sigma \in \Sigma$ be such that $\sigma[\phi] = \sigma$ and suppose $\sigma[\phi][\psi]$ exists. (Note that if we instead suppose $\sigma[\psi][\phi]$ exists, so does $\sigma[\psi]$, and hence, since $\sigma[\phi] = \sigma$, $\sigma[\phi][\psi]$ exists.) By hypothesis, $\sigma[\phi][\psi] = \sigma[\phi][\psi][\phi]$. Since $\sigma[\phi] = \sigma$, $\sigma[\phi][\psi][\phi] = \sigma[\psi][\phi]$, hence $\sigma[\phi][\psi] = \sigma[\psi][\phi]$ as required. □

The equality

$$[\phi][\phi_1]\ldots[\phi_n] = [\phi][\phi_1]\ldots[\phi_n][\phi]$$

intuitively means that the information that ϕ is persistent, in the sense that, once received, it still holds after updating in turn with $\phi_1, \ldots, \phi_n$. Proposition 32.2 then says that, taken together, idempotence and f-commutativity precisely amount to information always being persistent. It will come as no surprise that our two notions of consequence coincide on the class of idempotent and f-commutative abstract frames:[12]

Proposition 32.3 *An abstract frame $\mathscr{F}$ is such that $\vDash_{UT}^{\{\mathscr{F}\}} = \vDash_{TT}^{\{\mathscr{F}\}}$ iff $\mathscr{F}$ is idempotent and f-commutative.*

Proof If a frame $\mathscr{F}$ is such that $\vDash_{UT} = \vDash_{TT}$ (suppressing $\mathscr{F}$ in the notation), then it is idempotent and f-commutative: Since $\vDash_{TT}$ satisfies (R), so does $\vDash_{UT}$. Take $\phi, \phi_1, \ldots, \phi_n \in L$, $\sigma \in \Sigma$ and consider $\sigma[\phi][\phi_1]\ldots[\phi_n]$, assuming it is defined. By (R), $\phi, \phi_1, \ldots, \phi_n \vDash_{UT} \phi$. By definition of $\vDash_{UT}$, this means that $range([\phi_n] \circ \ldots \circ [\phi_1] \circ [\phi]) \subseteq fix([\phi])$. Hence $\sigma[\phi][\phi_1]\ldots[\phi_n] = \sigma[\phi][\phi_1]\ldots[\phi_n][\phi]$, and the result follows from Proposition 32.2.

If a frame is idempotent and f-commutative, then $\vDash_{UT} = \vDash_{TT}$: First, it is always the case that $\vDash_{UT} \subseteq \vDash_{TT}$, so all we need to prove is $\vDash_{UT} \supseteq \vDash_{TT}$. Assume $\phi_1, \ldots, \phi_n \vDash_{TT} \psi$, and let $\sigma' \in range([\phi_n] \circ \ldots \circ [\phi_1])$. There is $\sigma \in \Sigma$ such that $\sigma[\phi_1]\ldots[\phi_n] = \sigma'$. Using idempotence and f-commutativity we see that $\sigma[\phi_1]\ldots[\phi_n][\phi_i] = \sigma[\phi_1]\ldots[\phi_n]$, i.e. $\sigma[\phi_1]\ldots[\phi_n] \in fix([\phi_i])$, for every $i \in \{1,\ldots,n\}$. Since $\phi_1, \ldots, \phi_n \vDash_{TT} \psi$, this implies that $\sigma[\phi_1]\ldots[\phi_n] \in fix([\psi])$. Thus $\phi_1, \ldots, \phi_n \vDash_{UT} \psi$ as required. □

---

[12] This result generalizes Proposition 2.3 in [11] by providing necessary as well as sufficient conditions for the equivalence to hold.

32.4 Classic and Dynamic Consequence in PAL

Proposition 32.2 provides an *abstract* characterization of the properties of information that make classical and dynamic consequence coincide. This abstract perspective can be made *concrete* by looking at specific forms of information update expressed in specific logical languages. This is the representation step advocated by van Benthem in [8]. We take such a step in the present section, and instantiate the two consequence relations within the logic of public announcements, PAL, where information updates are truthful public announcements.[13]

Technically, PAL expresses updates by means of an announcement operator $[!\_]$, which goes together with epistemic modalities K_i and a common knowledge modality C. Sets of models for PAL provide us with concrete versions of the abstract frames we were considering. More precisely, a *concrete frame* $\mathscr{K}$ is a set of multi-modal pointed S5-models $\mathscr{M}, w$, which is closed under submodels and change of designated world. (Putting the two conditions together, if $\mathscr{M}, w \in \mathscr{K}$ and $\mathscr{M}' \subseteq \mathscr{M}$, then $\mathscr{M}', w' \in \mathscr{K}$, for $w \in |\mathscr{M}|$ and $w' \in |\mathscr{M}'|$.) Closure under submodels guarantees that updates can be performed. The necessity to allow for changes in the designated world will become clear later.

Any concrete frame $\mathscr{K}$ generates an abstract frame

$$\mathscr{F}^{\mathscr{K}} = (\mathscr{K}, \{[\phi]\}_{\phi \in L})$$

where $[\phi]$ records the effect of updating with ϕ. Thus, $[\phi](\mathscr{M}, w)$ is $\mathscr{M}|\phi, w$ if $\mathscr{M}, w \vDash \phi$ and undefined otherwise, where $\mathscr{M}|\phi$ is $\mathscr{M}$ restricted to the worlds in $|\mathscr{M}|$ in which ϕ is true. Given a class of concrete frames $\mathfrak{K}$, we write $\mathfrak{F}^{\mathfrak{K}}$ for the class of abstract frames generated by frames in $\mathfrak{K}$. A PAL formula ϕ is valid on a concrete frame $\mathscr{K}$ iff it is true in every pointed model in $\mathscr{K}$, and it is valid on a class $\mathfrak{K}$ of concrete frames (notation: $\vDash^{\mathfrak{K}}$) iff it is valid on every frame in $\mathfrak{K}$. Also, observe that, for each $\mathscr{M}, w \in \mathscr{K}$,

$$\mathscr{M}, w \in \mathit{fix}([\psi]) \Leftrightarrow \text{for all } w' \in |\mathscr{M}|, \ \mathscr{M}, w' \vDash \psi$$
$$\Leftrightarrow \mathscr{M} \vDash \psi$$

We can now see the interplay between consequence relations and their concrete representations through the following two equivalences. First, as the label has it, Update to Test consequence is the abstract version of testing after updating:

Proposition 32.4 $\vDash^{\mathfrak{K}} [!\phi_1] \dots [!\phi_n] C \psi$ *iff* $\phi_1, \dots, \phi_n \vDash^{\mathfrak{F}^{\mathfrak{K}}}_{UT} \psi$.

---

[13] Not all information updates are of this kind, e.g. because what we often get is 'soft' information that might be overridden. We leave a systematic investigation of the behavior of classical and dynamic consequence in these wider contexts to future research.

This correspondence is known,[14] but we give a detailed proof here so as to make fully explicit what the assumptions are.

Proof From Left to Right: Assume that $\vDash^{\mathfrak{K}} [!\phi_1]\ldots[!\phi_n]C\psi$. Let $\mathscr{F}^{\mathscr{K}} \in \mathfrak{F}^{\mathfrak{K}}$ and $\mathscr{M}, w \in \mathscr{K}$ be such that $\mathscr{M}, w \in range([\phi_n] \circ \ldots \circ [\phi_1])$. We need to show that $\mathscr{M}, w \in fix([\psi])$, that is, $\mathscr{M} \vDash \psi$. Let $w' \in |\mathscr{M}|$. Since $\mathscr{K}$ is closed under change of designated world, $\mathscr{M}, w' \in \mathscr{K}$. Moreover, because the effect of an update does not depend upon which world is designated, $\mathscr{M}, w' \in range([\phi_n] \circ \ldots \circ [\phi_1])$. By hypothesis, this guarantees that $\mathscr{M}, w' \vDash C\psi$, hence in particular that $\mathscr{M}, w' \vDash \psi$, as required.

From Right to Left: Assume that $\phi_1, \ldots, \phi_n \vDash^{\mathfrak{F}^{\mathfrak{K}}}_{UT} \psi$. Let $\mathscr{K} \in \mathfrak{K}$ and $\mathscr{M}, w \in \mathscr{K}$; we need to show that $\mathscr{M}, w \vDash [!\phi_1]\ldots[!\phi_n]C\psi$. That is, if the updates can be performed, ψ is common knowledge once they have been performed. So let $\mathscr{M}' = (\ldots(\mathscr{M}|\phi_1)\ldots)|\phi_n$. Thus, $\mathscr{M}', w \in \mathscr{K}$, and by definition of $\mathfrak{F}^{\mathfrak{K}}$, $\mathscr{M}', w \in range([\phi_n] \circ \ldots \circ [\phi_1])$. Therefore, by our initial assumption, $\mathscr{M}', w \in fix([\psi])$, so $\mathscr{M}' \vDash \psi$, which implies in turn that $\mathscr{M}', w \vDash C\psi$. $\square$

Note that the left to right direction of the proof requires that frames are closed under change of designated world. If that were not the case, there could be a $\neg\psi$ world in the model which is not reachable from the designated world so that $C\psi$ holds even though the model is not a fixed point for updating with ψ. Since that part of the proof only uses the facticity of C, this also shows that $\vDash^{\mathfrak{K}} [!\phi_1]\ldots[!\phi_n]C\psi$ implies $\vDash^{\mathfrak{K}} [!\phi_1]\ldots[!\phi_n]\psi$, which is a valid rule in PAL.

Second, Test to Test consequence is the abstract version of a classical notion of consequence:

Proposition 32.5 $\vDash^{\mathfrak{K}} (C\phi_1 \wedge \ldots \wedge C\phi_n) \rightarrow C\psi$ *iff* $\phi_1, \ldots, \phi_n \vDash^{\mathfrak{F}^{\mathfrak{K}}}_{TT} \psi$.

Proof The left to right direction is similar to the previous proof. Assume that $\vDash^{\mathfrak{K}} (C\phi_1 \wedge \ldots \wedge C\phi_n) \rightarrow C\psi$. Let $\mathscr{F}^{\mathscr{K}} \in \mathfrak{F}^{\mathfrak{K}}$ and $\mathscr{M}, w \in \mathscr{K}$ be such that $\mathscr{M}, w \in fix([\phi_i])$, $1 \leq i \leq n$. We need to show that $\mathscr{M}, w \in fix([\psi])$, i.e. $\mathscr{M} \vDash \psi$. Let $w' \in |\mathscr{M}|$. Since $\mathscr{K}$ is closed under change of designated world, $\mathscr{M}, w' \in \mathscr{K}$. And since $\mathscr{M} \vDash \phi_i$, we have $\mathscr{M}, w' \vDash C\phi_i$ for $1 \leq i \leq n$. So by hypothesis, $\mathscr{M}, w' \vDash C\psi$, from which it follows that $\mathscr{M}, w' \vDash \psi$.

Right to Left: Assume $\phi_1, \ldots, \phi_n \vDash^{\mathfrak{F}^{\mathfrak{K}}}_{TT} \psi$. Let $\mathscr{K} \in \mathfrak{K}$ and $\mathscr{M}, w \in \mathscr{K}$ be such that $\mathscr{M}, w \vDash C\phi_1 \wedge \ldots \wedge C\phi_n$. We need to show that $\mathscr{M}, w \vDash C\psi$. Let $\mathscr{M}^*$ be the submodel of $\mathscr{M}$ consisting of all those worlds that are connected to w. It is sufficient to show that $\mathscr{M}^*, w \vDash C\psi$. (Recall that the accessibility relations are equivalence relations.) Since $\mathscr{K}$ is closed under submodels, $\mathscr{M}^*, w \in \mathscr{K}$. Because $\mathscr{M}^* \vDash C\phi_1 \wedge \ldots \wedge C\phi_n$ and all worlds in $\mathscr{M}^*$ are connected to w, $\mathscr{M}^*, w \in fix([\phi_i])$, $1 \leq i \leq n$. Hence, by our initial assumption, $\mathscr{M}^*, w \in fix([\psi])$, which means that ψ is true everywhere in $\mathscr{M}^*$, so $\mathscr{M}^*, w \vDash C\psi$. $\square$

---

[14] In [7] van Benthem, states that "modulo a few details, dynamic validity amounts to PAL validity" (p. 192). We spell out these details here.

Let *Prop* be the set of purely propositional formulas. It is well known that for such formulas classical consequence and its dynamic version coincide.

Proposition 32.6 *If $\phi_1, \ldots, \phi_n, \psi \in Prop$, then $\vDash^{\Re} [!\phi_1] \ldots [!\phi_n] C\psi$ iff $\vDash^{\Re} (C\phi_1 \wedge \ldots \wedge C\phi_n) \rightarrow C\psi$.*

Viewed from our perspective, Proposition 32.6 holds because of the special properties of propositional formulas with respect to updates. Propositional formulas generate idempotent and f-commutative frames. In the light of Proposition 32.3, together with Propositions 32.4 and 32.5, this readily implies Proposition 32.6. But this also suggests a more general question. Proposition 32.6 describes sufficient conditions for the concrete version of Proposition 32.3, which gives sufficient *and necessary* conditions. So, more generally, for which sublanguages of full modal logic do we have for every $\Re$ that $\vDash^{\Re} [!\phi_1] \ldots [!\phi_n] C\psi$ iff $\vDash^{\Re} (C\phi_1 \wedge \ldots \wedge C\phi_n) \rightarrow C\psi$? Proposition 32.3 says that we need to characterize the class of modal formulas that generate idempotent and f-commutative frames.

We shall consider this question for formulas of modal logic without common knowledge. This covers PAL without common knowledge by virtue of the reduction axioms. Let us say that a formula ϕ is *persistent* iff, for all concrete frames $\mathcal{K}, \mathcal{F}^{\mathcal{K}}$ restricted to $Prop \cup \{\phi\}$ (i.e. $(\mathcal{K}, \{[\psi]\}_{\psi \in Prop \cup \{\phi\}})$) is idempotent and f-commutative. On the face of it, generating idempotent and f-commutative frames is a property of sets of formulas. But giving a definition for formulas rather than sets thereof is adequate, because a set $\Gamma \supseteq Prop$ of formulas generates idempotent and f-commutative frames iff every formula in Γ is persistent in the sense just defined. (Testing persistence against *propositional* formulas is sufficient because the effect of updating with a non-propositional formula can always be mimicked using a suitably interpreted propositional formula.) Our question about formulas for which classical and dynamic consequence coincide may then be thus phrased:

Question 32.1 What is the class of persistent modal formulas?

Persistence can be analyzed into two more familiar features, corresponding respectively to f-commutativity and idempotence. A formula ϕ is *globally preserved under submodels* iff for any $\mathcal{M}$ and $\mathcal{M}'$ with $\mathcal{M}' \subseteq \mathcal{M}$, if $\mathcal{M} \vDash \phi$, then $\mathcal{M}' \vDash \phi$. A formula ϕ is *successful* if for any $\mathcal{M}, w$, if $\mathcal{M}, w \vDash \phi$, then $\mathcal{M}|\phi, w \vDash \phi$. Global preservation under submodels is indeed equivalent to the fact that, for any ψ, if $\mathcal{M}, w \in fix([\phi])$ and $\mathcal{M}', w = [\psi](\mathcal{M}, w)$ then $\mathcal{M}', w \in fix([\phi])$. Successfulness is equivalent in turn to the fact that $range([\phi]) \subseteq fix([\phi])$. (By the reasoning used in the proof of Proposition 32.4 success at a world makes for success at every world.)

Thus, Question 32.1 actually asks for a characterization of the class of modal formulas that are both successful and globally preserved under submodels. By Corollary 6.4 in [17], a formula is globally preserved under submodels iff it is globally equivalent to a universal modal formula.[15] The long sought-after characterization of

---

[15] A formula ϕ is globally equivalent to a universal modal formula if there is a formula ψ constructed using only (negations of) atoms, conjunction, disjunction and K such that for all $\mathcal{M}$, $\mathcal{M} \vDash \phi$ iff $\mathcal{M} \vDash \psi$.

successful formulas was provided in [13] where it is shown that, in S5, all unsuccessful formulas are look-alikes of the infamous Moore formula $p \wedge \neg K p$. Note that if we were talking about local, instead of global, preservation under submodels,[16] we would be done. As shown in [1], a formula is locally preserved under submodels iff it is locally equivalent to a universal formula (we will take the liberty to say 'locally universal'), and universal formulas are always successful. However, matters are more complicated when global preservation is concerned: in S5, the Moore formula is trivially globally preserved under submodels, since it is not globally satisfiable, but it is not successful.

Limiting ourselves from now on to a modal language with only one epistemic modality (as we in effect did in the previous paragraph), we provide a partial answer to Question 32.1. In order to do so, Carnapian disjunctive normal forms for modal formulas prove useful. (This strategy is inspired by [13].)

Definition 32.3 A formula δ is in *normal form* iff it is a disjunction of conjunctions of the form $\delta = \alpha \wedge \Box \beta_1 \wedge \ldots \wedge \Box \beta_n \wedge \Diamond \gamma_1 \wedge \ldots \wedge \Diamond \gamma_m$, where α and each γ_i are conjunctions of literals and each β_j is a disjunction of literals.

For the limited class of modal formulas such that their disjunctive normal form consists of only one disjunct, Question 32.1 gets a rather satisfactory answer.

Proposition 32.7 *Let δ be a conjunction in normal form. In S5, δ is persistent iff δ is locally universal.*

Proof From Left to Right: We prove the contrapositive. Assume δ is not locally universal. Either δ is globally satisfiable (meaning that there is a $\mathcal{M}$ such that $\mathcal{M} \vDash \delta$) or it is not. If δ is not globally satisfiable, note first that it still is locally satisfiable (meaning that there is a $\mathcal{M}, w$ such that $\mathcal{M}, w \vDash \delta$), or δ would be locally equivalent to $p \wedge \neg p$, which is universal. Being locally but not globally satisfiable, it readily follows that δ is not successful, hence not persistent, and we are done. So we assume that δ is globally satisfiable. There has to be a γ_i such that $\alpha \wedge \Box \beta_1 \wedge \ldots \wedge \Box \beta_n \nvDash \gamma_i$, since otherwise δ would be equivalent to $\alpha \wedge \Box \beta_1 \wedge \ldots \wedge \Box \beta_n$, which is locally universal. It follows that

(*) $\alpha \wedge \beta_1 \wedge \ldots \wedge \beta_n \wedge \neg \gamma_i$ is satisfiable.

Take a model $\mathcal{M}$ such that $\mathcal{M} \vDash \delta$. By (*), it is possible to extend $\mathcal{M}$ to a model $\mathcal{M} \cup \{w\}$ such that $\mathcal{M} \cup \{w\}, w \vDash \alpha \wedge \beta_1 \wedge \ldots \wedge \beta_n \wedge \neg \gamma_i$ and $\mathcal{M} \cup \{w\} \vDash \delta$. But then $\mathcal{M} \cup \{w\} | (\alpha \wedge \beta_1 \wedge \ldots \wedge \beta_n \wedge \neg \gamma_i) \nvDash \delta$, so δ is not globally preserved under submodels.

From Right to Left: it is known that locally universal formulas are successful, see e.g. [6]. Moreover, by the result in [1], universal formulas are locally preserved under submodels, and *a fortiori* globally so.   □

---

[16] ϕ is locally preserved under submodels iff $\mathcal{M}' \subseteq \mathcal{M}$, $w \in |\mathcal{M}'|$, and $\mathcal{M}, w \vDash \phi$ implies $\mathcal{M}', w \vDash \phi$.

Thus, for conjunctions in normal form, classical and dynamic notions of consequence are equivalent exactly when the information which comes into play is stable, in the sense of being locally preserved under submodels. We leave it as an open question whether this result carries over: is it the case that, in S5, for any formula ϕ, ϕ is persistent iff ϕ is locally universal?

32.5 Classical Consequence *Versus* Classical Update

In the last two sections we studied the conditions under which abstract and concrete frames behave classically with respect to dynamic consequence. Rothschild and Yalcin in [18] recently asked similar questions, concerning the conditions under which abstract frames behave classically with respect to updates themselves. Comparing results will prove instructive. The main lesson of Proposition 32.7 is that some non-propositional formulas pass the classicality test for consequence. Could it be so for updates as well? The answer is not immediate. Intuitively, being classical with respect to updates is more demanding than being classical only with respect to the visible effects of these updates on the consequence relation.

Following Rothschild and Yalcin, being classical with respect to updates means that informational states and propositional contents can be represented by sets of worlds, in such a way that updating with ϕ is taking the intersection of the current informational state with the set of ϕ-worlds. When this is so, the abstract frame is said to be static:[17]

Definition 32.4 (Rothschild and Yalcin) An abstract frame $(\Sigma, \{[\phi]\}_{\phi \in L})$ is *static* iff there are functions $f : L \to \wp(\Sigma)$ and $g : \Sigma \to \wp(\Sigma)$ such that for any $\sigma \in \Sigma$, $g(\sigma[\phi]) = g(\sigma) \cap f(\phi)$.

A result similar to Proposition 32.3 ensues.

Proposition 32.8 (Rothschild and Yalcin) *An abstract frame is static iff it is idempotent and commutative.*

Clearly, commutativity implies f-commutativity, but the converse is not true, even assuming idempotence. Actually, idempotence and f-commutativity correspond to a weaker notion of being static where $g(\sigma[\phi]) = g(\sigma) \cap f(\phi)$ is replaced by $g(\sigma[\phi]) \subseteq g(\sigma) \cap f(\phi)$ in Definition 32.4.

Going concrete, we shall say that a modal formula ϕ is *strongly persistent* iff, for all concrete frames $\mathscr{K}$, $\mathscr{F}^{\mathscr{K}}$ restricted to *Prop* $\cup \{\phi\}$ is idempotent and commutative. (Just as with 'persistent', the definition of 'strongly persistent' can be given for formulas rather than sets of formulas, and for similar reasons.) Here is a full characterization of the strongly persistent formulas.

---

[17] This exact definition was to be found in an early version of [18]. The definition in the final version is slighly more complex, but the extra complexity is irrelevant to our present purpose.

Proposition 32.9 *Let ϕ be any modal formula. In S5, ϕ is strongly persistent iff ϕ is equivalent to a propositional formula.*

Proof The direction from right to left is immediate. We prove the other direction.[18] Consider an arbitrary strongly persistent formula ϕ. First, we note that it is sufficient to prove the following:

(*) If $\mathcal{M}, w \vDash \phi$ and $\mathcal{M}, w \equiv_{Prop} \mathcal{M}', w'$, then $\mathcal{M}', w' \vDash \phi$,

where $\equiv_{Prop}$ is elementary equivalence restricted to propositional formulas. To see this, let ϕ^{Prop} be the set of propositional consequences of ϕ. We claim that

(**) $\phi^{Prop} \vDash \phi$

By compactness, ϕ is then equivalent to a propositional formula. To prove (**), assume $\mathcal{M}', w' \vDash \phi^{Prop}$ and show $\mathcal{M}', w' \vDash \phi$. Let $Prop_{w'}$ be the set of propositional formulas that are true in $\mathcal{M}', w'$. $Prop_{w'} \cup \{\phi\}$ is consistent, since otherwise there would be propositional formulas $\psi_1, \ldots, \psi_n$ in $Prop_{w'}$ such that $\phi \vDash \neg(\psi_1 \wedge \ldots \wedge \psi_n)$, i.e. $\phi \vDash \neg\psi_1 \vee \ldots \vee \neg\psi_n$, contradicting the fact that $\mathcal{M}', w' \vDash \phi^{Prop}$. So there is $\mathcal{M}, w$ with $\mathcal{M}, w \vDash Prop_{w'}$ and $\mathcal{M}, w \vDash \phi$. $\mathcal{M}, w \vDash Prop_{w'}$ means that $\mathcal{M}, w \equiv_{Prop} \mathcal{M}', w'$ so (*) applies and we have $\mathcal{M}', w' \vDash \phi$, as required.[19]

We now prove (*), assuming that ϕ is strongly persistent. Consider two structures $\mathcal{M}, w$ and $\mathcal{M}', w'$ in a concrete frame $\mathcal{K}$ with $\mathcal{M}, w \vDash \phi$ and such that $\mathcal{M}, w \equiv_{Prop} \mathcal{M}', w'$. Since they are S5-models, we may also assume without loss of generality that there are no two different worlds satisfying exactly the same propositional formulas in $\mathcal{M}$ or $\mathcal{M}'$.

Since $\mathcal{M}, w \vDash \phi$, $[\phi]$ is defined on $\mathcal{M}, w$ in $\mathcal{F}^{\mathcal{K}}$. Let $Prop_w$ be the conjunction of propositional atoms and negations thereof that are true in $\mathcal{M}, w$. (The proof does not go through if there is an infinite number of atoms.) $[Prop_w]$ is also defined on $(\mathcal{M}, w)[\phi]$. By commutativity, $(\mathcal{M}, w)[\phi][Prop_w] = (\mathcal{M}, w)[Prop_w][\phi]$. Hence, by idempotence,

(***) $(\mathcal{M}, w)[\phi][Prop_w] = (\mathcal{M}, w)[\phi][Prop_w][\phi]$

Since $\mathcal{M}, w \equiv_{Prop} \mathcal{M}', w'$, $[Prop_w]$ is defined on $\mathcal{M}', w'$ too. Moreover, we have $(\mathcal{M}', w')[Prop_w] = (\mathcal{M}, w)[\phi][Prop_w]$, since the result of a successful announcement of $Prop_w$ is always the same one-world structure. Also, it follows from (***) that $(\mathcal{M}', w')[Prop_w] = (\mathcal{M}', w')[Prop_w][\phi]$. But then, by commutativity, $(\mathcal{M}', w')[\phi][Prop_w]$ is defined, which implies that $\mathcal{M}', w' \vDash \phi$. This completes the proof. $\square$

In the context of PAL, classicality *with respect to consequence only* and classicality *with respect to updates in general* end up being two very different things.

---

[18] Our proof of Proposition 32.9 is inspired by a simplified proof of Proposition 32.8 by Johan van Benthem (private correspondence). Whether there is deeper connection still remains to be seen. Is there a sense in which the possibility of a classical representation forces the equivalence to purely propositional formulas?

[19] The first part of the proof is well-known from model theory, relying only on compactness and the fact that ϕ^{Prop} is closed under disjunction; see [10], Lemma 3.2.1.

In the second case, only propositional formulas pass the test. Requiring classicality with respect to updates in general really means dealing with announcements that are deprived of epistemic content. By contrast, in the first case, universal formulas do pass the test. Requiring classicality with respect to consequence is compatible with information that has epistemic content, as long as this epistemic content is about knowledge rather than doubts.

32.6 Conclusion

What does logical dynamics tell us about logical consequence? Our focus has been on properties of content and how they should be represented. Such questions are familiar in formal semantics, regarding whether certain phenomena (anaphora, presuppositions, etc.) should be given a classical account in terms of propositions interpreted as sets of possible worlds or dynamically represented in terms of context change potentials. Debates regarding logical consequence typically take a different form. The focus is on the validity of inference rules, and the model-theoretic definition of logical consequence most often remains classical. But, in principle, there is no telling apart the question of content and the question of validity: some rules may be valid only with respect to some particular types of contents. From the dynamic perspective, this becomes clear with the splitting of logical consequence into a classical and a dynamic notion. The broadest notion is the dynamic one, which is about what follows from the information received. In this context, classicality for logical consequence emerges as a property of content: when information is persistent, dynamic consequence boils down to the standard semantic definition in terms of preservation of truth and satisfies the classical structural rules.

This new take on logical consequence is also a new take on the dynamic *versus* classical dispute in semantics. The dispute has several faces: dynamic or classical *what*? Asking the question about updates is not the same as asking the question about logical consequence. Rothschild and Yalcin conclude their paper [18] by asking about classes of semantic systems which would lie between the static systems and the information-sensitive systems. Our static systems (the idempotent and f-commutative frames) constitute such a class: they may be information-sensitive for updates, but they are information-insensitive for logical consequence. This suggests a fully parametric approach: given a certain kind of manifestation of content (through logical consequence, updates of common ground, but also possibly other manifestations, such as, say, presupposition accommodation), what are the properties of content that make a classical analysis possible or force a dynamic account?

Acknowledgments Thanks to Chris Barker, Johan van Benthem, Rosalie Iemhoff, Daniel Rothschild and Seth Yalcin for helpful comments and suggestions. Both authors were supported by the ESF-funded project Logic for Interaction (LINT; http://www.illc.uva.nl/lint/index.php), a Collaborative Research Project under the Eurocores program LogICCC, and by a grant from the Swedish Research Council.

References

1. Andréka H, van Benthem J, Németi I (1998) Modal languages and bounded fragments of predicate logic. J Philos Logic 27:217–274
2. Beall JC, Restall G (2006) Logical pluralism. Clarendon Press, Oxford
3. Burgess JP (2005) No requirement of relevance. In: Shapiro S (ed) The oxford handbook of philosophy of mathematics and logic. Oxford University Press, Oxford, pp 727–751
4. van Benthem J (1991) Language in action. Categories, lambdas, and dynamic logic. North-Holland/Elsevier, Amsterdam.
5. van Benthem J (1996) Exploring logical dynamics. CSLI Publications, Stanford
6. van Benthem J (2004) What one may come to know. Analysis 64:95–105
7. van Benthem J (2008) Logical dynamics meets logical pluralism? Australas J Logic 6:182–209
8. van Benthem J (2011) Logical dynamics of information and interaction. Cambridge University Press, Cambridge
9. Cantini A (2012) Paradoxes and contemporary logic. In: Zalta EN (ed) The stanford encyclopedia of philosophy. (Summer 2012 edn).
10. Chang CC, Keisler HJ (1973) Model theory. North-Holland, Amsterdam. (Studies in logic and the foundations of mathematics, vol 73, 3rd edn. 1990).
11. van der Does J, Groeneveld W, Veltman F (1997) An update on might. J Logic Lang Inform 6(4):361–380
12. Field H (2008) Saving truth from paradox. Oxford University Press, Oxford
13. Holliday WH, Icard TF (2010) Moorean phenomena in epistemic logic. Advances in modal logic. College, London, pp 178–199
14. Humberstone L (2011) The connectives. MIT Press, Cambridge, Mass
15. Łoś J, Suszko R (1958) Remarks on sentential logics. Indagationes Mathematicas 20:171–183
16. Plaza JA (1989) Logics of public communication. Methodologies for intelligent systems, vol 4. North-Holland, Amsterdam, pp 201–216.
17. de Rijke M, Sturm H (2000) Global definability in basic modal logic. In: Wansing H (ed) Essays on non-classical logic. King's College University Press, London, pp 111–135
18. Rothschild D, Yalcin S (2012) On the dynamics of conversation. Manuscript.
19. Scott D (1971) On engendering an illusion of understanding. J Philos 68(21):787–807
20. Shoesmith DJ, Smiley TJ (1978) Multiple-conclusion logic. Cambridge University Press, Cambridge
21. Tennant N (2005) Relevance in reasoning. In: Shapiro S (ed) The oxford handbook of philosophy of mathematics and logic. Oxford University Press, Oxford, pp 696–725
22. Zardini E (2011) Truth without contra(di)ction. Rev Symbolic Logic 4(04):498–535

Chapter 33
Dynamic Epistemic Logic as a Substructural Logic

Guillaume Aucher

Abstract Dynamic Epistemic Logic (DEL) is an influential logical framework for reasoning about the dynamics of beliefs and knowledge. It has been related to older and more established logical frameworks. Despite these connections, DEL remains, arguably, a rather isolated logic in the vast realm of non-classical logics and modal logics. This is problematic if logic is to be viewed ultimately as a unified and unifying field and if we want to avoid that DEL goes on "riding off madly in all directions" (a metaphor used by van Benthem about logic in general). In this article, we show that DEL can be redefined naturally and meaningfully as a two-sorted substructural logic. In fact, it is even one of the most primitive substructural logics since it does not preserve any of the structural rules. Moreover, the ternary semantics of DEL and its dynamic interpretation provides a conceptual foundation for the Routley & Meyer's semantics of substructural logics.

33.1 Introduction

Dynamic Epistemic Logic (DEL) is an influential logical framework for reasoning about the dynamics of beliefs and knowledge, which has drawn the attention of a number of researchers ever since the seminal publication of [11]. A number of contributions have linked DEL to older and more established logical frameworks: it has been embedded into (automata) PDL [34, 40], it has been given an algebraic semantics [8, 9], and it has been related to epistemic temporal logic [6, 32] and the situation calculus [31, 38]. Many of these links have been established by van Benthem himself. Despite these connections, DEL remains, arguably, a rather isolated logic in the vast realm of non-classical logics and modal logics. This is problematic if

G. Aucher (✉)
University of Rennes 1 – INRIA, 263, Avenue du Gnral Leclerc,
35042 Rennes, Cedex, France
e-mail: guillaume.aucher@irisa.fr

A. Baltag and S. Smets (eds.), *Johan van Benthem on Logic
and Information Dynamics*, Outstanding Contributions to Logic 5,
DOI: 10.1007/978-3-319-06025-5_33, © Springer International Publishing Switzerland 2014

logic is to be viewed ultimately as a unified and unifying field and if we want to avoid that DEL goes on "riding off madly in all directions" (a metaphor used by van Benthem [28, 30] about logic in general). In this article, we will show that DEL can be redefined naturally and meaningfully as a two-sorted substructural logic. In fact, it is even one of the most primitive substructural logics since it does not preserve any of the structural rules.

Substructural logics will also benefit from this interaction with DEL. The well-known semantics for substructural logics is based on a ternary relation introduced by Routley and Meyer for relevance logic in the 1970s [59–62]. However, the introduction of this ternary relation was originally motivated by technical reasons, and it turns out that providing a non-circular and conceptually grounded interpretation of this relation remains problematic [18]. As we shall see, the ternary semantics of DEL provides a conceptual foundation for Routley & Meyer's semantics. In fact, the dynamic interpretation induced by the DEL framework turns out to be not only meaningful, but also consistent with the interpretations of this ternary relation proposed in the substructural literature.

The article is structured as follows. In Sect. 33.2 we recall the core of DEL viewed from a semantic perspective. In Sect. 33.3 we briefly recall elementary notions of relevance and substructural logics and we observe that the ternary relation of relevance logic can be interpreted as a sort of update. In Sect. 33.4 we proceed further to define a substructural language based on this idea. This substructural language extends the DEL language with operators stemming from the Lambek calculus (a substructural logic), but we show that these different substructural operators actually correspond to the DEL operators of [3, 4]. This allows us to show that DEL *is* a (two-sorted) substructural logic. In this section we also formally relate these operators to the dynamic inferences introduced by van Benthem [23]. In Sect. 33.5 we conclude and give some personal views about the future of DEL and logical dynamics.

33.2 Dynamic Epistemic Logic

Dynamic epistemic logic (DEL) is a relatively recent non-classical logic [11] which extends ordinary modal epistemic logic [45] by the inclusion of *event/action models* (called $\mathscr{L}_\alpha$-models in this article) to describe actions, and a *product update* operator that defines how *epistemic models* (called $\mathscr{L}$-models in this article) are updated as the consequence of executing actions described through event models (see [10, 30, 37] for more details). So, the methodology of DEL is such that it splits the task of representing the agents' beliefs and knowledge into three parts: first, one represents their beliefs/knowledge about an initial situation; second, one represents their beliefs/knowledge about an event taking place in this situation; third, one represents the way the agents update their beliefs/knowledge about the situation after (or during) the occurrence of the event. Following this methodology, we also split the exposition of the DEL framework into three sections.

33.2.1 Representation of the Initial Situation: $\mathscr{L}$-Model

In the rest of this article, ATM is a countable set of propositional letters called *atomic facts* which describe static situations, and $AGT := \{1, \ldots, m\}$ is a finite set of agents.

Definition 33.1 (*Language $\mathscr{L}$ and $\mathscr{L}$-structure*) We define the language $\mathscr{L}$ inductively as follows:

$$\mathscr{L} : \varphi ::= p \mid \neg\varphi \mid \varphi \wedge \varphi \mid \varphi \vee \varphi \mid \Box_j\varphi$$

where p ranges over ATM and j over AGT. We define $\bot := p \wedge \neg p$ for a chosen $p \in ATM$ and we also define $\top := \neg\bot$. The formula $\Diamond_j\varphi$ is an abbreviation for $\neg\Box_j\neg\varphi$, the formula $\varphi \rightarrow \psi$ is an abbreviation for $\neg\varphi \vee \psi$, and the formula $\varphi \leftrightarrow \psi$ is an abbreviation for $(\varphi \rightarrow \psi) \wedge (\psi \rightarrow \varphi)$.

A $\mathscr{L}$-*structure* is defined inductively as follows, with φ ranging over $\mathscr{L}$:

$$X ::= \varphi \mid (X, X)$$

We abusively write $\varphi \in X$ when the formula $\varphi \in \mathscr{L}$ is a substructure of X.

A (pointed) $\mathscr{L}$-model $(\mathscr{M}, w)$ represents how the actual world represented by w is perceived by the agents. Atomic facts are used to state properties of this actual world.

Definition 33.2 (*$\mathscr{L}$-model*) A $\mathscr{L}$-model is a tuple $\mathscr{M} = (W, R_1, \ldots, R_m, I)$ where:

- W is a non-empty set of possible worlds,
- $R_j \subseteq W \times W$ is an accessibility relation on W, for each $j \in AGT$,
- $I : W \rightarrow 2^{ATM}$ is a function assigning to each possible world a subset of ATM. The function I is called an *interpretation*.

We write $w \in \mathscr{M}$ for $w \in W$, and $(\mathscr{M}, w)$ is called a pointed $\mathscr{L}$-model (w often represents the actual world). We denote by $\mathscr{C}$ the set of pointed $\mathscr{L}$-models. If $w, v \in W$, we write wR_jv or $(\mathscr{M}, w) R_j(\mathscr{M}, v)$ for $(w, v) \in R_j$, and $R_j(w)$ denotes the set $\{v \in W \mid wR_jv\}$.

Intuitively, wR_jv means that in world w agent j considers that world v might correspond to the actual world. Then, we define the following epistemic language that can be used to describe and state properties of $\mathscr{L}$-models:

Definition 33.3 (*Truth conditions of $\mathscr{L}$*) Let $\mathscr{M}$ be a $\mathscr{L}$-model, $w \in \mathscr{M}$ and $\varphi \in \mathscr{L}$. $\mathscr{M}, w \models \varphi$ is defined inductively as follows:

$$\begin{array}{lll}
\mathcal{M}, w \models p & \text{iff} & p \in I(w) \\
\mathcal{M}, w \models \neg \psi & \text{iff} & \text{not } \mathcal{M}, w \models \varphi \\
\mathcal{M}, w \models \varphi \wedge \psi & \text{iff} & \mathcal{M}, w \models \varphi \text{ and } \mathcal{M}, w \models \psi \\
\mathcal{M}, w \models \varphi \vee \psi & \text{iff} & \mathcal{M}, w \models \varphi \text{ or } \mathcal{M}, w \models \psi \\
\mathcal{M}, w \models \square_j \varphi & \text{iff} & \text{for all } v \in R_j(w), \mathcal{M}, v \models \varphi
\end{array}$$

We write $\mathcal{M} \models \varphi$ when $\mathcal{M}, w \models \varphi$ for all $w \in \mathcal{M}$, and $\models \varphi$ when for all $\mathcal{L}$-model $\mathcal{M}, \mathcal{M} \models \varphi$. A $\mathcal{L}$-formula φ is said to be *valid* if $\models \varphi$. We extend the scope of the relation $\models$ to also relate pointed $\mathcal{L}$-models to structures:

$$\mathcal{M}, w \models X, Y \quad \text{iff} \quad \mathcal{M}, w \models X \text{ and } \mathcal{M}, w \models Y$$

Let C be a class of pointed $\mathcal{L}$-models, let X, Y be $\mathcal{L}$-structures. We say that X *entails* Y *in the class* C, written $X \models_{\overline{C}} Y$, when the following holds:

$X \models_{\overline{C}} Y \quad$ iff $\quad$ for all pointed $\mathcal{L}$-model $(\mathcal{M}, w) \in C$, if for all $\varphi \in X \mathcal{M}, w \models \varphi$,
then there is $\psi \in Y$ such that $\mathcal{M}, w \models \psi$.

We also write $X \models Y$ for $X \models_{\overline{\mathcal{C}}} Y$, where $\mathcal{C}$ is the class of all pointed $\mathcal{L}$-models.

The formula $\square_j \varphi$ reads as "agent j believes φ". Its truth conditions are defined in such a way that agent j believes φ is true in a possible world when φ holds in all the worlds agent j considers possible.

Example 33.1 Assume that agents A, B and C play a card game with three cards: a white one, a red one and a blue one. Each of them has a single card but they do not know the cards of the other players. At each step of the game, some of the players show their/her/his card to another player or to both other players, either privately or publicly. We want to study and represent the dynamics of the agents' beliefs/knowledge in this game. The initial situation is represented by the pointed $\mathcal{L}$-model $(\mathcal{M}, w)$ of Fig. 33.1.

In this example, $AGT := \{A, B, C\}$ and $ATM := \{r_j, b_j, w_j \mid j \in AGT\}$ where r_j stands for 'agent j has the red card', b_j stands for 'agent j has the blue card' and w_j stands for 'agent j has the white card'. The boxed possible world corresponds to the actual world. The propositional letters not mentioned in the possible worlds do not hold in these possible worlds. The accessibility relations are represented by arrows indexed by agents between possible worlds. Reflexive arrows are omitted in the figure, which means that for all worlds $v \in \mathcal{M}$ and all agents $j \in AGT$, $v \in R_j(v)$. In this model, we have for example the following statement: $\mathcal{M}, w \models (w_B \wedge \neg \square_A w_B) \wedge \square_C \neg \square_A w_B$. It states that player A does not 'know' that player B has the white card and player C 'knows' it.

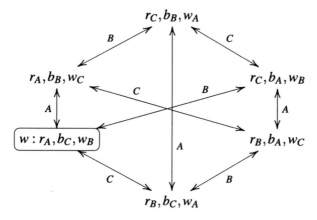

Fig. 33.1 Cards example

33.2.2 Representation of the Event: $\mathscr{L}_\alpha$-Model

The language $\mathscr{L}_\alpha$ was introduced in [12]. The propositional letters p_ψ describing events are called *atomic events* and range over $ATM_\alpha = \{p_\psi \mid \psi \text{ ranges over } \mathscr{L}\}$. The reading of p_ψ is "an event of precondition ψ is occurring".

Definition 33.4 (*Language $\mathscr{L}_\alpha$ and $\mathscr{L}_\alpha$-structure*) We define the language $\mathscr{L}_\alpha$ inductively as follows:

$$\mathscr{L}_\alpha : \alpha ::= p_\psi \mid \neg\alpha \mid \alpha \wedge \alpha \mid \alpha \vee \alpha \mid \Box_j \alpha$$

where ψ ranges over $\mathscr{L}$ and j over AGT. We define $\bot := p_\psi \wedge \neg p_\psi$ for a chosen $\psi \in \mathscr{L}$ and we define $\top := \neg\bot$. The formula $\Diamond_j \alpha$ is an abbreviation for $\neg\Box_j \neg\alpha$, the formula $\alpha \rightarrow \beta$ is an abbreviation for $\neg\alpha \vee \beta$, and the formula $\alpha \leftrightarrow \beta$ is an abbreviation for $(\alpha \rightarrow \beta) \wedge (\beta \rightarrow \alpha)$.

A $\mathscr{L}_\alpha$-*structure* is defined inductively as follows, with β ranging over $\mathscr{L}_\alpha$:

$$\mathscr{S}_\alpha : X_\alpha ::= \beta \mid (X_\alpha, X_\alpha)$$

We abusively write $\alpha \in X_\alpha$ when the formula $\alpha \in \mathscr{L}_\alpha$ is a substructure of X_α.

A pointed $\mathscr{L}_\alpha$-model $(\mathscr{E}, e)$ represents how the actual event represented by e is perceived by the agents. Intuitively, $f \in R_j(e)$ means that while the possible event represented by e is occurring, agent j considers possible that the possible event represented by f is actually occurring.

Definition 33.5 ($\mathscr{L}_\alpha$-*model*, [11]) A $\mathscr{L}_\alpha$-*model* is a tuple $\mathscr{E} = (W_\alpha, R_1, \ldots, R_m, I)$ where:

- W_α is a non-empty set of possible events,
- $R_j \subseteq W_\alpha \times W_\alpha$ is an accessibility relation on W_α, for each $j \in AGT$,
- $I : W_\alpha \to \mathcal{L}$ is a function assigning to each possible event a formula of $\mathcal{L}$. The function I is called the *precondition function*.

Let P be a subset of $\mathcal{L}$. A *P-complete $\mathcal{L}_\alpha$-model* is a $\mathcal{L}_\alpha$-model which satisfies moreover the following condition:

- $I(e) \in P$, for each $e \in W_\alpha$ *(P-complete)*

We write $e \in \mathcal{E}$ for $e \in W_\alpha$, and $(\mathcal{E}, e)$ is called a *pointed $\mathcal{L}_\alpha$-model* (e often represents the actual event). We denote by $\mathcal{C}_\alpha$ the set of pointed $\mathcal{L}_\alpha$-models, by $\mathcal{C}_\alpha^P$ the set of pointed P-complete event models. If $e, f \in W_\alpha$, we write eR_jf or $(\mathcal{E}, e)R_j(\mathcal{E}, f)$ for $(e, f) \in R_j$, and $R_j(e)$ denotes the set $\{f \in W_\alpha \mid eR_jf\}$.

The truth conditions of the language $\mathcal{L}_\alpha$ are identical to the truth conditions of the language $\mathcal{L}$:

Definition 33.6 (*Truth conditions of $\mathcal{L}_\alpha$*) Let $\mathcal{E}$ be a $\mathcal{L}_\alpha$-model, $e \in \mathcal{E}$ and $\alpha \in \mathcal{L}_\alpha$. $\mathcal{E}, e \models \alpha$ is defined inductively as follows:

$$
\begin{aligned}
\mathcal{E}, e &\models p_\psi & \text{iff}\quad & I(e) = \psi \\
\mathcal{E}, e &\models \neg\alpha & \text{iff}\quad & \text{not } \mathcal{E}, e \models \alpha \\
\mathcal{E}, e &\models \alpha \wedge \beta & \text{iff}\quad & \mathcal{E}, e \models \alpha \text{ and } \mathcal{E}, e \models \beta \\
\mathcal{E}, e &\models \alpha \vee \beta & \text{iff}\quad & \mathcal{E}, e \models \alpha \text{ or } \mathcal{E}, e \models \beta \\
\mathcal{E}, e &\models \Box_j\alpha & \text{iff}\quad & \text{for all } f \in R_j(e), \mathcal{E}, f \models \alpha
\end{aligned}
$$

Let C be a class of pointed $\mathcal{L}_\alpha$-models, let X_α, Y_α be $\mathcal{L}_\alpha$-structures. We say that X *entails* Y *in the class* C, written $X_\alpha \models_{\overline{C}} Y_\alpha$, when the following holds:

$$
\begin{aligned}
X_\alpha \models_{\overline{C}} Y_\alpha \quad \text{iff}\quad & \text{for all pointed } \mathcal{L}_\alpha\text{-model } (\mathcal{E}, e) \in C, \\
& \text{if for all } \alpha \in X_\alpha \mathcal{E}, e \models \alpha, \text{ then there is } \beta \in Y_\alpha \text{ such that } \mathcal{E}, e \models \beta.
\end{aligned}
$$

We also write $X_\alpha \models Y_\alpha$ for $X_\alpha \models_{\overline{\mathcal{C}_\alpha}} Y_\alpha$, where $\mathcal{C}_\alpha$ is the class of all pointed $\mathcal{L}_\alpha$-models.

Example 33.2 Let us resume Example 33.1 and assume that players A and B show their card to each other. As it turns out, C noticed that A showed her card to B but did not notice that B did so to A. Players A and B know this. This event is represented in the $\mathcal{L}_\alpha$-model $(\mathcal{E}, e)$ of Fig. 33.2. The boxed possible event e corresponds to the actual event 'players A and B show their *red* and *white* cards respectively to each other' (with precondition $r_A \wedge w_B$), f stands for the event 'player A shows her *white* card' (with precondition w_A) and g stands for the

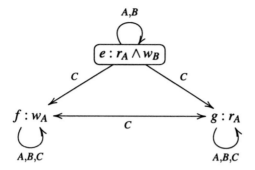

Fig. 33.2 Players A and B show their cards to each other in front of player C

atomic event 'players A shows her *red* card' (with precondition r_A). The following statement holds in the example of Fig. 33.2:

$$\mathcal{E}, e \models p_{r_A \wedge w_B} \wedge \left(\Diamond_A p_{r_A \wedge w_B} \wedge \Box_A p_{r_A \wedge w_B} \right) \wedge \left(\Diamond_B p_{r_A \wedge w_B} \wedge \Box_B p_{r_A \wedge w_B} \right)$$
$$\wedge \left(\Diamond_C p_{w_A} \wedge \Diamond_C p_{r_A} \wedge \Box_C \left(p_{w_A} \vee p_{r_A} \right) \right). \tag{33.1}$$

It states that players A and B show their cards to each other, players A and B 'know' this and consider it possible, while player C considers possible that player A shows her *white* card and also considers possible that player A shows her *red* card, since he does not know her card. In fact, that is all that player C considers possible since he believes that either player A shows her *red* card or her *white* card.

The $\mathscr{L}_\alpha$-model of Fig. 33.3 corresponds to a 'public announcement' or 'public display' of the fact that agent A has the red card. In particular, the following statement holds in the example of Fig. 33.3:

$$\mathcal{E}, e \models p_{r_A} \wedge \Box_A p_{r_A} \wedge \Box_B p_{r_A} \wedge \Box_C p_{r_A}$$
$$\wedge \Box_A \Box_A p_{r_A} \wedge \Box_A \Box_B p_{r_A} \wedge \Box_A \Box_C p_{r_A}$$
$$\wedge \Box_B \Box_A p_{r_A} \wedge \Box_B \Box_B p_{r_A} \wedge \Box_B \Box_C p_{r_A}$$
$$\wedge \Box_C \Box_A p_{r_A} \wedge \Box_C \Box_B p_{r_A} \wedge \Box_C \Box_C p_{r_A}$$
$$\wedge \dots$$

Fig. 33.3 Public announcement $\mathscr{L}_\alpha$-model of r_A

It states that player A shows her red card and that players A, B and C 'know' it, that players A, B and C 'know' that each of them 'know' it, etc…in other words, there is common knowledge among players A, B and C that player A shows her red card.[1]

$$\mathscr{E}, e \models p_{r_A} \wedge \square^*_{AGT} p_{r_A}.$$

33.2.3 Update of the Initial Situation by the Event: Product Update

The DEL product update of [11] is defined as follows. This update yields a new $\mathscr{L}$-model $(\mathscr{M}, w) \otimes (\mathscr{E}, e)$ representing how the new situation which was previously represented by $(\mathscr{M}, w)$ is perceived by the agents after the occurrence of the event represented by $(\mathscr{E}, e)$.

Definition 33.7 (*Product update*) Let $(\mathscr{M}, w) = (W, R_1, \ldots, R_m, I, w)$ be a pointed $\mathscr{L}$-model and let $(\mathscr{E}, e) = (W_\alpha, R_1, \ldots, R_m, I, e)$ be a pointed $\mathscr{L}_\alpha$-model such that $\mathscr{M}, w \models I(e)$. The *product update* of $(\mathscr{M}, w)$ and $(\mathscr{E}, e)$ is the pointed $\mathscr{L}$-model $(\mathscr{M} \otimes \mathscr{E}, (w, e)) = (W^\otimes, R_1^\otimes, \ldots, R_m^\otimes, I^\otimes, (w, e))$ defined as follows: for all $v \in W$ and all $f \in W_\alpha$,

- $W^\otimes = \{(v, f) \in W \times W_\alpha \mid \mathscr{M}, v \models I(f)\}$,
- $R_j^\otimes(v, f) = \{(u, g) \in W^\otimes \mid u \in R_j(v) \text{ and } g \in R_j(f)\}$,
- $I^\otimes(v, f) = I(v)$.

Example 33.3 As a result of the event described in Example 33.2, the agents update their beliefs. We get the situation represented in the $\mathscr{L}$-model $(\mathscr{M}, w) \otimes (\mathscr{E}, e)$ of Fig. 33.4. In this $\mathscr{L}$-model, we have for example the following statement:

$$(\mathscr{M}, w) \otimes (\mathscr{E}, e) \models (w_B \wedge \square_A w_B) \wedge \square_C \neg \square_A w_B.$$

---

[1] We write $\mathscr{E}, e \models \square^*_{AGT} \alpha$ when for all $f \in \left(\bigcup_{j \in AGT} R_j \right)^* (e)$, $\mathscr{E}, f \models \alpha$. See for example [41] for a detailed study of the operator $\square^*_{AGT}$ of common knowledge.

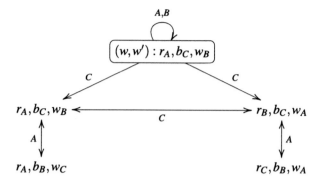

Fig. 33.4 Situation after the update of the situation represented in Fig. 33.1 by the event represented in Fig. 33.2

It states that player A 'knows' that player B has the white card but player C believes that it is not the case.

33.3 Substructural Logics

Substructural logics are a family of logics lacking some of the structural rules of classical logic. A structural rule is a rule of inference which is closed under substitution of formulas. The structural rules for classical logic are given in Fig. 33.5 (U, X, Y, Z denote $\mathscr{L}$-structures). While (*Weakening*) and (*Contraction*) are often dropped like in relevance logic and linear logic, the rule of (*Associativity*) is often preserved. We shall see in this article that DEL invalidates all of them.

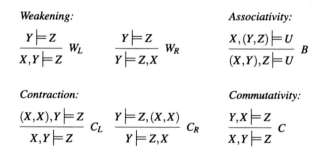

Fig. 33.5 Structural rules of classical logic

33.3.1 A Substructural Language

Our exposition of substructural logics is based on [39, 57, 58]. The logical frame-work presented in [57] is much more general and studies a wide range of substructural logics: relevance logic, linear logic, lambek calculus, display logic, etc…For what concerns us in this article, we will only introduce a fragment of this general frame-work. The semantics of this fragment is based on the ternary relation of the frame semantics for relevant logic originally introduced by Routley & Meyer [59–62]. Another semantics proposed independently by Urquhart [63–65] at about the same time will be discussed at the end of this section.

Definition 33.8 (*Language* $\mathscr{L}_{\mathsf{Sub}}$ *and* $\mathscr{L}_{\mathsf{Sub}}$-*structure*) The *language* $\mathscr{L}_{\mathsf{Sub}}$ is defined inductively as follows:

$$\mathscr{L}_{\mathsf{Sub}} : \quad \varphi ::= \top \mid \bot \mid p \mid \neg\varphi \mid \varphi \wedge \varphi \mid \varphi \vee \varphi \mid \Box\varphi \mid$$
$$\varphi \supset \varphi \mid \varphi \subset \varphi \mid \varphi \circ \varphi$$

where p ranges over ATM.

A $\mathscr{L}_{\mathsf{Sub}}$-*structure* is defined inductively as follows, with φ ranging over $\mathscr{L}_{\mathsf{Sub}}$:

$$X ::= \varphi \mid (X, X) \mid (X; X)$$

Definition 33.9 (*Point set, plump accessibility relation*) A *point set* $\mathscr{P} = (P, \sqsubseteq)$ is a set P together with a partial order $\sqsubseteq$ on P. The set $Prop(\mathscr{P})$ of *propositions* on $\mathscr{P}$ is the set of all subsets X of P which are *closed upwards*: that is, if $x \in X$ and $x \sqsubseteq x'$ then $x' \in X$. We abusively write $x \in \mathscr{P}$ for $x \in P$.

- A binary relation $\mathscr{S}$ is a *positive two–place accessibility relation* on the point set $\mathscr{P}$ iff for any $x, y \in \mathscr{P}$ where $x\mathscr{S}y$, if $x' \sqsubseteq x$ then there is a $y' \sqsupseteq y$, where $x'\mathscr{S}y'$. Similarly, if $x\mathscr{S}y$ and $y \sqsubseteq y'$ then there is some $x' \sqsubseteq x$, where $x'\mathscr{S}y'$.
- A ternary relation $\mathscr{R}$ is a *three-place accessibility relation* iff whenever $\mathscr{R}xyz$ and $z \sqsubseteq z'$ then there are $y' \sqsupseteq y$ and $x' \sqsubseteq x$, where $\mathscr{R}x'y'z'$. Similarly, if $x' \sqsubseteq x$ then there are $y' \sqsubseteq y$ and $z' \sqsupseteq z$, where $\mathscr{R}x'y'z'$, and if $y' \sqsubseteq y$ then there are $x' \sqsubseteq x$ and $z' \sqsupseteq z$, where $\mathscr{R}x'y'z'$.
- A ternary relation $\mathscr{R}$ is a *plump accessibility relation* on the point set $\mathscr{P}$ if and only if for any $x, y, z, x', y', z' \in \mathscr{P}$ such that $\mathscr{R}xyz$, if $x' \sqsubseteq x$, $y' \sqsubseteq y$ and $z \sqsubseteq z'$, then $\mathscr{R}x'y'z'$.

Our definition of $\mathscr{L}_{\mathsf{Sub}}$-model corresponds to the definition of a *model* in [57, Chap. 11] stripped out from all its truth sets. These other features are not needed for what concerns us here.

Definition 33.10 ($\mathscr{L}_{\mathsf{Sub}}$-*model*) A $\mathscr{L}_{\mathsf{Sub}}$-*model* is a tuple $\mathscr{M}_{\mathscr{R}} = (\mathscr{P}, \mathscr{S}, \mathscr{R}, \mathscr{I})$ where:

- $\mathscr{P} = (P, \sqsubseteq)$ is a point set;

- $\mathscr{S} \subseteq \mathscr{P} \times \mathscr{P}$ is a positive two–place accessibility relation on $\mathscr{P}$;
- $\mathscr{R} \subseteq \mathscr{P} \times \mathscr{P} \times \mathscr{P}$ is a three–place accessibility relation on $\mathscr{P}$;
- $\mathscr{I} : P \to 2^{ATM}$ is an interpretation function.

We abusively write $x \in \mathscr{M}_{\mathscr{R}}$ for $x \in \mathscr{P}$, and $(\mathscr{M}_{\mathscr{R}}, x)$ is called a *pointed* $\mathscr{L}_{\mathsf{Sub}}$-*model*.

Note that in the above definition, there could be multiple positive two–place accessibility relations $\mathscr{S}_1, \ldots, \mathscr{S}_n$ corresponding to multiple modalities $\Box_1, \ldots \Box_n$. We refrain from defining $\mathscr{L}_{\mathsf{Sub}}$-models in their full generality in order to ease the readability of the article.

Definition 33.11 (*Truth conditions of* $\mathscr{L}_{\mathsf{Sub}}$) Let $\mathscr{M}_{\mathscr{R}}$ be a $\mathscr{L}_{\mathsf{Sub}}$-model, $x \in \mathscr{M}_{\mathscr{R}}$ and $\varphi \in \mathscr{L}_{\mathsf{Sub}}$. The relation $\mathscr{M}_{\mathscr{R}}, x \Vdash \varphi$ is defined inductively as follows:

$$
\begin{array}{lll}
\mathscr{M}_{\mathscr{R}}, x \Vdash \top & & \text{always} \\
\mathscr{M}_{\mathscr{R}}, x \Vdash \bot & & \text{never} \\
\mathscr{M}_{\mathscr{R}}, x \Vdash p & \text{iff} & p \in \mathscr{I}(x) \\
\mathscr{M}_{\mathscr{R}}, x \Vdash \neg\varphi & \text{iff} & \text{not } \mathscr{M}_{\mathscr{R}}, x \Vdash \varphi \\
\mathscr{M}_{\mathscr{R}}, x \Vdash \varphi \wedge \psi & \text{iff} & \mathscr{M}_{\mathscr{R}}, x \Vdash \varphi \text{ and } \mathscr{M}_{\mathscr{R}}, x \Vdash \psi \\
\mathscr{M}_{\mathscr{R}}, x \Vdash \varphi \vee \psi & \text{iff} & \mathscr{M}_{\mathscr{R}}, x \Vdash \varphi \text{ or } \mathscr{M}_{\mathscr{R}}, x \Vdash \psi \\
\mathscr{M}_{\mathscr{R}}, x \Vdash \Box\varphi & \text{iff} & \text{for all } y \in \mathscr{M}_{\mathscr{R}}, \text{ where } x\mathscr{S}y, \mathscr{M}_{\mathscr{R}}, y \Vdash \varphi \\
\mathscr{M}_{\mathscr{R}}, x \Vdash \varphi \supset \psi & \text{iff} & \text{for all } y, z \in \mathscr{P} \text{ where } \mathscr{R}xyz, \text{ if } \mathscr{M}_{\mathscr{R}}, y \Vdash \varphi \text{ then } \mathscr{M}_{\mathscr{R}}, z \Vdash \psi \\
\mathscr{M}_{\mathscr{R}}, x \Vdash \psi \subset \varphi & \text{iff} & \text{for all } y, z \in \mathscr{P} \text{ where } \mathscr{R}yxz \text{ if } \mathscr{M}_{\mathscr{R}}, y \Vdash \varphi \text{ then } \mathscr{M}_{\mathscr{R}}, z \Vdash \psi \\
\mathscr{M}_{\mathscr{R}}, x \Vdash \varphi \circ \psi & \text{iff} & \text{there are } y, z \in \mathscr{P} \text{ such that } \mathscr{R}yzx, \mathscr{M}_{\mathscr{R}}, y \Vdash \varphi \text{ and } \mathscr{M}_{\mathscr{R}}, z \Vdash \psi
\end{array}
$$

We extend the scope of the relation $\Vdash$ to also relate points to $\mathscr{L}_{\mathsf{Sub}}$-structures:

$$
\begin{array}{lll}
\mathscr{M}_{\mathscr{R}}, x \Vdash X, Y & \text{iff} & \mathscr{M}_{\mathscr{R}}, x \Vdash X \text{ and } \mathscr{M}_{\mathscr{R}}, x \Vdash Y \\
\mathscr{M}_{\mathscr{R}}, x \Vdash X; Y & \text{iff} & \text{there are } y, z \in \mathscr{M}_{\mathscr{R}} \text{ such that } \mathscr{R}yzx, \mathscr{M}_{\mathscr{R}}, y \Vdash X \text{ and } \mathscr{M}_{\mathscr{R}}, z \Vdash Y
\end{array}
$$

We say that $\mathscr{M}_{\mathscr{R}}$ validates a $\mathscr{L}_{\mathsf{Sub}}$-structure X when for all $x \in \mathscr{M}_{\mathscr{R}}$, $\mathscr{M}_{\mathscr{R}}, x \Vdash X$. Let X be a structure and let $\varphi \in \mathscr{L}_{\mathsf{Sub}}$. We say that X *entails* φ, written $X \Vdash \varphi$, when the following holds:

$$X \Vdash \varphi \quad \text{iff} \quad \text{for all pointed } \mathscr{L}_{\mathsf{Sub}}\text{-model } (\mathscr{M}_{\mathscr{R}}, x), \text{ if } \mathscr{M}_{\mathscr{R}}, x \Vdash X, \text{ then } \mathscr{M}_{\mathscr{R}}, x \Vdash \varphi.$$

Note that unlike many substructural logics, we use a Boolean negation. We list below some key inferences of substructural logics, more precisely of the Lambek Calculus:

$$\varphi; \psi \Vdash \chi \quad \text{iff} \quad \varphi \Vdash \psi \supset \chi \tag{33.2}$$

$$\varphi \Vdash \psi \supset \chi \quad \text{iff} \quad \varphi \circ \psi \Vdash \chi \tag{33.3}$$

$$\varphi \circ \psi \Vdash \chi \quad \text{iff} \quad \psi \Vdash \chi \subset \varphi \tag{33.4}$$

$$\varphi \Vdash \psi \supset \chi \quad \text{iff} \quad \psi \Vdash \chi \subset \varphi \qquad (33.5)$$

33.3.1.1 Urquhart's Semantics

The Urquhart's semantics for relevance logic was developed independently from the Routley & Meyer's semantics in the early 1970s. An *operational frame* is a set of points $\mathscr{P}$ together with a function which gives us a new point from a pair of points:

$$\sqcup : \mathscr{P} \times \mathscr{P} \to \mathscr{P}. \qquad (33.6)$$

An *operational model* is then an operational frame together with a relation $\Vdash$ which indicates what formulas are true at what points. The truth conditions for the implication $\supset$ are defined as follows:

$$x \Vdash \varphi \supset \psi \text{ iff for each } y, \text{ if } y \Vdash \varphi \text{ then } x \sqcup y \Vdash \psi. \qquad (33.7)$$

As one can easily notice, an operational frame is a Routley and Meyer frame where $\mathscr{R}xyz$ holds if and only if $x \sqcup y = z$. Hence, the ternary relation $\mathscr{R}$ of the Routley and Meyer semantics is a generalization of the function $\sqcup$ of the Urquhart's semantics. Because it is a *relation*, it allows moreover to apply x to y and yield either a set of outcomes or no outcome at all.

33.3.2 Updates as Ternary Relations

The ternary relation $\mathscr{R}$ of the Routley & Meyer semantics was introduced originally for technical reasons: any 2-ary (n-ary) connective of a logical language can be given a semantics by resorting to a 3-ary (resp. $n + 1$-ary) relation on worlds. Subsequently, a number of philosophical interpretations of this ternary relation have been proposed and we will briefly recall some of them at the end of this section (see [18, 50, 58] for more details). However, one has to admit that providing a non-circular and conceptually grounded interpretation of this relation remains problematic. In this article, we propose a new *dynamic* interpretation of this relation, inspired by the ternary semantics of DEL.

First, one should observe that the DEL product update $\otimes$ of Definition 33.7 can be seen as a partial function $\mathscr{F}$ from a pair of pointed $\mathscr{L}$-model and pointed $\mathscr{L}_\alpha$-model to another pointed $\mathscr{L}$-model:

$$\mathscr{F} : \mathscr{C} \times \mathscr{C}_\alpha \to \mathscr{C} \qquad (33.8)$$

There is a formal similarity between this abstract definition of the DEL product update and the function $\sqcup$ of Eq. (33.6) introduced by Urquhart in the early 1970s for providing a semantics to the implication of relevance logic. This similarity is not only formal but also intuitively meaningful. Indeed, the intuitive interpretation of the DEL product update operator is very similar to the intuitive interpretation of the function ⊔ of Urquhart. Points are sometimes also called "worlds", "states", "situations", "set-ups", and as explained by Restall:

> We have a class of points (over which x and y vary), and a function ⊔ which gives us new points from old. The point $x \sqcup y$ is supposed, on Urquhart's interpretation, to be the body of information given by combining x with y.   [58, p. 363]

and also, keeping in mind the truth conditions for the connective ⊃ of Eq. (33.7):

> To be committed to $A \supset B$ is to be committed to B whenever we gain the information that A. To put it another way, a body of information warrants $A \supset B$ if and only if whenever you *update* that information with new information which warrants A, the resulting (perhaps new) body of information warrants B. (my emphasis)   [58, p. 362]

From these two quotes, it is natural to interpret the DEL product update ⊗ of Definition 33.7 as a specific kind of Urquhart's function ⊔ (Eq. (33.6)). Moreover, as explained by Restall, this substructural "update" can be nonmonotonic and may correspond to some sort of *revision*:

> [C]ombination is sometimes nonmonotonic in a natural sense. Sometimes when a body of information is combined with another body of information, some of the original body of information might be lost. This is simplest to see in the case motivating the failure of $A \vdash B \supset A$. A body of information might tell us that A. However, when we combine it with something which tells us B, the resulting body of information might no longer warrant A (as A might with B). Combination might not simply result in the addition of information. It may well warrant its *revision*. (my emphasis)   [58, p. 363]

Our dynamic interpretation of the ternary relation is consistent with the above considerations: sometimes, updating beliefs amounts to *revise* beliefs. As it turns out, belief revision has also been extensively studied within the DEL framework and DEL has been extended to deal with this phenomenon [1, 2, 14, 15, 26, 36, 46].

More generally, an update can be seen as a *partial* function $\mathscr{F}$ from a pair of pointed $\mathscr{L}$-model and pointed $\mathscr{L}_\alpha$-model to a *set* of pointed $\mathscr{L}$-model:

$$\mathscr{F} : \mathscr{C} \times \mathscr{C}_\alpha \to \mathscr{P}(\mathscr{C}). \tag{33.9}$$

Equivalently, an update can be seen as a ternary relation $\mathscr{R}$ defined on $\mathscr{C} \cup \mathscr{C}_\alpha$ between three pointed models $((\mathscr{M}, w), (\mathscr{E}, e), (\mathscr{M}_f, w_f))$ where $(\mathscr{M}, w)$ is a pointed $\mathscr{L}$-model, $(\mathscr{E}, e)$ is a pointed $\mathscr{L}_\alpha$-model and $(\mathscr{M}_f, w_f)$ is another pointed $\mathscr{L}$-model:

$$\mathscr{R} \subseteq \mathscr{C} \times \mathscr{C}_\alpha \times \mathscr{C}. \tag{33.10}$$

The ternary relation of Eq. (33.10) then resembles the ternary relation of the Routley & Meyer semantics. This is not surprising since the Routley & Meyer semantics generalizes the Urquhart semantics (they are essentially the same, since as we explained it in the previous section, an operational frame is a Routley & Meyer frame where $\mathscr{R}xyz$ holds if and only if $x \sqcup y = z$). Viewed from the perspective of DEL, the ternary relation then represents a particular sort of *update*. With this interpretation in mind, $\mathscr{R}xyz$ reads as 'the occurrence of event y in world x results in the world z' and the corresponding conditional $\alpha \supset \varphi$ reads as 'the occurrence in the current world of an event satisfying property α results in a world satisfying φ'.

The dynamic reading of the ternary relation and its corresponding conditional is very much in line with the so-called "Ramsey Test" of conditional logic. The Ramsey test can be viewed as the very first modern contribution to the logical study of conditionals and much of the contemporary work on conditional logic can be traced back to the famous footnote of Ramsey [55]. Roughly, it consists in defining a counterfactual conditional in terms of belief revision: an agent currently believes that φ would be true if ψ were true (i.e. $\psi \supset \varphi$) if and only if he should believe φ after *learning* ψ. A first attempt to provide truth conditions for conditionals, based on Ramsey's ideas, was proposed by Stalnaker. He defined his semantics by means of selection functions over possible worlds $f : W \times 2^W \to W$. As one can easily notice, Stalnaker's selection functions could also be considered from a formal point of view as a special kind of ternary relation, since a relation $\mathscr{R}_f \subseteq W \times 2^W \times W$ can be canonically associated to each selection function f. Moreover, like the ternary relation corresponding to a product update (Eq. (33.10)), this ternary relation is 'two-sorted': the antecedent of a conditional takes value in a set of worlds (instead of a single world).[2] So, the dynamic reading of the ternary semantics is consistent with the dynamic reading of conditionals proposed by Ramsey.

This dynamic reading was not really considered and investigated by substructural logicians when they connected the substructural ternary semantics with conditional logic [18]. On the other hand, the dynamic reading of inferences has been stressed to a large extent by van Benthem [27, 30] (we will come back to this point in Sect. 33.4.2), and also by Baltag and Smets who distinguished *dynamic* belief revision from *static* (standard) belief revision [13–15]. What distinguishes *dynamic* belief revision from *static* belief revision is that the latter is a revision of the agent's beliefs about the state of the world as it was before an event, and the former is a revision of the state of the world as it is after the event. Note, however, that this important distinction between static belief revision and dynamic belief revision collapses in the case of relevant logic, because in that case we only deal with propositional formulas. This shows again that a *dynamic* interpretation of the ternary semantics of substructural logic is consistent with the interpretations proposed by substructural logicians.

To summarize our discussion, the DEL product update provides substructural logics with an intuitive and consistent interpretation of its ternary relation. This interpretation is consistent in the sense that the intuitions underlying the definitions of the DEL framework are coherent with those underlying the ternary semantics of substructural logic, as witnessed by our quotes and citations from the substructural literature.

---

[2] Note that Burgess [35] already proposed a ternary semantics for conditionals, but his truth conditions and his interpretation of the ternary relation were quite different from ours.

33.3.2.1 Other interpretations of the ternary relation

One interpretation, due to Barwise [16] and developed by Restall [56], takes worlds to be 'sites' or 'channels', a site being possibly a channel and a channel being possibly a site. If x, y and z are sites, $\mathcal{R}xyz$ reads as 'x is a channel between y and z'. Hence, if $\varphi \supset \psi$ is true at channel x, it means that all sites y and z connected by channel x are such that if φ is information available in y, then ψ is information available in z. Another similar interpretation due to Mares [49] adapts Israel and Perry's theory of information [54] to the relational semantics. In this interpretation, worlds are situations in the sense of Barwise and Perry's situation semantics [17] and pieces of information—called *infons*—can carry information about other infons: an infon might carry the information that a red light on a mobile phone carries the information that the battery of the mobile phone is low. In this interpretation, the ternary relation $\mathcal{R}$ represents the informational links in situations: if there is an informational link in situation x that says that an infon σ carries the information that the infon π also holds, then if $\mathcal{R}xyz$ holds and y contains the infon σ, then z contains the infon π. Other interpretations of the ternary relation have been proposed in [18], with a particular focus on their relation to conditionality.

33.4 DEL is a Substructural Logic

In this section, we will extend the languages $\mathcal{L}$ and $\mathcal{L}_\alpha$ of Sect. 33.2 with the substructural operators $\circ$, $\supset$ and $\subset$. We will also provide a substructural semantics for this language based on the idea to view an update as a ternary relation of a substructural frame ($\mathcal{L}_{\mathsf{Sub}}$-model). This idea is motivated and intuitively grounded in the analysis of the previous section.

33.4.1 An Extended DEL Language

Our language extends both the language $\mathcal{L}$ and the language $\mathcal{L}_\alpha$ of Sect. 33.2. Like our semantics, it is two-sorted: it contains both formulas of $\mathcal{L}$ and formulas of $\mathcal{L}_\alpha$.

Definition 33.12 (*Language $\mathcal{L}_\mathcal{R}$*) The *language $\mathcal{L}_\mathcal{R}$* is two-sorted and is defined by a double induction as follows:

$$\mathcal{L}_\mathcal{R}^1: \quad \varphi ::= p \quad | \quad \neg\varphi \quad | \quad \varphi \wedge \varphi \quad | \quad \varphi \vee \varphi \quad | \quad \Box_j \varphi \quad | \quad \alpha \supset \varphi \quad | \quad \varphi \circ \alpha$$
$$\mathcal{L}_\mathcal{R}^2: \quad \alpha ::= p_\psi \quad | \quad \neg\alpha \quad | \quad \alpha \wedge \alpha \quad | \quad \alpha \vee \alpha \quad | \quad \Box_j \alpha \quad | \quad \varphi \subset \varphi$$

where p ranges over ATM, ψ ranges over $\mathcal{L}_\mathcal{R}^1$ and j over AGT. The abbreviations $\varphi \to \psi$, $\varphi \leftrightarrow \psi$ and $\alpha \to \beta$, $\alpha \leftrightarrow \beta$ are defined as in Definitions 33.1 and 33.4.

Definition 33.13 (*$\mathscr{L}_{\mathscr{R}}$-structure and $\mathscr{L}_{\mathscr{R}}$-sequent*) The $\mathscr{L}_{\mathscr{R}}$-structures are defined inductively as follows:

$$\begin{array}{llll}
\mathscr{S}_1: & X & ::= & \varphi \mid (X, X) \mid (X; X_\alpha) \\
\mathscr{S}_2: & X & ::= & \varphi \mid (X, X)
\end{array}$$

where φ ranges over $\mathscr{L}_{\mathscr{R}}$ and X_α ranges over $\mathscr{L}_\alpha$-structures. A $\mathscr{L}_{\mathscr{R}}$-sequent is a $\mathscr{L}_\alpha$-sequent or an expression of the form $X \vdash Y$, where $X \in \mathscr{S}_1$, $Y \in \mathscr{S}_2$.

Definition 33.14 (*DEL product update model*) The *DEL product update model* is the tuple $\mathscr{M}_\otimes = (\mathscr{P}, \mathscr{R}_1, \ldots, \mathscr{R}_m, \mathscr{R}_\otimes, \mathscr{I})$ where:

- $\mathscr{P} := (\mathscr{C} \cup \mathscr{C}_\alpha, \leftrightarrow)$ where $\leftrightarrow$ is the bisimilarity relation;
- $\mathscr{R}_j \subseteq \mathscr{P} \times \mathscr{P}$ is a positive two-place accessibility relation on $\mathscr{P}$ for each $j \in AGT$ such that for all $x, y \in \mathscr{P}$, where $x = (\mathscr{M}_x, w_x)$ and $y = (\mathscr{M}_y, w_y)$:

$$x \in \mathscr{R}_j(y) \text{ iff } \mathscr{M}_x = \mathscr{M}_y \text{ and } w_x \in R_j(w_y)$$

- $\mathscr{R}_\otimes := \{(x, y, z) \in \mathscr{C} \times \mathscr{C}_\alpha \times \mathscr{C} \mid x \otimes y = z\}$ is a plump ternary relation on $\mathscr{P}$;
- $\mathscr{I}(x) := I(x)$, for all $x \in \mathscr{C} \cup \mathscr{C}_\alpha$.

The DEL product update model is a $\mathscr{L}_{\mathsf{Sub}}$-model where points are pointed $\mathscr{L}$-models and pointed $\mathscr{L}_\alpha$-models. The ternary relation $\mathscr{R}_\otimes$ is defined and motivated by the explanations of the previous section. Note that the accessibility relations R_j of $\mathscr{L}$-models and $\mathscr{L}_\alpha$-models are seen in this definition as positive two-place accessibility relations $\mathscr{R}_j$. The truth conditions are the same as the ones for $\mathscr{L}_{\mathscr{R}}$-models:

Definition 33.15 (*Truth conditions of $\mathscr{L}_{\mathscr{R}}$*) Let $\mathscr{M}_\otimes$ be the DEL product update model, $x \in \mathscr{M}_\otimes$ and $\varphi \in \mathscr{L}_{\mathscr{R}}$. The relation $\mathscr{M}_\otimes, x \Vdash \varphi$ is defined inductively as follows:

$$\begin{array}{lll}
\mathscr{M}_\otimes, x \Vdash p & \text{iff} & p \in \mathscr{I}(x) \\
\mathscr{M}_\otimes, x \Vdash \neg\varphi & \text{iff} & \text{not } \mathscr{M}_\otimes, x \Vdash \varphi \\
\mathscr{M}_\otimes, x \Vdash \varphi \wedge \psi & \text{iff} & \mathscr{M}_\otimes, x \Vdash \varphi \text{ and } \mathscr{M}_\otimes, x \Vdash \psi \\
\mathscr{M}_\otimes, x \Vdash \varphi \vee \psi & \text{iff} & \mathscr{M}_\otimes, x \Vdash \varphi \text{ or } \mathscr{M}_\otimes, x \Vdash \psi \\
\mathscr{M}_\otimes, x \Vdash \Box_j\varphi & \text{iff} & \text{for all } y \in \mathscr{P} \text{ such that } x\mathscr{R}_j y, \mathscr{M}_\otimes, y \Vdash \varphi \\
\mathscr{M}_\otimes, x \Vdash \alpha \supset \psi & \text{iff} & \text{for all } y, z \in \mathscr{P} \text{ such that } \mathscr{R}_\otimes xyz, \text{ if } \mathscr{M}_\otimes, y \Vdash \alpha \text{ then } \mathscr{M}_\otimes, z \Vdash \psi \\
\mathscr{M}_\otimes, x \Vdash \psi \subset \varphi & \text{iff} & \text{for all } y, z \in \mathscr{P} \text{ such that } \mathscr{R}_\otimes yxz, \text{ if } \mathscr{M}_\otimes, y \Vdash \varphi \text{ then } \mathscr{M}_\otimes, z \Vdash \psi \\
\mathscr{M}_\otimes, x \Vdash \varphi \circ \alpha & \text{iff} & \text{there are } y, z \in \mathscr{P} \text{ such that } \mathscr{R}_\otimes yzx, \mathscr{M}_\otimes, y \Vdash \varphi \text{ and } \mathscr{M}_\otimes, z \Vdash \alpha
\end{array}$$

We extend the scope of the relation $\Vdash$ to also relate points to $\mathscr{L}_{\mathscr{R}}$-structures:

$$\begin{array}{lll}
\mathscr{M}_\otimes, x \Vdash X, Y & \text{iff} & \mathscr{M}_\otimes, x \Vdash X \text{ and } \mathscr{M}_\otimes, x \Vdash Y \\
\mathscr{M}_\otimes, x \Vdash X; Y & \text{iff} & \text{there are } y, z \in \mathscr{M}_{\mathscr{R}} \text{ such that } \mathscr{R} yzx, \mathscr{M}_\otimes, y \Vdash X \text{ and } \mathscr{M}_\otimes, z \Vdash Y
\end{array}$$

Let $C \subseteq \mathscr{C} \cup \mathscr{C}_\alpha$ be a class of pointed $\mathscr{L}$-models or $\mathscr{L}_\alpha$-models, and let $X \vdash \varphi$ be a $\mathscr{L}_\mathscr{R}$-sequent. We say that X *entails φ in the class* C, written $X \Vdash_C \varphi$, when the following holds:

$$X \Vdash_C \varphi \quad \text{iff} \quad \text{for all } x \in C, \text{ if } \mathscr{M}_\otimes, x \Vdash X \text{ then } \mathscr{M}_\otimes, x \Vdash \varphi.$$

We also write $X \Vdash \varphi$ for $X \Vdash_{\overline{\mathscr{C} \cup \mathscr{C}_\alpha}} \varphi$.

Naturally, the truth conditions for $\Vdash$ coincide with the truth conditions for $\models$ if we only consider epistemic or event formulas:

Proposition 33.1 *Let $\mathscr{M}_\otimes$ be the DEL product update model, $\varphi \in \mathscr{L}$ and $x \in \mathscr{M}_\otimes$ such that $x \in \mathscr{C}$. Then, $\mathscr{M}_\otimes, x \Vdash \varphi$ iff $x \models \varphi$. Let $\alpha \in \mathscr{L}_\alpha$ and let $y \in \mathscr{M}_\otimes$ such that $y \in \mathscr{C}_\alpha$. Then, $\mathscr{M}_\otimes, y \Vdash \alpha$ iff $y \models \alpha$.*

Remark 33.1 The frame semantics of substructural logic is very abstract and general and it provides a rich framework which captures a wide range of logics, such as arrow logic [23, Chap. 8], action frames and domain space (see [57, Example 11.12–11.15] for more details). But the epistemic temporal models of ETL [53] (which have been related to DEL in [32, 33]) can also be viewed as models of the ternary semantics of substructural logic [5].

33.4.2 DEL Operators are Substructural Operators

In this section, we will show that the DEL operators introduced in [3, 4] correspond to the substructural operators $\circ$, $\supset$ and $\subset$. We will also relate the work of van Benthem on dynamic inference with the DEL–sequents of [3, 4, 7].

33.4.2.1 Dynamic Inferences and DEL–sequents

Dynamic Inferences
In the so-called 'dynamic turn', van Benthem was interested in various dynamic styles of inference where propositions are procedures changing information states. These dynamic styles of inference differ greatly from the classical Tarskian valid inferences because the latter are supposed to transmit and preserve truth. Among various dynamic styles of inference (such as the so-called test-test, update-update or update-test consequence [21, 23, 51]), he studied the concrete following one, which can be defined within the DEL framework:

Definition 33.16 (*Dynamic inference* [25]) Let $\varphi_0, \varphi_1, \ldots, \varphi_n, \psi \in \mathscr{L}$. We define the *dynamic inference* $\varphi_0, \varphi_1, \ldots, \varphi_n \models \psi$ as follows:

$\varphi_1, \ldots, \varphi_n \models \varphi$ iff   for all pointed $\mathscr{L}$-model $(\mathscr{M}, w)$, and public announcement
$\mathscr{L}_\alpha$-models $(\mathscr{E}_1, e_1), \ldots, (\mathscr{E}_n, e_n)$ of $\varphi_1, \ldots, \varphi_n$
respectively, $(\mathscr{M}, w) \otimes (\mathscr{E}_1, e_1) \otimes \ldots \otimes (\mathscr{E}_n, e_n) \models \varphi$.

van Benthem noticed that various dynamic styles of inference obey structural rules of inference which are non-classical. For example, all the structural rules of classical logic of Fig. 33.5 fail for dynamic inference, but the structural rules below characterize completely the dynamic inference [25] (below, $\overrightarrow{\varphi}$ stands for $\varphi_1, \ldots, \varphi_n$ and $\overrightarrow{\psi}$ stands for $\psi_1, \ldots, \psi_n$, where $\varphi_1, \ldots, \varphi_n, \psi_1, \ldots, \psi_n \in \mathscr{L}$):

$\quad$ if $\overrightarrow{\varphi} \models \varphi$ then $\psi, \overrightarrow{\varphi} \models \varphi$ $\hfill$ (Left-Monotonicity)

$\quad$ if $\overrightarrow{\varphi} \models \varphi$ and $\overrightarrow{\varphi}, \varphi, \overrightarrow{\psi} \models \psi$ then $\overrightarrow{\varphi}, \overrightarrow{\psi} \models \psi$ $\hfill$ (Left-Cut)

$\quad$ if $\overrightarrow{\varphi} \models \varphi$ and $\overrightarrow{\varphi}, \overrightarrow{\psi} \models \psi$ then $\overrightarrow{\varphi}, \varphi, \overrightarrow{\psi} \models \psi$ $\hfill$ (Cautious Monotonicity)

DEL–sequents

In [3], I introduced what I called DEL–sequents. They are a particular sort of dynamic inference and are defined as follows:

Definition 33.17 (*DEL–sequent* [3]) Let $\varphi, \varphi_f \in \mathscr{L}$ and $\alpha \in \mathscr{L}_\alpha$. We define the logical consequence relation $\varphi, \alpha \models \varphi_f$ as follows:

$\varphi, \alpha \models \varphi_f$ iff   for all pointed $\mathscr{L}$-model $(\mathscr{M}, w)$, all $\mathscr{L}_\alpha$-model $(\mathscr{E}, e)$ such that
$\mathscr{M}, w \models I(e), \mathscr{M}, w \models \varphi$ and $\mathscr{E}, e \models \alpha$, it holds that $(\mathscr{M}, w) \otimes (\mathscr{E}, e) \models \varphi_f$.

In [7], DEL–sequents are generalized to take into account sequences of events and not only 'one-shot' occurrence of events. Several generalized DEL–sequents are introduced in [7] but they are all reducible to the following one:

Definition 33.18 (*Generalized DEL–sequent* [7]) Let $\varphi_0, \ldots, \varphi_n \in \mathscr{L}$, let $\alpha_1, \ldots, \alpha_n \in \mathscr{L}_\alpha$ and let $\psi \in \mathscr{L}$. Then,

$$\varphi_0, \alpha_1, \varphi_1, \ldots, \alpha_n, \varphi_n \models \psi$$
$$\text{iff}$$

if for all pointed $\mathscr{L}$-model $(\mathscr{M}, w)$, and $\mathscr{L}_\alpha$-models $(\mathscr{E}_1, e_1), \ldots, (\mathscr{E}_n, e_n)$ such that for all $i \in \{1, \ldots, n\}$, $\mathscr{E}_i, e_i \models \alpha_i$, $(\mathscr{M}, w) \otimes (\mathscr{E}_1, e_1) \otimes \ldots \otimes (\mathscr{E}_i, e_i)$ is defined and makes φ_i true, then it holds that $(\mathscr{M}, w) \otimes (\mathscr{E}_1, e_1) \otimes \ldots \otimes (\mathscr{E}_n, e_n) \models \psi$.

As one can easily notice, dynamic inferences can be translated into DEL–sequents if we resort to the common knowledge/belief operator $\square^*_{AGT} \varphi$ (see for example [41] for a definition and a detailed study of this operator):

Proposition 33.2 *Let* $\varphi_0, \varphi_1, \ldots, \varphi_n, \varphi \in \mathcal{L}$. *Then, the following holds:*

$$\varphi_1, \ldots, \varphi_n \vDash \varphi \quad iff \quad \top, p_{\varphi_1} \wedge \Box^*_{AGT} p_{\varphi_1}, \ldots, \top, p_{\varphi_n} \wedge \Box^*_{AGT} p_{\varphi_n}, \top \vDash \varphi \wedge \Box^*_{AGT} \varphi$$

Thus, DEL–sequents are more expressive than dynamic inferences, and also more abstract because they 'operate' at a deeper level, a semantical one. It is this more general and abstract approach towards dynamic styles of inference that will allow us to relate more precisely and closely DEL with substructural logics, and explain to a certain extent why the substructural phenomena occurring in dynamic inferences and observed by van Benthem arise.

33.4.2.2 DEL–sequents for Progression, Regression and Epistemic Planning

Substructural logics and dynamic logics of information flow are long-standing interests of van Benthem [22, 23, 25, 27, 29, 30]. Recently again in [29], he expressed some worries about interpreting the Lambek Calculus (the paradigmatic substructural logic) as a base logic of information flow while trying to connect the operators $\circ$, $\supset$ and $\subset$ of substructural logic to some sort of DEL operators. Indeed, the DEL operators usually rely on the regular algebra of sequential composition, choice and iteration which are of a quite different nature. Recently, I introduced some DEL operators called progression, regression and epistemic planning [3, 4], the operator of regression being a natural generalization of the standard and original action modality $[\mathcal{E}, e]\varphi$ of DEL [11]. It turns out that these operators can all be identified with connectives of the substructural language $\mathcal{L}_{\mathcal{R}}$. We first briefly recall their definitions below and then we give our correspondence results between the two kinds of operators.

Progression
The operator of *progression* is denoted $\otimes$ in [3]. In [4, Definition 41], a constructive definition of this operator is provided using characteristic formulas (called "Kit Fine" formulas). Here, we provide an alternative and non–constructive definition of the *progression* of φ by α, denoted $\varphi \otimes \alpha$:

Theorem 33.1 *Let* $(\mathcal{M}_f, w_f)$ *be a pointed* $\mathcal{L}$-*model and let* $\varphi \in \mathcal{L}$ *and* $\alpha \in \mathcal{L}_\alpha$. *Then,*

$$\mathcal{M}_f, w_f \vDash \varphi \otimes \alpha \quad iff \quad \text{there is a pointed } \mathcal{L}\text{-model } (\mathcal{M}, w) \text{ and a pointed}$$
$$\mathcal{L}_\alpha\text{-model } (\mathcal{E}, e) \text{ such that } (\mathcal{M}, w) \otimes (\mathcal{E}, e) \leftrightarrows (\mathcal{M}_f, w_f),$$
$$\mathcal{M}, w \vDash \varphi \text{ and } \mathcal{E}, e \vDash \alpha$$

Proof It follows from Lemmata 43 and 44 of [3].

Epistemic Planning
The operator of *epistemic planning* is denoted $\oslash_P$ in [4]. It is defined relatively to a finite set P of formulas/preconditions/atomic events. In [4, Def. 34.14–34.15], a constructive definition of this operator is provided using characteristic formulas

(called "Kit Fine" formulas). As it turns out, an alternative and non–constructive definition of the *epistemic planning from* φ *to* φ_f, denoted $\varphi \otimes_P \varphi_f$, exists as well:

Theorem 33.2 ([4]) *Let* $\varphi, \varphi_f \in \mathcal{L}$ *and let* P *be a finite subset of* $\mathcal{L}$. *Then, for all* P-*complete* $\mathcal{L}_\alpha$-*model* $(\mathcal{E}, e)$, *it holds that*

$$\mathcal{E}, e \models \varphi \otimes_P \varphi_f \quad iff \quad \begin{array}{l} \text{there is } (\mathcal{M}, w) \text{ such that } \mathcal{M}, w \models \varphi, \\ \mathcal{M}, w \models I(e) \text{ and } (\mathcal{M}, w) \otimes (\mathcal{E}, e) \models \varphi_f. \end{array}$$

The dual of the operator $\varphi \otimes_P \varphi_f$ is defined by:

$$\varphi[\otimes]_P \varphi_f := \neg(\varphi \otimes_P \neg \varphi_f). \tag{33.11}$$

Theorem 33.2 entails that $\varphi[\otimes]_P \varphi_f$ can be alternatively defined as follows: for all P-complete $\mathcal{L}_\alpha$-model $(\mathcal{E}, e)$, it holds that

$$\mathcal{E}, e \models \varphi[\otimes]_P \varphi_f \quad iff \quad \text{for all } (\mathcal{M}, w) \text{ such that } \mathcal{M}, w \models \varphi, \text{ if} \tag{33.12}$$
$$\mathcal{M}, w \models I(e) \text{ then } (\mathcal{M}, w) \otimes (\mathcal{E}, e) \models \varphi_f$$

Example 33.4 In the situation depicted in the $\mathcal{L}$-model of Fig. 33.1, agent B does not know that agent A has the red card and does not know that agent C has the blue card: $\mathcal{M}, w \models (\lozenge_B r_A \wedge \lozenge_B \neg r_A) \wedge (\lozenge_B b_C \wedge \lozenge_B \neg b_C)$. Our problem is therefore the following:

> What sufficient and necessary property (i.e. 'minimal' property) an event should fulfill so that its occurence in the initial situation $(\mathcal{M}, w)$ results in a situation where agent B knows the true state of the world, i.e. agent B knows that agent A has the red card and that agent C has the blue card?

The answer to this question obviously depends on the kind of atomic events we consider. In this example, the events $P = \{p_{b_C}, p_{r_A}, p_{w_B}\}$ under consideration are the following. First, agent C shows her blue card (p_{b_C}), second, agent A shows her red card (p_{r_A}), and third, agent B herself shows her white card (p_{w_B}). Answering this question amounts to compute the formula $(M, w) \otimes_P \square_B (r_A \wedge b_C \wedge w_B)$. Applying the algorithm of [4, Definition 34.15], we obtain that

$$(\mathcal{M}, w) \otimes_P \square_B (r_A \wedge b_C \wedge w_B) \leftrightarrow \square_B(p_{b_C} \vee p_{r_A}) \text{is valid.}$$

In other words, this result states that agent B should believe either that agent A shows her red card or that agent C shows her blue card in order to know the true state of the world. Indeed, since there are only three different cards which are known by the agents and agent B already knows her card, if she learns the card of (at least) one of the other agents, she will also be able to infer the card of the third agent.

Regression

The operator of *regression* is denoted $\oslash$ in [3]. In [4, Definition 41], a constructive definition of this operator is provided using characteristic formulas (called "Kit Fine" formulas) by adapting and translating the reduction axioms of [11]. As it turns out, an alternative and non–constructive definition of the *regression* of φ_f by α, denoted $\alpha \oslash \varphi_f$, exists as well:

Theorem 33.3 *Let* $\alpha \in \mathscr{L}_\alpha$ *and* $\varphi_f \in \mathscr{L}$. *Then, for all* $\mathscr{L}$-model $(\mathscr{M}, w)$, *it holds that*

$$\mathscr{M}, w \models \alpha \oslash \varphi_f \quad \text{iff} \quad \begin{array}{l} \text{there is } (\mathscr{E}, e) \text{ such that } \mathscr{E}, e \models \alpha, \\ \mathscr{M}, w \models I(e) \text{ and } (\mathscr{M}, w) \otimes (\mathscr{E}, e) \models \varphi_f \end{array}$$

Note that we could define a dual operator of $\alpha \oslash \varphi_f$ as follows:

$$\alpha[\oslash]\varphi_f = \neg \left(\alpha \oslash \neg \varphi_f \right) \tag{33.13}$$

Then, the counterpart of Theorem 33.3 for this dual operator is as follows:

$$\mathscr{M}, w \models \alpha[\oslash]\varphi_f \quad \text{iff} \quad \begin{array}{l} \text{for all } (\mathscr{E}, e) \text{ such that } \mathscr{E}, e \models \alpha, \\ \text{if } \mathscr{M}, w \models I(e) \text{ then } (\mathscr{M}, w) \otimes (\mathscr{E}, e) \models \varphi_f \end{array} \tag{33.14}$$

As shown in [4, Sect. 6], the operator $\alpha[\oslash]\varphi_f$ is a generalization of the original and more standard DEL operator $[\mathscr{E}, e]\varphi$ almost exclusively used in the DEL literature [11].

Correspondence between DEL and Substructural Operators

As one can easily notice, there is a strong similarity between the operations of progression, epistemic planning and regression and the operations of substructural logic, more precisely of the Lambek Calculus. In fact, there exists a rigorous mapping between them, as the following theorem shows:

Theorem 33.4 *Let* P *be a finite subset of* $\mathscr{L}$, *let* $x = (\mathscr{M}, w) \in \mathscr{C}$ *and let* $y = (\mathscr{E}, e) \in \mathscr{C}_\alpha^P$ *be a* P-complete pointed event model. *Let* $\varphi, \psi \in \mathscr{L}$ *and let* $\alpha \in \mathscr{L}_\alpha$. *Then,*

$$\begin{array}{llll} \mathscr{M}_\otimes, x \Vdash \varphi \circ \alpha & \text{iff} & x \models \varphi \otimes \alpha \\ \mathscr{M}_\otimes, x \Vdash \alpha \supset \varphi & \text{iff} & x \models \alpha[\oslash]\varphi \\ \mathscr{M}_\otimes, y \Vdash \psi \subset \varphi & \text{iff} & y \models \varphi[\otimes]_P \psi \end{array}$$

Moreover, for all $\alpha, \alpha_1, \ldots, \alpha_n \in \mathscr{L}_\alpha$, *for all* $\varphi, \psi, \varphi_0, \varphi_1, \ldots, \varphi_n \in \mathscr{L}$, *we have:*

$$\begin{array}{llll} \varphi; \alpha \Vdash \psi & & \text{iff} & \varphi, \alpha \models \psi \\ (((\varphi_0; \alpha_1), \varphi_1); \ldots; \alpha_n), \varphi_n \Vdash \psi & \text{iff} & \varphi_0, \alpha_1, \varphi_1, \ldots, \alpha_n, \varphi_n \models \psi \end{array}$$

The key Theorem 42 of [3] relates DEL–sequents and the operator of progression: for all $\varphi, \varphi_f \in \mathscr{L}$ and $\alpha \in \mathscr{L}_\alpha$, it holds that

$$\varphi, \alpha \models \varphi_f \text{ iff } \varphi \otimes \alpha \models \varphi_f. \tag{33.15}$$

Substructural operators	DEL operators
$\circ$	$\otimes$
$\supset$	$[\oslash]$
$\subset$	$[\oslash]$

Fig. 33.6 Correspondence between DEL and substructural operators

As it turns out, this theorem is also valid in any substructural logic: it corresponds to the theorem of Eq. (33.2). More generally, all the theorems of the non-associative Lambek calculus hold in our DEL setting if we use the translation given in Fig. 33.6. In particular, we have the following results which are the counterparts of Eqs. (33.3), (33.4) and (33.5) in our setting:

Corollary 33.1 *Let P be a finite subset of $\mathcal{L}$. For all $\varphi, \varphi_f \in \mathcal{L}$ and $\alpha \in \mathcal{L}_\alpha$, it holds that*

$$\varphi; \alpha \Vdash \varphi_f \quad \text{iff} \quad \varphi \Vdash \alpha[\oslash]\varphi_f \tag{33.16}$$

$$\varphi \Vdash \alpha[\oslash]\varphi_f \quad \text{iff} \quad \varphi \otimes \alpha \Vdash \varphi_f \tag{33.17}$$

$$\varphi \otimes \alpha \Vdash \varphi_f \quad \text{iff} \quad \alpha \Vdash_{\overline{\mathscr{C}_\alpha^P}} \varphi[\oslash]_P \varphi_f \tag{33.18}$$

$$\varphi \Vdash \alpha[\oslash]\varphi_f \quad \text{iff} \quad \alpha \Vdash_{\overline{\mathscr{C}_\alpha^P}} \varphi[\oslash]_P \varphi_f \tag{33.19}$$

33.5 Conclusion

We proved in this article that DEL is a two-sorted substructural logic. Also, we argued in Sect. 33.3.2 that our embedding of DEL within the framework of substructural logic is intuitively consistent, in the sense that in this embedding the intuitions underlying the DEL framework are coherent with the intuitive interpretations proposed for the ternary semantics of substructural logics. This may explain to a certain extent *why* some substructural phenomena arise in the dynamic inferences of Sect. 33.4.2.1. As observed by van Benthem, "it seemed that structural rules address mere *symptoms* of some underlying phenomenon" [30, p. 297]. I claim that these "symptoms" are caused at a deeper semantic level by the fact that an update, and in that case the DEL product update, can be represented by the ternary relation of substructural logics.

In a certain sense, this article is in line with the approach of van Benthem [28, 30] and contributes to relate even more closely the research programs of Logical Pluralism [19] and Logical Dynamics [30]. Roughly, the informal idea underpinning the connection between these two logical paradigms is to consider different reasoning styles and their corresponding consequence relations as the result of different sorts of updates induced by various informational tasks (such as observation, memory, questions and answers, dialogue, or general communication). We showed in this article that this approach is not only meaningful from an intuitive point of view, but it can also be realized at a formal level if the ternary relation of substructural logic is interpreted intuitively as a sort of update. So, we hope that our embedding will strengthen the connections between the two areas of research represented by Logical Pluralism (and substructural logics) on the one hand and Logical Dynamics on the other hand. In fact, our point of view is also very much in line with the claim of Gärdenfors and Makinson [44, 48] that non-monotonic reasoning and belief revision are "two sides of the same coin": as a matter of fact, non-monotonic reasoning is a reasoning style and belief revision is a sort of update. Likewise, the formal connection in this case also relies on a similar idea based on the Ramsey test.

In this article, we focused our attention on the DEL product update. It is, however, a particular kind of update operator and the ternary relation of substructural logics could actually be a representation of any sort of update, including the various revision and update operators which have been studied in the logics of "common sense reasoning" of artificial intelligence and philosophical logic, such as conditional logic [52], default and non-monotonic logics [42, 47], belief revision theory [43], etc... Different kinds of updates, induced by different informational tasks, define different kinds of reasoning styles. If one adheres to our interpretation of the ternary relation, the dynamic notion of update then becomes the foundational concept of substructural logics.

This observation gives rise, in turn, to a research thread where updates are the central objects of study and where we can (re-)analyze various updates within the generic and abstract logical framework of substructural logics. This research thread is of course very much in line with van Benthem's long standing interest in information and logical dynamics, but also with his interest in modal correspondence theory, the area of logic where he first contributed [20, 24]. For example, we could elicit a number of axioms and inference rules that define specific properties of updates, some being possibly satisfied by the DEL product update. In other words, we could develop a correspondence theory for analyzing and studying the notion of update similar to the correspondence theory developed by van Benthem [20, 24] for modal logic. A basic correspondence theory with a complete characterization of the DEL product update in terms of axioms and inference rules is given in [5].

Acknowledgments I thank Olivier Roy and Ole Hjortland for organizing and inviting me to an inspiring workshop on substructural epistemic logic in Munich in February 2013. Also, I thank Johan van Benthem and Igor Sedlar for comments on an earlier version of this article. Finally, I thank Sean Sedwards for checking the English of this article.

References

1. Aucher G (2004) A combined system for update logic and belief revision. In: Barley M, Kasabov NK (eds) PRIMA, vol 3371., Lecture Notes in Computer ScienceSpringer, Berlin, pp 1–17
2. Aucher G (2008) Perspectives on belief and change. Ph.D. thesis, University of Otago - University of Toulouse.

3. Aucher G (2011) Del-sequents for progression. J Appl Non-class Logics 21(3–4):289–321
4. Aucher G (2012) Del-sequents for regression and epistemic planning. J Appl Non-class Logics 22(4):337–367
5. Aucher G (2013) Update logic. Research report RR-8341, INRIA. http://hal.inria.fr/hal-00849856
6. Aucher G, Herzig A (2011) Exploring the power of converse events. In: Girard P, Roy O, Marion M (eds) Dynamic formal epistemology, vol 351. Springer, The Netherlands, pp 51–74
7. Aucher G, Maubert B, Schwarzentruber F (2012) Generalized DEL-sequents. In: del Cerro LF, Herzig A, Mengin J (eds) JELIA, vol 7519., Lecture Notes in Computer ScienceSpringer, New York, pp 54–66
8. Baltag A, Coecke B, Sadrzadeh M (2005) Algebra and sequent calculus for epistemic actions. Electron Notes Theoret Comput Sci 126:27–52
9. Baltag A, Coecke B, Sadrzadeh M (2007) Epistemic actions as resources. J Logic Comput 17(3):555–585
10. Baltag A, Moss L (2004) Logic for epistemic programs. Synthese 139(2):165–224
11. Baltag A, Moss LS, Solecki S (1998) The logic of public announcements and common knowledge and private suspicions. In: Gilboa I (ed) TARK. Morgan Kaufmann, San Francisco, pp 43–56
12. Baltag A, Moss L, Solecki S (1999) The logic of public announcements, common knowledge and private suspicions. Indiana University, Technical report
13. Baltag A, Smets S (2006) Conditional doxastic models: a qualitative approach to dynamic belief revision. Electron Notes Theoret Comput Sci 165:5–21
14. Baltag A, Smets S (2008) The logic of conditional doxastic actions. Texts in logic and games, vol 4. Amsterdam University Press, Amsterdam, pp 9–31
15. Baltag A, Smets S (2008) A qualitative theory of dynamic interactive belief revision. Texts in logic and games, vol 3. Amsterdam University Press, Amsterdam, pp 9–58
16. Barwise J (1993) Constraints, channels, and the flow of information. In: Cooper R, Barwise J, Mukai K (eds) Situation theory and its applications, vol 3. Center for the Study of Language and Information, US, pp 3–27
17. Barwise J, Perry J (1983) Situations and attitudes. MIT Press, Cambridge
18. Beall J, Brady R, Dunn JM, Hazen A, Mares E, Meyer RK, Priest G, Restall G, Ripley D, Slaney J et al (2012) On the ternary relation and conditionality. J philos logic 41(3):595–612
19. Beall JC, Restall G (2006) Logical pluralism. Oxford University Press, Oxford
20. van Benthem, J (1977) Modal correspondence theory. Ph.D. thesis, University of Amsterdam.
21. van Benthem J (1991) General dynamics. Theoret Linguis 17(1–3):159–202
22. van Benthem J (1991) Language in action: categories, lambdas and dynamic logic, vol 130. North Holland, Amsterdam
23. van Benthem J (1996) Exploring logical dynamics. CSLI publications, Stanford
24. van Benthem J (2001) Correspondence theory. In: Gabbay D, Guenthner F (eds) Handbook of philosophical logic, vol 3. Reidel, Dordrecht, pp 325–408
25. van Benthem J (2003) Structural properties of dynamic reasoning. In: Peregrin J (ed) Meaning: the dynamic turn. Elsevier, Amsterdam, pp 15–31
26. van Benthem J (2007) Dynamic logic for belief revision. J Appl Non-class Logics 17(2):129–155
27. van Benthem J (2007) Inference in action. Publications de l'Institut Mathématique-Nouvelle Série 82(96):3–16
28. van Benthem J (2008) Logical dynamics meets logical pluralism? Australas J Logic 6:182–209
29. van Benthem J (2010) Modal logic for open minds. CSLI publications, Stanford
30. van Benthem J (2011) Logical dynamics of information and interaction. Cambridge University Press, Cambridge
31. van Benthem J (2011) Mccarthy variations in a modal key. Artif intell 175(1):428–439
32. van Benthem J, Gerbrandy J, Hoshi T, Pacuit E (2009) Merging frameworks for interaction. J Philos Logic 38(5): 491–526. doi:10.1007/s10992-008-9099-x. http://dx.doi.org/10.1007/s10992-008-9099-x

33. van Benthem J, Gerbrandy J, Pacuit E (2007) Merging frameworks for interaction: DEL and ETL. In: Samet D (ed) Theoretical aspect of rationality and knowledge (TARK XI). ILLC, Brussels, pp 72–82
34. van Benthem J, Kooi B (2004) Reduction axioms for epistemic actions. In: Schmidt R, Pratt-Hartmann I, Reynolds M, Wansing H (eds) AiML-2004: advances in modal logic, number UMCS-04-9-1 in technical report series. University of Manchester, Manchester, pp 197–211
35. Burgess JP (1981) Quick completeness proofs for some logics of conditionals. Notre Dame J Formal Logic 22(1):76–84
36. van Ditmarsch H (2005), Prolegomena to dynamic logic for belief revision. Synthese 147:229–275.
37. van Ditmarsch H, van der Hoek W, Kooi B (2007) Synthese library. In: Dynamic epistemic logic, vol 337. Springer, New York.
38. van Ditmarsch HP, Herzig A, Lima TD (2009) From situation calculus to dynamic epistemic logic. J Logic Comput 21(2):179–204
39. Dunn JM, Restall G (2002) Relevance logic. In: Gabbay D, Guenthner F (eds) Handbook of philosophical logic, vol 6. Kluwer, Dordrecht, pp 1–128
40. van Eijck J (2004) Reducing dynamic epistemic logic to PDL by program transformation. Technical report SEN-E0423, CWI.
41. Fagin R, Halpern J, Moses Y, Vardi M (1995) Reasoning about knowledge. MIT Press, Cambridge
42. Gabbay DM, Hogger CJ, Robinson JA, Siekmann J, Nute D (1998 eds) Nonmonotonic reasoning and uncertain reasoning. In: Handbook of logic in artificial intelligence and logic programming, vol 3. Clarendon Press, Oxford.
43. Gärdenfors P (1988) Knowledge in flux (modeling the dynamics of epistemic states). Bradford/MIT Press, Cambridge
44. Gärdenfors P (1991) Belief revision and nonmonotonic logic: two sides of the same coin? Logics in AI. Springer, New york, pp 52–54
45. Hintikka J (1962) Knowledge and belief, an introduction to the logic of the two notions. Cornell University Press, Ithaca, London
46. Liu F (2008) Changing for the better: preference dynamics and agent diversity. Ph.D. thesis, ILLC, University of Amsterdam.
47. Makinson D (2005) Bridges from classical to nonmonotonic logic. King's College, London
48. Makinson D, Gärdenfors P (1989) Relations between the logic of theory change and nonmonotonic logic. In: Fuhrmann A, Morreau M (eds) The logic of theory change. Lecture notes in computer science, vol 465. Springer, Berlin, pp 185–205.
49. Mares ED (1996) Relevant logic and the theory of information. Synthese 109(3):345–360
50. Mares ED, Meyer RK (2001) Relevant Logics. In: Goble L (ed) The Blackwell guide to philosophical logic. Wiley-Blackwell, Oxford
51. Muskens R, van Benthem J, Visser A (2011) Dynamics. In: van Benthem JFAK, ter Meulen A (eds) Handbook of logic and language. Elsevier, Amsterdam, pp 607–670
52. Nute D, Cross CB (2001) Conditional logic. Handbook of philosophical logic, vol 4. Kluwer Academic Pub, Dordrecht, pp 1–98
53. Parikh R, Ramanujam R (2003) A knowledge based semantics of messages. J Logic Lang Inform 12(4):453–467
54. Perry J, Israel D (1990) What is information? In: Hanson PP (ed) Information, language, and cognition 1. Columbia Press, Vancouver
55. Ramsey F (1929) Philosophical papers. In: General propositions and causality. Cambridge University Press, Cambridge.
56. Restall G (1996) Information flow and relevant logics. Logic, language and computation: the 1994 Moraga proceedings. CSLI, Stanford, pp 463–477
57. Restall G (2000) An introduction to substructural logics. Routledge, London
58. Restall G (2006) Relevant and substructural logics. Handbook of the history of logic, vol 7. Elsevier, London, pp 289–398
59. Routley R, Meyer RK (1972) The semantics of entailment-ii. J Philos Logic 1(1):53–73

60. Routley R, Meyer RK (1972) The semantics of entailment-iii. J philos logic 1(2):192–208
61. Routley R, Meyer R (1973) The semantics of entailment. Stud Logic Found Math 68:199–243
62. Routley R, Plumwood V, Meyer RK (1982) Relevant logics and their rivals. Ridgeview Publishing Company, Atascadero
63. Urquhart AI (1971) Completeness of weak implication. Theoria 37(3):274–282
64. Urquhart A (1972) A general theory of implication. J Symbolic Logic 37(443):270
65. Urquhart A (1972) Semantics for relevant logics. J Symbolic Logic, pp 159–169.

Chapter 34
Arrows Pointing at Arrows: Arrow Logic, Relevance Logic, and Relation Algebras

J. Michael Dunn

> *Time flies like an arrow; fruit flies like a banana.*
>
> Groucho Marx

Abstract Richard Routley and Robert K. Meyer introduced a ternary relational semantics for various relevance logics in the early 1970s. Johan van Benthem and Yde Venema introduced "arrow logic" in the early 1990s and about the same time I showed how a variation of the Routley–Meyer semantics could be used to provide an interpretation of Tarski's axioms for relation algebras. In this paper I explore the relationships between the van Benthem–Venema semantics for arrow logics, and the Routley–Meyer semantics for relevance logic, and conclude with a comparison between van Benthem's version of the semantics for arrow logic aimed at relation algebras, and my own version of the Routley–Meyer semantics which I used to give a representation of relation algebras (but at a type level higher than Tarski's original intended interpretation of an element as a relation, for me it is a set of relations). In the process I show how van Benthem's semantics for arrow logic can be just slightly tweaked (just one additional constraint) so as to give a representation of relation algebras.

34.1 Introduction

Traditionally, logic is thought of as timeless, static, situated in Plato's world of the forms, unmoving like Rodin's statue of The Thinker. But thinkers such as Johan van Benthem have emphasized the temporal/dynamic aspects of logic.[1] This has

---

[1] van Benthem's work has been broadly influential and stands out among those working on temporal and dynamic aspects of logic. I won't try to mention many others but will content myself with Aristotle and his sea battle tomorrow, [28] (who I believe introduced the phrase "dynamic logic"), and Arthur Prior [30] for his ground breaking work on modality and temporal logic.

J. M. Dunn (✉)
School of Informatics and Computing, Department of Philosophy,
Indiana University, Bloomington, USA
e-mail: dunn@indiana.edu

A. Baltag and S. Smets (eds.), *Johan van Benthem on Logic
and Information Dynamics*, Outstanding Contributions to Logic 5,
DOI: 10.1007/978-3-319-06025-5_34, © Springer International Publishing Switzerland 2014

taken many different forms, even in van Benthem's own work, but one of the most interesting, to me at least, has been what he in [4] has labeled "arrow logic". My interest could be viewed as self-serving, because it stems from strong connections that I see between arrow logic and the Anderson and Belnap's relevance logic.

Arrow logic has a somewhat complex history (with many contributors) but Johan van Benthem and Yde Venema seem to be the originators. This paper will focus on [5], but it will not examine the iteration operator which is what is added to arrow logic to make it *dynamic* arrow logic.[2] Related material may also be found in Chap. 8 of [6]. I have also found [39] very useful. The first occurrence of arrow logic seems to be in Venema's dissertation [38]. His report [37] is highly relevant to 34.5 below and also to my [15], but I only became aware of this after the present paper was at press.

The semantics of both arrow logic and relevance logic find themselves within the general frame (pun intended) of modal logic. [21] used what has come to be called a Kripke frame to define the modal connectives of necessity $\Box$ and possibility $\Diamond$. The original idea of Kripke was to use a structure (G, K, R) where K is a nonempty set of "possible worlds," $G \in K$ is the "actual world," and R is the relation of "relative possibility" (or "accessibility") between worlds. Researchers soon saw that one did not need to pick out a distinguished world G in assessing the validity of a formula (or inference), but simply let each world take its turn, and so the idea of a frame developed as just a structure (K, R). Researchers also saw that there were various ways to interpret the components of a frame, so to give just one example K might be thought of as a set of moments in time, and R might be thought of as the temporal order. And they also saw that varying the requirements on R could become an "economic stimulus package" for modal logicians.[3]

In this paper I explore the relationships between the van Benthem–Venema semantics for arrow logics, and the Routley–Meyer semantics for relevance logic, and conclude with a comparison between van Benthem's version of the semantics for arrow logic aimed at relation algebras, and my own version of the Routley–Meyer semantics aimed at the same target.

34.2 Arrow Logic

Arrow logic arises from the very nice idea of constructing a logic that captures the "modal" properties of graph arrows (as opposed to worlds). As I have already suggested one of the astonishing things about the Kripke semantics for modal logic is that the "worlds" have taken on various forms: times, information states, spatial

---

[2] Bimbó and Dunn's [7], and Chaps. 4 and 7 of [8] might be useful for this purpose.

[3] Kripke's work grew in a kind of hothouse environment around 1960 when many researchers more or less independently came up with ideas closely related to what many of us still call the Kripke semantics for modal logic. See [11] and [18] for fascinating history.

temporal coordinates, even impossible and/or incomplete worlds, etc. Arrow logic is a multi-modal logic—i.e., a modal logic having several different modalities. It also is structurally different from basic modal logic in that while one of those modalities ("converse") is unary and can be modeled with a binary accessibility relation, another of those modalities uses a binary relation ("composition") which requires a ternary accessibility relation. And to complete the picture there is even a nullary operation ("identity") which requires a unary relation (property, or set). Arrow diagrams (directed graphs) might seem a somewhat narrow, even nerdish application. But arrow logic is not just the logic of white boards, it is also about what those diagrams stand for, and that can be essentially anything relational. In particular as [25, p. 5] aptly puts it, "Arrow logic is the modal logic of transitions," and transitions are certainly a key part of actions.

An arrow frame is a structure (A, C, F, I), where A is a non-empty set, $C \subseteq A^3$, $F \subseteq A^2$, $I \subseteq A$.[4] van Benthem calls the members of A "arrows," and thinks of $Cxyz$ as meaning "arrow x is the composition of y and z," Fxy as meaning "arrow y is the converse (flip) of x," and Ix as x is an "identity arrow". See the diagram:

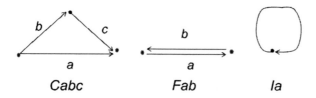

$$Cabc \qquad\qquad Fab \qquad\qquad Ia$$

An important point is that an arrow a is not just an ordered pair (γ, δ). There can be more than one arrow a that leads from point γ to point δ,[5] and not every ordered pair (γ, δ) is associated with an arrow. Arrow frames can be interpreted as sets of ordered pairs, but they are more abstract that that.

The basic language for an arrow logic contains the unary connectives of negation $\neg$ and $\smile$, and the binary connectives of conjunction $\wedge$ and $\bullet$. The connective $\smile$ is to be understood as a kind of converse, and the connective $\bullet$ is to be understood as a kind of multiplication of relations, as we shall explain. A model adds a valuation V to a frame which assigns a subset $V(p)$ of A to each propositional letter p. This then induces the following inductive definition as to when a sentence holds at an arrow x.

$x \models p$ iff $x \in V(p)$
$x \models \neg\varphi$ iff not $x \models \varphi$

---

[4] van Benthem actually uses C^3, R^2, and I^1. We do not bother to use the superscripts to denote degree, and we use F for "flip" instead of R ("reverse"?) because we do not want any confusion with the Routley–Meyer ternary relation R.

[5] A cautionary and picky note regarding the reading in abstract arrow logic of $Cxyz$: x is not *the* composition of y and z. There can be more than one arrow from the beginning of y to the end of z. And similarly for Fxy: y is not *the* converse of x—there can be more than one arrow from the end of y to the beginning of x.

$x \models \varphi^\smile$ iff $\exists y$ such that Fxy and $y \models \varphi$

$x \models \varphi \wedge \psi$ iff $x \models \varphi$ and $x \models \psi$

$x \models \varphi \bullet \psi$ iff $\exists y, z$ such that $Cxyz$ and $y \models \varphi$ and $z \models \psi$

$x \models Id$ iff Ix.

van Benthem [5] gives axioms for the minimal arrow logic (we call it MAL) for these frames:

$(\varphi_1 \vee \varphi_2) \bullet \psi \leftrightarrow (\varphi_1 \bullet \psi) \vee (\varphi_2 \bullet \psi)$

$\varphi \bullet (\psi_1 \vee \psi_2) \leftrightarrow (\varphi \bullet \psi_1) \vee (\varphi \bullet \psi_2)$

$(\varphi_1 \vee \varphi_2)^\smile \leftrightarrow \varphi_1^\smile \vee \varphi_2^\smile.$

Implicitly van Benthem is assuming the usual definitions of $\vee$, $\rightarrow$, and $\leftrightarrow$ in terms of conjunction and negation, and some usual complete set of axioms for classical propositional calculus, as well as the usual rule of modus ponens for $\rightarrow$. Completeness for MAL follows from Jónsson and Tarski's [19, 20] representation of Boolean algebras with operators, since identifying provably equivalent formulas in the usual way we obtain a Boolean algebra with operators corresponding to $\bullet$ and $\smile$.

van Benthem [5, p. 17] then goes on to note that:

(2.1) $\neg(\varphi^\smile) \rightarrow (\neg\varphi)^\smile$ corresponds to the frame condition $\forall x \exists y Fxy$,

(2.2) $(\neg\varphi)^\smile \rightarrow \neg(\varphi^\smile)$ corresponds to the frame condition $\forall x, y, z (Fxy$ and Fxz implies $y = z$).

He notes that these two conditions together make the binary relation F into "a unary function r of "reversal," and that

(2.3) $\varphi^{\smile\smile} \leftrightarrow \varphi$ corresponds to the idempotence of $r : r(r(x)) = x$.

He then says "Let us assume this much henceforth in our arrow frames." We shall call these standard arrow frames and the logic they determine standard arrow logic (SAL). van Benthem goes on to discuss some further correspondences and says: "Obviously, there are many further choices here, and 'Arrow Logic' really stands for a family of modal logics, whose selection may depend on intended applications." Certainly among the further choices one that cries out for attention is a set of choices that will characterize relation algebras. van Benthem in fact goes on to list 5 more correspondence principles that concern axioms for relation algebras:

	Axiom	Corresponding Frame Condition
(2.4)	$(\varphi \bullet \psi)^\smile = \psi^\smile \bullet \varphi^\smile$	$\forall xyz : Cxyz \Rightarrow Cr(x)r(z)r(y)$
(2.5)	$\varphi \bullet \neg(\varphi^\smile \vee \psi) \rightarrow \neg\varphi$	$\forall xyz : Cxyz \Rightarrow Czr(y)x$
(2.6)	$Id \rightarrow Id^\smile$	$\forall x : Ix \Rightarrow Ir(x)$
(2.7)	$Id \bullet \varphi \rightarrow \varphi$	$\forall xyz : (Iy$ and $Cxyz) \Rightarrow x = z$
(2.8)	$\varphi \bullet (\psi \bullet \chi) \leftrightarrow (\varphi \bullet \psi) \bullet \chi$	$\exists x(Cxyz$ and $Cyuv) \Leftrightarrow \exists w(Cxuw$ and $Cwvz)$.

van Benthem [5, p. 19] expresses some hesitation about this last axiom, and [6] does the same, but I think more clearly, or at least more quotably:

Correspondence analysis reveals a natural border line in what is expressed by principles like the above. Some of them are purely universal, making no demands on the supply of arrows, whereas others are existential. The former, rather than the latter, seem to form the true logical core of any field.

I will say more about this in the last section of the paper.

34.3 Routley and Meyer's Semantics for Relevance Logic

We will be comparing van Benthem's arrow logic with the Routley–Meyer semantics for relevance logic. Routley and Meyer published "Semantics of Entailment I, II, III" [33–35] in the years 1972 and 1973.[6] Routley and Meyer use a frame $(K, R, *, 0)$, K is a set, $0 \in K$, $R \subseteq K^3$, and $*$ is a unary operation on K of period two, i.e., for $a \in K$, $a^{**} = a$ a is called an "involution." Routley and Meyer call the members of K "set ups," and put various constraints on a frame, but we shall not explore these in detail now. We do though note that they defined a binary relation $a \le b$[7] as $R0ab$ and gave R properties that assure that $\le$ is a quasi-order (reflexive and transitive).

Routley and Meyer assume a basic language that contains the unary connective $\sim$ of De Morgan negation (which is weaker than the Boolean negation $\neg$ used in arrow logic), and the binary connectives of conjunction $\wedge$, disjunction $\vee$, and (relevant) implication $\rightarrow$. They take a valuation v to be a function that assigns to each pair (p, a) (p an atomic sentence) a member of $\{T, F\}$. From this they inductively define a function I that assigns to each pair (φ, a) (φ an arbitrary formula) a member of $\{T, F\}$.

But there is an important restriction. They require the Hereditary Condition on atomic sentences: if $a \le b$ and $v(p, a) = T$, then $v(p, b) = T$. It can then be shown by induction that it extends as well to compound formulas. The Hereditary Condition is needed to show that $v(\varphi \rightarrow \varphi, 0) = T$. For the sake of a ready comparison to the clauses for arrow logic above, we assign to each atomic sentence a set $V(p) \subseteq K$, and for an arbitrary formula φ we write $x \models \varphi$ rather than $I(\varphi, x) = T$. The Hereditary Condition then just amounts to requiring that $V(p)$ is a cone, i.e., if $a \in V(p)$ and $a \le b$, then $b \in V(p)$.

The Routley–Meyer evaluation clauses can now be stated as follows:

$x \models p$ iff $x \in V(p)$
$x \models \sim \varphi$ iff not $x^* \models \varphi$
$x \models \varphi \wedge \psi$ iff $x \models \varphi$ and $x \models \psi$
$x \models \varphi \vee \psi$ iff $x \models \varphi$ or $x \models \psi$
$x \models \varphi \rightarrow \psi$ iff $\forall y, z$, if $Rxyz$ and $y \models \varphi$ then $z \models \psi$.

---

[6] And "Semantics of Entailment IV" [36] written in 1972 but published in 1982. As with the "Kripke semantics," there were a lot of "competitors" in the early 1970s with essentially the same, or very similar ideas, including (in alphabetical order) Charlewood, Fine, Gabbay, Maksimova, and Urquhart. I believe the label "Routley–Meyer" has stuck because of their persistence and skill in exploring and promoting this framework.

[7] They actually use the notation $<$ but because the relation turns out to be reflexive it has become standard to use $\le$.

There has been some controversy and complaint over the abstract character of the Routley–Meyer semantics. See [10] and somewhat ironically [3, p. 995], which says:

> There appears to be an over-application of the Henkin method in intensional logic, generating facile possible-worlds semantics. For instance, could it be that the Routley semantics lacks explanatory power, due to lack of potential falsification?

I do not see the problem with very general mathematical theories, and in fact the Jónsson-Tarski [19, 20] theory of "Boolean algebras with operators" ends up having many different applications, some of which they foresaw, e.g., Tarski's cylindric algebras, and others of which were only seen with hindsight, e.g., the Kripke semantics for modal logic. Gaggle theory (see [8, 13]) was constructed intentionally to abstract both the Routley–Meyer semantics and the Kripke semantics (and the representation theorem for Boolean algebras with operators of [19, 20] so as to give a very general approach to the semantics of substructural logics (including relevance logics). And I think, and here is the irony, that van Benthem was careful not to embed any particular logic into arrow logic. But I completely agree with van Benthem if he is suggesting that there must be concrete interpretations of the abstract semantics in order to make the general theory interesting.

One way to view $Rxyz$ is as "relative relative possibility" (the repetition is not a typo). I in fact suggested this to Routley and Meyer after reading an early draft of their first paper on the semantics of entailment. They say ([35], p. 206):

> Consider a natural English rendering of Kripke's binary R. xRy "says that 'world' y is possible relative to world x." An interesting ternary generalization is to read $xRyz$ to say that 'worlds' y and z are compossible (better, maybe, compatible) relative to x. (The reading is suggested by Dunn.)

Another way to view the ternary relation grows out of an early suggestion by Peter Woodruff that it be interpreted as an indexed set of binary accessibility relations. We can replace the ternary relation R on a Routley–Meyer frame with a function assigning to each world x a binary accessibility relation R_x and if we have a multimodal logic with a necessity operator $\Box_x$ for each $x \in K$ we can define $y \models \Box_x \varphi$ iff $\forall z(yR_xz \Rightarrow z \models \varphi)$. This is very reminiscent of Pratt's dynamic logic except Pratt had two types: indices were programs, and worlds were worlds. But here we have only one type which can play the different roles of indexing the necessity operator or being a world (evaluation point).

I have more recently exploited Woodruff's construction as a duality between data (static) and computation (dynamic). The basic idea is that propositions can be viewed as either sets of states or as the set of actions these states index. See Dunn [16] for a general explanation, and [14, 15, 17] for concrete applications. Barwise [1] had a formally similar idea, which was of two 'sites' being connected by a 'channel.' He did not rule out the case where channels might also be sites. See Restall [31] for more on the relationship between Barwise's channels and the Routley–Meyer ternary accessibility relation. See [2] for a variety of other interpretations of the ternary relation.

34.4 The Obvious Similarities, and Apparent Differences

We start with the frames. Both (A, C, F, I) and $(K, R, *, 0)$ have a non-empty set A, K and a ternary relation C, R on that set. Both F and $*$ are binary relations—though the last is a function. But we saw that in a standard arrow frame F is required to be an involution and can be denoted by the function symbol τ. We shall say more about that. This leaves I and 0 to correspond to each other. We shall say more about that too. Perhaps the most striking similarity (and the biggest difference from the usual Kripke semantics for modal logic) is that both languages have a binary connective, and both use a ternary frame in the truth clause for the connective, though of course they use it in different ways.

There are a number of striking differences, both in the logical connectives, and in the frames. We shall examine these one by one and show, using known results, how in each case the difference can be reinterpreted so as to not make a difference at all. Incidentally, when I said "known results" I might have put it more clearly by saying "scattered results." In fact I think that there are very few of us who would know all of the "relevant" literature (pun intended). I was fortunate to have grown up in this environment, which is one of the reasons that I am writing this paper. I now present what the American talk show host David Letterman would call his "Top Ten List," and in each case provide a "fix."

Difference 1. van Benthem formulates arrow logic without disjunction, and with connectives for the material conditional and material equivalence not present in relevance logic.

Fix 1. $\varphi \vee \psi$ can be defined in the usual way as $\neg(\neg\varphi \wedge \neg\psi)$, and once (Fix 2) we add Boolean negation to the Routley–Meyer framework the material conditional and material equivalence can be defined in this expanded language of relevance logic in the usual way.

A crucial difference of vocabulary has to do with the two kinds of negation. The van Benthem negation is "Boolean" whereas the Routley–Meyer negation is merely "De Morgan" and utilizes the famous $*$ operator to do a kind of "bait and switch" requiring that φ be false, not at x, but rather at x^*.

Difference 2. The Routley–Meyer framework is missing a Boolean negation.

Fix 2. Fortunately there is a well-known way to add Boolean negation to relevance logic. Meyer [26] has shown how to convert an arbitrary one of their models into an equivalent "Boolean" model. The main trick is to add a new element $0'$ and define a new $*'$ that is like $*$ but with $0'^{*'} = 0'$, and a new relation R' that is like R but requiring $R'0'ab$ iff $R'a0'b$ iff $a = b$, and also $R'ab0'$ iff $a = b^*$. We can now add a Boolean negation $\neg$ with the evaluation clause

$x \models \neg\varphi$ iff not $x \models \varphi$.

We still have the Hereditary Condition since $\leq$ becomes simply $=$.

Difference 3. The van Benthem framework is missing a De Morgan negation.

Fix 3. Meyer has suggested adding in effect $*$ as a unary connective in relevance logic, defining $x \models \bigstar\varphi$ iff $x^* \models \varphi$, remarking that in effect it already exists in Boolean relevance logic since we can define it as $\neg \sim \varphi$. He then observes that

De Morgan negation need not be taken as a primitive but can be defined as $\sim \varphi = \neg \bigstar \varphi$. Star is much like converse, so the same trick allows us to define a De Morgan negation connective within arrow logic, $\sim \varphi = \neg(\varphi^{\smile})$.

Difference 4. The van Benthem framework is missing a relevant implication connective.

Fix 4. It has been known since my 1966 dissertation (cf. [12]) that instead of having $\rightarrow$ as primitive, one can instead take as primitive a binary connective o that I called "cotenability" at the time, but which has since been called "intensional conjunction" and now most commonly "fusion".[8] One can then define $\varphi \rightarrow \psi = \sim (\psi o \sim \varphi)$. One can do the same thing with van Benthem's • now using the De Morgan negation we found hidden in his framework with Fix 3. One can add a clause to the definition of a model that $y \models \varphi \rightarrow \psi$ iff $\forall z, x$, if $Cxyz$ and $z \models \varphi$ then $x \models \psi$, thus transliterating the Routley–Meyer clause.

Difference 5. The Routley–Meyer framework is missing the • connective.

Fix 5. You guessed it. We can do the reverse of Fix 4 and define $\varphi \bullet \psi = \sim (\psi \rightarrow \sim \varphi)$.

Difference 6. Routley–Meyer require that $V(p)$ is a cone whereas van Benthem does not.

Fix 6. Fix 2 has already taken care of this since it makes the Hereditary Condition vacuous.

Difference 7. A van Benthem frame contains the binary relation F whereas a Routley–Meyer frame does not.

Fix 7. Define Fxy iff $x^* = y$. It is easy to see that $x \models \bigstar \varphi$ iff $\exists y$ such that Fxy and $y \models \varphi$, i.e., $x^* = y$ and $y \models \varphi$. i.e., $x^* \models \varphi$.

Difference 8. A Routley–Meyer frame contains the operator * whereas a van Benthem frame does not.

Fix 8. This is in fact a real but small difference. As we saw above van Benthem seems to accept frame conditions which make F an idempotent operation, and these in effect make F an operation, which he denotes as r. More importantly it is well-known that the $*$ operator can be replaced with a binary "compatibility" relation (see [14, 24, 32]). We might as well denote this by F.

Difference 9. An arrow frame has the set $I \subseteq K$, whereas a Routley–Meyer frame has only the element $0 \in K$.

Fix 9. It has also become common to replace the single 0 with a set of "zeroes" Z.[9]

Difference 10. The language of arrow logic has a 0-ary connective Id whereas the language of relevance logic does not.

Fix 10. It has become common to consider relevance logics conservatively extended by the propositional constant t, which is interpreted as denoting the set Z.

---

[8] In the linear logic community it is "multiplicative conjunction."

[9] I don't know the when/where/who about how this originated, but I know that for me this was important in the representation of algebras of relevance logic, because the set Z corresponds to the identity element. See e.g., [15].

34.5 Relation to Relation Algebras, or Hindsight is Better than 20–20 Vision

Dunn [15] contains a representation of relation algebras using Routley–Meyer frames.[10] It appeared in an edited volume in memory of Alonzo Church, which began as a volume to be published in 1993 in honor of Alonzo Church on the occasion of his 90th birthday. Church was still alive back in 1992 when I turned in my contribution. I published it informally in the Indiana University Logic Group Preprint Series [14]. I mention these dates to show that I came up with the representation at about the same time that van Benthem and Venema were creating arrow logic. They had similar motivations in wanting to use the language of relation algebras but to interpret a variable as a set of frame elements, that can then themselves be viewed as relations. As I said in [15]:

> An ideal representation of relation algebras would send the elements into relations, but [22] has shown that such a representation is impossible. It turns out that we are in one sense close to such an ideal representation, but we are off a type level. Elements are **not** carried into relations, but into *sets of relations*. ... Peter Woodruff suggested early on that the ternary relation $R\alpha\beta\gamma$ that arises in the Routley–Meyer semantics for relevance logic should be viewed as an "indexed" binary accessibility relation ... An equivalent idea, with slightly different metaphysical overtones is to view β as itself a binary relation. More accurately we view β as something like the Fregean "object correlate" of a binary relation. By thus not literally identifying β with a set of ordered pairs, but rather thinking of it as determining a set of ordered pairs, we avoid unnecessary set theoretic problems about what happens when one has e.g., $R\beta\beta\beta$.

In the Introduction of [15] I cited [4] (as well as [29] and [1]) as a kind of advertisement for relation algebras, saying that they "have received renewed attention in recent years because of an interest in a more dynamic conception of logic, which incorporates 'actions', i.e., relations between states." But I admit to not having seen the close connection to arrow logic at the time.

But I want now to explain the relationship between the Routley–Meyer semantics I adapted for relation algebras, and the arrow-logic approach to relation algebras. I should mention that my aim was to develop the frames I needed by stages so as to make clear their relationship to Routley and Meyer's work. Having this different motivation led to frame requirements that do not always match the requirements that van Benthem suggests for an arrow logic valid in relation algebras. But as we shall see they are equivalent, with one very small difference. The frames I used were structures $(U, R, \smile, Z)$. One immediate difference then is that in my representation of relation algebras I used a unary operator for converse, whereas arrow logic uses a binary relation instead. But as I suggested above this can be finessed by redoing the Routley–Meyer semantics with a binary relation of compatibility. In [15] it was proved that every relation algebra is isomorphic to a relation algebra defined

---

[10] The reader might also want to look at [9, 23, 27] for other relationships between relevance logic and relation algebras.

on subsets of an *associative assertional frame* on which is defined an *involution* satisfying *tagging* and *antilogism*. I briefly explain the italicized terms:

Definition 34.1 (i) A frame is *associative* iff $\exists x(Rabx$ and $Rxcd) \Leftrightarrow \exists x(Raxd$ and $Rbcx)$. In a notation introduced by Routley–Meyer this is often written as $R^2(ab)cd \Leftrightarrow R^2a(bc)d$, which makes transparent why such a frame is called associative.
(ii) A frame is *assertional* iff $\exists z \in Z(Rzab)$ iff $a = b$ iff $\exists z \in Z(Razb)$.
(iii) An *involution* is a unary operation on U such that for all $a \in U, a^{\smile\smile} = a$.
(iv) An involution satisfies *tagging* iff $Rabc \Rightarrow Rb^{\smile}a^{\smile}c^{\smile}$.
(v) And finally an involution satisfies *antilogism* iff $Rabc \Leftrightarrow Rc^{\smile}ab^{\smile}$.

Theorem 34.1 *Relation algebra frames satisfying conditions* (i)–(v) *as defined above are equivalent to arrow frames satisfying van Benthem's conditions* (2.1–2.8) *plus* $\exists z(Iz$ *and* $Cxzx)$.

Proof The proof will take up the rest of this section. There are some small but notable differences between these frame conditions and those of van Benthem listed above. We have already discussed how van Benthem's (2.3) is equivalent to taking $\smile$ to be an involution. And it is an easy "transliteration" to see that Associativity is the same as van Benthem's frame condition (2.8), and that Tagging is the same as his frame condition (2.4).

So we are left with his frame conditions (2.5–2.7).

(2.5) $Cxyz \Rightarrow Czr(y)x$ transliterates with $Ryzx \Rightarrow Ry^{\smile}xz$. This last can be shown using first Tagging, and then Antilogism/Involution: $Ryzx$ implies $Rz^{\smile}y^{\smile}x^{\smile}$ implies $Ry^{\smile}xz$.

(2.6) $Ix \Rightarrow Ir(x)$ transliterates as $z \in Z \Rightarrow z^{\smile} \in Z$. Assume $z \in Z$. Since $z = z$, it follows from the Routley–Meyer frame being Assertional that $\exists i \in Z, Rizz$. By Tagging, $Rz^{\smile}i^{\smile}z^{\smile}$, and then by Antilogism (and Involution) $Rzz^{\smile}i$. So since the frame is Assertional, $z^{\smile} = i \in Z$.

(2.7) $(Iy$ and $Cxyz) \Rightarrow x = z$. This transliterates with $(y \in Z$ and $Ryzx) \Rightarrow x = z$. This clearly follows from a frame being Assertional.

We have thus obtained (after transliteration) all of the arrow frame conditions (2.1–2.8) from the Routley–Meyer relation algebra frame conditions (i)–(v). But now what about the reverse? We can assume from what has been said already that (i), (iii), and (iv) are taken care of.

We still have to consider (ii). There are really 4 parts to a frame being Assertional:

(1) $\exists z \in Z(Rzab) \Rightarrow a = b$
(2) $\exists z \in Z(Razb) \Rightarrow a = b$
(3) $a = b \Rightarrow \exists z \in Z(Rzab)$, or equivalently $(3')$ $\exists z \in Z(Rzaa)$
(4) $a = b \Rightarrow \exists z \in Z(Razb)$, or equivalently $(4')$ $\exists z \in Z(Raza)$

It is easy to show that we can do with just (1) and (3). We show that (1) implies (2). Suppose $Razb$ with $z \in Z$. By Tagging we get $Rz^{\smile}a^{\smile}b^{\smile}$. But we have already showed that Z is closed under converse, and so by (1) $a^{\smile} = b^{\smile}$, and so by (1) $a = b$. We next show that $(4')$ follows from $(3')$. If $(3')$, then as a special case we have $\exists z \in Z$, $Rza^{\smile}a^{\smile}$ then by Tagging (and Involution) $Raz^{\smile}a$. But since Z is closed under converse $z^{\smile}$ is the desired member of Z.

As already noted, (1) is essentially the same as the transliteration of (2.7) (change $x = z$ to $z = x$), and we have shown that (2) follows from (1).

But what about (3) and (4)? We have also shown that (3) implies (4). But there seems to be no way to derive (3) or (4) from transliterations of the arrow frame conditions (2.1–2.8). And when one looks at the axiom $Id \bullet \varphi \rightarrow \varphi$ that corresponds to (2.7) one sees immediately why. There is just $\rightarrow$ and not $\leftrightarrow$. The converse axiom $\varphi \rightarrow Id \bullet \varphi$ is also true of relation algebras, and it requires $\exists z \in Z(Rzaa)$, which can easily be seen to be equivalent to (3). It seems that van Benthem was perhaps just giving the flavor of the frame conditions that correspond to the axioms of a relation algebra, rather than giving the complete list.[11] Venema [37] is explicit about giving a complete list, and does have the transliteration of (3').

We next tackle (v) and show that Antilogism can be derived from (2.5). We must show $Rabc \Leftrightarrow Rc^{\smile}ab^{\smile}$. Left to right: (2.5) says $Rabc$ implies $Ra^{\smile}cb$, but then by Tagging and Involution $Rc^{\smile}ab^{\smile}$. Right to left: Tagging (with Involution) says that $Rc^{\smile}ab^{\smile}$ implies $Ra^{\smile}cb$, and (2.5) says that this last implies $Rabc$.

Remark Define $R_i = \{(x, y) : Rxiy\}$. [14, 15] points out that this can be seen as giving a derivative representation where the subsets I of the a frame are interpreted as sets of relations, i.e., relational databases. The operations are interpreted as operations on such databases. Complement, meet, and join are just the corresponding operations on sets (as with Stone's representation of a Boolean algebra), and converse and roughly speaking relative product are interpreted pointwise in terms of the concrete relations of converse and relative product on the concrete relations R_i. The reader is referred to [14, 15] for details. The only tricky part is that the sets of relations $\{R_i\}_{i \in I}$ are required to be closed downward under inclusion, so if $i \in I$ and $R \subseteq R_i$ then $\exists j \in I$ such $R = R_j$. This works out OK with converse but we must build it into "relative product."

34.6 Conclusion

van Benthem was wise in urging that there are various interpretations of abstract arrow models. The same is true of Routley–Meyer frames, though the aim of their first implementation was to give a semantics for relevant logic. van Benthem was rightly conservative in associating any particular postulates with arrow logic, whereas Routley–Meyer were targeted on frames that gave completeness theorems for the relevance logic **R** of relevant implication and **E** of entailment (though they also studied frame requirements for related logics). We have explored transliterating the respective frames and the models on them. Once one gets used to the transliterations, the main differences between the frames for the relevance logic **R** and the frames for

---

[11] He is also missing the frame conditions (2) and (4) corresponding to $\varphi \leftrightarrow \varphi \bullet Id$, but that is ok since as we have seen (2) follows from (1), and (4) from (2). This axiom is in fact redundant in relation algebras.

relation algebras have to do with the fact that unlike the frames for relation algebras, **R** requires $Raaa$ (Total Reflexivity) and $Rabc \Rightarrow Rbac$ (Commutativity), whereas on the other side the relation algebra frames require $Rabc \Rightarrow Rb^{\smile}a^{\smile}c^{\smile}$ (Tagging), which is not a requirement on a Routley–Meyer frame for **R**. But by taking a couple of steps forward in terms of requirements on an arrow frame, and a couple of steps backwards on the requirements on a Routley–Meyer frame, we have made them meet in the middle.

This answers a question implicitly left open by van Benthem. The question is: can we represent relation algebras using arrow frames? This question was not explicitly stated, but is clearly suggested by his giving correspondences between various axioms of relation algebras and requirements on an arrow frame. By our transliterations above we have shown that the answer is yes if we only add $\exists z(Iz\&Cxzx)$—the transliteration of $\exists z \in Z, Rzxx$. The proof requires showing the existence of a maximal filter, and van Benthem's preference for universally quantified formulas (no existential quantifiers) explains his reluctance to deal with postulates such as this.[12] The existence of such maximal filters is also required for Associativity, and it is made clear in [15] how the "Squeeze Lemma" implicit in Routley–Meyer (see [12]), makes more or less automatic the fulfillment of postulates such as these (including Associativity).

Acknowledgments I thank Katalin Bimbó for her helpful comments and suggestions, and also Johan van Benthem for his suggestions of ways to improve and expand the paper. I have not had time to take as much advantage of this as I wish, van Benthem and I envisage a joint follow up to this paper.

References

1. Barwise J (1993) In: Aczel P, Israel D, Peters S (eds) Constraints, channels, and the flow of information, Situation theory and its applications, CSLI Publications (CSLI Lecture Notes 37), Stanford, CA, pp 3–27
2. Beall JC, Brady R, Dunn JM, Hazen AP, Mares E, Meyer RK, Priest G, Restall G, Ripley R, Slaney J, Sylvan R (formerly Routley) (2012) On the ternary relation and conditionality. J Philos Logic 41:595–612
3. van Benthem J (1984) Review of 'on when a semantics is not a semantics' by Copeland BJ. J Symbolic Logic 49:994–995
4. van Benthem J (1991) Language in action: categories. Lambdas and Dynamic Logic, North Holland, Amsterdam
5. van Benthem J (1994) A note on dynamic arrow logic. In: van Eijck J, Visser A (eds) Logic and information flow. The MIT Press, Cambridge
6. van Benthem J (1996) Exploring logical dynamics. The European association for logic, language and information. CSLI Publications and FoLLI, Stanford, CA
7. Bimbó K, Dunn JM (2005) Relational semantics for Kleene logic and action logic. Notre Dame Journal of Formal Logic 46:461–490

---

[12] van Benthem has told me that this preference has to do with wanting to reduce computational complexity and moreover to avoid undecidability.

8. Bimbó K, Dunn JM (2008) Generalized Galois logics. Relational semantics of nonclassical logical calculi. CSLI Publications (CSLI Lecture, Notes, no. 188), Stanford, CA

9. Bimbó K, Dunn JM, Maddux R (2009) Relevance logic and relational algebras. The review of symbolic logic, vol 2, pp 102–131

10. Copeland BJ (1979) On when a semantics is not a semantics: some reasons for disliking the Routley-Meyer semantics for relevance logic. J Philos Logic 8:399–413

11. Copeland BJ (2002) The genesis of possible world semantics. J Philos Logic 31:99–137

12. Dunn JM (1985) Relevance logic and entailment. In: Gabbay D, Guenthner F, Reidel D (eds) Handbook of philosophical logic, vol 3, 1stedn. Kluwer Academic Publishers, Dordrecht, Holland, pp 117–224. Updated with joint author Restall G (2002) Relevance logic. In: Gabbay D, Guenthner F (eds) Handbook of philosophical logic, vol 6, 2nd edn. Kluwer Academic Publishers, Dordrecht, pp 1–128

13. Dunn JM (1990) Gaggle theory: an abstraction of Galois connections and residuation with applications to negation and various logical operators, In: Logics in AI, Proceedings of European workshop JELIA 1990. Lecture Notes in Computer Science, no. 478. Springer, Berlin, pp 31–51

14. Dunn JM (1993) Representation of relation algebras using Routley-Meyer frames, version of Dunn (2001). Informally published in Indiana University logic group preprint series, IULG-93-28

15. Dunn JM (2001a) Representation of relation algebras using Routley-Meyer frames. In: Anderson CA, Zelëny M (eds) Logic, meaning and computation: essays in memory of Alonzo Church. Kluwer, Dordrecht, pp 77–108 (Informally published as Dunn (1993))

16. Dunn JM (2001b) Ternary relational semantics and beyond. Logical Studies 7:1–20

17. Dunn JM, Meyer RK (1997) Combinators and structurally free logic. Logic J IGPL 5:505–537

18. Goldblatt R (2006) Mathematical modal logic: a view of its evolution. In: Gabbay D, Woods J (eds) Handbook of the history of logic, vol 6. Elsevier, Amsterdam, pp 1–98

19. Jónsson B, Tarski A (1951) Boolean algebras with operators part I. Am J Math 73:891–939

20. Jónsson B, Tarski A (1951) Boolean algebras with operators part II. Am J Math 74:127–162

21. Kripke S (1963) Semantical analysis of modal logic II. Zeitschrift für Mathematische Logik und Grundlagen der Mathematik 9:67–96

22. Lyndon RC (1950) The representation of relations algebras. Ann Math 51:707–729

23. Maddux R (2010) Relevance logic and relational algebras. The review of symbolic logic, vol 3, pp 41–70

24. Mares E(1995) A star-free semantics for R. J Symbolic Logic 60:579–90

25. Marx M (1995) Algebraic Relativization and Arrow Logic. Institute for Logic, Language and Computation (ILLC Dissertation Series), University of Amsterdam

26. Meyer RK (1974) New axiomatics for relevance logics I. J Philos Logic 3:53–86

27. Mikulas S (2009) Algebras of relations and relevance logic. J Logic Comput 19:305–321

28. Pratt V(1976) Semantical considerations on Floyd-Hoare logic. In: Proceedings of the 17th annual IEEE symposium on foundations of computer, science, pp 109–121

29. Pratt V (1991) Action logic and pure induction. In: van Eijck V (ed) Logics in AI, Proceedings of European workshop JELIA 1990, Lecture Notes in Computer Science, no. 478. Springer, Berlin, pp 31–51

30. Prior AN (1957) Time and modality. Clarendon Press, Oxford

31. Restall G (1995) Information flow and relevant logics. In: Seligman J, Westerståhl D (eds) Logic, language, and computation, CSLI Publications (CSLI Lecture Notes, no. 58), Stanford, CA, pp 463–477

32. Restall G (2000) Defining double negation elimination. Logic J IGPL 8:853–860

33. Routley R, Meyer RK (1972a) The semantics of entailment–II. J Philos Logic 1:53–73

34. Routley R, Meyer RK (1972b) The semantics of entailment–III. J Philos Logic 1:192–208

35. Routley R, Meyer RK (1973) The semantics of entailment. In: Leblanc H (ed) Truth, syntax and modality, Proceedings of the Temple University conference on alternative semantics. Amsterdam, North Holland, pp 199–243

36. Routley R, Meyer RK (1982) The semantics of entailment–IV: E, Π', and Π''. In: Routley R, Meyer RK, Plumwood V, Brady R (eds) Relevant logics and their rivals, Part I, The basic philosophical and semantical theory. Ridgeview Publishing Company, Atascadero, CA, Appendix 1:407–424
37. Venema Y (1989) Two-dimensional modal logics for relational algebras and temporal logic of intervals. ITLI Prepublication Series LP-89-03, Institute for Logic, Language, Information, University of Amsterdam
38. Venema Y (1991) Multi-dimensional modal logic. Doctoral dissertation, Institute for Logic, Language, Competition, University of Amsterdam
39. Venema Y (1996) A crash course in arrow logic. In: Marx M, Pólos L, Masuch M (eds) Arrow logic and multi-modal logic. The European association for logic, language and information. CSLI Publications and FoLLI, Stanford, CA, pp 3–61

Chapter 35
Situation Theory Reconsidered

Jeremy Seligman

Abstract We recall a largely forgotten intellectual project: that of providing a formal theory of situations that does justice to informal ideas about situations and information flow with the 'situation theory' community of the late 1980s and early 1990s. Instead of defending specific desiderata, and in the spirit of Barwise's 'Branch Points', we record some difficulties that defined the project by posing a series of twelve questions. Drawing on the theory of channels and information flow (Barwise and Seligman, late 1990s), with some modifications and extensions, we provide a version of situation theory that answers some of these questions. One of the main extensions is to allow probabilistic constraints. We also consider a more recent proposal by van Benthem to capture many of situation theory's insights using a modal logic closely related to dependency logic and use this as an alternative but comparable way of answering our questions.

> Zuang Zhou was wandering when he saw a peculiar kind of magpie. 'What kind of bird is that!' he exclaimed. 'Its wings are enormous but they get it nowhere; its eyes are huge but it can't even see where it's going!' Then he hitched up his robe, strode forward, cocked his crossbow and prepared to take aim. As he did so, he spied a cicada that had found a lovely spot of shade. Behind it, a praying mantis, stretching forth its claws, prepared to snatch the cicada. The peculiar magpie was close behind, ready to make off with the praying mantis. Zhuang Zhou, shuddering at the sight, said, 'Ah! - things do nothing but make trouble for each other—one creature calling down disaster on another!' He threw down his crossbow, turned about and hurried from the park, but the park keeper [taking him for a poacher] raced after him with shouts of accusation.   'Mountain Tree' *Zhuangzi*[1]

In the late 1980s and early 1990s, at the time I was a graduate student, a research project out of Stanford, specifically the newly created Centre of Studies in Language and Information (CSLI), inspired and promoted by Jon Barwise and John Perry,

---

[1] The translation from this third century B.C. Chinese text is abridged with modifications from *Thesaurus Linguae Sericae* http://tls.uni-hd.de/home_en.lasso

J. Seligman (✉)
The University of Auckland, Auckland, New Zealand
e-mail: j.seligman@auckland.ac.nz

A. Baltag and S. Smets (eds.), *Johan van Benthem on Logic*   895
and Information Dynamics, Outstanding Contributions to Logic 5,
DOI: 10.1007/978-3-319-06025-5_35, © Springer International Publishing Switzerland 2014

created a minor whirlpool of intellectual excitement that influenced me greatly and many of my peers. The project was that of Situation Theory: the attempt to provide a formal theory of information based on the concept of a 'situation'. The above quotation captures for me the essential insight that lay behind this work: when reasoning and acting in the world we do so within a limited context; when we change perspective different information is available and this makes a significant difference. Our conceptual boundaries can always be expanded, or even contracted, and since logic, inference and rational activity in general takes place within a conceptually bounded space, an appreciation of the fragility and versatility of our concepts is essential for an understanding of the capacity and application of reason.

van Benthem has had a persistent interest in the continuation of these ideas, has written on the subject on several occasions [13–15] and has prompted me, on many occasions, to do more to represent this line of thought to a wider community. I have not succeeded to meet his expectations, partly because my own interests have shifted, to some extent, but mostly because it is very hard. 'Situation Theory' (henceforth: ST) is somewhat of a misnomer. There never was a theory in any very precise sense of the word, only a more-or-less shared idea about what such as theory should achieve—a vague set of desiderata together with a conviction that there should be nothing vague about it. ST should provide a set of conceptual primitives on the basis of which the older absolute metaphysics and theory of language and logic, would be replaced by an alternative that took seriously the essential role that context, or 'situatedness' plays in our thinking. Jon Barwise, in particular, had a vision of ST in which 'theory' should be understood in the sense of 'set theory'. It would serve as a foundation for the information sciences in the way that set theory serves as a foundation for mathematics.

Yet there was also a revisionary aspect to the project. At the time the dominant approaches to providing a systematic semantic theory were those of Donald Davidson (mostly within the philosophical community) and Richard Montague (within the logico-linguistic community). While differing in important philosophical details, these approaches shared a global perspective on language and its relationship to non-linguistic reality, in which 'reference' and 'truth' play a central role. Language refers to certain aspects of reality in a largely context-independent way, allowing us to describe it truly or falsely, and logic is a matter of monitoring which inferences are sure to be truth-preserving, notwithstanding a serious interest in context by such philosophers as David Kaplan and John Perry, who drew our attention to those aspects of language, such as demonstratives and indexicals, that depend essentially on the context of use. The ST project aimed to push these insights further, inserting context into the picture at every point. Reference occurs in and is influenced by contextual factors, and what we take to be true descriptions are never true simpliciter but true *about* some specific part of reality.[2] Or to put it in the jargon of the time, the concepts

---

[2] John Barwise later came to associate this essential aboutness of descriptive language with the views of the English philosopher J. Austin, expressed mostly clearly in [3] and called the resulting notion of proposition 'austinian'. This was an essential ingredient of Barwise and Etchemendy's account on the Liar Paradox in [8].

of reference and truth need to be 'situated'. So, instead of propositions, which are true or false by definition, the focus of situation theory should be 'infons'—units of information - that may or may not be 'supported' by small parts of the world, known as 'situations'. Likewise, in place of universally valid laws of logic and natural science, ST aimed to make a place for 'constraints' that held locally but not necessarily everywhere. Even the basic division of the world into objects, properties and relations, was taken to be relative to a 'scheme of individuation' - something that we impose on the world—allowing for different ways of doing this.

There were several fully articulated theories that made some progress towards this goal. Perhaps the most well-known is the use of non-well-founded set theory, as developed by Peter Aczel [1] and others, to model situations as sets of 'infons': the basic units of information. The locality of a situation is thereby understood in a purely informational sense: the information supported by a situation is just the information it contains, and its identity is determined extensionally: distinct situations are distinguished by supporting different infons. The use of a universe of sets as a general modelling tool and, in particular, the use of the $\in$ relation to represent a 'part' or 'component' is unsurprising; the innovation here was to extend this representation of parts to allow for circularity. This enabled Barwise and Etchemendy, in [8], to give a very smooth situation-theoretic account of The Liar and other self-referential propositions. This was extended by Aczel, Lunnon and others [2, 21, 28] to a theory of 'structure objects' in which the relation of membership from set theory is generalised to that of a structural component, with associated operators for substitution and abstraction. The focus of the structured-object models of situation theory (henceforth: SOST) was on the provision of a highly 'intensional' account of situations, infons, propositions and properties, in which syntactic structure is mirrored in the semantics, allowing for very precise control over identities, such as whether or not the infon that $a = b$ is the same as the infon that $b = a$. As such, it was a victim of its own success. There were simply too many choices to be made, and little sense of having reached explanatory bedrock.

For me, another weakness of the SOST models was that they ignored what I find to be the most interesting part of the project: to account for the role of different representational perspectives in our reasoning and how shifting from one perspective to another opens up new vistas. For this one needs some story about how information in very different systems of representation can be related. A central example for researchers at the time, which has been pursued in great depth subsequently, is the relationship between diagrammatic and sentential reasoning. When we represent information diagrammatically, certain inferences are more easily made; others are more difficult. Likewise, the significance of signs (natural and artificial) depends on their particular context. A pile of stones on the bank of a river indicates where it is safe to ford; smoke rising from some point on the horizon means that there is a fire there. The provision of a systematic account of such regularities was also a desideratum of the situation theory project but was overlooked in the SOST models.

All of this was motivation for the development of the theory of classification and channels in [12], in which Barwise and I attempted to provide an account of information flow that is more or less independent of an account of information content. van Benthem, in [13], makes a useful distinction that will help to make this point clearer. The defining feature of *information-as-range* is the elimination of possibilities. When you pick a card from a pack without showing it to me, I know only that it is one of 54 possible cards. If I then learn that it is red, this number is cut in half. The information that I have gained can be identified with this reduction of possibilities. Identifying the informational content of a sign with the range of possibilities consistent with it was the approach taken by Bar-Hillel and Carnap in their seminal [4]. It is also implicit in the Tarskian approach to semantics and the ubiquitous concept of a 'Californian proposition': a set of possible worlds.

Information-as-range is to be distinguished from *information-as-correlation*, the second of van Benthem's categories, according to which the informational content of an event is given by other events with which it is correlated. The mercury aligned with '70°' on the thermometer indicates that the air temperature is 70°F because of the correlation between the temperature and the height of the mercury. The correlation, in this case, is due to certain laws of natural science, the particular construction of the thermometer with its helpful gradations, the physical proximity of the mercury bulb to the surrounding air, the relative kinetic stability of the air, the origin of the instrument (America, where the Fahrenheit scale is commonly used) and many other contingencies. In situation theoretic parlance, the correlation is due to a 'constraint' that holds in the situation being described. Science typically studied correlations from a probabilistic or statistical point of view. This is considered in Sect. 35.2.5, below.

From a slightly more abstract perspective, we can see that information-as-range occurs wherever there is a classification of things into types. By knowing the type or types of an unknown thing, or 'token', one has some, albeit incomplete, information by having eliminated those other things that are not of the same type. Information-as-correlation occurs when there is a systematic relationship between things of different types. In [12] this is captured by the concept of a 'channel', in which the types and tokens of one classification are related to those of another. The theory of classification and channels from [12] is therefore very much in alignment with van Benthem's distinction.

Yet despite these thematic connections and a number of common examples, much was left unsaid about the relationship between [12] and the earlier project of situation theory. This paper will explore some more explicit connections in more detail, including an elaboration of van Benthem's approach as an alternative interpretation of ST.

35.1 Twelve Questions About Situation Theory

Situation theory begins with the idea that reality is composed of *situations* which differ in the information they *support*. The smallest unit of information is called an *infon*, which is the situation theoretic analogue of an atomic proposition. From the very start, it is worth noting that situations, as constituents of reality, most naturally fall within the scope of metaphysics, whereas the issue of what information they do or do not support is at least partly epistemological. The development of the theory is characterised by conflicting intuitions from these two traditional domains of philosophy. We write

$$s \models \sigma$$

to mean that situation s supports infon σ. For example, suppose σ is the information that I am typing. This is supported by the situation s that I am currently part of, or rather one such situation because I participate in many simultaneously, some of which contain others. We write

$$s \trianglelefteq t$$

to mean that situation s is *part of* situation t, taking this to be a partial order (reflexive, transitive and antisymmetric). Situations are subject to certain *constraints* that bind together infons more or less strongly. Given infons σ and τ, we write

$$\sigma \Rightarrow \tau$$

for the constraint that σ *involves* τ. For example, the constraint

$$\langle\!\langle \text{I am typing slowly} \rangle\!\rangle \Rightarrow \langle\!\langle \text{I am typing} \rangle\!\rangle$$

would relate situations supporting the antecedent infon to those supporting the consequent infon quite strongly, whereas the constraint

$$\langle\!\langle \text{I am typing 'slowly'} \rangle\!\rangle \Rightarrow \langle\!\langle \text{you are reading 'slowly'} \rangle\!\rangle$$

is a little more tenuous. The relationship between the three relations $\trianglelefteq$, $\models$ and $\Rightarrow$ raises a number of questions, none of which were answered definitively by the early pioneers of ST.

35.1.1 Parts and Persistence

Question 1 Do situations support all the information supported by their parts?[3]

A positive answer to this question amounts to acceptance of the following principle:

> *Persistence*: if $s \trianglelefteq t$ and $s \models \sigma$ then $t \models \sigma$.

Resistance to Persistence comes from two sources. Firstly, if one takes 'support' in something like the ordinary epistemic sense, then it is surely defeasible. The infon

---

[3] Considered by Barwise in [6] as Choice 6.

《Emily ate the last cookie》 may be supported by a situation in the kitchen with biscuit crumbs, a red-faced girl and an empty cookie jar, but may no longer be supported by a larger situation, including her brother hiding in the pantry. Secondly, even setting aside issues of defeasibility, certain kinds of information do not seem to persists at all. The infon 《everyone is dancing》 may be supported by a situation at my Greek friend's wedding but not by a larger situation that includes the activities of, say, the whole human population at the time of the wedding.[4] A defence of Persistence against the first problem is available to those holding a firmly realist interpretation of 'support' who can insist that the situation in the kitchen, no matter what the weight of evidence, does not support the infon 《Emily ate the last cookie》 because Emily did not eat it. This interpretation was favoured by Barwise and Perry from [9] on. To defend Persistence against the second problem requires that the information that everyone is dancing in the two situations cannot be the same infon, the hidden ambiguity is a consequence of the change of meaning of 'everyone' in the two cases. There are a number of ways of implementing this idea. We could include the situation explicitly in the infon, so that 《everyone$_s$ is dancing》 is the information that everyone is dancing in s. The connection between 《everyone$_s$ is dancing》 and 《everyone$_t$ is dancing》 would then need to be explained, perhaps by a constraint: 《everyone$_s$ is dancing》 $\Rightarrow$ 《everyone$_t$ is dancing》 when $t \trianglelefteq s$ (but not vice versa).[5] Persistence can also be expressed in terms of the 'information containment' order, defined as $s \sqsubseteq t$ iff for all infons σ, if $s \models \sigma$ then $t \models \sigma$. With this, Persistence is the requirement that each part of a situation is informationally contained within it: that $\trianglelefteq$ implies $\sqsubseteq$.

Question 2 Can different situations support exactly the same infons?

A negative answer amounts to acceptance of the following principle:

Extensionality: $s = t$ if for every σ, $s \models \sigma$ iff $t \models \sigma$.

As its namesake in set theory suggests, the principle allows us to model a situation by the set of infons it supports. It is implied by Persistence together with its converse, which would then also permit modelling $\trianglelefteq$ as the subset relation. Such models became the main focus of attempts to develop a theory of situations in the late 1980s and early1990s. Perhaps surprisingly, Extensionality is not the last word on the identity of situations. The existence of situations that support information about themselves, such as the above-mentioned 《everyone$_s$ is dancing》 raises a problem:

---

[4] Persistence is studied in classical model theory as the question of which formulas are satisfied in a model whenever they are satisfied in a submodel.

[5] Alternatively, we could include a 'parameter' dom for the domain of quantification to give a *parametric* infon 《everyone$_{dom}$ is dancing》, allowing the value of the parameter to be 'anchored' in the context of a particular situation. Then s supports this infon iff it supports that everyone in dom[s] is dancing, where dom[s] is a set of individuals anchored to dom by s. This would allow persistence to hold for non-parametric infons but fail for parametric ones. Of course, one could also make similar a distinction between logically simple infons and complex ones, involving quantification, with varying persistence properties. These and other ideas were explored at the time.

if this is the only infon that s supports, and moreover, there is a situation t that only supports $\langle\!\langle$everyone$_t$ is dancing$\rangle\!\rangle$. Then both $s = t$ and $s \neq t$ are compatible with Extensionality.[6]

A negative to Question 2 implies that there is something more to the individuation of situations than merely the infons they support. One motivation for thinking this is to allow for different representational perspectives, or 'schemes of individuation', discussed most explicitly by Barwise in [7] and one of the main motivations for our [12]. Given that information presupposes a dividing-up of the world into types, objects, properties, and other ontological paraphernalia, there is at least the possibility that any one way of doing this is inadequate in some respects, perhaps even incoherent if extended beyond its normal area of application.[7] So if the infons are supplied by a particular representational perspective, but the situations and their division into parts transcends this perspective, one would not expect the $\trianglelefteq$ and $\sqsubseteq$ orders to line up perfectly.[8]

There are many more questions one could pose about the mereology of situations, which we will just summarise briefly as follows:

Question 3 What is the mereology of situations? Is there a largest situation? Is there a smallest one? Given any two situations, must there be a third that contains them both as parts? Or are there incompatible situations, which are not part of any common situation? If so, how can we interpret such incompatibility?

Instead of charting the many possible answers, let us distinguish two. On an *actualist* view of situations, the only situations that exist are the ones that make up

---

[6] The move to non-well-founded set theory is motivated by such considerations. If a situation s is modelled as the set of infons it supports and if those infons are modelled also as sets in such a way that the things they are about occur as hereditary members, related to the infon by the transitive closure $\in^*$ of the membership relation, then we have that $s \in^* \langle\!\langle$everyone$_s$ is dancing$\rangle\!\rangle \in s$ and so a counterexample to the well-foundedness of $\in$. This can easily be avoided, as pointed out by Paul King [20] and others, by choosing to model the component structure of infons in a way that does not require them to be hereditary members, but such a sidestep doesn't really help. The more important issue concerning the identity of situations, mentioned above, is independent of the way in which they are modelled. Nonetheless, the study of non-well-foundedness reveals various solutions in the form of strengthened principles of individuation, such as Peter Aczel's Anti-foundation Axiom in [1], which was subsequently used to obtain the SOST models of situation theory mentioned in the introduction.

[7] Barwise gives the example of the Cantorian conception of set. A more quotidian example is our system of directions: up, down, left, right. When extended around the surface of the earth, incoherence is less than a hemisphere away. See Gupta [18]. Schemes will be discussed more in Sect. 35.1.3, below.

[8] A line of reasoning that purports to undermine this attempt at conceptual relativism goes as follows. In making the objection to identifying $\sqsubseteq$ and $\trianglelefteq$ explicit we reveal its rather awkward presupposition: that situations are individuated and even ordered in a way that is independent of any scheme of individuation. This smacks of blind faith or incoherence. Yet this attack can be undermined by considering that the way of individuating situations and ordering them into parts need only be different from the way presupposed by $\models$ and $\sqsubseteq$. When considering multiple schemes of individuation one would have to distinguish between multiple $\sqsubseteq$ relations. Then the distinction between $\trianglelefteq$ and $\sqsubseteq$ is just that of the ordering given by two schemes: an internal one and an external one.

the actual world, not those that might have existed or ones that we imagine in works of fiction. On this view, it is difficult if not impossible to make sense of incompatible situations. Yet, there need not be a largest situation. The view of the world as open-ended, allowing ever bigger situations without limit is one that is familiar, by analogy with the universe of sets, and which forms the basis of Barwise's and Etchemendy's solution to the Liar paradox in [8].

On a *possibilist* view of situations, there is no sharp distinction drawn between situations that are part of the actual world and those that are not. Rather, actuality is an indexical notion, just as it is in David Lewis' metaphysics of possible worlds. On this view, there are many incompatible situations, corresponding to different eventualities, and so no largest situation. There may be (many) maximal situations: situations that are not part of any larger situation. These could be thought of as different possible worlds, although, as on the actualist view, such maximal situations are not required. Without maximal situations we would have a universe of possibilia without possible worlds. With a possibilist view of situations one can of course provide a reductionist account of necessity, as is done by Lewis and others with possible worlds, and this may be considered the principle motivation for the view.

35.1.2 Constraints and Information Flow

Some of the tension created by Persistence and its converse was relieved by distinguishing between information supported by a situation and information that is 'carried' by means of constraints. We have already seen the example of the constraint between my typing slowly and my typing: if I am doing one then I must be doing the other. This is represented by the constraint

$$\langle\!\langle \text{I am typing slowly} \rangle\!\rangle \Rightarrow \langle\!\langle \text{I am typing} \rangle\!\rangle$$

Yet the issue of how this constraint is related to the support of infons by situations is vexed. We will talk of a constraint 'holding' at or of a situation to mean that the constraint applies there. If the constraint $\sigma \Rightarrow \tau$ holds in situation s, which also supports the infon σ, then infon τ is said to be *carried* by s. Here we are presupposing that information carried by a situation (about another situation) is somehow explicitly indicated by a constraint, but it is helpful to frame this as a question, thus:

Question 4 Is all information carried by a situation indicated by a constraint? In other words, if a situation s carries the information τ about situation t, must there be a constraint $\sigma \Rightarrow \tau$ such that $s \models \sigma$ with the appropriate relationship between s and t?

We write $s \models_t \tau$ to mean that situation s carries the information τ about situation t, and $\sigma \underset{s,t}{\Rightarrow} \tau$ to mean that the constraint $\sigma \Rightarrow \tau$ relates situations s to t. Then a positive answer to Question 4 is given by the principle:

Indication: $s \models_t \tau$ iff there is a σ such that $s \models \sigma$ and $\sigma \underset{s,t}{\Rightarrow} \tau$

But the definition of carrying invites another obvious question:

Question 5 Is information that is carried by a situation also supported by it?

In the case of our example of typing slowly a positive answer is attractive. One cannot type slowly without typing and so, any situation in which there is some slow typing going on is also a situation in which there is some typing. More generally, constraints of this kind capture the internal logic of the infons themselves, independent of contingencies and context, and without offering a firm definition, we can call them *analytic* constraints. We will use $\sigma \underset{s}{\Rightarrow} \tau$ to mean that $\sigma \Rightarrow \tau$ is an analytic constraint that holds in situation s. So a positive answer to Question 5 for analytic constraints therefore amounts to the principle:

Analytic Involvement: if $s \models \sigma$ and $\sigma \underset{s}{\Rightarrow} \tau$ then $s \models \tau$

It is less clear that information carried by virtue of the inferences of classical logic are also supported by a situation. One central reason for denying this is the assumption that situations have bounded subject matter. The situation of my typing on my laptop has nothing to do with events in China in 223 B.C.E. so it would be strange indeed for it to support the information that the assassin Jing Ke either did or did not succeed in killing the king of Qin. But it is not only irrelevant tautologies that are problematic. My typing slowly also entails that I am typing either slowly or fast, and even that either I or you are typing. But it would be odd to insists that the situation that involves my typing must also include facts about you.[9] Since the ST's origins in Barwise's treatment of the naked infinitives [5], partial logic has been a close ally. Yet, no definitive answer as to the rules of such a logic has been given.

So, when asked about constraints in general, Question 5 should be answered negatively. Indeed, the distinction between information supported and information carried is a central tenet of situation theory, and its consequence is information *flow* between situations. The constraint that my typing the word 'slowly' involves your reading the same word relates two situations: one situation s in which I am typing and a geographically and temporally remote situation t in which you are reading. The holding of this constraint requires that if s supports $\langle\langle$I am typing 'slowly'$\rangle\rangle$ then t supports $\langle\langle$you are reading 'slowly'$\rangle\rangle$. Information about t is carried by s, by virtue of some sort of informational connection between the two situations. This is the phenomenon van Benthem calls 'information as correlation', which is captured by the principle:

Involvement: if $s \models \sigma$ and $\sigma \underset{s,t}{\Rightarrow} \tau$ then $t \models \tau$

---

[9] This sort of observation led to some parallels between research on situation theory and relevant logic, summarisd in [22].

Involvement follows from Carrying, as does Analytic Involvement, if in addition we accept the reduction of one to the other:

$$Analytic\ Reduction: \sigma \underset{s}{\Rightarrow} \tau \text{ iff } \sigma \underset{s,s}{\Rightarrow} \tau$$

To get some purchase on these principles, situation theorists aimed to understand not only the abstract properties of constraints but also the metaphysics needed to underwrite them. As they involve some sort of conditional relationship between infons, an obvious place to look is the possible-worlds semantics for conditions developed by Lewis and others. But ST was ideologically opposed to possible worlds, wanting an account closer to home, by appealing to the basic concepts of the theory: situations, infons and support.[10] In searching for an answer, the earlier situation theorists were inspired by classical information theory, as introduced by Claude Shannon, and appealed to the concept of a *communication channel* between the two situations. Such a channel is not a matter of logic or anything to do with the information content as such, but is rather a contingent correlation between two situations, by which an agent with access to one situation may thereby gain information about the other. The most fundamental question left open by early work in ST is the following:

Question 6 What is a channel? How are channels related to constraints? And what is the consequence for the relationship between information carried and information supported?

A rough answer is that a channel provides a connection between two situations s and t that somehow underwrites constraints so that they obey the Involvement principle. An agent who is 'attuned' to the constraint can thereby gain information about one situation from information accessed at another.[11] There is an apparent ambiguity in talk of channels, between a passive understanding, whereby a channel is an abstraction from the telegraph wire that links a sender to a receiver, which is a precondition for communication between them, and a more active understanding, whereby a channel is an abstraction of the message-sending activity than occurs in those wires. We will try to stick to the latter interpretation, recognising that the former is a part of it.

Channels, in the latter, more inclusive sense, were also the focus of Fred Dretske's more-or-less simultaneous attempt to naturalise epistemology in [16] which took channels to form part of the natural fabric of the world, created by ordinary nomic dependancies. Agents exploit these channels to gain information about the world around them and so to act in advantageous ways. Many of his examples are about microorganisms that show an astonishing ability to use the natural regularities of their environment to do things that one might otherwise have thought involve representation and inference.

---

[10] van Benthem's account, to be discussed in Sect. 35.3.3 has this flavour.

[11] Such talk of information access and processing was commonplace in discussions of situation theory at the time but never made formally precise. A good discussion is by Israel and Perry in [19].

Crucial to Dretske's project were two properties of channels. Firstly, that information does not degrade: that if a situation carries the information that σ and this in turn carries the information that τ, then the original situation also carries the information that τ. This he called the Xerox Principle. There are various ways of implementing or violating this idea in situation theory, depending on whether this is taken to be a principle about the existence of channels or of their reliability.

Question 7 If a channel connects s to t and another channel connects t to u, then is there a channel that connects s to u? And if so, what constraints does it licence? In particular, does $\sigma \underset{s,u}{\Rightarrow} \upsilon$ follow from $\sigma \underset{s,t}{\Rightarrow} \tau$ and $\tau \underset{t,u}{\Rightarrow} \upsilon$?

The second property of channels, is that they may in fact fail. Error is possible. Dretske likens the carrying of information to other concepts that are both absolute, in the sense of not admitting degrees, but also depend on some presupposed standard. His example is the concept 'flat' which implies that there are no deviations from the level: no 'bumps'. Of course every real surface, however flat, still has bumps when looked at closely, which is to say that whether or not a surface is flat depends on how closely you look. What varies is what counts as a 'bump' ([17], p. 366).

Struggling to deal with these two properties of channels and information flow led to my work with Barwise, especially [10–12], in which we considered various models for channels that accounted for the Xerox principle and yet also the possibility of error. As a stop-gap in the development of the theory, many early situation theorists appealed to the concept of a 'background condition', borrowed from experimental science, in which one expects observations to conform to the predictions of a certain law only when the conditions of the experiment are controlled to ensure that other factors that may influence the experiment are controlled. For example, an experimenter may try to ensure that the laboratory is at a constant temperature throughout the experiment, or that it is conducted in a vacuum.

We will adopt the notation $[B]\sigma \Rightarrow \tau$ for a constraint that depends on a background condition B, without settling the matter as to what such a condition is, and merely state our final question about constraints and channels:

Question 8 Can channels fail and if so how? How, if at all, is this dependent on the satisfaction of a 'background condition'?

35.1.3 Infons and Schemes of Individuation

The mainstay of research in situation theory left most of the questions we have considered to one side and focussed on developing the theory as a viable tool for formal semantics and other applications. This required a more detailed account of the structure of infons. The ontology most commonly used was similar, if not identical to that used in standard mathematical logic. One starts with a stock of individuals and relations and uses these to construct *basic* infons of a positive and negative type. For example, given a relation r and suitable objects a and b, we have the basic infons

$$\langle\!\langle r, a, b; +\rangle\!\rangle \quad \text{and} \quad \langle\!\langle r, a, b; -\rangle\!\rangle$$

meaning that a does stand in relation r to b and, respectively, that it does not. The two infons are said to have *polarity* $+$ and $-$ respectively, and infons that only differ in their polarity are said to be *dual* to each other.

Question 9 Can a situation support both an infon and its dual? Can it support neither?

Situation theorists were more or less in agreement on this one. Situations may support neither an infon nor its dual, but cannot support both. An explanation for this is that the notion of 'support' has a hidden modality: to say that $s \models \langle\!\langle I \text{ am typing}; +\rangle\!\rangle$ is to say that, given how things are in s, I *must* be typing, and to say that $s \models \langle\!\langle I \text{ am typing}; -\rangle\!\rangle$ is to say that, given how things are in s, I *cannot* be typing. The failure of s to support either infon of the pair is simply a result of the lack of information in s to determine which is the case; but to suppose that s supports both is an outright contradiction.[12] Yet the force of these modalities is similar to that of the 'analytic constraints' considered in the last subsection. And this raises the prospect that behind talk of polarities, constraints of some kind play a role.

Many other more complicated ways of constructing infons were discussed in the literature, resulting in a rich ontology of types, roles, abstract objects, and other paraphernalia. But similar such complications need to be introduced in any theory aiming to provide a useful tool to linguists or other semanticists wanting to express subtle grammatical and logical distinctions. These will not be the focus of this paper and so I will just package them in the following rather open-ended question:

Question 10 Are there general ways of combining infons into more complex infons? How is the structure of such compounds related to the constraints in which they participate? (In particular: are there negations and disjunctions of infons, quantified infons, etc. and what logic do they have?)

Information arises whenever there is some classification of things into different types: these are in and those are out. This was the basis of Carnap and Bar-Hillel's early discussion of information in the 'semantic' sense in [4]. With the logical atomist world of objects, properties and relations, it is an easy step to identify information content with a set of possible configurations of those parts. Carnap did this with 'state descriptions': conjunctions of atomic propositions or their negations, one for each atomic proposition. For him, the content of some piece of information was the set of state descriptions compatible with the information itself: 'information-as-range', in van Benthem's phrase. A drawback of this approach is that it assumes some particular division of the world into objects, properties and relations and uses this to interpret all others. Setting aside the question of how we arrive at such a fundamental ontology, there are worries as to whether this best reflects the wide variety of representational

---

[12] By 'outright' contradiction, I mean a contradiction in the metalanguage, in which we are explaining what 'support' means. However attractive impossible situations might be to someone of paraconsistent leanings, he or she must therefore come up with a different explanation for 'supports', at least when talking to a classical logician.

means of conveying information. Much of the information we process in daily life is not obviously in this form, and there is even some doubt as to whether the objects and properties we spend much of our time worrying about can meet the demands of logic.[13] We freely exchange information in the form of loosely worded remarks that resist being pinned down to propositional form and even with quite different forms of representation, such as diagrams, gestures and the like. Furthermore, if information-processing is taken to be present in all forms of life, and even non-life, as Dretske's examples suggest, we must be considerably more liberal in what counts as information content.

Thoughts in this direction led early situation theorist to propose that situations and infons are relative to a 'scheme of individuation'. The central insight was that the world and those who interact with it enter into a collaboration when it comes to the production of information. The same situation can be seen from different perspectives, using different conceptual, or rather 'informational' resources, and yet when those resources are applied, the facts are determined one way or the other. To say that information is supported is to presuppose one of those perspectives, but this should not thereby exclude others or require that they be interpretable using present resources. Contemporary logicians are familiar with the idea that a language is just something one concocts on the fly to model a particular scenario. We have come a long way from the 'universal science' of Carnap and now distance ourselves from the project of producing a universal language. Yet there is still a reluctance to address the consequences of this humility for the relationship between logic and metaphysics.

Question 11 What, then, is a 'scheme of individuation'? Can one situation be 'viewed' from the perspective of more than one scheme?

Barwise provided one answer to this question in [7], p. 252, by associating each situation with its own set of objects and relations, and consequently infons, which it may or may not support by it. By attributing the scheme to a situation, he removes the possibility of a positive answer to the second part of the question. Also, his proposal retains the division into objects and properties, which buys into a particular way of organising information content that we might want to resist. Instead, I will suggest only that a situation supplies a set of 'issues'. Each issue α gives rise to two infons: $\langle\langle\alpha; +\rangle\rangle$ and $\langle\langle\alpha; -\rangle\rangle$. And these may or may not be supported by the situation. The internal logic of these issues and how it relates to constraints and the flow of information will be left open for now.

Finally, in a self-reflective mode, one can ask about the perspective we are currently taking on situations, infons, constraints, and so forth. It would appear that we are using a 'theorists' scheme of individuation to talk about situation theory itself. The idea that situation theory could be its own metatheory is an attractive one and one that draws further on Barwise's analogy with set theory. There were some negative results, e.g. Plotkin's [23], which produced the equivalent of Russell's paradox for 'naive' situation theory. On the other had, Barwise and Etchemedies work on the

---

[13] Objects of interest to us, such as trees, cites and other people typically have rather unclear boundaries and their properties are vague.

Liar paradox in [8] gives a concrete model version of situation theory that models some (but perhaps not all) central semantic concepts of the theory itself.

Question 12 To what extent can situation theory express the facts and constraints of situation theory itself? For example, are there infons with content $s \models \sigma$ and which situations support it?

I intend to return to many of the questions posed here in Sect. 35.3. But before that, we will need to review and develop some theory from [12].

35.2 Logic and Information Flow in Classifications and Channels

The following is a rather dense and rapid development of the theory of information flow and channels from [12] so as to make it suitable for Sect. 35.3, in which the theory is used to interpret concepts from situation theory. It departs from [12] in a few minor but crucial ways. A major consideration motivating this newer approach was to be able to model information flow in systems described probabilistically, using the criterion that exceptions to a constraint occur with zero probability, which does not of course mean that they do not occur. The starting point is a general account of 'classification' and its relationship to logic.

35.2.1 Logic of Classification

A *classification* $\mathscr{A} = \langle \mathscr{A}^\wedge, \mathscr{A}^\vee, \models_\mathscr{A} \rangle$ consists of a set $A^\vee$ of *tokens*, a set $A^\wedge$ of *types* and a binary relation $\models_\mathscr{A}$ between them. We conceive of the types as classifying the tokens into kinds. Example abound but, as the notation suggests, a central motivating example comes from logical semantics, in which the types are formulas and the tokens are models. Another important example, is that of a topological space, in which the tokens are points and the types are open sets; here points are classified according to their proximity to each other in the sense of proximity given by the topology.[14] Yet the most important example for the interpretation of ST, is that of a system: some interacting coalition of physical, biological or sociological processes that has a determinate concept of 'state' and 'event'.[15] Any system event classifies the states into those that are compatible with the event's occurrence and those that are not. Every classification $\mathscr{A}$ has a 'natural' logic induced by the classification relation.

---

[14] The logical and topological interpretation may coincide when, for example, the space is the Stone space of the Boolean algebra of logically equivalent formulas. This was used in [26] to use classifications in the analysis of the logic of diagrammatic reasoning.

[15] We are restricted here to a classical notion of state; quantum processes are quite different.

We write $\Gamma \vdash_{\mathscr{A}} \Delta$ to mean that every token a that is of every type in Γ is of at least one type in Δ, giving a range of inference patterns:

$$\alpha \vdash_{\mathscr{A}} \beta \text{ if } a \models_{\mathscr{A}} \alpha \text{ then } a \models_{\mathscr{A}} \beta$$
$$\alpha, \beta \vdash_{\mathscr{A}} \text{ if } a \models_{\mathscr{A}} \alpha \text{ then } a \not\models_{\mathscr{A}} \beta$$
$$\vdash_{\mathscr{A}} \alpha, \beta \text{ if } a \not\models_{\mathscr{A}} \alpha \text{ then } a \models_{\mathscr{A}} \beta$$

Let $[\Gamma; \Delta]$ be the set of tokens a for which are described positively by Γ and negatively by Δ, i.e., $a \models_{\mathscr{A}} \alpha$ for every $\alpha \in \Gamma$ and $a \not\models_{\mathscr{A}} \alpha$ for every $\alpha \in \Delta$. The members of $[\Gamma; \Delta]$ are counterexamples to the inference from Γ to Δ and so:

Fact 1 $\Gamma \vdash_{\mathscr{A}} \Delta$ *iff* $[\Gamma; \Delta] = \emptyset$

If $\mathscr{A}$ is a semantic classification then $\vdash_{\mathscr{A}}$ is of course the standard (multiple conclusion) consequence relation. More generally, we define a *local logic* $L = \langle N_L, \vdash_L \rangle$ on $\mathscr{A}$ (henceforth: 'logic on $\mathscr{A}$') to consist of a set N_L of *normal* tokens and a *consequence* relation $\vdash_L$ between sets of types. Here we think of $\vdash_L$ being defined in some other way than $\vdash_{\mathscr{A}}$, e.g., by a proof system in the case that types are formulas, and N_L also being an independently specified class, e.g., some operant background condition in the case that tokens are states of a system. Now some familiar properties of logics can be formulated[16]:

L is *sound*   : If $\Gamma \vdash_L \Delta$ then $N_L \cap [\Gamma; \Delta] = \emptyset$.
L is *complete* : If $N_L \cap [\Gamma; \Delta] = \emptyset$ then $\Gamma \vdash_L \Delta$.

Fact 2 *The natural logic* $L_A = \langle A^{\vee}, \vdash_{\mathscr{A}} \rangle$ *on classification $\mathscr{A}$ is both sound and complete.*

So too is the *trivial logic* $T_{\mathscr{A}}$ with the empty set of normal tokens and $\Gamma \vdash_{T_{\mathscr{A}}} \Delta$ for all Γ, Δ. In fact, given any consequence relation $\vdash$, the logic $\langle \emptyset, \vdash \rangle$ is sound. This may seem a little strange, but we will show in Sect. 35.2.5 that there are interesting examples.

The logics on a classification $\mathscr{A}$ form a complete lattice $\mathsf{Log}(\mathscr{A})$ under the order: $L_1 \sqsubseteq L_2$ iff $\vdash_{L_1} \subseteq \vdash_{L_2}$ and $N_{L_2} \subseteq N_{L_1}$. The sound-and-complete logics also form a complete lattice $\mathsf{Log}^*(\mathscr{A})$ but is not in general a sub-lattice of $\mathsf{Log}(\mathscr{A})$. Its bottom element is $L_{\mathscr{A}}$ and its top element is $T_{\mathscr{A}}$.

The following principles are all satisfied by natural logics (where $\bar{\Sigma} = \mathscr{A}^{\wedge} \setminus \Sigma$):

---

[16] In [12] 'local logic' refers only to what we are calling sound local logics; there a logic L is 'sound' iff $N_L = \mathscr{A}^{\vee}$, so that a 'sound and complete' logic is just a the natural logic of a classification. Present purposes dictate a slightly more general approach, but many of the results of [12] carry through.

I	Identity	$\alpha \vdash_L \alpha$
W	Weakening	if $\Gamma \vdash_L \Delta$ then $\Gamma, \Gamma' \vdash_L \Delta, \Delta'$
C	Cut	if $\Gamma \vdash_L \alpha, \Delta$ and $\Gamma, \alpha \vdash_L \Delta$ then $\Gamma \vdash_L \Delta$
SC$_l$	Strong Cut (left)	if $\Gamma \vdash_L \alpha, \Delta$ for all $\alpha \in \Sigma$ and $\Gamma, \Sigma \vdash_L \Delta$ then $\Gamma \vdash_L \Delta$
SC$_r$	Strong Cut (right)	if $\Gamma \vdash_L \Sigma, \Delta$ and $\Gamma, \alpha \vdash_L \Delta$ for all $\alpha \in \Sigma$ then $\Gamma \vdash_L \Delta$
P	Partition	if $\Gamma, \Sigma \vdash_L \bar{\Sigma}, \Delta$ for all Σ then $\Gamma \vdash_L \Delta$.
SP	Strong Partition	if (for some Σ) $\Gamma, \Gamma' \vdash_L \Delta', \Delta$ for all $\Gamma' \cup \Delta' = \Sigma$ then $\Gamma \vdash_L \Delta$.

Fact 3 *If L satisfies* I, W *and* P *then it satisfies all of the rest.*

Many of these principles are familiar to the logician. The exceptions are Partition and Strong Partition, which are also related to familiar ideas in logic but are somewhat disguised. Note that the converse to Partition says if the set of tokens $[\Gamma; \Delta]$ is non-empty, so there are some tokens described positively by Γ and negatively by Δ, which is a form of consistency, then there is a way of extending Γ to Σ such that $[\Sigma; \bar{\Sigma}]$ is also non-empty. Or, in other words, the 'consistent' pairing of Γ and Δ can be extended to a 'maximally consistent' pairing. When applied to standard logic-based classifications, Partition is of course no more than the familiar Lindenbaum Lemma. Strong Partition is a generalisation of both Partition and Cut, showing their common structure. Taking Σ to be $\mathscr{A}^{\wedge}$ we get Partition; taking it to be a singleton, we get Cut.

We can also reason in a more interactive way by defining logics *between* classifications $\mathscr{A}$ and $\mathscr{B}$. For some mixed set Γ of types from $\mathscr{A}^{\wedge}$ and $\mathscr{B}^{\wedge}$ (assumed, for now to be disjoint), we write $\Gamma_{\mathscr{A}}$ for those that are in $\mathscr{A}^{\wedge}$ and $\Gamma_{\mathscr{B}}$ for those that are in $\mathscr{B}^{\wedge}$. Then the natural logic between $\mathscr{A}$ and $\mathscr{B}$ can be defined by $\Gamma \vdash_{\mathscr{A}+\mathscr{B}} \Delta$ iff for all $a \in \mathscr{A}^{\vee}$ and $b \in \mathscr{B}^{\vee}$, if a is of every type in $\Gamma_{\mathscr{A}}$ and b is of every type in $\Gamma_{\mathscr{B}}$ then either a is of one of the types in $\Delta_{\mathscr{A}}$ or b is of one of the types in $\Delta_{\mathscr{B}}$. Two observations are immediate. Firstly, $\vdash_{\mathscr{A}+\mathscr{B}}$, as the notation suggests, is just the natural logic of the sum $\mathscr{A} + \mathscr{B}$.[17] Secondly, it is not so interesting.

Fact 4 $\Gamma \vdash_{\mathscr{A}+\mathscr{B}} \Delta$ *iff* $\Gamma_{\mathscr{A}} \vdash_{\mathscr{A}} \Delta_{\mathscr{A}}$ *and* $\Gamma_{\mathscr{B}} \vdash_{\mathscr{B}} \Delta_{\mathscr{B}}$

In other words, there is no interaction between the two component logics. Other logics in $\mathrm{Log}(\mathscr{A} + \mathscr{B})$, which we call *distributed logics*, are potentially more interesting. For such a logic L, N_L is a non-trivial relation between tokens of $\mathscr{A}$ and tokens of $\mathscr{B}$, representing some informational connection between them. It is sound iff the following condition holds:

if $N_L(a, b)$ and $\Gamma \vdash_{\mathscr{A}+\mathscr{B}} \Delta$ and $a \models_{\mathscr{A}} \alpha$ for every $\alpha \in \Gamma_{\mathscr{A}}$ and $b \models_{\mathscr{A}} \beta$ for every $\alpha \in \Gamma_{\mathscr{B}}$ then either $a \models_{\mathscr{A}} \alpha$ for some $\alpha \in \Delta_{\mathscr{A}}$ or $b \models_{\mathscr{A}} \beta$ for some $\beta \in \Delta_{\mathscr{B}}$.

Such a logic exploits the connection represented by N_L to reason about the two classifications in parallel.

---

[17] Concretely, $\mathscr{A} + \mathscr{B}$ is the classification of pairs of tokens $\langle a, b \rangle$ with $a \in \mathscr{A}^{\vee}$ and $b \in \mathscr{B}^{\vee}$ of types from the disjoint union $\mathscr{A}^{\wedge} + \mathscr{B}^{\wedge}$ such that $\langle a, b \rangle \models_{\mathscr{A}+\mathscr{B}} \alpha$ iff $a \models_{\mathscr{A}} \alpha$, and $\langle a, b \rangle \models_{\mathscr{A}+\mathscr{B}} \beta$ iff $a \models_{\mathscr{A}} \beta$. More abstractly, it is the sum in the category of classifications and infomorphisms, to be introduced in Sect. 35.2.2.

Returning to the primary example of the classification of states by events, a local logic L can be understood as a local pattern of regularity in the system's behaviour. The constraints $\vdash_L$ represent dependencies between events which, if the logic is sound, are found whenever the system's state is in N_L. Unsound logics allow some departure from conformity, interesting examples of which will be considered in a little bit later (Sect. 35.2.5). A distributed logic represents local patterns of interaction between two systems: ways in which they are partly co-ordinated. This is a first step toward a model of information flow.

35.2.2 Logic and Information Flow

An *infomorphism* $f\colon\mathscr{A}\rightleftarrows\mathscr{B}$ from $\mathscr{A}$ to $\mathscr{B}$ is a pair of functions $f^\wedge\colon\mathscr{A}^\wedge\ \to\ \mathscr{B}^\wedge$ and $f^\vee\colon\mathscr{B}^\vee\ \to\ \mathscr{A}^\vee$ such that $f^\vee(b)\ \models_\mathscr{A}\ \alpha$ iff $b\models_\mathscr{B} f^\wedge\alpha$ for each $b\in\mathscr{B}^\vee$ and $\alpha\in\mathscr{A}^\wedge$. Infomorphisms arise naturally in both the logical (semantic) and topological examples of classification. If $\mathscr{A}$ and $\mathscr{B}$ are both classifications of models by formulas, of two different languages, say, then any interpretation of the language of $\mathscr{A}$ in that of $\mathscr{B}$ gives rise to an infomorphism $f\colon\mathscr{A}\rightleftarrows\mathscr{B}$. The image φ is of course the interpretation of $\mathscr{A}$-formula ϕ in the language of $\mathscr{B}$ and the image $f^\vee M$ of $\mathscr{B}$ model M is the model of $\mathscr{A}$ constructed by interpreting the primitives of the language in the way their images are interpreted in M.[18] Take, for example, the interpretation of arithmetic in set theory. φ is a sentence in the language of set theory that interprets the arithmetic sentence ϕ, and $f^\vee M$ is a model of arithmetic constructed from a particular model M of set theory. In the topological setting, infomorphisms are just continuous functions.

In our main example, infomorphisms represent dependancies between systems. If the system represented by $\mathscr{B}$ is in state b then the system represented by $\mathscr{A}$ will be in state $f^\vee b$, so that if event α occurs in system $\mathscr{A}$, event $f^\wedge\alpha$ will occur in system $\mathscr{B}$. Of course, the nature of the dependency may have many sources. The systems could be remotely located but more-or-less simultaneously evolving, with the infomorphism representing the instantaneous (or near-instantaneous) determination of system $\mathscr{A}$ by system $\mathscr{B}$. A particular case of this is where system $\mathscr{A}$ is a subsystem of system $\mathscr{B}$. Or the systems may be temporally distant, with the infomorphism representing the deterministic persistence of information over time. Each past state of the system b resulted in present system $f^\vee b$, whereas each present event α corresponds to a past event $f^\wedge\alpha$.

Classifications under infomorphisms, as abstract mathematical objects, have been extensively studied as the category of Chu spaces over Set with binary values, or Chu(Set,2).[19] Among other properties (typically more interesting to those working

---

[18] Not every infomorphism is an interpretation, in the usual sense, since there is no requirement of compositionally.

[19] A number of papers by Vaughan Pratt, such as [24], explore the state-event interpretation of Chu spaces.

$\overrightarrow{f}\,L$	The *right image* under f	$\Gamma \vdash_{\overrightarrow{f}_L} \Delta$ iff $f^{\wedge^{-1}}[\Gamma] \vdash_L f^{\wedge^{-1}}[\Delta]$
in $\mathsf{Log}(\mathscr{B})$	of logic L in $\mathsf{Log}(\mathscr{A})$	$N_{\overrightarrow{f}L} = f^{\vee^{-1}}[N_L]$.
$\overleftarrow{f}\,L$	The *left image* under f	$\Gamma \vdash_{\overleftarrow{f}_L} \Delta$ iff $f^{\wedge}[\Gamma] \vdash_L f^{\wedge}[\Delta]$
in $\mathsf{Log}(\mathscr{A})$	of logic L in $\mathsf{Log}(\mathscr{B})$	$N_{\overleftarrow{f}_L} = f^{\vee}[N_L]$.

on Chu spaces), this category is cartesian closed, and this simple fact is used to prove many of the properties we will be interested in. In particular, it gives us the existence of the limits and co-limits of diagrams.

Moreover, infomorphisms can be used to relate the logics on one classification to those on another. In fact, each infomorphism $f:\mathscr{A} \rightleftarrows \mathscr{B}$ induces order-homomorphisms $\overrightarrow{f} : \mathsf{Log}(\mathscr{A}) \rightarrow \mathsf{Log}(\mathscr{B})$ and $\overleftarrow{f} : \mathsf{Log}(\mathscr{B}) \rightarrow \mathsf{Log}(\mathscr{A})$, defined as follows.[20]

Fact 5 *Given $f:\mathscr{A} \rightleftarrows \mathscr{B}$, , and $L_1 \in \mathsf{Log}(\mathscr{A})$, $L_2 \in \mathsf{Log}(\mathscr{B})$,*

If L_1 is sound then $\overrightarrow{f}\,L_1$ is sound; conversely, if $f^{\wedge}$ surjective.

If L_1 is complete then $\overrightarrow{f}\,L_1$ is complete; conversely, if $f^{\wedge}$ surjective.

If L_2 is sound then $\overleftarrow{f}\,L_2$ is sound; conversely, if $f^{\wedge}$ injective and $f^{\vee}$

surjective.

If L_2 is complete then $\overleftarrow{f}\,L_2$ is complete, if both $f^{\wedge}$ and $f^{\vee}$ are surjective;

conversely, if $f^{\wedge}$ is injective.

In particular, when restricted to sound and complete logics, $\overleftarrow{f} : \mathsf{Log}^*(\mathscr{B}) \rightarrow \mathsf{Log}^*(\mathscr{A})$ is a homomorphism and $\overrightarrow{f} : \mathsf{Log}^*(\mathscr{A}) \rightarrow \mathsf{Log}^*(\mathscr{B})$ is also a homomorphism if $f^{\wedge}$ and $f^{\vee}$ are surjective.

This brief study of the properties of soundness and completeness, as logics are propagated across infomorphisms, is enough to provide a representation of sound and complete logics:

Fact 6 *A logic is sound and complete iff it is a left image of a natural logic.*

Proof If L on $\mathscr{A}$ is sound and complete, then let $\mathscr{B}$ be the restriction of $\mathscr{A}$ to normal tokens, i.e., such that $\mathscr{B}^{\wedge} = \mathscr{A}^{\wedge}$ and $\mathscr{B}^{\vee} = N_L$. Then the infomorphism $f:\mathscr{A} \rightleftarrows \mathscr{B}$ with $f^{\wedge}$ the identity and $f^{\vee}$ the inclusion, is such that $L = \overleftarrow{f}\,L_B$. The converse follows from Facts 2 and 5.

This shows that it would be possible to do without the extra layer of structure we added when moving from classifications to local logics, by paying sufficient attention to infomorphisms. However, for unsound and incomplete logics, a little more is needed.

---

[20] $\overleftarrow{f}$ and $\overrightarrow{f}$ may not preserve lattice joins and meets.

van Benthem [14] provides another perspective on the 'flow' properties of infomorphisms, by defining a two-sorted language for describing classifications and then asking which formulas in this language are preserved by infomorphisms. Using our present notation, van Benthem's language is the language of predicate logic with variables a for tokens, α for types, and a single binary relation $\models$. Of these formulas, the ones that are preserved by arbitrary infomorphisms, are what he calls *flow formulas*:

$$\phi \quad ::= \quad a \models \alpha \mid a \not\models \alpha \mid (\phi \wedge \phi) \mid \exists a\, \phi \mid \forall \alpha\, \phi$$

Notice in particular that the formula expressing that a (finite) constraint does not hold in the natural logic of a classification can be expressed by a flow formula, e.g. $\alpha_1, \alpha_2 \not\vdash \beta_1, \beta_2$ is expressed by

$$\exists a\, (a \models \alpha_1 \wedge a \models \alpha_2 \wedge a \not\models \beta_1 \wedge a \not\models \beta_2)$$

35.2.3 Information Flow Along a Channel

When two classifications have different complementary concerns because they are about different things or use very different ways of classifying similar things one should not expect that either one can be infomorphically related to the other. In the case of logical semantic classifications, we may have two very different languages and two very different classes of models. Nonetheless, one classification may convey information about the other, in the sense of information-as-correlation. To represent the correlation, one can interpret each in a more expressive theory using a pair of interpretations. Generalising, we define a *binary channel* $c: \mathscr{A} \overset{\rightarrow\;\leftarrow}{\;} \mathscr{B}$ between $\mathscr{A}$ and $\mathscr{B}$ to be a pair of infomorphisms $c_{\mathscr{A}}: \mathscr{A} \rightleftarrows [c]$ and $c_{\mathscr{B}}: \mathscr{B} \rightleftarrows [c]$. The classification $[c]$ is called the *core* of the channel.

When the classifications are concrete systems, the relationship between a subsystem and the system of which it is a part can be represented by an infomorphism. Likewise, the relationship between subsystems can be represented by a channel. Or, put conversely, to represent the informational connections between two systems we can see them as subsystems of something larger. The sense in which they are parts may be taken in a literal mereological sense or something more epistemic, in which the two systems are merely represented within a third system. More concretely still, if we consider a communication channel such as a telegraph wire, we can model events at the source and receiver within distinct classifications $\mathscr{S}$ and $\mathscr{R}$. The wire connecting the two then determines a correlation between them models as a channel $c: \mathscr{S} \overset{\rightarrow\;\leftarrow}{\;} \mathscr{R}$ in which $c_{\mathscr{S}}{}^{\vee}(a)$ and $c_{\mathscr{R}}{}^{\vee}(a)$ are the states of the source and receiver when the channel is in state a and events σ and ρ of the source and receiver correspond to channel events $c_{\mathscr{S}}{}^{\wedge}(\sigma)$ and $c_{\mathscr{R}}{}^{\wedge}(\rho)$.

Information flow along a channel can then be expressed by pushing logics between the connected classifications using the homomorphisms $\overrightarrow{c_{\mathscr{B}}}\overleftarrow{c_{\mathscr{A}}} : \mathsf{Log}(\mathscr{A}) \rightarrow \mathsf{Log}(\mathscr{B})$

and $\overrightarrow{c_{\mathscr{A}}}\overleftarrow{c_{\mathscr{B}}}:\mathsf{Log}(\mathscr{B}) \to \mathsf{Log}(\mathscr{A})$. The *image along channel* c of a logic L on $\mathscr{B}$ is the logic $\overrightarrow{c_{\mathscr{A}}}\overleftarrow{c_{\mathscr{B}}}L$ on $\mathscr{A}$.

Fact 7 *A logic is sound iff it is a channel image of a natural logic.*

Proof We construct a classification $\mathsf{Cla}(L)$ with the same types as $\mathscr{A}$ but with tokens $\langle \Gamma, \Delta \rangle$ for which $\Gamma \not\vdash_L \Delta$, and such that $\langle \Gamma, \Delta \rangle \models_{\mathsf{Cla}(L)} \alpha$ iff $\alpha \in \Gamma$ and $\alpha \notin \Delta$. We then define a channel $c_L:\mathscr{A} \overset{*}{\leftrightarrow} \mathsf{Cla}(L)$ whose core $[c_L]$ has again the same types as $\mathscr{A}$ and the same classification relation, but restricted to tokens in N_L. The maps $c_{L\mathscr{A}}{}^{\wedge}$ and $c_{L\mathsf{Cla}(L)}{}^{\wedge}$ are both identity and $c_{L\mathscr{A}}{}^{\vee}$ is the inclusion $N_L \subseteq \mathscr{A}^{\vee}$. Finally, $c_{L[c_L]}{}^{\vee}(a) = \langle a^+, a-\rangle$ where $a^+ = \{\alpha \mid a \models_{\mathscr{A}} \alpha\}$ and $a^- = \{\alpha \mid a \not\models_{\mathscr{A}} \alpha\}$. It can then be checked that if L in $\mathsf{Log}(\mathscr{A})$ is sound then $L = \overrightarrow{c_{L\mathscr{A}}}\overleftarrow{c_{L\mathsf{Cla}(L)}}L_{\mathsf{Cla}(L)}$. The converse follows from Facts 2 and 5.

Channels can be composed: given $c:\mathscr{A} \overset{*}{\leftrightarrow} \mathscr{B}$ and $d:\mathscr{B} \overset{*}{\leftrightarrow} \mathscr{C}$ the channel $cd:\mathscr{A} \overset{*}{\leftrightarrow} \mathscr{C}$ is defined using the push-out channel $p:[c] \overset{*}{\leftrightarrow} [d]$ of $c_{\mathscr{B}}:\mathscr{B} \rightleftarrows [c]$ and $d_{\mathscr{B}}:\mathscr{B} \rightleftarrows [c]$. In diagrams[21]:

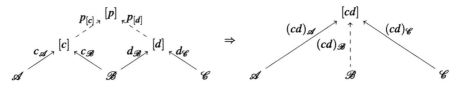

Note that the composition of two binary channels gives, in addition to a new binary channel, an infomorphism $(cd)_{\mathscr{B}}$ from the intermediate classification $\mathscr{B}$ to the core of the new channel. We can push the natural logic from this intermediate classification to the core, to get the logic $\overrightarrow{(cd)_{\mathscr{B}}}L_{\mathscr{B}}$ which can then be used for reasoning about the relationship between $\mathscr{A}$ and $\mathscr{C}$. Composition thus occurs by 'absorbing' into the new channel the informational content of the intermediate classification $\mathscr{B}$ and how it connects $\mathscr{A}$ and $\mathscr{C}$. For a concrete example, think again of the case of a telegraphic communication, in which both sender and receiver are represented as system classifications, as is the wire that connects them. One channel connects the sender to the wire and a second connects the wire to the receiver. When these are composed we get a channel connecting sender to receiver, in which all relevant information about the wire has been absorbed.

The construction can be generalised to combine any system of channels to produce a unique channel between any two classifications of the given channel. More precisely, given classifications $\mathscr{A}_0, \ldots, \mathscr{A}_n$ and a chain of channels $c_i:\mathscr{A}_i \overset{*}{\leftrightarrow} \mathscr{A}_{i+1}$ for $0 \le i < n$, we can repeat this step, pushing of the logics $L_{\mathscr{A}_1}, \ldots, L_{\mathscr{A}_{n-1}}$ on to the core and combine them with the lattice join in $\mathsf{Log}([c_0 \ldots c_{n-1}])$ to get

$$L_{cd} = \bigvee_{i=0}^{n-1} \overrightarrow{(c_i c_{i+1})_{\mathscr{A}_{i+1}}}L_{\mathscr{A}_{i+1}}$$

---

[21] Specifically, $(cd)_{\mathscr{A}} = p_{[c]}c_{\mathscr{A}}$ and $(cd)_{\mathscr{B}} = p_{[d]}c_{\mathscr{B}}$.

This core logic can be added to logics pushed from $\mathscr{A}$ to $\mathscr{C}$ or vice versa. So, for example, given $L \in \mathsf{Log}(\mathscr{C})$, we get $\overrightarrow{c_{\mathscr{A}}}(L_{cd} \vee \overleftarrow{c_{\mathscr{C}}}L)$ in $\mathsf{Log}(\mathscr{A})$. The logic represents those inferences that are possible within the system. For example, $\langle 1, \alpha \rangle, \langle 3, \beta \rangle \vdash \langle 4, \gamma \rangle$ in $\mathsf{Log}([c_0 \ldots c_{n-1}])$ represents the inference from α in $\mathscr{A}_1$ and β in $\mathscr{A}_3$ to γ in $\mathscr{A}_4$. The tokens of the core can be seen as sequences $a_1, \ldots, a_n$ of tokens in the component classifications, so, given sound component logics, the inference implies that if $a_1 \models \alpha$, $a_3 \models \beta$ then $a_4 \models \gamma$.

35.2.4 Logic in Channels: Core Logics and Distributed Logics

More generally, a *channel* between classifications $\{\mathscr{A}_i\}_{i \in I}$ is a family of infomorphisms $\{c_i : \mathscr{A}_i \rightleftarrows [c]\}_{i \in I}$ which we can think of as a model of informational relationships between the *component* classifications $\{\mathscr{A}_i\}_{i \in I}$. It can be composed with any other channel $\{d_i : \mathscr{A}_i \rightleftarrows [d]\}_{i \in J}$ to form $\{cd_i : \mathscr{A}_i \rightleftarrows [cd]\}_{i \in I+J}$[22] which models any interaction between the two systems. Logics on component classifications can be pushed from any one classification $\mathscr{A}_i$ to any other $\mathscr{A}_j$ using the homomorphism $\overleftarrow{c_j} \overrightarrow{c_i} : \mathsf{Log}(\mathscr{A}_i) \to \mathsf{Log}(\mathscr{A}_j)$. Thinking again of the case in which the classifications model concrete systems, an observation of one component $\mathscr{A}_i$ provides information about it, the simplest kind of which is that some event $\alpha \in \mathscr{A}_i^{\wedge}$ has occurred. Suppose for example that the lighting in a house is wired in a complicated way so that each light is controlled by a number of switches, some in different rooms, and that each room is classified according to simply observable features: the light being on or off, a switch being down or up, etc. Now suppose that a light turns on in the bedroom. What information is conveyed by this event? If it is represented by type α in classification $\mathscr{A}_i$, in which the events occurring in the bedroom are classified, then conclusions about the hallway, represented in classification $\mathscr{A}_j$, say, are given by pushing the logic L_α across to $\mathscr{A}_j$ to obtain $\overleftarrow{c_j} \overrightarrow{c_j} L_\alpha$. Furthermore, the natural logic $L_{[c]}$ of the channel core, which represents constraints relating the component classifications, such as are induced by the wiring of the house, can be factored into this process, so that information from the occurrence of α in $\mathscr{A}_i$ can be combined with $L_{[c]}$ as $\overrightarrow{c_j} L_\alpha \vee L_{[c]}$ and then pushed to $\mathscr{A}_j$ as $\overleftarrow{c_j}(\overrightarrow{c_j} L_\alpha \vee L_{[c]})$.[23]

Distributed logics can also be studied in this more general setting to obtain logics that explicitly represent the interaction between classifications. The sum $\mathscr{A} + \mathscr{B}$ is the core of the *sum channel* $\sigma : \mathscr{A} \overset{+}{\leftrightarrows} \mathscr{B}$ which involves minimal interaction between the two classifications, and the logic $L_{\mathscr{A}+\mathscr{B}}$ is just the join $(\overrightarrow{\sigma_{\mathscr{A}}} L_{\mathscr{A}} \vee \overrightarrow{\sigma_{\mathscr{B}}} L_{\mathscr{B}})$.[24] More generally, minimal interaction among a family of classifications, $\{\mathscr{A}_i\}_{i \in I}$ is

---

[22] $I + J$ is the disjoint union of I and J. The construction of cd is a generalisation of the binary case. Just take the co-limit of $\{c_j : \mathscr{A}_i \rightleftarrows [c]\}_{i \in I+J}$

[23] Indeed, any core logic $L \in \mathsf{Log}([c])$, such as that is given by the construction of a channel from a chain, can be factored into the pushing process in a similar way.

[24] $\sigma_{\mathscr{A}}^{\wedge}$ and $\sigma_{\mathscr{B}}^{\wedge}$ are the inclusions of $\mathscr{A}^{\wedge}$ and $\mathscr{B}^{\wedge}$ in the disjoint union $\mathscr{A}^{\wedge} + \mathscr{B}^{\wedge}$. And $\sigma_{\mathscr{A}}^{\vee}$ and $\sigma_{\mathscr{B}}^{\vee}$ are the projections of $\mathscr{A}^{\vee} \times \mathscr{B}^{\vee}$ to $\mathscr{A}^{\vee}$ and $\mathscr{B}^{\vee}$, respectively.

modelled by the sum channel with core $\sum_{i \in I} \mathscr{A}_i$. As in the binary case, we define $\mathsf{Log}(\sum_{i \in I} \mathscr{A}_i)$ to be the class of *distributed logics* over the family $\{\mathscr{A}_i\}_{i \in I}$.

Given channels c and d between classifications $\{\mathscr{A}_i\}_{i \in I}$, an infomorphism $f:[c] \rightleftarrows [d]$ is a *channel morphism* from c to d iff $fd_i = c_j$ for all $i \in I$. For any channel c over the family there is a unique channel morphism c_+ from sum $\sum_{i \in I} \mathscr{A}_i$ to c (by the limit properties of sums), and so a uniquely defined distributed logic $\overleftarrow{c_+}L_{[c]}$, which we call the *natural distributed logic* of c. Using Fact 7 it is then easy to show:

Fact 8 *Every distributed logic satisfying* WIP *is the natural distributed logic of some channel.*

35.2.5 Normality Reconsidered

An important class of logics fail to satisfy the Partition rule and so cannot be studied using the concept of a natural logic. Suppose $P = \langle \Omega, E, \mu \rangle$ is a probability space with outcomes Ω, events E and probability measure $\mu : E \to [0, 1]$. Let $\mathscr{P}$ be the a classification with $\mathscr{P}^\wedge = E$, $\mathscr{P}^\vee = \Omega$ and $\models_{\mathscr{P}} = \in$. Then we can define $\vdash_\mu$ as $\Gamma \vdash_\mu \Delta$ iff $\mu([\Gamma; \Delta]) = 0$, which states that the probability of a counterexample is 0. Since many systems are better modelled probabilistically, this class of consequence relations is perhaps more important than those considered so far. Moreover, to build a bridge between logical conceptions of information flow and ideas from communication theory, they are essential; see Seligman [27].

The relation $\vdash_\mu$ satisfies WIC and even a countable version of SC but may not satisfy P. If, for example, Ω is the real interval $[0, 1]$, E and the set of its measurable subsets and μ is any distribution that assigns singletons zero probability (e.g., the uniform distribution), then for all Γ', Δ' such that $\Gamma \cup \Delta = \mathscr{P}^\wedge$, either $[\Gamma; \Delta]$, if all singletons $\{x\}$ are in Δ, or there is a singleton in Γ', and in both cases $\mu([\Gamma; \Delta]) = 0$, which means that $\Gamma' \vdash_\mu \Delta'$ but not $\emptyset \vdash_\mu \emptyset$, so violating P. Moreover, there is no non-empty set N of reals with respect to which $\vdash_m u$ is sound. Although $\vdash_\mu [0, x)$, $(x, 1]$, x is in neither $[0, x)$ nor $(x, 1]$ and so is a counterexample. The only sound logic with this consequence relation is therefore the rather curious logic $\langle \emptyset, \vdash_\mu \rangle$, whose completion is the trivial logic $T_{\mathscr{A}}$.

To provide a more satisfactory treatment of $\vdash_\mu$, we will need to look elsewhere. The central idea is to shift to a classification of sets of tokens as either possible (in the sense of having positive probability) or not. So we define the *extension* classification $\delta \mathscr{A}$ to have the same types as $\mathscr{A}$, but sets of tokens from $\mathscr{A}$ as tokens, i.e. $\delta \mathscr{A}^\vee = \mathrm{pow} \mathscr{A}^\vee$. Then $X \models_{\delta \mathscr{A}} \alpha$ iff $a \models_{\mathscr{A}} \alpha$ for every $a \in X$. An *extended logic* on $\mathscr{A}$ is a logic in $\mathsf{Log}(\delta \mathscr{A})$. Now, letting $L_\mu = \langle N_\mu, \vdash_\mu \rangle$ where N_μ is the set of $X \subseteq \mathscr{P}^\vee$ such that $\mu(X) > 0$, we have that

Fact 9 L_μ *is a sound and complete extended logic on* $\mathscr{P}$.

The infomorphism $\delta_{\mathscr{A}}:\delta\mathscr{A}\rightleftarrows\mathscr{A}$ is defined by mapping each token a of $\mathscr{A}$ to the singleton $\{a\}$ and with the identity on types. It can be used to push any logic L on $\delta\mathscr{A}$ to the logic $\overrightarrow{\delta_{\mathscr{A}}}L$ on $\mathscr{A}$. In the case of L_μ the result is just the logic $\langle\emptyset,\vdash_\mu\rangle$ since no singleton has positive probability. We therefore have an example of a vacuously sound logic on $\mathscr{A}$, which is not complete, but which is the image of an interesting sound and complete logic on the extended classification $\delta\mathscr{A}$. In the opposite direction, $\overleftarrow{\delta_{\mathscr{A}}}$ is an embedding of the lattice $\mathsf{Log}(\mathscr{A})$ in $\mathsf{Log}(\delta\mathscr{A})$, showing that the concept of extended logics generalises that of a logic.

Each infomorphism $f:\mathscr{A}\rightleftarrows\mathscr{B}$ can also be extended to an infomorphism $\delta_f:\delta\mathscr{A}\rightleftarrows\delta\mathscr{B}$ called the *image extension* of f by defining $\delta_f{}^\vee X$ to be the image of X under $f^\vee$, i.e. the set $\{f^\vee x\mid x\in X\}$. ($\delta_f{}^\wedge$ is just $f^\wedge$ again.) Then the following diagram commutes:

$$
\begin{array}{ccc}
\delta\mathscr{A} & \xrightarrow{\ \delta_f\ } & \delta\mathscr{B}\\[2pt]
\delta_{\mathscr{A}}\Big\downarrow & & \Big\downarrow\delta_{\mathscr{B}}\\[2pt]
\mathscr{A} & \xrightarrow{\ \ f\ \ } & \mathscr{B}
\end{array}
$$

The logic homomorphisms $\overleftarrow{\delta_f}\overleftarrow{\delta_{\mathscr{B}}}:\mathsf{Log}(\mathscr{B})\to\mathsf{Log}(\delta\mathscr{A})$ and $\overrightarrow{\delta_f}\overleftarrow{\delta_{\mathscr{A}}}:\mathsf{Log}(\mathscr{A})\to\mathsf{Log}(\delta\mathscr{B})$ can be used to push logics on $\mathscr{A}$ and $\mathscr{B}$ to extended logics on $\mathscr{B}$ and $\mathscr{A}$.

The generalisation to extended logics preserves all of the nice features of logics and their infomorphisms studied so far, in that similar constructions are possible, allowing the metaphor of information flow within a network of infomorphisms to be realised also in the movement of probabilistic logics.[25]

35.2.6 The Natural Logic of a Channel

More extended logics can be defined without the help of a probability function, using any channel. For any channel $\{c_i:\mathscr{A}_i\rightleftarrows[c]\}_{i\in I}$ we define the *extension* channel $\{\delta c_i:\delta\mathscr{A}_i\rightleftarrows[\delta c]\}_{i\in I}$ by taking δc_i to be the image extension of c_i. This ensures that $c_i\delta_{\mathscr{A}_i}=\delta_{[c]}\delta c_i$, which is to say that the following diagram commutes:

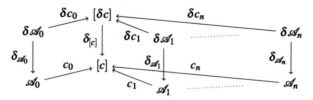

---

[25] There is a lot to investigate about extension logics, such as the relationship to structural properties. Here I conjecture that a logic satisfies WIC iff it is the image of a sound and complete extension logic perhaps with some additional properties.

The *natural logic* L_c of channel c is the result of pushing the natural logic of each component classification $\mathscr{A}_i$ first up to its extensions $\delta\mathscr{A}_i$ and then along the infomorphism δc_i to the core $[\delta c]$ of the channel extension and taking lattice meets, to produce, in $\mathsf{Log}(\delta c)$, the logic

$$L_c \;=\; \bigwedge_{i\in I} \overrightarrow{\delta c_i}\overleftarrow{\delta_{\mathscr{A}_i}} L_{\mathscr{A}_i}$$

It is worth looking at a concrete example of this construction in more detail. Consider again a probability classification $\mathscr{P}$ over $[0, 1]$ for which $\mu(X) = 0$ iff X is a null set,[26] and let $\mathscr{A}_n$ be the classification with the same types as $\mathscr{P}$ but whose tokens are the intervals $I_{m,n} = \left[\frac{m}{n+1}, \frac{m+1}{n+1}\right)$, for $m = 0, \ldots, n-1$, and $I_{n,n} = \left[\frac{n}{n+1}, 1\right]$ and with $\models_{\mathscr{A}_n} = \subseteq$. For example, $\mathscr{A}_2$ has tokens $[0, \frac{1}{3})$, $[\frac{1}{3}, \frac{2}{3})$ and $[\frac{2}{3}, 1)$ and, e.g., $[0, \frac{1}{3}) \models_{\mathscr{A}_2} \alpha$ iff $[0, \frac{1}{3}) \subseteq \alpha$. The classifications $\mathscr{A}_n$ provide an approximation to $\mathscr{P}$ in a sense that can be made precise by constructing a channel $\{c_i{:}\mathscr{A}_n \overrightarrow{\leftarrow} [c]\}_{n\in\mathbb{N}}$ whose natural logic L_μ. To define c, let $[c]$ be $\mathscr{P}$ itself and let each c_n be the identity on types and map each real x (in $\mathscr{P}^\vee$) to the unique interval $I_{m,n}$ that contains it. Now the natural logic L_c is defined so that $\Gamma \vdash_{L_{\mathscr{P}}} \Delta$ iff $\Gamma \vdash_{L_{\delta\mathscr{A}_n}} \Delta$ for all n. But $\vdash_{L_{\delta\mathscr{A}_n}} = \vdash_{L_{\mathscr{A}_n}}$, and so $\Gamma \vdash_{L_{\mathscr{P}}} \Delta$ iff there is no $I_{m,n}$ such that $I_{m,n} \subseteq [\Gamma; \Delta]$. And $\mu[\Gamma; \Delta] > 0$ iff there is some such interval. So $\vdash_{L_c} = \vdash_m u$. Moreover, by Fact 5, L_c is sound and complete, and so by Fact 9, $L_c = L_\mu$. Thus:

Fact 10 L_μ *is the natural logic of a channel.*

By calling the natural logic of a channel 'natural' we are promoting it in a way that deserves some justification. This comes from thinking of classifications as representing the space of events in a system and a channel as representing the way in which a number of systems are related. If we suppose that the states of the component systems are all possible, then the normal sets of tokens in the channel's core, are those that correspond to possible occurrences in its components. For an inference $\Gamma \vdash \Delta$ about the channel core to have a genuine counterexample, it must come from the existence of a violating state in one of the components and not merely one of its own connection states.

Reflecting on this explanation, we see that it involves an implicit assumption that the components are independent: that they can be in one or other state without any effect on the states of the other components. Stronger logics on the channel will arise from greater interaction between the components.

Another issue that arises is the relationship between the 'natural distributed logic' of a channel (defined at the end of Sect. 35.2.4) and its 'natural logic'. This can be explained with the following commutative diagram:

---

[26] Null in the underlying measure, which we can assume to be, e.g. Borel measure, so that the singletons and countable unions of them are all null. Typical probability measures, such as the uniform distribution, or any normal distribution are all of this kind.

$$\delta(\Sigma_{i\in I}\,\mathscr{A}_i) \xrightarrow{\;\;\delta_{c_+}\;\;} \delta[c]$$

$$\delta_+ \downarrow \qquad\qquad\qquad \downarrow \delta_{[c]}$$

$$\Sigma_{i\in I}\,\mathscr{A}_i \xrightarrow[\;\;c_+\;\;]{} [c]$$

The 'natural distributed logic' of c is defined as $\overleftarrow{c_+}L_{[c]}$, which is the result of pushing the natural logic of the channel's core to the sum, so as to distribute it for reasoning about the channel's components. What we are calling the 'natural logic' of c, the logic L_c, is an extended logic and so lives in $\mathsf{Log}(\delta[c])$. But it too can be distributed for reasoning about the components, by pushing it along the channel morphism δ_{c_+} to get the *distributed natural logic* $\overleftarrow{\delta_{c_+}}L_c$, which is an extended distributed logic, living in $\mathsf{Log}(\delta(\Sigma_{i\in I}\,\mathscr{A}_i))$.

The natural logic of a channel arises from regarding the components as defining what is possible in the core. Fixing the state of the component classifications does not always completely determine the state of the core, so allowing a degree of non-determinism, and this is reflected in the natural logic.

35.2.7 Logical Operators

The abstract setting we have been using makes it easy to give criteria for the existence of various logical operations. Given a logic L on $\mathscr{A}$, we say that unary operation $\neg$ and binary operations $\wedge$, $\vee$, and $\rightarrow$ on the set of types $\mathscr{A}^\wedge$, are *Gentzen negation, conjunction, disjunction* and *implication*, respectively, for L iff

$$
\begin{array}{lll}
\Gamma \vdash_L \Delta, \neg\alpha & \text{iff} & \alpha, \Gamma \vdash_L \Delta \\
\neg\alpha, \Gamma \vdash_L \Delta & \text{iff} & \Gamma \vdash_L \Delta, \alpha \\
\Gamma \vdash_L \Delta, (\alpha \wedge \beta) & \text{iff} & \Gamma \vdash_L \Delta, \alpha \text{ and } \Gamma \vdash_L \Delta, \beta \\
(\alpha \wedge \beta), \Gamma \vdash_L \Delta & \text{iff} & \alpha, \beta, \Gamma \vdash_L \Delta \\
\Gamma \vdash_L \Delta, (\alpha \vee \beta) & \text{iff} & \Gamma \vdash_L \Delta, \alpha, \beta \\
(\alpha \vee \beta), \Gamma \vdash_L \Delta & \text{iff} & \alpha, \Gamma \vdash_L \Delta \text{ and } \beta, \Gamma \vdash_L \Delta \\
\Gamma \vdash_L \Delta, (\alpha \rightarrow \beta) & \text{iff} & \alpha, \Gamma \vdash_L \Delta, \beta \\
(\alpha \rightarrow \beta), \Gamma \vdash_L \Delta & \text{iff} & \Gamma \vdash_L \Delta, \alpha \text{ and } \beta, \Gamma \vdash_L \Delta
\end{array}
$$

35.3 Reconstructing Situation Theory

In this final section, I will attempt to apply the theory developed above to restate some of the founding intuitions of situation theory in a new form, taking a stand on various issues raised in Sect. 35.1. I'll then do the same with van Benthem's model from [13].

35.3.1 Situations as Local Logics

The proposal is to model a situation s as a local logic. As such, we are considering the partiality of situations to be based on a degree of non-determinism. We can think of the tokens of the underlying classification $[s]$ of the situation as being determinate possibilities, representing possible ways in which the situation may be connected to other situations and thereby interact with them. The types represent potential information about the situation. As such, each type α models an issue that may or may not be settled, positively or negatively, by s. Then we interpret[27]

$$s \models \langle\!\langle \alpha; + \rangle\!\rangle :: \vdash_s \alpha$$
$$s \models \langle\!\langle \alpha; - \rangle\!\rangle :: \alpha \vdash_s$$

Consider our concrete example of the classification of states of a system by events that occur within it; a situation s is then a local pattern of regularity in the behaviour of the system when its state lies within the region N_s. If s is sound and complete, then $s \models \langle\!\langle \alpha; + \rangle\!\rangle$ just in case the occurrence of α is already determined: no state in N_s is compatible with α not occurring. Likewise, $s \models \langle\!\langle \alpha; - \rangle\!\rangle$ just in case it is already determined that α cannot occur: no state in N_s is compatible with α occurring. This interpretation is in line with the modal character of the $\models$ relation noted in Sect. 35.1.3. Question 9 is answered in the expected way: it is not possible for a situation to support both an infon and its dual but it may support neither.

The next step is to model analytic constraints in a similar way:

$$\langle\!\langle \alpha; + \rangle\!\rangle \underset{s}{\Rightarrow} \langle\!\langle \beta; + \rangle\!\rangle :: \quad \alpha \vdash_s \beta$$

$$\langle\!\langle \alpha; + \rangle\!\rangle \underset{s}{\Rightarrow} \langle\!\langle \beta; - \rangle\!\rangle :: \quad \alpha, \beta \vdash_s$$

$$\langle\!\langle \alpha; - \rangle\!\rangle \underset{s}{\Rightarrow} \langle\!\langle \beta; + \rangle\!\rangle :: \quad \vdash_s \alpha, \beta$$

$$\langle\!\langle \alpha; - \rangle\!\rangle \underset{s}{\Rightarrow} \langle\!\langle \beta; - \rangle\!\rangle :: \quad \alpha, \beta \vdash_s \alpha$$

These constraints, on our concrete interpretation, concern the relative co-occurrence of event. In fact, the situation s contains more information than is expressed by the binary involvement relation. As a simple example, the impossibility of the co-occurrence of α, β and γ, namely $\alpha, \beta, \gamma \vdash_s$ cannot be reduced to facts of the form $\sigma \underset{s}{\Rightarrow} \tau$. In answer to Question 5, we get as a consequence the Analytic Involvement principle: if $s \models \sigma$ and $\sigma \underset{s}{\Rightarrow} \tau$ then $s \models \tau$.

Incomplete situations are ones in which not all of the potentially available information about the co-occurrence of events is represented as constraints. This possibility is due to the epistemic character of situations, which may contain less than all the

---

[27] A potential for confusion here is to mistake the $\models$ of situation theory with the relation $\models_{[s]}$ of the classification $[s]$ but we trust that any difficulties can be resolved in context. Of course, the classification of situations by the infons they support is another classification, of which the situation theoretic $\models$ is the classification relation.

information. This is a *second* source of the lack of bivalence in ST: for example, even if the state of the system is totally determined, so that N_s is a singleton, it may still be that neither $s \models \langle\!\langle \alpha; + \rangle\!\rangle$ nor $s \models \langle\!\langle \alpha; - \rangle\!\rangle$.

Finally, unsound situations are also of interest. A lack of soundness breaks the modal force of the constraints. If s is unsound, then it may be that $s \models \langle\!\langle \alpha; + \rangle\!\rangle$ even though there is a possible state of the system in N_s that is not compatible with α's occurrence. What does this mean? I do not have a general answer to this question, but in some cases, we can make sense of it. Consider the examples of Sect. 35.2.5, in which constraints are defined to hold when the probability of a counterexample is 0. Since being infinitely improbably and being impossible are not the same[28] there is a gap of exactly this kind, so suggesting an interpretation of unsound situations: the tokens that lie in N_s but that are not of type α when $s \models \langle\!\langle \alpha; + \rangle\!\rangle$ are those that, while possible in some objective sense, are taken to be so unlikely that they are disregarded.

For constraints between situations s and t, we look to local logics on the classification $[s] + [t]$. Given such a logic / situation u, we can therefore interpret:

$$[u: \sigma \underset{s,t}{\Rightarrow} \tau] \quad :: \quad \sigma_s \underset{u}{\Rightarrow} \tau_t$$

where σ_s and τ_t are the obvious u-infon correlates of σ and τ (which are infons of s and t, respectively.)[29] So the primary definition of constraint in the current setting is that of a *conditional* constraint, in which the 'background' condition for constraints relating information in s to information in t is supplied by a third situation, u. Of course, the framework of local logics suggests an extension of the notation we introduced in Sect. 35.1 involving many situations and the full two-sided $\vdash$, of which $\underset{s,t}{\Rightarrow}$ is just a special case.

Much of in Sect. 35.3.1 was devoted to a discussion of how local logics can be pushed along infomorphisms in an interrelated web of classifications. This serves as a basis for a long and detailed answer to Question 6 about the existence of channels and how they underwrite constraints. We will focus here on the simplest case. For any binary channel $\{c_s:[s] \rightleftharpoons [c], c_t:[t] \rightleftharpoons [c]\}$ the logics s and t can be combined via c to get the natural distributed logic $\overleftarrow{c_+}L_{[c]}$ on the sum $[s] + [t]$. This, then, provides a direct answer to Question 6 which can be represented by:

$$[c: \sigma \underset{s,t}{\Rightarrow} \tau] \quad :: \quad [\overleftarrow{c_+}L_{[c]}: \sigma \underset{s,t}{\Rightarrow} \tau]$$

Moreover, by the composition of channels, we get a version of the Xerox principle, in answer to Question 7: If $[c: \sigma \underset{s,t}{\Rightarrow} \tau]$ and $[d: \tau \underset{t,u}{\Rightarrow} v]$ then $[cd: \sigma \underset{s,u}{\Rightarrow} v]$

Aside from answering Question 6 about the nature of channels, the theory of information flow articulated in [12] and recapitulated in Sect. 35.2 is intended to say

---

[28] as every reader of Douglas Adam's *Hitchhiker's Guide* knows.

[29] The sum $[s] + [t]$ is defined as a limit by infomorphisms $\iota_s:[s] \rightleftharpoons [s] + [t]$ and $\iota_t:[t] \rightleftharpoons [s] + [t]$ and these are used to find the 'correlates', e.g. $\langle\!\langle \alpha; + \rangle\!\rangle$ corresponds to $\langle\!\langle \iota_s^\wedge(\alpha); + \rangle\!\rangle$.

something about the possibility of error, our Question 8. In part, this is addressed by the concept of a conditional constraint, with its explicit dependence on the 'background condition' supplied by a channel. But this is only half the story. At the heart of our interpretation is the definition of support: $s \models \langle\!\langle \alpha; + \rangle\!\rangle$ iff $\vdash_s \alpha$. But the extent to which this gives accurate information about the tokens of s depends both on the normal tokens of s and whether it is sound. We must also consider carefully the application of reasoning about situations in the abstractly probabilistic setting, modelled by the 'natural logic' of a channel, from Sect. 35.2.6. It is difficult to reduce this level of analysis to concepts that we have taken from situation theory: support, involvement, and the like. Nonetheless, we have seen in Sect. 35.2 that a wide range of local logics can be defined by projecting the natural logic of a classification along infomorphisms, and so to this extent, explicit consideration of normality can be replaced by the algebra of these projections. To put this in the language of situation theory, the 'background condition' u of $[u: \sigma \Rightarrow \tau]$ can be derived by pushing the information from a possibly remote situation v along a network of connections. A simple example of this is the definition of the distributed logic of a channel c as $\overleftarrow{c_+} L_{[c]}$ that we just used to explain the role of a channel in conditionalising constraints. A suggestion then, is to augment the concepts of situation theory with those of these information-pushing operations.

35.3.2 Perspectives as Infomorphisms

By considering the source of the logic of a situation as coming from the information network in which it resides, we presuppose that there is such a network, that the world is to be modelled as such a network, and so that there are not only situations but the infomorphisms that link them. Or, to put it more accurately, that there are classifications and interpretations of them within other classifications. All of this has to be part of our model. Because of this the present framework goes a fair way beyond the tacit realism of early proponents of situation theory. Instead of a world of situations built from a scheme of individuation, with the possibility of different schemes of individuation as a promissory note, our model builds in the notion of a 'perspective' at the ground level. An infomorphism is just a model of the interpretation of one classification in another, or, in other words, the seeing of one classification in terms of another. Thus, in answer to Question 11, we have no need of an additional layer of relativism: it lies at the core.

That said, it would be good to have some account of infon structure: how more complex items of information are related to their parts. To gesture at an answer to Question 10, we can mention the way in which logic operators on types can be defined implicitly by their Gentzen conditions, as explained in Sect. 35.2.7. But this is only one of many ways in which information can be structured. Rather than providing a final answer to this question, the present approach to situation theory leaves it to more detailed consideration of specific systems of representation. The hope is that the concept of a local logic is sufficiently flexible to account for any such system or that it can be appropriately extended.[30]

Likewise, we have not yet touched on the mereology of situations and its relationship to information flow, as queried in Questions 1, 2, and 3. The present interpretation resists any absolute judgement as to whether one situation is part of another by focusing instead on the informational relationships between them. No global perspective is assumed from the vantage point of which one could make sense of parts and wholes. Nonetheless, when modelling particular systems that *are* conceptualised in this way, their mereological structure induces channels. Whenever situations s and t are parts of situation u, the way in which they are included can be represented by infomorphisms $:[s] \underset{\leftarrow}{\rightarrow} [u]$ and $:[t] \underset{\leftarrow}{\rightarrow} [u]$, , which is of course a channel, but not one that is distinguished from other channels in an informationally significant way.

On the matter of the individuation of situations, it is clear that by taking a situation to be a local logic, there is more to it than merely the infons it supports, which would give us a negative answer to Question 2, rejecting the principle of Extensionality. But from this negative answer a number of new questions emerge. Can two distinct situations *carry* the same information? Is it even *possible* for Extensionality to hold given a sufficiently rich and interconnected network of information? Is every situation the image of a natural logic? This last question is especially interesting as a positive answer suggests a complete reduction of situations to classifications and their interpretation by infomorphisms. This offers a potential answer also to Question 12, concerning the possibility of situation theory being its own metatheory. To make sense of this we would have to ask how to interpret the concepts of classification and infomorphism in terms of situations, infons and the like. Perhaps a more fruitful line would be to ask whether the theory of classification and infomorphisms can be *its* own metatheory. Here the example of Category Theory provides some insight but further consideration of this idea will have to be left for another occasion.

35.3.2.1 Some Alternatives/Additions

With the decision to model situations as local logics and perspectives as infomorphisms, we have seen that the pertinent question becomes: which logics and which infomorphisms? Each classification merely records what is classified as what and there are many ways of identifying logical structure within. One significant way of doing this, as we have seen, is to define a logic on one classification by means of an interpretation of it in another, that is to say, by pushing logical structure along infomorphisms. Suppose we have one situation s, based on a classification $[s]$ within which many other situations are interpreted. We can record these interpretations as infomorphisms, $t_s:[t] \underset{\leftarrow}{\rightarrow} [s]$ where t is a situation (logic) based on classification $[t]$. Take for example, the case of a faithful representation. Situation s is one in which

---

[30] It was the realisation that the theory of [12] was unable to account for probabilistic structure that led to the present variant.

we the doctor examining the X-rays of Jackie's leg and situation t is the situation of the leg itself, classified into the types 'broken' and 'unbroken'. These are mapped by t_s into visually distinct types of X-ray picture, with the possible X-rays mapped to possible states of the leg.

An alternative comes from thinking of how situations are identified from the outside, by means of informational connections, which we are modelling as info-morphisms. So suppose we have a situation s and another situation t, together with an informorphism $t_s : [t] \rightleftarrows [s]$. Situation t is represented inside s by means of this informorphism as the logic $\overleftarrow{t_s}\,s$, the left image of s along t_s. This is the information that s possesses about t, so we define

$$s \models \langle\!\langle t \models \sigma; + \rangle\!\rangle \quad \text{as} \quad \overleftarrow{t_s}\,s \models \sigma$$

$$s \models \langle\!\langle t \models \sigma; - \rangle\!\rangle \quad \text{as} \quad \overleftarrow{t_s}\,s \not\models \sigma$$

$$s \models (\sigma \Rightarrow \tau \text{ holds in } t) \quad \text{as} \quad \sigma \Rightarrow \tau \text{ holds in } \overleftarrow{t_s}\,s$$

The corresponding Indication Principle therefore also applies: if $s \models (t \models \sigma)$ and $s \models (\sigma \Rightarrow \tau \text{ holds in } t)$ then $s \models (t \models \tau)$. But the relationship between $s \models (t \models \sigma)$ and $t \models \sigma$ is more complicated, in general. What is at issue is the relationship between the two logics t and $\overleftarrow{t_s}\,s$. To model the requirement that t_s gives information about t, as opposed to any other logic that shares $[t]$, we suppose that $\overleftarrow{t_s}\,s \leq t$ and from this it follows that $s \models (t \models \sigma)$ implies $t \models \sigma$ (but not vice versa).[31]

A third perspective comes from thinking of the logic of a situation as determined by the informational relationships between a classification s and other classifications, as represented by a set $\mathbb{C}$ of infomorphisms. We define $\ln(\mathbb{C}, s)$ to be the set of infomorphisms in $\mathbb{C}$ that point to $\mathscr{A}$, i.e. those of the form $f : \mathscr{B} \rightleftarrows \mathscr{A}$ for some $\mathscr{B}$. This set forms a channel with core $\mathscr{A}$ and so determines the natural channel logic $L_{\ln(\mathbb{C},s)}$ on s. Thus we can say that

$$\ln \mathbb{C}, s \models \sigma \text{ iff } L_{\ln(\mathbb{C},s)} \models \sigma$$

All that changes in these various approaches to modelling situations, is the source of the local logic used to determine which infons are supported and which constraints are satisfied. As such, aspects of both the 'information as range' and 'information as correlation' are involved.

35.3.3 Using van Benthem's Constraint Logic

van Benthem in [13] introduces a model of situations and constraints, which we will now summarise, with some slight changes in notation and terminology. The

---

[31] Notice that in $s \models (t \models \sigma)$, the infon σ is an infon of the classification $[t]$ not of $[s]$. In fact, the subexpression $(t \models \sigma)$ does not denote an infon at all.

starting point is to think of a situation as something that can be in one of a range of 'local states'. Each issue is interpreted as a predicate of local states. A 'global state' is an assignment of local states to situations, so that a particular global state determines exactly how the issues are resolved by each situation. But not all global states are possible. This restriction permits a modal understanding of both support and information flow.

So, first define a *constraint frame* to be a structure $F = \langle S, L, G \rangle$ that consists of a set S of *situations*, a set L of *local states*, a set G of functions assigning a local state to each site, each of which is called a *global state*.[32] Each situation can be in one of a range of local states and a global state specifies the local state of each situation, but not all assignments of local states to situations are possible, only those in the set G. van Benthem defines the following language for describing global states:

$$\phi ::= \alpha(\mathbf{s}) \mid \neg\phi \mid (\phi \wedge \phi) \mid [\mathbf{s}]\phi \mid U\phi$$

where $\mathbf{s}$ is a tuple of *situation names* and α is a *predicate*. When two situation names st occur together, we can understand this as referring to the composite situation made up of the situations named by s and t.

The language is interpreted in a *constraint model* $M = \langle F, V \rangle$ where $F = \langle X, L, G \rangle$ is a constraint frame and V is a *valuation* function that assigns a subset of X to each situation name and a set of sets of local states to each predicate. Then, given any state g in G, a satisfaction relation $\Vdash$ can be defined by:

$$
\begin{aligned}
M, g &\Vdash \alpha(\mathbf{s}) &&\text{iff} && g(V(\mathbf{s})) \in V(\alpha) \\
M, g &\Vdash [\mathbf{s}]\phi &&\text{iff} && M, g' \Vdash \phi \text{ for all } h \sim_{\mathbf{s}} g \\
M, g &\Vdash U\phi &&\text{iff} && M, h \Vdash \phi \text{ for all } h \in G
\end{aligned}
$$

where the functions V and g are lifted to tuples in the obvious way,[33] so that $\alpha(\mathbf{s})$ states that the (composite) situation $\mathbf{s}$ whose local state (i.e. the set of local states of its consistent situations) is of type α. The relation $\sim_{\mathbf{s}}$ on G is defined by:

$$g \sim_{s_1,\dots,s_n} h \quad \text{iff} \quad gV(s_i) = hV(s_i) \text{ for each i.}$$

Thus the modal formula $[\mathbf{s}]\phi$ says that ϕ holds in all global states in which situation $\mathbf{s}$ has the same local state configuration as it does in the current global state, or, in other words, that $\mathbf{s}$ "settles the truth" of ϕ.

van Benthem comments that the resulting logic is decidable and is axiomatised as a normal model logic with $S5$ axioms for $[\mathbf{s}]$ and U together with $U\phi \rightarrow [\mathbf{s}]\phi$ and $[\mathbf{s}]\phi \rightarrow [\mathbf{t}]\phi$ whenever $\mathbf{t} \subseteq \mathbf{s}$.[34] Moreover, it is an equivalent to the first-order logic

---

[32] Compare Rosenschien's notion of 'physical information' in [25], which was designed for the purpose of comparing local and global description of information content when designing robots.

[33] More precisely, $V(s_1, \dots, s_n) = \langle V(s_1), \dots, V(s_n) \rangle$ and $g(x_1 \dots x_n) = \langle g(x_1), \dots, g(x_n) \rangle$.

[34] Here '$\mathbf{t} \subseteq \mathbf{s}$' is just the syntactic requirement that names in the string $\mathbf{t}$ also occur in the string $\mathbf{s}$.

of dependency in the sense that the two are mutually effectively interpretable (van Benthem [13], p.8).

Likewise, we can interpret many of the concepts of situation theory as follows. The first matter to decide is how to represent an infon. The 'predicates' α apply to local states of situations and so serve as the obvious candidates for issues and so we can give a contextual definition of σ_s, for each infon σ in the situation s:

$$\langle\!\langle \alpha; + \rangle\!\rangle_s \quad :: \quad \alpha(s)$$
$$\langle\!\langle \alpha; - \rangle\!\rangle_s \quad :: \quad \neg\alpha(s)$$

This is sufficient to interpret all statements about infons, concerning their support and participation in constraints:

$$s \models \sigma \quad :: \quad [s]\sigma_s$$
$$s \models_t \tau \quad :: \quad [s]\tau_t$$
$$\sigma \underset{s}{\Rightarrow} \tau \quad :: \quad U(\sigma_s \to \tau_s)$$
$$\sigma \underset{s,t}{\Rightarrow} \tau \quad :: \quad U(\sigma_s \to \tau_t)$$

With this interpretation, we can answer some of the questions of Sect. 35.1. Firstly, note that the interpretation takes a situation to support that information that it carries about itself. This is a nice connection between the two concepts. But it fact, since the issues are predicates of local states, this can be simplified. The following is valid:

$$s \models \sigma \quad \leftrightarrow \quad \sigma_s$$

And as a consequence, it is clear, in answer to Question 9 that the following are both valid:

$$\neg(s \models \langle\!\langle \alpha; + \rangle\!\rangle \wedge s \models \langle\!\langle \alpha; - \rangle\!\rangle) \qquad s \models \langle\!\langle \alpha; + \rangle\!\rangle \vee s \models \langle\!\langle \alpha; - \rangle\!\rangle$$

Interestingly, however, the possibility of a gap emerges with information 'carrying' in general as $s \models_t \langle\!\langle \alpha; + \rangle\!\rangle \vee s \models_t \langle\!\langle \alpha; - \rangle\!\rangle$ is invalid. And more generally still, there are many examples of formulas ϕ whose 'truth' is not guaranteed to be settled by a situation.

Next, our interpretation has taken a stance on Analytic Reduction, explicitly accepting it.[35] Question 5 for all constraints is answered by validities corresponding to one half of the Indication Principle and the Carrying Principle:

---

[35] There are two relevant differences between analytic and general constraints. One is that analytic constraints carry information about the situation itself, as captured by this equivalence. But this other is that there is a different source for their modal force of analytic constraints. A fuller treatment of them should replace U with another modal operator with a wider range of gobal state functions.

$$\sigma \underset{s,t}{\Rightarrow} \tau \wedge s \models \sigma \quad \rightarrow \quad s \models_t \tau \quad \text{and} \quad s \models_t \tau \quad \rightarrow \quad t \models \tau$$

Interestingly, again, the other half of the Indication Principle does *not* hold in the current interpretation. It is possible for $s \models_t \tau$ without there being any infon σ such that $\sigma \underset{s,t}{\Rightarrow} \tau$. This is because the carrying relation is determined only by the interaction of the global and local states for s and t, plus the interpretation of τ itself; what it is about s that 'settles the truth' of $t \models \tau$ need not be explicitly defined by some infon.

The Xerox principle of Question 7 is validated by

$$\sigma \underset{s,t}{\Rightarrow} \tau \wedge \tau \underset{t,u}{\Rightarrow} \upsilon \quad \rightarrow \quad \sigma \underset{s,u}{\Rightarrow} \upsilon$$

Or, to put it more vividly and in terms of information carrying:

$$s_1 \models \sigma_1 \wedge \sigma_1 \underset{s_1,s_2}{\Rightarrow} \sigma_2 \wedge \ldots \wedge \sigma_{n-1} \underset{s_{n-1},s_n}{\Rightarrow} \sigma_n \quad \rightarrow \quad s_1 \models_{s_n} \sigma_n$$

A background condition B can be incorporated as an arbitrary formula of the language, so allowing great flexibility in what can be expressed,

$$[B \colon \sigma \underset{s,t}{\Rightarrow} \tau] \quad :: \quad U(B \wedge \sigma_s \rightarrow \tau_t)$$

This provides one way of answering Question 8 on background conditions and how constraints may fail, given a modified relationship between constraints and information carried:

$$B \wedge [B \colon \sigma \underset{s,t}{\Rightarrow} \tau] \wedge s \models \sigma \quad \rightarrow \quad s \models_t \tau$$

Note that B need not describe the state of situation s itself; it could, for example, be the formula $\beta(u)$ where u is some other situation, appropriately related to s and t. This suggests a way of introducing channels into constraint models, namely as situations that provide background conditions for constraints:

$$[\beta(u) \colon \sigma \underset{s,t}{\Rightarrow} \tau]$$

If this constraint holds, we can consider u to be a channel between s and t that licences the constraint $\sigma \Rightarrow \tau$ when its local state is of type β. This is considerably more flexible than the model of channels given in Sect. 35.2, which requires a tighter relationship between the three situations. In particular, it requires that the local states of s and t are functionally dependant on that of u. Further exploration of the notion of functional dependence in this setting would therefore be useful in drawing further comparisons between the two models.[36] Another way of modelling channels and

---

[36] van Benthem notes (p. 6) that this is related to Beth's theorem in first-order logic that any implicit definition, e.g., of the local state of s in terms of the local state of u, relative to a theory can be given an explicit definition.

so addressing Question 6 is to augment the model with relations between situations (or sites), as discussed in [13], p.19., and adding modal operators for them. Further comparison between these two methods is needed.

Although constraint logic does not describe part-whole relationships between situations, there is a natural way of adapting it to do so. The language already has the ability to refer to tuples of situations, which we can regard as a composite of its components, and also a situation of sorts. This suggests the following slight change. Call the situations in the set X of a constraint frame 'basic situations' and consider composite situation names to be equivalent up to reordering and repetition, so that they can be regarded as denoting sets of basic situations. We therefore define an *unordered constraint model* to be a structure $M = \langle F, v \rangle$ where $F = \langle X, L, G \rangle$ is again a constraint frame but the valuation V assigns subsets of X to each situation name and sets of such subsets to each predicate. We add the symbol $\trianglelefteq$ to the language and define

$$M, g \Vdash \mathbf{s} \trianglelefteq \mathbf{t} \quad \text{iff} \quad V(\mathbf{s}) \subseteq V(\mathbf{t})$$

Where $V(\mathbf{s}) = \{V(s) \mid s \in \mathbf{s}\}$. When we add $\trianglelefteq$ to the language, we must strengthen the axioms to include

$$\mathbf{s} \trianglelefteq \mathbf{t} \quad \rightarrow \quad ([\mathbf{s}]\phi \rightarrow [\mathbf{t}]\phi)$$

and add axioms to express that $\trianglelefteq$ is reflexive, transitive and respects concatenation:

$$\mathbf{s} \trianglelefteq \mathbf{s}$$
$$\mathbf{s} \trianglelefteq \mathbf{t} \wedge \mathbf{t} \trianglelefteq \mathbf{u} \quad \rightarrow \quad \mathbf{s} \trianglelefteq \mathbf{u}$$
$$\mathbf{s} \trianglelefteq \mathbf{st} \wedge \mathbf{t} \trianglelefteq \mathbf{st}$$
$$\mathbf{s} \trianglelefteq \mathbf{u} \wedge \mathbf{t} \trianglelefteq \mathbf{u} \quad \rightarrow \quad \mathbf{st} \trianglelefteq \mathbf{u}$$

Since $\mathbf{s} \trianglelefteq \mathbf{t} \wedge \mathbf{t} \trianglelefteq \mathbf{s}$ is only true when $\mathbf{s}$ and $\mathbf{t}$ have the same denotation, we can abbreviate this as $\mathbf{s} = \mathbf{t}$ and add axioms to ensure that this is a congruence[37]:

$$\mathbf{s} = \mathbf{t} \quad \rightarrow \quad \phi(\mathbf{s}) \leftrightarrow \phi(\mathbf{t})$$

This provides a mereology for situations in answer to Question 3 but in a somewhat question-begging way. Any doubt about the mereological structure obtained can be translated into a doubt about which subsets of X are to be counted as situations. For example, if we wish to question the existence of the largest situation, which is modelled by X itself, we could restrict V accordingly. Nonetheless, the model of situation theory obtained is in all cases an 'actualist' metrology, since each situation is a part of the whole system, and its local state is merely the restriction of the global state to that part.

---

[37] It is not yet clear to me whether or not the translation into the guarded fragment of first-order predicate logic, given by van Benthem, can be extended to the language with $\trianglelefteq$.

Also since the valuation function V is unconstrained, nothing is implied about the relationship between $\trianglelefteq$ and $\sqsubseteq$, and so in answer to Questions 1 and 2 we get a failure of both Persistence and Extensionality.[38] Persistence is expressed by

$$s \trianglelefteq t \wedge s \models \sigma \quad \rightarrow \quad t \models \tau$$

which is only valid (for σ positive and negative) on trivial models in which V assigns each predicate to either all local states or none of them. This is a consequence of the bivalence of support mentioned above. Once again, the corresponding statement about carrying information is valid:

$$s \trianglelefteq t \wedge s \models_u \sigma \quad \rightarrow \quad t \models_u \sigma$$

And when we take $u = s$ we get the valid principle:

$$s \trianglelefteq t \wedge s \models \sigma \quad \rightarrow \quad t \models_s \sigma$$

For our earlier example of non-persistence, $\langle\!\langle everyone\ is\ dancing \rangle\!\rangle$, this seems to get things about right. The situation at the wedding supports that everyone is dancing and any larger situation, although it may not itself support that everyone is dancing, still *carries* this information about the situation at the wedding.

That Extensionality does not hold in general can be seen by considering two separate, perhaps even disjoint, but equinumerous subsets s and t of X that have been assigned the same local states by the current global state and all $\sim_s$- and $\sim_t$-similar ones. This would result in the satisfaction of σs, $t\sigma$ for all infons σ but also $s \neq t$.[39]

Finally, concerning schemes of individuation (Questions 11 and 12) constraint logic, like any logic is built from an inductively specified formal language, which represents a single way of representing information, no matter how expressively rich. There are ways in which we could incorporate changing perspectives, for example by dividing the basic vocabulary of predicates into subsets or looking at other ways of carving out 'fragments' of the language, or by considering different constraint logical languages with interpretations between them, perhaps even internalising this idea as is done in interpretability logic. These are all additions to the present theory that would require further work.

---

[38] One could add the relation $\sqsubseteq$ to the language in a fairly straightforward way, but it clearly has a second-order interpretation and I suspect that this would add greatly to the complexity of the logic.

[39] In fact there are two kinds of equivalence between distinct situations that may be considered in these models. The first is necessary local-state equivalence, whereby $g(s) = g(t)$ for all $g \in G$. Such situations have perfectly synchronised identical local states and so support the same infons. The second concerns the predicates used to classify the local states. Even if s and t have quite different local states, they may still satisfy the same predicates and so support the same infons.

35.4 Conclusion

The two proposals for reconstructing situation theory have many similar features as well as profound differences in orientation.

In addition to a largely overlapping series of answers to our twelve questions, there is a more direct mathematical relationship between the two in the form of a partial interpretation of each in the other.

A constraint model can be regarded as a channel in which the model as a whole serves as the core and the component classifications are specified by individual situations. More precisely, given $F = \langle X, L, G \rangle$ and $M = \langle F, V \rangle$, we defined a channel $c = \{c_s : \mathscr{A}_s \overset{\rightarrow}{\underset{\leftarrow}{}} [c]\}_{s \in S}$ as follows. Each situation name $s \in S$ determines a classification $\mathscr{A}_s$ whose tokens are local states (L) and whose types are predicates, with $w \models \alpha$ iff $w \in V(\alpha)$ (for $w \in L$). The core classification $[c]$ has global states (G) as tokens and formulas of the language of constraint logic as types. Then the equations $c_s^{\wedge}(\alpha) = \alpha(s)$ and $c_s^{\vee}(g) = g(V(s))$ ensure that c_s is an infomorphism. With this representation, the two interpretations of $\models$ and $\Rightarrow$ coincide. This can be extended to a representation of the part-whole structure of an unordered constraint model only if some of the above-mentioned issues concerning Persistence are first resolved.[40]

Conversely, we can take a limit of any network of classifications and infomorphisms to produce a channel c whose components $\{\mathscr{A}_x\}_{x \in X}$ are all the classifications and whose defining infomorphisms commute with all those in the network. This channel can then be represented as a constraint model, whose set of basic situations X is the index set of the channel, whose local state set L is the disjoint union of the token sets of all the classifications (i.e., the set of pairs $\langle x, a \rangle$ for each token a of $\mathscr{A}_x$), and whose global function set G is defined to be the set of functions $g_d : X \rightarrow L$ such that $g(x) = c_x^{\vee}(d)$ for each token d of $[c]$. For predicates, we can take the types of $[c]$ with $V(\alpha)$ the set of $\langle x, a \rangle$ such that $a \models_{\mathscr{A}_x} c^{\wedge}\alpha$. Since there is no part-whole structure to the information network, there is nothing here to represent. Indeed, this is implied by the above construction, according to which each classification is represented by a *basic* situation.

A full study of such representations would be of interest.

Beyond these connections, there lies an interesting difference in orientation. The model of situations as local logics, which in turn are induced by pushing the natural logics of classifications along infomorphisms takes differences in perspective as fundamental. Information content and flow is modelled as emerging from the process of interpretation. The account of situations based on constraint models takes the same phenomena to result from the interplay between the local and global states of a system. Much can be learned from further investigation into these differences.

---

[40] For example, if for each situation s there is a function P_s on the set of predicates and the valuation V is restricted so that for $x \subseteq y, y \in V(P_s\alpha)$ iff $x \in V(\alpha)$ then we recover Persistence in the form: $s \trianglelefteq t \wedge s \models \sigma \rightarrow t \models P_s\sigma$. And then we can represented the part-whole relationship by the informorphism $\iota_{s,t} : [s] \overset{\rightarrow}{\underset{\leftarrow}{}} [t]$ defined by $\iota_{s,t}^{\wedge}(\alpha) = P_s\alpha$ and $\iota_{s,t}^{\vee}(g) = g(V(s))$.

Moreover, van Benthem's [13] has many more features of interest. In particular, he shows how constraint models can be combined with the apparatus of dynamic epistemic logic to bring agents and actions into the picture. At one point (p. 6), he suggest that agents be viewed *as* situations. This echoes an old idea from the early days of situation theory, during which time *everything* was by default a situation. Yet there is now a chance that some good sense could be made of this fascinating idea.

《莊子.山木》莊周反入，三月不庭。藺且從而問之：「夫子何為頃間甚不庭乎？」
莊周曰：「吾守形而忘身，觀於濁水而迷於清淵。且吾聞諸夫子曰：『入其俗，從其
俗。』今吾遊於雕陵而忘吾身，異鵲感吾顙，遊於栗林而忘真，栗林虞人以吾為戮，
吾所以不庭也。

References

1. Aczel P (1988) Non-well-founded sets. CSLI Publications, Stanford, CSLI lecture notes
2. Aczel P (1990) Replacement systems and the aziomatisation of situation theory. In: Cooper R, Mukai K, Perry J (eds) Situation theory and its aplications, vol I. CSLI Publications, Stanford
3. Austin J (1950) Truth. Proc Aristotelian Soc Suppl 24:111–128
4. Bar-Hillel Y, Carnap R (1952) An outline of a theory of semantic information. Technical report 247, Research Laboratory for Electronics, MIT.
5. Barwise J (1981) Scenes and other situations. J Philos 78(7):369–397
6. Barwise J (1989) Notes on branch points in situation theory. In: The situation in logic, CSLI Lecture Notes, Chap 11, p 255277. CSLI Press, Stanford, CA.
7. Barwise J (1989) Situations, facts, and true propositions. In: The situation in logic, CSLI Lecture Notes, Chap 10, p 221254. CSLI Press, Stanford, CA.
8. Barwise J, Etchemendy J (1987) The liar: an essay in truth and circularity. Oxford University Press Inc, New York, NY
9. Barwise J, Perry J (1983) Situations and attitudes. MIT Press, Cambridge, MA
10. Barwise J, Seligman J (1993) Imperfect information flow. In: Eighth annual IEEE symposium on logic in computer, Science, pp 252–260.
11. Barwise J, Seligman J (1994) The rights and wrongs of natural regularity. Philos Perspect 8:331
12. Barwise J, Seligman J (1997) Information flow: the logic of distributed systems. Cambridge University Press, Cambridge
13. van Benthem J (2006) Information as correlation versus information as range. Technical report PP-2006-07, ILLC, University of Amsterdam
14. van Benthem J (2010) Information transfer across Chu spaces. Logic J IGPL 8(6):719–731
15. van Benthem J, Israel D (1999) Review of Information flow: the logic of distributed systems, Jon Barwise and Jerry Seligman. J Logic Lang Inf 8(3):390–397
16. Dretske F (1981) Knowledge and the flow of information. Basil Blackwell, Oxford
17. Dretske F (1981) The pragmatic dimension of knowledge. Philos Stud 40:363–378
18. Gupta A (1999) Meaning and misconceptions. In: Jackendoff R, Bloom P, Wynn K (eds) Language, logic, and concepts: essays in memory of John Macnamara. The MIT Press, Cambridge, pp 15–41
19. Israel D, Perry J (1991) Information and architecture. In: Barwise J, Gawron JM, Plotkin G, Tutiya S (eds) Situation theory and its applications, vol 2. CSLI Press, Stanford, CA
20. King PJ (1994) Reconciling Austinian and Russellian accounts of the Liar paradox. J Philos Logic 23:451–494
21. Lunnon R (1991) Many sorted universes, SRD's, and injective sums. In: Barwise J, Gawron P, Plotkin G, Tutiya S (eds) Situation theory and its applications, vol II. CSLI Publications, Stanford, pp 51–79

22. Mares E, Seligman J, Restall G (2011) Situations, constraints and channels. In: van Benthem J, ter Meulen A (eds) Handbook of logic and language. Eslevier, Burlington, pp 329–344
23. Plotkin G (1990) An illative theory of relations. In: Cooper R, Mukai K, Perry J (eds) Situation theory and its aplications, vol I. CSLI Publications, Stanford
24. Pratt V (2002) Event-state duality: the enriched case. In: Proceedings of the CONCUR'02, Brno.
25. Rosenschein SJ (1989) Synthesizing information-tracking automata from environment descriptions. In: The principles of knowledge representation and reasoning KR'89.
26. Seligman J (1993) An algebraic appreciation of diagrams. In: The 9th Amsterdam colloquium, pp 607–625.
27. Seligman J (2009) Channels: from logic to probability. In: Sommaruga G (ed) Formal theories of information: from Shannon to semantic information theory. Springer, Berlin, pp 193–233
28. Seligman J, Moss LS (2011) Situation theory. In: van Benthem J, ter Meulen A (eds) Handbook of logic and language. Elsevier, North Holland, pp 253–328

Chapter 36
Unified Correspondence

Willem Conradie, Silvio Ghilardi and Alessandra Palmigiano

> *Correspondence Theory may be applied to any kind of semantic entity.*
> *(J. van Benthem, Correspondence theory, Handbook of Philosophical Logic, p. 381).*

Abstract The present chapter is aimed at giving a conceptual exposition of the mathematical principles underlying Sahlqvist correspondence theory. These principles are argued to be inherently algebraic and order-theoretic. They translate naturally on relational structures thanks to Stone-type duality theory. The availability of this analysis in the setting of the algebras dual to relational models leads naturally to the definition of an *expanded* (object) language in which the well-known 'minimal valuation' meta-arguments can be encoded, and of a *calculus for correspondence* of a proof-theoretic style in the expanded language, mechanically computing the first-order correspondent of given propositional formulas. The main advantage brought about by this formal machinery is that correspondence theory can be ported in a uniform way to families of nonclassical logics, ranging from substructural logics to mu-calculi, and also to different semantics for the same logic, paving the way to a uniform correspondence theory.

W. Conradie (✉)
Department of Mathematics, University of Johannesburg, Johannesburg, South Africa
e-mail: wconradie@uj.ac.za

S. Ghilardi
Department of Mathematics, Università degli Studi di Milano, Milan, Italy
e-mail: silvio.ghilardi@unimi.it

A. Palmigiano
Faculty of Technology Policy and Management, Delft University of Technology, Delft, The Netherlands
e-mail: a.palmigiano@tudelft.nl

A. Baltag and S. Smets (eds.), *Johan van Benthem on Logic and Information Dynamics*, Outstanding Contributions to Logic 5, DOI: 10.1007/978-3-319-06025-5_36, © Springer International Publishing Switzerland 2014

36.1 Introduction

Correspondence theory has been one of van Benthem's core interests since early in his career, and is the field to which his most celebrated result in mathematical logic—the *van Benthem Characterization Theorem*—belongs. Throughout his subsequent career, he has been pointing out various correspondence phenomena embedded in his many research interests, which he collected e.g. in [5–7]. Most recently, [8] ties in with his current interests in information flow. van Benthem has always been eager to point out unexplored research directions, and these chapters are no exception. The correspondence phenomena he identified are often *fringe* phenomena, in the sense that they are clearly recognizable as *instances* of correspondence, but are not embedded in a systematic theory, see especially [5]. We are now in a position to bring the fringe to the core and build a unifying theory around these scattered instances. Clearly, such an encompassing theory cannot be unfolded in the scope of the present chapter; our objectives are more modest, and are:

(a) to give a conceptual exposition of the mathematical principles underlying the correspondence mechanism, and how these principles work uniformly across different logics and also across different semantics for the same logic;
(b) to give pointers to the recent literature, and to mention the most important directions in which correspondence theory has been extended;
(c) to give a second reading to van Benthem's fringe examples, to show how the general principles identified in item (a) are still at work in these examples, and to point at ways in which the general theory accounts for them.

36.2 Correspondence via Duality

Relational semantics for modal logic provides a very clear understanding of what modal axioms mean in many different contexts of application, and is the essential reason why modal logic has become the successful formalism it is. With the introduction of Kripke semantics in the early 1960s, modal logic found itself in a very special position among non-classical logic, thanks to the fact that relational structures can be used *both* as semantics for modal logic *and* for classical first-order logic. This common semantic ground immediately elicited a whole research programme in the model theory of modal logic, focusing on its *expressivity*. A high point of this programme was of course van Benthem's theorem characterizing modal logic as the bisimulation invariant fragment of first-order logic [4].

A host of simple but insightful connections started to pop up between modal axioms which have been previously and independently studied (e.g. in formal philosophy), and basic properties of relational structures, such as reflexivity or transitivity. These connections are established via the notion of *local validity* of a modal formula in a relational structure, i.e., of that formula being satisfied at a given state *for every* valuation. The style of argument used to establish each of these connections is fairly uniform, so let us briefly review how this is done by way of one such example.

Example 36.1 The following are equivalent for any relational structure $\mathcal{F} = (W, R)$ and any $w \in W$:

1. The modal formula $\Box p \rightarrow \Box\Box p$ is true in $\mathcal{F}$ at w under all assignments $V(p) \subseteq W$;

2. $\mathcal{F}$ satisfies the first-order formula $\alpha(x)$ expressing the inclusion $R[R[x]] \subseteq R[x]$ whenever the free variable x is interpreted as w.[1]

Proof For the interesting direction, i.e., assume 1 and prove 2, we need to assume that there are states v and v s.t. wRv and vRv, and show that wRv. Consider the assignment $V^*(p) := R[w]$; this is the smallest assignment of p under which the antecedent of $\Box p \rightarrow \Box\Box p$ is true. Hence, by modus ponens w must satisfy also the conclusion $\Box\Box p$ under the same assignment, which implies that v satisfies $\Box p$ under V^*, which implies that $v \in V^*(p) = R[w]$.

The Sahlqvist formulas, introduced by Sahlqvist [45] and further developed by van Benthem [4] and others, form the best known class of modal formulas whose syntactic shape makes it possible for similar proof arguments to succeed.

New perspective. So what is special about the 'Sahlqvist shape', and how can we systematically recognize and reproduce it in the syntax of other, non-modal logics? The aim of the present chapter is illustrating that the answers to these questions are inherently *algebraic* and *order-theoretic*. Taking this perspective has the advantage of endowing correspondence results with greater generality between logics and enhanced portability to different semantics. Such a claim of course requires elaborate justification, and it is our hope that the reader will be convinced of this by the end of the chapter. For now, let us say the following: modal logic, like all propositional logics, can be interpreted into *algebras* in a *canonical* way, in the same sense in which first-order logic is interpreted into relational structures in a canonical way. On the other hand, the interpretation of modal formulas into relational structures seems to offer some degrees of choice; for instance, one could use either the forward or backward direction of the relations to interpret the modal operators. The relational models alone do not seem to provide enough justification to establish that the usual interpretation of modal formulas into relational structures is canonical in the informal sense. This looks like a fundamental asymmetry between the algebraic and relational semantics of modal logic. *Symmetry* is restored, in a sense, if we allow *Stone duality* (between complete atomic modal algebras and Kripke frames) to enter the picture: indeed, the relational interpretation of modal formulas is uniquely identified as the *dual characterization* of its interpretation on algebras. Hence, its being canonical can be derived as a consequence of this strong link, and of the canonicity of the algebraic interpretation. This is pictured in Fig. 36.1(b).

This discussion provides a general illustration of how, thanks to duality, the advantages of the algebraic perspective on modal logic can be transferred to Kripke frames.

---

[1] For $x \in W$ we let $R[x] = \{v \in W \mid Rxv\}$, and for $X \subseteq W$ we let $R[X] = \bigcup\{R[x] \mid x \in X\}$.

(a)   **(b)**

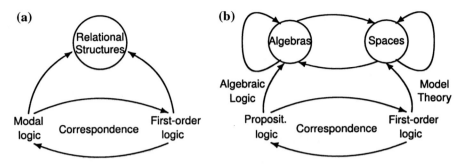

Fig. 36.1 From a model-theoretic (**a**) to a duality-based (**b**) approach to correspondence

But more specifically, the link between the relational and the algebraic interpretation of modal logic directly invests correspondence theory: indeed, thanks to it, we will be able to show that the Sahlqvist-style correspondence mechanism is driven by properties which naturally live in the *algebraic* side of diagram (b).

Finally, having been able to recognize modal correspondence theory as part of the logical fallout of the specific duality between the algebraic and the set-based semantics of modal logic, it will also become clear that correspondence theory is by no means unique to modal logic, and is uniformly available in great generality to all (classes of) propositional logics for which such dualities are available. Before being able to motivate these conclusions, let us take a step back, and resume the example we started with.

Example, continued. The proof in Example 36.1 gives an illustration of the so-called *minimal valuation argument*: assuming that a modal formula $\varphi \to \psi$ is locally valid at a given state w, we instantiate with the *minimal valuation* which satisfies φ at w. In fact, this argument is a special case of a more general reasoning pattern, which is typically employed when proving the equivalence of the following statements:

(1) for every assignment V, if $\mathcal{F}, V, w \Vdash \varphi$ then $\mathcal{F}, V, w \Vdash \psi$;
(2) for every assignment V^* ranging in a given subclass K, if $\mathcal{F}, V^*, w \Vdash \varphi$ then $\mathcal{F}, V^*, w \Vdash \psi$.

The equivalence between (1) and (2), for a suitable choice of K, is the crucial require-ment on which the local correspondence mechanism is grounded. Indeed, (1) is just a reformulation of $\varphi \to \psi$ being locally valid at a given state w. If K is a class of assignments V^* mapping each proposition variable to a subset of W which admits a *uniform description* (for instance, in the case of our example above, K can be taken as the set of all the assignments such that either $V(p) = \varnothing$ or $V(p) = R[v]$ for some $v \in W$), and if further there are only finitely many members of K rele-vant for any given state w, then (2) can be further manipulated[2] into an equivalent condition in the language to which this uniform description belongs. This is done by orderly substituting the predicate variables (ranging over *arbitrary* subsets) with

---

[2] This aspect of the story deserves a separate account, which will be given in Sect. 36.6.

formal descriptions, in the target language, of *definable* subsets. So, for instance, if these uniform descriptions are expressible in the first-order language of $\mathcal{F}$ (as is the case of the example above), then the equivalent condition will yield a local first-order correspondent of $\varphi \to \psi$; if the descriptions are expressible in the first-order language of $\mathcal{F}$ enriched with fixed points (as is the case of the Löb formula), then we will have local correspondence with first-order logic with least fixed points. Thus, in each context, this uniform descriptions for the assignments in K targets the language we want to establish modal correspondence with, which explains why we will refer to the class K which we choose in each particular context as the class of *domesticated* assignments, as opposed to the *arbitrary* assignments, which roam wildly and for which no such description is available.

Having indicated that the equivalence between (1) and (2) is the crux of the matter, let us take a closer look at it. It is immediate that (1) always implies (2). It is also clear that the converse direction is false in its full generality, and our being able to prove it depends on our being able to find, for a given arbitrary assignment V such that $w \in [\![\varphi]\!]_V$ (where $[\![\varphi]\!]_V$ denotes the extension of φ in $\mathcal{F}$ under the assignment V), a *domesticated* assignment V^* such that $w \in [\![\varphi]\!]_{V^*}$ and $[\![\psi]\!]_{V^*} \subseteq [\![\psi]\!]_V$. The latter requirement is typically achieved by assuming that the extension function induced by ψ is *monotone*, and defining V^* so that $V^*(p) \subseteq V(p)$ for all the relevant proposition variables. Therefore, the two sufficient requirements on V^* in order for the equivalence between (1) and (2) to go through are:

$$w \in [\![\varphi]\!]_{V^*} \quad \text{and} \quad V^*(p) \subseteq V(p). \tag{36.1}$$

In all the different contexts in which (both the scattered instances of and the systematic) correspondence results hold, the *general strategy* to find this domesticated assignment V^* can be described as follows[3]: for each relevant variable p and every $w \in [\![\varphi]\!]_V$, the required domesticated V^* is defined by stipulating $V^*(p) := [\![\alpha]\!]_V \subseteq V(p)$ for some suitable (modal) formula α. It is often the case that α does *not* belong to the original language. To fix ideas, let us review what happens in the proof that (2) implies (1) when $\varphi \to \psi$ is the formula $\Box p \to \Box\Box p$ in the Example 36.1: fix an arbitrary V such that $w \in [\![\Box p]\!]_V$; then the following chain of equivalences holds:

$$
\begin{aligned}
w \in [\![\Box p]\!]_V \quad &\text{iff} \quad \{w\} \subseteq \Box_R V(p) \\
&\text{iff} \quad \{w\} \subseteq (R^{-1}[V(p)^c])^c \\
&\text{iff} \quad \{w\} \cap R^{-1}[V(p)^c] = \varnothing \\
&\text{iff} \quad R[\{w\}] \cap V(p)^c = \varnothing \\
&\text{iff} \quad R[w] \subseteq V(p). \tag{36.2}
\end{aligned}
$$

---

[3] It is certainly not the only way to describe the correspondence mechanism, but it is useful for our purposes.

The above chain of equivalences effectively rewrites our assumptions into a workable choice of domesticated valuation: indeed, it says that a valuation satisfies the antecedent of our given formula at w iff it assigns p to a superset of $R[w]$. This immediately implies that the set of valuations satisfying the antecedent of our given formula at w (ordered pointwise) has a *minimum*, given by the valuation V^* assigning $V^*(p) := R[w]$ and $\varnothing$ to all the other variables. This valuation clearly satisfies both requirements in clause (36.1), which as discussed, are sufficient condition for the equivalence between (1) and (2) to be established.

But, more interestingly, for which α can we identify the set $R[w]$ as $[\![\alpha]\!]_V$? Or more precisely, how should we expand our base language (and hence also our original assignments V) so that we get $V^*(p) = [\![\alpha]\!]_V$? This example shows that we certainly need to expand our language with at least the following two types of syntactic ingredients:

(a) ingredients which enable us to speak about *singletons*;
(b) ingredients which enable us to speak about *direct R-images of subsets*.

As to (b), it is well known that, for every subset X (which might be in particular a singleton), the assignment $X \mapsto R[X]$ provides the interpretation for the *backward-looking diamond* $\blacklozenge$, which is interpreted by the semantic diamond associated with R^{-1}. This is a well known situation in modal *tense* logic, where the backward-looking modalities belong to the base language; for the modal languages in which this is not the case, the backward-looking modalities will be added to the expanded language.

As to (a), the most convenient way for us to speak about singletons is to introduce a special sort of variables $\mathbf{h}, \mathbf{i}, \mathbf{j}, \mathbf{k}, \dots$ in the extended modal language, which are to be interpreted as singletons; we call them *nominals*, after the analogous devices adopted in hybrid logic. In Sects. 36.3 and 36.4, the expanded language will be discussed more formally and generally. For the moment, we only remark that nominals, interpreted as singletons, make it possible to encode *local* satisfaction of modal formulas as *global* satisfaction of certain inequalities, as follows:

$$\mathcal{F}, V, w \Vdash \varphi \quad \text{iff} \quad \mathcal{F}, V_{\mathbf{j}:=w} \Vdash (\mathbf{j} \leq \varphi), \tag{36.3}$$

where $V_{\mathbf{j}:=w}$ is the extended $\mathbf{j}$-variant of V sending $\mathbf{j}$ to $\{w\}$, and for every valuation V and formulas ψ and χ we write $\mathcal{F}, V \Vdash \psi \leq \chi$ to indicate that $[\![\psi]\!]_V \subseteq [\![\chi]\!]_V$.

Given both types of syntactic ingredients, we can stipulate in the example above $V^*(p) := [\![\alpha]\!]_V$ for $\alpha = \blacklozenge\mathbf{j}$. Notice that the introduction of this language expansion is harmless w.r.t. our target language: the standard translation of formulas in the language expanded with both nominals and backward-looking modalities falls within the basic first-order frame language. But more interestingly, which advantages does this expanded language bring to us?

Firstly, we have gained a better calculus: for instance, the equivalence between the beginning and the end of the (rather clumsy) chain of set-theoretic equivalences (36.2) above can be justified in one line as the following instance of the well known tense axiomatics:

$$\mathbf{j} \leq \Box p \quad \text{iff} \quad \blacklozenge\mathbf{j} \leq p. \tag{36.4}$$

Secondly and more importantly, we have defined a formal setting in which the computation of the minimal valuation is internalized at the level of a suitable object language. Indeed, the left-to-right direction of (36.4) provides us with the minimal valuation $V^*(p)$ which is expressed in the extended language as $\blacklozenge\mathbf{j}$. This enables us to proceed to a full *mechanization of the minimal valuation argument*. Indeed, after computing the needed minimal valuation as above, the actual instantiation with this valuation is facilitated and justified by the following version of *Ackermann's lemma* [2]:

Lemma 36.1 *Let* α, $\beta(p)$, $\gamma(p)$ *be formulas of a (modal) language* $\mathcal{L}^+$ *over the set of variables* PROP; *let* $p \in$ PROP *such that* p *does not occur free in* α, β *is negative in* p *and* γ *is positive in* p, *then the following are equivalent for every* $\mathcal{L}^+$-*Kripke frame* $\mathcal{F}$:

(a) $\mathcal{F} \Vdash (\alpha \leq p \Rightarrow \beta(p) \leq \gamma(p))$;
(b) $\mathcal{F} \Vdash \beta(\alpha/p) \leq \gamma(\alpha/p)$.

Proof For every formula φ and valuation V, let $[\![\varphi[p]]\!]_V$ be the unary operation on $\mathcal{P}(W)$ sending $X \in \mathcal{P}(W)$ to $[\![\varphi]\!]_{V'}$ where V' is the p-variant of V sending p to X.

As to the direction from (a) to (b): assume contrapositively that $[\![\beta(\alpha/p)]\!]_V \not\subseteq [\![\gamma(\alpha/p)]\!]_V$ for some valuation V. Let V^* be the p-variant of V such that $V^*(p) := [\![\alpha]\!]_V$. Then, because the variable p does not occur in α, we have $[\![\alpha]\!]_{V^*} = [\![\alpha]\!]_V = V^*(p)$, which proves that $\mathcal{F}, V^* \Vdash \alpha \leq p$. However, for every formula ξ, the following chain of equalities holds: $[\![\xi(p)]\!]_{V^*} = [\![\xi[p]]\!]_{V^*}(V^*(p)) = [\![\xi[p]]\!]_{V^*}([\![\alpha]\!]_V) = [\![\xi(\alpha/p)]\!]_V$. This and the contrapositive assumption prove that $\mathcal{F}, V^* \not\Vdash \beta(p) \leq \gamma(p)$.

Conversely, assume that $\mathcal{F} \Vdash \beta(\alpha/p) \leq \gamma(\alpha/p)$, and let V be such that $[\![\alpha]\!]_V \subseteq V(p)$. Then, since β and γ are respectively negative and positive in p, we have: $[\![\beta(p)]\!]_V \subseteq [\![\beta(\alpha/p)]\!]_V \subseteq [\![\gamma(\alpha/p)]\!]_V \subseteq [\![\gamma(p)]\!]_V$, which proves that $\mathcal{F}, V \Vdash \beta(p) \leq \gamma(p)$.

The proof of the direction (a) $\Rightarrow$ (b) in the lemma above encodes the minimal valuation argument in a very general way. This provides us with a crucial step towards mechanizing the correspondence process via the elimination of variables, as we can now simply appeal to the lemma instead of making an ad hoc minimal valuation argument. Notice also that the lemma does not depend on the particular choice of language $\mathcal{L}^+$.[4] Besides the assumptions of monotonicity/antitonicity of the interpretation of formulas, the only requirement encoded in the proof is that the minimal valuation be defined in terms of the resources of $\mathcal{L}^+$.

Towards a calculus for correspondence. Using the resources of the expanded language, and the stipulations made in clauses (36.3) and (36.4), it is not difficult to check the soundness on Kripke frames of the following chain of equivalences:

---

[4] In fact, it works also when $\mathcal{F}$ is an ordered algebra where the operations interpret the $\mathcal{L}^+$-connectives.

$$\forall p[\Box p \leq \Box\Box p] \quad \text{iff} \quad \forall p \forall \mathbf{j}[\mathbf{j} \leq \Box p \Rightarrow \mathbf{j} \leq \Box\Box p]$$
$$\text{iff} \quad \forall p \forall \mathbf{j}[\blacklozenge \mathbf{j} \leq p \Rightarrow \mathbf{j} \leq \Box\Box p]$$
$$\text{iff} \quad \forall \mathbf{j}[\mathbf{j} \leq \Box\Box\blacklozenge \mathbf{j}] \qquad \text{(Lemma 36.1)}$$
$$\text{iff} \quad \forall \mathbf{j}[\blacklozenge\blacklozenge \mathbf{j} \leq \blacklozenge \mathbf{j}].$$

indeed, thanks to the above stipulations on the interpretation of nominals, the condition $\mathcal{F} \Vdash \Box p \leq \Box\Box p$ can be equivalently rewritten as $\mathcal{F} \Vdash \forall \mathbf{j}[\mathbf{j} \leq \Box p \Rightarrow \mathbf{j} \leq \Box\Box p]$; by the stipulations on the additional modal operators, this clause can equivalently be rewritten as $\mathcal{F} \Vdash \forall \mathbf{j}[\blacklozenge \mathbf{j} \leq p \Rightarrow \mathbf{j} \leq \Box\Box p]$. Hence, by Ackermann's lemma applied to $\alpha := \blacklozenge \mathbf{j}$, $\beta(p) := \mathbf{j}$, $\gamma(p) := \Box\Box p$, we get that $\mathcal{F} \Vdash \forall \mathbf{j}[\mathbf{j} \leq \Box\Box\blacklozenge \mathbf{j}]$, which can be rewritten as $\mathcal{F} \Vdash \forall \mathbf{j}[\blacklozenge\blacklozenge \mathbf{j} \leq \blacklozenge \mathbf{j}]$.

To sum up, what we are heading towards is introducing a formal (*object*) language and a *syntactic* machinery in which the *semantic* 'minimal valuation' *meta*-argument given in Example 36.1 can be encoded. This small copernican revolution can be traced back to [37]. As to the benefits it brings: once we are dealing with syntax, we are free to interpret these strings of symbols and transformation rules in all sorts of models which happen to soundly interpret them; for instance, *atomistic*[5] tense Boolean algebras, and more specifically, the *complex algebras*, i.e. the modal algebras dually associated with relational structures, are obvious sound models. In the latter, nominals would then be interpreted as atoms of the algebra, and it is easy to see that the first equivalence is sound precisely because of atomicity. In fact, thanks to duality, the soundness of the chain of equivalences above w.r.t. complex algebras is the equivalent counterpart of the soundness proof on frames. Interpreting $\forall \mathbf{j}[\blacklozenge\blacklozenge \mathbf{j} \leq \blacklozenge \mathbf{j}]$ on complex algebras, where $\mathbf{j}$ ranges over the singletons, we readily obtain the well known first-order condition

$$\forall x (R[R[x]] \subseteq R[x]),$$

which standardly abbreviates the usual transitivity condition.

But there is more. In fact, we can do just as well with much more general algebras than the complex algebras of Kripke frames. All we need of an algebraic model for this (very simple) proof to be sound is its being a poset endowed with a pair of adjoint operations $\blacklozenge \dashv \Box$, and its being join-generated by some designated subset J (which will provide the interpretation for nominals). Of course, for the sake of finding a suitable environment for *classes* of logics, we need to assume more: in particular, we want to assume the existence of a rich enough algebraic environment, able to provide interpretation to logical connectives; certain complete (distributive) lattice expansions which we will introduce below are adequate for most purposes. This enables us to explore the full domain of applicability of correspondence arguments, which turns out to be much wider than classical modal logic.

---

[5] A lattice is atomistic if every element is the supremum of a set of atoms.

This concludes the informal presentation of the view on correspondence theory pursued in the present chapter. In the following section, we will expand on some of the technical details supporting this perspective.

36.3 A Calculus for Correspondence

Let us start by formally introducing the expanded syntax we mentioned in the previous section: it will include the backward-looking box corresponding to the diamond taken as a primitive operator, as well as a denumerably infinite set of sorted variables NOM called *nominals*, the elements of which will be denoted with $\mathbf{i}, \mathbf{j}$, possibly indexed.

The *formulas* of $\mathcal{L}^+$ are given by the following recursive definition:

$$\varphi ::= \perp \mid p \mid \mathbf{j} \mid \varphi \vee \psi \mid \neg\varphi \mid \Diamond\varphi \mid \blacksquare\varphi,$$

where $p \in$ PROP and $\mathbf{j} \in$ NOM. The derived connectives $\wedge, \square, \rightarrow, -, \ldots$ are defined in the standard way. In order to formalize the correspondence arguments, we will have to expand $\mathcal{L}^+$ to accommodate inequalities and quasi-inequalities. To be precise, if $\varphi, \varphi_1, \ldots, \varphi_n, \psi, \psi_1, \ldots, \psi_n \in \mathcal{L}^+$ then $\varphi \leq \psi$ is an inequality and $\varphi_1 \leq \psi_1 \& \cdots \& \varphi_n \leq \psi_n \Rightarrow \varphi \leq \psi$ is a quasi-inequality. Disjunctions $\varphi \leq \psi \mathbin{⅋} \chi \leq \xi$ between inequalities will be sometimes considered.

Formulas, inequalities and quasi-inequalities not containing any propositional variables (but possibly containing nominals) will be called *pure*. As we will see next, these can be readily translated into the first-order frame correspondence language; hence we aim to introduce rules for a calculus of syntactic transformations of quasi-inequalities, by means of which quasi-inequalities in $\mathcal{L}^+$ can be transformed into *pure* ones, so as to preserve logical equivalence. In order to motivate this calculus, let us introduce the intended interpretation of the expanded language.

A *valuation for* $\mathcal{L}^+$ on a Kripke frame $\mathcal{F} = (W, R)$ is any map V from the set PROP $\cup$ NOM of propositional variables and nominals into the powerset $\mathcal{P}(W)$, such that each $\mathbf{i} \in$ NOM is assigned to the singleton subset $\{x\}$ for some $x \in W$. A *model* for $\mathcal{L}^+$ is a tuple $\mathcal{M} = (\mathcal{F}, V)$ such that $\mathcal{F}$ is a Kripke frame and V is a valuation for $\mathcal{L}^+$. For any such model, the satisfaction relation for formulas in $\mathcal{L}^+$ is recursively defined as follows (here we report only the new connectives):

$$\mathcal{M}, w \Vdash \mathbf{i} \quad \text{iff} \quad V(\mathbf{i}) = \{w\},$$
$$\mathcal{M}, w \Vdash \blacksquare\varphi \quad \text{iff} \quad \text{for every } v, \text{ if } vRw \text{ then } \mathcal{M}, v \Vdash \varphi.$$

The local satisfaction relation extends to inequalities and quasi-inequalities as follows:

$$\mathcal{M}, w \Vdash \varphi \leq \psi \quad \text{iff} \quad \text{if } \mathcal{M}, w \Vdash \varphi \text{ then } \mathcal{M}, w \Vdash \psi,$$
$$\mathcal{M}, w \Vdash (\&_{i=1}^{n} \varphi_i \leq \psi_i) \Rightarrow \varphi \leq \psi \quad \text{iff} \quad \text{if } \mathcal{M}, w \Vdash \varphi_i \leq \psi_i \text{ for } 1 \leq i \leq n$$
$$\text{then } \mathcal{M}, w \Vdash \varphi \leq \psi.$$

From the clauses above, the global satisfaction relation for inequalities and quasi-inequalities is defined in the usual way, by universally quantifying over w; namely,

$$\mathcal{M} \Vdash \varphi \leq \psi \qquad \text{iff} \quad \text{for any } w, \text{ if } \mathcal{M}, w \Vdash \varphi \text{ then } \mathcal{M}, w \Vdash \psi,$$
$$\mathcal{M} \Vdash (\&_{i=1}^{n} \varphi_i \leq \psi_i) \Rightarrow \varphi \leq \psi \quad \text{iff} \quad \text{for any } w, \text{ if } \mathcal{M}, w \Vdash \varphi_i \leq \psi_i \text{ for } 1 \leq i \leq n$$
$$\text{then } \mathcal{M}, w \Vdash \varphi \leq \psi.$$

For every model $\mathcal{M} = (\mathcal{F}, V)$ and every $\varphi \in \mathcal{L}^+$, the symbol $[\![\varphi]\!]_{\mathcal{M}}$ denotes as usual the set of states of $\mathcal{M}$ at which φ is satisfied. When there could be no confusion about $\mathcal{F}$, the symbol $[\![\varphi]\!]_V$ will alternatively be used.

As mentioned in Sect. 36.2, *local* satisfaction of formulas can be encoded as a special case of the *global* satisfaction of inequalities, as reported in the following proposition:

Proposition 36.1 *For any Kripke frame $\mathcal{F}$, any valuation V for $\mathcal{L}^+$ and any $\varphi \in \mathcal{L}$,*

$$\mathcal{F}, V, w \Vdash \varphi \quad \text{iff} \quad \mathcal{F}, V' \models \mathbf{j} \leq \varphi \text{ and } V'(\mathbf{j}) = \{w\},$$

with $V' \sim_{\mathbf{j}} V$ and $\mathbf{j}$ a nominal not occurring in φ.

The Ackermann lemma (Lemma 36.1) implies that the following rules are sound and invertible w.r.t. the standard Kripke semantics:

$$\frac{\forall p[(\alpha \leq p \,\&\, \&_{1 \leq i \leq n}(\gamma_i(p) \leq \delta_i(p))) \Rightarrow \varphi(p) \leq \psi(p)]}{\&_{1 \leq i \leq n}(\gamma_i(\alpha/p) \leq \delta_i(\alpha/p)) \Rightarrow \varphi(\alpha/p) \leq \psi(\alpha/p)} \,(\text{LA}) \qquad \frac{\forall p[\varphi(p) \leq \psi(p)]}{\varphi(\bot/p) \leq \psi(\bot/p)} \,(\bot)$$

subject to the restrictions that α is p-free, and that φ and the δ_i are negative in p, while ψ and the γ_i are positive in p. Notice that the rule $(\bot)$ can be regarded as the special case of (LA) in which $\alpha := \bot$. Likewise, a mirror-image version of Lemma 36.1 implies that the following rules are sound and invertible w.r.t. the standard Kripke semantics:

$$\frac{\forall p[(p \leq \alpha \,\&\, \&_{1 \leq i \leq n}(\gamma_i(p) \leq \delta_i(p))) \Rightarrow \varphi(p) \leq \psi(p)]}{[\&_{1 \leq i \leq n}(\gamma_i(\alpha/p) \leq \delta_i(\alpha/p)) \Rightarrow \varphi(\alpha/p) \leq \psi(\alpha/p)]} \,(\text{RA}) \qquad \frac{\forall p[\varphi(p) \leq \psi(p)]}{\varphi(\top/p) \leq \psi(\top/p)} \,(\top)$$

subject to the restrictions that α is p-free, and that φ and the δ_i are positive in p, while ψ and the γ_i are negative in p. In addition to this, the following proposition is an immediate consequence of the stipulations above:

Proposition 36.2 *For every model $\mathcal{M} = (\mathcal{F}, V)$ for $\mathcal{L}^+$, every $\mathbf{j} \in \mathsf{NOM}$, and all $\varphi, \psi, \chi \in \mathcal{L}^+$,*

1. $\mathcal{F}, V \Vdash \varphi \le \psi$ iff $\mathcal{F}, V \models \forall \mathbf{j}[\mathbf{j} \le \varphi \Rightarrow \mathbf{j} \le \psi]$, *for any nominal* $\mathbf{j}$ *not occurring in* $\varphi \le \psi$.
2. $\mathcal{F}, V \Vdash \varphi \vee \chi \le \psi$ iff $\mathcal{F}, V \Vdash \varphi \le \psi$ *and* $\mathcal{F}, V \Vdash \chi \le \psi$.
3. $\mathcal{F}, V \Vdash \varphi \le \chi \vee \psi$ iff $\mathcal{F}, V \Vdash \varphi - \chi \le \psi$, *where* $\varphi - \chi := \varphi \wedge \neg \chi = \neg(\neg\varphi \vee \chi)$.
4. $\mathcal{F}, V \Vdash \Diamond\varphi \le \psi$ iff $\mathcal{F}, V \Vdash \varphi \le \blacksquare\psi$.
5. $\mathcal{F}, V \Vdash \mathbf{j} \le \Diamond\psi$ iff $\mathcal{F}, V \models \exists \mathbf{i}[\mathbf{j} \le \Diamond \mathbf{i} \ \& \ \mathbf{i} \le \psi]$, *for any nominal* $\mathbf{i}$ *not occurring in* $\mathbf{j} \le \Diamond\psi$.
6. $\mathcal{F}, V \Vdash \psi \le \neg\varphi$ iff $\mathcal{F}, V \Vdash \varphi \le \neg\psi$.
7. $\mathcal{F}, V \Vdash \neg\varphi \le \psi$ iff $\mathcal{F}, V \Vdash \neg\psi \le \varphi$.

The proposition above essentially says that the following rules are sound and invertible w.r.t. the standard Kripke semantics:

$$\frac{\varphi \le \psi}{\forall \mathbf{j}[\mathbf{j} \le \varphi \Rightarrow \mathbf{j} \le \psi]} \ (\text{FA})^* \qquad \frac{\varphi \vee \chi \le \psi}{\varphi \le \psi \quad \chi \le \psi} \ (\vee\text{-}\Delta) \qquad \frac{\varphi \le \chi \vee \psi}{\varphi - \chi \le \psi} \ (\vee\text{RR})$$

$$\frac{\Diamond\varphi \le \psi}{\varphi \le \blacksquare\psi} \ (\Diamond\text{LA}) \qquad \frac{\mathbf{j} \le \Diamond\psi}{\exists \mathbf{i}(\mathbf{j} \le \Diamond\mathbf{i} \& \mathbf{i} \le \psi)} \ (\text{jCJP})^\dagger \qquad \frac{\varphi \le \neg\psi}{\psi \le \neg\varphi} \ (\neg\text{RGA}) \qquad \frac{\neg\varphi \le \psi}{\neg\psi \le \varphi} \ (\neg\text{LGA})$$

*where the introduced nominal $\mathbf{j}$ does not occur in derivation so far.
†where the introduced nominal $\mathbf{i}$ does not occur in derivation so far.

It is easy to show that the calculus admits derived invertible rules such as the following:

$$\frac{\varphi \le \chi \wedge \psi}{\varphi \le \chi \quad \varphi \le \psi} \ (\wedge\text{-}\Delta) \qquad \frac{\varphi \wedge \chi \le \psi}{\varphi \le \chi \to \psi} \ (\wedge\text{LR})$$

$$\frac{\varphi \le \Box\psi}{\blacklozenge\varphi \le \psi} \ (\Box\text{RA}) \qquad \frac{\Box\varphi \le \neg\mathbf{j}}{\exists \mathbf{i}(\Box\neg\mathbf{i} \le \neg\mathbf{j} \ \& \ \varphi \le \neg\mathbf{i})} \ (\neg\text{jCMP})^\dagger$$

The calculus introduced above can be used to derive first-order correspondents of formulas, inequalities, and quasi-inequalities; formal derivations in this calculus can be semantically interpreted as 'minimal valuation' meta-arguments, which justifies the statement that this calculus indeed mechanizes these meta-arguments. Several algorithms have been introduced in the literature (see, e.g., [18, 22, 30]) which specify how these derivations should proceed; these algorithms are also shown to be successful for classes of formulas which significantly extend the class of Sahlqvist formulas. Reporting in detail on these algorithms and their properties is certainly beyond the aims of this chapter; however we conclude the present section by discussing examples, since we believe that this, rather than the extensive theory, will give the reader a better idea on how to proceed in practice.

Example 36.2 In [39] Goranko and Vakarelov show that the formula $p \wedge \Box(\Diamond p \to \Box q) \to \Diamond\Box\Box q$, which falls in their class of Inductive formulas, has a first-order frame correspondent which does not correspond to any Sahlqvist formula in the basic modal language. For the sake of a smoother application of the rules introduced above, we rewrite this formula as an inequality and proceed as follows:

$$\forall p \forall q (p \wedge \Box(\Diamond p \to \Box q) \le \Diamond\Box\Box q)$$
$$\text{iff} \forall p \forall q \forall \mathbf{j}(\mathbf{j} \le p \wedge \Box(\Diamond p \to \Box q) \Rightarrow \mathbf{j} \le \Diamond\Box\Box q) \qquad \text{(FA)}$$
$$\text{iff} \forall p \forall q \forall \mathbf{j}(\mathbf{j} \le p \;\&\; \mathbf{j} \le \Box(\Diamond p \to \Box q) \Rightarrow \mathbf{j} \le \Diamond\Box\Box q) \qquad (\wedge\text{RA})$$
$$\text{iff} \forall q \forall \mathbf{j}(\mathbf{j} \le \Box(\Diamond \mathbf{j} \to \Box q) \Rightarrow \mathbf{j} \le \Diamond\Box\Box q) \qquad \text{(LA)}$$
$$\text{iff} \forall q \forall \mathbf{j}(\blacklozenge(\blacklozenge\mathbf{j} \wedge \Diamond\mathbf{j}) \le q \Rightarrow \mathbf{j} \le \Diamond\Box\Box q) \qquad (\Box\text{RA}), (\wedge\text{LR}), (\Box\text{RA})$$
$$\text{iff} \forall \mathbf{j}(\mathbf{j} \le \Diamond\Box\Box\blacklozenge(\blacklozenge\mathbf{j} \wedge \Diamond\mathbf{j})). \qquad \text{(LA)}$$

Note that the last application of (LA) yields an empty & in the antecedent. Now the last quasi-inequality is pure, and translates, after some slight simplification, into the expected first-order local frame condition $\exists y(Rxy \wedge \forall z(R^2 yz \to \exists v(Rvz \wedge Rvx \wedge Rxv)))$.

36.4 Algebraic Soundness of the Calculus for Correspondence

Discrete Stone duality for Kripke frames guarantees that the interpretation of the expanded language on Kripke frames systematically translates to complete atomic modal algebras.

$\mathcal{L}^+$-valuations on Kripke frames translate as assignments on the dual algebras, under which, nominals are interpreted as *atoms*. Inequalities and quasi-inequalities are interpreted in algebras using their natural lattice order, and satisfaction and validity naturally carry over to algebras as well. In particular, it is not difficult to show that both Lemma 36.1 and Proposition 36.2 hold if Kripke frames are replaced by complete atomic modal algebras, which again means that the calculus for correspondence defined in the previous section is sound w.r.t. the algebraic duals of Kripke frames. However, this is neither surprising nor does it give us anything more than we had before.

The algebraic perspective starts to become interesting when noticing that, as we had mentioned in Sect. 36.2, almost all the rules of the calculus for correspondence are sound w.r.t. a *significantly larger* class of algebras than complete atomic modal algebras:

Definition 36.1 A *perfect distributive lattice* (cf. [27, Definition 2.9]) is a complete lattice $\mathbb{C}$ such that the set $J^\infty(\mathbb{C})$ of the completely join-prime elements[6] is join-dense in $\mathbb{C}$ (meaning that $a = \bigvee\{j \in J^\infty(\mathbb{C}) \mid j \le a\}$ for every $a \in \mathbb{C}$) and the set $M^\infty(\mathbb{C})$ of the completely meet-prime elements is meet-dense in $\mathbb{C}$ (meaning that $a = \bigwedge\{m \in M^\infty(\mathbb{C}) \mid a \le m\}$ for every $a \in \mathbb{C}$).

---

[6] An element c of a complete lattice is *completely join-prime* if $c \ne \bot$ and, for every subset S of the lattice, $c \le \bigvee S$ iff $c \le s$ for some $s \in S$, and is *completely meet-prime* if $c \ne \top$ and, for every subset S of the lattice, $c \ge \bigwedge S$ iff $c \ge s$ for some $s \in S$.

Analogously to the duality between complete atomic Boolean algebras and sets, a Stone-type duality holds between perfect distributive lattices and *posets*, as a consequence of which, perfect distributive lattices can be equivalently characterized (cf. [33, Definition 2.14]) as those lattices each of which is isomorphic to the lattice $\mathcal{P}^\uparrow(X)$ of the upward-closed subsets of some poset X. In particular, the role atoms had in the Boolean algebra setting is taken over, in this generalized duality, by the completely join-prime elements.

Definition 36.2 *(Perfect distributive lattice with operators)* (cf. [32]) A distributive lattice with operators (DLO) $\mathbb{A}$ is *perfect* if its lattice reduct is a perfect distributive lattice and every additional operation is, in each coordinate, either completely join- or meet-preserving or completely join- or meet-reversing.

So for instance, the unary additional operations in a DLO need to satisfy at least one property in the following array: for every $S \subseteq A$,

$$\Diamond(\bigvee S) = \bigvee\{\Diamond s \mid s \in S\} \quad \Box(\bigwedge S) = \bigwedge\{\Box s \mid s \in S\}$$
$$\rhd(\bigvee S) = \bigwedge\{\rhd s \mid s \in S\} \quad \lhd(\bigwedge S) = \bigvee\{\lhd s \mid s \in S\}. \tag{36.5}$$

It is not difficult to show that both Lemma 36.1 and all items of Proposition 36.2, with the exception of item 7, hold if $\mathcal{F}$ is replaced by a suitable perfect DLO (suitable in the sense that it has the appropriate array of operations and in particular, in it, the connective $\neg$ is interpreted e.g. as intuitionistic negation), and $\mathcal{L}^+$-valuations on frames are replaced with $\mathcal{L}^+$-assignments on perfect DLOs which map nominals to completely join-prime elements.

For instance, item 1 of Proposition 36.2 is sound because, by definition, in a perfect DLO every element is the join of the set of completely join-prime elements below it; item 5 is sound because the following equivalence holds in every perfect DLO $(\mathbb{A}, \Diamond)$: for every $j \in J^\infty(\mathbb{A})$ and every $a \in \mathbb{A}$,

$$
\begin{aligned}
j &\leq \Diamond a \\
&= \Diamond(\vee\{i \in J^\infty(\mathbb{A}) \mid i \leq a\}) & \text{(definition of perfect DLO)} \\
&= \vee\{\Diamond i \in J^\infty(\mathbb{A}) \mid i \leq a\} & \text{($\Diamond$ is completely $\vee$ −preserving)} \\
\text{iff} \quad j &\leq \Diamond i \text{ for some } i \in J^\infty(\mathbb{A}) \text{s.t. } i \leq a. & \text{(j is completely join-prime)}
\end{aligned}
$$

By general order-theoretic facts (see e.g. [31]), all the operations of a perfect DLO admit right or left residuals in each coordinate, or are adjoints[7]; this immediately proves items 2, 3, 4 and 6. This means that all the rules of the calculus given in the previous section, with the exception of ($\neg$LGA), are sound and invertible w.r.t. perfect DLOs. In fact, soundness and invertibility w.r.t. perfect DLOs can be shown for a

---

[7] Notice for instance that the defining clause of the least upper bound, i.e. $a \vee b \leq c$ iff $a \leq c$ and $b \leq c$ for all $a, b, c \in \mathbb{A}$ can be equivalently restated by saying that $\vee : \mathbb{A} \times \mathbb{A} \to \mathbb{A}$ is left adjoint to the diagonal map $\Delta : \mathbb{A} \to \mathbb{A} \times \mathbb{A}$ defined by the assignment $a \mapsto (a, a)$. Likewise, $\wedge$ is the right adjoint of Δ. This is why we refer to the corresponding rules as Δ-*rules*. More on adjoints and residuals can be found in the appendix.

few more rules: for instance, thanks to the fact that in a perfect DLO every element is not only the join of the set of completely join-prime elements below it, but is also the meet of the set of completely meet-prime elements above it, the language $\mathcal{L}^+$ can be further expanded by adding a new sort of variables $\mathbf{l}, \mathbf{m}, \mathbf{n} \cdots \in \mathsf{CONOM}$, referred to as *co-nominals*, ranging over the completely meet-prime elements, and it can be easily shown that the following facts hold in every perfect DLO $\mathbb{A}$, which can be added to (the DLO-version of) the list of Proposition 36.2: for all $a, b \in \mathbb{A}$,

8. $a \leq b$ iff for every $m \in M^\infty(\mathbb{A})$, if $b \leq m$ then $a \leq m$;
9. $a \leq b$ iff for every $j \in J^\infty(\mathbb{A})$ and every $m \in M^\infty(\mathbb{A})$, if $j \leq a$ and $b \leq m$ then $j \leq m$;

these equivalences imply that the following rules are sound and invertible w.r.t. perfect DLOs:

$$\frac{\varphi \leq \psi}{\forall \mathbf{m}[\psi \leq \mathbf{m} \Rightarrow \varphi \leq \mathbf{m}]} \text{ (UA)} \qquad \frac{\varphi \leq \psi}{\forall \mathbf{j} \forall \mathbf{m}[(\mathbf{j} \leq \varphi \ \& \ \psi \leq \mathbf{m}) \Rightarrow \mathbf{j} \leq \mathbf{m}]} \text{ (ULA)}$$

It can also be shown that the *derived* rules $(\wedge \text{RA})$, $(\wedge \text{LR})$, $(\Box \text{RA})$, and $(\neg \mathbf{j} \text{CMP})$ introduced in the previous section are sound and invertible w.r.t. DLOs, except that they cannot be soundly derived anymore, but need to be added to the calculus as primitive rules, and their soundness and invertibility should be proved from first principles. Indeed, they can be shown to be sound and invertible for $\Box$ taken as a primitive connective, the implication $\rightarrow$ and subtraction $-$ respectively interpreted by means of the Heyting and the dual Heyting implications, and $\neg \mathbf{i}$ in $(\neg \mathbf{j} \text{CMP})$ replaced by $\mathbf{m} \in \mathsf{CONOM}$. In which case, the following rules can also be shown to be sound and invertible on perfect DLOs, using the fact that the Heyting implication is completely join-reversing in its first coordinate and completely meet-preserving in its second one, and the dual Heyting implication is completely join-preserving in its first coordinate and completely meet-reversing in its second one:

$$\frac{\varphi \rightarrow \chi \leq \mathbf{m}}{\exists \mathbf{j} \exists \mathbf{n}[\mathbf{j} \leq \varphi \ \& \ \chi \leq \mathbf{n} \ \& \ \mathbf{j} \rightarrow \mathbf{n} \leq \mathbf{m}]} \text{ ($\rightarrow$Appr)} \qquad \frac{\mathbf{j} \leq \chi - \psi}{\exists \mathbf{i} \exists \mathbf{m}[\mathbf{i} \leq \chi \ \& \ \psi \leq \mathbf{m} \ \& \ \mathbf{j} \leq \mathbf{i} - \mathbf{m}]} \text{ ($-$Appr)}$$

By now, the reader may have realized that the way rules are introduced easily and uniformly generalizes to any additional operation in a DLO, and applies also to the algebraic interpretation of logical languages outside the scope of modal logic, such as for instance the substructural logics, many-valued logics, and so on. For instance, the following rules for the substructural *fusion* $\circ$ and its two right residuals $/_\circ$ and $\backslash_\circ$, and for *fission* $\star$ and its two left residuals $/_\star$ and $\backslash_\star$ can be shown to be sound and invertible on DLOs:

$$\frac{\dfrac{\varphi \circ \chi \leq \psi}{\varphi \leq \psi /_\circ \chi}}{\chi \leq \varphi \backslash_\circ \psi} \text{ (}\circ\text{R)} \qquad \frac{\dfrac{\varphi \leq \chi \star \psi}{\varphi /_\star \chi \leq \psi}}{\psi \backslash_\star \varphi \leq \chi} \text{ (}\star\text{R)}$$

$$\frac{j \leq \chi \circ \psi}{\exists i \exists h[i \leq \chi \ \& \ h \leq \psi \ \& \ j \leq i \circ h]} \ (\circ\text{Appr})$$

$$\frac{\varphi \star \chi \leq m}{\exists n \exists l[\chi \leq n \ \& \ \varphi \leq l \ \& \ n \star l \leq m]} \ (\star\text{Appr})$$

$$\frac{\varphi /_\circ \chi \leq m}{\exists n \exists j[\varphi \leq n \ \& \ j \leq \chi \ \& \ n /_\circ j \leq m]} \ (/_\circ\text{Appr})$$

$$\frac{\varphi \backslash_\circ \chi \leq m}{\exists j \exists n[j \leq \varphi \ \& \ \chi \leq n \ \& \ j \backslash_\circ n \leq m]} \ (\backslash_\circ\text{Appr})$$

$$\frac{j \leq \chi /_\star \psi}{\exists i \exists m[i \leq \chi \ \& \ \psi \leq m \ \& \ j \leq i /_\star m]} \ (/_\star\text{Appr})$$

$$\frac{j \leq \chi \backslash_\star \psi}{\exists m \exists i[\chi \leq m \ \& \ i \leq \psi \ \& \ j \leq m \backslash_\star i]} \ (\backslash_\star\text{Appr})$$

Duality, relational structures and target correspondence language. Just in the same way in which the duality between complete atomic Boolean algebras and sets can be expanded to a duality between complete atomic modal algebras and relational structures consisting of *sets* endowed with arrays of relations, the duality between perfect distributive lattices and posets can be expanded to a duality between perfect DLOs and relational structures $\mathcal{F} = (W, \leq, \ldots)$, consisting of *posets* endowed with arrays of relations. Each relation in the array induces (and up to isomorphism is induced by) one additional operation in the usual way, i.e., n-ary operations correspond to $n+1$-ary relations. Examples of such structures can be found in Sect. 36.11, where more details and references are provided. The only important detail for the sake of the present discussion is that the *complex algebras* $\mathcal{F}^+$ for these frames can be defined as in the classical setting, with the notable difference that they are based on the (perfect distributive) lattice $\mathcal{P}^\uparrow(W)$ of the *upward-closed* subsets of $(W, \leq)$. This is unsurprising, and perfectly fits with the well-known fact that the valuations for e.g. intuitionistic logic are to be *persistent*. As in the case of classical modal logic, these relational structures are *both* models for the extended propositional (modal) language, *and* for the first-order language(s) which are naturally interpreted on them, and which will be our target correspondence languages. The only remaining open issue is then to establish a *standard translation* of pure formulas and quasi-inequalities of the extended propositional language $\mathcal{L}^+$ into these first-order correspondence languages. How? Because of space constraints we will not give full details, which are straightforward and can be found in [22]; instead, we restrict our attention to the interpretation of the variables in NOM and CONOM in the dual relational structures, and justify why this interpretation gives rise to first-order definable conditions on any structure $\mathcal{F} = (W, \leq, \ldots)$. Duality is crucial to establish this interpretation. Indeed, there is only one solution which takes all the following facts into account:

(a) on perfect distributive lattices, nominals and co-nominals are respectively interpreted as completely join- and meet-prime elements;
(b) the complex algebra of $\mathcal{F} = (W, \leq, \ldots)$ is based on the perfect distributive lattice $\mathcal{P}^\uparrow(W)$;
(c) the collections of all completely join- and meet-prime elements of $\mathcal{P}^\uparrow(W)$ are respectively[8]

$$\{x{\uparrow} \mid x \in W\} \quad \text{and} \quad \{W \setminus x{\downarrow} \mid x \in W\};$$

---

[8] As usual, $x{\uparrow}$ denotes the subset $\{y \mid y \in W \text{ and } x \leq y\}$, and $x{\downarrow}$ denotes the subset $\{y \mid y \in W \text{ and } y \leq x\}$.

(d) the unique homomorphic extension $\widehat{V}$ of each $\mathcal{L}^+$-valuation on $\mathcal{F}$ is to be an $\mathcal{L}^+$-valuation on $\mathcal{F}^+$;

(e) it should be that case that, for all models $(\mathcal{F}, V)$ and all $\varphi, \psi \in \mathcal{L}^+$,

$$\mathcal{F}, V \Vdash \varphi \leq \psi \quad \text{iff} \quad \mathcal{F}^+, \widehat{V} \Vdash \varphi \leq \psi.$$

The only way to define the interpretation of $\mathbf{j} \in \mathsf{NOM}$ and $\mathbf{m} \in \mathsf{CONOM}$ which takes all these facts into account is to stipulate that $\mathcal{L}^+$-valuations V on $\mathcal{F}$ assign variables $\mathbf{j} \in \mathsf{NOM}$ to elements in $\{x\uparrow \mid x \in W\}$ and variables $\mathbf{m} \in \mathsf{CONOM}$ to elements in $\{W \setminus x\downarrow \mid x \in W\}$. As was the case in the classical setting, the interpretations of nominals and co-nominals are clearly definable in the most restricted correspondence language which the structures $\mathcal{F}$ are models of.

Stepping back from this discussion, we note two points: duality was crucial in establishing the connection of clearest practical value to our current agenda, namely being able to translate the pure fragment of the extended language $\mathcal{L}^+$ into the target first-order correspondence language. However, the reasoning used in establishing this connection illustrates a *methodological* point about dualities, namely, that they can be used not only as a proof tool, but also as a *defining* tool. For instance, in more general settings than the ones presented so far, like lattice based logics, the algebraic semantics is clear but one might be in the dark as to what an appropriate relational semantics might be, both as regards an appropriate class of relational structures and as to the appropriate interpretation of the propositional language in such a class. This is where duality can be used as a defining tool: firstly, relational structures can be extracted, as it were, from perfect lattices [27]; secondly, the interpretation of the propositional formulas in algebras transfers via the duality to these extracted structures. To mention a related but different example, in [43, 44] duality is used to semantically identify the intuitionistic counterparts of public announcement logic and of the logic of epistemic actions and knowledge.

36.5 Four Conclusions and a Question

Conclusion 1: thanks to the algebraic insights facilitated by duality, correspondence theory can be developed uniformly for more than modal-like logics; as we have illustrated, also substructural logics, intuitionistic logic and its fragments, MV-logics, as well as distributive and intuitionistic modal logic, and more in general, all the logics the algebraic semantics of which is given by DLOs can be encompassed. Also, μ-calculus (see Sect. 36.8.3), monotone modal logic [23] and their lattice-based extensions are examples of logics which can be uniformly treated by this theory.

Conclusion 2: the algebraic and algorithmic developments for correspondence can and have been merged. This now allows for algebraic canonicity to be treated either independently from correspondence in the style of [36], or via correspondence as in [22]. And there is more: as discussed at the end of the previous subsection, even in vastly more general settings, concrete relational structures can be extracted from the algebras. Therefore, even in these rarified algebraic settings, speaking of correspondence theory does not amount to merely establishing an elaborate social convention, or a manner of speaking, by means of which we can pretend that relational models which are not really there virtually manifest themselves by means of their algebraic ghosts. On the contrary, the obtained correspondence theory makes sense, on the extracted relational structures, in the traditional way.

Conclusion 3: for the sake of the present chapter, we have distilled the main features of the algebraic-algorithmic approach into a more informal presentation of a calculus for correspondence, the set of rules of which can be modified, expanded or reduced, so that the calculus can be adapted to different logical languages, and so that it can be proven sound w.r.t. different semantics; however, the underlying mathematical principles which drive this calculus (as well as the algorithms, and more in general, all the Sahlqvist-style correspondence arguments) remain stable across the different settings, and are: the Ackermann lemma in any of its many forms, the residuation/adjunction properties of the operations interpreting the logical connectives, and the approximation properties of the 'states' (or co-states) of the relational semantics, which generate (and co-generate) their dual complex algebras.

Conclusion 4: The computation process of first-order correspondents can be neatly divided in two stages: a first stage, in which quasi-inequalities are transformed into *pure* ones, and a second stage, where pure quasi-inequalities are interpreted in the given classes of relational structures. Different relational semantics might then yield different interpretations of the same pure quasi inequality, and some instances of this will be discussed in Sect. 36.9. The definition of this syntactic calculus and the possibility of soundly interpreting it in a generalized algebraic environment (which can then be translated, in a second stage, into several concrete relational semantics) gives some mathematical flesh to van Benthem's insight that "Correspondence Theory may be applied to any kind of semantic entity".

Question: How powerful is this algebraic-algorithmic procedure? In the case of classical modal logic it is state of the art, and covers syntactically characterized classes of formulas which *significantly extend* the Sahqvist class (viz. Inductive, Recursive, see [39]). But can we claim that, in all the other (e.g. lattice-based) cases, the algorithmic procedure is just as powerful? The answer to this question requires being able to recognize Sahlqvist, Inductive, Recursive classes for each logical language to which the algorithmic correspondence applies. In Sect. 36.7 we suggest a way in which this can be done.

36.6 The van Benthem Formulas

One aspect of the discussion in Sect. 36.2 still needs to be justified, which concerns how to extract the correspondent of a given modal formula, provided the equivalence between clauses (1) and (2) holds (which are reported below); before moving on to what we have promised to do at the end of the previous section, in the present section we discuss this aspect briefly. As mentioned early on in Sect. 36.2, suppose that, for a certain subclass of valuations K, the following are equivalent:

(1) $\mathcal{F}, V, w \Vdash \varphi(p_1, \ldots, p_n)$ for every assignment V;
(2) $\mathcal{F}, V^*, w \Vdash \varphi(p_1, \ldots, p_n)$ for every assignment $V^* \in K$.

Suppose moreover that each member $V^* \in K$ and $1 \leq i \leq n$, the subset $V^*(p_i)$ can be defined (possibly parametrically) by a formula $\alpha_i(w, \overline{v})$ in some extension L' of the frame correspondence language L_0. Here we typically think of L' as L_0 itself or some language in between L_0 and L_2 such as first-order logic with least fixed points, or perhaps a first-order logic with branching quantifiers such as information friendly logic.

Let Σ be the set of all L'-formulas $\mathrm{ST}_x(\varphi)(\alpha_1(w, \overline{v}), \ldots, \alpha_n(w, \overline{v}))$ obtained by substituting in $\mathrm{ST}_x(\varphi)$ the predicate symbols $P_1, \ldots P_n$ with the L'-formulas $\alpha_1(w, \overline{v}), \ldots, \alpha_n(w, \overline{v})$ corresponding to the valuations in K. Clearly, $\forall \overline{P} \mathrm{ST}_x(\varphi) \models \Sigma[x := w]$, where $\overline{P}$ is the vector of all predicate symbols occurring in $\mathrm{ST}_x(\varphi)$. But also, because of the equivalence between (1) and (2) assumed above, $\Sigma \models \forall \overline{P} \mathrm{ST}_x(\varphi)[x := w]$. If Σ is finite, then $\bigwedge \Sigma$ is clearly an L' local frame correspondent for φ.

Even if Σ is infinite, we can still find an L' equivalent, provided L' is compact: Since $\Sigma \models \forall \overline{P} \mathrm{ST}_x(\varphi)[x := w]$ we have $\Sigma \models \mathrm{ST}_x(\varphi)[x := w]$, and we may then appeal to the compactness of L' to find some finite subset $\Sigma' \subseteq \Sigma$ such that $\Sigma' \models \mathrm{ST}_x(\varphi)[x := w]$.

We claim that $\Sigma' \models \forall \overline{P} \mathrm{ST}_x(\varphi)[x := w]$. Indeed, let $\mathcal{M}$ be any L_1-model such that $\mathcal{M} \models \Sigma'[x := w]$. Since the predicate symbols in $\overline{P}$ do not occur in Σ', every $\overline{P}$-variant of $\mathcal{M}$ also models Σ', and hence also $\mathrm{ST}_x(\varphi)$. It follows that $\mathcal{M} \models \forall \overline{P} \mathrm{ST}_x(\varphi)[x := w]$. Thus we may take $\bigwedge \Sigma'$ as a local first-order frame correspondent for φ.

The case in which $L' = L_0$ and K is the class of all parametrically L_0-definable valuations was studied by van Benthem in [4]. Under these assumptions, the class of formulas for which the equivalence between (1) and (2) holds was named the *van Benthem formulas* in [17]. All the well known syntactically characterized classes of first-order definable modal formulas (Sahlqvist, Inductive, etc.) are encompassed by the van Benthem formulas. However, in its full generality, the class of van Benthem formulas is of little practical use. Indeed, for infinite sets Σ, the above argument, relying on compactness as it does, does not enable us to explicitly calculate a correspondent for a given formula φ, or devise an algorithm which produces frame correspondents for each member of a given *class* of modal formulas. One therefore typically concentrates on cases in which the class K can be described by L'-formulas of uniform shape and hence of bounded complexity. In [23] an account of classical correspondence is given in terms of a hierarchy of such classes K.

36.7 Characterizing the Sahlqvist Formulas Across Different Logics

As discussed at the end of Sect. 36.5, being able to measure the effectiveness of the algebraic-algorithmic approach across different logics requires being able to recognize Sahlqvist, Inductive, Recursive classes for each logical language to which the algorithmic correspondence applies. In the following subsection, we give a very portable definition of Sahlqvist formulas, or rather inequalities, that is general enough to be applied unchanged across a wide variety of logics. In Sect. 36.7.2, we contrast this briefly with other definitions in the literature.

36.7.1 The Sahlqvist Inequalities: A General Purpose Definition

Given a logic with DLOs as algebraic semantics, what should 'morally' be the class of Sahlqvist formulas for this logic? As glimpsed above, the reduction strategy for Sahlqvist formulas is based on the order-theoretic properties of adjunction and residuation possessed by the operations interpreting the connectives. More specifically, it is the order of alternation of connectives with these properties over certain variable occurrences which is of crucial importance, since it enables the input clause to be transformed into an equivalent one satisfying the restrictions under which (LA) or (RA) can be applied. Our answer will accordingly be couched in these terms.

To fix ideas, let us consider a logical signature containing classical negation (like that of basic modal logic) but otherwise undefined. (Negated) Sahlqvist formulas in such a signature can be described in terms of their generation trees, as illustrated in Fig. 36.2. Namely, the nodes in the upper part are labelled with connectives interpreted by means of left residuals or Δ-adjoints. The lower parts of branches ending in positive variables are labelled with connectives interpreted by means of right adjoints. This is the basic Sahlqvist shape that we are going to reproduce across signatures.

To port this shape to signatures without classical negation, we will have to introduce some bookkeeping machinery and the following auxiliary definitions and notation: we work with the usual notion of a *generation tree* of a formula. A *signed generation tree* (see e.g., [33]) associates with each node in a generation tree a sign, $+$ or $-$, in such a way that children of nodes labelled with connectives which are order preserving (order reversing) in the appropriate coordinate have the same (opposite) sign as their parent. The *positive (negative) generation tree of* φ, denoted $+\varphi$ $(-\varphi)$, is thus obtained by signing the root in the generation tree of φ with $+$ $(-)$ and propagating the signs.

Fig. 36.2 The basic Sahlqvist shape

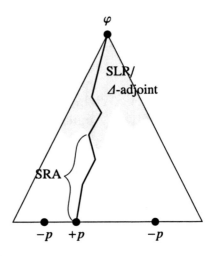

$$\varphi$$

SLR/
$\varDelta$-adjoint

SRA

$-p$ $+p$ $-p$

Definition 36.3 (*Order types and critical branches*) An *order type* over $n \in \mathbb{N}$ is an n-tuple $\epsilon \in \{1, \partial\}^n$. For any formula $\varphi(p_1, \ldots p_n)$, any order type ϵ over n, and any $1 \le i \le n$, an ϵ-*critical node* in a signed generation tree of φ is a (leaf) node $+p_i$ with $\epsilon_i = 1$ or $-p_i$ with $\epsilon_i = \partial$. An ϵ-*critical branch* in the tree is a branch from an ϵ-critical node.

We are now ready to reproduce the Sahlqvist shape in non-classical settings. For definiteness' sake we work in the distributive setting, and consider a signature which provides a representative sample of connectives commonly encountered in the literature, taken from intuitionistic logic, Distributive Modal logic (cf. [22, 33]), and substructural logic.

Definition 36.4 Nodes in generation trees are classified according to Table 36.1. A branch in a signed generation tree $*s$, $* \in \{+, -\}$, is *excellent* if it is the concatenation of two paths P_1 and P_2, one of which may possibly be of length 0, such that P_1 is a path from the leaf consisting (apart from variable nodes) only of SRA-nodes, and P_2 consists (apart from variable nodes) only of Δ-adjoint and SLR-nodes.

Definition 36.5 For any order type ϵ, the signed generation tree of a formula φ is ϵ-*Sahlqvist* if every ϵ-critical branch is excellent. An inequality $\varphi \le \psi$ is ϵ-*Sahlqvist* if the trees $+\varphi$ and $-\psi$ are both ϵ-Sahlqvist. An inequality is *Sahlqvist* if it is ϵ-Sahlqvist for some ϵ.

Notice that, according to Definition 36.5, the generation trees of the two sides of an ϵ-Sahlqvist inequality reproduce the pattern illustrated in Fig. 36.2, modulo the order type ϵ.

We wish to stress the methodology that Definition 36.5 aims at exemplifying. This definition is intended to serve as a template applicable to any signature via a classification of connectives such as the one of Table 36.1. The place of any given

Table 36.1 Classification of nodes

Δ-adjoints		Syntactically Right Adjoints (SRA)	
+	∨ ∧	+	∧ □ ▷
−	∧ ∨	−	∨ ◇ ◁
Syntactically Left Residuals (SLR)		**Syntactically Right Residuals (SRR)**	
+	◇ ◁ ∘	+	∨ ★ →
−	□ ▷ ★ →	−	∧ ∘

logical connective in this classification is *not inherent* to the connective; rather, it entirely depends on the order-theoretic properties of the interpretation of the given connectives, *relative to a specific semantics*, and is hence bound to change when switching to a different interpretation. For instance, the classification of Table 36.1 is relative to the usual interpretation of logical connectives in the setting of *distributive* lattices. When interpreted in general lattices, $+\vee$ and $-\wedge$ do not fall into the SRR category anymore, because their standard interpretations in general lattices are not residuated.

In essence, Definition 36.5 gives us a winning strategy which guarantees the success of our calculus, as well as of algorithms like ALBA, SQEMA and indeed the Sahlqvist-van Benthem algorithm. Success consists in eliminating all occurring variables by means of applications of the Ackermann rules (LA) or (RA). The Sahlqvist shape guarantees that, for every variable, input inequalities can be transformed into a shape to which an Ackermann rule is applicable. The order type ϵ tells us which occurrences of a given variable p we need to 'display', i.e., get to occur in inequalities of the form $p \leq \alpha$ or $\alpha \leq p$ as prescribed by (LA) or (RA). The Sahlqvist shape guarantees that this is always possible. Indeed, going down a critical branch, we can surface the subtree containing the SRA part of the critical branch, by applying *approximation rules*[9] to the SLR-nodes and Δ-rules (see footnote 7) to the Δ-adjoint nodes. Then the SRA-nodes on the remainder of this branch can be stripped off by means of the residuation/adjunction rules, thus surfacing the variable occurrence and simultaneously calculating the minimal valuation for it. Finally, notice that the remaining occurrences of p are of the opposite order type: this guarantees that they have the right polarity to receive the calculated minimal valuations, as prescribed by (LA) or (RA).

Example 36.3 The Dunn axioms for positive modal logic $\Box p \wedge \Diamond q \leq \Diamond(p \wedge q)$ and $\Box(p \vee q) \leq \Box p \vee \Diamond q$, as well their intuitionistic counterparts $\Diamond(p \rightarrow q) \leq \Box p \rightarrow \Diamond q$ and $\Diamond p \rightarrow \Box q \leq \Box(p \rightarrow q)$ are all Sahlqvist inequalities. Specifically, the first inequality is ϵ-Sahlqvist with $\epsilon(p) = 1$ and $\epsilon(q) = 1$, the second is ϵ-Sahlqvist with

---

[9] The approximation rules are those which introduce new nominals or co-nominals. All the other rules introduced so far, except (LA), (RA), (⊤), and (⊥), are collectively referred to as residuation/adjunction rules.

Table 36.2 Universal and choice nodes

	Choice		Universal	
+	∨ ◊ ◁	+	□ ▷	
−	∧ □ ▷	−	◊ ◁	

$\epsilon(p) = \partial$ and $\epsilon(q) = \partial$, and neither is ϵ-Sahlqvist for any other order type. The third and fourth inequalities are both ϵ-Sahlqvist with $\epsilon(p) = 1$ and $\epsilon(q) = \partial$, and again neither is ϵ-Sahlqvist for any other order type.

The Löb inequality $\Box(\Box p \rightarrow p) \leq \Box p$ is not ϵ-Sahlqvist for any order type, because in the positive generation tree of the left hand side both positive and negative occurrences of p have the properly SRR-node $+ \rightarrow$ as ancestor, making their corresponding branches non-excellent.

In similar way, the Frege inequality $p \rightarrow (q \rightarrow r) \leq (p \rightarrow q) \rightarrow (p \rightarrow r)$ is not ϵ-Sahlqvist for any order type, because both positive and negative occurrences of q have properly SRR-nodes $+ \rightarrow$ as ancestors, making their corresponding branches non-excellent.

36.7.2 Other Approaches to Syntactic Characterization

Definitions of Sahlqvist-like classes generally come in two flavours: positive, or constructive, definitions that tell one how the formulas in the class can be built up, and negative definitions which define a class by banning certain alternations of connectives. While not being explicitly constructive, the definition offered in the previous subsection is clearly positive. We would like to contrast it with the negative definition used in [33]. This definition classifies the connectives of Distributive Modal Logic (DML) as *Choice* and *Universal*, according to the Table 36.2. A signed generation tree is then declared to be ϵ-Sahlqvist if on no ϵ-critical branch there is a choice node with a universal node as ancestor.[10] The notion of a Sahlqvist inequality is then further defined exactly as in Definition 36.5. Comparing Tables 36.1 and 36.2 will also make it clear that, in terms of adjunction and residuation, Choice and Universal have the following meaning:

$$\text{Choice} = \text{Not a right adjoint}$$
$$\text{Universal} = \text{Neither a left residual nor a } \Delta\text{-adjoint.}$$

Thus, when restricted to the signature of DML, this definition and Definition 36.5 are equivalent. However, generalizing the Choice-Universal style definition does become problematic once binary connectives like the intuitionistic implication $\rightarrow$ is

---

[10] This has been slightly paraphrased in order to exploit the terminology already introduced above.

involved. Indeed, the Heyting implication is not a right adjoint, and is neither a left residual nor a Δ-adjoint, and hence $+ \rightarrow$ is both Choice and Universal. Now, applying the Choice-Universal style definition, inequalities such as $p \rightarrow \Box p \leq \Diamond \Box p$, which cannot be solved, would be classified as ϵ-Sahlqvist with $\epsilon(p) = 1$. One way to remedy this is to declare the ancestor relation to be reflexive, rendering an occurrence of a choice-and-universal node a violation of the rule prohibiting choice nodes in the scope of universal ones. A more elegant solution, we maintain and hope the reader would agree, would be to adopt a definition in the style of the previous subsection.

36.8 Three Moves Towards a Unified Correspondence Theory

The present section is aimed at discussing how three recent directions in correspondence theory can be encompassed in the algebraic-algorithmic approach, based on the recognition that all these directions are predicated on the same basic order-theoretic principles we have discussed in the previous sections. The first generalization, presented in Sect. 36.8.1, concerns correspondence settings in which the target language is first-order logic with *least* (or more in general *extremal*) *fixed points* (FO + LFP). In Sect. 36.8.2, various syntactic generalizations of the Sahlqvist class will be discussed, which are obtained by relaxing the requirements of Definition 36.5. Finally, Sect. 36.8.3 focuses on a recent research line in which van Benthem has been active, which extends algorithmic correspondence theory to propositional logics expanded with fixed points, such as the modal mu-calculus.

36.8.1 Expanding the Target Language with Fixed Points

When trying to reduce an inequality with the calculus of correspondence, one reason of failure is that it is not possible to obtain a form to which (LA) or (RA) is applicable, and particularly because any obtainable α (as in the formulation of these rules) is not p-free. Consider for example the Löb inequality $\Box(\Box p \rightarrow p) \leq \Box p$. Let us apply the calculus to it:

$$\forall p[\Box(\Box p \rightarrow p) \leq \Box p]$$
$$\text{iff} \quad \forall p \forall \mathbf{i} \forall \mathbf{m}[(\mathbf{i} \leq \Box(\Box p \rightarrow p) \ \& \ \Box p \leq \mathbf{m}) \Rightarrow \mathbf{i} \leq \mathbf{m}]$$
$$\text{iff} \quad \forall p \forall \mathbf{i} \forall \mathbf{m}[(\blacklozenge \mathbf{i} \leq \Box p \rightarrow p \ \& \ \Box p \leq \mathbf{m}) \Rightarrow \mathbf{i} \leq \mathbf{m}]$$
$$\text{iff} \quad \forall p \forall \mathbf{i} \forall \mathbf{m}[(\blacklozenge \mathbf{i} \wedge \Box p \leq p \ \& \ \Box p \leq \mathbf{m}) \Rightarrow \mathbf{i} \leq \mathbf{m}].$$

We would have been able to apply (LA), had it not been for the p occurring on the left hand side of $\blacklozenge \mathbf{i} \wedge \Box p \leq p$. So this is how far we can get and no further, and with good reason: the Löb inequality has no first-order frame correspondent, as is well known. However, the Ackermann lemma can be strengthened to the following version:

Lemma 36.2 *Let $\alpha(p)$, $\beta(p)$, and $\gamma(p)$ be formulas of a language $\mathcal{L}^+$ interpreted on perfect DLOs, with $\alpha(p)$ and $\beta(p)$ positive in p and $\gamma(p)$ negative in p. Then the following are equivalent for every perfect DLO $\mathbb{C}$ and variable assignment v:*

1. *$\mathbb{C}, v \models \beta(\mu p.\alpha(p)/p) \leq \gamma(\mu p.\alpha(p))/p)$;*
2. *there exists some $v' \sim_p v$ such that $\mathbb{C}, v' \models \alpha(p) \leq p$, and $\mathbb{C}, v' \models \beta(p) \leq \gamma(p)$,*

where $\mu p.\alpha(p)$ is the least fixed point of $\alpha(p)$.

Proof We begin by noting that, since we are working in a complete lattice, least fixed points of monotone (term) functions exist by the Knaster-Tarski theorem. As regards '1 $\Rightarrow$ 2', let $v'(p) := v(\mu p.\alpha(p))$. As regards '2 $\Rightarrow$ 1', $\mathbb{C}, v' \models \alpha(p) \leq p$ implies that $v'(p)$ is a pre-fixed point of $\alpha(\cdot)$,[11] and hence $\mu p.\alpha(p) \leq v'(p)$. Therefore, $\beta(\mu p.\alpha(p)/p) \leq \beta(v'(p)) \leq \gamma(v'(p)) \leq \gamma(\mu p.\alpha(p)/p)$.

Lemma 36.2 justifies the following rule:

$$\frac{\forall p[(\alpha(p) \leq p \ \& \ \&_{1 \leq i \leq n}\beta_i(p) \leq \gamma_i(p)) \Rightarrow \varphi \leq \psi]}{\&_{1 \leq i \leq n}\beta_i(\mu p.\alpha(p)/p) \leq \gamma_i(\mu p.\alpha(p)/p) \Rightarrow \varphi \leq \psi} \quad \text{(RLA)}$$

where α, β_i, and γ_i are as in the lemma, and φ and ψ are negative and positive in p, respectively. Back to the Löb inequality, we can now apply RLA to eliminate p:

$$\text{iff} \quad \forall i \forall m[\Box(\mu p.(\blacklozenge i \wedge \Box p)) \leq m \Rightarrow i \leq m]$$
$$\text{iff} \quad \forall i[i \leq \Box(\mu p.(\blacklozenge i \wedge \Box p))]$$
$$\text{iff} \quad \forall i[\blacklozenge i \leq \mu p.(\blacklozenge i \wedge \Box p)].$$

Under duality with Kripke frames, the condition above translates as $\forall w[R[w] \subseteq \mu X.(R[w] \cap (R^{-1}[X^c])^c)]$, which gives the expected condition of transitivity and converse well foundedness.

As another example, consider the van Benthem inequality $\Box \Diamond \top \leq \Box(\Box(\Box p \to p) \to p)$:

$$\forall p[\Box \Diamond \top \leq \Box(\Box(\Box p \to p) \to p)]$$
$$\text{iff} \quad \forall p \forall i \forall m[(i \leq \Box \Diamond \top \ \& \ \Box(\Box(\Box p \to p) \to p) \leq m) \Rightarrow i \leq m]$$
$$\text{iff} \quad \forall p \forall i \forall m \forall n[(i \leq \Box \Diamond \top \ \& \ \Box n \leq m \ \& \ \Box(\Box p \to p) \to p \leq n) \Rightarrow i \leq m]$$
$$\text{iff} \quad \forall p \forall i \forall j \forall m \forall n[(i \leq \Box \Diamond \top \ \& \ \Box n \leq m \ \& \ j \leq \Box(\Box p \to p) \ \& \ j \to p \leq n) \Rightarrow i \leq m]$$
$$\text{iff} \quad \forall p \forall i \forall j \forall m \forall n[(i \leq \Box \Diamond \top \ \& \ \Box n \leq m \ \& \ \blacklozenge j \leq \Box p \to p \ \& \ j \to p \leq n) \Rightarrow i \leq m]$$

---

[11] Here $\alpha(\cdot)$ is obtained from the term function α by leaving p free and fixing all other variables to the values prescribed by v.

$$\text{iff} \quad \forall p \forall i \forall j \forall m \forall n [(\mathbf{i} \leq \Box \Diamond \top \; \& \; \Box \mathbf{n} \leq \mathbf{m} \; \& \; \blacklozenge \mathbf{j} \wedge \Box p \leq p \; \& \; \mathbf{j} \to p \leq \mathbf{n}) \Rightarrow \mathbf{i} \leq \mathbf{m}]$$

$$(*) \; \text{iff} \quad \forall i \forall j \forall m \forall n [(\mathbf{i} \leq \Box \Diamond \top \; \& \; \Box \mathbf{n} \leq \mathbf{m} \; \& \; \mathbf{j} \to \mu p.(\blacklozenge \mathbf{j} \wedge \Box p) \leq \mathbf{n}) \Rightarrow \mathbf{i} \leq \mathbf{m}]$$

$$\text{iff} \quad \forall i \forall j \forall n [(\mathbf{i} \leq \Box \Diamond \top \; \& \; \mathbf{j} \to \mu p.(\blacklozenge \mathbf{j} \wedge \Box p) \leq \mathbf{n}) \Rightarrow \forall m [\Box \mathbf{n} \leq \mathbf{m} \Rightarrow \mathbf{i} \leq \mathbf{m}]]$$

$$\text{iff} \quad \forall i \forall j \forall n [(\mathbf{i} \leq \Box \Diamond \top \; \& \; \mathbf{j} \to \mu p.(\blacklozenge \mathbf{j} \wedge \Box p) \leq \mathbf{n}) \Rightarrow \blacklozenge \mathbf{i} \leq \mathbf{n}]$$

$$\text{iff} \quad \forall i \forall j [\mathbf{i} \leq \Box \Diamond \top \Rightarrow \forall n [\mathbf{j} \to \mu p.(\blacklozenge \mathbf{j} \wedge \Box p) \leq \mathbf{n} \Rightarrow \blacklozenge \mathbf{i} \leq \mathbf{n}]]$$

$$\text{iff} \quad \forall i \forall j [\mathbf{i} \leq \Box \Diamond \top \Rightarrow \blacklozenge \mathbf{i} \leq \mathbf{j} \to \mu p.(\blacklozenge \mathbf{j} \wedge \Box p)]$$

$$\text{iff} \quad \forall i \forall j [\mathbf{i} \leq \Box \Diamond \top \Rightarrow \mathbf{i} \leq \Box (\mathbf{j} \to \mu p.(\blacklozenge \mathbf{j} \wedge \Box p))]$$

$$\text{iff} \quad \forall j [\Box \Diamond \top \leq \Box (\mathbf{j} \to \mu p.(\blacklozenge \mathbf{j} \wedge \Box p))].$$

In the equivalence marked with (∗), the Right Ackermann lemma has been applied with $\alpha(p) := \blacklozenge \mathbf{j} \wedge \Box p$ and $\beta(p) := \mathbf{j} \to p$ being positive in p, and $\gamma(p) := \mathbf{n}$ being negative in p.

Correspondence with FO+LFP has been studied in [7, 8, 11, 20] and other chapters. It is not possible here to do justice to this work, but that is not the aim of the current chapter.

36.8.2 Syntactic Generalizations of the Sahlqvist Class

The class of Sahlqvist formulas is, quite rightly, considered to be the paradigmatic syntactically definable class of modal formulas admitting first-order correspondents. This pre-eminent status can, however, blind one to the fact that there is much interesting and systematic correspondence theory that can be done with formulas that lie strictly *outside* this class. There is indeed life beyond the Sahlqvist formulas. Some of this work is orthogonal to the Sahlqvist theme, in the sense that the arguments bear no obvious resemblance to the minimal valuation strategy: here we are thinking, for example, of the modal reduction principles interpreted over transitive frames, which all have first-order correspondents [3]. In the present section we will, however, be looking at classes of formulas that represent the natural generalization of the Sahlqvist formulas, in the sense that they are obtained by taking the order-theoretic insights underlying the Sahlqvist 'winning strategy' (see discussion following Definition 36.5) to their natural boundaries of applicability.

A very noticeable feature of Definition 36.5 is the fact that nodes lower down on critical branches need to be syntactically right adjoint, not, e.g., syntactically right residual. For unary connectives, residuation and adjunction are equivalent notions (see appendix), so this imposes no restriction, but for connectives of higher arity it does. For example, the Löb inequality $\Box(\Box p \to p) \leq \Box p$, considered in Sect. 36.8.1, is *not* Sahlqvist, for in the generation tree $+\Box(\Box p \to p)$ both the leaves $+p$ and $-p$ have the properly SRR node $+ \to$ as ancestor, and hence for no choice of order type ϵ will the ϵ-critical paths be excellent. The Löb formula belongs to the class of so-called *Recursive* or *Regular formulas*, introduced in [38], which all have frame correspondents in FO + LFP (see also [20]). We will not burden the reader with precise definitions here, but the intuition is that one firstly relaxes the definition of an excellent branch (Definition 36.4) to that of a *good branch* by also allowing the occurrence of SRR nodes (and not just of SRA nodes) in the lower part of critical branches. Merely substituting the "good" for "excellent" in the definition of the Sahlqvist inequalities (Definition 36.5) would be too liberal, however, for that would allow inequalities like $\Diamond(\Box p \star p) \leq \Diamond p$, on which the calculus of correspondence fails entirely, since we cannot bring them into a shape to which even the recursive Ackermann lemma is applicable. To ensure that the calculus works, it is enough to add the further requirement that at most one ϵ-critical branch may pass through any given properly SRR-node; this yields precisely the ϵ-*Recursive inequalities*.

But here the reader may very well protest that we have promised extensions of the Sahlqvist class for which first-order correspondence holds, while the Recursive formulas are only guaranteed to have correspondents in FO + LFP. Indeed, the Recursive inequalities as a generalization of the Sahlqvist class is still too liberal. In order to guarantee first-order correspondence, the ordinary non-recursive Ackermann lemma will have to be applicable for each variable elimination. In order to ensure this, one needs to impose upon the variables in Recursive inequalities a partial ordering, and demand not only that at most one ϵ-critical branch pass through any given properly SRR-node, but also that if an ϵ-critical branch passes through a properly SRR-node, all variables occurring on other branches passing through it have to be strictly less (according to the ordering) than the variable on the critical branch. This gives rise to the classes of *Inductive formulas* and *inequalities*, for formal definitions of which the reader is referred to [21, 22, 38]. As an example, the Frege inequality $p \to (q \to r) \leq (p \to q) \to (p \to r)$ from the implicative fragment of intuitionistic logic is Inductive; however, it is not Sahlqvist, as shown in Example 36.3. For an ALBA reduction of this inequality see [22, Example 7.5].

36.8.3 Correspondence for Propositional Logics with Fixed Points

In the generalized setting of Sect. 36.8.1, fixed points have been added to the target language so as to be able to extend the correspondence methodology up to classes of formulas, pre-eminently exemplified by the Löb's formula, for which minimal valuations exist but are not elementarily definable. However, once fixed points are brought into the correspondence picture on the *target* side, it is natural to extend the correspondence program to settings in which fixed points belong also to the *source* language, like the modal mu-calculus; the extra expressivity of the source language will be safely accommodated by the expanded target language. In this vein, in [8], van Benthem and his collaborators syntactically characterized a certain class of formulas in the language of modal mu-calculus as the counterpart of the Sahlqvist class (hence named the class of *Sahlqvist mu-formulas*), on the basis of the minimal valuation methodology, via an extension of the classical model-theoretic proof. In [15], a correspondence result theoretically independent from [8] has been given for logics with fixed points on a weaker than classical base (thus applicable e.g. also to intuitionistic modal mu-calculus, or to certain substructural logics expanded with fixed points). In [15], the results in [8] are encompassed into the algebraic-algorithmic unified correspondence theory, and Sahlqvist mu-formulas are recognized in essence as Recursive formulas (see Sect. 36.8.2) on the basis of the approach outlined in Sects. 36.7.1 and 36.8.2. The chapter [15] is rather technical; however, thanks to the insights developed so far in the present exposition, and particularly on the existing tight connection between the minimal valuation argument and (the recursive and non-recursive versions of) the Ackermann lemma, we are now in a position to give an informal account of these results, as well as of their relationship with results in [8].

Concretely, embedding the Sahlqvist-type theorem of [8] into the algebraic-algorithmic correspondence theory requires:

(a) extending (the distributive/intuitionistic/non-distributive versions of) the calculus for correspondence with dedicated approximation and adjunction/residuation rules (see footnote 9, p. 21) capable of transforming systems of mu-inequalities into equivalent systems of mu-inequalities in Ackermann shape[12];

(b) giving a (distributive/intuitionistic/non-distributive) counterpart of the class of Sahlqvist mu-formulas as defined in [8] in the style of Definition 36.5;

(c) motivating the definitions in (b) by giving surjective projections from the non-classical languages involved to the classical, which preserve and reflect Sahlqvist status. An analogous projection has been given in [22] between DML and classical modal logic.

Due to space constrains we will only address (a) and (b). As to (a), notice preliminarily that the calculus for correspondence introduced in Sect. 36.3 is already enough to perform the elimination of predicate variables on a *restricted* class of mu-formulas/inequalities,[13] as in the following example (cf. [8, Example 5.3]):

$$\forall p[\nu X. \Box(p \wedge X) \leq p]$$

$$\text{iff} \quad \forall p \forall \mathbf{i} \forall \mathbf{m}[(\mathbf{i} \leq \nu X.\Box(p \wedge X) \,\&\, p \leq \mathbf{m}) \Rightarrow \mathbf{i} \leq \mathbf{m}] \quad \text{(ULA)}$$

$$(\ast)\,\text{iff} \quad \forall \mathbf{i} \forall \mathbf{m}[(\mathbf{i} \leq \nu X.\Box(\mathbf{m} \wedge X) \Rightarrow \mathbf{i} \leq \mathbf{m}] \quad \text{(LA)}$$

$$\text{iff} \quad \forall \mathbf{m}[\nu X.\Box(\mathbf{m} \wedge X) \leq \mathbf{m}]. \quad \text{(FA inverse)}$$

Indeed, the application of the rule (LA) is sound because the term function $\gamma(p) = \nu X.\Box(p \wedge X)$ is monotone in p. However, this calculus is certainly not powerful enough to be successful over the whole class of (intuitionistic counterparts of) Sahlqvist mu-formulas in [8]. In [15], an enhancement of the calculus has been

---

[12] Notice that, thanks to the very general way in which the various versions of Ackemann's lemma have been stated, the corresponding Ackermann rules apply without changes to logical languages with fixed points.

[13] Namely, the one formed by those inequalities such that, for some order type ϵ, all ϵ-critical branches are excellent (cf. Definition 36.4) or good (cf. Sect. 36.8.2) according to the letter of these notions, and hence no fixed point binders occur in ϵ-critical branches.

defined by firstly adding approximation rules, of which the following are special instances (cf. [15] for the complete account):

$$\frac{\mathbf{i} \leq \mu X.\varphi(X, \psi/!x, \overline{z})}{\exists \mathbf{j}[\mathbf{i} \leq \mu X.\varphi(X, \mathbf{j}/!x, \overline{z}) \ \& \ \mathbf{j} \leq \psi]} \ (\mu^+\text{-A}) \qquad \frac{\nu X.\varphi(X, \psi/!x, \overline{z}) \leq \mathbf{m}}{\exists \mathbf{n}[\nu X.\varphi(X, \mathbf{n}/!x, \overline{z}) \leq \mathbf{m} \ \& \ \psi \leq \mathbf{n}]} \ (\nu^+\text{-A})$$

$$\frac{\mathbf{i} \leq \mu X.\varphi(X, \psi/!x, \overline{z})}{\exists \mathbf{n}[\mathbf{i} \leq \mu X.\varphi(X, \mathbf{n}/!x, \overline{z}) \ \& \ \psi \leq \mathbf{n}]} \ (\mu^-\text{-A}) \qquad \frac{\nu X.\varphi(X, \psi/!x, \overline{z}) \leq \mathbf{m}}{\exists \mathbf{j}[\nu X.\varphi(X, \mathbf{j}/!x, \overline{z}) \leq \mathbf{m} \ \& \ \mathbf{j} \leq \psi]} \ (\nu^-\text{-A})$$

where, in $(\mu^+\text{-A})$ (resp., $(\mu^-\text{-A})$) the associated term function of $\varphi(X, x, \overline{z})$ is completely $\bigvee$-preserving in $(X, x) \in \mathbb{C} \times \mathbb{C}$ (resp., in $(X, x) \in \mathbb{C} \times \mathbb{C}^\partial$), and in $(\nu^+\text{-A})$ (resp., $(\nu^-\text{-A})$) the associated term function of $\varphi(X, x, \overline{z})$ is completely $\bigwedge$-preserving in $(X, x) \in \mathbb{C} \times \mathbb{C}$ (resp., in $(X, x) \in \mathbb{C} \times \mathbb{C}^\partial$), for any perfect DLO $\mathbb{C}$ of the appropriate signature. Moreover, in each rule the variable x is assumed not to occur in ψ. The notation $\varphi(!x)$ means that the variable x has a unique occurrence in φ.

Some motivating intuitions and examples illustrating the functioning and applicability of these rules, as well as of the adjunction-rules below, are given in the ensuing discussion.

Secondly, adjunction rules for fixed point binders have been added, of which the following are special instances (cf. [15] for the complete account):

$$\frac{\mu X.(A(X) \vee B(p)) \leq \chi}{p \leq \nu X.(E(X) \wedge D(\chi/p))} \ (\mu\text{-Adj}) \qquad \frac{\chi \leq \nu X.(E(X) \wedge D(p))}{\mu X.(A(X) \vee B(\chi/p)) \leq p} \ (\nu\text{-Adj})$$

where, in each rule,

$$A(X) = \bigvee_{i \in I} \delta_i(X), \quad B(p) = \bigvee_{j \in J} \delta'_j(p), \quad E(X) = \bigwedge_{i \in I} \beta_i(X) \text{ and } D(p) = \bigwedge_{j \in J} \beta'_j(p)$$

with I and J finite sets of indexes, each δ_i and δ'_j interpreted as a unary left adjoint (typically, δ_i and δ'_j are concatenations of diamonds over a variable), and each β_i and β'_j interpreted as a unary right adjoint (typically, β_i and β'_j are boxed atoms). Finally, $\delta_i \dashv \beta_i$ and $\delta'_j \dashv \beta'_j$ for each i and j.

Notice that, unlike the rules for propositional connectives, the rules above are *contextual*, i.e., dependent on assumptions on the formulas in the scope of the fixed point binder. This reflects the fact that the semantic interpretations of fixed point binders do not have intrinsic order-theoretic properties, but at most preserve those of the term functions associated with the formulas in their scope.

In [15], rules generalizing the ones above are proven to be sound w.r.t. the natural algebraic/relational semantics of (intuitionistic) modal mu-calculus. Thanks to these rules, inequalities we could previously not treat, such as $p \leq \nu X[\Box(X \wedge (q \rightarrow \bot)) \vee (\Diamond p \wedge \Diamond q)]$ (cf. [8, Example 5.4]) can be reduced as follows:

$$\forall p \forall q [p \leq \nu X[\Box(X \wedge (q \to \bot)) \vee (\Diamond p \wedge \Diamond q)]]$$

iff $\forall p \forall q \forall \mathbf{i} \forall \mathbf{m}[(\mathbf{i} \leq p \ \& \ \nu X[\Box(X \wedge (q \to \bot)) \vee (\Diamond p \wedge \Diamond q)] \leq \mathbf{m}) \Rightarrow \mathbf{i} \leq \mathbf{m}]$ (ULA)

iff $\forall q \forall \mathbf{i} \forall \mathbf{m}[\nu X[\Box(X \wedge (q \to \bot)) \vee (\Diamond \mathbf{i} \wedge \Diamond q)] \leq \mathbf{m} \Rightarrow \mathbf{i} \leq \mathbf{m}]$ (RA)

iff $\forall q \forall \mathbf{i} \forall \mathbf{m} \forall \mathbf{j}[(\mathbf{j} \leq q \ \& \ \nu X[\Box(X \wedge (\mathbf{j} \to \bot)) \vee (\Diamond \mathbf{i} \wedge \Diamond q)] \leq \mathbf{m}) \Rightarrow \mathbf{i} \leq \mathbf{m}]$ $(\nu^- \text{-A})$

iff $\forall \mathbf{i} \forall \mathbf{m} \forall \mathbf{j}[\nu X[\Box(X \wedge (\mathbf{j} \to \bot)) \vee (\Diamond \mathbf{i} \wedge \Diamond \mathbf{j})] \leq \mathbf{m} \Rightarrow \mathbf{i} \leq \mathbf{m}]$ (RA)

iff $\forall \mathbf{i} \forall \mathbf{j}[\mathbf{i} \leq \nu X[\Box(X \wedge (\mathbf{j} \to \bot)) \vee (\Diamond \mathbf{i} \wedge \Diamond \mathbf{j})]]$.   (UA inverse)

In the application of $(\nu^- \text{-A})$ above, $\varphi(X, !x, z)$ is $\Box(X \wedge (x \to \bot)) \vee z$, and ψ is q.

Moreover, the following alternative reduction is now possible for the inequality $\nu X.\Box(p \wedge X) \leq p$, treated as the first example of the present subsection:

$$\forall p[\nu X.\Box(p \wedge \Box X) \leq p]$$

iff $\forall p \forall \mathbf{i} \forall \mathbf{m}[(\mathbf{i} \leq \nu X.\Box(p \wedge \Box X) \ \& \ p \leq \mathbf{m}) \Rightarrow \mathbf{i} \leq \mathbf{m}]$ (ULA)

iff $\forall p \forall \mathbf{i} \forall \mathbf{m}[\mu X.\blacklozenge(\blacklozenge X \vee \mathbf{i}) \leq p \ \& \ p \leq \mathbf{m}) \Rightarrow \mathbf{i} \leq \mathbf{m}]$ $(\nu - \text{Adj})$

iff $\forall \mathbf{i} \forall \mathbf{m}[\mu X.\blacklozenge(\blacklozenge X \vee \mathbf{i}) \leq \mathbf{m} \Rightarrow \mathbf{i} \leq \mathbf{m}]$ (RA)

iff $\forall \mathbf{i}[\mathbf{i} \leq \mu X.\blacklozenge(\blacklozenge X \vee \mathbf{i})]$.   (UA inverse)

The application of $(\nu\text{-Adj})$ is performed modulo distributing modal connectives. Notice that, by unfolding the least fixed point $\mu X.\blacklozenge(\blacklozenge X \vee \mathbf{i})$, the clause $\forall \mathbf{i}[\mathbf{i} \leq \mu X.\blacklozenge(\blacklozenge X \vee \mathbf{i})]$ can be rewritten as $\forall \mathbf{i}[\mathbf{i} \leq \bigvee_{\kappa \geq 1} \blacklozenge^\kappa \mathbf{i}]$, which immediately translates on Kripke frames into the well known condition expressing the reflexivity of the transitive closure of the relation interpreting $\Box$.

As to (b), the class of ϵ-*Recursive inequalities*, in the intuitionistic modal mu-language, has been syntactically defined in [15] closely following the approach of Definition 36.5; this class is the intuitionistic counterpart of the class of Sahlqvist mu-formulas defined in [8]. Analogously to Definition 36.5, the definition of ϵ-Recursive inequalities is grounded on a classification of the nodes in the signed generation trees of formulas similar to the specification given in Table 36.3. However, as was mentioned early on, the fixed point binders escape to some extent the order-theoretic classification, since their interpretation does not enjoy inherent order-theoretic properties, but rather preserves, in some cases, those of the term function in its scope. To take this fact into account, we firstly group nodes according to categories (we use the names *skeleton* and *PIA* for these categories, also appearing in [8], to explicitly establish a connection with the model-theoretic analysis conducted there), and secondly, we group nodes within each category according to their contextually relevant order-theoretic properties.

The shape of the ϵ-Recursive inequalities is in essence the Sahlqvist/Inductive/Recursive shape introduced and discussed in Sect. 36.7.1; as to the similarities, the outer skeleton is exactly the same as the outer part of a Sahlqvist formula; moreover the PIA part is defined in such a way that, when restricted to the binder-free fragment, it gives the inner part of the ϵ-Recursive formulas (cf. Sect. 36.8.2). The complete definition of the PIA part incorporates extra conditions regulating the relative positions of free fixed point variables and variables which we want to solve for;

Table 36.3 Skeleton and PIA nodes

Outer Skeleton	Inner Skeleton	PIA
Δ-adjoints	Binders	Binders
+ ∨ ∧	+ μ	+ ν
− ∧ ∨	− ν	− μ
SLR	SLA	SRA
+ ◇ ◁ ∘	+ ◇ ◁ ∨	+ □ ▷ ∧
− □ ▷ ⋆ →	− □ ▷ ∧	− ◇ ◁ ∨
	SLR	SRR
	+ ∧ ∘	+ ∨ ⋆ →
	− ∨ ⋆ →	− ∧ ∘

these conditions ensure that formulas in the scope of binders have the appropriate order-theoretic properties guaranteeing the applicability of the μ- and ν-adjunction rules. The inner skeleton essentially arises by the introduction of fixed point binders into the outer part of a Sahlqvist formula. As to the differences, this introduction blocks the application of Δ-rules (and more generally also the possibility of applying rules to single connectives), leaving us with only μ- and ν-approximation rules. Hence all the nodes are reclassified according to the properties which they enjoy and which are now relevant. Similar to the PIA formulas, inner skeletons incorporate extra conditions regulating the relative positions of free fixed point variables and variables which we want to solve for; these conditions ensure that formulas in the scope of binders have the appropriate order-theoretic properties guaranteeing the applicability of the μ- and ν-approximation rules. The shape of the inequalities in this class provides a winning strategy analogous to the one described for the Sahlqvist inequalities in Sect. 36.7.1. Again, the order type ϵ tells us which occurrences of a given variable we need to 'display'. The ϵ-Recursive shape guarantees that this is always possible. Indeed, going down a critical branch, we can surface the subtree containing the PIA part of the critical branch by applying *approximation rules* to the Skeleton nodes. Then adjunction/residuation rules such as (μ-Adj) and (ν-Adj) are applied to display the critical occurrences of variables in the subtrees containing the PIA parts, and to simultaneously calculate the minimal valuation for them. Finally, notice that the remaining occurrences of variables are of the opposite order type: this guarantees that they have the right polarity to receive the calculated minimal valuations, as prescribed by (LA), (RA) or their recursive counterparts.

The analysis of PIA-formulas conducted in [8] can be summarized in the slogan "PIA formulas provide minimal valuations". In this respect, the crucial model-theoretic property possessed by PIA-formulas is the *intersection property*, isolated by van Benthem in [6]. The order-theoretic import of this property is clear: if a formula has the intersection property then the term function associated with it is completely meet preserving. In the complete lattice setting in which we find ourselves, this is equivalent to it being a right adjoint; this is exactly the order-theoretic property guaranteeing the soundness of adjunction/residuation rules like (μ-Adj) and (ν-Adj).[14]

36.9 Correspondence Across Different Semantics

In Sect. 36.7 we saw that it is possible to uniformly implement the 'correspondence calculus' for different logics, and how, accordingly, the definitions of syntactic classes like the Sahlqvist class could be ported to these logics. In the current section we shift our focus to consider the related question of what happens when we keep the logical language and the order-theoretic properties of the connectives fixed, while varying the relational semantics. In terms of Fig. 36.1b this means that we maintain the algebraic interpretation of the logic while imposing different dualities. The key point we wish to illustrate is that the calculus of correspondence is sound in the setting of perfect distributive lattice expansions, and hence that the elimination of propositional variables can proceed largely independently of any considerations on the dual relational structures; the outcome of the reduction/elimination process can be then further translated so as to fit different relational environments.

We take as our running example Pierce's law $((p \to q) \to p) \to p$, which was considered in [5, Sect. 3.2] where correspondence is studied for this and other formulas belonging to the implicative fragment of intuitionistic logic. The calculus of Sect. 36.3 gives us the following reduction, which is sound on perfect Heyting algebras (i.e., perfect distributive lattices expanded with the right residual $\to$ of $\wedge$):

$$\forall p \forall q [(p \to q) \to p \leq p]$$
$$\text{iff} \quad \forall p [(p \to \bot) \to p \leq p]$$
$$\text{iff} \quad \forall p \forall \mathbf{j} \forall \mathbf{m} [(\mathbf{j} \leq (p \to \bot) \to p \,\&\, p \leq \mathbf{m}) \Rightarrow \mathbf{j} \leq \mathbf{m}] \qquad (36.6)$$
$$\text{iff} \quad \forall \mathbf{j} \forall \mathbf{m} [\mathbf{j} \leq (\mathbf{m} \to \bot) \to \mathbf{m} \Rightarrow \mathbf{j} \leq \mathbf{m}]$$
$$\text{iff} \quad \forall \mathbf{m} [(\mathbf{m} \to \bot) \to \mathbf{m} \leq \mathbf{m}].$$

Thus, the propositional variables have been eliminated, and we can interpret the result on intuitionistic frames (i.e. posets), via the well known duality between perfect Heyting algebras and intuitionistic frames. Recall that, in intuitionistic frames, variables (and consequently all formulas) are evaluated to upward-closed subsets (upsets), and that in particular $\llbracket \varphi \to \psi \rrbracket = (((\llbracket \varphi \rrbracket^c \cup \llbracket \psi \rrbracket)^c) \downarrow)^c = (\llbracket \varphi \rrbracket \cap \llbracket \psi \rrbracket^c) \downarrow^c$, where S^c and $S \downarrow$ denote the set theoretic complement and downward closure of the subset S, respectively. All this fits with the duality between perfect Heyting algebras and intuitionistic frames, according to which the algebra elements correspond to up-sets of frames, and in particular the meet prime elements correspond to the

---

[14] We must warn the reader that this account, and in particular the formulation of the additional rules, is slightly oversimplified. Complete details can be found in [15].

complements of principal down-sets, which we denote by $w{\downarrow}^c$. So, translating the outcome obtained above via this duality yields:

$\forall m[(m \to \bot) \to m \le m]$
iff $\forall w[w{\downarrow} \subseteq ((w{\downarrow}^c \cap \varnothing^c){\downarrow}^c \cap w{\downarrow}){\downarrow}]$
iff $\forall w[w \in ((w{\downarrow}^c){\downarrow}^c \cap w{\downarrow}){\downarrow}]$
iff $\forall w \exists v[w \le v \ \& \ v \le w \ \& \ v \in (w{\downarrow}^c){\downarrow}^c]$
iff $\forall w[w \notin (w{\downarrow}^c){\downarrow}]$
iff $\forall w \forall v[v \not\le w \ \Rightarrow \ w \not\le v]$

iff $\forall w[((w{\downarrow}^c \cap \varnothing^c){\downarrow}^c \cap w{\downarrow}^{cc}){\downarrow}^c \subseteq w{\downarrow}^c]$
iff $\forall w[w{\downarrow} \subseteq ((w{\downarrow}^c){\downarrow}^c \cap w{\downarrow}){\downarrow}]$
iff $\forall w \exists v[w \le v \ \& \ v \in (w{\downarrow}^c){\downarrow}^c \cap w{\downarrow})]$
iff $\forall w[w \in (w{\downarrow}^c){\downarrow}^c]$
iff $\forall w \forall v[v \notin w{\downarrow} \ \Rightarrow \ w \not\le v]$
iff $\forall w \forall v[w \le v \ \Rightarrow \ v \le w]$.

Thus, as discussed in [5, Example 78], we see that Pierce's law takes us to classical propositional logic, by constraining the ordering on intuitionistic frames to be discrete.

Pierce's law (as well as any other axiom in the implicative fragment of intuitionistic logic) can be alternatively interpreted on ternary frames, as they are defined e.g. in [42], where a Kripkean semantics is employed for the non-associative Lambek calculus, and a restricted Sahlqvist theorem is proven. A *ternary frame* (cf. [42, Definition 1]) is a structure (W, R) such that W is a nonempty set and R is a ternary relation on W. For all $X, Y \subseteq W$, let $R[Y, X] = \{z \mid \exists x \exists y[x \in Y \ \& \ y \in X \ \& \ R(xyz)]\}$. Implication can be interpreted on ternary frames as follows: for all $X, Y \subseteq W$,

$$X \implies Y = \{z \mid \forall x \forall y[(R(yxz) \ \& \ x \in X) \Rightarrow y \in Y]\} = R[Y^c, X]^c.$$

Valuations send proposition letters to arbitrary subsets of the universe of ternary frames. Thus, the complex algebra of the ternary frame (W, R) can be defined as the perfect algebra $(\mathcal{P}(W), \cup, \cap, W, \varnothing, \implies)$, and this assignment can be extended to a fully fledged discrete Stone-type duality for BAOs, in the style of e.g. [47]. In particular, $\implies$ as defined above is order-reversing (in fact, completely join-reversing) in its first coordinate and order-preserving (in fact, completely meet-preserving) in its second coordinate[15] (see Sect. 5 and references therein for more details). Thus, the very same reduction performed in (36.6) is sound also w.r.t. the complex algebras of ternary frames defined above, or equivalently, w.r.t. ternary frame semantics. Relying on this duality, the final clause of (36.6) can be interpreted on ternary frames as follows (we abuse notation and write w^c for $\{w\}^c = W \setminus \{w\}$):

$\forall m[(m \to \bot) \to m \le m]$
iff $\forall w[w \in R[\{w\}, R[W, w^c]^c]]$
iff $\forall w \exists y[R(wyw) \ \& \ y \in R[W, w^c]^c]$

iff $\forall w[R[w^{cc}, R[\varnothing^c, w^c]^c]^c \subseteq w^c]$
iff $\forall w \exists x \exists y[R(xyw) \ \ y \in R[W, w^c]^c \ \& \ x = w]$
iff $\forall w \exists y[R(wyw) \ \ \forall x \forall z[R(xzy) \Rightarrow z = w]]$.

Notice that, in the more familiar case in which the operation $\bullet$, uniquely identifying $\implies$, coincides with meet, the ternary relation which dually represents the binary map given by $(U, V) \mapsto U \cap V$ is $R = \{(x, x, x) \mid x \in W\}$; in this case, $X \implies Y$ reduces to the classical $X^c \cup Y$, and the first-order clause above is always true.

---

[15] In fact, $\implies$ can be uniquely identified as the right residual of $\bullet$ (*fusion*), given by $Y \bullet Z := \{x \mid \exists y \exists z[y \in Y \ \& \ z \in Z \ \& \ R(x, y, z)]\}$.

A second example. The inequality $(p \wedge q) \to r \leq (p \to r) \vee (q \to r)$, in the intuitionistic language, is reduced by the calculus as follows:

$\forall p \forall q \forall r [(p \wedge q) \to r \leq (p \to r) \vee (q \to r)]$

$\forall i \forall m \forall p \forall q \forall r [(i \leq (p \wedge q) \to r \ \& \ (p \to r) \vee (q \to r) \leq m) \Rightarrow i \leq m]$

$\forall i \forall m \forall p \forall q \forall r [(i \leq (p \wedge q) \to r \ \& \ (p \to r) \leq m \ \& \ (q \to r) \leq m) \Rightarrow i \leq m]$

$\forall i \forall j \forall k \forall m \forall p \forall q \forall r [(i \leq (p \wedge q) \to r \ \& \ (j \to r) \leq m \ \& \ j \leq p \ \& \ (k \to r) \leq m \ \& \ k \leq q)$
$\Rightarrow i \leq m]$

$\forall i \forall j \forall k \forall m \forall r [(i \leq (j \wedge k) \to r \ \& \ (j \to r) \leq m \ \& \ (k \to r) \leq m) \Rightarrow i \leq m]$

$\forall i \forall j \forall k \forall m \forall r [(i \wedge (j \wedge k) \leq r \ \& \ (j \to r) \leq m \ \& \ (k \to r) \leq m) \Rightarrow i \leq m]$

$\forall i \forall j \forall k \forall m [((j \to (i \wedge j \wedge k)) \leq m \ \& \ (k \to (i \wedge j \wedge k)) \leq m) \Rightarrow i \leq m]$

$\forall i \forall j \forall k \forall m [(j \to (i \wedge j \wedge k)) \vee (k \to (i \wedge j \wedge k)) \leq m \Rightarrow i \leq m]$

$\forall i \forall j \forall k [i \leq (j \to (i \wedge j \wedge k)) \vee (k \to (i \wedge j \wedge k))]$

$\forall i \forall j \forall k [i \leq (j \to (i \wedge j \wedge k)) \ \vartheta \ i \leq (k \to (i \wedge j \wedge k))]$

$\forall i \forall j \forall k [i \wedge j \leq k \ \vartheta \ i \wedge k \leq j].$

For reasons analogous to those discussed in the previous example, this reduction is sound w.r.t. several classes of algebras based on perfect distributive lattices (and hence w.r.t. the classes of set-based structures dual to each of these), which include, but are not limited to, perfect (i.e. complete and atomic) Boolean algebras (hence sets), perfect Heyting algebras (hence posets) and the perfect BAO of the previous example (hence ternary frames as in the previous example). When interpreted according to the first or third option, or equivalently on sets or ternary frames, the last line in the reduction above becomes:

$$\forall w \forall v \forall v [\{w\} \cap \{v\} \subseteq \{v\} \ \vartheta \ \{w\} \cap \{v\} \subseteq \{v\}],$$

which always holds, as was expected, since the inequality treated above is classically (but not intuitionistically) valid. When interpreted in perfect Heyting algebras, or equivalently on posets, the last line in the reduction above can be further translated into

$$\forall w \forall v \forall v [w{\uparrow} \cap v{\uparrow} \subseteq v{\uparrow} \ \vartheta \ w{\uparrow} \cap v{\uparrow} \subseteq v{\uparrow}],$$

which is equivalent to the condition that every principal up-set be linearly ordered. Indeed, it is clear that, if in a poset there are states w, v, v such that $w \leq v$ and $w \leq v$ but $v \not\leq v$ and $v \not\leq v$, then neither inclusion in the condition above holds for these states; conversely, reasoning by cases should convince the reader that if in a poset every principal up-set is linearly ordered, then the displayed condition holds. For instance, if $v \not\leq v$ and $v \not\leq v$, and $w{\uparrow} \cap v{\uparrow} \neq \varnothing \neq w{\uparrow} \cap v{\uparrow}$, let us assume that $w{\uparrow} \cap v{\uparrow} \not\subseteq v{\uparrow}$, i.e. that there exists some $x \in w{\uparrow} \cap v{\uparrow}$ such that $v \not\leq x$, and let $y \in w{\uparrow} \cap v{\uparrow}$. Then $y \not\leq x$, but since $x, y \in w{\uparrow}$, the assumption implies that $x \leq y$, and hence $y \in v{\uparrow}$, as desired.

Intuitionistic correspondence via Gödel translation. So far in the present section, we have seen that the purely syntactic encoding of correspondence arguments is particularly advantageous in those situations (common to many nonclassical logics) in which a given logical language is interpreted on more than one type of set-based structures; indeed, the soundness of a given algorithmic reduction depends exclusively on the order-theoretic properties of the interpretation of the logical connectives, and, provided these properties are satisfied in each interpretation, the same reduction will yield first-order correspondents in relational structures of different types. Sometimes, as in the case of intuitionistic modal logic, the availability of different relational semantics for a given logic reflects itself in the fact that the category of perfect algebras naturally associated with the logic in question is dually equivalent to each category of relational structures supporting the interpretation of that logic. However, in some cases, the roles of algebras and relational structures might be reversed, in the sense that more than one category of algebras might be dually associated with one and the same category of relational structures. This is the case in e.g. the category of posets and p-morphisms, which is dually equivalent to both the category of perfect Heyting algebras and complete homomorphisms via Birkhoff duality, and to a suitable full subcategory of perfect modal algebras and complete homomorphisms via the Jónsson-Tarski duality. Notice that, for every poset $(W, \leq)$, the inclusion map $\mathcal{P}^{\uparrow}(W) \hookrightarrow (\mathcal{P}(W), [\leq])$ satisfies the clauses of the Gödel assignment, i.e. $U \mapsto U = [\leq]U$ and $(U \rightarrow V) \mapsto (U \rightarrow V) = [\leq](U^c \cup V)$ for every $U, V \in \mathcal{P}^{\uparrow}(W)$, which implies the well known fact that an intuitionistic formula is valid on a given poset $(W, \leq)$ if its Gödel translation is. On the syntactic side, the Sahlqvist/Inductive shape of formulas in the language of intuitionistic logic is preserved under the Gödel translation. In the light of these observations it is natural to ask to what extent intuitionistic correspondence arguments can be subsumed by classical correspondence arguments via the Gödel translation. This question, formulated as vaguely as we have, can be reformulated more concretely in ways which—more importantly for our purposes here—lend themselves to be investigated with the tools of the unified correspondence theory outlined in the present chapter.

One such reformulation is: can the reduction steps for the intuitionistic language which are sound on perfect Heyting algebras be *simulated* by suitable reduction steps for the target modal language of the Gödel translation and which are sound on perfect BAOs? And is the Gödel translation itself, as it were, such a simulation? This would be the case, in a sense, if the minimal valuations calculated in performing the reduction steps on an intuitionistic inequality and on its Gödel translation were always semantically identical. In general, one cannot expect this to hold, as the minimal valuation provided by the calculus in the classical setting need not be *persistent*, as required by the intuitionistic notion of validity. However, running the calculus on the Gödel translation of the inequality in the example above proves instructive; below, $\square$ stands for $[\leq]$ and $\blacklozenge$ for $\langle \geq \rangle$.

$$\forall p \forall q \forall r [\square((\square p \wedge \square q) \rightarrow \square r) \leq \square(\square p \rightarrow \square r) \vee \square(\square q \rightarrow \square r)]$$

$$\forall \mathbf{i} \forall \mathbf{m} \forall p \forall q \forall r [(\mathbf{i} \leq \square((\square p \wedge \square q) \rightarrow \square r) \ \& \ \square(\square p \rightarrow \square r) \vee \square(\square q \rightarrow \square r) \leq \mathbf{m})$$
$$\Rightarrow \mathbf{i} \leq \mathbf{m}]$$

$$\forall i \forall m \forall p \forall q \forall r[(i \leq \Box((\Box p \wedge \Box q) \rightarrow \Box r) \,\&\, \Box(\Box p \rightarrow \Box r) \leq \mathbf{m} \,\&\, \Box(\Box q \rightarrow \Box r)$$
$$\leq \mathbf{m}) \Rightarrow i \leq \mathbf{m}]$$

$$\forall i \forall j \forall k \forall m \forall p \forall q \forall r[(i \leq \Box((\Box p \wedge \Box q) \rightarrow \Box r) \,\&\, \Box(\mathbf{j} \rightarrow r) \leq \mathbf{m}$$
$$\&\, \mathbf{j} \leq \Box p \,\&\, \Box(\mathbf{k} \rightarrow r) \leq \mathbf{m} \,\&\, \mathbf{k} \leq \Box q) \Rightarrow i \leq \mathbf{m}]$$

$$\forall i \forall j \forall k \forall m \forall p \forall q \forall r[(i \leq \Box((\Box p \wedge \Box q) \rightarrow \Box r) \,\&\, \Box(\mathbf{j} \rightarrow r) \leq \mathbf{m}$$
$$\&\, \blacklozenge \mathbf{j} \leq p \,\&\, \Box(\mathbf{k} \rightarrow r) \leq \mathbf{m} \,\&\, \blacklozenge \mathbf{k} \leq q) \Rightarrow i \leq \mathbf{m}]$$

$$\forall i \forall j \forall k \forall m \forall r[(i \leq \Box((\Box \blacklozenge \mathbf{j} \wedge \Box \blacklozenge \mathbf{k}) \rightarrow \Box r) \,\&\, \Box(\mathbf{j} \rightarrow r) \leq \mathbf{m} \,\&\, \Box(\mathbf{k} \rightarrow r) \leq \mathbf{m})$$
$$\Rightarrow i \leq \mathbf{m}]$$

$$\forall i \forall j \forall k \forall m \forall r[\blacklozenge(\blacklozenge \mathbf{i} \wedge (\Box \blacklozenge \mathbf{j} \wedge \Box \blacklozenge \mathbf{k})) \leq r \,\&\, \Box(\mathbf{j} \rightarrow r) \leq \mathbf{m} \,\&\, \Box(\mathbf{k} \rightarrow r) \leq \mathbf{m})$$
$$\Rightarrow i \leq \mathbf{m}]$$

$$\forall i \forall j \forall k \forall m[(\Box(\mathbf{j} \rightarrow \blacklozenge(\blacklozenge \mathbf{i} \wedge \Box \blacklozenge \mathbf{j} \wedge \Box \blacklozenge \mathbf{k})) \leq \mathbf{m} \,\&\, \Box(\mathbf{k} \rightarrow \blacklozenge(\blacklozenge \mathbf{i} \wedge \Box \blacklozenge \mathbf{j} \wedge \Box \blacklozenge \mathbf{k}))$$
$$\leq \mathbf{m}) \Rightarrow i \leq \mathbf{m}]$$

$$\forall i \forall j \forall k \forall m[(\Box(\mathbf{j} \rightarrow \blacklozenge(\blacklozenge \mathbf{i} \wedge \Box \blacklozenge \mathbf{j} \wedge \Box \blacklozenge \mathbf{k})) \vee \Box(\mathbf{k} \rightarrow \blacklozenge(\blacklozenge \mathbf{i} \wedge \Box \blacklozenge \mathbf{j} \wedge \Box \blacklozenge \mathbf{k})) \leq \mathbf{m})$$
$$\Rightarrow i \leq \mathbf{m}]$$

$$\forall i \forall j \forall k[i \leq \Box(\mathbf{j} \rightarrow \blacklozenge(\blacklozenge \mathbf{i} \wedge \Box \blacklozenge \mathbf{j} \wedge \Box \blacklozenge \mathbf{k})) \vee \Box(\mathbf{k} \rightarrow \blacklozenge(\blacklozenge \mathbf{i} \wedge \Box \blacklozenge \mathbf{j} \wedge \Box \blacklozenge \mathbf{k}))]$$

$$\forall i \forall j \forall k[i \leq \Box(\mathbf{j} \rightarrow \blacklozenge(\blacklozenge \mathbf{i} \wedge \Box \blacklozenge \mathbf{j} \wedge \Box \blacklozenge \mathbf{k})) \,\invamp\, i \leq \Box(\mathbf{k} \rightarrow \blacklozenge(\blacklozenge \mathbf{i} \wedge \Box \blacklozenge \mathbf{j} \wedge \Box \blacklozenge \mathbf{k}))]$$

$$\forall i \forall j \forall k[\blacklozenge \mathbf{i} \wedge \mathbf{j} \leq \blacklozenge(\blacklozenge \mathbf{i} \wedge \Box \blacklozenge \mathbf{j} \wedge \Box \blacklozenge \mathbf{k}) \,\invamp\, \blacklozenge \mathbf{i} \wedge \mathbf{k} \leq \blacklozenge(\blacklozenge \mathbf{i} \wedge \Box \blacklozenge \mathbf{j} \wedge \Box \blacklozenge \mathbf{k})].$$

The minimal valuation computed above assigns p to $\blacklozenge \mathbf{j} = \langle \geq \rangle \{w\} = w{\uparrow}$; analogously, q is mapped by the same valuation to $\blacklozenge \mathbf{k} = v{\uparrow}$, and r to $\blacklozenge(\blacklozenge \mathbf{i} \wedge (\Box \blacklozenge \mathbf{j} \wedge \Box \blacklozenge \mathbf{k}))$. The assignment for r can be rewritten as follows:

$$\blacklozenge(\blacklozenge \mathbf{i} \wedge (\Box \blacklozenge \mathbf{j} \wedge \Box \blacklozenge \mathbf{k})) = (w{\uparrow} \cap ((v{\uparrow}^c){\downarrow})^c \cap ((v{\uparrow}^c){\downarrow})^c){\uparrow} = (w{\uparrow} \cap v{\uparrow} \cap v{\uparrow}){\uparrow}$$
$$= w{\uparrow} \cap v{\uparrow} \cap v{\uparrow}.$$

So, in this case, the minimal valuation provided by the reduction of the Gödel translation in the boolean setting is *exactly the same* as that provided by the reduction of the original inequality in the intuitionistic setting. This example is of course not enough to justify any general claims, but it does suggest a line for further investigation, namely the identification of classes of intuitionistic formulas for which the correspondence arguments are subsumed by the correspondence arguments of their Gödel translations in the strongest sense, as discussed above. As an initial observation in this direction we note that whenever the algorithm solves for positive occurrences of variables (cf. discussion on signed generation trees before Definition 36.5), as in the example above, these variable occurrences will surface, if at all, on the right-hand side of inequalities; this, together with the fact that the Gödel translation prefixes all variables with a $\Box$, implies that the minimal valuations provided by the algorithm will be (the extensions of) finite disjunctions of $\blacklozenge$-terms. The latter are always upward closed, as required by the intuitionistic semantics.

Things do not work out so nicely for all intuitionistic Sahlqvist formulas, as revisiting our first example in the current section, the Pierce inequality, will show.

This inequality is ϵ-Sahlqvist for $\epsilon(p) = \partial$ and $\epsilon(q) = 1$ and for no other order type ϵ. Running the correspondence algorithm on its Gödel translation yields:

$$\forall p \forall q [\Box(\Box(\Box p \to \Box q) \to \Box p) \leq \Box p]$$

$$\forall p [\Box(\Box(\Box p \to \Box\bot) \to \Box p) \leq \Box p]$$

$$\forall \mathbf{i} \forall \mathbf{m} \forall p [(\mathbf{i} \leq \Box(\Box(\Box p \to \Box\bot) \to \Box p) \ \& \ \Box p \leq \mathbf{m}) \Rightarrow \mathbf{i} \leq \mathbf{m}]$$

$$\forall \mathbf{i} \forall \mathbf{m} \forall \mathbf{n} \forall p [(\mathbf{i} \leq \Box(\Box(\Box p \to \Box\bot) \to \Box p) \ \& \ \Box \mathbf{n} \leq \mathbf{m} \ \& \ p \leq \mathbf{n}) \Rightarrow \mathbf{i} \leq \mathbf{m}]$$

$$\forall \mathbf{i} \forall \mathbf{m} \forall \mathbf{n} [(\mathbf{i} \leq \Box(\Box(\Box \mathbf{n} \to \Box\bot) \to \Box \mathbf{n}) \ \& \ \Box \mathbf{n} \leq \mathbf{m}) \Rightarrow \mathbf{i} \leq \mathbf{m}]$$

$$\forall \mathbf{i} \forall \mathbf{n} [\mathbf{i} \leq \Box(\Box(\Box \mathbf{n} \to \Box\bot) \to \Box \mathbf{n}) \Rightarrow \forall \mathbf{m} [\Box \mathbf{n} \leq \mathbf{m} \Rightarrow \mathbf{i} \leq \mathbf{m}]]$$

$$\forall \mathbf{i} \forall \mathbf{n} [\mathbf{i} \leq \Box(\Box(\Box \mathbf{n} \to \Box\bot) \to \Box \mathbf{n}) \Rightarrow \mathbf{i} \leq \Box \mathbf{n}]$$

$$\forall \mathbf{n} [\Box(\Box(\Box \mathbf{n} \to \Box\bot) \to \Box \mathbf{n}) \leq \Box \mathbf{n}].$$

The minimal valuation provided by the above reduction assigns p to a co-atom, i.e. to a set of type $W \setminus \{w\}$, which need not be upward-closed. There are probably other, less naïve ways in which the intuitionistic correspondence argument for the Pierce axiom can be simulated classically via its Gödel translation, but we leave this question open.

Finally, notice that the preservation of the intuitionistic Sahlqvist or Inductive classes under the Gödel translation is a very restricted phenomenon. This preservation occurs thanks mainly to the lack of order-theoretic variety in the intuitionistic signature. Namely, the interpretation of each binary connective in the intuitionistic signature is either a right residual or a right adjoint; in other words, the intuitionistic signature does not include any 'pure diamond-type' connective. As soon as pure diamond-type connectives are added, this transfer breaks down: for instance, Sahlqvist inequalities in the language of intuitionistic modal logic are not preserved under the Gödel translation. Indeed, the Gödel-translation of the Sahlqvist inequality $\Box\Diamond p \leq \Diamond p$ yields $\Box\Diamond[\leq]p \leq \Diamond[\leq]p$, which is not Sahlqvist, and actually—by van Benthem's classification of the modal reduction principles [3]—it does not even have a first-order frame correspondent.

36.10 Conclusions

Unified correspondence. As van Benthem has aptly remarked, our "algebraic analysis is a combinatorial formalization of essentials of correspondence reasoning." Indeed, classical correspondence arguments have been mechanized, and transformed into chains of equivalent rewritings of quasi-inequalities in the extended language $\mathcal{L}^+$. The language $\mathcal{L}^+$ can be captured by the monadic second-order frame language. The chains of equivalent rewritings aim at transforming quasi-inequalities in $\mathcal{L}^+$ into equivalent quasi-inequalities in a *fragment* of $\mathcal{L}^+$ which can be captured by the first-order frame language. In this process, minimal valuation arguments, which are pivotal for *local* correspondence, are encoded as applications of the Ackermann rule. To support the claim that these rewritings encode correspondence arguments as desired, the soundness of the rewriting rules needs to be verified. This has been done, via *duality theory*, in an *algebraic* setting. This move to algebras, per se, is not indispensable as long as the classical setting is concerned. However, the algebraic setting brings about a crucial advantage: it makes it possible to identify the properties really underlying the correspondence mechanism. And it turns out that no property exclusive to the classical setting is needed. This observation paves the way for rolling out correspondence theory, in great uniformity, to a wide variety of logics, including e.g. classical and intuitionistic modal mu-calculus (see [15]), polyadic and hybrid modal logics (see [19, 24]), monotone modal logic [23], modal logics with propositional quantifiers [13] or graded modalities [28], and substructural logics (see also below). This is what we understand as *unified correspondence*. In this setting, it is possible, e.g., to give a general purpose definition of Sahlqvist formulas (cf. Sect. 36.7) simultaneously applicable to several languages, and purely based on the order-theoretic behaviour of the interpretations of logical connectives.

Dropping distributivity. We wish to stress that the soundness of the approximation rules introduced in section 4 depends on the perfect lattices being completely join-generated by the set of their completely join *prime* elements, which implies that the perfect lattices in which these rules are sound are necessarily *distributive*. However, more general approximation rules can be introduced, which are sound on (non-distributive) perfect lattices. Hence, correspondence theory in the style illustrated in the present chapter covers also logics with algebraic semantics based on general lattices, for instance *substructural logics* (cf. [21] for complete details).

Complexity. While we hope that the reader is convinced that the calculus of correspondence facilitates simple, perspicuous and uniform derivations, we do not claim that it improves upon the computational complexity of other methods like the traditional Sahlqvist-van Benthem algorithm. Still, a few remarks on complexity are perhaps in order. When restricted to the class of Sahlqvist formulas, or to any other class of formulas on which it is guaranteed to succeed, the calculus of correspondence yields an algorithm for computing the first-order correspondents of the members of this class. It is not difficult to see that this algorithm's runtime complexity is polynomial in the size of the input formula. More sophisticated versions of the calculus could involve more costly computations like testing for monotonicity of terms (as opposed to mere syntactic positivity), and can take us to the full complexity of the underlying logic or beyond (see e.g., [16]). When applied to arbitrary formulas, the calculus of correspondence is only a semi-algorithm, as is to be expected, since the question whether a formulas has a first-order frame correspondent is undecidable [14]. Some considerations relevant to implementation and computational optimization are treated in [34] and Chap. 13 of [30].

Constructive canonicity. Perhaps the most important classical applications of correspondence is its connection to canonicity. Indeed, it has been appropriately argued [46] that the correspondence machinery can be extended and made applicable also in the context of descriptive frames, where it leads to canonicity results. Such results are often stated as persistence results (validity can be moved from a descriptive frame to its underlying Kripke frame); however, when seen from the dual, algebraic side, they can be stated as *transfer* results, namely that validity transfers from an algebra to its canonical extension. Formulated in this way, canonicity requires a rich metatheory for which the ultrafilter theorem (depending on the axiom of choice) must be available. However, there is a method for building canonical extensions 'without ultrafilters' in a constructive way. The idea in [36] is to exploit a Galois connections induced by an abstract 'containment' relation between filters and ideals, and to define the canonical extension as the resulting algebra of Galois-stable subsets. Indeed, in the presence of the axiom of choice, this construction is isomorphic to the canonical extension defined via duality. However, the canonical extension defined in [36] has an autonomous life also in a constructive (topos-theoretically valid) metatheory, and moreover, it has a rich enough internal structure that the transfer results for Sahlqvist-type equations can be proved in two steps, without relying on any correspondence result. Thus, canonicity (the alter ego of correspondence) is meaningful also in a purely constructive context.

Inverse correspondence. We focused on the question of finding first-order (or FO + LFP) correspondents for modal formulas. In this way we 'cover' only a fragment of the first-order (or FO + LFP) correspondence language, so it is natural to reverse direction and ask which first-order (or FO + LFP) frame conditions are modally definable. The more specific question of characterizing the first-order formulas which are frame correspondents of Sahlqvist formulas was answered by Marcus Kracht [41] and more generally for the inductive formulas by Stanislav Kikot [40]. Analogous questions for intuitionist modal logic or when correspondence with FO + LFP is sought are still open.

Step-by-step construction of finitely generated free algebras, and correspondence methods. The step-by-step construction of finitely generated free algebras is gaining more and more attention, viz. [1, 12, 25, 35]. The case of equations of rank 1 has been thoroughly investigated in connection with research issues relevant to coalgebraic logic; interestingly, preliminary results show that, in order to extend these results beyond rank 1, the correspondence machinery is needed in the setting of the so-called *step frames* [10], two-sorted Kripke frames modelling partially defined modalities. Possible developments of this line of investigation invest proof-theoretic questions related to the subformula property [9].

36.11 Appendix

36.11.1 Distributive Complex Algebras and Frames

An element $c \neq \bot$ of a complete lattice $\mathbb{C}$ is *completely join-irreducible* iff $c = \bigvee S$ implies $c \in S$ for every $S \subseteq \mathbb{C}$; moreover, c is *completely join-prime* if $c \neq \bot$ and, for every subset S of the lattice, $c \leq \bigvee S$ iff $c \leq s$ for some $s \in S$. An element $c \neq \bot$ of a complete lattice is an *atom* if there is no element y in the lattice such that $\bot < y < c$. An element $c \neq \top$ of a complete lattice is *completely meet-irreducible* iff $c = \bigwedge S$ implies $c \in S$ for every $S \subseteq \mathbb{C}$; moreover, c is *completely meet-prime* if $c \neq \top$ and, for every subset S of the lattice, $c \geq \bigwedge S$ iff $c \geq s$ for some $s \in S$. An element $c \neq \top$ of a complete lattice is a *co-atom* if there is no element y in the lattice such that $c < y < \top$.

If c is an atom (resp. a co-atom), then c is completely join-prime (resp. meet-prime), and if c is completely join-prime (resp. meet-prime), then c is completely join-irreducible (resp. meet-irreducible). If $\mathbb{C}$ is *frame distributive* (i.e. finite meets distribute over arbitrary joins) then the completely join-irreducible elements are completely join-prime, and if $\mathbb{C}$ is a complete Boolean lattice, then the completely join-prime elements are atoms. The collections of all completely join- and meet-irreducible elements of $\mathbb{C}$ are respectively denoted by $J^\infty(\mathbb{C})$ and $M^\infty(\mathbb{C})$.

Definition 36.6 A *perfect* lattice is a complete lattice $\mathbb{C}$ such that $J^\infty(\mathbb{C})$ join-generates $\mathbb{C}$ (i.e. every element of $\mathbb{C}$ is the join of elements in $J^\infty(\mathbb{C})$) and $M^\infty(\mathbb{C})$ meet-generates $\mathbb{C}$ (i.e. every element of $\mathbb{C}$ is the meet of elements in $M^\infty(\mathbb{C})$). A *perfect distributive lattice* is a perfect lattice such that $J^\infty(\mathbb{C})$ coincides with the set of all completely join-prime elements of $\mathbb{C}$ and $M^\infty(\mathbb{C})$ coincides with the set of all completely meet-prime elements of $\mathbb{C}$; a *perfect Boolean lattice* is a perfect lattice such that $J^\infty(\mathbb{C})$ coincides with the set of all the atoms of $\mathbb{C}$ (or $M^\infty(\mathbb{C})$ coincides with the set of all the co-atoms of $\mathbb{C}$).

Complete atomic modal algebras are those modal algebras $\mathbb{A}$ the lattice reducts of which is a perfect Boolean lattice and moreover, their $\Diamond$ operation preserves arbitrary joins, i.e. $\Diamond(\bigvee S) = \bigvee_{s \in S} \Diamond s$ for every $S \subseteq \mathbb{A}$. Discrete Stone duality between complete atomic modal algebras and their complete homomorphisms and Kripke frames and their bounded morphisms is defined on objects by mapping any Kripke frame $\mathcal{F} = (W, R)$ to its *complex algebra* $\mathcal{F}^+ = (\mathcal{P}(W), \langle R \rangle)$, where $\langle R \rangle X = R^{-1}[X] = \{w \in W : \exists x(x \in X \,\&\, wRx)\}$ for every $X \in \mathcal{P}(W)$, and every complete atomic modal algebra $\mathbb{A} = (\mathbb{B}, \Diamond)$ to its *atom structure* $\mathbb{A}_+ = (J^\infty(\mathbb{B}), R)$, where xRy iff $x \leq \Diamond y$ for all atoms $x, y \in J^\infty(\mathbb{B})$. As a consequence of this duality, the Stone representation theorem holds for complete atomic modal algebras, which states that these can be equivalently characterized as the modal algebras each of which is isomorphic to the complex algebra of some Kripke frame.

Likewise, a Stone-type duality (extending the finite *Birkhoff* duality) holds between perfect distributive lattices and their complete homomorphisms and *posets* and monotone maps, which is defined on objects as follows: every poset X is associated with the lattice $\mathcal{P}^\uparrow(X)$ of the upward-closed subsets of X, and every perfect lattice $\mathbb{C}$ is associated with $(J^\infty(\mathbb{C}), \geq)$ where $\geq$ is the reverse lattice order in $\mathbb{C}$, restricted to $J^\infty(\mathbb{C})$. As a consequence of this duality, perfect distributive lattices can be equivalently characterized (see e.g. [32]) as those lattices each of which is isomorphic to the lattice $\mathcal{P}^\uparrow(X)$ of the upward-closed subsets of some poset X.

As was mentioned early on, just in the same way in which the duality between complete atomic Boolean algebras and sets can be expanded to a duality between complete atomic modal algebras and Kripke frames, the duality between perfect distributive lattices and posets can be expanded to a duality between perfect DLOs and posets endowed with arrays of relations, each of which dualizes one additional operation in the usual way, i.e., n-ary operations give rise to $n + 1$-ary relations, and the assignments between operations and relations are defined as in the classical setting. We are not going to report on this duality in full detail (we refer e.g. to [22, 33, 47]), but we limit ourselves to mention that, for instance, the DLOs endowed with four unary operators as in (36.5) are dual to the relational structures $\mathcal{F} = (W, \leq, R_\Diamond, R_\Box, R_\lhd, R_\rhd)$ such that $(W, \leq)$ is a nonempty poset, $R_\Diamond, R_\Box, R_\lhd, R_\rhd$ are binary relations on W and the following inclusions hold:

$$\geq \circ\, R_\Diamond \circ \geq\ \subseteq\ R_\Diamond \qquad \leq \circ\, R_\rhd \circ \geq\ \subseteq\ R_\rhd$$
$$\leq \circ\, R_\Box \circ \leq\ \subseteq\ R_\Box \qquad \geq \circ\, R_\lhd \circ \leq\ \subseteq\ R_\lhd.$$

The *complex algebra* of any such relational structure $\mathcal{F}$ (cf. [33, Sect. 2.3]) is

$$\mathcal{F}^+ = (\mathcal{P}^\uparrow(W), \cup, \cap, \varnothing, W, \langle R_\Diamond\rangle, [R_\Box], \langle R_\lhd\rangle, [R_\rhd]),$$

where, for every $X \subseteq W$,

$$[R_\Box]X := \{w \in W \mid R_\Box[w] \subseteq X\} = (R_\Box^{-1}[X^c])^c$$
$$\langle R_\Diamond\rangle X := \{w \in W \mid R_\Diamond[w] \cap X \neq \varnothing\} = R_\Diamond^{-1}[X]$$
$$[R_\rhd]X := \{w \in W \mid R_\rhd[w] \subseteq X^c\} = (R_\rhd^{-1}[X])^c$$
$$\langle R_\lhd\rangle X := \{w \in W \mid R_\lhd[w] \cap X^c \neq \varnothing\} = R_\lhd^{-1}[X^c].$$

Here $(\cdot)^c$ denotes the complement relative to W, while $R[x] = \{w \mid w \in W$ and $xRw\}$ and $R^{-1}[x] = \{w \mid w \in W$ and $wRx\}$. Moreover, $R[X] = \bigcup\{R[x] \mid x \in X\}$ and $R^{-1}[X] = \bigcup\{R^{-1}[x] \mid x \in X\}$.

Adjunction and Residuation

Let P and Q be partial orders. The maps $f : P \to Q$ and $g : Q \to P$ form an *adjoint pair* (notation: $f \dashv g$) iff for every $x \in P$ and $y \in Q$, $f(x) \leq y$ iff $x \leq g(y)$. Whenever $f \dashv g$, f is the *left adjoint* of g and g is *the right adjoint* of f. Adjoint maps are order-preserving. If a map admits a left (resp. right) adjoint, the adjoint is unique and can be computed pointwise from the map itself and the order.

Proposition 36.3 *1. Right adjoints (resp. left adjoints) between complete lattices are exactly the completely meet-preserving (resp. join-preserving) maps;*

2. right (resp. left) adjoints on powerset algebras $\mathcal{P}(W)$ are exactly the maps defined by assignments of type $X \mapsto [R]X = (R^{-1}[X^c])^c$ (resp. $X \mapsto \langle R\rangle X = R^{-1}[X]$) for some binary relation R on W.

3. *For any binary relation R on W, the left adjoint of $[R]$ is the map $\langle R^{-1} \rangle$, defined by the assignment $X \mapsto R[X]$.*

Proof 1. See [26, Proposition 7.34].
2. For a left adjoint $f : \mathcal{P}(W) \longrightarrow \mathcal{P}(W)$, define R as follows: for every $x, z \in W$, $x R z$ iff $x \in f(\{z\})$. For a right adjoint $g : \mathcal{P}(W) \longrightarrow \mathcal{P}(W)$, define R as follows: for every $x, z \in W$, $x R z$ iff $x \notin g(W \setminus \{z\})$.

The notion of adjunction can be made parametric and generalized to n-ary maps in a component-wise fashion: an n-ary map $f : P^n \to P$ on a poset P is *residuated* if there exists a collection of maps $\{g_i : P^n \to P \mid 1 \leq i \leq n\}$ s.t. for every $1 \leq i \leq n$ and for all $x_1, \ldots, x_n, y \in P$,

$$f(x_1, \ldots, x_n) \leq y \quad \text{iff} \quad x_i \leq g_i(x_1, \ldots, x_{i-1}, y, x_{i+1}, \ldots, x_n).$$

The map g_i is the *i-th residual* of f. Residuated maps are order preserving in each coordinate, and for each $1 \leq i \leq n$, the residual g_i is order-preserving in its ith coordinate and order-reversing in all other coordinates. The facts stated in the following example and proposition are well known in the literature in their binary instance (cf. [31, Sect. 3.1.3]):

Example 36.4 For every $(n + 1)$-ary relation S on W and every $(X_1, \ldots, X_n) \in \mathcal{P}(W)^n$, let

$$S[X_1, \ldots, X_n] := \{y \in W \mid \exists x_1 \cdots \exists x_n [\bigwedge_{i=1}^{n} x_i \in X_i \wedge S(x_1, \ldots, x_n, y)]\}.$$

The n-ary operation on $\mathcal{P}(W)$ defined by the assignment $(X_1, \ldots, X_n) \mapsto S[X_1, \ldots, X_n]$ is residuated and its i-th residual is the map $g_i : \mathcal{P}(W)^n \to \mathcal{P}(W)$ which maps every n-tuple $(X_1, \ldots, X_{i-1}, Y, X_{i+1}, \ldots, X_n)$ to the set $\{w \in W \mid \alpha_S^i(w)\}$, where $\alpha_S^i(w)$ is the following first-order formula:

$$\forall x_1 \cdots \forall y \cdots \forall x_n [(\bigwedge_{k \in \mathbf{n}_i} x_k \in X_k \& S(x_1, \ldots, w, \ldots, x_n, y)) \Rightarrow y \in Y],$$

and moreover $\mathbf{n}_i = \{1, \ldots, n\} \setminus \{i\}$.

Proposition 36.4 *If $f : P^n \to P$ is residuated and $\{g_i : P^n \to P \mid 1 \leq i \leq n\}$ is the collection of its residuals, then:*

1. *if P is a complete lattice, then f preserves arbitrary joins in each coordinate;*
2. *if P is a powerset algebra, f coincides with the map defined by the assignment $S[X_1, \ldots, X_n]$ as in Example 36.4, for some $(n + 1)$-ary relation S on W.*

References

1. Abramsky S (2005) A Cook's tour of the finitary non-well-founded sets. In: We will show them: essays in honour of Dov Gabbay, pp 1–18
2. Ackermann W (1935) Untersuchung über das Eliminationsproblem der mathematischen logic. Mathematische Annalen 110:390–413
3. van Benthem J (1976) Modal reduction principles. J Symbolic Logic 41(2):301–312
4. van Benthem J (1983) Modal logic and classical logic. Bibliopolis, Napoli
5. van Benthem J (2001) Correspondence theory. In: Gabbay DM, Guenthner F (eds) Handbook of philosophical logic, vol 3. Kluwer Academic, Dordrecht, pp 325–408
6. van Benthem J (2005) Minimal predicates, fixed-points, and definability. J Symbolic Logic 70(3):696–712
7. van Benthem J (2006) Modal frame correspondence and fixed-points. Studia Logica 83:133–155
8. van Benthem J, Bezhanishvili N, Hodkinson I (2012) Sahlqvist correspondence for modal mu-calculus. Studia Logica 100:31–60
9. Bezhanishvili N, Ghilardi S (2013) Bounded proofs and step frames. In: Logic group preprint series, no 306. Utrecht University
10. Bezhanishvili N, Ghilardi S, Jibladze M (2013) Free modal algebras revisited: the step-by-step method. In: Leo Esakia on duality in modal and intuitionistic logics. Outstanding contributions to Logic, vol 4, 2014. Springer
11. Bezhanishvili N, Ghilardi S (2013) Bounded proofs and step frames. In: Proceedings of Automated Reasoning with Analytic Tableaux and Related Methods (TABLEAUX 2013), pp 44–58
12. Bezhanishvili N, Kurz K (2007) Free modal algebras: a coalgebraic perspective. In: Proceedings of CALCO 2007, pp 143–157
13. Bull RA (1969) On modal logic with propositional quantifiers. J Symbolic Logic 34:257–263
14. Chagrov A, Chagrova LA (2006) The truth about algorithmic problems in correspondence theory. In: Governatori G, Hodkinson I, Venema Y (eds) Advances in modal logic, vol 6. College Publications, London, pp 121–138
15. Conradie W, Fomatati Y, Palmigiano A, Sourabh S Sahlqvist correspondence for intuitionistic modal mu-calculus (To appear in Theoretical Computer Science)
16. Conradie W, Goranko V (2008) Algorithmic correspondence and completeness in modal logic III: semantic extensions of the algorithm SQEMA. J Appl Non-class Logics 18:175–211
17. Conradie W, Goranko V, Vakarelov D (2005) Elementary canonical formulae: a survey on syntactic, algorithmic, and model-theoretic aspects. In: Schmidt R, Pratt-Hartmann I, Reynolds M, Wansing H (eds) Advances in modal logic, vol 5. Kings College, London, pp 17–51
18. Conradie W, Goranko V, Vakarelov D (2006) Algorithmic correspondence and completeness in modal logic I: the core algorithm SQEMA. Logical Methods Comput Sci 2(1:5):1–26
19. Conradie W, Goranko V, Vakarelov D (2006) Algorithmic correspondence and completeness in modal logic. II. Polyadic and hybrid extensions of the algorithm SQEMA. J Logic Comput 16:579–612
20. Conradie W, Goranko V, Vakarelov D (2010) Algorithmic correspondence and completeness in modal logic. V. Recursive extensions of SQEMA. J Appl Logic 8:319–333
21. Conradie W, Palmigiano A Algorithmic correspondence and canonicity for non-distributive logics (Submitted)
22. Conradie W, Palmigiano A (2012) Algorithmic correspondence and canonicity for distributive modal logic. Ann Pure Appl Logic 163:338–376
23. Conradie W, Palmigiano A, Sourabh S Algebraic modal correspondence: Sahlqvist and beyond (Submitted)
24. Conradie W, Robinson C An extended Sahlqvist theorem for hybrid logic (In preparation)
25. Coumans D, van Gool S On generalizing free algebras for a functor. J Logic Comput
26. Davey BA, Priestley HA (2002) Lattices and order. Cambridge Univerity Press, Cambridge
27. Dunn JM, Gehrke M, Palmigiano A (2005) Canonical extensions and relational completeness of some substructural logics. J Symbolic Logic 70(3):713–740

28. Fine K (1972) In so many possible worlds. Notre Dame J Formal Logic 13:516–520
29. Frittella S, Palmigiano A, Santocanale L Characterizing uniform upper bounds on the length of D-chains in finite lattices via correspondence theory for monotone modal logic (In preparation)
30. Gabbay DM, Schmidt RA, Szałas A (2008) Second-order quantifier elimination: foundations, computational aspects and applications. Vol 12 of studies in logic: mathematical logic and foundations. College Publications, London
31. Galatos N, Jipsen P, Kowalski T, Ono H (2007) Residuated lattices: an algebraic glimpse at substructural logics. Elsevier, Amsterdam
32. Gehrke M, Jónsson B (1994) Bounded distributive lattices with operators. Math Japon 40
33. Gehrke M, Nagahashi Y, Venema H (2005) A Sahlqvist theorem for distributive modal logic. Ann Pure Appl Logic 131:65–102
34. Georgiev D (2006) An implementation of the algorithm SQEMA for computing first-order correspondences of modal formulas, master's thesis. Sofia University, Faculty of mathematics and computer science
35. Ghilardi S (1995) An algebraic theory of normal forms. Ann Pure Appl Logic 71:189–245
36. Ghilardi S, Meloni G (1997) Constructive canonicity in non-classical logics. Ann Pure Appl Logic 86:1–32
37. Goranko V, Vakarelov D (2001) Sahlqvist formulas in hybrid polyadic modal logics. J Logic Comput 11:737–754
38. Goranko V, Vakarelov D (2002) Sahlqvist formulas unleashed in polyadic modal languages. In: Wolter F, Wansing H, de Rijke M, Zakharyaschev M (eds) Advances in modal logic, vol 3. World Scientific, Singapore, pp 221–240
39. Goranko V, Vakarelov D (2006) Elementary canonical formulae: extending Sahlqvist theorem. Ann Pure Appl Logic 141(1–2):180–217
40. Kikot S (2009) An extension of Kracht's theorem to generalized Sahlqvist formulas. J Appl Non-class Logics 19:227–251
41. Kracht M (1993) How completeness and correspondence theory got married. In: de Rijke M (ed) Diamonds and defaults. Kluwer Academic, Dordrecht, pp 175–214
42. Kurtonina N (1998) Categorical inference and modal logic. J Logic Lang Inform 7:399–411
43. Kurz A, Palmigiano A (2013) Epistemic updates on algebras. Logical Methods Comput Sci 9(4). doi:10.2168/LMCS-9(4:17)2013
44. Ma M, Palmigiano A, Sadrzadeh M (2013) Algebraic semantics and model completeness for intuitionistic public announcement logic. Ann Pure Appl Logic 165(4):963–995
45. Sahlqvist H (1975) Correspondence and completeness in the first and second-order semantics for modal logic. In: Kanger S (ed) Proceedings of the 3rd Scandinavian logic symposium, Uppsala 1973. Springer, Amsterdam, pp 110–143
46. Sambin G, Vaccaro V (1989) A new proof of Sahlqvist's theorem on modal definability and completeness. J Symbolic Logic 54:992–999
47. Sofronie-Stokkermans V (2000) Duality and canonical extensions of bounded distributive lattices with operators, and applications to the semantics of non-classical logics I. Studia Logica 64(1):93–132

ADDENDUM I
Reflections on the Contributions

Reading this book, one quickly realizes that the number of substantial contributions found in its 1,000 pages defies any attempt at integrative summary, let alone, detailed response on my part. Moreover, I do not see myself as the leader of this talented interdisciplinary flock, but rather as one of a swarm of birds exploring a landscape. Each bird may take turns at the head when the swarm is in full flight, but occasionally, we also perch on branches in a non-hierarchical easy comradery. This dynamic perching is what artist Nina Gierasimczuk has captured in her nice image of all the contributors that appears on the next page. I hope they will all find their perches with ease.

The authors and their topics do deserve serious responses, but this may well have to be my project for the year to come. Instead, for now, I have tried to also read the chapters in another intellectual mode that I have always found congenial, not that of agreeing or disagreeing with specific claims, but that of being inspired. The reflections that follow were what resulted, though the reader will see a clear continuity with the themes mentioned in my scientific autobiography. After all, minds as well as bodies can only resonate with their eigenfrequencies.

My reflections follow the sequence and grouping of chapters in the book, not because this is a unique natural order of things, but because, to me, the topics in this book are highly interconnected. One can start at any end, and find oneself in the midst of things instantaneously. In the passages to come, I will just start with one topic that occupies me, and then get the ball rolling naturally from one basic theme to another. This deep entanglement of themes also reflects another conviction of mine, namely, that there is no natural border line between approaching the topics underlying this book from the perspectives of mathematics, computation, philosophy, or even other disciplines such as linguistics or cognitive science. For instance, usually, I just do not feel a sharp transition in looking at a mathematical result about a fundamental theme like information and agency and a philosophical analysis of that same topic. The various stances form a natural continuum of intellectual pleasures, and sometimes even simple necessities, in seeing topics relevant to logical dynamics in their full intellectual breadth.

A. Baltag and S. Smets (eds.), *Johan van Benthem on Logic and Information Dynamics*, Outstanding Contributions to Logic 5, DOI: 10.1007/978-3-319-06025-5, © Springer International Publishing Switzerland 2014

Mathematical and Computational Perspectives

The chapters in this part may be seen as dealing with wide-ranging tools, but at the same time, they deal with broad issues connected to these tools, that run through the whole area of information and agency.

The first of these issues is *which notion (or notions) of process* lies behind games and interactive agency in general. My own work has mostly approached this issue indirectly, by looking at semantic levels of guarded process equivalence such as bisimulation and the associated invariances for modal languages.[1] Erich Grädel and Martin Otto show the power of this modal invariance approach, which extends much further than most people realize, namely, at least into the guarded fragment of first-order logic and its fixed point extensions. And a focus on structural 'sameness' returns in various chapters in other parts. But when one looks at the model theory in the Grädel and Otto chapter, one striking feature is its connections with combinatorial features of size and complexity, and also, with the use of concrete computing devices such as automata that represent constructive definitions of relevant processes. The encounter between modal logic and *automata theory* is something that was entirely absent from modal logic in my formative years, even though in hindsight, one can say that the invention of propositional dynamic logic in the 1970s heralded its arrival. Yde Venema demonstrates the elegant mixture of ideas that results when model theory meets automata theory, and presents a little gem: an (at least, to me) elegant and comprehensible proof of the Janin and Walukiewicz invariance theorem that the bisimulation-invariant fragment of monadic second-order logic is the modal μ-calculus.[2] Automata will also return in other parts of the book as concrete models for agents or their strategies.

Another broad topic involved with processes and automata is that of induction, recursion, and *fixed point logics*. While these systems have so far been mostly studied in computer science, and are not yet part of the classical mathematical foundations of modal logic, I feel that they should. Recursion is everywhere, from 'reflective equilibrium' in philosophy to strategic equilibria in game theory, or iterative patterns in language and cognitive behavior.[3] And recursion and process structure is just one aspect of taking a computational perspective. Another is the difficulty or *complexity of tasks*, since this is what determines whether the process makes sense in the realm of agency. Several chapters in this part have complexity issues in the background, that bring to light a fine-structure of the actual performance of logical systems that goes unnoticed with standard presentations. Many people flatter themselves that they can understand a formal system without understanding its actual complexity in performance. Do we understand a natural language when we grasp its semantic meaning system, but have no idea what it would take to actually use it? I do not think so, and I do not think so for logical reasoning systems either.

---

[1] This mathematical idea occurs just as well in philosophy, in thinking about 'identity criteria'.

[2] I hope that something similar is possible for the fixed-point extension of the guarded fragment.

[3] If I had the power of changing basic logic curricula, I would probably add fixed point logics to the standard predicate-logical fare that students get.

In a very stark practical form, issues of complexity are central to the treatment of database theory by Balder ten Cate and Phokion Kolaitis, the closer to polynomial the better, but I also enjoyed that chapter for its closeness of logical model theory and the syntax of databases relevant to today's practice. Personally, I have always felt that, beyond its practical uses, database theory is a concrete laboratory for process and information representation.[4] One thing that intrigues me here is the essential *role of syntax*. While I was brought up as a student to believe that syntax is just a nuisance or a concession to human frailty, while the real essences are semantic, this prejudice is under pressure. Syntax is close to important structure of information that gets washed out in semantic models, and also, syntax is close to specifying automata, so that it gives essential procedural information about what we mean, and what we do.[5] Another aspect of databases that is central to much of this book is the notion of *dependence*. I fully agree with Pietro Galliani and Jouko Väänänen that this deserves a sharp and separate focus in logic, since it is a notion with attractive formal structure that unifies across information, computation, games and interactive agency. Incidentally, their chapter also makes connections between dependence logic and dynamic-epistemic logic, providing one of the many instances of system rapprochement in this book.

But I was not done yet with my earlier issues of syntax and process structure. Samson Abramsky's inspiring analysis of computation in a rich higher-order type-theoretic and categorial framework reminds me of that other main tradition in logic, that of *proof theory and category theory*. Much of the work I have done is about model-theoretic logics *of* processes, whereas Abramsky's perspective might be described as logic *as* process. I hope that, when I have come to a full understanding of what he is achieving in his framework, I will also have a better view of the difference between the two process aspects of logic, which has long intrigued me, but also baffles me a bit. This distinction between "of" and "as" is another running thread through this book, and I will draw attention to it in several later parts.

Finally, let me turn to what to some readers may seem to be an outlier, Hajnal Andréka and Ístvan Németi's analysis of relativity theories in a logical framework. To me, this chapter about *theories in empirical science* represents many strands in the above. It is in the empirical sciences, rather than pure mathematics, that different conceptions of information meet, as well as associated notions of process. Just think for a moment about the notions of signal and measurement that are crucial to relativity theory, and you will quickly see the point. And there are further links, such as the discussion of relative interpretation (a notion on which I worked with David Pearce in the 1980s) which is about sameness of theories, and hence sameness of informational and process structure. This chapter also reflects a conviction of mine, arising from breaking with all that I was taught as a student in courses on the philosophy of science. I now think that there is no significant border line between the worlds of science and

---

[4] Already in my 1989 paper on 'Semantic Parallels', I pointed out connections between abstract data types and theories as structured bodies of knowledge in the philosophy of science.

[5] Syntax also triggers complexity in ever new ways, witness the large new decidable fragment of first-order logic discovered by ten Cate and Segoufin as reported by Grädel and Otto.

of common sense daily behavior. Indeed, if we view science as a very successful species of general human behavior, then we can profit from its providing a sort of restricted cognitive and conceptual lab for our general concerns about information-driven agency in logical dynamics.

This is a very long passage, and I have not even begun to summarize all topics in this part. Let me just say that many topics raised here will return later in perhaps unlikely places, witness the use of "semantic automata" in the part on language and cognition. Computational complexity, too, will return in later chapters on cognition. Likewise, automata models for fixed point logics are closely related to evaluation games and their winning strategies, which will return in the part on games. Vice versa, many later topics of this book would also make sense in this part. For instance, several later chapters will be concerned with non-monotonic reasoning, but as we all know from the case of logic programs, databases are an excellent testing ground for such logics too.

But let me end by returning to the first issue that I raised in connection with this part, the appropriate notion of process. One way of thinking about Abramsky's program is as a search for a stable notion of process that would be adequate to modern computing. Now this ties in with a general issue that has long occupied me. The logical dynamics program seems to presuppose some very broad notion of information-driven distributed *social interactive computation*. What sort of mathematical structure can we expect for this notion of distributed behavior, and will its universal model (if there is one) have anything like the classical elegance of Turing's original analysis of sequential computing?

Dynamics of Knowledge and Belief Over Time

The chapters in this second part show what has become of the simple epistemic and doxastic logics I used to know when starting in the 1970s. The current world of logics for knowledge acquisition, belief change, and learning is fast developing, and many things seem to be crystallizing out in the process, even though there are also important remaining choices of a most convenient stable formalism. One persistent framework exploits the natural fit between *epistemic logic and dynamic logic of actions* whose state of the art is exemplified in Jan van Eijck's chapter. This simple and perspicuous combination makes sense from a logical dynamics standpoint where information and action belong together, but it also shows up in more surprising places, such as the recent informational mechanisms that update process structure in the chapter by Patrick Girard and Hans Rott, or notions of group information like common knowledge where the actors in 'epistemic PDL' seem to become complex information processing agents. I wonder whether we will also need other natural systems, with stronger fixed point logics for both components.

The paper by Girard and Rott also addresses another issue that intrigues me: the duality between two approaches to understanding update and agency. One is the construction of dynamic-epistemic logics with explicit mechanisms for the relevant tasks, the other is the formulation of, often philosophically motivated, postulates on update or belief change, as in AGM belief revision theory. I would like to understand the interplay of these *constructive and postulational* approaches much better. The contrast seems a very general one having to do with the standards of adequacy for any formal analysis.

Several chapters in this part then take what is obviously a further natural step, also given my own past interests in temporal structure. Update mechanisms describe the next step in some dynamical system, but we also want to understand the *system evolution over time*. In our field, this is not just a matter of routine mathematical technique taking difference equations to solvable differential equations: there are serious connection problems. One technical issue is linking dynamic-epistemic logics to epistemic temporal logics of various sorts that have been proposed in philosophy, computer science, and game theory. Valentin Goranko and Eric Pacuit map out the candidates and the issues in a way that leaves me nothing to add, except for noting that many popular frameworks seem to be converging, from epistemic temporal logic to interpreted systems. But equally basic is seeing that temporal structure is not just a matter of rolling out local dynamics: there may be genuine procedural information about the evolution of a process that is sui generis. I do not think that we have a good general understanding of the logic of protocols that govern long-term behavior. One concrete area where this theme becomes especially conspicuous is *learning theory*, and Nina Gierasimczuk, Vincent Hendricks and Dick de Jongh show abundantly how learning theory both meshes with and enriches dynamic-epistemic logic. To me, learnability in the long run seems a very natural counterpart to our usual concerns about logical systems, including the earlier-mentioned complexity, and I hope that this theme will penetrate more into the consciousness of logicians. Kevin Kelly's chapter is along the same lines, but adds many surprising connections to temporal logics, and the philosophical literature on inquiry that I find fascinating. I had long admired his book "The Logic of Reliable Inquiry", and seeing it so close to my own concerns comes as a pleasant surprise, making me realize the importance to logic of the philosophy of science in completely new ways.

Moving from time to space, the chapter by Wiebe van der Hoek and Nick Bezhanishvili reminds us of the importance of generalized topological and neighborhood models for my enterprise. I have worked with these on many occasions, but an issue of general import remains. In topological models, there are two major families of examples: trees, close to process evolution, and Euclidean spaces. The *spatial view of information* is very powerful, too, and I still have not managed to absorb it organically into my thinking.

The final intriguing theme in this part is the combination of *logic and probability*. While these are often seen as rivals in formal philosophy or some areas of computer science, their combination seems increasingly natural to me. Sometimes this is just a matter of adding richer probabilistic fine-structure to qualitative logical notions, but sometimes, making our very universe of objects probabilistic is needed to get the deep results, such as the existence of equilibria in strategic games. Lorenz Demey and Barteld Kooi show how combinations of dynamic-epistemic logic and probability make sense both in theory and practice, distinguishing the subtly different notions of prior probability, observation probability, and occurrence probability (a notion of process information that also underlies learning theory). But even then issues remain about the border line between the two components. One thing that has occupied me for a while now is the manifold roles of numbers in probability theory. They perform what are in fact extremely different jobs: indicating strengths of belief, giving global summaries of past experience, but also, they help us weigh and glue factors in update or learning rules. I wonder whether one can find truly qualitative versions of all these roles that work well in non-numerical scenarios. In particular, update gluing may involve forms of relation merge that resemble aggregation procedures in social choice. One telling case study at the level of static attitudes is Hannes Leitgeb's paper on how qualitative belief and numerical probability can be related in a mathematically precise way based on a notion of stability with respect to making errors, and I am sure that any reader of his paper will be able to formulate the follow-up questions about the information dynamics.

Games

The chapters in the part on logic and game theory are a natural continuation of an interest in information dynamics and information-driven action. Games are a natural microcosm for about everything on the agenda of philosophical logic, and when viewing games as a model for modern computation, also for the agenda of computational logic. To me, they are also our best current model for the social intelligence that seems typical for human cognition. Logicians and game theorists seem congenial communities, and the chapter by Giacomo Bonanno and Cédric Dégremont shows many instances of that. There are many logics of games, but what would be a best logical framework for games seems a matter of continuing discussion, even in my new book "Logic in Games". I have come to think in terms of a *Theory of Play* that would combine features of a game as a global process structure with modeling players and their relevant habits, and it is surprising to see how game theorists often think the same way when charting what actually happens in the course of play. While I have tended to focus on individual players here, the chapter by Thomas Ågotnes and Hans van Ditmarsch provides a welcome extension to the realm of coalitions and groups, one of the most striking features of working in an interactive setting. Moreover, while most work on logics of games has been monopolized by semantic perspectives, proof theory has a natural role as well, and the chapter by Sergei Artemov re-examines basic intuitions of the founders of game theory in terms of the reasoning leading to strategic equilibrium. In doing so Artemov uses his framework of justification logic, itself a beautiful example of the virtues of combining features of epistemic logic with more fine-grained notions of evidence coming from proof theory. Not all different temperaments lead to happy marriages, but it is good to see that some do.

Many of the earlier computational themes return in the realm of games. The chapter by Ramaswamy Ramanujam shows how automata can serve as concrete models of players or their strategies, and his use of a logic for specifying player types may be just what is needed to tame the potential explosion of options in my proposed Theory of Play. I have the feeling that much of the current sophisticated work on automata, games, and strategies in computational logic (witness also the chapters by Abramsky and Venema) has a much broader significance for our understanding of intelligent interaction in general, even though, sadly, I am not the person who is able to write an authoritative book on this fast-growing line of research in the logical foundations of computing.

While all this is mainly about applying logic to games, there is also a converse direction entangled with it, the use of games to elucidate major features of logic. In terms of two prepositions, in addition to logic *of* games, there is logic *as* games. The chapter by Gabriel Sandu comes from the latter tradition, namely, the semantics of IF logics of independence (and, ex silentio, dependence) in terms of games with imperfect information caused by limited observation and memory. Sandu takes the game-theoretic connection seriously, and develops a link with signaling games, one of the most natural counterparts to our interest in information dynamics. But one can come to logic in other ways as well. Ågotnes and van Ditmarsch present their 'knowledge games' played over epistemic models that uncover a lot of procedural information lying hidden inside what look like static structures. This perspective may well affect our understanding of knowledge, not just as a static attitude, but also as a procedural notion. Overall, despite having written a 550 page book in the area, I am still intrigued by the entanglement of logic of games and logic as games, whose structure might well be the DNA of our field: logic gets applied to games and agency, but it may also get transformed in the process.

Agency

Games are one species of general agency, and I do not wish to claim that it is the only format for making sense of social behavior. The broader area of agency is again one where logicians, philosophers, and computer scientists meet naturally. Peter Millican and Michael Wooldridge raise the crucial issue of *what are agents*, merging computational and philosophical perspectives, and made me realize that I should think much harder on what I am doing. They contrast several models of agents, and tie logic to the intentional stance, asking where this makes sense and where not. As for technical frameworks, Hector Levesque and Yongmei Liu make a connection that had long eluded me, between the situation calculus, one of the major computational paradigms for agency in AI, and dynamic-epistemic logic. I wish this insight had reached me earlier, so that I could have discussed it with John McCarthy at Stanford, one of the most creative and open minds I have known over the years.

More in the philosophical tradition, several chapters in this part remind me of things that need to be done. Wesley Holliday and John Perry discuss epistemic predicate logic, and how to model *knowledge of objects* in a way that does justice to the philosophical state of the art. This has shaken my prejudice that adding predicate logic to the propositional systems in the bulk of my work is merely a routine issue of no particular priority in the larger scheme of things. I now see that these are highly natural issues, and much needs to be done. The authors also suggest that, in this richer setting, we need a dynamics of roles for agents, in addition to the dynamics of information. That might realize one of my longstanding desires, to find significant interfaces between my work and that of the Stanford situation theorists. Another philosophical-computational framework that has long intrigued me is STIT logic of agency, based on *choice and control* of options over time. Roberto Ciuni and Jon Horty explain STIT in a way that is totally accessible to people outside of the framework, and extend an incipient junction with logics of games that I had studied with Eric Pacuit. Thanks to them, I am now beginning to see how there are the makings of a solid bridge here, when focusing on higher level concepts such as choice, freedom, and knowledge.

Higher-level concepts also take me to what I see as the double role of logic, providing either more, or less *conceptual zoom*. We often think of logical analysis as providing more detail, perhaps even formalizing arguments down to a rock-bottom level where a computer could check all individual steps. But the opposite direction is often just as prominent, where logic provides high-level concepts that hover far above the details of some reasoning practice, and capture essential global features. In fact, this is how I see the common use of deontic concepts such as "may" or "must": they encapsulate complex considerations into one label that lends itself to reasoning. Oliver Roy, Albert Anglberger and Norbert Gratzl zoom out on the role of deontics, and discuss logics for best action in game-theoretic and social scenarios. I would hope that there is actually a stable qualitative level lying above the details of decision theory and game theory where we can make perfect sense of the way in which we really see all of our life's choices: in terms of entangled informational-evaluative notions of duty, best action, but also, I would think, of the powerful pervasive notions of hope and fear.

Finally, another arena of agency close to logic that comes with vivid intuitions and concrete experiences is that of *argumentation*. This area has long been the exclusive preserve of informal argumentation theorists, who were often inspired by severe critics of formal logic (in its pre-dynamic 1950s mode) such as Toulmin or Perelman. Davide Grossi and Dov Gabbay present modern mathematical network models for basic styles of argumentation, and show how these models fit very naturally with the methodology of bisimulation and modal logic used in much of this book. In doing so, they also address something that has long baffled me when comparing my own work with that of Dov: how to relate dynamic logics of information and action with the more dynamical-systems oriented approach of argumentation networks and their temporal evolution.

Language and Cognition

Over the years, many people have chided me for losing my interest in natural language that was so prominent in the 1980s. And indeed, I have strayed from that path since the mid-1990s, as described in my scientific autobiography. But it is becoming increasingly clear that, under the surface, there is continuity. In this part, Lawrence Moss discusses my early work on the monotonicity calculus and *natural logic*, which was an attempt at seeing tractable and cognitively independent subsystems of reasoning inside the monolith called first-order logic, and he also connects it to modern dynamic semantics. This reminds me that the logics of agency in my current work give theoretical laws at a meta-level, whereas there is a perfectly natural matching question how these things are expressed in natural language, and which reasoning modules then make sense.[6] One case study in making my past and current life coherent again is the analysis by Sven Ove Hansson and Fenrong Liu of my early work on comparatives based on context-dependent unary predicates. I suddenly see that my interest in preference aspects of agency may be connected to these basic ways in which we use evaluative terms in natural language, while their discussion of the dynamics of context shifting seems a natural complement to the pure information dynamics in most of the logics found in this book.

There are also more general philosophical questions to be asked about what is happening here. Natural language is not just another perspective on agency: it is an empirical phenomenon, and logic of natural language has a challenging status, being descended from the realm of pure thought to deal with the realities of life on Earth. Martin Stokhof engages in rethinking of what formal semantics can hope to achieve, and in particular, what is the role of its beloved tool of formal systems. In the process, he raises important issues about the status of various blends of dynamic semantics, making me realize, amongst other things that the approach found in dynamic-epistemic logics is closer to Stalnaker's pragmatics than that of my close Amsterdam colleagues.[7] Hans Kamp offers further reflection on the role of logic and semantics vis-à-vis natural language, focusing on the role played by inferential intuitions about valid consequences, and what our systems achieve. I cannot do full justice to this here, but one issue that struck me is his claim concerning beneficial bootstrapping: teaching logical structures can improve linguistic appreciation and performance. I feel this is very true, and it resonates with my feeling that we need to understand, not natural language and formal logic as separate systems, but precisely the creative process of *hybridization* that goes on between the two in areas such as scientific language, but really also, in everyday life.

The final two chapters in this part are about cognition in general. Alistair Isaac, Jakub Szymanik and Rineke Verbrugge discuss current research at the interface of

---

[6] Larry also discusses natural logic in mathematics, which may look like a strange excursion. It is not to me, given what I have said about the thin borderline between common sense and science.

[7] I have approached these issues in terms of a pervasive 'implicit/explicit' distinction in styles of logical modeling, since I feel that it occurs far beyond the dynamics arena, including the contrast between epistemic logics and another highlight of my Amsterdam home, intuitionistic logic.

logic, computation, and cognition. This sort of picture of what is happening on the ground is light-years away from unreflected distinctions between normative versus descriptive that only obscure what is in fact an intriguing and rich interface. Of course, the normative descriptive distinction continues to make sense, but the boundary may not lie conveniently in between disciplines. Reading the piece has clarified my somewhat timid thinking about empirical complexity and social interaction. The chapter also injects new ideas into the mainstream models of agency in this book, such as much sparser information representations than the lush epistemic models I normally work with.[8] I also see with interest that my early work on natural logic and semantic automata makes a return here, also in the work of my Stanford students.[9] Peter Gärdenfors takes a different perspective here, minimizing computation and logic in favor of rich situated environments that allow simple agents to thrive. This situated stance is close to what my Stanford colleagues like Barwise and Perry advocated in the early 1980s, and it reinforces my suspicion that the days of situation theory are not over yet.

Styles of Reasoning

The chapters in the final part on styles of reasoning address a theme that really plays throughout this book, the meeting of different logical paradigms. Denis Bonnay and Dag Westerståhl discuss what the logical dynamics program implies in terms of the core business of traditional logic, and determine when natural dynamic consequence relations behave classically.[10] This chapter made me realize that some features of the logical dynamics program are less clear than I had thought. In particular, while I extend the agenda of logic at an object-level, at a meta-level the laws that govern this extended repertoire of informational actions work entirely in a setting of classical consequence. Incidentally, I now think that there is also room for an architecture with a plurality of consequence relations from a cognitive point of view, where the dynamic notions may be closer to what happens in real life, while classical consequence governs the abstract information representations that we retain over time. Guillaume Aucher takes this interface much further, working at the abstract level of ternary models for information structure found in relevant logic, and in my own earlier work on categorial grammar. He finds abstract laws of dynamic-epistemic update that greatly generalize my own attempts at abstract correspondence analysis in this area. In doing so he makes a big step toward connecting up two ways of thinking about information and action that always seemed quite disjoint to me: the semantic dynamic logic approach and the resource-based proof-theoretic approach. Still in

---

[8] What also strikes me is the guiding role of mathematical results, such as the representation of neural networks by default logics due to Gärdenfors, Gabbay, van Lambalgen, and Leitgeb.

[9] I need to reevaluate the different boxes in which I have stored the fragments of my life!

[10] They relate this to interesting recent work by Rothschildt and Yalcin on what makes a system 'dynamic' which is in the line of my old work on abstract update operators in the 1980s.

this same sphere of ideas, Michel Dunn analyzes connections between relevance logic and my earlier work on 'arrow logics' capturing the essence of abstract state transitions in decidable fragments of the complete algebra of binary relations. I am sure that much more can and will be done along the lines he has drawn here. Relevant information and resource semantics has also been proposed as an appropriate level for connecting up with the basic insights in situation theory of distributed situations related by information channels. Jeremy Seligman revisits situation theory in the light of modal and dynamic logics, and through his eyes I can see once more, how suddenly things are moving and getting clearer here.

The final chapter by Willem Conradie, Silvio Ghilardi and Alessandra Palmigiano closes a circle in this book, and in my research, going back to the topic of my dissertation. It proposes a powerful algebraic generalization of modal correspondence techniques that promises to cover most of the systems in this book, including fixed point logics, unveiling their connections, and revealing new logical vocabulary that may transform our idea of the syntax-semantics interface in formal systems for logical dynamics. Once again, I see this as the indispensability of engaging in pure mathematical thinking behind the concrete system building in the area of agency.

This final part of the book may seem to illustrate the attitude that most authors ascribe to me (correctly): the search for unity and translation, or at least compatibility, between formal frameworks. In one sense, this is what I want, since I hope that the area and its logical structure are stable, and it would be nice to arrive at a sort of Church Thesis for agency telling us that we have found the stable essentials of the notion. But on the other hand, I am also aware of the dangers of unification and peace in terms of creative impulses and sheer intellectual vigor. Competition is as good for us as cooperation, and academic life definitely has features of both war and peace. I am not sure how to think of this book then, whether as a harbinger of world peace, or just of a ceasefire in the area. But I do know that my thinking on what I myself and others are doing has greatly changed after reading it.

In summary, I thank the authors for all that they have contributed. When the birds leave Nina's tree again, and spiral upward to the sky, I only hope to be part of their flight to the next foreign lands.

ADDENDUM II
The Life of Logic, a Scientific Autobiography

How does one end up as a logician? Choice problems have been a constant companion in my life, starting in my gymnasium days. I loved classical languages and history, but also mathematics and the sciences. In those days, one had to choose one type or the other eventually, but I managed to beat the system. I did my official school exam in the 'beta' science track, but with the help of extra lessons after classes, I also took a parallel national exam in the 'alpha' language track. With those two degrees in hand, I still found myself without any preference for a field of study at the university, so I took physics, since people told me it is the hardest discipline, and best for keeping your brain active while waiting for inspiration to strike. And then I ran into logic. A fellow student who had observed that I always managed to talk myself into a corner in discussions suggested I should read a logic book to find out what was wrong with me. As it happened, it was a 19th century text by William Stanley Jevons, which had been translated into a popular Dutch pocket book series. That chance encounter set my course: I was intrigued by the subject, and switched to studying mathematics and philosophy, as the two obvious companion disciplines.

Well, this is the official story. I did have one very specific burning ambition at age 18, to become a literary author. I had collected all my heartfelt short stories and sent them to a well-known Dutch publisher. The answer was that there was a little merit, and a lot of adolescent immaturity, and I was advised to submit again in some 20 years. I got the point. Disappointments guide our lives more firmly than fond hopes.

My interest in logic had some features of a spiritual conversion. I remember the feeling of enlightenment coming from realizing that there are mathematical patterns behind the daily stream of our language and reasoning. That feeling was much reinforced by the organized religion behind this spiritual experience. Reading Nagel and Newman's book *Gödel's Proof* was like entering a world of holy gospel.

My life as a student was at the intersection of philosophy and mathematics. The logic students and teachers formed a truly interdisciplinary team, and I was lucky to see a golden generation in action, with people such as Dick de Jongh, Hans Kamp, Anne Troelstra, Wim Blok, Peter van Emde Boas, and others, including my supervisor Martin Löb. And there was of course that mysterious thing called the international community. I still remember the feeling of anticipation and then fulfillment when

A. Baltag and S. Smets (eds.), *Johan van Benthem on Logic*   989
and Information Dynamics, Outstanding Contributions to Logic 5,
DOI: 10.1007/978-3-319-06025-5, © Springer International Publishing Switzerland 2014

picking up a blue airmail envelope from a logician in the United States or New Zealand. Vanished pleasures! For me, modal logic was the ideal bridge between philosophy and mathematics, combining the best of both: mathematical challenge and conceptual motivation. I well remember the excitement of those early days, with lots of new questions floating around each week. Of course, there was also the intense pressure of having to start doing creative work on a par with these formidable others. One such experience that I remember vividly concerns my proof (by proving theorems logicians are really trying to prove themselves) that the McKinsey Axiom is not first-order definable on modal frames. In that period, I once had to spend a week in an Amsterdam hospital in the aftermath of surgery, and as I was lying there in a somewhat depressing third-class ward, on a low-budget student insurance policy, thinking about the abstractions of modal logic was my escape. One evening, the crucial uncountable frame and the Löwenheim-Skolem argument that was the core of my first published JSL paper suddenly appeared before my eyes.

My student generation was equally remarkable. These were the days of Liberation in the air, the barriers of rank between students and professors were down, and we all expected a golden new age for the whole planet Earth, which was uniformly inhabited by kind and reasonable people anyway. Looking back, many things that would seem unusual now seemed perfectly normal then. I hitchhiked extensively, starting alone and picking up companions on the way, from Holland to lots of countries, on a minimal budget, including North Africa, Iran, Afghanistan, India, Nepal, and the Soviet Union. My parents wanted to know where I was going, but I told them I did not know my destination exactly, there was nothing for us to discuss anyway, but they could write me poste restante in Tehran, Kabul, or Kathmandu. And it worked: I found letters from my mother waiting for me, and sent terse postcards in return (one has to limit oneself to essentials when communicating with anxious parents). My current academic trips to what are considered exotic countries are very pale copies of these student travels, whose adventures (good and also bad) have formed me for life.

My dissertation topic of modal correspondence theory was also a reflection of my Amsterdam environment. I did not want to do proof theory or intuitionism (these were the old topics my professors did, I wanted to be myself in this new age), but I did want to bridge between the mathematics and the philosophy in my environment. Correspondence theory was a way of employing techniques from classical logic to understand modal logic, then still the paradigm of a philosophically motivated system. My experience in that work continues to determine my general attitudes in research: developing modal and classical logic in tandem, and in the same spirit, being wary of ideological choices between logical systems, but also, appreciating that small languages qua expressive power can be beautiful, and being able to analyze phenomena at different levels of zoom. I think it is such broad themes that define a field, rather than specific formal systems or subfields, and I was happy to see later that creative mathematicians and philosophers of my acquaintance feel the same way. My supervisor Löb was not very supportive in all this, since he disliked modal logic and constantly worried whether it was respectable. Still, I learnt a lot from him in many other ways, and what I did not get from him, I got through the support of Dick de Jongh, and at a crucial moment in writing my dissertation also Anne Troelstra.

Varied personal experiences with research continued of course. Lots of topics in the dissertation revealed their true sense only much later, such as the discovery of bisimulation and proving what is now called Van Benthem's Theorem, then a side comment on the modal language used on models rather than frames. There were also major disappointments, such as having proved the Sahlqvist correspondence theorem independently, but running into an anonymous JSL referee who happened to know that some Norwegian guy had an unpublished thesis with this result. End of the story. I felt an early urge to collect my work into a book, and was invited by a Polish colleague to publish it with Ossolineum around 1979. The book never appeared there, it came out in 1983 as *Modal Logic and Classical Logic* with Bibliopolis in Naples, but not every story needs to be told here. I did get an advance for the book in Poland which was deposited for me in a bank in Warsaw, and annual statements duly arrived. I may have been the only one in my generation to have a capitalist nest egg in a communist country.

After this period, I thought the modal phase should be closed. I looked around for new topics, and for a while, I tried the philosophy of science. I liked some things that I saw, especially the logical analysis of empirical theories, opening my eyes to the fact that there is more to science from a logical perspective than pure mathematics. But I found Sneed's work, the major formal paradigm at that time, largely definition-mongering without very exciting questions, so eventually, I gave up. By the way, interests fade in my life, not because I come to despise their topics or practitioners, but the initial love degenerates into a mild appreciation that is not enough for action.

My next enterprise was the logic of time, where I had become enamored of developing an alternative interval paradigm, rather than points, as primitive entities. At Jaakko Hintikka's invitation, I wrote up my lecture notes into a book *The Logic of Time*, which brought together structure theory of intervals with techniques from modal logic. This idea was in the air around 1980, and many people proposed it. I remember attending a colloquium by a speaker at Stanford, who was announced as a brilliant leader in Artificial Intelligence having revolutionized our understanding of time, and then telling us something that sounded much like my work. I considered speaking up, but did not: why be the European spoil-sport who points out in bad English that he already had these ideas in an obscure book in some insignificant country? But that evening, I decided to call the speaker in his hotel room, and he said he had just heard my name over dinner. Jon Barwise had told him that people should stop giving talks about temporal logic before they had read van Benthem's book. Sometimes (but do not get your hopes up too much) life deals us sweet surprises. *The Logic of Time* has been one of my most widely read publications, far beyond the impact of my modal logic, and I have heard back from readers in the most diverse walks of life, from Dutch high school students to Austrian architects. The book is out of fashion now, and major handbook articles on temporal logic do not even mention it. But I am sure that the interval paradigm will make a comeback: it always has, it is just too natural to die. By the way, my later interest in logics of space is a natural continuation of this work, and in particular, editing the *Handbook of Spatial Logics* in 2007 was a labor of love.

Most of my work in the 1980s was on logic and natural language. This connection was already in the air in my student days. A group of us physics students would go to the faculty of Humanities to take classes in Chomsky's new formal grammar, with the side benefit of being able to watch the gorgeous fashion show at lunchtime when the literary students took their break. Even hardcore scientists have occasional longings for a better, more beautiful life. One day I had learned that the Dutch language has infinitely many sentences, and I rushed to my landlady [I lived in a tiny student room under her wings] to tell her about this wonderful insight. When she heard the trivial proof by recursion on "(Mary thinks that John thinks that)* the weather is bad", she was very disappointed, and told me not to be silly. But the feeling that "there is gold in them there hills", as Austin said about natural language, persisted, and significantly, logicians that I admired such as Hans Kamp and Jon Barwise had moved in that direction. While my initial reaction to Montague grammar had been mainly like that to Sneed: a grand machine with too many definitions and too few real results, things were changing now, and I jumped in.

Topics that intrigued me were not formalizing fragments of natural language, but general themes such as the power of human languages for describing reality, with a focus on their quantifier repertoire. I joined the small band of logicians working on generalized quantifiers, and went for questions of expressive power in definability and semantic universals about shared conceptual structures across natural languages. Eventually, I developed an interest in natural logic of reasoning close to the linguistic surface, resulting in the 'monotonicity calculus', and procedural-computational views of linguistic interpretation, that led to my work on 'semantic automata'. You can find all these themes in my book *Essays in Logical Semantics* of 1986, written toward the end of my period in Groningen, and the informal start of the ILLC in Amsterdam.

Despite what I just said about Montague Grammar, the general machinery behind the complex syntax of natural language did come to intrigue me. I opted for categorial grammar in the elegant version proposed by Lambek in 1958, and really brought to the world's attention in the dissertation of Wojciech Buszkowski. One of my early discoveries turned out to be another disappointment. I found a truly beautiful correspondence between categorical derivations and special linear terms in the lambda calculus, but then learnt that it was a special case of the well-known Curry-Howard isomorphism. All that had been revealed was my ignorance of basic proof theory. I made up for this by entering a proof-theoretic phase concentrating on grammar, recognizing power, and related topics, and developed a wide-ranging theory of language in a categorical perspective, which you can see in my book *Language in Action. Categories, Lambdas and Dynamic Logic* of 1991. This work also sits at a cusp with the more general idea of resources and substructural rules, as occurring in relevant logic and linear logic, that are still so prominent in logic today. Some of my work even had some practical impact, such as a simple numerical invariant for pruning the search space of Lambek derivations that I once saw running on a TNO computer with a banner streaming on the screen computing successive 'van Benthem counts'. Pure theorists may brag about the virgin uselessness of their work, but the experience of having an actual use can be very powerful.

Herman Hendriks claims that each of my books has an odd chapter that predicts the next line. The dynamic logic in my title was certainly significant, and I shifted my interest away from natural language. Even so, I did edit the *Handbook of Logic and Language* with Alice ter Meulen in 1997, as a public service to the field at a time when the partisan fights of earlier phases were abating, and the true achievements became visible. By the way, historically, churches and sects have been very successful forms of organization, so I do not want to belittle the power of partisanship.

Moving from language, my interest became the general notion of information. I was struck by the many conceptual parallels in the study of natural language, computer science, AI, and philosophy, and my paper 'Semantic Parallels in Natural Language and Computation' at the 1987 Granada Logic Colloquium contains a host of these, many of which became separate research lines. These include the abstract analysis of intuitionistic and modal information models, substructural characterizations of styles of inference and update, and other things that still occupy me, such as the connection between proof-theoretic combination of pieces of evidence and model-theoretic views of information. These concerns return in the *Handbook of the Philosophy of Information* that I edited with Pieter Adriaans in 2008. Some occur in the editorials, and many more in the chapter on 'Logic and Information' with Maricarmen Martinez, where we try to come to grips with the variety of notions of information in logic, semantic, proof-theoretic, and also correlation- and channel-based as in the situation theory of Jon Barwise, John Perry, and their school.

But information should not be studied on its own. One powerful idea in computer science that has always appealed to me is the dictum of 'no representation without process'. One should know the process a representation is made for, a point that is still underappreciated in natural language semantics and large areas of philosophy. So, along with my interest in information came an interest in computation. By that time, the importance of bisimulation as a view of process equivalence (rediscovered independently in the early 1980s) had become clear to me, and so, around 1990, a return to modal logic made sense. I started doing work on dynamic logics of computation and action in general, and have kept working along these lines, taking my earlier modal work to the area of fixed point logics for induction and recursion.

One aspect of my taking computation seriously was an interest in computational complexity, the mathematics of difficulty of tasks. I have come to believe that complexity is an essential aspect in truly understanding the topics we usually study, and this interest led to a new look at the undecidability or decidability of logical systems. I became interested in the exact reasons for the usual commonplaces such as 'predicate logic is undecidable'. Does this really tell us that core reasoning with quantifiers is complex, or might there be historical accidents of formulation? This is the line that led to the discovery of decidable core logics of relational algebra ('arrow logic') and of predicate logic based on generalized semantics, but also, in another manifestation, the Guarded Fragment of first-order logic, a large decidable realm far beyond basic modal logic. My general feeling is that we should always distinguish between true contents of logical systems and 'wrappings', accidents of set-theoretic formulations or other fashions. Then there may be much more decidability and even lower complexity in logic than is usually thought. I once gave a talk on my new 'geometric' semantics for predicate logic at the Berkeley logic seminar, and Leon Henkin told me that they still had discussions in the 1950s about what should be the right formulation of first-order semantics. Over dinner, Henkin added that he would have loved to see me debate with Tarski. Well, we shall never know.

All this set the stage for the main theme of this book, that of logical dynamics as an integrated view of the nature of logic as a dual study of statics and dynamics. There is no need for me to repeat this here, since it has been explained in various pieces at the beginning of this book. One of the earliest moments I felt that I was on to something big occurred in 1991 when preparing for an invited lecture on logic and information flow at the Congress of Logic, Methodology and Philosophy of Science in Uppsala. However, I was quickly put back on the ground. On the eve of my lecture, there was a party at Dag Prawitz' house in Stockholm, and I managed to lose my way and miss the last train. There was of course no way I would go back to my distinguished colleague and confess that I could not even remember a few simple travel instructions. So I found a bench at the station and prepared for sleeping out, as I had done so often as a traveling student. All around me were somewhat shady characters, drunks and addicts, but I hung on to my spot. Around 1:30, I suddenly woke from my fitful slumbers: the police were sweeping the station clean, and turning us out into the street, with long subsequent hours of deep cold and discomfort. I found an early morning bus to Uppsala, and gave my talk, but the intellectual epiphany had disappeared.

The progress of my ideas on logical dynamics is easy to follow in books. *Exploring Logical Dynamics* collected many themes and results, with major developments coming out of collaborations with colleagues like Hajnal Andréka and Istvan Németi and Jan Bergstra, a new habit that I acquired in this period, perhaps in line with the logical dynamics idea. Another prominent feature was the work done by my Ph.D. students, who enabled me to see much further than I could have done on my own (perhaps they also did some of the more dangerous missions). You will see many of the themes I mentioned earlier, now as threads in one overarching endeavor.

Conspicuously missing, however, was the theme of multi-agent interaction, which only entered after I became influenced by students like Willem Groeneveld, Jelle Gerbrandy, Hans van Ditmarsch, and (though it is hard to think of him as having been a student) Alexandru Baltag. Dynamic-epistemic logic was born around 2000 (the current ascriptions to Plaza, whose work was totally unknown then, are a form of overblown courtesy that distorts the historical record), and I became an enthusiastic participant. I had a traveling talk called 'Update Delights' in 1999, and still remember an invited lecture at the ESSLLI Summer School in Birmingham where the chair pointed out that my title was a rare instance of a two-word expression in English that is three-way ambiguous. My book *Logical Dynamics of Information and Interaction* from 2011 tells the story as I see it now, with logic as a theory of agency where pure information and knowledge update based on observations, inferences and acts of communication such as questions needs to be in balance with agents' beliefs and how they correct themselves. Much of our quality resides in learning from errors, and the point is that logic can incorporate this essential feature. In addition, the book reflects another growing conviction of mine, that just dealing with pure information may not be a natural boundary. In all we do, information is in balance with how we evaluate the world, and again logic is up to the task of describing this.

Of course, there are also persistent technical strands from my earlier work in all this, such as the central role of modal logic and dynamic logic, and the use of mathematical notions in logic to demystify mysterious innovations. For instance, the down-to-earth analysis of update as relativization was one of my points at Jelle Gerbrandy's thesis defense. Another point at that defense were connections between dynamic-epistemic logic and Process Algebra which have not panned out yet as I hoped. And in recent years, I have taken up modal frame correspondence analysis of dynamic-epistemic logics, returning to themes and techniques from my dissertation.

My most recent book is *Logic in Games*, and I see its emphasis on the social process of intelligent interaction (a phrase with a nice Mozartesque ring that I once coined for a strategic European funding program) as a fitting ending to the logical dynamics trilogy. Games in logic had always been on my radar, ever since I read the Luce and Raiffa classic *Games and Decisions* as a student, and then started out as a young teacher in the 1970s telling my students about Lorenzen dialogues and Hintikka evaluation games. But a deeper interest only started at the time of my Spinoza Award project in 1996, a sort of oeuvre award of the Dutch national science organization that allowed me to pursue new lines by offering a substantial sum of money for 5 years that I was free to spend. I chose three: computational logic, didactic innovation in logic, and logic and games, where we first entered into serious contacts with game theorists, a congenial mathematical community. Incidentally, spending the money turned out to be not totally free. When the award was announced, I had quickly computed that it sufficed for buying one of the smaller Florida keys, and I felt that buying an island for logic in the Caribbean might be the best investment in perpetuity that anyone could make for our field. But that was one step too far for our national science foundation, who refused to think big like our seafaring ancestors.

Logic and games have been a natural match ever since people started thinking about argumentation in Greek and Chinese antiquity, and in my book, I show how dynamic-epistemic logics can analyze the structure of games in innovative ways, leading toward a love child of logic and game theory that might be called a Theory of Play. But I also study the manifold current uses of games to understand logic, and these two themes, 'logic of games' and 'logic as games' form two intertwined strands in my book, which also presents many hybrids between them. I now see this entanglement of strands as the DNA of logic, but how the duality works exactly is still a mystery to me.

What is next? One thing that just seems to be happening naturally these days is a return to philosophy. I feel that the sort of logics I am pursuing now might transform the logic-philosophy interface that has been a bit dormant after the roaring 1960s and 1970s, and one project is a book on epistemology called *The Music of Knowledge* with Alexandru Baltag and Sonja Smets. Another influence that I feel is one that was entirely absent in my student days: the importance of empirical facts about human reasoning, as they are coming to light these days in cognitive (neuro-)science. I feel that logical theorizing should balance 'intuitions' with reality checks, but facts still scare me a bit, and I am mainly content with admiring those of my current students who seem equally at home in mathematical logic and cognitive psychology. My 2006 paper 'Logic and Psychology: Do the Facts Matter?' shows my cautious, and hence ambiguous, enthusiasm in this realm. One way in which I may face the facts is in a return to my old interests in natural language, where the logical dynamics perspective suggests very different views of what we can, and perhaps should, study by way of key expressions and phenomena. I now believe that the usual emphasis on successful communication is too limited, and that there is much more to the dynamic stability of language with fallible users that has escaped our attention so far. But sometimes, there are even more brute facts than that. Nowadays, I often show students in logic classes frequency tables of words in English or Chinese text corpora, to see which expressions really occur a lot. Fortunately, many logical items score very well.

Finally, I am still intrigued by many technical issues, of which the interface of logic and probability is probably the most urgent right now. Looking at the realities of research in formal philosophy, but also many other fields adjoining logic, this combination seems inevitable, also for deep theoretical reasons. The way I see it now, the mind works on an analogy with the body. Our conscious span of bodily control is in a tiny physical zone, around one meter say, with the bulk behavior of atoms and molecules underneath, and that of astronomical constellations above us. Likewise, our neat little world of conscious deliberation, communication, and decision is just a tiny slice in between the statistics of neural nets in our bodies and the statistics of the crowds and societies of which we form part. I would love to understand these interfaces better, and it may involve deep connections between logic and probability beyond those we already know. To do this well, I may well have to go back to the physics studies of my early student days—something which my sons have been urging me to do anyway while I still have the brain power.

Was logic a good choice? An interviewer of the Dutch national radio once asked me, off the record after a public broadcast on logic, why someone like me had not gone into really interesting subjects like physics or literature instead of this very narrow topic that he found small-minded, being self-centered around our own thinking. But logic has been good for me. It fit with the needs of a young boy who could not choose between the humanities and the sciences, and it put me at an intellectual crossroads between disciplines that keeps opening new vistas, with congenial colleagues at Amsterdam, Stanford, and now also China. And in addition to the delights of research, it sometimes afforded moments of transcendence. I once gave a talk on logic in Ayacucho, high in the Andes, for an audience of mathematicians who only spoke Spanish (and perhaps Quechua), so I talked in Dutch and a friend of mine translated. And my friend told me that at one moment he felt his own personality had disappeared, since the audience was obviously understanding what I was saying through him while he did not. That is the power of resonance afforded by logic.

Of course, there are only few logicians, so I have always tried to work in environments where I would be the average rather than the exception, such as the

ILLC in Amsterdam, CSLI at Stanford, or the ESSLLI Summer Schools. Moreover, there have always been enough students sharing my constellation of interests, that circle around colleagues in different fields like electrons, hard to detect at first, but crucial to keeping the whole process working together. In fact, students have been an integral part of my intellectual development, and all my recent books testify to that role. As I said, they enable one to see and achieve much more, as a sort of extended eyes and ears (though not in the sense of the ancient Persian imperial court).

But I should not over-systematize or rationalize my life's choices. In addition to all the rational factors outlined here, I also owe an enormous debt of gratitude to mere chance, or at least circumstances beyond my control. I got my first university job because my professor Löb saw something in me, and I was recalled to Amsterdam in 1986 because my old teachers were willing to take a chance against prevailing currents of thought in mathematical logic. I met many people who influenced me in totally unpredictable productive ways, such as my Dutch high school friend Frans Zwarts or new friends like Jon Barwise or Dov Gabbay. I found highly creative students who chose to study with me though their talents would have taken them anywhere, and their opinions and needs often affected the course of my own work. Sometimes, I think that is all there is to life in general: a beautiful accident.

ADDENDUM III
Bibliography of Johan van Benthem

Books

1977   *Modal Correspondence Theory*, dissertation, Universiteit van Amsterdam, Instituut voor Logica en Grondslagenonderzoek van de Exacte Wetenschappen, 148 p.

1982   *Logica, Taal en Betekenis*, Spectrum, Utrecht. With authors' collective GAMUT.

1983   *The Logic of Time*, Reidel, Dordrecht (Synthese Library 156). Revised and expanded edition published in 1991.

1985   *Modal Logic and Classical Logic*, Bibliopolis, Napoli (Indices 3) and Humanities Press, Atlantic Heights.

1985   *A Manual of Intensional Logic*, CSLI Lecture Notes 1, Center for the Study of Language and Information, Stanford.

1986   *Essays in Logical Semantics*, Reidel, Dordrecht, Studies in Linguistics and Philosophy 29.

1987   *Situations, Language and Logic*, Reidel, Dordrecht, Studies in Linguistics and Philosophy 34. With J-E Fenstad, P-K Halvorsen and T. Langholm.

1988   *A Manual of Intensional Logic*, CSLI Publications, Stanford and The University of Chicago Press, Chicago.

1991   *Language in Action: Categories, Lambdas and Dynamic Logic*, North-Holland, Amsterdam (Studies in Logic 130). Paperback reprint with new Appendix, The MIT Press, 1995.

1991   *Logic, Language and Meaning*, The University of Chicago Press. Expansion and revision of GAMUT 1982.

1991   *Logica voor Informatici*, Addison-Wesley, Amsterdam. With co-authors H. van Ditmarsch, J. Ketting, W. Meyer Viol.

1996   *Exploring Logical Dynamics*, Studies in Logic, Language and Information, CSLI Publications and Cambridge University Press.

2001   *Logic in Games*, Lecture Notes and Book Preversion, ILLC Amsterdam.

2001   *Logic in Action*, Spinoza Project, ILLC Amsterdam. With Paul Dekker, Jan van Eijck, Maarten de Rijke, and Yde Venema.

A. Baltag and S. Smets (eds.), *Johan van Benthem on Logic and Information Dynamics*, Outstanding Contributions to Logic 5, DOI: 10.1007/978-3-319-06025-5, © Springer International Publishing Switzerland 2014

2002   *Introducción a la Lógica,* Eudeba, Buenos Aires, Spanish translation of GAMUT 1991.

2003   *Logica voor Informatica,* revised and expanded edition of LVI 1991. With Hans van Ditmarsch and Josje Lodder.

2005   *Hoe Wiskunde Werkt,* ILLC/KdV/IIS, with Robbert Dijkgraaf.

2009   *Logica in Actie,* Open University, Heerlen, SDU Academic Service: http:// www.sdu.nl/catalogus/25999. With Hans van Ditmarsch, Jan van Eijck and Jan Jaspars.

2009–2012   *A Door to Logic,* Chinese translation of selected papers, 4 volumes ("Modal Logic, Information and Computation", "Modal Correspondence Theory", "Logic, Language and Cognition", "Logic, Philosophy and Methodology"), Science Press, Beijing.

2010   *Modal Logic for Open Minds,* CSLI Publications, Stanford.

2011   *Logical Dynamics of Information and Interaction,* Cambridge University Press, Cambridge UK.

2012   *Logic in Action,* Open Source Textbook, http://www.logicinaction.org/ with Hans van Ditmarsch, Jan van Eijck, Jan Jaspars and collaborators in Amsterdam, Beijing, Seville, and Stanford.

2014   *Logic, Language and Meaning,* Chinese edition, Commercial Press, Beijing.

2014   *Logic in Games,* The MIT Press, Cambridge (Mass.).

To appear
The Music of Knowledge, Dynamic-Epistemic Logic and Epistemology. With co-authors Alexandru Baltag and Sonja Smets.

Edited Books

1985   *Generalized Quantifiers in Natural Language,* Foris, Dordrecht (GRASS, vol. 4); with A. ter Meulen.

1988   *Categories, Polymorphism and Unification,* Centre for Cognitive Science (University of Edinburgh) and Institute for Language, Logic and Information (University of Amsterdam); with E. Klein.

1988   *Categorial Grammar,* John Benjamins, Amsterdam and Philadelphia; with W. Buszkowski and W. Marciszewski.

1990   *Semantics and Contextual Expression,* Foris, Dordrecht, (GRASS, vol. 11); with R. Bartsch and P. van Emde Boas.

1996   *Proceedings 10th International Congress on Logic, Methodology and Philosophy of Science. Florence 1995,* volumes "Logic and Scientific Methods", "Structures and Norms in Science", Kluwer Academic Publishers, Dordrecht; with M. Dalla Chiara, K. Doets and D. Mundici.

1996   *Logic and Theory of Argumentation,* Royal Dutch Academy, Amsterdam; with F. van Eemeren, R. Grootendorst and F. Veltman.

1997 *Handbook of Logic and Language*, http://mitpress.mit.edu/catalog/item/ default.asp?tid=8458&ttype=2. North-Holland, Amsterdam and The MIT Press; with Alice ter Meulen.

2001 *Proceedings 8th Conference on Theoretical Aspects of Rationality and Knowledge*, Morgan Kaufmann Publishers, San Francisco.

2002 *Words, Proofs, and Diagrams*, CSLI Publications, Stanford; with D. Barker-Plummer, D. Beaver, D. Israel, P. Scotto di Luzio.

2006 *The Age of Alternative Logics,* Springer, Dordrecht; with G. Heinzmann, M. Rebuschi and H. Visser.

2006 *Handbook of Modal Logic,* http://www.lsc.liv.ac.uk/~wolter/hml/http:// www.elsevier.com/wps/find/bookdescription.cws_home/708884/description. Elsevier, Amsterdam; with Patrick Blackburn and Frank Wolter.

2007 *A Meeting of the Minds*, Proceedings LORI Workshop Beijing August 2007, King's College Publications, London; with Ju Shier and Frank Veltman.

2007 *Handbook of Spatial Logics*, http://www.dit.unitn.it/~aiellom/hsl. Springer, Dordrecht; with Marco Aiello and Ian Pratt-Hartmann.

2007 *Foundations of the Formal Sciences V: Infinite Games*. Studies in Logic, College Publications, London; with Stefan Bold, Benedikt Löwe, Thoralf Räsch.

2007 *Logic at a Cross-Roads: Logic and its Interdisciplinary Environment*, Allied Publishers, Mumbai; with A. Gupta and R. Parikh.

2007 *Logics for Interaction*, Augustus de Morgan Workshop 2005, Texts in Logic and Games, Amsterdam University Press; with Dov Gabbay and Benedikt Löwe.

2008 *Logic and Intelligent Interaction*, Proceedings Workshop *ESSLLI XV*, University of Hamburg; with Eric Pacuit, 153 p.

2008 *Handbook of the Philosophy of Information*, http://www.illc.uva.nl/HPI/. Elsevier Science Publishers, Amsterdam; with Pieter Adriaans.

2009 *Wat Cognitiewetenschappers Bezielt*; with B. Mols en P. Hagoort, NWO.

2010 *Handbook of Logic and Language, Second Edition*, Elsevier Science Direct, Amsterdam; with Alice ter Meulen.

2011 "Logic at the Cross-Roads", Vol. 1 + 2: *Proof, Computation, and Agency* and *Games, Norms and Reasons*, Springer; with A. Gupta and E. Pacuit.

2011 *Logic and Philosophy Today*, College Publications, London, Studies In Logic, Vol's. 29 and 30; with Amitabha Gupta.

2013 *Logic Across the University. Foundations and Applications*, College Publications, London, Studies in Logic, Vol. 47; with Fenrong Liu.

To appear

2014 *Reasoning with Strategies*; Birkhäuser Verlag Basel. (With Sujata Ghosh and Rineke Verbrugge.)

2014 *A Formal Epistemology Reader*, Springer, Dordrecht. (With Horacio Arló-Costa and Vincent Hendricks.)

Articles in Journals

1973

1 Bestaan Denkwetten? *Algemeen Nederlands Tijdschrift voor Wijsbegeerte* 65, 120–125.

1974

2 Hintikka on Analyticity. *Journal of Philosophical Logic* 3, 419–431.
3 Semantic Tableaus, *Nieuw Archief voor Wiskunde* 22, 44–59.

1975

4 A Set-Theoretical Equivalent of the Prime Ideal Theorem for Boolean Algebras, *Fundamenta Mathematicae* 89, 151–153.
5 A Note on Modal Formulae and Relational Properties, *Journal of Symbolic Logic* 40, 55–58.

1976

6 Modal Reduction Properties, *Journal of Symbolic Logic* 41, 301–312.
7 Modal Formulas are either Elementary or not $\Sigma - \Delta$-Elementary, *Journal of Symbolic Logic* 41, 436–438.
8 Enkele Opmerkingen over Zelfreferentie en Zelfweerlegging, *Algemeen Nederlands Tijdschrift voor Wijsbegeerte* 68, 250–270.

1977

9 Tense Logic and Standard Logic, *Logique et Analyse* 20, 41–83.

1978

10 Four Paradoxes, *Journal of Philosophical Logic* 7, 49–72.
11 Ramsey Eliminability, *Studia Logica* 37:4, 321–336.
12 Two Simple Incomplete Modal Logics, *Theoria* 44:1, 25–37.
13 Transitivity Follows from Dummett's Axiom, *Theoria* 44:2, 117–118. (With Willem Blok.)

1979

14 Canonical Modal Logics and Ultrafilter Extensions, *Journal of Symbolic Logic* 44:1, 1–8.
15 Minimal Deontic Logics, *Bulletin of the Section of Logic* 8:1, 36–42.
16 What is Dialectical Logic? *Erkenntnis* 14, 333–347.
17 Syntactic Aspects of Modal Incompleteness Theorems, *Theoria* 45:2, 63–77.

1980

18 Some Kinds of Modal Completeness, *Studia Logica* 39:2/3, 125–141.

1981

19 Historische Vergissingen? Kanttekeningen bij de Fregeaanse Revolutie in de Logica, *Kennis en Methode* 5:2, 94–116.
20 Fundering of Ondermijning? *Nieuw Archief voor Wiskunde* 29:3, 254–284.

1982

21 The Dynamics of Interpretation, *Journal of Semantics* 1, 3–20. (With Jan van Eijck.)
22 The Logical Study of Science, *Synthese* 51, 431–472.
23 Later than Late: on the Logical Origin of the Temporal Order, *Pacific Philosophical Quarterly* 63, 193–203.

1983

24 Logical Semantics as an Empirical Science, *Studia Logica* 42:2/3, 299–313.
25 Halldén Completeness by Glueing of Kripke Frames, *Notre Dame Journal of Formal Logic* 24, 426–430. (With Lloyd Humberstone.)
26 Determiners and Logic, *Linguistics and Philosophy* 6:4, 447–478.

1984

27 Questions about Quantifiers, *Journal of Symbolic Logic* 49:2, 443–466.
28 Tense Logic and Time, *Notre Dame Journal of Formal Logic* 25:1, 1–16.
29 Foundations of Conditional Logic, *Journal of Philosophical Logic* 13, 303–349.
30 A Mathematical Characterization of Interpretation between Theories, *Studia Logica* 43:3, 295–303. (With David Pearce.)
31 Possible Worlds Semantics: a Research Program that Cannot Fail? *Studia Logica* 43:4, 379–393.
32 Analytic/Synthetic: Sharpening a Philosophical Tool, *Theoria* 50:2/3, 106–137.

1985

33 Situations and Inference, *Linguistics and Philosophy* 8, 3–9.
34 The Variety of Consequence, According to Bernard Bolzano, *Studia Logica* 44:4, 389–403.

1986

35 Tenses in Real Time, *Zeitschrift für mathematische Logik und Grundlagen der Mathematik* 32, 61–72.
36 Partiality and Non-Monmotonicity in Classical Logic, *Logique et Analyse* 29, 225–247.
37 The Relational Theory of Meaning, *Logique et Analyse* 29, 251–273.

1987

38 Meaning: Interpretation and Inference, *Synthese* 73:3, 451–470.

1988

39 A Note on Jónsson's Theorem, *Algebra Universalis* 25, 391–393.

40 Vragen om Typen, *GLOT* 10:3, 333–352.
41 Logical Syntax, *Theoretical Linguistics* 14, 119–142.

1989

42 Notes on Modal Definability, *Notre Dame Journal of Formal Logic* 30:1, 20–35.
43 Polyadic Quantifiers, *Linguistics and Philosophy* 12:4, 437–464.
44 Logical Constants across Varying Types, *Notre Dame Journal of Formal Logic* 30:3, 315–342.

1990

45 Categorial Grammar and Type Theory, *Journal of Philosophical Logic* 19, 115–168.
46 Kunstmatige Intelligentie: Een Voortzetting van de Filosofie met Andere Middelen, *Algemeen Nederlands Tijdschrift voor Wijsbegeerte* 82, 83–100.
47 Computation versus Play as a Paradigm for Cognition, *Acta Philosophica Fennica* 49, 236–251.

1991

48 Editorial Information Sciences, *Journal of Logic, Language, and Information* 1, 1–4.
49 Language in Action, *Journal of Philosophical Logic* 20, 1–39.
50 General Dynamics, *Theoretical Linguistics* 17: 1/2/3, 151–201.

1992

51 Logic as Programming, *Fundamenta Informaticae* 17:4, 285–317.
52 Modeling the Kinematics of Meaning, *Proceedings Aristotelean Society* 1992, 105–122.

1993

53 Modal Frame Classes Revisited, *Fundamenta Informaticae* 18: 2/3/4, 307–317.
54 The Elusive Locus of Logicality, guest editorial, *Journal of Logic and Computation* 3:5, 451–453.
55 Reflections on Epistemic Logic, *Logique et Analyse* 34 (vol. 133–134), 5–14.

1994

56 Modal Logic, Transition Systems and Processes, *Journal of Logic and Computation* 4:5, 811–855. (With Jan van Eijck and Vera Stebletsova.)

1995

57 Logic of Transition Systems, *Journal of Logic, Language and Information* 3:4, 247–283. (With Jan Bergstra.)
58 Directions in Generalized Quantifier Theory, *Studia Logica* 55:3, 389–419. (With Dag Westerståhl.)

59 Back and Forth Between Modal Logic and Classical Logic, *Bulletin of the Interest Group in Pure and Applied Logics* 3, August 1995, 685–720. London and Saarbruecken.

1996

60 Logica in Beweging: de Dynamiek van Redeneren en Betekenis, *Handelingen KNAW*, 6.1.96, Afdeling Letterkunde, Amsterdam.
61 Wat is Mis met de Filosofie? *Ergo Cogito* 5, Historische Uitgeverij Groningen, 27–36.
62 Space, Time and Computation: Trends and Problems, editorial, special issue on Spatial and Temporal Reasoning, *International Journal of Applied Intelligence* 6:1, 5–9. (With Frank Anger, Rita Rodriguez and Hans Guesgen.)

1997

63 Logic, Language and Information: The Makings of a New Science? guest editorial, *Journal of Logic, Language and Information* 6:1, 1–3.
64 Cognitive Actions in Focus, guest editorial (with co-editor Yoav Shoham), special TARK issue, *Journal of Logic, Language and Information* 6:2, 119–121.
65 Modal Foundations for Predicate Logic, *Bulletin of the IGPL* 5:2, 259–286, London and Saarbruecken (R. de Queiroz, ed., *Proc's WoLLIC, Recife 1995*).
66 Modal Deduction in Second-Order Logic and Set Theory. Part I, *Logic and Computation* 7:2, 251–265. (With Giovanna d'Agostino, Angelo Montanari and Alberto Policriti.)

1998

67 Programming Operations that are Safe for Bisimulation, *Studia Logica* 60:2 (Logic Colloquium. Clermont-Ferrand 1994), 311–330.
68 Logische Dynamiek, Themanummer *ANTW* 90:1, 54–70.
69 Modal Logics and Bounded Fragments of Predicate Logic, *Journal of Philosophical Logic* 27:3, 1998, 217–274. (With H. Andréka and I. Németi.)
70 Modal Deduction in Second-Order Logic and Set Theory, II, *Studia Logica* 60, 387–420. (With Giovanna d'Agostino, Angelo Montanari and Alberto Policriti.)
71 Points on Time, discussion note, ENRAC Electronic Newsletter 4:10, *Reasoning about Action and Change*, Linköping: http://www.ida.liu.se/ext/etai/rac/notes/1998/03/debet.html.

1999

72 Wider Still and Wider: Resetting the Bounds of Logic, in A. Varzi, ed., *The European Review of Philosophy*, CSLI Publications, Stanford, 21–44.
73 Temporal Patterns and Modal Structure, in A. Montanari, A. Policriti and Y. Venema, eds., Special issue on Temporal Logic, *Logic Journal of the IGPL* 7:1, 7–26.
74 Modality, Bisimulation and Interpolation in Infinitary Logic, *Annals of Pure and Applied Logic* 96, 29–41.

75 Interpolation, Preservation, and Pebble Games, *Journal of Symbolic Logic* 64:2, 881–903. (With Jon Barwise.)
76 The Range of Modal Logic, *Journal of Applied Non-Classical Logics* 9:2/3 (issue in Memory of George Gargov), 407–442.

2000

77 Information Transfer Across Chu Spaces, *Logic Journal of the IGPL* 8:6, Nov 2000, 719–731.

2001

78 Games in Dynamic Epistemic Logic, in G. Bonanno and W. van der Hoek, eds., *Bulletin of Economic Research* 53:4, 219–248 (Proceedings LOFT-4, Torino).
79 Action and Procedure in Reasoning, *Cardozo Law Review* 22, 1575–1593.

2002

80 Logic in Action, 5 jaar later, *Alg. Ned's Tijdschrift voor Wijsbegeerte* 2, 146–150.
81 Extensive Games as Process Models, in M. Pauly and P. Dekker, eds., special issue of *Journal of Logic, Language and Information* 11, 289–313.

2003

82 A Modal Walk through Space, *Journal of Applied Non-Classical Logic* 12:3/4, 319–363. (With Marco Aiello.)
83 Connecting the Different Faces of Information, editorial guest-edited volume, *Journal of Logic, Language and Information* 12:4. (With Robert van Rooij.)
84 Conditional Probability Meets Update Logic, *Journal of Logic, Language and Information* 12:4, 409–421.
85 Logic Games are Complete for Game Logics, *Studia Logica* 75, 183–203.
86 Logic and the Dynamics of Information, *Minds and Machines* 13:4, special issue (L. Floridi, ed.), 503–519.
87 Euclidean Hierarchy in Modal Logic, *Studia Logica* 75, 327–344. (With G. Bezhanishvili and M. Gehrke.)
88 Reasoning About Space: The Modal Way, *Logic and Computation* 13:6, 889–920. (With Marco Aiello and Guram Bezhanishvili.)

2004

89 A Mini-Guide to Logic in Action, *Philosophical Researches 2004,* Suppl., 21–30, Beijing, Chinese Academy of Sciences.
90 What One May Come to Know, *Analysis* 64:282, 95–105.
91 Diversity of Logical Agents in Games, *Philosophia Scientiae* 8:2, 163–178. (With Fenrong Liu.)

2005

92 Minimal Predicates, Fixed-Points, and Definability, *Journal of Symbolic Logic* 70:3, 696–712.

93 Guards, Bounds, and Generalized Semantics, *Journal of Logic, Language and Information* 14, 263–279.

2006

94 Epistemic Logic and Epistemology: the State of their Affairs, *Philosophical Studies* 128, 49–76.
95 Modal Frame Correspondences and Fixed-Points, *Studia Logica* 83:1, 133–155.
96 Where is Logic Going, and Should It? *Topoi 25*, 117–122.
97 Logics of Communication and Change. *Information and Computation* 204:11, 1620–1662. (With Jan van Eijck and Barteld Kooi.)
98 Epistemic Logic and Epistemology: the State of their Affairs, Chinese translation of [94] by F. Liu, *World Philosophy* 6, pp 73–83.

2007

 99 Rationalizations and Promises in Games, *Philosophical Trends*, 'Supplement 2006' on logic, Chinese Academy of Social Sciences, Beijing, 1–6.
100 Introduction, *Topoi* 26:1–2, special issue on Logic and Psychology, with co-editors Helen and Wilfrid Hodges.
101 Multimodal Logics of Products of Topologies, *Studia Logica* 84, 369–392. (With Guram Bezhanishvili, Balder ten Cate, and Darko Sarenac.)
102 Rational Dynamics and Epistemic Logic in Games, *International Game Theory Review* 9:1, 13–45. Erratum reprint, Volume 9:2, 377–409.
103 Dynamic Logic of Belief Revision, *Journal of Applied Non-Classical Logics* 17:2, 129–155.
104 Dynamic Logic of Preference Upgrade, *Journal of Applied Non-Classical Logics* 17:2, 157–182. (With Fenrong Liu.)
105 A New Modal Lindström Theorem, *Logica Universalis* 1, 125–138.

2008

106 Editorial, *Synthese* 160:1, 1–4 (With Vincent Hendricks and John Symons).
107 Logic and Reasoning: do the Facts Matter? *Studia Logica* 88, 67–84.
108 Tell It Like It Is: Information Flow in Logic, *Journal of Peking University* (Humanities and Social Science Edition), No. 1, 80–90.
109 Inference in Action, *Publications de l'Institut Mathématique,* Nouvelle Série 82:96, Beograd, 3–16.
110 Abduction at the Interface of Logic and Philosophy of Science, *Theoria* 22/3, 271–273.
111 Merging Observation and Access in Dynamic Epistemic Logic, *Studies in Logic* 1:1, 1–16.
112 The Many Faces of Interpolation, *Synthese* 164:3, 451–460.
113 Modal Logic and Invariance, *Journal of Applied Non-Classical Logics* 18:2–3, 153–173. (With Denis Bonnay.)
114 Modeling Simultaneous Games in Dynamic Logic, *Synthese (KRA)* 165:2, 247–268. (With Sujata Ghosh and Fenrong Liu.)
115 Constanten, of Variabelen, van het Logisch Denken, *ANTW* 100:4, 296–304.

116 Logical Pluralism Meets Logical Dynamics? *The Australasian Journal of Logic* 6, 28 pages. December 17, 2008.

2009

117 Everything Else Being Equal: A Modal Logic for Ceteris Paribus Preferences, *Journal of Philosophical Logic* 38:1, 83–125. (With Patrick Girard and Olivier Roy.)
118 The Information in Intuitionistic Logic, *Synthese* 167:2, 251–270.
119 Merging Frameworks for Interaction, *Journal of Philosophical Logic* on-line, January 16, 2009. (With Jelle Gerbrandy, Tomohiro Hoshi and Eric Pacuit.) Paper version: *JPL* 38:5 (2009), 491–526.
120 Logic and Intelligent Interaction, Editorial, special issue of *Knowledge, Rationality and Action. Synthese* 169:2, 219–221. (With Thomas Ågotnes and Eric Pacuit.)
121 Lindström Theorems for Fragments of First-Order Logic, *Logical Methods in Computer Science* 5:3, 1–27. (With Balder ten Cate and Jouko Väänänen.)
122 Dynamic Update with Probabilities, *Studia Logica* 93:1, 67–96. (With Jelle Gerbrandy and Barteld Kooi.)

2010

123 Temporal Logics of Agency, Editorial, *Journal of Logic, Language and Information* 19:4, 1–5. (With Eric Pacuit.)
124 Game Solution, Epistemic Dynamics, and Fixed-Point Logics, *Fundamenta Informaticae* 100, 19–41. (With Amélie Gheerbrant.)
125 The Dynamics of Awareness, *Knowledge, Rationality and Action*, Springer on-line. (With Fernando Velazquez.) *Synthese* 177:1, 5–27.
126 A Logician Looks at Argumentation Theory, *Cogency* 1:2, Universidad Diego Portales, Santiago de Chile.
127 McCarthy Variations in a Modal Key, *Artificial Intelligence* 175:1, 428–439. John McCarthy's Legacy, Leora Morgenstern and Sheila A. McIlraith, eds.
128 Logic and Philosophy Today: Editorial Introduction, *Journal of Indian Council of Philosophical Research* XXVII:1 and 2, 1–7. (With Amitabha Gupta.)
129 Categorial versus Modal Information Theory, *Linguistic Analysis* 36:1–4, 533–544.
130 Joachim Lambek: the Beauty of Mathematics in Language, *Linguistic Analysis* 36: 1–4, i–viii (with Michael Moortgat).
131 Куда должна, и должна ли, двигаться логика?, Russian Translation of "Where is Logic Going, and Should It?" (*Topoi*, 2006, #96), *Vox 9*, electronic journal of philosophy, Moscow, http://vox-journal.org/html/issues/vox9.

2011

132 Toward a Theory of Play: A Logical Perspective on Games and Interaction, *GAMES* 2:1, 52–86; doi:10.3390/g2010052. (With Eric Pacuit and Olivier Roy.)
133 Introduction to 'Logic and Philosophy of Science: in the Footsteps of E.W. Beth', *Synthese* 179:2, 203–206. (With Theo Kuipers and Henk Visser.)

134 Логика в действии. Введение, Russian translation of 'A Miniguide to Logic in Action', *VOX 10* (http://vox-journal.org/html/issues/145, Philosophical Institute, Russian Academy of Sciences, Moscow.

135 Dynamic Logic of Evidence-Based Beliefs, *Studia Logica* 99:1, 61–92. (With Eric Pacuit.)

136 Logic in a Social Setting, *Episteme* 8:3, 227–247.

137 Models of Reasoning in Ancient China, *Studies in Logic* 4:3, 57–81. (With Fenrong Liu and Jeremy Seligman.)

2012

138 Toward a Dynamic Logic of Questions, *Journal of Philosophical Logic* 41:4, 633–669. (With Stefan Minica.)

139 The Logic of Empirical Theories Revisited, *Synthese* 186:3, 775–792.

140 Logic and Psychology: Do the Facts Matter? Chinese translation, *Journal of Hubei University* (*Philosophy and Social Sciences*) 39:3, 1–9.

141 Sahlqvist Correspondence for Modal Mu-Calculus. *Studia Logica* 100, 31–60. (With Nick Bezhanishvili and Ian Hodkinson.)

142 The Nets of Reason, *Argument and Computation* 3:2/3, 83–86.

2014

143 Evidence and Plausibility in Neighborhood Structures. *Annals of Pure and Applied Logic* 165:1, 106–133. (With David Fernandez Duque and Eric Pacuit.)

144 Natural Language and Logic of Agency, *Journal of Logic, Language and Information* 23:3, 367–382.

145 Modeling Reasoning in a Social Setting, *Studia Logica* 102:2, 235–265.

146 Priority Structures in Deontic Logic, *Theoria* 80:2, 116–152. (With Davide Grossi and Fenrong Liu.)

To Appear

Where is Logic Going?, preface to a collection of discussion pieces, *Studies in Logic*. (With Fenrong Liu.)

Articles in Books

1980

1 Points and Periods, in Ch. Rohrer, ed., *Time, Tense and Quantifiers*, Niemeyer Verlag, Tübingen, 39–58.

1981

2 Tense Logic, Second-Order Logic and Natural Language, in U. Mönnich, ed., *Aspects of Philosophical Logic*, Reidel, Dordrecht, 1–20.

3 Why is Semantics What? in J. Groenendijk, T. Janssen and M. Stokhof, eds., *Formal Methods in the Study of Language*, Math. Centre Tract 135, Amsterdam, 29–49.

1982

4 Recht en Redeneren, in A. Soeteman and P. W. Brouwer, eds., *Logica en Recht*, Tjeenk Willink, Zwolle, 61–70.

1983

5 Five Easy Pieces, in A. ter Meulen, ed., *Studies in Model-Theoretic Semantics*, Foris, Dordrecht (GRASS series, vol. 1), 1–17.
6 Higher-Order Logic, in D. Gabbay and F. Guenthner, eds., *Handbook of Philosophical Logic*, vol. I., Reidel, Dordrecht, 275–329. (With Kees Doets.)

1984

7 Correspondence Theory, in D. Gabbay and F. Guenthner, eds., *Handbook of Philosophical Logic*, Vol. II., Reidel, Dordrecht, 167–247.
8 The Logic of Semantics, in F. Landman and F. Veltman, eds., *Varieties of Formal Semantics*, Foris, Dordrecht (GRASS series, vol. 3), 55–80.

1985

9 Themes from a Workshop, in J. van Benthem and A. ter Meulen, eds., *Generalized Quantifiers in Natural Language*, Foris, Dordrecht (GRASS 4), 161–169.
10 Semantics of Time, in J. Jackson and J. Michon, eds., *Time, Mind and Behaviour*, Springer, Heidelberg, 266–278.

1986

11 A Linguistic Turn: New Directions in Logic, in R. Marcus et al., eds., *Proceedings 7th International Congress of Logic, Methodology and Philosophy of Science. Salzburg 1983*, North-Holland, Amsterdam, 205–240.
12 The Ubiquity of Logic in Natural Language, in W. Leinfellner and F. Wuketits, eds., *The Tasks of Contemporary Philosophy*, Hölder-Pichler-Tempsky Verlag, Wien, 177–186.
13 Het Categoriale Wereldbeeld, in C. Hoppenbrouwers et al., eds., *Proeven van Taalwetenschap*, TABU, Groningen, 1–18.

1987

14 Semantic Automata, in D. de Jongh et al., eds., *Studies in the Theory of Generalized Quantifiers and Discourse Representation*, Foris, Dordrecht (GRASS series, vol. 8), 1–25.
15 Towards a Computational Semantics, in P. Gärdenfors, ed., *Generalized Quantifiers: Linguistic and Logical Approaches*, Reidel, Dordrecht, 31–71.
16 Verisimilitude and Conditionals, in T. Kuipers, ed., *What is Closer-to-the-Truth?*, Rodopi, Amsterdam, 103–128.

17 Categorial Grammar and Lambda Calculus, in D. Skordev, ed., *Mathematical Logic and its Applications*, Plenum Press, New York, 39–60.

1988

18 Categorial Equations, in E. Klein and J. van Benthem, eds., *Categories, Polymorphism and Unification*, Edinburgh and Amsterdam, 1–17.
19 The Lambek Calculus, in R. Oehrle, E. Bach and D. Wheeler, eds., *Categorial Grammars and Natural Language Structures*, Reidel, Dordrecht, 35–68.
20 Strategies of Intensionalization, in I. Bodnar, A. Maté and L. Pólos, eds., *A Filozofiai Figyelo Kiskonyvtara*, Kezirat Gyanant, Budapest, 41–59.
21 Semantic Type Change and Syntactic Recognition, in G. Chierchia, B. Partee and R. Turner, eds., *Properties, Types and Meaning*, vol. I, Reidel, Dordrecht, 231–249.
22 Games in Logic, in J. Hoepelman, ed., *Representation and Reasoning*, Niemeyer Verlag, Tübingen, 3–15, 165–168.
23 New Trends in Categorial Grammar, in W. Buszkowski et al., eds., *Categorial Grammar*, John Benjamin, Amsterdam/Philadelphia, 23–33.
24 The Semantics of Variety in Categorial Grammar, in W. Buszkowski et al., eds., *Categorial Grammar*, John Benjamin, Amsterdam, 37–55.

1989

25 Time, Logic and Computation, in J. W. de Bakker, W.-P. de Roever and G. Rozenberg, eds., *Linear Time, Branching Time and Partial Order in Logics and Models for Concurrency*, Springer, Berlin, 1–49.
26 Semantic Parallels in Natural Language and Computation, in H-D Ebbinghaus et al., eds., *Logic Colloquium. Granada 1987*, North-Holland, Amsterdam, 331–375.
27 Reasoning and Cognition, in H. Schnelle and N-O Bernsen, eds., *Logic and Linguistics. Research Directions in Cognitive Science*, L. Erlbaum, Hove (UK), 185–208.
28 Logical Semantics, in H. Schnelle and N-O Bernsen, eds., *Logic and Linguistics. Research Directions in Cognitive Science*, L. Erlbaum, Hove (UK), 109–126.

1990

29 What is Extensionality? in J. Kelemen et al., eds., *Annales Universitatis Scientiarum Budapestinensis De Rolando Eötvös Nominatae* XXII–IIII, 213–220.

1991

30 Beyond Accessibility: New Semantics for Modal Predicate Logic, in M. de Rijke, ed., *Modal Logic Colloquium*, Dutch Network for Language, Logic and Information, 1–14.
31 Linguistic Universals in Logical Semantics, in D. Zaefferer, ed., *Semantic Universals and Universal Semantics*, Foris, Berlin (GRASS series), 17–36.

32 Generalized Quantifiers and Generalized Inference, in J. van der Does and J. van Eijck, eds., *Quantification in the Netherlands*, 'Semantic Parallels Project', CWI, Amsterdam.

1992

33 Fine-Structure in Categorial Semantics, in M. Rosner and R. Johnson, eds., *Computational Linguistics and Logical Semantics*, Cambridge UP, 127–157.

1993

34 Beyond Accessibility: Functional Models for Modal Logic, in M. de Rijke, ed., *Diamonds and Defaults*, Kluwer, Dordrecht, 1–18.

35 Quantifiers and Inference, in M. Krynicki, M. Mostowski and L. W. Szczerba, eds., *Quantifiers*, vol. II, Kluwer Academic Publishers, Dordrecht, 1–20.

36 Logic and the Flow of Information, in D. Prawitz, B. Skyrms and D. Westerståhl, eds., *Proceedings 9th International Congress of Logic, Methodology and Philosophy of Science. Uppsala 1991*, Elsevier Science Publishers, Amsterdam, 693–724.

1994

37 The Landscape of Deduction, in K. Dosen and P. Schröder-Heister, eds., *Substructural Logics*, Clarendon Press, Oxford, 357–376.

38 Dynamic Arrow Logic, in J. van Eijck and A. Visser, eds., *Logic and Information Flow*, The MIT Press, Cambridge (Mass.), 15–29.

39 General Dynamic Logic, in D. Gabbay, ed., *What is a Logical System?*, Oxford University Press, Oxford, 107–139.

40 A New World Underneath Standard Logic, in K. Apt, L. Schrijvers and N. Temme, eds., *From Universal Morphisms to Megabytes: A Baayen Space Odyssey*, Centre for Mathematics and Computer Science (CWI), Amsterdam, 179–186.

1995

41 Temporal Logic, in D. Gabbay, C. Hoggar and J. Robinson, eds., *Handbook of Logic in Artificial Intelligence and Logic Programming*, 4, Oxford University Press, Oxford, 241–350.

42 NNIL, A Study in Intuitionistic Propositional Logic, in A. Ponse, M. de Rijke and Y. Venema, eds., *Modal Logic and Process Algebra*, CSLI Lecture Notes, Stanford, 289–326. (With Albert Visser, Dick de Jongh and Gerard Renardel de Lavalette.)

43 Submodel Preservation Theorems in Finite-Variable Fragments, in A. Ponse, M. de Rijke and Y. Venema, eds., *Modal Logic and Process Algebra*, CSLI Lecture Notes, Cambridge UP, 1–11. (With Hajnal Andréka and Istvan Németi.)

1996

44 Quantifiers in the World of Types, in Jaap van der Does and Jan van Eijck, eds., *Quantifiers, Logic and Language*, CSLI Lecture Notes 54, Stanford, 47–62.

45 Complexity of Contents versus Complexity of Wrappings, in M. Marx, M. Masuch and L. Pólos, eds., *Arrow Logic and Multimodal Logic*, Studies in Logic, Language and Information, CSLI Publications, Stanford (and Cambridge University Press), 203–219.

46 Dynamics, a chapter in J. van Benthem and A. ter Meulen, eds., *Handbook of Logic and Language*, Elsevier Science Publishers, Amsterdam, 587–648. (With Reinhard Muskens and Albert Visser.)

47 Inference, Methodology and Semantics, in P. Bystrov and V. Sadofsky, eds., *Philosophical Logic and Logical Philosophy, Essays in Honour of Vladimir Smirnov*, Kluwer Academic Publishers, Dordrecht, 63–82.

48 Bisimulation: The Never-Ending Story, in J. Tromp, ed., *A Dynamic and Quick Intellect. Liber Amicorum Paul Vitányi*, CWI, Amsterdam, 23–27.

49 Modal Logic as a Theory of Information, in J. Copeland, ed., *Logic and Reality. Essays on the Legacy of Arthur Prior*, Clarendon Press, Oxford, 135–168.

50 Logic and Argumentation Theory, in F. van Eemeren, R. Grootendorst, J. van Benthem and F. Veltman, eds., *Proceedings Colloquium on Logic and Argumentation*, Royal Dutch Academy of Sciences, Amsterdam, 27–41.

1997

51 Modal Quantification over Structured Domains, in M. de Rijke, ed., *Advances in Intensional Logic*, Kluwer, Dordrecht, 1–28. (With Natasha Alyeshina.)

1998

52 Proofs, Labels, and Dynamics in Natural Language, in U. Reyle and H-J Ohlbach, eds., *Festschrift for Dov Gabbay*, 31–41, Kluwer Academic Publishers, Dordrecht.

53 Shifting Contexts and Changing Assertions, in A. Aliseda-Llera, R. van Glabbeek and D. Westerståhl, eds., *Computing Natural Language,* 51–65.

1999

54 Modal Foundations for Predicate Logic, in Ewa Orlowska, ed., *Logic at Work*, To the Memory of Elena Rasiowa, Physica Verlag, Heidelberg, 39–54.

55 Logical Constants, Computation and Simulation Invariance, in Th. Childers, ed., *The Logica 98 Yearbook*, Institute of Philosophy, Czech Academy of Sciences, 11–19.

2000

56 Linguistic Grammar as Dynamic Logic, in V. M. Abrusci and C. Casadio, eds., *Dynamic Perspectives in Logic and Linguistics*, Bulzoni, Roma, 7–17.

57 Reasoning in Reverse, Preface to "Abduction and Induction, their Relation and Integration", P. Flach and A. Kakas, eds., Kluwer, Dordrecht, ix-xi.

58 Explaining Language by Economic Behaviour, in A. Rubinstein, *Economics and Language*, Cambridge University Press, 93–107.

2001

59 Modal Logic in Two Gestalts, in M. de Rijke, H. Wansing and M. Zakharyashev, eds., *Advances in Modal Logic II*, Uppsala 1998, CSLI Publications, Stanford, 73–100.

60 Higher-Order Logic, reprint with addenda, in D. Gabbay, ed., *Handbook of Philosophical Logic*, vol. I., second edition, Kluwer, Dordrecht, 189–243.

61 Correspondence Theory, reprint with addenda, in D. Gabbay, ed., *Handbook of Philosophical Logic*, vol. III., second edition, Kluwer, Dordrecht, 325–408.

62 Nonstandard Reasoning, *International Encyclopedia of the Behavioral and Social Sciences*, volume 16, 10696–10699, Elsevier, Colchester, UK.

2002

63 Invariance and Definability: Two Faces of Logical Constants, in W. Sieg, R. Sommer, and C. Talcott, eds., *Reflections on the Foundations of Mathematics. Essays in Honor of Sol Feferman*, ASL Lecture Notes in Logic 15, 426–446.

64 Mathematical Logic and Natural Language, in B. Löwe, W. Malzkorn, and T. Rasch, eds., *Foundations of the Formal Sciences II: Applications of Mathematical Logic in Philosophy and Linguistics*, Kluwer, Dordrecht, 25–38.

65 Modal Logic, in D. Jacquette, ed., *A Companion to Philosophical Logic*, Blackwell, Oxford, 391–409.

66 Logical Patterns in Space, in D. Beaver et al., eds. *Words, Proofs, and Diagrams*, CSLI Publications, Stanford, 5–25. (With Marco Aiello.)

67 Action and Procedure in Reasoning, in M. MacCrimmon and P. Tillers, eds., *The Dynamics of Judicial Proof*, Physica Verlag, Heidelberg, 243–259.

2003

68 Rational Dynamics and Epistemic Logic in Games, in S. Vannucci, ed., *Logic, Game Theory and Social Choice III*, University of Siena, Department of Political Economy, 19–23.

69 *Logic for Concurrency and Synchronization*, guest preface, R. de Queiroz, ed., Trends in Logic, Kluwer, xvii.

70 Logic and Game Theory: Close Encounters of the Third Kind, in G. Mints and R. Muskens, eds., *Games, Logic, and Constructive Sets*, CSLI Publications, Stanford, 3–22.

71 Fifty Years: Changes and Constants in Logic, in V. Hendricks and J. Malinowski, eds., *Trends in Logic, 50 Years of Studia Logica*, Kluwer, Dordrecht, 35–56.

72 Categorial Grammar at a Cross-Roads, in G-J Kruijff and R. Oehrle, eds. *Resource-Sensitivity, Binding, Anaphora*, Kluwer, Dordrecht, 3–21.

73 Is there still Logic in Bolzano's Key? in E. Morscher, ed., *Bernard Bolzanos Leistungen in Logik, Mathematik und Physik*, Bd.16, Academia Verlag, Sankt Augustin 2003, 11–34.

74 Structural Properties of Dynamic Reasoning, in *Meaning: the Dynamic Turn*, J. Peregrin, ed., Elsevier, Amsterdam, 2003, 15–31.

2004

75 The Inexhaustible Content of Modal Boxes, in A. Troelstra, ed., *Liber Amicorum for Dick de Jongh*, 16 p. ILLC Amsterdam, ISBN 90 5776 1289.

76 De Kunst van het Vergaderen, in Wiebe van der Hoek, ed., *Liber Amicorum 'John-Jules Charles Meijer 50'*, 5–7, Onderzoeksschool SIKS, Utrecht.

77 Probabilistic Features in Logic Games, invited presentation, Open Court Symposium, APA Chicago. In D. Kolak and J. Symons, eds., *Quantifiers, Questions, and Quantum Physics*, Springer Verlag, New York, 189–194.

2005

78 The Categorial Fine-Structure of Natural Language, in C. Casadio, P.J. Scott, R.A.G. Seely, eds., *Language and Grammar: Studies in Mathematical Linguistics and Natural Language,* CSLI Publications Stanford (CSLI Lecture Notes 168), 3–29.

79 An Essay on Sabotage and Obstruction, in D. Hutter, ed., *Mechanizing Mathematical Reasoning, Essays in Honor of Jörg Siekmann on the Occasion of his 69th Birthday*, Springer Verlag, 268–276.

80 Formal Methods in Philosophy, in V. Hendricks and J. Symons, eds., *Formal Philosophy*, Automatic Press, New York and London.

81 Open Problems in Logic and Games, in S. Artemov, H. Barringer, A. d'Avila Garcez, L. Lamb and J. Woods, eds., *Essays in Honour of Dov Gabbay*, King's College Publications, London, 229–264.

82 A Note on Modeling Theories, in *Confirmation, Empirical Progress and Truth Approximation. Essays in Debate with Theo Kuipers*, R. Festa, A. Aliseda and J. Peijnenburg, eds., *Poznan Studies in the Philosophy of the Sciences and the Humanities*, Rodopi Amsterdam, 403–419.

83 The Geometry of Knowledge, in J-Y Béziau, A. Costa Leite and A. Facchini, eds., *Aspects of Universal Logic*, Centre de Recherches Sémiologiques, Université de Neuchatel, 1–31. (With Darko Sarenac.)

2006

84 Open Problems in Update Logic. D. Gabbay, S. Goncharov and M. Zakharyashev, eds., *Mathematical Problems from Applied Logic I*, Springer, 137–192.

85 The Epistemic Logic of IF Games. R. Auxier and L. Hahn, eds., *The Philosophy of Jaakko Hintikka*, Schilpp Series, Open Court Publishers, Chicago, 481–513.

86 A New Modal Lindström Theorem. H. Lagerlund, S. Lindström and R. Sliwinski, eds., *Modality Matters*, University of Uppsala, 55–60.

87 Preference Logic, Conditionals, and Solution Concepts in Games. H. Lagerlund, S. Lindström and R. Sliwinski, eds., *Modality Matters*, University of Uppsala, 61–76. (With Sieuwert van Otterloo and Olivier Roy.)

88 Logical Construction Games. *Acta Philosophica Fennica* 78, T. Aho and A-V Pietarinen, eds., Truth and Games, Essays in Honour of Gabriel Sandu, 123–138.

89 Alternative Logics and Classical Concerns, in J. van Benthem, G. Heinzmann, M. Rebuschi and H. Visser, eds., *The Age of Alternative Logics*, Springer, Dordrecht, 1–7.

90 A Mini-Guide to Logic in Action, updated version, in F. Stadler and M. Stöltzner, eds., *Time and History*, Ontos Verlag, Frankfurt, 419–440.

91 One is a Lonely Number: on the Logic of Communication, in Z. Chatzidakis, P. Koepke and W. Pohlers, eds., *Logic Colloquium '02*, ASL and A.K. Peters, Wellesley MA, 96–129.

92 Logic in Philosophy, in D. Jacquette, ed., *Handbook of the Philosophy of Logic*, Elsevier, Amsterdam, 65–99.

93 Modal Logic, a Semantic Perspective, in J. van Benthem, P. Blackburn and F. Wolter, eds., *Handbook of Modal Logic*, Elsevier, Amsterdam, 1–84.

2007

94 Cognition as Interaction, in G. Bouma, I. Krämer and J. Zwarts, eds., *Cognitive Foundations of Interpretation*, KNAW Amsterdam, 27–38.

95 Editorial, *A Meeting of the Minds*, Proceedings LORI Workshop Beijing August 2007, King's College Publications, London. (With Ju Shier and Frank Veltman.)

96 What is Spatial Logic?, editorial, *Handbook of Spatial Logics*, Springer, Dordrecht, 1–11. With co-editors Marco Aiello and Ian Pratt-Hartmann.

97 Modal Logics of Space, in M. Aiello et al., eds., *Handbook of Spatial Logics*, Springer, Dordrecht, 217–298. (With Guram Bezhanishvili.)

98 Five Questions on Philosophy of Mathematics, in V. Hendricks and H. Leitgeb, eds., *Philosophy of Mathematics, Five Questions*, Automatic Press, Copenhagen.

99 Five Questions on Games, in V. Hendricks and P. Guldborg Hansen, eds., *Game Theory, Five Questions*, Automatic Press, Copenhagen.

100 Logic Games, From Tools to Models of Interaction, in A. Gupta, R. Parikh and J. van Benthem, eds., *Logic at the Crossroads*, Allied Publishers, Mumbai, 283–317.

2008

101 Modal Fixed-Point Logic and Changing Models, in A. Avron, N. Dershowitz and Rabinovich, eds., *Pillars of Computer Science: Essays Dedicated to Boris (Boaz) Trakhtenbrot on the Occasion of his 85th Birthday*, Springer, Berlin, 146–165. (With Daisuke Ikegami.)

102 Games that Make Sense: Logic, Language and Multi-Agent Interaction, in K. Apt and R. van Rooij, eds., *Proceedings KNAW Colloquium on Games and Interactive Logic*, Texts in Logic and Games, Amsterdam University Press, 197–209.

103 Computation as Conversation, in S. Cooper, B. Löwe and A. Sorbi., eds., *New Computational Paradigms, Changing Conceptions of What is Computable*, Springer, New York, 35–58.

104 Information Is What Information Does, editorial, *Handbook of the Philosophy of Information*, Elsevier, Amsterdam. (With Pieter Adriaans.)

105 The Logical Stories of Information, in *Handbook of the Philosophy of Information*, Elsevier, Amsterdam, 217–280. (With Maricarmen Martinez.)

106 Man Muss Immer Umkehren, in C. Dégremont, L. Keiff, H. Rueckert, eds., *Dialogues, Logics and Other Strange Things: Essays in Honour of Shahid Rahman*, College Publications, London.

107 A Brief History of Natural Logic, in M. Chakraborty, B. Löwe, M. Nath Mitra, S. Sarukkai, eds., *Logic, Navya-Nyaya and Applications, Homage to Bimal Krishna Matilal*, College Publications, London 2008, 21–42.

2009

108 Logic and Philosophy in the Century That Was, in F. Stoutland, ed., *Philosophical Probings: Essays on Von Wright's Later Work*, Automatic Press, Roskilde, 163–167.

109 Preface to new printing of *E. W. Beth, Door Wetenschap tot Wijsheid*, Amsterdam Academic Archive. (With Henk Visser.)

110 Editor's Preface, *Wiebefest 2009*, Department of Computing, University of Liverpool. (With John-Jules Meijer, Cees Witteveen, and Mike Wooldridge.)

111 Preface, in L. Kurzen and F. Velazquez-Quesada, eds., *Logical Dynamics for Information and Preferences*, Seminar Yearbook, ILLC, Amsterdam, vii–viii.

112 Argumentation in the Lense of Artificial Intelligence, Preface, in I. Rahwan and G. Simari, eds., *Argumentation in Artificial Intelligence*, Springer, 2009, vii–viii.

113 Preface, in U. Wybraniec-Skardowska, 'Polish Logic, a few Lines from a Personal Perspective', ILLC prepublications, Amsterdam.

114 Actions that Make Us Know, in J. Salerno, ed., *New Essays on the Knowability Paradox*, Oxford University Press, Oxford, 129–146.

115 For Better of for Worse: Dynamic Logics of Preference, in T. Grüne-Yanoff and S-O Hansson, eds., *Preference Change*, Springer, Dordrecht, 57–84.

2010

116 A Logician Takes a Look at Argumentation Theory, Chinese translation of the same paper from *Cogency*, Yearbook, Philosophy Department, Tsinghua University.

117 A Brief History of Natural Logic, Chinese translation of the same paper from 2008, *Values and Culture*, Beijing Normal University Publications, 152–167.

118 Logic, Rational Agency, and Intelligent Interaction, in C. Glymour, W. Wei and D. Westerståhl, eds., *Logic, Methodology and Philosophy of Science XIII Beijing 2007*, College Publications, London, 137–161.

119 Logic, Mathematics, and General Agency, in P. E. Bour, M. Rebuschi and L. Rollet, eds., *Construction*, College Publications, London, 281–300.

120 Frame Correspondence in Modal Predicate Logic, in S. Feferman and W. Sieg, eds., *Proofs, Categories and Computations*, College Publications, London.

121 In Praise of Strategies, In J. van Eijck and R. Verbrugge, eds., *Foundations of Social Software*, Studies in Logic, College Publications, 283–317.

2011

122 Belief Update as Social Choice, in P. Girard, O. Roy and M. Marion, eds., *Dynamic Formal Epistemology*, Springer, Dordrecht, 151–160.

2012

123 In Praise of Strategies, reprint with addenda, J. van Eijck and R. Verbrugge, eds., *Games, Actions, and Social Software*, Lecture Notes in Computer Science 7010, Springer, Heidelberg, 96–116.

124 Dynamic Logic in Natural Language, in G. Russell and D. Graff Fara, eds., *The Routledge Companion to Philosophy of Language*, Routledge, New York and London, 652–666.

125 CRS and Guarded Logics, in H. Andréka, M. Ferenczi and I. Németi, eds., *Cylindric-like Algebras and Algebraic Logic*, Bolyai Society Mathematical Studies 22, Mathematical Institute, Hungarian Academy of Sciences, Budapest, 273–301.

2013

126 Reasoning about Strategies, in B. Coecke, L. Ong and P. Panangaden, eds., *Computation, Logic, Games, and Quantum Foundations. The Many Facets of Samson Abramsky*, Springer Lecture Notes in Computer Science 7860, 333–347.

127 Two Logical Faces of Belief Revision, in R. Trypuz, ed., *Krister Segerberg on Logic of Actions*, Outstanding Logicians Series, Springer, Dordrecht, 281–300.

2014

128 Connecting Logics of Choice and Change, in Th. Mueller, *Nuel Belnap on Indeterminism and Free Action*, Outstanding Logicians Series, Springer, Dordrecht. (With Eric Pacuit.)

129 Logica en Recht: Naar een Rijkere Relatie, in M. Groenhuijsen, E. Hondius and A. Soeteman, eds., *Het Recht in het Geding*, Boom Juridische Uitgevers, Den Haag, 71–83.

To Appear

Deontic Logic and Changing Preferences, in D. Gabbay, J. Horty, R. van der Meyden, X. Parent and L. van der Torrre, eds., *Handbook of Deontic Logic and Normative Systems,* (with Fenrong Liu).

Dynamic Logics of Belief Change, in H. van Ditmarsch, J. Halpern, W. van der Hoek and B. Kooi, eds., *Handbook of Logics for Knowledge and Belief*, College Publications (with Sonja Smets).

Modal Logic, *Internet Encyclopedia of Philosophy*.

Reviews

1978

1 Systems of Intensional and Higher-Order Modal Logic (D. Gallin) *Mededelingen Wiskundig Genootschap.*

1979

2 Filosofische Grondslagen van de Wiskunde (D. van Dalen) *Algemeen Nederlands Tijdschrift voor Wijsbegeerte* 71:1, 58–64.

3 Investigations in Modal and Tense Logics (D. Gabbay), Critical Notice, *Synthese* 40, 353–373.

4 Inleiding tot de Symbolische Logica (H. Hubbeling and H. de Swart) *Algemeen Nederlands Tijdschrift voor Wijsbegeerte* 71:2, 125–127.

5 Filosoferen (S. Ysseling and R. de Kwant), *Algemeen Nederlands Tijdschrift voor Wijsbegeerte* 71:2, 129–131.

1980

6 Analyticiteit (H. Perrick). *Algemeen Nederlands Tijdschrift voor Wijsbegeerte* 72:2, 119–122.

7 Complementarity in Mathematics (W. Kuyk), *Algemeen Nederlands Tijdschrift voor Wijsbegeerte* 72:3, 200–202.

8 Juristische Logik als Argumentationslehre (Ch. Perelman), *Algemeen Nederlands Tijdschrift voor Wijsbegeerte* 72:4, 269–273.

9 Checking Landau's "Grundlagen" in the AUTOMATH System (L. van Benthem-Jutting), *Mededelingen Wiskundig Genootschap.*

1981

10 Outline of a Nominalist Theory of Propositions (P. Gochet), *Algemeen Nederlands Tijdschrift voor Wijsbegeerte* 73:1, 61–63.

11 De Dialoog van Galilei (M. Finocchiaro), *Kennis en Methode* 5:3, 273–283.

1982

12 The Continuing Story of Conditionals ("Ifs"; W. Harper et al., eds.) *Studies in Language* 6:1, 125–136.

1983

13 The Logic of Natural Language (F. Sommers), *Philosophical Books* 24:2, 99–102.

14 Bernard Bolzano: Leben und Wirkung (C. Christian, ed.), *Archives Internationales d'Histoire des Sciences* 33, 360–361.

15 Argumentation. Approaches to Theory Formation (E. Barth and J. Martens, eds.), *Algemeen Nederlands Tijdschrift voor Wijsbegeerte* 75:4, 380–381.

1984

16 On When a Semantics is not a Good Semantics (J. Copeland), *Journal of Symbolic Logic* 49:3, 994–995.

17 The Lattice of Modal Logics (W. Blok), *Journal of Symbolic Logic* 49:4, 1419–1420.

1986

18 The Logic of Aspect (A. Galton), *The Philosophical Review* 95:3, 434–437.

19 A Companion to Modal Logic (G. Hughes and M. Creswell), *Journal of Symbolic Logic* 51:3, 824–826.

20 Boolean Semantics for Natural Language (E. Keenan and L. Faltz) *Language* 62:4, 112–115.

21 Building Models by Games (W. Hodges), *Mededelingen Wiskundig Genootschap* 1986, 368–370.

1989

22 Logicheskije Metodi Analiza Nauchnowo Znanija (V. A. Smirnov), *Studia Logica* 48:1, 135–136.

23 Logicheskaja Semantika i Filosofskije Osnovanija Logiki (E. D. Smirnova), *Studia Logica* 48:1, 136–137.

1990

24 Handbook of Philosophical Logic, vol. IV (D. Gabbay and F. Guenthner, eds.), *Language* 66:2 (1989), 396–400.

1991

25 Knowledge in Flux. Modelling the Dynamics of Epistemic States (P. Gärdenfors), *Studia Logica.* 49:4, 421–424.

1999

26 Information Flow (J. Barwise and J. Seligman), *Journal of Logic, Language and Information* 8:3, 390–397. (With David Israel.)

2007

27 Quantifiers in Language and Logic (S. Peters and D. Westerståhl), *Notre Dame Philosophical Reviews.*

2013

28 Bernard Bolzano's "Wissenschaftslehre", *Topoi* 3:2, 301–303.

General Publications

1977

1 De Logica van Zinovjev, *Rusland-Bulletin* 2:1, 3–9.

1978

2 Logica en Argumentatietheorie, *Spectator* 7:5/6, 263–276.

1979

3 In Alle Redelijkheid, inaugural lecture, Onderzoeksbulletin 4, Filosofisch Instituut, Rijksuniversiteit Groningen, 25 p.
4 Logische Theorie en Wiskundige Praktijk, *Euclides* 55:6, 249–254.

1981

5 Logische Lijnen in de Zeventiger Jaren, *Wijsgerig Perspectief* 22:1, 1–29.

1983

6 Tijd, Taal en Logica, *Wijsgerig Perspectief* 24:3, 78–81.

1984

7 Aan de Rede van Barbarije, *Kennis en Methode* 8:2, 125–136.

1986

8 Die linguistische Wende in der Logik, *Information Philosophie*, Mai 1986, 18–28.
9 Taal en Informatie: naar Nieuwe Toepassingen, *Forum der Letteren* 27, 174–187.
10 Rekenen met Taal, inaugural lecture, Mathematisch Instituut, University of Amsterdam.
11 Aan de Rede van Barbarije, in P. Meeuse, *Harmonie als Tegenspraak*, De Bezige Bij, A'dam, 219–235.

1989

12 The Lure of Information and Computation, *Universiteit en Hogeschool* 36:1, 65–74.

1992

13 Bouwwerk Zonder Funderingen, in: Redactie Wetenschappen NRC, *De Mond Vol Tanden*, Prometheus, Amsterdam, 119–126.

1993

14 Is Informatica een Wetenschap? *Automatiseringsgids* 27:10, 11–13.

1994

15 Wat is er Mis met de Filosofie? *Epimedium* 58, Filosofisch Instituut, Utrecht, 4–9.

16 Logica in Veelvoud, RADAR 1994, ARAMITH uitgevers, Bloemendaal, 364–373.

1996

17 Logica in Beweging, Lezingenboekje Uitreiking Spinoza Premie, NWO, Den Haag.

1997

18 Wat Cultuurbeleid van Wetenschapsbeleid Kan Leren, *Boekman Cahier* 31, 38–45, Boekman Stichting, Amsterdam.

1998

19 Van Grondslagenonderzoek naar Informatiewetenschap, KNAW, Afdeling Gees-teswetenscappen, Amsterdam.
20 De Boeken in Mijn Leven, *Amsterdamse Boekengids* 14, 37–38.
21 Internationalism is a State of Mind, *Reizigers*, UvA Yearbook 98, Vossius Press, Amsterdam.
22 E.W. Beth: het Geheim van Wetenschappelijk Succes, in P. Klein, ed., "Een Beeld van de Akademie: 1808–1998", KNAW, Amsterdam, 187–188.

1999

23 Boeken Top Drie, *Natuur en Techniek* 67:11, 48.

2000

24 Instroom Blues, *Nieuw Archief voor Wiskunde* 5:1/3, p. 263. Also *Euclides*, December 2000.

2001

25 Interdisciplinariteit en Monogamie, *Grensgevallen*, Instituut I$_2$O, Universiteit van Amsterdam.

2002

26 Science and Society in Flux, prize-winning essay at the Hollandsche Maatschap-pij van Wetenschappen, in A. Verrijn-Stuart et al., eds., *The Future of the Sciences and Humanities*, Amsterdam University Press, Amsterdam, 63–90.
27 De Multidisciplinaire Universiteit: Losse Gedachten van een Zwerver, in B. de Reuver and R. de Klerk, eds., *Feestbundel Karel van Dam*, SCO Kohnstamm Instituut, Amsterdam, 65–71.

2003

28 De Kunst van het Kennis Maken, Opening Academisch Jaar, Amsterdam University Press, Amsterdam.
29 Wat Drijft Informatica Onderzoekers? *Programmaboek ICT-Kennis Congres 2003*, NWO, Den Haag.

30 Het ABC van Communicatie, *Verslag KNAW* (Dutch Academy of Arts and Sciences), Afdeling Natuurkunde, 24 november 2003.

2004

31 Reidansende Eiwitten, *Boekman Cahier* 59/59, 165–166, Amsterdam.

2005

32 De Kunst van het Kennis Maken, BLIND, electronisch interdisciplinair tijdschrift, http://www.ziedaar.nl/editions/3/.
33 Echte versus Virtuele Realiteiten: de Speelse Logica van Kaarten en Vergaderen, *BRES Planète* 232, 24–28.
34 Leer het Brein Kennen, with J. Jolles, R. de Groot, H. Dekkers, C. de Glopper, H. Uijlings and A. Wolff-Albers, OECD and KNAW, Amsterdam.
35 Talentenkracht/Talent Power, Freudental Instituut, Utrecht. With Jan de Lange and Robbert Dijkgraaf.
36 'L'Art et la Logique de la Conversation. "Dossier Logique", Éditions *Pour la Science*, Paris, 68–73.

2006

37 Informatiestroom voor Oplettende Mensen, in B. Mols, ed., *Het Raadsel van Informatie*, Boom, Meppel, 11–26.
38 Een Man uit Eén Stuk, Liber Amicorum Professor Bob Hertzberger, 13–14, FNWI, UvA.
39 Verdriet en Vorm vanuit een Hoger Standpunt, Liber Amicorum for Sijbolt Noorda, Amsterdam University Press, UvA.
40 *Brain Lessons*, Neuropsych Publishers, Maastricht, J. Jolles, R. de Groot, H. Dekkers, C. de Glopper, H. Uijlings and A. Wolff-Albers.

2007

41 *Eenheid van Cognitie*, Boekje Talentenkracht Publieksdag, Ouwehands Dierenpark, Freudental Instituut Utrecht.
42 Interview, in *Game Theory: 5 Questions*, V. Hendricks and P. Hansen, eds., Automatic Press, Copenhagen, 9–19.
43 Adios a la Soledad, *Azafea*, University of Salamanca, 21–33.
44 Patterns of Intelligent Interaction: Games, Action, and Social Software, *NIAS Newsletter* May 2007, Wassenaar.
45 Intelligent Systems: The Man Inside the Machine, *Festschrift for Jaap van den Herik 60*, MICC, University of Maastricht, 164–167.
46 Een Stroom van Informatie: Logica op het Grensvlak van Alfa, Beta en Gamma', Natuurkundige Voordrachten, Nieuwe Reeks 85, Genootschap Diligentia, Den Haag, 115–123.

2008

47 De Beste Boeken volgens Johan van Benthem, Academische Boekengids 67, Maart.

48 Een Postzegel vol Logica, *De Gids*, Maart 2008, 191–205.
49 Interview, in *Philosophy of Computing and Information: 5 Questions*, L. Floridi, ed., Automatic Press, Copenhagen, 9–19.
50 Interview, in *Epistemology: 5 Questions*, V. Hendricks and D. Pritchard, eds., Automatic Press, Copenhagen, 39–46.
51 *E.W. Beth. A Centenary Celebration*, KNAW Amsterdam. (With Paul van Ulsen and Henk Visser.)

2009

52 Voorwoord, in B. Mols, *Geestdrift. Wat cognitiewetenschappers bezielt*, NWO, Den Haag, 2–3. (With Peter Hagoort.)
53 Alledaags Gesprek is Toppunt van Prestatie, interview, in B. Mols, *Geestdrift. Wat Cognitiewetenschappers Bezielt*, NWO, Den Haag, 36–41.
54 Interview: Logic, Language, Cognition, *The Reasoner* 3:12, December 2009.
55 A Stamp Full of Logic, Farsi translation, *Mathematics Education Journal*, Iran.

2010

56 The Journal *Synthese*, Chinese Social Sciences Today, 5 May, p. 17. (With Vincent Hendricks and John Symons.)
57 Dialogue on Logic, Language and Cognition–Interview with Johan van Benthem, Guo Meiyun and Liu Xinwen, *Philosophical Trends* 3, 62–71.
58 Computation, Conversation, and Celebration, in T. Janssen and L. Torenvliet, eds., *Lectori Salutem*, festschrift on the occasion of the retirement of Peter van Emde Boas, ILLC, University of Amsterdam, 1–5.
59 Preface, *Logic and Interactive Rationality* Yearbook 2009, D. Grossi, L. Kurzen and F. Velazquez-Quesada, eds., ILLC, Amsterdam, vii.
60 Interview, in *Epistemic Logic: 5 Questions*, V. Hendricks and O. Roy, eds., Automatic Press, Copenhagen, 35–46.
61 Opencourse Logic in Action, *De Nieuwe Wiskrant* 30:1, 41–48, with Jan Jaspars. Ook in *Wiskunde en Onderwijs* 37:145, 2011, 28–39.

2011

62 Intelligentna interakcja: trendy dynamiczne w dziesiejszej logice, Polish translation of #46, Group in Logic, University of Opole.
63 Preface, *Logic and Interactive Rationality* Yearbook 2010, D. Grossi, S. Minica, B. Rodenhauser and S. Smets, eds., ILLC, Amsterdam, vi.

2012

64 Preface, *Logic and Interactive Rationality* Yearbook 2011, A. Baltag, D. Grossi, A. Marcoci, B. Rodenhäuser and S. Smets, eds., ILLC, Amsterdam, 1.
65 Dynamic Logic in Natural Language, in Th. Graf, D. Paperno, A. Szabolcsi, and J. Tellings, eds., *Theories of Everything: In Honor of Ed Keenan*. UCLA Working Papers in Linguistics 17, 4 p., Creative Commons License http://creativecommons.org/licenses/by-nc/3.0/.

66 Johan Frederik Staal, *Levensberichten en Herdenkingen,* KNAW, Amsterdam, 71–75.

2013

67 *The UvA Meets China,* Amsterdam University Press. (With Anouk Tso.)

2014

68 Interview, in Tr. Lupher and Th. Adajian, eds., *Five Questions on the Philosophy of Logic,* Automatic Press, Copenhagen.
69 Preface, *Dynamics Yearbook 2013,* Amsterdam, Beijing and Berkeley.

Conference Proceedings

1979

1 Partial Logical Consequence, *Abstracts 6th International Congress of Logic, Methodology and Philosophy of Science,* Hannover, Section 5.

1991

2 Natural Language: from Knowledge to Cognition, in E. Klein and F. Veltman, eds., *ESPRIT Symposium,* Brussels, Springer Verlag, Berlin, 159–171.

1997

3 Two Dynamic Strategies, *Proceedings 11th Amsterdam Colloquium,* ILLC Amsterdam. (With A. ter Meulen.)

1998

4 Process Operations in Extended Dynamic Logic, invited lecture, *Proceedings LICS 98,* Indianapolis, IEEE Publications, Los Alamitos, 244–250.

2001

5 Preface of the General Chair, *Temporal Representation and Reasoning, TIME-2001,* IEEE Computer Society, Los Alamitos, p. ix.
6 Logics for Information Update, *Proceedings TARK VIII,* Morgan Kaufmann, Los Altos, 51–88.

2004

7 Reduction Axioms for Epistemic Actions, *Proceedings Advances in Modal Logic 2004,* Computer Science Department, University of Manchester. Report UMCS-04 9-1, Renate Schmidt, Ian Pratt-Hartmann, Mark Reynolds, Heinrich Wansing, eds., 197–211. (With Barteld Kooi.)

2005

8 Common Knowledge in Update Logics, in R. van der Meyden, ed., *Proceedings TARK 10*, Singapore, 253–261. (With Jan van Eijck and Barteld Kooi.)

2006

9 Dynamic Update with Probabilities, W. van der Hoek and M. Wooldridge, eds., *Proceedings LOFT 2006*, Liverpool. (With Jelle Gerbrandy and Barteld Kooi.)

10 The Tree of Knowledge in Action, *Proceedings AiML Melbourne 2006*. (With Eric Pacuit.)

2007

11 Dynamic Epistemic Logic and Epistemic Temporal Logic, *Proceedings TARK Namur*, 2007, 72–81. (With Jelle Gerbrandy and Barteld Kooi.)

12 Lindström Theorems for Fragments of First-order Logic, *Proceedings of LICS 2007*, 280–292. (With Balder ten Cate and Jouko Väänänen).

13 Modeling Simultaneous Games with Concurrent Dynamic Logic; in J. van Benthem, S. Ju and F. Veltman, eds., *A Meeting of the Minds*, Proceedings LORI Workshop Beijing, College Publications, London, 243–258. (With Fenrong Liu and Sujata Ghosh.)

2008

14 Decisions, Actions, and Games, a Logical Perspective; in *Proceedings of the Third Indian Conference on Logic and Applications ICLA 2009*, R. Ramanujam and Sundar Sarukkai, Eds., Springer Lecture Notes in AI 5378, 1–22.

15 Multi-Agent Belief Dynamics: Bridges between Dynamic Doxastic and Doxastic Temporal Logics, in G. Bonanno, W. van der Hoek and B. Löwe, eds., *Proceedings LOFT*. (With Cédric Dégremont).

2009

16 Toward a Dynamic Logic of Questions, in J. Horty and E. Pacuit, eds., *Proceedings LORI II Chongqing*, Springer Lecture Notes in AI 5834, 28–42. (With Stefan Minica.)

2010

17 Logic Between Expressivity and Complexity, in J. Giesl and R. Hähnle, eds., *Proceedings IJCAR 2010*, LNAI 6173, Springer, Heidelberg, 122–126.

18 Deontics = Betterness + Priority, in G. Governatori and G. Sartor, eds., *Proceedings Deontic Logic in Computer Science*, DEON 2010, Fiesole, Italy. Lecture Notes in Computer Science 6181, Springer, 50–65. (With Davide Grossi and Fenrong Liu.)

2011

19 Exploring a Theory of Play, invited lecture, in K. R. Apt, ed., *Proceedings TARK Groningen*, ACM Digital Library, 12–16.

2012

20 Foundational Issues in Logical Dynamics, invited lecture, in Th. Bolander et al., eds., *Advances in Modal Logic, Copenhagen 2012*, College Publications, London, 95–96.

21 Evidence Logic: A New Look at Neighborhood Structures, in Th. Bolander et al., eds., *Advances in Modal Logic, Copenhagen 2012*, College Publications, London, 97–118. (With David Fernandez Duque and Eric Pacuit.)

CPSIA information can be obtained at www.ICGtesting.com
Printed in the USA
BVOW06*1412100914

366183BV00007BB/36/P

9 783319 060248